Biostatistical Analysis

Biostatistical Analysis

Second Edition

JERROLD H. ZAR

Department of Biological Sciences
Northern Illinois University

PRENTICE-HALL, INC., *Englewood Cliffs, N.J.* 07632

Library of Congress Cataloging in Publication Data

Zar, Jerrold H., (date)
Biostatistical analysis.

Bibliography: p.
Includes index.
1. Biometry. I. Title.
QH323.5.Z37 1984 574′.01′5195 83–13884
ISBN 0–13–077925–3

Editorial/production supervision: *Bette Kurtz*
Interior design: *Ros Herion and Bette Kurtz*
Cover design: *Edsal Enterprises*
Manufacturing buyer: *John Hall*

Printed in the United States of America

10 9 8

ISBN 0-13-077925-3

PRENTICE-HALL INTERNATIONAL, INC., *London*
PRENTICE-HALL OF AUSTRALIA PTY. LIMITED, *Sydney*
EDITORA PRENTICE-HALL DO BRASIL, LTDA., *Rio de Janeiro*
PRENTICE-HALL CANADA INC., *Toronto*
PRENTICE-HALL OF INDIA PRIVATE LIMITED, *New Delhi*
PRENTICE-HALL OF JAPAN, INC., *Tokyo*
PRENTICE-HALL OF SOUTHEAST ASIA PTE. LTD., *Singapore*
WHITEHALL BOOKS LIMITED, *Wellington, New Zealand*

Contents

Preface xiii

1 Introduction 1

1.1 Types of Biological Data 2 1.2 Accuracy and Significant Figures 4
1.3 Frequency Distributions 5 1.4 Cumulative Frequency Distributions 12

2 Populations and Samples 14

2.1 Populations 14 2.2 Samples from Populations 15
2.3 Random Sampling 15 2.4 Parameters and Statistics 16

3 Measures of Central Tendency 18

3.1 The Arithmetic Mean 18 3.2 The Median and Other Quantiles 20
3.3 The Mode 23 3.4 Other Measures of Central Tendency 23
3.5 The Effect of Coding Data 24 Exercises 26

4 Measures of Dispersion and Variability 27

4.1 The Range 27 4.2 The Mean Deviation 29 4.3 The Variance 29
4.4 The Standard Deviation 31 4.5 The Coefficient of Variation 32
4.6 Indices of Diversity 32 4.7 Other Measures of Dispersion 36
4.8 The Effect of Coding Data 36 Exercises 38

5 Testing for Goodness of Fit 40

5.1 Chi-Square Goodness of Fit 40 **5.2** Statistical Significance 41
5.3 Statistical Errors in Hypothesis Testing 43
5.4 Chi-Square Goodness of Fit for More Than Two Categories 45
5.5 Subdividing Chi-Square Analyses 46
5.6 Chi-Square Correction for Continuity 48
5.7 Bias in Chi-Square Calculations 49 **5.8** Heterogeneity Chi-Square 49
5.9 The Log-Likelihood Ratio 52
5.10 Kolmogorov-Smirnov Goodness of Fit for Discrete Data 53
5.11 Kolmogorov-Smirnov Goodness of Fit for Continuous Data 55
5.12 Sample Size Required for Kolmogorov-Smirnov Goodness of Fit for Continuous Data 58 Exercises 58

6 Contingency Tables 61

6.1 Chi-Square Analysis 62 **6.2** The 2 × 2 Contingency Table 64
6.3 Heterogeneity Testing of 2 × 2 Tables 67
6.4 Subdividing Contingency Tables 69
6.5 Bias in Chi-Square Contingency Analysis 70
6.6 The Log-Likelihood Ratio for Contingency Tables 71
6.7 Three-Dimensional Contingency Tables 72
6.8 Log-Linear Models for Multidimensional Contingency Tables 77 Exercises 78

7 The Normal Distribution 79

7.1 Symmetry and Kurtosis 81 **7.2** Proportions of a Normal Distribution 83
7.3 The Distribution of Means 86 **7.4** Assessing Departures from Normality 88
Exercises 96

8 One-Sample Hypotheses 97

8.1 Two-Tailed Hypotheses Concerning the Mean 97
8.2 One-Tailed Hypotheses Concerning the Mean 101
8.3 Confidence Limits for the Population Mean 103
8.4 Reporting Variability about the Mean 104
8.5 Sample Size and Estimation of the Population Mean 108
8.6 Power and Sample Size in Tests Concerning the Mean 110
8.7 Sampling Finite Populations 112
8.8 Confidence Limits for the Population Median 113
8.9 One-Sample Hypotheses Concerning the Median 114
8.10 Confidence Limits for the Population Variance 115
8.11 Hypotheses Concerning the Variance 115
8.12 Power and Sample Size in Tests Concerning the Variance 116
8.13 Hypotheses Concerning Symmetry and Kurtosis 118
8.14 The Effect of Coding 120 Exercises 120

9 Two-Sample Hypotheses 122

9.1 Testing for Difference between Two Variances 122
9.2 Confidence Interval for Variance Ratio 125
9.3 Testing for Difference between Coefficients of Variation 125
9.4 Testing for Difference between Two Means 126
9.5 Confidence Limits for Means 131
9.6 Sample Size and Estimation of the Difference between Two Population Means 132
9.7 Power and Sample Size in Tests for Difference between Two Means 134
9.8 Nonparametric Statistical Methods 138
9.9 Two-Sample Rank Testing 138
9.10 Testing for Difference between Two Medians 145
9.11 The Effect of Coding 146
9.12 Two-Sample Testing of Nominal Scale Data 146
9.13 Testing for Difference between Two Diversity Indices 146 Exercises 148

10 Paired-Sample Hypotheses 150

10.1 The Paired-Sample *t* Test 150
10.2 Confidence Limits for the Population Mean Difference 152
10.3 Power and Sample Size in Paired-Sample Testing of Means 153
10.4 Paired-Sample Testing by Ranks 153
10.5 Confidence Limits for the Population Median Difference 156
10.6 Paired-Sample Testing of Nominal Scale Data 156 Exercises 160

11 Multisample Hypotheses: The Analysis of Variance 162

11.1 Single Factor Analysis of Variance 163
11.2 Confidence Limits for Means 170
11.3 Power and Sample Size in Analysis of Variance 171
11.4 Nonparametric Analysis of Variance 176
11.5 Testing for Difference among Several Medians 179
11.6 The Effect of Coding 180
11.7 Multisample Testing for Nominal Scale Data 181
11.8 Homogeneity of Variances 181 Exercises 183

12 Multiple Comparisons 185

12.1 The Tukey Test 186 **12.2** The Newman-Keuls Test 190
12.3 Confidence Intervals Following Multiple Comparisons 191
12.4 Comparison of a Control Mean to Each Other Group Mean 194
12.5 Scheffé's Multiple Contrasts 196
12.6 Nonparametric Multiple Comparisons 199
12.7 Nonparametric Multiple Contrasts 201
12.8 Multiple Comparisons among Medians 202
12.9 Multiple Comparisons among Variances 203 Exercises 205

13 Two-Factor Analysis of Variance 206

13.1 Two-Factor Analysis of Variance with Equal Replication 206
13.2 Two-Factor Analysis of Variance with Unequal Replication 214
13.3 Two-Factor Analysis of Variance without Replication 217
13.4 Nonparametric Two-Factor Analysis of Variance 219
13.5 The Randomized Block Experimental Design 222
13.6 Multiple Comparisons and Confidence Intervals 226
13.7 Power and Sample Size in Two-Factor Analysis of Variance 227
13.8 Nonparametric Randomized Block Analysis of Variance 228
13.9 Multiple Comparisons for Nonparametric Randomized Block Analysis of Variance 230
13.10 Dichotomous Nominal Scale Data in Randomized Blocks 231
13.11 Multiple Comparisons with Dichotomous Randomized Block Data 233
Exercises 234

14 Data Transformations 236

14.1 The Logarithmic Transformation 238
14.2 The Arcsine Transformation 239
14.3 The Square Root Transformation 241
14.4 Other Transformations 242
Exercises 243

15 Multiway Factorial Analysis of Variance 244

15.1 Three-Factor Analysis of Variance 245
15.2 The Latin Square Experimental Design 248
15.3 Higher Order Factorial Analysis of Variance 248
15.4 Nonparametric Multiway Factorial Analysis of Variance 249
15.5 Factorial Analysis of Variance with Unequal Replication 250
15.6 Multiple Comparisons and Confidence Intervals in Multiway Analysis of Variance 251
15.7 Power and Sample Size in Multiway Analysis of Variance 251
Exercises 252

16 Nested (Hierarchical) Analysis of Variance 253

16.1 Nesting within One Main Factor 255
16.2 Nesting in Factorial Experiments 258
16.3 Multiple Comparisons and Confidence Intervals 259
16.4 Power and Sample Size in Nested Analysis of Variance 260
Exercise 260

17 Simple Linear Regression 261

17.1 Regression vs. Correlation 261
17.2 The Simple Linear Regression Equation 263
17.3 Testing the Significance of a Regression 268
17.4 Confidence Intervals in Regression 272
17.5 Inverse Prediction 276

17.6 Interpretations of Regression Functions 278
17.7 Regression with Replication: Testing for Linearity 278
17.8 Power and Sample Size in Regression 284
17.9 Regression through the Origin 284
17.10 Data Transformations in Regression 286 **17.11** The Effect of Coding 289
Exercises 290

18 Comparing Simple Linear Regression Equations 292

18.1 Comparing Two Slopes 292 **18.2** Comparing Two Elevations 295
18.3 Comparing Points of Two Regression Lines 299
18.4 Comparing More than Two Slopes 300
18.5 Comparing More than Two Elevations 302
18.6 Multiple Comparisons among Slopes 302
18.7 Multiple Comparisons among Elevations 303
18.8 An Overall Test for Coincidental Regressions 304 Exercises 304

19 Simple Linear Correlation 306

19.1 The Correlation Coefficient 306
19.2 Hypotheses about the Correlation Coefficient 309
19.3 Confidence Intervals for the Correlation Coefficient 311
19.4 Power and Sample Size in Correlation 312
19.5 Comparing Two Correlation Coefficients 313
19.6 Power and Sample Size in Comparing Two Correlation Coefficients 314
19.7 Comparing More than Two Correlation Coefficients 315
19.8 Multiple Comparisons among Correlation Coefficients 316
19.9 Rank Correlation 318
19.10 Correlation for Dichotomous Nominal Scale Data 321
19.11 Intraclass Correlation 323 **19.12** The Effect of Coding 325 Exercises 326

20 Multiple Regression and Correlation 328

20.1 Intermediate Computational Steps 329
20.2 The Multiple Regression Equation 333
20.3 Analysis of Variance of Multiple Regression or Correlation 335
20.4 Hypotheses Concerning Partial Regression Coefficients 337
20.5 Standardized Partial Regression Coefficients 338 **20.6** Partial Correlation 339
20.7 Round-Off Error and Coding Data 340
20.8 Selection of Independent Variables 341 **20.9** Predicting *Y* Values 344
20.10 Testing Differences between Two Partial Regression Coefficients 346
20.11 "Dummy" Variables 346
20.12 Comparing Multiple Regression Equations 347
20.13 Multiple Regression through the Origin 349
20.14 Nonlinear Regression 349
20.15 Descriptive vs. Predictive Models 351
20.16 Concordance: Rank Correlation among Several Variables 352
20.17 Concordance between Groups 357 Exercises 359

21 Polynomial Regression 361

21.1 Polynomial Curve Fitting 361 **21.2** Round-Off Error and Coding Data 365
21.3 Quadratic Regression 366 Exercises 368

22 The Binomial Distribution 369

22.1 Binomial Probabilities 370 **22.2** Sampling a Binomial Population 375
22.3 Confidence Limits for Proportions 378
22.4 Goodness of Fit for the Binomial Distribution 380 **22.5** The Binomial Test 383
22.6 The Sign Test 386 **22.7** Power of the Binomial and Sign Tests 387
22.8 Confidence Interval for the Median 390 **22.9** The Fisher Exact Test 390
22.10 Comparing Two Proportions 395
22.11 Power and Sample Size in Comparing Two Proportions 397
22.12 Comparing More than Two Proportions 400
22.13 Multiple Comparisons for Proportions 401 Exercises 403

23 The Poisson Distribution and Randomness 406

23.1 Poisson Probabilities 406
23.2 Confidence Limits for the Poisson Parameter 408
23.3 Goodness of Fit of the Poisson Distribution 409
23.4 The Binomial Test Revisited 411 **23.5** Comparing Two Poisson Counts 415
23.6 Serial Randomness of Nominal Scale Categories 416
23.7 Serial Randomness of Measurements: Parametric Testing 418
23.8 Serial Randomness of Measurements: Nonparametric Testing 419
Exercises 420

24 Circular Distributions: Descriptive Statistics 422

24.1 Data on a Circular Scale 422 **24.2** Graphical Presentation of Data 424
24.3 Sines and Cosines of Circular Data 426 **24.4** The Mean Angle 428
24.5 Angular Dispersion 430 **24.6** The Median and Modal Angles 432
24.7 Confidence Limits for the Mean and Median Angles 432
24.8 Diametrically Bimodal Distributions 432
24.9 Second-Order Analysis: The Mean of Mean Angles 434
24.10 Confidence Limits for the Second-Order Mean Angle 436 Exercises 439

25 Circular Distributions: Hypothesis Testing 440

25.1 Goodness of Fit Testing 440
25.2 The Significance of the Mean and Median Angles 442
25.3 Parametric Two-Sample and Multisample Testing of Angles 446
25.4 Nonparametric Two-Sample and Multisample Testing of Angles 450
25.5 Two-Sample and Multisample Testing of Angular Distances 452
25.6 Two-Sample and Multisample Testing of Angular Dispersion 454
25.7 Parametric One-Sample Second-Order Analysis of Angles 455
25.8 Nonparametric One-Sample Second-Order Analysis of Angles 456

25.9 Parametric Two-Sample Second-Order Analysis of Angles 457
25.10 Nonparametric Two-Sample Second-Order Analysis of Angles 459
25.11 Parametric Paired-Sample Testing with Angles 461
25.12 Nonparametric Paired-Sample Testing with Angles 463
25.13 Angular Correlation and Regression 463
25.14 Serial Randomness of Nominal Scale Categories on a Circle 465 Exercises 467

APPENDIX A Analysis of Variance Hypothesis Testing 470

A.1 Determination of Appropriate F's and Degrees of Freedom 470
A.2 Two-Factor Analysis of Variance 473
A.3 Three-Factor Analysis of Variance 474
A.4 Nested Analysis of Variance 474

APPENDIX B Statistical Tables and Graphs 477

Interpolation 477
Table B.1 Critical Values of the Chi-Square Distribution 479
Table B.2 Proportions of the Normal Curve (One-Tailed) 483
Table B.3 Critical Values of the t Distribution 484
Table B.4 Critical Values of the F Distribution 486
Table B.5 Critical Values of the q Distribution 522
Table B.6 Critical Values of q' for the One-Tailed Dunnett's Test 538
Table B.7 Critical Values of q' for the Two-Tailed Dunnett's Test 539
Table B.8 Critical Values of d_{max} for the Kolmogorov-Smirnov Goodness of Fit for Discrete or Grouped Data 540
Table B.9 Critical Values of D for the Kolmogorov-Smirnov Goodness of Fit for Continuous Distributions 546
Table B.10 Critical Values of the Mann-Whitney U Distribution 550
Table B.11 Critical Values of the Wilcoxon T Distribution 563
Table B.12 Critical Values of the Kruskal-Wallis H Distribution 565
Table B.13 Critical Values of the Friedman χ_r^2 Distribution 567
Table B.14 Critical Values of Q for Nonparametric Multiple Comparison Testing 568
Table B.15 Critical Values of Q' for Nonparametric Multiple Comparison Testing with a Control 569
Table B.16 Critical Values of the Correlation Coefficient, r 570
Table B.17 Fisher's z Transformation for Correlation Coefficients, r 572
Table B.18 Correlation Coefficients, r, Corresponding to Fisher's z Transformation 574
Table B.19 Critical Values of the Spearman Rank Correlation Coefficient, r_s 577
Table B.20 Critical Values of g_1 579
Table B.21 Critical Values of D for the Kolmogorov-Smirnov Goodness of Fit of a Normal Distribution 583
Table B.22 Critical Values of D'Agostino's D for Normality Testing 585
Table B.23 The Arcsine Transformation 586
Table B.24 Proportions Corresponding to Arcsine Transformations, X' 589
Table B.25 Proportions of the Binomial Distribution for $p = q = 0.5$ 591
Table B.26 Critical Values of C for the Sign Test or the Binomial Test with $p = 0.5$ 592
Table B.27 Critical Values for Fisher's Exact Test 602
Table B.28 Critical Values for the Runs Test 627

Table B.29 Critical Values of C for the Mean Square Successive Difference Test 634
Table B.30 Angular Deviation, s, as a Function of Vector Length, r 636
Table B.31 Circular Standard Deviation, s', as a Function of r 638
Table B.32 Critical Values of Rayleigh's z 640
Table B.33 Critical Values of u for the V Test of Circular Uniformity 641
Table B.34 Correction Factor, K, for the Watson and Williams Test 642
Table B.35 Critical Values of Watson's U^2 644
Table B.36 Critical Values of R' for the Moore Test of Circular Uniformity 647
Table B.37 Critical Values for the Runs Test on a Circle 648
Table B.38 Common Logarithms of Factorials 652
Table B.39 Ten Thousand Random Digits 653

Figure B.1 Power and Sample Size in Analysis of Variance 657
Figure B.2 Confidence Limits for the Mean Angle 665

Answers to Exercises 667

Literature Cited 680

Index 697

Preface

The majority of contemporary biological investigations require a basic appreciation and knowledge of statistical techniques. This is becoming increasingly apparent to biological researchers, journal editors, and college curriculum planners. Owing to the magnificent diversity of scientific endeavors that can be found among the biological sciences, this book presents a broad collection of data analysis techniques, which will address the statistical needs of the great majority of contemporary biological researchers.

A book of this type is called upon to fulfill a dual purpose. First, it serves as an introductory textbook, assuming no prior knowledge of statistics and covering a sufficient enough variety of concepts and procedures to satisfy a large portion of the biological disciplines that require statistical analysis. Secondly, it is expected to serve as a reference work that will be repeatedly consulted long after formal instruction has ended.

The material in this book requires no mathematical competence beyond elementary algebra, although the discussions include some topics that appear seldom, if at all, in other general texts. Also, cognizance is taken of the increasing use of digital computer capability in academic and industrial biological institutions of all sizes. There are statistical procedures that are of importance but which involve computations so demanding that they practically, if not actually, preclude noncomputer execution. The principles of some of these are presented with the assumption that computer programs will perform the laborious computations, but with the realization that the biologist must enter into the interpretation of the results of the computer's calculations.

A final attempt at achieving a book with self-sufficiency for most biostatistical needs is the inclusion of a thorough set of statistical tables, the majority of which are more extensive than those found in other introductory or advanced texts, including many that are not found at all in other general texts.

A book of this nature requires and benefits from the assistance of many people. I am indebted to the library services of Northern Illinois University and the University of Illinois. I also gratefully acknowledge the cooperation of the computer services people at Northern Illinois University who assisted in running the many author-prepared computer programs used to generate the statistical appendix tables. For the tables taken from previously published sources, thanks are here offered for the permission given to reproduce them; full acknowledgement of each source is found immediately following the appearance of the reprinted material in the text.

Over the years, my teachers, students, and colleagues have aided in guiding me to the material that is presented in this volume. Available space precludes mention of all those providing input to this writing endeavor, but special note must be made of Edward Batschelet, University of Zurich, who, with enthusiasm, patience, and kindness, provided me with encouragement and inspiration from the time of manuscript reading for the first edition until the time of his death shortly before the manuscript for this second edition was completed. Finally, I acknowledge my wife, Carol, and my sons, David and Adam, for their prolonged patience during the preparation of both editions of this book.

J.H.Z.

DeKalb, Illinois

Biostatistical Analysis

1 Introduction

Many of the investigations in biological sciences have become quantitative, in that a great many types of biological observations consist of numerical facts called *data*. (One numerical fact is a *datum*.) As biological entities are counted or measured, it becomes apparent that some objective methods are necessary to aid the investigator in presenting and analyzing research data.

The word "statistics" is derived from the Latin for "state," indicating the historical importance of governmental data gathering, which related principally to census taking and tax collecting. The layman uses this term as a synonym for "data"; one hears of college enrollment statistics (how many freshman students, how many students from each state, etc.), statistics of a football game (how many first downs, how many completed passes, etc.), labor statistics (numbers unemployed, numbers employed in various occupations), and so on. Hereafter, this use of the word "statistics" will not appear in this book. Instead, "statistics" will be used to refer to the analysis and interpretation of data with a view toward objective evaluation of the reliability of the conclusions based on the data. Statistics applied to biological problems is simply called *biostatistics* or, sometimes, *biometry* (the latter term literally meaning "biological measurement").

Before data can be analyzed, they must be collected, and here statistical considerations can aid in the design of experiments and in the setting up of hypotheses to be tested. Many are the biologists who will attempt the analysis of their research data only to find that too few data were collected to enable reliable conclusions to be drawn, or that much extra effort was expended in collecting data that cannot be of ready aid in the analysis of the experiment. Thus, a knowledge of basic statistical principles and procedures is important even before an experiment is begun.

Once the data have been obtained, we may organize and summarize them in such a way as to arrive at their orderly presentation. Such procedures are often termed *descriptive statistics*. For example, a tabulation might be made of the heights of all members of a freshman English class, indicating an average height for each sex, or for each age. However, it might be desired to make some generalizations from these data. We might, for example, wish to make a reasonable estimate of the heights of all freshmen in the university. Or we might wish to conclude whether the males in the university are on the average taller than the females. The ability to make such generalized conclusions, inferring characteristics of the whole from characteristics of its parts, lies within the realm of *inferential statistics*.

1.1 TYPES OF BIOLOGICAL DATA

A characteristic that varies from one biological entity to another is termed a *variable* (or *variate*). Different sorts of variables may be encountered by biologists, and it is desirable to be able to distinguish among them. The classification used here is that which is standardly employed (Senders, 1958; Siegel, 1956; Stevens, 1946, 1968).

Data on a Ratio Scale. Consider that the heights of a group of plants constitute a variable of interest, and perhaps the number of leaves per plant is another variable under consideration. Thanks to measuring devices at the biologist's disposal, it is possible to assign a numerical value to the height of each plant, and simply counting the leaves allows a numerical value to be assigned to the number of leaves on each plant. Regardless of whether the height measurements are recorded in centimeters, inches, or any other units, and regardless of whether the leaves are counted in a number system using base 10 or any other base, there are two fundamentally important characteristics of these data.

First, there is a constant size interval between any adjacent units on the measurement scale. That is, the difference in height between a 36-cm and a 37-cm plant is the same as the difference between a 39-cm and a 40-cm plant, and the difference between 8 and 10 leaves is equal to the difference between 9 and 11 leaves. (This may seem simple minded, but it is very important, as we shall see on examining the other scales of measurement.)

Second, it is important that there exists a zero point on the measurement scale and that there is a physical significance to this zero. This enables us to say something meaningful about the ratio of measurements. We can say that a 30-cm (11.8-in.) tall plant is half as tall as a 60-cm (23.6-in.) plant, and that a plant with 45 leaves has 3 times as many leaves as a plant with 15.

Measurement scales having a constant interval size and a true zero point are said to be *ratio scales* of measurement. Besides lengths and numbers of items, ratio scales include weights (mg, lb, etc.), volumes (cc, cu ft, etc.), capacities (ml, qt, etc.), rates (cm/sec, mph, mg/min, etc.), and lengths of time (hr, yr, etc.).

Data on an Interval Scale. Some measurement scales possess a constant interval size but not a true zero; they are called *interval scales*. An outstanding example is that of the two common temperature scales: Celsius (C) and Fahrenheit (F). We can see that the same difference exists between 20°C (68°F) and 25°C (77°F) as between 5°C (41°F) and 10°C (50°F); i.e., the measurement scale is composed of equal-sized intervals. But it cannot be said that a temperature of 40°C (104°F) is twice as hot as a temperature of 20°C (68°F); i.e., the zero point is arbitrary. [Temperature measurements on the absolute, or Kelvin (K), scale can be referred to a physically meaningful zero and thus constitute a ratio scale.]

Some interval scales encountered in biological data collection are *circular scales*. Time of day and time of the year are examples of such scales. The interval between 2:00 PM (i.e., 1400 hr) and 3:30 PM (1530 hr) is the same as the interval between 8:00 AM (0800 hr) and 9:30 AM (0930 hr). But one cannot speak of ratios of times of day because the zero point (midnight) on the scale is arbitrary, in that one could just as well set up a scale for time of day which would have noon, or 3:00 PM, or any other time as the zero point. Circular biological data are occasionally compass points, as if one records the compass direction in which an animal or plant is oriented. Since the designation of north as 0° is arbitrary, this circular scale is a form of interval scale of measurement. Some special statistical procedures are available for circular data; these are discussed in Chapters 24 and 25.

Data on an Ordinal Scale. The preceding paragraphs on ratio and interval scales of measurement discussed data between which we know numerical differences. For example, if man *A* weighs 70 kg and man *B* weighs 80 kg, then man *A* is known to weigh 10 kg more than *B*. But our data may, instead, be a record only of the fact that man *A* weighs more than man *B* (with no indication of how much more). Thus, we may be dealing with relative differences rather than with quantitative differences. Such data consist of an ordering or ranking of measurements and are said to be on an *ordinal* scale of measurement ("ordinal" being from the Latin word for "order"). One may speak of one biological entity being shorter, darker, faster, or more active than another; the sizes of five cell types might be labelled 1, 2, 3, 4, and 5, to denote their magnitudes relative to each other; or success in learning to run a maze may be recorded as *A*, *B*, or *C*.

It is often true that biological data expressed on the ordinal scale could have been expressed on the interval or ratio scale had exact measurements been obtained (or obtainable). Sometimes data that were originally on interval or ratio scales will be changed to ranks; for example, examination grades of 99, 85, 73, and 66% (ratio scale) might be recorded as A, B, C, and D (ordinal scale), respectively.

Ordinal scale data contain and convey less information than ratio or interval data, for only relative magnitudes are known. Consequently, quantitative comparisons are impossible (e.g., we cannot speak of a grade of C being half as good as a grade of A, or of the difference between cell sizes 1 and 2 being the same as the difference between sizes 3 and 4). However, we will see that a great many statistical procedures are, in fact, applicable to ordinal data.

Data on a Nominal Scale. Sometimes the variable under study is classified by some quality it possesses rather than by a numerical measurement. In such cases the variable is called an *attribute*, and we are said to be using a *nominal scale* of measurement. Genetic phenotypes are commonly encountered biological attributes; the possible manifestation of an animal's eye color may be blue or brown, and if human hair color were the attribute of interest, we might record black, brown, blonde, or red. On a nominal scale ("nominal" is from the Latin word for "name"), animals might be classified as male or female, or as left- or right-handed. Or plants might be classified as dead or alive, or as with or without thorns. Taxonomic categories also form a nominal classification scheme (e.g., a plant might be classified as pine, spruce, or fir). Sometimes data from an ordinal, interval, or ratio scale of measurement may be recorded in nominal scale categories. For example, heights may be recorded as tall or short, or performances on an examination as pass or fail.

As will be seen, statistical methods useful with ratio, interval, or ordinal data generally are not applicable to nominal data, and we must, therefore, be able to identify such situations when they occur.

Continuous and Discrete Data. When we spoke above of plant heights, we were dealing with a variable that could be any conceivable value within any observed range; this is referred to as a *continuous variable*. That is, if we measure a height of 35 cm and a height of 36 cm, an infinite number of heights is possible in the range from 35 to 36 cm: a plant might be 35.07 cm tall or 35.988 cm tall, or 35.3263 cm tall, etc., although, of course, we do not have devices sensitive enough to detect this infinity of heights. A continuous variable is one for which there is a possible value between any other two possible values.

However, when speaking of the number of leaves on a plant, we are dealing with a variable that can take on only certain values. It might be possible to observe 27 leaves, or 28 leaves, but 27.43 leaves and 27.9 leaves are values of the variable that are impossible to obtain. Such a variable is termed a *discrete* or *discontinuous variable* (also known as a *meristic variable*). The number of white blood cells in 1 mm^3 of blood, the number of giraffes visiting a water hole, and the number of eggs laid by a grasshopper are all discrete variables. The possible values of a discrete variable generally are consecutive integers, but this is not necessarily so. If the leaves on our plants are always formed in pairs, then only even integers are possible values of the variable. And the ratio of number of wings to number of legs of insects is a discrete variable that may only have the value of 0, 0.3333. . . , or 0.6666 . . . (i.e., $\frac{0}{6}$, $\frac{2}{6}$, or $\frac{4}{6}$, respectively).*

1.2 ACCURACY AND SIGNIFICANT FIGURES

Accuracy is the nearness of a measurement to the actual value of the variable being measured. *Precision* is not a synonymous term, but refers to the closeness to each other of repeated measurements of the same quantity.

If we report that the hind leg of a frog is 8 cm long, we are stating the number 8 (a value of a continuous variable) as an estimate of the frog's true leg length. This estimate was made using some sort of a measuring device. Had the device been capable of more accuracy, we might have concluded that the leg was 8.3 cm long, or perhaps

*The ellipses (. . .) may be read as "and so on." Here, they indicate that $\frac{2}{6}$ and $\frac{4}{6}$ are repeating decimal fractions, which could just as well have been written as 0.3333333333333 . . . and 0.6666666666666 . . . , respectively.

8.32 cm long. When recording values of continuous variables, it is important to designate the accuracy with which the measurements have been made. By convention, the value 8 denotes a measurement in the range of 7.50000 . . . to 8.49999 . . . , the value 8.3 designates a range of 8.25000 . . . to 8.34999 . . . , and the value 8.32 implies that the true value lies within the range of 8.31500 . . . to 8.32499 . . . . That is, the reported value is the midpoint of the implied range, and the size of this range is designated by the last decimal place in the measurement. The value of 8 cm implies a range of accuracy of 1 cm, 8.3 cm implies a range of 0.1 cm, and 8.32 cm implies a range of 0.01 cm. Thus, to record a value of 8.0 implies greater accuracy of measurement than does the recording of a value of 8, for in the first instance the true value is said to lie between 7.95000 . . . and 8.049999 . . . (i.e., within a range of 0.1 cm), whereas 8 implies a value between 7.50000 . . . and 8.49999 . . . (i.e., within a range of 1 cm). To state 8.00 cm implies an accuracy in measurement which ascertains the frog's limb length to be between 7.99500 . . . and 8.00499 . . . cm (i.e., within a range of 0.01 cm). Those digits in a number that denote the accuracy of the measurement are referred to as *significant figures*. Thus, 8 has one significant figure, 8.0 and 8.3 each have two significant figures, and 8.00 and 8.32 each have three.

In working with exact values of discrete variables, the preceding considerations do not apply. That is, it is sufficient to state that our frog has 4 limbs or that its left lung contains 13 flukes. The use of 4.0 or 13.00 would be inappropriate, for since the numbers involved are exactly 4 and 13, there is no question of accuracy or significant figures.

But there are instances where significant figures and implied accuracy come into play with discrete data. An entomologist may report that there are 72,000 moths in a particular forest area. In doing so, it is probably not being claimed that this is the exact number but an estimate of the exact number, perhaps accurate to 2 significant figures. In such a case, 72,000 would imply a range of accuracy of 1000, so that the true value might lie anywhere from 71,500 to 72,500. If the entomologist wished to convey the fact that this estimate is believed to be accurate to the nearest 100 (i.e., to 3 significant figures), rather than to the nearest 1000, he had better present his data in the form of *scientific notation*, as follows: If the number 7.2×10^4 (= 72,000) is written, a range of accuracy of 0.1×10^4 (= 1000) is implied, and the true value is assumed to lie between 71,500 and 72,500. But if 7.20×10^4 were written, a range of accuracy of 0.01×10^4 (= 100) would be implied, and the true value would be assumed to be in the range of 71,950 to 72,050. Thus, the accuracy of large values (and this applies to continuous as well as discrete variables) can be expressed succinctly using scientific notation.

1.3 FREQUENCY DISTRIBUTIONS

When collecting and summarizing large amounts of data, it is often helpful to record the data in the form of a *frequency table*. Such a table simply involves a listing of all the observed values of the variable being studied and how many times each value is observed. Consider the tabulation of the frequency of occurrence of sparrow nests in each of several different locations. This is illustrated in Example 1.1, where the observed nest sites are listed, and for each site the number of nests observed is

Example 1.1

The location of sparrow nests. A frequency table of nominal data.

Nest site	*Number of nests observed*
A. Vines	56
B. Building eaves	60
C. Low tree branches	46
D. Tree and building cavities	49

recorded. The distribution of the total number of observations among the various categories is termed a *frequency distribution*. Example 1.1 is a frequency table for nominal data, and these data may also be presented graphically by means of a *bar graph* (Fig. 1.1), where the height of each bar is proportional to the frequency in the class represented. The widths of all bars in a bar graph should be equal so that the

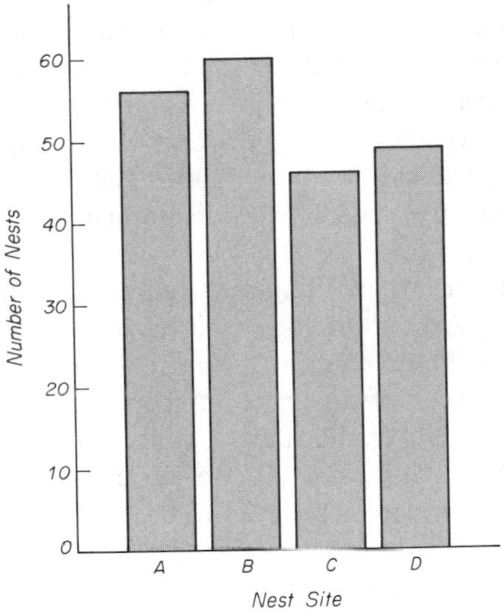

Figure 1.1 A bar graph of the sparrow nest data of Example 1.1. An example of a bar graph for nominal data.

eye of the reader is not distracted from the differences in bar heights; this also makes the area of each bar proportional to the frequency it represents.

A frequency tabulation of ordinal data might appear as in Example 1.2, which

Example 1.2

Numbers of sunfish, tabulated according to amount of black pigmentation. A frequency table of ordinal data.

Pigmentation class	*Amount of pigmentation*	*Number of fish*
0	No black pigmentation	13
1	Faintly speckled	68
2	Moderately speckled	44
3	Heavily speckled	21
4	Solid black pigmentation	8

presents the observed numbers of sunfish collected in each of five categories, each category being a degree of skin pigmentation. A bar graph (Fig. 1.2) can be prepared for this frequency distribution just as for nominal data.

In preparing frequency tables of interval and ratio scale data, we make a procedural distinction between discrete and continuous data. Example 1.3 shows discrete data which are frequencies of litter sizes in foxes, and Fig. 1.3 presents this frequency distribution graphically.

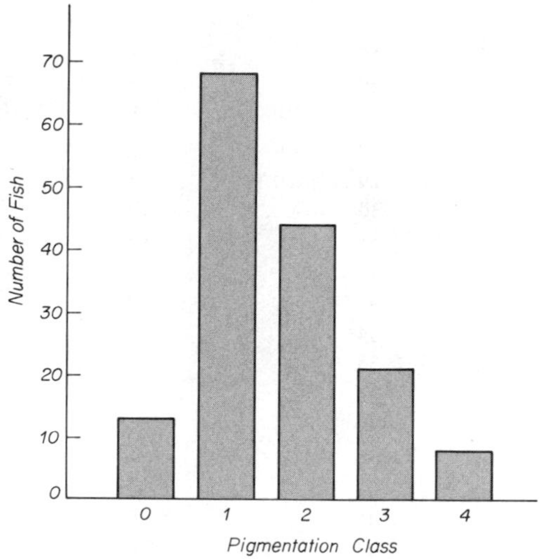

Figure 1.2 A bar graph of the sunfish pigmentation data of Example 1.2. An example of a bar graph for ordinal data.

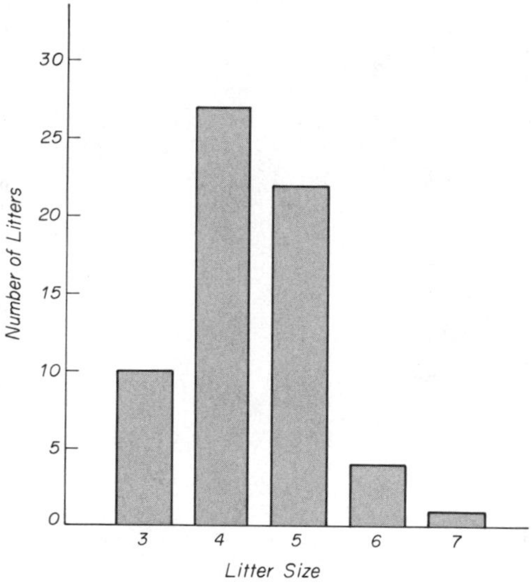

Figure 1.3 A bar graph of the fox litter data of Example 1.3. An example of a bar graph for discrete, ratio scale, data.

Example 1.3

Frequency of occurrence of various litter sizes in foxes. A frequency table of discrete data.

Litter size	*Frequency*
3	10
4	27
5	22
6	4
7	1

Example 1.4a shows discrete data which are the numbers of aphids found per clover plant. These data create quite a lengthy frequency table, and it is not difficult to imagine sets of data whose tabulation would result in an even longer list of frequencies. Thus, for purposes of preparing bar graphs, we often cast data into a frequency table by grouping them.

Example 1.4b is a table of the data from Example 1.4a, arranged by grouping the data into size classes. The bar graph for this distribution appears as Fig. 1.4. Such grouping results in the loss of some information and is generally utilized only to make frequency tables and bar graphs easier to read, and not for calculations performed

Example 1.4a

Number of aphids observed per clover plant. A frequency table of discrete data.

Number of aphids on a plant	*Number of plants observed*	*Number of aphids on a plant*	*Number of plants observed*
0	3	20	17
1	1	21	18
2	1	22	23
3	1	23	17
4	2	24	19
5	3	25	18
6	5	26	19
7	7	27	21
8	8	28	18
9	11	29	13
10	10	30	10
11	11	31	14
12	13	32	9
13	12	33	10
14	16	34	8
15	13	35	5
16	14	36	4
17	16	37	1
18	15	38	2
19	14	39	1
		40	0
		41	1

Total number of observations = 424

Example 1.4b

Number of aphids observed per clover plant. A frequency table grouping the data of Example 1.4a.

Number of aphids on a plant	*Number of plants observed*
0–3	6
4–7	17
8–11	40
12–15	54
16–19	59
20–23	75
24–27	77
28–31	55
32–35	32
36–39	8
40–43	1

Total number of observations = 424

on the data. There have been several "rules of thumb" proposed to aid in deciding into how many classes data might reasonably be grouped, for the use of too few groups will obscure the general shape of the distribution. But such "rules" or recom-

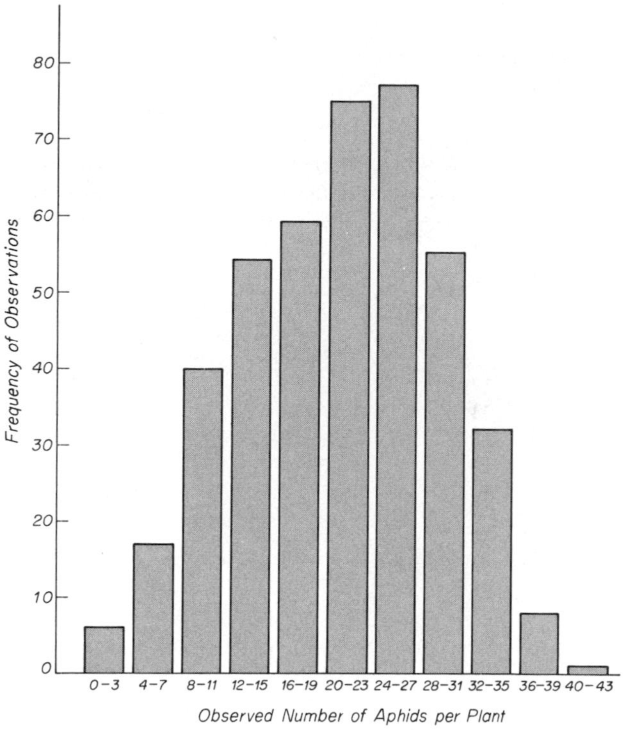

Figure 1.4 A bar graph of the aphid data of Example 1.4b. An example of a bar graph for grouped discrete data.

mendations are only rough guides, and the choice is generally left to good judgment, bearing in mind that from 10 to 20 groups are useful for most biological work. (See also Doane, 1976.) In general, groups should be established that are equal in the size interval of the variable being measured. (For example, the group size interval in Example 1.4b is 4 aphids per plant.)

Because continuous data, contrary to discrete data, can take on an infinity of values, one is essentially always dealing with a frequency distribution tabulated by groups. If the variable of interest were a weight, measured to the nearest 0.1 mg, a frequency table entry of the number of weights measured to be 48.6 mg would be interpreted to mean the number of weights grouped between 48.5500 . . . and 48.6499 . . . mg (although in a frequency table this class interval is usually written as 48.55–48.65). Example 1.5 presents a tabulation of 130 determinations of the amount of phosphorus, in milligrams per gram of dried leaf. (Ignore the last two columns of this table until Section 1.4).

In presenting this frequency distribution graphically, one can prepare a *histogram*, which is the name given to a bar graph based on continuous data. This is done in Fig. 1.5a; note that rather than indicating the range on the horizontal axis, we indicate only the midpoint of the range, a procedure that results in less crowded printing on the graph. Note also that adjacent bars in a histogram are often drawn touching each other, whereas in the other bar graphs discussed they generally are not.

Often a *frequency polygon* is drawn instead of a histogram. This is done by plotting the frequency of each class as a dot at the class midpoint and then connecting each adjacent pair of dots by a straight line (Fig. 1.5b). It is, of course, the same as

Example 1.5

Determinations of the amount of phosphorus in leaves. A frequency table of continuous data.

Phosphorus (mg/g of leaf)	*Frequency (i.e., number of determinations)*	**Cumulative frequency** *Starting with low values*	*Starting with high values*
8.15–8.25	2	2	130
8.25–8.35	6	8	128
8.35–8.45	8	16	122
8.45–8.55	11	27	114
8.55–8.65	17	44	103
8.65–8.75	17	61	86
8.75–8.85	24	85	69
8.85–8.95	18	103	45
8.95–9.05	13	116	27
9.05–9.15	10	126	14
9.15–9.25	4	130	4

Total frequency = 130

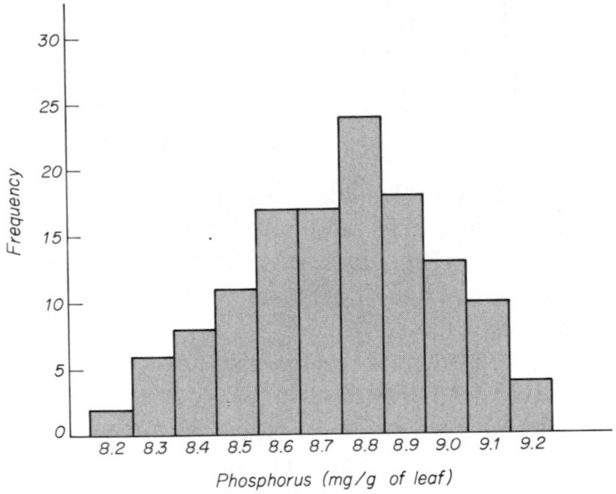

Figure 1.5a A histogram of the leaf phosphorus data of Example 1.5. An example of a histogram for continuous data.

if the midpoints of the tops of the histogram bars were connected by straight lines. Instead of plotting frequencies on the vertical axis, one can plot *relative frequencies*, or proportions of the total frequency. This enables different distributions to be readily compared and even plotted on the same axes. Sometimes, as in Fig. 1.5b, frequency is indicated on one vertical axis and the corresponding relative frequency on the other.

Frequency polygons are also commonly used for discrete distributions, but one can argue against their use when dealing with ordinal data, as the polygon implies to the reader a constant size interval horizontally between points on the polygon. Frequency polygons should not be employed for nominal scale data.

If we have a frequency distribution of values of a continuous variable that falls into a large number of class intervals, the data may be grouped as was demonstrated with discrete variables. This results in fewer intervals, but each interval is, of course,

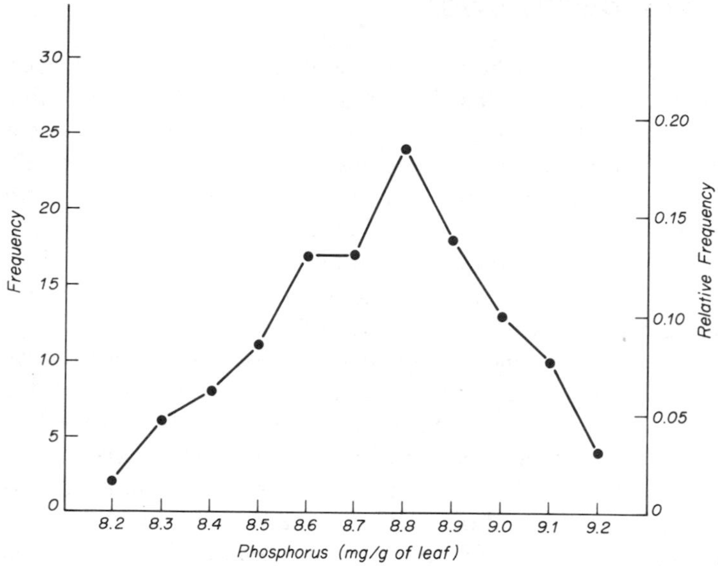

Figure 1.5b A frequency polygon for the leaf phosphorus data of Example 1.5.

larger. The midpoints of these intervals may then be used in the preparation of a histogram or frequency polygon. The user of frequency polygons is cautioned that such a graph is simply an aid to the eye in following trends in frequency distributions, and one should not attempt to read frequencies between points on the polygon. Also note that the method presented for the construction of histograms and frequency polygons requires that the class intervals be equal. Lastly, the vertical axis (i.e., the frequency scale) on frequency polygons and bar graphs generally should begin with zero, especially if graphs are to be compared with one another. If this is not done, the eye may be misled by the appearance of the graph. If, for example, a bar graph of the data of Example 1.1 were constructed the same size as Fig. 1.1, but with the vertical axis representing frequencies of 45 to 60 rather than 0 to 60, the results would appear as in Fig. 1.6. Huff (1954) illustrates other techniques that can mislead the readers of graphs.

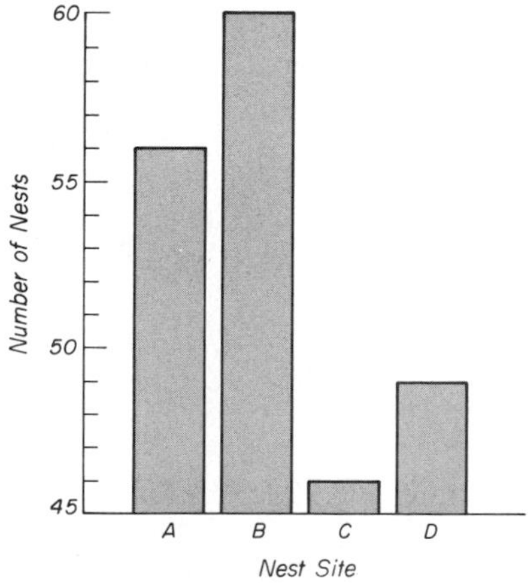

Figure 1.6 A bar graph of the sparrow nest data of Example 1.1, drawn with the vertical axis starting at 45. Compare this with Fig. 1.1, where the axis starts at 0.

1.4 CUMULATIVE FREQUENCY DISTRIBUTIONS

A frequency distribution informs us of how many observations occurred for each value (or group of values) of a variable. That is, examination of the frequency table of Example 1.3 (or its corresponding bar graph or frequency polygon) would yield information such as "how many fox litters of 4 were observed?", the answer being 27. But if it is desired to ask questions such as "how many litters of 4 or more were observed?", or "how many fox litters of 5 or less were observed?", we are speaking of *cumulative frequencies*. To answer the first question, we sum all frequencies for litter sizes 4 and up, and for the second question we sum all frequencies from the smallest litter size up through a size of 5. We arrive at answers of 54 and 59, respectively.

In Example 1.5, the phosphorus concentration data are cast into two cumulative frequency distributions, one with cumulation commencing at the low end of the measurement scale and one with cumulation being performed from the high values toward the low values. The choice of the direction of cumulation is immaterial, as can be demonstrated. If one desired to calculate the number of phosphorus determinations less than 8.55 mg/g, namely 27, a cumulation starting at the low end might be used, whereas the knowledge of the frequency of determinations greater than 8.55 mg/g, namely 103, can be readily obtained from the cumulation commencing from the high end of the scale. But one can easily calculate any frequency from a low-to-high cumulation (e.g., 27) from its complementary frequency from a high-to-low cumulation (e.g., 103), simply by knowing that the sum of these two frequencies is the total frequency (i.e., 130); therefore, in practice it is not necessary to calculate both sets of cumulations.

Cumulative frequency distributions are useful in determining medians, percentiles, and other quantiles, as discussed in Section 3.2. They are not often presented

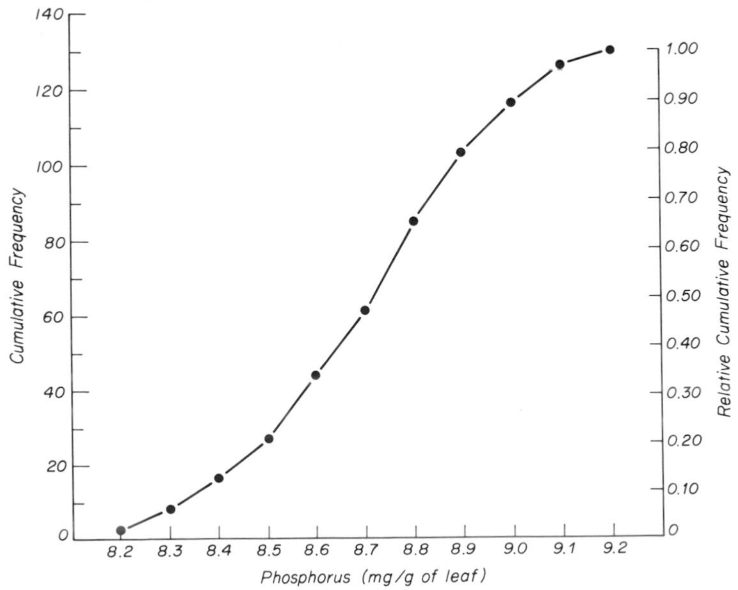

Figure 1.7 Cumulative frequency polygon of the leaf phosphorus data of Example 1.5, with cumulation commencing from the lowest to the highest values of the variable.

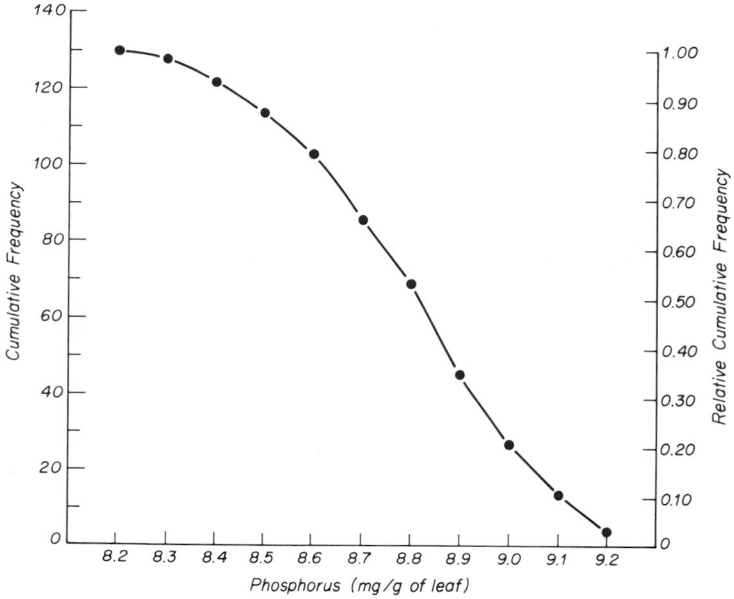

Figure 1.8 Cumulative frequency polygon of the leaf phosphorus data of Example 1.5, with cumulation commencing from the highest to the lowest values of the variable.

in bar graphs, but *cumulative frequency polygons* (sometimes called *ogives*) are not uncommon. (See Figs. 1.7 and 1.8.) Relative frequencies (proportions of the total frequency) can be plotted instead of (or, as in Figs. 1.7 and 1.8, in addition to) frequencies on the vertical axis of a cumulative frequency polygon. This enables different distributions to be readily compared and even plotted on the same axes.

2 Populations and Samples

The primary objective of statistical analysis is to infer characteristics of a group of data by analyzing the characteristics of a small sampling of the group. This generalization from the part to the whole requires the consideration of such important concepts as population, sample, parameter, statistic, and random sampling. These topics are discussed in this chapter.

2.1 POPULATIONS

Basic to statistical analysis is the desire to draw conclusions about a group of measurements of a variable being studied. This entire collection of measurements about which one wishes to draw conclusions is the *population* or *universe*. For example, an investigator may desire to draw conclusions about the tail lengths of bobcats in Montana. All Montana bobcat tail lengths are, therefore, the population under consideration. If a study is concerned with the blood glucose concentration in children of a certain age, then the blood glucose levels in all children of that age comprise the population of interest.

Populations are often very large, such as the antenna lengths of all grasshoppers in Kansas or the eye colors of all female New Zealanders, but occasionally populations of interest may be relatively small, such as the ages of men who have traveled to the moon or the heights of women who have swum the English Channel.

2.2 SAMPLES FROM POPULATIONS

If the population under study is very small, it might be practical to obtain all the measurements in the population. If one wishes to draw conclusions about the ages of all men who have traveled to the moon, it would not be unreasonable to attempt to collect the ages of the small number of individuals under consideration. Generally, however, populations of interest are so large as to render the obtaining of all the measurements unfeasible. For example, we could not reasonably expect to measure the antenna length of each grasshopper in Kansas. What can be done in such cases is to obtain a subset of all the measurements in the population. This subset of measurements comprises a *sample*, and from the characteristics of samples conclusions can be drawn about the characteristics of the populations from which the samples came.

Often one samples a population that does not actually exist. Suppose an experiment is performed in which a food supplement is administered to 40 guinea pigs, and the sample data consist of the growth rates of these 40 animals. Then the population about which conclusions might be drawn is the growth rates of all the guinea pigs that conceivably might have been administered the same food supplement under identical conditions. Such a population is said to be "imaginary" and is also referred to as "hypothetical" or "potential."

2.3 RANDOM SAMPLING

Samples from populations can be obtained in a number of ways; however, to reach valid conclusions about populations by induction from samples, statistical procedures typically assume that the samples are obtained in a *random* fashion. To sample a population randomly requires that each member of the population has an equal and independent chance of being selected. That is, not only must each measurement in the population have an equal chance of being chosen as a member of the sample, but the selection of any member of the population must in no way influence the selection of any other member. Throughout this book, "sample" will always imply "random sample."

It is sometimes possible to assign each member of a population a unique number and to draw a sample by choosing a set of such numbers at random. This is equivalent to having all members of a population in a hat and drawing a sample from them while blindfolded. Appendix Table B.39 provides 10,000 random digits for this purpose. In this table, each digit from 0 to 9 has an equal and independent chance of appearing anywhere in the table. Similarly, each combination of two digits, from 00 to 99, is found at random in the table, as is each three-digit combination, from 000 to 999, etc.

Assume that a random sample of 200 names is desired from a telephone directory having 274 pages, 3 columns of names per page, and 98 names per column. Entering Table B.39 at random (i.e., do not always enter the table at the same place), one might decide first to arrive at a random combination of three digits. If this three-digit number is 001 to 274, it can be taken as a randomly chosen page number (if it is 000 or larger than 274, simply skip it and choose another three-digit number, e.g., the next one on the table). Then one might examine the next digit in the table; if it is a 1, 2, or 3, let it denote a page column (if any digit besides 1, 2, or 3 is encountered, it is simply ignored). Then one could look at the next two-digit number in the table; if it is from 01 to 98, let it represent a randomly selected name within that column. This procedure would be performed a total of 200 times to obtain the desired random sample. One can proceed in any direction in the random number table: left to right, right to left, upward, downward, or diagonally. But the direction should be decided on before looking at the table.

Very often it is not possible to assign a number to each member of a population, and random sampling then involves biological, rather than simply mathematical, considerations. That is, the techniques for sampling Montana bobcats or Kansas grasshoppers require quite a bit of knowledge about the organism to insure that the sampling is random.

2.4 PARAMETERS AND STATISTICS

Several measures help to describe or characterize a population. For example, generally a preponderance of measurements occurs somewhere around the middle of the range of a population of measurements. Thus, some indication of a population "average" would express a useful bit of descriptive information. Such information is called a *measure of central tendency*, and several such measures (e.g., the mean and the median) will be discussed in Chapter 3.

It is also important to describe how dispersed the measurements are around the "average." That is, we can ask whether there is a wide spread of values in the population or whether the values are rather concentrated around the middle. Such a descriptive property is called a *measure of dispersion*, and several such measures (e.g., the range and the standard deviation) will be discussed in Chapter 4.

A quantity such as a measure of central tendency or a measure of dispersion is called a *parameter* when it describes or characterizes a population, and we shall be very interested in discussing parameters and drawing conclusions about them. Section 2.2 pointed out, however, that one seldom has data for entire populations, but nearly always has to rely on samples to arrive at conclusions about populations. Thus, one rarely is able to calculate parameters. However, by random sampling of populations, parameters can be estimated very well, as we shall see throughout this book. An estimate of a population parameter is called a *statistic*. It is biostatistical convention to represent population parameters by Greek letters and sample statistics by Latin letters; the following chapters will demonstrate this custom for specific examples.

The statistics one calculates will vary from sample to sample for samples taken from the same population. Since one uses sample statistics as estimates of population parameters, it behooves the researcher to arrive at the "best" estimates possible. As for what properties to desire in a "good" estimate, consider the following.

First, it is desirable that if we take an indefinitely large number of samples from a population, the long run average of the statistics obtained will equal the parameter being estimated. That is, for some samples a statistic may underestimate the parameter of interest, and for others it may overestimate that parameter; but in the long run the estimates that are too low and those that are too high will "average out." If such a property is exhibited by a statistic, we say that we have an *unbiased* statistic or an unbiased estimator.

Second, not only should the deviations of the statistic from the parameter cancel out in the long run, but it is also desirable that a statistic obtained from any single sample from a population be very close to the value of the parameter being estimated. This property of a statistic is referred to as *precision*,* *efficiency*, or *reliability*. As one frequently secures only one sample from a population, it is important to arrive at a close estimate of a parameter from a single sample.

Third, keep in mind that one can take larger and larger samples from a population (the largest sample being the population itself). As the sample size increases, a *consistent* statistic will become a better estimate of the parameter it is estimating. Indeed, if the sample were the size of the population, the best estimate would equal the parameter itself.

In the chapters that follow, the statistics recommended as estimates of parameters are "good" estimates in the sense that they possess a desirable combination of unbiasedness, efficiency, and consistency.

*The precision of a sample statistic, as defined here, should not be confused with the precision of a measurement, defined in Section 1.2.

3 Measures of Central Tendency

In samples, as well as in populations, one generally finds a preponderance of values somewhere around the middle of the range of observed values. The description of this concentration near the middle is an *average* to the layman, and a *measure of central tendency* to the statistician. It is also termed a *measure of location*, for it indicates where, among all the possible values of a variable, the sample or population is located.

Various measures of central tendency are useful parameters, in that they describe a property of populations. This chapter discusses the characteristics of these parameters and the sample statistics which are good estimates of them.

3.1 THE ARITHMETIC MEAN

The most widely used measure of central tendency is the *arithmetic mean*,* usually referred to simply as the *mean*.

Each measurement in a population may be referred to as an X_i (read "X sub i") value. Thus, one measurement might be denoted as X_1, another as X_2, another as X_3, and so on. The subscript i might be any integer value up through N, the total number of X_i values in the population. The mean of the population is denoted by the Greek letter μ (lowercase mu), and is calculated as the sum of all the X_i values divided by the size of the population.

The calculation of the population mean can be abbreviated concisely by the formula

*As an adjective, "arithmetic" is pronounced with the accent on the third syllable.

$$\mu = \frac{\sum_{i=1}^{N} X_i}{N}. \tag{3.1}$$

The Greek letter Σ (capital sigma) means "summation," and $\sum_{i=1}^{N} X_i$ means "summation of all X_i values from X_1 through X_N." Thus, for example, $\sum_{i=1}^{4} X_i = X_1 + X_2 + X_3 + X_4$ and $\sum_{i=3}^{5} X_i = X_3 + X_4 + X_5$. Since, in statistical computations, summations are nearly always performed over the entire set of X_i values, this book will assume $\sum X_i$ to mean "sum X_i's over all values of i," simply as a matter of printing convenience, and $\mu = \sum X_i/N$ would therefore designate the same calculation as would $\mu = \sum_{i=1}^{N} X_i/N$.

The most efficient, unbiased, and consistent estimate of the population mean, μ, is the sample mean, denoted as $\bar{X}$ (read as "X bar"). Whereas the size of the population (which we generally do not know) is denoted as N, the size of a sample is indicated by n, and $\bar{X}$ is calculated as

$$\bar{X} = \frac{\sum_{i=1}^{n} X_i}{n} \quad \text{or} \quad \bar{X} = \frac{\sum X_i}{n}, \tag{3.2}$$

which is read "the sample mean equals the sum of all measurements in the sample divided by the number of measurements in the sample." Example 3.1 demonstrates the calculation of the sample mean. Note that the mean has the same units of measurement as do the individual observations. The question of how many decimal places should be reported for the mean will be answered in Section 7.3; until then we shall simply record the mean with one more decimal place than the data.

Example 3.1

A sample from a population of butterfly wing lengths.

X_i (cm)	X_i (cm)
3.3	4.0
3.5	4.0
3.6	4.0
3.6	4.1
3.7	4.1
3.8	4.1
3.8	4.2
3.8	4.2
3.9	4.3
3.9	4.3
3.9	4.4
4.0	4.5

$$\sum X_i = 95.0 \text{ cm}$$
$$n = 24$$
$$\bar{X} = \frac{\sum X_i}{n} = \frac{95.0 \text{ cm}}{24} = 3.96 \text{ cm}$$

If, as in Example 3.1, a sample contains multiple identical data for several values of the variable, then it may be convenient to record the data in the form of a frequency table, as in Example 3.2. Then X_i can be said to denote each different measurement and f_i can denote the frequency with which that X_i occurs in the sample. The sample mean may then be calculated as

$$\bar{X} = \frac{\sum f_i X_i}{n}. \tag{3.3}$$

Example 3.2 demonstrates this calculation for the same data as in Example 3.1.

Example 3.2
The data from Example 3.1 recorded as a frequency table.

X_i (cm)	f_i	f_iX_i (cm)
3.3	1	3.3
3.4	0	0
3.5	1	3.5
3.6	2	7.2
3.7	1	3.7
3.8	3	11.4
3.9	3	11.7
4.0	4	16.0
4.1	3	12.3
4.2	2	8.4
4.3	2	8.6
4.4	1	4.4
4.5	1	4.5
	$\sum f_i = 24$	$\sum f_iX_i = 95.0$ cm

$$\sum f_i = n = 24$$

$$\bar{X} = \frac{\sum f_iX_i}{n} = \frac{95.0 \text{ cm}}{24} = 3.96 \text{ cm}$$

$$\begin{aligned} \text{median} &= 3.95 \text{ cm} + (\tfrac{1}{4})(0.1 \text{ cm}) \\ &= 3.95 \text{ cm} + 0.025 \text{ cm} \\ &= 3.975 \text{ cm} \end{aligned}$$

If data are plotted as a histogram (Fig. 3.1), the mean is the *center of gravity* of the histogram. That is, if the histogram were made of a solid material, it would balance horizontally with the fulcrum at $\bar{X}$. The mean generally is applicable only to ratio or interval scale data.

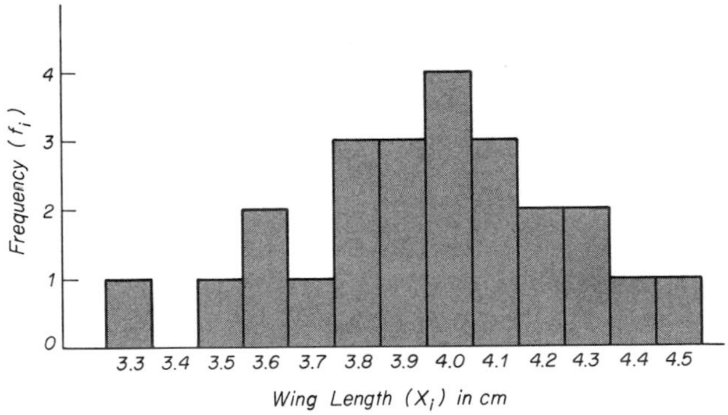

Figure 3.1 A histogram of the data in Example 3.2. The mean (3.96 cm) is the center of gravity of the histogram, and the median (3.975 cm) divides the histogram into two equal areas.

3.2 THE MEDIAN AND OTHER QUANTILES

The median is the middle measurement in a set of data. That is, there are just as many observations larger than the median as there are smaller. The sample median is the best estimate of the population median.* In a symmetrical distribution (Figs. 3.2a and 3.2b) the sample median is also an unbiased and consistent estimate of μ,

*There is no widely accepted symbol for the sample median. Later (Section 8.8), the symbol M (Greek capital mu) will be used to denote the population median.

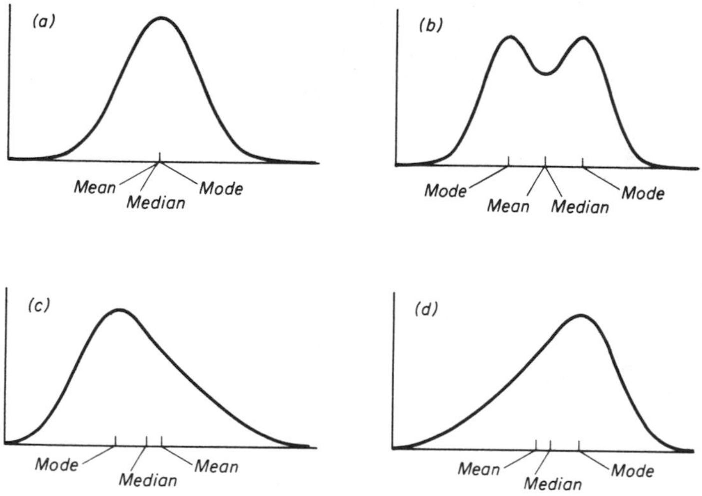

Figure 3.2 Frequency distributions showing measures of central tendency. Values of the variable are along the abscissa (horizontal axis), and the frequencies are along the ordinate (vertical axis). Distributions (a) and (b) are symmetrical, (c) is positively skewed, and (d) is negatively skewed. Distributions (a), (c), and (d) are unimodal, and distribution (b) is bimodal.

but it is not as efficient a statistic as $\bar{X}$, and should not be used as a substitute for $\bar{X}$. (If the frequency distribution is asymmetrical, the median is a poor estimate of the mean.)

The median of a set of data is readily found by first arranging the measurements in order of magnitude (either increasing or decreasing order), as is done with the samples in Examples 3.1, 3.2, and 3.3. Then,

$$\text{median} = X_{(n+1)/2} \quad \text{if } n \text{ is odd.} \tag{3.4}$$

Thus, for species A in Example 3.3, the median $= X_{(n+1)/2} = X_5 = 40$ mo. If n is even, then there are two middle measurements, and the median is defined as the midpoint

Example 3.3

Life expectancy of two hypothetical species of birds in captivity.

Species A X_i (mo)	*Species B* X_i (mo)
34	34
36	36
37	37
39	39
40	40
41	41
42	42
43	43
79	44
	45

Species A:

$n = 9$

median $= X_5 = 40$ mo

$\bar{X} = 43.4$ mo

Species B:

$n = 10$

$$\text{median} = \frac{X_5 + X_6}{2} = \frac{40 \text{ mo} + 41 \text{ mo}}{2} = 40.5 \text{ mo}$$

$\bar{X} = 40.1$ mo

between them. That is,

$$\text{median} = \frac{X_{n/2} + X_{(n/2)+1}}{2} \text{ if } n \text{ is even.} \tag{3.5}$$

For species B in Example 3.3, the median $= (X_{n/2} + X_{(n/2)+1})/2 = (X_5 + X_6)/2 =$ 40.5 mo. Note that the median has the same units as each individual measurement. If data are plotted as a frequency histogram (e.g., as in Fig. 3.1), the median is the value of X that divides the area of the histogram into two equal parts.

If the median falls among identical observations (referred to as *tied* values), as in Example 3.1 or 3.2, a special problem arises. If we utilize Equation 3.5, then we conclude the median to be 4.0 cm. But 11 measurements are less than 4.0 cm and 9 are greater; thus, 4.0 cm does not fit the definition of the median as that value having equal numbers of observed values greater than and less than itself.

When the median falls among tied observations, we may interpolate to estimate its exact value. Using the data of Example 3.2, we desire to calculate a value below which 50% of the observations lie. Fifty percent of the observations would be 12 observations. Since the first 7 classes in the frequency table include 11 observations and 4 observations are in class 4.0 cm, we know that the median lies within the range of 3.95 to 4.05 cm. Assuming that the 4 observations in class 4.0 cm are distributed evenly within the 0.1 cm range of 3.95 to 4.05 cm, then the median will be $(\frac{1}{4})(0.1 \text{ cm})$ $= 0.025$ cm into this class. Thus, the median $= 3.95 \text{ cm} + 0.025 \text{ cm} = 3.975$ cm. In general, for the median within a class interval containing tied observations,

$$\text{median} = \begin{pmatrix}\text{lower limit}\\ \text{of interval}\end{pmatrix} + \left(\frac{0.5n - \text{cum. freq.}}{\text{no. of observations in interval}}\right)\begin{pmatrix}\text{interval}\\ \text{size}\end{pmatrix}, \tag{3.6}$$

where "cum. freq." refers to the cumulative frequency of the previous classes. By using this procedure, the calculated median will be the value of X that divides the area of the histogram of the sample into two equal parts.

The median expresses less information than does the mean, for it does not take into account the actual value of each measurement, but only considers the rank of each measurement. Still, it may offer certain advantages in some situations. First, it is plain from the two samples in Example 3.3 that extremely high (or extremely low) measurements will not affect the median as much as they affect the mean (causing the sample median to be called a "resistant" statistic). Thus, when we deal with skewed populations, we may prefer the median to the mean to express central tendency.

Note that in Example 3.3 the researcher would have to wait 79 mo to compute a mean life expectancy for species A (45 mo for species B), whereas the median could be determined in only 40 mo (41 mo for species B). Also, to calculate a median one does not need to have accurate data for all members of the sample. If we did not have the first three data for species A accurately recorded, they could simply have been stated as "less than 39 mo," and the median could have been determined just as readily, although calculations of the mean would not have been possible. Lastly, the median can be used not only on interval and ratio scale data, but also on data on the ordinal scale, data for which the use of the mean usually would not be considered appropriate.

Just as the median is the value above and below which lie half the set of data, one can define measures above (or below) which lie other fractional parts of the data.

If the data are divided into four equal parts, for example, we speak of quartiles. The first quartile is the $(n + 1)/4$th measurement. One-fourth of all the ranked observations are less than the first quartile and three-fourths are less than the third quartile. The second quartile is the median. In general, the jth m-tile is the $j(n + 1)/m$th measurement. Besides the median and quartiles, *quintiles*, *deciles*, and *percentiles* (or "centiles") are sometimes used for dividing samples into fifths, tenths, and hundredths, respectively, and all measures of this type are sometimes collectively termed *quantiles*. The expression "LD_{50}," commonly used in some areas of biological research, is simply the 50th percentile of the lethal doses, or the median lethal dose. That is 50% of the experimental subjects survived this dose, whereas 50% did not. Likewise, "LD_{50}" is the median lethal concentration, or the 50th percentile of the lethal concentrations.

3.3 THE MODE

The *mode* is commonly defined as the most frequently occurring measurement in a set of data. In Example 3.2, the mode is 4.0 cm. But it is perhaps better to define a mode as a measurement of great concentration, for some frequency distributions may have more than one such point of concentration, even though these concentrations might not contain precisely the same frequencies. Thus, a sample consisting of the data: 6, 7, 7, 8, 8, 8, 8, 8, 8, 9, 9, 10, 11, 12, 12, 12, 12, 12, 13, 13, and 14 mm would be said to be bimodal, the two modes being at 8 mm and 12 mm (Some authors would refer to 8 mm as the "major mode" and call 12 mm the "minor mode.") A distribution in which each different measurement occurs with equal frequency is said to have no mode. If two consecutive values of X have frequencies great enough to declare the X values modes, the mode of the distribution is said to be the midpoint of these two X's; e.g., the mode of 3, 5, 7, 7, 7, 8, 8, 8, and 10 liters is 7.5 liters.

The sample mode is the best estimate of the population mode. When we sample a symmetrical unimodal population, the mode is an unbiased and consistent estimate of the mean and median (Fig. 3.2a), but it is relatively inefficient and should not be so used. As a measure of central tendency, the mode is affected by skewness less than is the mean or the median, but it is very much more affected by sampling than these other two measures. The mode, but neither the median nor the mean, may be used for data on the nominal, as well as the ordinal, interval, and ratio scales of measurement. In a unimodal asymmetric distribution (Figs. 3.2c and 3.2d), the median lies about one-third the distance between the mean and the mode.

The mode is not often used in biological research, although it is often interesting to report the number of modes detected in a population, if there are more than one.

3.4 OTHER MEASURES OF CENTRAL TENDENCY

The *range midpoint*, or *midrange*, is also a measure of central tendency, being half-way between the highest and lowest values in the set of data. It is not to be considered a good estimate of the mean and is a seldom-used measure, for it utilizes relatively little information from the data (although many times the so-called "mean" daily temperature reported is merely the mean of the minimum and maximum, and is thus a range midpoint). The mean of any two symmetrically located percentiles, such as the mean of the 1st and 3rd quartiles (i.e., the 25th and 75th percentiles), may be used in the same fashion as the range midpoint as a measure of central tendency (see Dixon and Massey, 1969: 133–134), but such a procedure is seldom encountered. Since such measures are based on quantiles, they may be applied to either ratio, interval, or ordinal data.

The *geometric mean* is the nth root of the product of the n data:

$$\text{geometric mean} = \sqrt[n]{X_1 X_2 X_3 \ldots X_n} = \sqrt[n]{\prod_{i=1}^{n} X_i}. \tag{3.7}$$

Capital Greek pi, Π, means "take the product" in an analogous fashion as Σ indicates "take the sum." The geometric mean may also be computed as the antilogarithm of the arithmetic mean of the logarithms of the data. It is appropriate only when all the data are positive values (and if the data are not all equal, the geometric mean is less than the arithmetic mean). This measure finds use in averaging ratios where it is desired to give each ratio equal weight, and in averaging percent changes, discussions of which are found in Croxton, Cowden, and Klein (1967:178–182).

The *harmonic mean* is the reciprocal of the arithmetic mean of the reciprocals of the data:

$$\text{harmonic mean} = \frac{1}{\frac{1}{n}\sum \frac{1}{X_i}} = \frac{n}{\sum \frac{1}{X_i}}. \tag{3.8}$$

It is occasionally used when dealing with averaging rates, as described by Croxton, Cowden, and Klein (1967: 182–188).

The geometric and harmonic means may be appropriately calculated only for ratio scale data. They are so rarely encountered that the term "mean" typically implies "arithmetic mean."

3.5 THE EFFECT OF CODING DATA

Oftentimes in the manipulation of data, considerable time and effort can be saved if *coding* is employed. Coding is the conversion of the original measurements into easier-to-work-with values by simple arithmetic operations. Generally coding employs a *linear transformation* of the data, such as multiplying (or dividing) or adding (or subtracting) a constant. The addition or subtraction of a constant is sometimes termed a translation of the data (i.e., changing the origin), whereas the multiplication or division by a constant causes an expansion or contraction of the scale of measurement. The first set of data in Example 3.4 are coded by subtracting a constant value of 840 g. Not only is each coded value equal to $X_i - 840$ g, but the mean of the coded values is equal to $\bar{X} - 840$ g. Thus, the easier-to-work-with coded values may be used to calculate a mean that then is readily converted to the mean of the original data, simply by adding back the coding constant. In sample 2 of Example 3.4, the observed

Example 3.4

Coding data to facilitate calculations.

Sample 1 (Coding by Subtraction: A = −840 g)		Sample 2 (Coding by Division: M = 0.001 liters/ml)	
X_i (g)	$[X_i] = X_i - 840$ g	X_i (ml)	$[X_i] = 0.001$ liters
842	2	8,000	8.000
844	4	9,000	9.000
846	6	9,500	9.500
846	6	11,000	11.000
847	7	12,500	12.500
848	8	13,000	13.000
849	9		
$\sum X_i = 5922$ g	$\sum [X_i] = 42$ g	$\sum X_i = 63{,}000$ ml	$\sum [X_i] = 63.000$ liters
$\bar{X} = \frac{5922 \text{ g}}{7}$	$[\bar{X}] = \frac{42 \text{ g}}{7}$	$\bar{X} = 10{,}500$ ml	$[\bar{X}] = 10.500$ liters
$= 846$ g	$= 6$ g		

Sample 1:

$$\bar{X} = [\bar{X}] - A = 6 \text{ g} - (-840 \text{ g}) = 846 \text{ g}$$

Sample 2:

$$\bar{X} = \frac{[\bar{X}]}{M} = \frac{10.500 \text{ liters}}{0.001 \text{ liters/ml}} = 10{,}500 \text{ ml}$$

data are coded by dividing each observation by 1000 (i.e., by multiplying by 0.001). The resultant mean only needs to be multiplied by the coding factor of 1000 (i.e., divided by 0.001) to arrive at the mean of the original data. Since the other measures of central tendency have the same units as the mean, they are affected by coding in exactly the same fashion. For calculations more involved than computing means, the advantages of coding will become more apparent.

In general, if we code X by addition of a constant, A, the coded X is

$$[X_i] = X_i + A. \tag{3.9}$$

In sample 1 of Example 3.4, $A = -840$ g. The mean of a set of data thus coded is

$$[\bar{X}] = \bar{X} + A; \tag{3.10}$$

so if one has calculated $[\bar{X}]$ using coded data, it is a simple matter to determine what the sample mean would have been if the data had not been coded, namely

$$\bar{X} = [\bar{X}] - A. \tag{3.11}$$

If one codes X by multiplying by a constant, M, then each coded datum is

$$[X_i] = MX_i. \tag{3.12}$$

In sample 2 of Example 3.4, $M = 1/1000 = 0.001$ liters/ml. The mean of the coded data is

$$[\bar{X}] = M\bar{X}. \tag{3.13}$$

Knowing $[\bar{X}]$, one can determine that the mean of the uncoded data is

$$\bar{X} = \frac{[\bar{X}]}{M}. \tag{3.14}$$

EXERCISES

3.1. If $X_1 = 3.1$ kg, $X_2 = 3.4$ kg, $X_3 = 3.6$ kg, $X_4 = 3.7$ kg, and $X_5 = 4.0$ kg, what is the value of (a) $\sum_{i=1}^{4} X_i$, (b) $\sum_{i=2}^{4} X_i$, (c) $\sum_{i=1}^{5} X_i$, (d) $\sum X_i$?

3.2. **(a)** Calculate the mean of the five weights in Exercise 3.1.
(b) Calculate the median of these weights.

3.3. The ages, in years, of the faculty members of a university department are 32.2, 37.5, 41.7, 53.8, 50.2, 48.2, 46.3, 65.0, and 44.8. **(a)** Calculate the mean age of these nine faculty members. **(b)** Calculate the median of the ages. **(c)** If the person 65.0 years of age retires and is replaced on the faculty with a person 46.5 years old, what is the new mean age? **(d)** What is the new median age?

3.4. Consider the following frequency tabulation of leaf weights (in grams):

X_i	f_i
1.85–1.95	2
1.95–2.05	1
2.05–2.15	2
2.15–2.25	3
2.25–2.35	5
2.35–2.45	6
2.45–2.55	4
2.55–2.65	3
2.65–2.75	1

Using the midpoints of the indicated ranges of X_i, (a) calculate the mean leaf weight using Equation 3.2, and (b) calculate the mean leaf weight using Equation 3.3. (c) Calculate the median leaf weight using either Equation 3.4 or 3.5, and (d) calculate the median using Equation 3.6. (e) Determine the mode of the frequency distribution.

4 Measures of Dispersion and Variability

In addition to a measure of central tendency, it is generally desirable to have a *measure of dispersion* of data. A measure of dispersion, or a *measure of variability*, as it is sometimes called, is an indication of the scatter of measurements around the center of the distribution, or the opposite of how clustered the measurements are around the center. Measures of dispersion of a population are parameters of the population, and the sample measures of dispersion that estimate them are statistics.

4.1 THE RANGE

The difference between the highest and lowest measurements in a group of data is termed the *range*. If sample measurements are arranged in increasing order of magnitude, as if the median were about to be determined, then

$$\text{sample range} = X_n - X_1. \tag{4.1}$$

Sample 1 in Example 4.1 is a hypothetical set of data in which $X_1 = 1.2$ g and $X_n = 2.4$ g. Thus, the range may be expressed as 1.2 to 2.4 g, or as 2.4 g − 1.2 g = 1.2 g. Note that the range has the same units as the individual measurements. Sample 2 in Example 4.1 has the same range as sample 1.

The range is a relatively crude measure of dispersion, inasmuch as it does not take into account any measurements except the highest and the lowest. Furthermore, as it is unlikely that a sample will contain both the highest and lowest values in the population, the sample range usually underestimates the population range; therefore, it is a biased and inefficient estimator. Nonetheless, it is considered useful by some to present the sample range as an estimate (although a poor one) of the population range.

Example 4.1

Calculation of measures of dispersion for two hypothetical samples.

Sample 1

X_i (g)	$X_i - \bar{X}$ (g)	$\lvert X_i - \bar{X} \rvert$ (g)	$(X_i - \bar{X})^2$ (g^2)
1.2	−0.6	0.6	0.36
1.4	−0.4	0.4	0.16
1.6	−0.2	0.2	0.04
1.8	0.0	0.0	0.00
2.0	0.2	0.2	0.04
2.2	0.4	0.4	0.16
2.4	0.6	0.6	0.36
$\sum X_i = 12.6$ g	$\sum (X_i - \bar{X}) = 0.0$ g	$\sum \lvert X_i - \bar{X} \rvert = 2.4$ g	$\sum (X_i - \bar{X})^2$ $= 1.12\ g^2$ = "sum of squares"

$$\bar{X} = \frac{12.6\text{ g}}{7} = 1.8\text{ g}$$

$$\text{range} = X_7 - X_1 = 2.4\text{ g} - 1.2\text{ g} = 1.2\text{ g}$$

$$\text{mean deviation} = \frac{\sum \lvert X_i - \bar{X} \rvert}{n} = \frac{2.4\text{ g}}{7} = 0.34\text{ g}$$

$$s^2 = \frac{\sum (X_i - \bar{X})^2}{n - 1} = \frac{1.12\text{ g}^2}{6} = 0.1867\text{ g}^2$$

$$s = \sqrt{0.1867\text{ g}^2} = 0.43\text{ g}$$

Sample 2

X_i (g)	$X_i - \bar{X}$ (g)	$\lvert X_i - \bar{X} \rvert$ (g)	$(X_i - \bar{X})^2$ (g^2)
1.2	−0.6	0.6	0.36
1.6	−0.2	0.2	0.04
1.7	−0.1	0.1	0.01
1.8	0.0	0.0	0.00
1.9	0.1	0.1	0.01
2.0	0.2	0.2	0.04
2.4	0.6	0.6	0.36
$\sum X_i = 12.6$ g	$\sum (X_i - \bar{X}) = 0.0$ g	$\sum \lvert X_i - \bar{X} \rvert = 1.8$ g	$\sum (X_i - \bar{X})^2$ $= 0.82\ g^2$ = "sum of squares"

$$\bar{X} = \frac{12.6\text{ g}}{7} = 1.8\text{ g}$$

$$\text{range} = X_7 - X_1 = 2.4\text{ g} - 1.2\text{ g} = 1.2\text{ g}$$

$$\text{mean deviation} = \frac{\sum \lvert X_i - \bar{X} \rvert}{n} = \frac{1.8\text{ g}}{7} = 0.26\text{ a}$$

$$s^2 = \frac{\sum (X_i - \bar{X})^2}{n - 1} = \frac{0.82\text{ g}^2}{6} = 0.1367\text{ g}^2$$

$$s = \sqrt{0.1367\text{ g}^2} = 0.37\text{ g}$$

For example, taxonomists are often concerned with having an estimate of what the highest and lowest values in a population are expected to be. Whenever the range is specified in reporting data, however, it is usually a good practice to report another measure of dispersion as well. The range is applicable to ordinal, interval, and ratio scale data.

4.2 THE MEAN DEVIATION

As is evident from the two samples in Example 4.1, the range conveys no information about how clustered about the middle of the distribution the measurements are. Since the mean is so useful a measure of central tendency, one might express dispersion terms of deviations from the mean. The sum of all deviations from the mean, i.e., $\sum (X_i - \bar{X})$, will always equal zero, however, so that such a summation would be useless as a measure of dispersion (as seen in Example 4.1).

Summing the absolute values of the deviations from the mean results in a quantity that is an expression of dispersion about the mean. Dividing this quantity by n yields a measure known as the *mean deviation*, or *mean absolute deviation* of the sample; this measure has the same units as do the data. In Example 4.1, sample 1 is more variable (or more dispersed, or less concentrated) than sample 2. Although the two samples have the same range, the mean deviation, calculated as

$$\text{sample mean deviation} = \frac{\sum |X_i - \bar{X}|}{n}, \tag{4.2}$$

expresses the differences in dispersion. Mean deviation can also be defined by using the sum of the absolute deviations from the median instead of from the mean.

4.3 THE VARIANCE

Another method of eliminating the signs of the deviations from the mean is to square the deviations. The sum of the squares of the deviations from the mean is called the *sum of squares*, abbreviated SS, and is defined as follows:

$$\text{population SS} = \sum (X_i - \mu)^2 \tag{4.3}$$

$$\text{sample SS} = \sum (X_i - \bar{X})^2. \tag{4.4}$$

The mean sum of squares is called the *variance* (or *mean square*, the latter being short for *mean squared deviation*), and for a population is denoted by σ^2 ("sigma squared," using the lowercase Greek letter):

$$\sigma^2 = \frac{\sum (X_i - \mu)^2}{N}. \tag{4.5}$$

The best estimate of the population variance, σ^2, is the sample variance, s^2:

$$s^2 = \frac{\sum (X_i - \bar{X})^2}{n - 1} \tag{4.6}$$

If, in Equation 4.5, we replace μ by $\bar{X}$ and N by n, the result is a quantity that is a biased estimate of σ^2. The dividing of the sample sum of squares by $n - 1$ (called the *degrees of freedom*, abbreviated DF) rather than by n, yields an unbiased estimate, and it is Equation 4.6 that should be used to calculate the sample variance.

If all observations are equal, then there is no variability and $s^2 = 0$; and s^2 becomes increasingly large as the amount of variability, or dispersion, increases. Since s^2 is a mean sum of squares, it can never be a negative quantity.

The variance expresses the same type of information as does the mean deviation, but it has certain very important properties relative to probability and hypothesis testing that make it distinctly superior. Thus, the mean deviation is very seldom encountered in biostatistical analysis.

The calculation of s^2 can be tedious for large samples, but it can be facilitated by the use of the equality

$$\text{sample SS} = \sum X_i^2 - \frac{(\sum X_i)^2}{n}. \tag{4.7}$$

This formula is much simpler to work with than is Equation 4.4. Example 4.2 demonstrates its use to obtain a sample sum of squares.

Example 4.2

"Machine formula" calculation of variance, standard deviation, and coefficient of variation.

Sample 1

X_i (g)	X_i^2 (g^2)
1.2	1.44
1.4	1.96
1.6	2.56
1.8	3.24
2.0	4.00
2.2	4.84
2.4	5.76
$\sum X_i = 12.6$ g	$\sum X_i^2 = 23.80\ g^2$

$n = 7$

$\bar{X} = \frac{12.6\ g}{7} = 1.8$ g

$SS = \sum X_i^2 - \frac{(\sum X_i)^2}{n}$

$= 23.80\ g^2 - \frac{(12.6\ g)^2}{7}$

$= 23.80\ g^2 - 22.68\ g^2$

$= 1.12\ g^2$

$s^2 = \frac{SS}{n-1}$

$= \frac{1.12\ g^2}{6} = 0.1867\ g^2$

$s = \sqrt{0.1867\ g^2} = 0.43$ g

$V = \frac{s}{\bar{X}} = \frac{0.43\ g}{1.8\ g}$

$= 0.24 = 24\%$

Sample 2

X_i (g)	X_i^2 (g^2)
1.2	1.44
1.6	2.56
1.7	2.89
1.8	3.24
1.9	3.61
2.0	4.00
2.4	5.76
$\sum X_i = 12.6$ g	$\sum X_i^2 = 23.50\ g^2$

$n = 7$

$\bar{X} = \frac{12.6\ g}{7} = 1.8$ g

$SS = 23.50\ g^2 - \frac{(12.6\ g)^2}{7}$

$= 0.82\ g^2$

$s^2 = \frac{0.82\ g^2}{6} = 0.1367\ g^2$

$s = \sqrt{0.1367\ g^2} = 0.37$ g

$V = \frac{0.37\ g}{1.8\ g}$

$= 0.21 = 21\%$

Since sample variance equals sample SS divided by DF,

$$s^2 = \frac{\sum X_i^2 - \frac{(\sum X_i)^2}{n}}{n-1}. \tag{4.8}$$

This last formula is often referred to as a "working formula," or "machine formula," because of its computational advantages. There are, in fact, two major advantages in calculating SS by Equation 4.7 rather than by Equation 4.4. First, fewer computational steps are involved, a fact that decreases chance of error. On a good desk calculator, the summed quantities, $\sum X_i$ and $\sum X_i^2$, can both be obtained with only one pass through the data, whereas Equation 4.4 requires one pass through the data to calculate $\bar{X}$, and at least one more pass to calculate and sum the squares of the deviations, $X_i - \bar{X}$. Second, there may be a good deal of rounding error in calculating each $X_i - \bar{X}$, a situation that leads to decreased accuracy in computation, but which is avoided by the use of Equation 4.7.*

For data recorded in frequency tables,

$$\text{sample SS} = \sum f_i X_i^2 - \frac{(\sum f_i X_i)^2}{n}, \tag{4.9}$$

where f_i is the frequency of observations with magnitude X_i. But with a desk calculator it is often faster to use Equation 4.7 for each individual observation, disregarding the class groupings.

The variance has square units. If measurements are in grams, their variance will be in grams squared, or if the measurements are in cubic centimeters, their variance will be in terms of cubic centimeters squared, even though such squared units have no physical interpretation. The question of how many decimal places to report for the variance will be considered in Section 7.3.

4.4 THE STANDARD DEVIATION

The *standard deviation*† is the positive square root of the variance; therefore, it has the same units as the original measurements. Thus, for a population,

$$\sigma = \sqrt{\frac{\sum X_i^2 - \frac{(\sum X_i)^2}{N}}{N}}, \tag{4.10}$$

and for a sample,

$$s = \sqrt{\frac{\sum X_i^2 - \frac{(\sum X_i)^2}{n}}{n-1}}. \tag{4.11}$$

Examples 4.1 and 4.2 demonstrate the calculation of s. This quantity frequently is abbreviated SD, and on rare occasions is called the *root mean square deviation*. Remember that the standard deviation is, by definition, always a nonnegative quantity.‡ Section 7.3 will explain how to determine the number of decimal places that may appropriately be recorded for the standard deviation.

*Computational formulas advantageous on desk calculators may not prove accurate on computers (Wilkinson and Dallal, 1977), largely because computers may use fewer significant figures. (Also see Ling, 1974.)

†It may have been the great American statistician Karl Pearson (1857–1936) who coined the term *standard deviation*, during the period 1895–1900 (Pearson, 1971).

‡The sample s is actually a slightly biased estimate of the population σ, in that on the average it estimates a trifle low, especially in small samples. But this fact is generally considered to be offset by the statistic's usefulness. Correction for this bias is sometimes possible (e.g., Bliss, 1967: 131; Dixon and Massey, 1969: 136; Gurland and Tripathi, 1971; Tolman, 1971), but it is rarely employed.

4.5 THE COEFFICIENT OF VARIATION

The *coefficient of variation** or *coefficient of variability*, is defined as

$$V = \frac{s}{\bar{X}} \quad \text{or} \quad V = \frac{s}{\bar{X}} \cdot 100\% \tag{4.12}$$

Since $s/\bar{X}$ is generally a small quantity, it is frequently multiplied by 100% in order to express V as a percentage.

As a measure of variability, the variance and standard deviation have magnitudes that are dependent on the magnitude of the data. Elephants have ears that are perhaps 100 times larger than those of mice. If elephant ears were no more variable, relative to their size, than mouse ears, relative to their size, the standard deviation of elephant ear lengths would be 100 times as great as the standard deviation of mouse ear lengths (and the variance of the former would be $100^2 = 10{,}000$ times the variance of the latter). The coefficient of variation expresses sample variability relative to the mean of the sample (and is on rare occasion referred to as the "relative standard deviation"). It is called a measure of *relative variability* or *relative dispersion*.

Since s and $\bar{X}$ have identical units, V has no units at all, a fact emphasizing that it is a relative measure, divorced from the actual magnitude or units of measurement of the data. Thus, had the data in Example 4.2 been measured in pounds, kilograms, or tons, instead of grams, the calculated V would have been the same. The coefficient of variability may be calculated only for ratio scale data; it is, for example, not valid to calculate coefficients of variation of temperature data measured on the Celsius or Fahrenheit temperature scales. Simpson, Roe, and Lewontin (1960: 89–95) present a good discussion of V and its biological application, especially with regard to zoomorphological measurements.

4.6 INDICES OF DIVERSITY

For nominal scale data there is no mean or median to serve as a reference for discussion of dispersion. Instead, we can invoke the concept of *diversity*, the distribution of observations among categories. Consider that sparrows are found to nest in four different types of location (vines, eaves, branches, and cavities). If, out of 20 nests observed, 5 are found at each of the four locations, then we would say that there was great diversity in nesting sites. If, however, 17 nests were found in cavities and only 1 in each of the other three locations, then we would consider the situation to be one of very low nest-site diversity. In other words, observations distributed evenly among categories result in high diversity, whereas a set of observations where the bulk of the data occurs in few of the categories is one exhibiting low diversity.

Of the quantitative descriptions of diversity available, those based on *information theory* have become popular. The underlying considerations of these measures can be visualized by considering *uncertainty* to be synonymous with diversity. If 17 out of

*The term *coefficient of variation* was introduced by the statistical giant, Karl Pearson, about whom more will be related later.

20 nest sites were to be found in cavities, then one would be relatively certain of being able to predict the location of a randomly encountered nest site. However, if nests were found to be distributed evenly among the various locations (a situation of high nest-site diversity), then there would be a good deal of uncertainty involved in predicting the location of a nest site selected at random. If a set of nominal scale data may be considered to be a random sample, then a quantitative expression appropriate as a measure of diversity is that of Shannon (1948):

$$H' = -\sum_{i=1}^{k} p_i \log p_i \tag{4.13}$$

(often referred to as the Shannon-Wiener diversity index or the Shannon-Weaver index). Here, k is the number of categories and p_i is the proportion of the observations found in category i. Denoting n to be the sample size, and f_i to be the number of observations in category i, then $p_i = f_i/n$. Some mathematical manipulation arrives at the equivalent function:

$$H' = \frac{n \log n - \sum_{i=1}^{k} f_i \log f_i}{n} \tag{4.14}$$

a formula that is easier to use than Equation 4.13 because it eliminates the necessity of calculating the proportions (p_i). Published tables of $n \log n$ and $f_i \log f_i$ are available (e.g., Brower and Zar, 1977: 139; Lloyd, Zar, and Karr, 1968). Any logarithmic base may be used to compute H'; bases 10, e, and 2 (in that order of commonness) are the most frequently encountered. A value of H' (or of any other measure of this section, except evenness measures) calculated using one logarithmic base may be converted to that of another base; Table 4.1 gives factors for doing this for bases 10, e, and 2. Unfortunately, H' is known to be an underestimate of the diversity in the sampled population (Bowman et al., 1971). However, this bias decreases with increasing sample size.

TABLE 4.1 MULTIPLICATION FACTORS FOR CONVERTING AMONG DIVERSITY MEASURES (H, H', H_{max}, OR H'_{max}) CALCULATED USING DIFFERENT LOGARITHMIC BASES*

To convert to:	To convert from: Base 2	Base e	Base 10
Base 2	1.0000	1.4427	3.3219
Base e	0.6931	1.0000	2.3026
Base 10	0.3010	0.4343	1.0000

For example, if $H' = 0.255$ using base 10; H' would be $(0.255)(3.3219) = 0.847$ using base 2.

*The measures J and J' are unaffected by change in logarithmic base.

The magnitude of H' is affected not only by the distribution of the data but also by the number of categories, for, theoretically, the maximum possible diversity for a set of data consisting of k categories is

$$H'_{\max} = \log k. \tag{4.15}$$

Therefore, many users of Shannon's index prefer to calculate

$$J' = \frac{H'}{H'_{\max}} \tag{4.16}$$

instead of, or in addition to, H', thus expressing the observed diversity as a proportion of the maximum possible diversity. The quantity J' has been termed *evenness* (Pielou, 1966) and may also be referred to as *homogeneity* or *relative diversity*. The measure $1 - J'$ may then be viewed as a measure of *heterogeneity* or *dominance*. Since k is typically an underestimate of the number of categories in the population, the sample evenness, J', is typically an overestimate of the population evenness. (That is, J' is a biased statistic.) Example 4.3 demonstrates the calculation of H' and J'.

If a set of data may not be considered a random sample, then Equation 4.13 (or 4.14) is not an appropriate diversity measure (Pielou, 1966). Examples of such situations may be when we have, in fact, data comprising an entire population, or data that are a sample obtained nonrandomly from a population. In such a case, one may use the information-theoretic diversity measure of Brillouin (1962: 7–8):

$$H = \frac{\log\left(\dfrac{n!}{\prod_{i=1}^{k} f_i!}\right)}{n}, \tag{4.17}$$

where Π (capital Greek pi) means to take the product, just as Σ means to take the sum.* Equation 4.17 may be written, equivalently, as

$$H = \frac{\log \dfrac{n!}{f_1!f_2!\ldots f_k!}}{n} \tag{4.18}$$

or as

$$H = \frac{(\log n! - \sum \log f_i!)}{n}. \tag{4.19}$$

Table B.38 gives logarithms of factorials to ease this calculation. Other such tables are available, as well (e.g., Brower and Zar, 1977: 141; Lloyd, Zar, and Karr, 1968; Pearson and Hartley, 1966: Table 51).†

The maximum possible Brillouin diversity for a set of n observations distributed among k categories is

$$H_{\max} = \frac{\log n! - (k - d)\log c! - d\log(c + 1)!}{n}, \tag{4.21}$$

*$n!$ is read as "n factorial" and implies the product, $(n)(n - 1)(n - 2)\ldots(2)(1)$.

†For moderate to large n (or f_i), "Stirling's approximation" is excellent:

$$\log n! = (n + 0.5)\log n - 0.4343n + 0.3991. \tag{4.20}$$

Example 4.3

Indices of diversity for nominal scale data. The nesting sites of sparrows.

	Observed Frequencies (f_i)		
Category (i)	*Sample 1*	*Sample 2*	*Sample 3*
Vines	5	1	2
Eaves	5	1	2
Branches	5	1	2
Cavities	5	17	34
$H' = \frac{n \log n - \sum f_i \log f_i}{n}$	$[20 \log 20 - (5 \log 5 + 5 \log 5 + 5 \log 5 + 5 \log 5)]/20$	$[20 \log 20 - (1 \log 1 + 1 \log 1 + 1 \log 1 + 17 \log 17)]/20$	$[40 \log 40 - (2 \log 2 + 2 \log 2 + 2 \log 2 + 34 \log 34)]/40$
	$= [26.0206 - (3.4949 + 3.4949 + 3.4949 + 3.4949)]/20$	$= [26.0206 - (0 + 0 + 0 + 20.9176)]/20$	$= [64.0824 - (0.6021 + 0.6021 + 0.6021 + 52.0703)]/40$
	$= 12.0410/20 = 0.602$	$= 5.1030/20 = 0.255$	$= 10.2058/40 = 0.255$
	$H'_{\max} = \log 4 = 0.602$	$H'_{\max} = \log 4 = 0.602$	$H'_{\max} = \log 4 = 0.602$
	$J' = \frac{0.602}{0.602} = 1.00$	$J' = \frac{0.255}{0.602} = 0.42$	$J' = \frac{0.255}{0.602} = 0.42$

where c is the integer portion of n/k, and d is the remainder. The Brillouin-based evenness measure is, therefore,

$$J = \frac{H}{H_{max}}, \tag{4.22}$$

with $1 - J$ being a dominance measure. Inasmuch as we consider that we have data from an entire population, k is a population measurement, rather than an estimate of one, and J is not a biased estimate as is J'.

For further considerations of these and other diversity measures, see Brower and Zar 1977: Chapter 5B.

4.7 OTHER MEASURES OF DISPERSION

Besides the preceding statistics, other measures are used, albeit infrequently, to express dispersion. The *semiquartile*, or *interquartile, range* is the distance between the first quartile (25th percentile) and the third quartile (75th percentile). Distances between other quartiles may be used similarly (see Dixon and Massey, 1969: 134–139), for either ratio, interval, or ordinal data.

4.8 THE EFFECT OF CODING DATA

In Section 3.5 it was shown how coding data may facilitate statistical computations. Such benefits are even more apparent when calculating SS, s^2, and s, because of the labor, and concomitant chances of error, involved in the squaring of unwieldy numbers.

When data are coded by adding or subtracting a constant, all the measures of dispersion except the coefficient of variation are not changed from what they were for the uncoded data, because these measures involve deviations, and deviations are not changed by translation. Sample 1 in Example 4.4 demonstrates these relationships. To arrive at the desired coefficient of variation, simply decode $\bar{X}$ and s before calculating V.

When coding by multiplication or division, however, the measures of dispersion are affected, for the magnitudes of the deviations will be changed. The standard deviation, range, and mean deviation are changed in the same manner as the measures of central tendency (Section 3.5). However, the variance changes as the square of the coding constant, whereas the coefficient of variation is unchanged, as is shown in sample 2 of Example 4.4.

When calculating information-theoretic diversity indices, coding by multiplication or division does not affect the results (see Example 4.3). Coding by addition or subtraction should not be employed.

Table 4.2 summarizes the effect of coding on sample statistics considered thus far. A coded datum may be defined as

$$[X_i] = MX_i + A, \tag{4.23}$$

Example 4.4

Coding data to facilitate the calculation of measures of dispersion.

Sample 1 (Coding by Subtraction: A = −840 g)

Without coding X_i		*Using coding* $[X_i]$	
X_i (g)	X_i^2 (g²)	$[X_i]$ (g)	$[X_i]^2$ (g²)
842	708,964	2	4
843	710,649	3	9
844	712,336	4	16
846	715,716	6	36
846	715,716	6	36
847	717,409	7	49
848	719,104	8	64
849	720,801	9	81
$\Sigma X_i = 6765$ g	$\Sigma X_i^2 = 5{,}720{,}695$ g²	$\Sigma [X_i] = 45$ g	$\Sigma [X_i]^2 = 295$ g²

Without coding:

$$s^2 = \frac{5720695 \text{ g}^2 - \dfrac{(6765 \text{ g})^2}{8}}{7}$$

$$= 5.98 \text{ g}^2$$

$$s = 2.45 \text{ g}$$

$$\bar{X} = 845.6 \text{ g}$$

$$V = \frac{s}{\bar{X}} = \frac{2.44 \text{ g}}{845.6 \text{ g}}$$

$$= 0.0029 = 0.29\%$$

Using coding:

$$[s^2] = \frac{295 \text{ g}^2 - \dfrac{(45 \text{ g})^2}{8}}{7}$$

$$= 5.98 \text{ g}^2$$

$$[s] = 2.44 \text{ g}$$

$$[\bar{X}] = 5.6 \text{ g}$$

Sample 1 (Coding by Division: M = 0.01)

Without coding X_i		*Using coding* $[X_i]$	
X_i (sec)	X_i^2 (sec²)	$[X_i]$ (sec)	$[X_i^2]^2$ (sec²)
800	640,000	8.00	64.00
900	810,000	9.00	81.00
950	902,500	9.50	90.25
1100	1,210,000	11.00	121.00
1250	1,562,500	12.50	156.25
1300	1,690,000	13.00	169.00
$\Sigma X_i = 6300$ sec	$\Sigma X_i^2 = 6{,}815{,}000$ sec²	$\Sigma [X_i] = 63.00$ sec	$\Sigma [X_i]^2 = 681.50$ sec²

Without coding:

$$s^2 = \frac{6815000 \text{ sec}^2 - \dfrac{(6300 \text{ sec})^2}{6}}{5}$$

$$= 40{,}000 \text{ sec}^2$$

$$s = 200 \text{ sec}$$

$$\bar{X} = 1050 \text{ sec}$$

$$V = 0.19 = 19\%$$

Using coding:

$$[s^2] = \frac{681.50 \text{ sec}^2 - \dfrac{(63.00 \text{ sec})^2}{6}}{5}$$

$$= 4 \text{ sec}^2$$

$$[s] = 2.00 \text{ sec}$$

$$[\bar{X}] = 10.50 \text{ sec}$$

$$[V] = 0.19 = 19\%$$

where M is a multiplication coding constant and A is an addition coding constant. (In Example 4.4, we see that $M = 1$ and $A = -840$ g in sample 1; $M = 0.01$ and $A = 0$ in sample 2.)

TABLE 4.2 THE EFFECT OF CODING DATA ON SAMPLE STATISTICS, WHERE $[X_i] = MX_i + A$

Statistic	Value without coding	Value using coding
Mean, $\bar{X}$	$\bar{X} = \frac{[\bar{X}] - A}{M}$	$[\bar{X}] = M\bar{X} + A$
Median and mode	similar to mean	
Sum of squares, SS	$\text{SS} = \frac{[\text{SS}]}{M^2}$	$[\text{SS}] = M^2\text{SS}$
Variance, s^2	$s^2 = \frac{[s^2]}{M^2}$	$[s^2] = M^2s^2$
Standard deviation, s	$s = \frac{[s]}{M}$	$[s] = Ms$
Range and mean deviation	similar to standard deviation	
Coefficient of variation, V (if $A = 0$)*	$V = [V]$	$[V] = V$
Shannon diversity index, H' (if $A = 0$)*	$H' = [H']$	$[H'] = H'$
Shannon evenness index, J' (if $A = 0$)*	$J' = [J']$	$[J'] = J'$

*If $A \neq 0$, one cannot convert between coded and uncoded statistics.

EXERCISES

4.1. The body weights, in grams, collected from a population of rodent body weights are:

66.1, 77.1, 74.6, 61.8, 71.5.

(a) Compute the "sum of squares" and the variance of these data using Equations 4.3 and 4.6, respectively.

(b) Compute the "sum of squares" and the variance of these data by using Equations 4.7 and 4.8, respectively.

4.2. Consider the following data, which are amino acid concentrations (mg/100 ml) in arthropod hemolymph:

240.6, 238.2, 236.4, 244.8, 240.7, 241.3, 237.9.

(a) Determine the range of the data.

(b) Calculate the "sum of squares" of the data.

(c) Calculate the variance of the data.

(d) Calculate the standard deviation of the data.

(e) Calculate the coefficient of variation of the data.

4.3. The following frequency distribution of tree species was observed in a random sample from a forest:

Species	*Frequency*
White oak	44
Red oak	3
Shagbark hickory	28
Black walnut	12
Basswood	2
Slippery elm	8

(a) Use the Shannon index to express the tree species diversity.
(b) Compute the maximum Shannon diversity possible for the given number of species and individuals.
(c) Calculate the Shannon evenness for these data.

4.4. Assume the data in Exercise 4.3 were an entire population (e.g., all the trees planted around a group of buildings).
(a) Use the Brillouin index to express the tree species diversity.
(b) Compute the maximum Brillouin diversity possible for the given number of species and individuals.
(c) Calculate the Brillouin evenness measure for these data.

5 Testing for Goodness of Fit

The concepts of statistical inference will be introduced in this chapter by using some statistical methods especially suited for the analysis of nominal scale data. As nominal data are counts of items or events in each of several classifications, methods for their analysis are referred to as *enumeration statistical methods*.

The widely used chi-square* statistic was introduced by Karl Pearson† in the years surrounding 1900, and its theory and application were subsequently expanded by him and R. A. Fisher (Lancaster, 1969: Chapter 1). Various aspects of chi-square analyses will be discussed in Chapters 5 and 6. The more recently developed maximum likelihood ratio approaches to enumeration data analysis are introduced in Sections 5.9 and 6.6. Goodness of fit for ordered data is best handled by the Kolmogorov-Smirnov test, considered in Sections 5.10 and 5.11, or by the Watson test of Section 25.1.

5.1 CHI-SQUARE GOODNESS OF FIT

It is frequently desired to obtain a sample of nominal scale data and to infer whether the population from which it came conforms to a certain theoretical distribution. For example, a plant geneticist may raise 100 progeny from a cross that is hypothesized to result in a 3: 1 phenotypic ratio of yellow-flowered to green-flowered plants. Perhaps a ratio of 84 yellow: 16 green is observed, although out of this total of 100

*The Greek letter "chi," χ, is pronounced as the "ky" in "sky."

†Karl Pearson (1857–1936), British mathematician; one of the founders of the field of statistics.

plants, the geneticist's hypothesis would predict a ratio of 75 yellow: 25 green. The question to be asked, then, is whether the observed frequencies (84 and 16) deviate significantly from the frequencies expected if the hypothesis were true (75 and 25).

The statistical procedure for attacking the question first involves the concise statement of the hypothesis to be tested. The hypothesis in this case is that the population which was sampled has a 3: 1 ratio of yellow-flowered to green-flowered plants. Statistically, this is referred to as a *null hypothesis* (abbreviated H_0), because it is a statement of "no difference"; in this instance, we are hypothesizing that the population flower color ratio is not different from 3: 1. If it is concluded that H_0 is false, then an *alternate hypothesis* (abbreviated H_A) will be assumed to be true. In this case, H_A would be that the population sampled has a flower color ratio which *is not* 3 yellow: 1 green. One states a null hypothesis and an alternate hypothesis for every statistical test performed, and all possible outcomes are accounted for by the two hypotheses.

The following calculation of a statistic called *chi-square* is used as a measure of how far a sample distribution deviates from a theoretical distribution:

$$\chi^2 = \sum_{i=1}^{k} \frac{(f_i - \hat{f}_i)^2}{\hat{f}_i}. \tag{5.1}$$

Here, f_i is the frequency, or number of counts, observed in class i, $\hat{f}_i$ is the frequency expected in class i if the null hypothesis is true,* and the summation is performed over all k categories of data; χ is lowercase Greek chi. Example 5.1 shows the chi-square calculation for the flower color data presented. In this sample, there are two categories of data (i.e., $k = 2$): yellow-flowered plants and green-flowered plants. The expected frequency, $\hat{f}_i$, of each class is calculated by multiplying the total number of observations, n, by the proportion of the total that the null hypothesis predicts for the class. Therefore, for the two classes in the example, $\hat{f}_1 = (100)(\frac{3}{4}) = 75$ and $\hat{f}_2 = (100)(\frac{1}{4}) = 25$.

It should be apparent, by examining Equation 5.1, that larger disagreement between observed and expected frequencies (i.e., larger $f_i - \hat{f}_i$ values) will result in a larger χ^2 value. Thus, this type of calculation is referred to as a measure of *goodness of fit* (although it might better have been named a measure of "poorness of fit"). A calculated χ^2 value can be as small as zero, in the case of a perfect fit (i.e., each f_i value equals its corresponding $\hat{f}_i$), or very large if the fit is very bad; it can never be a negative value.

It is fundamentally important to appreciate that the chi-square statistic is calculated using the actual frequencies observed. It is not valid to convert the data to percentages and to attempt to submit the percentages to Equation 5.1.

5.2 STATISTICAL SIGNIFICANCE

Now that we have seen how to calculate a chi-square value as a measure of disagreement between observed and expected frequencies, we need to be able to draw conclusions about the significance of the disagreement.

*The symbol $\hat{f}$ is pronounced "f hat."

Example 5.1

Calculation of chi-square goodness of fit of data consisting of 100 flower colors to a hypothesized color ratio of 3: 1.

H_0: The sample data came from a population having a 3: 1 ratio of yellow to green flowers.
H_A: The sample data came from a population not having a 3: 1 ratio of yellow to green flowers.

The data recorded are the 100 observed frequencies, f_i, in each of the two flower color categories, with the frequencies expected under the null hypothesis, $\hat{f}_i$, in parentheses.

	Category (Flower Color)		
	Yellow	*Green*	*n*
f_i	84	16	100
($\hat{f}_i$)	(75)	(25)	

$$\text{degrees of freedom} = \nu = k - 1 = 2 - 1 = 1$$

$$\chi^2 = \sum \frac{(f_i - \hat{f}_i)^2}{\hat{f}_i} = \frac{(84 - 75)^2}{75} + \frac{(16 - 25)^2}{25}$$
$$= \frac{9^2}{75} + \frac{9^2}{25}$$
$$= 1.080 + 3.240$$
$$= 4.320$$
$$0.025 < P < 0.05$$

Therefore, reject H_0.

What is meant by statistical significance is derived from considerations of probability. Consider that the null hypothesis is true, i.e., the geneticist sampled a population of plants in which the yellow-to-green ratio is indeed 3 to 1. What we wish to ask is if it is likely to obtain from such a population a random sample of plants having an 84: 16 flower color ratio. If such a sample ratio can occur reasonably often, then we have no cause to reject H_0. If, however, there is little probability of obtaining 84 yellow-flowered and 16 green-flowered plants in a random sample from a population with a 3: 1 ratio, then we may infer that the null hypothesis is false and that the alternate hypothesis is true (i.e., the sample came from a population with a color ratio that is not 3: 1).

The computation of the probabilities that we require involves such complex mathematics that we are fortunate in having available tables of chi-square probabilities to aid in hypothesis testing. Below is reproduced a portion of the first row of Appendix Table B.1, which is a table of χ^2 values having certain probabilities of occurrence. As will be explained in Section 5.3, α (lowercase Greek alpha) refers to probability.

$\alpha = 0.10$	0.05	0.025	0.01	0.005	0.001
2.706	3.841	5.024	6.635	7.879	10.828

By means of this table, we can see, for example, that the probability (P) of a χ^2 equal to or greater than 2.706 is 0.10 (i.e., 10%); this statement can be written concisely as $P(\chi^2 \geq 2.706) = 0.10$. As another example, we see that $P(\chi^2 \geq 3.841) = 0.05$. Now, in Example 5.1 we obtained $\chi^2 = 4.320$. By consulting Table B.1, or the small portion of it here reproduced, we see that this value has associated with it a probability somewhere between 0.025 and 0.05, for $P(\chi^2 \geq 5.024) = 0.025$ and $P(\chi^2 \geq 3.841) = 0.05$. Thus, for this example, one can state that $0.025 < P(\chi^2 \geq 4.320) < 0.05$, or, simply, $0.025 < P < 0.05$. What this table tells us is that if H_0 were true, and if we repeated this same experiment a very large number of times, we could expect to get results that deviate at least this much from the hypothetical frequencies from 2.5 to 5% of the time.*

As explained in Section 5.3, biostatisticians often specify that if the magnitude of a calculated test statistic (such as χ^2) has an associated probability of 5% or less, its occurrence is so unlikely to be due to random sampling alone that we may reasonably conclude that the null hypothesis is false. This is the case for the data in Example 5.1; therefore, we reject H_0 and accept H_A, concluding that the population sampled has a flower color ratio other than 3 yellow: 1 green.

5.3 STATISTICAL ERRORS IN HYPOTHESIS TESTING

One needs an objective criterion for rejecting or not rejecting the null hypothesis for a particular statistical test. Theoretically, very large χ^2 values might be calculated even when the hypothesis is true, although the larger the chi-square the smaller is the probability that H_0 is true. But how small a probability (i.e., how large a χ^2) shall we require to reject the null hypothesis? As explained below, a probability of 5% or less is commonly used as the criterion for rejection of H_0. The probability used as the criterion for rejection is termed the *significance level*, denoted by α, and the value of the test statistic corresponding to this probability (e.g., $\chi^2 = 3.841$, for the 5% significance level) is the *critical value* of the statistic.

It is very important to realize that a true null hypothesis occasionally will be rejected, which of course means that we have committed an error. Moreover, this error will be committed with a frequency of α. That is, if H_0 is in fact a true statement about a statistical population, it will be concluded (erroneously) to be false 5% of the time. The rejection of a null hypothesis when it is in fact true is a *Type I error* (also called an α error, or an error of the first kind). On the other hand, if H_0 is in fact false, our test may occasionally not detect this fact, and we shall have reached an erroneous conclusion by not rejecting H_0. This error, of not rejecting the null hypothesis when it is in fact false, is a *Type II error* (also called a β error, or an error of the second kind). The *power* of a statistical test is defined as $1 - \beta$; i.e, power is the probability of rejecting the null hypothesis when it is in fact false and should be rejected.

Whereas the probability of committing a Type I error is α, the specified significance level, the probability of committing a Type II error is β, a value that generally

*Some desk calculators and computer programs have the capability of determining the exact probability of a given χ^2 (Guenther, 1977). For the present example, we would thereby find that $P(\chi^2 \geq 4.320) = 0.038$.

we neither specify nor know. What we do know is that for a given sample size, n, the value of α is related inversely to the value of β. That is, lower probabilities of committing a Type I error are associated with higher probabilities of committing a Type II error, and the only way to reduce both types of error simultaneously is to increase n. Thus, for a given α, larger samples will result in statistical test with greater power $(1 - \beta)$.

Table 5.1 summarizes these two types of statistical errors. Since, for a given n, one cannot minimize both of them, it is important to ask what the acceptable combination of the two might be. By experience, and thence by convention, an α of 0.05 is usually considered to be a "small enough" chance of committing a Type I error, while not being so small as to result in "too large a chance" of a Type II error. But there is nothing sacrosanct about the 0.05 level. Although it is the most widely used and perhaps the most widely useful significance level, experimenters may decide for themselves whether it is more important to minimize one type of error or the other.

TABLE 5.1 THE TWO TYPES OF ERRORS IN HYPOTHESIS TESTING

	If H_0 is true	*If H_0 is false*
If H_0 is rejected:	Type I error	No error
If H_0 is not rejected:	No error	Type II error

In some situations, for example, a 5% chance of an incorrect rejection of H_0 is unacceptably high, so the 1% level of significance is sometimes used. It is necessary, of course, to state the significance level used when reporting the results of a statistical test. Indeed, rather than simply stating whether the null hypothesis is rejected, it is good practice to state also the test statistic itself and the best estimation of its exact probability. (For Example 5.1, we could write $\chi^2 = 4.320$ and $0.025 < P < 0.05$.) In this way, readers of the research results may draw their own conclusions, even if their choice of significance level is different from yours. Bear in mind, however, that the choice of α is to be made before ever seeing the data. Otherwise there is a great risk of having the choice influenced by the examination of the data, introducing bias rather than objectivity into the proceedings. The best practice generally is to decide on the null hypothesis, alternate hypothesis, and significance level before commencing the experiment. It is conventional to refer to rejection of H_0 at the 5% level of significance as denoting a "significant" difference (indicated by *; for Example 5.1, one might write $\chi^2 = 4.320^*$) and rejection at the 1% level as denoting a "highly significant" difference (indicated by **).

Note throughout this book that a null hypothesis (a hypothesis expressing "no difference") is the only type of hypothesis that can be tested statistically, and that only when it is rejected can one routinely express a specific level of confidence in the conclusion. That is, rejection of a null hypothesis is done with a stated probability of error (α), but acceptance of a null hypothesis is done with a chance of error (β) that is unknown (although for some statistical testing procedures,* there are methods of estimating β, given α and n). Also bear in mind that failing to reject a null hypothesis is not "proof" that the hypothesis is true. It denotes only that there is not sufficient evidence to conclude that it is false.

5.4 CHI-SQUARE GOODNESS OF FIT FOR MORE THAN TWO CATEGORIES

Example 5.1 considered chi-square testing for goodness of fit when there are two categories of data (i.e., $k = 2$). This analysis may be extended readily to any larger number of classes, as Example 5.2 exemplifies. Here, $k = 4$, and the calculated χ^2, using Equation 5.1, is 8.972. (We shall routinely express a calculated chi-square to three decimal places, because that is the accuracy of the table of critical values, Appendix Table B.1. Therefore, to avoid rounding error, we shall perform all intermediate computations, including those of $\hat{f}_i$, to four decimal places.)

Example 5.2

Chi-square goodness of fit for $k = 4$.

H_0: The sample comes from a population having a 9: 3: 3: 1 ratio of yellow-smooth to yellow-wrinkled to green-smooth to green-wrinkled seeds.

H_A: The sample comes from a population not having a 9: 3: 3: 1 ratio of the above four seed phenotypes.

The sample data are recorded as observed frequencies, f_i, with the frequencies expected under the null hypothesis, $\hat{f}_i$, in parentheses.

	Yellow smooth	*Yellow wrinkled*	*Green smooth*	*Green wrinkled*	n
f_i	152	39	53	6	250
$(\hat{f}_i)$	*(140.6250)*	*(46.8750)*	*(46.8750)*	*(15.6250)*	

$$\nu = k - 1 = 3$$

$$\chi^2 = \frac{11.3750^2}{140.6250} + \frac{7.8750^2}{46.8750} + \frac{6.1250^2}{46.8750} + \frac{9.6250^2}{15.6250}$$

$$= 0.9201 + 1.3230 + 0.8003 + 5.9290$$

$$= 8.972$$

$0.025 < P < 0.05$

Therefore, reject H_0.

It has already been pointed out that larger χ^2 values will result from larger differences between f_i and $\hat{f}_i$, but large calculated χ^2 values may also simply be the result of a large number of classes of data, because the calculation involves the summing over all classes. Thus, in considering the significance of a calculated χ^2, the value of k must in some way be taken into account. What is done is to consider a quantity known as *degrees of freedom* (abbreviated DF, or by lowercase Greek nu, ν).

*For example, Section 22.7 introduces a method for estimating power in testing for a hypothesized proportion of observations in each of two categories, providing n is large enough.

For the goodness of fit testing discussed in this chapter, DF $= k - 1$. Thus, for Example 5.2, DF $= 4 - 1 = 3$, while the calculated $\chi^2 = 8.972$. Entering Appendix Table B.1 in the row for 3 DF, we see that $P(\chi^2 \geq 7.815) = 0.05$ and $P(\chi^2 \geq 9.348) = 0.025$. Therefore, $0.025 < P(\chi^2 \geq 8.972) < 0.05$, and we would reject the null hypothesis which states that the sample came from a population having a 9: 3: 3: 1 phenotypic ratio of yellow-smooth: yellow-wrinkled: green-smooth: green-wrinkled seeds. Tabled critical values are frequently denoted as $\chi^2_{\alpha,\nu}$, so that we could write $\chi^2_{0.05,3} = 7.815$, $\chi^2_{0.10,3} = 6.251$, etc.

When we say, in a goodness of fit problem such as Example 5.1 or 5.2, that DF $= k - 1$, we are stating that, given the frequencies in any $k - 1$ of the categories, we can readily calculate the frequency in the remaining category. This is true because n is known, and the sum of the frequencies in all k categories equals n. (In other words, one has "freedom" in assigning frequencies to only $k - 1$ of the categories.) Another way of looking at chi-square degrees of freedom is to note that DF equals k minus the number of sample constants used to calculate the expected frequencies. In the present examples, only one constant, n, was so used, so $\nu = k - 1$. Degrees of freedom other than $k - 1$ will be encountered in later chapters.

5.5 SUBDIVIDING CHI-SQUARE ANALYSES

In Example 5.2, the chi-square analysis detected a difference between the observed and expected frequencies too great to be attributed to chance, and the null hypothesis was rejected. This conclusion may be satisfactory in some instances, but in many cases the investigator will wish to perform further analysis.

For the example under consideration, the null hypothesis is that the sample came from a population having a 9: 3: 3: 1 phenotypic ratio. If the chi-square analysis had not led to a rejection of the hypothesis, we would proceed no further. But since H_0 was rejected, we may wish to ask whether the significant disagreement between observed and expected frequencies was concentrated in certain of the classes, or whether the difference was due to the effects of the data in all of the classes. Of the four individual contributions to the chi-square value—0.9201, 1.3230, 0.8003, and 5.9290—that resulting from the last class (the green-wrinkled seeds) contributes a relatively large amount to the size of the calculated χ^2. Thus we see that the nonconformity of the sample frequencies to those expected from a population with a 9: 3: 3: 1 ratio is due largely to the magnitude of the discrepancy between f_4 and $\hat{f}_4$.

This line of thought can be examined as shown in Example 5.3. First, we test H_0: f_1, f_2, and f_3 came from a population having a 9: 3: 3 ratio. (H_A: the frequencies in the first three categories came from a population having a phenotypic ratio other than 9: 3: 3.) This null hypothesis is not rejected, indicating that the frequencies in the first three categories conform acceptably well to those predicted by H_0. Then we test the frequency of green-wrinkled seeds against the combined frequencies for the other three phenotypes, under the null hypothesis of a 1: 15 ratio. The calculated χ^2 value causes us to reject this hypothesis, however, and we draw the conclusion that the nonconformity of the data in Example 5.2 to the hypothesized frequencies is due primarily to the observed frequency of green-wrinkled seeds.

Example 5.3

Chi-square goodness of fit, subdividing the chi-square analysis of Example 5.2.

H_0: The sample came from a population with a 9: 3: 3 ratio of the first three phenotypes in Example 5.2.

H_A: The sample came from a population not having a 9: 3: 3 ratio of the first three phenotypes in Example 5.2.

Seed Characteristics

	Yellow smooth	*Yellow wrinkled*	*Green smooth*	n
f_i	152	39	53	244
$(\hat{f}_i)$	*(146.4000)*	*(48.8000)*	*(48.8000)*	

$$\nu = k - 1 = 2$$

$$\chi^2 = \frac{5.6000^2}{146.4000} + \frac{9.8000^2}{48.8000} + \frac{4.2000^2}{48.8000}$$

$$= 0.2142 + 1.9680 + 0.3615$$

$$= 2.544$$

$0.25 < P < 0.50$

Therefore, do not reject H_0.

H_0: The sample came from a population with a 1: 15 ratio of green-wrinkled to other seed phenotypes.

H_A: The sample came from a population not having the 1: 15 ratio stated in H_0.

Seed Characteristics

	Green wrinkled	*Others*	n
f_i	6	244	250
$(\hat{f}_i)$	*(15.625)*	*(234.375)*	

$$\nu = k - 1 = 1$$

$$\chi^2 = \frac{9.6250^2}{15.6250} + \frac{9.6250^2}{234.3750}$$

$$= 5.9290 + 0.3953$$

$$= 6.324$$

$0.01 < P < 0.025$

Therefore, reject H_0.

Equation 5.1 can be rewritten as

$$\chi^2 = \sum_{i=1}^{k} \frac{f_i^2}{\hat{f}_i} - n, \tag{5.2}$$

where n is the sum of all the f_i's, namely the total number of observations in the sample. Although this formula renders the calculation of χ^2 a little easier, it has one big disadvantage. It does not enable us to examine each contribution to χ^2 [i.e., each $(f_i - \hat{f}_i)^2/\hat{f}_i$], and, as shown in this section, such an examination is an aid in determining how one might subdivide an overall chi-square analysis into component chi-square analyses. Thus, Equation 5.2 is seldom encountered.

Strictly speaking, it is not proper to test statistical hypotheses developed after examining the data to be tested. Therefore, the analyses of this section should be considered only a guide to developing additional hypotheses, hypotheses that should then be stated in advance of their being tested with a new set of data.

5.6 CHI-SQUARE CORRECTION FOR CONTINUITY

Chi-square values obtained from actual data, using Equation 5.1, belong to a discrete, or discontinuous, distribution, in that they can take on only certain values. For instance, in Example 5.1 we calculated a chi-square value of 4.320 for $f_1 = 84$, $f_2 = 16$, $\hat{f}_1 = 75$, and $\hat{f}_2 = 25$. If we had observed $f_1 = 83$ and $f_2 = 17$, the calculated chi-square value would have been $(83 - 75)^2/75 + (17 - 25)^2/25 = 0.8533 + 2.5600 = 3.413$; for $f_1 = 82$ and $f_2 = 18$, $\chi^2 = 2.613$; etc. These chi-square values obviously form a discrete distribution, for values between 4.320 and 3.413 or between 3.413 and 2.613 are not possible with the given $\hat{f}_i$ values. However, the theoretical χ^2 distribution, from which Table B.1 is derived, is a continuous distribution. That is, for a given ν, all values of χ^2 between 2.613 and 4.320 are possible. Thus, the results from chi-square analyses are only approximations to the theoretical distribution, and our conclusions are not really taking place exactly at the level of α which we set. This situation would be most unfortunate were it not for the fact that the approximation is a very good one except when $\nu = 1$ (and in the instances described in Section 5.7). In the case of $\nu = 1$, it is usually recommended to use the *Yates correction for continuity* (Yates, 1934), where the absolute value of each deviation of f_i from $\hat{f}_i$ is reduced by 0.5 units. That is,

$$\chi_c^2 = \sum_{i=1}^{2} \frac{(|f_i - \hat{f}_i| - 0.5)^2}{\hat{f}_i}, \tag{5.3}$$

where χ_c^2 denotes the chi-square value calculated with the correction for continuity.

Thus, for Example 5.1 (an analysis where $\nu = 1$) $\chi_c^2 = (|84 - 75| - 0.5)^2/75 + (|16 - 25| - 0.5)^2/25 = 72.2500/75 + 72.2500/25 = 0.9633 + 2.8900 = 3.853$. For this calculated χ_c^2, we find by Table B.1 that $0.025 < P < 0.05$. Although this is the same conclusion as is arrived at without the correction for continuity, this will not always be the case. Without the continuity correction, the calculated χ^2 may be inflated enough to cause us to reject H_0, whereas the corrected χ_c^2 value might not. In other words, not correcting for continuity may cause us to commit the Type I error with a probability greater than the stated α. The Yates correction should routinely be used when $\nu = 1$; it is not applicable for $\nu > 1$. For very large n, the effect of discontinuity is small, even for $\nu = 1$, and in such cases the Yates correction will change the calculated chi-square very little.

As another example of the use of the correction for continuity, we can reexamine the second null hypothesis of Example 5.3 and calculate $\chi_c^2 = (9.6250 - 0.5)^2/15.6250 + (9.6250 - 0.5)^2/234.3750 = 5.3290 + 0.3553 = 5.684$.

For $k = 2$, if H_0 involves a 1:1 ratio,

$$\chi^2 = \frac{(f_1 - f_2)^2}{n} \tag{5.4}$$

may be used in place of Equation 5.1, and

$$\chi_c^2 = \frac{(|f_1 - f_2| - 1)^2}{n} \tag{5.5}$$

may be used instead of Equation 5.3. Note that in these two shortcut equations, $\hat{f}_1$ and $\hat{f}_2$ need not be calculated, thus avoiding the concomitant rounding errors.

5.7 BIAS IN CHI-SQUARE CALCULATIONS

Section 5.6 explained that when one calculates the chi-square statistic, the theoretical chi-square distribution is being approximated. This approximation is a very acceptable one, except when $\nu = 1$ (in which case the Yates correction for continuity usually should be employed), and in the following circumstances.

If $\hat{f}_i$ values are very small, the calculated χ^2 is biased in that it is larger than the theoretical χ^2 value it is supposed to estimate, and we shall tend to reject the null hypothesis with a probability greater than α. This situation is clearly undesirable, and although other authors have suggested different "rules," a thorough analysis by Cochran (1954) concludes that *no expected frequency, $\hat{f}_i$, should be less than 1.0 and that no more than 20% of the expected frequencies should be less than 5.0*. More recent analysis (Roscoe and Byars, 1971) shows that this rule of thumb is good, except that it may be conservative at significance levels less than 5%, especially when the $\hat{f}_i$'s are all equal.

If one is faced with a sample size not sufficiently large for an unbiased chi-square calculation, one may use the log-likelihood method described in Section 5.9 for assessing goodness of fit. The chi-square calculation could be made, however, if the data for the classes with the offensively low $\hat{f}_i$ values were simply eliminated from H_0 and the subsequent analysis. Or, certain of the classes of data might be meaningfully combined so as to result in all $\hat{f}_i$ values being large enough to proceed with the analysis. Such modified procedures are not to be recommended as routine practice. Rather, the experimenter should strive to obtain a sufficiently large n for the analysis to be performed. When $k = 2$ and each f_i is small, the use of the binomial test (Section 22.5) is preferable to chi-square analysis. [Similarly, use of the multinomial, rather than the binomial, distribution is appropriate when $k > 2$ and the f_i's are small; however, this is a tedious procedure and will not be demonstrated here (Radlow and Alf, 1975).]

5.8 HETEROGENEITY CHI-SQUARE

It is frequently the case that a number of sets of data are being tested against the same null hypothesis, and we wish to decide whether we may combine all of the sets in order to perform one overall chi-square analysis. As an example, let us examine some of the classic data of Gregor Mendel (1933: 322). In one series of 10 experiments, Mendel obtained yellow and green seeds in the frequencies shown in Example 5.4. The data from each of the 10 samples are tested against the null hypothesis that there is a 3: 1 ratio of yellow-to-green seeds in the population from which the sample came. Since H_0 is not rejected in any of the 10 cases, it is reasonable to test another null

Example 5.4

An example of heterogeneity chi-square analysis. The seed color frequencies resulting from pea breeding experiments of Gregor Mendel (1933: 322).

The null hypothesis for each experiment is that the population sampled has a 3:1 ratio of yellow-to-green seeds.

The null hypothesis for heterogeneity chi-square testing is that all 10 samples could have come from the same population.

For each experiment, the observed frequencies, f_i, are given, with the frequencies predicted by the null hypothesis, $\hat{f}_i$, in parentheses.

Experiment	*Yellow seeds*	*Green seeds*	*Total seeds* (n)	*Chi-square*	DF
1	25 (*27.0000*)	11 (*9.0000*)	36	0.5926	1
2	32 (*29.2500*)	7 (*9.7500*)	39	1.0342	1
3	14 (*14.2500*)	5 (*4.7500*)	19	0.0175	1
4	70 (*72.7500*)	27 (*24.2500*)	97	0.4158	1
5	24 (*27.7500*)	13 (*9.2500*)	37	2.0270	1
6	20 (*19.5000*)	6 (*6.5000*)	26	0.0513	1
7	32 (*33.7500*)	13 (*11.2500*)	45	0.3630	1
8	44 (*39.7500*)	9 (*13.2500*)	53	1.8176	1
9	50 (*48.0000*)	14 (*16.0000*)	64	0.3333	1
10	44 (*46.5000*)	18 (*15.5000*)	62	0.5376	1
Total of chi-squares				7.1899	10
Chi-square of totals (i.e., pooled chi-square)	355 (*358.5000*)	123 (*119.5000*)	478	0.1367	1
Heterogeneity chi-square				7.0532 ($0.50 < P < 0.75$)	9

hypothesis, that all 10 samples could have, in fact, come from the same population. This new hypothesis may be tested by the procedure called *heterogeneity chi-square* analysis (sometimes referred to as "interaction" chi-square analysis). In addition to performing the 10 separate chi-square tests, we total all 10 f_i values and 10 $\hat{f}_i$ values and perform a chi-square test on these totals. But in totaling these values, commonly called *pooling* them, we must assume that all 10 samples came from the same population (or from populations having identical seed color ratios). If this assumption is true, we say that the samples are *homogeneous*. If this assumption is false, the samples are said to be *heterogeneous*, and the chi-square analysis on the pooled data would not be justified. So we are faced with the desirability of testing for heterogeneity, using the null hypothesis that the samples could have come from the same population (i.e., H_0: the samples are homogeneous).

Testing for heterogeneity among replicated goodness of fit tests is based on the fact that the sum of chi-square values is itself a chi-square value. If the samples are indeed homogeneous, then the total of the individual chi-square values should closely approximate the chi-square for the total frequencies. In Example 5.4, the total chi-square is 7.1899, with a total of 10 DF; and the chi-square of the totals is 0.1367, with 1 DF. The difference between these two chi-squares is itself a chi-square (called the *heterogeneity chi-square*), 7.053, with DF $= 10 - 1 = 9$.

Consulting Table B.1, we see that for the heterogeneity chi-square, $0.50 < P < 0.75$, so H_0 is not rejected. Thus we conclude that the 10 samples could have come from the same population and that their frequencies might justifiably be pooled. The Yates correction for continuity may not be applied in a heterogeneity chi-square analysis. But if we conclude that the sample data may be pooled, we should then analyze these pooled data using the correction for continuity. Thus, for Example 5.4, $\chi_c^2 = 0.128$, rather than $\chi^2 = 0.137$, should be used, once it has been determined that the data may be pooled.

Example 5.5 demonstrates how one can be misled by pooling heterogeneous samples. If the six samples shown were pooled and a chi-square computed ($\chi^2 = 0.234$), one would not reject the null hypothesis. But such a procedure would have ignored the strong indication ($P < 0.001$ for heterogeneity chi-square) that the

Example 5.5

Hypothetical data for heterogeneity chi-square analysis, demonstrating misleading results from the pooling of heterogeneous samples.

H_0: The sampled population has a 1:1 ratio of right- to left-handed men.
H_A: The sampled population does not have a 1:1 ratio of right- to left-handed men.
Sample frequencies observed, f_i, are listed, with the frequencies predicted by H_0 ($\hat{f}_i$) in parentheses.

Sample	*Right-handed*	*Left-handed*	*n*	*Chi-square*	DF
1	3 (*7.0000*)	11 (*7.0000*)	14	4.5714*	1
2	4 (*8.0000*)	12 (*8.0000*)	16	4.0000*	1
3	5 (*10.0000*)	15 (*10.0000*)	20	5.0000*	1
4	14 (*9.0000*)	4 (*9.0000*)	18	5.5556*	1
5	13 (*8.5000*)	4 (*8.5000*)	17	4.7647*	1
6	17 (*11.0000*)	5 (*11.0000*)	22	6.5455*	1

*Statistically significant.

Total of chi-squares				30.4372	6
Chi-square of totals (i.e., pooled chi-square)	56 (*53.5000*)	51 (*53.5000*)	107	0.2336	1
Heterogeneity chi-square				30.2036* $P < 0.001$	5

Therefore, one is not justified in performing a goodness of fit analysis on the pooled data.

samples came from at least two different populations. A heterogeneity chi-square analysis on the first three samples, and another such analysis on the last three samples, appear to be called for.

It is also important to realize that the pooling of homogeneous data may result in a more powerful analysis. Example 5.6 presents hypothetical data for four replicate chi-square analyses. None of the individual chi-square tests detects a significant deviation from the null hypothesis, but on pooling them, the chi-square test performed on the larger number of data is able to reject H_0. The nonsignificant heterogeneity chi-square shows that we are justified in pooling the replicates in order to analyze a single set of data with a large n.

Example 5.6

Hypothetical data for heterogeneity chi-square analysis, demonstrating how nonsignificant sample frequencies can result in significant pooled frequencies.

For each sample, and for the pooled sample:
H_0: The sampled population has equal frequencies of right- and left-handed men.
H_A: The sampled population does not have equal frequencies of right- and left-handed men.
For heterogeneity testing:
H_0: All the samples came from the same population.
H_A: The samples came from at least two different populations.
For each sample, the observed frequencies, f_i, are given, together with the expected frequencies, $\hat{f}_i$, in parentheses.

Sample	*Right-handed*	*Left-handed*	n	*Chi-square*	DF
1	15	7	22	2.9091	1
	(11.0000)	(11.0000)			
2	16	8	24	2.6667	1
	(12.0000)	(12.0000)			
3	12	5	17	2.8824	1
	(8.5000)	(8.5000)			
4	13	5	18	3.5556	1
	(9.0000)	(9.0000)			
Total of chi-squares				12.0138	4
Chi-square of totals (pooled chi-square)					
	56	25	81	11.8642	1
	(40.5000)	(40.5000)			
Heterogeneity chi-square				0.1496	3
				$0.975 < P < 0.99$	

Therefore, one is justified in pooling the four sets of data. On doing so, $\chi_c^2 = 11.111$, DF = 1.

5.9 THE LOG-LIKELIHOOD RATIO

A relatively recent statistical development is the use of the *log-likelihood ratio* for analyses of the sort for which one would employ chi-square. The log-likelihood ratio,*

*Proposed by Wilks (1935).

$\sum f_i \ln(f_i/\hat{f}_i)$, which may also be written as $\sum f_i \ln f_i - \sum f_i \ln \hat{f}_i$, has advantages and disadvantages compared with chi-square. One of the disadvantages is its somewhat more tedious computation. Another is that its theoretical distribution is poorly known; however, twice this quantity, a value called G, approximates the χ^2 distribution. Thus,

$$G = 2\sum f_i \ln \frac{f_i}{\hat{f}_i} = 4.60517 \sum f_i \log_{10} \frac{f_i}{\hat{f}_i}, \tag{5.6}$$

or, equivalently,

$$G = 2[\sum f_i \ln f_i - \sum f_i \ln \hat{f}_i] = 4.60517[\sum f_i \log_{10} f_i - \sum f_i \log_{10} \hat{f}_i] \tag{5.7}$$

is applicable as a test for goodness of fit, utilizing Table B.1 with the same number of degrees of freedom as would be used for chi-square testing. The Yates correction for continuity and heterogeneity analyses are performed in a fashion analogous to the chi-square analyses described earlier in this chapter.

Williams (1976) recommends G be used in preference to χ^2 whenever any $|f_i - \hat{f}_i| < \hat{f}_i$. The two methods often yield the same conclusions; when they do not, many statisticians prefer G and therefore recommend its routine use.

Example 5.7 demonstrates the G test for the data of Example 5.2.

Example 5.7

Calculation of the G statistic for the log-likelihood ratio goodness of fit test. The data and the hypotheses are the same as in Example 5.2.

	Yellow smooth	*Yellow wrinkled*	*Green smooth*	*Green wrinkled*	n
f_i	152	39	53	6	250
$(\hat{f}_i)$	*(140.6250)*	*(46.8750)*	*(46.8750)*	*(15.6250)*	

$\nu = k - 1 = 3$

$$\begin{aligned} G &= 4.60517[\sum f_i \log f_i - \sum f_i \log \hat{f}_i] \\ &= 4.60517[(152)(2.18184) + (39)(1.59106) + (53)(1.72428) \\ &\quad + (6)(0.77815) - (152)(2.14806) - (39)(1.67094) \\ &\quad - (53)(1.67094) - (6)(1.19382)] \\ &= 4.60517[331.63968 + 62.05134 + 91.38684 + 4.66890 \\ &\quad - 326.50512 - 65.16666 - 88.55982 - 7.16292] \\ &= 4.60517[2.35224] \\ &= 10.832 \end{aligned}$$

$0.01 < P < 0.025$

Therefore, reject H_0.

5.10 KOLMOGOROV-SMIRNOV GOODNESS OF FIT FOR DISCRETE DATA

The preceding sections deal with goodness of fit tests applicable to nominal scale data. Sections 5.10 and 5.11 present goodness of fit testing for data in ordered categories.

In Example 5.8, an experiment was performed in which 35 cats were given a choice of five identical containers, each holding a food differing from the others only in moisture content. (The cats were tested one at a time, so that no individual's actions

Example 5.8

Kolmogorov-Smirnov goodness of fit for discrete ordinal scale data.

H_0: Cats show no preference among the five food moisture contents (i.e., cats prefer all five equally).

H_A: Cats do show preference among the five food moisture contents (i.e., cats do not prefer all five equally).

Moisture class (X_i)

	(*Moist*) 1	2	3	4	(*Dry*) 5	n
f_i	2	18	10	4	1	35
$\hat{f}_i$	7	7	7	7	7	35
F_i	2	20	30	34	35	
$\hat{F}_i$	7	14	21	28	35	
$\lvert d_i \rvert$	5	6	9	6	0	

$d_{\max}$ = maximum $\lvert d_i \rvert = 9$

$(d_{\max})_{0.05,5,35} = 7$

Therefore, reject H_0.

$0.002 < P < 0.005$

would influence another's.) The data are observed frequencies, f_i, the numbers of animals choosing each of the five food moistures. Food moisture is recorded on an ordinal scale, for although we can say that food 1 is moister than food 2, and food 2 is moister than food 3, we cannot say that the difference in moisture between foods 2 and 3 is quantitatively equal to the difference between 1 and 2. That is, we can only speak of relative magnitudes, and not of quantitative measurements, of the foods' moisture contents.

The Kolmogorov-Smirnov goodness of fit test* (Kolmogorov, 1933; Smirnov, 1939a, 1939b), also called the Kolmogorov-Smirnov one-sample test, is suitable for assessing goodness of fit of an observed to an expected cumulative frequency distribution. (Section 1.4 introduces the concept of a cumulative frequency distribution.) In Example 5.8, each observed frequency (f_i) is listed with the frequency expected ($\hat{f}_i$) under the null hypothesis that the two distributions are the same, the expected frequencies ($\hat{f}_i$) being calculated just as they would be in a chi-square goodness of fit analysis (Sections 5.1–5.5). The cumulative observed frequencies (F_i) and cumulative expected frequencies ($\hat{F}_i$) are then calculated. The cumulative frequency for category i is the sum of all frequencies from categories 1 through i.

For each category, i, one determines the absolute difference between the two cumulative frequency distributions:

$$\lvert d_i \rvert = \lvert F_i - \hat{F}_i \rvert. \tag{5.8}$$

The largest $\lvert d_i \rvert$, let us call it $d_{\max}$, is the test statistic.

Critical values of $d_{\max}$ are found in Appendix Table B.8, the use of which requires

*The name of the test honors the two Russian mathematicians who worked on its development: Andrei Nikolaevich Kolmogorov (1903–1973) and Nikolai Vasil'evich Smirnov (b. 1900).

that in the experiment n, the total number of data, is an even multiple of k, the number of categories.* It should also be noted that the tabled critical values are for the case where all $\hat{f}_i$ are equal; but the table also works well for unequal $\hat{f}_i$, as long as the inequality is not great (Pettitt and Stephens, 1977).

When applicable (i.e., when the categories are ordered), the Kolmogorov-Smirnov test is more powerful than the chi-square test when n is small or when $\hat{f}_i$ values are small, and likely in other cases as well, for the order of the categories is taken into consideration by the former test but not by the latter. Thus, for example, if the f_i values had been 2, 1, 4, 18, and 10, respectively, the same χ^2 would have been calculated as would be for the data in Example 5.8, whereas the calculated d_{max} would be different.

5.11 KOLMOGOROV-SMIRNOV GOODNESS OF FIT FOR CONTINUOUS DATA

The Kolmogorov-Smirnov goodness of fit procedure was originally developed for use with continuous data (which may be on ratio, interval, or ordinal scales of measurement), rather than with discrete data. Example 5.9 presents the results of data collection where the vertical locations of 15 moths on a 25-meter tree trunk were recorded as heights above the ground. The vertical distances above the ground are the measurements, X_i, that come from a continuous distribution. For each X_i we record the observed frequency, f_i (i.e., the number of moths at that height). The cumulative observed frequencies, F_i, are then determined, from which the cumulative relative frequencies are obtained as

$$\text{rel } F_i = \frac{F_i}{n}, \tag{5.9}$$

where n is the number of measurements taken; rel F_i is simply the proportion of the sample being measurements $\leq X_i$. Then, for each X_i we determine the cumulative relative expected frequency, rel $\hat{F}_i$. In the present example, since H_0 proposes a uniform distribution over the heights 0 to 25 meters, rel $\hat{F}_i = X_i/25$ m. [If H_0 had referred to a uniform distribution over heights 1 to 25 m from the ground, then rel $\hat{F}_i$ would have been $(X_i - 1 \text{ m})/24$ m.]

To find the test statistic for the Kolmogorov-Smirnov goodness of fit for continuous data, we must calculate both

$$D_i = |\text{rel } F_i - \text{rel } \hat{F}_i| \tag{5.10}$$

and

$$D'_i = |\text{rel } F_{i-1} - \text{rel } \hat{F}_i| \tag{5.11}$$

for each i. For Equation 5.11, it is important to know that $F_0 = 0$, so $D'_1 = \text{rel } \hat{F}_1$. The test statistic is

$$D = \max[(\max D_i), (\max D'_i)], \tag{5.12}$$

which means "D is the largest D_i or the largest D'_i, whichever is larger." Critical

*If n/k is not a whole number, then we can consult the critical values for the n above and the n below the n in the experiment and, conservatively, use the larger of the two critical values.

Example 5.9

Kolmogorov-Smirnov goodness of fit for continuous ratio scale data, vertical distribution of moths on tree trunks.

H_0: Moths are distributed uniformly from ground level to height of 25 m.
H_A: Moths are not distributed uniformly from ground level to height of 25 m.
Each X_i is a height (in meters) at which a moth was observed on a tree trunk.

i	X_i	f_i	F_i	rel F_i	rel $\hat{F}_i$	D_i	D'_i
1	1.4	1	1	0.0667	0.0560	0.0107	0.0560
2	2.6	1	2	0.1337	0.1040	0.0297	0.0373
3	3.3	1	3	0.2000	0.1320	0.0680	0.0017
4	4.2	1	4	0.2667	0.1680	0.0987	0.0320
5	4.7	1	5	0.3333	0.1880	0.1453	0.0787
6	5.6	2	7	0.4667	0.2240	0.2427	0.1093
7	6.4	1	8	0.5333	0.2560	0.2773	0.2107
8	7.7	1	9	0.6000	0.3080	0.2920	0.2253
9	9.3	1	10	0.6667	0.3720	0.2947	0.2280
10	10.6	1	11	0.7333	0.4240	0.3093	0.2427
11	11.5	1	12	0.8000	0.4600	0.3400	0.2733
12	12.4	1	13	0.8667	0.4960	0.3707	0.3040
13	18.6	1	14	0.9333	0.7440	0.1893	0.1227
14	22.3	1	15	1.0000	0.8920	0.1080	0.0413

$n = 15$

max $D_i = D_{12} = 0.3707$
max $D'_i = D'_{12} = 0.3040$
$D = 0.3707$
$D_{0.05,15} = 0.33760$
Therefore, reject H_0.
$0.02 < P < 0.05$

values for this test statistic are referred to as $D_{\alpha,n}$ and are found in Table B.9. If $D \geq D_{\alpha,n}$, the H_0 is rejected at the α level of significance.

Figure 5.1 demonstrates why both D_i and D'_i are necessarily examined in comparing an observed with a hypothesized cumulative frequency distribution, when the measurement scale is continuous. (See also D'Agostino and Noether, 1973; Fisz, 1963: 12.5A; Gibbons, 1976: 3.1). What is required is the maximum deviation of the observed distribution, F (which looks like a staircase when graphed), and the hypothesized distribution, $\hat{F}$. For each $\hat{F}_i$, we must examine the vertical distance, $D_i = |F_i - \hat{F}_i|$, at the left-hand end of each step, as well as the vertical distance, $D'_i = |F_{i-1} - \hat{F}_i|$, at the right-hand end of each step.

A lesser-known, but quite good, test may be used as an alternative to the Kolmogorov-Smirnov test for goodness of fit with continuous data. That is the Watson goodness of fit test; it is discussed in Section 25.1 as being especially suited for data on a circular scale, but it is applicable as well to data such as in the present section.

Grouped Data. If continuous data are collected such that $f_i > 1$, then the Kolmogorov-Smirnov test becomes conservative (meaning that the testing is occurring at an α smaller than that which we state; and the probability of a Type II error is

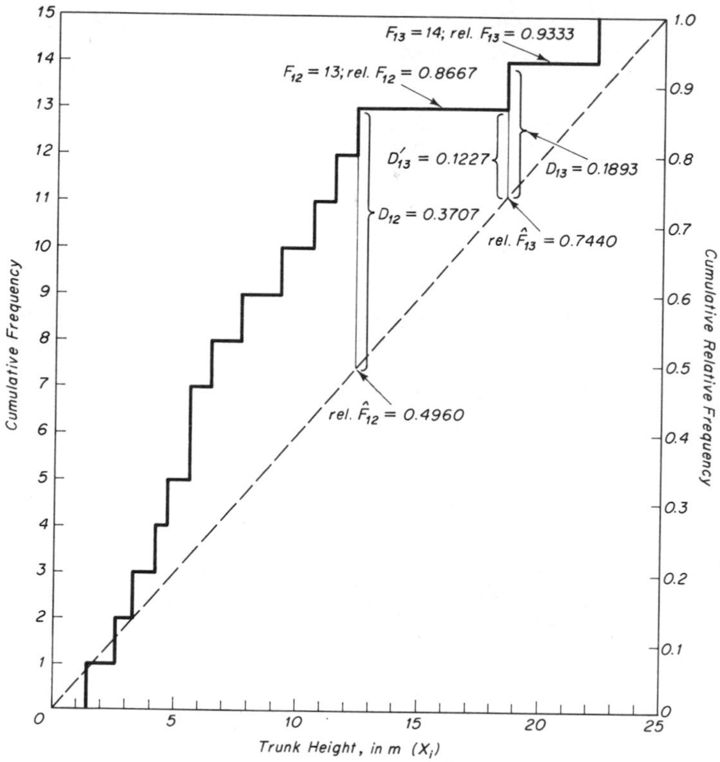

Figure 5.1 Graphical representation of Example 5.9, Kolmogorov-Smirnov goodness of fit testing for continuous data.

inflated). In such cases, the appropriate analysis resembles that for discrete data and the procedures of Section 5.10 should be used. Example 5.10 shows how the data of Example 5.9 would look had the investigator recorded them in 5-meter ranges of

Example 5.10

Kolmogorov-Smirnov goodness of fit for continuous, but grouped, data.

The data of Example 5.9, recorded in 5-meter segments of tree trunk height. H_0 and H_A as in Example 5.9.

	Trunk Height					
X_i	0–5 m	5–10 m	10–15 m	15–20 m	20–25 m	n
f_i	5	5	3	1	1	15
$\hat{f}_i$	3	3	3	3	3	15
F_i	5	10	13	14	15	
$\hat{F}_i$	3	6	9	12	15	
$\|d_i\|$	2	4	4	2	0	

$d_{max} = \text{maximum } |d_i| = 4$
$(d_{max})_{0.05,5,15} = 5$
Do not reject H_0.
$0.05 < P < 0.10$

trunk heights. Note that power is lost (and H_0 is not rejected) by grouping the data, and grouping should be avoided whenever possible.

5.12 SAMPLE SIZE REQUIRED IN KOLMOGOROV-SMIRNOV GOODNESS OF FIT FOR CONTINUOUS DATA

When dealing with continuous data, we may ask how large a sample is needed to be able to detect a difference of a given magnitude between an observed and a hypothesized cumulative frequency distribution. All that need be done is to seek the desired minimum detectable difference in the body of the table of critical values of D (Table B.9), for the selected significance level, α. For example, to be able to detect a difference as small as 0.30 between an observed and a hypothesized cumulative relative frequency distribution, at a significance level of 0.05, we need a sample of at least 20 (for $D_{0.05,19} = 0.30143$, which is larger than 0.30; and $D_{0.05,20} = 0.29516$, which is smaller than 0.30; and if the desired difference is not on the table we use the nearest smaller one). Thus, in Example 5.9, we estimate that at least 20 moths would have had to be observed to have been able to detect a difference—either D_i or D'_i—as small as 0.30. And, with $n = 20$. this means that (0.30)(20) = 0.60 is the smallest difference between cumulative frequency distributions—$|d_i|$ or $|d'_i|$—that could be detected.

If the desired detectable difference is beyond the $D_{\alpha,n}$ values in Table B.9 (i.e. $< D_{\alpha,160}$), then we know that the required sample size is greater than 160. This sample size may be estimated by employing the values of d_α at the end of Table B.9.* If we wish to detect a difference as small as Δ, then the sample size should be at least

$$n = \frac{d_\alpha^2}{\Delta^2}. \tag{5.14}$$

For example, if the collector of data in Example 5.9 had desired to be able to detect a difference, D_i or D'_i, as small as 0.10, a sample size of at least 185 moth observations should have been obtained, for

$$n = \frac{(1.35810)^2}{(0.10)^2} = 184.4.$$

EXERCISES

5.1. Consult Appendix Table B.1.

(a) What is the probability of computing a χ^2 at least as large as 3.452, if DF = 2 and the null hypothesis is true?

(b) What is $P(\chi^2 \geq 8.668)$ if $\nu = 5$?

*These values at the end of Table B.9 are

$$d_\alpha = D_{\alpha,n}\sqrt{n}, \tag{5.13}$$

which are asymptotic as n becomes very large.

(c) What is $\chi^2_{0.05,4}$?
(d) What is $\chi^2_{0.01,8}$?

5.2. Each of 126 individuals of a certain mammal species was placed in an enclosure containing equal amounts of each of six different foods. The frequency with which the animals chose each of the foods was:

Food item (i)	f_i
1	13
2	26
3	31
4	14
5	28
6	14

(a) Test the hypothesis that there is no preference among the food items.
(b) If the null hypothesis is rejected, determine if the analysis may be subdivided to ascertain which of the foods were preferred.

5.3. A sample of hibernating bats consisted of 44 males and 54 females. Test the hypothesis that the hibernating population consists of equal numbers of males and females.

5.4. In attempting to determine whether there is a 1:1 sex ratio among hibernating bats, samples were taken from four different locations:

Location	*Males*	*Females*
1	44	54
2	31	40
3	12	18
4	15	16

By performing a heterogeneity chi-square analysis, determine whether the four samples may justifiably be pooled. If they may, pool them and retest the null hypothesis of equal sex frequencies.

5.5. Test the hypothesis and data of Exercise 5.2 using the log-likelihood *G*.

5.6. A bird feeder is placed at each of 6 different heights. It is recorded which feeder was selected by each of 18 cardinals. Using the Kolmogorov-Smirnov procedure for discrete data, test the null hypothesis that each feeder is equally acceptable to cardinals.

Feeder height	*Number observed*
1 (lowest)	2
2	3
3	3
4	4
5	4
6 (highest)	2

5.7. A straight line is drawn on the ground perpendicular to the shore of a body of water. Then the locations of ground arthropods of a certain species are measured along a 1-meter-wide band on either side of the line. Use the Kolmogorov-Smirnov procedure on the following data to test the null hypothesis of uniform distribution of this species from the water's edge to a distance of 10 meters inland.

Distance from water (m)	*Numbers observed*	*Distance from water* (m)	*Numbers observed*
0.3	1	3.4	1
0.6	1	4.8	1
1.0	1	4.9	1
1.1	1	4.1	1
1.2	1	4.6	1
1.4	1	4.7	1
1.6	1	4.9	1
1.9	1	5.3	1
2.1	1	5.8	1
2.2	1	6.4	1
2.4	1	6.8	1
2.6	1	7.5	1
2.8	1	7.7	1
3.0	1	8.8	1
3.1	1	9.4	1

5.8. For a Kolmogorov-Smirnov goodness of fit test with continuous data at the 5% level of significance, how large a sample is necessary to detect a difference as small as 0.25 between cumulative relative frequency distributions?

6 Contingency Tables

In many situations, enumeration data are collected simultaneously for two variables, and it is desired to test the hypothesis that the frequencies of occurrence in the various categories of one variable are independent of the frequencies in the second variable. In Example 6.1, such data are tabulated for frequencies of observations of each of four hair colors and each of two sexes. These data are said to be arranged in a *contingency table*.

The number of columns in a contingency table is denoted by c and the number of rows by r, and there are said to be $r \times c$ "cells" in the table. In Example 6.1, $r = 2$ and $c = 4$, so we are said to be considering a 2×4 (read as "two by four") contingency table (i.e., a table with 8 cells). Note that the data might have been tabulated just as well as a 4×2 table, having the categories of hair color being the rows and the categories of sex being the columns. But this would in no way change the statistical hypotheses, tests, or conclusions that follow.

The null hypothesis for contingency table testing is that the frequencies of observations found in the rows are independent of the frequencies of observations found in the columns (or, that the column frequencies are independent of the row frequencies). For Example 6.1, H_0 could be stated "hair color relative frequencies are the same for both sexes," which is the same as saying "relative frequencies of males and females (i.e., sex ratios) are the same for each hair color." Then H_A could be stated "the hair color relative frequencies are not the same for males and females," or "the sex ratios are not the same for persons of all hair colors."

Sections 6.7 and 6.8 will introduce procedures for analyzing contingency tables of more than two dimensions, where frequencies are tabulated simultaneously for more than two variables.

Example 6.1

A 2 × 4 contingency table for testing the independence of hair color and sex in humans.

H_0: Human hair color is independent of sex in the population sampled.
H_A: Human hair color is not independent of sex in the population sampled.
$\alpha = 0.05$
The observed frequency, f_{ij}, in each cell is shown, with the frequency expected if H_0 is true (i.e., $\hat{f}_{ij}$) in parentheses.

	Hair Color				
Sex	*Black*	*Brown*	*Blond*	*Red*	**Total**
Male	32	43	16	9	100 (= R_1)
	(*29.0000*)	(*36.0000*)	(*26.6667*)	(*8.3333*)	
Female	55	65	64	16	200 (= R_2)
	(*58.0000*)	(*72.0000*)	(*53.3333*)	(*16.6667*)	
Total	87	108	80	25	300 (= n)
	(= C_1)	(= C_2)	(= C_3)	(= C_4)	

$$\chi^2 = \sum\sum \frac{(f_{ij} - \hat{f}_{ij})^2}{\hat{f}_{ij}}$$

$$= \frac{(32 - 29.0000)^2}{29.0000} + \frac{(43 - 36.0000)^2}{36.0000} + \frac{(16 - 26.6667)^2}{26.6667} + \frac{(9 - 8.3333)^2}{8.3333}$$

$$+ \frac{(55 - 58.0000)^2}{58.0000} + \frac{(65 - 72.0000)^2}{72.0000} + \frac{(64 - 53.3333)^2}{53.3333}$$

$$+ \frac{(16 - 16.6667)^2}{16.6667}$$

$$= 0.3103 + 1.3611 + 4.2667 + 0.0533 + 0.1552 + 0.6806 + 2.1333 + 0.0267 = 8.987$$

$\nu = (r - 1)(c - 1) = (2 - 1)(4 - 1) = 3$
$\chi^2_{0.05,3} = 7.815$
Therefore, reject H_0.
$0.025 < P < 0.05$

6.1 CHI-SQUARE ANALYSIS

The most common procedure for analyzing contingency table data is by using the chi-square statistic.* Recall that for the computation of chi-square one utilizes observed and expected frequencies (and never proportions or percentages). For the goodness of fit analysis introduced in Section 5.1, f_i denoted the frequency observed in category i of the variable under study. In a contingency table, we have two variables under consideration, and we denote an observed frequency as f_{ij}. By means of the double subscript, f_{ij} refers to the frequency observed in row i and column j of the contingency table. In Example 6.1, the value in row 1 column 1 is denoted as f_{11}, that in row 2 column 3 as f_{23}, and so on. Thus, $f_{11} = 32, f_{12} = 43, f_{13} = 16, \ldots, f_{23} = 64$, and $f_{24} = 16$.

*The early development of chi-square analysis of contingency tables is credited to Pearson (1904) and Fisher (1922).

The total frequency in row i of the table is denoted as R_i and is obtained as $R_i = \sum_{j=1}^{c} f_{ij}$. Thus, $R_1 = f_{11} + f_{12} + f_{13} + f_{14} = 100$, which is the total number of males in the sample, and $R_2 = f_{21} + f_{22} + f_{23} + f_{24} = 200$, which is the total number of females in the sample. The column totals, C_j, are obtained by analogous summations: $C_j = \sum_{i=1}^{r} f_{ij}$. For example, the total number of blonds in the sample data is $C_3 = \sum_{i=1}^{2} f_{i3} = f_{13} + f_{23} = 80$, the total number of redheads is $C_4 = \sum_{i=1}^{2} f_{i4} = 25$, and so on. The total number of observations in all cells of the table is called the grand total, and is $\sum_{i=1}^{r} \sum_{j=1}^{c} f_{ij} = f_{11} + f_{12} + f_{13} + \cdots + f_{23} + f_{24} = 300$, which is simply n, the size of our sample. The computation of the grand total may be written in several other notations: $\sum_i \sum_j f_{ij}$, or $\sum_{i,j} f_{ij}$, or simply $\sum \sum f_{ij}$. When no indices are given on the summation signs, we assume that the summation of all values in the sample is desired.

For chi-square analysis of contingency tables, one uses the formula

$$\chi^2 = \sum \sum \frac{(f_{ij} - \hat{f}_{ij})^2}{\hat{f}_{ij}} \tag{6.1}$$

In this formula,* similar to Equation 5.1 for chi-square goodness of fit, $\hat{f}_{ij}$ refers to the frequency expected in row i and column j if the null hypothesis is true. If, in Example 6.1, hair color is in fact independent of sex, then $\frac{100}{300} = \frac{1}{3}$ of all black-haired people would be expected to be males and $\frac{200}{300} = \frac{2}{3}$ would be expected to be females. That is, $\hat{f}_{11} = \frac{100}{300}(87) = 29$ (the expected number of black-haired males), $\hat{f}_{21} = \frac{200}{300}(87) = 58$ (the expected number of black-haired females), $\hat{f}_{12} = \frac{100}{300}(108) = 36$ (the expected number of brown-haired males), etc. The general formula for obtaining expected frequencies in a contingency table is

$$\hat{f}_{ij} = \frac{(R_i)(C_j)}{n}, \tag{6.4}$$

and it is in this way that the $\hat{f}_{ij}$ values in Example 6.1 were obtained. Note that one can check for arithmetic errors in one's calculations by observing that $R_i = \sum_{j=1}^{c} f_{ij} = \sum_{j=1}^{c} \hat{f}_{ij}$ and $C_j = \sum_{i=1}^{r} f_{ij} = \sum_{i=1}^{r} \hat{f}_{ij}$. That is, the row totals of the expected frequencies equal the row totals of the observed frequencies, and the column totals of the expected frequencies equal the column totals of the observed frequencies.

Once χ^2 has been calculated, its significance can be ascertained from Table B.1, but to do so, one must determine the degrees of freedom of the contingency table.

*Just as Equation 5.2 is equivalent to Equation 5.1 for chi-square goodness of fit, the following are mathematically equivalent to Equation 6.1 for contingency tables:

$$\chi^2 = \sum \sum \frac{f_{ij}^2}{\hat{f}_{ij}} - n \tag{6.2}$$

or

$$\chi^2 = n\left(\sum \sum \frac{f_{ij}^2}{R_i C_j} - 1\right). \tag{6.3}$$

These formulas are computationally simpler than Equation 6.1, the latter not even requiring the calculation of expected frequencies; however, they do not allow for the examination of the contributions to the computed chi-square, the utility of which will be seen in Section 6.4.

In Section 5.4, DF was described as the number of categories over which the calculation of χ^2 is summed minus the number of sample constants used to calculate the expected frequencies. In contingency table analyses, the χ^2 calculation is performed over $r \times c$ cells; to calculate all the $\hat{f}_{ij}$ values, we need to know n and at least $r - 1$ of the row totals and $c - 1$ of the column totals. So DF $= rc - 1 - (r - 1) - (c - 1)$, which is written more simply as

$$\text{DF} = (r - 1)(c - 1). \tag{6.5}$$

In Example 6.1, a 2×4 table, $\nu = (2 - 1)(4 - 1) = 3$. Since the calculated χ^2 statistic is 8.994, and the critical value, $\chi^2_{0.05,3} = 7.815$, we have a significant χ^2 test, and H_0 is rejected; we conclude, with 95% confidence, that the proportions of the hair colors in the population we sampled are not the same for both sexes.

6.2 THE 2 × 2 CONTINGENCY TABLE

The smallest possible contingency table is that consisting of 2 rows and 2 columns. It is referred to as a 2×2 table or a fourfold table, and it is commonly encountered in biological research. For such tables, the computing formula

$$\chi^2 = \frac{n(f_{11}f_{22} - f_{12}f_{21})^2}{(C_1)(C_2)(R_1)(R_2)} \tag{6.6}$$

may be used as an alternative to Equation 6.1. Although the two equations are mathematically equivalent, Equation 6.6 may yield more accurate results, for as it requires neither the calculation of $\hat{f}_{ij}$ nor that of $f_{ij} - \hat{f}_{ij}$ values, the associated rounding errors are avoided.

In a 2×2 table, $\nu = (r - 1)(c - 1) = 1$, and, as introduced in Section 5.6, we are faced with the desirability of correcting for continuity. The Yates correction for continuity applied to Equation 6.1 would result in the formula

$$\chi^2_c = \sum\sum \frac{(|f_{ij} - \hat{f}_{ij}| - 0.5)^2}{\hat{f}_{ij}} \tag{6.7}$$

or, applied to Equation 6.6,

$$\chi^2_c = \frac{n(|f_{11}f_{22} - f_{12}f_{21}| - n/2)^2}{(C_1)(C_2)(R_1)(R_2)}. \tag{6.8}$$

Crow (1952) points out that if $|f_{11}f_{22} - f_{12}f_{21}| \leq n/2$, Yates' correction will increase, rather than decrease, χ^2, and in such cases Equation 6.6, not 6.8, should be used.

Example 6.2 presents data from an experiment in which diseased guinea pigs were divided into two groups, one of which was treated with a drug and the other was not. The null hypothesis states that there is no difference in survival between the treated and untreated guinea pigs. Since the calculated χ^2_c value is less than the critical value, H_0 is not rejected, and survival is assumed to be independent of the drug treatment.

The Fisher exact test (Section 22.9) may also be used for the analysis of 2 × 2 contingency tables. It is preferable, in fact, when cell frequencies are small (see Section 6.5). A method for testing differences between proportions is in some cases a more powerful test than χ^2 (Section 22.10). Section 22.11 discusses the estimation of statistical power and of required sample size relevant to 2 × 2 contingency tables.

Dunnett and Gent (1977) describe procedures for testing whether proportions in a 2 × 2 table differ by a hypothesized amount. For example, we could have used the data of Example 6.2 to ask whether the proportion of animals surviving with the drug treatment was a specified amount different from or greater than the proportion surviving without the treatment.

Example 6.2

A 2 × 2 (fourfold) contingency table.

H_0: The survival of the animals is independent of whether the drug is administered.
H_A: The survival of the animals is associated with the administration of the drug.
$\alpha = 0.05$

	Dead	*Alive*	*Total*
Treated	9	15	24
Not treated	15	10	25
Total	24	25	49

$$\chi^2_c = \frac{n(|f_{11}f_{22} - f_{12}f_{21}| - n/2)^2}{(C_1)(C_2)(R_1)(R_2)}$$

$$= \frac{49[|(9)(10) - (15)(15)| - 24.5]^2}{(24)(25)(24)(25)}$$

$$= 1.662$$

$\nu = 1$

$\chi^2_{0.05,1} = 3.841$

Therefore, do not reject H_0.

$0.1 < P < 0.25$

The Continuity Correction Question. There has been considerable controversy among statisticians over when the use of a continuity correction is warranted and what form that correction should take (e.g., Conover, 1974; Mantel, 1976). Conover (1974) concludes that the Yates correction is justified when both the row totals and the column totals are set in advance of the experiment, especially if either $R_1 = R_2$ or $C_1 = C_2$. In Example 6.2 the row totals (but not the column totals) were set by the experimenter which is a common situation. If both the row and the column totals are not set in advance, the Yates correction may be appropriate in some cases but not in others, and we cannot readily determine which cases are which. In order to routinely employ this correction we will sometimes engage in conservative statistical testing. This means that we are actually testing at a more stringent (i.e., lower) significance level than the stated α, running an increased risk of committing a Type II error. If H_0 is rejected, we may depend on our conclusion, but with a doubtful statement of the probability that H_0 is true; if H_0 is not rejected, this conclusion may have a greater probability of being in error than we would like. To routinely neglect a continuity correction would avoid the possibility of this inflated Type II error by increasing the probability of a Type I error.

Haber (1980) has shown that a continuity correction due to Cochran (1942, 1952) gives decidedly better results when routinely employed than does either the Yates-corrected or noncorrected chi-square calculation. He provides this simple procedure, shown in Example 6.3, for computing the Cochran-corrected chi-square (which we shall indicate as $\chi^2_{c'}$). First we determine each of the four expected frequencies (using Equation 6.4), denoting the smallest as $\hat{f}$. Then, the absolute difference between the smallest expected frequency and its corresponding observed frequency is $d = |f - \hat{f}|$; and

$$\text{if } f \leq 2\hat{f},$$

then define $D =$ the largest multiple of 0.5 that is $< d$;

$$\text{if } f > 2\hat{f},$$

then define $D = d - 0.5$. The chi-square value with the Cochran correction for continuity is

$$\chi^2_{c'} = \frac{n^3 D^2}{R_1 R_2 C_1 C_2}. \tag{6.9}$$

Example 6.3

Cochran's corrected chi-square statistic for the 2×2 table of Example 6.2.

H_0: The survival of the animals is independent of whether the drug is administered.
H_A: The survival of the animals is associated with the administration of the drug.

$$\alpha = 0.05$$

The observed frequency, f_{ij}, in each cell is shown, with the frequency expected if H_0 is true (i.e., $\hat{f}_{ij}$) in parentheses.

	Dead	*Alive*	*Total*
Treated	9 *(11.7551)*	15 *(12.2449)*	24
Not treated	15 *(12.2449)*	10 *(12.7551)*	25
Total	24	25	49

The smallest expected frequency is $\hat{f}_{11} = 11.7551$; call it $\hat{f}$.

$$d = |f_{11} - \hat{f}_{11}| = |9 - 11.7551| = 2.7551$$
$$2\hat{f} = 2(11.7551) = 23.5102$$

As $f_{11} = 9 < 2\hat{f}$, define $D = 2.5$.

$$\chi^2_{c'} = \frac{n^3 D^2}{R_1 R_2 C_1 C_2} = \frac{(49)^3(2.5)^2}{(24)(25)(24)(25)} = 2.043$$

$\chi^2_{0.05,1} = 3.841$
Therefore, do not reject H_0.
$0.10 < P < 0.25$

6.3 HETEROGENEITY TESTING OF 2 × 2 TABLES

Just as one may test for heterogeneity of replications of a goodness of fit analysis (Section 5.8), it is frequently desirable to test for heterogeneity of 2 × 2 contingency tables. Suppose that in addition to the experiment summarized in Example 6.2, it is found that the same experiment had been performed three other times. It would then be meaningful to determine whether all four of the experiments were performed with samples that could have come from the same population (i.e., H_0: the four samples are homogeneous; and H_A: the four samples are heterogeneous). To test for heterogeneity, one first calculates the χ^2 for each of the four samples (recalling from Section 5.8 that the Yates correction for continuity may not be used in a heterogeneity chi-square analysis). The four contingency tables and their uncorrected chi-square values are presented in Example 6.4a, along with the contingency table formed from the pooled data and the completion of the heterogeneity analysis.

As the heterogeneity χ^2 is 0.181 and $\chi^2_{0.05,3} = 7.815$, the null hypothesis is not rejected (i.e., the four samples are concluded to be homogeneous). This tells us that the four samples may be pooled justifiably. Such pooling, if justified, is generally

Example 6.4a

A heterogeneity chi-square analysis of 2 × 2 contingency tables.

H_0: The four samples are homogeneous.
H_A: The four samples are heterogeneous.
$\alpha = 0.05$

Experiment 1

	Dead	*Alive*	*Total*	
Treated	9	15	24	$\chi^2 = 2.4806$
Not treated	15	10	25	DF = 1
Total	24	25	49	

Experiment 2

	Dead	*Alive*	*Total*	
Treated	13	12	25	$\chi^2 = 2.1222$
Not treated	18	7	25	DF = 1
Total	31	19	50	

Experiment 3

	Dead	*Alive*	*Total*	
Treated	12	13	25	$\chi^2 = 2.0525$
Not treated	17	8	25	DF = 1
Total	29	21	50	

Experiment 4

	Dead	*Alive*	*Total*	
Treated	10	14	24	$\chi^2 = 2.4522$
Not treated	16	9	25	DF = 1
Total	26	23	49	

Pooled Data for Experiments 1–4

	Dead	*Alive*	*Total*	
Treated	44	54	98	$\chi^2 = 8.9262$
Not treated	66	34	100	DF = 1
Total	110	88	198	

Total chi-square	9.1075	DF = 4
Chi-square of totals	8.9262	DF = 1
Heterogeneity chi-square	0.1813	DF = 3

$\chi^2_{0.05,3} = 7.815$
Therefore, do not reject H_0.
$0.975 < P < 0.99$

advantageous, as it provides for chi-square testing with a relatively large n. In Example 6.4b, the pooled data are analyzed, and it is concluded that there is a significant difference between the survival of the treated and untreated guinea pigs, a difference which was not detected using any of the four smaller samples originally recorded. Note that the Yates correction or the Cochran correction is used on the pooled data, after the heterogeneity analysis has been performed.

Example 6.4b

The 2 × 2 contingency table analysis (with correction for continuity) for the pooled data of Example 6.4a.

H_0: Survival is independent of the administration of the drug.
H_A: Survival is not independent of the administration of the drug.
$\alpha = 0.05$

	Dead	*Alive*	*Total*
Treated	44	54	98
Not treated	66	34	100
Total	110	88	198

$\chi^2_c = 8.092$
$\chi^2_{c'} = 8.183$
DF = 1
$\chi^2_{0.05,1} = 3.841$
Therefore, reject H_0.
$0.001 < P < 0.005$

6.4 SUBDIVIDING CONTINGENCY TABLES

In Example 6.1, a 2 × 4 contingency table, it was concluded that there was a significant difference in human hair color frequencies between males and females. Examination of the f_{ij} values in this table reveals that the proportion of males in the blond column is decidedly less than in the other columns. (To express the percent males and females in each column, as in Example 6.5a, helps to elucidate this fact, although such percentages may not be used in any chi-square calculation.) Thus, we might suspect that the significant χ^2 calculated in Example 6.1 was due largely to column 3 in the table. We might momentarily ignore the data for column 3 and consider the remaining 2 × 3 table (Example 6.5b). The nonsignificant χ^2 for this table supports the null hypothesis

Example 6.5a

The data of Example 6.1, where for each hair color the percent males and females is indicated.

	Hair Color				
Sex	*Black*	*Brown*	*Blond*	*Red*	**Total**
Male	32 (37%)	43 (40%)	16 (20%)	9 (36%)	100
Female	55 (63%)	65 (60%)	64 (80%)	16 (64%)	200
Total	87	108	80	25	300

In Example 6.1, the null hypothesis that the 4 hair colors are independent of sex was rejected at $\alpha = 0.05$.

Example 6.5b

The 2 × 3 contingency table formed from columns 1, 2, and 4 of the original 2 × 4 table. $\hat{f}_{ij}$ values for the cells of the 2 × 3 table are shown in parentheses.

H_0: The occurrence of black, brown, and red hair is independent of sex.
H_A: The occurrence of black, brown, and red hair is not independent of sex.
$\alpha = 0.05$

	Hair Color			
Sex	*Black*	*Brown*	*Red*	**Total**
Male	32 (*33.2182*)	43 (*41.2364*)	9 (*9.5455*)	84
Female	55 (*53.7818*)	65 (*66.7636*)	16 (*15.4545*)	136
Total	87	108	25	220

$\chi^2 = 0.245$ with DF $= 2$
$\chi^2_{0.05,2} = 5.991$
Therefore, do not reject H_0.
$0.75 < P < 0.9$

that these three hair colors are independent of sex. In Example 6.5c, a 2 × 2 table is formed by considering blond versus all other hair colors. Here, the null hypothesis of independence is rejected. By the described series of manipulations of the original contingency table, we have confirmed our suspicion that among the four hair colors tested, blond occurs among the sexes with relative frequencies different from those of the other colors. Snee (1974) describes graphical methods for determining how to subdivide contingency tables.

Example 6.5c

The 2 × 2 contingency table formed by combining columns 1, 2, and 4 of the original table.

H_0: Occurrence of blond and nonblond hair color is independent of sex.
H_A: Occurrence of blond and nonblond hair color is not independent of sex.
$\alpha = 0.05$

	Hair Color		
Sex	*Blond*	*Nonblond*	**Total**
Male	16	84	100
Female	64	136	200
Total	80	220	300

$\chi_c^2 = 7.928$
$\chi_{c'}^2 = 8.457$
DF = 1
$\chi_{0.05,1}^2 = 3.841$
Therefore, reject H_0.
$0.001 < P < 0.005$

Strictly speaking, it is not proper to test statistical hypotheses developed after examining the data to be tested. Therefore, the analysis of a subdivided contingency table should be considered only a guide to developing additional hypotheses; these hypotheses should then be stated in advance of their being tested with a new set of data.

Other considerations of chi-square contingency table analyses can be found in Bliss (1967: Chapters 3 and 4), Simpson, Roe, and Lewontin (1960: Chapter 13), and Snedecor and Cochran (1980: Chapter 11).

6.5 BIAS IN CHI-SQUARE CONTINGENCY ANALYSES

Section 5.7 recommended that in a chi-square goodness of fit analysis no $\hat{f}_i$ should be less than 1.0 and no more than 20% of the $\hat{f}_i$'s should be less than 5.0. Similarly, in chi-square analyses of contingency tables, it is recommended that no $\hat{f}_{ij}$ be less than 1.0 and no more than 20% less than 5.0 (Cochran, 1954). If a contingency table has an expected frequency < 1 in any cell and/or expected frequencies < 5 in more than one-fifth of its cells, the resulting chi-square value will be biased and may not be compared legitimately with the values in Table B.1.

If a 2 × 2 table has insufficiently large frequencies for a chi-square analysis, then the Fisher exact test (Section 22.9) might profitably be employed. If a contingency table with such frequencies has more than two rows and/or columns, one might simply discard the rows and/or columns with the offensively low $\hat{f}_{ij}$'s, or combine rows and/or columns for the same purpose, but such procedures are not recommended as routine practice; when possible, it would be better to repeat the experiment with a sufficiently large n.

6.6 THE LOG-LIKELIHOOD RATIO FOR CONTINGENCY TABLES

The log-likelihood ratio was introduced in Section 5.9, where the G statistic was presented as an alternative to chi-square for goodness of fit testing. The G test may also be applied to contingency tables (Wilks, 1935), where

$$G = 2[\sum_i \sum_j f_{ij} \ln f_{ij} - \sum_i R_i \ln R_i - \sum_j C_j \ln C_j + n \ln n], \tag{6.10}$$

or

$$G = 4.60517[\sum \sum f_{ij} \log f_{ij} - \sum R_i \log R_i - \sum C_j \log C_j + n \log n]. \tag{6.11}$$

Since G is approximately distributed as χ^2, Table B.1 may be used with $(r - 1)(c - 1)$ degrees of freedom. In Example 6.6, the contingency table of Example 6.1 is analyzed using the G statistic.

Example 6.6

The G test for the contingency table data of Example 6.1.

H_0: Hair color is independent of sex.
H_A: Hair color is not independent of sex.
$\alpha = 0.05$

	Hair Color				
Sex	*Black*	*Brown*	*Blond*	*Red*	**Total**
Male	32	43	16	9	100
Female	55	65	64	16	200
Total	87	108	80	25	300

$$\begin{aligned} G &= 4.60517[\sum \sum f_{ij} \log f_{ij} - \sum R_i \log R_i - \sum C_j \log C_j + n \log n] \\ &= 4.60517[(32)(1.50515) + (43)(1.63347) + (16)(1.20412) \\ &\quad + (9)(0.95424) + (55)(1.74036) + (65)(1.81291) \\ &\quad + (64)(1.80618) + (16)(1.20412) - (100)(2.00000) \\ &\quad - (200)(2.30103) - (87)(1.93952) - (108)(2.03342) \\ &\quad - (80)(1.90309) - (25)(1.39794) + (300)(2.47712)] \\ &= 4.60517(2.06518) \\ &= 9.510 \text{ with DF} = 3 \end{aligned}$$

$\chi^2_{0.05,3} = 7.815$
Therefore, reject H_0.
$0.01 < P < 0.025$

In the case of a 2 × 2 table, the Yates correction for continuity is applied as follows: If $f_{11}f_{22} - f_{12}f_{21}$ is negative, add 0.5 to f_{11} and f_{22} and subtract 0.5 from f_{12} and f_{21}; if $f_{11}f_{22} - f_{12}f_{21}$ is positive, subtract 0.5 from f_{11} and f_{22} and add 0.5 to f_{12} and f_{21}. Then Equation 6.10 or 6.11 may be used to calculate G_c, the corrected G statistic.

Williams (1976) recommends G be used in preference to χ^2 whenever $|f_{ij} - \hat{f}_{ij}| < \hat{f}_{ij}$ for any cell. Both methods commonly result in the same conclusions. When they do not, many statisticians prefer G and recommend its routine use.

6.7 THREE-DIMENSIONAL CONTINGENCY TABLES

In this chapter, thus far, we have considered two-dimensional contingency tables (that is, tables with rows and columns as the two dimensions); each of the two dimensions (row and column) represented a nominal scale variable. However, we may collect and tabulate enumeration data with respect to three or more variables and thus have what are referred to as multidimensional contingency tables (i.e., tables with three or more dimensions). Such tables have been the subject of considerable investigation in recent years (e.g., see Bishop, Fienberg, and Holland, 1975; Everitt, 1977; Fienberg, 1970, 1980; Goodman, 1970; Lancaster, 1969; Upton, 1978), and computer program libraries often include provision for their analysis.

Figure 6.1 shows a three-dimensional contingency table. The three "rows" are species, the four "columns" are geographic locations, and the two "tiers"* are presence and absence of a disease. If a sample is obtained containing individuals of these species, from these locations, with and without the disease in question, then observed frequencies can be recorded in the 24 cells of this 3 × 4 × 2 contingency table. We

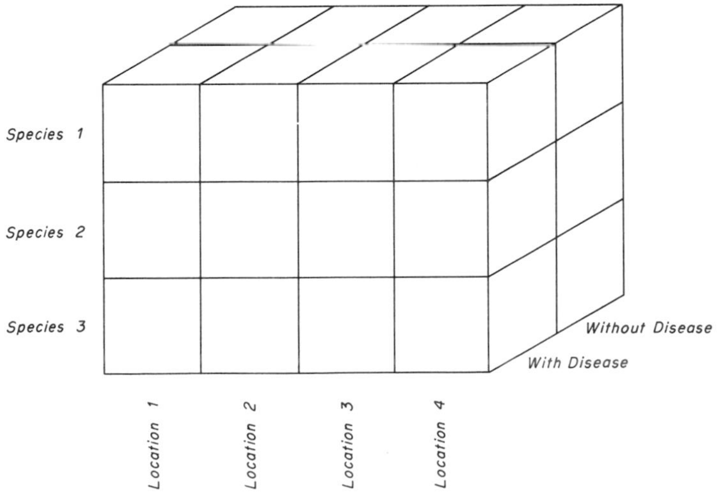

Figure 6.1 A three-dimensional contingency table, where the three rows are species, the four columns are locations, and the two tiers are occurrence of a disease. An observed frequency, f_{ijl}, is recorded in each combination of row, column, and tier.

*Some authors use the term "layer," rather than "tier" for the third contingency table dimension.

shall refer to the observed frequency in row i, column j, and tier l as f_{ijl}. We shall refer to the number of rows, columns, and tiers as r, c, and t, respectively. The sum of the frequencies in row i will be designated R_i, the sum in column j as C_j, and the sum in tier l as T_l.

Example 6.7 presents a $2 \times 2 \times 2$ contingency table where data (f_{ijl}) are collected as described above, but only for two species and two locations. Note that throughout the following discussions the sum of the expected frequencies for a given row, column, or tier equals the sum of the observed frequencies for that row, column, or tier.

Example 6.7

Test for mutual independence in a $2 \times 2 \times 2$ contingency table.

H_0: Disease occurrence, species, and location are mutually independent in the population sampled.

H_A: Disease occurrence, species, and location are not all mutually independent in the population sampled.

The observed frequencies (f_{ijl}):

	Disease Present		**Disease Absent**		**Species Totals**
	Location 1	*Location 2*	*Location 1*	*Location 2*	($r = 2$)
Species 1	44	12	38	10	$R_1 = 104$
Species 2	28	22	20	18	$R_2 = 88$

Disease totals: ($t = 2$) $T_1 = 106$ $T_2 = 86$

Grand total: $n = 192$

Location totals ($c = 2$): $C_1 = 130$, $C_2 = 62$

The expected frequencies ($\hat{f}_{ijl}$):

	Disease Present		**Disease Absent**		
	Location 1	*Location 2*	*Location 1*	*Location 2*	**Species Totals**
Species 1	38.8759	18.5408	31.5408	15.0425	$R_1 = 104$
Species 2	32.8950	15.6884	26.6884	12.7283	$R_2 = 88$

Disease totals: $T_1 = 106$ $T_2 = 86$

Grand total: $n = 192$

Location totals: $C_1 = 130$, $C_2 = 60$

$$\chi^2 = \sum\sum\sum \frac{(f_{ijl} - \hat{f}_{ijl})^2}{\hat{f}_{ijl}}$$

$$\chi^2 = \frac{(44 - 38.8759)^2}{38.8759} + \frac{(12 - 18.5408)^2}{18.5408} + \frac{(38 - 31.5408)^2}{31.5408} + \frac{(10 - 15.0425)^2}{15.0425}$$
$$+ \frac{(28 - 32.8950)^2}{32.8950} + \frac{(22 - 15.6884)^2}{15.6884} + \frac{(20 - 26.6884)^2}{26.6884} + \frac{(18 - 12.7283)^2}{12.7283}$$
$$= 0.6754 + 2.3075 + 1.3228 + 1.6903 + 0.7284 + 2.5392 + 1.6762 + 2.1834$$
$$= 13.123$$

$\nu = rct - r - c - t + 2 = (2)(2)(2) - 2 - 2 - 2 + 2 = 4$

$\chi^2_{0.05,4} = 9.488$

Reject H_0.

$0.01 < P < 0.025$

Mutual Independence. We can test more than one hypothesis using multidimensional contingency table data. An overall kind of hypothesis is that which states mutual independence among all the variables. Another way of expressing this H_0 is that there are no interactions (either three-way or two-way) among any of the variables. For this hypothesis, the expected frequency in row i, column j, and tier l is

$$\hat{f}_{ijl} = \frac{R_i C_j T_l}{n^2}, \tag{6.12}$$

where n is the total of all the frequencies in the entire contingency table.

In Example 6.7 this null hypothesis would imply that presence or absence of the disease occurred independently of species and location. For three dimensions, this null hypothesis is tested by computing

$$\chi^2 = \sum_{i=1}^{r} \sum_{j=1}^{c} \sum_{l=1}^{t} \frac{(f_{ijl} - \hat{f}_{ijl})^2}{\hat{f}_{ijl}}, \tag{6.13}$$

which is a simple extension of the chi-square calculation for a two-dimensional table (by Equation 6.1). The degrees of freedom for this test are the sums of the degrees of freedom for all interactions:

$$\nu = (r-1)(c-1)(t-1) + (r-1)(c-1) + (r-1)(t-1) + (c-1)(t-1), \tag{6.14}$$

which is equivalent to

$$\nu = rct - r - c - t + 2. \tag{6.15}$$

Partial Independence. If the above null hypothesis is not rejected, then we conclude that all three variables are mutually independent and the analysis proceeds no further. If, however, H_0 is rejected, then we may test further to determine between which variables dependencies and independencies exist. For example, we may test whether one of the three variables is independent of the other two, a situation known as *partial independence*.*

For the hypothesis of rows being independent of columns and tiers, we need total frequencies for rows and total frequencies for combinations of columns and tiers. We designate the total frequency in column j and tier l as $(CT)_{jl}$. Expected frequencies are then computed as

$$\hat{f}_{ijl} = \frac{R_i (CT)_{jl}}{n}, \tag{6.16}$$

and Equation 6.13 is used with degrees of freedom

$$\nu = (r-1)(c-1)(t-1) + (r-1)(c-1) + (r-1)(t-1), \tag{6.17}$$

which is equivalent to

$$\nu = rct - ct - r + 1. \tag{6.18}$$

*A different hypothesis is that of *conditional independence*, where two of the variables are said to be independent in each level of the third (but each may have dependence on the third). This is discussed in the references cited at the beginning of this section.

For the null hypothesis of columns being independent of rows and tiers, we compute expected frequencies using column totals, C_j, and the totals for row and tier combinations, $(RT)_{il}$:

$$\hat{f}_{ijl} = \frac{C_j(RT)_{il}}{n}, \tag{6.19}$$

and

$$\nu = rct - rt - c + 1. \tag{6.20}$$

And, for the null hypothesis of tiers being independent of rows and columns, we use tier totals, T_l, and the totals for row and column combinations, $(RC)_{ij}$:

$$\hat{f}_{ijl} = \frac{T_l(RC)_{ij}}{n}; \tag{6.21}$$

$$\nu = rct - rc - t + 1. \tag{6.22}$$

In Example 6.8, all three hypotheses for partial dependence are tested. In one of the three (the last), H_0 is not rejected; thus we conclude that presence of disease is independent of species and location. However, the hypothesis test of Example 6.7 concluded that all three variables are not independent of each other. Therefore, we suspect that species and location are not independent. The independence of these two variables may be tested using a two-dimensional contingency table, as described earlier in this chapter and demonstrated in Example 6.9. In the present case, the species-location interaction is tested by way of a 2×2 contingency table and we conclude that these two factors are not independent (i.e., species occurrence depends on geographic location).

Example 6.8

Test for partial independence in a $2 \times 2 \times 2$ contingency table. Since the H_0 of overall independence was rejected in Example 6.7, we may test the following three pairs of hypotheses.

H_0: Species is independent of location and disease.
H_A: Species is not independent of location and disease.
The expected frequencies ($\hat{f}_{ijl}$):

	Disease Present		**Disease Absent**		
	Location 1	*Location 2*	*Location 1*	*Location 2*	**Species Totals**
Species 1	39.0000	18.4167	31.4167	15.1667	$R_1 = 104$
Species 2	33.0000	15.5833	26.5833	12.8333	$R_2 = 88$
Location and Disease Totals:	$(CT)_{11} = 72$	$(CT)_{12} = 34$	$(CT)_{21} = 58$	$(CT)_{22} = 28$	*Grand total*: $n = 192$

$$\chi^2 = \frac{(44 - 39.0000)^2}{39.0000} + \frac{(12 - 18.4167)^2}{18.4167} + \frac{(38 - 31.4167)^2}{31.4167} + \cdots + \frac{(18 - 12.8333)^2}{12.8333}$$
$$= 0.6410 + 2.2357 + 1.3795 + 1.7601 + 0.7576 + 2.6422 + 1.6303 + 2.0801$$
$$= 13.126$$

$$\nu = rct - ct - r + 1 = (2)(2)(2) - (2)(2) - 2 + 1 = 3$$

$\chi^2_{0.05,3} = 7.815$
Reject H_0.
Species is not independent of location and presence of disease.
$0.01 < P < 0.025$

H_0: Location is independent of species and disease.
H_A: Location is not independent of species and disease.
The expected frequencies ($\hat{f}_{ijl}$):

	Disease Present		**Disease Absent**		
	Species 1	*Species 2*	*Species 1*	*Species 2*	**Location Totals**
Location 1	37.9167	33.8542	32.5000	25.7292	$C_1 = 130$
Location 2	18.0833	16.1458	15.5000	12.2708	$C_2 = 62$
Species and disease totals:	$(RT)_{11} = 56$	$(RT)_{12} = 50$	$(RT)_{21} = 48$	$(RT)_{22} = 38$	*Grand total:* $n = 192$

$$\chi^2 = \frac{(44 - 37.9167)^2}{37.9167} + \frac{(28 - 33.8542)^2}{33.8542} + \cdots + \frac{(18 - 12.2708)^2}{12.2708}$$
$$= 0.9760 + 1.0123 + 0.9308 + 1.2757 + 2.0464 + 2.1226 + 1.9516 + 2.6749$$
$$= 12.990$$

$\nu = rct - rt - c + 1 = (2)(2)(2) - (2)(2) - 2 + 1 = 3$
$\chi^2_{0.05,3} = 7.815$
Reject H_0.
Location is not independent of species and presence of disease.
$P < 0.001$

H_0: Presence of disease is independent of species and location.
H_A: Presence of disease is not independent of species and location.
The expected frequencies ($\hat{f}_{ijl}$):

	Species 1		**Species 2**		
	Location 1	*Location 2*	*Location 1*	*Location 2*	**Disease Totals**
Disease present	45.2708	12.1458	26.5000	22.0833	$T_1 = 106$
Disease absent	36.7292	9.8542	21.5000	17.9167	$T_2 = 86$
Species and location totals:	$(RC)_{11} = 82$	$(RC)_{12} = 22$	$(RC)_{21} = 48$	$(RC)_{22} = 40$	*Grand total:* $n = 192$

$$\chi^2 = \frac{(44 - 45.2708)^2}{45.2708} + \frac{(12 - 12.1458)^2}{12.1458} + \cdots + \frac{(18 - 17.9167)^2}{17.9167}$$
$$= 0.0357 + 0.0018 + 0.0849 + 0.0003 + 0.0440 + 0.0022 + 0.1047 + 0.0004$$
$$= 0.274$$

$\nu = rct - rc - t + 1 = (2)(2)(2) - (2)(2) - 2 + 1 = 3$
$\chi^2_{0.05,3} = 7.815$
Do not reject H_0.
$0.95 < P < 0.975$

Example 6.9.

Test for independence of two variables, following tests for partial dependence.

The hypothesis test of Example 6.7 concluded that all three variables are not mutually independent, while the last test in Example 6.8 concluded that presence of disease is independent of species and location. Therefore, it is desirable (and permissible) to test the following two-dimensional contingency table:

H_0: Species occurrence is independent of location.
H_A: Species occurrence is not independent of location.

	Location 1	*Location 2*	*Total*
Species 1	82	22	104
Species 2	48	40	88
Total	130	62	192

$\chi_c^2 = 11.787$
$\chi_{c'}^2 = 12.690$
$\nu = (r - 1)(c - 1) = 1$
$\chi^2_{0.05,1} = 3.841$
Reject H_0.
$P < 0.001$

The Log-Likelihood Ratio. In the hypothesis testing described above, the log-likelihood ratio (see Section 6.6) may be used in lieu of chi-square. Some authors prefer it and many multidimensional contingency table computer programs employ it.

6.8 LOG-LINEAR MODELS FOR MULTIDIMENSIONAL CONTINGENCY TABLES

The analysis of contingency tables with three or more dimensions (i.e., with three or more variables) is often accomplished (especially by computer) by using what are known as "log-linear models." A "model" in statistical analysis is an expression of how observed data (in this case observed frequencies, f_{ijl}) are affected by variables and combinations of variables. "Log-linear" refers to a procedure whereby a multiplicative relationship (e.g., $\hat{f}_{ijl} = R_i C_j T_l / n^2$) is transformed to a linear relationship by the use of logarithms (e.g., $\log \hat{f}_{ijl} = \log R_i + \log C_j + \log T_l - 2 \log n$).

In the terminology of log-linear models, one tests interactions of variables. A null hypothesis that states no interactions of two or more variables is one that implies that all the variables are independent. Hypotheses of partial independence and of conditional independence may be tested using appropriate log-linear models. Of the references cited at the beginning of Section 6.7, those of Fienberg (1970 and 1980) and Everitt (1977) contain accounts of log-linear models especially readable for non-mathematicians, and the first of them is oriented specifically toward biological data.

EXERCISES

6.1. Consider the following data for the abundance of a certain species of bird. Using chi-square, test the null hypothesis that the ratio of numbers of males to females was the same in all four seasons.

Sex	*Spring*	*Summer*	*Fall*	*Winter*
Males	163	135	71	43
Females	86	77	40	38

6.2. The following data are frequencies of skunks found with and without rabies in two different geographic areas. Using chi-square, test the null hypothesis that the incidence of rabies in skunks is the same in both areas.

Area	*With rabies*	*Without rabies*
E	14	29
W	12	38

6.3. Using the data in Exercise 6.1, use the G test to test the hypothesis called for.

6.4. Use the G test to test the appropriate hypothesis for the data in Exercise 6.2.

6.5. Data were collected as in Exercise 6.2, but with the additional tabulation of the sex of each skunk recorded, as follows. Test for mutual independence; and, if H_0 is rejected, test for partial independence.

	With Rabies		**Without Rabies**	
Area	*Male*	*Female*	*Male*	*Female*
E	42	33	55	63
W	84	51	34	48

7 The Normal Distribution

Most frequency distributions of interval or ratio scale data are observed to have a preponderance of values around the mean with progressively fewer observations toward the extremes of the range of values (see, e.g., Fig. 1.5). If n is large, the frequency polygons of many biological data distributions are "bell-shaped" and look something like Fig. 7.1.

Figure 7.1 is a frequency curve for a *normal distribution.** Not all bell-shaped curves are normal, however, for a *normal distribution* is defined as one in which the frequency (f_i) of an observation of size X_i is as expressed by the relation:

$$f_i = \frac{1}{\sigma\sqrt{2\pi}} e^{-(X_i-\mu)^2/2\sigma^2}. \tag{7.1}$$

In this equation, there are two mathematical constants: π (lowercase Greek pi),†

*The normal distribution is sometimes called the *Gaussian distribution,* because of its development by the German mathematician and physicist, Karl Fredrich Gauss (1777–1855), in the early nineteenth century. However, it was mathematician Abraham de Moivre (1667–1754), born in France and transplanted to England, who first announced the equation for the distribution, in 1733, and the very influential French mathematician-physicist Pierre-Simon Marquis de Laplace (1749–1827) rediscovered it about the same time as did Gauss. Karl Pearson called the distribution "normal" to avoid "an international question of priority" (Pearson, 1920).

†π, the ratio between the circumference and the diameter of a circle, is a symbol introduced in 1706 by William Jones (Beckman, 1971: 141). Pi is an irrational number, meaning it cannot be expressed as the ratio between two integers. (See also the introduction to Chapter 24.) To 20 decimal places its value is 3.14159 26535 89792 33846 (although it may be noted that 10 places are sufficient to obtain the circumference of a circle as large as the earth's equator to within about a centimeter). Beckman (1971) presents a fascinating history of π and its calculation.

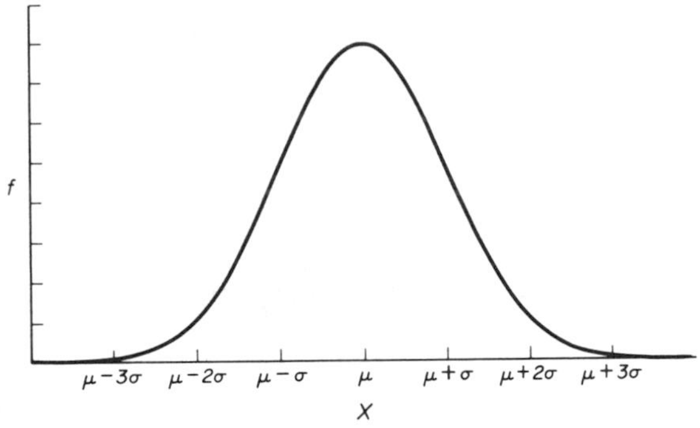

Figure 7.1 A normal distribution.

which equals 3.14159 . . . ; and e (the base of Naperian, or natural, logarithms),* which equals 2.71828 . . . . There are also two parameters (μ and σ^2) in the equation. Thus, for any given standard deviation, σ, there are an infinite number of normal curves possible, depending on μ. Figure 7.2a shows normal curves for $\sigma = 1$ and $\mu = 0$, 1, and 2. Likewise, for any given mean, μ, an infinity of normal curves is possible, each with a different value of σ. Figure 7.2b shows normal curves for $\mu = 0$ and $\sigma = 1$, 1.5, and 2.

A normal curve with $\mu = 0$ and $\sigma = 1$ is said to be a *standardized normal curve.*

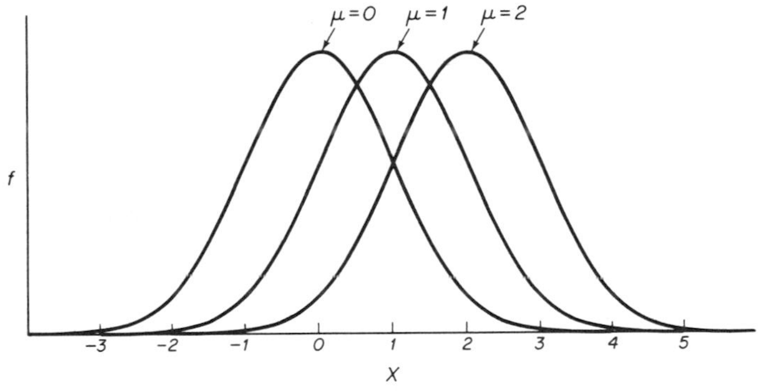

Figure 7.2a Normal distributions with $\sigma = 1$, varying in location with different means.

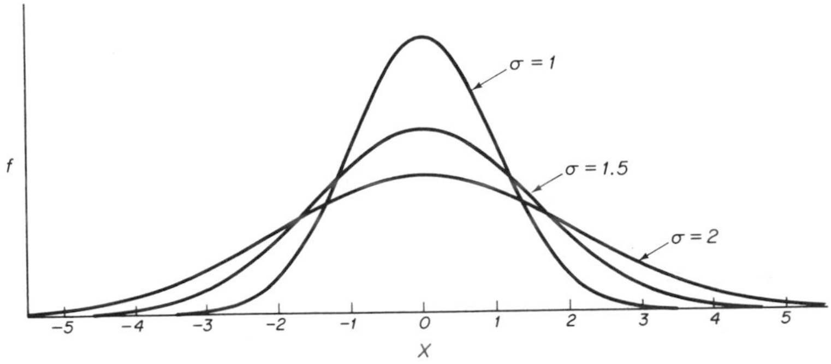

Figure 7.2b Normal distributions with $\mu = 0$, varying in spread with different standard deviations.

*e is also an irrational number, which to 20 decimal places is 2.71828 18284 59045 23536.

Thus, for a standardized normal distribution,

$$f_i = \frac{1}{\sqrt{2\pi}} e^{-X_i^2/2}. \tag{7.2}$$

7.1 SYMMETRY AND KURTOSIS

Mathematicians refer to $\sum (X_i - \mu)^p/N$ as the "pth moment about the mean," and it is often denoted as κ_p (using lowercase Greek kappa). For any distribution, the first moment about the mean,

$$\kappa_1 = \frac{\sum (X_i - \mu)}{N}, \tag{7.3}$$

will always be zero, since $\sum (X_i - \mu)$ is zero. The second moment about the mean,

$$\kappa_2 = \frac{\sum (X_i - \mu)^2}{N}, \tag{7.4}$$

has already been defined as the population variance, σ^2 (Section 4.3). Using the cube of the deviations from the mean, we obtain

$$\kappa_3 = \frac{\sum (X_i - \mu)^3}{N}, \tag{7.5}$$

a quantity that equals zero only if the data are from a symmetrical distribution. Since κ_3 has cubed units, the parameter

$$\gamma_1 = \frac{\kappa_3}{\sigma^3} = \frac{\sum (X_i - \mu)^3}{N\sigma^3} \tag{7.6}$$

is the more generally used measure of symmetry. This measure (called "gamma one") has no units; thus, coding the data has no effect on its value.

A symmetrical distribution is one in which the mean and the median are identical (see Fig. 3.2), and the portion of the frequency polygon to the left of the mean is a mirror image of the portion to the right of the mean. For a symmetrical distribution (normal or not), $\gamma_1 = 0$, whereas $\gamma_1 < 0$ indicates a negatively skewed distribution, and $\gamma_1 > 0$ indicates one that is positively skewed (see Fig. 3.2). The population parameters κ_3 and γ_1 are estimated from a sample by the statistics

$$k_3 = \frac{n \sum (X_i - \bar{X})^3}{(n-1)(n-2)} \tag{7.7}$$

and

$$g_1 = \frac{k_3}{s^3}, \tag{7.8}$$

respectively; and k_3 may also be calculated by the "machine formula":

$$k_3 = \frac{n \sum X_i^3 - 3 \sum X_i \sum X_i^2 + 2(\sum X_i)^3/n}{(n-1)(n-2)}. \tag{7.9}$$

A $g_1 < 0$ indicates that the sample comes from a population where the mean is less than the median, and a $g_1 > 0$ indicates sampling from a population whose distribution is skewed to the right. Testing for statistical significance of asymmetry will be described in Section 8.13.

As not all symmetrical distributions are normal, statisticians have desired ways of assessing the *kurtosis* (shape) of a distribution. Considering the fourth power of the deviations from the mean,

$$\kappa_4 = \frac{\sum (X_i - \mu)^4}{N} \tag{7.10}$$

is a measure of kurtosis. This population parameter is estimated by the sample statistic

$$k_4 = \frac{\sum (X_i - \bar{X})^4 n(n+1)/(n-1) - 3[\sum (X_i - \bar{X})^2]^2}{(n-2)(n-3)}, \tag{7.11}$$

which may be computed by either "machine formula"

$$k_4 = \frac{\sum X^4 - 4\sum X \sum X^3/n + 6(\sum X)^2 \sum X^2/n^2 - 3(\sum X)^4/n^3 n(n+1)/(n-1) - 3(SS)^2}{(n-2)(n-3)} \tag{7.12}$$

(from Bliss, 1967: 144), with the sample SS introduced as Equation 4.7, or

$$k_4 = \frac{(n^3 + n^2)\sum X^4 - 4(n^2 + n)\sum X^3 \sum X - 3(n^2 - n)(\sum X^2)^2 + 12n \sum X^2(\sum X)^2 - 6(\sum X)^4}{n(n-1)(n-2)(n-3)} \tag{7.13}$$

(Bennett and Franklin, 1954: 81). Because κ_4 has units to the fourth power, the quantity κ_4/σ^4 is a preferred measure (for it, as is the case with γ_1, has no units). For a normal (*mesokurtic*) distribution, $\kappa_4/\sigma^4 = 3$; therefore, the measure of kurtosis is often expressed as

$$\gamma_2 = \frac{\kappa_4}{\sigma^4} - 3, \tag{7.14}$$

so that a mesokurtic distribution will have a $\gamma_2 = 0$. The sample estimate of γ_2 is, then,

$$g_2 = \frac{k_4}{s^4} - 3. \tag{7.15}$$

A distribution having many values around the mean and in the "tails," far from the mean (see Fig. 7.3b), is called *leptokurtic* and will have a $\gamma_2 > 0$. Such a distribution

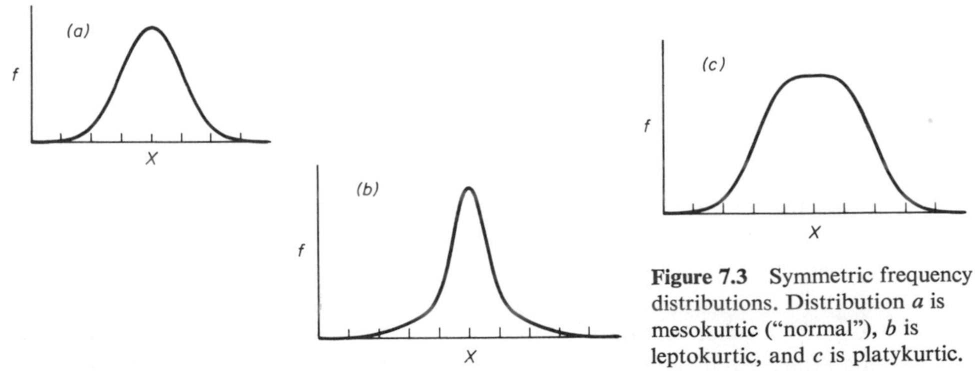

Figure 7.3 Symmetric frequency distributions. Distribution *a* is mesokurtic ("normal"), *b* is leptokurtic, and *c* is platykurtic.

might be the composite of two normal populations with the same μ but with different σ's. Those distributions exhibiting a $\gamma_2 < 0$ are said to be *platykurtic* (see Fig. 7.3c), and they have more values between the mean and tails than do normal (*mesokurtic*) distributions (see Fig. 7.3a). Such a distribution might be the composite of two normal populations with the same variance but different means.

7.2 PROPORTIONS OF A NORMAL DISTRIBUTION

If a normal population of 1000 body weights has a mean, μ, of 70 kg, one-half of the population (500 weights) is larger than 70 kg and one-half is smaller. This is true simply because the normal distribution is symmetrical. But if we desire to ask what portion of the population is larger than 80 kg, we need to know σ, the standard deviation of the population. If $\sigma = 10$ kg, then 80 kg is one standard deviation larger than the mean, and the portion of the population in question is the shaded area in Fig. 7.4a. If, however, $\sigma = 5$ kg, then 80 kg is two standard deviations above μ, and we are referring to a relatively small portion of the population, as shown in Fig. 7.4b.

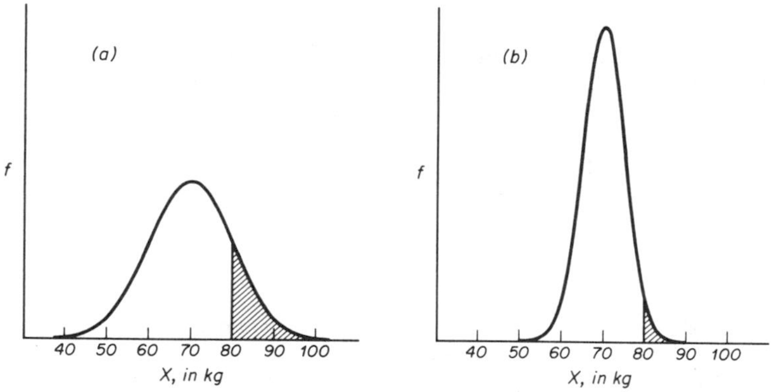

Figure 7.4 Two normal distributions with $\mu = 70$ kg. The shaded areas are the portions of the curves that lie above $X = 80$ kg. For distribution *a*, $\mu = 70$ kg and $\sigma = 10$ kg; for distribution *b*, $\mu = 70$ kg and $\sigma = 5$ kg.

Appendix Table B.2 enables us to determine proportions of normal distributions. For any X_i value from a normal population with mean μ, and standard deviation σ, the value

$$Z = \frac{X_i - \mu}{\sigma} \tag{7.16}$$

tells us how many standard deviations from the mean the X_i value is located. Carrying out the calculation of Equation 7.16 is known as *normalizing*, or *standardizing*, X_i; and Z is known as a *normal deviate*, or a *standard score*. The mean of a set of normal scores is 0, and the variance is 1. The quantity $Z + 5$ is called a *probit*.

Table B.2 tells us what proportion of a normal distribution lies beyond a given value of Z. If $\mu = 70$ kg, $\sigma = 10$ kg, and $X_i = 70$ kg, then $Z = (70\text{ kg} - 70\text{ kg})/10\text{ kg} = 0$, and by consulting Table B.2 we see that $P(X_i > 70\text{ kg}) = P(Z > 0) = 0.5000$.* That is, 0.5000 (or 50.00%) of the distribution is larger than 70 kg. To deter-

*Read $P(X_i > 70\text{ kg})$ as "the probability of an X_i greater than 70 kg"; $P(Z > 0)$ is read as "the probability of a Z greater than 0."

mine the proportion of the distribution that is greater than 80 kg in weight, $Z = (80 \text{ kg} - 70 \text{ kg})/10 \text{ kg} = 1$, and $P(X_i > 80 \text{ kg}) = P(Z > 1) = 0.1587$ (or 15.87%). This could be stated as being the probability of drawing at random a measurement, X_i, greater than 80 kg from a population with $\mu = 70$ kg and $\sigma = 10$ kg. What, then, is the probability of obtaining, at random, a measurement, X_i, which is less than 80 kg? Since $P(X_i > 80 \text{ kg}) = 0.1587$, then $P(X_i < 80 \text{ kg}) = 1.0000 - 0.1587 = 0.8413$; that is, if 15.87% of the population is greater than X_i, then 100% − 15.87% of the population is less than X_i.* Example 7.1a presents calculations for determining proportions of a normal distribution lying between a variety of limits.

Example 7.1a

Calculating proportions of a normal distribution of bone lengths, where $\mu = 60$ mm, $\sigma = 10$ mm, and $N = 2000$.

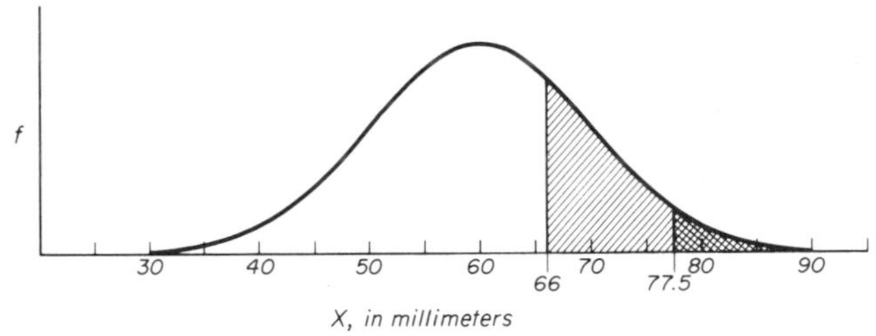

Example 7.1a Calculating proportions of a normal distribution of bone lengths, where $\mu = 60$ mm, $\sigma = 10$ mm, and $N = 2000$.

1. What proportion of the population of bone lengths is larger than 66 mm?

$$Z = \frac{X_i - \mu}{\sigma} = \frac{66 \text{ mm} - 60 \text{ mm}}{10 \text{ mm}} = 0.6$$

$$P(X_i > 66 \text{ mm}) = P(Z > 0.6) = 0.2743 \quad \text{or} \quad 27.43\%$$

2. What is the probability of picking, at random from this population, a bone larger than 66 mm?
   This is simply another way of stating the quantity calculated in part (1). The answer is 0.2743.
3. How many bone lengths in this population are greater than 66 mm?

$$(0.2743)(2000) = 549$$

4. What proportion of the population is smaller than 66 mm?

$$P(X_i < 66 \text{ mm}) = 1.0000 - P(X_i > 66 \text{ mm}) = 1.0000 - 0.2743 = 0.7257$$

5. What proportion of this population lies between 60 and 66 mm?
   Of the total population, 0.5000 is larger than 60 mm and 0.2743 is larger than 66 mm. Therefore, $0.5000 - 0.2743 = 0.2257$ of the population lies between 60 and 66 mm. That is, $P(60 \text{ mm} < X_i < 66 \text{ mm}) = 0.5000 - 0.2743 = 0.2257$.

*The statement that "$P(X_i > 80 \text{ kg}) = 0.1587$, therefore $P(X_i < 80) = 1.0000 - 0.1587$" is not mathematically accurate, because the case of $X_i = 80$ kg is not taken into account. But, as we are considering the distribution at hand to be a continuous one, the probability of X_i being *exactly* 80 kg (or being *exactly* any other value) is practically nil, so these types of probability statements offer no practical difficulties.

6. What portion of the area under the normal curve lies to the right of 77.5 mm?

$$Z = \frac{77.5 \text{ mm} - 60 \text{ mm}}{10 \text{ mm}} = 1.75$$

$$P(X_i > 77.5 \text{ mm}) = P(Z > 1.75) = 0.0401 \quad \text{or} \quad 4.01\%$$

7. How many bone lengths in the population are larger than 77.5 mm?

$$(0.0401)(2000) = 80$$

8. What is the probability of selecting at random from this population a bone measuring between 66 and 77.5 mm in length?

$$P(66 \text{ mm} < X_i < 77.5 \text{ mm}) = P(0.6 < Z < 1.75) = 0.2743 - 0.0401 = 0.2342$$

Note that Table B.2 contains no negative values of Z. However, if we are concerned with proportions in the left half of the distribution, we are simply dealing with areas of the curve that are mirror images of those present in the table. This is demonstrated in Example 7.1b.

Example 7.1b

Calculating proportions of a normal distribution of sucrose concentrations, where $\mu = 65$ mg/100 ml and $\sigma = 25$ mg/100 ml.

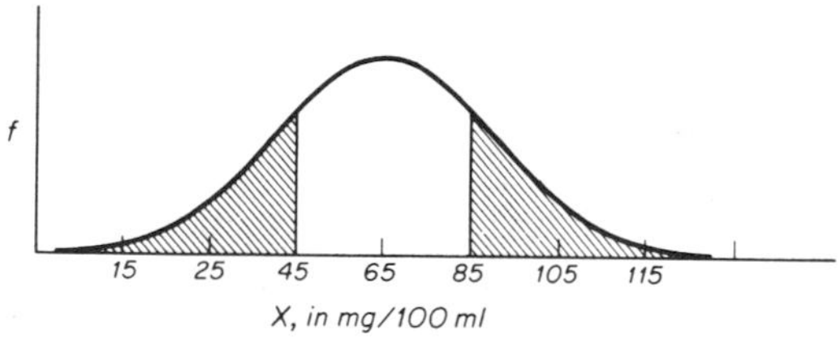

Example 7.1b Calculating proportions of a normal distribution of sucrose concentrations, where $\mu = 65$ mg/100 ml and $\sigma = 25$ mg/100 ml.

1. What proportion of the population is greater than 85 mg/100 ml?

$$Z = \frac{(X_i - \mu)}{\sigma} = \frac{85 \text{ mg/ 100 ml} - 65 \text{ mg/ 100 ml}}{25 \text{ mg/100 ml}} = 0.8$$

$$P(X_i > 85 \text{ mg/100 ml}) = P(Z > 0.8) = 0.2119 \quad \text{or} \quad 21.19\%$$

2. What proportion of the population is less than 45 mg/100 ml?

$$Z = \frac{45 \text{ mg/100 ml} - 65 \text{ mg/100 ml}}{25 \text{ mg/100 ml}} = -0.8$$

$$P(X_i < 45 \text{ mg/100 ml}) = P(Z < -0.8) = P(Z > 0.8) = 0.2119$$

That is, the probability of selecting from this population an observation less than 0.8 standard deviations below the mean is equal to the probability of obtaining an observation greater than 0.8 standard deviations above the mean.

3. What proportion of the population lies between 45 and 85 mg/100 ml?

$$\begin{aligned} P(45 \text{ mg/100 ml} < X_i < 85 \text{ mg/100 ml}) &= P(-0.8 < Z < 0.8) \\ &= 1.0000 - P(Z < -0.8 \quad \text{or} \quad Z > 0.8) \\ &= 1.0000 - (0.2119 + 0.2119) \\ &= 1.0000 - 0.4238 \\ &= 0.5762 \end{aligned}$$

Using the preceding considerations of the table of normal deviates (Table B.2), we can obtain the following formation:

68.26% of the measurements in a normal population
lie within the range of $\mu \pm \sigma$,
95.46% lie within $\mu \pm 2\sigma$,
99.73% lie within $\mu \pm 3\sigma$,
50% lie within $\mu \pm 0.67\sigma$,
95% lie within $\mu \pm 1.96\sigma$,
99% lie within $\mu \pm 2.58\sigma$,
99.5% lie within $\mu \pm 2.81\sigma$,
99.9% lie within $\mu \pm 3.31\sigma$.

7.3 THE DISTRIBUTION OF MEANS

If random samples of size n are drawn from a normal population, the means of these samples will form a normal distribution. The distribution of means from a nonnormal population will not be normal but will tend toward normality as n increases in size. Furthermore, the variance of the distribution of means will decrease as n increases; in fact, the variance of the population of all possible means of samples of size n from a population with variance σ^2 is

$$\sigma_{\bar{X}}^2 = \frac{\sigma^2}{n}. \tag{7.17}$$

The quantity $\sigma_{\bar{X}}^2$ is called the *variance of the mean*, and the preceding comments on the distribution of means comes from a very important mathematical theorem, known as the *central limit theorem*. A distribution of sample statistics is called a *sampling distribution*; therefore, we are discussing the sampling distribution of means.

Since $\sigma_{\bar{X}}^2$ has square units, its square root, $\sigma_{\bar{X}}$, will have the same units as the original measurements (and, therefore, the same units as the mean, μ, and the standard deviation, σ). This value, $\sigma_{\bar{X}}$, is the *standard deviation of the mean*. The standard deviation of a parameter or of a statistic is referred to as a *standard error*; thus, $\sigma_{\bar{X}}$ is frequently called the *standard error of the mean* (sometimes abbreviated SEM), or simply the *standard error* (sometimes abbreviated SE):

$$\sigma_{\bar{X}} = \sqrt{\frac{\sigma^2}{n}} \quad \text{or} \quad \sigma_{\bar{X}} = \frac{\sigma}{\sqrt{n}}. \tag{7.18}$$

Just as $Z = (X_i - \mu)/\sigma$ (Equation 7.16) is a normal deviate that refers to the normal distribution of X_i values,

$$Z = \frac{\bar{X} - \mu}{\sigma_{\bar{X}}} \tag{7.19}$$

is a normal deviate referring to the normal distribution of means ($\bar{X}$ values). Thus, we can ask questions such as: What is the probability of obtaining a random sample of nine measurements with a mean larger than 50.0 mm from a population having a mean of 47.0 mm and a standard deviation of 12.0 mm? This and other examples of the use of normal deviates for the sampling distribution of means are presented in Example 7.2.

As seen from Equation 7.18, to determine $\sigma_{\bar{X}}$ one must know σ^2 (or σ), which is

Example 7.2

Proportions of a sampling distribution of means.

1. If a population has $\mu = 47.0$ mm and $\sigma = 12.0$ mm, what is the probability of drawing from it a random sample of 9 measurements that has a mean larger than 50.0 mm?

$$\sigma_{\bar{X}} = \frac{12.0 \text{ mm}}{\sqrt{9}} = 4.0 \text{ mm}$$

$$Z = \frac{\bar{X} - \mu}{\sigma_{\bar{X}}} = \frac{50.0 \text{ mm} - 47.0 \text{ mm}}{4.0 \text{ mm}} = 0.75$$

$$P(\bar{X} > 50.0 \text{ mm}) = P(Z > 0.75) = 0.2266$$

2. What is the probability of drawing a sample of 25 measurements from the preceding population and finding that the mean of this sample is less than 40.0 mm?

$$\sigma_{\bar{X}} = \frac{12.0 \text{ mm}}{\sqrt{25}} = 2.4 \text{ mm}$$

$$Z = \frac{40.0 \text{ mm} - 47.0 \text{ mm}}{2.4 \text{ mm}} = -2.92$$

$$P(\bar{X} < 40.0 \text{ mm}) = P(Z < -2.92) = P(Z > 2.92) = 0.0018$$

3. If 500 random samples of size 25 are taken from the preceding population, how many of them would have means larger than 50.0 mm?

$$\sigma_{\bar{X}} = \frac{12.0 \text{ mm}}{\sqrt{25}} = 2.4 \text{ mm}$$

$$Z = \frac{50.0 \text{ mm} - 47.0 \text{ mm}}{2.4 \text{ mm}} = 1.25$$

$$P(\bar{X} > 50.0 \text{ mm}) = P(Z > 1.25) = 0.1056$$

Therefore, (0.1056)(500) = 53 samples would be expected to have means larger than 50.0 mm.

a population parameter. Since we very seldom can calculate population parameters, we must rely on estimating them from random samples taken from the population. The best estimate of $\sigma_{\bar{X}}^2$, the population variance of the mean, is

$$s_{\bar{X}}^2 = \frac{s^2}{n}, \tag{7.20}$$

the sample variance of the mean. Thus,

$$s_{\bar{X}} = \sqrt{\frac{s^2}{n}} \quad \text{or} \quad s_{\bar{X}} = \frac{s}{\sqrt{n}} \tag{7.21}$$

is an estimate of $\sigma_{\bar{X}}$ and is the sample standard error of the mean. Example 7.3 demonstrates the calculation of $s_{\bar{X}}$.

The importance of the standard error in hypothesis testing and related procedures will be evident in Chapter 8. At this point, however, it can be noted that the magnitude of $s_{\bar{X}}$ is helpful in determining the precision to which the mean and some measures of variability may be reported. Although different practices have been followed by many, we shall employ the following (Eisenhart, 1968). We shall state the standard error to two significant figures (e.g., 2.7 mm in Example 7.3). Then the standard deviation and the mean will be reported with the same number of decimal places (e.g.,

$\bar{X} = 137.6$ mm in Example 7.3*). The variance may be reported with twice the number of decimal places as the standard deviation.

Example 7.3

The calculation of the standard error of the mean, $s_{\bar{X}}$. The following are data for systolic blood pressure, in mm of mercury.

121	$n = 12$
125	$\bar{X} = \dfrac{1651 \text{ mm}}{12} = 137.6 \text{ mm}$
128	
134	$\text{SS} = 228{,}111 \text{ mm}^2 - \dfrac{(1651 \text{ mm})^2}{12}$
136	
138	$= 960.9167 \text{ mm}^2$
139	$s^2 = \dfrac{960.9167 \text{ mm}^2}{11} = 87.3561 \text{ mm}^2$
141	
144	$s = \sqrt{87.3561 \text{ mm}^2} = 9.35 \text{ mm}$
145	$s_{\bar{X}} = \dfrac{s}{\sqrt{n}} = \dfrac{9.35 \text{ mm}}{\sqrt{12}} = 2.7 \text{ mm}$ *or*
149	
151	$s_{\bar{X}} = \sqrt{\dfrac{s^2}{n}} = \sqrt{\dfrac{87.3561 \text{ mm}^2}{12}}$
$\sum X = 1651$ mm	
$\sum X^2 = 228{,}111$ mm²	$= \sqrt{7.2797 \text{ mm}^2} = 2.7 \text{ mm}$

7.4 ASSESSING DEPARTURES FROM NORMALITY

It is sometimes desired to test the hypothesis that a sample came from a population whose members follow a normal distribution. Methods for testing this hypothesis might be placed into four categories: use of the g statistics, goodness of fit testing, graphical methods, and other methods.

The testing of hypotheses utilizing g_1 and g_2 will be discussed in Section 8.13, but these procedures have relatively unsatisfactory performance in assessing departures from normality. Preferred methods will be discussed in the remainder of this chapter.

Goodness of Fit Testing of Normality. Goodness of fit procedures for testing the null hypothesis that a sample came from a normal population are based on the methods discussed in Chapter 5. Recall that for chi-square tests one needs to know both the observed frequency in each class or grouping of observations (f_i) and the frequency expected if the null hypothesis is true ($\hat{f}_i$). Example 7.4 demonstrates the procedure for calculating the expected frequencies and then presents the chi-square analysis for goodness of fit. Figure 7.5 shows the observed and expected frequency distributions. First, the sample mean and standard deviation ($\bar{X}$ and s) are calculated as estimates of the mean and standard deviation (μ and σ) of the population from which the sample came. Note that we may do this by using Equations 3.3 and 4.9, respectively, as the data are grouped. (Or, Equations 3.2 and 4.8, respectively, may be

*In Example 7.3, s is written with more decimal places than the Eisenhart recommendations indicate because it is an intermediate, rather than a final, result, and rounding off intermediate computations may lead to serious rounding error. Indeed, some authors routinely report extra decimal places, even in final results, with the consideration that readers of the results may use them as intermediates in additional calculations.

Example 7.4

The heights of the first 70 graduate students to enroll in my biostatistics course. Chi-square goodness of fit of a normal distribution.

H_0: This sample came from a normal population.
H_A: This sample did not come from a normal population.

Height class (in.)	*Height* (in.)(X_i)	*Observed frequency* (f_i)	f_iX_i (in.)	$f_iX_i^2$ (in.^{2})	$P(X_i)$	*Expected frequency* ($\hat{f}_i$)	$\frac{(f_i - \hat{f}_i)^2}{\hat{f}_i}$
< 62.5		0 }2			0.0102	0.7140 }1.5190	0.1523
62.5–63.5	63	2	126	7938	0.0115	0.8050	
63.5–64.5	64	2	128	8192	0.0219	1.5330	0.1423
64.5–65.5	65	3	195	12675	0.0357	2.4990	0.1004
65.5–66.5	66	5	330	21780	0.0542	3.7940	0.3834
66.5–67.5	67	4	268	17956	0.0755	5.2850	0.3127
67.5–68.5	68	6	408	27744	0.0995	6.9650	0.1337
68.5–69.5	69	5	345	23805	0.1122	7.8540	1.0371
69.5–70.5	70	8	560	39200	0.1191	8.3370	0.0136
70.5–71.5	71	7	497	35287	0.1156	8.0920	0.1474
71.5–72.5	72	7	504	36288	0.1026	7.1820	0.0046
72.5–73.5	73	10	730	53290	0.0858	6.0060	2.6560
73.5–74.5	74	6	444	32856	0.0611	4.2770	0.6941
74.5–75.5	75	3	225	16875	0.0414	2.8980	0.0036
75.5–76.5	76	2	152	11552	0.0256	1.7920	0.0241
76.5–77.5	77	0 }0	0	0	0.0145	1.0150 }1.9670	1.9670
> 77.5		0			0.0136	0.9520	
		$\Sigma f_i = 70$ $(= n)$	$\Sigma f_iX_i = 4912$	$\Sigma f_iX_i^2 = 345{,}438$	$\Sigma P(X_i) = 1.0000$	$\Sigma \hat{f}_i = 70.0000$	$\chi^2 = 7.7723$

$$\bar{X} = \frac{\Sigma f_iX_i}{n} = \frac{4912}{70}\text{ in.} = 70.17\text{ in.}$$

$$\text{SS} = \Sigma f_iX_i^2 - \frac{(\Sigma f_iX_i)^2}{n} = 345{,}438\text{ in.}^2 - \frac{(4912\text{ in.})^2}{70} = 755.9429\text{ in.}^2$$

$$s^2 = \frac{SS}{n-1} = \frac{755.94\text{ in.}^2}{69} = 10.9557\text{ in.}^2$$

$s = 3.31$ in. $\qquad \nu = 15 - 3 = 12$

$\chi^2 = 7.772 \qquad \chi^2_{0.05,12} = 21.026$

Therefore, do not reject H_0.

$(0.75 < P < 0.90)$

employed.) Then we can estimate the proportion of the total population in each size class, using $Z = (X_i - \bar{X})/s$ as an estimate of $Z = (X_i - \mu)/\sigma$, and consult Table B.2.

Let us begin with the largest size class, in this case those measurements > 77.5 in.* Here we ask what proportion of a normal population (and thus what proportion of a random sample from that normal population) with $\mu = 70.17$ in. and $\sigma = 3.31$ in. will be expected to be > 77.5 in. tall. Since $Z = (77.5\text{ in.} - 70.17\text{ in.})/3.31\text{ in.} = 2.21$, this proportion, read from Table B.2, is 0.0136. Thus, if our sample had

*Metric units of measurements are used throughout this book, with the sole exception of the data in Example 7.4, which have their English units retained for historical reasons.

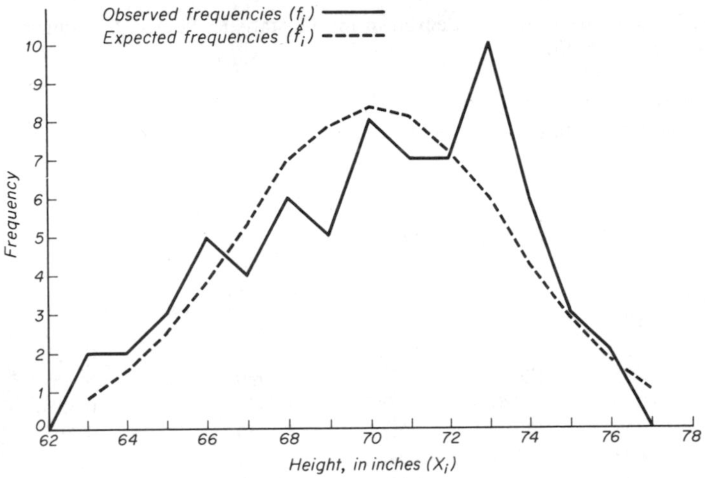

Figure 7.5 The frequency curve for the student height data in Example 7.4, showing also the frequency curve that would be expected if the data followed a normal distribution.

indeed come from a normal population, as the null hypothesis proposes, then (0.0136)(70) = 0.9520 individuals out of the 70 would be expected to be greater than 77.5 in. tall. Similarly, by calculating $Z = (76.5 \text{ in.} - 70.17 \text{ in.})/3.31 \text{ in.} = 1.91$, Table B.2 informs us that 0.0281 of the population would have a height greater than 76.5 in. Thus, $0.0281 - 0.0136 = 0.0145$ of the observations, or $(0.0145)(70) = 1.0150$ of the individuals, would be between 76.5 and 77.5 in.

To determine the proportion of the population > 75.5 in., we calculate $Z = (75.5 \text{ in.} - 70.17 \text{ in.})/3.31 \text{ in.} = 1.61$, and read from the table that $P(Z > 1.61) = 0.0537$. Therefore, the proportion of the population that is between 75.5 and 76.5 inches would be $P(1.61 < Z < 1.91) = 0.0537 - 0.0281 = 0.0256$.

We continue computing $P(X_i)$ in this fashion for all other height classes greater than that height class containing $\bar{X} = 70.17$ in. Then, similar probabilities are obtained, beginning with the smallest height class and proceeding until reaching the height class that contains $\bar{X}$. That is, for heights < 62.5 in., $Z = (62.5 \text{ in.} - 70.17 \text{ in.})/3.31 \text{ in.} = -2.32$, and $P(X_i < 62.5 \text{ in.}) = P(Z < -2.32) = 0.0102$. For heights < 63.5 in., $Z = -2.02$, so $P(62.5 \text{ in.} < X_i < 63.5 \text{ in.}) = P(-2.32 < Z < -2.02) = 0.0217 - 0.0102 = 0.0115$. For heights < 64.5 in., $Z = -1.71$, so $P(-2.02 < Z < -1.71) = 0.0436 - 0.0217 = 0.0219$.

Finally we determine the proportion of the population in that height class containing $\bar{X}$. This is

$$\begin{aligned} P(69.5 \text{ in.} < X_i < 70.5 \text{ in.}) &= 1.0000 - P(X_i < 69.5 \text{ in.}) - P(X_i > 70.5 \text{ in.}) \\ &= 1.0000 - P(Z < -0.20) - P(Z > 0.10) \\ &= 1.0000 - 0.4207 - 0.4602 \\ &= 0.1191. \end{aligned}$$

After obtaining all values of $P(X_i)$, the proportions of the normal curve lying within each of the 17 height classes, the expected frequencies for each height class are calculated as

$$\hat{f}_i = [P(X_i)][n]. \tag{7.22}$$

Thus, $\hat{f}_1 = (0.0102)(70) = 0.7140$, $\hat{f}_2 = (0.0115)(70) = 0.8050$, and so on.

Watson (1957) demonstrated that it is desirable to have at least 10 class intervals of X_i (i.e., $k \geq 10$) when performing a chi-square analysis for goodness of fit to normal distributions. Cochran (1954) suggests that bias in the chi-square analysis can be reduced adequately by assuring that no $\hat{f}_i$ values are less than 1.0. Thus, height classes in the tails of the distribution are combined; in our example, it is necessary to combine the two most extreme classes in the lower tail and the two most extreme classes in the upper tail. After we tabulate the f_i and $\hat{f}_i$ for each class, and assure that no $\hat{f}_i$ is < 1.0, we may perform the chi-square calculation. The number of degrees of freedom for the test is the number of classes of data ($k = 15$) minus the number of constants, determined from the sample, that were necessary to calculate the $\hat{f}_i$ values. As we required three sample constants, namely n, $\bar{X}$, and s, $\nu = k - 3 = 12$ for the chi-square goodness of fit of a normal distribution.

The log-likelihood goodness of fit procedure (Section 5.9) may be employed as an alternative to the chi-square analysis, in which case we need not be concerned with the magnitude of the $\hat{f}_i$'s. If a distribution's departure from normality occurs in one or both tails (a common situation), then the chi-square procedure of combining tail frequencies will destroy the test's power. The use of the log-likelihood test, or one of the tests below, is thus preferred.

Occasionally it may be the case that the $\hat{f}_i$ values are to be calculated using *a priori* values for μ and σ, rather than the sample statistics $\bar{X}$ and s. For example, the null hypothesis might have stated that the sample came from a normal population having a mean of 72 in. and a standard deviation of 3 in. By using these values of μ and σ, instead of the $\bar{X}$ and s calculated from the samples, we may calculate the $\hat{f}_i$ values in the fashion previously described, followed by the calculation of χ^2 or the log-likehood G. In this case, however, the only sample constant used in the computation of the $\hat{f}_i$'s would be n, and therefore $\nu = k - 1$.

Goodness of fit may also be tested by the Kolmogorov-Smirnov procedure (Section 5.11); indeed, this test is generally preferable to chi-square when dealing with continuous distributions such as the normal. Example 7.5 tabulates the observed frequencies, f_i, as in Example 7.4. (The expected frequencies, $\hat{f}_i$, are not required.) The column labelled "$P(X_i)$" in Example 7.4 contains the expected relative frequencies, which could be cumulated to yield the expected cumulative relative frequencies tabulated in Example 7.5. However, it is simpler, and reduces rounding error, if these cumulative relative frequencies are determined directly from a table of proportions of the normal curve (Table B.2).

With the values of $\bar{X} = 70.17$ in. and $s = 3.31$ in. computed from the sample (as shown in Example 7.4), we determine the relative cumulative expected frequencies using $Z = (X_i - \bar{X})/s$ as an estimate of $Z = (X_i - \mu)/\sigma$. For example, to find $P(X_i < 62.5$ in.) we compute $Z = (62.5$ in. $- 70.17$ in.)/3.31 in. $= -2.32$, and $P(Z < -2.32) = 0.0102$ (from Table B. 2). Similarly, cum. $\hat{F}_2 = P(X_i < 63.5$ in.) $= P(Z < -2.02) = 0.0217$, and so on. Keep in mind that Table B.2 gives proportions in the right-hand tail of the normal curve [i.e., $P(Z <$ a stated value)]; but, since the normal curve is symmetrical, the left-hand tail contains the same proportions. So $P(Z < -2.32) = P(Z > 2.32)$, and the latter is read directly from the table. If our computed Z is positive, however, then the cumulative relative expected frequency is 1.0000 minus the tabled proportion. For example, $P(X_i < 72.5 \text{ in.}) = P(Z < 0.70) = 1.0000 - P(Z > 0.70) = 1.0000 - 0.2420 = 0.7580$. When all values of rel F_i and rel $\hat{F}_i$ have been obtained, then D_i and D'_i may be determined, as shown in Example 7.5. Figure 7.6 shows rel F_i and rel $\hat{F}_i$ for Example 7.5.

Example 7.5

Kolmogorov-Smirnov goodness of fit of a normal distribution. The data are as in Example 7.4.

H_0: The sampled population is normally distributed.
H_A: The sampled population is not normally distributed.

i	Height class (in.)	Observed frequency (f_i)	Cumulative observed frequency (F_i)	Cumulative relative observed frequency (rel F_i)	Cumulative relative expected frequency (rel $\hat{F}_i$)	D_i	D'_i
1	< 62.5	0	0	0.0000	0.0102	0.0102	0.0102
2	62.5–63.5	2	2	0.0286	0.0217	0.0092	0.0217
3	63.5–64.5	2	4	0.0571	0.0436	0.0135	0.0150
4	64.5–65.5	3	7	0.1000	0.0793	0.0207	0.0220
5	65.5–66.5	5	12	0.1714	0.1335	0.0379	0.0335
6	66.5–67.5	4	16	0.2286	0.2090	0.0196	0.0376
7	67.5–68.5	6	22	0.3143	0.3085	0.0058	0.0799
8	68.5–69.5	5	27	0.3857	0.4207	0.0350	0.1064
9	69.5–70.5	8	35	0.5000	0.5398	0.0398	0.1541
10	70.5–71.5	7	42	0.6000	0.6554	0.0554	0.1554
11	71.5–72.5	7	49	0.7000	0.7580	0.0580	0.1580
12	72.5–73.5	10	59	0.8429	0.8438	0.0009	0.1438
13	73.5–74.5	6	65	0.9286	0.9049	0.0237	0.0620
14	74.5–75.5	3	68	0.9714	0.9463	0.0251	0.0177
15	75.5–76.5	2	70	1.0000	0.9719	0.0281	0.0005
16	76.5–77.5	0	70	1.0000	0.9864	0.0136	0.0136
17	> 77.5	0	70	1.0000	1.0000	0.0000	0.0000

maximum $D_i = 0.0580$; maximum $D'_i = 0.1580$; $D = 0.1580$
For normality testing, $D_{0.05,70} = 0.106$
Reject H_0.
$P < 0.01$

Typically, as in Example 7.5, Kolmogorov-Smirov goodness of fit testing for normality employs the sample statistics $\bar{X}$ and s as estimates of the population parameters μ and σ, respectively. This being the case, the use of the critical values in Table B.9 would provide for a very conservative test (i.e., one in which the probability of a Type II error is much inflated). It is much more preferable to employ the critical values in Table B.21 for such a test (Stephens, 1974), as demonstrated in Example 7.5.

The Watson goodness of fit test of Section 25.1 may also be used for normality testing. For appropriate critical values (not those in Table B.35), see Stephens (1974).

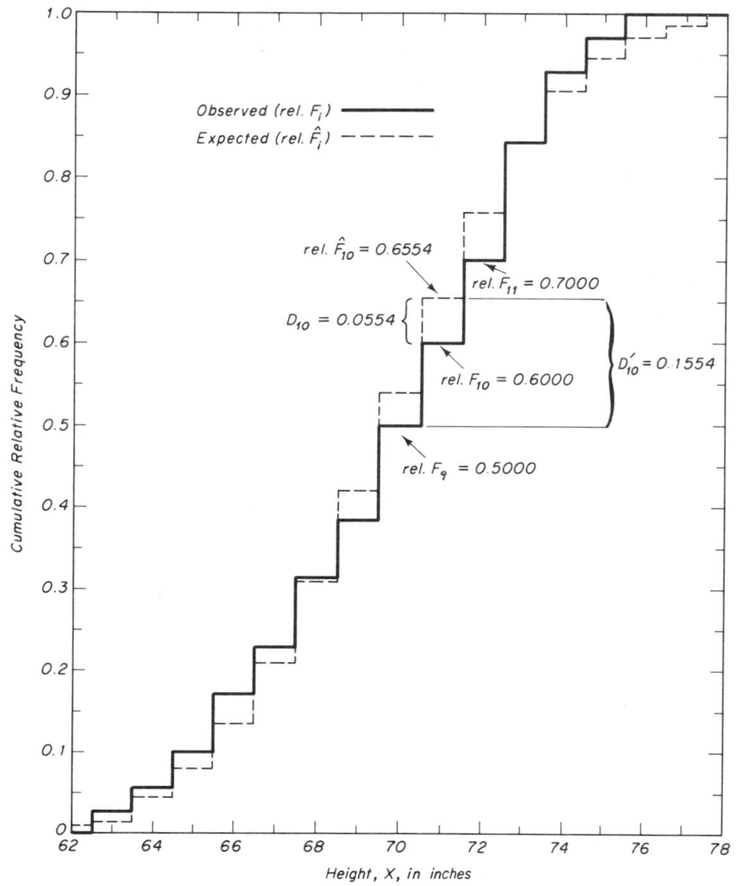

Figure 7.6 The observed and expected cumulative relative frequency distributions for the data in Example 7.5.

Graphical Assessment of Normality. The graphical representation of a normal distribution as a frequency curve is presented in Fig. 7.1. Plotting a cumulative frequency distribution for normally distributed data results in an S-shaped, or sigmoid, curve, as shown in Fig. 7.7, using the data of Example 7.4.

Graphical examination of the relative cumulative frequency distribution is aided greatly by the use of the *normal probability scale*, rather than a linear scale. As Fig. 7.7 shows, a given increment in X_i (on the abscissa) near the median is associated with a much larger change in relative frequency (on the ordinate) than is the same increment in X_i at very high or low percentiles. Using the normal probability scale on the ordinate expands the scale for high and low percentiles and compresses it for percentiles toward the median, and the resulting cumulative frequency plot is a straight line for a normal distribution. Figure 7.8 shows the data of Example 7.4 plotted as a cumulative distribution on such a scale. Graph paper with the normal probability scale on the ordinate is available commercially.

Leptokurtic distributions will appear sigmoid (S-shaped) rather than straight on such a plot, and a platykurtic distribution will appear as a reverse sigmoid curve. (The data in Fig. 7.8 lean slightly toward platykurtosis.) A negatively skewed distribu-

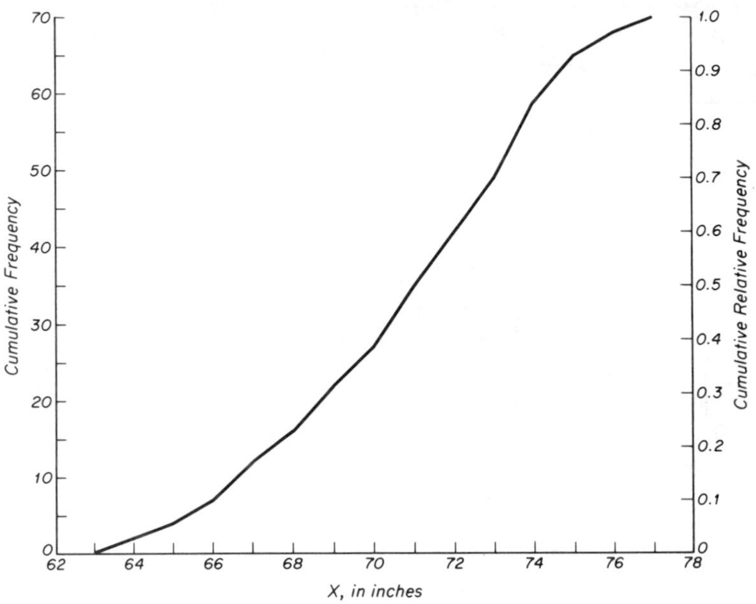

Figure 7.7 The cumulative frequency polygon of the student height data from Example 7.4.

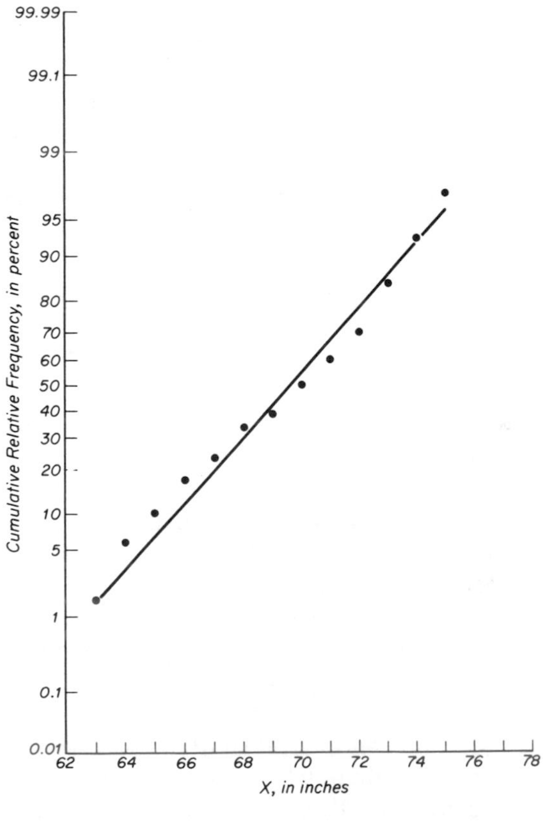

Figure 7.8 The cumulative relative frequency distribution for the data of Example 7.4, plotted with the normal probability scale as the ordinate. The expected frequencies (i.e., the frequencies from a normal distribution) would fall on the straight line shown.

tion will show an upward curve, as the lower portion of a sigmoid curve, and a positively skewed distribution will result in a shape resembling the upper portion of a sigmoid curve.

Further discussion of graphical methods of examining departures from normality may be found in Bliss (1967: 101–112).

Other Methods of Assessing Normality. Shapiro and Wilk (1965) presented a test for normality involving the calculation of a statistic they called W. The computation requires an extensive table of constants, however, because a different set of $n/2$ constants is required for each sample size, n. The authors provide a table of these constants (for n through 50) and also a table of critical values of W. The power of W has been shown to be excellent (Shapiro, Wilk, and Chen, 1968) when testing for departures from normality.

The computation of W for normality testing, especially when $n > 50$, can be very cumbersome. An alternative procedure is that of D'Agostino (1971a, 1971b), which involves the computation of a statistic he calls D, applicable as a powerful test for departure from normality. As Example 7.6 shows, the n observations in the sample are first arranged in ascending (or descending) order. Then,

$$D = \frac{T}{\sqrt{n^3 \text{SS}}}, \tag{7.23}$$

where SS is the "sum of squares," as defined by Equation 4.3 and computed by either Equation 4.7 or 4.9, and

$$T = \sum_{i=1}^{n} \left(i - \frac{n+1}{2} \right) X_i. \tag{7.24}$$

Example 7.6

D'Agostino's test for departure from normality applied to the data of Example 7.4.

H_0: This sample came from a normal population.
H_A: This sample did not come from a normal population.

X_i	f_i	i
63	2	1–2
64	2	3–4
65	3	5–7
66	5	8–12
67	4	13–16
68	6	17–22
69	5	23–27
70	8	28–35
71	7	36–42
72	7	43–49
73	10	50–59
74	6	60–65
75	3	66–68
76	2	69–70

$$SS = 755.94290 \text{ in.}^2 \text{ (computed as in Example 7.4)}$$

$$\frac{n+1}{2} = \frac{70+1}{2} = 35.5$$

$$\begin{aligned} T &= \Sigma\left(i - \frac{n+1}{2}\right)X_i \\ &= (1 - 35.5)(63 \text{ in.}) + (2 - 35.5)(63 \text{ in.}) + (3 - 35.5)(64 \text{ in.}) \\ &\quad + (4 - 35.5)(64 \text{ in.}) + (5 - 35.5)(65 \text{ in.}) + (6 - 35.5)(65 \text{ in.}) \\ &\quad + (7 - 35.5)(65 \text{ in.}) + \ldots + (66 - 35.5)(75 \text{ in.}) \\ &\quad + (67 - 35.5)(75 \text{ in.}) + (68 - 35.5)(75 \text{ in.}) + (69 - 35.5)(76 \text{ in.}) \\ &\quad + (70 - 35.5)(76 \text{ in.}) \\ &= 4579.00000 \text{ in.} \end{aligned}$$

$$D = \frac{T}{\sqrt{n^3 \text{ SS}}} = \frac{4579.00000 \text{ in.}}{\sqrt{(70^3)(755.94290 \text{ in.}^2)}} = 0.2844$$

Since D is neither ≤ 0.2726 nor ≥ 0.2864 (from Table B.22), do not reject H_0.

Table B.22 gives the upper and lower critical values of D'Agostino's D. Note that since the range of D values is small, at least 5 decimal place accuracy should be used in the computations. In his original papers on this test, D'Agostino (1971a, 1971b, 1972) converted D to another quantity, Y, and provided critical values of Y.

EXERCISES

7.1. A normally distributed population of lemming body weights has a mean of 63.5 g and a standard deviation of 12.2 g.

(a) What proportion of this population is 78.0 g or larger?

(b) What proportion of this population is 78.0 g or smaller?

(c) If there are 1000 weights in the population, how many of them are 78.0 g or larger?

(d) What is the probability of choosing at random from this population a weight smaller than 41.0 g?

7.2. (a) Considering the population of Exercise 7.1, what is the probability of selecting at random a body weight between 60.0 and 70.0 g?

(b) What is the probability of a body weight between 50.0 and 60.0 g?

7.3. (a) What is the standard deviation of all possible means of samples of size 10 which could be drawn from the population in Exercise 7.1?

(b) What is the probability of selecting at random from this population a sample of 10 weights that has a mean greater than 65.0 g?

(c) What is the probability of the mean of a sample of 10 being between 60.0 and 62.0 g?

8 One-Sample Hypotheses

This chapter will discuss how we can draw inferences about population parameters by examining appropriate sample estimates. In doing so, we shall be introduced to the *t* distribution. This was first presented by W. S. Gosset*, who published it under the pseudonym "Student" (1908), hence the common reference to "Student's *t* distribution" or "Student's *t* test." Fisher (e.g., 1925) helped outline the potential uses of this distribution, and it has since become one of the most important in statistical analysis. This chapter will also introduce procedures for expressing the confidence one can have in estimating parameters from sample statistics.

8.1 TWO-TAILED HYPOTHESES CONCERNING THE MEAN

Let us assume that the mean life span of horses is known to be 22 years. Suppose that a new breed of horse has been developed, and the breeders would like to know if a change in longevity has resulted. The life spans of 25 of these horses are found to have a mean, $\bar{X}$, of 24.23 years and a standard deviation, s, of 4.25 years. We can now set forth the hypothesis that these 25 life spans came from a population having a mean of 22 years. This would be the *null hypothesis* (abbreviated H_0: $\mu = 22$ yr), for it hypothesizes "no difference" between the population parameter μ and the quantity 22 years. The *alternate hypothesis* (H_A: $\mu \neq 22$ yr) states that the mean longevity of all horses of this new breed is some value other than 22 years.

*William Sealy Gosset (1876–1937), an English chemist with the title "Brewer" in a Dublin brewery. He had other noteworthy accomplishments, as well, in the development of statistical theory.

Recall that it is the null hypothesis, not the alternative hypothesis, that may be tested statistically. The statistical procedure is to ask the question, "If H_0 were true (i.e., if the population mean were indeed 22 years), would it be reasonably likely to obtain from the population a random sample of 25 with a mean of 24.23 years?" If we determine that the probability of obtaining a sample with a mean as extreme as 24.23 years from a population with a mean of 22 years is very small, then we may reasonably conclude that H_0 is false and assume H_A to be true. The question is, how small a probability should be considered so small as to have us conclude that H_0 is false? Section 5.2 introduced 0.05 (i.e., 5%) as the probability conventionally used in biological statistics, and this is referred to as α, the *level of significance*, of the statistical test. (Although $\alpha = 0.05$ is most commonly employed, $\alpha = 0.01$, or, less frequently, $\alpha = 0.10$, may also be encountered.)

In Section 7.3 (Equation 7.19), $Z = (\bar{X} - \mu)/\sigma_{\bar{X}}$ was introduced as a normal deviate, and it was shown how one could determine the probability of obtaining a sample with mean $\bar{X}$ from a population with mean μ. Note, however, that the calculation of Z requires the knowledge of $\sigma_{\bar{X}}$, which we typically do not have. The best we can do is to calculate $s_{\bar{X}}$ as an estimate of $\sigma_{\bar{X}}$, and for the present example $s_{\bar{X}} = s/\sqrt{n} = 4.25 \text{ yr}/\sqrt{25} = 0.85 \text{ yr}$. If n is large, then $s_{\bar{X}}$ is a very good estimate of $\sigma_{\bar{X}}$, and one can calculate Z using this estimate. However, for most biological situations, n is insufficiently large to consider $s_{\bar{X}}$ as an accurate estimate of $\sigma_{\bar{X}}$, and we resort to the *t distribution* and calculate

$$t = \frac{\bar{X} - \mu}{s_{\bar{X}}}. \tag{8.1}$$

As does the χ^2 distribution, introduced in Chapter 5, the t distribution has different shapes for different degrees of freedom (ν). For hypotheses concerning the mean, $\nu = n - 1$; the influence of ν on the shape of the t distribution is shown in Fig. 8.1. This distribution is leptokurtic (see Section 7.1), having a greater concentration of values around the mean and in the tails than does the normal distribution; but as n increases, the t distribution tends to resemble the normal distribution more closely, and for $\nu = \infty$ (i.e., for an infinitely large sample), the two distributions are identical.*

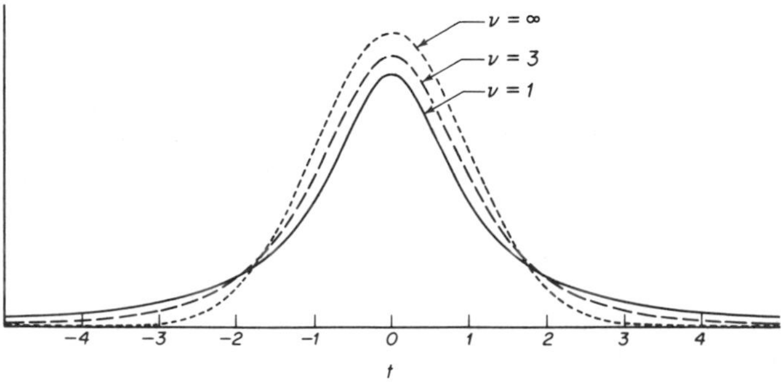

Figure 8.1 The t distribution for various degrees of freedom, ν. For $\nu = \infty$, the t distribution is identical to the normal distribution.

*Indeed, the t, normal, and chi-squared distributions are related as

$$t_{\alpha(2),\infty} = Z_{\alpha(2)} = \sqrt{\chi^2_{\alpha,1}}. \tag{8.2}$$

For our horse longevity data, $n = 25$, so $\nu = 24$, and the t distribution (which is the sampling distribution of all possible means of samples of size 25 that could be taken from the population) appears as in Fig. 8.2. In this figure, the mean of the

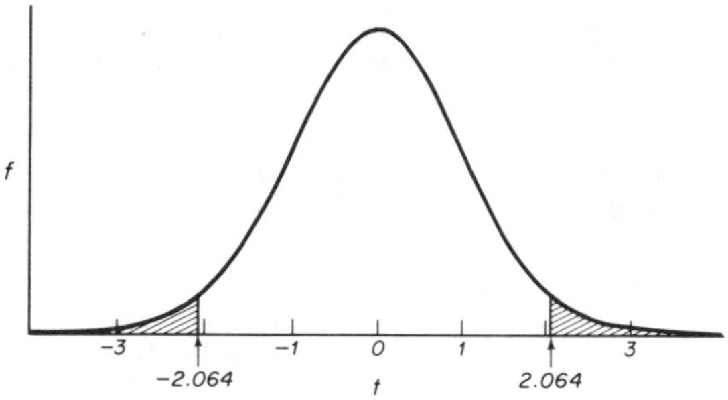

Figure 8.2 The t distribution for $\nu = 24$, showing the critical region (shaded area) for a two-tailed test using $\alpha = 0.05$. (The critical value of t is 2.064.)

distribution (i.e., $t = 0$) represents the mean hypothesized in H_0 (i.e., 22 yr), for, by Equation 8.1, $t = 0$ when $\bar{X} = 22$ yr; and the shaded areas represent the extreme 5% of the total area under the curve (2.5% in each tail). Thus, an $\bar{X}$ so far from μ that it lies in either of the shaded areas has a probability of less than 5% of occurring by chance alone, and we assume that it occurred because H_0 is, in fact, false. Since an extreme t value in either direction from μ will cause us to reject H_0, we are said to be considering a "two-tailed" (or "two-sided") test.

For $\nu = 24$ and $\alpha = 0.05$, we can consult Table B.3 to find that the shaded areas of the curve begin at 2.064 t units on either side of μ. That is, 2.064 is the *critical value* of t, and if t, calculated from Equation 8.1, is equal to or greater than 2.064 or is equal to or less than −2.064, that will be considered reasonable cause to reject H_0 and accept H_A. That portion of the distribution beyond the critical value (i.e., the shaded areas in the figure) is called the *critical region.* From the sample of 25 horses, $t = (\bar{X} - \mu)/s_{\bar{X}} = (24.23 \text{ yr} - 22 \text{ yr})/0.85 \text{ yr} = 2.624$. Since 2.624 lies within the critical region (i.e., $2.624 > 2.064$), H_0 is rejected, and we conclude that the new breed of horses does not have a life span of 22 years.

To summarize, the hypotheses for the two-tailed test are

$$H_0\text{:} \quad \mu = \mu_0 \text{ and } H_A\text{:} \quad \mu \neq \mu_0,$$

where μ_0 denotes the hypothesized value to which we are comparing the population mean. (In the above example, $\mu_0 = 22$ yr) The test statistic is calculated as $t = (\bar{X} - \mu)/s_{\bar{X}}$, and if its absolute value is larger than the critical value of t from Table B.3, one rejects H_0 and accepts H_A. The critical value of t can be abbreviated as $t_{\alpha(2),\nu}$, where $\alpha(2)$ refers to the two-tailed probability of α. Thus, for the preceding example, we could write $t_{0.05(2),24} = 2.064$. In general, for a two-tailed t test,

$$\text{if } |t| \geq t_{\alpha(2),\nu}, \quad \text{then reject } H_0.$$

Example 8.1 presents the computations for the analysis of the horse longevity data. A t of 2.624 is calculated, which for 24 degrees of freedom lies between the tabled critical values of $t_{0.02(2),24} = 2.492$ and $t_{0.01(2),24} = 2.797$. Therefore, if the null hypothesis, H_0, is a true statement about the population we sampled, the probability of $\bar{X}$ being at least this far from μ is between 0.01 and 0.02; that is, $0.01 < P(|t|$

$\geq 2.624) < 0.02$.* As this probability is less than 0.05, we reject H_0 and declare it is not a true statement. For a consideration of the types of errors involved in rejecting or accepting the null hypothesis, refer to Section 5.3.

Example 8.1

The two-tailed t test for significant difference between a mean and a hypothesized population mean of $\mu_0 = 22$ yr.

Age at death (in yr) of 25 horses of a particular breed: 17.2, 18.0, 18.7, 19.8, 20.3, 20.9, 21.0, 21.7, 22.3, 22.6, 23.1, 23.4, 23.8, 24.2, 24.6, 25.8, 26.0, 26.3, 27.2, 27.6, 28.1, 28.6, 29.3, 30.1, 35.1.

H_0: $\mu = 22$ yr

H_A: $\mu \neq 22$ yr

$\alpha = 0.05$

$n = 25$

$\bar{X} = 24.23$ yr

$s^2 = 18.0388$ yr

$$s_{\bar{X}} = \sqrt{\frac{18.0388\ \text{yr}^2}{25}} = 0.85\ \text{yr}$$

$$t = \frac{\bar{X} - \mu}{s_{\bar{X}}} = \frac{24.23\ \text{yr} - 22\ \text{yr}}{0.85\ \text{yr}} = 2.624$$

$\nu = n - 1 = 25 - 1 = 24$

$t_{0.05(2),24} = 2.064$

Since $|t| > t_{0.05(2),24}$, reject H_0 and conclude that the sample of 25 horse life spans came from a population whose mean, μ, is not 22 yr.

$$0.01 < P(|t| \geq 2.624) < 0.02$$

Frequently, the hypothesized value in the null and alternate hypotheses is zero. For example, the weights of 12 rats might be measured before and after the animals are placed on a regime of forced exercise for one week. The change in weight of the animals (i.e., weight after minus weight before) could be recorded, and it might have been found that the mean weight change was -0.65 g (i.e., the mean weight change is a 0.65-g weight loss). If we wished to infer whether such exercise causes any significant change in rat weight, we could state H_0: $\mu = 0$ and H_A: $\mu \neq 0$; Example 8.2 summarizes the t test for this H_0 and H_A. This test is two tailed, for a large $\bar{X} - \mu$ difference in either direction will constitute grounds for rejecting the veracity of H_0.

The theoretical basis of the t testing utilized throughout this chapter assumes that sample data came from a normal population, assuring that the mean at hand came from a normal distribution of means. Fortunately the t test is *robust*, meaning that its validity is not seriously affected by moderate deviations from the underlying assumption.†

A common situation in which one is dealing with a population known to be nonnormal is the case where the data are percentages or proportions. Such data are known to be binomial, rather than normal, and the treatment of such data is discussed in Section 14.2.

*Some desk calculators and computer programs have the capability of determining the probability of a given t (Guenther, 1977). For the present example, we would thereby find that $P(|t| \geq 2.624) = 0.015$.

†The effect of nonnormality is greater for smaller α, and the effect decreases as n increases (Ractliffe, 1968).

Example 8.2

A two-tailed test for significant difference between a sample mean and a hypothesized population mean of zero.

Weight change of 12 rats after being subjected to a regime of forced exercise. Each weight change (in g) is the weight after exercise minus the weight before.

1.7	H_0: $\mu = 0$
0.7	H_A: $\mu \neq 0$
−0.4	$\alpha = 0.05$
−1.8	$n = 12$
0.2	$\bar{X} = -0.65$ g
0.9	$s^2 = 1.5682$ g^2
−1.2	
−0.9	
−1.8	
−1.4	$s_{\bar{X}} = \sqrt{\frac{1.5682 \text{ g}^2}{12}} = 0.36$ g
−1.8	
−2.0	

$$t = \frac{\bar{X} - \mu}{s_{\bar{X}}} = \frac{-0.65 \text{ g}}{0.36 \text{ g}} = -1.81$$

$\nu = n - 1 = 11$

$t_{0.05(2),11} = 2.201$

Since $|t| < t_{0.05(2),11}$, do not reject H_0.

$0.05 < P < 0.10$

8.2 ONE-TAILED HYPOTHESES CONCERNING THE MEAN

In Section 8.1, we spoke of the hypotheses H_0: $\mu = \mu_0$ and H_A: $\mu \neq \mu_0$, because we were willing to consider a large deviation of $\bar{X}$ in either direction from μ as grounds for rejecting H_0. However, in many instances, our interest lies only in whether $\bar{X}$ is significantly larger (or significantly smaller) than μ_0, and this is termed a "one-tailed" (or "one-sided") test situation. For example, one might be testing a drug hypothesized to cause weight reduction in humans. The investigator is interested only in whether a weight loss occurs after the drug is taken. (In Example 8.2, using a two-sided test, we were interested in determining whether either weight loss or weight gain had occurred.) If there is either weight gain or no weight change, the drug will be considered a failure. Therefore, for this one-sided test, we should state H_0: $\mu \geq 0$ and H_A: $\mu < 0$. Here, the null hypothesis states that there is no mean weight loss (i.e., the weight change is greater than or equal to zero), and the alternate hypothesis states that there is a mean weight loss (i.e., the mean weight change is less than zero). By examining the alternate hypothesis, H_A, we see that H_0 will be rejected if t is in the left-hand critical region of the t distribution. In general,

$$\text{for } H_A\text{: } \mu < \mu_0,$$
$$\text{if } t \leq -t_{\alpha(1),\nu}, \text{ then reject } H_0.*$$

*For one-tailed testing of this H_0, probabilities of t up to 0.25 are indicated in Table B.3. If $t = 0$, then $P = 0.50$; so if $-t_{0.25(1),\nu} < t < 0$, then $0.25 < P < 0.50$; and if $t > 0$, then $P > 0.50$.

Example 8.3 summarizes such a set of 12 weight change data tested against this pair of hypotheses. From Table B.3 we find that $t_{0.05(1),11} = 1.796$, and the critical region for this test is shown in Fig. 8.3. From this figure, and by examining Table B.3, we see that $t_{\alpha(1),\nu} = t_{2\alpha(2),\nu}$, or $t_{\alpha(2),\nu} = t_{\alpha/2(1),\nu}$; that is, for example, the critical value of t for a one-sided test at $\alpha = 0.05$ is the same as the critical value of t for a two-sided test at $\alpha = 0.10$.

Example 8.3

A one-tailed t test for the hypotheses H_0: $\mu \geq 0$ and $H_A < 0$.

The data are weight changes of humans, tabulated after administration of a drug proposed to result in weight loss. Each weight change (in kg) is the weight after minus the weight before drug administration.

0.2
−0.5
−1.3
−1.6
−0.7
0.4
−0.1
0.0
−0.6
−1.1
−1.2
−0.8

$n = 12$
$\bar{X} = -0.61$ kg
$s^2 = 0.4008$ kg^2

$$s_{\bar{X}} = \sqrt{\frac{0.4008 \text{ kg}^2}{12}} = 0.18 \text{ kg}$$

$$t = \frac{\bar{X} - \mu}{s_{\bar{X}}} = \frac{-0.61 \text{ kg}}{0.18 \text{ kg}} = -3.389$$

$\nu = n - 1 = 11$

$t_{0.05(1),11} = 1.796$
If $t \leq -t_{0.05(1),11}$, reject H_0.
Conclusion: reject H_0.
$0.0025 < P(t \leq -3.389) < 0.005$

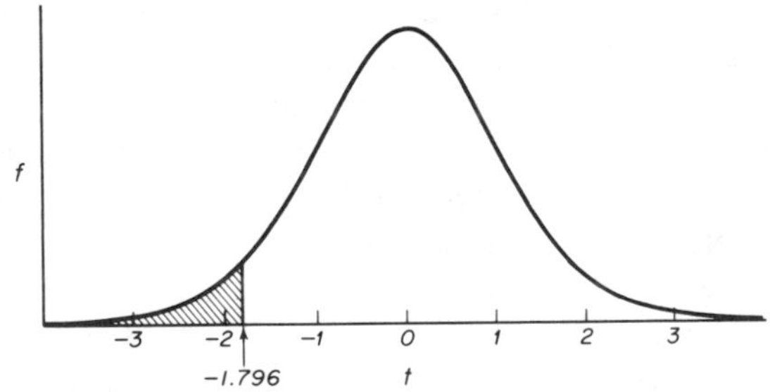

Figure 8.3 The distribution of t for $\nu = 11$, showing the critical region (shaded area) for a one-tailed test using $\alpha = 0.05$. (The critical value of t is -1.796.)

If we are interested in whether $\bar{X}$ is significantly greater than some value, μ_0, the hypotheses for the one-tailed test are H_0: $\mu \leq \mu_0$ and H_A: $\mu > \mu_0$. For example, a drug manufacturer might advertise that a product dissolves completely in gastric juice within 45 sec. The hypotheses appropriate for testing this claim are H_0: $\mu \leq 45$ sec and H_A: $\mu > 45$ sec, because we are not particularly interested in the possibility that the product dissolves faster than is claimed, but we wish to determine whether its dissolving time is longer than advertised. Thus, the rejection region would be in the right-hand tail, rather than in the left-hand tail (the latter being the case in Example

8.3.) The details of such a test are shown in Example 8.4. In general,

$$\text{for } H_A\text{: } \mu > \mu_0,$$
$$\text{if } t \geq t_{\alpha(1),\nu}, \text{ then reject } H_0.\text{*}$$

Example 8.4

The one-tailed t test for the hypotheses H_0: $\mu \leq 45$ sec and H_A: $\mu > 45$ sec.

Dissolving times (in sec) of a drug in agitated gastric juice: 42.7, 43.4, 44.6, 45.1, 45.6, 45.9, 46.8, 47.6.

H_0: $\mu \leq 45$ sec	$\bar{X} = 45.21$ sec
H_A: $\mu > 45$ sec	SS $= 18.8288$ sec^2
$\alpha = 0.05$	$s^2 = 2.6898$ sec^2
$n = 8$	$s_{\bar{X}} = 0.58$ sec

$$t = \frac{45.21 \text{ sec} - 45 \text{ sec}}{0.58 \text{ sec}} = 0.36$$

$\nu = 7$

$t_{0.05(1),7} = 1.895$

If $t \geq t_{0.05(1),7}$, reject H_0.

Conclusion: do not reject H_0.

$P(t \geq 0.36) > 0.50$

8.3 CONFIDENCE LIMITS FOR THE POPULATION MEAN

We learned in Section 8.1 that 5% of all possible sample means from a population with mean μ will yield t values—where $t = (\bar{X} - \mu)/s_{\bar{X}}$—that are either larger than $t_{0.05(2),\nu}$, or smaller than $-t_{0.05(2),\nu}$ (i.e., $|t| > t_{0.05(2),\nu}$). This means that 95% of all t values obtainable lie between the limits of $-t_{0.05(2),\nu}$ and $t_{0.05(2),\nu}$; that is,

$$P\left[-t_{0.05(2),\nu} < \frac{\bar{X} - \mu}{s_{\bar{X}}} < t_{0.05(2),\nu}\right] = 0.95. \tag{8.3}$$

It follows from this that

$$P[\bar{X} - t_{0.05(2),\nu}\, s_{\bar{X}} < \mu < \bar{X} + t_{0.05(2),\nu}\, s_{\bar{X}}] = 0.95. \tag{8.4}$$

Equation 8.4 is read as "the probability that the interval between $\bar{X} - t_{0.05(2),\nu}\, s_{\bar{X}}$ and $\bar{X} + t_{0.05(2),\nu}\, s_{\bar{X}}$ includes μ is 0.95." The expression within the brackets is called the *confidence interval* (abbreviated CI) for μ. What it tells us is that we can, knowing $\bar{X}$, $s_{\bar{X}}$, and ν, say that we are 95% confident that μ lies within the interval stated. In general, the confidence interval for μ can be stated as

$$\bar{X} - t_{\alpha(2),\nu}\, s_{\bar{X}} < \mu < \bar{X} + t_{\alpha(2),\nu}\, s_{\bar{X}}. \tag{8.5}$$

The quantity $\bar{X} - t_{\alpha(2),\nu}\, s_{\bar{X}}$ is called the *lower confidence limit* (abbreviated L_1), and $\bar{X} + t_{\alpha(2),\nu}\, s_{\bar{X}}$ is the *upper confidence limit* (abbreviated L_2); therefore, the two confi-

*For this H_0, if $t = 0$, then $P = 0.50$; therefore, if $0 < t < t_{0.25(1),\nu}$, then $0.25 < P < 0.50$, and if $t < 0$, then $P > 0.50$.

dence limits can be stated concisely as

$$\bar{X} \pm t_{\alpha(2),\nu}\, s_{\bar{X}} \tag{8.6}$$

(reading "±" to be "plus or minus"). When referring to a confidence interval, we call the quantity $1 - \alpha$ ($1 - 0.05 = 0.95$ in the present example) the *confidence level* (or, occasionally, the *confidence coefficient*).*

Although $\bar{X}$ is the best estimate of μ, it is still only an estimate, and the calculation of the confidence interval for μ allows us to express the precision of the estimate. Referring to Example 8.1, we would calculate the 95% confidence interval for μ as follows. Since $\bar{X} = 24.23$ yr, $s_{\bar{X}} = 0.85$ yr, $\nu = 24$, and $t_{0.05(2),24} = 2.064$, the confidence limits would be 24.23 yr ± (2.064)(0.85 yr) = 24.23 yr ± 1.75 yr, or $L_1 = 22.48$ yr and $L_2 = 25.98$ yr. Therefore, the confidence interval could be stated as 22.48 yr $< \mu <$ 25.98 yr, meaning that, based on the sample statistics, we are 95% confident that μ lies within the range of L_1 to L_2. This implies that if all possible samples of size n (= 25) were taken from the population, and a 95% confidence interval were calculated from each sample, 95% of all these intervals would contain μ.

Note that the smaller $s_{\bar{X}}$ is, the smaller will be the confidence interval, meaning that we have estimated μ more precisely. Also, it is obvious from the calculation of $s_{\bar{X}}$ (Equation 7.20), that a large n will result in a small $s_{\bar{X}}$. Thus, a parameter estimate from a large sample is more precise than an estimate of the same parameter from a small sample.

If, instead of the 95% confidence interval, we wished to state an interval having a 99% probability of containing μ, the value of $t_{0.01(2),24}$ would have been used. Since $t_{0.01(2),24} = 2.797$, the confidence limits would have been 24.23 yr ± (2.797) (0.85) = 24.23 yr ± 2.38 yr and the confidence interval would have been 21.85 yr $< \mu <$ 26.61 yr. Note that if we increase our confidence we concomitantly increase the width of the confidence interval. (Indeed, if we increase the confidence to 100%, then the confidence interval would be $-\infty < \mu < \infty$.) Note that it is the two-tailed t value that is utilized for confidence interval computation, as we are setting limits on both sides of μ.

8.4 REPORTING VARIABILITY ABOUT THE MEAN

It is very important to provide the reader of a research paper with some information concerning the variability of the data reported. But authors of such papers are often unsure of appropriate ways of doing so, and not infrequently do so improperly.

If we wish to describe the population that has been sampled, then the sample mean ($\bar{X}$) and the standard deviation (s) may be reported. The range might also be reported, but in general it should not be stated without being accompanied by another measure of variability, such as s. Such statistics are frequently presented as in Table 8.1.

*On rare occasion, the biologist sees reference to "fiducial intervals" or "fiducial limits," and although there are theoretical differences between fiducial limits and confidence limits, some authors use them synonymously.

TABLE 8.1 TAIL LENGTHS (IN mm) OF FIELD MICE FROM DIFFERENT LOCALITIES.

Location	*n*	$\bar{X} \pm$ SD (range in parentheses)
Bedford, Indiana	18	56.22 ± 1.33 (44.8 to 68.9)
Rochester, Minnesota	12	59.61 ± 0.82 (43.9 to 69.8)
Fairfield, Iowa	16	60.20 ± 0.92 (52.4 to 69.2)
Pratt, Kansas	16	53.93 ± 1.24 (46.1 to 63.6)
Mount Pleasant, Michigan	13	55.85 ± 0.90 (46.7 to 64.8)

If it is the author's intention to provide the reader with a statement about the precision of estimation of the population mean, the use of the standard error ($s_{\bar{X}}$) is appropriate. A typical presentation is shown in Table 8.2a. This table might instead be set up to show confidence intervals, rather than standard errors, as shown in Table 8.2b.

TABLE 8.2a ENZYME ACTIVITIES IN THE MUSCLE OF VARIOUS ANIMALS. DATA ARE $\bar{X} \pm$ SE, WITH *n* IN PARENTHESES.

	Enzyme Activity (μmole/min/g of tissue)	
Animal	*Isomerase*	*Transketolase*
Mouse	0.76 ± 0.09 (4)	0.39 ± 0.04 (4)
Frog	1.53 ± 0.08 (4)	0.18 ± 0.02 (4)
Trout	1.06 ± 0.12 (4)	0.24 ± 0.04 (4)
Crayfish	4.22 ± 0.30 (4)	0.26 ± 0.05 (4)

TABLE 8.2b ENZYME ACTIVITIES IN THE MUSCLE OF VARIOUS ANIMALS. DATA ARE $\bar{X} \pm$ 95% CONFIDENCE LIMITS.

		Enzyme Activity (μmole/min/g of tissue)	
Animal	*n*	*Isomerase*	*Transketolase*
Mouse	4	0.76 ± 0.28	0.39 ± 0.13
Frog	4	1.53 ± 0.25	0.18 ± 0.05
Trout	4	1.06 ± 0.38	0.24 ± 0.11
Crayfish	4	4.22 ± 0.98	0.26 ± 0.15

There are three very important points to note about Tables 8.1 and 8.2. First, n should be stated somewhere in the table, either in the caption or in the body of the table. (Thus, the reader has the needed information to convert from SD to SE or from SE to SD, if so desired.) One should *always* state n when presenting sample statistics ($\bar{X}$, s, $s_{\bar{X}}$, range, etc.), and if a tabular presentation is prepared it is very good practice to include n somewhere in the table, even if it is mentioned elsewhere in the paper.

Second, the measure of variability is clearly indicated. Not infrequently, an author will state something such as "the mean is 54.2 ± 2.7 g," with no explanation of what "± 2.7" denotes. This renders the statement worthless to the reader, because "± 2.7" will be assumed by some to indicate ± SD, by others to indicate ± SE, by others to indicate the 95% (or 99%, or other) confidence interval, and by others to indicate the range.* There is no widely accepted convention; one *must* state explicitly what quantity is meant by this type of statement. If such statements of "±" values appear in a table, then the explanation is best included somewhere in the table (either in the caption or in the body of the table), even if it is stated elsewhere in the paper.

Third, the units of measurement of the variable must be clear. There is little information conveyed by stating that the tail lengths of 24 birds have a mean of 8.42 and a standard error of 0.86, if the reader does not know whether the tail lengths were measured in centimeters, or inches, or some other unit. Whenever data appear in tables, the units of measurement should be stated somewhere in the table. Keep in mind that a table should be self-explanatory; one should not have to refer back and forth from the table to the text to determine what the tabled values represent.

Frequently, the types of information given in Tables 8.1, 8.2, and 8.3 are presented in graphs, rather than in tables. In such cases, the measurement scale is

TABLE 8.3 EVAPORATIVE WATER LOSS AND RESPIRATORY RATE OF A SMALL MAMMAL AT VARIOUS AIR TEMPERATURES. SAMPLE STATISTICS ARE MEAN ± STANDARD DEVIATION, WITH RANGE IN PARENTHESES.

	Air Temperature (°C)				
	16.2	24.8	30.7	36.8	40.9
Sample size	10	13	10	8	9
Evaporative water loss (mg/g/hr)	0.611 ± 0.164 (0.49 to 0.88)	0.643 ± 0.194 (0.38 to 1.13)	0.890 ± 0.212 (0.64 to 1.39)	1.981 ± 0.230 (1.50 to 2.36)	3.762 ± 0.641 (3.16 to 5.35)

typically indicated on the vertical axis, and the mean is indicated in the body of the graph by a short horizontal line or some other symbol. The standard deviation, standard error, or a confidence interval for the mean is commonly indicated on such graphs via a vertical line or rectangle. Often the range is also included, and in such instances the SD or SE may be indicated by a vertical rectangle and the range by a vertical line. Some authors will indicate a confidence interval (generally 95%) in addition to the range and either SD or SE. Figures 8.4, 8.5, and 8.6 demonstrate how vari-

*In older literature the ± symbol referred to yet another measure, known as the "probable error" (which has fallen into disuse). In a normal curve, the probable error (PE) is 0.6745 times the standard error, so that $\bar{X}$ ± PE includes 50% of the distribution.

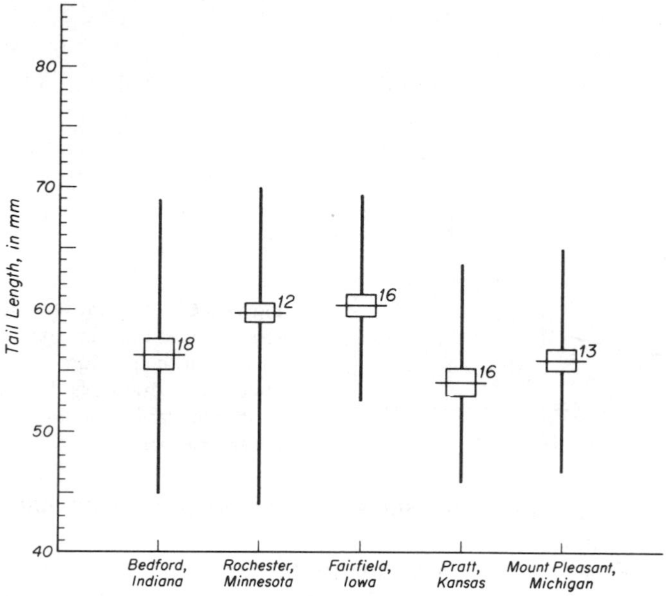

Figure 8.4 Tail lengths of male field mice from different localities, indicating the mean, the mean ± standard deviation (vertical rectangle), and the range (vertical line), with the sample size indicated for each location.

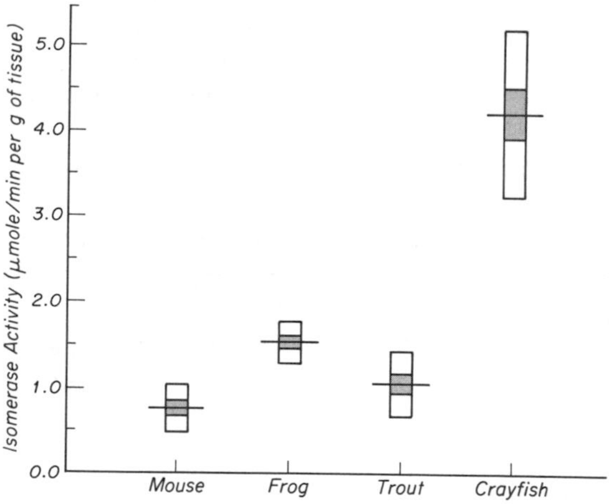

Figure 8.5 Levels of muscle isomerase in various animals. Shown is the mean ± standard error (shaded rectangle), and ± the 95% confidence interval (open rectangle). For each sample, $n = 4$.

ous combinations of these statistics may be presented graphically.* Note that when the horizontal axis on the graph represents an interval or ratio scale variable (as in Fig. 8.6), adjacent means may be connected by straight lines to aid in the recognition of trends.

In graphical presentation of data, as in tabular presentation, care must be taken to indicate clearly the following either on the graph or in the caption: the sample size

*Instead of the mean and a measure of variability based on the variance, one may present tabular or graphical descriptions of samples using the median and quartiles (e.g., McGill, Tukey, and Larsen, 1978), or the median and its confidence interval. Thus, a graphical presentation such as in Fig. 8.4 could have the range indicated by the vertical line, the median by the horizontal line, and the semiquartile range (Section 4.7) by the vertical rectangle.

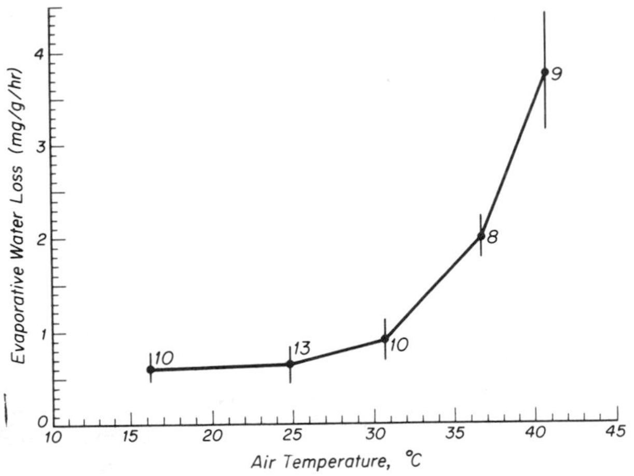

Figure 8.6 Evaporative water loss of a small mammal at various air temperatures. Shown at each temperature is the mean $\pm$ the standard deviation, and the sample size.

(n), the units of measurement, and what measures of variability (if any) are indicated (e.g., SD, SE, range, 95% confidence interval).

Some authors present $\bar{X} + 2s_{\bar{X}}$ in their graphs. An examination of the t table (Table B.3) will show that, except for small samples, this expression will approximate the 95% confidence interval for the mean. But for small samples, the true confidence interval is, in fact, greater than $\bar{X} \pm 2s_{\bar{X}}$. Thus, the general use of this expression is not to be encouraged, and the calculation of the accurate confidence interval is the wiser practice.

A word of caution is in order for those who set up confidence limits for two or more means and, by observing whether or not the limits overlap, attempt to determine whether there are differences among the population means. Such a procedure is not generally valid (see Simpson, Roe, and Lewontin, 1960: 350–354; Browne, 1979). The proper methods for testing for differences between means are discussed in the next several chapters. Andrews, Snee, and Sarner, 1980 explain how graphical display of means can be used validly to demonstrate significant differences among means.

8.5 SAMPLE SIZE AND ESTIMATION OF THE POPULATION MEAN

A commonly asked question is, "How large a sample must be taken to achieve a desired precision* in estimating the mean of a population?" The answer is obviously related to the concept of a confidence interval; for a confidence interval expresses the precision of a sample statistic, and the precision increases (i.e., the confidence interval becomes narrower) as the sample size increases.

Let us write Equation 8.6 as $\bar{X} + d$, which is to say that $d = t_{\alpha(2),\nu} s_{\bar{X}}$. We shall refer to d as the half-width of the confidence interval, which means that μ is estimated

*Recall from Section 2.4 that the precision of a sample statistic is the closeness with which it estimates the population parameter; it is not to be confused with the concept of the precision of a measurement, (defined in Section 1.2), which is the nearness of repeated measurements to each other.

to within $\pm d$. Now, the number of data we must collect to calculate a confidence interval of specified width depends upon: (1) the width desired (for narrower confidence intervals—i.e., more precision in estimating μ—requires a larger sample; (2) the variability in the population (which is estimated by s^2, and larger variability requires larger sample size); (3) the confidence level specified (for greater confidence—e.g., 99% vs. 95%—requires a larger sample size); and (4) the assurance that the confidence interval will be no larger than that specified (with greater assurance requiring larger samples).

If we have a sample estimate (s^2) of the variance of a normal population, then we can estimate the required sample size as

$$n = \frac{s^2 t^2_{\alpha(2),(n-1)} F_{\beta(1),(n-1,\nu)}}{d^2} \tag{8.7}$$

(Harris, Horvitz, and Mood, 1948). In this equation, s^2 is the sample variance, estimated with ν degrees of freedom, d is the half-width of the desired confidence interval, $1 - \alpha$ is the confidence level for the confidence interval, and $1 - \beta$ is the assurance that the confidence interval will be no larger than specified. Two-tailed critical values of Student's t, with $n - 1$ degrees of freedom, are found in Table B.3. One-tailed critical values for the F distribution,* with $n - 1$ and ν degrees of freedom, are found in Table B.4.

There is a basic difficulty in solving Equation 8.7, however; the values of $t_{\alpha(2),(n-1)}$ and $F_{\beta(1),(n-1,\nu)}$ depend upon n, the unknown sample size. The solution may be achieved by iteration—a process of trial and error with progressively more accurate approximations—as shown in Example 8.5. We begin the iterative process of estimation with an initial guess; the closer this initial guess is to the finally determined n, the faster we will arrive at the final estimate. Fortunately, the procedure works well even if this initial guess is far from the final n (although the process is faster if it is a high, rather than a low, guess).

Example 8.5

Determination of sample size needed to achieve a stated precision in estimating a population mean, using the data of Example 8.1.

If we specify that we wish to estimate μ by having a 90% probability that the 95% confidence interval will be no wider than 3.0 years, then $d = 1.5$ yr, $1 - \beta = 0.90$, $\beta = 0.10$, $1 - \alpha = 0.95$, and $\alpha = 0.05$. From Example 8.1 we have an estimate of the population variance: $s^2 = 18.0388\ \text{yr}^2$ with $\nu = 24$.

Let us guess that a sample of 50 is necessary; then,

$$t_{0.05(2),49} = 2.010 \quad \text{and} \quad F_{0.10(1),49,24} \approx F_{0.10(1),40,24} = 1.64.$$

As numerator degrees of freedom of 49 does not appear in Table B.4, we may either interpolate (as explained in the introduction to Appendix B) or (conservatively) use the next lower degrees of freedom that are in the table.

*This important distribution will be presented more thoroughly in Section 9.1 and in following chapters. It is sufficient for our present needs to note that in Appendix Table B.4, $n - 1$ is what is referred to as "Numerator DF," ν is "Denominator DF," and the desired probability, $\beta(1)$, corresponds to the column headings "$\alpha(1)$."

So we estimate (by Equation 8.7)

$$n = \frac{(18.0388)(2.010)^2(1.64)}{(1.5)^2} = 53.1.$$

Next, we might estimate $n = 53$, for which $t_{0.05(2),52} = 2.007$ and $F_{0.10(1),52,24} \approx F_{0.10(1),50,24} = 1.62$, and we calculate

$$n = \frac{(18.0388)(2.007)^2(1.62)}{(1.5)^2} = 52.3.$$

Therefore, we conclude that a sample size greater than 52 is required to achieve the specified confidence interval.

8.6 POWER AND SAMPLE SIZE IN TESTS CONCERNING THE MEAN

Sample Size Required. If we are to perform a one-sample test as described in Section 8.1 or 8.2, then it is desirable to know how many data should be collected. We can estimate the required sample size, n, if we have an estimate, s^2, of the population variance, σ^2.

We may specify that we wish to perform a t test with a probability of α of committing a Type I error, and a probability of β of committing a Type II error; and we can state that we want to be able to detect a difference between μ and μ_0 as small as δ (where μ is the actual population mean and μ_0 is the mean specified in the null hypothesis). To test at the α significance level with $1 - \beta$ power, the minimum sample size required to detect δ is

$$n = \frac{s^2}{\delta^2}(t_{\alpha,\nu} + t_{\beta(1),\nu})^2, \tag{8.8}$$

where α can be either $\alpha(1)$ or $\alpha(2)$, respectively, depending on whether a one-tailed or two-tailed test is to be used. However, ν depends on n, so n cannot be calculated directly but must be obtained by iteration* (i.e., by a series of estimations, each estimation coming closer to the answer than that preceding). This is demonstrated in Example 8.6.

Example 8.6

Estimation of required sample size to test H_0: $\mu = \mu_0$.

How large a sample is needed to reject the null hypothesis of Example 8.2, when sampling from the population in that example? We wish to test at the 0.05 level of significance with a 90% chance of detecting a mean significantly different from $\mu_0 = 0$ by as little as 1.0 g. In Example 8.2, $s^2 = 1.5682\ \text{g}^2$.

Let us guess that a sample size of 20 would be required. Then, $\nu = 19$, $t_{0.05(2),19} = 2.093$, $\beta = 1 - 0.90 = 0.10$, $t_{0.10(1),19} = 1.328$, and we use Equation 8.8 to calculate

$$n = \frac{1.5682}{(1.0)^2}(2.093 + 1.328)^2 = 18.4.$$

*If the population variance, σ^2, were actually known, rather than estimated by s^2, then Z_α would be substituted for t_α in this and the other computations in this section, and n would be determined in one step instead of iteratively.

We now use $n = 19$ as an estimate, in which case $\nu = 18$, $t_{0.05(2),18} = 2.101$, $t_{0.10(1),18} = 1.330$, and

$$n = \frac{1.5682}{(1.0)^2}(2.101 + 1.330)^2 = 18.5.$$

Thus, we conclude that a sample size of at least 19 is required.

Minimum Detectable Difference. By rearranging Equation 8.8, we can ask how small a δ (the difference between μ and μ_0) can be detected by the t test with $1 - \beta$ power, at the α level of significance, using a sample of specified size n:

$$\delta = \sqrt{\frac{s^2}{n}}(t_{\alpha,\nu} + t_{\beta(1),\nu}), \tag{8.9}$$

where $t_{\alpha,\nu}$ can be either $t_{\alpha(1),\nu}$ or $t_{\alpha(2),\nu}$, depending on whether a one-tailed or two-tailed test is to be performed. The estimation of δ is demonstrated in Example 8.7.

Example 8.7

Estimation of minimum detectable difference in a one-sample t test.

In the two-tailed test of Example 8.2, what is the smallest difference (i.e., difference between μ and μ_0) that is detectable 90% of the time using a sample of 25 data and a significance level of 0.05?

Using Equation 8.9:

$$\begin{aligned}\delta &= \sqrt{\frac{1.5682}{25}}(t_{0.05(2),24} + t_{0.10(1),24})\\ &= (0.25)(2.064 + 1.318)\\ &= 0.85 \text{ g}\end{aligned}$$

Power of One-Sample Testing. If our desire is to express the probability of correctly rejecting a false H_0, then we seek to estimate the power of the t test. Equation 8.8 can be rearranged to give

$$t_{\beta(1),\nu} = \frac{\delta}{\sqrt{\frac{s^2}{n}}} - t_{\alpha,\nu} \tag{8.10}$$

As shown in Example 8.8, for a stipulated δ, α, σ^2, and sample size, we can express $t_{\beta(1),\nu}$. Consulting Table B.3 allows us to convert $t_{\beta(1),\nu}$ to β, but only roughly (e.g., $\beta > 0.25$ in Example 8.8). However, $t_{\beta(1),\nu}$ may be considered to be approximated by $Z_{\beta(1)}$, so Table B.2 may be used to determine β.* Then, the power of the test is $1 - \beta$, as shown in Example 8.8.

Example 8.8

Estimation of power of a one-sample t test.

What is the probability of detecting a true difference (i.e., a difference between μ and μ_0) of at least 1.0 g for the experiment of Example 8.2?

*Approximating $t_{\beta(1),\nu}$ by $Z_{\beta(1)}$ apparently yields a β that is an underestimate (and a power that is an overestimate) of no more than 0.01 for ν at least 11 and no more than 0.02 for ν at least 7.

For $n = 12$, $\nu = 11$, $t_{0.05(2),11} = 2.201$, and $s^2 = 1.5682\ \text{g}^2$, and we use Equation 8.10 to find

$$t_{\beta(1),11} = \frac{1.0}{\sqrt{\frac{1.5682}{12}}} - 2.201$$
$$= 0.57.$$

Consulting Table B.3 tells us that, for $\nu = 11$, $\beta > 0.25$, so we can say the power is $1 - \beta < 0.75$. By considering 0.57 to be a normal deviate and consulting Table B.2, we conclude $\beta = 0.28$ and that the power of the test $1 - \beta = 0.72$.

8.7 SAMPLING FINITE POPULATIONS

In general we assume that a sample from a population is a very small portion of the totality of data in that population. Essentially, we consider that the population is infinite in size, so that the removal of a relatively small number of data from the population does not noticeably affect the probability of selecting further data.

However, if the sample size, n, is an appreciable portion of the population size, N, (say, at least 5%), then we are said to be sampling a *finite population*. In such a case, $\bar{X}$ is a substantially better estimate of μ the closer n is to N; specifically,

$$s_{\bar{X}} = \sqrt{\frac{s^2}{n}}\sqrt{1 - \frac{n}{N}} = \sqrt{\left(\frac{s^2}{n}\right)\left(1 - \frac{n}{N}\right)}, \tag{8.11}$$

where n/N is the *sampling fraction* and $1 - n/N$ is referred to as the *finite population correction.**

Obviously, from Equation 8.11, when n is very small compared to N, then the sampling fraction is almost zero, the finite population correction will be nearly one, and $s_{\bar{X}}$ will be nearly $\sqrt{s^2/n}$, just as we have used (Equation 7.21) when assuming the population size, N, to be infinite. As n becomes closer to N, the correction becomes smaller, and $s_{\bar{X}}$ becomes smaller, which makes sense intuitively. If $n = N$, then $1 - n/N = 0$ and $s_{\bar{X}} = 0$, meaning there is no error at all in estimating μ if the sample consists of the entire population; $\bar{X} = \mu$ if $n = N$. In computing confidence intervals when sampling finite populations (i.e., when n is not a negligibly small fraction of N), Equation 8.11 should be used instead of Equation 7.21.

If we are determining the sample size required to estimate the population mean with a stated precision (Section 8.5), and the sample size is an appreciable fraction of the population size, then the required sample size is calculated as

$$m = \frac{n}{1 + (n - 1)/N} \tag{8.12}$$

(Cochran, 1977: 77–78), where n is from Equation 8.7.

*One may also calculate $1 - n/N$ as $(N - n)/N$.

8.8 CONFIDENCE LIMITS FOR THE POPULATION MEDIAN

The sample median is used as the best estimate of M, the population median.* The $1 - \alpha$ confidence limits for M may be determined by considering the binomial distribution, which is discussed in some detail in Chapter 22. This determination is made simple by Appendix Table B.26. We shall refer to the two-tailed value read from the body of this table for a given n and α as $C_{\alpha(2),n}$, or simply C. The lower confidence limit is

$$L_1 = X_{C+1}, \tag{8.13}$$

and the upper confidence limit is

$$L_2 = X_{n-C}. \tag{8.14}$$

Confidence limits obtained using the binomial distribution are not typically associated with exactly $1 - \alpha$ confidence (because this is a discrete, rather than a continuous, distribution). Rather, the confidence should be stated as $\geq 1 - \alpha$. Example 8.9 demonstrates this procedure.

Example 8.9

Confidence limits for the median of the population from which the sample of Example 8.2 came.

The 12 data (in grams) are ranked: −2.0, −1.8, −1.8, −1.8, −1.4, −1.2, −0.9, −0.4, 0.2, 0.7, 0.9, 1.7.

For 95% confidence limits, $\alpha = 0.05$, and from Table B.26: $C = C_{0.05(2),12} = 2$. Then,

$$L_1 = X_{C+1} = X_{2+1} = X_3 = -1.8 \text{ g}$$

$$L_2 = X_{n-C} = X_{12-2} = X_{10} = 0.7 \text{ g}$$

Therefore, the confidence interval with confidence $\geq 95\%$ is $-1.8 \text{ g} < M < 0.7 \text{ g}$. That is, $P(-1.8 \text{ g} < M < 0.7 \text{ g}) \geq 95\%$.

A Large-Sample Approximation. For large samples (say, $n \geq 25$), an approximation of the lower confidence limit (based on Hollander and Wolfe, 1973: 40, derived from the normal distribution), is

$$L_1 = X_C, \tag{8.15}$$

where

$$C = n/2 - Z_{\alpha(2)} \sqrt{n/4} \tag{8.16}$$

rounded to the next lower integer, and $Z_{\alpha(2)}$ is the two-tailed normal deviate, read from Table B.2. (Recall that $Z_{\alpha(2)} = t_{\alpha(2),\infty}$, and so may be read from the last line of Table B.3). The upper confidence limit is

$$L_2 = X_{n-C+1}. \tag{8.17}$$

By this method we approximate a confidence interval for M with confidence $\geq 1 - \alpha$.

*Here M represents the Greek capital letter mu.

For example, if we have a sample of 40 data, the lower 95% confidence limit for the population median may be estimated as follows:

$$C = 40/2 - 1.9600\sqrt{40/4}$$
$$= 13.8, \text{ rounded down to } 13$$

and

$$L_1 = X_{13}.$$

The upper confidence limit would be

$$L_2 = X_{40-13+1} = X_{28}.$$

8.9 ONE-SAMPLE HYPOTHESES CONCERNING THE MEDIAN

In Example 8.2 we examined a sample of weight change data in order to ask whether the mean change in the sampled population was different from zero. Analogously, one may test H_0: $M = M_0$ against H_A: $M \neq M_0$, where M_0 can be zero or any other hypothesized population median.*

A simple method for testing this two-tailed hypothesis is to determine the confidence limits of Section 8.8 and reject H_0 (with probability $\leq \alpha$ of a Type I error) if $M_0 \leq L_1$ or $M_0 \geq L_2$. This is essentially a binomial test (Section 22.5), where we consider the number of data $< M_0$ as being in one category and the number of data $> M_0$ being in the second category. If either of these two numbers is less than or equal to the critical value in Table B.26, then H_0 is rejected. (Data equal to M_0 are ignored in this test.) For example, if we test H_0: $M = 0$ versus H_A: $M \neq 0$ for the data of Example 8.9, then by noting that M_0 (namely zero) lies within the confidence interval, we do not reject H_0. Alternatively, we may read the critical value from Table B.26 as $C_{0.05(2),12} = 2$; and, as neither 8 (the number of data less than zero) nor 4 (the number of data greater than zero) is less than or equal to 2, H_0 is not rejected.

For one-tailed hypotheses about the median, the binomial test may also be employed. For H_0: $M \geq M_0$ vs. H_A: $M < M_0$, H_0 is rejected if the number of data less than M_0 is $\leq$ the one-tailed critical value, $C_{\alpha(1),n}$. For H_0: $M \leq M_0$ vs. H_A: $M \geq M_0$, H_0 is rejected if the number of data greater than M_0 is $\leq C_{\alpha(1),n}$.

As an alternative to the binomial test, for either two-tailed or one-tailed hypotheses, one may use the more powerful Wilcoxon signed-rank test. The Wilcoxon procedure is applied as a one-sample median test by ranking the data as described in Section 10.4 and assigning a minus sign to each rank associated with a datum $< M_0$ and a plus sign to each associated with a datum $> M_0$. The sum of the ranks with a plus sign is called T_+ and the sum of the ranks with a minus sign is T_-, with the test then proceeding as described in Section 10.4. The Wilcoxon test assumes that the sampled population is symmetric (in which case the median and mean are identical and this procedure becomes a hypothesis test about the mean as well as about the median; but the one-sample t test is typically a more powerful test about the mean).

*Recall, from Section 8.8, that M, denoting the population median, is the capital Greek mu.

8.10 CONFIDENCE LIMITS FOR THE POPULATION VARIANCE

The sampling distribution of means is a symmetrical distribution, approaching the normal distribution as n increases. But the sampling distribution of variances is not symmetrical, and neither the normal nor the t distribution may be employed to set confidence limits around σ^2 or to test hypotheses about σ^2. However, theory states that

$$\chi^2 = \frac{\nu s^2}{\sigma^2}, \tag{8.18}$$

so by employing the χ^2 distribution we can define an interval within which there is a $1 - \alpha$ chance of including σ^2. Recall that the chi-square table (Table B.1) tells us the probability of a calculated χ^2 being greater than that in the table. If we desire to know the two χ^2 values that enclose $1 - \alpha$ of the chi-square curve, we want the portion of the curve between $\chi^2_{(1-\alpha/2),\nu}$ and $\chi^2_{\alpha/2,\nu}$ (for a 95% confidence interval, this would mean the area between $\chi^2_{0.975,\nu}$ and $\chi^2_{0.025,\nu}$). It follows from Expression (8.18) that

$$\chi^2_{(1-\alpha/2),\nu} \leq \frac{\nu s^2}{\sigma^2} \leq \chi^2_{\alpha/2,\nu}, \tag{8.19}$$

and

$$\frac{\nu s^2}{\chi^2_{\alpha/2,\nu}} \leq \sigma^2 \leq \frac{\nu s^2}{\chi^2_{(1-\alpha/2),\nu}}. \tag{8.20}$$

Since $\nu s^2 = \text{SS}$, we can also write Expression 8.20 as

$$\frac{\text{SS}}{\chi^2_{\alpha/2,\nu}} \leq \sigma^2 \leq \frac{\text{SS}}{\chi^2_{(1-\alpha/2),\nu}}. \tag{8.21}$$

Referring back to the data of Example 8.1, we would calculate the 95% confidence interval for σ^2 as follows. Since $\nu = 24$ and $s^2 = 18.0388$ yr², SS $= \nu s^2 = 432.9312$ yr². From Table B.1, we find $\chi^2_{0.025,24} = 39.364$ and $\chi^2_{0.975,24} = 12.401$. Therefore, $L_1 = \text{SS}/\chi^2_{\alpha/2,\nu} = 432.9312\ \text{yr}^2/39.364 = 11.00\ \text{yr}^2$ and $L_2 = \text{SS}/\chi^2_{(1-\alpha/2),\nu} = 432.9312\ \text{yr}^2/12.401 = 34.91\ \text{yr}^2$. Note that the confidence interval, $11.00\ \text{yr}^2 \leq \sigma^2 \leq 34.91\ \text{yr}^2$, is not symmetrical; that is, the distance from L_1 to s^2 is not the same as that from s^2 to L_2.

To obtain the $1 - \alpha$ confidence interval for the population standard deviation, simply use the square roots of the confidence limits for σ^2, so that

$$\sqrt{\frac{\text{SS}}{\chi^2_{\alpha/2,\nu}}} \leq \sigma \leq \sqrt{\frac{\text{SS}}{\chi^2_{(1-\alpha/2),\nu}}}. \tag{8.22}$$

For the preceding example, the 95% confidence interval for σ would be $\sqrt{11.00\ \text{yr}^2} \leq \sigma \leq \sqrt{34.91\ \text{yr}^2}$, or $3.32\ \text{yr} \leq \sigma \leq 5.91\ \text{yr}$.

8.11 HYPOTHESES CONCERNING THE VARIANCE

The procedures for testing hypotheses about the population variance come from the consideration of $\nu s^2/\sigma^2$ as a chi-square value (where $\nu = n - 1$), as introduced in Section 8.10. Consider the two-tailed pair of hypotheses, H_0: $\sigma^2 = \sigma_0^2$ and H_A: σ^2

$\neq \sigma_0^2$, where σ_0^2 may be any hypothesized population variance. Then, simply calculate

$$\chi^2 = \frac{\nu s^2}{\sigma_0^2} \quad \text{or, equivalently,} \quad \chi^2 = \frac{\text{SS}}{\sigma_0^2}, \tag{8.23}$$

and if the calculated χ^2 is $> \chi^2_{\alpha/2,\nu}$ or $< \chi^2_{(1-\alpha/2),\nu}$, then H_0 is rejected at the α level of significance. For example, if we wished to test $H_0: \sigma^2 = 10\ \text{yr}^2$ and $H_A: \sigma^2 \neq 10\ \text{yr}^2$ for the data of Example 8.1, with $\alpha = 0.05$, we would first calculate $\chi^2 = \text{SS}/\sigma^2$. Since, in the example, $\nu = 24$ and $s^2 = 18.0388\ \text{yr}^2$, $\text{SS} = \nu s^2 = 432.9312\ \text{yr}^2$. Also, since σ^2 is being hypothesized to be $10\ \text{yr}^2$, $\chi^2 = \text{SS}/\sigma_0^2 = 432.9312\ \text{yr}^2/10\ \text{yr}^2 = 43.293$. Two critical values are to be obtained from the chi-square table (Table B.1): $\chi^2_{(0.05/2),24} = \chi^2_{0.025,24} = 39.364$ and $\chi^2_{(1-0.05/2),24} = \chi^2_{0.975,24} = 12.401$. Since the calculated χ^2 is more extreme than one of these critical values (i.e., the calculated χ^2 is > 39.364), H_0 is rejected, and we conclude that the sample of data was obtained from a population having a variance different from $10\ \text{yr}^2$.

It is more common to consider one-tailed hypotheses concerning variances. For the hypotheses $H_0: \sigma^2 \leq \sigma_0^2$ and $H_A: \sigma^2 > \sigma_0^2$, H_0 is rejected if the χ^2 calculated from Equation 8.23 is $\geq \chi^2_{\alpha,\nu}$. For $H_0: \sigma^2 \geq \sigma_0^2$ and $H_A: \sigma^2 < \sigma_0^2$, a calculated χ^2 that is $\leq \chi^2_{(1-\alpha),\nu}$ is grounds for rejecting H_0. For the data of Example 8.4, a manufacturer might be interested in whether the variability in the dissolving times of the drug is greater than a certain value—say, 1.5 sec. Thus, $H_0: \sigma^2 \leq 1.5\ \text{sec}^2$ and $H_A: \sigma^2 > 1.5\ \text{sec}^2$ might be tested, as shown in Example 8.10.

Example 8.10

A one-tailed test for the hypotheses $H_0: \sigma^2 \leq 1.5\ \text{sec}^2$ and $H_A: \sigma^2 > 1.5\ \text{sec}^2$, using the data of Example 8.4.

$\text{SS} = 18.8288\ \text{sec}^2$

$\nu = 7$

$s^2 = 2.6898\ \text{sec}^2$

$$\chi^2 = \frac{\text{SS}}{\sigma_0^2} = \frac{18.8288\ \text{sec}^2}{1.5\ \text{sec}^2} = 12.553$$

$\chi^2_{0.05,7} = 14.067$

Since $12.553 < 14.067$, H_0 is not rejected.

$0.05 < P < 0.10$

8.12 POWER AND SAMPLE SIZE IN TESTS CONCERNING THE VARIANCE

Sample Size Required. We may ask how large a sample is required to perform the hypothesis tests of Section 8.11 at a specified power. For the hypotheses $H_0: \sigma^2 \leq \sigma_0^2$ vs. $H_A: \sigma^2 > \sigma_0^2$, the minimum sample size is that for which

$$\frac{\chi^2_{\beta,\nu}}{\chi^2_{\alpha,\nu}} = \frac{\sigma_0^2}{s^2}, \tag{8.24}$$

and this sample size, n, may be found by iteration (i.e., by a directed trial and error), as shown in Example 8.11. The ratio constituting the left side of Equation 8.24 increases in magnitude as n increases.

Example 8.11

Estimation of required sample size to test $H_0: \sigma^2 \leq \sigma_0^2$ vs. $H_A: \sigma^2 > \sigma_0^2$.

How large a sample is needed to reject $H_0: \sigma^2 \leq 2.5$ sec², using the data of Example 8.10, if we test at the 0.05 level of significance and with a power of 0.90? (Therefore, $\alpha = 0.05$ and $\beta = 0.10$.)

From Example 8.10, $s^2 = 2.6898$ sec². Since we have specified $\sigma_0^2 = 2.5$ sec, $\sigma_0^2/s^2 = 0.929$.

To begin the iterative process of estimating n, let us guess that a sample size of 30 is required. Then,

$$\frac{\chi^2_{0.10,29}}{\chi^2_{0.05,29}} = \frac{39.087}{42.557} = 0.918.$$

Since $0.918 < 0.929$, our estimate of n is too low. So we might guess that $n = 50$ is required:

$$\frac{\chi^2_{0.10,49}}{\chi^2_{0.05,49}} = \frac{62.038}{66.339} = 0.935.$$

Since $0.935 > 0.929$, $n = 50$ is too high an estimate and we might guess $n = 40$, for which $\chi^2_{0.10,39}/\chi^2_{0.05,39} = 50.660/54.572 = 0.928$.

Since $0.928 < 0.929$, our estimate is low (but only by a small amount), so we try $n = 41$: $\chi^2_{0.10,40}/\chi^2_{0.05,40} = 51.805/55.758 = 0.929$.

Therefore, we estimate that a sample size of at least 41 is required to perform a hypothesis test with the specified characteristics.

For the hypotheses $H_0: \sigma^2 \geq \sigma_0^2$ vs. $H_A: \sigma^2 < \sigma_0^2$, the minimum sample size is that for which

$$\frac{\chi^2_{\beta,\nu}}{\chi^2_{1-\alpha,\nu}} = \frac{\sigma_0^2}{s^2}. \tag{8.25}$$

Power of the Test. If we test $H_0: \sigma^2 \leq \sigma_0^2$ vs. $H_A: \sigma^2 > \sigma_0^2$, using significance level α and a sample size of n, then the power of the test is

$$1 - \beta = P(\chi^2 \geq \chi^2_{\alpha,\nu}\sigma_0^2/s^2). \tag{8.26}$$

Thus, in Example 8.10, $\alpha = 0.05$, $n = 8$, $\nu = 7$, $\chi^2_{0.05,7} = 14.067$, $s^2 = 2.6898$ sec², $\sigma_0^2 = 1.5$ sec², and the power of the test is

$$P[\chi^2 \geq (14.067)(1.5)/2.6898]$$
$$= P(\chi^2 \geq 7.845).$$

From Table B.1 we see that for $\chi^2 \geq 7.845$ with $\nu = 7$, $0.25 < P < 0.50$. By interpolation* between $\chi^2_{0.25,7}$ and $\chi^2_{0.50,7}$, we estimate that the power of the test was 0.36. This is low power. We could ask what power would result if we had a sample of 40 with the above s^2. Since $\chi^2_{0.05,39} = 54.572$, the power would be estimated as

$$P[\chi^2 \geq (54.572)(1.5)/2.6898]$$
$$= P(\chi^2 \geq 30.433).$$

Consulting Table B.1 for $\nu = 39$, $0.75 < P(\chi^2 \geq 30.433) < 0.90$. Or, by interpolation, $P = 0.83$.†

*See the beginning of Appendix B for a discussion of interpolation. In this example, the actual probability is 0.35, so the interpolation yielded an excellent result.

†The actual value of $P(\chi^2 \geq 30.433)$ with $\nu = 39$ is 0.84, so the interpolation worked very well.

For the test of H_0: $\sigma^2 \geq \sigma_0^2$ against H_A: $\sigma^2 < \sigma_0^2$, the power is

$$1 - \beta = P(\chi^2 \leq \chi^2_{1-\alpha,\nu}\sigma_0^2/s^2)$$
$$= 1 - P(\chi^2 \geq \chi^2_{1-\alpha,\nu}\sigma_0^2/s^2). \tag{8.27}$$

For the two-tailed test, H_0: $\sigma^2 = \sigma_0^2$ vs. H_A: $\sigma^2 \neq \sigma_0^2$, the power is

$$1 - \beta = 1 - P(\chi^2 \geq \chi^2_{1-\alpha/2,\nu}\sigma_0^2/s^2)$$
$$+ P(\chi^2 \geq \chi^2_{\alpha/2,\nu}\sigma_0^2/s^2). \tag{8.28}$$

8.13 HYPOTHESES CONCERNING SYMMETRY AND KURTOSIS

In Sections 8.1 and 8.2, the t statistic, $t = (\bar{X} - \mu)/s_{\bar{X}}$, was introduced to test hypotheses concerning the mean. In several later instances, we shall see that one can test hypotheses about parameters other than μ by using sample estimates other than $\bar{X}$ and standard errors other than $s_{\bar{X}}$. In general,

$$t = \frac{\text{estimator} - \text{parameter in } H_0}{\text{standard error of estimator}} \tag{8.29}$$

is a valid calculation when the sampling distribution of the estimator (i.e., of the sample statistic) is normal or nearly normal.

Testing Symmetry. Recall from Section 7.1 that a symmetrical population will have $\gamma_1 = 0$, and populations that are negatively or positively skewed will exhibit $\gamma_1 < 0$ and $\gamma_1 > 0$, respectively. Thus, one may propose the two-tailed hypotheses, H_0: $\gamma_1 = 0$ and H_A: $\gamma_1 \neq 0$, to determine whether a sampled population is symmetrical, and

$$t = \frac{g_1 - \gamma_1}{s_{g_1}} = \frac{g_1}{s_{g_1}}. \tag{8.30}$$

The calculation of g_1 was given in Section 7.1, and the standard error of g_1 is

$$s_{g_1} = \sqrt{\frac{6n(n-1)}{(n-2)(n+1)(n+3)}} \tag{8.31}$$

The use of the t statistic is valid only when the sampling distribution of g_1 approaches normality; unfortunately, this condition is not obtained when n is less than about 150. If the sample size is at least this large, then the preceding t value may be calculated and the associated degrees of freedom may be assumed to be ∞. However Table B.20 gives critical values of g_1, thus obviating the necessity of computing s_{g_1} and making possible hypothesis testing with small n.

Consider a sample of size 180, with $g_1 = 0.41$. For these data,

$$s_{g_1} = \sqrt{\frac{6(180)(180-1)}{(180-2)(180+1)(180+3)}} = \sqrt{0.0328} = 0.18.$$

If we wish to test H_0: $\gamma_1 = 0$ and H_A: $\gamma_1 \neq 0$ at the 5% level of significance, we calculate $t = (0.41 - 0)/0.18 = 2.28$ and compare its absolute value with $t_{0.05(2),\infty} = 1.960$. Since $|2.28| > 1.960$, we reject H_0 and conclude that the sampled population is asymmetrical. Using Table B.20 instead of Equations 8.30 and 8.31, we see that, at the 5% significance level, a g_1 with an absolute value ≥ 0.354 would cause us to reject H_0. Since $|0.41| > 0.354$, H_0 is rejected, and $0.02 < P < 0.05$.

If, from a sample having $g_1 = -0.5$ and $n = 30$, we are interested in testing whether there is a significant negative asymmetry in the population distribution, then H_0: $\gamma_1 \geq 0$ and H_A: $\gamma_1 < 0$. Table B.20 gives the critical values of g_1 at $\alpha = 0.05$ as 0.662; since $|-0.5| < 0.662$, we do not reject H_0, concluding that there is not a negative skew to the population distribution; there is a probability > 0.20 that H_0 is true.

Testing Kurtosis. Recall from Section 7.1 that a mesokurtic population will have $\gamma_2 = 0$, and populations that are platykurtic or leptokurtic will have $\gamma_2 < 0$ and $\gamma_2 > 0$, respectively. For very large samples (say, $n > 1000$), we may test the two-tailed hypotheses, H_0: $\gamma_2 = 0$ and H_A: $\gamma_2 \neq 0$, to determine whether the distribution of a sampled population departs from mesokurtic (i.e., "normal"). For such a test,

$$t = \frac{g_2 - \gamma_2}{s_{g_2}} = \frac{g_2}{s_{g_2}}, \tag{8.32}$$

where the standard error of g_2 is

$$s_{g_2} = \sqrt{\frac{24n(n-1)^2}{(n-3)(n-2)(n+3)(n+5)}} \tag{8.33}$$

Here, as with Equation 8.30, n is so large when the equation is used that $t_{\alpha(2),\infty}$ may be considered the critical value.

One-tailed hypotheses can also be tested. If the concern were strictly with the presence of platykurtosis, then we would employ Equation 8.32 to test H_0: $\gamma_2 \geq 0$ against H_A: $\gamma_2 < 0$, using $t_{\alpha(1),\infty}$ as the critical value. Testing H_0: $\gamma_2 \leq 0$ versus H_A: $\gamma_2 > 0$ looks specifically for leptokurtosis. Hypothesis testing for γ_2 is discussed further by D'Agostino and Tietjen (1971), Pearson and Hartley (1966: Table 34C), Shapiro, Wilk, and Chen (1968), and Snedecor and Cochran (1980: Section 5.14).

Testing Normality. It is often assumed that testing both symmetry and kurtosis simultaneously affords a test for normality of a data distribution. Madow (1940) proposed the following test:

$$\chi^2 = \left(\frac{g_1}{s_{g_1}}\right)^2 + \left(\frac{g_2}{s_{g_2}}\right)^2 \quad \text{or} \quad \chi^2 = \frac{g_1^2}{s_{g_1}^2} + \frac{g_2^2}{s_{g_2}^2}, \tag{8.34}$$

which has 2 degrees of freedom, to test H_0: The population is both symmetric and mesokurtic, against H_A: The population is not both symmetric and mesokurtic.

Since the above tests require very large samples, and since better tests for normality are available (Section 7.4), hypothesis testing for symmetry and kurtosis (especially the latter) is seldom encountered.

Confidence Limits. Knowing g_1 and s_{g_1} (or g_2 and s_{g_2}) and $t_{\alpha(2),\infty}$ when samples are large, one can set confidence limits on γ_1 (or γ_2), just as the knowledge of $\bar{X}$, $s_{\bar{X}}$, and $t_{\alpha(2),\nu}$ enables us to set confidence limits on μ; but there is very little call for such confidence intervals.

8.14 THE EFFECT OF CODING

In calculating a t value, the numerator and the denominator have identical units, so t is unitless. Coding of data will in no way affect the value of t in any one-sample hypothesis testing. Bear in mind that the constant, μ_0, in the hypotheses must be coded in the same fashion as are the data.

When performing the χ^2 calculation required in Section 8.11, note that both the numerator and denominator have the same units; the resultant χ^2 and the conclusions drawn therefrom will not be affected by the coding of the raw data.

EXERCISES

8.1. The following data are the lengths of the menstrual cycle in a random sample of 15 human females. Test the hypothesis that the mean length of human menstrual cycles is equal to a lunar month (a lunar month is 29.5 days).
The data are 26, 24, 29, 33, 25, 26, 23, 30, 31, 30, 28, 27, 29, 26, and 28 days.

8.2. Body temperatures were obtained from a sample of 8 intertidal crabs exposed to air at 26.2°C. Test the hypothesis that the mean body temperature of this species of crab under these conditions is less than 26.2°C.
The data are 25.8, 24.6, 26.1, 24.9, 25.1, 25.3, 24.0, and 24.5°C.

8.3. Present the following data in a graph that shows the mean, standard error, 95% confidence interval, range, and number of observations for each month.

Table of Caloric Intake (kcal/g of body weight) **of Squirrels**

Month	*No. of Data*	*Mean*	*Standard Error*	*Range*
January	13	0.458	0.026	0.289–0.612
February	12	0.413	0.027	0.279–0.598
March	17	0.327	0.018	0.194–0.461

8.4. A sample of size 18 has a mean of 13.55 cm and a variance of 6.4512 cm^2.

(a) Calculate the 95% confidence interval for the population mean.

(b) How large a sample would have to be taken from this population to estimate μ to within 1.00 cm, with 95% confidence? 2.00 cm with 95% confidence? 2.00 cm with 99% confidence?

8.5. We want to sample a population of lengths and to perform a test of $H_0: \mu = \mu_0$ vs. $H_A: \mu \neq \mu_0$, at the 5% significance level, with a 95% probability of rejecting H_0 when $\mu - \mu_0$ is at least 2.0 cm. The estimate of the population variance, σ^2, is $s^2 = 8.44$ cm^2.

(a) What minimum sample size should be used?

(b) What minimum sample size would be required if α were to be 0.01?

(c) What minimum sample size would be required if $\alpha = 0.05$ and power $= 0.99$?

(d) If $n = 25$ and $\alpha = 0.05$, what is the smallest difference $(\mu - \mu_0)$ that can be detected with 95% probability?

(e) If $n = 25$ and $\alpha = 0.05$, what is the probability of detecting a difference $(\mu - \mu_0)$ as small as 2.0 cm?

8.6. There are 200 members of a state legislature. The ages of a random sample of 50 of them are obtained, and it is found that $\bar{X} = 53.87$ yr and $s = 9.87$ yr.

(a) Calculate the 95% confidence interval for the mean age of all members of the legislature.

(b) If the above $\bar{X}$ and s had been obtained from a random sample of 100 from this population, what would the 95% confidence interval for the population mean have been?

9 Two-Sample Hypotheses

Among the most commonly employed biostatistical procedures is the comparison of two samples to infer whether differences exist between the two populations sampled. This chapter will introduce another very important sampling distribution, the F distribution—named for its discoverer, R. A. Fisher* (Snedecor, 1934: 15)—and will demonstrate further use of Student's t distribution.

The objective of many two-sample hypotheses is to make inferences about population parameters by examining sample statistics. Other hypothesis-testing procedures, however, draw inferences about populations without referring to parameters. Such procedures are called *nonparametric* methods, and several will be discussed in this and following chapters.

9.1 TESTING FOR DIFFERENCE BETWEEN TWO VARIANCES

If we have two samples of measurements, each sample taken at random from a normal population, we might ask if the variances of the two populations are equal. Consider the data of Example 9.1, where s_1^2, the estimate of σ_1^2, is 21.87 moths^2, and s_2^2, the estimate of σ_2^2, is 15.36 moths^2. The two-tailed hypotheses can be stated as H_0: $\sigma_1^2 = \sigma_2^2$ and H_A: $\sigma_1^2 \neq \sigma_2^2$, and we can ask, what is the probability of taking two samples from two populations having identical variances and having the two sample variances be as different as are s_1^2 and s_2^2? If this probability is rather low (say, $\alpha \leq 0.05$, as in previous chapters), then we reject the veracity of H_0 and conclude that the

*Sir Ronald Alymer Fisher (1890–1962).

two samples came from populations having unequal variances. If the probability is greater than α, we state that there is insufficient evidence to conclude that the variances of the two populations are not the same.

Example 9.1

A two-tailed variance ratio test for the hypotheses $H_0: \sigma_1^2 = \sigma_2^2$ and $H_A: \sigma_1^2 \neq \sigma_2^2$. The data are the numbers of moths caught during the night by 11 traps of one style and 8 traps of a second style.

H_0: $\sigma_1^2 = \sigma_2^2$
H_A: $\sigma_1^2 \neq \sigma_2^2$
$\alpha = 0.05$

Trap type 1	*Trap type 2*
41	52
34	57
33	62
36	55
40	64
25	57
31	56
37	55
34	
30	
38	

$n_1 = 11$	$n_2 = 8$
$\nu_1 = 10$	$\nu_2 = 7$
$SS_1 = 218.73$ moths²	$SS_2 = 107.50$ moths²
$s_1^2 = 21.87$ moths²	$s_2^2 = 15.36$ moths²

$$F = \frac{s_1^2}{s_2^2} = \frac{21.87}{15.36} = 1.42$$

$F_{0.05(2),10,7} = 4.76$

Therefore, do not reject H_0.

$P(F \geq 1.42) > 0.50$

$$s_p^2 = \frac{218.73 \text{ moths}^2 + 107.50 \text{ moths}^2}{10 + 7} = 19.19 \text{ moths}^2$$

The hypotheses may be submitted to the *variance ratio test, for* which one calculates

$$F = \frac{s_1^2}{s_2^2} \quad \text{or} \quad F = \frac{s_2^2}{s_1^2}, \text{ whichever is larger.} \tag{9.1}$$

We then ask whether the calculated ratio of sample variances (i.e., F) deviates so far from 1.0 as to enable us to reject H_0 at the α level of significance. For the data in Example 9.1, the calculated F is 1.42. The critical value, $F_{0.05(2),10,7}$, is obtained from Table B.4 and is found to be 4.76. As $1.42 < 4.76$, we do not reject H_0. The calculated F (namely 1.42) is smaller than the smallest tabled critical value (namely $F_{0.50(2),10,7} = 1.69$), so the probability of an F at least this large if H_0 is a true statement about the populations is greater than 0.50; i.e., $P(F \geq 1.42) > 0.50$.*

*Some desk calculators and computer programs have the capability of determining the probability of a given F (Guenther, 1977). For the present example, we would thereby find that $P(F \geq 1.42) = 0.66$.

Note that we consider degrees of freedom associated with the variances in both the numerator and denominator of the variance ratio. Furthermore, it is important to realize that F_{α,ν_1,ν_2} and F_{α,ν_2,ν_1} are not the same value, so one must be careful to refer to the numerator and denominator degrees of freedom in the correct order.

If $H_0: \sigma_1^2 = \sigma_2^2$ is not rejected, then s_1^2 and s_2^2 are assumed to be estimates of the same population variance, σ^2. The best estimate of this σ^2 that underlies both samples is what is called the *pooled variance*:

$$s_p^2 = \frac{SS_1 + SS_2}{\nu_1 + \nu_2} = \frac{\nu_1 s_1^2 + \nu_2 s_2^2}{\nu_1 + \nu_2}. \qquad (9.2)$$

One-tailed hypotheses may also be submitted to the variance ratio test. For $H_0: \sigma_1^2 \geq \sigma_2^2$ and $H_A: \sigma_1^2 < \sigma_2^2$, s_2^2 is always used as the numerator of the variance ratio; for $H_0: \sigma_1^2 \leq \sigma_2^2$ and $H_A: \sigma_1^2 > \sigma_2^2$, s_1^2 is always used as the numerator. (A look at the alternate hypothesis tells which variance belongs in the numerator of F.) Example 9.2 presents data submitted to the hypothesis test for whether ducks raised in captivity have less variability in clutch size than those breeding in the wild.

Example 9.2

A one-tailed variance ratio test for the hypothesis that duck clutch size is less variable in captive than in wild birds.

H_0: $\sigma_1^2 \geq \sigma_2^2$
H_A: $\sigma_1^2 < \sigma_2^2$
$\alpha = 0.05$

Clutch Size of Ducks

Captive	*Wild*
10	9
11	8
12	11
11	12
10	10
11	13
11	11
	10
	12

$n_1 = 7$ $\qquad n_2 = 9$
$\nu_1 = 6$ $\qquad \nu_2 = 8$
$SS_1 = 2.86 \text{ eggs}^2$ $\qquad SS_2 = 20.00 \text{ eggs}^2$
$s_1^2 = 0.48 \text{ eggs}^2$ $\qquad s_2^2 = 2.50 \text{ eggs}^2$

$$F = \frac{2.50}{0.48} = 5.21$$

$F_{0.05(1),8,6} = 4.15$
Therefore, reject H_0.
$0.025 < P(F \geq 5.21) < 0.05$

The variance ratio test is severely and adversely affected by sampling nonnormal populations. Thus, it must be employed with caution. There are nonparametric tests for the difference between the dispersions of two populations (e.g., Daniel, 1978: Section 3.2). While these do not have the assumption of normality, they have other serious drawbacks and are seldom used.

9.2 CONFIDENCE INTERVAL FOR VARIANCE RATIO

A $1 - \alpha$ confidence interval for the variance ratio, σ_1^2/σ_2^2, is defined by its lower confidence limit,

$$L_1 = \left(\frac{s_1^2}{s_2^2}\right)\left(\frac{1}{F_{\alpha(2),\nu_1,\nu_2}}\right), \tag{9.3}$$

and its upper confidence limit,

$$L_2 = \left(\frac{s_1^2}{s_2^2}\right)F_{\alpha(2),\nu_2,\nu_1}. \tag{9.4}$$

In Example 9.1, $s_1^2/s_2^2 = 1.42$, $F_{0.05(2),10,7} = 4.76$, and $F_{0,05(2),7,10} = 3.95$. Therefore, we would calculate $L_1 = 0.298$ and $L_2 = 5.61$, and we could state

$$P\left(0.298 \leq \frac{\sigma_1^2}{\sigma_2^2} \leq 5.61\right) = 0.95.$$

9.3 TESTING FOR DIFFERENCE BETWEEN COEFFICIENTS OF VARIATION

Some authors have suggested approximate procedures for testing the null hypothesis that two samples came from populations with identical coefficients of variation. However, Lewontin (1966) has shown that the variance ratio

$$F = \frac{(s^2_{\log})_1}{(s^2_{\log})_2} \tag{9.5}$$

can be used analogously to Equation 9.1. In Equation (9.5), $(s^2_{\log})_i$ refers to the variance of the logarithms of the data in sample i, where logarithms to any base may be employed.

A very useful property of coefficients of variation is that they have no units. Thus, V's may be compared even if they are calculated from measurements obtained in different units. Example 9.3 demonstrates a two-tailed test where this is the case. One-tailed testing is also possible, in the same fashion as discussed in Section 9.1.

Unfortunately, we are faced with the requirement of the variance ratio test that the two underlying distributions be normal (or nearly normal). Thus, this test must be applied with caution, for if the two sets of sample data are, in fact, from normal populations, the logarithms of the data *will not* be normally distributed; and the requirement here is that the *logarithms* be normally distributed.

Example 9.3

A two-tailed variance ratio test for the difference between two coefficients of variation.

H_0: The intrinsic variability of male weights is the same as the intrinsic variability of male heights (i.e., the population coefficients of variation of weight and height are the same).

H_A: The instrinsic variability of male weights is not the same as the intrinsic variability of male heights (i.e., the population coefficients of variation of weight and height are not the same).

Weight (kg)	*Log of weight*	*Height* (cm)	*Log of height*
72.5	1.86034	183.0	2.26245
71.7	1.85552	172.3	2.23629
60.8	1.78390	180.1	2.25551
63.2	1.80072	190.2	2.27921
71.4	1.85370	191.4	2.28194
73.1	1.86392	169.6	2.22943
77.9	1.89154	166.4	2.22115
75.7	1.87910	177.6	2.24944
72.0	1.85733		
69.0	1.83885		

$n_1 = 10$ $\qquad$ $n_2 = 8$

$\nu_1 = 9$ $\qquad$ $\nu_2 = 7$

$\bar{X}_1 = 70.73$ kg $\qquad$ $\bar{X}_2 = 178.82$ cm

$\text{SS}_1 = 246.1610 \text{ kg}^2$ $\qquad$ $\text{SS}_2 = 590.1350 \text{ cm}^2$

$s_1^2 = 27.3512 \text{ kg}^2$ $\qquad$ $s_2^2 = 84.3050 \text{ cm}^2$

$s_1 = 5.23$ kg $\qquad$ $s_2 = 9.18$ cm

$V_1 = 0.0739$ $\qquad$ $V_2 = 0.0513$

$(\text{SS}_{\log})_1 = 0.0098702632$ $\qquad$ $(\text{SS}_{\log})_2 = 0.0034727534$

$(s^2_{\log})_1 = 0.0010966959$ $\qquad$ $(s^2_{\log})_2 = 0.0004961076$

$$F = \frac{0.0010966959}{0.0004961076} = 2.21$$

$F_{0.05(2),9,7} = 4.82$

Therefore, do not reject H_0.

$0.20 < P(F \geq 2.21) < 0.50$

9.4 TESTING FOR DIFFERENCE BETWEEN TWO MEANS

Example 9.4 presents as data the time it takes blood to clot, as measured for each of 13 persons, 6 given one kind of drug and 7 given another. The two-tailed hypotheses, H_0: $\mu_1 - \mu_2 = 0$ and H_A: $\mu_1 - \mu_2 \neq 0$, can be proposed to ask whether, in the populations sampled, blood of persons treated with drug B has the same mean clotting time as does blood from persons treated with drug G. These hypotheses are commonly expressed in their equivalent forms: H_0: $\mu_1 = \mu_2$ and H_A: $\mu_1 \neq \mu_2$.

If the two samples came from normal populations, and if the two populations have equal variances, then a t value may be calculated in a manner analogous to the t test introduced in Section 8.1. The t value for testing the preceding hypotheses concerning the difference between two means is

$$t = \frac{\bar{X}_1 - \bar{X}_2}{s_{\bar{X}_1 - \bar{X}_2}}. \qquad (9.6a)$$

Example 9.4

A two-sample t test for the two-tailed hypotheses, $H_0: \mu_1 = \mu_2$ and $H_A: \mu_1 = \mu_2$ (which could also be stated as $H_0: \mu_1 - \mu_2 = 0$ and $H_2: \mu_1 - \mu_2 \neq 0$). The data are human blood-clotting times (in minutes) of individuals given one of two different drugs.

$H_0: \quad \mu_1 = \mu_2$
$H_A: \quad \mu_1 \neq \mu_2$

Given drug B	*Given drug G*
8.8	9.9
8.4	9.0
7.9	11.1
8.7	9.6
9.1	8.7
9.6	10.4
	9.5

$n_1 = 6 \qquad n_2 = 7$

$\nu_1 = 5 \qquad \nu_2 = 6$

$\bar{X}_1 = 8.75$ min $\qquad \bar{X}_2 = 9.74$ min

$SS_1 = 1.6950$ min $\qquad SS_2 = 4.0171$ min^2

$$s_p^2 = \frac{SS_1 + SS_2}{\nu_1 + \nu_2} = \frac{1.6950 + 4.0171}{5 + 6} = \frac{5.7121}{11} = 0.5193 \text{ min}^2$$

$$s_{\bar{X}_1 - \bar{X}_2} = \sqrt{\frac{s_p^2}{n_1} + \frac{s_p^2}{n_2}} = \sqrt{\frac{0.5193}{6} + \frac{0.5193}{7}} = \sqrt{0.0866 + 0.0742}$$

$$= \sqrt{0.1608} = 0.40 \text{ min}$$

$$t = \frac{\bar{X}_1 - \bar{X}_2}{s_{\bar{X}_1 - \bar{X}_2}} = \frac{8.75 - 9.74}{0.40} = \frac{-0.90}{0.40} = -2.475$$

$t_{0.05(2),\nu} = t_{0.05(2),11} = 2.201$

Therefore, reject H_0.

$0.02 < P(|t| \geq 2.475) < 0.05$

The quantity $\bar{X}_1 - \bar{X}_2$ is simply the difference between the two means, and $s_{\bar{X}_1 - \bar{X}_2}$ is the standard error of the difference between the means.

The quantity $s_{\bar{X}_1 - \bar{X}_2}$, along with $s^2_{\bar{X}_1 - \bar{X}_2}$, the variance of the difference between the means, is new to us, and we need to consider it further. Both $s^2_{\bar{X}_1 - \bar{X}_2}$, and $s_{\bar{X}_1 - \bar{X}_2}$ are statistics that can be calculated from the sample data and are estimates of the population parameters, $\sigma^2_{\bar{X}_1 - \bar{X}_2}$ and $\sigma_{\bar{X}_1 - \bar{X}_2}$, respectively. It can be shown mathematically that the variance of the difference between two variables is equal to the sum of the variances of the two variables,* so that $\sigma^2_{\bar{X}_1 - \bar{X}_2} = \sigma^2_{\bar{X}_1} + \sigma^2_{\bar{X}_2}$. Since $\sigma^2_{\bar{X}} = \sigma^2/n$, we can write

$$\sigma^2_{\bar{X}_1 - \bar{X}_2} = \frac{\sigma_1^2}{n_1} + \frac{\sigma_2^2}{n_2}. \tag{9.7}$$

As the two-sample t test requires that we assume $\sigma_1^2 = \sigma_2^2$, we can write

$$\sigma^2_{\bar{X}_1 - \bar{X}_1} = \frac{\sigma^2}{n_1} + \frac{\sigma^2}{n_2}. \tag{9.8}$$

*This is so if the two variables are independent, i.e., uncorrelated. If there is correlation between them, then see Chapter 10.

Thus, to calculate the estimate of $\sigma^2_{\bar{X}_1-\bar{X}_2}$, we must have an estimate of σ^2. Since both s_1^2 and s_2^2 are assumed to estimate σ^2, we compute the pooled variance, s_p^2, which is then used as the best estimate of σ^2.

$$s_p^2 = \frac{SS_1 + SS_2}{\nu_1 + \nu_2}. \tag{9.2}$$

and

$$s^2_{\bar{X}_1-\bar{X}_2} = \frac{s_p^2}{n_1} + \frac{s_p^2}{n_2}. \tag{9.9}$$

Thus,

$$s_{\bar{X}_1-\bar{X}_2} = \sqrt{\frac{s_p^2}{n_1} + \frac{s_p^2}{n_2}}, \tag{9.10}$$

and Equation 9.6a becomes

$$t = \frac{\bar{X}_1 - \bar{X}_2}{\sqrt{\frac{s_p^2}{n_1} + \frac{s_p^2}{n_2}}} \tag{9.6b}$$

Example 9.4 summarizes the procedure for testing the hypotheses under consideration. The critical value to be obtained from Table B.3 is $t_{\alpha(2),(\nu_1+\nu_2)}$, the two-tailed t value for the α significance level, with $\nu_1 + \nu_2$ degrees of freedom. We shall also write this as $t_{\alpha(2),\nu}$, defining the pooled degrees of freedom to be

$$\nu = \nu_1 + \nu_2 \quad \text{or, equivalently,} \quad \nu = n_1 + n_2 - 2. \tag{9.11}$$

In the two-tailed test, H_0 will be rejected if either $t \geq t_{\alpha(2),\nu}$ or $t \leq -t_{\alpha(2),\nu}$. Another way of stating this is that H_0 will be rejected if $|t| \geq t_{\alpha(2),\nu}$.

One-tailed hypotheses can be tested in situations where the investigator is interested in detecting a difference in only one direction. For example, a gardener may use a particular fertilizer for a particular kind of plant, and a new fertilizer is advertised as being an improvement. Let us say that plant height at maturity is an important characteristic of this kind of plant, with taller plants being preferable. An experiment was run, raising 10 plants on the present fertilizer and 8 on the new one, with the resultant 18 plant heights shown in Example 9.5. If the new fertilizer produces plants that are shorter than, or the same height as, plants grown with the present fertilizer, then we shall decide that the advertising claims are unfounded; therefore, the statements of $\mu_1 > \mu_2$ and $\mu_1 = \mu_2$ belong in the same hypothesis, namely the null hypothesis, H_0. If, however, mean plant height is indeed greater with the newer fertilizer, then it shall be declared to be distinctly better, with the alternate hypothesis (H_A: $\mu_1 < \mu_2$) concluded to be the true statement. The t statistic is calculated by Equation 9.6, just as for the two-tailed test. But this calculated t is then compared with the critical value $t_{\alpha(1),\nu}$, rather than with $t_{\alpha(2),\nu}$.

In other cases, the one-tailed hypotheses, H_0: $\mu_1 \leq \mu_2$ and H_A: $\mu_1 > \mu_2$, may be appropriate. Just as introduced in the one-sample testing of Sections 8.1 and 8.2, the following summary of procedures applies to two-tailed t testing:

For H_A: $\mu_1 \neq \mu_2$, if $|t| \geq t_{\alpha(2),\nu}$, then reject H_0.

For H_A: $\mu_1 < \mu_2$, if $t \leq -t_{\alpha(1),\nu}$, then reject H_0.*

For H_A: $\mu_1 > \mu_2$, if $t \geq t_{\alpha(1),\nu}$, then reject H_0.†

Example 9.5

A two-sample t test for the one-tailed hypotheses, H_0: $\mu_1 \geq \mu_2$ and H_A: $\mu_1 < \mu_2$ (which could also be stated as H_0: $\mu_1 - \mu_2 \geq 0$ and H_A: $\mu_1 - \mu_2 < 0$). The data are heights of plants, each grown with one of two different fertilizers.

H_0: $\mu_1 \geq \mu_2$
H_A: $\mu_1 < \mu_2$

Present fertilizer	*Newer fertilizer*
48.2 cm	52.3 cm
54.6	57.4
58.3	55.6
47.8	53.2
51.4	61.3
52.0	58.0
55.2	59.8
49.1	54.8
49.9	
52.6	

$n_1 = 10$	$n_2 = 8$
$\nu_1 = 9$	$\nu_2 = 7$
$\bar{X}_1 = 51.91$ cm	$\bar{X}_2 = 56.55$ cm
$SS_1 = 102.23$ cm^2	$SS_2 = 69.20$ cm^2

$$s_p^2 = \frac{102.23 + 69.20}{9 + 7} = \frac{171.43}{16} = 10.71 \text{ cm}^2$$

$$s_{\bar{X}_1-\bar{X}_2} = \sqrt{\frac{10.71}{10} + \frac{10.71}{8}} = \sqrt{2.41} = 1.55 \text{ cm}$$

$$t = \frac{\bar{X}_1 - \bar{X}_2}{s_{\bar{X}_1-\bar{X}_2}} = \frac{51.91 - 56.55}{1.55} = \frac{-4.64}{1.55} = -2.99$$

$t_{0.05(1),16} = 1.746$

Since t of -2.99 is less than -1.746, H_0 is rejected.

$0.0025 < P < 0.005$

Note that H_0: $\mu_1 = \mu_2$ may be written H_0: $\mu_1 - \mu_2 = 0$ and H_A: $\mu_1 \neq \mu_2$ as H_A: $\mu_1 - \mu_2 \neq 0$; the generalized two-sample hypotheses are H_0: $\mu_1 - \mu_2 = \mu_0$ and H_A: $\mu_1 - \mu_2 \neq \mu_0$, tested as

$$t = \frac{|\bar{X}_1 - \bar{X}_2| - \mu_0}{s_{\bar{X}_1-\bar{X}_2}}, \tag{9.12}$$

*For this one-tailed hypothesis test, probabilities of t up to 0.25 are indicated in Table B.3. If $t = 0$, then $P = 0.50$; so if $-t_{0.25(1),\nu} < t < 0$, then $0.25 < P < 0.50$; and if $t > 0$, then $P >$ 0.50.

†For this one-tailed hypothesis test, $t = 0$ indicates $P = 0.50$; therefore, if $0 < t < t_{0.25(1),\nu}$, then $0.25 < P < 0.50$; and if $t < 0$, then $P > 0.50$.

where μ_0 may be any hypothesized difference between population means. Stating $\mu_0 = 0$ (as in Example 9.4) is the most common situation, in which case Equations 9.2 and 9.6 are identical.

Also, H_0: $\mu_1 \leq \mu_2$ and H_A: $\mu_1 > \mu_2$ may be written as H_0: $\mu_1 - \mu_2 \leq 0$ and H_A: $\mu_1 - \mu_2 > 0$, respectfully. The generalized hypotheses for this type of one-tailed test are H_0: $\mu_1 - \mu_2 \leq \mu_0$ and H_A: $\mu_1 - \mu_2 > \mu_0$, for which the t is

$$t = \frac{\bar{X}_1 - \bar{X}_2 - \mu_0}{s_{\bar{X}_1-\bar{X}_2}}, \tag{9.13}$$

and μ_0 may be any desired value of $\mu_1 - \mu_2$.

Lastly, H_0: $\mu_1 \geq \mu_2$ and H_A: $\mu_1 < \mu_2$ may be written as H_0: $\mu_1 - \mu_2 \geq 0$ and H_A: $\mu_1 - \mu_2 < 0$, and the generalized one-tailed hypotheses of this type are H_0: $\mu_1 - \mu_2 \geq \mu_0$ and H_A: $\mu_1 - \mu_2 < \mu_0$, with the appropriate t statistic being that of Equation 9.13. For example, the gardener collecting the data of Example 9.5 may have decided, because the newer fertilizer is more expensive than the other, that it should be used only if the plants grown with it averaged at least 5.0 cm taller than plants grown with the present fertilizer. Then, $\mu_0 = \mu_1 - \mu_2 = -5.0$ cm and, by Equation 9.13, we would calculate $t = (51.91 - 56.55 + 5.0)/1.55 = 0.36/1.55 = 0.232$, which is not $\geq$ the critical value shown in Example 9.5; so H_0: $\mu_1 - \mu_2 \geq -5.0$ cm is not rejected.

By the procedure of Section 9.8, one can test whether the measurements in one population are a specified amount as large as those in a second population.

Violations of the Two-Sample *t* Test Assumptions. The two-sample t test assumes, by dint of its underlying theory, that both samples come at random from normal populations with equal variances. The biological researcher cannot, however, always be assured that these assumptions are correct. Fortunately, numerous studies have shown that the t test is robust enough to stand considerable departures from its theoretical assumptions, especially if the sample sizes are equal or nearly equal, and especially when two-tailed hypotheses are considered (e.g., Boneau, 1960; Box, 1953; Cochran, 1947, Srivastava, 1958). If the underlying populations are markedly skewed, then one should be wary of one-tailed testing, and if there is considerable nonnormality in the populations, then very small significance levels (say, $\alpha < 0.01$) should not be depended upon.

The larger the samples, the more robust the test. If the sample sizes are equal, the t test is remarkably robust. If unequal, then the probability of a Type I error will be less than α if the larger σ^2 is associated with the larger sample, and this probability will be greater than α if the smaller sample came from the population with the larger variance (Kohr and Games, 1974).

The power of the two-tailed t test is very little affected by skewness in the sampled populations, but there can be a serious effect on one-tailed tests. The actual power of the test is less than that discussed in Section 9.7 when the sampled populations are platykurtic and greater when the populations are leptokurtic, especially for small sample sizes (Glass, Peckham, and Sanders, 1972).

The comparison of two means from normal populations with unequal variances is known as the "Behrens-Fisher problem," and numerous solutions have been offered (e.g., Cochran, 1964; Cochran and Cox, 1957: 100–102; Dixon and Massey, 1969: 119; Fisher and Yates, 1963: 3–4, 60–61;* Gill, 1971; Lee and Gurland, 1975; Satterthwaite, 1946). One of the easiest, yet reliable, of such procedures is that attributed to Smith (1936) and also known as "Welch's approximate t" (Davenport and Webster, 1975; Mehta and Srinivasan, 1970; Scheffé, 1970; Wang, 1971). The test statistic is

$$t = \frac{\bar{X}_1 - \bar{X}_2}{\sqrt{\dfrac{s_1^2}{n_1} + \dfrac{s_2^2}{n_2}}}, \tag{9.14}$$

and the critical value is Student's t with degrees of freedom of

$$\nu = \frac{\left(\dfrac{s_1^2}{n_1} + \dfrac{s_2^2}{n_2}\right)^2}{\dfrac{\left(\dfrac{s_1^2}{n_1}\right)^2}{n_1 - 1} + \dfrac{\left(\dfrac{s_2^2}{n_2}\right)^2}{n_2 - 1}}. \tag{9.15}$$

(The degrees of freedom thus computed may be noninteger, in which case the next smaller integer should be used.)

If there are severe deviations from the normality and/or equality of variance assumptions, the nonparametric test of Section 9.8 could be employed, as it is not at all adversely affected by violations of these assumptions.

9.5 CONFIDENCE LIMITS FOR MEANS

In Section 8.3, we defined the confidence interval for a population mean as $\bar{X} \pm t_{\alpha(2),\nu}s_{\bar{X}}$, where $s_{\bar{X}}$ is the best estimate of $\sigma_{\bar{X}}$ and is calculated as $\sqrt{s^2/n}$. For the two-sample situation, where we assume that $\sigma_1^2 = \sigma_2^2$, the confidence interval for either μ_1 or μ_2 is calculated using s_p^2 (rather than either s_1^2 or s_2^2) as the best estimate of σ^2; and we use the two-tailed tabled t value with $\nu = \nu_1 + \nu_2$ degrees of freedom. Thus, for μ_i (where i is either 1 or 2, referring to either of the two samples), the $1 - \alpha$ confidence interval is

$$\bar{X}_i \pm t_{\alpha(2),\nu}\sqrt{\frac{s_p^2}{n_i}}. \tag{9.16}$$

For the data of Example 9.4, $\sqrt{s_p/n_2} = \sqrt{0.5193 \text{ min}^2/7} = 0.27$ min. Thus, the 95% confidence interval for μ_2 would be 9.74 min $\pm$ (2.201)(0.27 min) = 9.74 min $\pm$ 0.59 min, so that L_1 (the lower confidence limit) = 9.15 min and L_2 (the upper confidence limit) = 10.33 min. Thus, we can be 95% confident that the interval of 9.15 to 10.33 minutes includes the population mean, μ_2. This may be written as $P(9.15 \text{ min} \leq \mu_2 \leq 10.33 \text{ min}) = 0.95$. The confidence interval for μ_1 would be 8.75 min $\pm$ $(2.201)\sqrt{0.5193 \text{ min}^2/6}$ = 8.75 min $\pm$ (2.201)(0.29 min) = 8.75 min $\pm$ 0.64 min, so that L_1 = 8.11 min and L_2 = 9.39 min.

*In Fisher and Yates (1963), s refers to the standard error, not the standard deviation.

Confidence limits for the difference between the two population means can also be computed. The $1 - \alpha$ confidence interval for $\mu_1 - \mu_2$ is

$$\bar{X}_1 - \bar{X}_2 \pm t_{\alpha(2),\nu} s_{\bar{X}_1 - \bar{X}_2}. \tag{9.17}$$

Thus, for Example 9.4, the 95% confidence interval for $\mu_1 - \mu_2$ is (8.75 min − 9.74 min) ± (2.201)(0.40 min) = −0.99 min ± 0.88 min. Thus, $L_1 = -1.87$ min and $L_2 = -0.11$ min, and we can write $P(-1.87 \text{ min} \leq \mu_1 - \mu_2 \leq -0.11 \text{ min}) = 0.95$. This statement also implies that if one took from the two populations all possible two-sample combinations with $n_1 = 6$ and $n_2 = 7$, and computed a confidence interval from each one, 95% of all the confidence intervals would encompass $\mu_1 - \mu_2$.

If H_0: $\mu_1 = \mu_2$ is not rejected, then both samples are concluded to have come from populations having identical means, the common mean being denoted as μ. The best estimate of μ is the "pooled" or "weighted" or "common" mean:

$$\bar{X}_p = \frac{n_1 \bar{X}_1 + n_2 \bar{X}_2}{n_1 + n_2}. \tag{9.18}$$

Then the $1 - \alpha$ confidence interval for μ is

$$\bar{X}_p \pm t_{\alpha(2),\nu} \sqrt{\frac{s_p^2}{n_1 + n_2}}. \tag{9.19}$$

If H_0 is not rejected it is the confidence interval of Equation 9.19, rather than those of Equations 9.16 and 9.17, that one would calculate.

If $\sigma_1^2 \neq \sigma_2^2$. If the two population variances are not equal, then Equation 9.14 might be used to test H_0: $\mu_1 = \mu_2$. If this test rejects the null hypothesis, then confidence intervals for the two population means should be computed as

$$\bar{X}_i \pm t_{\alpha(2),\nu_i} \sqrt{\frac{s_i^2}{n_i}}, \tag{9.20}$$

rather than by Equation 9.16. A confidence interval for the difference between the population means is obtained, then, as follows:

$$\bar{X}_1 - \bar{X}_2 \pm t_{\alpha(2),\nu} \sqrt{\frac{s_1^2}{n_1} + \frac{s_2^2}{n_2}}, \tag{9.21}$$

rather than by Equation 9.17; ν is obtained from Equation 9.15. If H_0: $\mu_1 = \mu_2$ is not rejected, then Equations 9.18 and 9.19 can be employed to arrive at the confidence interval for the common mean.

9.6 SAMPLE SIZE AND ESTIMATION OF THE DIFFERENCE BETWEEN TWO POPULATION MEANS

In Section 8.5 it was shown how to determine the size of the sample that is needed to estimate a population mean, by obtaining with stated assurance a confidence interval of specified width. The same type of procedure may be employed to determine the sample size, n, required from each of two populations in order to estimate the difference between the two population means with specified precision. Just as with the one-sample case of Section 8.5, the estimation of sample size is an interative procedure, employing a series of successively improving estimates of the required n. As shown in Example 9.6, we use

$$n = \frac{2s_p^2(t_{\alpha(2),2(n-1)})^2 F_{\beta(1),(n-1),\nu}}{d^2} \tag{9.22}$$

(Harris, Horvitz, and Mood, 1948), where s_p^2 is the pooled variance, with ν degrees of freedom, $1 - \alpha$ is the confidence level of the desired confidence interval, critical values are from Table B.3 where the degrees of freedom are $2(n - 1)$, $1 - \beta$ is the probability that the half-width of the confidence interval will not exceed d, and critical values of F are found in Table B.4 for numerator DF of $2(n - 1)$ and denominator DF of ν. In general it takes fewer iterations (i.e., it is more efficient) to employ an initial guess too high rather than too low.

Example 9.6

Determination of sample size needed to achieve a stated precision in estimating the difference between two population means, using the data of Example 9.4.

If we specify that we wish to estimate $\mu_1 - \mu_2$ by having a 90% probability that the 95% confidence interval will be no wider than 1.0 min, then $d = 0.5$ min, $1 - \beta = 0.90$, $\beta = 0.10$, $1 - \alpha = 0.95$, and $\alpha = 0.05$. From Example 9.4 we have an estimate of the population variance: $s_p^2 = 0.5193$ min² with $\nu = 11$.

Let us guess that a sample size of 50 is necessary; then, $t_{0.05(2),2(50-1)} = t_{0.05(2),98} = 1.984$ and $F_{0.10(1),2(50-1),11} = F_{0.10(1),98,11} = 2.01$, so we estimate (by Equation 9.22)

$$n = \frac{2(0.5193)(1.984)^2(2.01)}{(0.5)^2} = 32.9.$$

Next, we might estimate $n = 33$, for which $t_{0.05(2),64} = 1.998$ and $F_{0.10(1),64,11} = 2.03$, and calculate

$$n = \frac{2(0.5193)(1.998)^2(2.03)}{(0.5)^2} = 33.7.$$

Now, estimating $n = 34$, we have $t_{0.05(2),66} = 1.997$ and $F_{0.10(1),66,11} = 2.03$, and we calculate

$$n = \frac{2(0.5193)(1.997)^2(2.03)}{(0.5)^2} = 33.6.$$

Therefore, we conclude that a sample of at least 34 (i.e., more than 33) should be taken from each of the two populations in order to achieve the specified confidence interval.

If, for some reason (say, there is a limited amount of the first drug available), n_1 is constrained to be no larger than 25, then the necessary n_2 would be determined, from Equation 19.23, to be

$$n_2 = \frac{(34)(25)}{2(25) - 34} = 53.1,$$

meaning that we should use n_2 of at least 54.

It is best to have equal sample sizes (i.e., $n_1 = n_2$) when estimating $\mu_1 - \mu_2$, but occasionally this is impractical. If sample 1 is constrained to have a size n_1, we can, after using the above procedure to calculate n, arrive at the required n_2 by

$$n_2 = \frac{nn_1}{2n_1 - n}, \tag{9.23}$$

as shown in Example 9.6. (If $2n_1 - n \leq 0$, then n_1 must be increased and/or α and/or β and/or d must be altered to obtain a positive n_2.) Note the efficiency in having equal sample sizes: If $n_1 = n_2$, a total of 68 data need to be collected in Example 9.6; but if $n_1 = 25$, then a total of $25 + 54 = 79$ data are to be obtained.

9.7 POWER AND SAMPLE SIZE IN TESTS FOR DIFFERENCE BETWEEN TWO MEANS

Sample Size Required. Prior to performing a two-sample test for difference between means (Section 9.4), an investigator may ask what size samples to collect. Assuming each sample comes from a normal population, we can estimate the minimum sample size to use to achieve desired test characteristics:

$$n \geq \frac{2s_p^2}{\delta^2}(t_{\alpha,\nu} + t_{\beta(1),\nu})^2 \tag{9.24}$$

(Cochran and Cox, 1957: 19–21).* Here, δ is the smallest population difference we wish to detect: $\delta = \mu_1 - \mu_2$ for the hypothesis test for which Equation 9.6 is used; $\delta = |\mu_1 - \mu_2| - \mu_0$ when Equation 9.12 is appropriate; $\delta = \mu_1 - \mu_2 - \mu_0$ when performing a test using Equation 9.13. Also in Equation 9.24 is $t_{\alpha,\nu}$, which may be either $t_{\alpha(1),\nu}$ or $t_{\alpha(2),\nu}$, depending, respectively, on whether a one-tailed or two-tailed test is to be performed.

Note that the required sample size depends on the following four factors:

1. δ, the minimum detectable difference between population means.† If we desire to detect a very small difference between means, then we shall need a larger sample than if we wished to detect only large differences.
2. σ^2, the population variance. If the variability within samples is great, then a larger sample size is required to achieve a given ability of the test to detect differences between means. We need to know the variability to expect among the data; assuming the variance is the same in each of the two populations sampled, σ^2 is estimated by the pooled variance, s_p^2, obtained from similar studies.
3. The significance level, α. If we perform the t test at a low α, then the critical value, $t_{\alpha,\nu}$, will be large and a large n is required to achieve a given ability to detect differences between means. That is, if we desire a low probability of committing a Type I error (i.e., of falsely rejecting H_0: $\mu_1 = \mu_2$), then we need large sample sizes.
4. The power of the test, $1 - \beta$. If we desire a test with a high probability of

*The method of Section 11.3 may also be used for estimation of sample size, but it offers no substantial advantage over the present procedure.

†δ is lowercase Greek delta. If μ_0 in the statistical hypotheses is not zero (see discussion surrounding Equations 9.12 and 9.13), then δ is the amount by which the absolute value of the difference between the population means differs from μ_0.

detecting a difference between population means (i.e., a low probability of committing a Type II error), then $\beta(1)$ will be small, $t_{\beta(1),\nu}$ will be large, and large sample sizes are required.

Example 9.7 shows how sample size is estimated. As $t_{\alpha(2),\nu}$ and $t_{\beta(1),\nu}$ depend on n, which is not yet known, Equation 9.24 must be solved iteratively, as we did with Equation 8.8. It matters little if the initial guess for n is inaccurate. Each iterative step will bring the estimate of n closer to the final result (which is declared when two successive iterations fail to change the value of n rounded to the next highest integer). In general, however, fewer iterations are required (i.e., the process is quicker) if one guesses high instead of low.

Example 9.7

Estimation of required sample size for a two-sample t test.

We desire to test for significant difference between the mean blood-clotting times of persons using two different drugs. We wish to test at the 0.05 level of significance, with a 90% chance of detecting a true difference between population means as small as 0.5 min. The within-population variability, based on a previous study of this type (Example 9.4), is estimated to be 0.52 min^2.

Let us guess that sample sizes of 100 will be required. Then, $\nu = 2(n-1) = 198$, $t_{0.05(2),198} = 1.972$, $\beta = 1 - 0.90 = 0.10$, $t_{0.10(1),198} = 1.286$, and we calculate (by Equation 9.24):

$$n \geq \frac{2(0.52)}{(0.5)^2}(1.972 + 1.286)^2 = 44.2.$$

Let us now use $n = 45$ to determine $\nu = 2(n-1) = 88$, $t_{0.05(2),88} = 1.987$, $t_{0.10(1),88} = 1.291$, and

$$n \geq \frac{2(0.52)}{(0.5)^2}(1.987 + 1.291)^2 = 44.7.$$

Therefore, we conclude that each of the two samples should contain at least 45 data.

If n_1 were constrained to be 30, then, using Equation 9.23, the required n_2 would be

$$n_2 = \frac{(44.7)(30)}{2(30) - 44.7} = 88.$$

For a given number of data ($n_1 + n_2$), maximum test power and robustness occurs when $n_1 = n_2$ (i.e., the sample sizes are equal). There are occasions, however, when equal sample sizes are impossible or impractical. If, for example, n_1 were fixed, then we would first determine n by Equation 9.24 and then find the required size of the second sample by Equation 9.23, as shown in Example 9.7. Note, from this example, that a total of $45 + 45 = 90$ data are required in the two equal-sized samples to achieve the desired power, whereas a total of $30 + 88 = 118$ data are needed if the two samples are as unequal as in this example.

Minimum Detectable Difference. Equation 9.24 can be rearranged to ask how small a population difference (δ, defined above) is detectable with a given sample size:

$$\delta \geq \sqrt{\frac{2s_p^2}{n}}(t_{\alpha,\nu} + t_{\beta(1),\nu}). \qquad (9.25)$$

The estimation of δ is demonstrated in Example 9.8.

Example 9.8

Estimation of minimum detectable difference in a two-sample t test.

In two-tailed testing for significant difference between mean blood-clotting times of persons using two different drugs, we desire to use the 0.05 level of significance and sample sizes of 20. What size difference between means do we have a 90% chance of detecting?

Using Equation 9.25 and the sample variance of Example 9.4, we calculate:

$$\delta = \sqrt{\frac{2(0.5193)}{20}}(t_{0.05(2),38} + t_{0.10(1),38})$$

$$= (0.2279)(2.024 + 1.304) = 0.76 \text{ min.}$$

Power of the Test. Further rearrangement of Equation 9.24 results in

$$t_{\beta(1),\nu} \geq \frac{\delta}{\sqrt{\frac{2s_p^2}{n}}} - t_{\alpha,\nu}, \tag{9.26}$$

which is analogous to Equation 8.10 in Section 8.6. On computing $t_{\beta(1),\nu}$, one can consult Table B.3 to determine $\beta(1)$, whereupon $1 - \beta(1)$ is the power. But this generally will only result in declaring a range of power (e.g., $0.75 < \text{power} < 0.90$). We may, with only slight overestimation of power (as noted in the footnote in Section 8.6) consider $t_{\beta(1)}$ to be approximated by a normal deviate and may thus employ Table B.2.

The above procedure for estimating power is demonstrated in Example 9.9, along with the following method (which is preferable when performing two-tailed testing at $\alpha = 0.01$ or $\alpha = 0.05$ and will be expanded on in the chapters on analysis of variance). We calculate

$$\phi = \sqrt{\frac{n\delta^2}{4s_p^2}} \tag{9.27}$$

Example 9.9

Estimation of the power of a two-sample t test.

What would have been the probability of detecting a true differencet of 1.0 min between mean blood-clotting times of persons using the two drugs of Example 9.4, if $n_1 = n_2 = 15$, and $\alpha(2) = 0.05$?

For $n = 15$, $\nu = 2(n - 1) = 28$, and $t_{0.05(2),28} = 2.048$. Using Equation 9.26:

$$t_{\beta(1),28} \leq \frac{1.0}{\sqrt{\frac{2(0.5193)}{15}}} - 2.048 = 1.752$$

Consulting Table B.3, we see that, for one-tailed probabilities and $\nu = 28$: $0.025 < P(t \geq 1.752) < 0.05$, so $0.025 < \beta < 0.05$.

Power $= 1 - \beta$, so $0.95 < \text{power} < 0.975$.

Or, by the normal approximation, we can estimate β by $P(Z \geq 1.752) = 0.04$. So power $= 0.96$.

To use Appendix Fig. B.1, we calculate

$$\phi = \sqrt{\frac{n\delta^2}{4s_p^2}} = \sqrt{\frac{(15)(1.0)}{4(0.5193)}} = 2.69.$$

In Fig. B.1, we find that $\phi = 2.69$ and $\nu = 28$ are associated with a power of about 0.96.

(from Kirk, 1968: 107–110), and ϕ (lowercase Greek phi—pronounced to rhyme with "sky") is then located on the first page of Appendix Fig. B.1, along the lower axis (taking care to distinguish between ϕ's for $\alpha = 0.01$ and $\alpha = 0.05$). Along the top margin of the graph are indicated pooled degrees of freedom, ν, for α of either 0.01 or 0.05 (although the symbol ν_2 is used on the graph for a reason that will be apparent in later chapters). By noting where ϕ vertically intersects the curve for the appropriate ν, one can read across to either the left or right axis to find the estimate of power.

Unequal Sample Sizes. For a given total number of data, $n_1 + n_2$, the two-sample t test has maximum power and robustness when $n_1 = n_2$. However, if $n_1 \neq n_2$, the above procedure for determining minimum detectable difference (Equation 9.25) and power (Equations 9.26 and 9.27) can be performed using the harmonic mean of the two sample sizes (Cohen, 1977: 42):

$$n = \frac{2n_1n_2}{n_1 + n_2}. \tag{9.28}$$

Thus, for example, if $n_1 = 6$ and $n_2 = 7$, then

$$n = \frac{2(6)(7)}{6 + 7} = 6.46.$$

Power of a Performed Test. What we estimated above was the power we would expect from a test performed with specified sample size, n, population variance, σ^2, and minimum detectable difference, δ. This sort of estimation is appropriate prior to collecting data for hypothesis testing. If a hypothesis test is performed, and the null hypothesis rejected, we know the probability of committing a Type I error (e.g., $0.02 < P < 0.05$ in Example 9.4). However, if H_0 is not rejected, we should desire to estimate the probability of having committed a Type II error. This may be done using Fig. B.1 by estimating ϕ to be

$$\phi = \sqrt{\frac{nd^2 - 2s_p^2}{4s_p^2}} \tag{9.29}$$

(from Kirk, 1968: 108), where d is the difference in sample means: $d = \bar{X}_1 - \bar{X}_2$. This computation is demonstrated in Example 9.10, where, even though H_0 was

Example 9.10

Estimation of the power of a two-sample t test after it has been performed.

What was the power of the hypothesis test in Example 9.4?

Since $n_1 \neq n_2$, we must calculate n by Equation 9.28:

$$n = \frac{2(6)(7)}{6 + 7} = 6.46.$$

Then, since $d = -0.99$ and $s_p^2 = 0.5193$,

$$\phi = \sqrt{\frac{nd^2 - 2s_p^2}{4s_p^2}} = \sqrt{\frac{(6.46)(-0.99)^2 - (2)(0.5193)}{(4)(0.5193)}} = 1.60$$

and, by consulting Fig. B.1 for $\nu = 11$, the power is estimated to be 0.55. (That is, the probability of committing a Type II error was 0.45.)

rejected, we might ask what the power of the test was. This procedure can be used only for two-tailed tests, where α is either 0.05 or 0.01 (simply because that is the limitation of the available graphs—Fig. B.1—that are needed).

9.8 NONPARAMETRIC STATISTICAL METHODS

The theory upon which the two-sample t test is based requires that the two sampled populations be normal and have equal variances. Many other common statistical procedures (e.g., analysis of variance and regression, topics to be discussed in later chapters) have similar underlying assumptions. Fortunately, as pointed out in Section 9.4, most of the commonly employed tests are sufficiently robust to allow us to disregard all but severe deviations from the theoretical assumptions.

However, a large body of statistical methods is available that comprises procedures not requiring the estimation of the population variance or mean and not stating hypotheses about parameters. These testing procedures are termed *nonparametric tests*. As these methods also typically do not make any assumptions about the distribution (e.g., normality) of the sampled populations, they are sometimes referred to as *distribution-free tests*.

Nonparametric tests, such as the two-sample testing procedure described in Section 9.9, may be applied in any situation where we would be justified in employing a parametric test, such as the two-sample t test, as well as in instances when the assumptions of the latter are untenable. However, if either the parametric or nonparametric approach is applicable, then the former will always be more powerful than the latter (i.e., the nonparametric method will have a greater probability of committing a Type II error), although often the difference in power is not great.

It is sometimes declared that only nonparametric testing may be employed when dealing with ordinal scale data, but this is not so. There is nothing in the theoretical basis of parametric hypothesis testing that requires interval or ratio scale data. (It might be argued, however, that a population of ordinal scale data is more likely to deviate unacceptably far from normality than is a population of interval or ratio data.) This point is discussed thoroughly by Anderson (1961), Gaito (1959, 1960), Savage (1957), and Stevens (1968).

Some general textbooks include good introductory coverage of nonparametric statistical methods, and there are a number of modern mathematical monographs on this important branch of statistical procedure. In this book, some nonparametric testing has already been discussed (chi-square and Kolmogorov-Smirnov testing), and others will be considered in this and following chapters.

9.9 TWO-SAMPLE RANK TESTING

Although nonparametric procedures have been proposed for testing differences between the dispersion, or variability, of two populations, none has achieved widespread acceptance. However, nonparametric analogues to the two-sample t test are commonly employed. We shall refer to the test originally proposed by Wilcoxon

(1945) and later enlarged upon by Mann and Whitney (1947). This test procedure is thus called the Wilcoxon-Mann-Whitney test, or, more commonly, the Mann-Whitney test [see Kruskal (1957) for additional history]. Watson's test (Section 25.5) may also be employed when the Mann-Whitney test is applicable, but the latter is easier to perform.

The Mann-Whitney Test. For this test, as for many other nonparametric procedures, the actual measurements are not employed, but we use instead the ranks of the measurements. The data may be ranked either from the highest to lowest or from the lowest to highest values. Example 9.11 ranks the measurements from highest to lowest: The greatest height in either of the two groups is given rank 1, the second greatest height is assigned rank 2, and so on, with the shortest height being assigned rank N, where

$$N = n_1 + n_2. \tag{9.30}$$

We then calculated the Mann-Whitney statistic,

$$U = n_1 n_2 + \frac{n_1(n_1 + 1)}{2} - R_1, \tag{9.31}$$

Example 9.11

The Mann-Whitney test for nonparametric testing of the two-tailed null hypothesis that there is no difference between the heights of male and female students.

H_0: Male and female students are the same height.
H_A: Male and female students are not the same height.
$\alpha = 0.05$

Heights of males	*Heights of females*	*Rank of male heights*	*Rank of female heights*
193 cm	175 cm	1	7
188	173	2	8
185	168	3	10
183	165	4	11
180	163	5	12
178		6	
170		9	
$n_1 = 7$	$n_2 = 5$	$R_1 = 30$	$R_2 = 48$

$$\begin{aligned} U &= n_1 n_2 + \frac{n_1(n_1 + 1)}{2} - R_1 \\ &= (7)(5) + \frac{(7)(8)}{2} - 30 \\ &= 35 + 28 - 30 \\ &= 33 \\ U' &= n_1 n_2 - U \\ &= (7)(5) - 33 \\ &= 2 \end{aligned}$$

$U_{0.05(2),7,5} = U_{0.05(2),5,7} = 30$
Since $33 > 30$, H_0 is rejected.
$0.01 < P(U \geq 33 \text{ or } U \leq 2) < 0.02$

where n_1 and n_2 are the number of observations in samples 1 and 2, respectively, and R_1 is the sum of the ranks of the observations in sample 1. For the two-tailed hypotheses, H_0: male and female students are the same height and H_A: male and female students are not the same height, the calculated U is compared with the two-tailed value of $U_{\alpha(2),n_1,n_2}$ found in Table B.10. This table assumes that $n_1 < n_2$; if $n_1 > n_2$, simply use $U_{\alpha(2),n_2,n_1}$ as the critical value. The Mann-Whitney statistic can also be calculated as

$$U' = n_2 n_1 + \frac{n_2(n_2 + 1)}{2} - R_2, \qquad (9.32)$$

(where R_2 is the sum of the ranks of the observations in sample 2), because the labelling of the two samples as 1 and 2 is purely arbitrary. For a two-tailed test we must compute both U and U', and the larger of the two is compared to the critical value, $U_{\alpha(2),n_1,n_2}$. If Equation 9.31 has been used to calculate U, then U' can be found quickly as

$$U' = n_1 n_2 - U. \qquad (9.33)$$

If Equation 9.32 has been used to compute U', then U is obtainable as

$$U = n_1 n_2 - U'. \qquad (9.34)$$

Then, if either U or U' is as great as or greater than $U_{\alpha(2),n_1,n_2}$, H_0 is rejected at the α level of significance. Note that neither parameters nor parameter estimates are employed in the statistical hypotheses or in the calculations of U or U'.

We may assign ranks either from large to small data (as in Example 9.11), or from small to large, calling the smallest datum rank 1, the next largest rank 2, and so on. The value of U obtained using one ranking procedure will be the same as the value of U' using the other procedure. Since for a two-tailed test both U and U' are employed, it makes no difference from which direction the ranks are assigned.

In summary, we note that after ranking the data of the two samples, we calculate U and U' using either Equations 9.31 and 9.33, which requires the determination of R_1, or Equations 9.32 and 9.34, which requires R_2. That is, the sum of the ranks for only one sum of the samples is needed. However, we may wish to compute both R_1 and R_2 in order to perform the following check on the assignment of ranks (which is especially desirable in the somewhat more complex case of assigning ranks to tied data, as will be shown below):

$$R_1 + R_2 = \frac{N(N + 1)}{2}. \qquad (9.35)$$

Thus, in Example 9.11,

$$R_1 + R_2 = 30 + 48 = 78$$

should equal

$$\frac{N(N + 1)}{2} = \frac{12(12 + 1)}{2} = 78.$$

This provides a check on (although it does not guarantee the accuracy of) the assignment of ranks.

If the underlying assumptions of the parametric counterpart of a nonparametric test are met, then the parametric procedure will be the more powerful. The Mann-Whitney test is one of the most powerful of nonparametric tests, however; when either the Mann-Whitney test or the two-sample t test is applicable, the former is about 95% as powerful as the latter (Mood, 1954); and when the assumptions of the t test are seriously violated, the Mann-Whitney test may be very much more powerful (Hodges and Lehman, 1956).

The Mann-Whitney Test with Tied Ranks. Example 9.12 demonstrates an important consideration encountered in tests requiring the ranking of observations. When two or more observations have exactly the same value, they are said to be *tied*. The rank assigned to each of the tied ranks is the mean of the ranks that would have been assigned to these ranks had they not been tied. For example, in the present set of data, which are ranked from low to high, the third and fourth lowest values are tied at 32 words per minute, so they are each assigned the rank of $(3 + 4)/2 = 3.5$. The eighth, ninth, and tenth observations are tied at 44 words per minute, so each of them receives the rank of $(8 + 9 + 10)/3 = 9$. Once the ranks have been assigned by this procedure, U and U' are calculated as previously described.

Example 9.12

The one-tailed Mann-Whitney test used to determine the effectiveness of high school training on the typing speed of college students. This example also demonstrates the assignment of ranks to tied data.

H_0: Typing speed is not greater in college students having had high school typing training.
H_A: Typing speed is greater in college students having had high school typing training.
$\alpha = 0.05$

Typing Speed (words per minute)

With training (*rank in parentheses*)	*Without training* (*rank in parentheses*)
44 (9)	32 (3.5)
48 (12)	40 (7)
36 (6)	44 (9)
32 (3.5)	44 (9)
51 (13)	34 (5)
45 (11)	30 (2)
54 (14)	26 (1)
56 (15)	
$n_1 = 8$	$n_2 = 7$
$R_1 = 83.5$	$R_2 = 36.5$

Since ranking was done from low to high and the alternate hypothesis states that the data of group 1 are larger than the data of group 2, use U' as the test statistic (as indicated in Table 9.1).

$$U' = n_2 n_1 + \frac{n_2(n_2 + 1)}{2} - R_2$$

$$= (7)(8) + \frac{(7)(8)}{2} - 36.5$$

$$= 56 + 28 - 36.5$$

$$= 47.5$$

$$U_{0.05(1),8,7} = U_{0.05(1),7,8} = 43$$

Since $47.5 > 43$, reject H_0.

$0.01 < P < 0.025$

The One-Tailed Mann-Whitney Test. For one-tailed hypotheses we need to declare which tail of the Mann-Whitney distribution is of interest, as this will determine whether U or U' must be calculated. This consideration is presented in Table 9.1. In Example 9.12 we have data that were ranked from lowest to highest and the alternate hypothesis states that the data in group 1 are greater in magnitude than those in group 2. Therefore, we need to compute U' and compare it to the one-tailed critical value, $U_{\alpha(1),n_1,n_2}$, from Table B.10.

TABLE 9.1 THE APPROPRIATE TEST STATISTIC FOR THE ONE-TAILED MANN-WHITNEY TEST

	H_0: *Group 1* $\geqq$ *Group 2* H_A: *Group 1* $<$ *Group 2*	H_0: *Group 1* $\leq$ *Group 2* H_A: *Group 1* $>$ *Group 2*
Ranking done from low to high	U	U'
Ranking done from high to low	U'	U

The Normal Approximation to the Mann-Whitney Test. Note that Table B.10 can be used only if the size of the smaller sample does not exceed 20 and the size of the larger sample does not exceed 40. Fortunately, the distribution of U approaches the normal distribution for larger samples. For large n_1 and n_2 we use the fact that the U distribution has a mean of

$$\mu_U = \frac{n_1 n_2}{2} \tag{9.36}$$

and a standard error of

$$\sigma_U = \sqrt{\frac{n_1 n_2 (N+1)}{12}}, \tag{9.37}$$

where $N = n_1 + n_2$, as used earlier. Thus, if a U, or a U', is calculated from data where either n_1 or n_2 is greater than that in Table B.10, its significance can be determined by computing

$$Z = \frac{|U - \mu_U| - 0.5}{\sigma_U}. \tag{9.38}$$

(The 0.5 in the numerator is called a correction for continuity. It is desirable in order to account for the fact that Z is a continuous distribution, but U is a discrete distribution.)

Recalling that the t distribution with $\nu = \infty$ is identical to the normal distribution, the critical value, Z_α, is equal to the critical value, $t_{\alpha,\infty}$. For example, if the hypotheses of Example 9.11 were being tested with sample sizes of 22 and 46, and if U were computed to be 282, we could then calculate

$$\mu_U = \frac{(22)(46)}{2} = 506,$$

$$\sigma_U = \sqrt{\frac{(22)(46)(68+1)}{12}} = 76.28,$$

and

$$Z = \frac{|282 - 506| - 0.5}{76.28} = \frac{223.5}{76.28} = 2.93.$$

For a two-tailed test at $\alpha = 0.05$, $t_{0.05(2),\infty} = 1.9600$. Since $Z > t_{0.05(2),\infty}$ (i.e., $2.93 > 1.9600$), we reject the null hypothesis that male and female student heights are the same in the sampled populations. When using the normal approximation for two-tailed testing, only U *or* U' (not both) need be calculated. If U' is computed instead of U, then U' is simply substituted for U in Equation 9.38, the rest of the testing procedure remaining the same.

One-tailed testing may also be performed using the normal approximation. Here one computes either U or U', in accordance with Table 9.1 and uses it in either Equation 9.39 or 9.40, respectively:

$$Z = \frac{U - \mu_U - 0.5}{\sigma_U}; \tag{9.39}$$

$$Z = \frac{U' - \mu_U - 0.5}{\sigma_U}. \tag{9.40}$$

The resultant Z is then compared to the one-tailed critical value, $Z_{\alpha(1)}$, or, equivalently, $t_{\alpha(1),\infty}$; and if $Z \geq$ the critical value, then Z is rejected.*

If tied ranks exist and the normal approximation is utilized, the computations are slightly modified as follows. One should calculate the quantity

$$\sum T = \sum (t_i^3 - t_i), \tag{9.41}$$

where t_i is the number of ties in a group of tied values, and the summation is performed over all groups of ties. Then,

$$\sigma_U = \sqrt{\frac{n_1 n_2}{N^2 - N} \cdot \frac{N^3 - N - \sum T}{12}}, \tag{9.42}$$

and this value is used in place of that from Equation 9.37. The computation of T is demonstrated, in another context, in Example 11.10.

The normal approximation is excellent for $\alpha(2) = 0.10$ or 0.05 [or for $\alpha(1) = 0.05$ and 0.025] and is also good for $\alpha(2) = 0.20$ and 0.02 [or for $\alpha(1) = 0.10$ or 0.01]. But for more extreme significance levels it is not as reliable, especially if n_1 and n_2 are dissimilar. Buckle, Kraft, and van Eeden (1969) propose another distribution, which they refer to as the "uniform approximation." They show it to be a good deal more accurate for $n_1 \neq n_2$, especially when the difference between n_1 and n_2 is great, and especially for small α. Fix and Hodges (1955) describe another approximation that is more accurate than the normal, but the normal remains the easiest to use and is indeed accurate enough for most data encountered.

The Mann-Whitney Test for Ordinal Data. In general, a nonparametric test that is applicable to ratio or interval data may also be used for ordinal data. Example 9.13 demonstrates this procedure. In this example, 25 undergraduate students were enrolled in an invertebrate zoology course. Each student was guided through

*By this procedure, Z must be positive in order to reject H_0. If it is negative, then the probability of H_0 being true is $P > 0.50$.

the course by one of two teaching assistants. On the basis of the final grades, we wish to test the null hypothesis that students perform equally well in the course, under both teaching assistants. The variable measured (i.e., the final examination grades) results in ordinal data, and the hypothesis is amenable to testing by the Mann-Whitney test.

Example 9.13

The Mann-Whitney test for ordinal data.

H_0: The performance of students is the same under the two teaching assistants.
H_A: Students do not perform equally well under the two teaching assistants.
$\alpha = 0.05$.

Teaching Assistant A		Teaching Assistant B	
Grade	*Rank of grade*	*Grade*	*Rank of grade*
A	3	A	3
A	3	A	3
A	3	B+	7.5
A−	6	B+	7.5
B	10	B	10
B	10	B−	12
C+	13.5	C	16.5
C+	13.5	C	16.5
C	16.5	C−	19.5
C	16.5	D	22.5
C−	19.5	D	22.5
		D	22.5
		D	22.5
		D−	25

$n_1 = 11$ $\quad$ $n_2 = 14$
$R_1 = 114.5$ $\quad$ $R_2 = 210.5$

$$\begin{aligned} U &= n_1 n_2 + \frac{n_1(n_1+1)}{2} - R_1 \\ &= (11)(14) + \frac{(11)(12)}{2} - 114.5 \\ &= 154 + 66 - 114.5 \\ &= 105.5 \\ U' &= n_1 n_2 - U \\ &= (11)(14) - 105.5 \\ &= 48.5 \end{aligned}$$

$U_{0.05(2),\,11,\,14} = 114$
Since $105.5 < 114$, do not reject H_0.
$0.10 < P(U \geq 105.5 \text{ or } U \leq 48.5) < 0.20$

Hypotheses Employing a Specified Difference Other Than Zero. Using the two-sample t test, one can examine hypotheses such as H_0: $\mu_1 - \mu_2 = \mu_0$, where μ_0 is not zero. Similarly, the Mann-Whitney test can be applied to hypotheses such as H_0: males are more than 5 cm taller than females (a one-tailed hypothesis with data such as those in Example 9.11) or H_0: the letter grades of students in one course are one grade higher than those of students in a second course (a one-tailed hypothesis with data such as those in Example 9.13). In the first hypothesis, one would list all the male heights but list all the female heights after increasing each of them by 5 cm. Then these listed heights would be ranked and the Mann-Whitney analysis would proceed as usual. For testing the second hypothesis, the letter grades for the students in the first course could be listed unchanged, with the grades for the second course increased by one letter grade before listing. Then all the listed grades would be ranked and subjected to the Mann-Whitney test.

When dealing with ratio or interval scale data, it is also possible to propose hypotheses employing a multiplication, rather than an addition, constant. Consider the hypothesis H_0: the wings of one species of insect are two times the length of the wings of a second species. We could test it by listing the wing lengths of the first species, listing each of the wing lengths of the second species multiplied by two, and then ranking the data and subjecting the ranks to the Mann-Whitney test. (The parametric t testing procedure ordinarily would be inapplicable for such a hypothesis, because multiplying the data by a constant changes the variance of the data by the square of the constant.)

9.10 TESTING FOR DIFFERENCE BETWEEN TWO MEDIANS

One can test the null hypothesis that two samples came from a population having the same median by a method called the *median test* (Mood, 1950: 394–395). The procedure is to determine the grand median for all the data in both samples and then to set up a 2 × 2 contingency table, as shown in Example 9.14. This contingency table can then be analyzed by chi-square (as in Section 6.2) or G (Section 6.6), or, if some frequencies are small (see Section 6.5), the Fisher exact test may be applied (Section 22.9). The median test is only about 64% as powerful as the two-sample t

Example 9.14

The two-sample median test, using the data of Example 9.13.

H_0: The two samples came from populations with identical medians (i.e., the median performance is the same under the two teaching assistants).
H_A: The medians of the two sampled populations are not equal.
$\alpha = 0.05$

The median of all N measurements, where $N = 25$, is $X_{(n+1)/2} = X_{12}$ = grade of C+. The following 2 × 2 contingency table is analyzed by chi-square:

Number	*Sample 1*	*Sample 2*	*Total*
Above median	6	6	12
Not above median	5	8	13
Total	11	14	25

$$\chi_c^2 = \frac{n\left(|f_{11}f_{22} - f_{12}f_{21}| - \frac{n}{2}\right)^2}{(C_1)(C_2)(R_1)(R_2)} \qquad \text{[Equation (6.8)]}$$

$= 0.031$

$\chi^2_{0.05,1} = 3.841$

Therefore, do not reject H_0.

$0.75 < P < 0.90$

test when applied to data where the latter is applicable (Mood, 1954), and only about 67% as powerful as the Mann-Whitney test of the preceeding section.

9.11 THE EFFECT OF CODING

Coding the raw data will not alter the test statistics or conclusions of the preceding sections. In the case of calculating F or t values, this is because the numerator and denominator will be unchanged by coding that employs the addition or subtraction of a constant, whereas coding by multiplication or division will change the numerator and denominator proportionately the same. As coding of any sort will not change the ranks of the data, the Mann-Whitney and median tests remain unaffected.

9.12 TWO-SAMPLE TESTING OF NOMINAL SCALE DATA

We may compare two samples of nominal data simply by arranging the data in a $2 \times C$ contingency table and proceeding as described in Chapter 6 and demonstrated in Example 6.1.

9.13 TESTING FOR DIFFERENCE BETWEEN TWO DIVERSITY INDICES

If the Shannon index of diversity, H', is obtained for each of two samples, it may be desired to test the null hypothesis that the diversities of the two sampled populations are equal. Hutcheson (1970) proposes a t test for this purpose:

$$t = \frac{H'_1 - H'_2}{s_{H'_1 - H'_2}}, \tag{9.43}$$

where

$$s_{H'_1 - H'_2} = \sqrt{s^2_{H'_1} + s^2_{H'_2}}. \tag{9.44}$$

The variance of H' may be approximated by

$$s^2_{H'} = \frac{\sum f_i \log^2 f_i - (\sum f_i \log f_i)^2/n}{n^2} \tag{9.45}$$

(Basharin, 1959; Lloyd, Zar, and Karr, 1968),* where s, f_i, and n are as defined in Section 4.6. Logarithms to any base may be utilized for this calculation, but those to base 10 are most commonly employed. The DF associated with the preceding t are approximated by

$$\nu = \frac{(s^2_{H'_1} + s^2_{H'_2})^2}{\frac{(s^2_{H'_1})^2}{n_1} + \frac{(s^2_{H'_2})^2}{n_2}} \tag{9.46}$$

(Hutcheson, 1970).

Example 9.15 demonstrates these computations. If one is faced with many calculations of $s^2_{H'}$, the tables of $f_i \log^2 f_i$ provided by Lloyd, Zar, and Karr (1968) will be

*Bowman *et al.* (1971) give an approximation [their Equation (11b)] that is more accurate for very small n.

Example 9.15

Comparing two indices of diversity.

H_0: The diversity of plant food items in the diet of Michigan blue jays is the same as the diversity of plant food items in the diet of Louisiana blue jays.

H_A: The diversity of plant food items in the diet of Michigan blue jays is not the same as in the diet of Louisiana blue jays.

$\alpha = 0.05$

Michigan Blue Jays

Diet item	f_i	$f_i \log f_i$	$f_i \log^2 f_i$
Oak	47	78.5886	131.4078
Corn	35	54.0424	83.4452
Blackberry	7	5.9157	4.9994
Beech	5	3.4949	2.4429
Cherry	3	1.4314	0.6830
Other	2	0.6021	0.1812

$s_1 = 6$	$n_1 = \sum f_i = 99$	$\sum f_i \log f_i = 144.0751$	$\sum f_i \log^2 f_i = 223.1595$

$$H'_1 = \frac{n \log n - \sum f_i \log f_i}{n} = \frac{197.5679 - 144.0751}{99} = 0.5403$$

$$s^2_{H'_1} = \frac{\sum f_i \log^2 f_i - (\sum f_i \log f_i)^2/n}{n^2} = 0.00137602$$

Louisiana Blue Jays

Diet item	f_i	$f_i \log f_i$	$f_i \log^2 f_i$
Oak	48	80.6996	135.6755
Pine	23	31.3197	42.6489
Grape	11	11.4553	11.9294
Corn	13	14.4813	16.1313
Blueberry	8	7.2247	6.5246
Other	2	0.6021	0.1812

$s_2 = 6$	$n_2 = \sum f_i = 105$	$\sum f_i \log f_i = 145.7827$	$\sum f_i \log^2 f_i = 213.0909$

$$H'_2 = \frac{n \log n - \sum f_i \log f_i}{n} = \frac{212.2249 - 145.7827}{105} = 0.6328$$

$$s^2_{H'_2} = \frac{\sum f_i \log^2 f_i - (\sum f_i \log f_i)^2/n}{n^2} = 0.00096918$$

$$s_{H'_1 - H'_2} = \sqrt{s^2_{H'_1} + s^2_{H'_2}} = \sqrt{0.00137602 + 0.00096918} = 0.0484$$

$$t = \frac{H'_1 - H'_2}{s_{H'_1 - H'_2}} = \frac{-0.0925}{0.0484} = -1.911$$

$$\nu = \frac{(s^2_{H'_1} + s^2_{H'_2})^2}{\frac{(s^2_{H'_1})^2}{n_1} + \frac{(s^2_{H'_2})^2}{n_2}} = \frac{(0.00137602 + 0.00096918)^2}{\frac{(0.00137602)^2}{99} + \frac{(0.00096918)^2}{105}}$$

$$= \frac{0.000005499963}{0.000000028071} = 196$$

$t_{0.05(2),196} = 1.972$

Therefore, do not reject H_0.

$0.05 < P < 0.10$

helpful. One-tailed as well as two-tailed hypotheses may be tested by this procedure. Also, the population diversity indices may be hypothesized to differ by some value, μ_0, other than zero, in which case the numerator of t would be $|H'_1 - H'_2| - \mu_0$.

EXERCISES

9.1. A sample of 29 plant heights of members of a certain species had $s^2 = 14.62\ \text{cm}^2$, and the heights of a sample of 25 from a second species had $s^2 = 8.45\ \text{cm}^2$. Test the null hypothesis that the variances of the two sampled populations are the same.

9.2. If $s_1^2 = 324.46\ \text{sec}^2$, $s_2^2 = 158.95\ \text{sec}^2$, $n_1 = 41$, and $n_2 = 36$, test the hypotheses H_0: $\sigma_1^2 \leq \sigma_2^2$ and H_A: $\sigma_1^2 > \sigma_2^2$.

9.3. Using the following data, test the null hypothesis that male and female turtles have the same mean serum cholesterol concentrations.

Serum Cholesterol (mg/100 ml)

Male	*Female*
220.1	223.4
218.6	221.5
229.6	230.2
228.8	224.3
222.0	223.8
224.1	230.8
226.5	

9.4. It is hypothesized that animals with a northerly distribution have shorter appendages than animals from a southerly distribution. Test this hypothesis (by computing t), using the following wing length data for birds (data are in millimeters).

Northern	*Southern*
120	116
113	117
125	121
118	114
116	116
114	118
119	123
	120

9.5. If $\bar{X}_1 = 4.6$ kg, $s_1^2 = 3.88\ \text{kg}^2$, $n_1 = 18$, $\bar{X}_2 = 6.0$ kg, $s_2^2 = 4.35\ \text{kg}^2$, and $n_2 = 26$, test the hypotheses H_0: $\mu_1 \geq \mu_2$ and H_A: $\mu_1 < \mu_2$.

9.6. If $\bar{X}_1 = 334.6$ g, $\bar{X}_2 = 349.8$ g, $SS_1 = 364.34\ \text{g}^2$, $SS_2 = 286.78\ \text{g}^2$, $n_1 = 19$, and $n_2 = 24$, test the hypothesis that the mean weight of population 2 is more than 10 g greater than the mean weight of population 1.

9.7. If the null hypothesis in Exercise 9.3 is rejected, compute the 95% confidence limits for μ_1, μ_2, and $\mu_1 - \mu_2$. If H_0 is not rejected, calculate the 95% confidence limits for the common population mean.

9.8. A sample is to be taken from each of two populations from which previous samples of size 14 have had $SS_1 = 244.66\ (\text{km/hr})^2$ and $SS_2 = 289.18\ (\text{km/hr})^2$. What size sample should be taken from each population in order to estimate $\mu_1 - \mu_2$ to within 2.0 km/hr, with 95% confidence?

9.9. Consider the populations described in Exercise 9.8.

(a) How large a sample should we take from each population if we wish to detect a difference between μ_1 and μ_2 of at least 5.0 km/hr, using a 5% significance level and a t test with 90% power?

(b) If we take a sample of 20 from one population and 22 from the other, what is the smallest difference between μ_1 and μ_2 that we have a 90% probability of detecting with a t test using $\alpha = 0.05$?

(c) If $n_1 = n_2 = 50$, and $\alpha = 0.05$, what is the probability of rejecting $H_0: \mu_1 = \mu_2$ when $\mu_1 - \mu_2$ is as small as 2.0 km/hr?

9.10. Using the Mann-Whitney test, test the appropriate hypotheses for the data in Exercise 9.3.

9.11. Using the Mann-Whitney procedure, test the appropriate hypotheses for the data in Exercise 9.4.

9.12. The following data are volumes (in cubic microns) of avian erythrocytes taken from normal (diploid) and intersex (triploid) individuals. Test the hypothesis (using the Mann-Whitney test) that the volume of intersex cells is 1.5 times the volume of normal cells.

Normal	*Intersex*
248	380
236	391
269	377
254	392
249	398
251	374
260	
245	
239	
255	

9.13. The following data are the frequencies of plant species being chosen for food by each of two species of insects. Test the null hypothesis that both insect species have the same food preference.

	Food 1	*Food 2*	*Food 3*	*Food 4*
Insect Species 1	13	21	5	8
Insect Species 2	8	16	12	20

10 Paired-Sample Hypotheses

The two-sample testing procedures discussed in Chapter 9 apply when the two samples are independent, independence implying that each datum in one sample is in no way associated with any specific datum in the other sample. However, there are instances when each observation in sample 1 is in some way correlated with an observation in sample 2, so that the data may be said to occur in pairs.

For example, a mammalogist might wish to test the null hypothesis that the left foreleg and left hindleg lengths of deer are equal. He could make these two measurements on a number of deer, but he would have to remember that the variation among the data might be owing to two possible factors. First, the null hypothesis might be false, there being, in fact, a difference between foreleg and hindleg length. Second, deer are of different sizes, and for each deer the hindleg length is correlated with the foreleg length (i.e., a deer with a large front leg is likely to have a large hind leg). Thus, as Example 10.1 shows, the data can be tabulated in pairs, one pair (i.e., one hindleg measurement and one foreleg measurement) per animal.

10.1 THE PAIRED-SAMPLE *t* TEST

The two-tailed hypotheses implied by Example 10.1 are H_0: $\mu_1 - \mu_2 = 0$ and H_A: $\mu_1 - \mu_2 \neq 0$ (which, as pointed out in Section 9.4, could also be stated H_0: $\mu_1 = \mu_2$ and H_A: $\mu_1 \neq \mu_2$). However, we can define a mean population difference, μ_d, as $\mu_1 - \mu_2$, and write the hypotheses as H_0: $\mu_d = 0$ and H_A: $\mu_d \neq 0$. Although the use of either μ_d or $\mu_1 - \mu_2$ is correct, the former will be used hereafter when it implies the paired-sample situation.

Example 10.1

The two-tailed paired-sample t test.

H_0: $\mu_d = 0$.
H_A: $\mu_d \neq 0$.
$\alpha = 0.05$

Deer (j)	*Hindleg length* (cm) (X_{1j})	*Foreleg length* (cm) (X_{2j})	*Difference* (cm) ($d_j = X_{1j} - X_{2j}$)
1	142	138	4
2	140	136	4
3	144	147	−3
4	144	139	5
5	142	143	−1
6	146	141	5
7	149	143	6
8	150	145	5
9	142	136	6
10	148	146	2

$n = 10$ $\qquad \bar{d} = 3.3$ cm
$\nu = n - 1 = 9$ $\qquad s_d^2 = 9.3444$ cm²
$s_{\bar{d}} = 0.97$ cm

$$t = \frac{\bar{d}}{s_{\bar{d}}} = \frac{3.3}{0.97} = 3.402$$

$t_{0.05(2),9} = 2.262$
Therefore, reject H_0.
$0.005 < P(|t| \geq 3.402) < 0.01$

The test statistic for the null hypothesis is

$$t = \frac{\bar{d}}{s_{\bar{d}}}. \tag{10.1}$$

Therefore, we do not use the original measurements for the two samples, but only the difference within each pair of measurements. One deals, then, with a sample of d_i values, whose mean is $\bar{d}$ and whose variance, standard deviation, and standard error are denoted as s_d^2, s_d, and $s_{\bar{d}}$, respectively. Thus, the *paired-sample t test*, as this procedure may be called, is essentially a one-sample t test, analogous to that described in Sections 8.1 and 8.2. In the paired-sample t test, n is the number of differences (i.e., the number of pairs of data), and $\nu = n - 1$. Note that the hypotheses used in Example 10.1 are special cases of the general hypotheses H_0: $\mu_d = \mu_0$ and H_A: $\mu_d \neq \mu_0$, where μ_0 is often, but not always, zero.

For one-tailed hypotheses with paired samples, one can test either H_0: $\mu_d \geq \mu_0$ and H_A: $\mu_d < \mu_0$, or H_0: $\mu_d \leq \mu_0$ and H_A: $\mu_d > \mu_0$, depending on the question to be asked. Example 10.2 presents data from an experiment designed to test whether a new fertilizer results in an increase of more than 250 kg/ha in crop yield over the old fertilizer. For testing this hypothesis, 18 test plots of the crop were set up. It is probably unlikely to find 18 field plots having exactly the same conditions of soil, moisture, wind, etc., but it should be possible to set up two plots with similar environmental conditions. If so, then the experimenter would be wise to set up 9 pairs of plots,

Example 10.2
A one-tailed paired-sample t test.

H_0: $\mu_d \leq 250$ kg/ha
H_A: $\mu_d > 250$ kg/ha
$\alpha = 0.05$

Crop Yield (kg/ha)

Plot (j)	With new fertilizer (X_{1j})	With old fertilizer (X_{2j})	d_j
1	2250	1920	330
2	2410	2020	390
3	2260	2060	200
4	2200	1960	240
5	2360	1960	400
6	2320	2140	180
7	2240	1980	260
8	2300	1940	360
9	2090	1790	300

$n = 9$ $\quad\quad \bar{d} = 295.6$ kg/ha
$\nu = n - 1 = 8$ $\quad\quad s_d = 80.6$ kg/ha
$s_{\bar{d}} = 26.9$ kg/ha

$$t = \frac{\bar{d} - 250}{s_{\bar{d}}} = 1.695$$

$t_{0.05(1),8} = 1.860$
Therefore, do not reject H_0
$0.05 < P < 0.10$

applying the new fertilizer to one plot of each pair and the old fertilizer to the other plot of that pair. As Example 10.2 shows, the statistical hypotheses to be tested are H_0: $\mu_d \leq 250$ kg/ha and H_A: $\mu_d > 250$ kg/ha.

The paired-sample t test does not have the normality and equality of variances assumptions of the two-sample t test, but assumes only that the differences, d_i, come from a normally distributed population of differences. If there is, in fact, pairwise correlation of data from the two samples, then the paired-sample t test will be more powerful than the two-sample t test. If no such correlation exists, then the two-sample t test will be the more powerful procedure. (If the data from Example 10.1 were subjected (inappropriately) to the two-sample t test, rather than to the paired-sample t test, a difference would not have been concluded, and a Type II error would have been committed.)

10.2 CONFIDENCE LIMITS FOR THE POPULATION MEAN DIFFERENCE

In paired-sample testing we deal with a sample of differences, d_i, so confidence limits for the mean of a population of differences, μ_d, may be determined as in Section 8.3. In the manner of Equation 8.6, the $1 - \alpha$ confidence interval for μ_d is

$$\bar{d} \pm t_{\alpha(2),\nu} s_{\bar{d}}. \tag{10.2}$$

For example, for the data in Example 10.1, we can compute the 95% confidence interval for μ_d to be 3.3 cm $\pm$ (2.262)(0.97 cm) = 3.3 cm $\pm$ 2.2 cm; the 95% confidence limits are L_1 = 1.1 cm and L_2 = 5.5 cm.

Furthermore, we may ask, as in Section 8.5, how large a sample is required to be $1 - \alpha$ confident in estimating μ_d to within $\pm d$ (using Equation 8.7).

10.3 POWER AND SAMPLE SIZE IN PAIRED-SAMPLE TESTING OF MEANS

By considering the paired-sample t test to be a one-sample t test for a sample of differences, d_i, we may employ the procedures of Section 8.6 to address questions of required sample size, minimum detectable difference, and power (Equations 8.8, 8.9, and 8.10, respectively). For this purpose we simply substitute $\bar{d}$ for $\bar{X}$, and s_d^2 for s^2.

10.4 PAIRED-SAMPLE TESTING BY RANKS

The *Wilcoxon paired-sample test* (Wilcoxon, 1945; Wilcoxon and Wilcox, 1964: 9) is a nonparametric analogue to the paired-sample t test, just as the Mann-Whitney test is a nonparametric procedure analogous to the two-sample t test. The literature refers to the test by a variety of names, but usually in conjunction with Wilcoxon's name* and some wording such as "paired sample" or "matched pairs," sometimes together with a phrase like "rank sum" or "signed rank."

Whenever the paired-sample t test is applicable, the Wilcoxon paired-sample test is also applicable. If, however, the d_i values are from a normal distribution, then the latter test has only 95% of the power in detecting differences as the former (Mood, 1954). However, there are instances when the Wilcoxon paired-sample test is applicable and the parametric paired-sample t test is not, as when one can not assume that the d_i's are from a normal distribution. Section 8.9 introduced the Wilcoxon procedure as a nonparametric one-sample test, but its greatest application is for paired-sample testing. The sign test (Section 22.6) could also be used for this purpose, but it is considerably less powerful.

Example 10.3 demonstrates the use of the Wilcoxon paired-sample test with the ratio scale data of Example 10.1, although it is commonly used with ordinal scale data. The testing procedure involves the calculation of differences, as does the paired-sample t test. Then one ranks the absolute values of the differences, from low to high, and affixes the sign of each difference to the corresponding rank. As introduced in Section 9.9, the rank assigned to tied observations is the mean of the ranks that would have been assigned to the observations had they not been tied.

Then we sum the ranks with a plus sign (we shall call this sum T_+) and the ranks with a minus sign (calling this sum T_-). For a two-tailed test (as in Example 10.4), we reject H_0 if either T_+ or T_- *is less than or equal to* the critical value, $T_{\alpha(2),n}$, from Table B.11.

*Frank Wilcoxon (1892–1965), American chemist and statistician.

Example 10.3
The Wilcoxon paired-sample test applied to the data of Example 10.1

H_0: Deer hindleg length is the same as foreleg length.
H_A: Deer hindleg length is not the same as foreleg length.
$\alpha = 0.05$

Deer (j)	*Hindleg length* (cm) (X_{1j})	*Foreleg length* (cm) (X_{2j})	*Difference* ($d_j = X_{1j} - X_{2j}$)	*Rank of* $\lvert d_j \rvert$	*Signed rank of* $\lvert d_j \rvert$
1	142	138	4	4.5	4.5
2	140	136	4	4.5	4.5
3	144	147	−3	3	−3
4	144	139	5	7	7
5	142	143	−1	1	−1
6	146	141	5	7	7
7	149	143	6	9.5	9.5
8	150	145	5	7	7
9	142	136	6	9.5	9.5
10	148	146	2	2	2

$n = 10$
$T_+ = 4.5 + 4.5 + 7 + 7 + 9.5 + 7 + 9.5 + 2 = 51$
$T_- = 3 + 1 = 4$
$T_{0.05(2),10} = 8$
Since $T_- < T_{0.05(2),10}$, H_0 is rejected.
$0.01 < P(T_- \text{ or } T_+ \leq 4) < 0.02$

Having calculated either T_+ or T_-, the other can be determined as

$$T_- = \frac{n(n+1)}{2} - T_+ \tag{10.3}$$

or

$$T_+ = \frac{n(n+1)}{2} - T_-. \tag{10.4}$$

For one-tailed testing we use one-tailed critical values from Table B.11 and either T_+ or T_- as follows. For the hypotheses

H_0: Measurements in population 1 $\leq$ measurements in population 2

and

H_A: Measurements in population 1 $>$ measurements in population 2

H_0 is rejected if $T_- \leq T_{\alpha(1),n}$. For the opposite hypotheses:

H_0: Measurements in population 1 $\geq$ measurements in population 2

and

H_A: Measurements in population 1 $<$ measurements in population 2

reject H_0 if $T_+ \leq T_{\alpha(1),n}$.

For the one-sample median test of Section 8.9, the two-tailed H_0: $M = M_0$ is rejected if favor of H_A: $M \neq M_0$ if either T_+ or $T_- \leq T_{\alpha(2),n}$. For H_0: $M \leq M_0$ vs.

H_A: $M > M_0$, reject H_0 if $T_- \leq T_{\alpha(1),n}$; and for H_0: $M \geq M_0$ vs. H_A: $M < M_0$, H_0 is rejected if $T_+ \leq T_{\alpha(1),n}$.

One will obtain a different value of T_+ (call it T'_+) or T_- (call it T'_-) if rank 1 is assigned to the largest, rather than the smallest, d_i (i.e., if the absolute values of the d_i's are ranked from high to low). If this is done, the test statistics are obtainable as

$$T_+ = m(n + 1) - T'_+ \tag{10.5}$$

and

$$T_- = m(n + 1) - T'_-, \tag{10.6}$$

where m is the number of ranks with the sign being considered.

Differences of zero are ignored in the analysis (i.e., they are discarded). Pratt (1959) recommended maintaining differences of zero until after ranking, and thereafter ignoring the ranks assigned to the zeros. This procedure may yield slightly better results in some circumstances, though worse results in others (Conover, 1973). If used, then the critical values of Rahe (1974) should be consulted or the normal approximation employed (see the following section) instead of using critical values of T from Table B.11.

If data are paired, the use of the Mann-Whitney test, instead of the Wilcoxon paired-sample test, may lead to the commission of a Type II error, with the concomitant inability to detect actual population differences.

The Wilcoxon paired-sample test has an underlying assumption that the sampled population is symmetrical about the median. Another, but less powerful, nonparametric test for paired samples is the sign test (described in Section 22.6), which does not have this assumption.

The Normal Approximation to the Wilcoxon Paired-Sample Test. For data consisting of more than 100 pairs (the limit of Table B.11), the significance of T (where either T_+ or T_- may be used for T) may be determined by considering that for such large samples the distribution of T is closely approximated by a normal distribution with a mean of

$$\mu_T = \frac{n(n + 1)}{4} \tag{10.7}$$

and a standard error of

$$\sigma_T = \sqrt{\frac{n(n + 1)(2n + 1)}{24}}. \tag{10.8}$$

Thus, we can calculate

$$Z = \frac{|T - \mu_T| - 0.5}{\sigma_T}, \tag{10.9}$$

where for T we may use, with identical results, either T_+ or T_-. (The 0.5 is a correction for continuity, which Claypool and Holbert (1974) found desirable for $\alpha > 0.035$.) Then, for a two-tailed test, Z is compared to the critical value, $Z_{\alpha(2)}$, or, equivalently, $t_{\alpha(2),\infty}$ (which for $\alpha = 0.05$ is 1.9600); if Z is greater than or equal to $Z_{\alpha(2)}$, then H_0 is rejected. The normal approximation is an excellent one, except at very small α, as shown at the end of Table B.11. If there are tied ranks, then use

$$\sigma_T = \sqrt{\frac{n(n+1)(2n+1) - \frac{\sum T}{2}}{24}}, \tag{10.10}$$

where

$$\sum T = \sum (t_i^3 - t_i) \tag{10.11}$$

is the correction for ties introduced in using the normal approximation to the Mann-Whitney test (Equation 9.41), applied here to ties of nonzero differences.

If we employ the Pratt procedure for handling differences of zero (described above), then the normal approximation is

$$Z = \frac{\left| T - \frac{n(n+1) - m'(m'+1)}{4} \right| - 0.5}{\sqrt{\frac{n(n+1)(2n+1) - m'(m'+1)(2m+1) - \frac{\sum T}{2}}{24}}} \tag{10.12}$$

(Cureton, 1967), where n is the total number of differences (including zero differences), and m' is the number of zero differences; $\sum T$ is as in Equation 10.11, applied to ties other than those of zero differences. We calculate T_+ or T_- by including the zero differences in the ranking and then deleting from consideration both the zero d_i's and the ranks assigned to them. For T in Equation 10.12, either T_+ or T_- may be used. If neither tied ranks nor zero d_i's are present, then Equation 10.12 becomes Equation 10.9.

One-tailed testing may also be performed using the normal approximation (Equation 10.9) or Cureton's procedure (Equation 10.12). The calculated Z is compared to $Z_{\alpha(1)}$ (or $t_{\alpha(1),\infty}$), and the direction of the arrow in the alternate hypothesis must be examined. If the arrow points to the left ("<"), then H_0 is rejected if $Z \geq Z_{\alpha(1)}$ *and* $T_+ < T_-$; if it points to the right (">"), then reject H_0 if $Z \geq Z_{\alpha(1)}$ *and* $T_+ > T_-$.

10.5 CONFIDENCE LIMITS FOR THE POPULATION MEDIAN DIFFERENCE

In Section 10.2 confidence limits were obtained for the mean of a population of differences. Given a population of differences, one can also determine confidence limits for the population median. This is done exactly as shown in Section 8.8; simply consider the observed differences between members of pairs (d_i) as a sample from a population of such differences.

10.6 PAIRED-SAMPLE TESTING OF NOMINAL SCALE DATA

Data in a 2 × 2 Table. Enumeration data may be collected from paired samples. If the data are dichotomous (i.e., the nominal scale variable has two possible values), then *McNemar's test* (McNemar, 1947) is applicable.

For example, assume that we wish to test whether two skin lotions are equally effective at relieving a poison ivy rash. The two lotions might be tested on 51 persons with poison ivy rash on both arms, applying one lotion to one arm and the other lotion to the second arm. The results of the experiment can be summarized in a 2×2 table, as shown in Example 10.4. (As introduced in Sections 6.1 and 6.2, f_{ij} denotes frequency observed in row i and column j of this table which has 2 rows and 2 columns.) Eleven of the patients responded to both lotions and 24 responded to neither lotion (i.e., $f_{11} = 11$ and $f_{22} = 24$). The null hypothesis states that the same proportion of persons receive relief with lotion 1 as with lotion 2; this is to say that the sample proportion $(f_{11} + f_{12})/n$ is an estimate of the same population proportion as is $(f_{11} + f_{21})/n$. H_0 is tested by considering a goodness of fit test to a 1:1 ratio, using the observed frequencies f_{12} and f_{21}, as shown in Example 10.6. As long as n is not less than 10, chi-square goodness of fit testing may be employed (Section 5.1); otherwise the binomial test (Section 22.5) is called for. For the chi-square goodness of fit to a 1:1 ratio, $\hat{f}_{12} = \hat{f}_{21} = (f_{12} + f_{21})/2$, and, because there is only one degree of freedom, Yates' correction for continuity usually is applied.

Example 10.4

McNemar's test for paired-sample nominal scale data.

H_0: The proportion of persons experiencing relief is the same with both lotions.
H_A: The proportion of persons experiencing relief is not the same with both lotions.
$\alpha = 0.05$

	Lotion 1	
Lotion 2	*Relief*	*No relief*
Relief	11	6
No relief	10	24

$$\chi_c^2 = \frac{(|f_{12} - f_{21}| - 1)^2}{f_{12} + f_{21}} = \frac{(|6 - 10| - 1)^2}{6 + 10} = 0.562$$

$\chi^2_{0.05,1} = 3.841$

Therefore, do not reject H_0.

$0.25 < P < 0.50$

The chi-square may also be computed as

$$\chi^2 = \frac{(f_{12} - f_{21})^2}{f_{12} + f_{21}}, \qquad (10.13)$$

and the contingency-corrected chi-square as

$$\chi_c^2 = \frac{(|f_{12} - f_{21}| - 1)^2}{f_{12} + f_{21}}, \qquad (10.14)$$

with the latter demonstrated in Example 10.6.

Another common type of data amenable to McNemar testing is the situation where there are two experimental responses recorded before and after some event, in which case one hears reference to this procedure as the McNemar test for significant changes.

Do not confuse this procedure with that of a 2×2 contingency table analysis (Section 6.2). Contingency table data are analyzed using a null hypothesis of independence between rows and columns, whereas in the case of data amenable to the McNemar test, there is intentional association between the row and column data.

Data in Larger Tables. The McNemar test may be extended to square tables larger than 2×2 (Bowker, 1948; Maxwell, 1970). What we test is whether the upper right corner of the table is symmetrical with the lower left corner. This is done by ignoring the data along the diagonal containing f_{ii} (i.e., row 1, column 1; row 2, column 2; etc.). We compute

$$\chi^2 = \sum_{i=1}^{r} \sum_{j>i} \frac{(f_{ij} - f_{ji})^2}{f_{ij} + f_{ji}}, \tag{10.15}$$

with degrees of freedom of

$$\nu = \frac{r(r-1)}{2}, \tag{10.16}$$

where r is the number of rows (or, equivalently, the number of columns) in the table of data. This is demonstrated in Example 10.5. For data in a 2×2 table, Equation 10.15 becomes Equation 10.13, and Equation 10.16 yields $\nu = 1$.

Example 10.5

Extending McNemar's test to a 3×3 table of nominal scale data.

H_0: Of men who adopt a religion different from that of their fathers, a change from religion A to B is as likely as a change from B to A.

H_A: Of men who adopt a religion different from that of their fathers, a change from religion A to B has a likelihood different than does a change from B to A.

Man's Religion	Man's Father's Religion		
	Protestant	*Catholic*	*Jewish*
Protestant	173	20	7
Catholic	15	51	2
Jewish	5	3	24

$r = 3$

$$\chi^2 = \sum_{i=1}^{r} \sum_{j>i} \frac{(f_{ij} - f_{ji})^2}{f_{ij} + f_{ji}}$$

$$= \frac{(20-15)^2}{20+15} + \frac{(7-5)^2}{7+5} + \frac{(2-3)^2}{2+3}$$

$$= 0.7143 + 0.3333 + 0.2000$$

$$= 1.248$$

$$\nu = \frac{r(r-1)}{2} = \frac{3(2)}{2} = 3$$

$\chi^2_{0.05,3} = 7.815$

Do not reject H_0.

$0.50 < P < 0.75$

Testing for Effect of Treatment Order. If two treatments are applied sequentially to a group of subjects, we might ask whether the response to each treatment depended on the order in which the treatments were administered. For example, suppose we have two medications for the treatment of poison ivy rash, but, instead of the situation in Example 10.4, they are to be administered orally rather than by external application to the skin; thus both arms receive a given medication at the same time, and the medications must be given at different times.

Gart (1969a) provides the following procedure to test for the difference in response between two sequentially applied treatments and to test whether the order of application had an effect on the response. The following 2×2 contingency table (see Section 6.2) is set up to test for a treatment effect:

	Order of Application of Treatments A and B		
	A, then B	*B, then A*	*Total*
Response with first treatment	f_{11}	f_{12}	R_1
Response with second treatment	f_{21}	f_{22}	R_2
Total	C_1	C_2	n

By rearranging the above data, the following 2×2 table may be used to test the null hypothesis of no difference in response due to order of treatment application:

	Order of Application of Treatments A and B		
	A, then B	*B, then A*	*Total*
Response with treatment A	f_{11}	f_{12}	R_1
Response with treatment B	f_{21}	f_{22}	R_2
Total	C_1	C_2	n

These two contingency tables may be tested by the chi-square test (Section 6.2) or the Fisher exact test (22.9). (If a one-tailed test is desired, the latter procedure is called for.) Example 10.6 demonstrates Gart's test.

Example 10.6

Gart's test for effect of treatment and treatment order.

H_0: The two oral medications have the same effect on relieving poison ivy rash.
H_A: The two oral medications do not have the same effect on relieving poison ivy rash.

	Order of Application of Medications A and B		
	A, then B	*B, then A*	*Total*
Response with 1st medication	14	8	22
Response with 2nd medication	4	15	19
Total	18	23	41

Using Equation 6.8:

$$\chi_c^2 = \frac{41\left[|(14)(15) - (8)(4)| - \frac{41}{2}\right]^2}{(18)(23)(22)(19)} = 5.877$$

$\chi^2_{0.05,1} = 3.841$; reject H_0.
That is, there is a difference in response to the two medications, regardless of the order in which they are administered.
$0.01 < P < 0.025$

H_0: The order of administration of the two oral medications does not affect their abilities to relieve poison ivy rash.
H_A: The order of administration of the two oral medications does affect their abilities to relieve poison ivy rash.

	Order of Application of Medications A and B		
	A, then B	*B, then A*	*Total*
Response with medication A	14	15	29
Response with medication B	4	8	12
Total	18	23	41

Using Equation 6.8:

$$\chi_c^2 = \frac{41\left[|(14)(8) - (15)(4)| - \frac{41}{2}\right]^2}{(18)(23)(29)(12)} = 0.282$$

$\chi^2_{0.05,1} = 3.841$; do not reject H_0.
That is, the effects of the two medications are not affected by the order in which they are administered.
$0.50 < P < 0.75$

EXERCISES

10.1. Concentrations of nitrogen oxides and of hydrocarbons were determined in a certain urban area (recorded in $\mu g/m^3$).

(a) Test the hypothesis that both classes of air pollutants were present in the same concentration.

Day	*Nitrogen oxides*	*Hydrocarbons*
1	104	108
2	116	118
3	84	89
4	77	71
5	61	66
6	84	83
7	81	88
8	72	76
9	61	68
10	97	96
11	84	81

(b) Calculate the 95% confidence interval for μ_d.

10.2. Using the data of Exercise 10.1, test the appropriate hypotheses with Wilcoxon's paired-sample test.

10.3. One hundred twenty-two pairs of brothers, one member of each pair overweight and the other of normal weight, were examined for presence of varicose veins. Use the McNemar test for the data below to test the hypothesis that there is no relationship between being overweight and developing varicose veins (i.e., that the same proportion of overweight men as normal weight men possess varicose veins). In the following data tabulation, "v.v." stands for "varicose veins."

	Overweight	
Normal Weight	*With v.v.*	*Without v.v.*
With v.v.	19	5
Without v.v.	12	86

$n = 122$

11 Multisample Hypotheses: The Analysis of Variance

When measurements of a variable are obtained for each of two samples, hypotheses such as those described in Chapter 9 are appropriate. However, biologists often collect measurements of a variable from three or more samples, a situation calling for multisample analyses, as introduced in this chapter.

It is tempting to some to attempt the testing of multisample hypotheses by applying two-sample tests to all possible pairs of samples. In this manner, for example, one might proceed to test the null hypothesis $H_0: \mu_1 = \mu_2 = \mu_3$ by testing all the following null hypotheses: $H_0: \mu_1 = \mu_2$, $H_0: \mu_1 = \mu_3$, and $H_0: \mu_2 = \mu_3$. But such a procedure of employing a series of two-sample tests to attack a multisample hypothesis is *invalid*. The calculated test statistic, t, and the critical values we find in the t table, are designed to test whether two sample statistics, $\bar{X}_1$ and $\bar{X}_2$, are likely to have come from the same population (or from two populations with identical means). In employing this test, we could randomly draw two sample means from the same population and wrongly conclude that they are estimates of two different populations' means; but we know that the probability of this error will be no greater than α. However, consider that three random samples were taken from a single population. As previously stated, three possible t tests can be performed, and the probability of wrongly concluding that two of the means estimate different parameters is considerably greater than α. In fact, if α is set at 5% and three means are tested, two at a time, by the two-sample t test, there is a 13% chance of wrongly concluding a difference between the two most extreme means. Using the critical values for t at $\alpha = 5\%$ for two-at-a-time comparisons of 10 means, there is a 63% chance of committing a Type I error; if there are 20 means being so tested, the probability of a Type I error is 92%. As the number of means increases, it approaches certainty that the t test will conclude the two extreme sample means to estimate different values of μ, even though they may have come from the same population (see Table 11.1). Two-sample tests, it must be emphasized, cannot be utilized validly to test multisample hypotheses. The appropriate procedures are introduced in the following sections.

TABLE 11.1 PROBABILITY OF COMMITTING A TYPE I ERROR BY USING MULTIPLE t TESTS TO SEEK DIFFERENCES BETWEEN ALL PAIRS OF k MEANS

	Level of Significance Used in the t Tests					
Number of means (k)	0.20	0.10	0.05	0.02	0.01	0.001
2	0.20	0.10	0.05	0.02	0.01	0.001
3	0.41	0.23	0.13	0.05	0.03	0.003
4	0.58	0.36	0.21	0.09	0.05	0.006
5	0.71	0.47	0.23	0.13	0.07	0.009
10	0.96	0.83	0.63	0.37	0.23	0.034
20	1.00	0.98	0.92	0.71	0.52	0.109
∞	1.00	1.00	1.00	1.00	1.00	1.000

Note: The particular values were derived from a table by Pearson (1942) by assuming equal population variances and large samples.

11.1 SINGLE FACTOR ANALYSIS OF VARIANCE

To test the null hypothesis H_0: $\mu_1 = \mu_2 = \ldots = \mu_k$, where k is the number of experimental groups, or samples, we need to become familiar with the topic of *analysis of variance*, often abbreviated ANOVA (or ANOV or AOV). Analysis of variance is a large area of statistical methods, owing its name and much of its early development to R. A. Fisher;* in fact, the F statistic was named in his honor by G. W. Snedecor† (1934: 15). There are many ramifications of analysis of variance considerations, the most common of which will be discussed in this and subsequent chapters. More complex applications and greater theoretical coverage are to be found in Cochran and Cox (1957), Cox (1958), Scheffé (1959), Guenther (1964), and others. At this point, it may appear strange that a procedure used for testing the equality of means should be named analysis of variance, but the reason for this terminology soon will become apparent.

Let us assume that we wish to test whether four different feeds result in different body weights in pigs. Since we are to test for the effect of only one factor (namely feed type) on the variable in question (namely body weight), the appropriate analysis is termed a single factor (or "single criterion" or "single classification" or "one-way") analysis of variance. Furthermore, each type of feed is said to be a *level* of the factor. The design of this experiment should have each experimental animal being assigned at random to receive one of the four feeds, with approximately equal numbers of pigs receiving each feed. (Although equal sample sizes are not required for the single

*Sir Ronald Alymer Fisher (1890–1962), British statistician.

†George W. Snedecor (1881–1974), American statistician.

factor ANOVA, such a situation is statistically desirable.) Because the pigs are assigned to the feed groups at random, the single factor ANOVA is said to represent a *completely randomized experimental design.*

Example 11.1 shows the weights of 19 pigs subjected to this feed experiment, and the null hypothesis to be tested would be H_0: $\mu_1 = \mu_2 = \mu_3 = \mu_4$. Each datum in the experiment may be uniquely represented by the double subscript notation, where X_{ij} denotes datum j in experimental group i. For example, X_{23} denotes the third pig weight in feed group 2; that is, $X_{23} = 74.0$ kg. Similarly, $X_{34} = 96.5$ kg, $X_{41} = 87.9$ kg, etc. We shall let the mean of group i be denoted by $\bar{X}_i$, and the grand mean of all observations will be designated by $\bar{X}$. Furthermore, n_i will represent the size of sample i, and $N = \sum_{i=1}^{k} n_i$ will be the total number of data in the experiment.

Example 11.1
A single factor analysis of variance (Model I).

Nineteen pigs are assigned at random among four experimental groups. Each group is fed a different diet. The data are pig body weights, in kilograms, after being raised on these diets. We wish to ask whether pig weights are the same for all four diets.

H_0: $\mu_1 = \mu_2 = \mu_3 = \mu_4$.
H_A: The mean weights of pigs on the four diets are not all equal.
$\alpha = 0.05$

	Feed 1	*Feed 2*	*Feed 3*	*Feed 4*	
	60.8	68.7	102.6	87.9	
	57.0	67.7	102.1	84.2	
	65.0	74.0	100.2	83.1	
	58.6	66.3	96.5	85.7	
	61.7	69.8		90.3	
n_i	5	5	4	5	$N = \sum_{i=1}^{k} n_i = 19$
$\sum_{j=1}^{n_i} X_{ij}$	303.1	346.5	401.4	431.2	
$\bar{X}_i$	60.62	69.30	100.35	86.24	
$\dfrac{\left(\sum_{j=1}^{n_i} X_{ij}\right)^2}{n_i}$	18373.922	24012.450	40280.490	37186.688	$\sum_{i=1}^{k} \dfrac{\left(\sum_{j=1}^{n_i} X_{ij}\right)^2}{n_i} = 119853.550$

$\sum_i \sum_j X_{ij} = 1482.2$

$\sum_i \sum_j X_{ij}^2 = 119981.900$

total DF $= N - 1 = 19 - 1 = 18$
groups DF $= k - 1 = 4 - 1 = 3$
error DF $= N - k = 19 - 4 = 15$

$$C = \frac{(\sum_i \sum_j X_{ij})^2}{N} = \frac{2196916.84}{19} = 115627.202$$

$$\text{total sum of squares} = \sum_i \sum_j X_{ij}^2 - C = 119981.900 - 115627.202 = 4354.698$$

$$\text{groups sum of squares} = \sum_i \frac{(\sum_j X_{ij})^2}{n_i} - C = 119853.550 - 115627.202 = 4226.348$$

$$\text{error sum of squares} = \text{total SS} - \text{groups SS} = 4354.698 - 4226.348 = 128.350$$

Summary of the Analysis of Variance

Source of variation	*SS*	*DF*	*MS*
Total	4354.698	18	
Groups	4226.348	3	1408.783
Error	128.350	15	8.557

$$F = \frac{\text{groups MS}}{\text{error MS}} = \frac{1408.783}{8.557} = 165$$

$F_{0.05(1),3,15} = 3.29$

Reject H_0.

$P < 0.0005$

Note: All sums of squares and mean squares have $(\text{kg})^2$ as units. For typographic convenience and ease in reading, however, the units for ANOVA computations will routinely be omitted.

Sources of Variation. In the two-sample t test (Section 9.4), where one assumes equality of the variances of the two sampled populations, the common population variance, σ^2, is estimated by the pooled variance, $s_p^2 = (\text{SS}_1 + \text{SS}_2)/(\nu_1 + \nu_2) = \sum_{i=1}^{2} [\sum_{j=1}^{n_i} (X_{ij} - \bar{X}_i)^2]/\sum_{i=1}^{2} (n_i - 1) =$ pooled SS/pooled DF (as in Equation 9.2). Similarly, in the ANOVA under discussion, we assume that $\sigma_1^2 = \sigma_2^2 = \sigma_3^2 = \sigma_4^2$, and we estimate the population variance assumed common to all k groups by a variance obtained using the pooled sum of squares and the pooled degrees of freedom for all the groups:

$$\text{within groups SS} = \sum_{i=1}^{k} \left[\sum_{j=1}^{n_i} (X_{ij} - \bar{X}_i)^2 \right] \tag{11.1}$$

and

$$\text{within groups DF} = \sum_{i=1}^{k} (n_i - 1) = N - k. \tag{11.2}$$

These two quantities are often referred to as the *error sum of squares* and the *error degrees of freedom*, respectively. The former divided by the latter is a statistic that is the best estimate of the variance, σ^2, common to all k populations.

The amount of variability among the k groups is important to our hypothesis testing. This can be denoted as

$$\text{among groups SS} = \sum_{i=1}^{k} n_i(\bar{X}_i - \bar{X})^2, \tag{11.3}$$

or, simply, the *groups sum of squares*, and

$$\text{among groups DF} = k - 1, \tag{11.4}$$

commonly called the *groups degrees of freedom*.

We also consider the variability present among all N data; that is,

$$\text{total SS} = \sum_{i=1}^{k} \sum_{j=1}^{n_i} (X_{ij} - \bar{X})^2 \tag{11.5}$$

and

$$\text{total DF} = N - 1. \tag{11.6}$$

In summary, each deviation of an observed datum from the grand mean of all data is due to a deviation of that datum from its group mean plus the deviation of that group mean from the grand mean; i.e.,

$$(X_{ij} - \bar{X}) = (X_{ij} - \bar{X}_i) + (\bar{X}_i - \bar{X}). \tag{11.7}$$

Furthermore, sums of squares and degrees of freedom are additive, so

$$\text{total SS} = \text{groups SS} + \text{error SS} \tag{11.8}$$

and

$$\text{total DF} = \text{groups DF} + \text{error DF}. \tag{11.9}$$

The total sum of squares may be calculated readily by a machine formula analogous to Equation 4.7:

$$\text{total SS} = \sum_{i=1}^{k} \sum_{j=1}^{n_i} X_{ij}^2 - C, \tag{11.10}$$

where

$$C = \frac{(\sum \sum X_{ij})^2}{N}, \tag{11.11}$$

a quantity referred to as the "correction term." Example 11.1 demonstrates these calculations.

A machine formula for the groups sum of squares is

$$\text{groups SS} = \sum_{i=1}^{k} \frac{\left(\sum_{j=1}^{n_i} X_{ij}\right)^2}{n_i} - C, \tag{11.12}$$

where $\sum_{j=1}^{n_i} X_{ij}$ is the sum of the n_i data from group i.

The error SS may be calculated as

$$\text{error SS} = \sum_{i=1}^{k} \sum_{j=1}^{n_i} X_{ij}^2 - \sum_{i=1}^{k} \frac{\left(\sum_{j=1}^{n_i} X_{ij}\right)^2}{n_i}, \tag{11.13}$$

which is the machine formula for Equation 11.1, but this quantity is most commonly computed by difference:

$$\text{error SS} = \text{total SS} - \text{groups SS}. \tag{11.14}$$

Similarly,

$$\text{error DF} = \text{total DF} - \text{groups DF}, \tag{11.15}$$

which yields the same result as Equation 11.2.

Dividing the groups SS or the error SS by the respective degrees of freedom results in a variance, referred to in ANOVA terminology as a *mean square* (abbreviated MS and short for *mean squared deviations from the mean*.) Thus,

$$\text{groups MS} = \frac{\text{groups SS}}{\text{groups DF}} \tag{11.16}$$

and

$$\text{error MS} = \frac{\text{error SS}}{\text{error DF}}, \tag{11.17}$$

the latter quantity occasionally being abbreviated as MSE. A total mean square could also be calculated (as total SS/total DF), but it is not needed in the ANOVA.

Table 11.2 summarizes the single factor ANOVA calculations.

TABLE 11.2 SUMMARY OF THE CALCULATIONS FOR A SINGLE FACTOR ANALYSIS OF VARIANCE

Source of variation	*Sum of squares* (SS)	*Degrees of freedom* (DF)	*Mean square* (MS)
Total $[X_{ij} - \bar{X}]$	$\sum_{i=1}^{k}\sum_{j=1}^{n_i} X_{ij}^2 - C$	$N - 1$	
Groups (i.e., among groups) $[\bar{X}_i - \bar{X}]$	$\sum_{i=1}^{k} \frac{\left(\sum_{j=1}^{n_i} X_{ij}\right)^2}{n_i} - C$	$k - 1$	$\frac{\text{groups SS}}{\text{groups DF}}$
Error (i.e., within groups) $[X_{ij} - \bar{X}_i]$	total SS − groups SS	total DF − groups DF (or $N - k$)	$\frac{\text{error SS}}{\text{error DF}}$

Note: $C = \left(\sum_{i=1}^{k}\sum_{j=1}^{n_i} X_{ij}\right)^2 \Big/ N$; $N = \sum_{i=1}^{k} n_i$; k is the number of groups; n_i is the number of data in group i.

Testing the Null Hypothesis. Statistical theory informs us that if the null hypothesis is true, then the groups MS, as well as the error MS, will be an estimate of σ^2, the variance common to all k populations. But if the k population means are not equal, then the groups MS will be greater than the error MS. Therefore, the test for the equality of means is a one-sided variance-ratio test:

$$F = \frac{\text{groups MS}}{\text{error MS}}, \tag{11.18}$$

with the groups DF being associated with the numerator of the ratio and the Error DF associated with the denominator. Thus, the critical value for this test is $F_{\alpha(1),(k-1),(N-k)}$. If the calculated F is at least as large as the critical value, then we reject H_0. But remember that all we conclude in such a case is that all the k population means are not equal. To determine between which means the equalities or inequalities lie, we must turn to the procedures of Chapter 12.

The Case Where $k = 2$. If $k = 2$, then H_0: $\mu_1 = \mu_2$, and either the two-sample t test (Section 9.4) or the single factor ANOVA may be applied; the conclusions obtained from these two procedures will be identical. The error MS will, in fact, be identical to the pooled variance in the t test; the groups DF will be 1; and the resultant F value from the analysis of variance will be the square of the resultant t value, while $F_{\alpha(1),1,(N-2)} = (t_{\alpha(2),(N-2)})^2$. If $k = 2$, and a one-tailed test between means is required, or if the hypothesis H_0: $\mu_1 - \mu_2 = \mu_0$ is desired for a μ_0 not equal to zero, then the t test is applicable, whereas the ANOVA is not.

ANOVA Using Means and Variances The above discussion assumes that all the data from the experiment to be analyzed are in hand. It may occur, however, that all we have are the means for each group and some measure of variability based on the variances of each group. That is, we may have $\bar{X}_i$ and either SS_i, s_i^2, s_i, or $s_{\bar{X}_i}$ for each group, rather than all the individual values of X_{ij}. If the sample sizes, n_i, are also known, then the single factor analysis of variance may still be performed, in the following manner.

First, determine the sample variance for each group; recall that

$$s_i^2 = (s_i)^2 = n_i(s_{\bar{X}_i})^2. \tag{11.19}$$

Then, calculate

$$\text{error SS} = \sum_{i=1}^{k} SS_i = \sum_{i=1}^{k} (n_i - 1)s_i^2 \tag{11.20}$$

and

$$\text{groups SS} = \sum_{i=1}^{k} n_i\bar{X}_i^2 - \frac{\left(\sum_{i=1}^{k} n_i\bar{X}_i\right)^2}{\sum_{i=1}^{k} n_i}. \tag{11.21}$$

Knowing the groups SS and error SS, the ANOVA can proceed in the usual fashion.

Two Types of ANOVA. In Example 11.1, the biologist designing the experiment was interested in whether all of these particular four feeds have the same effect on pig weight. That is, these four feeds were not randomly selected from a feed catalog but were specifically chosen. When the levels of a factor are specifically chosen, one is said to have designed a *fixed effects model*, or a *Model I, ANOVA*. In such a case, the null hypothesis H_0: $\mu_1 = \mu_2 = \mu_3 = \ldots = \mu_k$ is appropriate.

However, there are instances where the levels of a factor to be tested are indeed chosen at random. For example, we might have been interested in the effect of geographic location of the pigs, rather than the effect of their feed. It is possible that our concern might be with certain specific locations, in which case we would be employing a fixed effects model ANOVA. But we might, instead, be interested in testing the statement that in general there is a difference in pig weights in animals from different locations. That is, instead of being concerned with only the particular locations used in the study, the intent might be to generalize, considering the locations in our study to be a random sample from all possible locations. In this *random effects model*, or *Model II, ANOVA*,* all the calculations are identical to those for the fixed effects model, but the null hypothesis is better stated as H_0: There is no difference in pig weight among geographic locations (or H_0: There is no variability in weights among locations). Examination of Equation 11.18 shows that what the analysis asks is whether the variability among locations is greater than the variability within locations. Example 11.2 demonstrates the ANOVA for a random effects model. Although the biologist will find that Model I analyses are far more commonly encountered than are Model II situations, when dealing with more than one experimental factor (as in Chapters 13 and 15) the distinction between the two models becomes essential.

*Also referred to as a "components of variance model." The terms "Model I" and "Model II" for analysis of variance were introduced by Eisenhart (1947).

Example 11.2

A single factor analysis of variance for a random effects model (i.e., Model II) experimental design.

A laboratory employs a certain technique for determining the phosphorus content of hay. The question arises: "Do phosphorus determinations differ with the technician performing the analysis?" To answer this question, each of four randomly selected technicians was given five samples from the same batch of hay. The results of the 20 phosphorus determinations (in mg phosphorus/g of hay) are shown.

H_0: Determinations of phosphorus content do not differ among technicians.
H_A: Determinations of phosphorus content do differ among technicians.
$\alpha = 0.05$

	Technician 1	2	3	4
	34	37	34	36
	36	36	37	34
	34	35	35	37
	35	37	37	34
	34	37	36	35
Group sums:	173	182	179	176

$$\sum_i \sum_j X_{ij} = 710$$
$$\sum_i \sum_j X_{ij}^2 = 25234$$

$N = 20$

$C = \frac{(710)^2}{20} = 25205.00$

total SS $= 25234 - 25205.00 = 29.00$

groups (i.e., technicians) SS $= \frac{(173)^2}{5} + \frac{(182)^2}{5} + \frac{(179)^2}{5} + \frac{(176)^2}{5} - 25205.00$

$= 25214.00 - 25205.00 = 9.00$

error SS $= 29.00 - 9.00 = 20.00$

Source of variation	SS	DF	MS
Total	29.00	19	
Groups (technicians)	9.00	3	3.00
Error	20.00	16	1.25

$F = \frac{3.00}{1.25} = 2.40$

$F_{0.05(1),3,16} = 3.24$

Do not reject H_0.

$0.10 < P < 0.25$

Underlying Assumptions. Recall from Section 9.4 that to test H_0: $\mu_1 = \mu_2$ by the two-sample t test, we had to assume that $\sigma_1^2 = \sigma_2^2$ and that the two samples came from normal populations. Similarly, $\sigma_1^2 = \sigma_2^2 = \sigma_3^2 = \sigma_4^2$ should be true in order to apply the analysis of variance to H_0: $\mu_1 = \mu_2 = \mu_3 = \mu_4$, and each of the k samples should have come from a normal population.* Bartlett's test for homogeneity (Section 11.8) might be used to determine whether the assumption of equal variances is met. However, since Bartlett's test is not very efficient and is badly affected by nonnormality, it is generally not worthwhile to use it in conjunction with analyses of variance.

Fortunately, the analysis of variance is robust, operating well even with considerable heterogeneity of variances, as long as all n_i are equal or nearly equal (Glass, Peckham, and Sanders, 1972). If the n_i are quite different, then the probability of a Type I error will depart markedly from α, to a degree dependent on the magnitude of the heterogeneity (Box, 1954); if larger variances are associated with the larger samples, the probability of a Type I error will be $< \alpha$, and if they are associated with the smaller samples this probability will be $> \alpha$ (Kohr and Games, 1974).

The analysis of variance is also robust with respect to the assumption of the underlying populations' normality. The validity of the analysis is affected only slightly by even considerable deviations from normality (in skewness and/or kurtosis), especially as n increases (Box and Anderson, 1955; Srivastava, 1959; Tiku, 1971). If the underlying populations are very much platykurtic, and sample sizes are small, the actual power of the test will be less than that discussed in Section 11.3; if the populations are greatly leptokurtic, and sample sizes are small, the actual power will be greater than that calculated by the method of that section (Glass, Peckham, and Sanders, 1972).

Thus, the analysis of variance typically may be depended upon unless the data deviate severely from the underlying assumptions. In the latter case, the nonparametric procedure of Section 11.4 is appropriate, as it depends not at all on these assumptions. Or, one might consider alternate parametric analyses, which have been described (Gill, 1971; Sokal and Rohlf, 1981: 409–410; Welch, 1951) but not widely embraced.

11.2 CONFIDENCE LIMITS FOR MEANS

When $k > 2$, confidence limits for μ_i may be computed in a fashion analogous to that for the case where $k = 2$ [Section 9.5, Equation 9.16]. The $1 - \alpha$ confidence interval for μ_i is

$$\bar{X}_i \pm t_{\alpha(2),\nu}\sqrt{\frac{s^2}{n_i}}, \qquad (11.22)$$

where s^2 is the error mean square and ν is the error degrees of freedom from the analysis of variance. For example, let us consider the 95% confidence interval for μ_4 in Example 11.1. Here, $\bar{X}_4 = 86.24$ kg, $s^2 = 8.557\ \text{kg}^2$, $n_4 = 5$, and $t_{0.05(2),15} = 2.131$. Therefore, the lower 95% confidence limit, L_1, is $86.24\ \text{kg} - 2.131\sqrt{8.557\ \text{kg}^2/5} = 83.45$ kg, and the upper 95% confidence limit, L_2, is $86.24\ \text{kg} + 2.131\sqrt{8.557\ \text{kg}^2/5} = 89.03$ kg.

*In factorial analyses of variance, to be discussed in Chapters 13 and 15, the underlying assumptions are that the data in each cell (i.e., in each combination of factors) come from a normal population and that all of these populations have the same variance.

Computing a confidence interval for μ_i would only be warranted if that population mean were concluded to be different from each other population mean. And calculation of a confidence interval for each of the k μ's should be performed only if it is concluded that $\mu_1 \neq \mu_2 \neq \ldots \neq \mu_k$. However, the analysis of variance does not enable conclusions as to which population means are different from which. Therefore, we must first perform multiple comparison testing (Chapter 12), after which confidence intervals may be determined for each different population mean. Confidence intervals for differences between means should be calculated as shown in Section 12.3.

11.3 POWER AND SAMPLE SIZE IN ANALYSIS OF VARIANCE

In Section 9.6, dealing with the difference between two means, we saw how to estimate the sample size required to predict such a population difference with a specified level of confidence. When dealing with more than two means, we may also wish to determine the sample size necessary to estimate difference between population means, and the appropriate procedure will be found in Section 12.3.

In Section 9.7, methods were presented for estimating the power of the two-sample t test, the minimum sample size required for such a test, and the minimum difference between population means that is detectable by such a test. Let us now examine such procedures for analysis of variance situations, namely for dealing with more than two means. (The following discussion refers only to Model I—fixed effects model—analysis of variance, and it assumes normality of the sampled populations. For power of the random effects—Model II—analysis of variance, see Guenther, 1964: 61.)

If H_0 is true for an analysis of variance, then the variance ratio of Equation 11.18 follows the F distribution, this distribution being defined by the numerator and denominator degrees of freedom (ν_1 and ν_2, respectively). If, however, H_0 is false, then the ratio of groups MS to error MS follows instead what is known as the *noncentral F distribution*, which is defined by ν_1, ν_2, and a third quantity known as the *noncentrality parameter*. As power refers to probabilities of detecting a false null hypothesis, statistical discussions of the power of ANOVA testing depend upon the noncentral F distribution.

A number of authors have described procedures for estimating the power of an ANOVA, or the required sample size, or the detectable difference among means (e.g., Cohen, 1977: Ch. 8; Fox, 1956; Odeh and Fox, 1975; Patniak, 1949; Tang, 1938; Tiku, 1967, 1972), but the charts prepared by Pearson and Hartley (1951) provide one of the best of the methods and will be described below.

Power of the Test. Prior to performing an experiment, and collecting data from it, it is appropriate and desirable to investigate the power of the proposed test. (Indeed, it is possible that on doing so one would conclude that the power is so low that the experiment needs to be run with many more data or, perhaps, not run at all.)

Let us specify that an ANOVA involving k groups will be performed at the α significance level, with n data (i.e., replications) per group. We can then estimate the power of the test if we have an estimate of σ^2, the variability within the k populations (e.g., this estimate typically is s^2 from similar experiments), and an estimate of the variability among the populations. From this information we may calculate a quantity called ϕ (lowercase Greek phi*), which is related to the noncentrality parameter.

Our estimate of variability among populations might be in the form of mean squares from an ANOVA from similar experiments, in which case

$$\phi = \sqrt{\frac{(k-1)(\text{groups MS})}{ks^2}}. \tag{11.23}$$

(Recall that s^2 is error MS.) Or (in a related fashion), the variability among populations might be expressed in terms of deviations of the k population means, μ_i, from the overall mean of all populations, μ, in which case

$$\phi = \frac{n\sum_{i=1}^{k}(\mu_i - \mu)^2}{ks^2} \tag{11.24}$$

(e.g., Guenther, 1964: 47–49; Kirk, 1968: 107–110). The grand population mean is

$$\mu = \frac{\sum_{i=1}^{k}\mu_i}{k} \tag{11.25}$$

if all the samples are the same size.

Once ϕ has been obtained, we consult Appendix Fig. B.1. This figure consists of several pages, each with a different ν_1 (i.e., groups DF) indicated at the upper left of the graph. Values of ϕ are indicated on the lower axis of the graph for both $\alpha = 0.01$ and $\alpha = 0.05$. Each of the curves on the graph is for a different ν_2 (i.e., error DF), for both $\alpha = 0.01$ and 0.05, identified on the top margin of the graph. After turning to the graph for the ν_1 at hand, one locates the point at which the calculated ϕ intersects the curve for the given ν_2 and reads horizontally to either the right or left axis to determine the power of the test. This procedure is demonstrated in Example 11.3.

Example 11.3

Estimating the power of an analysis of variance when variability among population means is specified.

A proposed analysis of variance of plant root elongations is to comprise 10 roots at each of 4 chemical treatments. From previous experiments, we estimate σ^2 to be 7.5888 mm² and estimate that two of the population means are 8.0 mm, one is 9.0 mm, and one is 12.0 mm. What will be the power of the ANOVA if we test at the 0.05 level of significance?

$$k = 4$$
$$n = 10$$
$$\nu_1 = k - 1 = 3$$
$$\nu_2 = k(n-1) = 4(9) = 36$$
$$\mu = \frac{8.0 + 8.0 + 9.0 + 12.0}{4} = 9.25$$

*"Phi" rhymes with "sky."

$$\phi = \sqrt{\frac{n\sum(\mu_i - \mu)^2}{ks^2}}$$
$$= \sqrt{\frac{10[(8.0 - 9.25)^2 + (8.0 - 9.25)^2 + (9.0 - 9.25)^2 + (12.0 - 9.25)^2]}{4(7.5888)}}$$
$$= \sqrt{\frac{10(10.75)}{4(7.5888)}}$$
$$= \sqrt{3.5414}$$
$$= 1.88$$

In Fig. B.1, we enter the graph for $\nu_1 = 3$ with $\phi = 1.88$, $\alpha = 0.05$, and $\nu_2 = 36$ and read a power of about 0.86.

An alternative, and common, way to determine power is to specify the smallest difference we wish to detect between the two most different population means. Calling this minimum detectable difference δ, we compute

$$\phi = \sqrt{\frac{n\delta^2}{2ks^2}} \tag{11.26}$$

(Guenther, 1964: 48–49; Kirk, 1968: 109–110) and proceed to consult Fig. B.1 as above, and as demonstrated in Example 11.4. This procedure leads us to the statement that the power will be at least that determined from Fig. B.1 (and, indeed, it typically is greater).

Example 11.4

Estimating the power of an analysis of variance when minimum detectable difference is specified.

For the ANOVA proposed in Example 11.3, we do not estimate the population means, but rather specify that, using 10 data per sample, we wish to detect a difference between population means of at least 4.0 mm.

$k = 4$ $\quad\quad \delta = 4.0$ mm
$\nu_1 = 3$ $\quad\quad s^2 = 7.5888$ mm^2
$n = 10$
$\nu_2 = 36$

$$\phi = \sqrt{\frac{n\,\delta^2}{2ks^2}}$$
$$= \sqrt{\frac{10(4.0)^2}{2(4)(7.5888)}}$$
$$= \sqrt{2.6355}$$
$$= 1.62$$

In Fig. B.1, we enter the graph for $\nu_1 = 3$ with $\phi = 1.62$ and $\nu_2 = 36$ and read a power of about 0.72.

It is readily apparent from Fig. B.1 that, for a given α, ν_1, and ν_2, greater values of ϕ are associated with greater power; and, from Equations 11.23, 11.24, and 11.26, that ϕ increases with

1. increased sample size, n;
2. increased difference among population means (as measured either by groups MS, by $\sum(\mu_i - \mu)^2$, or by the minimum detectable difference, δ);
3. a fewer number of groups, k;
4. decreased variability within populations, σ^2, estimated by s^2.

Furthermore, recall that power increases with larger significance levels, α.

Estimating the power of a proposed ANOVA may effect considerable savings in time, effect, and expense. For example, such an estimation might conclude that the power is so very low that the experiment, as planned, ought not to be performed. The proposed experiment might be revised, perhaps by increasing n, so as to render the results more likely to be conclusive.

Power of a Performed Test. While it is desirable to estimate power prior to engaging in analysis of variance testing (as described above), it also is useful to ask with what power an ANOVA has been performed. This is especially useful if the null hypothesis is not rejected, for then we should wish to know how likely the test was to detect true difference among the population means.

As shown in Example 11.5, we compute

$$\phi = \sqrt{\frac{(k-1)(\text{groups MS} - s^2)}{ks^2}} \tag{11.27}$$

(Kirk, 1968: 108) and then proceed as above and as in Examples 11.3 and 11.4.

Example 11.5

Estimating the power of an analysis of variance after it has been performed.

In an experiment to determine whether the development time for insect embryos (measured as time elapsed from egg laying to hatching) is the same at three different experimental temperatures, four eggs were placed at each of two temperatures, and five eggs were placed at the third temperature. (Therefore, $k = 3$, $n_1 = n_2 = 4$, and $n_3 = 5$.) A one-way analysis of variance yields the following results:

H_0: $\mu_1 = \mu_2 = \mu_3$
H_A: The mean development time is not the same at all three temperatures.

Source of variation	SS	DF	MS
Total	26.9231	12	
Among groups	10.3731	2	5.1866
Error	16.5500	10	1.6550

$F = 3.13$
$F_{0.05(1),2,10} = 4.10$
Do not reject H_0; $0.05 < P < 0.10$

$$\phi = \sqrt{\frac{(k-1)(\text{groups MS} - s^2)}{ks^2}} = \sqrt{\frac{(3-1)(5.1866 - 1.6550)}{3(1.6550)}} = 1.19$$

On consulting Fig. B.1, with $\nu_1 = 2$, $\nu_2 = 10$, and $\phi = 1.19$, the power of the ANOVA is estimated to be 0.33; that is, there was a 67% chance of having committed a Type II error in this analysis.

Sample Size Required. Prior to performing an analysis of variance, we might ask how many data need to be obtained in order to achieve a desired power. We can specify the power with which we wish to detect a particular difference among the population means and then ask how large the sample from each population must be. This is done, with Equation 11.26, by iteration (i.e., by making an initial guess and repeatedly refining that estimate), as shown in Example 11.6.

Example 11.6

Estimation of required sample size for a one-way analysis of variance.

Let us propose an experiment such as that described in Example 11.5. How many replicate data should be collected so as to have an 80% probability of detecting a difference between population means as small as 2 days, testing at the 0.05 level of significance? We shall assume that $s^2 = 1.6550$ is a good estimate of σ^2.

Let us guess that $n = 15$ is required; then,

$$\nu_2 = 3(15 - 1) = 42,$$

and, by Equation 11.26,

$$\phi = \sqrt{\frac{(15)(2)^2}{2(3)(1.6550)}} = 2.46.$$

Consulting Fig. B.1, power ≈ 0.96, which is higher than we specified. Therefore, we can lower our guess, saying perhaps that $n = 10$; then,

$$\nu_2 = 3(10 - 1) = 27,$$

and

$$\phi = \sqrt{\frac{(10)(2)^2}{2(3)(1.6550)}} = 2.01,$$

which estimates a power of about 84%, a little higher than what we specified. Trying $n = 9$, we have

$$\nu_2 = 3(9 - 1) = 24,$$

and

$$\phi = \sqrt{\frac{(9)(2)^2}{2(3)(1.6550)}} = 1.90,$$

estimating a power of 0.79, which is lower than that specified.

Thus, we estimate that using sample sizes of at least 10 should result in an ANOVA with a power of at least 0.84.

Minimum Detectable Difference. If we specify the significance level and sample size for an ANOVA, and the power that we desire it to have, and if we have an estimate of σ^2, then we can ask what the smallest detectable difference between population means will be. By entering on Fig. B.1 the specified α, ν_1, and power, we can read a value of ϕ on the bottom axis. Then, by rearrangement of Equation 11.26, the minimum detectable difference is

$$\delta = \sqrt{\frac{2ks^2\phi^2}{n}}. \qquad (11.28)$$

Example 11.7 demonstrates this estimation procedure.

Example 11.7

Estimation of minimum detectable difference in a one-way analysis of variance.

In an experiment as that proposed in Example 11.5, assuming that $s^2 = 1.6550$ (days)^2 is a good estimate of σ^2, how small a difference between μ's can we have 90% confidence of detecting if $n = 20$ and $\alpha = 0.05$ are used?

Since $k = 3$ and $n = 20$, $\nu_2 = 3(20 - 1) = 57$. For $\nu_1 = 2$, $\nu_2 = 57$, $1 - \beta = 0.90$, and $\alpha = 0.05$, Fig. B.1 gives a ϕ of about 2.1, from which we compute an estimate of

$$\delta = \sqrt{\frac{2ks^2\phi^2}{n}} = \sqrt{\frac{2(3)(1.6550)(2.1)^2}{20}} = 1.48 \text{ days}.$$

Maximum Number of Groups Testable. For a given α, n, δ, and σ^2, power will decrease as k increases. It may occur that the total number of observations, N, will be limited, and for given ANOVA specifications the number of experimental groups, k, may have to be limited. As Example 11.8 illustrates, the maximum k can be determined by trial-and-error estimation of power, using Equation 11.26.

Example 11.8

Determination of maximum number of groups to be used in a one-way analysis of variance.

Consider an experiment such as that in Example 11.5. Perhaps we have six chemicals that might be tested, but we have only space and equipment to grow a total of 50 plants. Let us specify that we wish to test, with $\alpha = 0.05$ and $\beta \leq 0.20$ (i.e., power of at least 80%), and to detect a difference as small as 2 mm between population means.

If $k = 6$ were used, then $n = 50/6 = 8.3$ (call it 8), $\nu_1 = 5$, $\nu_2 = 6(8 - 1) = 42$, and (by Equation 11.26)

$$\phi = \sqrt{\frac{(8)(2)^2}{2(6)(1.6550)}} = 1.27,$$

for which Fig. B.1 indicates a power of about 0.60.

If $k = 5$ were used, $n = 50/5 = 10$, $\nu_1 = 4$, $\nu_2 = 5(10 - 1) = 45$, and

$$\phi = \sqrt{\frac{(10)(2)^2}{2(5)(1.6550)}} = 1.55,$$

for which Fig. B.1 indicates a power of about 0.74.

If $k = 4$ were used, $n = 50/4 = 12.5$ (call it 12), $\nu_1 = 3$, $\nu_2 = 4(12 - 1) = 44$, and

$$\phi = \sqrt{\frac{(12)(2)^2}{2(4)(1.6550)}} = 1.90,$$

for which Fig. B.1 indicates a power of about 0.87.

Therefore, we conclude that no more than four of the chemicals should be tested in an analysis of variance if we are limited to a total of 50 experimental plants.

11.4 NONPARAMETRIC ANALYSIS OF VARIANCE

If a set of data is collected according to a completely randomized design where $k > 2$, it is possible to test nonparametrically for intergroup differences. This is done by the *Kruskal-Wallis test* (Kruskal and Wallis, 1952), often called an analysis of variance by ranks. This test can be used in any situation where the parametric single factor ANOVA of Section 11.1 is applicable, and it will be $3/\pi = 95\%$ as powerful as the latter (Andrews, 1954); and it may also be employed in instances where the latter is not applicable, in which cases it may in fact be the more powerful test. The nonparametric analysis may be applied when the k samples do not come from normal populations and/or when the k population variances are heterogeneous. If $k = 2$, then the Mann-Whitney test (Section 9.9) is the appropriate nonparametric method.

Example 11.9 demonstrates the Kruskal-Wallis test procedure. Recall that in a nonparametric test, we do not use population parameters in statements of hypotheses, and neither parameters nor sample statistics are used in the test calculations. The Kruskal-Wallis test statistic, H, is calculated as

$$H = \frac{12}{N(N+1)} \sum_{i=1}^{k} \frac{R_i^2}{n_i} - 3(N+1), \tag{11.29}$$

where n_i is the number of observations in group i, $N = \sum_{i=1}^{k} n_i$ (the total number of observations in all k groups), and R_i is the sum of the ranks of the n_i observations in group i.* The procedure for ranking data is as presented in Section 9.9 for the Mann-Whitney test. A good check (but not a guarantee) of whether ranks have been assigned correctly is to see whether the sum of all the ranks equals $N(N+1)/2$.

Example 11.9

The Kruskal-Wallis single factor analysis of variance by ranks.

An entomologist is studying the vertical distribution of a fly species in a deciduous forest and obtains five collections of the flies from each of three different vegetation layers: herb, shrub, and tree.

H_0: The abundance of the flies is the same in all three vegetation layers.
H_A: The abundance of the flies is not the same in all three vegetation layers.
$\alpha = 0.05$
The data are as follows (with ranks of the data in parentheses):†

Numbers of Flies/m³ of Foliage

Herbs	*Shrubs*	*Trees*
14.0 (15)	8.4 (11)	6.9 (8)
12.1 (14)	5.1 (2)	7.3 (9)
9.6 (12)	5.5 (4)	5.8 (5)
8.2 (10)	6.6 (7)	4.1 (1)
10.2 (13)	6.3 (6)	5.4 (3)
$n_1 = 5$	$n_2 = 5$	$n_3 = 5$
$R_1 = 64$	$R_2 = 30$	$R_3 = 26$

$N = 5 + 5 + 5 = 15$

$$H = \frac{12}{N(N+1)} \sum_{i=1}^{k} \frac{R_i^2}{n_1} - 3(N+1)$$
$$= \frac{12}{15(16)}\left[\frac{64^2}{5} + \frac{30^2}{5} + \frac{26^2}{5}\right] - 3(16)$$
$$= \frac{12}{240}[1134.400] - 48$$
$$= 56.720 - 48$$
$$= 8.720$$

$H_{0.05,5,5,5} = 5.780$
Reject H_0.
$0.005 < P < 0.01$

*Interestingly, H (or H_c of Equation 11.32) could also be computed as

$$H = \frac{\text{groups SS}}{\text{total MS}}, \tag{11.30}$$

applying the procedures of Section 11.1 to the ranks of the data in order to obtain the groups SS and total MS. This relationship becomes important in Sections 13.4 and 15.4.

†To check whether ranks were assigned correctly, the sum of the ranks (or sum of the rank sums: $64 + 30 + 26 = 120$) is compared to $N(N+1)/2 = 15(16)/2 = 120$. This check will not guarantee that the ranks were assigned properly, but it will often catch errors of doing so.

Critical values of H for small sample sizes where $k \leq 5$ are given in Table B.12. For larger samples and/or for $k > 5$, H may be considered to be approximated by χ^2 with $k - 1$ degrees of freedom. [Kruskall and Wallis (1952) give two approximations that are better than chi-square when the n_i's are small or when significance levels less than 1% are to be used. But they are relatively complicated to use.]

If there are tied ranks, as in Example 11.10, H is a little lower than it should be,

Example 11.10

The Kruskal-Wallis test with tied ranks.

A limnologist obtained 8 containers of water from each of 4 ponds. The pH of each water sample was measured. The data are arranged in ascending order within each pond. (One of the containers from pond 3 was lost, so $n_3 = 7$, instead of 8; but the test procedure does not require equal numbers of data in each group.)

H_0: pH is the same in all four ponds.
H_A: pH is not the same in all four ponds.
$\alpha = 0.05$

Pond 1	*Pond 2*	*Pond 3*	*Pond 4*
Rank	Rank	Rank	Rank
7.68 (1)	7.71 (6*)	7.74 (13.5*)	7.71 (6*)
7.69 (2)	7.73 (10*)	7.75 (16)	7.71 (6*)
7.70 (3.5*)	7.74 (13.5*)	7.77 (18)	7.74 (13.5*)
7.70 (3.5*)	7.74 (13.5*)	7.78 (20*)	7.79 (22)
7.72 (8)	7.78 (20*)	7.80 (23.5*)	7.81 (26*)
7.73 (10*)	7.78 (20*)	7.81 (26*)	7.85 (29)
7.73 (10*)	7.80 (23.5*)	7.84 (28)	7.87 (30)
7.76 (17)	7.81 (26*)		7.91 (31)

*Tied ranks.

$n_1 = 8$	$n_2 = 8$	$n_3 = 7$	$n_4 = 8$
$R_1 = 55$	$R_2 = 132.5$	$R_3 = 145$	$R_4 = 163.5$

$N = 8 + 8 + 7 + 8 = 31$

$$H = \frac{12}{N(N+1)} \sum_{i=1}^{k} \frac{R_i^2}{n_i} - 3(N+1)$$

$$= \frac{12}{31(32)}\left[\frac{55^2}{8} + \frac{132.5^2}{8} + \frac{145^2}{7} + \frac{163.5^2}{8}\right] - 3(32)$$

$$= 11.876$$

Number of groups of tied ranks $= m = 7$.

$$\sum T = \sum (t_i^3 - t_i)$$

$$= (2^3 - 2) + (3^3 - 3) + (3^3 - 3) + (4^3 - 4) + (3^3 - 3) + (2^3 - 2) + (3^3 - 3)$$

$$= 168$$

$$C = 1 - \frac{\sum T}{N^3 - N} = 1 - \frac{168}{31^3 - 31} = 1 - \frac{168}{29760} = 0.9944$$

$$H_c = \frac{H}{C} = \frac{11.876}{0.9944} = 11.943$$

$\nu = k - 1 = 3$

$\chi^2_{0.05,3} = 7.815$

Reject H_0.

$0.005 < P < 0.01$

and a correction factor may be computed as

$$C = 1 - \frac{\sum T}{N^3 - N}, \tag{11.31}$$

and the corrected value of H is

$$H_c = \frac{H}{C}. \tag{11.32}$$

Here,

$$\sum T = \sum_{i=1}^{m} (t_i^3 - t_i), \tag{11.33}$$

where t_i is the number of ties in the ith group of ties, and m is the number of groups of tied ranks.

As in the parametric ANOVA, if we reject H_0 by the Kruskal-Wallis test, we do not know which groups differ from which other groups; all we know is that at least one difference among the k groups does exist. To locate the difference(s), a method such as that given in Section 12.6 would have to be employed.

11.5 TESTING FOR DIFFERENCE AMONG SEVERAL MEDIANS

Section 9.10 presented the *median test* for the two-sample case. This procedure may be expanded to multisample considerations (Mood, 1950: 398–399). The method requires the determination of the grand median of all observations in all k samples considered together. The number of data in each sample that are above and not above this median are tabulated, and the significance of the resultant $2 \times k$ contingency table is then analyzed, generally by chi-square (Section 6.1), alternatively by the G test of Section 6.6. For example, if there were four populations being compared, the statistical hypotheses would be H_0: All four populations have the same median, and H_A: All four populations do not have the same median. The median test would be the testing of the following contingency table:

	Sample 1	*Sample 2*	*Sample 3*	*Sample 4*	*Total*
Above median	f_{11}	f_{12}	f_{13}	f_{14}	R_1
Not above median	f_{21}	f_{22}	f_{23}	f_{24}	R_2
Total	C_1	C_2	C_3	C_4	n

This multisample median test is demonstrated in Example 11.11. Hays (1963: 621) recommends the test as applicable when $n \geq 20$ and $C_j \geq 5$. If H_0 is rejected, then the method of Section 12.8 can be used to attempt to discern which population medians are different from which.

Example 11.11

The multisample median test.

H_0: Median elm tree height is the same on all four sides of a building.
H_A: Median elm tree height is not the same on all four sides of a building.

A total of 48 elm seedlings were planted, 12 on each of a building's four sides. The heights, after several years of growth, were as follows:

	North	East	South	West
	7.1 m	6.9 m	7.8 m	6.4 m
	7.2	7.0	7.9	6.6
	7.4	7.1	8.1	6.7
	7.6	7.2	8.3	7.1
	7.6	7.3	8.3	7.6
	7.7	7.3	8.4	7.8
	7.7	7.4	8.4	8.2
	7.9	7.6	8.4	8.4
	8.1	7.8	8.6	8.6
	8.4	8.1	8.9	8.7
	8.5	8.3	9.2	8.8
	8.8	8.5	9.4	8.9
medians:	7.7 m	7.35 m	8.4 m	8.0 m

grand median = 8.1 m

The 2 × 4 contingency table (with expected frequencies—see Section 6.1—in parentheses):

	North	East	South	West	
Above median	3 (5.0000)	2 (5.0000)	9 (5.0000)	6 (5.0000)	20
Not above median	9 (7.0000)	10 (7.0000)	3 (7.0000)	6 (7.0000)	28
Total	12	12	12	12	48

$\chi^2 = 10.286$
$\chi^2_{0.05,3} = 7.815$
Reject H_0.
$0.01 < P < 0.025$

11.6 THE EFFECT OF CODING

In the parametric ANOVA, coding the data by addition or subtraction causes no change in any of the sums of squares or mean squares, so the resultant F and the ensuing conclusions are not affected at all. If the coding is performed by multiplying or dividing all the data by a constant, the sums of squares and the mean squares in the ANOVA each will be altered by an amount equal to the square of that constant, but the F value and the associated conclusions will remain unchanged.

A test utilizing ranks (such as the Kruskal-Wallis procedure) will not be affected at all by coding of the raw data. Thus, the coding of data for analysis of variance, either parametric or nonparametric, may be employed with impunity, and coding frequently renders data easier to manipulate. Neither will coding of data alter the conclusions from the hypothesis tests in Chapter 12 (multiple comparisons) or Chapters 13–16 (further analysis of variance procedures).

11.7 MULTISAMPLE TESTING FOR NOMINAL SCALE DATA

A $2 \times C$ contingency table may be analyzed to compare frequency distributions of nominal data for two samples. In a like fashion, an $R \times C$ contingency table may be set up to compare frequency distributions of nominal scale data from R samples. Contingency table procedures are discussed in Chapter 6.

Other procedures have been proposed for multisample analysis of nominal scale data (e.g., Light and Margolin, 1971; Windsor, 1948).

11.8 HOMOGENEITY OF VARIANCES

If we have three or more samples, and we compute a variance for each, then we can test the hypothesis that all the samples came from populations with identical variances (i.e., all the sample variances estimate the same population variance). This statement of the equality, or *homogeneity*, of variances can be written H_0: $\sigma_1^2 = \sigma_2^2 = \ldots = \sigma_k^2$, where k is the number of samples. *Homoscedasticity* is used as a synonym for homogeneity of variances; variance heterogeneity is then called *heteroscedasticity*.

The most common method employed to test for homogeneity of variances is Bartlett's test (Bartlett, 1937). In this procedure, the test statistic is

$$B = (\ln s_p^2)\left(\sum_{i=1}^{k} \nu_i\right) - \sum_{i=1}^{k} \nu_i \ln s_i^2, \qquad (11.34)$$

where $\nu_i = n_i - 1$ and n_i is the size of sample i. The pooled variance, s_p^2, is calculated as before as $\sum_{i=1}^{k} \text{SS}_i / \sum_{i=1}^{k} \nu_i$. Many biologists prefer to operate with common logarithms (base 10), rather than with natural logarithms (base e);* so Equation (11.34) may be written as

$$B = 2.30259\left[(\log s_p^2)\left(\sum_{i=1}^{k} \nu_i\right) - \sum_{i=1}^{k} \nu_i \log s_i^2\right]. \qquad (11.35)$$

The distribution of B is approximated by the chi-square distribution, with $k - 1$ degrees of freedom, but a more accurate chi-square approximation is obtained by computing a correction factor,

$$C = 1 + \frac{1}{3(k-1)}\left(\sum_{i=1}^{k} \frac{1}{\nu_i} - \frac{1}{\sum_{i=1}^{k} \nu_i}\right), \qquad (11.36)$$

with the corrected test statistic being

$$B_c = \frac{B}{C}. \qquad (11.37)$$

Example 11.12 demonstrates these calculations. The null hypothesis for testing

*In this book *ln* will denote the natural, or Naperian, logarithm, and *log* will denote the common, or Briggsian, logarithm.

the homogeneity of the variances of four populations may be written symbolically as H_0: $\sigma_1^2 = \sigma_2^2 = \sigma_3^2 = \sigma_4^2$, or, in words, as "the four population variances are homogeneous (i.e., are equal)." The alternate hypothesis can be stated as "the four population variances are not homogeneous (i.e., they are not all equal)," or "there is difference (or heterogeneity) among the four population variances." If H_0 is rejected, the further testing of Section 12.9 will allow us to ask which population variances are different from which.

Example 11.12

Bartlett's test for homogeneity of variances.

Nineteen pigs were divided into four groups, and each group was given a different feed. The data are weights, in kilograms, and we wish to test whether the variance of weights is the same for pigs on all four feeds. (These are the same data as in Example 11.1.)

H_0: $\sigma_1^2 = \sigma_2^2 = \sigma_3^2 = \sigma_4^2$
H_A: The four population variances are heterogeneous (i.e., are not all equal).
$\alpha = 0.05$

	Feed 1	*Feed 2*	*Feed 3*	*Feed 4*	
	60.8	68.7	102.6	87.9	
	57.0	67.7	102.1	84.2	
	65.0	74.0	100.2	83.1	
	58.6	66.3	96.5	85.7	
	61.7	69.8		90.3	
SS_i	37.57	34.26	22.97	33.55	$\sum SS_i = 128.35$
ν_i	4	4	3	4	$\sum \nu_i = 15$
s_i^2	9.39	8.56	7.66	8.39	
$\log s_i^2$	0.9727	0.9325	0.8842	0.9238	
$\nu_i \log s_i^2$	3.8908	3.7300	2.6526	3.6952	$\sum \nu_i \log s_i^2 = 13.9686$
$\frac{1}{\nu_i}$	0.250	0.250	0.333	0.250	$\sum \frac{1}{\nu_i} = 1.083$

$$s_p^2 = \frac{\sum SS_i}{\sum \nu_i} = \frac{128.35}{15} = 8.56 \qquad \log s_p^2 = 0.9325$$

$$\begin{aligned} B &= 2.30259[(\log s_p^2)(\textstyle\sum \nu_i) - \textstyle\sum \nu_i \log s_i^2] \\ &= 2.30259[(0.9325)(15) - 13.9686] \\ &= 2.30259(0.0189) \\ &= 0.0435 \end{aligned}$$

$$\begin{aligned} C &= 1 + \frac{1}{3(k-1)}\left(\sum \frac{1}{\nu_i} - \frac{1}{\sum \nu_i}\right) \\ &= 1 + \frac{1}{3(3)}\left(1.83 - \frac{1}{15}\right) \\ &= 1.113 \end{aligned}$$

$$B_c = \frac{B}{C} = \frac{0.0435}{1.113} = 0.0391$$

$\chi^2_{0.05,3} = 7.815$
Do not reject H_0.
$0.995 < P < 0.999$

Bartlett's test is powerful, but it is badly affected by nonnormal populations (Box, 1953; Box and Anderson, 1955; Gartside, 1972). If the population distribution is platykurtic, the true α is less than the stated α (i.e., the test is conservative and the probability of a Type II error is increased); if it is leptokurtic, the true α is greater than the stated α (i.e., the probability of a Type I error is increased). Many other variance homogeneity tests have been proposed, but they are all adversely affected by nonnormality, or are very low in power, or have other serious drawbacks (Brown and Forsythe, 1974; Games, Winkler, and Probert, 1972; Keselman, Games, and Clinch, 1979; Seber, 1977: 147–148). Thus, the Bartlett test remains commendable when the sampled populations are normal, and no testing procedure is especially good for hypotheses of variance homogeneity when they are not.

Because of the poor performance of tests for variance homogeneity, and the robustness of analysis of variance for multisample testing among means (Section 11.1), it is not recommended that the former be performed as a test of the underlying assumptions of the latter.

EXERCISES

11.1. The following data are weights of food (in kilograms) consumed per day by adult deer collected at different times of the year. Test the null hypothesis that food consumption is the same for all the months tested.

Feb.	*May*	*Aug.*	*Nov.*
4.7	4.6	4.8	4.9
4.9	4.4	4.7	5.2
5.0	4.3	4.6	5.4
4.8	4.4	4.4	5.1
4.7	4.1	4.7	5.6
	4.2	4.8	

11.2. An experiment is to have its results examined by analysis of variance. The variable is temperature (in degrees Celcius), with 12 measurements to be taken in each of 5 experimental groups. From previous experiments, we estimate the within-groups variability, σ^2, to be 1.54 (°C)^2. If the 5% level of significance is employed, what is the probability of the ANOVA detecting a difference between population means as small as 2.0°C?

11.3. For the experiment of Exercise 11.2, how many replicates are needed in each of the 5 groups to detect a difference between population means as small as 2.0°C with 95% power?

11.4. For the experiment of Exercise 11.2, what is the smallest difference between population means that we are 95% likely to detect with an ANOVA using 10 replicates per group?

11.5. Using the Kruskal-Wallis test, test nonparametrically the appropriate hypotheses for the data of Exercise 11.1.

11.6. Three different methods were used to determine the dissolved oxygen content of lake water. Each of the three methods was applied to a sample of water six times, with the following results. Test the null hypothesis that the three methods yield equally variable results.

Method 1 (mg/kg)	*Method 2* (mg/kg)	*Method 3* (mg/kg)
10.96	10.88	10.73
10.77	10.75	10.79
10.90	10.80	10.78
10.69	10.81	10.82
10.87	10.70	10.88
10.60	10.82	10.81

12 Multiple Comparisons

Using a single factor analysis of variance, we may test the null hypothesis H_0: $\mu_1 = \mu_2 = \ldots = \mu_k$. However, the rejection of H_0 does not imply that all k means are different from one another, and we know neither how many differences there are nor where differences are located among the k population means. For example, if $k = 3$, and H_0: $\mu_1 = \mu_2 = \mu_3$ is rejected, then usually we desire to ask whether H_A: $\mu_1 \neq \mu_2 = \mu_3$, or H_A: $\mu_1 = \mu_2 \neq \mu_3$, or H_A: $\mu_1 \neq \mu_2 \neq \mu_3$ is the appropriate alternate hypothesis.

As explained in the introduction to Chapter 11, it is generally invalid to employ multiple t tests to examine the differences between all possible pairs of means. Chapter 12 presents statistical procedures that may be used for such a purpose; they are called *multiple comparison tests*. Although multiple comparisons are very often desired after a Model I analysis of variance, they generally are not applied in Model II situations.

The multiple comparison problem has received much attention in the statistical literature, yet there is no agreement as to the "best" procedure to routinely employ. Among the most widely accepted and commonly used methods are the Tukey test (Tukey, 1953) and the Newman-Keuls test (Newman, 1939; Keuls, 1952), which will be discussed in Sections 12.1 and 12.2, respectively. The Duncan test (Duncan, 1955), often referred to as the "Duncan new multiple range test" because it succeeds an earlier procedure (Duncan, 1951), is also encountered; but it has a different theoretical basis, one that is not as widely accepted as that of the Tukey and Newman-Keuls procedures. (It is performed as is the Newman-Keuls test, except that different critical value tables are required.) We may also encounter a procedure called the "least significant difference" test (LSD), but in general it is not to be recommended for multiple comparisons. Yet another technique, Scheffé's test (Scheffé, 1953; 1959: Sections 3.4, 3.5) will be discussed in Section 12.5; it is especially suited for a kind of comparison called a multiple contrast. The relative performances of these, as well as other, multiple comparison procedures are discussed by many authors.* Although not actually required by theory, multiple comparison testing is most commonly performed only if an analysis of variance first rejects a multisample hypothesis of equal means.† (Thus, these tests are referred to as "a posteriori" tests.)

In general, the multiple comparison tests for means have the same underlying assumptions as does the analysis of variance: population normality and homogeneity of variance. Although the Tukey test appears to be robust with respect to departures from these assumptions (Keselman, 1976), the robustness of each of the above procedures is not well-known, with adverse effects on both Type I and Type II errors possible if the assumptions are greatly violated. The homogeneity of variance assumption apparently is the more serious, and parametric multiple comparison testing should not be performed if heteroscedasticity is pronounced. If it is not comfortable to assume normality or homoscedasticity, then a nonparametric analysis of variance (Section 11.4) may be employed, followed by nonparametric multiple comparison testing (Section 12.6 or 12.7).‡ Sections 12.7 and 12.8 deal with multiple comparisons among medians and variances, respectively.

In all multiple comparison testing, equal sample sizes are desirable for maximum power and robustness, but procedures are presented for analysis with unequal n.

This chapter considers multiple comparisons for the one-way analysis of variance experimental design. Other multiple comparison procedures are found in Sections 13.6 (for two-factor ANOVA), 13.9 (for nonparametric randomized block ANOVA), 13.10 (for dichotomous data in randomized blocks), 16.6 (for multiway ANOVA), 18.6 and 18.7 (for regression), and 19.8 (for correlation).

12.1 THE TUKEY TEST

The multiple comparison procedure exemplified by the Tukey test (also known as the "honestly significant difference test" and the "wholly significant difference test") considers the null hypothesis H_0: $\mu_B = \mu_A$ versus H_A: $\mu_B \neq \mu_A$, where the subscripts denote any possible pair of groups. For k groups, $k(k-1)/2$ different pairwise comparisons can be made.§ Example 12.1 demonstrates the testing procedure, utilizing

*For example, Bancroft (1968: Chapter 8), Dunnett (1970), Federer (1955: Section II-1), Harter (1957; 1970: Section 2), Steel and Torrie (1980: Chapter 8), Waldo (1976), and those listed in the reviews of Miller (1966, 1977).

†This results (except for Scheffé's test) in the probability of Type I error tending to be less than α (Bernhardson, 1975).

‡Howell and Games (1974) and Games and Howell (1976) discuss Behrens-Fisher procedures for when variances are unequal.

§The number of combinations of k groups taken 2 at a time is ${}_kC_2 = k!/[2!(k-2)!]$ (as explained in Section 22.1). This is equivalent to

$$ {}_kC_2 = \frac{k(k-1)(k-2)!}{2!(k-2)!} = \frac{k(k-1)}{2}. \tag{12.1} $$

Example 12.1

Tukey multiple comparison test with equal sample sizes. The data are strontium concentrations (mg/ml) in five different bodies of water.

First an analysis of variance is performed.

H_0: $\mu_1 = \mu_2 = \mu_3 = \mu_4 = \mu_5$

H_A: Mean strontium concentrations are not the same in all five bodies of water.

$\alpha = 0.05$.

Grayson's Pond	*Beaver Lake*	*Angler's Cove*	*Appletree Lake*	*Rock River*
28.2	39.6	46.3	41.0	56.3
33.2	40.8	42.1	44.1	54.1
36.4	37.9	43.5	46.4	59.4
34.6	37.1	48.8	40.2	62.7
29.1	43.6	43.7	38.6	60.0
31.0	42.4	40.1	36.3	57.3
$\bar{X}_1 = 32.1$ mg/ml	$\bar{X}_2 = 40.2$ mg/ml	$\bar{X}_3 = 44.1$ mg/ml	$\bar{X}_4 = 41.1$ mg/ml	$\bar{X}_5 = 58.3$ mg/ml
$n_1 = 6$	$n_2 = 6$	$n_3 = 6$	$n_4 = 6$	$n_5 = 6$

Source of variation	SS	DF	MS
Total	2437.5720	29	
Groups	2193.4420	4	548.3605
Error	244.1300	25	9.7652

$$F = \frac{548.3605}{9.7652} = 56.2$$

Since $F_{0.05(1),4,25} = 2.76$, H_0 is rejected.

Since a significant F resulted from the analysis of variance, the Tukey test is now applied on the means ranked in order of magnitude:

Samples ranked by mean (i)	1	2	4	3	5
Ranked sample means ($\bar{X}_i$, in mg/ml)	32.1	40.2	41.1	44.1	58.3

To test each H_0: $\mu_B = \mu_A$,

$$\mathrm{SE} = \sqrt{\frac{9.7652}{6}} = \sqrt{1.6275} = 1.28$$

Since $q_{0.05,25,k}$ does not appear in Table B.5, $q_{0.05,24,k}$ is utilized (as it is the critical value with the next lower DF).

Comparison (B vs. A)	*Difference* ($\bar{X}_B - \bar{X}_A$)	SE	q	$q_{0.05,24,5}$	*Conclusion*
5 vs. 1	58.3 − 32.1 = 26.2	1.28	20.47	4.166	Reject H_0: $\mu_5 = \mu_1$
5 vs. 2	58.3 − 40.2 = 18.1	1.28	14.14	4.166	Reject H_0: $\mu_5 = \mu_2$
5 vs. 4	58.3 − 41.1 = 17.2	1.28	13.44	4.166	Reject H_0: $\mu_5 = \mu_4$
5 vs. 3	58.3 − 44.1 = 14.2	1.28	11.09	4.166	Reject H_0: $\mu_5 = \mu_3$
3 vs. 1	44.1 − 32.1 = 12.0	1.28	9.38	4.166	Reject H_0: $\mu_3 = \mu_1$
3 vs. 2	44.1 − 40.2 = 3.9	1.28	3.05	4.166	Accept H_0: $\mu_3 = \mu_2$
3 vs. 4	Do not test				
4 vs. 1	41.1 − 32.1 = 9.0	1.28	7.03	4.166	Reject H_0: $\mu_4 = \mu_1$
4 vs. 2	Do not test				
2 vs. 1	40.2 − 32.1 = 8.1	1.28	6.33	4.166	Reject H_0: $\mu_2 = \mu_1$

Overall conclusion: $\mu_1 \neq \mu_2 = \mu_4 = \mu_3 \neq \mu_5$.

data similar to those in Example 11.1, except that all groups have equal numbers of data (i.e., all the n_i's are equal). Since the single factor analysis of variance rejects H_0: $\mu_1 = \mu_2 = \mu_3 = \mu_4 = \mu_5$, a multiple comparison test is called for to determine between which population means differences exist. The first step in the analysis is to arrange and number all five sample means in order of increasing magnitude. Then pairwise differences, $\bar{X}_B - \bar{X}_A$, are tabulated. Just as a difference between means divided by the appropriate standard error yields a t value (Section 9.6), a q value in the Tukey test is calculated by dividing a difference between means by

$$\text{SE} = \sqrt{\frac{s^2}{n}}, \tag{12.2}$$

where s^2 is the error mean square from the analysis of variance and n is the number of data in each of groups A and B. If this calculated q value,

$$q = \frac{\bar{X}_B - \bar{X}_A}{\text{SE}} \tag{12.3}$$

is equal to or greater than the critical value, $q_{\alpha,\nu,k}$, from Table B.5, then H_0: $\mu_B = \mu_A$ is rejected. The critical value in this test is known as a "Studentized range," abbreviated q, and is dependent upon α (the significance level), ν (the error DF for the analysis of variance), and k (the total number of means being tested).

The significance level, α, is the probability of encountering at least one Type I error (i.e., the probability of falsely rejecting at least one H_0) during the course of comparing all the pairs of means; it is not the probability of committing a Type I error for a single comparison. (That is, α for these multiple comparison tests is what statisticians refer to as "experimentwise error rate," rather than "comparisonwise error rate.")*

The conclusions reached by multiple comparison testing are dependent upon the order in which the pairwise comparisons are considered. The proper procedure is to compare first the largest mean against the smallest, then the largest against the next smallest, and so on, until the largest has been compared with the second largest. Then one compares the second largest with the smallest, the second largest with the next smallest, and so on. For example, after ranking four means in ascending order, the sequence of comparisons is as follows: means 4 vs. 1, 4 vs. 2, 4 vs. 3, 3 vs. 1, 3 vs. 2, 2 vs. 1. Another important procedural rule is that if no difference is found between two means, then it is concluded that no difference exists between any means enclosed by these two, and such differences are not tested for. Thus, in Example 12.1, since we conclude no difference to exist between population means 2 and 4, no tests are performed to judge the differences between means 3 and 4, or between means 2 and 3. Our conclusions in Example 12.1 are that sample 1 came from a population having a mean different from any of the other sampled populations; likewise, the population mean from which sample 5 came is different from any of the other population means; and samples 2, 4, and 3 came from populations having identical

*The Duncan's test, referred to at the start of this chapter, employs the comparisonwise error rate.

means. Therefore, we can conclude that $\mu_1 \neq \mu_2 = \mu_4 = \mu_3 \neq \mu_5$. In Example 12.1, each time a null hypothesis was accepted, a single line was drawn beneath the appropriate sample means to visualize the similarity of the population means that they estimate.

If the k group sizes are not equal, a slight modification in the Tukey procedure is necessary. For each comparison involving unequal n, the standard error* is calculated by the following approximation (Kramer, 1956; Tukey, 1953):

$$SE = \sqrt{\frac{s^2}{2}\left(\frac{1}{n_A} + \frac{1}{n_B}\right)}. \tag{12.4}$$

This is shown in Example 12.2, using the analysis of variance data of Example 11.1.

Example 12.2

The Tukey test with unequal sample sizes.

The data (in kg) are those from Example 11.1, where H_0: $\mu_1 = \mu_2 = \mu_3 = \mu_4$ was rejected.
$k = 4$
s^2 = error MS = 8.557
error DF = 15

Samples ranked by mean (i)	1	2	4	3
Ranked sample means ($\bar{X}_i$)	60.62	69.30	86.24	100.35
Sizes of samples (n_i)	5	5	5	4

Comparison (B vs. A)	*Difference* ($\bar{X}_B - \bar{X}_A$)	SE	q	$q_{0.05,15,4}$	*Conclusion*
3 vs. 1	100.35 − 60.62 = 39.73	1.39*	28.58	4.076	Reject H_0: $\mu_3 = \mu_1$
3 vs. 2	100.35 − 69.30 = 31.05	1.39	22.34	4.076	Reject H_0: $\mu_3 = \mu_2$
3 vs. 4	100.35 − 86.24 = 14.11	1.39	10.15	4.076	Reject H_0: $\mu_3 = \mu_4$
4 vs. 1	86.24 − 60.62 = 25.62	1.31†	19.56	4.076	Reject H_0: $\mu_4 = \mu_1$
4 vs. 2	86.24 − 69.30 = 16.94	1.31	12.93	4.076	Reject H_0: $\mu_4 = \mu_2$
2 vs. 1	69.30 − 60.62 = 8.68	1.31	6.63	4.076	Reject H_0: $\mu_2 = \mu_1$

*Since $n_3 \neq n_1$:

$$SE = \sqrt{\left(\frac{8.557}{2}\right)\left(\frac{1}{4} + \frac{1}{5}\right)} = \sqrt{(4.2785)(0.25 + 0.20)} = \sqrt{1.9253} = 1.39$$

†Since $n_4 = n_1$:

$$SE = \sqrt{\frac{8.557}{5}} = \sqrt{1.7114} = 1.31$$

Thus, we conclude that $\mu_1 \neq \mu_2 \neq \mu_4 \neq \mu_3$.

*This procedure has been shown to be excellent (e.g., Dunnett, 1980a; Keselman, Murray, and Rogan, 1976; Smith, 1971), often resulting in a probability of Type I error less than the stated α. Some authors have used the harmonic mean of all k sample sizes, which usually results in a probability of Type I error differing from the stated α more than does the use of Equation 12.4, and this probability is at times greater than α (Dunnett, 1980a). Multiple comparison testing procedures have been proposed (and compared by Dunnett, 1980b) for cases where population variances are unequal, but in such a situation we should consider using both nonparametric ANOVA and nonparametric multiple comparison testing.

The null hypothesis $H_0: \mu_B = \mu_A$ may, of course, also be written $H_0: \mu_B - \mu_A = 0$. The hypothesis $H_0: \mu_B - \mu_A = \mu_0$, where $\mu_0 \neq 0$, may also be tested; this is accomplished by replacing $\bar{X}_B - \bar{X}_A$ with $|\bar{X}_B - \bar{X}_A| - \mu_0$ in the numerator of Equation 12.3.

Not infrequently, a multiple comparison test will yield ambiguous results, in the form of conclusions of overlapping sets of similarities. For example, one might arrive at the following:

$$\bar{X}_1 \quad \bar{X}_2 \quad \bar{X}_3 \quad \bar{X}_4$$

for an experimental design consisting of four groups of data. Here the four samples appear to have come from populations among which there were two different population means: samples 1 and 2 were taken from one population and samples 2, 3, and 4 came from the second population. This is clearly impossible, for we have assigned sample 2 to both populations. The reason behind such a conclusion is that the multiple range test was not able to determine accurately from which population sample 2 came (i.e., at least one Type II error has been committed). Therefore, we can state that $\mu_1 \neq \mu_3 = \mu_4$, but we cannot conclude how μ_2 is related to the other population means. Repeating the analysis with a larger number of data would tend to yield more decisive conclusions, as the test would then have more power.

It is also possible for $H_0: \mu_1 = \mu_2 = \ldots = \mu_k$ to be rejected by an analysis of variance and for the subsequent Tukey test to fail to detect differences between any pair of means. This occurrence will not be encountered commonly, but it reflects the fact that the analysis of variance is a more powerful test than is the multiple comparison test (i.e., Type II errors are more likely to occur in multiple comparison testing than in performing an analysis of variance). Repeating the experiment with larger sample sizes would tend to result in a multiple comparison analysis more capable of locating differences among means.

12.2 THE NEWMAN-KEULS TEST

The Newman-Keuls test, also commonly referred to as the "Student-Newman-Keuls test" or "SNK test," is performed exactly as is Tukey's test, with one exception. The sample means are ranked, pairwise differences between means are determined, and a standard error is computed, just as in Section 12.1. Then the test statistic, q, is calculated just as is q for Tukey's test (i.e., by Equation 12.3). The difference in the Newman-Keuls procedure lies in the determination of the critical value, which is $q_{\alpha,\nu,p}$, where p is the number of means in the range of means being tested. To compare ranked means $\bar{X}_5$ and $\bar{X}_1$, for example, we are considering a range of 5 means, so $p = 5$; to test mean 5 vs. mean 2, $p = 4$, and so on; this is shown in Example 12.3.

A multiple comparison test of the type that employs different critical values for different ranges of means is called a *multiple range test*. Example 12.3 analyzes the data of Example 12.1 using the Newman-Keuls procedure.

If all k sample sizes are not equal, then SE is calculated using Equation 12.4, rather than Equation 12.2, just as for the Tukey test.

Example 12.3

The Newman-Keuls multiple range test applied to the data (in mg/ml) of Example 12.1.

First, an analysis of variance is performed and the sample means are ranked, just as in Example 12.1:

Samples ranked by mean (i)	1	2	4	3	5
Ranked sample means ($\bar{X}_i$)	32.1	40.2	41.1	44.1	58.3

SE = 1.28

Comparison (B vs. A)	*Difference* ($\bar{X}_B - \bar{X}_A$)	SE	q	p	$q_{0.05,24,p}$	*Conclusion*
5 vs. 1	26.2	1.28	20.47	5	4.166	Reject $H_0: \mu_5 = \mu_1$
5 vs. 2	18.1	1.28	14.14	4	3.901	Reject $H_0: \mu_5 = \mu_2$
5 vs. 4	17.2	1.28	13.44	3	3.532	Reject $H_0: \mu_5 = \mu_4$
5 vs. 3	14.2	1.28	11.09	2	2.919	Reject $H_0: \mu_5 = \mu_3$
3 vs. 1	12.0	1.28	9.38	4	3.901	Reject $H_0: \mu_3 = \mu_1$
3 vs. 2	3.9	1.28	3.05	3	3.532	Accept $H_0: \mu_3 = \mu_2$
3 vs. 4	Do not test	1.28				
4 vs. 1	9.0	1.28	7.03	3	3.532	Reject $H_0: \mu_4 = \mu_1$
4 vs. 2	Do not test	1.28				
2 vs. 1	8.1	1.28	6.33	2	2.919	Reject $H_0: \mu_2 = \mu_1$

Overall conclusion: $\mu_1 \neq \mu_2 = \mu_4 = \mu_3 \neq \mu_5$

In Examples 12.1 and 12.3, the Tukey and Newman-Keuls tests arrived at the same conclusions. This will not always be the case. The Newman-Keuls test tends to conclude more significant differences than does Tukey's procedure. (That is, the former is more powerful.) While some authors (e.g., Miller, 1966: 44, 88) have criticized the Tukey test as being unnecessarily conservative (i.e., too few significant differences are concluded), others (e.g., Einot and Gabriel, 1975; Ramsey, 1978) recommend against the Newman-Keuls test because it may falsely declare significant differences with a probability greater than α. While the lack of agreement among statisticians continues, I tend to favor the latter argument and shall recommend (although not vigorously) the Tukey test in preference to the Newman-Keuls.

12.3 CONFIDENCE INTERVALS FOLLOWING MULTIPLE COMPARISONS

Once it has been concluded which of three or more sample means are significantly different from which others, we can calculate confidence intervals for each different population mean. If one mean is declared different from all others, then we use Equation 11.22, introduced in Section 11.2:

$$\bar{X}_i \pm t_{\alpha(2),\nu}\sqrt{\frac{s^2}{n_i}}. \tag{11.22}$$

If two or more means are concluded to be the same, then a pooled sample mean is the best estimate of the underlying population mean:

$$\bar{X}_p = \frac{\sum n_i \bar{X}_i}{\sum n_i}, \tag{12.5}$$

where the summation is over all samples concluded to have come from the same population, and the confidence interval is, then,

$$\bar{X}_p \pm t_{\alpha(2),\nu}\sqrt{\frac{s^2}{\sum n_i}}, \tag{12.6}$$

again summing over all samples whose means are declared indistinguishable. This is analogous to the two-sample situation handled by Equation 9.16, and it is demonstrated in Example 12.4.

Example 12.4

Confidence intervals (CI) for the population means from Example 12.1.

It was concluded in Example 12.1 that $\mu_1 \neq \mu_2 = \mu_4 = \mu_3 \neq \mu_5$. Therefore, we may calculate confidence intervals for μ_1, for $\mu_{2,4,3}$, and for μ_5 (where $\mu_{2,4,3}$ indicates the mean of the common population from which samples 2, 4, and 3 came).

Using Equation 11.22:

$$95\%\text{ CI for }\mu_1 = \bar{X}_1 \pm t_{0.05(2),25}\sqrt{\frac{s^2}{n_1}} = 32.1 \pm (2.060)\sqrt{\frac{9.7652}{6}}$$
$$= 32.1\text{ mg/ml} \pm 2.6\text{ mg/ml}$$

Again using Equation 11.22:

$$95\%\text{ CI for }\mu_5 = \bar{X}_5 \pm t_{0.05(2),25}\sqrt{\frac{s^2}{n_5}} = 58.3\text{ mg/ml} \pm 2.6\text{ mg/ml}$$

Using Equation 12.5:

$$\bar{X}_p = \bar{X}_{2,4,3} = \frac{n_2\bar{X}_2 + n_3\bar{X}_3 + n_4\bar{X}_4}{n_2 + n_3 + n_4} = \frac{(6)(40.2) + (6)(41.1) + (6)(44.1)}{6 + 6 + 6} = 41.8\text{ mg/ml}$$

Using Equation 12.6:

$$95\%\text{ CI for }\mu_{2,4,3} = \bar{X}_{2,4,3} \pm t_{0.05(2),25}\sqrt{\frac{s^2}{6 + 6 + 6}} = 41.8\text{ mg/ml} \pm 1.5\text{ mg/ml}$$

Using Equation 12.7:

$$95\%\text{ CI for }\mu_5 - \mu_{2,4,3} = \bar{X}_5 - \bar{X}_{2,4,3} \pm q_{0.05,25,5}\sqrt{\frac{s^2}{2}\left(\frac{1}{n_5} + \frac{1}{n_2 + n_4 + n_3}\right)}$$
$$= 58.3 - 41.8 \pm (4.166)(1.04)$$
$$= 16.5\text{ mg/ml} \pm 4.3\text{ mg/ml}$$

Using Equation 12.7:

$$95\%\text{ CI for }\mu_{2,4,3} - \mu_1 = \bar{X}_{2,4,3} - \bar{X}_1 \pm q_{0.05,25,5}\sqrt{\frac{s^2}{2}\left(\frac{1}{n_2 + n_4 + n_3} + \frac{1}{n_1}\right)}$$
$$= 41.8 - 32.1 \pm (4.166)(1.04)$$
$$= 9.7\text{ mg/ml} \pm 4.3\text{ mg/ml}$$

If μ_A and μ_B are concluded to be different, the $1 - \alpha$ confidence interval for $\mu_B - \mu_A$ may be computed by Tukey's procedure as

$$(\bar{X}_B - \bar{X}_A) \pm (q_{\alpha,\nu,k})(\text{SE}). \tag{12.7}$$

Here, as in Section 12.1, ν is the error degrees of freedom in the analysis of variance, k is the total number of means, and SE is calculated as in either Equation 12.2 or 12.4, depending on whether or not n_A and n_B are equal. Thus, for example, the 95% confidence interval for $\mu_4 - \mu_1$ in Example 12.2 would be calculated as $(86.24 - 60.62) \pm (4.076)(1.31) = 25.62$ kg $\pm$ 5.34 kg and $L_1 = 20.28$ kg and $L_2 = 30.96$ kg. This procedure is further demonstrated in Example 12.4.

The Tukey procedure for computing confidence intervals for differences between means should be used regardless of whether the Tukey or the Newman-Keuls test was used to determine which means are different.

Sample Size and Estimation of the Difference between Two Population Means. In Section 9.6 it was shown how to estimate the sample size required in order to obtain a confidence interval of specified width for a difference between the population means associated with a two-sample test. In the multisample situation, we speak of all differences between pairs of population means, using a similar procedure but one that employs the q, rather than the t statistic. As in Section 9.6, iteration is necessary, whereby we determine n such that

$$n = \frac{s^2(q_{\alpha,\mathrm{DF},k})^2 F_{\beta(1),\mathrm{DF},\nu}}{d^2} \tag{12.8}$$

(Steel and Torrie, 1980: 232–233; Tukey, 1953). Here, d is the half-width of the $1 - \alpha$ confidence interval, $1 - \beta$ is the probability that the confidence interval is no wider than specified, s^2 is the estimate of error variance, with ν degrees of freedom, and k is the total number of means. DF is the error degrees of freedom that the experiment would have with the estimated n (i.e., $\mathrm{DF} = k(n - 1)$ for a single-factor ANOVA). This procedure is demonstrated in Example 12.5.

Example 12.5

Estimation of sample size needed to achieve a stated precision in estimating difference between pairs of means in an ANOVA experimental design, using the data of Example 12.1.

If we specify that we wish to estimate each $\mu_A - \mu_B$ (where it has been concluded that $\mu_A \neq \mu_B$) by having a 95% probability that the 99% confidence interval will be no wider than 10.0 mg/ml, then $d = 5.0$ mg/ml, $1 - \beta = 0.95$, $\beta = 0.05$, $1 - \alpha = 0.99$, and $\alpha = 0.01$. From Example 12.1, we have an estimate of the population variance, namely $s^2 = 9.7652\ (\mathrm{mg/ml})^2$ with $\nu = 25$; also $k = 4$.

Let us guess that a sample size of 20 is necessary; then, $q_{0.01,4(20-1),4} = q_{0.01,76,4} \approx q_{0.01,60,4} = 4.595$* and $F_{0.05(1),4(20-1),25} = F_{0.05(1),76,25} \approx F_{0.05(1),70,25} = 1.81$,* so we estimate (by Equation 12.8):

$$n = \frac{(9.7652)(4.595)^2(1.81)}{(5.0)^2} = 14.9.$$

Next, we might guess $n = 15$, for which $q_{0.01,4(15-1),4} = q_{0.01,56,4} \approx q_{0.01,40,4} = 4.696$ and $F_{0.05(1),4(15-1),25} = F_{0.05(1),56,25} \approx F_{0.05(1),50,25} = 1.84$, and calculate:

$$n = \frac{(9.7652)(4.696)^2(1.84)}{(5.0)^2} = 15.8.$$

Now, guessing $n = 16$, we have $q_{0.01,60,4} = 4.595$ and $F_{0.05(1),60,25} = 1.82$, and we calculate:

$$n = \frac{(9.7652)(4.595)^2(1.82)}{(5.0)^2} = 15.0.$$

Therefore, we have concluded that the desired n lies in the neighborhood of 15 to 16, so at least 16 data should be collected from each of the four populations in order to achieve the specified confidence interval.

*In Tables B.5–B.7, degrees of freedom of 76 are not to be found. In cases such as this, we may interpolate (as explained at the beginning of Appendix B), or, conservatively, we may employ the next lower degrees of freedom, as shown in this example.

12.4 COMPARISON OF A CONTROL MEAN TO EACH OTHER GROUP MEAN

Sometimes the objective of multisample experiments with k samples, or groups, is to determine whether the mean of one group, designated as a "control," differs significantly from each of the means of the $k - 1$ other groups. Dunnett (1955) has provided a procedure for such testing, which differs from the multiple comparison approaches in Sections 12.1 and 12.2 in that the investigator is here not interested in all possible comparisons of pairs of group means, but only in those $k - 1$ comparisons involving the "control" group. Knowing k, the total number of groups in the experiment, and ν, the error degrees of freedom from the analysis of variance for H_0: $\mu_1 = \mu_2 = \ldots = \mu_k$, one obtains critical values from either Table B.6 or Table B.7, depending on whether the hypotheses are to be one-tailed or two-tailed, respectively. We shall refer to these tabled values as $q'_{\alpha,\nu,p}$, for they are used in a manner similar to that of the $q_{\alpha,\nu,p}$ values employed in the Newman-Keuls test. As in the SNK procedure, the error rate, α, denotes the probability of committing a Type I error somewhere among all of the pairwise comparisons made. The standard error for Dunnett's test is

$$\mathrm{SE} = \sqrt{\frac{2s^2}{n}} \tag{12.9}$$

where group sizes are equal, or

$$\mathrm{SE} = \sqrt{s^2\left(\frac{1}{n_A} + \frac{1}{n_{\text{control}}}\right)} \tag{12.10}$$

when group sizes are not equal. The test statistic, analogous to that of Equation 12.3, is

$$q = \frac{\bar{X}_{\text{control}} - \bar{X}_A}{\mathrm{SE}}. \tag{12.11}$$

The Dunnett testing procedure is demonstrated in Example 12.6 for one-tailed hypotheses. The critical values, $q'_{\alpha(1),\nu,p}$, are in Table B.6. For a two-tailed test, if $|q'| \geq q'_{\alpha(2),\nu,p}$, then H_0: $\mu_{\text{control}} = \mu_A$ is rejected. In a one-tailed test H_0: $\mu_{\text{control}} \leq \mu_A$ would be rejected if $q \geq q'_{\alpha(1),\nu,p}$; and H_0: $\mu_{\text{control}} \geq \mu_A$ would be rejected if $|q| \geq q'_{\alpha(1),\nu,p}$ and $\bar{X}_{\text{control}} < \bar{X}_A$ (i.e., if $q < -q'_{\alpha(1),\nu,p}$).

Example 12.6

Dunnett's test for comparing a control mean to each other group mean.

The yield (in metric tons per hectare) of each of several plots of potatoes has been determined after a season's application of a standard fertilizer. Likewise, the potato yields from several plots were determined for each of four new fertilizers. A manufacturer wishes to sell us one of the four new fertilizers, claiming they will increase potato crop yield. A total of 80 plots is available for use in this experiment.

Optimum allocation of plots among the 5 fertilizer groups will be such that the control group (let us say it is group 2) has a little less than $\sqrt{k-1} = \sqrt{4} = 2$ times as many data as each of the other groups. Therefore, we use $n_2 = 24$ and $n_1 = n_3 = n_4 = n_5 = 14$.

An analysis of variance was performed and H_0: $\mu_1 = \mu_2 = \mu_3 = \mu_4 = \mu_5$ was rejected; for this analysis, error MS = 10.42 (metric tons/ha)^2 and error DF = 75.

$$\mathrm{SE} = \sqrt{10.42\left(\frac{1}{14} + \frac{1}{24}\right)} = 1.1 \text{ metric tons/ha}$$

Groups ranked by mean (i)	1	2	3	4	5
Ranked group means ($\bar{X}_i$)	17.3	21.7	22.1	23.6	27.8

As the control group (i.e., the group with the standard fertilizer) is group 2, we wish to test each H_0: $\mu_2 \geq \mu_A$ against H_A: $\mu_2 < \mu_A$.

Comparison (2 vs. A)	*Difference* ($\bar{X}_2 - \bar{X}_A$)	SE	$\lvert q \rvert$	p	$q'_{0.05(1),75,p}$	*Conclusion*
2 vs. 1	21.7 − 17.3 = 4.4			2		Since $\bar{X}_2 > \bar{X}_1$, accept H_0: $\mu_2 \geq \mu_1$.
2 vs. 5	27.8 − 21.7 = 6.1	1.1	5.55	4	2.10	Reject H_0: $\mu_2 \geq \mu_5$.
2 vs. 4	23.6 − 21.7 = 1.9	1.1	1.73	3	1.95	Accept H_0: $\mu_2 \geq \mu_4$.
2 vs. 3	Do not test					

Conclusion: Only fertilizer 5 produces a yield greater than the yield from the control fertilizer (fertilizer 2).

The null hypothesis H_0: $\mu_{\text{control}} = \mu_A$ is, of course, a special case of H_0: $\mu_{\text{control}} - \mu_A = \mu_0$, where $\mu_0 = 0$. Other values of μ_0 may appear in the hypothesis, however, and such hypotheses may be tested by placing $\lvert \bar{X}_{\text{control}} - \bar{X}_A \rvert - \mu_0$ in the numerator of the q calculation. In a similar manner, H_0: $\mu_{\text{control}} - \mu_A \leq \mu_0$ or H_0: $\mu_{\text{control}} - \mu_A \geq \mu_0$ may be tested.

When comparisons of group means to a control mean is an investigator's principal desire, the control group ought to contain more observations than the other groups. Dunnett (1955) showed that the optimal size of the control group typically should be a little less than $\sqrt{k-1}$ times the size of each other group.

Sample Size and Estimation of the Difference between One Population Mean and the Mean of a Control Population. This situation is similar to that discussed at the end of Section 12.3 but it pertains to one of the two means of the population being designated as the control. The procedure would use Equation 12.8 substituting the Dunnett's test critical value, $q'_{\alpha(2),\text{DF},p}$, for the Tukey critical value, $q_{\alpha(2),\text{DF},k}$.

Confidence Intervals for Differences between Control and Other Group Means. Using Dunnett's q' statistic, and the SE of Equation 12.9 or 12.10, we may calculate confidence limits for the difference between the control mean and each of the other group means:

$$1 - \alpha \text{ CI for } \mu_{\text{control}} - \mu_A = (\bar{X}_{\text{control}} - \bar{X}_A) \pm (q'_{\alpha(2),\nu,p})(\text{SE}). \qquad (12.12)$$

One-tailed confidence limits are also possible. One can express 95% confidence that a difference, $\mu_{\text{control}} - \mu_A$ is no less than (i.e., is at least as large as)

$$(\bar{X}_{\text{control}} - \bar{X}_A) - (q'_{\alpha(1),\nu,p})(\text{SE}), \qquad (12.13)$$

or it might be desired to state that the difference is no greater than

$$(\bar{X}_{\text{control}} - \bar{X}_A) + (q'_{\alpha(1),\nu,p})(\text{SE}). \qquad (12.14)$$

12.5 SCHEFFÉ'S MULTIPLE CONTRASTS

Scheffé (1953; 1959: Section 3.4, 3.5) has provided a multiple comparison procedure (sometimes called the S test) that can be used to test null hypotheses of the form $H_0\colon \mu_B - \mu_A = 0$, just as can the Tukey or Newman-Keuls tests (but with less power); it can also be used for the special considerations that will be shown presently.

For $H_0\colon \mu_B - \mu_A = 0$, one computes

$$S = \frac{|\bar{X}_B - \bar{X}_A|}{\text{SE}}, \tag{12.15}$$

where

$$\text{SE} = \sqrt{s^2\left(\frac{1}{n_B} + \frac{1}{n_A}\right)}; \tag{12.16}$$

and the critical value is

$$S_\alpha = \sqrt{(k-1)F_{\alpha(1),k-1,N-k}}. \tag{12.17}$$

In the preceding equations, s^2 is the error mean square, and $k - 1$ and $N - k$ are the groups DF and error DF, respectively, from the analysis of variance. Note that this test does not require equal sample sizes.

Example 12.7 demonstrates Scheffé's procedure for the same data as analyzed in Example 12.1. In this case, the use of this test yields the same conlusions as does the Tukey test. Such will not always be true; in fact, it is common to have the former procedure detect fewer differences than the latter. That is, Scheffé's test is more apt to have us commit Type II errors. Because of its relative lack of power it is not generally recommended for the types of multiple comparisons thus far described.

Example 12.7

Scheffé's test for the hypotheses and data of Example 12.1.

For $\alpha = 0.05$, the critical value, S_α, for each comparison, is

$$\sqrt{(k-1)F_{0.05(1),k-1,N-k}} = \sqrt{(4)(2.76)} = \sqrt{11.04} = 3.32.$$

Since $n_B = n_A$ for each comparison, the SE for each comparison is

$$\sqrt{s^2\left(\frac{1}{n_B} + \frac{1}{n_A}\right)} = \sqrt{(9.7652)(0.33)} = \sqrt{3.2547} = 1.80.$$

Comparison (B vs. A)	*Difference* $\bar{X}_B - \bar{X}_A$	SE	S	S_α	*Conclusion*
5 vs. 1	26.2	1.80	14.56	3.32	Reject $H_0\colon \mu_5 - \mu_1 = 0$
5 vs. 2	18.1	1.80	10.06	3.32	Reject $H_0\colon \mu_5 - \mu_2 = 0$
5 vs. 4	17.2	1.80	9.56	3.32	Reject $H_0\colon \mu_5 - \mu_4 = 0$
5 vs. 3	14.2	1.80	7.89	3.32	Reject $H_0\colon \mu_5 - \mu_3 = 0$
3 vs. 1	12.0	1.80	6.67	3.32	Reject $H_0\colon \mu_3 - \mu_1 = 0$
3 vs. 2	3.9	1.80	2.17	3.32	Accept $H_0\colon \mu_3 - \mu_2 = 0$
3 vs. 4	Do not test				
4 vs. 1	9.0	1.80	5.00	3.32	Reject $H_0\colon \mu_4 - \mu_1 = 0$
4 vs. 2	Do not test				
2 vs. 1	8.1	1.80	4.50	3.32	Reject $H_0\colon \mu_2 - \mu_1 = 0$

Overall conclusion: $\mu_1 \neq \mu_2 = \mu_4 = \mu_3 \neq \mu_5$.

However, there are hypotheses for which the S test is better qualified than other procedures. These refer to considerations spoken of as "multiple contrasts." Utilizing once again the data of Example 12.1, one might have proposed the hypothesis H_0: the mean strontium concentration in areas 2, 4, and 3 is no different from the mean concentration in area 5. This hypothesis could be stated concisely as H_0: $(\mu_2 + \mu_4 + \mu_3)/3 - \mu_5 = 0$. The S test then considers that $(\mu_2 + \mu_4 + \mu_3)/3 - \mu_5$ can be expressed as $\mu_2/3 + \mu_4/3 + \mu_3/3 - \mu_5$, so that μ_2, μ_4, μ_3, and μ_5 are preceded by coefficients, c, of $c_2 = \frac{1}{3}$, $c_4 = \frac{1}{3}$, $c_3 = \frac{1}{3}$, and $c_5 = -1$, respectively. The test statistic, S, is calculated as

$$S = \frac{|\sum c_i \bar{X}_i|}{\text{SE}}, \tag{12.18}$$

where

$$\text{SE} = \sqrt{s^2\left(\sum \frac{c_i^2}{n_i}\right)}. \tag{12.19}$$

Note that the sum of the coefficients is always zero; i.e., $\sum c_i = 0$.

With $k = 5$ we could consider H_0: $\mu_1 - (\mu_2 + \mu_3 + \mu_4)/3 = 0$, or H_0: $(\mu_1 + \mu_5)/2 - (\mu_2 + \mu_3 + \mu_4)/3 = 0$, or H_0: $(\mu_1 + \mu_4)/2 - (\mu_2 + \mu_3)/2 = 0$, or other contrasts. These several hypotheses are tested in Example 12.8. In grouping means to

Example 12.8

Scheffé's test for multiple contrasts, using the data of Example 12.1.

For $\alpha = 0.05$, the critical value, S_α, for each contrast is

$$\sqrt{(k-1)F_{0.05(1),k-1,N-k}} = \sqrt{(5-1)F_{0.05(1),4,25}} = \sqrt{4(2.76)} = 3.32$$

Contrast	SE	S	*Conclusion*
$\frac{\bar{X}_2 + \bar{X}_3 + \bar{X}_4}{3} - \bar{X}_5 = 41.8 - 58.3 = -16.5$	$\sqrt{9.7652\left[\frac{(\frac{1}{3})^2}{6} + \frac{(\frac{1}{3})^2}{6} + \frac{(\frac{1}{3})^2}{6} + \frac{(1)^2}{6}\right]} = 1.47$	11.22	Reject H_0: $\frac{\mu_2 + \mu_3 + \mu_4}{3} - \mu_5 = 0$
$\bar{X}_1 - \frac{\bar{X}_2 + \bar{X}_3 + \bar{X}_4}{3} = 32.1 - 41.8 = -9.7$	$\sqrt{9.7652\left[\frac{(1)^2}{6} + \frac{(\frac{1}{3})^2}{6} + \frac{(\frac{1}{3})^2}{6} + \frac{(\frac{1}{3})^2}{6}\right]} = 1.47$	6.60	Reject H_0: $\mu_1 - \frac{\mu_2 + \mu_3 + \mu_4}{3} = 0$
$\frac{\bar{X}_1 + \bar{X}_5}{2} - \frac{\bar{X}_2 + \bar{X}_3 + \bar{X}_4}{3} = 45.2 - 41.8 = 3.4$	$\sqrt{9.7652\left[\frac{(\frac{1}{2})^2}{6} + \frac{(\frac{1}{2})^2}{6} + \frac{(\frac{1}{3})^2}{6} + \frac{(\frac{1}{3})^2}{6} + \frac{(\frac{1}{3})^2}{6}\right]} = 1.16$	2.93	Accept H_0: $\frac{\mu_1 + \mu_5}{2} - \frac{\mu_2 + \mu_3 + \mu_4}{3} = 0$
$\frac{\bar{X}_1 + \bar{X}_4}{2} - \frac{\bar{X}_2 + \bar{X}_3}{2} = 36.6 - 42.15 = -5.55$	$\sqrt{9.7652\left[\frac{(\frac{1}{2})^2}{6} + \frac{(\frac{1}{2})^2}{6} + \frac{(\frac{1}{2})^2}{6} + \frac{(\frac{1}{2})^2}{6}\right]} = 1.28$	4.34	Reject H_0: $\frac{\mu_1 + \mu_4}{2} - \frac{\mu_2 + \mu_3}{2} = 0$

be used as contrasts, we should have some reasonable basis for defining the groups. For example, the first body of water in Example 12.1 might be rural and the others urban; or the first four might be standing water and the fifth a river.

If one group of data has been designated a "control" to which all other groups are to be compared, we may compare each group to that control by Dunnett's test (Section 12.4). Combinations of groups may be compared to the control using Scheffé's test, although the test proposed by Shaffer (1977) may be more powerful for simple contrasts and large k.

Confidence Intervals for Contrasts. For the difference between two means, we may calculate a $1 - \alpha$ confidence interval by Tukey's method (Equation 12.7) or by the Scheffé procedure:

$$(\bar{X}_B - \bar{X}_A) \pm S_\alpha \sqrt{s^2\left(\frac{1}{n_A} + \frac{1}{n_B}\right)}, \tag{12.20}$$

but the former will result in a narrower confidence interval and is, therefore, preferable.

However, the Scheffé procedure enables us to establish confidence limits for a contrast. In general, the $1 - \alpha$ confidence interval for a contrast is

$$\sum c_i \bar{X}_i + S_\alpha \mathrm{SE}, \tag{12.21}$$

(of which Equation 12.20 is a special case). SE is as in Equation 12.19. Example 12.9 demonstrates this for the significantly different contrasts of Example 12.8.

Example 12.9

Confidence intervals for the significantly different contrasts of Example 12.8.

The critical value, S_α, for each confidence interval is $\sqrt{(k-1)F_{\alpha(1),k-1,N-k}}$. This is the same critical value as found in Example 12.8; that is, $S_{0.05} = 3.32$.

95% confidence interval for $\frac{\mu_2 + \mu_3 + \mu_4}{3} - \mu_5$

$$= \left(\frac{\bar{X}_2 + \bar{X}_3 + \bar{X}_4}{3} - \bar{X}_5\right) \pm S_\alpha \mathrm{SE}$$

$$= -16.5 \pm (3.32)(1.47)$$

$$= -16.5 \text{ mg/ml} \pm 4.9 \text{ mg/ml}$$

$L_1 = -21.4$ mg/ml

$L_2 = -11.6$ mg/ml

95% confidence interval for $\mu_1 - \frac{\mu_2 + \mu_3 + \mu_4}{3}$

$$= \left(\bar{X}_1 - \frac{\bar{X}_2 + \bar{X}_3 + \bar{X}_4}{3}\right) \pm S_\alpha \mathrm{SE}$$

$$= -9.7 \pm (3.32)(1.47)$$

$$= -9.7 \text{ mg/ml} \pm 4.9 \text{ mg/ml}$$

$L_1 = -14.6$ mg/ml

$L_2 = -4.8$ mg/ml

12.6 NONPARAMETRIC MULTIPLE COMPARISONS

Let us consider the multisample situation where the nonparametric Kruskal-Wallis test is applied and the null hypothesis is rejected. In such a case, the experimenter usually will desire to determine between which of the samples significant differences occur.

Nonparametric multiple comparisons may be effected in a fashion paralleling the Tukey test (Section 12.1), by using rank sums instead of means. This is demonstrated in Example 12.10. The rank sums from the Kruskal-Wallis test are first arranged in increasing order of magnitude. Pairwise differences between rank sums are then tabulated, starting with the difference between the largest and smallest rank sums, and proceeding in the same sequence as described in Section 12.1. The standard error is calculated as

$$SE = \sqrt{\frac{n(nk)(nk+1)}{12}} \tag{12.22}$$

(Nemenyi, 1963; Wilcoxon and Wilcox, 1964: 10),* and the tabled Studentized range to be used is $q_{\alpha,\infty,k}$.

Example 12.10

Nonparametric Tukey-type multiple comparisons. The data are those from Example 11.9, where the Kuskal-Wallis test rejected the null hypothesis that fly abundance is the same for the three different vegetation heights sampled.

$$SE = \sqrt{\frac{n(nk)(nk+1)}{12}} = \sqrt{\frac{5(15)(16)}{12}} = \sqrt{100} = 10.00$$

Samples ranked by rank sums (i)	3	2	1
Rank sums (R_i)	26	30	64

Comparison (B vs. A)	*Difference* ($R_B - R_A$)	SE	q	$q_{0.05,\infty,3}$	*Conclusion*
1 vs. 3	64 − 26 = 38	10.00	3.80	3.314	Reject H_0: Fly abundance is the same at vegetation heights 3 and 1.
1 vs. 2	64 − 30 = 34	10.00	3.40	3.314	Reject H_0: Fly abundance is the same at vegetation heights 2 and 1.
2 vs. 3	30 − 26 = 4	10.00	0.40	3.314	Accept H_0: Fly abundance is the same at vegetation heights 3 and 2.

Overall conclusion: Fly abundance is the same at vegetation heights 3 and 2 but is different at height 1.

*Some authors (e.g., Miller, 1966: 166) perform this test in an equivalent fashion by considering the difference between mean ranks ($\bar{R}_A$ and $\bar{R}_B$) rather than rank sums (R_A and R_B), in which case the appropriate standard error would be

$$SE = \sqrt{\frac{k(nk+1)}{12}}. \tag{12.23}$$

The above is a Tukey-type multiple comparison test, a nonparametric analog to the procedure of Section 12.1. If an analog is preferred to the Student-Newman-Keuls multiple range test of Section 12.2, the standard error would be

$$\mathrm{SE} = \sqrt{\frac{n(np)(np+1)}{12}} \qquad (12.24)$$

and $q_{\alpha,\infty,p}$ would be the critical value.*

The above multiple comparison testing requires that there be equal numbers of data in each of the k groups. If such is not the case, then we may use the procedure of Section 12.7; but a more powerful test is the following.

Dunn (1964; Hollander and Wolfe, 1973: 125) proposed using a standard error of

$$\mathrm{SE} = \sqrt{\frac{N(N+1)}{12}\left(\frac{1}{n_A} + \frac{1}{n_B}\right)} \qquad (12.26)$$

for a test statistic we shall call

$$Q = \frac{\bar{R}_B - \bar{R}_A}{\mathrm{SE}}, \qquad (12.27)$$

where $\bar{R}$ indicates a mean rank (i.e., $\bar{R}_A = R_A/n_A$ and $\bar{R}_B = R_B/n_B$). If tied ranks are present, then the following is preferred to Equation 12.26 (Dunn, 1964):

$$\mathrm{SE} = \sqrt{\left(\frac{N(N+1)}{12} - \frac{\sum T}{12(N-1)}\right)\left(\frac{1}{n_A} + \frac{1}{n_B}\right)}. \qquad (12.28)$$

In the latter equation, $\sum T$ is as used in the Kruskal-Wallis test when ties are present and is defined in Equation 11.33. Critical values for this test, $Q_{\alpha,k}$, are given in Table B.14. The testing procedure is demonstrated in Example 12.11; note that it is the mean ranks ($\bar{R}_i$), rather than the rank sums (R_i), that are arranged in order of magnitude.

Example 12.11

Nonparametric multiple comparisons with unequal sample sizes. The data are those from Example 11.10, where the Kruskall-Wallis test rejected the null hypothesis that water pH was the same in all four ponds examined.

$\sum T = 168$, as in Example 11.10.

For $n_A = 8$ and $n_B = 8$,

$$\mathrm{SE} = \sqrt{\left(\frac{N(N+1)}{12} - \frac{\sum T}{12(N-1)}\right)\left(\frac{1}{n_A} + \frac{1}{n_B}\right)}$$

$$= \sqrt{\left(\frac{31(32)}{12} - \frac{168}{12(30)}\right)\left(\frac{1}{8} + \frac{1}{8}\right)}$$

$$= \sqrt{20.5500} = 4.53$$

For $n_A = 7$ and $n_B = 8$,

$$\mathrm{SE} = \sqrt{\left(\frac{31(32)}{12} - \frac{168}{12(30)}\right)\left(\frac{1}{7} + \frac{1}{8}\right)} = \sqrt{22.0179} = 4.69$$

*If mean ranks, rather than rank sums, are used, then

$$\mathrm{SE} = \sqrt{\frac{p(np+1)}{12}}. \qquad (12.25)$$

Samples ranked by mean ranks (i)	1	2	4	3
Rank sums (R_i)	55	132.5	163.5	145
Sample sizes (n_i)	8	8	8	7
Mean ranks ($\bar{R}_i$)	6.88	16.56	20.44	20.71

Comparison (B vs. A)	*Difference* ($\bar{R}_B - \bar{R}_A$)	SE	Q	$Q_{0.05,4}$	*Conclusion*
3 vs. 1	20.71 − 6.88 = 13.83	4.69	2.95	2.639	Reject H_0: Water pH is the same in ponds 1 and 3.
3 vs. 2	20.71 − 16.56 = 4.15	4.69	0.88	2.639	Accept H_0: Water pH is the same in ponds 2 and 3.
3 vs. 4	Do not test				
4 vs. 1	20.44 - 6.88 = 13.56	4.53	2.99	2.639	Reject H_0
4 vs. 2	Do not test				
2 vs. 1	16.56 - 6.88 = 9.68	4.53	2.14	2.639	Accept H_0

Overall conclusion: Water pH is the same in ponds 2, 4, and 3 but is different in pond 1.

Nonparametric Comparisons of a Control to Other Groups. Subsequent to a Kruskal-Wallis test in which H_0 is rejected, a nonparametric analysis may be performed to seek either one-tailed or two-tailed significant differences between one group (designated as the "control") and each of the other groups of data. This is done in a manner paralleling that of the procedure of Section 12.4, but using group rank sums instead of group means. The standard error to be calculated is

$$SE = \sqrt{\frac{n(np)(np+1)}{6}} \tag{12.29}$$

(Wilcoxon and Wilcox, 1964: 11), and one uses critical values of either $q'_{\alpha(1),\infty,p}$ or $q'_{\alpha(2),\infty,p}$ (from Table B.6 or Table B.7, respectively) for one-tailed or two-tailed hypotheses, respectively.

The preceding nonparametric test requires equal sample sizes. [If the n's are not all equal, then the procedure suggested by Dunn (1964; Hollander and Wolfe, 1973: 131) may be employed.] By this method group B is considered to be the control and uses Equation 12.27, where the appropriate standard error is that of Equation 12.26 or 12.28, depending on whether there are ties or no ties, respectively. We shall refer to critical values for this test, which may be two-tailed or one-tailed, as $Q'_{\alpha,k}$; and they are given in Table B.15.

12.7 NONPARAMETRIC MULTIPLE CONTRASTS

Multiple contrasts, introduced in Section 12.5, can be tested nonparametrically using the Kruskal-Wallis H statistic rather than F. As an analog of Equation 12.18, we compute

$$S = \frac{|\sum c_i \bar{R}_i|}{SE}, \tag{12.30}$$

where

$$\mathrm{SE} = \sqrt{\left(\frac{N(N+1)}{12}\right)\left(\sum \frac{c_i^2}{n_i}\right)}, \qquad (12.31)$$

unless there are tied ranks, in which cases we use

$$\mathrm{SE} = \sqrt{\left(\frac{N(N+1)}{12} - \frac{\sum T}{12(N-1)}\right)\left(\sum \frac{c_i^2}{n_i}\right)}, \qquad (12.32)$$

where $\sum T$ is as in Equation 11.33. The critical value for these multiple contrasts is $\sqrt{H_{\alpha,n_1,n_2,\ldots}}$, using Table B.12 to obtain the critical values of H. If the needed critical value of H is not on that table, then χ^2_{k-1} may be used.

12.8 MULTIPLE COMPARISONS AMONG MEDIANS

If the null hypothesis is rejected in a multisample median test (Section 11.5), then it is usually desirable to ascertain between which groups significant differences exist. A Tukey-type multiple comparison test has been provided by Levy (1979), using

$$q = \frac{f_{1B} - f_{1A}}{\mathrm{SE}}. \qquad (12.33)$$

As shown in Example 12.12, we employ the values of f_{1j} for each group, where f_{1j} is the number of data in group j that are greater than the grand median. (The values of f_{1j} are the observed frequencies in the first row in the contingency table used in the multisample median test of Section 11.5.) The values of f_{1j} are ranked and pairwise differences between the ranks are examined as in other Tukey-type tests. The appropriate standard error is

$$\mathrm{SE} = \sqrt{\frac{n(N+1)}{4N}} \qquad (12.34)$$

if N, the total number of data in all groups, is odd, or

$$\mathrm{SE} = \sqrt{\frac{nN}{4(N-1)}} \qquad (12.35)$$

if N is even. The critical values to be used are $q_{\alpha,k,\infty}$. This multiple comparison procedure, demonstrated in Example 12.12, appears to possess low statistical power.

The above procedure is for equal sample sizes (n). If the sample sizes are slightly unequal the test can be used by employing the harmonic mean sample size,

$$n = \frac{k}{\sum_{j=1}^{k} \frac{1}{n_j}} \qquad (12.36)$$

for an approximate result.

Example 12.12

Tukey-type multiple comparison for differences among medians, using the data of Example 11.11.

Samples ranked by f_{1j} (j)	2	1	4	3
Ranked f_{1j}	2	3	6	9
Sample sizes (n_j)	12	12	12	12

$k = 4$
$N = 12 + 12 + 12 + 12 = 48$
$n = 12$
By Equation 12.35,

$$\text{SE} = \sqrt{\frac{(12)(48)}{4(48-1)}} = 1.750$$

H_0: Median of population B = Median of population A
H_A: Median of population $B \neq$ Median of population A

Comparison	$f_{1B} - f_{1A}$	SE	q	$q_{0.05,4,\infty}$	*Conclusion*
3 vs. 2	$9 - 2 = 7$	1.750	4.000	3.633	Reject H_0.
3 vs. 1	$9 - 3 = 6$	1.750	3.429	3.633	Accept H_0.
3 vs. 4	Do not test				
4 vs. 2	$6 - 2 = 4$	1.750	2.286	3.633	Accept H_0.
4 vs. 1	Do not test				
1 vs. 2	Do not test				

Overall conclusion: The medians of populations 3 and 2 (i.e., south and east—see Example 11.11) are not the same; but the test lacks the power to allow clear conclusions about the medians of populations 4 and 1.

12.9 MULTIPLE COMPARISONS AMONG VARIANCES

If the null hypothesis that k population variances are all equal (see Section 11.8) is rejected, then we may wish to determine which of the variances differ from which others. Levy (1975a, 1975c) suggests multiple comparison procedures for this purpose based on a logarithmic transformation of sample variances.

A test analogous to the Tukey test of Section 12.1 is performed by calculating

$$q = \frac{\ln s_B^2 - \ln s_A^2}{\text{SE}}, \tag{12.37}$$

where

$$\text{SE} = \sqrt{\frac{2}{\nu}}, \tag{12.38}$$

if both samples being compared are of equal size. If $\nu_A \neq \nu_B$, then the above procedure is not usable, but we can employ

$$\text{SE} = \sqrt{\frac{1}{\nu_A} + \frac{1}{\nu_B}}. \tag{12.39}$$

Just as in Sections 12.1 and 12.2, the subscripts A and B refer to the pair of groups being compared; and the sequence of pairwise comparisons must follow that given in those sections. This is demonstrated in Example 12.13.* The critical value for this test is $q_{\alpha,\infty,k}$ (from Table B.5).

*Recall (as in Section 11.8) that "ln" refers to natural logarithms (i.e., logarithms using base e). If one prefers using common logarithms ("log"; logarithms in base 10), then

$$q = \frac{2.30259(\log s_B^2 - \log s_A^2)}{\text{SE}}. \tag{12.40}$$

Example 12.13

Tukey-type multiple comparison test for differences among four variances (i.e., $k = 4$).

i	s_i^2	n_i	ν_i	$\ln s_i^2$
1	2.74 g^2	50	49	1.0080
2	2.83 g^2	48	47	1.0403
3	2.20 g^2	50	49	0.7885
4	6.42 g^2	50	49	1.8594

Samples ranked by variances (i)	3	1	2	4
Ranked logarithms of sample variances ($\ln s_i^2$)	0.7885	1.0080	1.0403	1.8594
Sample degrees of freedom (ν_i)	49	49	47	49

Comparison (B vs. A)	*Difference* ($\ln s_B^2 - \ln s_A^2$)	SE	q	$q_{0.05,\infty,4}$	*Conclusions*
4 vs. 3	1.8594 − 0.7885 = 1.0709	0.202*	5.301	3.633	Reject H_0: $\sigma_4^2 = \sigma_3^2$
4 vs. 1	1.8594 − 1.0080 = 0.8514	0.202	4.215	3.633	Reject H_0: $\sigma_4^2 = \sigma_1^2$
4 vs. 2	1.8594 − 1.0403 = 0.8191	0.204†	4.015	3.633	Reject H_0: $\sigma_4^2 = \sigma_2^2$
2 vs. 3	1.0403 − 0.7885 = 0.2518	0.204	1.234	3.633	Accept H_0: $\sigma_2^2 = \sigma_3^2$
2 vs. 1	Do not test				
1 vs. 3	Do not test				

*Since $\nu_4 = \nu_3$: $\text{SE} = \sqrt{\frac{2}{\nu}} = \sqrt{\frac{2}{49}} = 0.202$.

†Since $\nu_4 \neq \nu_2$: $\text{SE} = \sqrt{\frac{1}{\nu_4} + \frac{1}{\nu_2}} = \sqrt{\frac{1}{49} + \frac{1}{47}} = 0.204$.

Overall conclusion: $\sigma_3^2 = \sigma_1^2 = \sigma_2^2 \neq \sigma_4^2$

A Newman-Keuls-type test can also be performed using the logarithmic transformation. For this test, we calculate q using Equation 12.37; but the critical value, $q_{\alpha,\infty,p}$, depends on p, the range of variances being tested (just as p is the range of means being tested in Section 12.2).

It must be pointed out that the methods of this section, as well as those of Section 11.8, are valid only if the sampled populations are normal or very close to normal, and are severely affected if this assumption is not satisfied.

Comparing a Control Variance to Each Other Group Variance. If the investigator's intent in multiple comparison testing is to compare each possible pair of variances, then the procedures above are applicable. If, however, it is desired to stipulate that one of the variances (call it the "control," or sample B, variance) is to be compared with each other variance (but the others are not to be compared with each other), then the Dunnett-type test of Levy (1975b) may be employed for more powerful testing.

Here, in a fashion analogous to that of Section 12.4 for means, we calculate

$$q = \frac{\ln s_{\text{control}}^2 - \ln s_A^2}{\text{SE}}, \tag{12.41}$$

where

$$\mathrm{SE} = \sqrt{\frac{4}{\nu}} \tag{12.42}$$

if the control sample and sample A are of equal size. If $\nu_A \neq \nu_{\text{control}}$, then use

$$\mathrm{SE} = \sqrt{\frac{2}{\nu_A} + \frac{2}{\nu_{\text{control}}}}. \tag{12.43}$$

The appropriate critical value for the two-tailed test (i.e., H_0: $\sigma^2_{\text{control}} = \sigma^2_A$) is $q'_{\alpha(2),\infty,p}$ (from Table B.7) and for the one-tailed test (i.e., either H_0: $\sigma^2_{\text{control}} \leq \sigma^2_A$ or H_0: $\sigma^2_{\text{control}} \geq \sigma^2_A$) is $q'_{\alpha(1),\infty,p}$ (from Table B.6). This testing should be used only if it can be assumed that the underlying populations are normally distributed (or very nearly so).

EXERCISES

12.1. **(a)** Apply the Tukey test procedure to the following results from an analysis of variance: $k = 3$ and the three sample means are 14.8, 20.2, and 16.2; the error mean square and degrees of freedom are 8.46 and 21, respectively; there are eight data in each of the three groups. **(b)** Employ the Student-Newman-Keuls test for the same data. **(c)** Calculate the 95% confidence interval for each different population mean and for each difference between means.

12.2. Assume that in the experiment described in Example 11.2, group 1 was set up to be a control. Use a two-tailed Dunnett's test to compare the control mean with each other mean.

12.3. Use Scheffé's S test on the data of Example 11.2. **(a)** Test the hypothesis that the means of the populations represented by groups 1 and 4 are the same as the means of groups 2 and 3. **(b)** Test the hypothesis that the means of groups 2 and 4 are the same as the mean of group 3.

12.4. The following ranks result in a significant Kruskal-Wallis test. Employ nonparametric multiple range testing to determine between which of the three groups population differences exist.

Group 1	*Group 2*	*Group 3*
8	10	14
4	6	13
3	9	7
5	11	12
1	2	15

13 Two-Factor Analysis of Variance

Section 11.1 introduced methods for the analysis of the effect of a single factor (such as the type of feed) on a variable (such as the weight of pigs). Chapter 13 will show how the effects on a variable of two factors can be assessed simultaneously.

A simultaneous analysis of the effect of more than one factor on population means is termed a *factorial analysis of variance*, and there are important advantages to such an experimental design. Among them is the simple fact that one experiment can suffice for the analysis, and it is not necessary to perform a one-way ANOVA for each factor. Thus, we may economize with respect to time, effort, and often money. Also, factorial analysis of variance procedures can test for interaction among factors.

The two-factor analysis of variance is introduced in this chapter. Analysis of variance for experiments consisting of more than two factors will be discussed in Chapter 15. Underlying assumptions of such analyses are those discussed in Section 11.1.

13.1 TWO-FACTOR ANALYSIS OF VARIANCE WITH EQUAL REPLICATION

Example 13.1 presents data from a two-way analysis of variance, in which the variable under consideration is blood calcium concentration, and the two factors being simultaneously tested are hormone treatment and sex. Since there are two levels in the first factor (hormone treated and nontreated) and two levels in the second factor (female and male), this experimental design is termed a 2×2 (or 2^2) factorial. The two factors are said to be "crossed" because each level of one factor is found in combination with each level of the second factor.* There are 5 replicate observations (i.e., calcium determinations on each of 5 birds) for each of the $2 \times 2 = 4$ combinations of the two factors; therefore, there are a total on $N = 2 \times 2 \times 5 = 20$ data in this experiment. In general, it is advantageous to have equal replication (what is sometimes called a "balanced experimental design"), but Section 13.2 will consider cases with unequal numbers of data per cell, and Section 13.3 will discuss analyses with only one datum per combination of factors.

For the general case of the two-way factorial analysis of variance, we can refer to one factor as A and to the other as B. Furthermore, let us have a represent the number of levels in factor A, b the number of levels in factor B, and n the number of replicates. A triple subscript on the variable, as X_{ijl}, will enable us to identify uniquely the value that is replicate l of the combination of level i of factor A and level j of factor B. In Example 13.1, $X_{213} = 21.3$ mg/100 ml, $X_{115} = 12.8$ mg/100 ml, etc.

Example 13.1

A Model I two-way analysis of variance with equal replication.

The data are plasma calcium concentrations (in mg/100 ml) of birds of both sexes, half of each sex being treated with a hormone and half remaining untreated.

H_0: There is no effect of hormone treatment on the mean plasma calcium concentration of the birds (i.e., $\mu_1 = \mu_2$).

H_A: There is an effect of hormone treatment on the mean plasma calcium concentration of the birds (i.e., $\mu_1 \neq \mu_2$).

H_0: There is no difference in mean plasma calcium concentration between male and female birds (i.e., $\mu_{♂} = \mu_{♀}$).

H_A: The mean plasma calcium concentration is different in male and female birds (i.e., $\mu_{♂} \neq \mu_{♀}$).

H_0: There is no interaction of sex and hormone treatment on the mean plasma calcium concentration of the birds.

H_A: There is interaction of sex and hormone treatment on the mean plasma calcium concentration of the birds.

$\alpha = 0.05$

	No Hormone Treatment		Hormone Treatment	
	Female	*Male*	*Female*	*Male*
	16.5	14.5	39.1	32.0
	18.4	11.0	26.2	23.8
	12.7	10.8	21.3	28.8
	14.0	14.3	35.8	25.0
	12.8	10.0	40.2	29.3
Cell totals:	$\sum_{l=1}^{5} X_{11l}$ $= 74.4$	$\sum_{l=1}^{5} X_{12l}$ $= 60.6$	$\sum_{l=1}^{5} X_{21l}$ $= 162.6$	$\sum_{l=1}^{5} X_{22l}$ $= 138.9$

*Two or more factors can exist in an ANOVA without being crossed; see Chapter 16.

$$\sum_{i=1}^{2}\sum_{j=1}^{2}\sum_{l=1}^{5} X_{ijl} = 436.5$$

$$\sum_{i=1}^{2}\sum_{j=1}^{2}\sum_{l=1}^{5} X_{ijl}^2 = 11354.31$$

a = number of hormone groups = 2

b = number of sexes = 2

n = number of replicates per cell = 5

$N = abn = (2)(2)(5) = 20$

$$\text{Total for no hormone} = \sum_{j=1}^{2}\sum_{l=1}^{5} X_{1jl} = 74.4 + 60.6 = 135.0$$

$$\text{Total for hormone} = \sum_{j=1}^{2}\sum_{l=1}^{5} X_{2jl} = 162.6 + 138.9 = 301.5$$

$$\text{Total for females} = \sum_{i=1}^{2}\sum_{l=1}^{5} X_{i1l} = 74.4 + 162.6 = 237.0$$

$$\text{Total for males} = \sum_{i=1}^{2}\sum_{l=1}^{5} X_{i2l} = 60.6 + 138.9 = 199.5$$

$$C = \frac{\left(\sum_{i=1}^{a}\sum_{j=1}^{b}\sum_{l=1}^{n} X_{ijl}\right)^2}{N} = \frac{(436.5)^2}{20} = 9526.6125$$

$$\text{total SS} = \sum_{i=1}^{a}\sum_{j=1}^{b}\sum_{l=1}^{n} X_{ijl}^2 - C = 11354.31 - 9526.6125 = 1827.6975$$

$$\text{total DF} = N - 1 = 19$$

$$\text{cells SS} = \sum_{i=1}^{a}\sum_{j=1}^{b} \frac{\left(\sum_{l=1}^{n} X_{ijl}\right)^2}{n} - C$$

$$= \frac{(74.4)^2 + (60.6)^2 + (162.6)^2 + (138.9)^2}{5} - 9526.6125$$

$$= 1461.3255$$

$$\text{cells DF} = ab - 1 = (2)(2) - 1 = 3$$

$$\text{within cells (i.e., error) SS} = \text{total SS} - \text{cells SS}$$
$$= 1827.6975 - 1461.3255 = 366.3720$$

$$\text{within cells (error) DF} = ab(n-1) = (2)(2)(4) = 16$$

$$\text{factor } A \text{ (hormone group) SS} = \frac{\sum_{i=1}^{a}\left(\sum_{j=1}^{b}\sum_{l=1}^{n} X_{ijl}\right)^2}{bn} - C$$

$$= \frac{(\text{sum without hormone})^2 + (\text{sum with hormone})^2}{\text{number of data per hormone group}} - C$$

$$= \frac{(135.0)^2 + (301.5)^2}{(2)(5)} - 9526.6125$$

$$= 1386.1125$$

$$\text{factor } A \text{ DF} = a - 1 = 1$$

$$\text{factor } B \text{ (sex) SS} = \frac{\sum_{j=1}^{b}\left(\sum_{i=1}^{a}\sum_{l=1}^{n} X_{ijl}\right)^2}{an} - C$$

$$= \frac{(\text{sum for females})^2 + (\text{sum for males})^2}{\text{number of data per sex}} - C$$

$$= \frac{(237.0)^2 + (199.5)^2}{(2)(5)} - 9526.6125 = 70.3125$$

$$\text{factor } B \text{ DF} = b - 1 = 1$$

$$A \times B \text{ interaction SS} = \text{cells SS} - \text{Factor } A \text{ SS} - \text{Factor } B \text{ SS}$$
$$= 1461.3255 - 1386.1125 - 70.3125 = 4.9005$$

$$A \times B \text{ interaction DF} = (\text{factor } A \text{ DF})(\text{factor } B \text{ DF}) = (1)(1) = 1$$

Analysis of Variance Summary Table

Source of variation	SS	DF	MS
Total	1827.6975	19	
Cells	1461.3255	3	
Factor *A* (hormone)	1386.1125	1	1386.1125
Factor *B* (sex)	70.3125	1	70.3125
$A \times B$	4.9005	1	4.9005
Within cells (error)	366.3720	16	22.8982

For H_0: There is no effect of hormone treatment on the plasma calcium concentration of the birds.

$$F = \frac{\text{hormone MS}}{\text{within cells MS}} = \frac{1386.1125}{22.8982} = 60.5$$

$F_{0.05(1),1,16} = 4.49$

Therefore, reject H_0.

$P < 0.0005$

For H_0: There is no difference in plasma calcium concentration between male and female birds.

$$F = \frac{\text{sex MS}}{\text{within cells MS}} = \frac{70.3125}{22.8982} = 3.07$$

$F_{0.05(1),1,16} = 4.49$

Therefore, do not reject H_0.

$0.05 < P < 0.10$

For H_0: There is no interaction of sex and hormone treatment affecting the plasma calcium concentration of the birds.

$$F = \frac{\text{hormone} \times \text{sex interaction MS}}{\text{within cells MS}} = \frac{4.9005}{22.8982} = 0.214$$

$F_{0.05(1)\ 1,16} = 4.49$

Therefore, do not reject H_0.

$P > 0.25$

For the two-factor ANOVA we calculate:

$$\text{total SS} = \sum_{i=1}^{a} \sum_{j=1}^{b} \sum_{l=1}^{n} X_{ijl}^2 - C, \tag{13.1}$$

or $11354.31 - C$, where the correction term is

$$C = \frac{\left(\sum_{i=1}^{a} \sum_{j=1}^{b} \sum_{l=1}^{n} Y_{ijl} \right)^2}{N}. \tag{13.2}$$

Since

$$N = abn, \tag{13.3}$$

or $N = (2)(2)(5) = 20$, in our example, $C = (436.5)^2/20 = 9526.6125$ and total SS $= 11354.31 - 9526.6125 = 1827.6975$. The associated degrees of freedom are

$$\text{total DF} = N - 1; \tag{13.4}$$

or, total DF $= 20 - 1 = 19$ in Example 13.1.

Next, we can consider each combination of factor A and factor B, each combination being called a *cell*; and we treat each cell as if it were a group in a single factor ("one-way") analysis of variance:

$$\text{cells SS} = \frac{\sum_{i=1}^{a}\sum_{j=1}^{b}\left(\sum_{l=1}^{n} X_{ijl}\right)^2}{n} - C. \tag{13.5}$$

Since the number of cells is ab,

$$\text{cells DF} = ab - 1. \tag{13.6}$$

Furthermore,

$$\text{within cells SS} = \text{total SS} - \text{cells SS} \tag{13.7}$$

and

$$\text{within cells DF} = \text{total DF} - \text{cells DF}. \tag{13.8}$$

The latter quantity could also be calculated as

$$\text{within cells DF} = ab(n - 1). \tag{13.9}$$

The terms *error SS* and *error DF* are very commonly used for within cells SS and DF, respectively.

All the calculations to this point are analogous to those for the one-way ANOVA (Section 11.1); but our desire in the two-factor ANOVA is not to consider the aforementioned cells, but to consider the effects of each of the two factors independently of the other. For factor A this is done as follows:

$$\text{factor } A \text{ SS} = \frac{\sum_{i=1}^{a}\left(\sum_{j=1}^{b}\sum_{l=1}^{n} X_{ijl}\right)^2}{bn} - C, \tag{13.10}$$

and

$$\text{factor } A \text{ DF} = a - 1. \tag{13.11}$$

Simply put, the factor A SS is calculated by considering factor A to be the sole factor in a single-factor analysis of variance of the data. That is, we obtain the sum for each level of factor A (ignoring the fact that the data are also categorized into levels of factor B); the sum of a level is what is in parentheses in Equation 13.10. Then we square each of these level sums and divide the sum of these squares by the number of data per level (i.e., bn). On subtracting the "correction term," C, we arrive at the factor A SS. (If the data were in fact analyzed by a single factor ANOVA, then the groups SS would indeed be the same as the factor A SS just described, and the groups DF would be the same as the factor A DF; but the error SS in the one-way ANOVA would be the within cells SS plus the factor B and the interaction sums of squares described below (and the error DF would be the sum of the within cells, factor B, and interaction degrees of freedom).

For factor B computations, we simply ignore the division of the data into levels of factor A and proceed as if factor B were the single factor in a one-way ANOVA:

$$\text{factor } B \text{ SS} = \frac{\sum_{j=1}^{b}\left(\sum_{i=1}^{a}\sum_{l=1}^{n} X_{ijl}\right)^2}{an} - C, \tag{13.12}$$

and

$$\text{factor } B \text{ DF} = b - 1. \tag{13.13}$$

In Example 13.1, the variability among cells (from Equation 13.5) is not equal to the variability between levels of factor A plus the variability between levels of factor B (from Equation 13.10 plus Equation 13.12). The amount of variability left unaccounted for is that which is owing to the effect of interaction between factors A and B. This is referred to as the $A \times B$ interaction and is readily calculated by difference:

$$A \times B \text{ interaction SS} = \text{cells SS} - \text{factor } A \text{ SS} - \text{factor } B \text{ SS}. \tag{13.14}$$

The interaction degrees of freedom may also be calculated by difference, or as

$$\begin{aligned} A \times B \text{ interaction DF} &= (\text{factor } A \text{ DF})(\text{factor } B \text{ DF}) \\ &= (a - 1)(b - 1). \end{aligned} \tag{13.15}$$

An interaction between two factors means that the effect of one factor is not independent of the presence of a particular level of the other factor. In Example 13.1 no interaction implies that the effect (if any) of hormone treatment on plasma calcium is the same in both sexes, or, equivalently, that any difference in plasma calcium between males and females is the same under both hormone treatments.* Therefore, interaction (if any) between factors is an effect on the variable (e.g., plasma calcium) in addition to the effects of each factor considered separately.

Table 13.1 summarizes the computing formulas for the two-factor analysis of variance with equal replication.

The Model I ANOVA. Recall from Section 11.1 the distinction between fixed and random factors. Example 13.1 is an ANOVA where the levels of both factors are fixed; we did not simply pick these levels at random. A factorial analysis of variance in which all (in this case both) factors are fixed effects is termed a *Model I ANOVA*. In such a model, the null hypothesis of no difference among the levels of a factor is tested using $F =$ factor MS/error MS. As shown in Example 13.1, the appropriate F tests conclude that there is a highly significant effect of the hormone treatment on the plasma calcium content, and that there is not a significantly different mean plasma calcium concentration between males and females.

In addition, we can test for significant interaction in a Model I ANOVA by $F =$ interaction MS/error MS and find, in our present example, that there is no significant interaction between sex and hormone treatment. This is interpreted to mean that the effect on calcium of the hormone treatment is not different in males and females; i.e., the effect of the hormone is not dependent on sex. This concept of interaction (or its converse, independence) is analogous to that employed in the analysis of contingency tables (see Chapter 6).

If, in a two-factor analysis of variance, the effects of one or both factors are significant, the interaction effect may or may not be significant. In fact, it is possible

*Symbolically, the null hypothesis for interaction effect could be stated as $H_0\colon \mu_{11} - \mu_{12} = \mu_{21} - \mu_{22}$ or $H_0\colon \mu_{11} - \mu_{21} = \mu_{12} - \mu_{22}$, where μ_{ij} is the population mean of the variable in the presence of level i of factor A and level j of factor B.

TABLE 13.1 TWO-WAY FACTORIAL ANOVA WITH EQUAL REPLICATION

Source of variation	*Sum of squares* (SS)	*Degrees of freedom* (DF)	*Mean square* (MS)
Total	$\sum_{i=1}^{a}\sum_{j=1}^{b}\sum_{l=1}^{n} X_{ijl}^2 - C$	$N-1$	
Cells	$\dfrac{\sum_{i=1}^{a}\sum_{j=1}^{b}\left(\sum_{l=1}^{n} X_{ijl}\right)^2}{n} - C$	$ab-1$	
Factor A	$\dfrac{\sum_{i=1}^{a}\left(\sum_{j=1}^{b}\sum_{l=1}^{n} X_{ijl}\right)^2}{bn} - C$	$a-1$	$\dfrac{\text{SS}}{\text{DF}}$
Factor B	$\dfrac{\sum_{j=1}^{b}\left(\sum_{i=1}^{a}\sum_{l=1}^{n} X_{ijl}\right)^2}{an} - C$	$b-1$	$\dfrac{\text{SS}}{\text{DF}}$
$A \times B$ interaction	Cells SS − factor A SS − factor B SS	$(a-1)(b-1)$	$\dfrac{\text{SS}}{\text{DF}}$
Within cells (error)	Total SS − cells SS	$ab(n-1)$ or total DF − cells DF	$\dfrac{\text{SS}}{\text{DF}}$

Note: Here, a is the number of levels in factor A, b is the number of levels in factor B, and n is the number of replicates per cell.

$$C = \frac{\left(\sum_{i=1}^{a}\sum_{j=1}^{b}\sum_{l=1}^{n} X_{ijl}\right)^2}{N}$$

and

$$N = abn.$$

to encounter situations where there is a significant interaction even though each of the individual factor effects is judged to be insignificant.

The Model II ANOVA. If a factorial design is composed only of factors with random levels, then we are said to be employing a *Model II ANOVA* (a relatively uncommon situation). In such a case, where two factors are involved, the appropriate hypothesis testing for significant factor effects is accomplished by calculating $F =$ factor MS/interaction MS (see Table 13.2). We test for the interaction effect, as before, by $F =$ interaction MS/error MS. The Model II ANOVA for designs with more than two factors will be discussed in Chapter 15.

The Model III ANOVA. If a factorial design has both a fixed effect and a random effect factor, then it is said to be a *mixed model*, or a *Model III*, ANOVA. The appropriate F statistics are calculated as shown in Table 13.2.

TABLE 13.2 COMPUTATION OF THE *F* STATISTIC FOR TESTS OF SIGNIFICANCE IN A TWO-FACTOR ANOVA WITH REPLICATION

Hypothesized effect	*Model I (factors A and B both fixed)*	*Model II (factors A and B both random)*	*Model III (factor A fixed; factor B random)*
Factor *A*	$\frac{\text{factor } A \text{ MS}}{\text{error MS}}$	$\frac{\text{factor } A \text{ MS}}{A \times B \text{ MS}}$	$\frac{\text{factor } A \text{ MS}}{A \times B \text{ MS}}$
Factor *B*	$\frac{\text{factor } B \text{ MS}}{\text{error MS}}$	$\frac{\text{factor } B \text{ MS}}{A \times B \text{ MS}}$	$\frac{\text{factor } B \text{ MS}}{\text{error MS}}$
$A \times B$ interaction	$\frac{A \times B \text{ MS}}{\text{error MS}}$	$\frac{A \times B \text{ MS}}{\text{error MS}}$	$\frac{A \times B \text{ MS}}{\text{error MS}}$

Pooling Mean Squares. If one does not conclude there to be a significant interaction effect, then the interaction MS and the within cells (i.e., the error) MS are theoretically estimates of the same population variance. Because of this, some authors suggest the pooling of the interaction and within cells sums of squares and degrees of freedom in such cases. From these pooled SS and DF values, one can obtain a pooled mean square, which then should be a better estimate of the population random error (i.e., within cell variability) than either the error MS or the interaction MS alone.

The conservative researcher who does not engage in such pooling can be assured that the probability of a Type I error is at the stated α level. But the probability of a Type II error may be greater than is acceptable to some. The chance of the latter type of error is reduced by the pooling previously described, but confidence in stating the probability of committing a Type I error may be reduced (Brownlee, 1965: 509). Rules of thumb for deciding when to pool have been proposed (e.g., Paull, 1950), but statistical advice beyond this book should be obtained if such pooling is contemplated. The analyses in this text will proceed according to the conservative, nonpooling approach.

Multiple Comparisons. If significant differences are concluded among the levels of a factor, then the multiple comparison procedures of Sections 12.1, 12.2, 12.3, 12.4, or 12.5 may be employed. For such purposes, s^2 is the within cells MS, ν is the within cells DF, and the n of Chapter 12 is replaced in the present situation with the total number of data per level of the factor being tested (i.e., what we have noted in this section as bn data per level of factor A and an data per level of factor B).

Confidence Limits for Means. We may compute confidence intervals for population means of levels of a fixed factor by the methods in Section 11.2. That section's error mean square, s^2, is the within cells MS of the present discussion; the error degrees of freedom, ν, is the within cells DF; and n in Section 11.2 is replaced in the present context by the total number of data in the level being examined. Confidence intervals for differences between population means are obtained by the procedures of Section 12.3. This is demonstrated in Example 13.2.

Example 13.2

Confidence limits for the results of Example 13.1.

We concluded that mean plasma calcium concentration is different between birds with the hormone treatment and those without.

$$\bar{X}_1 = \frac{\text{total for nonhormone group}}{\text{number in nonhormone group}} = \frac{135.0 \text{ mg/100 ml}}{10} = 13.50 \text{ mg/100 ml}$$

$$\bar{X}_2 = \frac{\text{total for hormone group}}{\text{number in hormone group}} = \frac{301.5 \text{ mg/100 ml}}{10} = 30.15 \text{ mg/100 ml}$$

$$95\% \text{ CI for } \mu_1 = \bar{X}_1 \pm t_{0.05(2),\nu}\sqrt{\frac{s^2}{bn}}$$

$$= 13.50 \pm t_{0.05(2),16}\sqrt{\frac{22.8982}{(2)(5)}}$$

$$= 13.50 \pm (2.120)(1.513)$$

$$= 13.50 \text{ mg/100 ml} \pm 3.21 \text{ mg/100 ml}$$

$L_1 = 10.29$ mg/100 ml; $L = 16.71$ mg/100 ml

$$95\% \text{ CI for } \mu_2 = \bar{X}_2 \pm t_{0.05(2),\nu}\sqrt{\frac{s^2}{bn}}$$

$$= 30.15 \text{ mg/100 ml} \pm 3.21 \text{ mg/100 ml}$$

$L_1 = 26.94$ mg/100 ml; $L_2 = 33.36$ mg/100 ml

$$95\% \text{ CI for } \mu_1 - \mu_2 = \bar{X}_1 - \bar{X}_2 \pm q_{0.05,\nu,k}\sqrt{\frac{s^2}{bn}}$$

$$= 13.50 - 30.15 \pm q_{0.05,16,2}\sqrt{\frac{22.8982}{(2)(5)}}$$

$$= -16.65 \pm (2.998)(1.513)$$

$$= -16.65 \text{ mg/100 ml} \pm 4.54 \text{ mg/100 ml}$$

$L_1 = -21.19$ mg/100 ml; $L_2 = -12.11$ mg/100 ml

We concluded that mean calcium concentration is not different in males and females. Therefore, $\mu_♂ = \mu_♀$, and we would not speak of confidence intervals for each of these two means or for the difference between the means. If we desired, we could pool the means and speak of a confidence interval for the pooled population mean, μ_p:

$$\text{pooled } \bar{X} = \bar{X}_p = \frac{\text{total for females} + \text{total for males}}{\text{number of females} + \text{number of males}}$$

$$= \frac{237.0 + 199.5}{10 + 10} = 21.82 \text{ mg/100 ml}$$

$$95\% \text{ CI for } \mu_p = \bar{X}_p \pm t_{0.05(2),16}\sqrt{\frac{s^2}{20}}$$

$$= 21.82 \pm (2.120)(1.070)$$

$$= 21.82 \text{ mg/100 ml} \pm 2.27 \text{ mg /100 ml}$$

$L_1 = 19.55$ mg/100 ml; $L_2 = 24.09$ mg/100 ml

13.2 TWO-FACTOR ANALYSIS OF VARIANCE WITH UNEQUAL REPLICATION

The procedures outlined in Section 13.1 for two-factor analysis of variance require that n, the number of replicates per cell, be the same in all cells. In general it is desirable, for optimum power, to design experiments with equal cell sizes, but occasionally this is impossible or impractical. Figure 13.1 shows two-factor experimental designs with various kinds of replication.

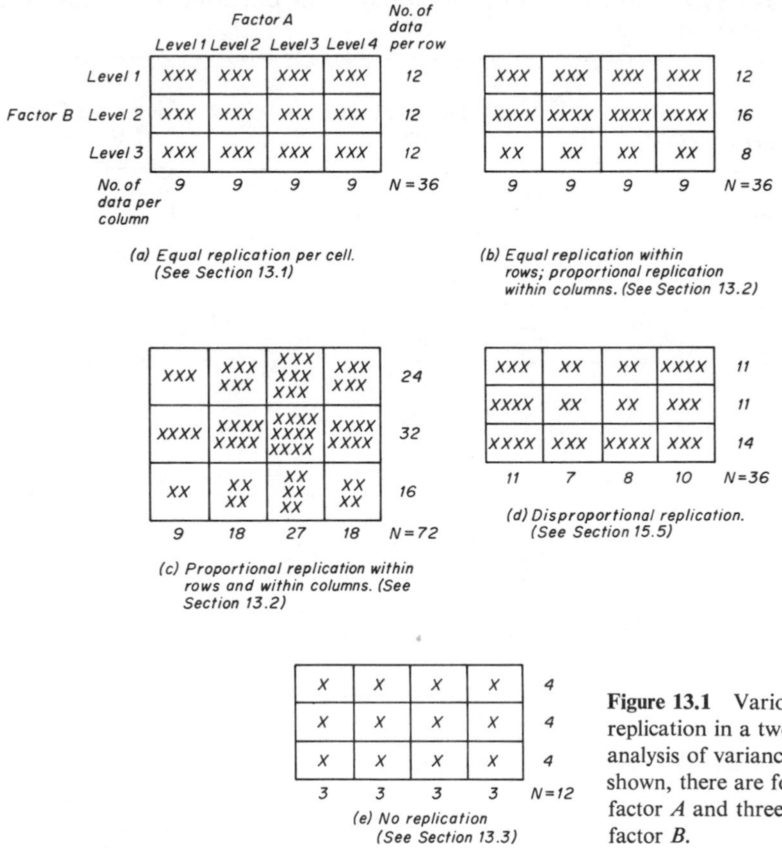

Figure 13.1 Various kinds of replication in a two-factor analysis of variance. In all cases shown, there are four levels of factor *A* and three levels of factor *B*.

Proportion Replication. A two-factor experimental design exhibits proportional replication if the number of data in the cell in row *i* and column *j* is

$$n_{ij} = \frac{(\text{number of data in row } i)(\text{number of data in column } j)}{N}, \qquad (13.16)$$

where *N* is the total number of data in all cells.* (For example, in Fig. 13.1c, there are two data in row 3, column 1; and (16)(9)/72 = 2. The proper analysis of variance computes the total SS in a fashion similar to that in Equation 13.1. The appropriate computations of sums of squares, degrees of freedom, and mean squares are shown in Table 13.3.† Once these quantities are obtained, the hypothesis testing proceeds just as in Section 13.1. (See Table 13.2.) Some factorial analysis of variance computer programs accommodate proportionally replicated data.

*The number of replicates in each of the *ab* cells need not be checked against Equation 13.16 to determine whether proportional replication is present. One need check only one cell in each of $a - 1$ levels of factor *A* and one in each of $b - 1$ levels of factor *B* (Huck and Layne, 1974).

†If replication is equal, the computations of Table 13.3 are identical to those in Table 13.1.

TABLE 13.3 TWO-WAY FACTORIAL ANOVA WITH UNEQUAL, BUT PROPORTIONAL, REPLICATION

Source of variation	*Sum of squares* (SS)	*Degrees of freedom* (DF)	*Mean square* (MS)
Total	$\sum_{i=1}^{a}\sum_{j=1}^{b}\sum_{l=1}^{n_{ij}} X_{ijl}^2 - C$	$N - 1$	$\frac{\text{SS}}{\text{DF}}$
Cells	$\sum_{i=1}^{a}\sum_{j=1}^{b}\frac{\left(\sum_{l=1}^{n_{ij}} X_{ijl}\right)^2}{n_{ij}} - C$	$ab - 1$	$\frac{\text{SS}}{\text{DF}}$
Factor A	$\sum_{i=1}^{a}\frac{\left(\sum_{j=1}^{b}\sum_{l=1}^{n_{ij}} X_{ijl}\right)^2}{\sum_{j=1}^{b} n_{ij}} - C$	$a - 1$	$\frac{\text{SS}}{\text{DF}}$
Factor B	$\sum_{j=1}^{b}\frac{\left(\sum_{i=1}^{a}\sum_{l=1}^{n_{ij}} X_{ijl}\right)^2}{\sum_{i=1}^{a} n_{ij}} - C$	$b - 1$	$\frac{\text{SS}}{\text{DF}}$
$A \times B$ Interaction	cells SS − factor A SS − factor B SS	$(a-1)(b-1)$	$\frac{\text{SS}}{\text{DF}}$
Within cells (error)	total SS − cells SS	$\sum_{i=1}^{a}\sum_{j=1}^{b}(n_{ij} - 1)$ or total DF − cells DF	$\frac{\text{SS}}{\text{DF}}$

Note: Here, a is the number of levels in factor A, b is the number of levels in factor B, and n_{ij} is the number of replicates in the cell formed by level i of factor A and level j of factor B;

$$C = \frac{\left(\sum_{i=1}^{a}\sum_{j=1}^{b}\sum_{l=1}^{n_{ij}} X_{ijl}\right)^2}{N}, \quad \text{and} \quad N = \sum_{i=1}^{a}\sum_{j=1}^{b} n_{ij}.$$

Disproportional Replication. We should strive to collect data so that there is equal or proportional replication in the cells. If this is not the case, we may employ a computer program capable of performing factorial analysis of variance with unequal and disproportional replication (Section 15.5). Alternatively, if a few cells have numbers of data in excess of those needed for equal or proportional replication, then data may be deleted, at random, within such cells, so that equality or proportionality is achieved. Then the ANOVA can proceed as usual, as described above or as in Section 13.1.

If one cell is one datum short of the number required for equal or proportional replication, a value may be estimated for inclusion in place of the missing datum, as follows (Shearer, 1973):

$$\hat{X}_{ijl} = \frac{aA_i + bB_j - \sum_{i=1}^{a}\sum_{j=1}^{b}\sum_{l=1}^{n_{ij}} X_{ijl}}{N + 1 - a - b}, \tag{13.17}$$

where $\hat{X}_{ijl}$ is the estimated value for replicate l in level i of factor A and level j of factor B; A_i is the sum of the other data in level i of factor A, B_j is the sum of the other data in level j of factor B, $\sum\sum\sum X_{ijl}$ is the sum of all the data, and N is the total number of data (including the missing datum) in the experimental design. For example, if datum X_{124} had been missing in Example 13.1, it could have had a quantity inserted in its place, estimated by Equation 13.17, where $a = 2$, $b = 2$, $N = 20$, $A_i = A_1 =$ the sum of all known data from animals receiving no hormone treatment, $B_j = B_2$ = the sum of all known data from males, and $\sum\sum\sum X_{ijl}$ = the sum of all 19 known data from both hormone treatments and both sexes. After the missing datum has been estimated, it is inserted into the data set and the ANOVA computations may proceed, with the provision that a missing datum is not counted in determining total and within cells degrees of freedom. (Therefore, if a datum were missing in Example 13.1, the total DF would have been 18 and the within cells DF would have been 15.)

If more than one datum is missing (but neither more than 10% of the total number of data nor more data than the number of levels of any factor), then Equation 13.17 could be used iteratively to derive estimates of the missing data (e.g., using cell means as initial estimates). The number of such estimates would not enter into the total or within cells degrees of freedom determinations.

If only a few cells (say, no more than the number of levels in either factor) are each one datum short of the numbers required for equal or proportional replication, then the mean of the data in each such cell may be inserted as an additional datum in that cell. In the latter situation, the analysis proceeds as usual but with the total DF and the within cells DF each being determined without counting such additional inserted data. Instead of employing these cell means themselves, however, they could be used as starting values for employing Equation 13.17 in the iterative fashion of Example 13.6.

13.3 TWO-FACTOR ANALYSIS OF VARIANCE WITHOUT REPLICATION

Occasionally we might encounter a two-factor experimental design in which there is only one datum for each combination of factors (i.e., $n = 1$ for all cells). In such a situation of no replication, each datum may be denoted uniquely by a double subscript, as X_{ij}, where i refers to one of the levels of factor A and j refers to a level of factor B.

For a levels of factor A and b levels of factor B, the appropriate computations of sums of squares, degrees of freedom, and mean squares are shown in Table 13.4. The total SS and DF are obtained just as when there is replication (Section 13.1 and Table 13.1), as are the SS and DF for factor A and for factor B. It should be noted that when $n = 1$, the cells SS of Section 13.1 is identical to the total SS (and the cells DF would be the same as the total DF). Consequently, the within groups (or error) SS and DF are both zero.

That part of the total variability not accounted for by the effect of the two factors is:

$$\text{remainder SS} = \text{total SS} - \text{factor } A \text{ SS} - \text{factor } B \text{ SS} \tag{13.18}$$

with

$$\text{remainder DF} = \text{total DF} - \text{factor } A \text{ DF} - \text{factor } B \text{ DF}. \tag{13.19}$$

TABLE 13.4 TWO-WAY FACTORIAL ANOVA WITHOUT REPLICATION

Source of variation	*Sum of squares* (SS)	*Degrees of freedom* (DF)	*Mean square* (MS)
Total	$\sum_{i=1}^{a}\sum_{j=1}^{b} X_{ij}^2 - C$	$N - 1$	
Factor A	$\dfrac{\sum_{i=1}^{a}\left(\sum_{j=1}^{b} X_{ij}\right)^2}{b} - C$	$a - 1$	$\dfrac{\text{SS}}{\text{DF}}$
Factor B	$\dfrac{\sum_{j=1}^{b}\left(\sum_{i=1}^{a} X_{ij}\right)^2}{a} - C$	$b - 1$	$\dfrac{\text{SS}}{\text{DF}}$
Remainder	total SS − factor A SS − factor B SS	$(a - 1)(b - 1)$	$\dfrac{\text{SS}}{\text{DF}}$

Note: Here, a is the number of levels in factor A, and b is the number of levels in factor B.

$$C = \frac{\left(\sum_{i=1}^{a}\sum_{j=1}^{b} X_{ij}\right)^2}{N}$$

$$N = ab$$

Note that quantities 13.18 and 13.19 are what are referred to as "interaction" quantities when replication is present.

Table 13.4 summarizes the ANOVA calculations just described, and Table 13.5 summarizes the significance tests that may be performed to test hypotheses about

TABLE 13.5 COMPUTATION OF THE F STATISTIC FOR TESTS OF SIGNIFICANCE IN A TWO-FACTOR ANOVA WITHOUT REPLICATION

If It Is Assumed That There May Be a Significant Interaction Effect

Hypothesized effect	*Model I (factors A and B both fixed)*	*Model II (factors A and B both random)*	*Model III (factor A fixed; factor B random)*
Factor A	Test with caution*	$\dfrac{\text{factor } A \text{ MS}}{\text{remainder MS}}$	$\dfrac{\text{factor } A \text{ MS}}{\text{remainder MS}}$
Factor B	Test with caution*	$\dfrac{\text{factor } B \text{ MS}}{\text{remainder MS}}$	Test with caution*
$A \times B$ interaction	No test possible	No test possible	No test possible

*Analysis can be performed as in Model II, but with increased chance of Type II error.

If It Is Correctly Assumed That There Is No Significant Interaction Effect

Hypothesized effect	*Model I*	*Model II*	*Model III*
Factor A	$\dfrac{\text{factor } A \text{ MS}}{\text{remainder MS}}$	$\dfrac{\text{factor } A \text{ MS}}{\text{remainder MS}}$	$\dfrac{\text{factor } A \text{ MS}}{\text{remainder MS}}$
Factor B	$\dfrac{\text{factor } B \text{ MS}}{\text{remainder MS}}$	$\dfrac{\text{factor } B \text{ MS}}{\text{remainder MS}}$	$\dfrac{\text{factor } B \text{ MS}}{\text{remainder MS}}$
$A \times B$ interaction	No test possible	No test possible	No test possible

each of the factors. Since we do not have measures of both interaction and error variability, a test for the significance of interaction is not possible. Testing for the effect of each of the two factors in a Model I analysis (or testing for the effect of the random factor in a Model III design) is not advisable, because there may, in fact, be interaction between the two factors (and there will be decreased test power); but if a significant difference is concluded, then that conclusion may be accepted.

13.4 NONPARAMETRIC TWO-FACTOR ANALYSIS OF VARIANCE

Data obtained from a two-factor experimental design may be analyzed nonparametrically by an extension of the Kruskal-Wallis test introduced in Section 11.4 for single-factor analysis (Scheirer, Ray, and Hare, 1976). The procedure is demonstrated in Example 13.4. First, all N data are ranked, just as in the one-factor case.* Then the ranks are summed for each cell, and the rank sums are obtained for each level

Example 13.4

A nonparametric two-way analysis of variance.

The data are fat concentrations (in percent) in the milk of rats. The two experimental factors are diet (three levels) and radiation (two levels).

H_0: There is no effect of radiation treatment on amount of milk fat in rats.
H_A: There is an effect of radiation treatment on amount of milk fat in rats.

H_0: There is no difference in milk fat concentration among rats on the three diets.
H_A: There is difference in milk fat concentration among rats on the three diets.

H_0: There is no interaction of radiation and diet on the milk fat concentration in rats.
H_A: There is interaction of radiation and diet on the milk fat concentration in rats.

$\alpha = 0.05$

The data are as follows, with ranks shown in parentheses.

	No Radiation Treatment			**Radiation Treatment**		
	Diet L	*Diet M*	*Diet H*	*Diet L*	*Diet M*	*Diet H*
	12.9 (10)	14.5 (21)	13.4 (16)	12.1 (4)	12.2 (5.5)	13.1 (13)
	13.6 (17)	15.0 (24)	13.8 (18)	11.9 (1)	12.3 (7)	13.3 (15)
	12.7 (9)	14.8 (22.5)	14.3 (20)	12.6 (8)	13.0 (11)	13.1 (13)
	14.0 (19)	14.8 (22.5)	12.2 (5.5)	12.0 (2.5)	13.1 (13)	12.0 (2.5)
Sum of ranks†	55	90	59.5	15.5	36.5	43.5

$N = 24$.

*Although not a guarantee that ranks have been assigned correctly, it is good practice to check to see whether the sum of all the ranks equals $N(N + 1)/2$. Errors are common in assigning ranks when there are tied data.)

†Let us check whether the sum of the ranks (or the sum of the rank sums: $55 + 90 + 59.5 + 15.5 + 36.5 + 43.5 = 300$) is equal to $N(N + 1)/2 = 24(25)/2 = 300$. If it does, the ranking is not necessarily correct; but if it does not, then we know there is error in assigning the ranks.

Rank sum for no radiation = 55 + 90 + 59.5 = 204.5
Rank sum for radiation = 15.5 + 36.5 + 43.5 = 95.5
Rank sum for diet L = 55 + 15.5 = 70.5
Rank sum for diet M = 90 + 36.5 = 126.5
Rank sum for diet H = 59.5 + 43.5 = 103

$$C = \frac{N(N+1)^2}{4} = \frac{24(25)^2}{4} = 3750$$

$$\text{cells SS} = \frac{55^2 + 90^2 + 59.5^2 + 15.5^2 + 36.5^2 + 43.5^2}{4} - C = 782.5000$$

cells DF = 6 − 1 = 5

$$\text{factor } A \text{ (radiation) SS} = \frac{204.5^2 + 95.5^2}{12} - C = 495.0417$$

factor A DF = 2 − 1 = 1

$$\text{factor } B \text{ (diets) SS} = \frac{70.5^2 + 126.5^2 + 103^2}{8} - C = 197.6875$$

factor B DF = 3 − 1 = 2
$A \times B$ interaction SS = 782.5000 − 495.0417 − 197.6875 = 89.7708
$A \times B$ interaction DF = (factor A DF)(factor B DF) = (1)(2) = 2

$$\text{total MS} = \frac{N(N+1)}{12} = \frac{24(25)}{12} = 50.0000^*$$

Analysis of Variance Summary Table

Source of variation	SS	DF
Cells	782.5000	5
Factor A (radiation)	495.0417	1
Factor B (diets)	197.6875	2
$A \times B$	89.7708	2

For the null hypothesis concerning the effect of factor A (radiation):

$$H = \frac{\text{factor } A \text{ SS}}{\text{total MS}} = \frac{495.0417}{50.0000} = 9.901$$

$\chi^2_{0.05,1} = 3.841$
Therefore, reject H_0; we conclude that the fat content of milk is different in irradiated and non-irradiated rats.
$0.001 < P < 0.005$

For the null hypothesis concerning the effect of factor B (diets):

$$H = \frac{\text{factor } B \text{ SS}}{\text{total MS}} = \frac{197.6875}{50.0000} = 3.954$$

$\chi^2_{0.05,2} = 5.991$
Therefore, do not reject H_0; we conclude that the three diets have no different effects on the fat content of rat milk.
$0.10 < P < 0.25$

For the null hypothesis concerning interaction between factors A and B:

$$H = \frac{A \times B \text{ SS}}{\text{total MS}} = \frac{89.7708}{50.0000} = 1.795$$

$\chi^2_{0.05,2} = 5.991$
Therefore, do not reject H_0; we conclude that there is no interactive effect of radiation and diet on the fat content of rat milk.
$0.25 < P < 0.50$

*Even though there are several tied data in this example, this computation of total MS is very close to that which would be obtained using Equation 11.10 (49.8478).

of each factor. The sums of squares of the ranks are calculated in the same manner as the sums of squares of raw data are computed in Section 13.1. That is, consider X_{ijl} to be rank l in level i of factor A and level j of factor B; the needed sums of squares and degrees of freedom are calculated as follows:

cells SS by Equation 13.5
cells DF by Equation 13.6
factor A SS by Equation 13.10
factor A DF by Equation 13.11
factor B SS by Equation 13.12
factor B DF by Equation 13.13
$A \times B$ interaction SS by Equation 13.14
$A \times B$ interaction DF by Equation 13.15

In the above sums of squares computations, the "correction term" may be obtained as

$$C = \frac{N(N+1)^2}{4}, \tag{13.20}$$

instead of by Equation 13.2*

The total mean square of the ranks (i.e., the variance of all N ranks) is†

$$\text{total MS} = \frac{N(N+1)}{12}. \tag{13.21}$$

The above computations are demonstrated in Example 13.4. Once the above have been calculated, one may test for significance of any source of variation with the Kruskal-Wallis statistic,

$$H = \frac{\text{source SS}}{\text{total MS}}. \tag{13.22}$$

H is closely approximated by χ^2 with the degrees of freedom appropriate to the source of variation being tested.

*Equation 13.20 is equivalent to Equation 13.2 because the sum of the consecutive integers 1 through N is

$$\sum_{i=1}^{N} i = N(N+1)/2; \quad \text{therefore,} \quad C = [N(N+1)/2]^2/N$$
$$= N(N+1)^2/4.$$

†Equation 13.21 is equivalent to Equation 13.1 divided by Equation 13.4 owing to the sum of the squares of the consecutive integers 1 through N being

$$\sum_{i=1}^{N} i^2 = \frac{N(N+1)(2N+1)}{6};$$

therefore,

$$\text{total MS} = \frac{(\text{total SS})}{N-1} = \frac{\dfrac{N(N+1)(2N+1)}{6} - \dfrac{N(N+1)^2}{4}}{N-1},$$

which reduces to

$$\frac{N(N+1)}{12}.$$

The above procedure may be used when there are equal cell sizes. Equation 13.21 is valid if there are no tied data. If ties are present, this equation gives a total MS that is too high, resulting in conservative testing. The inaccuracy is not great unless many ties are present, in which case Equation 11.10 can be applied to the ranks to obtain the total SS; that quantity divided by $N - 1$ is the total MS.

13.5 THE RANDOMIZED BLOCK EXPERIMENTAL DESIGN

The statistical term "block" is conceptually an extension of the term "pair" introduced in Chapter 10. In that chapter, it was shown how members of two groups might be paired and that the resultant paired-sample testing could be more powerful than the nonpaired two-sample test. We shall now see that the randomized block ANOVA is an extension of the paired t test just as the one-way ANOVA is an extension of the two-sample t test, and also that the data from such an experimental design can be analyzed by Model III ANOVA considerations.*

Consider the experiment represented by the data in Example 13.5. The variable in question is weight gain in guinea pigs; the fixed effect factor whose effect is to be tested is diet. Twenty animals are to be used in this experiment, five on each of four diets. However, the experimenter believes there are some environmental factors that likely would affect weight gain and does not feel that all 20 animal cages can be kept in the laboratory at identical conditions of temperature, light, etc. Therefore, five blocks of experimental units are established, i.e., five groups of guinea pigs. Each block of animals consists of four guinea pigs, one on each of the experimental diets. All members of a block are considered to be at identical conditions (except, of course, for diet).

The assignment of an experimental diet to each of the individuals in a block is done at random; for this purpose, Table B.50 may be consulted. A random series of digits allows us to assign diets 1 through 4 to the four animals in the first block. Then a second random series of four digits determines the assignment of the four diets to the individuals in the second block, etc. After the data are collected, they are commonly arranged in an array, such as shown in Example 13.5.

The analysis of the randomized block data is performed as a mixed model two-way ANOVA, without replication, by consulting Tables 13.4 and 13.5. In the present example, diet is the fixed factor, and block is always the random factor. The null hypothesis of equal weight gain on all diets is readily tested. A null hypothesis of equal weight gain among blocks is usually of no particular interest; furthermore, it is not advisable, because it would require knowledge about an interaction effect. Thus, it is generally not tested.

*A common example of paired data is observation of the same subjects at two different times (e.g., before and after some treatment); similarly, blocks may be subjects from which data are collected at three or more times. Thus, in such situations some researchers refer to the randomized block ANOVA as an "ANOVA with repeated measures" or an "ANOVA of treatments by subjects experimental design."

Example 13.5

A randomized block analysis of variance (Model III two-factor analysis of variance without replication).

H_0: The mean weight gain of guinea pigs is the same on each of four specified diets (i.e., $\mu_1 = \mu_2 = \mu_3 = \mu_4$).

H_A: The mean weight gain of guinea pigs is not the same on each of the four specified diets.

$\alpha = 0.05$

Each guinea pig is housed in a separate cage. A block consists of a group of four animals that we can be reasonably assured will experience identical environmental conditions (light, temperature, draft, noise, etc.). Each block has each of its four animals assigned at random to one of the four experimental diets, so that each animal in a given block is to receive a different diet. By consulting a table of random numbers (as Table B.50), only the digits 1, 2, 3, or 4 need be considered; any others, and any repeats of 1, 2, 3, or 4, are simply ignored. The assignment of diets to animals within blocks is shown, where the assignment in the first block may have resulted from a random digit sequence from Table B.50 of 34102, the assignment in the second block by a random digit sequence of 153924, the assignment in block 3 by 7352431, etc. Shown in parentheses is the weight gain (in grams) of each of the 20 animals.

Block 1	Diet 3 (2.1)	Diet 4 (1.3)	Diet 1 (1.5)	Diet 2 (2.7)
Block 2	Diet 1 (1.4)	Diet 3 (2.2)	Diet 2 (2.9)	Diet 4 (1.0)
Block 3	Diet 3 (2.4)	Diet 2 (2.1)	Diet 4 (1.1)	Diet 1 (1.4)
Block 4	Diet 4 (1.3)	Diet 2 (3.0)	Diet 1 (1.2)	Diet 3 (2.0)
Block 5	Diet 1 (1.4)	Diet 4 (1.5)	Diet 3 (2.5)	Diet 2 (3.3)

The preceding data (weight gains, in grams) are summarized in the following table.

	Diets					
Blocks	1	2	3	4	*Totals* (B_j)	
1	1.5	2.7	2.1	1.3	7.6	
2	1.4	2.9	2.2	1.0	7.5	$a = 4$
3	1.4	2.1	2.4	1.1	7.0	$b = 5$
4	1.2	3.0	2.0	1.3	7.5	$N = ab = 20$
5	1.4	3.3	2.5	1.5	8.7	
Totals:					*Grand total:*	
(G_i)	6.9	14.0	11.2	6.2	$\sum_{i=1}^{a}\sum_{j=1}^{b} X_{ij} = 38.3$	

$$\sum_{i=1}^{a}\sum_{j=1}^{b} X_{ij}^2 = 82.67$$

$$C = \frac{\left(\sum_{i=1}^{a}\sum_{j=1}^{b} X_{ij}\right)^2}{N} = 73.3445$$

$$\text{total SS} = \sum_{i=1}^{a}\sum_{j=1}^{b} X_{ij}^2 - C = 82.67 - 73.3445 = 9.3255$$

$$\text{diets SS} = \frac{\sum_{i=1}^{a} G_i^2}{b} - C = \frac{(6.9)^2 + (14.0)^2 + (11.2)^2 + (6.2)^2}{5} - 73.3445 = 8.1535$$

$$\text{blocks SS} = \frac{\sum_{j=1}^{b} B_j^2}{a} - C = \frac{(7.6)^2 + (7.5)^2 + (7.0)^2 + (7.5)^2 + (8.7)^2}{4} - 73.3445$$
$$= 0.3930$$
$$\text{remainder SS} = \text{total SS} - \text{diets SS} - \text{blocks SS} = 9.3255 - 8.1535 - 0.3930$$
$$= 0.7790$$

Analysis of Variance Summary Table

Source of variation	SS	DF	MS
Total	9.3255	19	
Diets	8.1535	3	2.7178
Blocks	0.3930	4	0.0982
Remainder	0.7790	12	0.0649

To test H_0: $F = \frac{\text{diets MS}}{\text{remainder MS}} = \frac{2.7178}{0.0649} = 41.9$

$F_{0.05(1),3,12} = 3.49$

Therefore, reject H_0.

$P < 0.0005$

Example 13.5 introduces some simplified symbols:

$$G_i = \text{sum (over all blocks) of all data in group } i$$
$$= \sum_{i=1}^{a} X_{ij} \tag{13.23}$$
$$B_j = \text{sum (over all groups) of all data in block } j$$
$$= \sum_{j=1}^{b} X_{ij} \tag{13.24}$$

Therefore, we can write

$$\text{groups SS} = \frac{\sum_{i=1}^{a} G_i^2}{b} - C \tag{13.25}$$

and

$$\text{blocks SS} = \frac{\sum_{j=1}^{b} B_j^2}{a} - C, \tag{13.26}$$

where the total SS and the quantity C are as in Table 13.4. The sums of squares, degrees of freedom, and mean squares for the randomized block ANOVA are also as in Table 13.4, where factor A is groups and factor B is blocks.

Missing Data in the Randomized Block Design. On occasion, a datum in a randomized block design may be missing due to no fault of the experimental conditions under study. An experimental mouse may be ingested by a pet hawk, a culture tube may accidentally be shattered on a desk top, an experimental plant may be chewed by a hungry guinea pig, etc. In such a situation, one may discard the entire block of data in which the missing datum belonged, but in general it is undesirable to discard data. Alternatively, we may employ the following procedure to estimate what the missing datum would have been (C. C. Li, 1964: 228–233).

A datum which is missing from block j in experimental group i may be estimated as

$$\hat{X}_{ij} = \frac{aG_i + bB_j + \sum_{i=1}^{a} \sum_{j=1}^{b} X_{ij}}{(a-1)(b-1)}. \tag{13.27}$$

where G_i is the sum of the known data in group i, B_j is the sum of the known data in block j, and $\sum_{i=1}^{a} \sum_{j=1}^{b} X_{ij}$ is the sum of all known data. The total, groups, and blocks sums of squares are then calculated using this estimated value in place of the missing datum, except that the groups SS is slightly biased, a situation corrected by subtracting the quantity

$$\text{bias} = \frac{[B_i - (a-1)\hat{X}_{ij}]^2}{a(a-1)}. \tag{13.28}$$

Also, 1 must be subtracted from the groups DF, which means that the remainder DF is also reduced by 1.

If there are more than one missing data, but not too many (say, no more than 10% of all data are missing, but in no case more than $a - 1$ data missing), then the following procedure may be used. Insert into the data array estimates of all but one of the missing data. Such estimates may be guesses; one might use as a guess, the mean of the known data for the group in which the missing datum would be located, or even the mean of all known data. Then the remaining missing datum is estimated by Equation 13.27, where the group and block sums include all data except that one being estimated. Then that estimate is placed in the data array and one of the other estimates is recalculated using Equation 13.27. This is done for each missing datum in turn. Then, the process is repeated for each of the missing data, thus arriving at somewhat better estimates of all of them. This iterative, or repetitive, process is engaged in until there is no further change in the estimated values. Typically, only two (or sometimes three) cycles of the process are required.

The total SS, blocks SS, and groups SS are then computed using all data, both known and estimated. But this groups SS is biased and should be corrected in the following manner (Glen and Kramer, 1958). Use Equation 13.28 to determine the bias for each block having one missing datum. Then, use Equation 13.29 to determine the bias for each block containing two or more data:

$$\text{bias} = \sum^{m_j} \hat{X}_{ij}^2 + \frac{B_j^2}{a - m_j} - \frac{(B_j + \sum^{m_j} \hat{X}_{ij})^2}{a}, \tag{13.29}$$

where m_j is the number of missing data in block j (m_j must be less than a), and each summation is performed over all m_j missing data in that block.

Finally, the results of all the above bias calculations are added and this sum is subtracted from the groups SS. Using the total SS, the blocks SS, and the corrected groups SS, the remainder SS is calculated. The total DF and the remainder DF each must be reduced by the number of missing data estimated; the groups DF and blocks DF are as usual (i.e., $a - 1$ and $b - 1$, respectively).

13.6 MULTIPLE COMPARISONS AND CONFIDENCE INTERVALS

If a two-factor analysis of variance reveals a significant effect among levels of a fixed effects factor having more than two levels, then we can determine between which levels the difference(s) occur(s). The may be done using the Tukey test (Section 12.1) or the Newman-Keuls test (Section 12.2). The appropriate SE is calculated by Equation 12.2, substituting for n the number of data in each level (i.e., there are bn data in each level of factor A and an data in levels of factor B); s^2 is the within cells MS; and ν is the within cells degrees of freedom. If there is no replication in the experiment, then we are obliged to use the remainder MS in place of the within cells MS, and to use the remainder DF as ν.

The calculation of confidence limits for the population mean estimated by each significantly different level mean can be performed by the procedures of Section 12.3, as can the computation of confidence limits for differences between pairs of significantly different level means.

If it is desired to compare a control level mean to each of the other level means, Dunnett's test, described in Section 12.4, may be used. Section 12.4 also shows how to calculate confidence limits for the difference between such means. Scheffé's procedure for multiple contrasts (Section 12.5) may also be applied to the levels of a factor, where the critical value in Equation 12.17 employs either a or b in place of k (depending, respectively, on whether the levels of factor A or B are being examined), and the within cells DF is used in place of $N - k$. In all references to Chapter 12, n in the standard error computation is to be replaced by the number of data per level, and s^2 and ν are the within cells MS and DF, respectively.

Multiple comparison testing and confidence interval determination are appropriate for levels of a fixed-effects factor but are not used with random-effects factors.

If Interaction Is Significant. On concluding that there is a significant interaction between factors A and B, it may be desired to perform multiple comparison testing to seek significant differences among cell means. This can be done with any of the above-mentioned procedures, using the equations from Chapter 12, where n (the number of data per cell) is appropriate instead of the number of data per level. For the Scheffé test critical value (Equation 12.17), k is the number of cells (i.e., $k = ab$) and $N - k$ is the within cells DF.

Randomized Block Analysis of Variance. For a randomized block experimental design (Section 13.5), we may perform multiple comparison testing among, and calculate confidence intervals for, the means of the levels of the fixed-effects factor, using any of the procedures of Sections 12.1 through 12.5. (This may not be done for block means, unless it is assumed that there is no interaction effect, but even then it is rarely desirable.) For such tests, s^2 refers to the remainder MS and ν refers to the remainder DF.

If there were any missing data that had to be estimated, then the standard error for the difference between two group means may be corrected upward (e.g., C. C. Li, 1964: 236), but the amount of correction is usually trivial.

13.7 POWER AND SAMPLE SIZE IN TWO-FACTOR ANALYSIS OF VARIANCE

If the reader is familiar with the concepts and procedures of estimating power, sample size, and minimum detectable difference for a single-factor ANOVA, as discussed in Section 11.3, then it will be seen how the same considerations can be applied to fixed-effects factors in a two-factor analysis of variance. (The handling of the fixed factor in a mixed model ANOVA will be explained at the end of this section.)

We can consider either factor A or factor B (or both, but one at a time). Let us say k' is the number of levels of the factor being examined. (That is, $k' = a$ for factor A; $k' = b$ for factor B.) Let us define n' as the number of data in each level. (That is, $n' = bn$ for factor A; $n' = an$ for factor B.) We shall also have s^2 refer to the within cells MS.

Power of the Test. We can now generalize Equation 11.23 as

$$\phi = \sqrt{\frac{(k' - 1)(\text{factor MS})}{k's^2}}, \tag{13.30}$$

Equation 11.24 as

$$\phi = \sqrt{\frac{n' \sum_{m=1}^{k'} (\mu_m - \mu)^2}{k's^2}}, \tag{13.31}$$

Equation 11.25 as

$$\mu = \frac{\sum_{m=1}^{k'} \mu_m}{k'}, \tag{13.32}$$

and Equation 11.26 as

$$\phi = \sqrt{\frac{n'\delta^2}{2k's^2}}, \tag{13.33}$$

in order to estimate the power of the analysis of variance in detecting differences among the population means of the levels of the factor under consideration.

In addition, we may estimate the power of the test in detecting an interaction effect by using

$$\phi = \sqrt{\frac{A \times B\ \text{SS}}{[(A \times B\ \text{DF}) + 1]s^2}} \tag{13.34}$$

(e.g., Guenther, 1964: 108; Kirk, 1968: 179). Here, $\nu_1 = A \times B$ DF and $\nu_2 =$ error DF.

After any of the computations of ϕ have taken place, either as above, or as below, then we proceed to employ Appendix Fig. B.1 just as we did in Section 11.3, with ν_1 being the factor DF (i.e., $k' - 1$), and ν_2 referring to the within cells (i.e., error) DF.

Power of a Performed Test. As demonstrated in Section 11.3, we can estimate the power of an ANOVA that has been performed, an especially desirable capability when H_0 has not been rejected. Using the above notation, with s^2 being the within cells MS for the analyzed data, Equation 11.27 can be generalized as

$$\phi = \sqrt{\frac{(k' - 1)(\text{factor MS} - s^2)}{k's^2}}. \tag{13.35}$$

Sample Size Required. By using Equation 13.33 with a specified power, significance level, and detectable difference between means, we can determine the necessary minimum number of data per level, n', needed in the experiment. This is done iteratively, as it was in Example 11.6.

Minimum Detectable Difference. In Example 11.7 we estimated the smallest detectable difference between population means, given the significance level, sample size, and power of a one-way ANOVA. We can pose the same question in the two-factor experiment, generalizing Equation 11.28 as

$$\delta = \sqrt{\frac{2k's^2\phi^2}{n'}}. \tag{13.36}$$

Maximum Number of Levels Testable. The considerations of Example 11.8 can be applied to the two-factor case by using Equation 13.33 instead of Equation 11.26.

The Mixed Model ANOVA. All the preceding considerations of this section can be applied to the fixed factor in a mixed model (Model III) two-factor analysis of variance with the following modifications.

For factor A fixed, with replication within cells, substitute the interaction MS for the within cells MS, and use the interaction DF for ν_2.

For factor A fixed, with no replication (i.e., a randomized block experimental design), substitute the remainder MS for the within cells MS, and use the remainder DF for ν_2. If there is no replication, then $n = 1$, and $n' = b$. (Recall that if there is no replication, we do not perform testing for interaction effect.)

13.8 NONPARAMETRIC RANDOMIZED BLOCK ANALYSIS OF VARIANCE

Friedman's test (1937, 1940) is a nonparametric analysis that may be performed on a randomized block experimental design, and it is especially useful with data that do not meet the parametric analysis of variance assumptions of normality and homoscedasticity, namely that the k samples (i.e., the k levels of the fixed-effect factor) come from populations that are each normally distributed and have the same variance. If these assumptions of the parametric ANOVA are met, the Friedman test will be $3k/[\pi(k + 1)]$ as powerful as the parametric method (van Elteren and Noether, 1959). (That is, the power of the latter compared to the former ranges from 64%, when $k = 2$*; to 72% when $k = 3$; to 95% when $k = \infty$.) If the assumptions are seriously violated, the former should not be used, but the Friedman test is valid.

*Where $k = 2$, the Friedman test is equivalent to the sign test (Section 22.6).

In Example 13.6, Friedman's test is applied to the data of Example 13.5. The data within each of the b blocks are assigned ranks. The ranks are then summed for each of the a groups, each rank sum being denoted as R_i. The test statistic, χ_r^2, is calculated as

$$\chi_r^2 = \frac{12}{ba(a+1)} \sum_{i=1}^{a} R_i^2 - 3b(a+1). \tag{13.37}$$

Critical values for some combinations of a and b are given in Table B.13. For other values of b and a, the distribution of χ_r^2 may be considered to approximate the χ^2 distribution (Table B.1), with $a - 1$ degrees of freedom.*

Example 13.6

Friedman's analysis of variance by ranks applied to the randomized block data of Example 13.5.

H_0: The weight gain of guinea pigs is the same on each of four specified diets.
H_A: The weight gain of guinea pigs is not the same on each of the four specified diets.
$\alpha = 0.05$.
The following data (weight gain, in grams) have been ranked within each block. The ranks are shown in parentheses.

	Diets (i)			
Blocks	1	2	3	4
1	1.5	2.7	2.1	1.3
	(2)	(4)	(3)	(1)
2	1.4	2.9	2.2	1.0
	(2)	(4)	(3)	(1)
3	1.4	2.1	2.4	1.1
	(2)	(3)	(4)	(1)
4	1.2	3.0	2.0	1.3
	(1)	(4)	(3)	(2)
5	1.4	3.3	2.5	1.5
	(1)	(4)	(3)	(2)
Rank sum (R_i)	8	19	16	7

$$\chi_r^2 = \frac{12}{ba(a+1)} \sum R_i^2 - 3b(a+1)$$
$$= \frac{12}{(5)(4)(4+1)}[(8)^2 + (19)^2 + (16)^2 + (7)^2] - (3)(5)(4+1)$$
$$= 0.12(730) - 75 = 87.60 - 75 = 12.60$$

$\nu = a - 1 = 3$
$\chi^2_{0.05,3} = 7.815$
Therefore, reject H_0.
$0.005 < P < 0.01$

*The F distribution also may be utilized to assess the significance of χ_r^2 (Kendall and Babington–Smith, 1939). Although it works somewhat better than does the χ^2 distribution (Friedman, 1940), it is much more difficult to employ.

If tied ranks are present, they may be taken into consideration by computing

$$(\chi_r^2)_c = \frac{\sum_{i=1}^{a} R_i^2 - \frac{\left(\sum_{i=1}^{a} R_i\right)^2}{a}}{\frac{ba(a+1)}{12} - \frac{\sum T}{a-1}} \tag{13.38}$$

(Kendall, 1962: Chapter 6), where the correction for ties is

$$\sum T = \frac{\sum_{i=1}^{m} (t_i^3 - t_i)}{12}, \tag{13.39}$$

and t_i is the number of ties in the ith group of ties, and m is the number of groups of tied ranks.

The *Kendall coefficient of concordance* (W) is another form of Friedman's χ_r^2:

$$W = \frac{\chi_r^2}{b(a-1)}. \tag{13.40}$$

It is used as a measure of the agreement of rankings within blocks and is considered further in Section 20.16.

Multiple Observations per Cell. In the experiment of Example 13.6 there is one datum for each combination of diet and block. Although this is the typical situation, one might also encounter an experimental design in which there are multiple observations recorded for each combination of block and experimental group. Recall that each combination of level of factor A (experimental group) and level of factor B (block) is called a cell; for n replicate data per cell,

$$\chi_r^2 = \frac{12}{ban^2(na+1)} \sum_{i=1}^{a} R_i^2 - 3b(na+1) \tag{13.41}$$

(Marascuilo and McSweeney, 1977: 376–377), with the critical values again being $\chi^2_{\alpha,k-1}$. Note that if $n = 1$, Equation 13.41 reduces to Equation 13.37.

13.9 MULTIPLE COMPARISONS FOR NONPARAMETRIC RANDOMIZED BLOCK ANALYSIS OF VARIANCE

A multiple comparison analysis applicable to ranked data in a randomized block is similar to the Tukey procedure for ranked data in a one-way ANOVA design (Section 12.6). In this case, Equation 12.3 is used with the difference between rank sums, i.e., $R_B - R_A$, in the numerator and

$$\mathrm{SE} = \sqrt{\frac{ba(a+1)}{12}} \tag{13.42}$$

in the denominator;* and this is used in conjunction with the appropriate critical values of $q_{\alpha,\infty,k}$.

If the various groups are to be compared one at a time with a control group, then

$$SE = \sqrt{\frac{ba(a+1)}{6}} \tag{13.44}$$

may be used in Dunnett's procedure, in a fashion similar to that explained at the end of Section 12.6.

The preceding multiple comparisons are applicable to the levels of the fixed-effect factor, not to the blocks (levels of the random-effect factor).

Multiple contrasts, as introduced in Sections 12.5 and 12.7, may be performed using rank sums.† We employ Equation 12.30, with

$$SE = \sqrt{\frac{ba(a+1)}{12}\left(\sum_i c_i^2\right)}, \tag{13.46}$$

unless there are tied ranks, in which case‡

$$SE = \sqrt{\frac{\left(ba(a+1) - \frac{\sum T}{(a-1)}\right)\left(\sum_i c_i^2\right)}{12}} \tag{13.48}$$

(Marascuilo and McSweeney, 1967). The critical value for the multiple contrasts is $\sqrt{(\chi_r^2)_{\alpha,a,b}}$, using Table B.13 to obtain $(\chi_r^2)_{\alpha,a,b}$. If the needed critical value is not on that table, then $\sqrt{\chi^2_{\alpha,a-1}}$ may be used.

Multiple Observations per Cell. If there is replication per cell (as in Equation 13.41), the standard errors of this section are modified by replacing a with an and b with bn wherever they appear. (See Marascuilo and McSweeney, 1977: 378.)

13.10 DICHOTOMOUS NOMINAL SCALE DATA IN RANDOMIZED BLOCKS

A randomized block experimental design may contain values of a dichotomous variable (i.e., a variable with two possible values: e.g., "present" or "absent," "dead" or "alive," "true" or "false," "left" or "right," "male" or "female," etc.), in which

*If desired, mean ranks ($\bar{R}_A = R_A/b$ and $\bar{R}_B = R_B/a$) can be used in the numerator of Equation 12.3, in which case the denominator will be

$$SE = \sqrt{\frac{a(a+1)}{12b}}. \tag{13.43}$$

†If mean ranks are used,

$$SE = \sqrt{\frac{a(a+1)}{12}\left(\frac{\sum_i c_i^2}{b}\right)}. \tag{13.45}$$

‡If mean ranks are used,

$$SE = \sqrt{\frac{\left(a(a+1) - \frac{\sum T}{b(a-1)}\right)\left(\frac{\sum_i c_i^2}{b}\right)}{12}}. \tag{13.47}$$

case Cochran's Q test† (Cochran, 1950) may be applied. For such an analysis, one value of the attribute is recorded with a "1," and the other with a "0." In Example 13.7, the data are the occurrence or absence of mosquito attacks on humans wearing one of several types of clothing. The null hypothesis is that the proportion of people attacked is the same for each type of clothing worn. (It is common in blocked experimental designs to make repeated observations on the same individuals, each individual being a block.)

Example 13.7

Cochran's Q test.

H_0: The proportion of humans attacked by mosquitoes is the same for all 5 clothing types.
H_A: The proportion of humans attacked by mosquitoes is not the same for all 5 clothing types.
$\alpha = 0.05$
A person attacked is scored as a "1"; a person not attacked is scored as a "0."

	Clothing Type					
Person (block)	*Light, loose*	*Light, tight*	*Dark, long*	*Dark, short*	*None*	*Totals* (B_j)
1	0	0	0	1	0	1
2*	1	1	1	1	1	*
3	0	0	0	1	1	2
4	1	1	0	1	0	3
5	0	1	1	1	1	4
6	0	1	0	0	1	2
7	0	0	1	1	1	3
8	0	0	1	1	0	2
*Totals** (G_i)	1	3	3	6	4	$\sum_{i=1}^{a} G_i = \sum_{j=1}^{b} B_j = 17$

$a = 5$
$b = 7^*$

*The data for block 2 are deleted from the analysis, because 1's occur for all clothing. (See test discussion in Section 13.10.)

$$Q = \frac{(a-1)\left[\sum_{i=1}^{a} G_i^2 - \frac{\left(\sum_{i=1}^{a} G_i\right)^2}{a}\right]}{\sum_{j=1}^{b} B_j - \frac{\sum_{j=1}^{b} B_j^2}{a}} = \frac{(5-1)\left[1+9+9+36+16-\frac{17^2}{5}\right]}{17 - \frac{(1+4+9+16+4+9+4)}{5}}$$

$$= \frac{52.8}{7.6} = 6.947$$

$\nu = a - 1 = 4$
$\chi^2_{0.05,4} = 9.488$
Therefore, do not reject H_0.
$0.10 < P < 0.25$

†William G. Cochran, influential American statistician (1909–1980).

For a groups and b blocks, where G_i is the sum of the 1's in group i and B_j is the sum of the 1's in block j,

$$Q = \frac{(a-1)\left[\sum_{i=1}^{a} G_i^2 - \frac{\left(\sum_{i=1}^{a} G_i\right)^2}{a}\right]}{\sum_{j=1}^{b} B_j - \frac{\sum_{j=1}^{b} B_j^2}{a}} \tag{13.49}$$

Note, as shown in Example 13.7, that $\sum B = \sum G$, which is the total number of 1's in the set of data. This test statistic, Q, is distributed approximately as chi-square, with $a - 1$ degrees of freedom. Tate and Brown (1970) explain that the value of Q is unaffected by having blocks containing either all 0's or all 1's. Thus, any such block may be disregarded in the calculations. They further point out that the approximation of Q to χ^2 is a satisfactory one only if the number of data is large. These authors suggest as a rule of thumb that a should be at least 4 and ba should be at least 24, where b is the number of blocks remaining after all those containing either all 0's or all 1's are disregarded. For sets of data smaller than these suggestions allow, the analysis may proceed but with caution exercised if Q is near a borderline of significance. In these cases it would be better to use the tables of Tate and Brown (1964) or Patil (1975).

If $a = 2$, then Cochran's test is identical to McNemar's test (Section 10.4), except that the latter employs a correction for continuity.

13.11 MULTIPLE COMPARISONS WITH DICHOTOMOUS RANDOMIZED BLOCK DATA

Marascuilo and McSweeney (1967) present a multiple comparison procedure that may be used for multiple contrasts as well as for pairwise comparisons for data subjected to the Cochran Q test of Section 13.10. It may be performed using group means, $\bar{R}_i = G_i/b$.

For pairwise comparisons, the test statistic is

$$S = \frac{|\bar{R}_B - \bar{R}_A|}{\text{SE}} \tag{13.50}$$

(which parallels Equation 12.15), where

$$\text{SE} = \sqrt{2\left(\frac{a\sum_j B_j - \sum_j B_j^2}{ab^2(a-1)}\right)}. \tag{13.51}$$

For multiple contrasts, the test statistic is that of Equation 12.18, where $\bar{R}_i$ replaces $\bar{X}_i$ and

$$\text{SE} = \sqrt{\left(\frac{a\sum_j B_j - \sum_j B_j^2}{ab^2(a-1)}\right)\sum_i c_i^2} \tag{13.52}$$

The critical value for such multiple comparisons is $\sqrt{\chi^2_{\alpha,a-1}}$.

EXERCISES

13.1. Fish of each sex were given one of three different hormone treatments, after which the blood calcium was measured (in mg Ca/100 ml blood). The data are as follows.

	Treatment 1	*Treatment 2*	*Treatment 3*
Male	16.87	19.07	32.45
	16.18	18.77	28.71
	17.12	17.63	34.65
	16.83	16.99	28.79
	17.19	18.04	24.46
Female	15.86	17.20	30.54
	14.92	17.64	32.41
	15.63	17.89	28.97
	15.24	16.78	28.46
	14.80	16.92	29.65

(a) Test the hypothesis that there is no difference in mean blood calcium concentration among the three hormone treatments.
(b) Test the hypothesis that there is no difference in mean blood calcium concentration between males and females.
(c) Test the hypothesis that there is no interaction between sex and hormone treatment in affecting mean blood calcium concentration (i.e., that the effect of hormone treatment is independent of sex).

13.2. Using the data from Exercise 13.1, test nonparametrically:
(a) for difference in blood calcium concentration among the three hormone treatments:
(b) for difference between males and females in blood calcium concentration;
(c) for interaction between sex and hormone treatment in affecting blood calcium concentration.

13.3. Six greenhouse benches were set up as blocks. Within each block one of each of four varieties of house plants was planted. The plant heights (in centimeters) obtained are tabulated as follows. Test the hypothesis that all four varieties of plants reach the same maximum height.

Block	*Variety 1*	*Variety 2*	*Variety 3*	*Variety 4*
1	19.8	21.9	16.4	14.7
2	16.7	19.8	15.4	13.5
3	17.7	21.0	14.8	12.8
4	18.2	21.4	15.6	13.7
5	20.3	22.1	16.4	14.6
6	15.5	20.8	14.6	12.9

13.4. Consider the data of Exercise 13.3. Nonparametrically test the hypothesis that all four varieties of plants reach the same maximum height.

13.5. A textbook distributor wishes to assess potential acceptance of four general biology textbooks. He asks each of 15 biology professors to examine the books and to respond as to which ones they would seriously consider for their courses. In the table, a positive response is recorded as 1 and a negative response as a 0. Test the hypothesis that there is no difference in potential acceptance among the four textbooks.

Professor	*Textbook 1*	*Textbook 2*	*Textbook 3*	*Textbook 4*
1	1	1	0	0
2	1	1	0	1
3	1	0	0	0
4	1	1	1	1
5	1	1	0	1
6	0	1	0	0
7	0	1	1	0
8	1	1	1	0
9	0	0	1	0
10	1	0	1	0
11	0	0	0	0
12	1	1	0	1
13	1	0	0	1
14	0	1	1	0
15	1	1	0	0

14 Data Transformations

For the valid application of parametric analyses of variance and related procedures, such as those using Student's t, certain basic assumptions must be met. First, we must be able to assume that the data for each group were obtained randomly from a normal population. Second, the sampled populations must all have equal variances (in which case it can be said that the variances are *homoscedastic*, the opposite situation being referred to as *heteroscedastic*). Third, the effects of the factor levels must be assumed to be *additive*.

The requirements of normality and homoscedasticity have been stated previously, but the assumption of additivity (sometimes referred to as linearity) needs elaboration. Consider the two-way analysis of variance data in Example 14.1A. In this example, the effect of factor A is said to be additive, for each datum in level 2 differs from the corresponding datum in level 1 by the addition of the same amount (10 g). Similarly, each datum in level 3 differs from its corresponding level 2 datum by a constant value (5 g). Examining the effect of the two levels of factor B, we also observe additivity (by a constant value of 10 g). What we are assuming is that in the population, on the average, the effects of the factor levels are additive. Occasionally, however, the effect of a factor may not be additive; multiplicative effects are such an instance. In Example 14.1B, we can see that factor A has an effect such that each datum in level 3 differs from its corresponding member in level 2 by a multiplication factor of 2, and each member of level 2 differs from that in level 1 by a factor of 3. Examining the two levels of factor B, we can see that the data are affected by a multiplication factor of 2. The importance of Example 14.1C will be discussed shortly.

Experience has shown that analyses of variance and t tests are usually robust enough to perform well even if the data deviate somewhat from the requirements of

Example 14.1

Additive and multiplicative effects.

A. A hypothetical two-way analysis of variance design, where the effects of the factors are additive. (Data are in grams.)

	Factor A		
Factor B	*Level 1*	*Level 2*	*Level 3*
Level 1	10	20	25
Level 2	20	30	35

B. A hypothetical two-way analysis of variance design, where the effects of the factors are multiplicative. (Data are in grams.)

	Factor A		
Factor B	*Level 1*	*Level 2*	*Level 3*
Level 1	10	30	60
Level 2	20	60	120

C. The two-way analysis of variance design of Example 14.1B, showing the logarithms (rounded to two decimal places) of the data.

	Factor A		
Factor B	*Level 1*	*Level 2*	*Level 3*
Level 1	1.00	1.48	1.78
Level 2	1.30	1.78	2.08

normality, homoscedasticity, and additivity. (See Section 11.1.) But severe deviations can lead to spurious conclusions.

In cases of heteroscedastic data, the inequality of variances often may be "corrected," and this correction will usually result simultaneously in data that do not deviate intolerably from the normality and additivity assumptions.

Also, there are types of data for which it is known on theoretical grounds that the population sampled is not normal. Such a problem can often be "corrected," and such correction procedures generally will result in data that have acceptable homoscedasticity and additivity characteristics.

Situations of nonadditivity exist, as Example 14.1B demonstrates, and they may often be "corrected," the resultant data generally also being acceptably homoscedastic and normal.

These "corrections" for heteroscedasticity, nonnormality, and nonadditivity are frequently possible by means of changing, or transforming, the data from their original form (X values) to a different form (let us call them X' values). Many statistical researchers have addressed themselves to general questions of data transformations (e.g., Bartlett, 1947; Box and Cox, 1964; Kendall and Stuart, 1966: 87–96), and an extraordinarily thorough monograph on transformation methodology has been prepared by Thöni (1967). The most commonly employed transformations are described in the following sections. The transformation of data used in regression analysis will be discussed in a later chapter (Section 17.10).

14.1 THE LOGARITHMIC TRANSFORMATION

If the factor effects in an analysis of variance are, in fact, multiplicative rather than additive, then the logarithms of the data will exhibit additivity (as Example 14.1C shows). Instead of the transformation, $X' = \log X$, however,

$$X' = \log(X + 1) \qquad (14.1)$$

is preferred on theoretical grounds and is especially preferable when some of the observed values are small numbers (particularly zero) (Bartlett, 1947). Logarithms in base 10 are generally utilized, but any base would be satisfactory.

A second instance when the logarithmic transformation is applicable is when there is heteroscedasticity and the standard deviations are proportional to the means (i.e., there is a constant coefficient of variation). Such a case is shown in Example 14.2.

Example 14.2

Data in which there is heterogeneity of variance and the standard deviations are proportional to the means (i.e., the coefficients of variation are the same). Primes identify statistics obtained using the transformed data (e.g., $\bar{X}'$, s', L').

The original data:

Group 1 (cm)	*Group 2* (cm)
3.1	7.6
2.9	6.4
3.3	7.5
3.6	6.9
3.5	6.3
$\bar{X}_1 = 3.28$ cm	$\bar{X}_2 = 6.94$ cm
$s_1^2 = 0.0820$ cm^2	$s_2^2 = 0.3630$ cm^2
$s_1 = 0.29$ cm	$s_2 = 0.60$ cm
$V_1 = 0.09$	$V_2 = 0.09$

The logarithmically transformed data, using Equation (14.1):

Group 1	*Group 2*
0.61278	0.93450
0.59106	0.86923
0.63347	0.92942
0.66276	0.89763
0.65321	0.86332
$\bar{X}'_1 = 0.63066$	$\bar{X}'_2 = 0.89882$
$s_1^{2\prime} = 0.0008586657$	$s_2^{2\prime} = 0.0010866641$
$s'_1 = 0.02930$	$s'_2 = 0.03296$
$V'_1 = 0.04646$	$V'_2 = 0.03667$
$s'_{\bar{X}_1} = 0.01310$	$s'_{\bar{X}_2} = 0.01474$

Calculating confidence limits:

$$95\%\ \text{confidence interval for } \mu'_1 = \bar{X}'_1 \pm (t_{0.05(2),\,4})(0.01310)$$
$$= 0.63066 \pm (2.776)(0.01310)$$
$$= 0.63066 \pm 0.03637$$
$$L'_1 = 0.59429 \quad \text{and} \quad L'_2 = 0.66703$$

95% confidence limits for μ_1, in the original units:

$$L_1 = \text{antilog } 0.59429 - 1 = 3.93 - 1 = 2.93 \text{ cm}$$
$$L_2 = \text{antilog } 0.66703 - 1 = 4.65 - 1 = 3.65 \text{ cm}$$

This transformation may also convert a positively skewed distribution into a symmetrical one.

Using the transformed data, we can now proceed with parametric testing. In Example 14.2, we calculate a confidence interval for the mean of population 1. By finding the antilogarithm and subtracting 1, we can express the confidence limits in the units of the original data. But the confidence limits in the original units will not be symmetrical about the mean.*

If the distribution of X' is normal, then the distribution of X is said to be *lognormal.*

14.2 THE ARCSINE TRANSFORMATION

It is known from statistical theory that percentages or proportions form a binomial, rather than a normal, distribution, the deviation from normality being great for small or large percentages (0 to 30% and 70 to 100%). If the square root of each proportion, p, in a binomial distribution is transformed to its arcsine (i.e., the angle whose sine is $\sqrt{p}$), then the resultant data will have an underlying distribution that is nearly normal. This transformation,

$$p' = \arcsin \sqrt{p}, \tag{14.2}$$

is performed easily with the aid of Table B.23. For proportions of 0 to 1.00 (i.e., percentages of 0 to 100%), the transformed values will range between 0 and 90 degrees (although some authors' tables present the transformation in terms of radians†).

The arcsine transformation ("arcsine" is abbreviated "arcsin") frequently is referred to as the "angular transformation," and "inverse sine" or "$\sin^{-1}$" is sometimes written to denote "arcsine."

Example 14.3 demonstrates calculations using data submitted to the arcsine transformation. Transformed values (such as means or confidence limits) may be transformed back to proportions, as

$$p = (\sin p')^2; \tag{14.3}$$

and Table B.24 is useful for this purpose. But confidence limits generally will not be symmetrical about the mean when expressed in proportions.‡

*Also, an antilogarithmic transformation to obtain $\bar{X}$ in terms of the original units is known to result in a somewhat biased estimator of μ, the estimator being more unbiased for larger variances of $\bar{X}'$ values.

†A radian is $180°/\pi = 57.29577951308232\ldots$ degrees.

‡The mean, $\bar{p}$, that is obtained from $\bar{p}'$ by consulting Table B.24 is, however, slightly biased. Quenouille (1950) suggests correcting for the bias by adding to $\bar{p}$ the quantity $0.5 \cos(2\bar{p}')(1 - e^{-2s^2})$, where s^2 is the variance of the p' values.

Example 14.3

The arcsine transformation for percentage data.

Original data (p):

Group 1 (%)	*Group 2* (%)
84.2	92.3
88.9	95.1
89.2	90.3
83.4	88.6
80.1	92.6
81.3	96.0
85.8	93.7
$\bar{p}_1 = 84.7\%$	$\bar{p}_2 = 92.7\%$
$s_1^2 = 12.29\ (\%)^2$	$s_2^2 = 6.73\ (\%)^2$
$s_1 = 3.5\%$	$s_2 = 2.6\%$

Transformed data (by using Equation 14.2 or Table B.23) (p'):

Group 1	*Group 2*
66.58	73.89
70.54	77.21
70.81	71.85
65.96	70.27
63.51	74.21
64.38	78.46
67.86	75.46
$\bar{p}'_1 = 67.09$	$\bar{p}'_2 = 74.48$
$(s_1^2)' = 8.0052$	$(s_2^2)' = 8.2193$
$s'_1 = 2.83$	$s'_2 = 2.87$
$s_{\bar{x}_1}{}' = 1.07$	$s_{\bar{x}_2}{}' = 1.08$

Calculating confidence limits:

95% confidence interval for μ'_1: $\bar{p}_1 \pm (t_{0.05(2),6})(1.07) = 67.09 \pm 2.62$

$L'_1 = 64.47$ and $L'_2 = 69.71$

By using Table B.24 to transform backward from L'_1, L'_2, and p'_1:

95% confidence limits for μ_1: $L_1 = 81.5\%$ and $L_2 = 88.0\%$

$\bar{p}_1 = 84.9\%$

This transformation is not as good at the extreme ends of the range of possible values (i.e., near 0 and 100%) as it is elsewhere. If, rather than simply having percentages as one's data, one knows the actual proportions, X/n, then the arcsine transformation is improved by replacing $0/n$ with $1/4n$ and n/n with $1 - 1/4n$ (Bartlett, 1937). Anscombe (1948) proposed an even better transformation:

$$p' = \sqrt{(n + \tfrac{1}{2})} \arcsin \sqrt{\frac{X + \frac{3}{8}}{n + \frac{3}{4}}}. \tag{14.4}$$

And a slight modification of the Freeman and Tukey (1950) transformation, namely

$$p' = \frac{1}{2}\left[\arcsin \sqrt{\frac{X}{n+1}} + \arcsin \sqrt{\frac{X+1}{n+1}}\right], \tag{14.5}$$

yields very similar results, except for small and large proportions where it appears to be preferable. The use of the latter transformation will be shown in Section 22.13. Determination of transformed proportions, p', by either Equation 14.4 or 14.5 is facilitated by using Table B.23.

14.3 THE SQUARE ROOT TRANSFORMATION

The square root transformation is applicable when the group variances are proportional to the means. This most often occurs in biological data when samples are taken from a Poisson distribution (i.e., when the data consist of counts of randomly occurring objects or events; see Chapter 21 for discussion of the Poisson distribution). Transforming such data by utilizing their square roots results in a sample whose underlying distribution is normal. However, Bartlett (1936) proposed that

$$X' = \sqrt{X + 0.5} \tag{14.6}$$

is preferable to $X' = \sqrt{X}$, especially when there are very small data and/or when some of the observations are zero (see Example 14.4). Actually,

$$X' = \sqrt{X + \frac{3}{8}} \tag{14.7}$$

has even better variance-stabilizing qualities than Equation 14.6 (Anscombe, 1948; Kihlberg, Herson, and Schutz, 1972), and Freeman and Tukey (1950) show

$$X' = \sqrt{X} + \sqrt{X + 1} \tag{14.8}$$

to yield similar results but to be preferable for $X \leq 2$.

Equation 14.6 is most commonly employed. Statistical computation may then be performed on the transformed data. The mean can be expressed in terms of the original data by squaring it and then subtracting 0.5, although the resultant statistic is slightly biased.*

Example 14.4
The square root transformation for Poisson data.

Original data:

	Group 1	*Group 2*	*Group 3*	*Group 4*
	2	6	9	2
	0	4	5	4
	2	8	6	1
	3	2	5	0
	0	4	11	2
$\bar{X}_i$	1.4	4.8	7.2	1.8
s_i^2	1.8	5.2	7.2	2.2

*Thöni (1967: 16) has shown that an unbiased estimate of μ would be obtained by adding $(1 - 1/n)s^2$ to the $\bar{X}$ derived by untransforming $\bar{X}'$, where s^2 is the variance of the transformed data.

Transformed data; by Equation (14.6):

	Group 1	*Group 2*	*Group 3*	*Group 4*
	1.581	2.550	3.082	1.581
	0.707	2.121	2.345	2.121
	1.581	2.915	2.550	1.225
	1.871	1.581	2.345	0.707
	0.707	2.121	3.391	1.581
$\bar{X}'_i$	1.289	2.258	2.743	1.443
$(s_i^2)'$	0.297	0.253	0.222	0.272
$s'_{\bar{X}_i}$	0.244	0.225	0.211	0.233
$(L'_1)_i$	0.612	1.633	2.157	0.796
$(L'_2)_i$	1.966	2.883	3.329	2.090

On transforming back to original units [e.g., $\bar{X} = (\bar{X}')^2 - 0.5$]:

	Group 1	*Group 2*	*Group 3*	*Group 4*
$\bar{X}_i$	1.2	4.6	7.0	1.6
$(L_1)_i$	−0.1	2.2	4.2	0.1
$(L_2)_i$	3.4	7.8	10.6	3.9

14.4 OTHER TRANSFORMATIONS

The logarithmic, arcsine, and squre root transformations are those most commonly required to handle nonnormal, heteroscedastic, or nonadditive data. Other transformations are only rarely called for.

If the standard deviations of groups of data are proportional to the square of the means of the groups, then the *reciprocal transformation*,

$$X' = \frac{1}{X}, \tag{14.9}$$

may be employed. (If counts are being transformed, then

$$X' = \frac{1}{X + 1} \tag{14.10}$$

may be used to allow for observations of zero.) See Thöni (1967: 32) for further discussion of the use of this transformation.

If the standard deviations decrease as the group means increase, and/or if the distribution is skewed to the left, then

$$X' = X^2 \tag{14.11}$$

might prove useful.

If the data come from a population with what is termed a "negative binomial distribution," then the use of inverse hyperbolic sines may be called for (see Anscombe, 1948; Bartlett, 1947; Beall, 1940, 1942; Thöni, 1967: 20–24).

Thöni (1967) mentions other, infrequently employed, transformations.

EXERCISES

14.1. Perform the logarithmic transformation on the following data [using Equation (14.1)] and calculate the 95% confidence interval for μ. Express the confidence limits in terms of the original units (i.e., ml). The data are 3.67, 4.01, 3.85, 3.92, 3.71, 3.88, 3.74, and 3.82 ml.

14.2. Transform the following proportions by the arcsine transformation(using Table B.23) and calculate the 95% confidence interval for μ. Express the confidence limits in terms of proportions (using Table B.24).

0.733, 0.804, 0.746, 0.781, 0.772, and 0.793

14.3. Apply the square root transformation to the following data [using Equation (14.6)] and calculate the 95% confidence interval for μ. Transform the confidence limits back to the units of the original data. The data are 4, 6, 3, 8, 10, 3.

15 Multiway Factorial Analysis of Variance

Chapter 13 discussed the analysis of the effects on a variable of two factors acting simultaneously. In such a procedure—a two-way, or two-factor, analysis of variance—one can determine whether either of the factors has a significant effect on the magnitude of the variable and also whether the interaction of the two factors significantly affects the variable. By expanding the considerations of the two-way analysis of variance, we can assess the effects on a variable of the simultaneous application of three or more factors, this being done by what we may refer to as a multiway analysis of variance.

It is not unreasonable to expect a researcher to perform a one-way or two-way analysis by hand, although computer programs are frequently employed, especially when the experiment consists of a large number of observations. However, it has become uncommon for analyses of variance with more than two factors to be analyzed other than on a digital computer, owing to considerations of time, ease, and accuracy. Therefore, this chapter will not instruct the reader in the computational mechanics of multiway analyses of variance. Rather, it will presume that established computer programs will be used to perform the necessary mathematical manipulations. We shall consider only the subsequent examination and interpretation of the numerical results of the computer's labor.

This chapter begins with a discussion of analysis of variance with three factors and proceeds to expand this consideration to analyses with more than three factors. We shall also introduce nonparametric ANOVA for three or more factors, multiway ANOVA with unequal replication, the performance of multiple comparisons, and the computation of confidence intervals for such analyses of variance. Lastly, we shall consider power and sample size determinations for multiway analyses of variance.

15.1 THREE-FACTOR ANALYSIS OF VARIANCE

For a particular variable, we may wish to assess the effects of three factors; let us refer to them as factors A, B, and C. For example, we might desire to determine what effect the following three factors have on the rate of oxygen consumption of crabs: species, temperature, and sex. Example 15.1 shows experimental data collected for crabs of both sexes, representing three species, and measured at three temperatures. For each cell (i.e., each combination of species, temperature, and sex) there was an oxygen consumption datum for each of four crabs (i.e., there were four replicates); therefore, 72 animals were used in the experiment.

Example 15.1

A three-factor analysis of variance (Model I), where the variable is respiratory rate of crabs (in ml O_2/hr).

1 H_0: Mean respiratory rate is the same in all three crab species (i.e., $\mu_1 = \mu_2 = \mu_3$).
H_A: Mean respiratory rate is not the same in all three crab species.

2 H_0: Mean respiratory rate is the same at all three experimental temperatures (i.e., $\mu_{\text{low}} = \mu_{\text{med}} = \mu_{\text{high}}$).
H_A: Mean respiratory rate is not the same at all three experimental temperatures.

3 H_0: Mean respiratory rate is the same for males and females (i.e., $\mu_♂ = \mu_♀$).
H_A: Mean respiratory rate is not the same for males and females (i.e., $\mu_♂ \neq \mu_♀$).

4 H_0: Differences in mean respiratory rate among the three species are independent of (i.e., are the same at) the three experimental temperatures; or, differences in mean respiratory rate among the three temperatures are independent of (i.e., are the same in) the three species. (Testing for $A \times B$ interaction.)
H_A: Differences in mean respiratory rate among the species are not independent of the experimental temperatures.

5 H_0: Differences in mean respiratory rate among the three species are independent of sex (i.e., are the same for both sexes); or, differences in mean respiratory rate between males and females are independent of (i.e., are the same in) the three species. (Testing for $A \times C$ interaction.)
H_A: Differences in mean respiratory rate among the species are not independent of sex.

6 H_0: Differences in mean respiratory rate among the three experimental temperatures are independent of (i.e., are the same in) the two sexes; or, differences in mean respiration rate between the sexes are independent of (i.e., are the same at) the three temperatures. (Testing for $B \times C$ interaction.)
H_A: Differences in mean respiratory rate among the three temperatures are not independent of sex.

7 H_0: Differences in mean respiratory rate among the species (or temperatures, or sexes) are independent of the other two factors. (Testing for $A \times B \times C$ interaction.)
H_A: Differences in mean respiratory rate among the species (or temperatures, or sexes) are not independent of the other two factors.

Species 1

Low temp.		*Med. temp.*		*High temp.*	
♂	♀	♂	♀	♂	♀
1.9	1.8	2.3	2.4	2.9	3.0
1.8	1.7	2.1	2.7	2.8	3.1
1.6	1.4	2.0	2.4	3.4	3.0
1.4	1.5	2.6	2.6	3.2	2.7

Species 2

Low temp.		Med. temp.		High temp.	
♂	♀	♂	♀	♂	♀
2.1	2.3	2.4	2.0	3.6	3.1
2.0	2.0	2.6	2.3	3.1	3.0
1.8	1.9	2.7	2.1	3.4	2.8
2.2	1.7	2.3	2.4	3.2	3.2

Species 3

Low temp.		Med. temp.		High temp.	
♂	♀	♂	♀	♂	♀
1.1	1.4	2.0	2.4	2.9	3.2
1.2	1.0	2.1	2.6	2.8	2.9
1.0	1.3	1.9	2.3	3.0	2.8
1.4	1.2	2.2	2.2	3.1	2.9

By consulting the computer output of Table 15.1, we can prepare the following summary:

Effect in hypothesis	*Calculated F*	*Critical F**	*Conclusion*	*P*
1. Species (factor A)	24.475	$F_{0.05(1),2,54} \cong 3.18$	Reject H_0	$P \ll 0.0005$
2. Temperature (factor B)	332.024	$F_{0.05(1),2,54} \cong 3.18$	Reject H_0	$P \ll 0.0005$
3. Sex (factor C)	0.239	$F_{0.05(1),1,54} \cong 4.03$	Accept H_0	$P > 0.50$
4. $A \times B$	7.418	$F_{0.05(1),4,54} \cong 2.56$	Reject H_0	$P < 0.0005$
5. $A \times C$	4.986	$F_{0.05(1),2,54} \cong 3.18$	Reject H_0	$0.01 < P < 0.025$
6. $B \times C$	2.360	$F_{0.05(1),2,54} \cong 3.18$	Accept H_0	$0.10 < P < 0.25$
7. $A \times B \times C$	1.485	$F_{0.05(1),4,54} \cong 2.56$	Accept H_0	$0.10 < P < 0.25$

*There are no critical values in Table B.4 for $\nu_2 = 54$, so the values for the next lower DF, $\nu_2 = 50$, were utilized.

We conclude that the respiratory rates of the three species are not all the same, and that the respiratory rates at the three temperatures are not all the same. Furthermore, it is concluded that the respiratory rate of a species is dependent on temperature, that the respiratory rate of a species is dependent on sex, and that the respiratory rate at a given temperature is dependent on sex.

Table 15.1 presents the computer output for the analysis of these experimental results, such output typically giving the sums of squares, degrees of freedom, and mean squares pertaining to the hypotheses to be tested. Some computer programs also give the F values calculated, assuming that the experiment calls for a Model I analysis, which is the case with most biological data. The major work that the computer has performed for us is the calculation of the sums of squares. We could easily have arrived at the degrees of freedom for each factor as follows: number of levels -1 (so, for factor A, DF $= 3 - 1 = 2$; for factor B, DF $= 3 - 1 = 2$; and for factor C, DF $= 2 - 1 = 1$). The degrees of freedom for each interaction are $A \times B$ DF $=$

TABLE 15.1 COMPUTER OUTPUT FROM A THREE-FACTOR ANALYSIS OF VARIANCE OF THE DATA PRESENTED IN EXAMPLE 15.1

SOURCE OF VARIATION	SUM OF SQUARES	DF	MEAN SQUARE
FACTOR A	1.81750	2	0.90875
FACTOR B	24.65583	2	12.32791
FACTOR C	0.00889	1	0.00889
A × B	1.10167	4	0.27542
A × C	0.37028	2	0.18514
B × C	0.17528	2	0.08764
A × B × C	0.22056	4	0.05514
ERROR	2.00500	54	0.03713

factor A DF $\times$ factor B DF $= 2 \times 2 = 4$; $A \times C$ DF $=$ factor A DF $\times$ factor C DF $= 2 \times 1 = 2$; $B \times C$ DF $=$ factor B DF $\times$ factor C DF $= 2 \times 1 = 2$; and $A \times B \times C$ DF $=$ factor A DF $\times$ factor B DF $\times$ factor C DF $= 2 \times 2 \times 1 = 4$. The error DF is, then, the total DF (i.e., $N - 1$) minus all the other degrees of freedom. Each needed mean square is then obtained by dividing the appropriate sum of squares by its associated DF. Since we are dealing with a Model I (fixed effects model) ANOVA, the computation of each F value consists of dividing a mean square by the error MS.

We may now consider the testing of the hypotheses stated in Example 15.1. To test whether oxygen consumption is the same among all three species, we compare the species F (i.e., 24.475) with the critical value, $F_{0.05(1),2,54} \cong 3.18$; since the former exceeds the latter, the null hypothesis is rejected.* In a similar fashion, we test the hypothesis concerning each of the other two factors, as well as each of the four hypotheses regarding the interactions of factors, by comparing the calculated F values with the critical values from Table B.4. The procedure and conclusions are given in Example 15.1.

Recall from Chapter 13 that the test for a two-way interaction asks whether differences in the variable among levels of one factor are the same at all levels of the second factor. A test for a three-factor interaction may be thought of as asking if the interaction between any two of the factors is the same at all levels of the third factor.

It is only for a factorial ANOVA with all factors fixed that we compute all F values utilizing the error MS. If any of the factors are random effects, then the analysis becomes more complicated. The proper F calculations for such situations appear in Appendix A.

If there are not equal numbers of replicates in each cell of a factorial analysis of variance design, then the usual ANOVA computations are invalid (see Section 15.5).

A factorial ANOVA experimental design may also include nesting (see chapter 16 for a discussion of nesting). For example, in Example 15.1 we might have performed two or more respiratory rate determinations on each of the four animals per

*Since Table B.4 does not contain critical values of F for $\nu_2 = 54$, the critical values for the next lower ν_2 (namely, 50) are employed. (Or, the critical value for $\nu_2 = 54$ could be estimated by interpolation, as explained in the introduction to Appendix B.)

cell. Some of the available computer programs for factorial analysis of variance also provide for nested (also called hierarchical) experimental designs.

15.2 THE LATIN SQUARE EXPERIMENTAL DESIGN

The randomized block design, described in Section 13.5, is a two factor analysis of variance in which there is one fixed effects factor, one random effects factor, and no replication. The levels of the random effects factor are termed "blocks," and it is assumed that no variation exists within blocks except that due to differences among levels of the fixed effects factor.

There are times when two sources of variability may be accounted for by blocking, and such a situation may be described by a *Latin square* experimental design, shown diagrammatically:

II	I	III
III	II	I
I	III	II

Here, there are not only three horizontal blocks (rows), but also three vertical blocks (columns) of data. There are three levels of the fixed effects factor (designated I, II, and III), and they are assigned to each row and column at random, except that each level is represented exactly once in each column and in each row. In the Latin square design, the number of vertical blocks, the number of horizontal blocks, and the number of levels in the fixed effects factor must be equal. The design shown is referred to as a 3×3 Latin square and is the smallest possible Latin square experimental design. Twelve different configurations are possible for a 3×3 Latin square analysis, there being 576 possible 4×4 squares, 161,280 5×5 squares, and 812,851,200 6×6 squares; several sets of configurations are listed by Fisher and Yates (1963: 86–89) to facilitate the setting up of Latin square experimental designs.

The only hypothesis generally of interest in a Latin square analysis is the one concerning equality among the levels of the fixed effects factor. We test this hypothesis by considering the experimental design to be a three-factor ANOVA with one fixed and two random factors, with no replication. Appendix A.3 summarizes the procedure used for such a test.

The Latin square analysis of variance is rarely applied with biological data, with the occasional exception of those from certain agricultural field plot studies and other, uncommonly encountered, situations. This experimental design is more fully discussed elsewhere (e.g., Snedecor and Cochran, 1980: Section 14.10; Steel and Torrie, 1980: 221ff; Woolf, 1968: Chapter 10). A good discussion of the handling of missing data in this design is that by C. C. Li (1964: 235–237).

15.3 HIGHER ORDER FACTORIAL ANALYSIS OF VARIANCE

In theory, there is no limit to the number of factors that might be analyzed simultaneously, but the numbers of possible interactions to be dealt with in computation and interpretation soon become unwieldy as larger analyses are considered (see Table 15.2). Consequently, most computer programs available for factorial analysis of variance are capable of considering a maximum of from five to ten factors.

TABLE 15.2 NUMBER OF HYPOTHESES POTENTIALLY TESTABLE IN FACTORIAL ANALYSES OF VARIANCE

	Number of factors			
	2	3	4	5
Main factors	2	3	4	5
2-way interactions	1	3	6	10
3-way interactions		1	4	10
4-way interactions			1	5
5-way interactions				1

Note: The number of mth-order interactions in a k-factor ANOVA is the number of ways k factors can be combined m at a time:

$$_kC_m = \frac{k!}{m!(k-m)!}.$$

Once the sums of squares for all factors and interactions have been computed, the degrees of freedom and mean squares are readily determined, as Section 15.1 describes. However, most available computer programs also provide the degrees of freedom and mean squares. If all factors are fixed, then the F required to test each null hypothesis is obtained by dividing the appropriate mean square by the error MS. If, however, any of the factors represent random effects, then the analysis becomes considerably more complex, and in some cases impossible. Appendix A presents the procedures applicable to hypothesis testing in such cases. See Section 15.5 for consideration of analyses with unequal replication.

15.4 NONPARAMETRIC MULTIWAY FACTORIAL ANALYSIS OF VARIANCE

As an expansion of the procedure of Section 13.4, we can perform factorial analysis of variance by ranks for three or more factors. We simply use the ranks of the data, rather than the raw data, and subject these ranks to the same computations as in Section 15.1 or 15.3 in order to obtain sums of squares. Knowing these sums of squares, and the degrees of freedom for each source of variation (as determined in Section 15.1 or 15.3, also using Equation 13.20), we can obtain the Kruskal-Wallis H (by Equation 13.22) and test its significance as shown in Example 13.4.

15.5 FACTORIAL ANALYSIS OF VARIANCE WITH UNEQUAL REPLICATION

Although equal replication is always desirable for optimum power and ease of computation in analysis of variance, it is not essential for the performance of the computation in a single-factor ANOVA (Section 11.1). However, all the techniques thus far discussed for ANOVA designs consisting of two or more factors require equal numbers of data per cell (with the exception of the case of proportional replication described in Section 13.2). For example, the data in Example 15.1 are composed of four replicates in each combination of species, temperature, and sex. If there were five or more replicates in a very small number of cells, then it is not highly criticizable to discard (at random within a cell) those few data necessary to arrive at equal numbers of replicate data. However, a more general approach is available, a procedure by which data suffering from replication inequality can be analyzed and interpreted by analysis of variance considerations. The mathematical manipulations involved are sufficiently complex as to be attempted reasonably only by a computer, but it is worthwhile for the biologist to be aware of the fact that programs for such an analysis are available. (These procedures may employ a type of multiple linear regression—see Section 20.11—and may be referred to as "general linear models.")

If inequality is due to one or a few cells containing one less datum than the others, then a factorial analysis of variance may be performed after inserting an estimate of each missing datum. If one datum is missing, an estimate of its value may be found as follows (Shearer, 1973):

$$\hat{X} = \frac{aA_i + bB_j + cC_l + \cdots - (k - 1)\Sigma X}{N + k - 1 - a - b - c - \cdots}, \tag{15.1}$$

where $\hat{X}$ is the estimated value for a missing datum in level i of factor A, level j of factor B, level l of factor C, etc.; a, b, c, etc. are the numbers of levels in factors A, B, C, etc., respectively; A_i is the sum of all the other data in level i of factor A, B_j is the sum of the other data in level j of factor B, etc.; the summation of $aA_i + bB_j + cC_l + \ldots$ is over all factors; k is the number of factors; ΣX is the sum of all the other data in all levels of all factors; and N is the total number of data (including the missing one) in the experimental design. This estimated value may then be inserted with the other data in the analysis of variance computations.

An alternative method of handling experimental designs with one or a few cells containing one less datum than the other is much simpler, but not as desirable, as the procedure above. For each small cell the mean of the cell's observed data can be inserted as an additional datum. The analysis of variance is then performed as usual, but with the total DF and within cells DF calculated without including the number of such additional data. (That is, the total and within cells DF are those appropriate to the set of original observations.) The best estimation procedure for missing data is to use the cell means as starting values for employing Equation 15.1 iteratively, just as Equation 13.17 is used iteratively in Example 13.6.

Nonparametric factorial analysis of variance can also be performed if cell sizes are unequal. This is done by subjecting the ranks of the data, rather than the raw data, to a computer program of the kind referred to in this section. With the sums of squares determined in this manner, the Kruskal-Wallis H can be determined using Equation 13.22, and tests of significance performed as in Example 13.4.

15.6 MULTIPLE COMPARISONS AND CONFIDENCE INTERVALS IN MULTIWAY ANALYSIS OF VARIANCE

As we have seen, for each factor a hypothesis may be tested concerning the equality of the level means in that factor. If the null hypothesis of equality is rejected, then it may be desirable to ascertain between which levels the difference(s) lie(s). This can be done by the multiple comparison procedures prescribed for two-way analyses of variance in Section 13.6. Also mentioned in that section is the calculation of confidence intervals with respect to level means in a two-factor analysis of variance; those considerations also apply to an ANOVA with more than two factors. It should be remembered that the sample size, n, referred to in Chapter 12 is replaced in the present context by the total number of data per level (i.e., the number of data used to calculate a level mean); k is replaced by the number of levels of the factor being tested; s^2 will be replaced by the MS appropriate in the denominator of the F ratio used to test for significance of the factor being examined; the degrees of freedom, ν (in q, q', and t) is the DF associated with this MS; and F in Scheffé's test is the same as in the ANOVA.

15.7 POWER AND SAMPLE SIZE IN MULTIWAY ANALYSIS OF VARIANCE

The principles and procedures of Section 13.7 (for two-way ANOVA) may be readily expanded to multifactor analysis of variance. In section 13.7, k' is the number of levels of the factor under consideration, n' is the total number of data in each level of that factor, s^2 is the within cells MS, ν_2 is the within cells DF, and $\nu_1 = k' - 1$. Then, the power of the ANOVA in detecting differences among level means may be estimated using Equations 13.30–13.33 and 13.35; and Equation 13.34 can be applied to any interaction (two-factor or otherwise) simply by employing the appropriate SS and DF.

Equation 13.33 may be used to estimate the minimum number of data per level that would be needed to achieve a specified power, given the significance level and detectable difference desired among means.

Equation 13.36 enables us to estimate the smallest difference among level means detectable with the ANOVA. As shown in Section 13.7, we can also use Equation 13.3 to estimate the maximum number of levels testable.

The Mixed Model ANOVA. The above procedures are applicable when all the factors are fixed effects (i.e., we have a Model I ANOVA). They may also be applied to any fixed-effects factor in a mixed-model ANOVA, but in such cases we must modify our method as follows.

Consider the appropriate denominator for the F calculated to test for the significance of the factor in question (see Appendix A.) Then, substitute this denominator for the within cells MS (s^2); and substitute this denominator DF for ν_2.

EXERCISES

15.1. Test for all factor and interaction effects in the following $4 \times 3 \times 2$ Model I analysis of variance, where a_i is a level of factor A, b_i is a level of factor B, and c_i is a level of factor C.

	a_1			a_2			a_3			a_4		
	b_1	b_2	b_3	b_1	b_2	b_3	b_1	b_2	b_3	b_1	b_2	b_3
	4.1	4.6	3.7	4.9	5.2	4.7	5.0	6.1	5.5	3.9	4.4	3.7
c_1	4.3	4.9	3.9	4.6	5.6	4.7	5.4	6.2	5.9	3.3	4.3	3.9
	4.5	4.2	4.1	5.3	5.8	5.0	5.7	6.5	5.6	3.4	4.7	4.0
	3.8	4.5	4.5	5.0	5.4	4.5	5.3	5.7	5.0	3.7	4.1	4.4
	4.8	5.6	5.0	4.9	5.9	5.0	6.0	6.0	6.1	4.1	4.9	4.3
c_2	4.5	5.8	5.2	5.5	5.3	5.4	5.7	6.3	5.3	3.9	4.7	4.1
	5.0	5.4	4.6	5.5	5.5	4.7	5.5	5.7	5.5	4.3	4.9	3.8
	4.6	6.1	4.9	5.3	5.7	5.1	5.7	5.9	5.8	4.0	5.3	4.7

15.2. Test for the effects of all factors and interactions in the following $2 \times 2 \times 2 \times 3$ Model I analysis of variance design, where a_i is a level of factor A, b_i is a level of factor B, c_i is a level of factor C, and d_i is a factor of level D.

	a_1				a_2			
	b_1		b_2		b_1		b_2	
	c_1	c_2	c_1	c_2	c_1	c_2	c_1	c_2
	12.2	13.4	12.2	13.1	10.9	12.1	10.1	11.2
d_1	12.6	13.1	12.4	13.0	11.3	12.0	10.2	10.8
	12.5	13.5	12.3	13.4	11.2	11.7	9.8	10.7
	11.9	12.8	11.8	12.7	10.6	11.3	10.0	10.9
d_2	11.8	12.6	11.9	12.5	10.4	11.1	9.8	10.6
	12.1	12.4	11.6	12.3	10.3	11.2	9.8	10.7
	12.6	13.0	12.5	13.0	11.1	11.9	10.0	10.9
d_3	12.8	12.9	12.7	12.7	11.1	11.8	10.4	10.5
	12.9	13.1	12.4	13.2	11.4	11.7	10.1	10.8

15.3. Test for all factor and interaction effects in the following Model I 3×2 analysis of variance with unequal replication.

a_1		a_2		a_3	
b_1	b_2	b_1	b_2	b_1	b_2
34.1	35.6	38.6	40.3	41.0	42.1
36.9	36.3	39.1	41.3	41.4	42.7
33.2	34.7	41.3	42.7	43.0	43.1
35.1	35.8	41.4	41.9	43.4	44.8
			40.8		44.5

16 Nested (Hierarchical) Analysis of Variance

In Chapters 13 and 15 we dealt with analysis of variance experimental designs that the statistician refers to as *crossed.* A crossed experiment is one where all possible combinations of levels of the factors exist; the cells of data are formed by each level of one factor being in combination with each level of every other factor. Thus, Example 13.1 is a two-factor crossed experimental design, since each sex is found in combination with each hormone treatment. In Example 15.1, each sex is found in combination with each experimental temperature and each species.

In some experimental designs, however, we may have some different levels of one factor occurring in combination with the levels of one or more other factors, and other distinctly different levels occurring in combination with others. In Example 16.1, where blood cholesterol concentration is the variable, there are two factors, drug type and drug source. Each drug was obtained from two sources, but the two sources are not the same for all the drugs. Thus, the experimental design is not crossed; rather, we say it is *nested* (or *hierarchical*), with one factor, drug source, being nested within another, drug type. The nested factor, as in the present example, is typically random, so the experiment may be considered to be a kind of one-way ANOVA, where the levels of this variable (drug sources) are the samples, and the individual cholesterol determinations are called subsamples.

Sometimes experiments are designed with nesting in order to test hypotheses about the samples. More typical, however, is the inclusion of a random-effects nested factor in order to account for some within group variability and thus make the hypothesis testing for one or more factors (usually fixed effects) of primary interest.

Example 16.1

A hierarchical (nested) analysis of variance.

The variable is human female blood cholesterol concentration (in mg/100 ml of plasma). This variable was measured after the administration of one of three different drugs, each drug having been obtained from two sources.

	Drug 1		**Drug 2**		**Drug 3**		
	Source A	*Source Q*	*Source D*	*Source B*	*Source L*	*Source S*	
	102	103	108	109	104	105	
	104	104	110	108	106	107	
n_{ij}	2	2	2	2	2	2	
$\sum_{l=1}^{n_{ij}} X_{ijl}$	206	207	218	217	210	212	
$\frac{\left(\sum_{l=1}^{n_{ij}} X_{ijl}\right)^2}{n_{ij}}$	21218.0	21424.5	23762.0	23544.5	22050.0	22472.0	$\sum_{i=1}^{a}\sum_{j=1}^{b}\frac{\left(\sum_{l=1}^{n_{ij}} X_{ijl}\right)^2}{n_{ij}} = 134471.0$
n_i	4		4		4		$N = 12$
$\sum_{j=1}^{b}\sum_{l=1}^{n_{ij}} X_{ijl}$	413		435		422		
$\frac{\left(\sum_{j=1}^{b}\sum_{l=1}^{n_{ij}} X_{ijl}\right)^2}{n_i}$	42642.25		47306.25		44521.00		$\sum_{i=1}^{a}\frac{\left(\sum_{j=1}^{b}\sum_{l=1}^{n_{ij}} X_{ijl}\right)^2}{n_i} = 134469.50$

$$C = \frac{\left(\sum_{i=1}^{a}\sum_{j=1}^{b}\sum_{l=1}^{n_{ij}} X_{ijl}\right)^2}{N} = \frac{(1270)^2}{12} = 134408.33$$

$$\text{total SS} = \sum_{i=1}^{a}\sum_{j=1}^{b}\sum_{l=1}^{n_{ij}} X_{ijl}^2 - C = 134480.00 - 134408.33 = 71.67$$

$$\text{among all subgroups SS} = \sum_{i=1}^{a}\sum_{j=1}^{b}\frac{\left(\sum_{l=1}^{n_{ij}} X_{ijl}\right)^2}{n_{ij}} - C = 134471.00 - 134408.33 = 62.67$$

$$\text{error SS} = \text{total SS} - \text{among all subgroups SS} = 71.67 - 62.67 = 9.00$$

$$\text{groups SS} = \sum_{i=1}^{a}\frac{\left(\sum_{j=1}^{b}\sum_{l=1}^{n_{ij}} X_{ijl}\right)^2}{n_i} - C = 134469.50 - 134408.33 = 61.17$$

$$\text{subgroups SS} = \text{among all subgroups SS} - \text{groups SS} = 62.67 - 61.17 = 1.50$$

Source of variation	SS	DF	MS
Total	71.67	11	
Among all subgroups	62.67	5	
Groups	61.17	2	30.58
Subgroups	1.50	3	0.50
Error	9.00	6	1.50

H_0: There is no difference among the drug sources in affecting blood cholesterol concentration.

H_A: There is difference among the drug sources in affecting blood cholesterol concentration.

$F = \frac{0.50}{1.50} = 0.33.$ $F_{0.05(1),3,6} = 4.76.$ Do not reject H_0. $P > 0.50.$

H_0: There is no difference in cholesterol concentrations owing to the three drugs.
H_A: There is difference in cholesterol concentrations owing to the three drugs.

$F = \frac{30.58}{0.50} = 61.16$. $F_{0.05(1),2,3} = 9.55$. Reject H_0. $0.0025 < P < 0.005$.

16.1 NESTING WITHIN ONE MAIN FACTOR

In the experimental design such as in Example 16.1, the primary concern is to detect true differences among levels of the fixed-effects factor (drug type). But we can often employ a more powerful test by nesting a random-effects factor that can account for some of the variability within the groups of interest. The partitioning of the variability in a nested ANOVA may be observed in this example.

Calculations in the Nested ANOVA. In the hierarchical design described, we can uniquely designate each datum by using a triple subscript notation, where X_{ijl} indicates the lth datum in subgroup j of group i. Thus, in Example 11.7, $X_{222} =$ 108 mg/100 ml, $X_{311} =$ 104 mg/100 ml, etc. For the general case, there are a groups, numbered 1 through a, and b is the number of subgroups in each group. For Example 16.1, there are three levels of factor A (drug type), and b (the number of levels of factor B, i.e., sources for each drug) is 2. The number of data in subgroup j of group i may be denoted by n_{ij}, and the total number of data in group i is n_i. The total number of observations in the entire experiment is $N = \sum_{i=1}^{k} n_i$ (which could also be computed as $N = \sum_{i=1}^{a} \sum_{j=1}^{b} n_{ij}$). The sum of the data in subgroup j of group i is calculated as $\sum_{l=1}^{n_{ij}} X_{ijl}$; the sum of the data in group i is $\sum_{j=1}^{b} \sum_{l=1}^{n_{ij}} X_{ijl}$; and the mean of group i is

$$\bar{X}_i = \frac{\sum_{j=1}^{b} \sum_{l=1}^{n_{ij}} X_{ijl}}{n_i}. \tag{16.1}$$

The grand mean of all the data is

$$\bar{X} = \frac{\sum_{i=1}^{a} \sum_{j=1}^{b} \sum_{l=1}^{n_{ij}} X_{ijl}}{N}. \tag{16.2}$$

The total sum of squares for this ANOVA design considers the deviations of all the $\bar{X}_{ijl}$ from $\bar{X}$ and is calculated

$$\text{total SS} = \sum_{i=1}^{a} \sum_{j=1}^{b} \sum_{l=1}^{n_{ij}} X_{ijl}^2 - C, \tag{16.3}$$

where the "correction term" is

$$C = \frac{\left(\sum_{i=1}^{a} \sum_{j=1}^{b} \sum_{l=1}^{n_{ij}} X_{ijl} \right)^2}{N}, \tag{16.4}$$

and

$$\text{total DF} = N - 1. \tag{16.5}$$

The variability among groups, i.e., the indication of the deviations $\bar{X}_i - \bar{X}$, is expressed as

$$\text{groups SS} = \sum_{i=1}^{a} \frac{\left(\sum_{j=1}^{b} \sum_{l=1}^{n_{ij}} X_{ijl} \right)^2}{n_i} - C, \tag{16.6}$$

and

$$\text{groups DF} = a - 1 \tag{16.7}$$

There is a total of *ab* subgroups in the design, and, considering them as if they were groups in a one-way ANOVA, we can calculate a measure of the deviations $X_{ij} - \bar{X}$ as

$$\text{among all subgroups SS} = \sum_{i=1}^{a} \sum_{j=1}^{b} \frac{\left(\sum_{l=1}^{n_{ij}} X_{ijl} \right)^2}{n_{ij}} - C, \tag{16.8}$$

and

$$\text{among all subgroups DF} = ab - 1. \tag{16.9}$$

The variability due to the subgrouping within groups causes the deviation of a subgroup from its group mean, $X_{ij} - \bar{X}_i$, and the appropriate sum of squares is the "among subgroups within groups" SS, which will be referred to as

$$\text{subgroups SS} = \text{among all subgroups SS} - \text{groups SS}, \tag{16.10}$$

and

$$\text{subgroups DF} = \text{among all subgroups DF} - \text{groups DF} = a(b - 1). \tag{16.11}$$

The within subgroups, or error, variability measures the deviations $X_{ijl} - \bar{X}_{ij}$; it is essentially the within cells variability encountered in Chapters 13 and 15. The appropriate sum of squares is obtained by difference:

$$\text{error SS} = \text{total SS} - \text{among all subgroups SS}, \tag{16.12}$$

with

$$\text{error DF} = \text{total DF} - \text{among all subgroups DF} = N - ab \tag{16.13}$$

The summary of this hierarchical analysis of variance is presented in Table 16.1. Recall that MS = SS/DF. Some similarities may be noted between Tables 13.1 and 16.1, but in the latter we cannot speak of interaction between the two factors.

Hypothesis Testing in the Hierarchical ANOVA. For the data in Example 16.1, we can test the null hypothesis that no difference in cholesterol occurs among subgroups (i.e., the source of the drugs has no effect on the concentration of blood cholesterol). We do this by examining

$$F = \frac{\text{subgroups MS}}{\text{error MS}}. \tag{16.14}$$

For our example, this is $F = 0.50/1.50 = 0.33$; since $F_{0.05(1),3,6} = 4.76$, H_0 is not rejected.

Next, we can test the null hypothesis that there is no difference in cholesterol with the administration of the three different drugs. We calculate

$$F = \frac{\text{groups MS}}{\text{subgroups MS}}, \tag{16.15}$$

which in the present example is $F = 30.58/0.50 = 61.16$. Since $F_{0.05(1),2,3} = 9.55$, H_0 is rejected.

TABLE 16.1 SUMMARY OF HIERARCHICAL (NESTED) SINGLE FACTOR ANALYSIS OF VARIANCE CALCULATIONS

Source of variation	SS	DF
Total $[X_{ijl} - \bar{X}]$	$\sum_{i=1}^{a}\sum_{j=1}^{b}\sum_{l=1}^{n_{ij}} X_{ijl}^2 - C$	$N - 1$
Among all subgroups $[\bar{X}_{ij} - \bar{X}]$	$\sum_{i=1}^{a}\sum_{j=1}^{b}\frac{\left(\sum_{l=1}^{n_{ij}} X_{ijl}\right)^2}{n_{ij}} - C$	$ab - 1$
Groups (i.e., among groups) $[\bar{X}_i - \bar{X}]$	$\sum_{i=1}^{a}\frac{\left(\sum_{j=1}^{b}\sum_{l=1}^{n_{ij}} X_{ijl}\right)^2}{n_i} - C$	$a - 1$
Subgroups (i.e., among subgroups within groups) $[\bar{X}_{ij} - \bar{X}_i]$	among all subgroups SS − groups SS	$a(b - 1)$
Error (i.e., within subgroups) $[\bar{X}_{ijl} - \bar{X}_{ij}]$	total SS − among all subgroups SS	$N - ab$

Note: $C = \frac{\left(\sum_{i=1}^{a}\sum_{j=1}^{b}\sum_{l=1}^{n_{ij}} X_{ijl}\right)^2}{N}$

a = number of groups; b = number of subgroups within each group; n_i = number of data in group i; n_{ij} = number of data in subgroup j for group i; N = total number of data in entire experiment.

If one does not reject the null hypothesis of no difference among subgroups within groups, then the subgroups MS might be considered to estimate the same population variance as does the error MS. Thus, some statisticians suggest that in such cases we calculate a pooled mean square by pooling the sums of squares and pooling the degrees of freedom for the subgroups and the error variability, for this will theoretically provide the ability to perform a more powerful test for differences between groups. However, there is no widespread agreement on this matter, so it is usually suggested that the biologist be conservative and not engage in such pooling, at least not without consulting a statistician.

If there are unequal numbers of subgroups in each group, then the analysis becomes more complex, and the preceding calculations are not applicable. This situation is generally submitted to computer analysis, often by the procedure referred to in Section 15.5.

A hierarchical experimental design might have each subgroup composed of sub-subgroups, thus involving an additional step in the hierarchy. For instance, for the data of Example 16.1, the different drugs define the groups, the different sources define the subgroups, and if different technicians or different instruments were used to perform the cholesterol analyses within each subgroup, then these technicians or instruments would define the sub-subgroups. Sokal and Rohlf (1981: 288) describe the necessary calculations for a design with sub-subgroups, although one generally resorts to computer calculation for hierarchical designs with more than the two steps in the hierarchy discussed in the preceding paragraphs. See Appendix A for assistance in hypothesis testing for such nested designs, and designs where the groups are not fixed effects and/or the subsamples are not random effects.

16.2 NESTING IN FACTORIAL EXPERIMENTS

Experimental designs are encountered where there are two or more crossed factors as well as one or more nested factors. For example, in Example 13.1 the two crossed factors are sex and hormone treatment, and five animals of each sex were given each hormone treatment. In addition, the experimenter might have obtained three blood samples (statistically, subsamples) from each animal so that individual birds would be samples and the triplicate blood collections would be subsamples (The animals represent a nested, rather than a crossed factor, because the same animal is not found at each combination of the other factors.) The analysis of variance table would then look like that in Example 16.2. The computation of sums of squares could be obtained by computer, and the appropriate hypothesis testing will be that indicated in Appendix A.4. Some available computer programs can operate with data where there is not equal replication, while others cannot.

The concept of nesting could be extended by considering that each subsample of blood in Example 16.2 was subjected to two or more (i.e., replicate) chemical analyses. Then we would have chemical analysis nested within animal, and animal nested within the two crossed factors.

Example 16.2

An analysis of variance with a random-effects factor (blood collection) nested within the two-factor crossed experimental design of Example 13.1.

For each of the 4 combinations of sex and hormone treatment ($a = 2$ and $b = 2$), there are 5 animals ($c = 5$), from each of which 3 blood collections are taken ($n = 3$). Therefore, the total number of data collected is $N = abcn = 60$.

Source of variation	SS	DF	MS
Total		$N - 1 = 59$	
Cells		$ab - 1 = 3$	
Hormone treatment (factor A)	*	$a - 1 = 1$	†
Sex (factor B)	*	$b - 1 = 1$	†
$A \times B$	*	$(a - 1)(b - 1) = 1$	†
Among all animals		$abc - 1 = 19$	
Cells		$ab - 1 = 3$	
Animals (within cells) (factor C)	*	$ab(c - 1) = 16$	†
Error (within animals)	*	$abc(n - 1) = 40$	†

*These sums of squares can be obtained from an appropriate computer program; the other sums of squares in the table may not be given by such a program, or MS might be given but not SS.

†The mean squares can be obtained from an appropriate computer program. Or, they may be obtained from the sums of squares and degrees of freedom (as MS = SS/DF). The degrees of freedom might appear in the computer output, or they may have to be determined by hand. The appropriate F statistics are those indicated in Appendix A.4.

H_0: There is no difference in mean blood calcium concentration between males and females.
H_A: There is a difference in mean blood calcium concentration between males and females.

$$F = \frac{\text{factor } A \text{ MS}}{\text{factor } C \text{ MS}} \qquad F_{0.05(1),\,1,\,16} = 4.49$$

H_0: The mean blood calcium concentration is the same in birds receiving and not receiving the hormone treatment.
H_A: The mean blood calcium concentration is not the same in birds receiving and not receiving the hormone treatment.

$$F = \frac{\text{factor } B \text{ MS}}{\text{factor } C \text{ MS}} \qquad F_{0.05(1),\,1,\,16} = 4.49$$

H_0: There is no interactive effect of sex and hormone treatment on mean blood calcium concentration.
H_A: There is interaction between sex and hormone treatment in affecting mean blood calcium concentration.

$$F = \frac{A \times B \text{ MS}}{\text{factor } C \text{ MS}} \qquad F_{0.05(1),\,1,\,16} = 4.49$$

H_0: There is no difference in blood calcium concentration among animals within combinations of sex and hormone treatment.
H_A: There is difference in blood calcium concentration among animals within combinations of sex and hormone treatment.

$$F = \frac{\text{factor } C \text{ MS}}{\text{error MS}} \qquad F_{0.05(1),\,16,\,40} = 1.90$$

16.3 MULTIPLE COMPARISONS AND CONFIDENCE INTERVALS

Whenever a fixed-effects factor is concluded by an ANOVA to have a significant effect on the variable, we may turn to the question of which of the factor's levels are different from which others. If there are only two levels of the factor, then of course we have concluded that their population means are different by the ANOVA. But if there are more than two levels, then a multiple comparison test must be employed.

The multiple comparison procedures usable in nested experimental designs are found in Chapter 12, with slight modifications as those we saw in Sections 13.6 and 15.6. Simply keep the following in mind when employing the tests of Sections 12.1, 12.2, 12.4, and 12.5:

1. k refers to the number of levels being compared. (In Example 16.1, $k = a$, the number of levels in factor A. In Example 16.2, $k = a$ when comparing levels of factor A, and $k = b$ when testing levels of factor B.)
2. The sample size, n, refers to the total number of data from which a level mean is calculated. (In Example 16.1, the sample size $bn = 4$ would be used in place of n. In Example 16.2 we would use $bcn = 30$ to compare level means for factor A and $acn = 30$ for factor B.)
3. The mean square, s^2, refers to the MS in the denominator of the F ratio appropriate to testing the factor in question in the ANOVA. (In Example 16.1, the subgroups (sources) MS would be used. In Example 16.2, the factor C MS would be used.)

4. The degrees of freedom, ν, for the critical value of q or q' are the degrees of freedom associated with the mean square above. (In Examples 16.1 and 16.2, these would be 3 and 16, respectively.)
5. The critical value of F in the Scheffé test has the same degrees of freedom as it does in the ANOVA for the factor under consideration. (In Example 16.1, these are 2 and 3. In Example 16.2 they are 1 and 16.)

Once a multiple comparison test has determined where differences lie among level means, we can express a confidence interval for each different mean, as was done in Sections 12.3, 13.6, and 15.6, keeping in mind the sample sizes, mean squares, and degrees of freedom defined above.

16.4 POWER AND SAMPLE SIZE IN NESTED ANALYSIS OF VARIANCE

In Sections 13.7 and 15.7, considerations of power and sample size for factorial analyses of variance were discussed. The same types of procedures may be employed for a fixed-effects factor within which nesting occurs. As previously used, k' is the number of levels of the factor, n' is the total number of data in each level, and $\nu = k' - 1$. The appropriate mean square, s^2, is that appearing in the denominator of the F ratio used to test that factor in the ANOVA, and ν_2 is the degrees of freedom associated with s^2.

Referring back to Section 13.7, then, the power of the nested ANOVA to detect differences among level means may be estimated using Equations 13.30–13.33 and 13.35. In addition, Equation 13.34 enables us to estimate (using the appropriate SS and DF) the power in detecting an interaction of crossed factors in a nested experimental design. Equation 13.33 may be used to esitmate the minimum number of data per level that would be needed to achieve a specified power, and Equation 13.36 allows the estimation of the smallest detectable difference among level means. Section 13.7 also shows how to estimate the maximum number of levels testable.

EXERCISE

16.1. Three water samples were taken from each of three locations. Two determinations of fluoride content were performed on each of the nine samples. Test the hypothesis that there was no difference in mean fluoride content among samples at a location. Then test the hypothesis that there is no difference in mean fluoride concentration among the locations. Data are in milligrams fluoride per liter of water.

Locations	1			2			3		
Samples	1	2	3	1	2	3	1	2	3
	1.1	1.3	1.2	1.3	1.3	1.4	1.8	2.1	2.2
	1.2	1.1	1.0	1.4	1.5	1.2	2.0	2.0	1.9

17 Simple Linear Regression

Previous chapters have discussed various methods of statistical analysis that deal with a single variable. Techniques that consider relationships between two variables are described in this and the following two chapters. Chapter 20 presents the expansion of such techniques to analyze situations where more than two variables may be related to each other.

17.1 REGRESSION VS. CORRELATION

The relationship between two variables may be one of functional dependence of one on the other. That is, the magnitude of one of the variables (the *dependent variable*) is assumed to be determined by—i.e., is a function of—the magnitude of the second variable (the *independent variable*), whereas the reverse is not true. For example, in the relationship between blood pressure and age in humans, blood pressure may be considered the dependent variable and age the independent variable; we may reasonably assume that although the magnitude of a person's blood pressure might be a function of age, age is not determined by blood pressure. This not to say that age is the only biological determinant of blood pressure, but we do consider it to be one determining factor.

Such a dependent relationship is termed a *regression*; the term *simple regression* refers to the fact that only two variables are being considered. In the case of simple regression, the adjective *linear* is used to refer to the relationship between the two variables being a straight line. Data amenable to simple regression analysis will consist of a dependent variable that is a random effect factor and an independent variable that is either a fixed-effect or a random-effect factor. (See the end of Section 11.1 to review these concepts.)

It is very convenient to graph simple regression data, using the ordinate (Y axis) for the dependent variable (conventionally termed Y) and the abscissa (X axis) for the independent variable (X). Thus, as shown in Fig. 17.1, the data of Example 17.1 appear as a scatter of points, each point representing a pair of X and Y values. One pair of X and Y data may be denoted as (X_1, Y_1), another as (X_2, Y_2), another as (X_3, Y_3), etc. (The line in this figure will be explained shortly.)

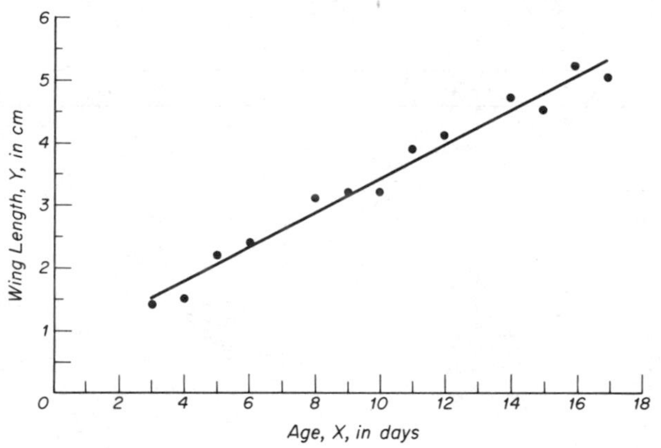

Figure 17.1 Sparrow wing length as a function of age. The data are from Example 17.1.

Example 17.1

Sparrow wing lengths at various times after hatching. The data are plotted in Fig. 17.1.

Age (days) (X)	*Wing length* (cm) (Y)
3.0	1.4
4.0	1.5
5.0	2.2
6.0	2.4
8.0	3.1
9.0	3.2
10.0	3.2
11.0	3.9
12.0	4.1
14.0	4.7
15.0	4.5
16.0	5.2
17.0	5.0

$n = 13$

In many kinds of biological data, however, the relationship between two variables is not one of dependence. In such cases, the magnitude of one of the variables changes as the magnitude of the second variable changes, but it is not reasonable to consider there to be an independent and a dependent variable. In such situations, *correlation*, rather than regression, analyses are called for, and both variables are theoretically to be random-effects factors, An example of data suitable for correlation analysis would be measurements of human arm and leg lengths. It might be found that an individual with long arms will in general possess long legs, so a relationship may be describable; but there is no justification in stating that the length of one limb is mathematically dependent upon the length of the other. Correlation techniques involving two variables will be discussed in Chapter 19; if more than two variables are being considered, then the appropriate procedures are found in Chapter 20.

17.2 THE SIMPLE LINEAR REGRESSION EQUATION

The simplest functional relationship of one variable to another in a population is the *simple linear regression*,

$$Y_i = \alpha + \beta X_i. \tag{17.1}$$

Here, α and β are population parameters (and, therefore, constants), and this expression will be recognized as the general equation for a straight line.*

Consider the data in Example 17.1, where wing length is the dependent variable and age is the independent variable. From a scatter plot of these data (Fig. 17.1), it appears that our sample of measurements from 13 birds represents a population of data in which wing length is linearly related to age. Thus, we would like to know the values of α and β that would uniquely describe the functional relationship existing in the population.

If all the data in a scatter diagram such as Fig. 17.1 occurred in a straight line, it would be an unusual situation. Generally, as is shown in this figure, there is considerable variability of data around any straight line we might draw through them. What we seek to define is what is commonly termed the "best fit" line through the data. The criterion for "best fit" that is generally employed utilizes the concept of *least squares*. Figure 17.2 is an enlarged portion of Fig. 17.1. Each value of X will have a corresponding value of Y lying on the line that we might draw through the scatter of data points. This value of Y is represented as $\hat{Y}$ to distinguish it from the Y value actually observed in our sample.† Thus, as Fig. 17.2 illustrates, an observed data point is denoted as (X_i, Y_i), and a point on the regression line is $(X_i, \hat{Y}_i)$.

The criterion of least squares considers the vertical deviation of each point from the line (i.e., the deviation describable as $Y_i - \hat{Y}_i$), and defines the best fit line as that which results in the smallest value for the sum of the squares of these deviations for all values of Y_i and $\hat{Y}_i$. That is, $\sum_{i=1}^{n} (Y_i - \hat{Y}_i)^2$ is to be a minimum,‡ where n is the number of data points comprising the sample. The sum of squares of these deviations is called the *residual sum of squares* (or, sometimes, the *error sum of squares*) and will be discussed later in this chapter.

*α and β have become standard symbols for these population parameters, and as such should not be confused with the standard use of the same Greek letters to denote the probabilities of a Type I and a Type II error, respectively (see Section 5.3).

†Statisticians refer to $\hat{Y}$ as "Y hat."

‡Another way to express this is to say that the correlation between Y_i's and $\hat{Y}_i$'s is to be maximum.

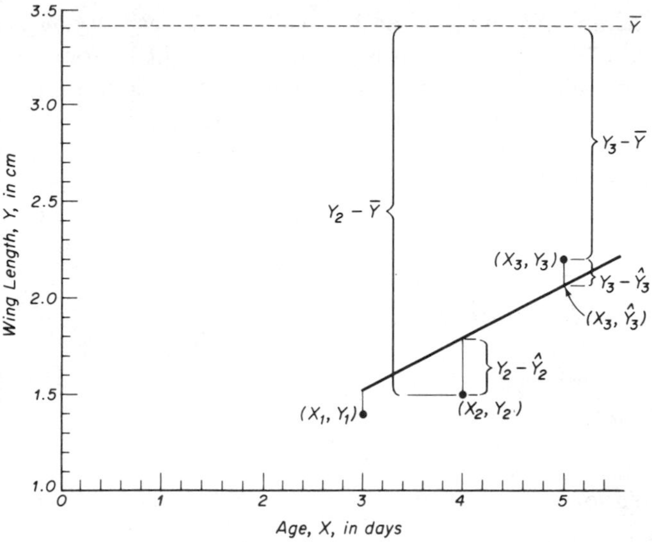

Figure 17.2 An enlarged portion of Fig. 17.1, showing the partitioning of Y deviations.

The only way to determine the population parameters α and β with complete confidence and accuracy would be to possess all the data for the entire population. Since this is nearly always impossible, we have to estimate these parameters from a sample of n data, where n is the number of pairs of X and Y values. The calculations required to arrive at such estimates, as well as to execute the testing of a variety of important hypotheses, involve the computation of sums of squared deviations from the mean, just as has been encountered before. Recall that the sum of squares of X_i values is defined as $\sum (X_i - \bar{X})^2$, which is more easily obtained on a calculator as $\sum X_i^2 - (\sum X_i)^2/n$. It will be convenient to define $x_i = X_i - \bar{X}$, so that this sum of squares can be abbreviated as $\sum x_i^2$, or, more simply, as $\sum x^2$.

We shall also be required to calculate a quantity referred to as the *sum of the crossproducts* of deviations from the mean:

$$\sum xy = \sum (X_i - \bar{X})(Y_i - \bar{Y}), \tag{17.2}$$

where y denotes a deviation of a Y value from the mean of all Y's just as x denotes a deviation of an X value from the mean of all X's. The sum of the crossproducts, analogously to the sum of squares, has a simple-to-use machine formula:

$$\sum xy = \sum X_i Y_i - \frac{(\sum X_i)(\sum Y_i)}{n}, \tag{17.3}$$

and it is recommended that the latter formula be employed.

The Regression Coefficient. The parameter β is termed the *regression coefficient*, or the *slope* of the best fit regression line. The best estimate of β is

$$b = \frac{\sum xy}{\sum x^2} = \frac{\sum X_i Y_i - \dfrac{(\sum X_i)(\sum Y_i)}{n}}{\sum X_i^2 - \dfrac{(\sum X_i)^2}{n}}. \tag{17.4}$$

Although the denominator in this calculation is always positive, the numerator may be either positive, negative, or zero, and the value of b theoretically can range from $-\infty$ to $+\infty$, including zero (see Fig. 17.3).

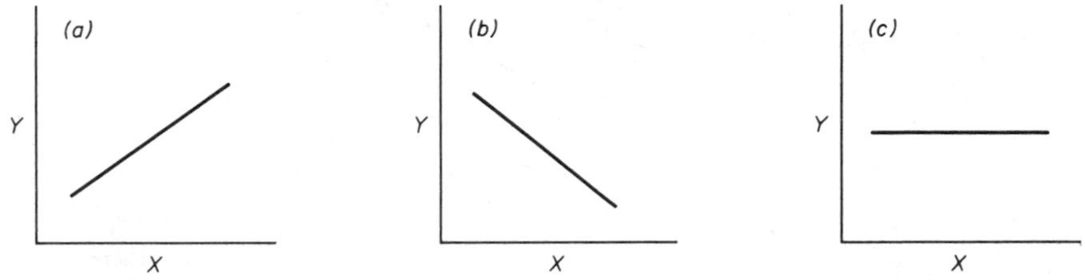

Figure 17.3 The slope of a linear regression line may be (a) positive, (b) negative, or (c) zero.

Example 17.2 demonstrates the calculation of b for the data of Example 17.1. Note that the units of b are the units of Y divided by the units of X. The regression coefficient expresses what change in Y is associated, on the average, with a unit change in X. In the present example, $b = 0.270$ cm/day indicates that there is a mean wing growth of 0.270 cm each day. Determination of the precision of b will be considered in Section 17.4.

Example 17.2

The simple linear regression equation calculated by the method of least squares for the data of Example 17.1.

$$n = 13 \qquad \sum X = 130.0 \qquad \bar{X} = 10.0 \qquad \sum X^2 = 1562.00$$

$$\sum x^2 = 1562.00 - \frac{(130.0)^2}{13} = 1562.00 - 1300.00 = 262.00$$

$$\sum Y = 44.4 \qquad \bar{Y} = 3.415 \qquad \sum XY = 514.80$$

$$\sum xy = 514.80 - \frac{(130.0)(44.4)}{13} = 514.80 - 444.00 = 70.80$$

$$b = \frac{\sum xy}{\sum x^2} = \frac{70.80}{262.00} = 0.270 \text{ cm/day}$$

$$a = \bar{Y} - b\bar{X} = 3.415 \text{ cm} - (0.270 \text{ cm/day})(10.0 \text{ day}) = 3.415 \text{ cm} - 2.700 \text{ cm} = 0.715 \text{ cm}$$

So, the simple linear regression equation is $\hat{Y} = 0.715 + 0.270X$.

The *Y* Intercept. An infinite number of lines possess any stated slope, all of them parallel (see Fig. 17.4). However, a line can be defined uniquely by stating, in addition to β, any one point on the line—i.e., any pair of coordinates, $(X_i, \hat{Y}_i)$. The point conventionally chosen is the point on the line where $X = 0$. The value of Y in the population at this point is the parameter α, which is called the *Y intercept*.

It can be shown mathematically that the point $(\bar{X}, \bar{Y})$ always lies on the best fit regression line. Thus, substituting $\bar{X}$ and $\bar{Y}$ in Equation (17.1), we find that:

$$\bar{Y} = \alpha + \beta\bar{X} \qquad (17.5)$$

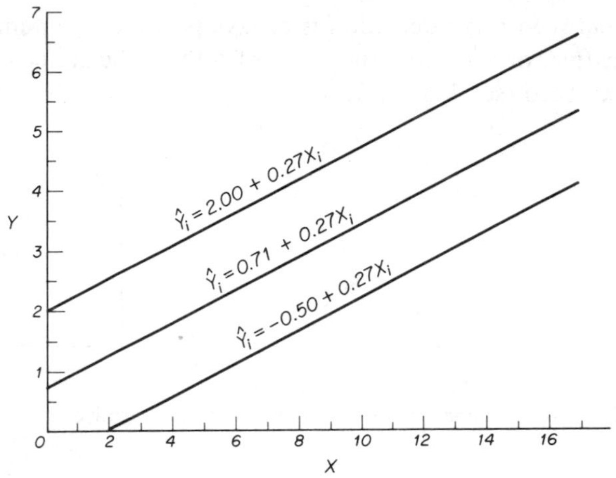

Figure 17.4 For any given slope, there exist an infinite number of possible regression lines, each with a different Y intercept. Three of this infinite number are shown here.

and

$$\alpha = \bar{Y} - \beta\bar{X}. \tag{17.6}$$

The best estimate of α is

$$a = \bar{Y} - b\bar{X}. \tag{17.7}$$

The calculation of a is shown in Example 17.2. Note that the Y intercept has the same units as any other Y value. (The precision of a is considered in Section 17.4.) The sample regression equation (which estimates the population relationship between Y and X stated in Equation 17.1) is written as

$$\hat{Y}_i = a + bX_i, \tag{17.8}$$

although some authors write

$$\hat{Y}_i = \bar{Y} + b(X_i - \bar{X}), \tag{17.9}$$

which is equivalent.

Figures 17.4 and 17.5 demonstrate that the knowledge of either a or b allows only an incomplete description of a regression function. But by specifying both a and b, one uniquely defines a line. Also, because a and b were calculated using the criterion of least squares, the residual sum of squares from this line is smaller than the residual sum of squares that would result from any other line (i.e., a line with any other a or b) that could be drawn through the data points.

Predicting Values of *Y*. Knowing the parameter estimates a and b for the linear regression equation, one can predict the value of the dependent variable expected at a stated value of X_i. For the regression in Example 17.2, the wing length of a sparrow at 13.0 days of age would be predicted to be

$$\begin{aligned}\hat{Y}_i &= a + bX_i \\ &= 0.715 \text{ cm} + (0.270 \text{ cm/day})(13.0 \text{ day}) = 4.225 \text{ cm}.\end{aligned}$$

The wing length in the population at 7.0 days of age would be estimated to be $\hat{Y}_i = 0.715 \text{ cm} + (0.270 \text{ cm/day})(7.0 \text{ day}) = 2.605 \text{ cm}$, and so on.

To plot a linear regression line graphically, we need to know only two points that lie on the line. We already know two points, namely $(\bar{X}, \bar{Y})$ and $(0, a)$; however, for ease and accuracy in drawing the line, two points that lie near extreme ends of the

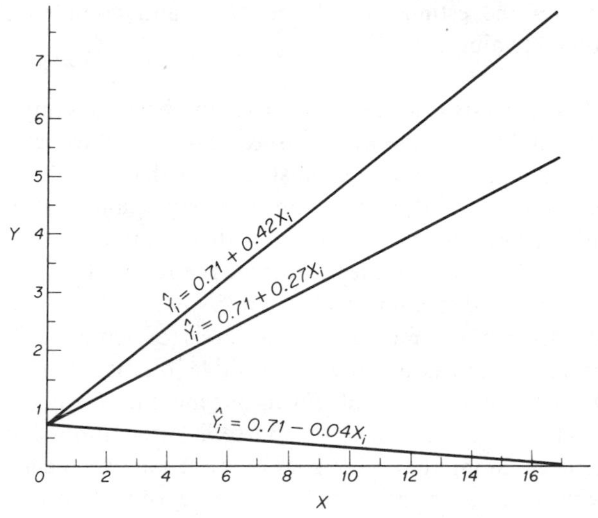

Figure 17.5 For any given Y intercept, there exist an infinite number of possible regression lines, each with a different slope. Three of this infinite number are shown here.

observed range of X are useful. For drawing the line in Fig. 17.1, the values of $\hat{Y}_i$ for $X_i = 3.0$ cm and $X_i = 17.0$ cm were used. These were found to be $\hat{Y}_i = 1.525$ and 5.305 cm, respectively.

A word of caution is in order concerning predicting $\hat{Y}_i$ values from a regression equation. Generally, it is an unsafe procedure to extrapolate from regression equations —that is, to predict $\hat{Y}_i$ values for X_i values outside the observed range of X_i. It would, for example, be unjustifiable to attempt to predict the wing length of a 20-day-old sparrow, or a 1-day-old sparrow, using the regression calculated for birds ranging from 3.0 to 17.0 days in age. What the linear regression describes is Y as a function of X *within the range of observed values of* X. For values of X above or below this range, the function may not be the same (i.e., α and/or β may be different); indeed, the relationship may not even be linear in such ranges, even though it is linear within the observed range. If there is good reason to believe that the described function holds for X values outside the range of those observed, then we may cautiously extrapolate. Otherwise, beware.

A classic example of nonsensical extrapolation was provided in 1874 by Mark Twain (1950: 156):

> In the space of one hundred and seventy-six years the Lower Mississippi has shortened itself two hundred and forty-two miles. That is an average of a trifle over one mile and a third per year [i.e., a slope of -1.375 mi/yr]. Therefore, any calm person, who is not blind or idiotic, can see that in the Old Oölitic Silurian Period, just a million years ago next November, the Lower Mississippi River was upward of one million three hundred thousand miles long, and stuck out over the Gulf of Mexico like a fishing-rod. And by the same token any person can see that seven hundred and forty-two years from now, the lower Mississippi will be only a mile and three-quarters long, and Cairo [Illinois] and New Orleans [Louisiana] will have joined their streets together, and be plodding comfortably along under a single mayor and a mutual board of aldermen.*

*Author Twain concludes by noting that "There is something fascinating about science. One gets such a wholesale return of conjecture out of a trifling investment of fact."

Section 17.4 discusses the estimation of the error and confidence intervals associated with predicting $\hat{Y}_i$ values.

Assumptions of Regression Analysis. Certain basic assumptions must be met in order to test validly hypotheses about regressions or to set confidence intervals for regression parameters. First, we must assume that for any value of X there exists in the population a normal distribution of Y values and that we sampled this distribution at random. Second, the variances of these population distributions of Y values must all be equal to one another. (Indeed, the residual mean square, to be described shortly, estimates this common variance, just as the error variance estimates the common variance assumed in the analysis of variance in previous chapters.) Third, the errors in Y are assumed to be additive (discussed in Chapter 14). Fourth, the values of Y are to be independent. Fifth, our measurements of X must be obtained without error. This last requirement, of course, is often impossible; so what we are doing in practice is assuming that the errors in the X data are negligible, or at least small compared with the measurement errors in the dependent variable. Regression statistics are known to be robust with respect to at least some of these underlying assumptions (e.g., Jacques and Norusis, 1973), so violations of them are not of concern unless they are severe. These assumptions will be discussed further in Section 17.10. Some nonparametric regression hypothesis testing is discussed by Daniel (1978: Chapter 10) and others.

17.3 TESTING THE SIGNIFICANCE OF A REGRESSION

The slope, b, of the regression line computed from the sample data expresses quantitatively the straight-line dependence of Y on X in the sample. But what is really desired is information about the functional relationship (if any) in the population from which the sample came. Indeed, the finding of a dependence of Y on X in the sample (i.e., $b \neq 0$) does not necessarily mean that there is a dependence in the population (i.e., $\beta \neq 0$). Consider Fig. 17.6, a scatter plot representing a population of data points with no dependence of Y on X; the best fit regression line would be parallel to the X axis (i.e., the slope, β, would be zero). However, it is possible, by random sampling, to obtain a sample of five data points having the values circled in the figure. By

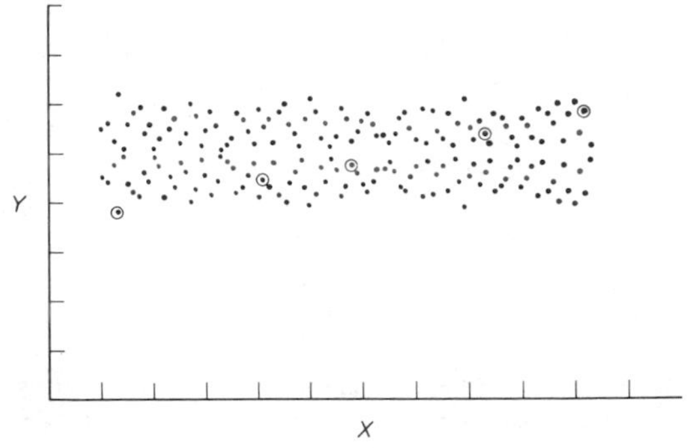

Figure 17.6 A hypothetical population of data points, having a regression coefficient, β, of zero. The circled points are a possible sample of five.

calculating b for this sample of five we would estimate that β was positive, even though it is, in fact, zero.

Now, it is not likely to obtain five such points out of this population, but we should desire to determine just how likely it is; therefore, we can set up the null hypothesis, H_0: $\beta = 0$, and the alternate hypothesis, H_A: $\beta \neq 0$. If we conclude that there is a reasonable probability (i.e., a probability greater than the chosen level of significance—say, 5%) that the calculated b could have come from sampling a population with a $\beta = 0$, the H_0 is not rejected. If the probability of obtaining the calculated b is small (say, 5% or less), then H_0 is rejected, and H_A is assumed to be true.

Analysis of Variance Testing. The preceding H_0 may be tested by an analysis of variance procedure. First, the overall variability of the dependent variable is calculated by computing the sum of squares of deviations of Y_i values from $\bar{Y}$, a quantity termed the *total sum of squares*:

$$\text{total SS} = \sum (Y_i - \bar{Y})^2 = \sum y^2 = \sum Y_i^2 - \frac{(\sum Y_i)^2}{n}. \tag{17.10}$$

Then, one determines the amount of variability among the Y_i values that results from there being a linear regression; this is termed the *linear regression sum of squares*:

$$\text{regression SS} = \sum (\hat{Y}_i - \bar{Y})^2 = \frac{(\sum xy)^2}{\sum x^2} = \frac{\left(\sum X_iY_i - \frac{\sum X_i \sum Y_i}{n}\right)^2}{\sum X_i^2 - \frac{(\sum X_i)^2}{n}}; \tag{17.11}$$

since $b = \sum xy / \sum x^2$, this can also be calculated as

$$\text{regression SS} = b \sum xy. \tag{17.12}$$

The value of the regression SS will be equal to that of the total SS only if each data point falls exactly on the regression line, a very unlikely situation. The scatter of data points around the regression line has already been alluded to, and the residual, or error, sum of squares has been defined. Knowing the total and the linear regression sums of squares, we may, by difference, obtain

$$\text{residual SS} = \sum (Y_i - \hat{Y}_i)^2 = \text{total SS} - \text{regression SS}. \tag{17.13}$$

Table 17.1 presents the analysis of variance summary for testing the hypothesis H_0: $\beta = 0$ against H_A: $\beta \neq 0$. Example 17.3 performs such an analysis for the data from Examples 17.1 and 17.2. The degrees of freedom associated with the total variability of Y_i values are $n - 1$. The degrees of freedom associated with the variability among Y_i's due to regression is always 1 in a simple linear regression.* The residual degrees of freedom are calculable as residual DF = total DF − regression DF = $n - 2$. Once the regression and residual mean squares are calculated (MS = SS/DF, as usual), H_0 may be tested by determining

$$F = \frac{\text{regression MS}}{\text{residual MS}}, \tag{17.14}$$

*In general, the regression DF is the number of parameters being estimated minus 1. Since we are here estimating two parameters (α and β), the regression DF = 2 − 1 = 1.

which is then compared to the critical value, $F_{\alpha(1),\nu_1,\nu_2}$, where ν_1 = regression DF = 1, and ν_2 = residual DF = $n - 2$.

The residual mean square is often written as $s^2_{Y \cdot X}$, a representation denoting that it is the variance of Y after taking into account the dependence of Y on X. The square root of this quantity, i.e., $s_{Y \cdot X}$, is called the *standard error of estimate* (occasionally termed the "standard error of the regression"). In Example 17.3, $s_{Y \cdot X} = \sqrt{0.047701 \text{ cm}^2}$

TABLE 17.1 SUMMARY OF THE CALCULATIONS FOR TESTING H_0: $\beta = 0$ AGAINST H_A: $\beta \neq 0$ BY AN ANALYSIS OF VARIANCE

Source of variation	*Sum of squares* (SS)	DF	*Mean square* (MS)
Total $[Y_i - \bar{Y}]$	$\sum y^2$	$n - 1$	
Linear regression $[\hat{Y}_i - \bar{Y}]$	$\frac{(\sum xy)^2}{\sum x^2}$	1	$\frac{\text{regression SS}}{\text{regression DF}}$
Residual $[Y_i - \hat{Y}_i]$	total SS − regression SS	$n - 2$	$\frac{\text{residual SS}}{\text{residual DF}}$

Note: To test the null hypothesis, we compute F = regression MS/residual MS. The critical value for the test is $F_{\alpha(1),1,(n-2)}$.

Example 17.3

Analysis of variance testing of H_0: $\beta = 0$ against H_A: $\beta \neq 0$, using the data of Examples 17.1 and 17.2.

$$n = 13$$
$$\sum Y = 44.4$$
$$\sum Y^2 = 171.30$$
$$\text{total SS} = \sum y^2 = 171.30 - \frac{(44.4)^2}{13}$$
$$= 171.30 - 151.6431$$
$$= 19.656923$$
$$\text{total DF} = n - 1 = 12$$
$$\sum xy = 70.80 \quad \text{(from Example 17.2)}$$
$$\sum x^2 = 262.00 \quad \text{(from Example 17.2)}$$
$$\text{regression SS} = \frac{(\sum xy)^2}{\sum x^2} = \frac{(70.80)^2}{262.00}$$
$$= \frac{5012.64}{262.00}$$
$$= 19.132214$$

Source of variation	SS	DF	MS
Total	19.656923	12	
Linear regression	19.132214	1	19.132214
Residual	0.524709	11	0.047701

$$F = \frac{19.132214}{0.047701} = 401.1$$

$F_{0.05(1),1,11} = 4.84$

Therefore, reject H_0.

$P \ll 0.0005$

$$r^2 = \frac{19.132214}{19.656923} = 0.97$$

$$s_{Y \cdot X} = \sqrt{0.047701} = 0.218 \text{ cm}$$

$= 0.218$ cm. The standard error of estimate is an overall indication of the accuracy with which the fitted regression function predicts the dependence of Y on X. Some researchers express their sample linear regression equations in the form:

$$\hat{Y}_i = a + bX_i \pm s_{Y\cdot X}, \tag{17.15}$$

so that for the regression in Example 17.2 one might write

$$\hat{Y}_i = 0.72 + 0.270X_i \pm 0.22.$$

If this notation is utilized (and it seldom is*), the quantity following the "$\pm$" symbol must be defined for the reader to be the standard error of estimate. The magnitude of $s_{Y\cdot X}$ is proportional to the magnitude of the dependent variable, Y, making examination of $s_{Y\cdot X}$ a poor method for comparing regressions. Thus, Dapson (1980) recommends using $s_{Y\cdot X}/\bar{Y}$ (a unitless measure) to judge regression fits.

The proportion (or percentage) of the total variation in Y that is explained or accounted for by the fitted regression is termed the *coefficient of determination*, r^2, which may be thought of as a measure of the strength of the straight-line relationship:

$$r^2 = \frac{\text{regression SS}}{\text{total SS}}. \tag{17.16}$$

From Example 17.3, $r^2 = 0.97$, or 97%. That portion of the total variation not explained by the regression is, of course, $1 - r^2$, or residual SS/total SS, and this is called the *coefficient of nondetermination*, a quantity seldom referred to.† In Example 17.3, $1 - r^2 = 1.00 - 0.97 = 0.03$, or 3%. (The quantity r is the correlation coefficient, to be introduced in Chapter 19.)

Another good way to express the accuracy of a regression, or to compare accuracies of several regressions, is to compute confidence intervals for predicted values of Y, as described in Section 17.4.

***t* Testing.** The preceding null hypothesis concerning β can also be tested by using Student's t statistic. In fact, the more general two-tailed hypotheses, H_0: $\beta = \beta_0$ and H_A: $\beta \neq \beta_0$, can be tested in this fashion. Most frequently, β_0 is zero in these hypotheses, in which case either the analysis of variance or the t test may be employed. But if any value of β other than zero is hypothesized, then the following procedure is applicable, whereas the analysis of variance is not. Also, the t-testing procedure allows for the testing of one-tailed hypotheses: either H_0: $\beta \leq \beta_0$ and H_A: $\beta > \beta_0$, or H_0: $\beta \geq \beta_0$ and H_A: $\beta < \beta_0$.

Since the t statistic is in general calculated as

$$t = \frac{(\text{parameter estimate}) - (\text{parameter value hypothesized})}{\text{standard error of parameter estimate}}, \tag{17.18}$$

we need to compute s_b, the standard error of the regression coefficient.

*One reason to avoid using the "$\pm$" notation in Equation 17.15 is to keep the reader from inferring that this is an equation for determining a confidence interval for $\hat{Y}$; the appropriate procedure for obtaining such confidence intervals is shown in Section 17.4.

†The standard error of estimate is directly related to the coefficient of nondetermination and to the variability of Y as

$$s_{Y\cdot X} = s_Y\sqrt{1 - r^2}. \tag{17.17}$$

The variance of b is calculated as

$$s_b^2 = \frac{s_{Y \cdot X}^2}{\sum x^2}. \tag{17.19}$$

Therefore,

$$s_b = \sqrt{\frac{s_{Y \cdot X}^2}{\sum x^2}}, \tag{17.20}$$

and

$$t = \frac{b - \beta_0}{s_b}. \tag{17.21}$$

To test H_0: $\beta = 0$ against H_A: $\beta \neq 0$ in Example 17.4, $s_b = 0.0135$ cm/day, so $t = (b - \beta_0)/s_b = (0.270 - 0)/0.0135 = 20.000$. The degrees of freedom for this testing procedure are $n - 2$; thus, the critical value of t, at the 5% significance level, is $t_{0.05(2),11} = 2.201$, and H_0 is rejected. In a case such as this (i.e., when $\beta_0 = 0$), where either the analysis of variance or the t test may be employed, $t^2 = F$ and $t^2_{\alpha(2),(n-2)} = F_{\alpha(1),1,(n-2)}$. The reader may refer to Section 8.2 to review the concepts and procedures involved in one-tailed t testing.

Example 17.4

Use of Student's t to test H_0: $\beta = 0$ against H_A: $\beta \neq 0$, employing the data of Examples 17.1 and 17.2.

$n = 13$

$b = 0.270$ cm/day

$$s_b = \sqrt{\frac{s_{Y \cdot X}^2}{\sum x^2}} = \sqrt{\frac{0.047701}{262.00}} = \sqrt{0.00018206} = 0.0135 \text{ cm/day}$$

$$t = \frac{b - 0}{s_b} = \frac{0.270}{0.0135} = 20.000$$

$t_{0.05(2),11} = 2.201$

Therefore, reject H_0.

$P \ll 0.001$

17.4 CONFIDENCE INTERVALS IN REGRESSION

In many cases, knowing the standard error of a statistic allows us to calculate confidence intervals for the parameter being estimated, as

$$\text{confidence interval} = \text{statistic} \pm (t)(\text{SE of statistic}). \tag{17.22}$$

This was first demonstrated in Section 8.3 for confidence intervals for means, and it has been used repeatedly in succeeding chapters. In addition, the second significant figure of the standard error of a statistic may be used as an indicator of the precision to which that statistic should be reported (as done with the mean in Section 8.4). The standard error of b has been given in Equation 17.20. For the data in Examples 17.1 and 17.2, the second significant figure of $s_b = 0.0135$ cm/day enables us to express b to the third decimal place (i.e., $b = 0.270$ cm/day).

Confidence Interval for the Regression Coefficient. For the $(1 - \alpha)$ confidence limits of β, we calculate

$$b \pm t_{\alpha(2),(n-2)} s_b. \tag{17.23}$$

Therefore,

$$L_1 = b - t_{\alpha(2),(n-2)} s_b \tag{17.24}$$

and

$$L_2 = b + t_{\alpha(2),(n-2)} s_b. \tag{17.25}$$

For Example 17.2, the 95% confidence interval for β would be $b \pm t_{0.05(2),11} s_b = 0.270 \pm (2.201)(0.0135) = 0.270 \pm 0.030$ cm/day. Thus, $L_1 = 0.270 - 0.030 = 0.240$ cm/day, $L_2 = 0.270 + 0.030 = 0.300$ cm/day, and we can state, with 95% confidence (i.e., we state that there is no greater than a 5% chance that we are wrong), that the population regression coefficient, β, lies between 0.240 and 0.300 cm/day. Figure 17.7 shows, by the broken lines, these confidence limits for the slope of the regression line. Within these limits, the various possible b values rotate the line about the point $(\bar{X}, \bar{Y})$.

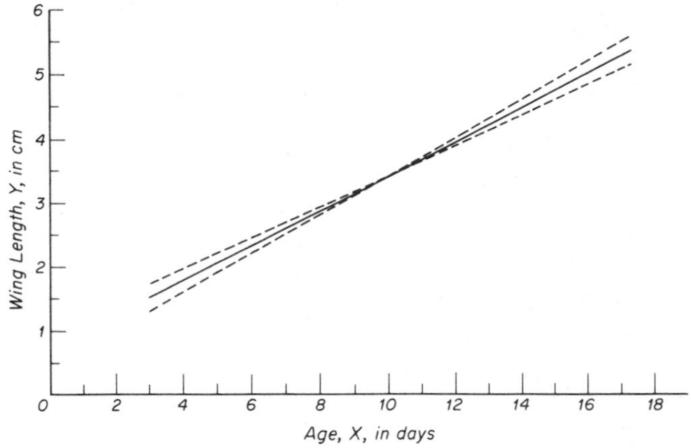

Figure 17.7 The regression line from Fig. 17.1, showing, by broken lines, the lines with slopes equal to the upper and lower 95% confidence limits for β.

Confidence Interval for an Estimated *Y*. As shown in Section 17.1, a regression equation allows one to estimate the value of $\hat{Y}_i$ existing in the population at a given value of X_i. The standard error of such a population value is

$$s_{\hat{Y}_i} = \sqrt{s^2_{Y \cdot X}\left[\frac{1}{n} + \frac{(X_i - \bar{X})^2}{\sum x^2}\right]}. \tag{17.26}$$

Example 17.5A shows how $s_{\hat{Y}_i}$ can be used in Equation 17.22 to calculate confidence intervals. It is apparent from Equation 17.26 that the standard error is a minimum for $X_i = \bar{X}$, and that it increases as estimates are made at values of X_i farther from the mean. If confidence limits were calculated for all points on the regression line, the result would be the *confidence bands* shown in Fig. 17.8.

Example 17.5

Standard errors of predicted values of Y.

The regression equation derived in Example 17.2 is used for the following considerations. For this regression, $a = 0.72$ cm, $b = 0.270$ cm/day, $\bar{X} = 10.0$ days, $\sum x^2 = 262.00$ days^2, $n = 13$, $s^2_{Y \cdot X} = 0.0477$ cm^2, and $t_{0.05(2),11} = 2.201$.

A. Equation 17.26 is used when one wishes to predict the mean value of $\hat{Y}_i$, given X_i, in the entire population. For example, we could ask: "What is the mean wing length of all 13.0-day-old birds in the population under study?"

$$\begin{aligned}\hat{Y}_i &= a + bX_i\\ &= 0.715 + (0.270)(13.0)\\ &= 0.715 + 3.510\\ &= 4.225 \text{ cm}\end{aligned}$$

$$s_{\hat{Y}_i} = \sqrt{s^2_{Y \cdot X}\left[\frac{1}{n} + \frac{(X_i - \bar{X})^2}{\sum x^2}\right]} \tag{17.26}$$

$$\begin{aligned}&= \sqrt{0.047701\left[\frac{1}{13} + \frac{(13.0 - 10.0)^2}{262.00}\right]}\\ &= \sqrt{(0.047701)(0.111274)}\\ &= 0.073 \text{ cm}\end{aligned}$$

$$\begin{aligned}95\% \text{ confidence interval} &= Y_i \pm t_{0.05(2),\,11} s_{\hat{Y}_i}\\ &= 4.225 \pm (2.201)(0.073)\\ &= 4.225 \pm 0.161 \text{ cm}\end{aligned}$$

$$L_1 = 4.064 \text{ cm}$$
$$L_2 = 4.386 \text{ cm}$$

B. Equation 17.28 is used when one proposes taking an additional sample of m individuals from the population and wishes to predict the mean Y value, at a given X, for these m new data. For example, one might ask: "If 10 13.0-day-old birds were taken from the population, what would be their mean wing length?"

$$\hat{Y}_i = 0.715 + (0.270)(13.0) = 4.225 \text{ cm}$$

$$(s_{\hat{Y}_i})_m = \sqrt{0.047701\left[\frac{1}{10} + \frac{1}{13} + \frac{(13.0 - 10.0)^2}{262.00}\right]} \tag{17.28}$$

$$\begin{aligned}&= \sqrt{(0.047701)(0.211274)}\\ &= 0.100 \text{ cm}\end{aligned}$$

$$\begin{aligned}95\% \text{ confidence interval} &= \hat{Y}_i \pm t_{0.05(2),\,11}(s_{\hat{Y}_i})_m\\ &= 4.225 \pm (2.201)(0.100)\\ &= 4.225 \pm 0.220 \text{ cm}\end{aligned}$$

$$L_1 = 4.005 \text{ cm}$$
$$L_2 = 4.445 \text{ cm}$$

C. Equation 17.29 is used when one wishes to predict the Y value of a single observation taken from the population at a specified X. For example, one could ask: "If one 13.0-day-old bird were taken from the population, what would be its wing length?"

$$\hat{Y}_i = 0.715 + (0.270)(13.0) = 4.225 \text{ cm}$$

$$(s_{\hat{Y}_i})_1 = \sqrt{0.047701\left[1 + \frac{1}{13} + \frac{(13.0 - 10.0)^2}{262.00}\right]} \tag{17.29}$$

$$\begin{aligned}&= \sqrt{(0.047701)(1.111274)}\\ &= 0.230 \text{ cm}\end{aligned}$$

$$\begin{aligned}95\% \text{ confidence interval} &= \hat{Y}_i \pm t_{0.05(2),\,11}(s_{\hat{Y}_i})_1\\ &= 4.225 \pm (2.201)(0.230)\\ &= 4.225 \pm 0.506 \text{ cm}\end{aligned}$$

$$L_1 = 3.719 \text{ cm}$$
$$L_2 = 4.731 \text{ cm}$$

Note from these three examples that the accuracy of prediction increases as does the number of data upon which the prediction is based. That is, for example, predictions about a mean for the entire population will be more accurate than a prediction about a mean from 10 members of the population, which is more accurate than a prediction about a single member of the population.

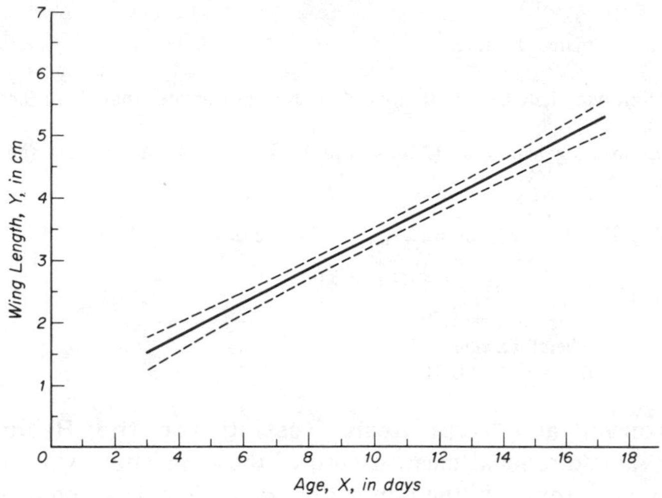

Figure 17.8 The 95% confidence bands (broken lines) for the regression line from Fig. 17.1 (the regression of Example 17.2).

If $X_i = 0$, then $\hat{Y}_i = a$, the Y intercept. Therefore,

$$s_a = \sqrt{s_{Y \cdot X}^2 \left[\frac{1}{n} + \frac{\bar{X}^2}{\sum x^2} \right]}. \tag{17.27}$$

If we predict a value of $\hat{Y}_i$ that is the mean of m additional measurements at X_i, its standard error would be

$$(s_{\hat{Y}_i})_m = \sqrt{s_{Y \cdot X}^2 \left[\frac{1}{m} + \frac{1}{n} + \frac{(X_i - \bar{X})^2}{\sum x^2} \right]}. \tag{17.28}$$

A special case of Equation 17.28 exists when we ask for the standard error involved in estimating $\hat{Y}_i$ for a single additional measurement at X_i:

$$(s_{\hat{Y}_i})_1 = \sqrt{s_{Y \cdot X}^2 \left[1 + \frac{1}{n} + \frac{(X_i - \bar{X})^2}{\sum x^2} \right]}. \tag{17.29}$$

Equation 17.26 is the special case of Equation 17.28 when m approaches infinity. Confidence intervals using Equations 17.28 and 17.29 are sometimes called *prediction intervals*. Examples 17.5B and C demonstrate the uses of the standard errors of predictions.

Testing Hypotheses About Estimated *Y* Values. Once we have computed the standard error of a predicted Y, we can test hypotheses about that prediction. For example, we might ask whether the mean population wing length of 13.0-day-old sparrows, call it $\mu_{\hat{Y}_{13.0}}$, is equal to some specified value (two-tailed test) or is greater than (or less than) some specified value (one-tailed test). We simply refer to Equation 17.18, as Example 17.6 demonstrates.

Example 17.6

Hypothesis testing with an estimated Y value.

H_0: The mean population wing length of 13.0-day-old birds is not greater than 4 cm (i.e., H_0: $\mu_{\hat{Y}_{13.0}} \leq 4$ cm).

H_A: The mean population wing length of 13.0-day-old birds is greater than 4 cm (i.e., H_A: $\mu_{\hat{Y}_{13.0}} > 4$ cm).

Since, from Example 17.5, $Y_{13.0} = 4.225$ cm and $s_{\hat{Y}_{13.0}} = 0.073$ cm,

$$t = \frac{4.225 - 4}{0.073} = \frac{0.225}{0.073} = 3.082$$

$t_{0.05(1),11} = 1.796$

Therefore, reject H_0.

$0.005 < P < 0.01$

Confidence Interval and Hypothesis Testing for the Residual Mean Square. The sample residual mean square of the population, $s^2_{Y \cdot X}$, is an estimate of the residual mean square in the population, $\sigma^2_{Y \cdot X}$. Confidence limits may be calculated for $\sigma^2_{Y \cdot X}$ just as they are for the population variance, σ^2, in Section 8.10. Simply use $\nu = n - 2$, rather than $\nu = n - 1$, and replace σ^2 with $\sigma^2_{Y \cdot X}$ and SS with residual SS in Equation 8.20 or 8.21. Also, a confidence interval for the population standard error of estimate, $\sigma_{Y \cdot X}$, may be obtained by analogy to Equation 8.22. Hypothesis testing for $\sigma^2_{Y \cdot X}$ or $\sigma_{Y \cdot X}$ may be performed by analogy to the procedures of Section 8.11.

17.5 INVERSE PREDICTION

Situations exist where the biologist wishes to predict the value of the independent variable (X_i) that is to be expected in the population at a specified value of the dependent variable (Y_i), a procedure known as *inverse prediction.* In Example 17.1, for instance, we might ask "How old is a bird that has a wing 4.5 cm long?" By simple algebraic rearrangement of the linear regression equation (Equation 17.8), we obtain

$$\hat{X}_i = \frac{Y_i - a}{b}. \tag{17.30}$$

From Fig. 17.8, it is clear that, although confidence limits calculated around the predicted $\hat{Y}_i$ are symmetrical above and below $\hat{Y}_i$, confidence limits associated with the predicted $\hat{X}_i$ are not symmetrical to the left and to the right of $\hat{X}_i$. The $1 - \alpha$ confidence limits for the X predicted at a given Y may be calculated as follows, which is demonstrated in Example 17.7:

$$\bar{X} + \frac{b(Y_i - \bar{Y})}{K} \pm \frac{t}{K}\sqrt{s^2_{Y \cdot X}\left[\frac{(Y_i - \bar{Y})^2}{\sum x^2} + K\left(1 + \frac{1}{n}\right)\right]}. \tag{17.31}$$

where* $K = b^2 - t^2 s^2_b$.

*It may be recalled that $F_{\alpha(1),1,\nu} = t^2_{\alpha(2),\nu}$. Therefore, we could compute $K = b^2 - Fs^2_b$, where $F = t^2_{\alpha(2),(n-2)} = F_{\alpha(1),1,(n-2)}$.

Example 17.7

Inverse prediction.

We wish to estimate, with 95% confidence, the age of a bird with a wing length of 4.5 cm.

Predicted age:

$$\hat{X} = \frac{Y_i - a}{b}$$
$$= \frac{4.5 - 0.715}{0.270}$$
$$= 14.019 \text{ days}$$

To compute 95% confidence interval:

$$t = t_{0.05(2),\,11} = 2.201$$
$$K = b^2 - t^2 s_b^2$$
$$= 0.270^2 - (2.201)^2(0.0135)^2$$
$$= 0.0720$$

95% confidence interval:

$$\bar{X} + \frac{b(Y_i - \bar{Y})}{K} \pm \frac{t}{K}\sqrt{s_{Y\cdot X}^2\left[\frac{(Y_i - \bar{Y})^2}{\sum x^2} + K\left(1 + \frac{1}{n}\right)\right]} \tag{17.31}$$

$$= 10.0 + \frac{0.270(4.5 - 3.415)}{0.0720}$$
$$\pm \frac{2.201}{0.0720}\sqrt{0.047701\left[\frac{(4.5 - 3.415)^2}{262.00} + 0.0720\left(1 + \frac{1}{13}\right)\right]}$$
$$= 10.0 + 4.069 \pm 30.569\sqrt{0.003913}$$
$$= 14.069 \pm 1.912 \text{ days}$$
$$L_1 = 12.157 \text{ days}$$
$$L_2 = 15.981 \text{ days}$$

This computation is a special case of the prediction of the $\hat{X}$ associated with multiple values of Y at that X. Such a situation would be where, for data as in Example 17.1, we had wing length measurements from m birds of the same age and wished to estimate that age. The predicted X would be

$$\hat{X}_i = \frac{\bar{Y}_i - a}{b}, \tag{17.32}$$

where $\bar{Y}_i$ is the mean of the m values of Y_i; and the confidence limits would be calculated as

$$\bar{X} + \frac{b(\bar{Y}_i - \bar{Y})}{K} \pm \frac{t}{K}\sqrt{(s_{Y\cdot X}^2)_*\left[\frac{(\bar{Y}_i - \bar{Y})^2}{\sum x^2} + K\left(\frac{1}{m} + \frac{1}{n}\right)\right]}, \tag{17.33}$$

where* $t = t_{\alpha(2),\,(n+m-3)}$, $K = b^2 - t^2(s_b^2)_*$,

$$(s_b^2)_* = \frac{(s_{Y\cdot X}^2)_*}{\sum x^2}, \tag{17.34}$$

and

$$(s_{Y\cdot X}^2)_* = \text{residual SS} + \sum_{j=1}^{m}(Y_{ij} - \bar{Y}_i)^2/(n + m - 3) \tag{17.35}$$

(Ostle and Mensing, 1975: 180–181; Seber, 1977: 190–191).

*Alternatively, one may compute $K = b^2 - F(s_b^2)_*$, where $F = t^2_{\alpha(2),\,(n+m-3)} = F_{\alpha(1),\,1,\,(n+m+3)}$.

17.6 INTERPRETATIONS OF REGRESSION FUNCTIONS

If we calculate the two constants, a and b, that define a linear regression equation, then we have quantitatively described the rate of change of Y with a change in X. However, although we have assumed a mathematical dependence of Y on X, we must not automatically assume that there is a biological cause and effect relationship. Causal relationships are concluded only with some insight into the natural phenomenon being investigated and may not be declared by statistical testing alone. Indeed, it is often necessary to determine the interrelationships between the two variables under study and other variables, for an observed dependence may, in fact, be due to the influence of one or more additional variables. (The methods in Chapter 20 are often used in this regard.)

We must also remember that a linear regression function is mathematically nothing more than a straight line forced to fit through a set of data points, and it may not at all describe a natural phenomenon. The biologist may frequently be chagrined when attempting to explain why the observed relationship is well described by a linear function or what biological insights are to be unfolded by the consideration of a particular slope or a particular magnitude of a Y intercept. In other words, although an empirically derived regression function often provides a satisfactory and satisfying description of a natural system, sometimes it does not. Section 20.15 further discusses the fitting of regression models.

Even if a regression function does not help us to explain the functional anatomy of a natural system, it may still be useful in its ability to predict Y, given X. In the sciences, equations may inaccurately represent natural processes yet may be employed advantageously to predict the value of one variable given the value of an associated variable. Thus, predicting $\hat{Y}$ (or $\hat{X}$) values and their standard errors is frequently a useful end in itself.

17.7 REGRESSION WITH REPLICATION: TESTING FOR LINEARITY

If, in Example 17.1, we had wing measurements for more than one bird for at least some of the recorded ages, then we could test the null hypothesis that the population regression is linear.* (Note that true replication requires that there are multiple birds at a given age, not that there are multiple wing measurements on the same bird.) Figure 17.9 presents the data of Example 17.8. A least squares, best fit, linear regression equation can be calculated for any set of at least two data, but neither the equation itself nor the testing for a significant slope (which requires at least three data) indicates whether Y is, in fact, a straight line function of X in the population sampled. We will occasionally encounter the suggestion that for data such as those in Fig. 17.9 the mean Y at each X be utilized for a regression analysis. However, to do so would be to discard information, and such a procedure is not recommended (Freund, 1971).

*Thornby (1972) presents a procedure to test the hypothesis of linearity even when there are not multiple observations of Y. But the computation is rather tedious.

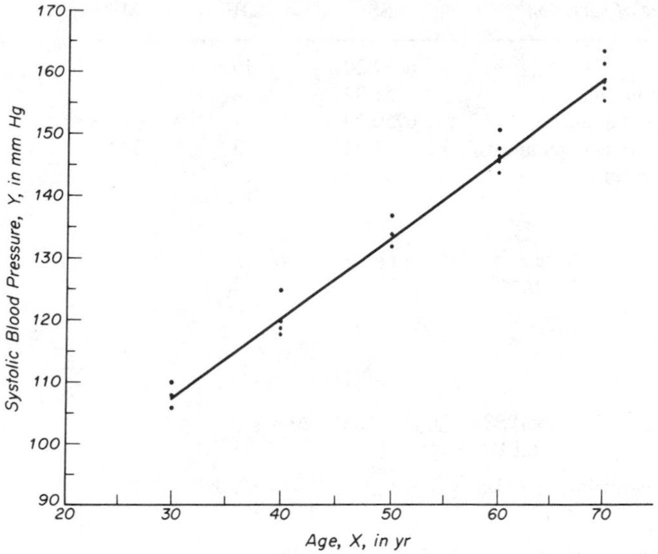

Figure 17.9 A regression where there are multiple values of Y for each value of X.

Example 17.8

Hypothesis testing for a regression where there are multiple values of Y for each value of X.

Age (yr) X	*Systolic blood pressure* (mm Hg) Y	n_i
30	108, 110, 106	3
40	125, 120, 118, 119	4
50	132, 137, 134	3
60	148, 151, 146, 147, 144	5
70	162, 156, 164, 158, 159	5

$N = 20$

$\sum\sum X_{ij} = 1050$ $\quad\sum\sum Y_{ij} = 2744$

$\sum\sum X_{ij}^2 = 59{,}100$ $\quad\sum\sum Y_{ij}^2 = 383{,}346$ $\quad\sum\sum X_{ij}Y_{ij} = 149{,}240$

$\sum x^2 = 3975.00$ $\quad\sum y^2 = 6869.20$ $\quad\sum xy = 5180.00$

$\bar{X} = 52.5$ $\quad\bar{Y} = 137.2$

$$b = \frac{\sum xy}{\sum x^2} = \frac{5180.00}{3975.00} = 1.303 \text{ mm Hg/yr}$$

$$a = \bar{Y} - b\bar{X} = 137.2 - (1.303)(52.5) = 68.4 \text{ mm Hg}$$

Therefore, the least squares regression line is $\hat{Y}_{ij} = 68.4 + 1.303X_{ij}$.

H_0: The population regression is linear.
H_A: The population regression is not linear.

$$\text{among groups SS} = \sum_{i=1}^{k}\frac{\left(\sum_{j=1}^{n_i} Y_{ij}\right)^2}{n_i} - \frac{\left(\sum_{i=1}^{k}\sum_{j=1}^{n_i} Y_{ij}\right)^2}{N} = 383{,}228.73 - 376{,}476.80$$
$$= 6751.93$$

among groups DF $= k - 1 = 4$
within groups SS = total SS − among groups SS = 6869.20 − 6751.93 = 117.27
within groups DF = total DF − among groups DF = 19 − 4 = 15
deviations from linearity SS = among groups SS − regression SS
= 6751.93 − 6750.29 = 1.64
deviations from linearity DF = among groups DF − regression DF = 4 − 1 = 3

Source of variation	SS	DF	MS
Total	6869.20	19	
Among groups	6751.93	4	
Linear regression	6750.29	1	
Deviations from linearity	1.64	3	0.55
Within groups	117.27	15	7.82

$$F = \frac{0.55}{7.82}$$

Since $F < 1.00$, do not reject H_0.

$P > 0.25$

H_0: $\beta = 0$.
H_A: $\beta \neq 0$.

$$\text{total SS} = \sum y^2 = 6869.20$$

$$\text{total DF} = N - 1 = 19$$

$$\text{regression SS} = \frac{(\sum xy)^2}{\sum x^2} = \frac{(5180.00)^2}{3975.00} = 6750.29$$

Source of variation	SS	DF	MS
Total	6869.20	19	
Linear regression	6750.29	1	6750.29
Residual	118.91	18	6.61

$$F = \frac{6750.29}{6.61} = 1021.2$$

$F_{0.05(1),1,18} = 4.41$

Therefore, reject H_0.

$P \ll 0.0005$

$$r^2 = \frac{6750.29}{6869.20} = 0.98$$

$$s_{Y \cdot X} = \sqrt{6.61} = 2.57 \text{ mm Hg}$$

Example 17.8 analyzes data consisting of multiple Y values at each X value, and Fig. 17.9 presents the data graphically. For each of the k unique X_i values we can speak of each of n_i values of Y_{ij}, using the double subscript exactly as in the one-way analysis of variance (Section 11.1); in Example 17.8, $n_1 = 3$, $n_2 = 4$, $n_3 = 3$, etc.; and $Y_{11} = 108$ mm, $Y_{12} = 110$ mm, $Y_{32} = 137$ mm, etc. Therefore,

$$\sum xy = \sum_{i=1}^{k} \sum_{j=1}^{n_i} X_{ij} Y_{ij} - \frac{(\sum \sum X_{ij})(\sum \sum Y_{ij})}{N}, \tag{17.36}$$

where $N = \sum_{i=1}^{k} n_i$, the total number of pairs of data. Also,

$$\sum x^2 = \sum_{i=1}^{k} \sum_{j=1}^{n_i} X_{ij}^2 - \frac{(\sum \sum X_{ij})^2}{N}, \tag{17.37}$$

and

$$\text{total SS} = \sum y^2 = \sum_{i=1}^{k} \sum_{j=1}^{n_i} Y_{ij}^2 - C, \tag{17.38}$$

where

$$C = \frac{(\sum \sum Y_{ij})^2}{N}, \quad \text{and} \quad N = \sum n_i. \tag{17.39}$$

It is then a simple matter, as Example 17.8 shows, to calculate the regression coefficient b, the Y intercept a, and the regression and residual sums of squares, using Equations 17.4, 17.7, 17.11, and 17.13, respectively. The total, regression, and residual degrees of freedom are $N - 1$, 1, and $N - 2$, respectively.

As shown in Section 17.3, the analysis of variance for significant slope involves the partitioning of the total variability of Y (i.e., $Y_{ij} - \bar{Y}$) into that variability due to regression ($\hat{Y}_i - \bar{Y}$) and that variability remaining (i.e., residual) after the regression line is fitted ($Y_{ij} - \hat{Y}_i$). However, by considering the k groups of Y values, we can also partition the total variability exactly as we did in the one-way analysis of variance (Section 11.1), by describing variability among groups ($\bar{Y}_i - \bar{Y}$) and within groups ($Y_{ij} - \bar{Y}_i$):

$$\text{among groups SS} = \sum_{i=1}^{k} \frac{\left(\sum_{j=1}^{n_i} Y_{ij}\right)^2}{n_i} - \text{C}, \tag{17.40}$$

$$\text{among groups DF} = k - 1, \tag{17.41}$$

$$\text{within groups SS} = \text{total SS} - \text{among groups SS}, \tag{17.42}$$

$$\text{within groups DF} = \text{total DF} - \text{among groups DF} = N - k. \tag{17.43}$$

The variability among groups can, in turn, be partitioned. Part of this variability ($\hat{Y}_i - \bar{Y}$) results from the linear regression being fitted, whereas the remainder is due to the deviation of each group of data from the regression line ($\bar{Y}_i - \hat{Y}_i$), as shown in Fig. 17.10.

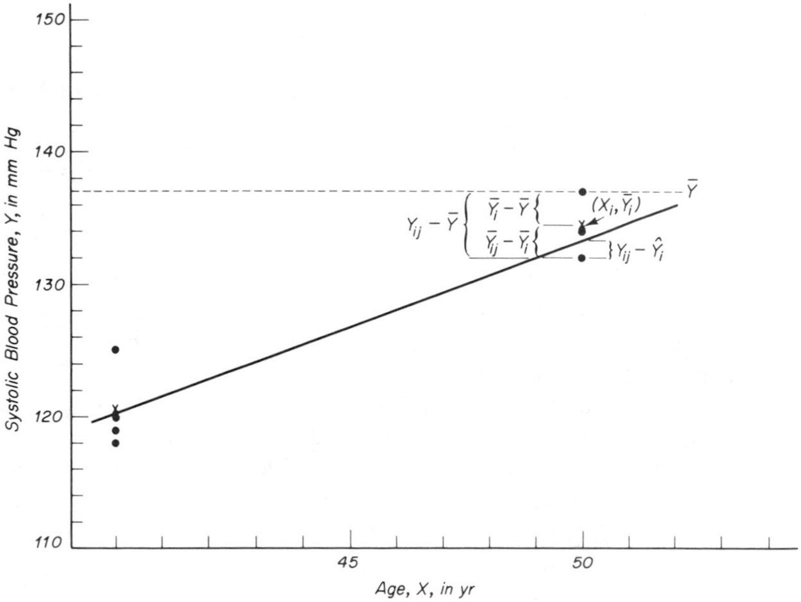

Figure 17.10 An enlarged portion of Fig. 17.9, showing the partitioning of Y deviations.

Therefore,

$$\text{deviations from linearity SS} = \text{among groups SS} - \text{regression SS} \quad (17.44)$$

and

$$\text{deviations from linearity DF} = \text{among groups DF} - \text{regression DF}$$
$$= k - 2. \quad (17.45)$$

Table 17.2 summarizes this partitioning of sums of squares.

TABLE 17.2 SUMMARY OF THE ANALYSES OF VARIANCE CALCULATIONS FOR TESTING H_0: THE POPULATION REGRESSION IS LINEAR, AND FOR TESTING H_0: $\beta = 0$

Source of variation	*Sum of squares* (SS)	DF	*Mean square* (MS)
Total $[Y_{ij} - \bar{Y}]$	$\sum y^2$	$N - 1$	
Linear regression $[\hat{Y}_i - \bar{Y}]$	$\frac{(\sum xy)^2}{\sum x^2}$	1	$\frac{\text{regression SS}}{\text{regression DF}}$
Residual $[Y_{ij} - \hat{Y}_i]$	total SS − regression SS	$N - 2$	$\frac{\text{residual SS}}{\text{residual DF}}$
Among groups $[\bar{Y}_i - \bar{Y}]$	$\sum_{i=1}^{k} \frac{\left(\sum_{j=1}^{n_i} Y_{ij}\right)^2}{n_i} - \frac{\left(\sum_{i=1}^{k}\sum_{j=1}^{n_i} Y_{ij}\right)^2}{N}$	$k - 1$	
Linear regression $[\hat{Y}_i - \bar{Y}]$	$\frac{(\sum xy)^2}{\sum x^2}$	1	
Deviations from linearity $[\bar{Y}_i - \hat{Y}_i]$	among groups SS − regression SS	$k - 2$	$\frac{\text{deviations SS}}{\text{deviations DF}}$
Within groups $[Y_{ij} - \bar{Y}_i]$	total SS − among groups SS	$N - k$	$\frac{\text{within groups SS}}{\text{within groups DF}}$

Note: To test H_0: the population regression is linear, we use $F =$ deviations MS/within groups MS, with a critical value of $F_{\alpha(1), (k-2), (N-k)}$. If that null hypothesis is not rejected, then H_0: $\beta = 0$ is tested using $F =$ regression MS/residual MS, with a critical value of $F_{\alpha(1), 1, (N-1)}$; if the null hypothesis of linearity is rejected, then $H_0: \beta = 0$ is tested using $F =$ regression MS/within groups MS, with a critical value of $F_{\alpha(1), 1, (N-k)}$.

Alternatively, and with identical results, we may consider the residual variability $(Y_{ij} - \hat{Y}_i)$ to be divisible into two components: within groups variability $(Y_{ij} - \bar{Y}_i)$ and deviations from linearity $(\bar{Y}_i - \hat{Y}_i)$. This partitioning of sums of squares and degrees of freedom is summarized in Table 17.3.*

If the population relationship between Y and X is a straight line (i.e., "H_0: The population regression is linear" is a true statement), then the deviations from linearity MS and the within groups MS will be estimates of the same variance; if the relationship is not a straight line (the H_0 is false), then the deviations from linearity MS will be significantly greater than the within groups MS. Thus, as demonstrated in Example 17.8,

*Some authors refer to deviations from linearity as "lack of fit" and to within groups variability as "pure error."

TABLE 17.3 SUMMARY OF ANALYSIS OF VARIANCE PARTITIONING OF SOURCES OF VARIATION FOR TESTING LINEARITY, AS AN ALTERNATIVE TO THAT IN TABLE 17.2

Source of variation	DF
Total $[Y_{ij} - \bar{Y}]$	$N - 1$
Among groups $[\bar{Y}_i - \bar{Y}]$	$k - 1$
Within groups $[Y_{ij} - \bar{Y}_i]$	$N - k$
Linear regression $[\hat{Y}_i - \bar{Y}]$	1
Residual $[Y_{ij} - \hat{Y}_i]$	$N - 2$
Within groups $[Y_{ij} - \bar{Y}_i]$	$N - k$
Deviations from linearity $[\bar{Y}_i - \hat{Y}_i]$	$k - 2$

Note: Sums of squares and mean squares are as in Table 17.2.

$$F = \frac{\text{deviations from linearity MS}}{\text{within groups MS}} \tag{17.46}$$

provides a one-tailed test of the null hypothesis of linearity.

If the null hypothesis of linearity is not rejected, then deviations from linearity MS and the within groups MS may be considered to be estimates of the same population variance. The latter will be the better estimate, as it is based on more degrees of freedom; but an even better estimate is the residual MS, $s^2_{Y \cdot X}$, for it constitutes a pooling of the deviations MS and the within groups MS. Therefore, if a regression is assumed to be linear, $s^2_{Y \cdot X}$ is the appropriate variance to use in the computation of standard errors (e.g., by Equations 17.20, 17.26–17.29) and confidence intervals resulting from them, and this residual mean square ($s^2_{Y \cdot X}$) is also appropriate in testing the hypothesis H_0: $\beta = 0$ (by either Equation 17.14 or 17.21), as demonstrated in Example 17.8.

If the population regression is concluded not to be linear, then the investigator would do well to consider the procedures of Section 17.10 or 20.14 or of Chapter 21, and to consider the linear regression analysis no further. If, however, we desire to test H_0: $\beta = 0$, then the within groups MS should be substituted for the residual MS ($s^2_{Y \cdot X}$). It would not be advisable to engage in predictions with the linear regression equation.

17.8 POWER AND SAMPLE SIZE IN REGRESSION

Although there are basic differences between regression and correlation (Section 17.1), a set of data for which there is a statistically significant regression coefficient (i.e., one rejects H_0: $\beta = 0$, as explained in Section 17.3) would also yield a statistically significant correlation coefficient (i.e., one would reject H_0: $\rho = 0$, to be discussed in Section 19.2). In addition, conclusions about the power of a significance test for a regression coefficient can be obtained by estimating power associated with the significance test for the correlation coefficient that would have been obtained from the same set of data.

After performing a regression analysis for a set of data, we may obtain the sample correlation coefficient, r, either from Equation 19.1, or, more simply, as

$$r = b\sqrt{\frac{\sum x^2}{\sum y^2}}, \tag{17.47}$$

or we may take the square root of the coefficient of determination, r^2, assigning to it the sign of b. Then, with r in hand, the procedures of Section 19.4 may be employed (Cohen, 1977: 76–77) to estimate power and minimum required sample size for the hypothesis test for the regression coefficient, H_0: $\beta = 0$.

17.9 REGRESSION THROUGH THE ORIGIN

Although not of common biological importance, a special type of regression procedure is called for when we are faced with sets of data for which we know, a priori, that in the population Y will be zero when X is zero; i.e., the population Y intercept is known to be zero. Since the point on the graph with coordinates (0, 0) is termed the *origin*, this regression situation is known as regression through the origin.

For regression through the origin, the linear regression equation would be

$$\hat{Y}_i = bX_i, \tag{17.48}$$

and some of the calculations unique to such a regression are as follows:

$$b = \frac{\sum X_i Y_i}{\sum X_i^2}, \tag{17.49}$$

$$\text{total SS} = \sum Y_i^2, \quad \text{with total DF} = n, \tag{17.50}$$

$$\text{regression SS} = \frac{(\sum X_i Y_i)^2}{\sum X_i^2}, \quad \text{with regression DF} = 1, \tag{17.51}$$

$$s_{Y\cdot X}^2 = \text{residual MS} = \frac{\sum Y_i^2 - \dfrac{(\sum X_i Y_i)^2}{\sum X_i^2}}{n-1}, \tag{17.52}$$

$$s_b^2 = \frac{s_{Y\cdot X}^2}{\sum X_i^2}. \tag{17.53}$$

The statistic r^2 is calculated and tests of hypotheses about the slope of the line are performed, as explained earlier in this chapter, with the exception that the above values are used; $n - 1$ is used as degrees of freedom whenever $n - 2$ is used for regressions not assumed to pass through the origin. Interestingly, a regression line forced through the origin does not necessarily pass through point $(\bar{X}, \bar{Y})$.

Confidence Intervals. For regressions passing through the origin, confidence intervals may be obtained in ways analogous to the procedures in Section 17.4. That is, a confidence interval for the population regression coefficient, β, is calculated using Equation 17.53 for s_b^2 and $n - 1$ degrees of freedom in place of $n - 2$. A confidence interval for an estimated $\hat{Y}$ is

$$s_{\hat{Y}} = \sqrt{s_{Y\cdot X}^2\left(\frac{X_i^2}{\sum X^2}\right)}, \tag{17.54}$$

using the $s^2_{Y \cdot X}$ of Equation 17.52; a confidence interval for $\hat{Y}_i$ predicted as the mean of m additional measurements at X_i is

$$(s_{\hat{Y}})_m = \sqrt{s^2_{Y \cdot X}\left(\frac{1}{m} + \frac{X_i^2}{\sum X^2}\right)} \tag{17.55}$$

and a confidence interval for the $\hat{Y}_i$ predicted for one additional measurement of X_i is

$$(s_{\hat{Y}})_m = \sqrt{s^2_{Y \cdot X}\left(1 + \frac{X_i^2}{\sum X^2}\right)} \tag{17.56}$$

(Seber, 1977: 192).

Inverse Prediction. For inverse prediction (see Section 17.5) with a regression passing through the origin,

$$\hat{X}_i = \frac{Y_i}{b}, \tag{17.57}$$

and the confidence interval for the X_i predicted at a given Y is

$$\bar{X} + \frac{bY_i}{K} \pm \frac{t}{K}\sqrt{s^2_{Y \cdot X}\left(\frac{Y_i^2}{\sum X_i^2} + K\right)}, \tag{17.58}$$

where $t = t_{\alpha(2),(n-1)}$ and* $K = b^2 - t^2 s_b^2$ (Seber, 1977: 192)

If X is to be predicted for multiple values of Y at that X, then

$$\hat{X}_i = \frac{\bar{Y}_i}{b}, \tag{17.59}$$

where $\bar{Y}_i$ is the mean of m values of Y; and the confidence limits would be calculated as

$$\bar{X} + \frac{b\bar{Y}_i}{K} \pm \frac{t}{K}\sqrt{(s^2_{Y \cdot X})_*\left(\frac{\bar{Y}_i^2}{\sum X^2} + \frac{K}{m}\right)}, \tag{17.60}$$

where $t = t_{\alpha(2),(n+m-2)}$ and† $K = b^2 - t^2(s_b^2)_*$;

$$(s_b^2)_* = \frac{(s^2_{Y \cdot X})_*}{\sum X^2}; \tag{17.61}$$

and

$$(s^2_{Y \cdot X})_* = \frac{\text{residual SS} + \sum_{j=1}^{m}(Y_{ij} - \bar{Y}_i)^2}{n + m - 2} \tag{17.62}$$

(Seber, 1977: 192).

17.10 DATA TRANSFORMATIONS IN REGRESSION

As previously mentioned, the described regression analyses considered in this book depend upon the assumptions of normality and homoscedasticity, with regard to the values of Y, the dependent variable. Chapter 14 discussed the logarithmic, square root, and arcsine transformations of data to achieve closer approximations to these assumptions. By consciously striving to satisfy one of the assumptions one

*Alternatively, $K = b^2 - Fs_b^2$ where $F = t^2_{\alpha(2),(n-1)} = F_{\alpha(1),1,(n-1)}$.

†Alternatively, $K = b^2 - F(s_b^2)_*$, where $F = t^2_{\alpha(2),(n+m-2)} = F_{\alpha(1),1,(n+m-2)}$.

often (but without guaranty) appeases the others. The same considerations are applicable to regression data.

Transformation of the independent variable will not affect the distribution of Y, so transformations of X generally may be made with impunity, and sometimes they conveniently convert a curved line into a straight line. However, transformations of Y do affect least squares considerations and will therefore be discussed. Acton (1966: Chapter 8) presents a readable discussion of transformations in regression.

If the values of Y are from a Poisson distribution (i.e., the data are counts, especially small counts), then the square root transformation is usually desirable:

$$Y' = \sqrt{Y + 0.5}, \tag{17.63}$$

where the values of the variable after transformation (Y') are then submitted to regression analysis. (Also refer to Section 14.3.)

If the Y values are from a binomial distribution (e.g., they are proportions or percentages), then the arcsine transformation is called for:

$$Y' = \arcsin \sqrt{Y} \tag{17.64}$$

(See also Section 14.2.) Table B.23 allows for ready use of this transformation.

The most commonly used transformation in regression is the logarithmic transformation (see also Section 14.1), although it is sometimes employed for the wrong reasons. This transformation,

$$Y' = \log Y, \tag{17.65}$$

or

$$Y' = \log (Y + 1), \tag{17.66}$$

is appropriate when there is heteroscedasticity owing to the variance of Y at any X increasing in proportion to the value of X. When this situation exists, it implies that values of Y can be measured more accurately at low than at high values of X. Figure 17.11 shows such data (from Example 17.9) before and after the transformation.

Example 17.9

Regression data before and after logarithmic transformation of Y.

Original data (as plotted in Fig. 17.11a), indicating the variance of Y (s_Y^2) at each X:

X	Y	s_Y^2
5	10.72, 11.22, 11.75, 12.31	0.4685
10	14.13, 14.79, 15.49, 16.22	0.8101
15	18.61, 19.50, 20.40, 21.37	1.4051
20	24.55, 25.70, 26.92, 28.18	2.4452
25	32.36, 33.88, 35.48, 37.15	4.2526

Transformed data (as plotted in Fig. 17.11b), indicating the variance of log Y ($s_{\log Y}^2$) at each X:

X	$\log Y$	$s_{\log Y}^2$
5	1.03019, 1.04999, 1.07004, 1.09026	0.000668
10	1.15014, 1.16997, 1.19005, 1.21005	0.000665
15	1.26975, 1.29003, 1.30963, 1.32980	0.000665
20	1.39005, 1.40993, 1.43008, 1.44994	0.000665
25	1.51001, 1.52994, 1.54998, 1.56996	0.000666

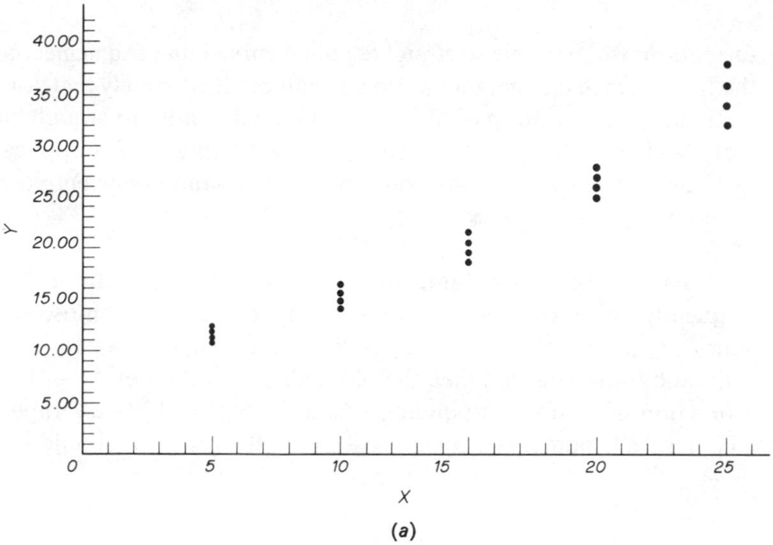

(a)

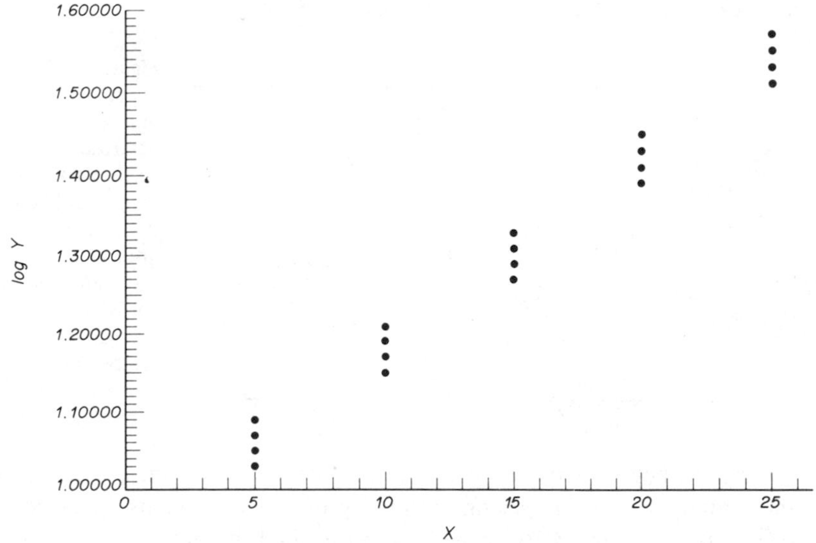

(b)

Figure 17.11 Regression data (of Example 17.9), exhibiting an increasing variability of Y with increasing magnitude of X. (a) The original data. (b) The data after logarithmic transformation of Y.

Many scatter plots of data imply a curved, rather than a straight line dependence of Y on X (e.g., Fig. 17.11a). Often, logarithmic or other transformations of the values of Y and/or X will result in a straight line relationship (as Fig. 17.11b) amenable to linear regression techniques. However, if original, nontransformed values of Y agree with our assumptions of normality, homoscedasticity, and additivity, then the data resulting from any of the preceding transformations will *not* abide by these assumptions. This is often not considered, and many biologists employing transformations do so simply to straighten out a curved line and neglect to consider whether the transformed data might indeed be analyzed legitimately by least squares regression methods. If a transformation may not be used validly to straighten out a curvilinear regression, then Section 20.14 (or, perhaps Chapter 21) is applicable.

Section 14.4 mentions some other, less commonly employed, data transformations.

Examination of Residuals. Since the logarithmic transformation is frequently proposed and employed to try to achieve homoscedasticity, we should consider how we might obtain a justification for such a transformation. If a regression is fitted by least squares, then the residuals (i.e., values of $Y_i - \hat{Y}_i$) may be plotted as a function of their corresponding X's, as in Fig. 17.12 (see Draper and Smith, 1981: Chapter 3). If homoscedasticity exists, then the residuals should be distributed evenly

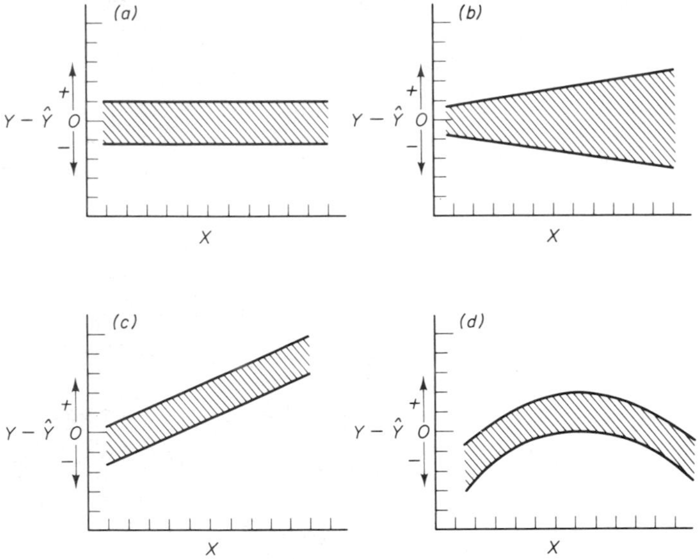

Figure 17.12 The plotting of residuals. (a) Data exhibiting homoscedasticity. (b) Data with heteroscedasticity of the sort in Example 17.9 (c) Data for which there was likely an error in the regression calculations, or an additional variable is needed in the regression model. (d) Data for which a linear regression does not accurately describe the relationship between Y and X, and a curvilinear relationship should be considered.

above and below zero (i.e., within the shaded area in Fig. 17.12a). If there is heteroscedasticity due to increasing variability in Y with increasing values of X, then the residuals will form a pattern such as in Fig. 17.12b, and a logarithmic transformation might be attempted. If the residuals form a pattern such as in Fig. 17.12c, we should suspect a calculation error or suspect that an additional important variable should be added to the regression model (see Chapter 20). Then pattern in Fig. 17.12d indicates that a *linear* regression is an improper model to describe the data; e.g., a quadratic regression (see Section 21.3) might be employed. Glejser (1969) suggests the fitting of the simple linear regression

$$E_i = a + bX_i, \tag{17.67}$$

where $E_i = | Y_i - \hat{Y}_i |$. A statistically significant b greater than zero indicates Fig. 17.12b to be the case, and the logarithmic transformation may be attempted. Then, after the application of the transformation, a plot of the new residuals (i.e., $\log Y_i - \widehat{\log Y_i}$) should be examined and Equation 17.67 fitted, where $E_i = |\log Y_i -$

$|\widehat{\log Y_i}|$. If this regression has a b not significantly different from zero, then we may assume that the transformation was justified.

Tests for normality in the distribution of residuals may be made by using the methods of Section 7.4 (employing $Y_i - \hat{Y}_i$ in place of X_i in that section); graphical examination of normality (as in Fig. 7.8) is often convenient.

17.11 THE EFFECT OF CODING

Either X or Y data, or both, may be coded prior to the application of regression analysis, and coding may facilitate computations, especially when the data are very large or very small in magnitude. As shown in Sections 3.5 and 4.8, coding may consist of adding a constant to (or subtracting it from) X, or multiplying (or dividing) X by a constant; or both addition (or subtraction) and multiplication (or division) may be applied simultaneously. Values of Y may be coded in the same fashion, indeed simultaneously with the coding of X values, using either the same or different coding constants. If we let M_X and M_Y represent constants by which X and Y, respectively, are to be multiplied, and let A_X and A_Y be constants then to be added to $M_X X$ and $M_Y Y$, respectively, then the transformed variables, $[X]$ and $[Y]$, are

$$[X] = M_X X + A_X \tag{17.68}$$

and

$$[Y] = M_Y Y + A_Y. \tag{17.69}$$

The slope, b, will not be changed by adding constants to X and/or Y, for such transformations have the effect of simply sliding the scale of one or both axes. But if multiplication factors are used in coding, then the resultant slope, $[b]$, will be equal to $(b)(M_Y/M_X)$, and the slope that would have been calculated had coding not been employed would be $b = [b](M_X/M_Y)$. The effects of coding on other regression statistics are shown in Table 17.4, where bracketed statistics are those resulting from analysis of data coded as in Equations 17.68 and 17.69. Note that coding in no way alters the value of r^2 or the t or F statistics calculated for hypothesis testing (except as noted in the table).

A common situation involving multiplicative coding factors is one where the variables were recorded using certain units of measurement, and it is desired to determine what regression statistics would have resulted if other units of measurement had been used.

For the data in Examples 17.1, 17.2, and 17.3, $a = 0.715$ cm, $b = 0.270$ cm/day, and $s_{Y \cdot X} = 0.218$ cm. If the wing length data were measured in inches, rather than in centimeters, there would have to be a coding by multiplying by 0.3937 in./cm (there are 0.3937 inches in one centimeter). By consulting Table 17.4, with $M_Y = 0.3937$ in./cm, $A_Y = 0$, $M_X = 1$, and $A_X = 0$, we can calculate that if a regression analysis were run on these data where X was recorded in inches, the slope would be $[b] = (0.270$ cm/day)(0.3937 in./cm) $= 0.106$ in./day; the Y intercept would be $[a] = (0.715$ cm) (0.3937 in./cm) $= 0.281$ in.; and the standard error of estimate would be $s_{Y \cdot X} =$ (0.3937 in./cm)(0.218 cm) $= 0.086$.

A relatively common biological case where A_X and/or A_Y are not zero is when

TABLE 17.4 THE EFFECT OF CODING ON REGRESSION STATISTICS

Statistic	*Value without coding*	*Value using coding*
Regression coefficient, b	$b = [b]M_X/M_Y$	$[b] = bM_Y/M_X$
Y intercept, a	$a = ([a] + [b]A_X - A_Y)/M_Y$	$[a] = aM_Y - [b]A_X + A_Y$
Standard error of estimate, $s_{Y \cdot X}$	$s_{Y \cdot X} = [s_{Y \cdot X}]/M_Y$	$[s_{X \cdot Y}] = M_Y s_{Y \cdot X}$
Coefficient of determination, r^2	$r^2 = [r^2]$	$[r^2] = r^2$
Test statistics,* t and F	$t = [t]$; $F = [F]$	$[t] = t$; $[F] = F$
Standard error of b, s_b	$s_b = [s_b]M_X/M_Y$	$[s_b] = s_b M_Y/M_X$
Standard error of a, s_a*	$s_a = [s_a]/M_Y$	$[s_a] = s_a M_Y$
Mean of X, $\bar{X}$	$\bar{X} = ([\bar{X}] - A_X)/M_X$	$[\bar{X}] = M_Y\bar{X} + A_X$
Mean of Y, $\bar{Y}$	$\bar{Y} = ([\bar{Y}] - A_Y)/M_Y$	$[\bar{Y}] = M_Y\bar{Y} + A_Y$
Sum of squares, SS	$\text{SS} = [\text{SS}]/M_Y^2$	$[\text{SS}] = (\text{SS})(M_Y^2)$
Mean square, MS	$\text{MS} = [\text{MS}]/M_Y^2$	$[\text{MS}] = (\text{MS})(M_Y^2)$

Note: If a regression is fit through the origin (Section 17.9), then coding by addition of a constant may not be used (i.e., $A_X = 0$ and $A_Y = 0$).

*A_X must be zero if s_a is desired or if it is desired to test hypotheses about the Y intercept.

we have temperature measurements in degrees Celsius (or Fahrenheit) and wish to determine the regression equation that would have resulted had the data been recorded in degrees Fahrenheit (or Celsius). The appropriate coding constants in Equations 17.68, and 17.69 may be determined by knowing that Celsius and Fahrenheit temperatures are related as follows:

$$\text{degrees Celsius} = \left(\frac{5}{9}\right)(\text{degrees Fahrenheit}) - \left(\frac{5}{9}\right)(32)$$

$$\text{degrees Fahrenheit} = \left(\frac{9}{5}\right)(\text{degrees Celsius}) + 32$$

This is summarized elsewhere (Zar, 1968a), as are the effects of multiplicative coding on logarithmically transformed data (Zar, 1967).

EXERCISES

17.1. The following data are the rates of oxygen consumption of birds, measured at different environmental temperatures:

Oxygen consumption (ml/g/hr)	*Temperature* (°C)
5.2	−18
4.7	−15
4.5	−10
3.6	−5
3.4	0
3.1	5
2.7	10
1.8	19

(a) Calculate a and b for the regression of oxygen consumption rate on temperature. (b) Test, by analysis of variance, the hypothesis $H_0: \beta = 0$. (c) Test, by the t test, the hypothesis $H_0: \beta = 0$. (d) Calculate the standard error of estimate of the regression. (e) Calculate the coefficient of determination of the regression. (f) Calculate the 95% confidence limits for β.

17.2. Utilize the regression equation computed for the data of Exercise 17.1. **(a)** What is the mean rate of oxygen consumption in the population for birds at 15°C? **(b)** What is the 95% confidence interval for this mean rate? **(c)** If we randomly chose one additional bird at 15°C from the population, what would its rate of oxygen consumption be estimated to be? **(d)** We can be 95% confident of this value lying between what limits?

17.3. The frequency of electrical impulses emitted from electric fish is measured from fish at different temperatures. The resultant data are as follows:

Temperature (°C)	*Impulse frequency* (number/sec)
20	225, 230, 239
22	251, 259, 265
23	266, 273, 280
25	287, 295, 302
27	301, 310, 317
28	307, 313, 325
30	324, 330, 338

(a) Compute a and b for the linear regression equation relating impulse frequency to temperature. **(b)** Test, by analysis of variance, $H_0: \beta = 0$. **(c)** Calculate the standard error of estimate of the regression. **(d)** Calculate the coefficient of determination of the regression. **(e)** Test H_0: The population regression is linear.

18 Comparing Simple Linear Regression Equations

It is common to possess more than one set of data and to have calculated a regression equation for each set. We might then ask whether the slopes of these lines are significantly different or whether they might be estimating the same population value of β. Furthermore, if we conclude that the slopes of the several lines are not significantly different, then we might wish to determine whether the several sets of data are likely from the same population (i.e., whether the population Y intercepts, as well as the slopes, are the same). In this chapter, procedures for testing differences among regression lines will be presented as summarized in Fig. 18.1.

18.1 COMPARING TWO SLOPES

A simple method for testing hypotheses about equality of two population regression coefficients involves the use of Student's t in a fashion analogous to that of testing for differences between two population means (Section 9.4). The test statistic is

$$t = \frac{b_1 - b_2}{s_{b_1 - b_2}}, \tag{18.1}$$

where the standard error of the difference between regression coefficients is

$$s_{b_1 - b_2} = \sqrt{\frac{(s^2_{Y \cdot X})_p}{(\sum x^2)_1} + \frac{(s^2_{Y \cdot X})_p}{(\sum x^2)_2}}, \tag{18.2}$$

and the pooled residual mean square is calculated as

$$(s^2_{Y \cdot X})_p = \frac{(\text{residual SS})_1 + (\text{residual SS})_2}{(\text{residual DF})_1 + (\text{residual DF})_2}, \tag{18.3}$$

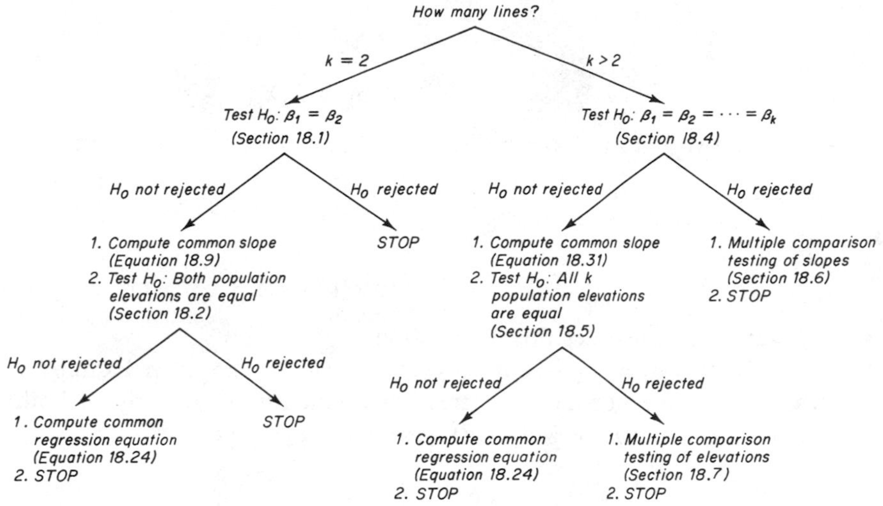

Figure 18.1 Flow chart for the comparison of regression lines.

the subscripts 1 and 2 referring to the two regression lines being analyzed. The critical value of t for this test has $(n_1 - 2) + (n_2 - 2)$ degrees of freedom (i.e., the sum of the two residual degrees of freedom), which is

$$\nu = n_1 + n_2 - 4. \tag{18.4}$$

Example 18.1 demonstrates these calculations for testing H_0: $\beta_1 = \beta_2$ against H_A: $\beta_1 \neq \beta_2$. The two regression lines are shown in Fig. 18.2.

Just as the t test for difference between means assumes that $\sigma_1^2 = \sigma_2^2$, the above t test assumes that $(\sigma_{Y \cdot X}^2)_1 = (\sigma_{Y \cdot X}^2)_2$. The presence of the latter condition can be tested by the variance ratio test, $F = (s_{Y \cdot X}^2)_{\text{larger}}/(s_{Y \cdot X}^2)_{\text{smaller}}$; but this is usually not done due to the limitations of this test (Section 9.1).

The $1 - \alpha$ confidence interval for the difference between two slopes, β_1 and β_2, is

$$(b_1 - b_2) \pm t_{\alpha(2),\nu} s_{b_1 - b_2}, \tag{18.5}$$

where ν is as in Equation 18.4. Thus, for Example 18.1,

$$\begin{aligned} 95\% \text{ confidence interval for } \beta_1 - \beta_2 &= (2.97 - 2.17) \pm (t_{0.05(2),52})(0.1165) \\ &= 0.80 \pm (2.007)(0.1165) \\ &= 0.80 \text{ ml/°C} \pm 0.23 \text{ ml/°C}; \end{aligned}$$

and the lower and upper 95% confidence limits for $\beta_1 - \beta_2$ are $L_1 = 0.57$ ml/°C and $L_2 = 1.03$ ml/°C, respectively.

If H_0: $\beta_1 = \beta_2$ is rejected (as in Example 18.1), we may wish to calculate the point where the two lines intersect. The intersection is at

$$X_I = \frac{a_2 - a_1}{b_1 - b_2}, \tag{18.6}$$

at which the value of $\hat{Y}$ may be computed either as

$$\hat{Y}_I = a_1 + b_1 X_I \tag{18.7}$$

or

$$\hat{Y}_I = a_2 + b_2 X_I. \tag{18.8}$$

For the two lines in Example 18.1, the point of intersection is at

$$X_I = \frac{24.91 - 10.57}{2.97 - 2.17} = 17.92°\text{C}$$

and

$$\hat{Y}_I = 10.57 + (2.97)(17.92) = 63.79 \text{ ml.}$$

(Figure 18.2 illustrates this intersection.)

If H_0: $\beta_1 = \beta_2$ is not rejected (as will be shown in Example 18.2), then an estimate of the population regression coefficient, β, underlying both b_1 and b_2 is called the *common* (or *weighted*) *regression coefficient*:

$$b_c = \frac{(\sum xy)_1 + (\sum xy)_2}{(\sum x^2)_1 + (\sum x^2)_2} \tag{18.9}$$

or equivalently (but with more chance of rounding error):

$$b_c = \frac{(\sum x^2)_1 b_1 + (\sum x^2)_2 b_2}{(\sum x^2)_1 + (\sum x^2)_2}. \tag{18.10}$$

Equation 18.1 is a special case of

$$t = \frac{|b_1 - b_2| - \beta_0}{s_{b_1 - b_2}}, \tag{18.11}$$

namely when $\beta_0 = 0$. By using Equation 18.11 we may test the hypothesis that the difference between two population regression coefficients is a specified magnitude; that is, H_0: $\beta_1 - \beta_2 = \beta_0$ may be tested against H_A: $\beta_1 - \beta_2 \neq \beta_0$.

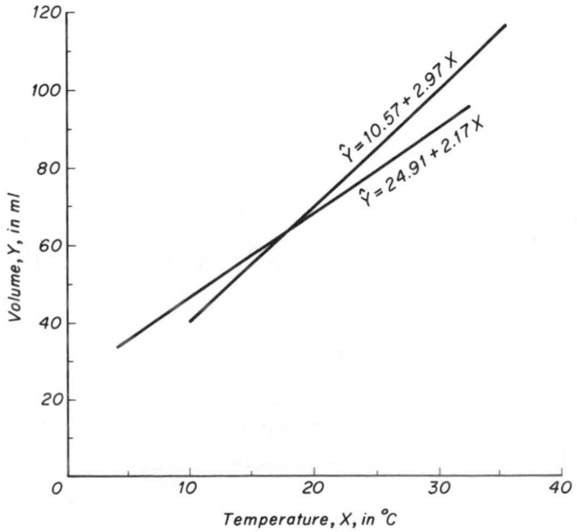

Figure 18.2 The two regression lines of Example 18.1. The two slopes are concluded significantly different and the two lines are found to intersect at X_I = 17.92°C and $\hat{Y}_I$ = 63.79 ml.

Example 18.1

Testing for difference between two population regression coefficients.

H_0: $\beta_1 = \beta_2$
H_A: $\beta_1 \neq \beta_2$

For Sample 1:

$\sum x^2 = 1470.8712$
$\sum xy = 4363.1627$
$\sum y^2 = 13299.5296$
$n = 26$

$$b = \frac{4363.1627}{1470.8712} = 2.97$$

$$\text{residual SS} = 13299.5296 - \frac{(4363.1627)^2}{1470.8712} = 356.7317$$

residual DF $= 26 - 2 = 24$

For Sample 2:

$\sum x^2 = 2272.4750$
$\sum xy = 4928.8100$
$\sum y^2 = 10964.0947$
$n = 30$

$$b = \frac{4928.8100}{2272.4750} = 2.17$$

$$\text{residual SS} = 10964.0947 - \frac{(4928.8100)^2}{2272.4750} = 273.9142$$

residual DF $= 30 - 2 = 28$

$$(s^2_{Y \cdot X})_p = \frac{356.7317 + 273.9142}{24 + 28} = 12.1278$$

$$s_{b_1 - b_2} = \sqrt{\frac{12.1278}{1470.8712} + \frac{12.1278}{2272.4750}} = 0.1165$$

$$t = \frac{2.97 - 2.17}{0.1165} = 6.867$$

$\nu = 24 + 28 = 52$
Reject H_0 if $|t| \geq t_{\alpha(2),\nu}$
$t_{0.05(2),52} = 2.007$
Reject H_0.
$P < 0.001$

One-tailed testing is also possible, asking whether one population regression coefficient is greater than the other. If we test H_0: $\beta_1 \geq \beta_2$ and H_A: $\beta_1 < \beta_2$, or H_0: $\beta_1 - \beta_2 \geq \beta_0$ vs. H_A: $\beta_1 - \beta_2 < \beta_0$, then H_0 is rejected if $t \leq -t_{\alpha(1),\nu}$; if we test H_0: $\beta_1 \leq \beta_2$ and H_A: $\beta_1 > \beta_2$, or H_0: $\beta_1 - \beta_2 \leq \beta_0$ vs. H_A: $\beta_1 - \beta_2 > \beta_0$, then we reject H_0 if $t \geq t_{\alpha(1),\nu}$. In either case, t is computed by Equation 18.1.

An alternative method of testing H_0: $\beta_1 = \beta_2$ is by the analysis of covariance procedure of Section 18.4. However, the preceding t test generally involves less computational effort.

18.2 COMPARING TWO ELEVATIONS

If H_0: $\beta_1 = \beta_2$ is rejected, we may assume that two different populations have been sampled. However, if two population regression lines are not concluded to have different slopes (i.e., H_0: $\beta_1 = \beta_2$ is not rejected), then the two lines are assumed to be parallel. In the latter case, we often wish to determine whether the two population regressions have the same elevation (i.e., the same vertical position on a graph) and thus coincide.

To test the null hypothesis that the elevations of the two population regression lines are the same, we define the following quantities for use in a t test:

Sum of squares of X for "common regression"

$$= A_c = (\sum x^2)_1 + (\sum x^2)_2, \tag{18.12}$$

Sum of crossproducts for "common regression"

$$= B_c = (\sum xy)_1 + (\sum xy)_2, \tag{18.13}$$

Sum of squares of Y for "common regression"

$$= C_c = (\sum y^2)_1 + (\sum y^2)_2, \tag{18.14}$$

Residual SS for "common regression"

$$= \mathrm{SS}_c = C_c - \frac{B_c^2}{A_c}, \tag{18.15}$$

Residual DF for "common regression"

$$= \mathrm{DF}_c = n_1 + n_2 - 3, \tag{18.16}$$

and

Residual MS for "common regression"

$$= (s_{Y\cdot X}^2)_c = \frac{\mathrm{SS}_c}{\mathrm{DF}_c}. \tag{18.17}$$

Then, the appropriate test statistic is

$$t = \frac{(\bar{Y}_1 - \bar{Y}_2) - b_c(\bar{X}_1 - \bar{X}_2)}{\sqrt{(s_{Y\cdot X}^2)_c\left[\frac{1}{n_1} + \frac{1}{n_2} + \frac{(\bar{X}_1 - \bar{X}_2)^2}{A_c}\right]}} \tag{18.18}$$

and the appropriate critical value of t is that for $\nu = \mathrm{DF}_c$. Example 18.2 and Fig. 18.3 consider the regression of human systolic blood pressure on age for men over 40 years old. A regression was fitted for data for men in each of two different occupations. The two-tailed null hypothesis is that in the two sampled populations the regression elevations are the same. This also says that blood pressure is the same in both groups, after accounting for the effect of age. In the example, H_0 is rejected, so we conclude that men in these two occupations do not have the same blood pressure. As an alternative to this t-testing procedure, the analysis of covariance of Section 18.4 may be used to test this hypothesis, but it generally involves more computational effort.

Example 18.2

Testing for difference between two population regression coefficients and elevations.

For Sample 1:	For Sample 2:
$n = 13$	$n = 15$
$\bar{X} = 54.65$ yr	$\bar{X} = 56.93$ yr
$\bar{Y} = 170.23$ mm Hg	$\bar{Y} = 162.93$ mm Hg
$\sum x^2 = 1012.1923$	$\sum x^2 = 1659.4333$
$\sum xy = 1585.3385$	$\sum xy = 2475.4333$
$\sum y^2 = 2618.3077$	$\sum y^2 = 3848.9333$
$b = 1.57$ mm Hg/yr	$b = 1.49$ mm Hg/yr
$a = 84.6$ mm Hg	$a = 78.0$ mm Hg
residual SS $= 136.2230$	residual SS $= 156.2449$
residual DF $= 11$	residual DF $= 13$

H_0: $\beta_1 = \beta_2$

H_A: $\beta_1 \neq \beta_2$

$$(s^2_{Y \cdot X})_p = \frac{136.2230 + 156.2449}{11 + 13} = 12.1862$$
$$\nu = 11 + 13 = 24$$
$$s_{b_1 - b_2} = 0.1392$$
$$t = \frac{1.57 - 1.49}{0.1392} = 0.575$$
$t_{0.05(2),24} = 2.064$; do not reject H_0.
$P > 0.50$

H_0: The two population regression lines have the same elevation.
H_A: The two population regression lines do not have the same elevation.

$$A_c = 1012.1923 + 1659.4333 = 2671.6256$$
$$B_c = 1585.3385 + 2475.4333 = 4060.7718$$
$$C_c = 2618.3077 + 3848.9333 = 6467.2410$$
$$b_c = \frac{4060.7718}{2671.6256} = 1.520 \text{ mm Hg/yr}$$
$$\text{SS}_c = 6467.2410 - \frac{(4060.7718)^2}{2671.6256} = 295.0185$$
$$\text{DF}_c = 13 + 15 - 3 = 25$$
$$(s^2_{Y \cdot X})_c = \frac{295.0185}{25} = 11.8007$$
$$t = \frac{(170.23 - 162.93) - 1.520(54.65 - 56.93)}{\sqrt{11.8007\left[\frac{1}{13} + \frac{1}{15} + \frac{(54.65 - 56.93)^2}{2671.6256}\right]}}$$
$$= \frac{10.77}{1.3105} = 8.218$$
$t_{0.05(2),25} = 2.060$; reject H_0.
$P < 0.001$

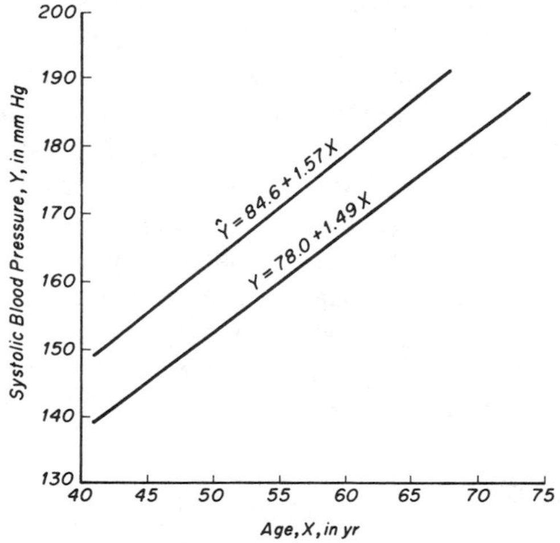

Figure 18.3 The two regression lines of Example 18.2.

If it is concluded that two population regressions do not have different slopes but do have different elevations, then the slopes computed from the two samples are both estimates of the common population regression coefficient, and the two regression equations should be written as

$$\hat{Y}_i = a_1 + b_c X_i \tag{18.19}$$

and

$$\hat{Y}_i = a_2 + b_c X_i, \tag{18.20}$$

which for the two lines in Example 18.2 and Fig. 18.3 would be

$$\hat{Y}_i = 84.6 + 1.52X_i$$

and

$$\hat{Y}_i = 78.0 + 1.52X_i.$$

If it is concluded that two population regressions have neither different slopes nor different elevations, then both sample regressions estimate the same population regression, and we must use a common regression coefficient, b_c, as well as a common Y intercept,

$$a_c = \bar{Y}_p - b_c\bar{X}_p, \tag{18.21}$$

where the pooled sample means of the two variables may be obtained as

$$\bar{X}_p = \frac{n_1\bar{X}_1 + n_2\bar{X}_2}{n_1 + n_2} \tag{18.22}$$

and

$$\bar{Y}_p = \frac{n_1\bar{Y}_1 + n_2\bar{Y}_2}{n_1 + n_2}. \tag{18.23}$$

Thus, when two samples have been concluded to estimate the same population regression, the regression equation would be

$$\hat{Y}_i = a_c + b_cX_i. \tag{18.24}$$

The above test of difference between elevations may be considered the same as asking whether the two population Y intercepts are different. However, it is not advisable to test $H_0: \alpha_1 = \alpha_2$ using sample estimates a_1 and a_2; the latter test would consider a point on each line that may lie far from the observed range of X's. There are many regressions for which the Y intercept has no importance beyond helping to define the line and in fact may be a sample statistic prone to misleading interpretation. In Fig. 18.3, for example, discussion of the Y intercepts (and testing hypotheses about them) would require a risky extrapolation of the regression lines far below the range of X for which data were obtained. This would assume that the linear relationship that was determined also holds between $X = 40$ yr and $X = 0$, a gravely incorrect assumption in the present case. Additionally, since the Y intercepts are so far from the mean values of X, their standard errors would be very large, and a test of $H_0: \alpha_1 = \alpha_2$ would lack statistical power. For such a test, Equations 18.25 and 18.26 are a special case of Equations 18.27 and 18.28 which follow:

$$t = \frac{a_1 - a_2}{s_{a_1 - a_2}}, \tag{18.25}$$

where

$$s_{a_1 - a_2} = \sqrt{(s^2_{Y\cdot X})_p\left[\frac{1}{n_1} + \frac{1}{n_2} + \frac{\bar{X}_1^2}{(\sum x^2)_1} + \frac{\bar{X}_2^2}{(\sum x^2)_2}\right]}. \tag{18.26}$$

We may also test one-tailed hypotheses about elevations. For data such as those in Example 18.2 and Fig. 18.3, it might have been the case that one occupation was considered to be more emotionally stressful, and the interest was to determine whether men in that occupation had higher blood pressure than men in the second occupation.

18.3 COMPARING POINTS ON TWO REGRESSION LINES

If the slopes of two regression lines and the elevations of the two lines have not been concluded to be different, then the two lines are estimates of the same population regression line. If the slopes of two lines are not concluded to be different, but their elevations are declared different, then the lines are assumed to be parallel, and for a given X_i the corresponding $\hat{Y}_i$ on one line is different from that on the other line.

If the slopes of two population regression lines are concluded different, then the lines are intersecting, rather than parallel. In such cases we may wish to test whether a $\hat{Y}$ on one line is the same as the $\hat{Y}$ on the second line at a particular X. For a two-tailed test we can state the null hypothesis as H_0: $(\mu_{\hat{Y},X})_1 = (\mu_{\hat{Y},X})_2$ and the alternate as H_A: $(\mu_{\hat{Y},X})_1 \neq (\mu_{\hat{Y},X})_2$. The test statistic is

$$t = \frac{\hat{Y}_1 - \hat{Y}_2}{s_{\hat{Y}_1 - \hat{Y}_2}}, \tag{18.27}$$

where

$$s_{\hat{Y}_1 - \hat{Y}_2} = \sqrt{(s^2_{Y\cdot X})_p \left[\frac{1}{n_1} + \frac{1}{n_2} + \frac{(X - \bar{X}_1)^2}{(\sum x^2)_1} + \frac{(X - \bar{X}_2)^2}{(\sum x^2)_2}\right]} \tag{18.28}$$

and the degrees of freedom are the pooled degrees of freedom,

$$\nu = n_1 + n_2 - 4. \tag{18.29}$$

Such a test is demonstrated in Example 18.3. One-tailed testing is also possible. The test should be applied with caution, however, as it assumes that each of the two predicted $\hat{Y}$'s has associated with it the same variance. Therefore, the test works best when the two lines have the same $\bar{X}$, the same $\sum X^2$, and the same n.

Example 18.3

Testing for difference between points on the two nonparallel regression lines of Example 18.1 and Fig. 18.2. We are testing whether the volumes (Y) are different in the two groups at $X = 12°\text{C}$.

H_0: $(\mu_{\hat{Y},12})_1 = (\mu_{\hat{Y},12})_2$
H_A: $(\mu_{\hat{Y},12})_1 \neq (\mu_{\hat{Y},12})_2$

Beyond the statistics given in Example 18.1 we need to know the following:

$$a_1 = 10.57 \text{ ml} \quad \text{and} \quad a_2 = 24.91 \text{ ml};$$
$$\bar{X}_1 = 22.93°\text{C} \quad \text{and} \quad \bar{X}_2 = 18.95°\text{C}.$$

We then compute:

$$\hat{Y}_1 = 10.57 + (2.97)(12) = 46.21 \text{ ml}$$
$$\hat{Y}_2 = 24.91 + (2.17)(12) = 50.95 \text{ ml}$$
$$s_{\hat{Y}_1 - \hat{Y}_2} = \sqrt{12.1278\left[\frac{1}{26} + \frac{1}{30} + \frac{(12 - 22.93)^2}{1470.8712} + \frac{(12 - 18.95)^2}{2272.4750}\right]}$$
$$= \sqrt{2.1135}$$
$$= 1.45 \text{ ml}$$
$$t = \frac{46.21 - 50.95}{1.45} = -3.269$$
$$\nu = 26 + 30 - 4 = 52$$
$$t_{0.05(2),52} = 2.007$$

Since $|t| > t_{0.05(2),52}$, reject H_0.

$0.001 < P < 0.002$

18.4 COMPARING MORE THAN TWO SLOPES

One can test H_0: $\beta_1 = \beta_2 = \cdots = \beta_k$, with the alternate hypothesis being that the k regression lines were not derived from samples estimating populations among which the slopes (β) were all equal. This calls for the utilization of a procedure known as *analysis of covariance.*

Analysis of covariance encompasses a large body of statistical methodology too extensive to be covered in this book, but the following procedures suffice to test for the homogeneity (i.e., equality) of regression coefficients. Just as an analysis of variance for H_0: $\mu_1 = \mu_2 = \cdots = \mu_k$ assumes that $\sigma_1^2 = \sigma_2^2 = \cdots = \sigma_k^2$, the testing of H_0: $\beta_1 = \beta_2 = \cdots = \beta_k$ proceeds with the assumption that $(\sigma_{Y\cdot X}^2)_1 = (\sigma_{Y\cdot X}^2)_2 = \cdots = (\sigma_{Y\cdot X}^2)_k$. Heterogeneity of the k residual mean squares can be tested by means of Bartlett's test (Section 11.1), but this generally is not done for the same reasons that the test is not often employed as a prelude to analysis of variance procedures (Section 11.1).

The basic calculations necessary to compare k regression lines require quantities already computed: $\sum x^2$, $\sum xy$, $\sum y^2$ (i.e., total SS), and the residual SS and DF for each computed line (Table 18.1). The values of the k residual sums of squares may then be summed, yielding what we shall call the "pooled" residual sum of squares, SS_p; and the sum of the k residual degrees of freedom is the "pooled" residual degrees of freedom, DF_p. The values of $\sum x^2$, $\sum xy$, and $\sum y^2$ for the regressions may also be summed, and from these sums a residual sum of squares may be calculated. This quantity will be termed the "common" residual sum of squares, SS_c.

TABLE 18.1 CALCULATIONS FOR TESTING FOR SIGNIFICANT DIFFERENCES BETWEEN SLOPES AND ELEVATIONS OF k SIMPLE LINEAR REGRESSION LINES

	$\sum x^2$	$\sum xy$	$\sum y^2$	Residual SS	Residual DF
Regression 1	A_1	B_1	C_1	$SS_1 = C_1 - \frac{B_1^2}{A_1}$	$DF_1 = n_1 - 2$
Regression 2	A_2	B_2	C_2	$SS_2 = C_2 - \frac{B_2^2}{A_2}$	$DF_2 = n_2 - 2$
⋮	⋮	⋮	⋮	⋮	⋮
Regression k	A_k	B_k	C_k	$SS_k = C_k - \frac{B_k^2}{A_k}$	$DF_k = n_k - 2$
"Pooled" regression				$SS_p = \sum_{i=1}^{k} SS_i$	$DF_p = \sum_{i=1}^{k} (n_i - 2) = \sum_{i=1}^{k} n - 2k$
"Common" regression	$A_c = \sum_{i=1}^{k} A_i$	$B_c = \sum_{i=1}^{k} B_i$	$C_c = \sum_{i=1}^{k} C_i$	$SS_c = C_c - \frac{B_c^2}{A_c}$	$DF_c = \sum_{i=1}^{k} n_i - k - 1$
"Total" regression*	A_t	B_t	C_t	$SS_t = C_t - \frac{B_t^2}{A_t}$	$DF_t = \sum_{i=1}^{k} n_i - 2$

*See Section 18.4 for explanation.

To test $H_0: \beta_1 = \beta_2 = \cdots = \beta_k$, one may calculate

$$F = \frac{\left(\frac{SS_c - SS_p}{k - 1}\right)}{\frac{SS_p}{DF_p}}, \tag{18.30}$$

a statistic with numerator and denominator degrees of freedom of $k - 1$ and DF_p, respectively.* Example 18.4 demonstrates this testing procedure for three regression lines calculated from three sets of data (i.e., $k = 3$).

Example 18.4

Testing for difference between three regression functions.*

	$\sum x^2$	$\sum xy$	$\sum y^2$	n	b	*Residual* SS	*Residual* DF
Regression 1	*430.14*	*648.97*	*1065.34*	*24*	1.51	86.21	22
Regression 2	*448.65*	*694.36*	*1184.12*	*29*	1.55	109.48	27
Regression 3	*502.31*	*714.33*	*1186.52*	*30*	1.42	170.68	28
"Pooled" regression						366.37	77
"Common" regression	1381.10	2057.66	3435.98		1.49	370.33	79
"Total" regression	*2144.06*	*3196.78*	*5193.48*	*83*		427.10	81

*The italicized values are those computed from the raw data; all other values are derived from them.

To test for differences between slopes: $H_0: \beta_1 = \beta_2 = \beta_3$; H_A: All three β's are not equal.

$$F = \frac{\frac{370.33 - 366.37}{3 - 1}}{\frac{366.37}{77}} = 0.42$$

Since $F_{0.05(1),2,77} \cong 3.13$, do not reject H_0.

$$b_c = \frac{2057.66}{1381.10} = 1.49$$

$P > 0.25$

To test for differences between elevations:

H_0: The three population regression lines have the same elevation.

H_A: The three lines do not have the same elevation.

$$F = \frac{\frac{427.10 - 370.33}{3 - 1}}{\frac{370.33}{79}} = 6.06$$

Since $F_{0.05(1),2,79} \cong 3.13$, reject H_0.

$0.0025 < P < 0.005$

If $H_0: \beta_1 = \beta_2 = \cdots = \beta_k$ is rejected, then one may wish to employ a multiple comparison test to determine which of the k population slopes differ from which others. This is analogous to the multiple comparison testing employed after rejecting H_0: $\mu_1 = \mu_2 = \cdots = \mu_k$ (Chapter 12), and it is presented in Section 18.6.

If $H_0: \beta_1 = \beta_2 = \cdots \beta_k$ is not rejected, then the common regression coefficient,

*The quantity $SS_c - SS_p$ is an expression of variability between the k regression coefficients; hence, it is associated with $k - 1$ degrees of freedom.

b_c, may be used as an estimate of the β underlying all k samples:

$$b_c = \frac{\sum_{i=1}^{k} (\sum xy)_i}{\sum_{i=1}^{k} (\sum x^2)_i}. \tag{18.31}$$

For Example 18.4, this is $b_c = 2057.66/1381.10 = 1.49$.

18.5 COMPARING MORE THAN TWO ELEVATIONS

Consider the case where it has been concluded that all k population slopes underlying our k samples of data are equal (i.e., H_0: $\beta_1 = \beta_2 = \cdots = \beta_k$ is accepted). In this situation, it is reasonable to ask whether all k population regressions are, in fact, identical, i.e., whether they have equal elevations as well as slopes, and thus the lines all coincide.

The null hypothesis of equality of elevations may be tested by a continuation of the analysis of covariance considerations outlined in Section 18.4. We can combine the data from all k samples, and from this compute $\sum x^2$, $\sum xy$, $\sum y^2$, a residual sum of squares, and residual degrees of freedom; the latter will be called the "total" residual sum of squares, SS_t, and "total" residual degrees of freedom, DF_t. (See Table 18.1.) The null hypothesis is tested with the test statistic

$$F = \frac{\dfrac{\text{SS}_t - \text{SS}_c}{k - 1}}{\dfrac{\text{SS}_c}{\text{DF}_c}}, \tag{18.32}$$

with $k - 1$ and DF_c degrees of freedom. An example of this procedure is offered in Example 18.4.

If the null hypothesis is rejected, we can then employ multiple comparisons to determine the location of significant differences among the elevations, as described in Section 18.6. If it is not rejected, then all k sample regressions are estimates of the same population regression, and the best estimate of that underlying population regression is given by Equation 18.24 using Equation 18.21.

Power and Sample Size in Comparing Regressions. In Section 17.8 it was explained that the procedure for consideration of power in correlation analysis (Section 19.4) could be used to estimate power and sample size in a regression analysis. Section 19.6 presents power and sample size estimation when testing for difference between two correlation coefficients. Unfortunately, utilization of that procedure for the case of comparing two regression coefficients is not valid—unless one has the rare case of $(\sum x^2)_1 = (\sum x^2)_2$ and $(\sum y^2)_1 = (\sum y^2)_2$ (Cohen, 1977: 110).

18.6 MULTIPLE COMPARISONS AMONG SLOPES

If an analysis of covariance concludes that k population slopes are not all equal, we may employ a multiple comparison procedure (Chapter 12) to determine which β's are different from which others. For example, the Tukey test (Section 12.1) may be employed to test for differences between each pair of β values, by H_0: $\beta_A = \beta_B$ and H_A: $\beta_A \neq \beta_B$, where A and B can represent any two of the k regression lines.

The test statistic is

$$q = \frac{b_B - b_A}{\text{SE}}. \tag{18.33}$$

If $\sum x^2$ is the same for lines A and B, then the standard error to be used is

$$\text{SE} = \sqrt{\frac{(s^2_{Y \cdot X})_p}{\sum x^2}}. \tag{18.34}$$

If $\sum x^2$ is different for lines A and B, then use

$$\text{SE} = \sqrt{\frac{(s^2_{Y \cdot X})_p}{2}\left[\frac{1}{(\sum x^2)_A} + \frac{1}{(\sum x^2)_B}\right]}. \tag{18.35}$$

The degrees of freedom for determining the critical value of q are the pooled residual DF (i.e., DF_p in Table 18.1). Although it is not mandatory to have first performed the analysis of covariance before applying the multiple range test, such a procedure is commonly followed.

The confidence interval for the difference between the slopes of population regressions A and B is

$$(b_B - b_A) \pm (q_{\alpha,\nu,p}) \quad (\text{SE}), \tag{18.36}$$

where ν is the pooled residual DF (i.e., DF_p in Table 18.1).

If one of several regression lines is considered to be a control with which each of the other lines is to be compared, then the procedures of Dunnett's test (introduced in Section 12.4) are appropriate. Here,

$$\text{SE} = \sqrt{\frac{2(s^2_{Y \cdot X})_p}{\sum x^2}} \tag{18.37}$$

if $\sum x^2$ is the same for line A (the control line) and line B, and

$$\text{SE} = \sqrt{(s^2_{Y \cdot X})_p\left[\frac{1}{(\sum x^2)_A} + \frac{1}{(\sum x^2)_B}\right]} \tag{18.38}$$

if it is not. Either two-tailed or one-tailed hypotheses may be thus tested.

The $1 - \alpha$ confidence interval for the difference between the slopes of the control line (line A) and another line (line B) are

$$(b_B - b_A) \pm (q'_{\alpha(2),\nu,p})(\text{SE}). \tag{18.39}$$

To apply Scheffé's procedure (Section 12.5), we would calculate SE as Equations 18.37 or 18.38.

18.7 MULTIPLE COMPARISONS AMONG ELEVATIONS

If H_0: $\beta_1 = \beta_2 = \cdots = \beta_k$ has been accepted and the null hypothesis of all k elevations being equal has been rejected, then one may apply multiple comparison procedures (similar to those of Chapter 12) to determine between which elevations differences occur in the populations sampled. The test statistic for the Tukey test (introduced in Section 12.1) is

$$q = \frac{|(\bar{Y}_A - \bar{Y}_B) - b_c(\bar{X}_A - \bar{X}_B)|}{\text{SE}}, \tag{18.40}$$

with DF_p degrees of freedom (see Table 17.1), where the subscripts A and B refer to the two lines the elevations of which are being compared, b_c is from Equation 18.31, and

$$\text{SE} = \sqrt{\frac{(s^2_{Y \cdot X})_c}{2}\left[\frac{1}{n_A} + \frac{1}{n_B} + \frac{(\bar{X}_A - \bar{X}_B)^2}{(\sum x^2)_A + (\sum x^2)_B}\right]} \tag{18.41}$$

If we use Dunnett's test to compare the elevation of a control regression line (let's call it line A) with that of another line (line B),

$$\text{SE} = \sqrt{(s^2_{Y \cdot X})_c\left[\frac{1}{n_A} + \frac{1}{n_B} + \frac{(\bar{X}_A - \bar{X}_B)^2}{(\sum x^2)_A + (\sum x^2)_B}\right]}. \tag{18.42}$$

Equation (18.42) would also be employed if Scheffé's test were being performed on elevations.

18.8 AN OVERALL TEST FOR COINCIDENTAL REGRESSIONS

It is possible to test the null hypothesis that k population regressions are coincident, i.e., that the β's are all identical *and* the α's are all identical. Here we would calculate

$$F = \frac{\dfrac{\text{SS}_t - \text{SS}_p}{2(k-1)}}{\dfrac{\text{SS}_p}{\text{DF}_p}} \tag{18.43}$$

with $2(k-1)$ and DF_p degrees of freedom. If this F is not significant, then all k sample regressions are assumed to estimate the same population regression, and the best estimate of that population regression is that given by Equation 18.24.

Some statistical workers prefer this test to those of the preceding sections in this chapter. However, if the null hypothesis is rejected, it is still necessary to employ the procedures of the previous sections if we wish to determine whether the differences within the regression are due to differences among slopes or elevations.

EXERCISES

18.1. Given:

For sample 1: $n = 28$, $\sum x^2 = 142.35$, $\sum xy = 69.47$, $\sum y^2 = 108.77$, $\bar{X} = 14.7$, $\bar{Y} = 32.0$.

For sample 2: $n = 30$, $\sum x^2 = 181.32$, $\sum xy = 97.40$, $\sum y^2 = 153.59$, $\bar{X} = 15.8$, $\bar{Y} = 27.4$.

(a) Test H_0: $\beta_1 = \beta_2$ vs. H_0: $\beta_1 \neq \beta_2$.

(b) If H_0 in part (a) is not rejected, test H_0: The elevations of the two population regressions are the same, vs. H_A: The two elevations are not the same.

18.2. Given:

For sample 1: $n = 33$, $\sum x^2 = 744.32$, $\sum xy = 2341.37$, $\sum y^2 = 7498.91$.

For sample 2: $n = 34$, $\sum x^2 = 973.14$, $\sum xy = 3147.68$, $\sum y^2 = 10366.97$.

For sample 3: $n = 29$, $\sum x^2 = 664.42$, $\sum xy = 2047.73$, $\sum y^2 = 6503.32$.

For the total of all 3 samples: $n = 96$, $\sum x^2 = 3146.72$, $\sum xy = 7938.25$, $\sum y^2 = 20599.33$.

(a) Test H_0: $\beta_1 = \beta_2 = \beta_3$, vs. H_0: All three β's are not equal.

(b) If H_0: in part (a) is not rejected, test H_0: The three population regression lines have the same elevation, vs. H_A: The lines do not have the same elevation.

19 *Simple Linear Correlation*

Chapter 17 introduced simple linear regression, the linear dependence of one variable (termed the dependent variable, Y) on a second variable (called the independent variable, X). In simple linear correlation, we also consider the linear relationship between two variables, but neither is assumed to be functionally dependent upon the other. Recall that the adjective "simple" refers to the fact that only two variables are considered simultaneously.

In regression we assume that for each X the Y values have come at random from a normal population. However, in correlation, not only are the Y's at each X assumed to be normal, but also the X values at each Y are assumed to have come at random from a normal population. This situation is referred to as sampling from a "bivariate normal distribution." The effect of deviations from the assumption of bivariate normality appears unimportant when there is, in fact, only slight correlation in the population; but if there is substantial population correlation, then there may be a marked adverse effect of such nonnormality, this effect not being diminished by increasing sample size (Norris and Hjelm, 1961). If data are to be transformed in correlation analysis, the considerations of Section 17.10 are applicable to X as well as to Y.

19.1 THE CORRELATION COEFFICIENT

Some authors refer to the two variables in a simple correlation analysis as X_1 and X_2. We shall here employ the more common designation of X and Y, which does not, however, imply dependence of Y on X as it does in regression. The *correlation coeffi-*

cient (sometimes called the "simple correlation coefficient" or the "product-moment correlation coefficient"*) is calculated as

$$r = \frac{\sum xy}{\sqrt{\sum x^2 \sum y^2}}. \tag{19.1}$$

Although the denominator of Equation 19.1 is always positive, the numerator may be positive, zero, or negative, thus enabling r to be either positive zero, or negative, respectively. A positive correlation implies that for an increase in the value of one of the variables, the other variable also increases in value; a negative correlation indicates that an increase in value of one of the variables is accompanied by a decrease in value of the other variable. If $\sum xy = 0$, then $r = 0$, and one has a zero correlation, denoting that there is no linear association between the magnitudes of the two variables; that is, a change in magnitude of one does not imply a change in magnitude of the other. Figure 19.1 presents these considerations graphically.

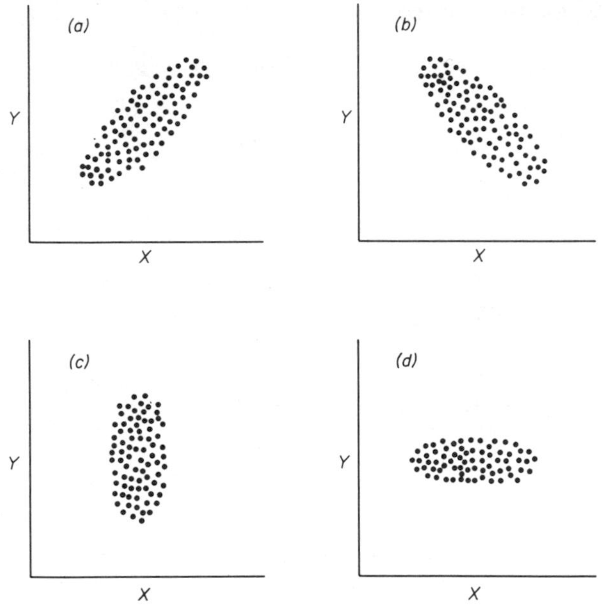

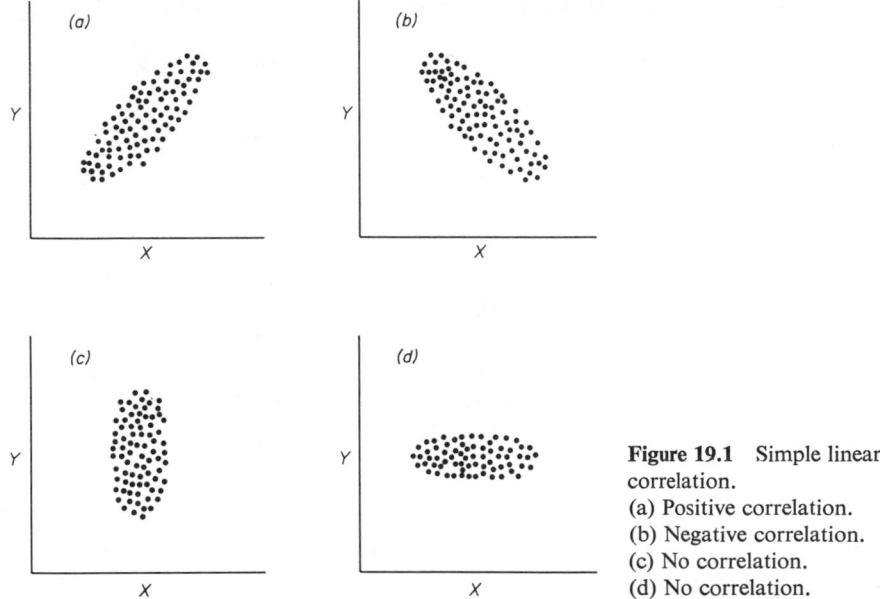

Figure 19.1 Simple linear correlation.
(a) Positive correlation.
(b) Negative correlation.
(c) No correlation.
(d) No correlation.

Also important is the fact that the absolute value of the numerator of Equation 19.1 can never be larger than the denominator. Thus, r can never be greater than 1.0 nor less than -1.0. Inspection of this equation further will reveal also that r has no units of measurement, for the units of both X and Y appear in both the numerator and denominator and thus cancel out arithmetically. A regression coefficient, b, may lie in the range of $-\infty \leq b \leq \infty$, and it expresses the magnitude of a change in Y associated with a unit change in X. But a correlation coefficient is unitless and $-1 \leq r \leq 1$.

*It is also referred to as the "Pearson product-moment correlation coefficient" because of the direct computation of the coefficient, and other pioneering work, by Karl Pearson (1857–1936) in 1896 (Pearson, 1920). The symbol, r, for the correlation coefficient can be traced to Sir Francis Galton (1822–1911) in 1877, who spoke of "reversion" in heredity studies (and who later also used r to indicate the slope of a regression line); indeed, in the early history of correlation, correlation coefficients were called "Galton functions" (Pearson, 1920).

Thus, the correlation coefficient is not a measure of quantitative change of one variable with respect to the other, but it is a measure of intensity of association between the two variables.

The coefficient of determination, r^2, was introduced in Section 17.3 as a measure of how much of the total variability in Y is accounted for by regressing Y on X. In a correlation analysis, r^2 (occasionally called the "correlation index") may be calculated most simply by squaring the correlation coefficient, r. It may be described as the amount of variability in one of the variables (either Y or X) accounted for by correlating that variable with the second variable. As in regression analysis, r^2 may be considered to be a measure of the strength of the straight-line relationship. The calculation of r and r^2 is demonstrated in Example 19.1.

Example 19.1

Calculation of the simple correlation coefficient, and testing $H_0: \rho = 0$. The data are wing and tail lengths among birds of a particular species.

Wing length (cm) (X)	*Tail length* (cm) (Y)
10.4	7.4
10.8	7.6
11.1	7.9
10.2	7.2
10.3	7.4
10.2	7.1
10.7	7.4
10.5	7.2
10.8	7.8
11.2	7.7
10.6	7.8
11.4	8.3

$n = 12$

$\sum X = 128.2$ cm $\qquad \sum Y = 90.8$ cm

$\sum X^2 = 1371.32$ cm² $\qquad \sum Y^2 = 688.40$ cm² $\qquad \sum XY = 971.37$ cm²

$\sum x^2 = 1.72$ cm² $\qquad \sum y^2 = 1.35$ cm² $\qquad \sum xy = 1.32$ cm²

$$\text{correlation coefficient} = r = \frac{1.32 \text{ cm}^2}{\sqrt{(1.72 \text{ cm}^2)(1.35 \text{ cm}^2)}} = 0.866$$

$$\text{coefficient of determination} = r^2 = 0.750$$

$$\text{standard error of } r = s_r = \sqrt{\frac{1 - (0.866)^2}{12 - 2}} = 0.158$$

To test $H_0: \rho = 0$, $H_A: \rho \neq 0$,

$r_{0.05(2),10} = 0.576$ (from Table B.16)

Therefore, reject H_0.

$P < 0.001$

Or:

$$t = \frac{r}{s_r} = \frac{0.866}{0.158} = 5.48$$

$t_{0.05(2),10} = 2.228$

Therefore, reject H_0.

$P < 0.001$

Or:

$$F = \frac{1 + |r|}{1 - |r|} = \frac{1.866}{0.134} = 13.93$$

$F_{0.05(2),10,10} = 3.72$

Therefore, reject H_0.

$P < 0.001$

The standard error of the correlation coefficient may be computed as

$$s_r = \sqrt{\frac{1 - r^2}{n - 2}}. \tag{19.2}$$

(The location of the decimal place at which the second significant digit of s_r is located may be noted and r may be expressed rounded off to that decimal place.)

19.2 HYPOTHESES ABOUT THE CORRELATION COEFFICIENT

The correlation coefficient, r, which we calculate from a sample, is an estimate of a population parameter, namely the correlation coefficient in the population that was sampled. This parameter is denoted by ρ, lowercase Greek rho. If we wish to ask whether there is, in fact, a correlation between Y and X in the population, we can test $H_0: \rho = 0$ against $H_A: \rho \neq 0$. We do this, as in Example 19.1 by the familiar Student's t considerations:

$$t = \frac{r}{s_r}, \tag{19.3}$$

where the standard error of r is calculated by Equation 19.2, and the degrees of freedom are $\nu = n - 2$. The null hypothesis is rejected if

$$|t| \geq t_{\alpha(2),\nu}.$$

Alternatively, this two-tailed hypothesis may be tested using

$$F = \frac{1 + |r|}{1 - |r|} \tag{19.4}$$

(Cacoullos, 1965), where the critical value is $F_{\alpha(2),\nu,\nu}$. (See Example 19.1.) Or, critical values of $|r|$ (namely, $r_{\alpha(2),\nu}$) may be read directly from Appendix Table B.16.*

One-tailed hypotheses about the population correlation coefficient may also be tested by the above procedures. For the hypotheses H_0: $\rho \leq 0$ and H_A: $\rho > 0$, compute either t or F (Equations 19.3 or 19.4, respectively) and reject H_0 if r is positive

*Critical values of r are, by rearranging Equation 19.3,

$$r_{\alpha,\nu} = \sqrt{\frac{t^2_{\alpha,\nu}}{t^2_{\alpha,\nu} + \nu}}, \tag{19.5}$$

where α may be either one-tailed or two-tailed, and $\nu = n - 2$. If a regression analysis is performed, rather than a correlation analysis, the probability of rejection of $H_0: \beta = 0$ is identical to the probability of rejecting $H_0: \rho = 0$. Also, r is related to b as

$$r = \frac{s_X}{s_Y} b, \tag{19.6}$$

where s_X and s_Y are the standard deviations of X and Y, respectively.

and either $t \geq t_{\alpha(1),\nu}$, or $F \geq F_{\alpha(1),\nu,\nu}$, or $r \geq r_{\alpha(1),\nu}$. To test H_0: $\rho \geq 0$ vs. H_A: $\rho < 0$, reject H_0 if r was negative and either $|t| \geq t_{\alpha(1),\nu}$, or $F \geq F_{\alpha(1),\nu,\nu}$, or $|r| \geq r_{\alpha(1),\nu}$.

If we wish to test H_0: $\rho = \rho_0$ for any ρ_0 other than zero, however, Equations 19.3 and 19.4 and Table B.16 are not applicable. Only for $\rho_0 = 0$ can r be considered to have come from a distribution approximated by the normal, and if the distribution of r is not normal, then the t and F statistics may not be validly calculated. Fisher (1915, 1921) dealt with this problem when he proposed a transformation enabling r to be converted to a value, called z, which estimates a population parameter, ζ (lowercase Greek zeta), that *is* normally distributed. The transformation* is

$$z = 0.5 \ln \left(\frac{1 + r}{1 - r} \right). \tag{19.7}$$

For values of r between 0 and 1, the corresponding values of Fisher's z will lie between 0 and ∞; and for r's from 0 to -1, the corresponding z's will fall between 0 and $-\infty$. For convenience, we may utilize Table B.17 to avoid having to perform the computation of Equation 19.7 to transform r to z.†

For the hypothesis H_0: $\rho = \rho_0$, then, we calculate a normal deviate, as

$$Z = \frac{z - \zeta_0}{\sigma_z} \tag{19.8}$$

where z is the transform of r, ζ_0 is the transform of the hypothesized coefficient, ρ_0, and the standard error of z is approximated by

$$\sigma_z = \sqrt{\frac{1}{n - 3}}. \tag{19.9}$$

In Example 19.1, $r = 0.866$ was calculated. If we had desired to test H_0: $\rho = 0.750$, we would have proceeded as shown in Example 19.2. Recall that the critical value of a

Example 19.2

Testing H_0: $\rho = \rho_0$, where $\rho_0 \neq 0$.

$r = 0.866$
$n = 12$
H_0: $\rho = 0.750$; H_A: $\rho \neq 0.750$.
$z = 1.3169$
$z_c = z_{0.750} = 0.9730$

$$Z = \frac{z - z_c}{\sqrt{\frac{1}{n-3}}} = \frac{1.3169 - 0.9730}{\sqrt{\frac{1}{9}}} = \frac{0.3439}{0.3333} = 1.0317$$

$Z_{0.05(2)} = t_{0.05(2),\infty} = 1.960$
Therefore, do not reject H_0.
$0.20 < P < 0.50$

*z is also equal to $r + r^3/3 + r^5/5 \ldots$ and is a quantity that mathematicians recognize as the "inverse hyperbolic tangent of r," namely $z = \tanh^{-1} r$.

†Fisher (1958: 205) has noted that there is a slight bias in z. This is approximately correctable by subtracting $r/2(n - 1)$ from z. Unless n is very small, however, this correction will be insignificant and may be ignored.

normal deviate may be obtained readily from the bottom line of the t table (Table B.3), for $Z_{\alpha(2)} = t_{\alpha(2),\infty}$. One-tailed hypotheses, H_0: $\rho \leq \rho_0$ and H_A: $\rho > \rho_0$, or H_0: $\rho \geq \rho_0$ and H_A: $\rho < \rho_0$, may also be tested using Equation 19.8 in which case the critical value of the normal deviate would, of course, be $Z_{\alpha(1)} = t_{\alpha(1),\infty}$.

If the variables in correlation analysis have come from a bivariate normal distribution, as generally may be assumed, then we may invoke the aforementioned procedures, as well as those that follow. Occasionally, only one of the two variables may be assumed to have been obtained randomly from a normal population. It may be possible to employ a data transformation (see Section 17.10) to remedy this situation. If that cannot be done, then the hypothesis H_0: $\rho = 0$ (or its associated one-tailed hypotheses) may be tested, but none of the other testing procedures of this chapter (except for the methods of Section 19.9) is valid. If neither variable came from a normal population, and data transformations do not improve this condition, then we may turn to the procedures of Section 19.9.

19.3 CONFIDENCE INTERVALS FOR THE CORRELATION COEFFICIENT

Fisher's transformation is also used for setting confidence limits on ρ. One converts r to z (as with the aid of Table B.17), and then the $1 - \alpha$ confidence limits may be computed for ζ:

$$z \pm Z_{\alpha(2)}\sigma_z \tag{19.10}$$

or, equivalently,

$$z \pm t_{\alpha(2),\infty}\sigma_z. \tag{19.11}$$

The lower and upper confidence limits, L_1 and L_2, are both z values and may be transformed to r values (Table B.17 being useful for this purpose). Example 19.3 demonstrates this procedure. Note that, although the confidence limits for ζ are symmetrical, the confidence limits for ρ are not.

Example 19.3

Setting confidence limits for a correlation coefficient. This example uses the data of Example 19.1.

$r = 0.866$; therefore, $z = 1.3169$ (from Table B.17).

$$\sigma_z = \sqrt{\frac{1}{n-3}} = 0.333$$

$$\begin{aligned} 95\% \text{ confidence interval for } \zeta &= z \pm Z_{0.05(2)}\sigma_z \\ &= z \pm t_{0.05(2),\infty}\sigma_z \\ &= 1.3169 \pm (1.9600)(0.3333) \\ &= 1.3169 \pm 0.6533 \\ L_1 &= 0.664 \\ L_2 &= 1.970 \end{aligned}$$

For the 95% confidence limits for ρ, transform L_1 and L_2 computed for ζ (using Table B.18): $L_1 = 0.581$; $L_2 = 0.962$.

19.4 POWER AND SAMPLE SIZE IN CORRELATION

If we test $H_0: \rho = 0$ at the α significance level, with a sample size of n, then we may determine the probability of correctly rejecting H_0 when ρ_0 is in fact a specified value other than zero. This is done (Cohen, 1977: 458) by using the Fisher z transformation for the critical value of r and for the sample r (from Table B.17 or Equation 19.7); let us call these two transformed values z_α and z, respectively. Then, the power of the test for $H_0: \rho = 0$ is $1 - \beta(1)$, where $\beta(1)$ is the one-tailed probability of the normal deviate

$$Z_{\beta(1)} = (z - z_\alpha)\sqrt{n-3}, \tag{19.12}$$

as demonstrated in Example 19.4. This procedure may be used for one-tailed as well as two-tailed hypotheses, so α may be either $\alpha(1)$ or $\alpha(2)$, respectively.

Example 19.4

Determination of power of the test of $H_0: \rho = 0$ in Example 19.1.

$$n = 12;\ \nu = 10$$

$$r = 0.866, \text{ so } z = 1.3169$$

$$r_{0.05(2),10} = 0.576, \text{ so } z_{0.05} = 0.6565$$

$$Z_{\beta(1)} = (1.3169 - 0.6565)\sqrt{12-3}$$

$$= 1.98$$

From Table B.2, $P(Z \geq 1.98) = 0.0239 = \beta$. Therefore, the power of the test is $1 - \beta = 0.98$.

If the desired power is stated, then we can ask how large a sample is required to reject $H_0: \rho = 0$ if it is truly false with a specified $\rho_0 \neq 0$. This can be estimated (Cohen, 1977: 458) by calculating

$$n = \left(\frac{Z_{\beta(1)} + Z_\alpha}{\zeta_0}\right)^2 + 3, \tag{19.13}$$

where ζ_0 is the Fisher transformation of the ρ_0 specified, and the significance level, α, can be either one-tailed or two-tailed. This procedure is shown in Example 19.5.

Example 19.5

Determination of required sample size in testing $H_0: \rho = 0$.

We desire to reject $H_0: \rho = 0.99\%$ of the time when $|\rho| \geq 0.5$ and the hypothesis is tested at the 0.05 level of significance. Therefore, $\beta(1) = 0.01$ and (from the last line of Table B.3) $Z_{\beta(1)} = 2.3263$; $\alpha(2) = 0.05$ and $Z_{\alpha(2)} = 1.9600$; and, for $r = 0.5$, $z = 0.5493$.

Then,

$$n = \left(\frac{2.3263 + 1.9600}{0.5493}\right)^2 + 3 = 63.9,$$

so a sample of size at least 64 should be used.

Hypothesizing ρ Other than 0. For the two-tailed hypothesis $H_0: \rho = \rho_0$, where $\rho_0 \neq 0$, the power of the test is determined from

$$Z_{\beta(1)} = (|z - z_0| - z_\alpha)\sqrt{n-3}, \tag{19.14}$$

instead of from Equation 19.12; here, z_0 is the Fisher transformation of ρ_0. Either two-tailed or one-tailed hypotheses may be considered, using $\alpha(2)$ or $\alpha(1)$, respectively. Sample size estimation is from Equation 19.13.

19.5 COMPARING TWO CORRELATION COEFFICIENTS

Hypotheses (either one-tailed or two-tailed) about two correlation coefficients may be tested by the use of

$$Z = \frac{z_1 - z_2}{\sigma_{z_1 - z_2}} \tag{19.15}$$

where

$$\sigma_{z_1 - z_2} = \sqrt{\frac{1}{n_1 - 3} + \frac{1}{n_2 - 3}} \tag{19.16}$$

If $n_1 = n_2$, then Equation 19.16 reduces to

$$\sigma_{z_1 - z_2} = \sqrt{\frac{2}{n - 3}}, \tag{19.17}$$

where n is the size of each sample.

As in Example 19.6, a conclusion that $\rho_1 = \rho_2$ would lead us to say that both our samples came from the same population, or from populations with identical correlation coefficients. In such a case, we may combine the information from the two samples to calculate a better estimate of the single underlying ρ. Let us call this estimate the "common," or "weighted," correlation coefficient. We obtain it by converting

$$z_w = \frac{(n_1 - 3)z_1 + (n_2 - 3)z_2}{(n_1 - 3) + (n_2 - 3)} \tag{19.18}$$

to its corresponding r value, r_w, as shown in Example 19.6. If both samples are of equal size (i.e., $n_1 = n_2$), then the previous equation reduces to

$$z_w = \frac{z_1 + z_2}{2}. \tag{19.19}$$

Example 19.6

Testing the hypothesis $H_0: \rho_1 = \rho_2$.

For a sample of 95 bird wing and tail lengths, a correlation coefficient of 0.84 was calculated. A sample of 98 such measurements from a second bird species yielded a correlation coefficient of 0.78. Let us test for the equality of the two population correlation coefficients.

$H_0: \rho_1 = \rho_2$; $H_A: \rho_1 \neq \rho_2$.

$r_1 = 0.84$ $\quad$ $r_2 = 0.78$

$z_1 = 1.2212$ $\quad$ $z_2 = 1.0454$

$n_1 = 95$ $\quad$ $n_2 = 98$

$$Z = \frac{1.2212 - 1.0454}{\sqrt{\frac{1}{n_1 - 3} + \frac{1}{n_2 - 3}}} = \frac{0.1758}{0.1463} = 1.202$$

$Z_{0.05(2)} = t_{0.05(2),\infty} = 1.960$

Therefore, do not reject H_0.

$0.20 < P < 0.50$

The common correlation coefficient may then be computed as

$$z_w = \frac{(n_1 - 3)z_1 + (n_2 - 3)z_2}{(n_1 - 3) + (n_2 - 3)} = \frac{(92)(1.2212) + (95)(1.0454)}{92 + 95} = 1.1319$$

$r_w = 0.81$

Testing for equality of two correlation coefficients works best if the two sample sizes are equal, or nearly so, or if neither n_1 nor n_2 is small (Rao, 1973: 434).

A null hypothesis such as $H_0: \rho_1 - \rho_2 = \rho_0$, where $\rho_0 \neq 0$, might be tested by substituting $|z_1 - z_2| - z_0$ for the numerator in Equation 19.5, but no utility for such a test is apparent.

19.6 POWER AND SAMPLE SIZE IN COMPARING TWO CORRELATION COEFFICIENTS

The power of the above test for difference between two correlation coefficients is estimated as $1 - \beta$, where β is the one-tailed probability of the normal deviate calculated as

$$Z_{\beta(1)} = \frac{(z_1 - z_2)}{\sigma_{z_1 - z_2}} - Z_\alpha \qquad (19.20)$$

(Cohen, 1977: 459), where α may be either one-tailed or two-tailed and where $Z_{\alpha(1)}$ or $Z_{\alpha(2)}$ is most easily read from the last line of Table B.3. Example 19.7 demonstrates this calculation for the data of Example 19.6.

Example 19.7

Determination of the power of the test of $H_0: \rho_1 = \rho_2$ in Example 19.6.

$$z_1 = 1.2212 \qquad z_2 = 1.0454$$
$$\sigma_{z_1 - z_2} = 0.1463$$
$$Z_\alpha = Z_{0.05(2)} = 1.960$$
$$Z_{\beta(1)} = \frac{1.2212 - 1.0454}{0.1463} - 1.960$$
$$= 1.202 - 1.960$$
$$= -0.76$$

From Table B.2,

$$\beta = P(Z \geq -0.76) = 1 - P(Z \leq -0.76) = 1 - 0.2236 = 0.78$$

Therefore,

$$\text{power} = 1 - \beta = 1 - 0.78 = 0.22.$$

If we state a desired power to detect a specified difference between transformed correlation coefficients, then the sample size required to reject H_0 when testing at the α level of significance is

$$n = 2\left(\frac{Z_\alpha + Z_{\beta(1)}}{z_1 - z_2}\right)^2 + 3 \qquad (19.21)$$

(Cohen, 1977: 459). This is shown in Example 19.8.

Example 19.8

Estimating the sample size necessary for the test of $H_0: \rho_1 = \rho_2$.

Let us say we wish to be 90% confident of detecting a difference, $z_1 - z_2$, as small as 0.5000 when testing $H_0: \rho_1 = \rho_2$ at the 5% significance level. Then $\beta(1) = 0.10$, $\alpha(2) = 0.05$, and

$$n = 2\left(\frac{1.9600 + 1.2816}{0.5000}\right)^2 + 3$$
$$= 45.0$$

So, sample sizes of at least 45 should be used.

The test for significance between correlation coefficients is most powerful for $n_1 = n_2$, and the above estimation is for a sample size of n in both samples. Sometimes the size of one sample is fixed and cannot be manipulated, and we then ask how large the second sample must be to achieve the desired power. If n_1 is fixed, and n is determined by Equation 19.21, then (by considering n to be the harmonic mean of n_1 and n_2),

$$n_2 = \frac{nn_1 + 3n_1 - 6n}{2n_1 - n - 3} \tag{19.22}$$

(Cohen, 1977: 137).*

19.7 COMPARING MORE THAN TWO CORRELATION COEFFICIENTS

If k samples have been obtained and an r has been calculated for each, it is often desirable to determine whether or not all the samples came from populations having identical ρ's. If H_0: $\rho_1 = \rho_2 = \cdots = \rho_k$ is not rejected, then all the samples might be combined and one value of r calculated to estimate the single population ρ. As Example 19.9 shows, the testing of this hypothesis involves the transforming of each r to a z value. One then may calculate

$$\chi^2 = \sum_{i=1}^{k} (n_i - 3)z_i^2 - \frac{\left[\sum_{i=1}^{k} (n_i - 3)z_i\right]^2}{\sum_{i=1}^{k} (n_i - 3)} \tag{19.23}$$

which may be considered to be a chi-square value with $k - 1$ degrees of freedom.

Example 19.9

Testing a three-sample hypothesis concerning correlation coefficients.

Given the following:

$n_1 = 24$ $n_2 = 29$ $n_3 = 32$

$r_1 = 0.52$ $r_2 = 0.56$ $r_3 = 0.87$

To test: H_0: $\rho_1 = \rho_2 = \rho_3$.

H_A: All three population correlation coefficients are not equal.

i	r_i	z_i	z_i^2	n_i	$n_i - 3$	$(n_i - 3)z_i$	$(n_i - 3)z_i^2$
1	0.52	0.5763	0.3321	24	21	12.1023	6.9741
2	0.56	0.6328	0.4004	29	26	16.4528	10.4104
3	0.87	1.3331	1.7772	32	29	38.6599	51.5388
Sums:					76	67.2150	68.9233

$$\chi^2 = \sum (n_i - 3)z_i^2 - \frac{[\sum (n_i - 3)z_i]^2}{\sum (n_i - 3)}$$

$$= 68.9233 - \frac{(67.2150)^2}{76}$$

$$= 9.478$$

*If the denominator in Equation 19.22 is ≤ 0, then we must either increase n_1 or change the desired power, significance level, or detectable difference in order to solve for n_2.

$$\nu = k - 1 = 2$$
$$\chi^2_{0.05,2} = 5.991$$
Therefore, reject H_0.
$$0.005 < P < 0.01$$

If H_0 had not been rejected, it would have been appropriate to calculate the common correlation coefficient:

$$z_w = \frac{\sum (n_i - 3)z_i}{\sum (n_i - 3)} = \frac{67.2150}{76} = 0.884$$
$$r_w = 0.71$$

If H_0 is not rejected, then all k sample correlation coefficients are assumed to estimate a common population ρ. A "common r" (or "weighted mean of r") may be obtained from transforming the weighted mean z value,

$$z_w = \frac{\sum_{i=1}^{k} (n_i - 3)z_i}{\sum_{i=1}^{k} (n_i - 3)} \tag{19.24}$$

to its corresponding r value (let's call it r_w), as is shown in Example 19.9.

Correcting for Bias. Fisher (1958: 205) pointed out that the z transformation is slightly biased, in that each z value will be a little inflated. This minor systematic error is likely to have only negligible effects on our previous considerations, but it might adversely affect the present hypothesis testing since several values of z, and therefore several small errors, are being summed. Thus, it is good practice when dealing with multisample hypotheses of correlation coefficients to first compute r_w (by Equation 19.24) and then for each sample subtract

$$\frac{r_w}{2(n_i - 1)}$$

from z_i prior to using Equation 19.23 (Snedecor and Cochran, 1980: 188). If the null hypothesis of equal population correlation coefficients is not rejected, then a new r_w is calculated. Neither the hypothesis test nor the common correlation coefficient is appreciably changed by correcting for bias in z unless sample sizes are small and/or there are many (instead of only three) samples being compared.

19.8 MULTIPLE COMPARISONS AMONG CORRELATION COEFFICIENTS

If the null hypothesis of the previous section (H_0: $\rho_1 = \rho_2 = \ldots = \rho_k$) is rejected, it is typically of interest to determine which of the k correlation coefficients are different from which others. This can be done, again using Fisher's z transformation (Levy, 1976).

In the fashion of Section 12.1 (where multiple comparisons were made among means), we can test each pair of correlation coefficients, r_B and r_A, by a Tukey-type test, if $n_B = n_A$:

$$q = \frac{z_B - z_A}{\text{SE}}, \tag{19.25}$$

where

$$\text{SE} = \sqrt{\frac{1}{n-3}} \tag{19.26}$$

and n is the size of each sample. If the sizes of the two samples, A and B, are not equal, then we can use

$$\text{SE} = \sqrt{\frac{1}{2}\left(\frac{1}{n_A - 3} + \frac{1}{n_B - 3}\right)}. \tag{19.27}$$

The appropriate critical value for this test is $q_{\alpha,\infty,k}$ (from Table B.5). This test is demonstrated in Example 19.10.

Example 19.10

Tukey-type multiple comparison testing among the three correlation coefficients in Example 19.9.

Samples ranked by correlation coefficient (i)	1	2	3
Ranked correlation coefficients (r_i)	0.52	0.56	0.87
Ranked transformed coefficients (z_i)	0.5763	0.6328	1.3331
Sample size (n_i)	24	29	32

Comparison B vs. A	*Difference* $z_B - z_A$	SE	q	$q_{0.05,\infty,3}$	*Conclusion*
3 vs. 1	1.3331 − 0.5763 = 0.7568	0.203	3.728	3.314	Reject $H_0: \rho_3 = \rho_1$
3 vs. 2	1.3331 − 0.6328 = 0.7003	0.191	3.667	3.314	Reject $H_0: \rho_3 = \rho_2$
2 vs. 1	0.6328 − 0.5763 = 0.0565	0.207	0.273	3.314	Accept $H_0: \rho_2 = \rho_1$

Overall conclusion: $\rho_1 = \rho_2 \neq \rho_3$

In a manner analogous to that in Section 12.2, Newman-Keuls-type multiple range testing could be performed instead of the Tukey-type procedure. This would involve computing q as in Equation 19.25 and using as critical values $q_{\alpha,\infty,p}$, where p is the range of correlation coefficients being compared (just as p is the range of means being compared in Section 12.2). It is typically unnecessary in multiple comparison testing to employ the correction for bias described at the end of Section 19.7.

Comparison of a Control Correlation Coefficient to Each Other Correlation Coefficient. The methods above enable us to compare each correlation coefficient with each other coefficient. If, instead, we desire only to compare each coefficient to one particular coefficient (call it the correlation coefficient of the "control" set of data), then a procedure analogous to the Dunnett test of Section 12.4 may be employed (Huitema, 1974).

Let us designate the control set of data as B, and each other group of data, in turn, as A. Then, we compute

$$q = \frac{z_B - z_A}{\text{SE}}, \tag{19.28}$$

for each A, in the same sequence as described in Section 12.4. The appropriate standard error is

$$\text{SE} = \sqrt{\frac{2}{n-3}} \tag{19.29}$$

if the samples A and B are of the same size, or

$$\text{SE} = \sqrt{\frac{1}{n_A - 3} + \frac{1}{n_B - 3}} \tag{19.30}$$

if $n_A \neq n_B$. The critical value is $q'_{\alpha(1),\infty,p}$ (from Table B.6) or $q'_{\alpha(2),\infty,p}$ (from Table B.7) for the one-tailed or two-tailed test, respectively.

Multiple Contrasts Among Correlation Coefficients. Section 12.5 introduces the concepts and procedures of multiple contrasts among means; these are multiple comparisons involving groups of means. In a similar fashion, multiple contrasts may be examined among correlation coefficients (Marascuilo, 1971: 454–455). We again employ the z-transformation and calculate, for each contrast, the test statistic

$$S = \frac{\left|\sum_i c_i z_i\right|}{\text{SE}}, \tag{19.31}$$

where

$$\text{SE} = \sqrt{\sum_i c_i^2 \sigma_{z_i}^2} \tag{19.32}$$

and c_i is a contrast coefficient, as described in Section 12.5. (For example, if we wished to test the hypothesis H_0: $(\rho_1 + \rho_2)/2 - \rho_3 = 0$, then $c_1 = \frac{1}{2}$, $c_2 = \frac{1}{2}$, and $c_3 = -1$.) The critical value for this test is*

$$S_\alpha = \sqrt{\chi^2_{\alpha,(k-1)}}. \tag{19.34}$$

19.9 RANK CORRELATION

If we have data obtained from a bivariate population that is far from normal, then the correlation procedures discussed thus far are generally inapplicable. Instead, we may operate with the ranks of the measurements for each variable. Two different *rank correlation* methods are commonly encountered, that proposed by Spearman† (1904), and that of Kendall‡ (1962).

Example 19.11 demonstrates Spearman's rank correlation procedure. After each measurement of a variable is ranked, as done in previously described nonparametric

*Since $\chi^2_{\alpha,\nu} = \nu F_{\alpha(1),\nu,\infty}$, it is equivalent to write

$$S_\alpha = \sqrt{(k-1)F_{\alpha(1),(k-1),\infty}} \tag{19.33}$$

but Equation 19.34 is preferable, because it engenders less rounding error in the calculations.

†Charles Edward Spearman (1863–1949), English psychologist and statistician.

‡Maurice George Kendall (1907–1983), English statistician.

Example 19.11

The Spearman rank correlation coefficient, computed for the data of Example 19.1.

X	Rank of X	Y	Rank of Y	d_i	d_i^2
10.4	4	7.4	5	−1	1
10.8	8.5	7.6	7	1.5	2.25
11.1	10	7.9	11	−1	1
10.2	1.5	7.2	2.5	−1	1
10.3	3	7.4	5	−2	4
10.2	1.5	7.1	1	0.5	0.25
10.7	7	7.4	5	2	4
10.5	5	7.2	2.5	2.5	6.25
10.8	8.5	7.8	9.5	−1	1
11.2	11	7.7	8	3	9
10.6	6	7.8	9.5	−3.5	12.25
11.4	12	8.3	12	0	0

$n = 12$ $\qquad$ $\sum d_i^2 = 42.00$

$$\begin{aligned} r_s &= 1 - \frac{6 \sum d_i^2}{n^3 - n} \\ &= 1 - \frac{6(42.00)}{1716} \\ &= 1 - 0.147 \\ &= 0.853 \end{aligned}$$

To test H_0: $\rho_s = 0$, H_A: $\rho_s \neq 0$,

$(r_s)_{0.05(2),12} = 0.587$ (from Table B.19)
Therefore, reject H_0.
$P < 0.001$

To employ the correction for ties (see Equation 19.37):

among the X's there are 2 measurements of 10.2 cm and 2 of 10.8 cm, so

$$\sum T_X = \frac{(2^3 - 2) + (2^3 - 2)}{12} = 1;$$

among the Y's there are 2 measurements tied at 7.2 cm and 2 at 7.8 cm, so

$$\sum T_Y = \frac{(2^3 - 2) + (2^3 - 2)}{12} = 1;$$

therefore,

$$r_s = \frac{(12^3 - 12)/6 - 42.00 - 1 - 1}{\sqrt{[(12^3 - 12)/6 - 2(1)][(12^3 - 12)/6 - 2(1)]}} = \frac{242}{284} = 0.852;$$

and the hypothesis test proceeds exactly as above.

testing procedures, Equation 19.1 can be applied to the ranks to obtain the *Spearman rank correlation coefficient*, r_s. However, a simpler computation is

$$r_s = 1 - \frac{6 \sum_{i=1}^{n} d_i^2}{n^3 - n}, \qquad (19.35)$$

where d_i is a difference between X and Y ranks.

The value of r_s, as an estimate of the population rank correlation coefficient, ρ_s, may range from −1 to +1, and it has no units; however, its value is not to be expected to be the same as the value of r that might have been calculated for the original data, rather than their ranks.*

Use Appendix Table B.19 to assess the significance of r_s. If n is greater than that provided for in this table, then r_s may be used in place of r in the hypothesis testing procedures of Section 19.2. If either the Spearman or the parametric correlation analysis (Section 19.2) is applicable, the former is $9/\pi^2 = 0.91$ as powerful as the latter (Daniel, 1978: 304; Hotelling and Pabst, 1936).

Correction for Tied Data. If there are tied data, then r_s is better calculated as

$$r_s = \frac{(n^3 - n)/6 - \sum d_i^2 - \sum T_X - \sum T_Y}{\sqrt{[(n^3 - n)/6 - 2\sum T_X][(n^3 - n)/6 - 2\sum T_Y]}} \tag{19.37}$$

(Kendall, 1962: 38). Here,

$$\sum T_X = \frac{\sum (t_i^3 - t_i)}{12}, \tag{19.38}$$

where t_i is the number of tied values of X in a group of ties, and

$$\sum T_Y = \frac{\sum (t_i^3 - t_i)}{12}, \tag{19.39}$$

where t_i is the number of tied Y's in a group of ties. If $\sum T_X$ and $\sum T_Y$ are zero, then Equation 19.37 is identical to Equation 19.35. Indeed, the two equations differ appreciably only if there are numerous tied data.

Other Hypotheses, Confidence Limits, and Power. If $n \geq 10$ and $\rho_s \leq 0.9$, then the Fisher z transformation may be used for Spearman coefficients, just as it was in Sections 19.2 through 19.6, for testing several additional kinds of hypotheses (including multiple comparisons), estimating power, and setting confidence limits around ρ_s. But in doing so it is recommended that $1.060/(n - 3)$ be used instead of $1/(n - 3)$ for the variance of z (Fieller, Hartley, and Pearson, 1957, 1961). That is,

$$\sigma_z = \sqrt{\frac{1.060}{n - 3}} \tag{19.40}$$

should be used for the standard error of z.

The Kendall Rank Correlation Coefficient. The Kendall rank correlation procedure will not be explained here. (See Daniel, 1978: 306–321; Kendall, 1962; Siegel, 1956: 213–223.) In this procedure, the sample correlation coefficient is usually designated τ (lowercase Greek tau, an exceptional use of a Greek letter to denote a sample statistic). The value of τ for a particular set of data will not necessarily resemble the value of r_s calculated for the same data. The performances of the Spearman and Kendall coefficients are very similar, but the former may be a little better, especially when n is large (Chow, Miller, and Dickinson, 1974), and for large n the Spearman measure is also easier to calculate than the Kendall.

*This statistic is related to the coefficient of concordance, W, from Section 20.16. Within two groups of ranks,

$$W = (r_s + 1)/2. \tag{19.36}$$

19.10 CORRELATION FOR DICHOTOMOUS NOMINAL SCALE DATA

Dichotomous nominal scale data are common in biology (e.g., observations may be recorded as male or female, dead or alive, thorned or thornless), and Chapter 22 is devoted to several aspects of the analysis of such data. If data are recorded for two dichotomous variables, they may be presented in the form of a 2 × 2 contingency table (as introduced in Section 6.2). The data of Example 19.12, for instance, might be

Example 19.12

Correlation for dichotomous nominal scale data. Data are collected to determine the degree of association, or correlation, between the presence of a plant disease and the presence of a certain species of insect.

Case	*Presence of plant disease*	*Presence of insect*
1	+	+
2	+	+
3	−	−
4	−	+
5	+	+
6	−	+
7	−	−
8	+	+
9	−	+
10	−	−
11	+	+
12	−	−
13	+	+
14	−	+

The data may be tabulated in the following 2 × 2 contingency table:

Insect	**Plant Disease** *Present*	*Absent*	*Total*
Present	6	4	10
Absent	0	4	4
Total	6	8	14

$$\phi_2 = \frac{f_{11}f_{22} - f_{12}f_{21}}{\sqrt{C_1C_2R_1R_2}}$$

$$= \frac{(6)(4) - (4)(0)}{\sqrt{(6)(8)(10)(4)}}$$

$$= 0.55$$

$$Q = \frac{f_{11}f_{22} - f_{12}f_{21}}{f_{11}f_{22} + f_{12}f_{21}} = \frac{(6)(4) - (4)(0)}{(6)(4) + (4)(0)} = 1.00$$

$$r_n = \frac{(f_{11} + f_{22}) - (f_{12} + f_{21})}{(f_{11} + f_{22}) + (f_{12} + f_{21})} = \frac{(6 + 4) - (4 + 0)}{(6 + 4) + (4 + 0)} = \frac{10 - 4}{10 + 4} = 0.43$$

cast into a 2 × 2 table, as shown. We shall set up such tables by having f_{11} and f_{22} be the frequencies of agreement between the two variables.

Various measures of association of two dichotomous variables have been suggested (Conover, 1980: Section 4.4; Everitt, 1977: Section 3.7; Gibbons, 1976: Section 7.4.2). So-called *contingency coefficients*, such as

$$\sqrt{\frac{\chi^2}{\chi^2 + n}} \tag{19.41}^*$$

and

$$\sqrt{\frac{\dfrac{\chi^2}{n}}{1 + \dfrac{\chi^2}{n}}} \tag{19.42}$$

employ the χ^2 statistic of Section 6.2. However, they have drawbacks, among them the lack of the desirable property of ranging between 0 and 1. [They are indeed zero when $\chi^2 = 0$ (i.e., when there is no association between the two variables), but the coefficients can never reach 1, even if there is complete agreement between the two variables.]

The Cramér, or phi, coefficient† (Cramér, 1946: 44),

$$\phi_1 = \sqrt{\frac{\chi^2}{n}}, \tag{19.43}$$

does range from 0 to 1 (as does ϕ_1^2, which may also be used as a measure of association). It is based upon χ^2 (uncorrected for continuity), as obtained from Equation 6.1, or, more readily, from Equation 6.6. Therefore, we can write

$$\phi_2 = \frac{f_{11}f_{22} - f_{12}f_{21}}{\sqrt{C_1C_2R_1R_2}}. \tag{19.44}$$

This measure is preferable to Equation 19.43 because it can range from -1 to $+1$, thus expressing not only the strength of an association between variables, but also the direction of the association. If $\phi_2 = 1$, this means that all the data in the contingency table lie in the upper left and lower right cells (i.e., $f_{12} = f_{21}$). In Example 19.12 this would mean there was complete agreement between the presence of both the disease and the insect; either both were always present or both were always absent. If $\phi_2 = -1$, $f_{11} = f_{22} = 0$, and all the data lie in the upper right and lower left cells of the contingency table.

The statistic ϕ_2 is also preferred over the previous coefficients of this section because it is amenable to hypothesis testing. The significance of ϕ_2 (i.e., whether it indicates that an association exists in the sampled population) can be assessed by considering the significance of the contingency table. If the frequencies are sufficiently large (see Section 6.5), the significance of χ_c^2 (chi-square with the correction for

*This measure has also been called the "Pearson coefficient of mean square contingency" (Yule, 1917: 64–65).

†This measure is commonly symbolized by the lowercase Greek phi (ϕ)—pronounced "fy" as in "simplify" (or, sometimes, "fee")—and is a sample statistic, not a population parameter as Greek letters typically designate. (It should not, of course, be confused with the quantity used in estimating the power of a statistical test, which appears elsewhere in this book.)

continuity) may be determined. If not, the Fisher exact test (Section 22.9) or the log-likelihood ratio (Section 6.6) may be employed.

The Yule coefficient of association* (Yule, 1900, 1912, 1917: 38):

$$Q = \frac{f_{11}f_{22} - f_{12}f_{21}}{f_{11}f_{22} + f_{12}f_{21}}, \tag{19.45}$$

ranges from -1 to $+1$ and is interpreted as is ϕ_2. As can be seen in Example 19.12, it declares "perfect agreement" ($Q = 1.00$) in cases where agreement is less than complete. A better measure is that of Ives and Gibbons (1967). It may be expressed as a correlation coefficient,

$$r_n = \frac{(f_{11} + f_{22}) - (f_{12} + f_{21})}{(f_{11} + f_{22}) + (f_{12} + f_{21})}. \tag{19.46}$$

The interpretation of positive and negative values of r_n (which can range from -1 to $+1$) is just as for ϕ_2.

The expression of significance of r_n involves statistical testing which will be described in Chapter 22. The binomial test (Section 22.5) may be utilized, with a null hypothesis of $H_0: p = 0.5$, using cases of perfect agreement and cases of disagreement as the two categories. Alternatively, the sign test (Section 22.6), the Fisher exact test (Section 22.9), or the chi-square contingency test (Section 6.2) could be applied to the data.

19.11 INTRACLASS CORRELATION

Correlation situations exist where it is impossible to designate one variable as X and one as Y. Consider the data in Example 19.13 where we wish to determine whether there is any relationship between the weights of identical twins. Although we clearly have pairs of data, it is not possible to say that the data in the first column have something in common which is different from what all the data in the second column have in common. Thus, we invoke the use of *intraclass correlation*, a concept generally approached by means of analysis of variance considerations (Model II single factor ANOVA). Aside from assuming random sampling from a bivariate normal distribution, this procedure also assumes that the population variances are equal.

If we consider each of the pairs in our example as groups in an ANOVA (i.e., $k = 7$), with each group containing two observations (i.e., $n = 2$), then we may calculate mean squares to express variability both between and within the k groups (see Section 11.1). Then the *intraclass correlation coefficient* is defined as

$$r_I = \frac{\text{groups MS} - \text{error MS}}{\text{groups MS} + \text{error MS}}, \tag{19.47}$$

this statistic being an estimate of the population intraclass correlation coefficient, ρ_I. To test H_0: $\rho_I = 0$ vs. H_0: $\rho_I \neq 0$, one may utilize

$$F = \frac{\text{groups MS}}{\text{error MS}}, \tag{19.48}$$

*Yule called this coefficient Q in honor of Lambert-Adolphe-Jacques Quetelet (1796–1874), a pioneering Belgian statistician.

Example 19.13
Intraclass correlation.

Testing for correlation between weights of members of pairs of human twins.

Group (*i.e., twin*)	*Weight of one member of group* (kg)	*Weight of other member of group* (kg)
1	70.4	71.3
2	68.2	67.4
3	77.3	75.2
4	61.2	66.7
5	72.3	74.2
6	74.1	72.9
7	71.1	69.5

Source of variation	SS	DF	MS
Total	220.17	13	
Groups	198.31	6	33.05
Error	21.86	7	3.12

(See Section 11.1 for the computation procedures.)

$$r_I = \frac{\text{groups MS} - \text{error MS}}{\text{groups MS} + \text{error MS}} = \frac{29.93}{36.17} = 0.827$$

To test $H_0: \rho_I = 0$, $H_A: \rho_I \neq 0$,

$$F = \frac{\text{groups MS}}{\text{error MS}} = \frac{33.05}{3.12} = 10.6$$

$F_{0.05(1),6,7} = 3.87$
Therefore, reject H_0.
$0.0025 < P < 0.005$

a statistic associated with Groups DF and error DF for the numerator and denominator, respectively.* If the measurements are equal within each group, then error MS $= 0$, and $r_I = 1$ (a perfect positive correlation). If there is more variability within groups than there is between groups, then r_I will be negative. The smallest it may be, however, is $-1/(n-1)$; therefore, only if $n = 2$ (as in Example 19.13) can r_I be as small as -1.

Theoretically, we are not limited to pairs of data (i.e., situations where $n = 2$) to speak of intraclass correlation. Consider, for instance, expanding the considerations of Example 19.13 into a study of weight correspondence among triplets instead of twins. Indeed, n need not even be equal for all groups. We might, for example, ask whether there is any concordance among adult weights of brothers; here, some families might consist of two brothers, some of three brothers, etc. If n is not 2 for all k groups, then

$$r_I = \frac{\text{groups MS} - \text{error MS}}{\text{groups MS} + (n-1)\text{ error MS}}, \qquad (19.50)$$

*If desired, F may be calculated first, followed by computing

$$r_I = (F-1)/(F+1). \qquad (19.49)$$

a calculation for which Equation 19.47 is a special case. If all n's are not equal, then the appropriate n for use in Equation 19.50 may be obtained as

$$n = \frac{\sum_{i=1}^{k} n_i - \frac{\sum_{i=1}^{k} n_i^2}{\sum_{i=1}^{k} n_i}}{k - 1}. \tag{19.51}$$

Equation 19.48 is applicable for hypothesis testing in all the preceding cases.

If $n = 2$ in all groups, then we may set confidence limits and test hypotheses as would be done with r, by utilizing the z transformation.* (See Sections 19.3 through 19.7.) However, the standard error of the resultant z_I will be

$$\sigma_{z_I} = \sqrt{\frac{1}{k - \frac{3}{2}}} \tag{19.52}$$

(Fisher 1958: 215). If $n > 2$ in any groups, then the Fisher transformation is computed as

$$z_I = 0.5 \ln \frac{1 + (n - 1)r_I}{1 - r_I} \tag{19.53}$$

(Fisher 1958: 219), using Equation 19.51 if the n's are unequal. Table B.17 may be used, rather than Equation 19.53 if r_I is first converted to

$$r' = \frac{nr_I}{2 + (n - 2)r_I} \tag{19.54}$$

(Rao, Mitra, and Matthai, 1966: 87), and using r' to enter the table. If Table B.18 is used to convert z_I to r', then r_I is obtainable as

$$r_I = \frac{-2r'}{(n - 2)r' - n}. \tag{19.55}$$

Also, if $n > 2$, then

$$\sigma_{z_I} = \sqrt{\frac{n}{2(n - 1)(k - 2)}} \tag{19.56}$$

(Fisher, 1958: 219).

Rothery (1979) has developed a nonparametric measure of intraclass coerrelation.

19.12 THE EFFECT OF CODING

Coding of the raw data will have no effect at all on any of the correlation coefficients presented in this chapter. Likewise, there will be no effect of coding on any z transformations or on any hypotheses concerning correlation coefficients or z transformations, nor will the coefficients of determination be altered.

*The correction for bias in z involves adding $1/(2k - 1)$ to z (Fisher, 1958: 216); it is typically negligible unless several z's are being summed, as in the procedure of Section 19.7.

EXERCISES

19.1. Measurements of serum cholesterol (mg/100 ml) and arterial calcium deposition (mg/100 g dry weight of tissue) were made on 12 animals. The data are as follows:

Calcium (X)	*Cholesterol* (Y)
59	298
52	303
42	233
59	287
24	236
24	245
40	265
32	233
63	286
57	290
36	264
24	239

(a) Calculate the correlation coefficient.
(b) Calculate the coefficient of determination.
(c) Test $H_0: \rho = 0$, vs. $H_A: \rho \neq 0$.
(d) Set 95% confidence limits on the correlation coefficient.

19.2. Using the data from Example 19.1:
(a) Test $H_0: \rho \leq 0$; $H_A: \rho > 0$.
(b) Test $H_0: \rho = 0.50$; $H_A: \rho \neq 0.50$.

19.3. Given: $r_1 = -0.44$, $n_1 = 24$, $r_2 = -0.40$, $n_2 = 30$.
(a) Test $H_0: \rho_1 = \rho_2$; $H_A: \rho_1 \neq \rho_2$.
(b) If H_0 in part (a) is not rejected, compute the common correlation coefficient.

19.4. Given: $r_1 = 0.45$, $n_1 = 18$, $r_2 = 0.56$, $n_2 = 16$. Test $H_0: \rho_1 \geq \rho_2$; $H_A: \rho_1 < \rho_2$.

19.5. Given: $r_1 = 0.85$, $n_1 = 24$, $r_2 = 0.78$, $n_2 = 32$, $r_3 = 0.86$, $n_3 = 31$.
(a) Test $H_0: \rho_1 = \rho_2 = \rho_3$, stating the appropriate alternate hypothesis.
(b) If H_0 in part (a) is not rejected, compute the common correlation coefficient.

19.6. **(a)** Calculate the Spearman rank correlation coefficient for the data of Example 19.1.
(b) Test $H_0: \rho_s = 0$ against $H_A: \rho_s \neq 0$.

19.7. To examine the proposition that the type of school a college president chooses to lead correlates with the type of school from which he received his undergraduate education, the following data were collected and arranged in a contingency table:

College as President	**Undergraduate College**	
	Public	*Private*
Public	9	5
Private	2	7

(a) Compute the coefficient, r_n, expressing this correlation.
(b) Using the Fisher exact test (Section 22.9), test the significance of the correlation.

19.8. Two samples of blood plasma from the same animal were submitted to each of four testing laboratories. The corticosterone concentrations, in grams per 100 ml, were determined as follows:

Laboratory	*Sample 1*	*Sample 2*
1	1.14	1.08
2	1.10	1.07
3	1.04	1.08
4	1.07	1.13

(a) Compute the intraclass correlation coefficient.

(b) Test whether there is a significant correlation between corticosterone determinations from the same laboratory.

20 Multiple Regression and Correlation

The previous three chapters have discussed the analysis of relationships between two variables. We shall now expand these considerations to interrelationship among three or more variables using the procedures of *multiple regression* and *multiple correlation*. If we have simultaneous measurements for more than two variables, and one of the variables is assumed to be dependent upon the others, then we are dealing with a *multiple regression* situation. Here, the observed values of the dependent variable (Y) are assumed to have come at random from a normal distribution of Y values existing in the population at the particular observed combination of independent variables. Furthermore, all such normal distributions, at all combinations of values for the dependent variables, are assumed to have the same variance. If none of the variables is assumed to be functionally dependent on any other, then we are dealing with *multiple correlation*, a situation requiring the assumption that each of the variables exhibits a normal distribution at each combination of all the other variables.* When dealing with two variables in simple correlation, we spoke of sampling a bivariate normal population; with samples dealing with more than two variables, we speak of a *multivariate normal distribution*. In multiple regression, the data transformation discussions of Section 17.10 apply. In multiple correlation, all variables must be treated as is Y in that discussion.

The computational procedures required for most multiple regression and correlation analyses are difficult enough to preclude hand calculation, but it is an

*Much development in multiple correlation theory began in the late nineteenth century by several pioneers, including Karl Pearson (1857–1936) and his colleague, George Udny Yule (1871–1951) (Pearson, 1967). (Pearson first called partial regression coefficients "double regression coefficients," and Yule later called them "net regression coefficients.")

uncommon science-oriented digital computer facility today that does not possess a program capable of performing the necessary operations (e.g., Dixon, 1975; Dixon and Brown, 1979; Nie et al., 1975; SAS Institute, 1979). Because of this, we will not dwell on the multiple regression and correlation calculations, but will concentrate instead on the interpretation of numerical results of the kind that a computer program would provide. Those readers desiring discussion of the details of multiple regression computations may consult works such as those by Bliss (1970: Chapter 18), Daniel and Wood (1971), Draper and Smith (1981), Dunn and Clark (1974: Chapters 11, 12), Efroymson (1960), Ezekial and Fox (1959), Jenrich (1977), J. C. R. Li (1964: Chapter 28), Neter and Wasserman (1974: Chapters 6–12), Seber (1977), and Williams (1959).

20.1 INTERMEDIATE COMPUTATIONAL STEPS

There are certain quantities that a computer program for multiple regression and/or correlation must calculate. Although we shall not concern ourselves with the mechanics of computation, we shall present certain intermediate steps in the calculating procedures here so that the user will not be a complete stranger to them if they appear on the printed computer output. Among the many different programs available for multiple regression and correlation, many do not print all the following intermediate results, or often they will do so only if the user specifically asks for them to appear in the output.

Consider n sets of M variables (the variables being referred to as X_I through X_M). If we desire to consider one of the M variables as being dependent on the others, then we may eventually designate that variable as Y, but the program will perform most of its computations simply considering all M variables as X's numbered 1 through M (as in Example 20.1a–e).

The sums of the observations of each of the M variables are calculated as

$$\sum_{j=1}^{n} X_{1j} \quad \sum_{j=1}^{n} X_{2j} \quad \cdots \quad \sum_{j=1}^{n} X_{Mj}. \tag{20.1}$$

For simplicity, let us refrain from indexing the $\sum$'s and assume that summations are always performed over all n sets of data. Thus, the sums of the variables could be denoted as:

$$\sum X_1 \quad \sum X_2 \quad \cdots \quad \sum X_M. \tag{20.2}$$

Sums of squares and sums of crossproducts are calculated just as for simple regression, or correlation, for each of the M variables. These sums, generally referred to as *raw sums of squares* and *raw sums of crossproducts*, are presented in computer output in the form of a *matrix*, or two-dimensional array (Example 20.1b):

$$\begin{matrix} \sum X_1^2 & \sum X_1X_2 & \sum X_1X_3 & \cdots & \sum X_1X_M \\ \sum X_2X_1 & \sum X_2^2 & \sum X_2X_3 & \cdots & \sum X_2X_M \\ \sum X_3X_1 & \sum X_3X_2 & \sum X_3^2 & \cdots & \sum X_3X_M \\ \vdots & \vdots & \vdots & & \vdots \\ \sum X_MX_1 & \sum X_MX_2 & \sum X_MX_3 & \cdots & \sum X_M^2 \end{matrix} \tag{20.3}$$

Example 20.1a

The $M \times n$ data matrix for a hypothetical multiple regression or correlation ($M = 5; n = 33$).

	Variable (i)				
j	1 (°C)	2 (cm)	3 (mm)	4 (min)	5 (ml)
1	6	9.9	5.7	1.6	2.12
2	1	9.3	6.4	3.0	3.39
3	−2	9.4	5.7	3.4	3.61
4	11	9.1	6.1	3.4	1.72
5	−1	6.9	6.0	3.0	1.80
6	2	9.3	5.7	4.4	3.21
7	5	7.9	5.9	2.2	2.59
8	1	7.4	6.2	2.2	3.25
9	1	7.3	5.5	1.9	2.86
10	3	8.8	5.2	0.2	2.32
11	11	9.8	5.7	4.2	1.57
12	9	10.5	6.1	2.4	1.50
13	5	9.1	6.4	3.4	2.69
14	−3	10.1	5.5	3.0	4.06
15	1	7.2	5.5	0.2	1.98
16	8	11.7	6.0	3.9	2.29
17	−2	8.7	5.5	2.2	3.55
18	3	7.6	6.2	4.4	3.31
19	6	8.6	5.9	0.2	1.83
20	10	10.9	5.6	2.4	1.69
21	4	7.6	5.8	2.4	2.42
22	5	7.3	5.8	4.4	2.98
23	5	9.2	5.2	1.6	1.84
24	3	7.0	6.0	1.9	2.48
25	8	7.2	5.5	1.6	2.83
26	8	7.0	6.4	4.1	2.41
27	6	8.8	6.2	1.9	1.78
28	6	10.1	5.4	2.2	2.22
29	3	12.1	5.4	4.1	2.72
30	5	7.7	6.2	1.6	2.36
31	1	7.8	6.8	2.4	2.81
32	8	11.5	6.2	1.9	1.64
33	10	10.4	6.4	2.2	1.82

Example 20.1b

A matrix of raw sums of squares and sums of cross-products, as it might occur as computer output (from the data of Example 20.1a).

	1	2	3	4	5
1	0.11270E 04	0.13661E 04	0.87270E 03	0.38130E 03	0.30273E 03
2	0.13661E 04	0.26757E 04	0.17218E 04	0.75560E 03	0.71860E 03
3	0.87270E 03	0.17218E 04	0.11466E 04	0.49707E 03	0.47978E 03
4	0.38130E 03	0.75560E 03	0.49707E 03	0.25827E 03	0.21564E 03
5	0.30273E 03	0.71860E 03	0.47978E 03	0.21564E 03	0.21677E 03

Note that in this and in following examples of computer output, numerical results are printed in scientific, or exponential, notation. Here, for example, 0.11270E 04 would indicate 0.11270 $\times 10^4$, or 1127.0. In many instances, exponential notation on computer output will be of the form 0.11270D 04, rather than 0.11270E 04. Both of these notations would be read the same way, but the "D" indicates that the computer program employed "double precision" for its computations, meaning that the calculations were performed with considerably more (although not necessarily double) significant figures than are used when the results are printed with the "E" notation. Also, it is very important to realize that the number of significant figures printed by a computer in no way indicates how many significant figures were used in the calculations.

Since $\sum X_i X_k = \sum X_k X_i$, this matrix is said to be symmetrical about the diagonal running from upper left to lower right.* Thus some computer programs print out only a half-matrix—such as,

$$\begin{matrix} \sum X_1^2 & & & & \\ \sum X_2 X_1 & \sum X_2^2 & & & \\ \sum X_3 X_1 & \sum X_3 X_2 & \sum X_3^2 & & \\ \vdots & \vdots & \vdots & & \\ \sum X_M X_1 & \sum X_M X_2 & \sum X_M X_3 & \cdots & \sum X_M^2 \end{matrix} \tag{20.4}$$

for this matrix as well as for those matrices that follow.

If a raw sum of squares, $\sum X_i^2$, is reduced by $(\sum X_i)^2/n$, we are left with a sum of squares that we have previously symbolized as $\sum x_i^2$ (see Section 17.2). Similarly, a raw sum of crossproducts, $\sum X_i X_k$, if diminished by $\sum X_i \sum X_k/n$, yields $\sum x_i x_k$. These quantities are known as *corrected* sums of squares and sums of crossproducts, respectively, and they may be presented as the following matrix (Example 20.1c):

$$\begin{matrix} \sum x_1^2 & \sum x_1 x_2 & \sum x_1 x_3 & \cdots & \sum x_1 x_M \\ \sum x_2 x_1 & \sum x_2^2 & \sum x_2 x_3 & \cdots & \sum x_2 x_M \\ \sum x_3 x_1 & \sum x_3 x_2 & \sum x_3^2 & \cdots & \sum x_3 x_M \\ \vdots & \vdots & \vdots & & \vdots \\ \sum x_M x_1 & \sum x_M x_2 & \sum x_M x_3 & \cdots & \sum x_M^2 \end{matrix} \tag{20.5}$$

Example 20.1c

A matrix of corrected sums of squares and sums of crossproducts, as it might appear as computer output (from the data of Example 20.1a).

	1	2	3	4	5
1	0.47218E 03	0.60027E 02	0.80727E 01	0.75636E 01	−0.60984E 02
2	0.60027E 02	0.70622E 02	−0.27091E 01	0.10161E 02	−0.68429E 01
3	0.80727E 01	−0.27091E 01	0.49091E 01	0.35854E 01	−0.47146E 00
4	0.75636E 01	0.10161E 02	0.35854E 01	0.44961E 02	0.80511E 01
5	−0.60984E 02	−0.68429E 01	−0.47146E 00	0.80511E 01	0.14747E 02

From Matrix 20.5, it is simple to calculate a matrix of simple correlation coefficients, for r_{ik} (representing the correlation between variables i and k) $= \sum x_i x_k / \sqrt{\sum x_i^2 \sum x_k^2}$.

*Hereafter, we shall refer to the values of a pair of variables as X_i and X_k.

$$
\begin{matrix}
r_{11} & r_{12} & r_{13} & \cdots & r_{1M} \\
r_{21} & r_{22} & r_{23} & \cdots & r_{2M} \\
r_{31} & r_{32} & r_{33} & \cdots & r_{3M} \\
\cdot & \cdot & \cdot & & \cdot \\
\cdot & \cdot & \cdot & & \cdot \\
\cdot & \cdot & \cdot & & \cdot \\
r_{M1} & r_{M2} & r_{M3} & \cdots & r_{MM}
\end{matrix} \tag{20.6}
$$

Each element in the diagonal of this matrix, i.e., r_{ii}, is equal to 1.0, for there will always be a perfect positive correlation between a variable and itself. (See Example 20.1d.)

Example 20.1d

A matrix of simple correlation coefficients, as it might appear as computer output (from the data of Example 20.1a).

	1	2	3	4	5
1	1.00000	0.32872	0.16767	0.05191	−0.73081
2	0.32872	1.00000	−0.14550	0.18033	−0.21204
3	0.16767	−0.14550	1.00000	0.24134	−0.05541
4	0.05191	0.18033	0.24134	1.00000	0.31267
5	−0.73081	−0.21204	−0.05541	0.31267	1.00000

The final major manipulation necessary before the important regression or correlation statistics of the following sections can be obtained is the computation of the *inverse* of a matrix. Inverting a matrix is to two-dimensional algebra what taking the reciprocal is to ordinary, one-dimensional algebra (e.g., see Batschelet, 1976: Section 14.7; Eason, et al., 1980: Section 18.5; Kleinbaum and Kupper, 1978: Appendix B; J. C. R. Li, 1964: Chapter 27; Searle, 1966: Chapter 9). Although the process of inverting a matrix of moderate size is too cumbersome to be performed readily by hand, digital computers can execute such manipulations with ease. A multiple regression or correlation program may invert the corrected sum of squares and crossproducts matrix, Matrix 20.5, resulting in a symmetrical matrix conventionally symbolized

$$
\begin{matrix}
c_{11} & c_{12} & c_{13} & \cdots & c_{1M} \\
c_{21} & c_{22} & c_{23} & \cdots & c_{2M} \\
\cdot & \cdot & \cdot & & \cdot \\
\cdot & \cdot & \cdot & & \cdot \\
\cdot & \cdot & \cdot & & \cdot \\
c_{M1} & c_{M2} & c_{M3} & \cdots & c_{MM}
\end{matrix} \tag{20.7}
$$

Or the correlation matrix, Matrix 20.6, may be inverted, yielding a different array of values, which we may designate

$$
\begin{matrix}
d_{11} & d_{12} & d_{13} & \cdots & d_{1M} \\
d_{21} & d_{22} & d_{23} & \cdots & d_{2M} \\
d_{31} & d_{32} & d_{33} & \cdots & d_{3M} \\
\cdot & \cdot & \cdot & & \cdot \\
\cdot & \cdot & \cdot & & \cdot \\
\cdot & \cdot & \cdot & & \cdot \\
d_{M1} & d_{M2} & d_{M3} & \cdots & d_{MM}
\end{matrix} \tag{20.8}
$$

The program utilized for the computer output presented in this chapter proceeded along the latter route, the inverse correlation matrix being the output shown in Example 20.1e. A program might compute either Matrix 20.7 or 20.8, the choice being irrelevant to our considerations, as the two are interconvertible:

$$c_{ik} = \frac{d_{ik}}{\sqrt{\sum x_i^2 \sum x_k^2}}, \tag{20.9}$$

or, equivalently,

$$c_{ik} = \frac{r_{ik} d_{ik}}{\sum x_i x_k}. \tag{20.10}$$

Example 20.1e

The inverse matrix of the simple correlation matrix of Example 20.1d.

	1	2	3	4	5
1	0.27610E 01	−0.36091E 00	−0.22915E 00	−0.69375E 00	0.21454E 01
2	−0.36091E 00	0.12501E 01	0.32710E 00	−0.32332E 00	0.12053E 00
3	−0.22915E 00	0.32710E 00	0.11761E 01	−0.35538E 00	0.78178E–01
4	−0.69375E 00	−0.32332E 00	−0.35538E 00	0.15142E 01	−0.10687E 01
5	0.21454E 01	0.12053E 00	0.78178E–01	−0.10687E 01	0.29319E 01

It is from manipulations of these types of arrays that a computer program can derive the sample statistics and components of analysis of variance described in the following sections. If partial correlation coefficients are desired (Section 20.6), the matrix inversion takes place as shown. If partial regression analysis is desired (Sections 20.2–20.4), then inversion is performed only on the $M - 1$ rows and $M - 1$ columns corresponding to the independent variables in either Matrix 20.5 or 20.6 (see Example 20.1f).

Example 20.1f

The inverse of the independent variable correlation matrix (i.e., if variable 5 is the dependent variable in Example 20.1a, this is the inverse of columns 1 through 4 and rows 1 through 4 of the correlation matrix of Example 20.1d).

	1	2	3	4
1	0.11911E 01	−0.44910E 00	−0.28635E 00	0.88264E–01
2	−0.44910E 00	0.12451E 01	0.32389E 00	−0.27938E 00
3	−0.28635E 00	0.32389E 00	0.11740E 01	−0.32688E 00
4	0.88264E–01	−0.27938E 00	−0.32688E 00	0.11247E 01

20.2 THE MULTIPLE REGRESSION EQUATION

Recall, from Section 17.2, that a simple linear regression for a population of paired variables is the relationship

$$\hat{Y}_i = \alpha + \beta X_i. \tag{17.1}$$

In this relationship, Y and X represent the dependent and independent variables, respectively; β is the regression coefficient in the population; and α (the "Y intercept") is the value of Y when X is zero.

In many situations, however, Y may be considered dependent upon more than one variable. Thus,

$$\hat{Y}_j = \alpha + \beta_1 X_{1j} + \beta_2 X_{2j} \tag{20.11}$$

may be proposed, implying that one variable (Y) is linearly dependent upon a second variable (X_1), and that Y is also linearly dependent upon a third variable (X_2). (The double subscript notation, X_{ij}, implies the jth observation of variable X_i.) In this particular multiple regression model we have one dependent variable and two independent variables.* The two population parameters, β_1 and β_2, are termed *partial regression coefficients*; β_1 expresses how much Y would change for a unit change in X_1, if X_2 were held constant. It is sometimes said that β_1 is a measure of the relationship of Y to X_1 after "controlling for" X_2; that is, it is a measure of the extent to which Y is related to X_1 after removing the effect of X_2 on Y and X_1. Similarly, β_2 describes the rate of change of Y as X_2 changes, with X_1 being held constant. They are called partial regression coefficients, then, because each expresses only part of the dependence relationship. The Y intercept, α, is the value of Y when *both* X_1 and X_2 are zero. Whereas Equation 17.1 mathematically represents a line (a two-dimensional figure), Equation 20.11 defines a plane (a three-dimensional figure); one seldom attempts to present a three- (or more-) variable relationship graphically. A regression with M variables defines an M-dimensional surface, sometimes referred to as a "response surface."

If we sample the population containing the three variables (Y, X_1, and X_2) in Equation 20.11, we can compute sample statistics to estimate the population parameters in the model. The multiple regression function derived from a sample of data would be

$$\hat{Y}_j = a + b_1 X_{1j} + b_2 X_{2j}. \tag{20.12}$$

Theoretically, there is no limit to the number of independent variables, X_j, which can be proposed to influence the dependent variable, Y. The general population model, of which Equation 20.11 is a special case, would be

$$\hat{Y}_j = \alpha + \beta_1 X_{1j} + \beta_2 X_{2j} + \beta_3 X_{3j} + \cdots + \beta_m X_{mj}, \tag{20.13}$$

or, more succinctly,

$$\hat{Y}_j = \alpha + \sum_{i=1}^{m} \beta_i Y_{ij}, \tag{20.14}$$

where m is the number of independent variables. The sample regression equation, containing the statistics used to estimate the population parameters when there are m independent variables would be

$$\hat{Y}_j = a + b_1 X_{1j} + b_2 X_{2j} + b_3 X_{3j} + \cdots + b_m X_{mj}, \tag{20.15}$$

or

$$\hat{Y}_j = a + \sum_{i=1}^{m} b_i X_{ij}. \tag{20.16}$$

At least $m + 2$ data points are required to perform a multiple regression analysis, where m is the number of independent variables determining each data point.

From the analysis shown in Example 20.1g, we arrive at a regression function having partial regression coefficients of $b_1 = -0.129$ ml/°C, $b_2 = -0.019$ ml/cm,

*Sometimes the independent variables are called "predictor" or "regressor" variables; the dependent variable may be referred to as the "response" or "criterion" variable.

Example 20.1g

Regression statistics computed for the regression model $\hat{Y} = \alpha + \beta_1 X_1 + \beta_2 X_2 + \beta_3 X_3 + \beta_4 X_4$ fit to the data of Example 20.1a. Here, B is a partial regression coefficient estimating β, SE is its standard error, T is the t for testing H_0: $\beta_i = 0$ against H_A: $\beta_i \neq 0$, and DF is the degrees of freedom for this t test. Note that common computer outputting procedure, as below, places subscripts in parentheses, so that $X(1)$ denotes X_1, etc. Also, computer output is typically all in capital letters.

VARIABLE	B	SE	T	DF
X(1)	−0.12932E 00	0.21287E−01	−0.60750E 01	28
X(2)	−0.18785E−01	0.56278E−01	−0.33379E 00	28
X(3)	−0.46215E−01	0.20727E 00	−0.22297E 00	28
X(4)	0.20876E 00	0.67034E−01	0.31141E 01	28

Y INTERCEPT = 0.29583E 01
COEFFICIENT OF DETERMINATION = 0.65893
STANDARD ERROR OF ESTIMATE = 0.42384E 00

ANALYSIS OF VARIANCE TABLE

SOURCE OF VARIATION	SUM OF SQUARES	DF	MEAN SQUARE
TOTAL	0.14747E 02	32	
MULTIPLE REGRESSION	0.97174E 01	4	0.24293E 01
RESIDUAL	0.50299E 01	28	0.17964E 00

FOR THE ANALYSIS OF VARIANCE, F = 0.13524E 02, WITH DF OF 4 AND 28

$b_3 = -0.05$ ml/mm, $b_4 = 0.209$ ml/min, and a Y intercept of $a = 2.96$ ml.* Therefore, we can write the regression function as $\hat{Y} = 2.96 - 0.129X_1 - 0.019X_2 - 0.05X_3 + 0.209X_4$. Therefore, b_1 is an estimate of the relationship between Y and X_1 after removing the effects of both X_2 and X_3.

Section 20.4 will explain that, if independent variables are highly correlated with each other, then the interpretation of partial regression coefficients becomes questionable, as does the testing of hypotheses about the coefficients.

20.3 ANALYSIS OF VARIANCE OF MULTIPLE REGRESSION OR CORRELATION

A computer program for multiple regression and correlation analysis will typically include an analysis of variance of the regression (see Example 20.1g). This analysis of variance is analogous to that in the case of simple regression, consisting of total, multiple regression, and residual sums of squares and degrees of freedom, as well as regression and residual mean squares. The total sum of squares is an expression of the total amount of variability among the Y values $[Y_j - \bar{Y}]$; the regression sum of squares expresses the variability among the Y values attributable to the regression being fit $[\hat{Y}_j - \bar{Y}]$; and the residual sum of squares tells us about the amount of variability of Y still remaining after fitting the regression $[Y_j - \hat{Y}_j]$. The necessary sums of squares, degrees of freedom, and mean squares are summarized in Table 20.1.

*By examining the magnitude of the standard errors of the four partial regression coefficients (namely, 0.021287, 0.056278, 0.20727, and 0.067034), we observe that their second significant figures are at the third, third, second, and third decimal places, respectively, making it appropriate to state the four coefficients to those precisions.

TABLE 20.1 DEFINITIONS OF THE APPROPRIATE SUMS OF SQUARES, DEGREES OF FREEDOM, AND MEAN SQUARES USED IN MULTIPLE REGRESSION OR MULTIPLE CORRELATION ANALYSIS OF VARIANCE

Source of variation	*Sum of squares* (SS)	DF*	*Mean square* (MS)
Total	$\sum (Y_i - \bar{Y})^2$	$n - 1$	
Regression	$\sum (\hat{Y}_i - \bar{Y}_i)^2$	m	$\frac{\text{Regression SS}}{m}$
Residual	$\sum (Y_i - \hat{Y}_i)^2$	$n - m - 1$	$\frac{\text{Residual SS}}{n - m - 1}$

*n = total number of data points (i.e., total number of Y values); m = number of independent variables in the regression model.

Remember that the expressions given for sums of squares are the defining equations; the actual computations of these quantities may involve the use of more manageable "machine formulas."

Note how a simple linear regression (i.e., Table 17.1) would conform to Table 20.1 as the special case where $m = 1$.

If we assume Y to be functionally dependent on each of the X's, then we are dealing with multiple regression. If no such dependence is implied, then any of the $M = m + 1$ variables could be designated as Y for the purposes of utilizing the computer program, this being a case of *multiple correlation.* In either situation, we can test the hypothesis that there is no interrelationship among the variables, as

$$F = \frac{\text{regression MS}}{\text{residual MS}}. \tag{20.17}$$

The numerator and denominator degrees of freedom for this variance ratio are the regression DF and the residual DF, respectively.

The ratio,

$$R^2 = \frac{\text{regression SS}}{\text{total SS}} \tag{20.18}$$

is the *coefficient of determination* for a multiple regression or correlation, or the *coefficient of multiple determination.* In a regression situation, it expresses the proportion of the total variability in Y attributable to the dependence of Y on all the X_i, as defined by the regression model fit to the data.* In the case of correlation, R^2 may be considered to be the amount of variability in any one of the M variables that is accounted for by correlating it with the other $M - 1$ variables. The positive square root of this quantity is referred to as the *multiple correlation coefficient:*

$$R = \sqrt{\frac{\text{regression SS}}{\text{total SS}}}. \tag{20.20}$$

*R^2 is actually a biased estimator of the population parameter ρ^2, the proportion of variance in Y in the population accounted for by the multiple regression equation. An unbiased estimate is

$$\hat{\rho}^2 = R^2 - \frac{m - 1}{n - m - 1}(1 - R^2) \tag{20.19}$$

(Bliss, 1970: 305).

In a multiple correlation analysis, Equation 20.17 provides the test for whether the multiple correlation coefficient is zero in the sampled population. In the case of a multiple regression analysis, Equation 20.17 tests the null hypothesis of no dependence of Y on any of the independent variables, X_i; i.e., H_0: $\beta_1 = \beta_2 = \cdots = \beta_m = 0$ (vs. H_A: All m population partial regression coefficients are not equal to zero). Once R^2 has been calculated, the following computation of F may be used as an alternative to Equation 20.17:

$$F = \left(\frac{R^2}{1 - R^2}\right)\left(\frac{\text{residual DF}}{\text{regression DF}}\right), \tag{20.21}$$

and F (from either Equation 20.17 or 20.21) provides a test of $H_0: \rho^2 = 0$ vs. $H_A: \rho^2 \neq 0$.

The square root of the residual mean square is the *standard error of estimate* for the multiple regression:

$$s_{Y \cdot 1,2,\ldots,m} = \sqrt{\text{residual MS}}. \tag{20.22}$$

20.4 HYPOTHESES CONCERNING PARTIAL REGRESSION COEFFICIENTS

In simple regression, we generally desire to test H_0: $\beta = \beta_0$, a two-tailed hypothesis where β_0 is most often zero. If Equation 20.17 yields a significant F (i.e., H_0: $\beta_1 = \beta_2 = \cdots = \beta_m$ is rejected), then each of the partial regression coefficients in a multiple regression equation may be submitted to an analogous hypothesis, H_0: $\beta_i = \beta_0$, where, again, the test is usually two-tailed and the constant is most frequently zero. For H_0: $\beta_i = 0$, Student's t may be computed as

$$t = \frac{b_i}{s_{b_i}}, \tag{20.23}$$

and it should be apparent, from considerations of Section 8.2 and the latter part of Section 17.3, how one-tailed tests and cases where $\beta_0 \neq 0$ would be handled.

We may obtain both b_i and s_{b_i} from the computer output shown in Example 20.1g. In the particular computer program employed for this example, the t value is also calculated for each b_i. If it had not been, then Equation 20.23 would have been applied. (Some computer programs present the square of this t value and call it a "partial F value.") The residual degrees of freedom are used for this test.

If the standard errors are not given by the computer program being utilized, then they may be calculated as

$$s_{b_i} = \sqrt{s^2_{Y \cdot 1,2,\ldots,m} c_{ii}}, \tag{20.24}$$

where $s^2_{Y \cdot 1,2,\ldots,m}$ is, of course, the square of the standard error of estimate, which is simply the residual mean square, and where c_{ii} is defined in Section 20.1. Knowing s_{b_i}, one can obtain a $1 - \alpha$ confidence interval for a partial regression coefficient, β_i, as

$$b_i \pm t_{\alpha(2),\nu} s_{b_i}, \tag{20.25}$$

where ν is the residual degrees of freedom.

In general, a significant F value in testing for dependence of Y on all X_i's (by Equation 20.17) will be associated with significance of some of the β_i's being concluded by the t test; but it is possible to have a significant F without any significant

t's, or even significant t's without a significant F (Geary and Leser, 1968). The latter situations often indicate a high degree of correlation between the several independent variables. Ordinarily, hypotheses about β_i's will not be appropriate if F is nonsignificant (Cramer, 1972).

If any of the partial regression coefficients are found to be nonsignificant (i.e., at least one H_0: $\beta_i = 0$ is not rejected), then the best procedure to follow is that of Section 20.8. By such an analysis, the least significant variable is dropped from the regression and a new regression equation is computed before deleting any other variables.

If independent variables, say X_1 and X_2, are correlated, then the partial regression coefficients associated with them (b_1 and b_2) may not be assumed to reflect the dependence of Y on X_1 or Y on X_2 that exists in the population. In practice, such *intercorrelation* among independent variables (also called *multicollinearity*, *collinearity*, or "ill conditioning among variables") is ignored if it is of low magnitude; but if the intercorrelation is high, conclusions regarding the significance of the correlated X_i's are likely to be spurious (Neter and Wasserman, 1974: Sections 7.9, 10.1).* When intercorrelation is present, standard errors of partial regression coefficients (s_{b_i}'s) are often large, meaning that the b_i's are imprecise estimates of the relationships in the population. As a consequence, a b_i may not be declared statistically significant (as by the above t test), even when Y and X_i are related in the population. An additional deleterious effect of intercorrelation is that it may lead to increased roundoff error in the computation of regression statistics. After examining the correlation Matrix 20.6, it may be wise to delete one or more of the intercorrelated independent variables from the regression model and reanalyze the remaining data.

20.5 STANDARDIZED PARTIAL REGRESSION COEFFICIENTS

The quantity

$$b'_i = b_i\left(\frac{s_{X_i}}{s_Y}\right), \quad \text{or, equivalently, } b'_i = b_i\sqrt{\frac{\sum x_i^2}{\sum y^2}}, \tag{20.26}$$

is termed a *standardized* (or *standard*) *partial regression coefficient* (occasionally called a "beta coefficient"). Standard partial regression coefficients are sometimes used as indications of relative importance of the various X_i's in determining the value of Y. The standardized coefficients are unitless; thus, a b'_i with a high absolute value is indicative of its associated X_i having a high degree of influence on Y. Many multiple regression computer programs include these standardized coefficients, and some include their standard errors, $s_{b'_i}$, as well. Tests of hypotheses concerning β'_i are not typically performed, for they would tell the user no more than those tests performed on hypotheses about β_i. Standardized partial regression coefficients suffer from the same problems with intercorrelation as do partial regression coefficients (see Section 20.4).

*If the intercorrelation is great, we may be unable to calculate the partial regression coefficients at all, since it may not be possible to perform the matrix inversion described in Section 20.1.

20.6 PARTIAL CORRELATION

The multiple correlation coefficient, R, reflects the overall interrelationship of all M variables. But we may desire to examine the variables two at a time. We could calculate a simple correlation coefficient, r, for each pair of variables (i.e., what Example 20.1d presents to us). But the problem with considering simple correlations of all variables, two at a time, is that such correlations will fail to take into account the interactions of any of the other variables on the two in question. *Partial correlation* solves this problem because it considers the correlation between each pair of variables while holding constant the value of each of the other variables. Symbolically, a partial correlation coefficient for a situation considering three variables (sometimes called a first-order partial correlation coefficient) would be $r_{ik \cdot l}$, which refers to the correlation between variables i and k, considering that variable l does not change its value (i.e., we have eliminated any effect of the interaction of variable l on the relationship between variables i and k). For four variables, a partial correlation coefficient, $r_{ik \cdot lp}$ (sometimes called a second-order partial correlation coefficient), expresses the correlation between variables i and k, assuming that variables l and p were held at constant values. In general, a partial correlation coefficient might be referred to as $r_{ik \cdots}$, meaning "the correlation between variables i and k, holding all other variables constant." A computer program for multiple regression and correlation providing partial correlation coefficients will generally do so in the form of a matrix, such as in Example 20.1h:

$$\begin{matrix} 1.00 & r_{12\cdots} & r_{13\cdots} & \cdots & r_{1M\cdots} \\ r_{21\cdots} & 1.00 & r_{23\cdots} & \cdots & r_{2M\cdots} \\ r_{31\cdots} & r_{32\cdots} & 1.00 & \cdots & r_{3M\cdots} \\ \vdots & \vdots & \vdots & & \vdots \\ r_{M1\cdots} & r_{M2\cdots} & r_{M3\cdots} & \cdots & 1.00 \end{matrix} \tag{20.27}$$

To test $H_0: \rho_{ik \cdots} = 0$, we may employ

$$t = \frac{r_{ik\cdots}}{s_{r_{ik\cdots}}}, \tag{20.28}$$

where

$$s_{r_{ik\cdots}} = \sqrt{\frac{1 - r^2_{ik\cdots}}{n - M}} \tag{20.29}$$

Example 20.1h

A matrix of partial correlation coefficients, as it might appear as computer output (from the data of Example 20.1a).

	1	2	3	4	5
1	1.00000	0.19426	0.12716	0.33929	−0.75406
2	0.19426	1.00000	−0.26977	0.23500	−0.06296
3	0.12716	−0.26977	1.00000	0.26630	−0.04210
4	0.33929	0.23500	0.26630	1.00000	0.50720
5	−0.75406	−0.06296	−0.04210	0.50720	1.00000

and M is the total number of variables in the multiple correlation. The statistical significance of a partial correlation coefficient (i.e., the test of H_0: $\rho_{ik\cdots} = 0$) may also be determined by employing Table B.16 for $n - M$ degrees of freedom. One-tailed hypotheses may also be performed as for simple correlation coefficients (Section 19.2). If a multiple regression and a multiple correlation analysis were performed on the same data, we would find that the test for H_0: $\beta_i = 0$ would be identical to the test for H_0: $\rho_{ik\cdots} = 0$ (by either t testing or "partial F" testing), where variable k is the dependent variable. Hypotheses such as H_0: $\rho_{ik\cdots} = \rho_0$, or similar one-tailed hypotheses, where $\rho_0 \neq 0$, may be testing using the z transformation (Section 19.2).

20.7 ROUND-OFF ERROR AND CODING DATA

It is important to realize that there may be rounding error associated with regression calculations even when they are performed on a computer (Freund, 1963; Healy, 1963; Longley, 1967; Wampler, 1970). Indeed, the number of significant figures utilized in the calculations by many computers and computer programs is less than the number possible on a good desk calculator. At each of the many steps summarized in Section 20.1, the computer rounds-off results, and round-off error at an early computational stage will be exacerbated during successive calculations. Such errors may be especially severe if variables have greatly different magnitudes, or if there is considerable intercorrelation (i.e., multicollinearity) among independent variables. Some multiple regression computer programs automatically code input data by subtracting the mean of each variable from each observed value of that variable or by transforming input data to be standardized, or normalized, scores (using the sample mean and standard deviation of each variable instead of μ and σ, respectively, in Equation 7.16). Some programs reduce round-off error by employing extra significant figures in arithmetic computations (often called "double precision"; see footnote in Example 20.1b).

Coding of X_i and/or Y affects multiple regression statistics in a fashion similar to the effects of coding X and/or Y in simple regression (Section 17.11). Each value of X_i might be coded by multiplying by a constant, M_{X_i}, and adding a constant, A_{X_i}, so that the coded datum is

$$[X_i] = M_{X_i} + A_{X_i}. \tag{20.30}$$

Similarly, each value of Y might be coded by multiplying by M_Y and adding A_Y, so that a coded value of Y would be

$$[Y] = M_Y Y + A_Y. \tag{20.31}$$

The effect of coding on multiple regression and multiple correlation statistics is shown in Table 20.2, where the bracketed statistics are those resulting from analysis of data coded as in Equations 20.30 and 20.31. Note that coding will not affect the values of R, R^2, $r_{ik\cdots}$, or any calculated t or F values thus far presented.

TABLE 20.2 THE EFFECT OF CODING ON MULTIPLE REGRESSION AND CORRELATION STATISTICS, WITH CODING PERFORMED AS IN EQUATIONS 20.30 AND 20.31

Statistic	*Value without coding*	*Value using coding*
Partial regression coefficient, b_i	$b_i = [b_i]M_{X_i}/M_Y$	$[b_i] = b_i M_Y/M_{X_i}$
Standard error of b_i, s_{b_i}	$s_{b_i} = [s_{b_i}]M_{X_i}/M_Y$	$[s_{b_i}] = s_{b_i} M_Y/M_{X_i}$
Y intercept, a	$a = \dfrac{[a] + \sum [b_i]A_{X_i} - A_Y}{M_Y}$	$[a] = aM_Y - \sum [b_i]A_{X_i} + A_Y$
Standard error* of a	$s_a = [s_a]/M_Y$	$[s_a] = s_a M_Y$
Mean of X_i, $\bar{X}_i$	$\bar{X}_i = \dfrac{[\bar{X}_i] - A_{X_i}}{M_{X_i}}$	$[\bar{X}] = M_{X_i}\bar{X}_i + A_{X_i}$
Mean of Y, $\bar{Y}$	$\bar{Y} = \dfrac{[\bar{Y}] - A_Y}{M_Y}$	$[\bar{Y}] = M_Y \bar{Y} + A_Y$
Standard error of estimate, $s_{Y \cdot 1,2,\ldots,m}$	$s_{Y \cdot 1,2,\ldots,m} = [s_{Y \cdot 1,2,\ldots,m}]/M_Y$	$[s_{Y \cdot 1,2,\ldots,m}] = M_Y s_{Y \cdot 1,2,\ldots,m}$
Coefficient of determination, R^2	$R^2 = [R^2]$	$[R^2] = R^2$
Multiple correlation coefficient, R	$R = [R]$	$[R] = R$
Test statistics,* t and F	$t = [t]$; $F = [F]$	$[t] = t$; $[F] = F$
Standardized partial regression coefficient, b_i'	$b_i' = [b_i']$	$[b_i'] = b_i'$
Standard error of b_i', $s_{b_i'}$	$s_{b_i'} = [s_{b_i'}]$	$[s_{b_i'}] = s_{b_i'}$
Correlation coefficient, r_{ik}	$r_{ik} = [r_{ik}]$	$[r_{ik}] = r_{ik}$
Partial correlation coefficient, $r_{ik\cdots}$	$r_{ik\cdots} = [r_{ik\cdots}]$	$[r_{ik\cdots}] = r_{ik\cdots}$
Corrected sum of squares, $\sum x_i^2$	$\sum x_i^2 = [\sum x_i^2]/M_{X_i}^2$	$[x_i^2] = (\sum x_i^2)(M_{X_i}^2)$
Corrected sum of crossproducts, $\sum x_i x_k$	$\sum x_i x_k = [\sum x_i x_k]/M_{X_i}M_{X_k}$	$[\sum x_i x_k] = (\sum x_i x_k)(M_{X_i}M_{X_k})$
Element of inverse corrected sum of squares and sum of crossproducts matrix, c_{ik}	$c_{ik} = [c_{ik}]/M_{X_i}M_{X_k}$	$[c_{ik}] = M_{X_i}M_{X_k}$
Element of inverse correlation matrix, d_{ik}	$d_{ik} = [d_{ik}]$	$[d_{ik}] = d_{ik}$
Sum of squares, SS	$SS = [SS]/M_Y^2$	$[SS] = (SS)(M_Y^2)$
Mean square, MS	$MS = [MS]/M_Y^2$	$[MS] = (MS)(M_Y^2)$

Note: If a regression is fitted through the origin (Section 20.13), then coding by addition of a constant may not be used (i.e., $A_{X_i} = 0$ and $A_Y = 0$).

*Each A_{X_i} must be zero if s_a is desired or if it desired to test hypotheses about the Y intercept.

20.8 SELECTION OF INDEPENDENT VARIABLES

In Example 20.1g, we obtained the statistics for the least squares best fit equation for our data. However, the fact that the data we considered consisted of four independent variables does not automatically imply that all four have a significant effect on the magnitude of the dependent variable. The problems facing the user of multiple regression analysis include the determination of which of the independent variables have significant effect on Y in the population sampled. A number of procedures can be used to conclude which is in some objective way the "best" regression model; even though the various methods do not necessarily arrive at the same conclusions (Hamaker, 1962; Draper and Smith, 1981: Chapter 6), there is no universal agreement among statisticians as to which *one* is the most advantageous method.

One procedure would be to fit a regression equation containing all four independent variables, then fit all four different equations consisting of three of the independent variables, then fit each of the possible six equations consisting of only two of the independent variables, and finally to perform a simple regression analysis utilizing each of the four independent variables separately. After fitting all 15 of these regression equations, we might choose the one resulting in the lowest residual mean square, or, perhaps, the largest R^2. There are at least two drawbacks to such a procedure, however. First, a very large number of regression equations have to be calculated, the number being $2^m - 1$. (The reader can study Section 22.1 to see where this number comes from.) Thus, if $m = 5$, there would be a total of 31 regressions to be fitted; if $m = 8$, 255 regressions would be required; if $m = 10$, we would find ourselves trying to choose among 1023 regression equations; etc. Even with the aid of a computer, the consideration of such large numbers of equations becomes prohibitively formidable. The second problem with considering all possible regressions is that of declaring an objective statistical method for determining which one of the many regression equations is to be considered the "best."

The following method offers us the ability to express whether or not one equation is preferable to another. The aim is to utilize as many variables as are required to provide a reliable description of what variables in the population effect a significant change in Y, yet to employ as few variables as are necessary for this purpose so as to minimize the time, energy, and finances to be expended in collecting further data or performing further calculations with the regression equation, and, we hope, to simplify the interpretation of the resultant regression equation.

If a multiple regression is fitted using all m independent variables (as we did in Example 20.1g), then we might ask whether the presence of any variable has insignificant influence in the sampled population and thus may be eliminated from the equation. We can examine $H_0: \beta_i = 0$ for each partial regression coefficient; if the t values for such tests are not provided in our computer output (as they are in Example 20.1g), then Equation 20.23 may be applied to calculate them.

If all m of the t values are equal to or greater than the critical value ($t_{\alpha(2),(n-m-1)}$), then we conclude that all the X_i's have a significant effect on Y and none of them should be deleted from our model. However, if some t values are less than the critical value, then the independent variable associated with the t with the lowest absolute value may be deleted from the model and a new multiple regression equation may be fitted, utilizing the remaining $m - 1$ independent variables. The null hypothesis H_0: $\beta_i = 0$ may then be tested for each partial regression coefficient in this new model, and if some t values are less than the critical value, then one more variable may be deleted and a new multiple regression analysis performed.

As demonstrated in Example 20.2, this procedure is repeated until all b_i's in the

Example 20.2

Stepwise regression analysis, utilizing the data from Example 20.1a. As shown in Example 20.1g, the multiple regression analysis for the model $\hat{Y} = \alpha + \beta_1 X_1 + \beta_2 X_2 + \beta_3 X_3 + \beta_4 X_4$ yields the following statistics:

Variable	b	s_b	t	ν
X_1	−0.12932	0.021287	−6.075	28
X_2	−0.018785	0.056278	−0.334	28
X_3	−0.046215	0.20727	−0.223	28
X_4	0.20876	0.067034	3.114	28

$a = 2.9583$

The critical value for testing $H_0: \beta_j = 0$ against $H_A: \beta_j \neq 0$ is $t_{0.05(2),28} = 2.048$. Therefore, H_0 would be rejected for β_1 and β_4, but not for β_2 or β_3. Of the t tests for the latter two, the t for testing the significance of β_3 has the smaller absolute value. Therefore, $\beta_3 X_3$ is deleted from the model, leaving $\hat{Y} = \alpha + \beta_1 X_1 + \beta_2 X_2 + \beta_4 X_4$. For this model with three independent variables, we obtain the following statistics:

Variable	b	s_b	t	ν
X_1	−0.13047	0.020312	−6.423	29
X_2	−0.015424	0.053325	−0.289	29
X_4	0.20450	0.063203	3.236	29

$a = 2.6725$

The critical value for testing the significance of the partial regression coefficients is $t_{0.05(2),29} = 2.045$. Therefore $H_0: \beta_j = 0$ would be rejected for β_1 and for β_4, but not for β_2. Therefore, $\beta_2 X_2$ is deleted from the regression model, leaving $\hat{Y} = \alpha + \beta_1 X_1 + \beta_4 X_4$. For this model with two independent variables, we obtain the following statistics:

Variable	b	s_b	t	ν
X_1	−0.13238	0.018913	−6.999	30
X_4	0.20134	0.061291	3.285	30

$a = 2.5520$

The critical value for testing $H_0: \beta_j = 0$ against $H_0: \beta_j \neq 0$ is $t_{0.05(2),30} = 2.042$. Therefore, both β_1 and β_4 are concluded to be different from zero, and $\hat{Y} = 2.552 - 0.132X_1 + 0.201X_4$ is our final model.

equation are concluded to estimate β_i's that are different from zero. (Each time a variable is thus deleted the multiple regression MS decreases slightly and the residual MS increases slightly.)

Another common stepwise procedure is to begin with the smallest possible regression model (i.e., one with only one independent variable; in other words, a simple regression) and gradually work up to the multiple regression model incorporating the largest number of significantly important independent variables. The problem here is first to determine which is the "best" simple regression model for the data, then which is the "best" model containing two independent variables, then which is the "best" one containing three X's, etc. Mantel (1970) describes how such a "step-up" procedure can involve more computational effort, and is fraught with more theoretical deficiencies, than is the "step-down" procedure previously described. Regarding computational effort, the step-up procedure might require as many as $2^m - 1$ regressions to be fit, whereas the step-down method will never involve the fitting of more than m regressions.

Finally, there is a stepwise regression procedure that employs both the addition and the elimination of independent variables. It begins just as the step-up procedure; but each time a variable is added all variables in the model are examined to see if any should be eliminated at that step.

Some computer program libraries contain routines for automatically performing addition and/or elimination of variables. However, as long as a program for multiple regression is available, data can be repeatedly submitted to the program with the user selecting a variable for deletion or addition each time as described above. Therefore, no special computer routine for stepwise multiple regression analysis is absolutely necessary.

Section 20.4 discussed the problem of *intercorrelation* among independent variables (multicollinearity). If substantial intercorrelation is present, then the partial regression coefficient, b_i, associated with an independent variable X_i, depends upon which other independent variables are in the regression model. In the stepwise regression procedure this often means that the magnitude (and even the sign) of a b_i changes as variables are deleted from the model. (Such is not the case in Example 20.2, where b_1, b_2, and b_4 change very little when X_3 is dropped from the model; and b_1 and b_4 change very little when b_2 is dropped.)

20.9 PREDICTING Y VALUES

Having fit a multiple regression equation to a set of data, we may desire to calculate the Y value to be expected at a particular combination of X_i values. Consider the a and b_i values determined in Example 20.2 for an equation of the form $\hat{Y} = a + b_1X_1 + b_2X_2$. Then the predicted value at $X_1 = 7°\text{C}$ and $X_2 = 2.0$ min, for example, would be $\hat{Y} = 2.552 - (0.132)(7) + (0.201)(2.0) = 2.03$ ml. Such predictions may be done routinely, as long as there is, in fact, a significant regression (i.e., the F from Equation 20.17 is significant).

In the consideration of the standard error of such a predicted Y, the reader might wish to refer to Section 17.4 for the calculations appropriate when $m = 1$. The following is the standard error of a mean Y predicted from a multiple regression equation:

$$s_{\hat{Y}} = \sqrt{s^2_{Y\cdot 1,2,\ldots,m}\left[\frac{1}{n} + \sum_{i=1}^{m}\sum_{k=1}^{m} c_{ik}x_ix_k\right]}. \tag{20.32}$$

In this equation, $x_i = X_i - \bar{X}_i$, where X_i is the value of independent variable i at which Y is to be predicted, and $\bar{X}_i$ is the mean of the observed values of variable i that were used to calculate the regression equation.

Thus, for the value of Y just predicted, we can solve Equation 20.32, as shown in Example 20.3.

Example 20.3

The standard error of a predicted Y. For the equation $\hat{Y} = 2.552 - 0.132X_1 + 0.201X_2$, derived from the data of Example 20.1a, where X_1 is the variable in column 1 of the data matrix, X_2 is the variable in column 4, and Y is the variable in column 5, one obtains the following quantities needed to solve Equation 20.32:

$$s^2_{Y\cdot 1,2} = 0.16844, \quad n = 33, \quad \bar{X}_1 = 4.4546,$$
$$\bar{X}_2 = 2.5424, \quad \sum x_1^2 = 472.18, \quad \sum x_2^2 = 44.961,$$
$$d_{11} = 1.0027, \quad d_{12} = -0.052051, \quad d_{21} = -0.052051, \quad d_{22} = 1.0027.$$

By employing Equation 20.9 each d_{ik} is converted to a c_{ik}, resulting in:

$$c_{11} = 0.0021236, \quad c_{12} = -0.00035724, \quad c_{21} = -0.00035724, \quad c_{22} = 0.022302.$$

What is the mean population value of Y at $X_1 = 7°C$ and $X_4 = 2.0$ min?

$$\hat{Y} = 2.552 - (0.132)(7) + (0.201)(2.0) = 2.030 \text{ ml}$$

What is the standard error of the mean population value of Y at $X_1 = 7°C$ and $X_4 = 2.0$ min? [Equation 20.32 is used.]

$$\begin{aligned} s^2_{\hat{Y}} &= 0.16844[\tfrac{1}{33} + (0.0021236)(7 - 4.4546)^2 \\ &\quad + (-0.00035724)(7 - 4.4546)(2.0 - 2.5424) \\ &\quad + (-0.00035724)(2.0 - 2.5424)(7 - 4.4546) \\ &\quad + (0.022302)(2.0 - 2.5424)^2] \\ &= 0.16844(\tfrac{1}{33} + 0.0213066) \\ &= 0.008693 \text{ ml}^2 \\ s_{\hat{Y}} &= \sqrt{0.008693 \text{ ml}^2} = 0.093 \text{ ml} \end{aligned}$$

Since $t_{0.05(2),30} = 2.042$, the 95% confidence interval for the predicted Y is $2.030 \pm (2.042)(0.093)$ ml $= 2.030 \pm 0.190$ ml.

What is the predicted value of one additional Y value taken from the population at $X_1 = 7°C$ and $X_4 = 2.0$ min?

$$\hat{Y} = 2.552 - (0.132)(7) + (0.201)(2.0) = 2.030 \text{ ml}$$

What is the standard error of the predicted value of one additional Y value taken from the population at $X_1 = 7°C$ and $X_4 = 2.0$ min? [Equation 20.34 is used.]

$$\begin{aligned} s_{\hat{Y}} &= \sqrt{0.16844[1 + \tfrac{1}{33} + 0.0213066]} \\ &= 0.421 \text{ ml} \end{aligned}$$

Since $t_{0.05(2),30} = 2.042$, the 95% confidence interval for the preceding predicted $\hat{Y}$ is $2.03 \pm (2.042)(0.421)$ ml $= 2.03 \pm 0.86$ ml.

What is the predicted value of the mean of 10 additional values of Y taken from the population at $X_1 = 7°C$ and $X_4 = 2.0$ min? [Equation 20.34 is used, substituting $\frac{1}{10}$ for 1.]

$$\hat{Y} = 2.552 - (0.132)(7) + (0.201)(2.0) = 2.030 \text{ ml}$$

What is the standard error of the predicted value of the mean of 10 additional values of Y taken from the population at $X_1 = 7°C$ and $X_4 = 2.0$ min?

$$\begin{aligned} s_{\hat{Y}} &= \sqrt{0.16844[\tfrac{1}{10} + \tfrac{1}{33} + 0.0213066]} \\ &= 0.16 \text{ ml} \end{aligned}$$

Since $t_{0.05(2),30} = 2.042$, the 95% confidence interval for the predicted Y is $2.03 \pm (2.042)(0.16)$ ml $= 2.03 \pm 0.33$ ml.

A special case of Equation 20.32 is where each $X_i = 0$. The Y in question is then the Y intercept, a, and

$$s_a = \sqrt{s^2_{Y\cdot 1,2,\ldots,m}\left[\frac{1}{n} + \sum_{i=1}^{m}\sum_{k=1}^{m} c_{ik}\bar{X}_i\bar{X}_k\right]}. \tag{20.33}$$

To predict the value of Y that would be expected if one additional set of X_i were obtained, we may use Equation 20.15, and the standard error of this prediction is

$$(s_{\hat{Y}})_1 = \sqrt{s^2_{Y\cdot 1,2,\ldots,m}\left[1 + \frac{1}{n} + \sum_{i=1}^{m}\sum_{k=1}^{m} c_{ik}x_i x_k\right]}, \tag{20.34}$$

as Example 20.3 shows. This situation is a special case of predicting the mean Y to be expected from obtaining p sets of X_i, where the X_1's in each set are equal, the X_2's in each set are equal, etc. Such a calculation is performed in Example 20.3, using

$$(s_{\hat{Y}})_p = \sqrt{s^2_{Y\cdot 1,2,\ldots,m}\left[\frac{1}{p} + \frac{1}{n} + \sum_{i=1}^{m}\sum_{k=1}^{m} c_{ik}x_i x_k\right]}. \tag{20.35}$$

20.10 TESTING DIFFERENCES BETWEEN TWO PARTIAL REGRESSION COEFFICIENTS

It is occasionally of interest to test $H_0: \beta_i - \beta_k = \beta_0$. This can be done by using

$$t = \frac{|b_i - b_k| - \beta_0}{s_{b_i - b_k}}, \tag{20.36}$$

which becomes

$$t = \frac{b_i - b_k}{s_{b_i - b_k}} \tag{20.37}$$

when $\beta_0 = 0$, in which case the null hypothesis is usually written as $H_0: \beta_i = \beta_k$. The standard error of the difference between two partial regression coefficients is

$$s_{b_i - b_k} = \sqrt{s^2_{Y\cdot 1,2,\ldots,m}[c_{ii} + c_{kk} - 2c_{ik}]} \tag{20.38}$$

and the degrees of freedom for this test are $n - m - 1$.

20.11 "DUMMY" VARIABLES

It is sometimes useful to introduce into a multiple regression model one or more additional variables, in order to account for the effects of one or more nominal scale variables on the dependent variable, Y. For example, we might be considering fitting the model $\hat{Y}_j = a + b_1X_{1j} + b_2X_{2j}$, where Y is human diastolic blood pressure, X_1 is age, and X_2 is body weight. In addition, we might be interested in determining the effect (if any) of sex on blood pressure. Our regression model could then be expanded to $\hat{Y}_j = a + b_1X_{1j} + b_2X_{2j} + b_3X_{3j}$, where X_3 is a "dummy variable," with one of two possible values: e.g., set $X = 0$ if the data are for a male and $X = 1$ if the data are for a female. By using this dummy variable, we can determine whether or not sex is a significant determinant of blood pressure (by the considerations of Section 20.4 for testing $H_0: \beta_3 = 0$). If it is, then the use of the model with all three independent variables will yield more accurate Y values than the preceding model with only two independent variables, if the regression equation is used for predicting blood pressure.

If there are three levels of the nominal scale variable, then two dummies would be needed in the regression model. For example, if we were considering the blood pressure of both sexes and of three races of humans, then we might fit the model $\hat{Y}_j = a + b_1X_{1j} + b_2X_{2j} + b_3X_{3j} + b_4X_{4j} + b_5X_{5j}$, where X_1, X_2, and X_3 are as before and X_4 and X_5 are used to denote race: e.g., if race 1, then $X_4 = 0$ and $X_5 = 0$; if race 2, then $X_4 = 0$ and $X_5 = 1$; and if race 3, then $X_4 = 1$ and $X_5 = 0$. In general, $L - 1$ dummies are required, where L is the number of levels of the variable to be represented by them.

20.12 COMPARING MULTIPLE REGRESSION EQUATIONS

It is not uncommon to desire to determine whether two or more multiple regressions, all containing the same variables, are estimating the same population regression function. We may test the null hypothesis that all the sample regression equations estimate the same population regression model by an extension of the considerations of Section 18.7. For a total of k regressions, the pooled residual sum of squares, SS_p, is simply the sum of all k residual sums of squares; and the pooled residual degrees of freedom, DF_p, is the sum of all k residual degrees of freedom. We then lump together all data from all k regressions and calculate a regression for this totality of data. The resulting total residual sum of squares and total degrees of freedom will be referred to as SS_t and DF_t, respectively.

The test of the null hypothesis (that there is a single population underlying all k sample regressions) is

$$F = \frac{\dfrac{\text{SS}_t - \text{SS}_p}{(m+1)(k-1)}}{\dfrac{\text{SS}_p}{\text{DF}_p}}, \qquad (20.39)$$

a statistic with $(m + 1)(k - 1)$ and DF_p degrees of freedom. Example 20.4 demonstrates this procedure.

Example 20.4

Comparing multiple regressions.

Let us say we have three multiple regressions, each fitted to a different sample of data, and each containing the same dependent variable and the same four independent variables. (Therefore, $m = 4$ and $k = 3$.) The residual sums of squares from each of the regressions are 437.8824, 449.2417, and 411.3548, respectively.

If the residual degrees of freedom for each of the regressions are 41, 32, and 38, respectively (that is, the three sample sizes were 46, 37, and 43, respectively), then the pooled residual sum of squares, SS_p, is 1298.4789, and the pooled residual degrees of freedom, DF_p, is 111.

Then, we combine all 126 data from all three samples and fit to these data a multiple regression having the same variables as the three individual regressions fitted previously. From this multiple regression let us say we have a total residual sum of squares, SS_t, of 1577.3106. The total residual degrees of freedom, DF_t, is 121.

Then we test H_0: All three sample regression functions estimate the same population regression, against H_A: All three sample regression functions do not estimate the same population regression:

$$F = \frac{\dfrac{SS_t - SS_p}{(m+1)(k-1)}}{\dfrac{SS_p}{DF_p}}$$

$$= \frac{\dfrac{1577.3106 - 1298.4789}{(5)(2)}}{\dfrac{1298.4789}{111}}$$

$$= 2.38$$

The degrees of freedom associated with F are 10 and 111.
Since $F_{0.05(1),10,111} \simeq 1.93$, reject H_0.
$0.01 < P < 0.025$

We may also employ the concept of parallelism in multiple regression as we did in simple regression. A simple linear regression may be represented as a line on a two-dimensional graph, and two such lines are said to be parallel (i.e., have the same slopes) if the vertical distance between them is constant for all values of the independent variable, meaning that the regression coefficients (i.e., slopes) of the two lines are the same. A multiple regression with two independent variables may be visualized as a plane in three-dimensional space. Two planes are parallel if the vertical distance between them is the same for all combinations of the independent variables, in which case all the partial regression coefficients for one regression are equal to those of the second, with only the Y intercepts possibly differing.

In general, two or more multiple regressions are said to be parallel if they all have the same β_1, β_2, β_3, etc. This may be tested by a straightforward extension of the procedure in Section 18.4. The residual sums of squares for all k regressions are summed to give the pooled residual sum of squares, SS_p; the pooled residual degrees of freedom are

$$DF_p = \sum_{i=1}^{k} n_i - k(m+1). \tag{20.40}$$

Additionally, we calculate a residual sum of squares for the "combined" regression in the following manner. Each element in a corrected sum of squares and sum of crossproducts matrix (Matrix 20.5) is formed by summing all those elements from the k regressions. For example, element $\sum x_1^2$ is formed as $(\sum x_1^2)_1 + (\sum x_1^2)_2 + (\sum x_1^2)_3 + \cdots + (\sum x_1^2)_k$, and element $\sum x_1x_2$ is formed as $(\sum x_1x_2)_1 + (\sum x_1x_2)_2 + \cdots + (\sum x_1x_2)_k$. The residual sum of squares obtained from the multiple regression analysis using the resulting matrix is the "common" residual sum of squares, SS_c; the degrees of freedom associated with it are

$$DF_c = \sum_{i=1}^{k} n_i - k - m. \tag{20.41}$$

Then the null hypothesis of all k regressions being parallel is tested by

$$F = \frac{\dfrac{SS_c - SS_p}{k-1}}{\dfrac{SS_p}{DF_p}}, \tag{20.42}$$

with $k - 1$ and DF_p degrees of freedom.

If the null hypothesis is not rejected, we then conclude that the independent variables affect the dependent variable in the same manner in all k regressions; we also conclude that all k regressions are parallel. Now we may ask whether the elevations of the k regressions are all the same. Here we proceed by an extension of the method in Section 18.5. All data for all k regressions are pooled together and one overall regression is fitted. The residual sum of square of this regression is the total residual sum of squares, SS_t, which is associated with degrees of freedom of

$$DF_t = \sum_{i=1}^{k} n_i - m - 1. \tag{20.43}$$

Then, the hypothesis of no difference among the k elevations is tested by

$$F = \frac{\dfrac{SS_t - SS_c}{k - 1}}{\dfrac{SS_c}{DF_c}}, \tag{20.44}$$

with $k - 1$ and DF_c degrees of freedom.

20.13 MULTIPLE REGRESSION THROUGH THE ORIGIN

As an expansion of the simple linear regression model presented in Section 17.5, we might propose a multiple regression model where $\alpha = 0$; that is, when all $X_i = 0$, then $Y = 0$:

$$\hat{Y}_j = \beta_1 X_{1j} + \beta_2 X_{2j} + \cdots + \beta_m X_{mj}. \tag{20.45}$$

This will be encountered only rarely in biological work, but it is worth noting that some multiple regression computer programs are capable of handling this model.* Striking differences in the computer output will be that total DF $= n$, regression DF $=$ m (the number of parameters in the model), and residual DF $= n - m$. Also, an *inverse pseudocorrelation matrix* may appear in the computer output in place of an inverse correlation or inverse sum of squares and crossproducts matrix.

20.14 NONLINEAR REGRESSION

Regression models such as

$$\hat{Y}_i = \alpha + \beta X_i, \tag{17.1}$$

$$\hat{Y}_j = \alpha + \beta_1 X_{1j} + \beta_2 X_{2j} + \cdots + \beta_m X_{mj}, \tag{20.13}$$

or

$$\hat{Y}_i = \alpha + \beta_1 X_i + \beta_2 X_i^2 + \cdots + \beta_m X_i^m \tag{21.2}$$

*Hawkins (1980) explains how a regression can be fit through the origin using the output from a computer program for fitting a regression not assumed to pass through the origin.

are more completely symbolized as

$$\hat{Y}_i = \alpha + \beta X_i + \epsilon_i, \tag{20.46}$$

$$\hat{Y}_j = \alpha + \beta_1 X_{1j} + \beta_2 X_{2j} + \cdots + \beta_m X_{mj} + \epsilon_j, \tag{20.47}$$

or

$$\hat{Y}_i = \alpha + \beta_1 X_i + \beta_2 X_i^2 + \cdots + \beta_m X_i^m + \epsilon_i, \tag{20.48}$$

respectively, where ϵ is the *residual*, the difference between the value of Y predicted from the equation (i.e., $\hat{Y}$) and the true value of Y in the population. All three of the preceding regression models are termed *linear* models because their parameters (i.e., α, β, and ϵ) appear in an additive fashion. However, cases do arise where the investigator wishes to fit to the data a model that is nonlinear with regard to its parameters. Such models might be those such as "exponential growth,"

$$\hat{Y}_i = \alpha\beta^{X_i} + \epsilon_i, \tag{20.49}$$

or

$$\hat{Y} = \alpha e^{\gamma X_i} + \epsilon_i; \tag{20.50}$$

"exponential decay,"

$$\hat{Y}_i = \alpha\beta^{-X_i} + \epsilon_i, \tag{20.51}$$

or

$$\hat{Y}_i = \alpha e^{-\gamma X_i} + \epsilon_i; \tag{20.52}$$

"asymptotic regression,"

$$\hat{Y}_i = \alpha - \beta\delta^{X_i} + \epsilon_i \tag{20.53}$$

or

$$\hat{Y}_i = \alpha - \beta(e^{-\gamma X_i}) + \epsilon_i; \tag{20.54}$$

or "logistic growth,"

$$\hat{Y}_i = \frac{\alpha}{1 + \beta\delta^{X_i}} + \epsilon_i; \tag{20.55}$$

where the various Greek letters are parameters in the model (see Snedecor and Cochran, 1980: 395, for graphs of such functions). Other nonlinear models would be those in which the residuals were not additive, but, for example, might be multiplicative:

$$\hat{Y}_i = \beta X_i \epsilon_i. \tag{20.56}$$

Sometimes a nonlinear model may be transformed into a linear one. For example, we may transform

$$\hat{Y}_i = \alpha X_i^{\beta} \epsilon_i \tag{20.57}$$

by taking the logarithm of each side of the equation, acquiring a model that is linear in its parameters:

$$\log Y_i = \log \alpha + \beta \log X_i + \log \epsilon_i. \tag{20.58}$$

Transformations must be employed with careful consideration, however, so that assumptions about the homoscedasticity of the residuals are not violated.

Biologists at times wish to fit nonlinear equations, some much more complex than the examples given, and computer programs are available for many of them. Such programs fall into two general groups. First are programs written to fit a particular model or a family of models, and the use of the program is little if any more complicated than the use of a multiple linear regression program (e.g., Zar, 1969; Dixon, 1975: 541–557). Second are general programs that can handle any of a very wide variety of models (e.g., Dixon, 1975: 541–572). To use the latter type of program, however, requires the user to submit a good deal of information, usually including the partial derivatives of the regression function with respect to each parameter in the model (thus, recalling differential calculus and/or consulting with a statistician would be in order).

Nonlinear regression programs typically involve some sort of an iterative procedure, *iteration* being the utilization of a set of parameter estimates to arrive at a set of somewhat better parameter estimates, using the new estimates to derive better estimates, etc. Thus, many of these programs require the user to submit initial estimates of (i.e., to guess the values of) the parameters in the model being fitted.

The program output for a nonlinear regression analysis is basically similar to much of the output from multiple linear regression analyses. Most importantly, the program should provide estimates of the parameters in the model (i.e., the statistics in the regression equation), the standard error of each of these statistics, and an analysis of variance summary including at least the regression and residual SS and DF. If regression and residual MS are not presented in the output, they may be calculated by dividing the appropriate SS by its associated DF. An F test of significance of the entire regression (or correlation) and the coefficient of determination may be obtained by means of Equations 20.17 and 20.18, respectively. Testing whether any of the parameters in the model is equal to any hypothesized value may be effected by t tests similar to those previously used for simple and partial regression coefficients (e.g., Section 20.4).

Further discussions of nonlinear regression are found in Draper and Smith (1981: Chapter 10) and Snedecor and Cochran (1980: Chapter 19). With the increasing availability of computer programs for such analyses, biologists will likely consider nonlinear models more frequently than they have previously.

20.15 DESCRIPTIVE VS. PREDICTIVE MODELS

Ideally, it is hoped that a regression model implies a biological dependence (i.e., a cause and effect) in nature, and that this dependence confirms the mathematical relationship described by the regression equation. However, regression equations are often useful solely as a means of predicting the value of a variable, if the values of a number of associated variables are known. For example, we may desire to be able to predict the weight (call it variable Y) of a mammal, given the length of the femur (variable X). Perhaps a polynomial regression such as

$$\hat{Y}_i = a + b_1X_i + b_2X_i^2 + b_3X_i^3 + b_4X_i^4 \quad (20.59)$$

might be found to fit the data rather well. (See Chapter 21 for details of polynomial regression.) Or perhaps we wish to be able to predict a man's blood pressure (call it variable P) as accurately as we can by using measurements of his weight (variable W), his age (variable A), and his height (variable H). By deriving additional regression terms composed of combinations and powers of the three measured independent variables, we might conclude statistical significance of each term in an equation such as

$$\hat{P}_i = a + b_1 W_i + b_2 A_i + b_3 H_i + b_4 W_i^2 + b_5 H_i^2 + b_6 W_i^3 + b_7 W_i A_i + b_8 H_i A_i + b_9 W_i^3 A_i. \quad (20.60)$$

Now, equations such as 20.59 and 20.60 might have statistically significant partial regression coefficients. They might also have associated with them small standard errors of estimate, meaning that the standard error of predicted Y_i's or P_i's would be small. Thus, we would have good regression equations for purposes of prediction; but we are not implying that the fourth power of femur length has any natural significance in determining mammal weights, or that terms such as $H_i A_i$ or $W_i^3 A_i$ have any biological significance relative to human blood pressure.

To realize a regression function that describes underlying biological phenomena, the investigator must possess a good deal of knowledge about the interrelationships in nature among the variables in the model. Is it indeed reasonable to assume underlying relationships to be linear, or is there a logical basis for seeking to define a particular nonlinear relationship? (For example, forcing a linear model to fit a set of data in no way "proves" that the underlying biological relationships are, in fact, linear.) Are the variables included in the model meaningful choices? (For example, we might find a significant regression of variable A on variable B, whereas a third variable, C, is actually causing the changes in both A and B.) Statistical analysis is only a tool; it can not substitute for incomplete or fallacious biological information.

20.16 CONCORDANCE: RANK CORRELATION AMONG SEVERAL VARIABLES

The concept of correlation between two variables can be expanded to consider association among more than two, as shown in earlier discussion of multiple correlation. Such association is readily measured nonparametrically by a statistic known as *Kendall's coefficient of concordance** (Kendall and Babington-Smith, 1939; Kendall, 1962: Chapter 6).† To demonstrate, let us expand the considerations of Example 19.1 to examine whether there is concordance (i.e., association) among the magnitudes of wing, tail, and bill lengths in birds of a particular species. Example 20.5 shows such data, for which we determine the ranks for each of the three variables (just as we did for each of the two variables in Example 19.11).

*Maurice George Kendall (1907–1983), English statistician.

†Wallis (1939) introduced this statistic independently, calling it the "correlation ratio," and designating it by η_r^2 (where η is the lowercase Greek eta).

Example 20.5

Kendall's coefficient of concordance.

Birds (i)	**Wing Length** (cm) Data	Ranks	**Tail Length** (cm) Data	Ranks	**Bill Length** (mm) Data	Ranks	*Sums of ranks* (R_i)
1	10.4	4	7.4	5	17	5.5	14.5
2	10.8	8.5	7.6	7	17	5.5	21
3	11.1	10	7.9	11	20	9.5	30.5
4	10.2	1.5	7.2	2.5	14.5	2	6
5	10.3	3	7.4	5	15.5	3	11
6	10.2	1.5	7.1	1	13	1	3.5
7	10.7	7	7.4	5	19.5	8	20
8	10.5	5	7.2	2.5	16	4	11.5
9	10.8	8.5	7.8	9.5	21	11	29
10	11.2	11	7.7	8	20	9.5	28.5
11	10.6	6	7.8	9.5	18	7	22.5
12	11.4	12	8.3	12	22	12	36

$M = 3$

$n = 12$

Without correction for ties:

$$W = \frac{\sum R_i^2 - \frac{(\sum R_i)^2}{n}}{\frac{M^2(n^3 - n)}{12}}$$

$$= \frac{(14.5^2 + 21^2 + 30.5^2 + \cdots + 36^2) - \frac{(14.5 + 21 + 30.5 + \cdots + 36)^2}{12}}{\frac{3^2(12^3 - 12)}{12}}$$

$$= \frac{5738.5 - \frac{(234)^2}{12}}{\frac{15444}{12}}$$

$$= \frac{1175.5}{1287} = 0.913$$

H_0: There is no association among the three variables.

H_A: There is association among the three variables.

$\chi_r^2 = M(n - 1)W$

$= (3)(12 - 1)(0.913)$

$= 30.129$

Since $n >$ that in Table B.13, we use Table B.1.

$\nu = n - 1 = 11$

$\chi^2_{0.05,11} = 19.675$

Reject H_0.

$0.001 < P < 0.005$

Incorporating the correction for ties (preferable):

In group 1 (wing length): there are 2 data tied at 10.2 cm (i.e., $t_1 = 2$); there are 2 data tied at 10.8 cm (i.e., $t_2 = 2$).

In group 2 (tail length): there are 2 data tied at 7.2 cm (i.e., $t_3 = 2$); there are 3 data tied at 7.4 cm (i.e., $t_4 = 3$); there are 2 data tied at 7.8 cm (i.e., $t_5 = 2$).

In group 3 (bill length): there are 2 data tied at 17 mm (i.e., $t_6 = 2$); there are 2 data tied at 20 mm (i.e., $t_7 = 2$).

Considering all seven groups of ties,

$$\begin{aligned} T &= \sum_{i=1}^{7} (t_i^3 - t_i) \\ &= (2^3 - 2) + (2^3 - 2) + (2^3 - 2) + (3^3 - 3) + (2^3 - 2) + (2^3 - 2) + (2^3 - 2) \\ &= 60 \end{aligned}$$

and

$$\begin{aligned} W_c &= \frac{1175.5}{\dfrac{15444 - 3(60)}{12}} \\ &= \frac{1175.5}{1272} = 0.924 \end{aligned}$$

Then, to test the significance of W_c:

$$\begin{aligned} (\chi_r^2)_c &= M(n-1)W_c \\ &= (3)(12-1)(0.924) \\ &= 30.492 \end{aligned}$$

The same conclusion is reached in this case with W_c as with W, namely: Reject H_0; $0.001 < P < 0.005$.

Several equivalent computational formulas for the coefficient of concordance are found in various texts. One that is easy to use is

$$W = \frac{\sum R_i^2 - \dfrac{(\sum R_i)^2}{n}}{\dfrac{M^2(n^3 - n)}{12}}, \tag{20.61}$$

where M is the number of variables being correlated, and n is the number of data per variable.* The numerator is simply the sum of squares of the n rank sums (analogous to Equation 4.7.)

The value of W may range from 0 (when there is no association and, consequently the R_i's are equal and the sum of squares of R_i is zero) to 1 (when there is complete agreement among the ranking of all groups). In Example 20.5 there is a very high level of concordance ($W = 0.913$), indicating that a bird with a large measurement for one of the variables is likely to have a large measurement for each of the other two variables.

We can ask whether a calculated sample W is significant; that is, whether it represents an association different from zero in the population that was sampled. A simple way to do this involves the relationship between the Kendall coefficient of concordance, W, and the Friedman chi-square, χ_r^2. Using the notation from this section,

$$\chi_r^2 = M(n-1)W. \tag{20.63}$$

*Another convenient formula is

$$W = \frac{12 \sum R_i^2 - 3M^2n(n+1)^2}{M^2(n^3 - n)}. \tag{20.62}$$

Therefore, we can convert a calculated W to its equivalent χ_r^2 and then employ our table of critical values of χ_r^2 (Table B.13). If either n or M is larger than that found in this table, then χ_r^2 may be assumed to be approximated by χ^2 with $n - 1$ degrees of freedom, and Table B.1 is used. This is demonstrated in Example 20.5.

The Coefficient of Concordance with Tied Ranks. If there are tied ranks within any of the M groups, then mean ranks are assigned as in previous discussions (e.g., Section 9.6, Example 9.6). Then W is computed with a correction for ties,

$$W_c = \frac{\sum R_i^2 - \frac{(\sum R_i)^2}{n}}{\frac{M^2(n^3 - n) - M \sum T}{12}}, \tag{20.64}$$

where

$$T = \sum_{i=1}^{m} (t_i^3 - t_i), \tag{20.65}$$

t_i is the number of ties in the ith group of ties, and m is the number of groups of tied ranks.* This computation of W_c is demonstrated in Example 20.5. W_c will not differ appreciably from W unless the numbers of tied data are great.

The Coefficient of Concordance for Assessing Agreement. A common use of Kendall's coefficient of concordance is to express the intensity of agreement among several rankings. In Example 20.6, each of three children has been asked to rank the palatability of six flavors of ice cream. We wish to ask whether the three evaluators arrive at the same rankings. The conclusion is that there is concordance (i.e., agreement) among the three.

Example 20.6

Kendall's coefficient of concordance used to assess agreement by three children in ranking palatability of six different ice cream flavors.

H_0: There is no agreement among the three rankings.
H_A: There is agreement among the three rankings.

	Flavors (i)						
Child	1	2	3	4	5	6	
1	5	1	3	2	4	6	
2	6	2	3	1	5	4	
3	6	3	2	1	4	5	
Rank sum (R_i)	17	6	8	4	13	15	$\sum R_i = 63$

*Analogous to Equation 20.62 is

$$W_c = \frac{12 \sum R_i^2 - 3M^2 n(n+1)^2}{M^2(n^3 - n) - M \sum T} \tag{20.66}$$

when ties are present.

$$M = 3$$
$$n = 6$$
$$W = \frac{\sum R_i^2 - \frac{(\sum R_i)^2}{n}}{\frac{M^2(n^3 - n)}{12}}$$
$$= \frac{17^2 + 6^2 + 8^2 + 4^2 + 13^2 + 15^2 - \frac{63^2}{6}}{\frac{3^2(6^3 - 6)}{12}}$$
$$= \frac{137.50}{157.50} = 0.873$$
$$\chi_r^2 = M(n - 1)W$$
$$= (3)(6 - 1)(0.873)$$
$$= 13.095$$

Using Table B.13, $(\chi_r^2)_{0.05,3,6} = 7.000$.
Therefore, reject H_0.
$P < 0.001$

The Relationship Between W and r_s. It is interesting to note that W is related to the mean value of all possible Spearman rank correlation coefficients that could be obtained from all possible pairs of variables. These correlation coefficients may be listed in a matrix array:

$$\begin{matrix} (r_s)_{11} & (r_s)_{12} & (r_s)_{13} & \cdots & (r_s)_{1M} \\ (r_s)_{21} & (r_s)_{22} & (r_s)_{23} & \cdots & (r_s)_{2M} \\ (r_s)_{31} & (r_s)_{32} & (r_s)_{33} & \cdots & (r_s)_{3M} \\ \cdot & \cdot & \cdot & & \cdot \\ \cdot & \cdot & \cdot & & \cdot \\ \cdot & \cdot & \cdot & & \cdot \\ (r_s)_{M1} & (r_s)_{M2} & (r_s)_{M3} & \cdots & (r_s)_{MM} \end{matrix} \qquad (20.67)$$

a form similar to that of Matrix 20.6. As in Matrix 20.6, each element of the diagonal, $(r_s)_{ii}$, is equal to 1.0, and each element below the diagonal is duplicated above the diagonal, as $(r_s)_{ik} = (r_s)_{ki}$. There are $M!/[2(M - 2)!]$ different r_s's possible for M variables.*

In Example 20.5, we are speaking of three r_s's: $(r_s)_{12}$, the r_s for wing length and tail length; $(r_s)_{13}$, the r_s for wing and bill lengths; and $(r_s)_{23}$, the r_s for tail and bill lengths. The Spearman rank correlation coefficient matrix would be

$$\begin{matrix} 1.000 & & \\ 0.853 & 1.000 & \\ 0.918 & 0.892 & 1.000 \end{matrix}$$

For Example 20.6, the r_s matrix would be

$$\begin{matrix} 1.000 & & \\ 0.771 & 1.000 & \\ 0.771 & 0.886 & 1.000 \end{matrix}$$

Denoting the mean r_s as $\overline{r_s}$, the relationship with W is

*That is, M things taken two at a time. (See Section 22.1.)

$$W_c = \frac{(M-1)\overline{r_s} + 1}{M}; \tag{20.68}$$

therefore,

$$\overline{r_s} = \frac{MW_c - 1}{M - 1}. \tag{20.69}$$

(If there are no ties, then $W_c = W$.) While the possible range of W is 0 to 1, $\overline{r_s}$ may range from $-1/(M-1)$ to 1. For Example 20.5, $\overline{r_s} = (0.853 + 0.918 + 0.892)/3 = 0.888$; for Example 20.6, $\overline{r_s} = 0.809$; and the reader can verify that Equations 20.68 and 20.69 hold.

If $M = 2$ (i.e., there are only two variables, or rankings, being correlated, as in Example 19.11), then either r_s or W might be computed; and

$$W_c = \frac{\overline{r_s} + 1}{2}, \tag{20.70}$$

and

$$r_s = 2W_c - 1. \tag{20.71}$$

When $M = 2$, the use of r_s is preferable, as it ranges from -1 to 1 and there are more thorough tables of critical values available.

20.17 CONCORDANCE BETWEEN GROUPS

If significant concordance is concluded for each of two groups of data, we may wish to ask if the agreement within each group is the same for both groups. For example, the data in Example 20.6 are for ice cream flavor preference as assessed by three children, while the data tabulated in Example 20.7 are evaluations of the same ice cream flavors by four adults. Since there is both significant concordance among the children and significant agreement among the adults, we may ask whether the consensus in the former group is the same as that in the latter. That is, the children agree among themselves, and the adults agree among themselves, but do the children concur with the adults in ranking the six flavors?

Example 20.7

Comparison of the concordance within the children data of Example 20.6 and data for the group of adults given below.

H_0: There is no agreement among the four rankings.
H_A: There is agreement among the four rankings.

	Flavors (i)						
Adult	1	2	3	4	5	6	
1	3	5	4	6	2	1	
2	2	6	5	4	3	1	
3	1	5	3	6	4	2	
4	1	4	6	5	2	3	
R_{i2}	7	20	18	21	11	7	$\sum R_{i2} = 84$

$M_2 = 4$
$n = 6$

$$W = \frac{7^2 + 20^2 + 18^2 + 21^2 + 11^2 + 7^2 - \dfrac{84^2}{6}}{\dfrac{4^2(6^3 - 6)}{12}}$$

$$= \frac{208}{280} = 0.743$$

$$\chi_r^2 = (4)(6 - 1)(0.743)$$
$$= 14.860$$
$$(\chi_r^2)_{4,6} = 7.600$$
Reject H_0.
$P < 0.001$

H_0: There is no agreement both within and between adults and children.
H_A: There is agreement both within and between adults and children.

$$\mathcal{W} = \frac{12 \sum R_{i1}R_{i2} - 3M_1M_2n(n+1)^2}{M_1M_2(n^3 - n)}$$

$$= \frac{12[(17)(7) + (6)(20) + (8)(18) + (4)(21) + (13)(11) + (15)(7)] - 3(3)(4)(6)(6+1)^2}{(3)(4)(6^3 - 6)}$$

$$= \frac{-2004}{2520} = -0.795$$

$$Z = \mathcal{W}\sqrt{M_1M_2(n-1)}$$
$$= -0.795\sqrt{(3)(4)(6-1)}$$
$$= -6.158$$
$Z_{0.05(1)} = 1.6449$
As $-6.158 < 1.6449$, do not reject H_0.
$P \gg 0.25$

Example 20.7 performs such an analysis, using the procedure of Schucany and Frawley (1973). We shall designate the rank sums from group 1 (the rankings by the children) as R_{i1} and the rank sums from group 2 (the adults' rankings) as R_{i2}. M_1 is the number of times the n items are ranked in group 1 (i.e., M_1 is the number of children), and M_2 is the number of rankings in group 2 (i.e., the number of adults). Then, we compute a multigroup coefficient of concordance:

$$\mathcal{W} = \frac{12 \sum_{i=1}^{n} R_{i1}R_{i2} - 3M_1M_2n(n+1)^2}{M_1M_2(n^3 - n)}. \tag{20.72}$$

This coefficient can range from 1 (when there is both perfect agreement among the rankings within each of the two groups and perfect agreement between the two groups) to -1 (when there is perfect agreement within each group, but the rank orders within the two groups are opposite each other); any lack of concordance within or between groups will drive $\mathcal{W}$ closer to zero.

The significance of $\mathcal{W}$ may be determined using

$$Z = \mathcal{W}\sqrt{M_1M_2(n-1)}, \tag{20.73}$$

where the critical value (from Table B.2) is $Z_{\alpha(1)}$. This is a test of the null hypothesis that there is no agreement in ranking both within each group and between the two groups, against the alternative that there is concordance both within and between groups. If $n \leq 3$, then the use of the critical values given by Frawley and Schucany (1972) is preferred to the preceding normal approximation.

Schucany and Frawley (1973) briefly discuss concordance among more than two groups.

EXERCISES

20.1. Given the following data:

Y (g)	X_1 (m)	X_2 (cm)	X_3 (m^2)	X_4 (cm)
51.4	0.2	17.8	24.6	18.9
72.0	1.9	29.4	20.7	8.0
53.2	0.2	17.0	18.5	22.6
83.2	10.7	30.2	10.6	7.1
57.4	6.8	15.3	8.9	27.3
66.5	10.6	17.6	11.1	20.8
98.3	9.6	35.6	10.6	5.6
74.8	6.3	28.2	8.8	13.1
92.2	10.8	34.7	11.9	5.9
97.9	9.6	35.8	10.8	5.5
88.1	10.5	29.6	11.7	7.8
94.8	20.5	26.3	6.7	10.0
62.8	0.4	22.3	26.5	14.3
81.6	2.3	37.9	20.0	0.5

(a) Fit the multiple regression model $\hat{Y} = \alpha + \beta_1 X_1 + \beta_2 X_2 + \beta_3 X_3 + \beta_4 X_4$ to the data, computing the partial regression coefficients and the Y intercept.
(b) By analysis of variance, test the hypothesis that there is no significant multiple regression relationship.
(c) If H_0 is rejected in part (b), compute the standard error of each partial regression coefficient and test each H_0: $\beta_i = 0$.
(d) Calculate the standard error of estimate and the coefficient of determination.
(e) What is the predicted mean population value of Y at $X_1 = 5.2$ m, $X_2 = 21.3$ cm, $X_3 = 19.7$ m^2, and $X_4 = 12.2$ cm?
(f) What are the 95% confidence limits for the $\hat{Y}$ of part (e)?
(g) Test the hypothesis that the mean population value of Y at the X_i's stated in part (e) is greater than 50.0 g.

20.2. Subject the data of Exercise 20.1 to a stepwise regression analysis.

20.3. Analyze the five variables in Exercise 20.1 as a multiple correlation.
(a) Compute the multiple correlation coefficient.
(b) Test the null hypothesis that the population multiple correlation coefficient is zero.
(c) Compute the partial correlation coefficient for each pair of variables.
(d) Determine which of the partial correlation coefficients estimate population partial correlation coefficients different from zero.

20.4. The following values were obtained for three multiple regressions of the form $\hat{Y} = a + b_1X_1 + b_2X_2 + b_3X_3$. Test the null hypothesis that each of the three sample regressions estimates the same population regression function.

Regression	*Residual sum of squares*	*Residual degrees of freedom*
1	44.1253	24
2	56.7851	27
3	54.4288	21
All data combined	171.1372	

20.5. Each of five research papers was read by each of four faculty reviewers. Each reviewer then ranked the quality of the five papers, as follows:

	Papers 1	2	3	4	5
Reviewer 1	5	4	3	1	2
Reviewer 2	4	5	3	2	1
Reviewer 3	5	4	1	2	3
Reviewer 4	5	3	2	4	1

(a) Calculate the Kendall coefficient of concordance.
(b) Test whether the rankings by the four reviewers are in agreement.
(c) Calculate the coefficient of concordance for the following four student reviewers of the same five research papers:

	Papers 1	2	3	4	5
Reviewer 1	5	3	1	2	4
Reviewer 2	4	5	2	1	3
Reviewer 3	4	5	1	2	3
Reviewer 4	5	4	2	3	1

(d) Test for statistical significance of the coefficient of concordance.
(e) If the coefficients of concordance of parts (a) and (c) are concluded significant in parts (b) and (d), respectively, then test for concordance between student and faculty reviewers.

21 Polynomial Regression

A special type of multiple regression is that concerning a *polynomial* expression

$$\hat{Y}_i = \alpha + \beta_1 X_i + \beta_2 X_i^2 + \beta_3 X_i^3 + \cdots + \beta_m X_i^m, \quad (21.1)$$

a model with parameters estimated in the expression

$$\hat{Y}_i = a + b_1 X_i + b_2 X_i^2 + b_3 X_i^3 + \cdots + b_m X_i^m, \quad (21.2)$$

for which a more concise symbolism is possible:

$$\hat{Y}_i = a + \sum_{j=1}^{m} b_j X_i^j. \quad (21.3)$$

If $m = 1$, then the polynomial regression reduces to a simple linear regression (with Equations 21.1 and 21.2 becoming Equations 17.1 and 17.8, respectively.)

21.1 POLYNOMIAL CURVE FITTING

As shown in Example 21.1, the data for the fitting of this model really only consist of two variables: the dependent variable, Y, and the independent variable, X. The remaining variables in the model are derived from X; the second independent variable in the model is the square of X, the third is the cube of X, etc.* If Y_i, X_i, X_i^2, $X_i^3, \ldots,$

*The transformation of X and/or Y in polynomial regression follows the considerations of Section 17.10.

X_i^m are submitted to a multiple regression computer program, then the statistics, b_j, in Equation 21.2 are readily obtained. Alternatively, there are programs specifically for the analysis of polynomial regression, to which one need only submit Y_i and X_i (e.g., Dixon, 1975: 593–620; Zar, 1968b; also see Nie et al., 1975: 371–372).

Example 21.1

Stepwise polynomial regression.

The following data are submitted to a stepwise polynomial regression analysis:

X (kg)	Y (hr)
1.22	40.9
1.34	41.8
1.51	42.4
1.66	43.0
1.72	43.4
1.93	43.9
2.14	44.3
2.39	44.7
2.51	45.0
2.78	45.1
2.97	45.4
3.17	46.2
3.32	47.0
3.50	48.6
3.53	49.0
3.85	49.7
3.95	50.0
4.11	50.8
4.18	51.1

First, a linear regression is fit to the data, resulting in:

$a = 37.389, \quad b = 3.1269, \quad \text{and } s_b = 0.15099$

To test H_0: $\beta = 0$ against H_A: $\beta \neq 0$, $t = \dfrac{b}{s_b} = 20.709$, with $\nu = 17$.

Since $t_{0.05(2),17} = 2.110$, H_0 is rejected.

Then, a quadratic (second-power) regression is fit to the data, resulting in:

$a = 40.302, \quad b_1 = 0.66658, \quad s_{b_1} = 0.91352$

$b_2 = 0.45397, \quad s_{b_2} = 0.16688$

To test H_0: $\beta_2 = 0$ against H_A: $\beta_2 \neq 0$, $t = 2.720$, with $\nu = 16$.

Since $t_{0.05(2),16} = 2.120$, H_0 is rejected.

Then, a cubic (third-power) regression is fit to the data, resulting in

$a = 32.767, \quad b_1 = 10.411, \quad s_{b_1} = 3.9030$

$b_2 = -3.3868, \quad s_{b_2} = 1.5136$

$b_3 = 0.47011, \quad s_{b_3} = 0.18442$

To test H_0: $\beta_3 = 0$ against H_A: $\beta_3 \neq 0$, $t = 2.549$, with $\nu = 15$.

Since $t_{0.05(2),15} = 2.131$, H_0 is rejected.

Then, a quartic (fourth-power) regression is fit to the data, resulting in

$a = 6.9265, \quad b_1 = 55.835, \quad s_{b_1} = 12.495$

$b_2 = -31.487, \quad s_{b_2} = 7.6054$

$$b_3 = 7.7625, \quad s_{b_3} = 1.9573$$
$$b_4 = -0.67507, \quad s_{b_4} = 0.18076$$

To test $H_0: \beta_4 = 0$ against $H_A: \beta_4 \neq 0$, $t = 3.735$, with $\nu = 14$.
Since $t_{0.05(2),14} = 2.145$, H_0 is rejected.

Then, a quintic (fifth-power) regression is fit to the data, resulting in

$$a = 36.239, \quad b_1 = -9.1615, \quad s_{b_1} = 49.564$$
$$b_2 = 23.387, \quad s_{b_2} = 41.238$$
$$b_3 = -14.346, \quad s_{b_3} = 16.456$$
$$b_4 = 3.5936, \quad s_{b_4} = 3.1609$$
$$b_5 = -0.31740, \quad s_{b_5} = 0.23467$$

To test $H_0: \beta_5 = 0$ against $H_A: \beta_5 \neq 0$, $t = 1.353$, with $\nu = 13$.
Since $t_{0.05(2),13} = 2.160$, do not reject H_0.

Therefore, it appears that a quartic polynomial is the optimum regression function for the data. But to be more certain, we add one more term beyond the quintic to the model (i.e., a sextic, or sixth-power, polynomial regression is fit to the data), resulting in

$$a = 157.88, \quad b_1 = -330.98, \quad s_{b_1} = 192.28$$
$$b_2 = 364.04, \quad s_{b_2} = 201.29$$
$$b_3 = -199.36, \quad s_{b_3} = 108.40$$
$$b_4 = 58.113, \quad s_{b_4} = 31.759$$
$$b_5 = -8.6070, \quad s_{b_5} = 4.8130$$
$$b_6 = 0.50964, \quad s_{b_6} = 0.29560$$

To test $H_0: \beta_6 = 0$ against $H_A: \beta_6 \neq 0$, $t = 1.724$, with $\nu = 12$.
Since $t_{0.05(2),12} = 2.179$, do not reject H_0.

In concluding that the quartic regression is that of optimum fit to the data, we have: $\hat{Y} = 6.9265 + 55.835X - 31.487X^2 + 7.7625X^3 - 0.67507X^4$. See Fig. 21.1 for graphical presentation of the preceding polynomial equations.

Regardless of how we perform the fitting of a polynomial regression, we need to determine the maximum power of the polynomial that has statistical significance. To do this, we shall employ a stepwise regression procedure, remembering that the maximum power, m, may be no greater than $n - 1$ if a polynomial is to be fit to the data, and, more practically, no greater than $n - 2$ if statistical analysis is to be performed on the resulting polynomial fit.

We may use the stepwise procedure detailed in Section 19.8, beginning with the fitting of a polynomial with a larger m than it is felt will be needed. Then we would examine the significance of the b_j associated with the largest power in the model. If this term represents a β_j significantly different from zero, the analysis will be terminated (or, we would likely be tempted to try a higher power). If not, the term will be deleted from the model and the new model fitted to the data, with the procedure repeated until the b_j associated with the highest power of X is significantly different from zero.

A second and probably more common method for fitting the polynomial model is to proceed from a small to a large model, rather than from large to small. We begin by fitting a linear regression to the data: $Y_i = \alpha + \beta X_i$ (as shown in Example 21.1 and Fig. 21.1a). Then we fit a second-degree polynomial (called a *quadratic* equation) to the data. This model, $\hat{Y}_i = \alpha + \beta_1 X_i + \beta_2 X_i^2$ (Fig. 21.1b), simply adds one term (called the quadratic term) to the simple regression. To determine whether the addition of this term significantly improves the accuracy of the prediction of Y values, we may

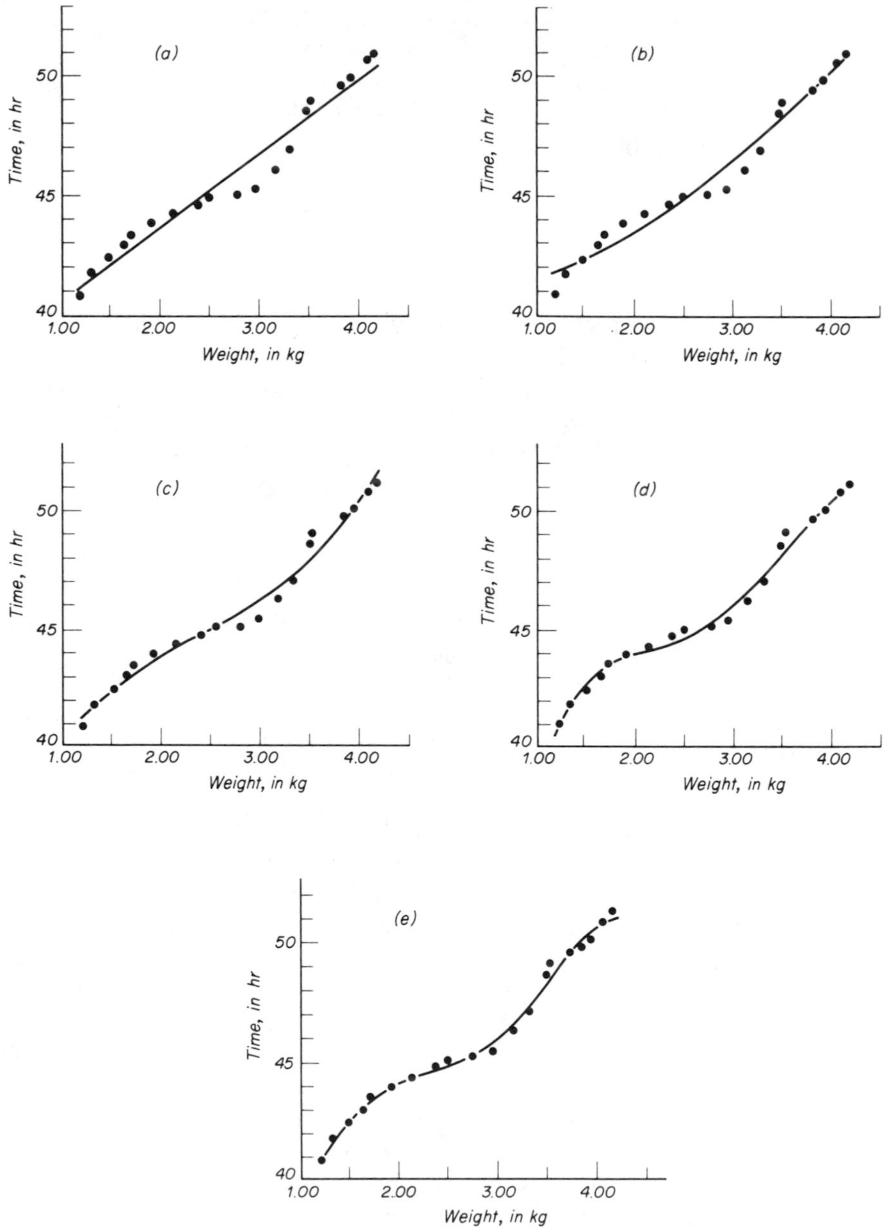

Figure 21.1 Fitting polynomial regression models. Each of the following regressions is fit to the 19 data points of Example 21.1. (a) Linear: $\hat{Y} = 37.389 + 3.1269X$. (b) Quadratic: $\hat{Y} = 40.302 + 0.66658X + 0.45397X^2$. (c) Cubic: $\hat{Y} = 32.767 + 10.411X - 3.3868X^2 + 0.47011X^3$. (d) Quartic: $\hat{Y} = 6.9265 + 55.835X - 31.487X^2 + 7.7625X^3 - 0.67507X^4$. (e) Quintic: $\hat{Y} = 36.239 - 9.1615X + 23.387X^2 - 14.346X^3 + 3.5936X^4 - 0.31740X^5$. The stepwise analysis of Example 21.1 concludes that the quartic equation provides the optimum fit; that is, the quintic expression does not provide a significant improvement in fit over the quartic.

use either of two testing procedures. One is to apply the t test to H_0: $\beta_2 = 0$ (see Section 20.4), with failure to reject the null hypothesis indicating that the simple regression model describes the dependence of Y on X sufficiently well and the quadratic term contributes insignificantly to this description. We may reach the same conclusion by calculating

$$F = \frac{\text{(regression SS for higher degree model)} - \text{(regression SS for lower degree model)}}{\text{residual MS for higher degree model}} \tag{21.4}$$

with a numerator DF of 1 and a denominator DF that is the residual DF for the higher degree model. This F test is equivalent to the preceding t test, because it also considers H_0: $\beta_j = 0$, where j is the last term in the higher degree model.

If it is concluded that $\beta_2 \neq 0$, then we may fit a third-degree polynomial (called a *cubic* equation) to the data by adding a cubic term to the model (see Example 21.1 and Fig. 21.1c). Then, H_0: $\beta_3 = 0$ may be tested by either the t or F procedure described above, and we can conclude whether the addition of the term $\beta_3 X_i^3$ significantly improves the model. If it does not, then the quadratic equation is assumed to be an appropriate description of the relationship between Y and X. If we conclude that $\beta_3 \neq 0$, then a fourth-degree polynomial (called a *quartic* equation) is fitted to the data (Fig. 21.1d), we test H_0: $\beta_4 = 0$, and so on.

Once we have not rejected H_0: $\beta_j = 0$ for the last term in the polynomial, we may cease adding further terms and conclude that the polynomial with $j - 1$ terms is the best model. However, as done in Example 21.1, it is good practice to carry the analysis one or two terms beyond the point where the preceding H_0 is not rejected, to try to make certain that significant terms are not being neglected inadvertently. For example, it is possible to not reject H_0: $\beta_3 = 0$, but by testing further to reject H_0: $\beta_4 = 0$. A polynomial regression may be fit through the origin using the considerations of Section 20.4.

After arriving at a final equation in a polynomial regression analysis it may be desired to predict values of Y at a given value of X. This can be done by the procedures or Section 20.9, by which the precision of a predicted $\hat{Y}$ (expressed by a standard error or confidence interval) may also be computed.

21.2 ROUND-OFF ERROR AND CODING DATA

In a polynomial regression analysis it is very often desirable to code X to avoid extremely large or extremely small ($\ll 1.0$) numbers resulting from raising X to the various powers. Although R^2, t, and F will not be affected by the coding shown below, Table 21.1 shows how other statistics are affected.

Let us code X by multiplying it by a constant, M_X, and code Y by multiplying it by M_Y. If desired, we may code Y by adding a constant, A_Y, whether or not M_Y is employed. Therefore, the coded variables are

$$[X] = M_X X \tag{21.5}$$

and

$$[Y] = M_Y Y + A_Y. \tag{21.6}$$

In Table 21.1 the bracketed statistics are those resulting from a polynomial regression analysis using values of X and Y that have been coded, and Example 21.5 demonstrates the use of coding in polynomial regression. Note that in polynomial regression we should not employ an addition constant, A_X, to code X, as may be done with other kinds of regression (see Sections 17.11 and 20.7).

TABLE 21.1 THE EFFECT OF CODING ON POLYNOMIAL REGRESSION STATISTICS, WITH CODING PERFORMED AS IN EQUATIONS 21.5 AND 21.6

Statistic	*Value without coding*	*Value with coding*
Partial regression coefficient, b_i	$b_i = [b_i]M_X^i/M_Y$	$[b_i] = b_i M_Y/M_X^i$
Standard error of b_i, s_{b_i}	$s_{b_i} = [s_{b_i}]M_X^i/M_Y$	$[s_{b_i}] = s_{b_i} M_Y/M_X^i$
Y intercept, a	$a = \dfrac{[a] - A_Y}{M_Y}$	$[a] = M_Y a + A_Y$
Standard error of a, s_a	$s_a = [s_a]/M_Y$	$[s_a] = M_Y s_a$
Mean of X^i, $\overline{X^i}$	$\overline{X^i} = [\overline{X^i}]/M_X^i$	$[\overline{X^i}] = M_X^i \overline{X^i}$
Mean of Y, $\bar{Y}$	$\bar{Y} = \dfrac{[Y] - A_Y}{M_Y}$	$[\bar{Y}] = M_Y \bar{Y} + A_Y$
Standard error of estimate, $s_{Y\cdot 1,2,\ldots,m}$	$s_{Y\cdot 1,2,\ldots,m} = [s_{Y\cdot 1,2,\ldots,m}]/M_Y$	$[s_{Y\cdot 1,2,\ldots,m}] = M_Y s_{Y\cdot 1,2,\ldots,m}$
Coefficient of determination, R^2	$R^2 = [R^2]$	$[R^2] = R^2$
Test statistics, t and F	$t = [t]$; $F = [F]$	$[t] = t$; $[F] = F$
Sum of squares, SS	$\text{SS} = [\text{SS}]/M_Y^2$	$[\text{SS}] = (\text{SS})(M_Y^2)$
Mean square, MS	$\text{MS} = [\text{MS}]/M_Y^2$	$[\text{MS}] = (\text{MS})(M_Y^2)$

See Table 20.2 for the effect of coding on multiple regression matrices and on correlation coefficients.

Note: If a regression is fit through the origin, then A_Y as well as A_X must be zero.

21.3 QUADRATIC REGRESSION

A common polynomial regression is the second-order, or *quadratic*, regression:

$$\hat{Y}_i = \alpha + \beta_1 X_i + \beta_2 X_i^2, \tag{21.7}$$

with three population parameters, α, β_1, and β_2 to be estimated by three regression statistics, a, b_1, and b_2 respectively, in the quadratic equation

$$\hat{Y}_i = a + b_1 X_i + b_2 X_i^2. \tag{21.8}$$

The geometric shape of the curve represented by Equation 21.7 is a *parabola*. An example of a quadratic regression line is shown in Fig. 21.2. If b_2 is negative as shown in Fig. 21.2, the parabola will be concave downward. If b_2 is positive (as shown in Fig. 21.1b), the curve will be concave upward.

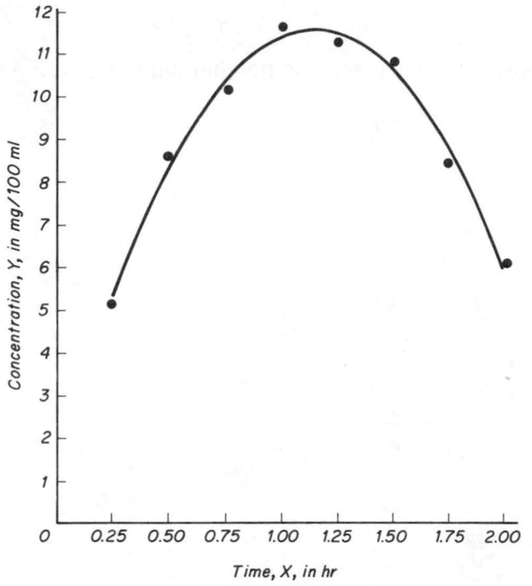

Figure 21.2 Quadratic fit to eight data points resulting in the equation $\hat{Y}_i = 1.39 + 17.769X_i - 7.74286X_i^2$.

Maximum and Minimum Values of Y_i. A common interest in polynomial regression analysis, especially where $m = 2$ (quadratic), is the determination of a maximum or minimum value of Y_i (Bliss, 1970: Section 14.4; Studier et al., 1975). A maximum value of Y_i is defined as one that is greater than those Y_i's which are very close to it: and a minimum Y_i is one that is less than those Y_i's very close to it. If, in a quadratic regression (Equation 21.8), the coefficient b_2 is negative, then there will be a maximum, as shown in Fig. 21.2. If b_2 is positive, there will be a minimum, (as is implied in Fig. 21.1b). It may be desired to determine what the maximum or minimum value of Y_i is and what the corresponding value of X_i is.

The maximum or minimum of a quadratic equation is at the following value of the independent variable:

$$\hat{X}_0 = \frac{-b_1}{2b_2}. \tag{21.9}$$

Placing $\hat{X}_0$ in the quadratic equation (Equation 21.8), we find that

$$\hat{Y}_0 = a - \frac{b_1^2}{4b_2}. \tag{21.10}$$

Thus, in Fig. 21.2, the maximum is at

$$\hat{X}_0 = \frac{-17.769}{2(-7.74286)} = 1.15 \text{ hr},$$

at which

$$\hat{Y}_0 = 1.39 - \frac{(17.769)^2}{4(-7.74286)} = 11.58 \text{ mg/100 ml}.$$

A confidence interval for a maximum or minimum $\hat{Y}_0$ may be computed by the procedures of Section 20.9.

EXERCISES

21.1. Subject the following data to a stepwise polynomial regression analysis:

Y (g)	X (cm)
4.5	21.4
4.4	21.7
4.6	22.3
4.7	22.9
4.5	23.2
4.4	23.8
4.5	24.8
4.2	25.4
4.4	25.9
4.2	27.2
3.8	27.4
3.4	28.0
3.1	28.9
3.2	29.2
3.0	29.8

21.2. Consider the following data, where X is temperature (in °C) and Y is numbers of eggs per cm^2:

X	Y
3.0	2.8
5.0	4.9
8.0	6.7
14.0	7.6
21.0	7.2
25.0	6.1
28.0	4.7

(a) Fit a quadratic equation to these data.
(b) Test for significance of the quadratic term.
(c) Estimate the mean population value of $\hat{Y}_i$ at $X_i = 10.0$°C and compute the 95% confidence interval for the estimate.
(d) Determine the values of X and Y at which the quadratic function is maximum.

22 The Binomial Distribution

In this chapter we will consider nominal scale data that come from a population with only two categories. As examples, members of a mammal litter might be classified as male or female, victims of an epidemic as dead or alive, trees in an area as spruce or fir, or progeny of a *Drosophila* cross as white-eyed or red-eyed.

The proportion of the population belonging to one of the two categories is denoted as p (here departing from the convention of using Greek letters for population parameters). Therefore, the proportion of the population belonging to the second class is $1 - p$, and the notation $q = 1 - p$ is commonly employed. For example, if 0.5 (i.e., 50%) of a population were male, then we would know that 0.5 (i.e., $1 - 0.5$) of the population were female, and we could write $p = 0.5$ and $q = 0.5$; if 0.4 (i.e., 40%) of a population were male, then 0.6 (i.e., 60%) of the population were female, and we could write $p = 0.4$ and $q = 0.6$.

If we took a random sample of 10 from a population where $p = q = 0.5$, then we might expect that the sample would consist of 5 males and 5 females. However, we should not be too surprised to find such a sample consisting of 6 males and 4 females, or 4 males and 6 females, although neither of these combinations would be expected with as great a frequency as samples possessing the population sex ratio of 5: 5. It would, in fact, be possible to obtain a sample of 10 with 9 males and 1 female, or even one consisting of all males, but the probabilities of such samples being encountered by random chance are relatively low.

If we were to obtain a large number of samples from the population under consideration, the frequency of samples consisting of no males, 1 male, 2 males, etc., would be described by the *binomial distribution* (sometimes referred to as the "Bernoulli distribution"*). Let us now examine binomial probabilities.

22.1 BINOMIAL PROBABILITIES

Consider a population consisting of two categories, where p is the proportion of individuals in one of the categories and $q = 1 - p$ is the proportion in the other. Then the probability of selecting at random from this population a member of the first category is p, and the probability of selecting a member of the second categoty is q.

For example, let us say we have a population of female and male animals, in proportions of $p = 0.4$ and $q = 0.6$, respectively, and we take a random sample of two individuals from the population. The probability of the first being a female is p (i.e., 0.4) and the probability of the second being a female is also p. Since *the probability of two independent events both occurring is the product of the probabilities of the two separate events*, the probability of having two females in a sample of two is $(p)(p) = p^2 = 0.16$; the probability of the sample of two consisting of two males is $(q)(q) = q^2 = 0.36$.

What is the probability of the sample of two consisting of one male and one female? This could occur by the first individual being a female and the second a male (with a probability of pq) or by the first being a male and the second a female (which would occur with a probability of qp). *The probability of either of two independent events is the sum of the probabilities of each event*, so the probability of one female and one male in the sample is $pq + qp = 2pq = 2(0.4)(0.6) = 0.48$. Note that $0.16 + 0.36 + 0.48 = 1.00$.

Now consider another sample from this population, where $n = 3$. The probability of all three individuals being female is $(p)(p)(p) = p^3 = (0.4)^3 = 0.064$. The probability of two females and one male is ppq (for a sequence of ♀♀♂) + pqp (for ♀♂♀) + qpp (for ♂♀♀), or: $3p^2q = 3(0.4)^2(0.6) = 0.288$. The probability of one female and two males is pqq (for ♀♂♂) + qpq (for ♂♀♂) + qqp (for ♂♂♀), or $3pq^2 = 3(0.4)(0.6)^2 = 0.432$. And, finally, the probability of all three being males is $qqq = q^3 = (0.6)^3 = 0.216$. Note that $p^3 + 3p^2q + 3pq^2 + q^3 = 0.064 + 0.288 + 0.432 + 0.216 = 1.000$ (meaning that there is a 100% probability that the three animals will be in one of these three combinations of sexes).

If we performed the same exercise with $n = 4$, we would find that the probability of four females is $p^4 = (0.4)^4 = 0.0256$, the probability of three females (and one male) is $4p^3q = 4(0.4)^3(0.6) = 0.1536$, the probability of two females is $6p^2q^2 = 0.3456$, the probability of one female is $4pq^3 = 0.3456$, and the probability of no females (i.e., all four are male) is $q^4 = 0.1296$. (The sum of these five terms is 1.0000, a good arithmetic check.)

If a random sample of size n is taken from a binomial population, then the

*The binomial formula discussed in the following section was first described in 1676 by Sir Isaac Newton (1642–1727); and its first proof, for positive integer exponents, was given by Jakob Bernoulli (1654–1705) in a 1713 publication (Cajori, 1954).

probability of X individuals being in one category (and, therefore, $n - X$ individuals in the second category) is

$$P(X) = \frac{n!}{X!(n-X)!} p^X q^{n-X}. \quad (22.1)$$

In this equation, $p^X q^{n-X}$ refers to the probability of a sample consisting of a particular arrangement of X items, each having a probability p, and $n - X$ items each with probability q; $n!/[X!(n - X)!]$, which is called the *binomial coefficient*, expresses the number of ways X items of one kind can be arranged with $n - X$ items of a second kind.* The binomial coefficient is often written in combinatorial notation, ${}_nC_X$, for

$$ {}_nC_X = \frac{n!}{X!(n-X)!} \quad (22.2)$$

is the number of possible combinations of n items divided into groups of X and $n - X$ items. For example, the number of ways a sample of 5 could be formed containing 2 females is ${}_5C_2 = 5!/[2!(5-2)!] = 10$:† ♀♀♂♂♂, ♀♂♀♂♂, ♀♂♂♀♂, ♀♂♂♂♀, ♂♀♀♂♂, ♂♀♂♀♂, ♂♀♂♂♀, ♂♂♀♀♂, ♂♂♀♂♀, ♂♂♂♀♀.

Therefore, ${}_nC_X p^X q^{n-X}$ is the Xth term in the expansion of $(p + q)^n$, and Table 22.1 shows this expansion for powers up through 6. Note that for any power, n, the sum of the two exponents in any term is n. Furthermore, the first term will always be p^n, the second will always contain $p^{n-1}q$, the third will always contain $p^{n-2}q^2$, etc.,

TABLE 22.1 EXPANSION OF THE BINOMIAL, $(p + q)^n$

n	$(p + q)^n$
1	$p + q$
2	$p^2 + 2pq + q^2$
3	$p^3 + 3p^2q + 3pq^2 + q^3$
4	$p^4 + 4p^3q + 6p^2q^2 + 4pq^3 + q^4$
5	$p^5 + 5p^4q + 10p^3q^2 + 10p^2q^3 + 5pq^4 + q^5$
6	$p^6 + 6p^5q + 15p^4q^2 + 20p^3q^3 + 15p^2q^4 + 6pq^5 + q^6$

*$a! = (a)(a-1)(a-2)(a-3)\cdots(2)(1)$; e.g., $6! = (6)(5)(4)(3)(2)(1) = 720$. It is also necessary to know that $0!$ is defined as 1.

†The order of the X items among themselves is not important in combinatorial counting, nor is the order of the $n - X$ items within the second category. If different orderings of the X times, or of the $n - X$, are to be counted, then we are dealing with permutations, rather than combinations. The number of permutations of X items taken from among n items is ${}_nP_X$:

$$ {}_nP_X = \frac{n!}{(n-X)!}. \quad (22.3)$$

In the sample of $n = 5$ and $X = 2$, if the two females were Martha (M) and Abigail (A), and the three males were George (G), John (J), and Thomas (T), then the number of ways of arranging the 5, considering permutations of females, is ${}_5P_2 = 5!/(5-2)! = 20$: M A ♂ ♂ ♂, A M ♂ ♂ ♂, M ♂ A ♂ ♂, A ♂ M ♂ ♂, M ♂ ♂ A ♂, A ♂ ♂ M ♂, and so on. The number of ways of arranging the 5 individuals if the order of the 3 males is important is ${}_5P_3 = 5!/(5-3)! = 60$; and the number of possible arrangements of the 5 individuals, counting all permutations of females and of males, is $n! = 120$.

with the last term always being q^n. The sum of all the terms in a binomial expansion will always be 1.0, for $p + q = 1$, and $(p + q)^n = 1^n = 1$.

As for the coefficients of these terms in the binomial expansion, the Xth term of the nth power expansion can be calculated by Equation 22.1. Furthermore, the examination of these coefficients as shown in Table 22.2 has been deemed interesting for centuries. This arrangement is known as *Pascal's triangle.** We can see from this triangular array that any binomial coefficient is the sum of two coefficients on the line above it, namely,

$$_nC_X = {}_{n-1}C_{X-1} + {}_{n-1}C_X. \tag{22.4}$$

TABLE 22.2 BINOMIAL COEFFICIENTS, $_nC_X$

n	$X = 1$	2	3	4	5	6	7	8	9	10	11	*Sum of coefficients*
1	1	1										2
2	1	2	1									4
3	1	3	3	1								8
4	1	4	6	4	1							16
5	1	5	10	10	5	1						32
6	1	6	15	20	15	6	1					64
7	1	7	21	35	35	21	7	1				128
8	1	8	28	56	70	56	28	8	1			256
9	1	9	36	84	126	126	84	36	9	1		512
10	1	10	45	120	210	252	210	120	45	10	1	1024

This can be more readily observed if we display the triangular array as follows:

```
                    1
                1       1
            1       2       1
        1       3       3       1
    1       4       6       4       1
1       5      10      10       5       1
```

Also note that the sum of all coefficients for the nth power binomial expansion is 2^n.

Thus, we can calculate probabilities of category frequencies occurring in random samples from binomial populations. If, for example, a sample of five (i.e., $n = 5$) is taken from a population composed of 50% males and 50% females (i.e., $p = 0.5$ and $q = 0.5$), then Example 22.1 shows how Equation 22.1 is used to determine the probability of the sample containing 0 males, 1 male, 2 males, 3 males, 4 males, and 5 males. These probabilities are found to be 0.03125, 0.15625, 0.31250, 0.31250, 0.15625,

*Blaise Pascal (1623–1662), one of the founding fathers of probability theory, had his triangular binomial coefficient derivation published in 1665, although knowledge of the triangular properties appears in Chinese writings as early as 1303 (Cajori, 1954; Struik, 1967: 79).

and 0.03125, respectively. This enables us to state that if we took 100 random samples of 5 from the population, 3 of them would be expected to contain all females, 16 to contain 1 male and 4 females, 31 to consist of 2 males and 3 females, etc. If we took 1400 random samples of 5, (0.03125)(1400) = 44 of them would be expected to contain all females, etc. Figure 22.1a shows graphically the binomial distribution for $p = q = 0.5$, for $n = 5$. Note, from Fig. 22.1a and Example 22.1, that when $p = q = 0.5$

Example 22.1
Computing binomial probabilities, $P(X)$, where $n = 5$, $p = 0.5$, and $q = 0.5$ (following Equation 22.1).

X	$P(X)$
0	$\frac{5!}{0!5!}(0.5^0)(0.5^5) = (1)(1.0)(0.03125) = 0.03125$
1	$\frac{5!}{1!4!}(0.5^1)(0.5^4) = (5)(0.5)(0.0625) = 0.15625$
2	$\frac{5!}{2!3!}(0.5^2)(0.5^3) = (10)(0.25)(0.125) = 0.31250$
3	$\frac{5!}{3!2!}(0.5^3)(0.5^2) = (10)(0.125)(0.25) = 0.31250$
4	$\frac{5!}{4!1!}(0.5^4)(0.5^1) = (5)(0.0625)(0.5) = 0.15625$
5	$\frac{5!}{5!0!}(0.5^5)(0.5^0) = (1)(0.03125)(1.0) = 0.03125$

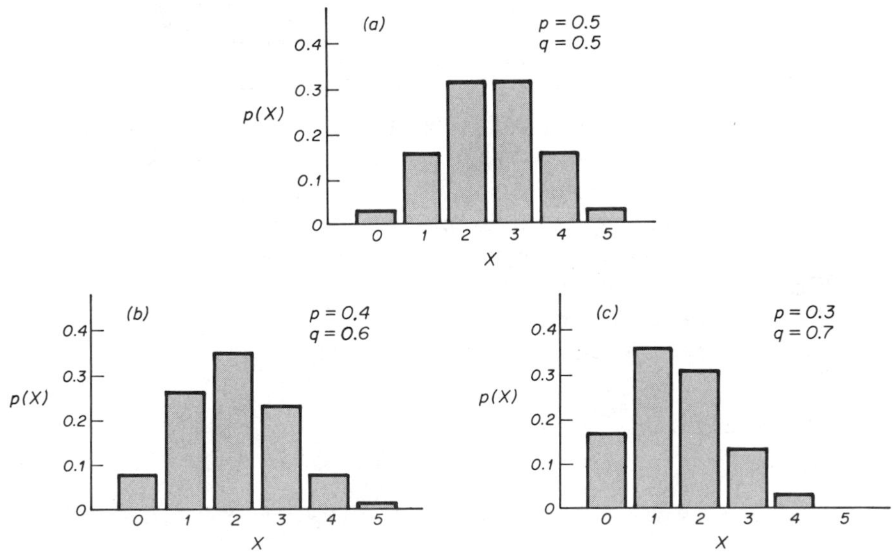

Figure 22.1 The binomial distribution, for $n = 5$. (a) $p = q = 0.5$. (b) $p = 0.4$, $q = 0.6$. (c) $p = 0.3$, $q = 0.7$. These graphs were drawn utilizing the proportions given by Equation 22.1.

the distribution is symmetrical (i.e., $P(0) = P(n)$, $P(1) = P(n-1)$, etc.),* and Equation 21.1 becomes

$$P(X) = \frac{n!}{X!(n-X)!}0.5^n. \tag{22.6}$$

Appendix Table B.25 gives binomial probabilities for $n = 2$ to $n = 20$, for $p = 0.5$.

Example 22.2 presents the calculation of binomial probabilities for the case where $n = 5$, $p = 0.4$, and $q = 1 - 0.4 = 0.6$. Thus, if one were sampling a population consisting of 40% males and 60% females, 0.07776 (i.e., 7.776%) of the samples would be expected to contain no males, 0.25920 to contain 1 male and 4 females, etc. Fig. 22.1b presents this binomial distribution graphically, whereas Fig. 22.1c shows the distribution where $p = 0.3$ and $q = 0.7$.

Example 22.2

Computing binomial probabilities, $P(X)$, where $n = 5$, $p = 0.4$, and $q = 0.6$ (following Equation 22.1).

X	$P(X)$
0	$\frac{5!}{0!5!}(0.4^0)(0.6^5) = (1)(1.0)(0.07776) = 0.07776$
1	$\frac{5!}{1!4!}(0.4^1)(0.6^4) = (5)(0.4)(0.1296) = 0.25920$
2	$\frac{5!}{2!3!}(0.4^2)(0.6^3) = (10)(0.16)(0.216) = 0.34560$
3	$\frac{5!}{3!2!}(0.4^3)(0.6^2) = (10)(0.064)(0.36) = 0.23040$
4	$\frac{5!}{4!1!}(0.4^4)(0.6^1) = (5)(0.0256)(0.6) = 0.07680$
5	$\frac{5!}{5!0!}(0.4^5)(0.6^0) = (1)(0.01024)(1.0) = 0.01024$

For calculating binomial probabilities for large n, it is often convenient to employ logarithms. For this reason, Table B.38, a table of logarithms of factorials, is provided. Alternatively, it is useful to note that the denominator of Equation 22.1 cancels out much of the numerator, so that it is possible to simplify the computation of $P(X)$, especially in the tails of the distribution (i.e., for low X and for high X), as shown in Example 22.3. If p is very small, then the use of the Poisson distribution (Section 23.1) should be considered.

The mean of a binomial distribution of counts, X, is

$$\mu_X = np, \tag{22.7}$$

the variance is

$$\sigma_X^2 = npq, \tag{22.8}$$

*The measure of symmetry (see Section 7.1) for a binomial distribution is

$$\gamma_1 = \frac{q-p}{\sqrt{npq}}, \tag{22.5}$$

so it can be seen that $\gamma_1 = 0$ only when $p = q = 0.50$, $\gamma_1 > 0$ implies a distribution skewed to the right (as in Figs. 22.1b and 22.1c), and $\gamma_1 < 0$ indicates a distribution skewed to the left.

Example 22.3

Computing binomial probabilities, $P(X)$, with $n = 400$, $p = 0.02$, and $q = 0.98$. (Many calculators can operate with large powers of numbers; otherwise, logarithms may be used.)

X	$P(X)$
0	$\frac{n!}{0!(n-0)!}p^0q^{n-0} = q^n = 0.98^{400} = 0.00031$
1	$\frac{n!}{1!(n-1)!}p^1q^{n-1} = npq^{n-1} = (400)(0.02)(0.98^{399}) = 0.00253$
2	$\frac{n!}{2!(n-2)!}p^2q^{n-2} = \frac{n(n-1)}{2!}p^2q^{n-2} = \frac{(400)(399)}{2}(0.02^2)(0.98^{398}) = 0.01028$
3	$\frac{n!}{3!(n-3)!}p^3q^{n-3} = \frac{n(n-1)(n-2)}{3!}p^3q^{n-3} = \frac{(400)(399)(398)}{(3)(2)}(0.02^3)(0.98^{397}) = 0.02784$

and so on.

and the standard deviation (also called standard error) of X is

$$\sigma_X = \sqrt{npq}. \tag{22.9}$$

Thus, if we have a binomially distributed population where p (e.g., the proportion of males) $= 0.5$ and q (e.g., the proportion of females) $= 0.5$, and we take 10 samples from that population, the mean of the 10 X's (i.e., the mean number of males per sample) would be $np = (10)(0.5) = 5$ and the standard error of the 10 X's would be $\sqrt{npq} = \sqrt{(10)(0.5)(0.5)} = 1.58$. Our concern typically is with the distribution of p rather than X, as will be explained in Section 22.2.

22.2 SAMPLING A BINOMIAL POPULATION

Let us consider a population of N observations: Y observations in one category and $N - Y$ in the second category. Then the proportion of observations in the first category is

$$p = \frac{Y}{N} \tag{22.10}$$

and the proportion in the second is

$$q = 1 - p \quad \text{or} \quad q = \frac{N - Y}{N}. \tag{22.11}$$

If a sample of n observations is taken from this population and X observations are in one category and $n - X$ are in the other, then the population parameter p is estimated by the sample statistic

$$\hat{p} = \frac{X}{n}, \tag{22.12}$$

which is the proportion of the sample that is in the first category.* The estimate of q is

*Placing the symbol "$\wedge$" above a letter is statistical convention for denoting an estimate of the quantity which that letter denotes. Thus, $\hat{p}$ refers to an estimate of p, and the statistic $\hat{q}$ is an estimate of the population parameter q.

$$\hat{q} = 1 - p \quad \text{or} \quad \hat{q} = \frac{n - X}{n}, \tag{22.13}$$

which is the proportion of the sample occurring in the second category. In Example 22.4 we have $X = 4$ and $n = 20$, so $\hat{p} = 4/20 = 0.20$ and $\hat{q} = 1 - p = 0.80$.

Example 22.4
Sampling a binomial population.

From a population of male and female spiders, a sample of 20 is taken, which contains 4 males and 16 females.

$$n = 20$$

$$X = 4$$

By Equation 22.12,

$$\hat{p} = \frac{X}{n} = \frac{4}{20} = 0.20.$$

Therefore we estimate that 20% of the population are males and, by Equation 22.13,

$$\hat{q} = 1 - \hat{p} = 1 - 0.20 = 0.80$$

or

$$\hat{q} = \frac{n - X}{n} = \frac{20 - 4}{20} = \frac{16}{20} = 0.80,$$

so we estimate that 80% of the population are females.

The variance of the estimate, $\hat{p}$ (or of $\hat{q}$) is, by Equation 22.15,

$$s^2_{\hat{p}} = \frac{\hat{p}\hat{q}}{n - 1} = \frac{(0.20)(0.80)}{20 - 1} = 0.008421.$$

If we consider that the sample consists of four 1's and sixteen 0's, then the variance of the twenty 1's and 0's is, by Equation 4.6, $s^2 = (4 - 4^2/20)/(20 - 1) = 0.168420$, and the variance of the mean, by Equation 7.20, is $s^2_{\bar{X}} = 0.168420/20 = 0.008421$.

The standard error (or standard deviation) of $\hat{p}$ (or of $\hat{q}$) is, by Equation 22.19, $s_{\hat{p}} = \sqrt{0.008421} = 0.092$.

If our sample of 20 were returned to the population (or if the population were extremely large), and we then took another sample of 20, and repeated this multiple sampling procedure many times, we could obtain many calculations of $\hat{p}$, each estimating the population parameter p. If, in the population, $p = 0$, then obviously any sample from that population would have $\hat{p} = 0$; and if $p = 1.0$, then each and every $\hat{p}$ would be 1.0. However, if p is neither 0 nor 1.0, then all the many samples from the population would not have the same values of $\hat{p}$. The variance of all possible $\hat{p}$'s is

$$\sigma^2_{\hat{p}} = \frac{pq}{n}, \tag{22.14}$$

which can be estimated from our sample as

$$s^2_{\hat{p}} = \frac{\hat{p}\hat{q}}{n - 1}. \tag{22.15}$$

This variance is essentially a variance of means, so Equation 22.14 is analogous to Equation 7.17, and Equation 22.15 to Equation 7.20. In Example 22.4 it is shown that the latter is true. The variance of $\hat{q}$ is the same as the variance of $\hat{p}$; i.e.,

$$\sigma_{\hat{q}}^2 = \sigma_{\hat{p}}^2 \tag{22.16}$$

and

$$s_{\hat{q}}^2 = s_{\hat{p}}^2. \tag{22.17}$$

The standard error of $\hat{p}$ (or of $\hat{q}$), also called the standard deviation, is

$$\sigma_{\hat{p}} = \sqrt{\frac{pq}{n}}, \tag{22.18}$$

which is estimated* from a sample as

$$s_{\hat{p}} = \sqrt{\frac{\hat{p}\hat{q}}{n-1}}. \tag{22.19}$$

The possible values of $\sigma_{\hat{p}}^2$, $\sigma_{\hat{q}}^2$, $\sigma_{\hat{p}}$, and $\sigma_{\hat{q}}$ range from a minimum of zero when either p or q is zero, to a maximum when $p = q = 0.5$; and $s_{\hat{p}}^2$, $s_{\hat{q}}^2$, $s_{\hat{p}}$, and $s_{\hat{q}}$ can range from a minimum of zero when either $\hat{p}$ or $\hat{q}$ is zero, to a maximum when $\hat{p} = \hat{q} = 0.5$.

Sampling Finite Populations.† If n is a substantial portion of the entire population of size N, then a finite population correction is called for (just like that found in Section 8.7) in estimating $\sigma_{\hat{p}}^2$ or $\sigma_{\hat{p}}$:

$$s_{\hat{p}}^2 = \frac{\hat{p}\hat{q}}{n-1}\left(1 - \frac{n}{N}\right) \tag{22.21}$$

and

$$s_{\hat{p}} = \sqrt{\frac{\hat{p}\hat{q}}{n-1}\left(1 - \frac{n}{N}\right)}, \tag{22.22}$$

where n/N is called the *sampling fraction*, and $1 - n/N$ is the finite population correction, the latter also being written as $(N - n)/N$. As N becomes very large compared to n, Equation 22.21 approaches Equation 22.15 and Equation 22.22 approaches 22.19.

We can estimate Y, the total number of occurrences in the population in the first category, as

$$\hat{Y} = \hat{p}N; \tag{22.23}$$

and the variance and standard error of this estimate are

$$s_{\hat{Y}}^2 = \frac{N(N-n)\hat{p}\hat{q}}{n-1} \tag{22.24}$$

and

$$s_{\hat{Y}} = \sqrt{\frac{N(N-n)\hat{p}\hat{q}}{n-1}}, \tag{22.25}$$

respectively.

*One often sees

$$s_{\hat{p}} = \sqrt{\frac{\hat{p}\hat{q}}{n}} \tag{22.20}$$

used to estimate $\sigma_{\hat{p}}$. Although it is inaccurate (being an underestimate), when n is large the difference between Equations 22.19 and 22.20 is slight.

†These procedures are from Cochran (1977: 52). They give results for sampling from finite populations, in which case the data follow what is known as the hypergeometric, rather than the binomial, distribution.

22.3 CONFIDENCE LIMITS FOR PROPORTIONS

Using a relationship between the F distribution and the binomial distribution (Fisher and Yates, 1963: 3), a confidence interval may be computed for the binomial parameter p (Bliss, 1967: 199–201; Brownlee, 1965: 148–149). As demonstrated in Example 22.5, the lower confidence limit is

$$L_1 = \frac{X}{X + (n - X + 1)F_{\alpha(2),\nu_1,\nu_2}}, \quad (22.26)$$

where

$$\nu_1 = 2(n - X + 1)$$

and

$$\nu_2 = 2X.$$

The upper confidence limit is

$$L_2 = \frac{(X + 1)F_{\alpha(2),\nu_1'\nu_2'}}{n - X + (X + 1)F_{\alpha(2),\nu_1',\nu_2'}}, \quad (22.27)$$

with

$$\nu_1' = 2(X + 1) = \nu_2 + 2$$

and

$$\nu_2' = 2(n - X) = \nu_1 - 2.$$

Since Table B.4 gives values of F to only three significant figures, the confidence limits may be in slight error in the third significant figure. (For large n the required critical values of F will not be found in Table B.4 and interpolation, as explained at the beginning of Appendix B, may be desirable.) The useful handbook by Burstein (1971) allows the determination of binomial confidence limits to an accuracy of 0.1% of the exact limits.

The confidence limits for q may be obtained by subtracting each of the confidence limits for p from 1. Thus, for Example 22.5, $q = 1 - p = 0.9800$, and the 95% confidence limits for q are: $L_1 = 1 - 0.0506 = 0.9494$ and $L_2 = 1 - 0.00550 = 0.99450$. Burstein (1971) gives tables and procedures for determining the sample size necessary to estimate p with a desired precision.

Example 22.5

Determination of the 95% confidence interval for the binomial population parameter, p.

A random sample of 200 persons contains 4 with immunity to a certain viral disease. What proportion of the population possesses such immunity?

$$n = 200$$

$$X = 4$$

$$\hat{p} = \frac{X}{n} = \frac{4}{200} = 0.0200$$

For the lower 95% confidence limit:

$\nu_1 = 2(n - X + 1) = 2(200 - 4 + 1) = 394$

$\nu_2 = 2X = 2(4) = 8$

$F_{0.05(2),394,8} \approx F_{0.05(2),\infty,8} = 3.67$

$$L_1 = \frac{X}{X + (n - X + 1)F_{0.05(2),394,8}} = \frac{4}{4 + (200 - 4 + 1)(3.67)} = 0.00550$$

For the upper 95% confidence interval:

$$\nu_1' = 2(X + 1) = 2(4 + 1) = 10$$
$$\text{or, } \nu_1' = \nu_2 + 2 = 8 + 2 = 10$$
$$\nu_2' = 2(n - X) = 2(200 - 4) = 392$$
$$\text{or, } \nu_2' = \nu_1 - 2 = 394 - 2 = 392$$
$$F_{0.05(2),10,392} \approx F_{0.05(2),10,300} = 2.09$$
$$L_2 = \frac{(X + 1)F_{0.05(2),10,392}}{n - X + (X + 1)F_{0.05(2),10,392}} = \frac{(4 + 1)(2.09)}{200 - 4 + (4 + 1)(2.09)} = 0.0506$$

Therefore, we can state:

$$P(0.00550 \leq p \leq 0.0506) = 0.95$$

Note: By harmonic interpolation (see beginning of Appendix B) we can find that $F_{0.05(2),394,8} = 3.69$ and $F_{0.05(2),10,392} = 2.10$, which when used in the computation of the above confidence limits yields $L_1 = 0.00547$ and $L_2 = 0.0506$. (It is worth noting that had the normal approximation of Equation 22.31 been used, we would have computed, inaccurately, confidence limits of 0.0200 ± 0.0220, namely $L_1 = -0.0020$ (but since $p < 0$ is not possible, we would state $L_1 = 0$) and $L_2 = 0.0420$.

Confidence Limits with Finite Populations. If n is appreciably large relative to N, then we are said to have sampled a finite population, and the accuracy of estimating p improves greatly as n approaches N. In such cases the lower confidence limit for p is

$$(L_1)_c = \frac{X - 0.5}{n} - \left(\frac{X - 0.5}{n} - L_1\right)\sqrt{1 - \frac{n}{N}} \tag{22.28}$$

and the upper confidence limit is

$$(L_2)_c = \frac{X'}{n} + \left(L_2 - \frac{X'}{n}\right)\sqrt{1 - \frac{n}{N}}, \tag{22.29}$$

where

$$X' = X + \frac{X}{n} \tag{22.30}$$

(see Burstein, 1975). The confidence interval is shorter with the finite population correction applied.

The Normal Approximation. We can obtain approximate confidence limits for p (or q) using the normal approximation to the binomial distribution, which employs the standard error of $\hat{p}$:

$$(1 - \alpha) \text{ confidence limits for } p = \hat{p} \pm \left(Z_{\alpha(2)} s_{\hat{p}} + \frac{1}{2n}\right), \tag{22.31}$$

and

$$(1 - \alpha) \text{ confidence limits for } q = \hat{q} \pm \left(Z_{\alpha(2)} s_{\hat{p}} + \frac{1}{2n}\right), \tag{22.32}$$

where for $s_{\hat{p}}$ one uses either 22.19 or 22.22, whichever is appropriate. The term $1/2n$ is a correction for continuity (Cochran, 1977: 57–58) which for large n is negligible.

The normal approximation usually does not give accurate confidence limits, it being especially poor when np or nq is less than 5 or when p is near 0 or 1 (say $p <$ 0.20 or $p < 0.80$). Also, the normal approximation gives confidence limits symmetrical around p, which can result in the computation of a nonsensical $L_1 < 0$ or $L_2 > 1$. Ghosh (1979) discusses another normal approximation, one that does not suffer these defects. However, with the availability of the exact method of Equations 22.26 and

22.27, using the normal approximation results in unnecessary inaccuracy and should be avoided. (At the end of Example 22.5 it is noted how undesirable the approximation is for the data of that example.)

Sample Size Requirements. If p and q are of sufficient magnitude to use the normal approximation, then one can estimate how large a sample is required to estimate p (and q) with an error no greater than δ. This sample size is

$$n = \frac{Z^2_{\alpha(2)} pq}{\delta^2} \tag{22.33}$$

(Cochran, 1977: 75–76). If the sample size, n, is not negligibly small compared to the population size, N, then the required sample size is estimated as

$$m = \frac{n}{1 + \frac{n-1}{N}}, \tag{22.34}$$

where n is as in Equation 22.33 (Cochran, 1977: 75–76). Recall that $t_{\alpha(2),\infty} = Z_{\alpha(2)}$. This sample size is also one which is required to perform a binomial test (Section 22.5) that is capable of detecting a population difference as small as δ from the hypothesized p. Burstein (1971) gives tables allowing for estimation of sample size for p or n not appropriate to the normal approximation.

If one does not have an estimate of p and q, but desires to estimate them with an error of δ, then a sample of size n or more should be taken, where

$$n = \frac{Z^2_{\alpha(2)}}{4\delta^2}. \tag{22.35}$$

If n/N is not a negligible fraction, then use the sample size indicated by Equation 22.34, with the n of Equation 22.35.

22.4 GOODNESS OF FIT FOR THE BINOMIAL DISTRIBUTION

When p Is Hypothesized to Be Known. In some biological instances, the population proportions, p and q, might be assumed to be known. For example, theory might tell us that 50% of mammalian sperm contain an X chromosome, whereas 50% contain a Y chromosome, and we can expect a 1: 1 sex ratio among the offspring. We may wish to test the hypothesis that our sample came from a binomially distributed population with equal sex frequencies. We may do this as follows, by using the goodness of fit testing introduced in Chapter 5.

Let us suppose that we have tabulated the sexes of the offspring from 54 litters of 5 animals each (Example 22.6). Assuming $p = q = 0.5$, the proportion of each possible litter composition can be computed by the procedures of Example 22.1, or they can be read directly from Table B.25. From these proportions, we can tabulate expected frequencies, and then we can subject observed and expected frequencies of each type of litter to a chi-square goodness of fit analysis (see Chapter 5), with $k - 1$ degrees of freedom (k being the number of classes of X).

Example 22.6

Goodness of fit of a binomial distribution, when p is known.

The data consist of observed frequencies of females in 54 litters of 5 offspring per litter. $X = 0$ denotes a litter having no females, $X = 1$ a litter having one female, etc.; f is the observed number of litters, and $\hat{f}$ is the number of litters expected if the null hypothesis is true. Computation of the values of $\hat{f}$ requires the values of $P(X)$, as obtained in Example 22.1.

H_0: The sex of the offspring are from a binomial distribution with $p = q = 0.5$.
H_A: The sex of the offspring are not from a binomial distribution with $p = q = 0.5$.

X_i	f_i	$\hat{f}_i$
0	3	(0.03125)(54) = 1.6875
1	10	(0.15625)(54) = 8.4375
2	14	(0.31250)(54) = 16.8750
3	17	(0.31250)(54) = 16.8750
4	9	(0.15625)(54) = 8.4375
5	1	(0.03125)(54) = 1.6875

$\chi^2 = 1.0208 + 0.2894 + 0.4898 + 0.0009 + 0.0375 + 0.2801 = 2.1185$
$\nu = k - 1 = 6 - 1 = 5$
$\chi^2_{0.05,5} = 11.070$
Therefore, do not reject H_0.
$0.75 < P < 0.90$

In Example 22.6, we do not reject the null hypothesis, and therefore conclude that the sampled population is binomial with $p = 0.5$.

To avoid bias in the chi-square computation, no expected frequency should be less than 1.0 (Cochran, 1954). If such small frequencies occur, then frequencies in the appropriate extreme classes of X may be pooled to arrive at sufficiently large F_i values. Such pooling was not necessary in Example 22.6, as no F_i was less than 1.0. But it will be shown in Example 22.7.

The G statistic (Section 5.9) may be calculated in lieu of chi-square, with the summation being executed over all classes except those where not only $f_i = 0$, but also all more extreme f_i's are zero. The Kolmogorov-Smirnov statistic of Section 5.10 could also have been used to determine the goodness of fit. Heterogeneity testing (Section 5.8) may be performed for several sets of data hypothesized to have come from a binomial distribution.

If the preceding null hypothesis had been rejected, we might have looked in a couple of directions for a biological explanation. The rejection of H_0 might have indicated that the population p was, in fact, not 0.5. Or, it might have indicated that the underlying distribution was not binomial. The latter possibility may occur when membership of an individual in one of the two possible categories is dependent upon another individual in the sample. In Example 22.6, for instance, identical twins (or other multiple identical births) might have been a common occurrence in the species in question. In that case, if one member of a litter was found to be female, then there would be a greater-than-expected likelihood of a second member of the litter being female.

Example 22.7

Goodness of fit of a binomial distribution, when p is not known but is estimated from the sample data.

The data consist of observed frequencies of left-handed persons in 75 samples of 8 persons each. $X = 0$ denotes a sample with no left-handed persons, $X = 1$ a sample with one left-handed person, etc.; f is the observed number of samples, and $\hat{f}$ is the number of samples expected if the null hypothesis is true. Each $\hat{f}$ is computed by multiplying 75 by $P(X)$, where $P(X)$ is obtained from Equation 22.1 by substituting $\hat{p}$ and $\hat{q}$ for p and q, respectively.

H_0: The frequencies of left- and right-handed persons in the population follow a binomial distribution.

H_A: The frequencies of left- and right-handed persons in the population do not follow a binomial distribution.

$$\bar{X} = \frac{\sum f_i X_i}{\sum f_i} = \frac{96}{75} = 1.2800$$

$$\hat{p} = \frac{\bar{X}}{n} = \frac{1.2800}{8} = 0.16 = \text{probability of a person being left-handed}$$

$$\hat{q} = 1 - \hat{p} = 0.84 = \text{probability of a person being right-handed}$$

X_i	f_i		f_iX_i	$\hat{f}_i$	
0	21		0	$\frac{8!}{0!8!}(0.16^0)(0.84^8)(75) = (0.24788)(75) = 18.59$	
1	26		26	$(0.37772)(75) = 28.33$	
2	19		38	$(0.25181)(75) = 18.89$	
3	6		18	$(0.09593)(75) = 7.19$	
4	2	3	8	$(0.02284)(75) = 1.71$	1.99
5	0		0	$(0.00348)(75) = 0.26$	
6	1		6	$(0.00033)(75) = 0.02$	
7	0		0	$(0.00002)(75) = 0.00$	
8	0		0	$(0.00000)(75) = 0.00$	

$\sum f_i = 75$

$\sum f_i X_i = 96$

$$\chi^2 = \frac{(21 - 18.59)^2}{18.59} + \frac{(26 - 28.33)^2}{28.33} + \frac{(19 - 18.89)^2}{18.89} + \frac{(6 - 7.19)^2}{7.19} + \frac{(3 - 1.99)^2}{1.99}$$

$= 1.214$

$\nu = k - 2 = 5 - 2 = 3$

$\chi^2_{0.05,3} = 7.815$

Therefore, do not reject H_0.

$0.50 < P < 0.75$

When p Is Not Assumed to Be Known. Commonly, we do not presume to know the value of p in the population but estimate it from a sample of data. As shown in Example 22.7, we may do this by calculating

$$\hat{p} = \frac{\sum_{i=1}^{k} f_i X_i \Big/ \sum_{i=1}^{k} f_i}{n}. \tag{22.36}$$

It then follows that $\hat{q} = 1 - \hat{p}$.

The values of $\hat{p}$ and $\hat{q}$ may be substituted in Equation 22.1 in place of p and q, respectively. Thus, expected frequencies may be calculated for each X, and a chi-square goodness of fit analysis may be performed just as it was in Example 22.6. In such a procedure, however, DF $= k - 2$, rather than $k - 1$, because two constants (n and $\hat{p}$) must be obtained from the sample, and DF is in general determined as k minus the number of such constants. The G statistic (Section 5.9) may be employed when p is not known, but the Kolmogorov-Smirnov test (Section 5.10) is very conservative in such cases and should be avoided.

The null hypothesis for such a test would be that the sampled population was distributed binomially. This implies that the members of the population are distributed among two categories and that the members occur independently of one another.

22.5 THE BINOMIAL TEST

With the ability to determine binomial probabilities, a simple procedure may be employed for goodness of fit testing of nominal data distributed among two categories. This method is especially welcome as an alternative to chi-square goodness of fit where the expected frequencies are small (See Section 5.7.) If p is very small, then the Poisson distribution (Section 23.1) may be used; it is simpler to employ when n is very large.

One-Tailed Testing. An animal might be introduced into a passageway, at the end of which it has a choice of turning either to the right or to the left. Food is placed out of sight to the left. We might state a null hypothesis that there is no preference for left turns, against the alternative that the animal prefers to turn toward the food. If we consider p to be the probability of turning toward the food, then the hypotheses (one tailed) would be stated as $H_0: p \leq 0.5$ and $H_A: p > 0.5$, and such an experiment might be utilized, for example, to determine the ability of the animal to smell the food. We may test H_0 as shown in Example 22.8. In this testing procedure, we determine the probability of obtaining, at random, a distribution of data deviating as much as, or more than, the observed data. In Example 22.8, the most likely distribution of data in a sample of 12 from a population where p, in fact, was 0.5, would be 6 left and 6 right. The samples deviating from a 6 : 6 ratio even more than our observed sample having a 10 : 2 ratio would be those possessing 11 left, 1 right, and 12 left, 0 right. The general one-tailed hypotheses are H_0: $p \leq p_0$ and H_A: $p > p_0$, or H_0: $p \geq p_0$ and $H_A: p < p_0$, where p_0 need not be 0.5. The determination of the probability of $\hat{p}$ as extreme as, or more extreme than, that observed is shown in Example 22.8, where the expected frequencies, $P(X)$, are obtained either from Table B.25 or by Equation 22.1. If the resultant probability is less than or equal to α, then H_0 is rejected.

Alternatively, a critical value for the one-tailed binomial test may be found by the confidence limit determinations of Section 22.3, using $X = pn$ and one-tailed values of F. If $H_A > 0.5$ (as in Example 22.8), then use Equation 22.27, so that the upper confidence limit is the critical value. If H_A: $p < 0.5$, then use Equation 22.26, so that the lower confidence limit is the critical value. This method is advantageous in that it renders unnecessary the calculation of $P(X)$ for each of several X, but it lacks the ability to declare the exact probability of H_0.

Example 22.8

A one-tailed binomial test.

Twelve animals were introduced into a passageway at the end of which they could turn to the left (where food was placed out of sight) or to the right. We wish to determine if animals choose the left more often than the right. Then, $n = 12$; and, if p is the probability of turning left, $H_0: p \leq 0.5$ and $H_A: p > 0.5$. In the following, $P(X)$ is obtained either from Table B.25 or by Equation 22.1.

X	$P(X)$
0	0.00024
1	0.00293
2	0.01611
3	0.05371
4	0.12085
5	0.19336
6	0.22559
7	0.19336
8	0.12085
9	0.05371
10	0.01611
11	0.00293
12	0.00024

On performing the experiment, 10 of the 12 animals turned to the left and 2 turned to the right. If H_0 is true, $P(X \geq 10) = 0.01611 + 0.00293 + 0.00024 = 0.01928$. Since this probability is less than 0.05, reject H_0.

Alternatively, this test could be performed by using the upper confidence limits as a critical value. By Equation 22.27, with a one-tailed F, we have

$$X = pn = (0.5)(12) = 6,$$
$$\nu_1' = 2(6 + 1) = 14,$$
$$\nu_2' = 2(12 - 6) = 12,$$
$$F_{0.05(1),14,12} = 2.64, \text{ and}$$
$$L_2 = \frac{(6 + 1)(2.64)}{12 - 6 + (6 + 1)(2.64)} = 0.755.$$

Since the observed $\hat{p}$ (namely $X/n = 10/12 = 0.833$) exceeds the critical value (0.755), we reject H_0.

Or, as a simpler alternative, consult Table B.26 for $n = 12$ and $\alpha(1) = 0.05$, and find a critical value of 2. As $10 > n - 2$, H_0 is rejected; and $0.01 < P(X \geq 10) < 0.025$.

An even simpler alternative is to consult Table B.26, which gives the lower confidence limit, $C_{\alpha(1),n}$, for one-tailed probabilities, $\alpha(1)$. If H_A: $p < 0.5$, then H_0 is rejected if $X < C_{\alpha(1),n}$, where $X = np$. If H_A: $p > 0.5$ (as in Example 22.8), then H_0 is rejected if $X > n - C_{\alpha(1),n}$.

Two-Tailed Testing. If no food were placed at the end of one of the passage turns, and, therefore, if there were no reason for considering a preference in only one of the two possible directions, then we would be dealing with two-tailed hypotheses, H_0: $p = 0.5$ and H_A: $p \neq 0.5$. The testing procedure would be identical to that in Example 22.8, except that we desire to know $P(X \leq 2 \text{ or } X \geq 10)$. This is the probability of a set of data deviating *in either direction* from the expected as much as or more than those data observed. This is shown in Example 22.9. The general two-tailed hypotheses are H_0: $p = p_0$ and H_A: $p \neq p_0$.

Example 22.9

A two-tailed binomial test.

The experiment is as described in Example 22.8, except that we have no a priori reason to hypothesize a particular direction of turning.

H_0: $p = 0.5$.
H_A: $p \neq 0.5$.
$n = 12$

The probabilities of X, for $X = 0$ through $X = 12$, are given in Example 22.8.

$$P(X \geq 10 \text{ or } X \leq 2) = 0.01611 + 0.00293 + 0.00024 + 0.01611 + 0.00293 + 0.00024 = 0.03856$$

Since this probability is less than 0.05, reject H_0.

Alternatively, this test could be performed by using the confidence limits as critical values. By Equations 22.26 and 22.27 we have

$$X = pn = (0.5)(12) = 6,$$

and for L_1 we have

$$\nu_1 = 2(12 - 6 + 1) = 14$$
$$\nu_2 = 2(6) = 12$$
$$F_{0.05(2),14,12} = 3.21$$
$$L_1 = \frac{6}{6 + (12 - 6 + 1)(3.21)} = 0.211,$$

and for L_2 we have

$$\nu_1' = 2(6 + 1) = 14$$
$$\nu_2' = 2(12 - 6) = 12$$
$$F_{0.05(2),14,12} = 3.21$$
$$L_2 = \frac{(6 + 1)(3.21)}{12 - 6 + (6 + 1)(3.21)} = 0.789.$$

As the observed $\hat{p}$ (namely $X/n = 10/12 = 0.833$) lies outside the range of 0.211 to 0.789, we reject H_0.

A simpler procedure uses Table B.26 to obtain critical values of $C_{0.05(2),6} = 2$ and $n - C_{0.05(2),6} = 12 - 2 = 10$. Since $X = 10$, H_0 is rejected; and $0.02 < P(X \leq 2$ or $X \geq 10) < 0.05$.

Instead of enumerating the several values of $P(X)$ required, one could determine critical values for the two-tailed binomial test using Equations 22.26 and 22.27. If the observed $\hat{p}$ lies outside the interval formed by L_1 and L_2, then H_0 is rejected, as shown in Example 22.9. Note, as in Example 22.9, that if we hypothesize $p = 0.50$, then ν_1 is the same for both L_1 and L_2, and ν_2 is the same for L_1 and L_2; therefore, $F_{\alpha(2)\nu_1,\nu_2}$ is the same for both critical values. Using the critical value approach to the binomial test may be computationally easier than the direct examination of the tails of the binomial distribution, but the latter procedure has the advantage of declaring an exact probability in stating conclusions from the hypothesis test.

A much simpler two-tailed binomial test is possible using Table B.26, as demonstrated in Example 22.9. If the observed count, $X = pn$, is either $\leq C_{\alpha(2),n}$ or $\geq n - C_{\alpha(2),n}$, then H_0 is rejected.

The Normal Approximation. If n is large (say, greater than 25), then the procedure just described is very burdensome indeed. If p is neither very small (i.e., close to 0) nor very large (close to 1), then a normal approximation for the binomial test may be used. As to what is meant by p being "very small" or "very large", Cochran (1977: 58) offers the following sample size recommendations for different magnitudes of $\hat{p}$:

$\hat{p}$	n
0.5	≥ 30
0.4 or 0.6	≥ 50
0.3 or 0.7	≥ 80
0.2 or 0.8	≥ 200
0.1 or 0.9	≥ 600
0.05 or 0.95	≥ 1400

This procedure makes use of estimates of the mean and standard deviation of a binomial distribution (Equations 22.7 and 22.9, respectively); the null hypothesis H_0: $p = p_0$ may be tested by

$$Z = \frac{\bar{X} - np_0}{s_X} = \frac{n\hat{p} - np_0}{\sqrt{n\hat{p}\hat{q}}}. \tag{22.37}$$

Since we are using a continuous distribution (Z) to work with discrete data, a correction for continuity is desirable. The test statistic with such a correction is

$$Z_c = \frac{|n\hat{p} - np_0| - 0.5}{\sqrt{n\hat{p}\hat{q}}} \tag{22.38}$$

(see Example 22.10). Recall that the critical value for a normal deviate (Z) is obtained readily by the equivalence $Z_\alpha = t_{\alpha,\infty}$. One-tailed hypotheses can also be subjected to the normal approximation to the binomial test, in which case $Z_{\alpha(1)}$ would be utilized as the critical value, rather than $Z_{\alpha(2)}$. The end of Section 22.3 discusses procedures for estimating the sample size required for a binomial test.

Example 22.10

The normal approximation to the binomial test.

H_0: $p = 0.5$.
H_A: $p \neq 0.5$.

Here, p is the proportion of females in a population of insects. A sample of 30 insects is taken (i.e., $n = 30$), and the sample contains 18 females (i.e., $n\hat{p} = 18$, $\hat{p} = \frac{18}{30} = 0.60$, and $\hat{q} = 1.00 - 0.60 = 0.40$).

We test the null hypothesis by computing

$$Z = \frac{|18 - (30)(0.5)| - 0.5}{\sqrt{(18)(0.40)(0.60)}} = 1.203$$

$Z_{0.05(2)} = t_{0.05(2),\infty} = 1.960$

Therefore, do not reject H_0.

$0.20 < P < 0.50$

22.6 THE SIGN TEST

The concept underlying the binomial test can be developed into a nonparametric paired-sample test. This procedure, the *sign test*,* may be employed whenever the Wilcoxon paired-sample test (Section 10.4) is appropriate, although it is not as powerful as the latter. The actual differences between members of a pair are not utilized in the sign test; only the direction (or sign) of each difference is tabulated.

*The sign test was reported more than two and a half centuries ago, making it one of the oldest statistical procedures. (Arbuthnott, 1710).

In Example 22.11, all that we need to record is whether each hindleg length is greater than, equal to, or less than its corresponding foreleg length; we do this by recording $+$, 0, or $-$, respectively. We then ask what the probability is of the observed distribution, or a more extreme distribution, of $+$ and $-$ signs if the null hypothesis is true. (A difference of zero is deleted from the analysis, so n is here defined as the number of differences having a sign.) The analysis proceeds as a binomial test with H_0: $p = 0.5$, and the null hypothesis tested is essentially that the median difference is zero (i.e., the population frequencies of positive differences and negative differences are the same). For large samples (say, $n \geq 30$), the *normal approximation*, described in Section 22.5, may be employed. The confidence interval approaches to the binomial test (Section 22.5) are also applicable (for any sample size) and are preferable to the normal approximation. An extension of the sign test to nonparametric testing for blocked data from more than two groups is found in the form of the Friedman test (Section 13.8).

Example 22.11

The sign test for the paired-sample data of Examples 10.1 and 10.4.

Deer	*Hindleg length* (cm)	*Foreleg length* (cm)	*Difference*
1	142	138	+
2	140	136	+
3	144	147	−
4	144	139	+
5	142	143	−
6	146	141	+
7	149	143	+
8	150	145	+
9	142	136	+
10	148	146	+

H_0: There is no difference between hindleg and foreleg length in deer.
H_A: There is a difference between hindleg and foreleg length in deer.
$n = 10$, and there are 8 positive differences and 2 negative differences.

Using Table B.25 for $n = 10$ and $p = 0.50$,

$$P(X \leq 2 \quad \text{or} \quad X \geq 8)$$
$$= 0.04395 + 0.00977 + 0.00098 + 0.00098 + 0.00977 + 0.04395$$
$$= 0.10940$$

Since the probability is greater than 0.05, do not reject H_0.

Using Table B.26 for $n = 10$, the critical values are $C_{0.05(2),10} = 1$ and $n - C_{0.05(2),\ 10} = 10 - 1$. Since neither $X = 2$ nor $X = 8$ is as small as 1 or as large as 9, H_0 is not rejected; and by consulting Table B.26 we state $0.10 < P < 0.20$.

22.7 POWER OF THE BINOMIAL AND SIGN TESTS

The power of and required sample size for the binomial test may be determined by examining the cumulative binomial distribution. As the sign test is essentially a binomial test with p hypothesized to be 0.5 (see Section 22.6), its power and sample size may be assessed in the same manner.

Power of the Test. If a binomial test is performed at a significance level α and with a sample size n, we can determine the power of the test (i.e., the probability of correctly rejecting H_0) as follows. First we determine the critical value(s) of X for the test. For a one-tailed test of $H_0: p \leq p_0$ vs. $H_A: p > p_0$, the critical value is the smallest value of X for which the probability of that X or a larger X is $\leq \alpha$. (In Example 22.8 this is found to be $X = 10$.) For a one-tailed test of $H_0: p \geq p_0$ vs. $H_A: p < p_0$, the critical value is the largest X for which the probability of that X or of a smaller X is $\leq \alpha$. Then we examine the binomial distribution for the observed proportion, $\hat{p}$, from our sample. The power of the test is $\geq$ the probability of an X at least as extreme as the critical value referred to above.* This is demonstrated in Example 22.12.

Example 22.12

Determination of the power of the one-tailed binomial test of Example 22.8.

$\hat{p} = X/n = 10/12 = 0.83$, $H_0: p \leq 0.5$, and $X = 10$ is the critical value for the test.

Then $P(X)$ for $X = 10$ through 12 for the binomial distribution having $p = 0.83$ and $n = 12$ is obtained from Equation 22.1 as

X	$P(X)$
12	0.107
11	0.263
10	0.296

Thus, the power of the test is $> 0.107 + 0.263 + 0.296 = 0.67$.

For a two-tailed test of $H_0: p = p_0$ vs. $H_A: p \neq p_0$, there are two critical values of X, one that cuts off $\alpha/2$ of the binomial distribution in each tail. Knowing these two X's we examine the binomial distribution for $p = \hat{p}$, and the power of the test is the probability in the latter distribution that X is at least as extreme as the critical values. This is demonstrated in Example 22.13.

Example 22.13

Determination of the power of the two-tailed sign test of Example 22.11.

$\hat{p} = X/n = 8/10 = 0.80$, and 1 and 9 are the critical values of the test of $H_0: p = 0.50$.

Then $P(X)$ for $X = 0, 1, 2, 8, 9$, and 10 for the binomial distribution having $p = 0.80$ and $n = 10$ is obtained from Equation 22.1 or from Table B.25 (as the reverse of the distribution for $p = 0.20$), as

X	$P(X)$
0	0.00000
1	0.00000
10	0.10737

Therefore, the power of the test is $\geq 0.00000 + 0.00000 + 0.26844 + 0.00007 + 0.30199$.

*If the critical X delineates a probability of exactly α in the tail of the distribution, then the power = that computed; if the critical value defines a tail $< \alpha$, then the power is $>$ that calculated.

Cohen (1977: Section 5.4) presents tables to estimate sample size requirements in the sign test.

The Normal Approximation. If n and p are of magnitudes conducive to the normal approximation (as indicated in Section 22.5), then the power of a binomial or sign test may be estimated as

$$\text{power} = P\left[Z < \frac{p_0 - p}{\sqrt{\frac{pq}{n}}} - Z_{\alpha(2)}\sqrt{\frac{p_0q_0}{pq}}\right] + P\left[Z > \frac{p_0 - p}{\sqrt{\frac{pq}{n}}} + Z_{\alpha(2)}\sqrt{\frac{p_0q_0}{pq}}\right] \tag{22.39}$$

(Marascuilo and McSweeney, 1977: 62). Here p_0 is the population proportion in the hypothesis to be tested, $q_0 = 1 - p_0$, p is the true population proportion, $q = 1 - p$, $Z_{\alpha(2)} = t_{\alpha(2),\infty}$, and the probabilities of Z are found from Table B.2, using the considerations of Section 7.2. This is demonstrated in Example 22.14.

Example 22.14

Estimation of power in a two-tailed binomial test, using the normal approximation.

We wish to test $H_0: p = 0.5$ vs. $H_A: p \neq 0.5$, using $\alpha = 0.05$, with a sample size of 50, when p is really 0.4.

To employ Equation 22.39 we determine

$$\frac{p_0 - p}{\sqrt{\frac{pq}{n}}} = \frac{0.5 - 0.4}{\sqrt{\frac{(0.4)(0.6)}{50}}} = 1.4434;$$

$$Z_{0.05(2)} = 1.960;$$

$$\sqrt{\frac{p_0q_0}{pq}} = \sqrt{\frac{(0.5)(0.5)}{(0.4)(0.6)}} = 1.0206;$$

and

$$\begin{aligned}\text{power} &= P[Z < 1.4434 - (1.960)(1.0206)] + P[Z > 1.4434 + (1.960)(1.0206)]\\ &= P(Z < -0.56) + P(Z > 3.44)\\ &= P(Z > 0.56) + P(Z > 3.44)\\ &= 0.2877 + 0.0003\\ &= 0.29\end{aligned}$$

For the one-tailed test $H_0: p \leq p_0$ vs. $H_A: p > p_0$, the estimated power is

$$\text{power} = P\left[Z > \frac{p_0 - p}{\sqrt{\frac{pq}{n}}} + Z_{\alpha(1)}\sqrt{\frac{p_0q_0}{pq}}\right]; \tag{22.40}$$

and for the one-tailed hypotheses $H_0\ p \geq p_0$ vs. $H_A: p < p_0$,

$$\text{power} = P\left[Z < \frac{p_0 - p}{\sqrt{\frac{pq}{n}}} - Z_{\alpha(1)}\sqrt{\frac{p_0q_0}{pq}}\right]. \tag{22.41}$$

Sample Size and Minimum Detectable Difference. The normal approximation can further be used in the following ways. We may specify a minimum detectable difference that is of interest by stating p to be a specific distance from the p_0 in the null hypothesis. Then, for a given α, the above procedure may be used to determine the power of the test. If the power thus calculated is below that desired, then the above computations can be effected using a larger value of n. If the power is greater than that desired, then reduce n and recalculate. In this fashion one can, by trial and error, estimate the sample size necessary to achieve a given power, given the minimum detectable difference and significance level.

In a similar fashion, we may state α, p_0, and n, and, by varying p, estimate (by trial and error) the value of p that will result in a desired power.

22.8 CONFIDENCE INTERVAL FOR THE MEDIAN

The confidence limits for a population median may be obtained by considering a binomial distribution with $p = 0.5$. The procedure thus is related to the binomial and sign tests in earlier sections of this chapter and may conveniently use Table B.26. That table gives $C_{\alpha,n}$, and from this we can state the confidence interval for a median to be

$$P(X_i \leq \text{population median} \leq X_j) > 1 - \alpha, \qquad (22.42)$$

where $i = C_{\alpha(2),n} + 1$ and $j = n - C_{\alpha(2),n}$ (e.g., MacKinnon, 1964). (Because of the discreteness of the binomial distribution, the confidence typically will be a little greater than the $1 - \alpha$ specified.) The confidence limits, therefore, are $L_1 = X_i$ and $L_2 = X_j$, with i and j as above. In these considerations, the data are assumed to be arranged in order of magnitude, so that X_i is the ith smallest measurement and X_j is the jth smallest. The above procedure is demonstrated in Example 22.15.

Example 22.15

A confidence interval for a median.

Let us determine a 95% confidence interval for the median of each of the two sets of data in Example 3.3, where the population median was estimated to be 40 mo for species A and 40.5 mo for species B.

For species A, $n = 9$, so (from Table B.26) $C_{0.05(2),9} = 1$ and $n - C_{0.05(2),9} = 9 - 1 = 8$. The confidence limits are, therefore, X_i and X_j, where $i = 1 + 1 = 2$ and $j = 8$; and we can state

$$P(X_2 \leq \text{population median} \leq X_8) > 0.95$$

or

$$P(36 \text{ mo} \leq \text{population median} \leq 43 \text{ mo}) > 0.95.$$

For species B, $n = 10$, and Table B.26 informs us that $C_{0.05(2),10} = 1$; therefore, $n - C_{0.05(2),10} = 10 - 1 = 9$. The confidence limits are X_i and X_j, where $i = 1 + 1 = 2$ and $j = 9$; thus,

$$P(X_2 \leq \text{population median} \leq X_9) > 0.95$$

or

$$P(36 \text{ mo} \leq \text{population median} \leq 44 \text{ mo}) > 0.95.$$

22.9 THE FISHER EXACT TEST

The analysis of 2×2 contingency tables may be considered in the following manner (Fisher, 1958: 96–97), and this procedure is applicable to either one-tailed or two-tailed hypotheses.

Recall from Section 6.2 the configuration of a 2×2 contingency table:

f_{11}	f_{12}	R_1
f_{21}	f_{22}	R_2
C_1	C_2	n

where f_{ij} denotes the frequency observed in row i and column j, and R_i and C_j are row and column totals, respectively. The *Fisher exact test** considers simultaneously two binomial probabilities: the probability of f_{11} coming at random from a total of R_1 counts and the probability of f_{12} coming at random from a total of R_2 counts. The probability of a particular table can be shown to be

$$P = \frac{\dfrac{R_1!R_2!C_1!C_2!}{n!}}{f_{11}!f_{12}!f_{21}!f_{22}!}. \tag{22.43}$$

The One-Tailed Test. Example 22.16 shows a 2×2 table where the data are the numbers of a certain beetle species and of a certain bug species, both of which were found on the upper and lower surfaces of leaves. If we suppose that the beetle might be more likely than the bug to be found on the upper rather than the lower surface (e.g., by virtue of its coloration or its resistance to insolation), then we are dealing with the following one-tailed hypotheses: H_0: The proportion of beetles on the upper side of the leaves is not greater than (i.e., is less than or equal to) the proportion of bugs on the upper surfaces, and H_A: The proportion of beetles on the upper side of the leaves is greater than the proportion of bugs on the upper surfaces.

The probability of the observed table occurring by random chance may be computed using Equation 22.43. Then we tabulate each possible table having an f_{11} value more extreme than that observed and apply Equation 22.43 to each such table. As shown in Example 22.16, the null hypothesis is tested by examining the sum of the probabilities of all these tables considered. This procedure yields the exact probability of obtaining these tables by chance if the null hypothesis were true; if this probability is less than or equal to the significance level, α, then H_0 is rejected.

Note that the quantity $R_1!R_2!C_1!C_2!/n!$ appears in each of the probability calculations and therefore need only be computed once. It is only the value of $f_{11}!f_{12}!f_{21}!f_{22}!$ that needs to be calculated anew for each table. The use of logarithms is, of course, advised for all but the very smallest tables, and Table B.38 provides logarithms of factorials. Ghent (1972), Leslie (1955), Leyton (1968), and Sakoda and Cohen (1957) have shown how the use of binomial coefficients can eliminate much of the laboriousness of Fisher exact test computations, and Ghent (1972) has expanded these considerations to tables with more than two rows and/or columns. Computer programs have been developed (e.g., Fleishman, 1977) to expand exact testing to $r \times c$ tables, where r and/or c are greater than 2.

*Named for Sir Ronald Alymer Fisher (1890–1962), a monumental English statistician with extremely strong influence in many areas of biostatistics and in statistical theory and method in general; it is also referred to on occasion as the Fisher-Irwin test or the Fisher-Yates test.

Example 22.16

A one-tailed Fisher exact test.

H_0: The proportion of beetles is not greater than the proportion of bugs on the upper side of leaves (i.e., beetles do not have greater affinity than bugs for upper leaf surfaces).

H_A: The proportion of beetles is greater than the proportion of bugs on the upper surface of leaves (i.e., beetles have a greater affinity than bugs for upper leaf surfaces).

The data are the four cells in the following table. The marginal totals are then calculated from the four data.

	Beetles	Bugs	
Upper leaf	12	7	19
Lower leaf	2	8	10
	14	15	29

$$P = \frac{\frac{R_1!R_2!C_1!C_2!}{n!}}{f_{11}!f_{12}!f_{21}!f_{22}!}$$

$$= \frac{\frac{19!10!14!15!}{29!}}{12!7!2!8!}$$

$= \text{antilog}\,[(\log 19! + \log 10! + \log 14! + \log 15! - \log 29!) - (\log 12! + \log 7! + \log 2! + \log 8)]$

$= \text{antilog}\,[15.75522 - 17.28932]$

$= \text{antilog}\,[-1.53410]$

$= \text{antilog}\,[0.46590 - 2.00000]$

$= 0.02923$

The tables more extreme than that observed are as follows:

13	6	19
1	9	10
14	15	29

$$P = \frac{\frac{19!10!14!15!}{29!}}{13!6!1!9!}$$

$= \text{antilog}\,[15.75522 - (\log 13! + \log 6! + \log 1! + \log 9!)]$

$= \text{antilog}\,[-2.46515]$

$= 0.003498$

14	5	19
0	10	10
14	15	29

$$P = \frac{\frac{19!10!14!15!}{29!}}{14!5!0!10!}$$

$= \text{antilog}\,[15.75522 - (\log 14! + \log 5! + \log 0! + \log 10!)]$

$= \text{antilog}\,[-3.82413]$

$= 0.0001499$

Therefore, the probability of H_0 being true is $0.02923 + 0.003498 + 0.0001499 = 0.03288$. Since this probability is less than 0.05, H_0 is rejected.

Instead of computing this exact probability of H_0, however, one may consult Table B.27, where $n = 29$, $m_1 = 10$, $m_2 = 14$, and the one-tailed critical values of f are 2 and 8 (see explanation preceding Table B.27). Since the observed f in the cell corresponding to $m_1 = 10$ and $m_2 = 14$ is 2, H_0 may be rejected.

Strictly speaking, the Fisher exact test is applicable to contingency tables where both the row totals and column totals are set in advance of data collection (an uncommon situation). Fortunately, the testing procedure appears to work with other contingency tables as well.

Feldman and Kluger (1963) have demonstrated a simpler computational procedure for obtaining the probabilities of the tables more extreme than the observed table. Let us designate the smallest of the four cell frequencies of a table as f_a and the cell frequency located diagonally from it as f_d. Let f_b be the remaining cell frequency in row 1 and f_c be the remaining cell frequency in row 2. In the next more extreme table f_b will change to $f'_b = f_b + 1$ and f_c will change to $f'_c = f_c + 1$. If P is the probability of one table, the probability of the next more extreme table is

$$P' = \frac{f_a f_d}{f'_b f'_c} P, \tag{22.44}$$

as demonstrated in Example 22.17.

An alternative to computing the exact probability is to consult Table B.27 to obtain a critical value with which to test the hypothesis.

Example 22.17

Simplified computations in a one-tailed Fisher exact test.

For the original table in Example 22.16,

$$f_a = 2, \quad f_d = 7, \quad f_b = 8, \quad f_c = 12, \quad \text{and} \quad P = 0.02923.$$

Therefore, $f'_b = 8 + 1 = 9$ and $f'_c = 12 + 1 = 13$
and the probability for the next more extreme table is

$$P' = \frac{f_a f_d}{f'_b f'_c} P = \frac{(2)(7)}{(9)(13)}(0.02923) = 0.003498.$$

For the new table with the above probability,

$$f_a = 1, \quad f_d = 6, \quad f_b = 9, \quad f_c = 13, \quad P = 0.003498,$$

Therefore, the next $f'_b = 9 + 1 = 10$ and the next $f'_c = 13 + 1 = 14$
and the probability for the next more extreme table is

$$P' = \frac{f_a f_d}{f'_b f'_c} = \frac{(1)(6)}{(10)(14)}(0.003498) = 0.0001499.$$

The Two-Tailed Test. Data such as those in Example 22.16 might have been subjected to two-tailed, instead of one-tailed, hypothesis testing. In that case, we would have been proposing: H_0: The proportion of beetles on the upper surfaces is the same as the proportion of bugs on the upper surfaces; and H_A: The proportion of beetles on the upper sides of the leaves is not the same as the proportion of bugs on the upper surfaces. This is the type of hypothesis that the chi-square contingency table analysis considers (Section 6.2), and it therefore asks whether the row and column categories are independent of one another.

The two-tailed probability is computed as follows. First, the probability of frequencies at least as extreme as those observed, and extreme in the same direction as the observed table is extreme, is calculated as in Example 22.16. This is the probability in one tail of the distribution. If either $R_1 = R_2$ or $C_1 = C_2$, then the two-tailed probability is twice the one-tailed probability. Otherwise, it is not, and the probability in the other tail is computed as follows. In the most extreme table thus far considered, we can identify the smallest of the four marginal frequencies and call it m_1. The smaller frequency from the opposite margin we call m_2. In our example, $m_1 = R_2 = 10$ and $m_2 = C_1 = 14$. Then, the frequency, f, associated with m_1 and m_2 is subtracted from m_1, and a new table is formed. Example 22.18 does this, where the f_{21} of the most extreme table from Example 22.16 is 0, and $m_1 - 0 = 10$. This newly formed table is the most extreme table in the second tail. If the probability of this table is no greater

Example 22.18

A two-tailed Fisher exact test.

Consider the data presented in Example 22.16 as being subjected to the following two-tailed hypotheses:

H_0: The proportion of beetles is the same as the proportion of bugs on upper leaf surfaces (i.e., both kinds of insects have the same affinity for upper leaf surfaces).

H_A: The proportion of beetles is not the same as the proportion of bugs on the upper leaf surfaces (i.e., both kinds of insects do not have the same affinity for upper leaf surfaces).

The observed table of data, and the marginal totals, are as follows:

	Beetles	Bugs	
Upper leaf	12	7	19
Lower leaf	2	8	10
	14	15	29

The probability of this table was found, in Example 22.16, to be 0.02923. Also in Example 22.16, the probabilities of tables more extreme, in the same direction in which the observed table is extreme, were found to be 0.003498 and 0.0001499.

Then, to examine the other tail of the distribution of insects, we consider the value of f associated with the smaller of the R_i's and the smaller of the C_j's in the most extreme table analyzed thus far. In our example, $R_2 = 10$ and $C_1 = 14$; therefore, the f in quesion is f_{21}. Since $R_2 < C_1$, we set f_{21} equal to $R_2 - f_{21}$ and obtain the following table:

4	15	19
10	0	10
14	15	29

$$P = \frac{\dfrac{19!10!14!15!}{29!}}{4!15!10!0!}$$
$$= \text{antilog}\,[15.75522 - (\log 4! + \log 15! + \log 10! + \log 0!)]$$
$$= \text{antilog}\,[-4.30125]$$
$$= 0.00004997$$

Then, f_{21} is reduced by one, and the process is repeated as long as the resultant probability is no greater than that of the observed table:

5	14	19
9	1	10
14	15	29

$$P = \frac{\dfrac{19!10!14!15!}{29!}}{5!14!9!1!} = 0.001499$$

6	13	19
8	2	10
14	15	29

$$P = \frac{\dfrac{19!10!14!15!}{29!}}{6!13!8!2!} = 0.01574$$

7	12	19
7	3	10
14	15	29

$$P = \frac{\dfrac{19!10!14!15!}{29!}}{7!12!7!3!} = 0.07796 \text{ (which is greater than } P \text{ for the observed table)}$$

The probability of a table being as extreme or more extreme as the observed table is $P = 0.02923 + 0.003498 + 0.0001499 + 0.00004997 + 0.001499 + 0.01574 = 0.05017$.

Since this probability is greater than 0.05, H_0 is not rejected.

As an alternative to performing all the preceding computations, consult Table B.27 for $n = 29$, $m_1 = 10$, and $m_2 = 14$; the two-tailed critical values of f are 1 and 8. Since the observed f corresponding to $m_1 = 10$ and $m_2 = 14$ is 2, H_0 is not rejected.

than that of the observed table, then the f being manipulated (in our example, f_{21}) is changed by 1 to make it less extreme (i.e., to make it closer to the f in the same column or in the same row). Then the probability of this new table is computed. This process is repeated as often as necessary to determine the sum of the probabilities of all tables having probabilities at least as small as that of the table originally observed. This procedure is performed in Example 22.17.

The simplified computational procedure of Feldman and Kluger (1963) may also be used for probabilities in the second tail. The most extreme table is identified, and its probability calculated, just as in Example 22.18. Let us say f'_a is the smallest cell frequency in this table, and f'_d is the cell frequency diagonally opposite it. The remaining cell frequency in row 1 is f'_b, and the remaining cell frequency in row 2 is f'_c. In the next less extreme table, $f_b = f'_b - 1$ and $f_c = f'_c - 1$. If P' is the probability of a given table, then the probability of the next less extreme table is

$$P = \frac{f'_b f'_c}{f_a f_d} P' \tag{22.45}$$

as demonstrated in Example 22.19.

Example 22.19
Simplified computations for a two-tailed Fisher exact test.

For the most extreme table in the second tail in Example 22.18,

$$f'_a = 0, \quad f'_d = 4, \quad f'_b = 10, \quad f'_c = 15, \quad P' = 0.00004997.$$

Therefore, $f_a = 0 + 1 = 1$ and $f_d = 4 + 1 = 5$
and the probability for the next less extreme table is

$$P = \frac{f'_b f'_c}{f_a f_b} P' = \frac{(10)(15)}{(1)(5)}(0.00004997) = 0.001499.$$

For the new table with the above probability:

$$f'_a = 1, \quad f'_d = 5, \quad f'_b = 9, \quad f'_c = 14, \quad P' = 0.001499.$$

Therefore, $f_a = 1 + 1 = 2$ and $f_d = 5 + 1 = 6$
and the probability for the next less extreme table is

$$P = \frac{f'_b f'_c}{f_a f_d} P' = \frac{(9)(14)}{(2)(6)}(0.001499) = 0.01574.$$

This process is repeated until the computed P is greater than 0.02923.

Power of the Test. If sample sizes are large enough to employ the normal approximation, then the power of the one-tailed or two-tailed Fisher exact test may be approximated by the computations of Section 22.11

22.10 COMPARING TWO PROPORTIONS

Ideally, the testing of differences among proportions can be expressed as a contingency table (Chapter 6). For the data of Example 6.2, the null hypothesis could be "The proportion of animals surviving is the same in both the experimental (drug administered) and control (no drug treatment) groups." If we are concerned with the difference in proportions between only two groups, the Fisher exact test might be used (either one-tailed or two-tailed, as demonstrated in Examples 22.16 and 22.18, respectively).

An approximate testing procedure for differences between two proportions is as follows. If $\hat{p}_1 = X_1/n_1$ is from one sample, and $\hat{p}_2 = X_2/n_2$ is from a second, then we may desire to test $H_0: p_1 - p_2 = p_0$ (or its one-tailed analogue). We can do this by calculating

$$Z = \frac{|\hat{p}_1 - \hat{p}_2| - p_0}{\sqrt{\frac{\bar{p}\bar{q}}{n_1} + \frac{\bar{p}\bar{q}}{n_2}}}, \tag{22.46}$$

where

$$\bar{p} = \frac{X_1 + X_2}{n_1 + n_2}, \tag{22.47}$$

or

$$\bar{p} = \frac{n_1\hat{p}_1 + n_2\hat{p}_2}{n_1 + n_2}, \tag{22.48}$$

and

$$\bar{q} = 1 - \bar{p}. \tag{22.49}$$

This is a reasonably good approximate test as long as n is large and neither np nor nq is very small (as noted for the "normal approximation" in Section 22.5).

The most common case is where $p_0 = 0$ and, therefore, $H_0: p_1 - p_2 = 0$, or, equivalently, $H_0: p_1 = p_2$; then

$$Z = \frac{\hat{p}_1 - \hat{p}_2}{\sqrt{\frac{\bar{p}\bar{q}}{n_1} + \frac{\bar{p}\bar{q}}{n_2}}}. \tag{22.50}$$

If the preceding null hypothesis is rejected, approximate confidence intervals for $p_1 - p_2$ might be computed as

$$(\hat{p}_1 - \hat{p}_2) \pm Z_{\alpha(2)}\sqrt{\frac{\bar{p}\bar{q}}{n_1} + \frac{\bar{p}\bar{q}}{n_2}}. \tag{22.51}$$

Equation 22.50 performs a test equivalent to χ^2 testing of a 2×2 contingency table without correcting for continuity (and $Z^2 = \chi^2$).* In both Equations 22.46 and 22.50, however, it is generally desirable to introduce a correction for continuity. Let us consider that $\hat{p}_1 = X_1/n_1$ is the larger of the two proportions and $\hat{p}_2 = X_2/n_2$ is the smaller. Then, in the numerator of Z replace $\hat{p}_1$ with $(X_1 - 0.5)/n_1$ and $\hat{p}_2$ with $(X_2 + 0.5)/n_2$. Example 22.20 tests $H_0: p_1 = p_2$ for the 2×2 contingency table data analyzed in Example 6.2 by chi-square. (Except for rounding error in Z, Z equals the square root of the χ^2_c in Example 6.2.)

If the two proportions to be compared are the means from two sets of proportions, then the two-sample t tests of Section 9.3 could be used, in which case the data should be transformed as indicated in Section 14.2, preferably using Equation 14.4 or 14.5. Or, the nonparametric two-sample testing of Sections 9.9 or 9.10 could be employed.

*Some authors have used $\sqrt{\hat{p}_1\hat{q}_1/n_1 + \hat{p}_2\hat{q}_2/n_2}$ as the denominator in Equations 22.46 and 22.50, resulting in what is essentially a large-sample approximation to the Fisher exact test. Eberhardt and Fligner (1977) have shown that this provides a more powerful test when $n_1 = n_2$; however, the probability of a Type I error may be greater than the stated α. Using Equation 22.50 for two-tailed hypotheses results in a Z that is (aside from rounding errors) equal to the square root of the chi-square that would be calculated for the 2×2 contingency table.

Example 22.20

Analysis of the 2 × 2 contingency table of Example 6.2 by considering proportions.

H_0: The same proportion of treated and untreated animals survive.
H_A: The proportion surviving is not the same for both treated and untreated animals.

	Alive	*Dead*	*Total*	*Proportion alive*	*Proportion dead*
Treated	15	9	24	$\hat{p}_1 = 0.625$	$\hat{q}_1 = 0.375$
Not treated	10	15	25	$\hat{p}_2 = 0.400$	$\hat{q}_2 = 0.600$
Total:	25	24	49	$\bar{p} = 0.510$	$\bar{q} = 0.490$

Using Equation 22.50:

$$Z = \frac{0.625 - 0.400}{\sqrt{\frac{(0.510)(0.490)}{24} + \frac{(0.510)(0.490)}{25}}}$$

$$= \frac{0.225}{\sqrt{0.010413 + 0.009996}}$$

$$= \frac{0.225}{0.143}$$

$$= 1.573$$

$Z_{0.05(2)} = t_{0.05(2),\infty} = 1.960$

Therefore, do not reject H_0.

$0.10 < P < 0.20$

Using the correction for continuity:

$$Z = \frac{\frac{15 - 0.5}{24} - \frac{10 + 0.5}{25}}{0.143}$$

$$= \frac{0.604 - 0.420}{0.143}$$

$$= \frac{0.184}{0.143}$$

$$= 1.287$$

$Z_{0.05(2)} = 1.960$

Therefore, do not reject H_0.

$0.10 < P < 0.20$

22.11 POWER AND SAMPLE SIZE IN COMPARING TWO PROPORTIONS

Here we present procedures for estimating power and sample size in comparing two proportions (or performing a test of a 2 × 2 contingency table), provided that the data are amenable to the normal approximation (as explained in Section 22.10).

Estimating Power. If the test of H_0: $p_1 = p_2$ vs. H_A: $p_1 \neq p_2$ is to be performed at the α significance level, with n_1 data in sample 1 and n_2 data in sample 2; and if the two samples come from populations actually having proportions of p_1 and p_2, respectively; then an estimate of power is

$$\text{power} = P\left[Z < \frac{-Z_{\alpha(2)}\sqrt{\bar{p}\bar{q}/n_1 + \bar{p}\bar{q}/n_2} - (p_1 - p_2)}{\sqrt{p_1q_1/n_1 + p_2q_2/n_2}}\right]$$
$$+ P\left[Z > \frac{Z_{\alpha(2)}\sqrt{\bar{p}\bar{q}/n_1 + \bar{p}\bar{q}/n_2} - (p_1 - p_2)}{\sqrt{p_1q_1/n_1 + p_2q_2/n_2}}\right] \quad (22.52)$$

(Marascuilo and McSweeney, 1977: 111), where

$$\bar{p} = \frac{n_1p_1 + n_2p_2}{n_1 + n_2}, \quad (22.53)$$

$$q_1 = 1 - p_1, \quad (22.54)$$

$$q_2 = 1 - p_2, \quad (22.55)$$

and

$$\bar{q} = 1 - \bar{p}. \quad (22.56)$$

The calculation utilizes Table B.2 and the considerations of Section 7.2 and is demonstrated in Example 22.21.

For the one-tailed test of $H_0: p_1 \geq p_2$ vs. $H_A: p_1 < p_2$, the estimated power is

$$\text{power} = P\left[Z < \frac{-Z_{\alpha(1)}\sqrt{\bar{p}\bar{q}/n_1 + \bar{p}\bar{q}/n_2} - (p_1 - p_2)}{\sqrt{p_1q_1/n_1 + p_2q_2/n_2}}\right]; \quad (22.57)$$

and for the one-tailed hypotheses $H_0: p_1 \leq p_2$ vs. $H_A: p_1 > p_2$,

$$\text{power} = P\left[Z > \frac{Z_{\alpha(1)}\sqrt{\bar{p}\bar{q}/n_1 + \bar{p}\bar{q}/n_2} - (p_1 - p_2)}{\sqrt{p_1q_1/n_1 + p_2q_2/n_2}}\right]. \quad (22.58)$$

Example 22.21

Estimation of power in a two-tailed test comparing two proportions.

We propose to test H_0: $p_1 = p_2$ vs. H_A: $p_1 \neq p_2$, with $\alpha = 0.05$, $n_1 = 50$, and $n_2 = 45$, where in the sampled populations $p_1 = 0.75$ and $p_2 = 0.50$. The power of the test can be estimated as follows.

We first compute (by Equation 22.53):

$$\bar{p} = \frac{(50)(0.75) + (45)(0.50)}{50 + 45} = 0.6316$$

and

$$\bar{q} = 1 - \bar{p} = 0.3684.$$

Then, using Equation 22.52,

$$\frac{\bar{p}\bar{q}}{n_1} = \frac{(0.6316)(0.3684)}{50} = 0.0047; \quad \frac{\bar{p}\bar{q}}{n_2} = 0.0052;$$

$$\frac{p_1q_1}{n_1} = \frac{(0.75)(0.25)}{50} = 0.0038; \quad \frac{p_2q_2}{n_2} = \frac{(0.50)(0.50)}{45} = 0.0056;$$

$$Z_{0.05(2)} = 1.9600;$$

and

$$\text{power} = P\left[Z < \frac{-1.9600\sqrt{0.0047 + 0.0052} - (0.75 - 0.50)}{\sqrt{0.0038 + 0.0056}}\right]$$
$$+ P\left[Z > \frac{1.9600\sqrt{0.0047 + 0.0052} - (0.75 - 0.50)}{\sqrt{0.0038 + 0.0056}}\right]$$
$$= P(Z < -4.59) + P(Z > -0.57)$$
$$= P(Z > -4.59) + [1 - P(Z > 0.57)]$$
$$= 0.0000 + 1.0000 - 0.2843$$
$$= 0.72$$

Estimating Necessary Sample Size for Equal Samples. Several procedures have been proposed in the statistical literature to estimate the sample sizes that would be required to detect a true population difference between proportions, δ, when testing for difference between proportions with specified probabilities of Type I and Type II errors. Casagrande, Pike, and Smith (1978) review some of these and propose the following as a superior method. If we specify α, β (where $\beta = 1 -$ power), δ, p_1, and p_2, then the size of each of the two samples should be at least

$$n = \frac{A\left[1 + \sqrt{1 + \frac{4\delta}{A}}\right]^2}{4\delta^2}, \tag{22.59}$$

where

$$A = [Z_\alpha\sqrt{2\bar{p}\bar{q}} + Z_{\beta(1)}\sqrt{p_1q_1 + p_2q_2}]^2 \tag{22.60}$$

$$\bar{p} = \frac{p_1 + p_2}{2}, \tag{22.61}$$

$\bar{q} = 1 - \bar{p}$, and Z_α is either $Z_{\alpha(2)} = t_{\alpha(2),\infty}$ or $Z_{\alpha(1)} = t_{\alpha(1),\infty}$ (depending on whether one is test a two-tailed or a one-tailed null hypothesis, respectively). These calculations are demonstrated in Example 22.22. For two-tailed testing, $\delta = |p_1 - p_2|$; for one-tailed testing, $\delta = p_1 - p_2$ if H_A: $p_1 > p_2$, or $\delta = p_2 - p_1$ if H_A: $p_2 > p_1$.

Example 22.22

Estimation of sample size required to detect a specified difference between population proportions, when testing with specified α and β.

We wish to test H_0: $p_1 = p_2$ vs. H_A: $p_1 \neq p_2$, at $\alpha = 0.05$, with 90% power (i.e., $\beta = 0.10$). If, in the sampled populations, $p_1 = 0.45$ and $p_2 = 0.25$, how large a sample should be collected from each population?

$\bar{p} = \frac{0.45 + 0.25}{2} = 0.35$; $\bar{q} = 1 - 0.35 = 0.65$;

$Z_{\alpha(2)} = Z_{0.05(2)} = t_{0.05(2),\infty} = 1.9600$;

$Z_{\beta(1)} = Z_{0.10(1)} = t_{0.10(1),\infty} = 1.2816$;

$A = [1.9600\sqrt{(2)(0.35)(0.65)} + 1.2816\sqrt{(0.45)(0.55) + (0.25)(0.75)}]^2 = 4.6975$;

$\delta = |0.45 - 0.25| = 0.20$; and

$n = \frac{A[1 + \sqrt{1 + 4(0.20)/A}]^2}{4(0.20)^2} = 127.2$

So the sample sizes should each be at least 128.

Estimating Sample Sizes for Unequal Samples. It is desirable to plan data collection so as to obtain equal sample sizes (i.e., $n_1 = n_2$). However, this is not always practical, so Fleiss, Tytun, and Ury (1980) showed how we may specify that if the ratio between two samples is

$$r = \frac{n_2}{n_1} \tag{22.62}$$

(i.e., n_2 is r times as large as n_1), then the required size of the first sample is

$$n_1 = \frac{B\left[1 + \sqrt{1 + \frac{2(r+1)\delta}{B}}\right]^2}{4r\delta^2}, \tag{22.63}$$

where

$$B = [Z_\alpha\sqrt{(r+1)p'_0q'_0} + Z_{\beta(1)}\sqrt{rp_1q_1 + p_2q_2}]^2 \quad (22.64)$$

$$p'_0 = \frac{p_1 + rp_2}{r+1}, \quad (22.65)$$

$$q'_0 = 1 - p'_0, \quad (22.66)$$

and Z_α and $Z_{\beta(1)}$ are as in Equation 22.60. Note that if $n_1 = n_2$, then $r = 1$, and Equation 22.64 becomes Equation 22.60, Equation 22.65 becomes Equation 22.61, and Equation 22.63 is identical to Equation 22.59.

Minimum Detectable Difference and Sample Size. If the smallest difference between proportions (i.e., $p_1 - p_2$) to be detected with a desired power by testing $H_0: p_1 - p_2$ is specified, then the above procedure can estimate the power of the test. If the power thus computed is less than that desired, the calculations can be repeated using larger sample sizes, a process that can be repeated until the desired power is achieved. (If, during these computations, the estimated power is greater than that desired, then the calculations can be redone with smaller sample sizes.)

Similarly, one may state α and the sample sizes, and by varying p_1 and p_2 estimate (by trial and error) the value of p_1 and of p_2 (and, therefore, of $p_1 - p_2$) that will result in a desired power.

22.12 COMPARING MORE THAN TWO PROPORTIONS

As indicated at the beginning of Section 22.9, it is best to test for differences among proportions by contingency table analysis. For example, the null hypothesis of Example 6.1 could be stated as "The proportions of males and females are the same among individuals of each of the four hair colors."

Alternatively, the normal approximation is applicable (if n and p are large enough to use the approximation, as indicated previously in this chapter). Using this approximation, one tests $H_0: p_1 = p_2 = \ldots = p_k$ against the alternative hypothesis that all k proportions are not the same, as

$$\chi^2 = \sum_{i=1}^{k} \frac{(X_i - n_i\bar{p})^2}{n_i\bar{p}\bar{q}} \quad (22.67)$$

(Pazer and Swanson, 1972: 187–190). Here,

$$\bar{p} = \frac{\sum_{i=1}^{k} X_i}{\sum_{i=1}^{k} n_i} \quad (22.68)$$

is a pooled proportion, $\bar{q}$ is as in Equation 22.49, and χ^2 has $k - 1$ degrees of freedom. Example 22.23 demonstrates this procedure, which is equivalent to χ^2 testing of a contingency table with two rows (or two columns).

Example 22.23

Comparing four proportions, using the data of Example 6.1.

$n_1 = 87,\ X_1 = 32,\ \hat{p}_1 = \frac{32}{87} = 0.368,\ \hat{q}_1 = 0.632$

$n_2 = 108,\ X_2 = 43,\ \hat{p}_2 = \frac{43}{108} = 0.398,\ \hat{q}_2 = 0.602$

$n_3 = 80,\ X_3 = 16,\ \hat{p}_2 = \frac{16}{80} = 0.200,\ \hat{q}_3 = 0.800$

$n_4 = 25,\ X_4 = 9,\ \hat{p}_2 = \frac{9}{25} = 0.360,\ \hat{q}_4 = 0.640$

$\bar{p} = \frac{\sum X_i}{\sum n_i} = \frac{32 + 43 + 16 + 9}{87 + 108 + 80 + 25} = \frac{100}{300} = \frac{1}{3} = 0.3333\ldots$

$\bar{q} = 1 - \bar{p} = \frac{2}{3}$

$$\chi^2 = \sum \frac{(X_i - n_i\bar{p})^2}{n_i\bar{p}\bar{q}}$$

$$= \frac{[32 - (87)(\frac{1}{3})]^2}{(87)(\frac{1}{3})(\frac{2}{3})} + \frac{[43 - (108)(\frac{1}{3})]^2}{(108)(\frac{1}{3})(\frac{2}{3})} + \frac{[16 - (80)(\frac{1}{3})]^2}{(80)(\frac{1}{3})(\frac{2}{3})} + \frac{[9 - (25)(\frac{1}{3})]^2}{(25)(\frac{1}{3})(\frac{2}{3})}$$

$$= 0.4655 + 2.0417 + 6.4000 + 0.0800$$

$$= 8.987$$

$\nu = k - 1 = 4 - 1 = 3$

$\chi^2_{0.05,3} = 7.815$

Therefore, reject H_0.

$0.025 < P < 0.05$

Note how the calculated χ^2 compares with that in Example 6.1.

We can, instead, test whether k p's are equal not only to each other but to a specified constant, p_0(i.e., $H_0: p_1 = p_2 = \ldots = p_k = p_0$). This is done by computing

$$\chi^2 = \sum_{i=1}^{k} \frac{(X_i - n_i p_0)^2}{n_i p_0(1 - p_0)}, \tag{22.69}$$

which is then compared to the critical value of χ^2 for k (rather than $k - 1$) degrees of freedom.

If each of the several proportions to be compared to each other is the mean of a set of proportions, then we can use the multisample testing procedures of Chapters 11, 12, 13, 15, and 16. To do so, the individual data should be transformed as suggested in Section 14.2, preferably by Equation 14.4 or 14.5, if possible.

Finally, it should be noted that comparing several p's yields the same results as if one compared the associated q's.

22.13 MULTIPLE COMPARISONS FOR PROPORTIONS

If the null hypothesis $H_0: p_1 = p_2 = \ldots p_k$ (see Section 22.12) is rejected, then we may desire to determine specifically which population proportions are different from which others. The following procedure (similar to that of Levy, 1975a) allows for testing analogous to the Tukey or Student-Newman-Keuls tests introduced in Chapter 12. An angular transformation (Section 14.2) of each sample proportion is to be used. If $\hat{p}$, but not X and n, is known, then Equation 14.2 may be used. If, however, X and n are also known, then either Equation 14.4 or 14.5 is preferable. (The latter two equations give similar results, except for small or large $\hat{p}$, where Equation 14.5 is probably better.)

As shown in Example 22.24, the multiple comparison procedure is similar to that in Chapter 12 (the Tukey test being in Section 12.1). The standard error for each comparison is, in degrees,*

$$\text{SE} = \sqrt{\frac{205.18}{n + 0.5}} \tag{22.70}$$

if the two samples being compared are the same size, or

$$\text{SE} = \sqrt{\frac{102.59}{n_A + 0.5} + \frac{102.59}{n_B + 0.5}} \tag{22.71}$$

if they are not; the critical value is $q_{\alpha,\infty,k}$ (from Table B.5).

Use of the normal approximation to the binomial is possible in multiple comparison testing (e.g., Marascuilo, 1971: 380–382), but the above procedure is preferable.

Example 22.24

Tukey-type multiple comparison testing among the four proportions of Example 22.23.

Samples ranked by proportion (i)	3	4	1	2
Ranked sample proportions ($p_i = X_i/n_i$)	16/80 = 0.200	9/25 = 0.360	32/87 = 0.368	43/108 = 0.398
Ranked transformed proportions (p'_i, in degrees)	26.84	37.19	37.44	39.18

Comparison B vs. A	*Difference* $p'_B - p'_A$	SE	q	$q_{0.05,\infty,4}$	*Conclusion*
2 vs. 3	39.18 − 26.84 = 12.34	1.49	8.28	3.633	Reject $H_0: p_2 = p_3$
2 vs. 4	39.18 − 37.19 = 1.99	2.23	0.89	3.633	Accept $H_0: p_2 = p_4$
2 vs. 1	Do not test				
1 vs. 3	37.44 − 26.84 = 10.60	1.56	6.79	3.633	Reject $H_0: p_1 = p_3$
1 vs. 4	Do not test				
4 vs. 3	37.19 − 26.84 = 10.35	2.30	4.50	3.633	Reject $H_0: p_4 = p_3$

Overall conclusion: $p_3 \neq p_4 = p_1 = p_2$ (which is the same conclusion reached in Example 6.4).

Equation 14.5 is used for the transformations. For sample 3, for example, $X/(n + 1) = 16/81 = 0.198$ and $(X + 1)/(n + 1) = 17/81 = 0.210$, so $p' = \frac{1}{2}[\arcsin \sqrt{0.198} + \arcsin \sqrt{0.210}] = 26.84$. Using Table B.23 to obtain the two needed arcsines, we have $p' = \frac{1}{2}[26.42 + 27.27] = 26.84$.

Comparison of a Control Proportion to Each Other Proportion. A procedure analogous to the Dunnett test of Section 12.4 may be used as a multiple comparison test where instead of comparing all pairs of proportions we desire to compare one proportion (designated as the "control") to each of the others. Calling

*The constant 205.18 square degrees results from $(180°/4\pi)^2$, which follows from the variances reported by Anscombe (1948) and Freeman and Tukey (1950).

the control group B, and each other group, in turn, A, we compute the Dunnett test statistic:

$$q = \frac{p'_B - p'_A}{\text{SE}}. \quad (22.72)$$

Here, the proportions have been transformed as earlier in this section, and the appropriate standard error is

$$\text{SE} = \sqrt{\frac{410.35}{n + 0.5}} \quad (22.73)$$

if samples A and B are the same size, or

$$\text{SE} = \sqrt{\frac{205.18}{n_A + 0.5} + \frac{205.18}{n_B + 0.5}} \quad (22.74)$$

if $n_A \neq n_B$. The critical value is $q'_{\alpha(1),\infty,p}$ (from Table B.6) or $q'_{\alpha(2),\infty,p}$ (from Table B.7) for one-tailed or two-tailed testing, respectively.

Multiple Contrasts among Proportions. The Scheffé procedure for multiple contrasts among means (Section 12.5) may be adapted to proportions by using angular transformations as done earlier in this section. For each contrast, we calculate

$$S = \frac{\left|\sum_i c_i p'_i\right|}{\text{SE}}, \quad (22.75)$$

where

$$\text{SE} = \sqrt{\sum_i c_i^2 \left(\frac{205.18}{n_i + 0.5}\right)^2}, \quad (22.76)$$

and c_i is a contrast coefficient as described in Section 12.5. (For example, if we wished to test the hypothesis $H_0: (p_1 + p_2 + p_4)/3 - p_3 = 0$, then $c_1 = \frac{1}{3}$, $c_2 = \frac{1}{3}$, $c_3 = -1$, and $c_4 = \frac{1}{3}$.)

EXERCISES

22.1. If, in a binomial population, $p = 0.3$ and $n = 6$, what proportion of the population does $X = 2$ represent?

22.2. If, in a binomial population, $p = 0.22$ and $n = 5$, what is the probability of $X = 4$?

22.3. Determine whether the following data, where $n = 4$, are likely to have come from a binomial population with $p = 0.25$:

X	f
0	30
1	51
2	33
3	10
4	2

22.4. Determine whether the following data, where $n = 4$, are likely to have come from a binomial population:

X	f
0	20
1	41
2	33
3	11
4	4

22.5. A randomly selected male mouse of a certain species was placed in a cage with a randomly selected male mouse of a second species, and it was recorded which animal exhibited dominance over the other. The experimental procedure was performed, with different pairs of animals, a total of 20 times, with individuals from species 1 being dominant 6 times and those from species 2 being dominant 14 times. Test the null hypothesis that there is no difference in the ability of either species to dominate.

22.6. A hospital treated 412 skin cancer patients over a period of time. Of these, 197 were female. Using the normal approximation to the binomial test, test the hypothesis that equal numbers of males and females seek treatment for skin cancer.

22.7. Test the null hypothesis of Exercise 5.3, using the binomial test normal approximation.

22.8. Ten students were given a mathematics aptitude test in a quiet room. The same students were given a similar test in a room with background music. Their performances were as follows. Using the sign test, test the hypothesis that the music has no effect on test performance.

Student	*Score without music*	*Score with music*
1	114	112
2	121	122
3	136	141
4	102	107
5	99	96
6	114	109
7	127	121
8	150	146
9	129	127
10	130	128

22.9. In a random sample of 1215 animals, 62 exhibited a certain genetic defect. Determine the 95% confidence interval for the proportion of the population displaying this defect.

22.10. Estimate the power of the hypothesis test of Exercise 22.5 if $\alpha = 0.05$.

22.11. Using the normal approximation, estimate the power of the hypothesis test of Exercise 22.6 if $\alpha = 0.05$.

22.12. Eight mature spruce and nine mature pine trees have been grown in a row to protect some buildings from the wind. After a severe windstorm, four spruce and one pine suffered considerable damage. Test the hypothesis that both species are equally resistant to wind damage. Compute the exact probability for this test and compare it with the conclusion reached by using Table B.27.

22.13. The study of Exercise 6.2 was repeated with larger numbers of data, so that the normal approximation may be used in the binomial test of the hypothesis that the proportion of skunks with rabies is the same in both geographic areas: 31 of the 87 animals collected from area "E" had rabies, as did 43 of the 158 taken from area "W."

22.14. For an analysis such as that in Exercise 22.13, using a significance level of 0.05 and sample sizes of $n_1 = n_2 = 300$, what would the power of the test be if the population proportions were $p_1 = 0.333$ and $p_2 = 0.250$?

22.15. If an analysis such as that in Exercise 22.13 were desired to be performed at the 5% significance level, with 90% power to detect a difference between population proportions of 0.333 and 0.250, what is the minimum size sample that should be taken from each population?

22.16. Using the data of Exercise 6.1, test the null hypothesis that there is the same proportion of males in all four seasons.

22.17. For the same data used in Exercise 22.16, test the hypothesis H_0: $p_1 = p_2 = p_3 = p_4 = 0.5$.

22.18. If the null hypothesis in Exercise 22.17 is rejected, perform a Tukey-type multiple comparison test to conclude which population proportions are different from which.

23 The Poisson Distribution and Randomness

The *Poisson distribution** is important in describing *random* occurrences, these occurrances being either objects in space or events in time. A random distribution of objects in space is one in which each portion of the space has the same probability of containing an object and the occurrence of an object in any portion of the space in no way influences the occurrence of any other of the objects in any portion of space. A biological example might be the distribution of bacteria in a liquid medium. A random distribution of events in time is one in which each time period has an equal chance of witnessing an event, and the occurrence of any event is independent of the occurrence of any other event. An example might be the firing of certain nerve fibers.

Testing for sequential randomness is described in Sections 23.6 to 23.8.

23.1 POISSON PROBABILITIES

The terms of the Poisson distribution are

$$P(X) = \frac{e^{-\mu}\mu^X}{X!} \tag{23.1}$$

or, equivalently,

$$P(X) = \frac{\mu^X}{e^{\mu}X!}, \tag{23.2}$$

*Described in 1837 by Simeon-Denis Poisson (1781–1840) (Struik, 1967: 108, 147), a French mathematician and physicist, although Abraham de Moivre (1667–1754) apparently described it previously in 1718 (David, 1962: 168). It was also described independently by others, including "Student" (W. S. Gosset, 1876–1937) in 1909 (Haight, 1967: 117).

where $P(X)$ is the probability of X occurrences in a unit of space (or time) and μ is the population mean number of occurrences per unit of space (or time). Thus,

$$P(X = 0) = e^{-\mu}, \tag{23.3}$$

$$P(X = 1) = e^{-\mu}\mu, \tag{23.4}$$

$$P(X = 2) = \frac{e^{-\mu}\mu^2}{2}, \tag{23.5}$$

$$P(X = 3) = \frac{e^{-\mu}\mu^3}{(3)(2)}, \tag{23.6}$$

$$P(X = 4) = \frac{e^{-\mu}\mu^4}{(4)(3)(2)}, \tag{23.7}$$

etc., where $P(X = 0)$ is the probability of no occurrences, $P(X = 1)$ is the probability of one occurrence, etc. Figure 23.1 presents some Poisson probabilities graphically.

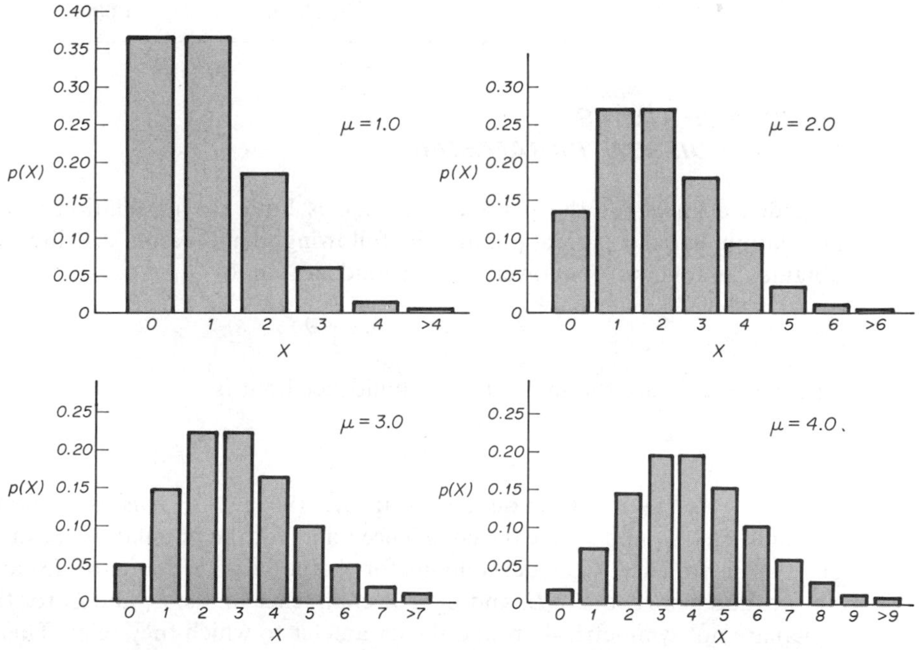

Figure 23.1 The Poisson distribution for various values of μ. These graphs were prepared by using Equation 23.1.

It is noteworthy that the Poisson distribution approaches the binomial distribution when n is large and p is very small. For example, Table 23.1 compares the Poisson distribution where $\mu = 1$ with the binomial distribution where $n = 100$ and $p = 0.01$ (and, therefore, $\mu = np = 1$). Therefore, the Poisson distribution has importance in describing binomially distributed events having low probability. Another interesting property of the Poisson distribution is that $\sigma^2 = \mu$; that is, the variance and the mean are equal.

TABLE 23.1 THE POISSON DISTRIBUTION WHERE $\mu = 1$ COMPARED WITH THE BINOMIAL DISTRIBUTION WHERE $n = 100$ AND $p = 0.01$ (I.E., WITH $\mu = 1$) AND THE BINOMIAL DISTRIBUTION WHERE $n = 10$ AND $p = 0.1$ (I.E., WITH $\mu = 1$)

X	$P(X)$ for Poisson: $\mu = 1$	$P(X)$ for binomial: $n = 100, p = 0.01$	$P(X)$ for binomial: $n = 10, p = 0.1$
0	0.36788	0.36603	0.34868
1	0.36788	0.36973	0.38742
2	0.18394	0.18486	0.19371
3	0.06131	0.06100	0.05740
4	0.01533	0.01494	0.01116
5	0.00307	0.00290	0.00149
6	0.00050	0.00046	0.00014
7	0.00007	0.00006	0.00001
>7	0.00001	0.00002	0.00000
Total	1.00000	1.00000	1.00001

23.2 CONFIDENCE LIMITS FOR THE POISSON PARAMETER

Confidence limits for the parameter (which is both the population mean and the population variance) of a population following the Poisson distribution may be obtained as follows. The lower $1 - \alpha$ confidence limit is

$$L_1 = \frac{\chi^2_{(1-\alpha/2),\nu}}{2}, \tag{23.8}$$

where $\nu = 2X$; and the upper $1 - \alpha$ confidence limit is

$$L_2 = \frac{\chi^2_{\alpha/2,\nu}}{2}, \tag{23.9}$$

where $\nu = 2(X + 1)$ (Pearson and Hartley, 1966: 81). This is demonstrated in Example 23.1. L_1 and L_2 are the confidence limits for the population mean and for the population variance. Confidence limits for the population standard deviation, σ, are simply the square roots of L_1 and L_2. The confidence limits, L_1 and L_2 (or their square roots) are not symmetrical around the parameter to which they refer. This procedure is a fairly good approximation. A more accurate (and tedious) method is the examination of the tails of the Poisson distribution (e.g., Example 23.3) to determine the value of X that cuts off $\leq \alpha/2$ of each tail.

Example 23.1
Confidence limits for the Poisson parameter.

An oak leaf contains 4 galls. Assuming that there is a random occurrence of galls on oak leaves in the population, estimate with 95% confidence the mean number of galls per leaf in the population.

The population mean, μ, is estimated as $X = 4$ galls/leaf.

The 95% confidence limits for μ are

$$L_1 = \frac{\chi^2_{(1-\alpha/2),\nu}}{2}, \text{ where } \nu = 2X = 2(4) = 8$$

$$L_1 = \frac{\chi^2_{0.975,8}}{2} = \frac{2.180}{2} = 1.1 \text{ galls/leaf}$$

$$L_2 = \frac{\chi^2_{\alpha/2,\nu}}{2}, \text{ where } \nu = 2(X+1) = 2(4+1) = 10$$

$$L_2 = \chi^2_{0.025,10} = \frac{20.483}{2} = 10.2 \text{ galls/leaf}$$

Therefore, we can state

$$P(1.1 \text{ galls/leaf} \leq \mu \leq 10.2 \text{ galls/leaf}) \geq 0.95$$

or

$$P(1.1 \text{ galls/leaf} \leq \sigma^2 \leq 10.2 \text{ galls/leaf}) \geq 0.95;$$

and, using the square roots of L_1 and L_2,

$$P(1.0 \text{ galls/leaf} \leq \sigma \leq 3.2 \text{ galls/leaf}) \geq 0.95.$$

23.3 GOODNESS OF FIT OF THE POISSON DISTRIBUTION

The goodness of fit of the Poisson distribution to a set of observed data may be tested by chi-square, just as was done with the binomial distribution in Section 22.4. The frequencies in the tails of the distribution should be pooled to arrive at a tabulation having no expected frequencies less than 1.0 (Cochran, 1954), and the degrees of freedom are $k - 2$ (k being the number of categories of X remaining after such pooling). Example 23.2 fits the Poisson distribution to a set of biological data. Here, μ is estimated by $\bar{X}$, and either Equation 23.1 or 23.2 is employed. The G statistic (Section 5.9) may be used for goodness of fit analysis instead of chi-square. It will give equivalent results when n/k is large; if n/k is very small, G is preferable to χ^2 (Rao and Chakravarti, 1956).

Example 23.2

Fitting the Poisson distribution.

The data are the number of sparrow nests in an area of given size. The area used is 1 hectare. Then, X_i is the number of nests in a hectare; f_i is the frequency of X_i nests per hectare; and $P(X_i)$ is the probability of X_i nests per hectare, if the nests are distributed randomly (as computed from Equation 23.1 or 23.2).

H_0: The population of sparrow nests is distributed randomly.
H_A: The population of sparrow nests is not distributed randomly.

X_i	f_i		$P(X_i)$	$\hat{f}_i = [P(X_i)][n]$	
0	9		0.332871	13.315	
1	22		0.366158	14.646	
2	6		0.201387	8.055	
3	2		0.073842	2.954	
4	1	3	0.020307	0.812	3.983
≥ 5	0		0.005435	0.217	

$n = \sum f_i = 40$

$\bar{X} = \frac{\sum f_i X_i}{\sum f_i} = \frac{44}{40} = 1.10$

$\chi^2 = 1.398 + 3.693 + 0.524 + 0.243 = 5.858$

$\nu = k - 2 = 2$

$\chi^2_{0.05,2} = 5.991$

Therefore, do not reject H_0.

$0.05 < P < 0.10$

If μ were known for the particular population sampled, or if it were desired to assume a certain value of μ, then the parameter would not have to be estimated by $\bar{X}$, and the degrees of freedom for χ^2 or G goodness of fit testing would be $k - 1$. It is only when this parameter is specified that the Kolmogorov-Smirnov goodness of fit procedure (Section 5.10) may be applied (Massey, 1951).

The null hypothesis in Poisson goodness of fit testing is one of a random distribution of entities in space or time. Rejection of the hypothesis of randomness may result from one of two situations. First, the population distribution may be *uniform*; that is, each unit of space (or time) has the same number of entities. Second, the population may be arranged in a *clustered*, or *contagious*, distribution. Figures 23.2 and 23.3 present these possibilities diagrammatically. If a population has a random distribution, $\sigma^2 = \mu$, and $\sigma^2/\mu = 1.0$. If the population distribution is more uniform than random, $\sigma^2 < \mu$, and $\sigma^2/\mu < 1.0$. And if the population is distributed contagiously, $\sigma^2 > \mu$, and $\sigma^2/\mu > 1.0$.*

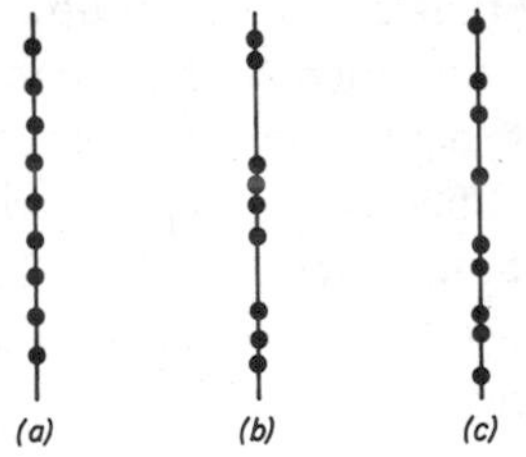

Figure 23.2 Distributions in one-dimensional space: (a) uniform, in which $\sigma^2 < \mu$; (b) contagious, or clustered, in which $\sigma^2 > \mu$; (c) random (Poisson), in which $\sigma^2 = \mu$.

The investigator generally has some control over the size of the space or the length of the time interval that is considered to be the unit of observation. Thus, in Example 23.2, we might have chosen areas twice the size of those used, in which case each f_i would have been twice the size as in our example, and $\bar{X}$ would have been 2.20 instead of 1.10. If we wish to consider analyses involving the Poisson distribution, it is desirable to use a sample distribution with a fairly small mean—let us say certainly below 10, preferably below 5, and ideally in the neighborhood of 1. If the mean is too large, then the Poisson too closely resembles the binomial, as well as the normal, distribution. If it is too small, however, then the number of categories, k, with appreciable frequencies will be too small for sensitive analysis.

*An alternative test for goodness of fit of the Poisson distribution utilizes the test statistic $\chi^2 = \mathrm{SS}/\bar{X}$ with $n - 1$ degrees of freedom (or $\chi^2 = \mathrm{SS}/\mu$ with $\nu = n$, if μ is known). This procedure is discussed briefly by Simpson, Roe, and Lewontin (1960: 311–312) and Steel and Torrie (1980: 530–531); it may yield different conclusions than the method of fitting expected to observed frequencies.

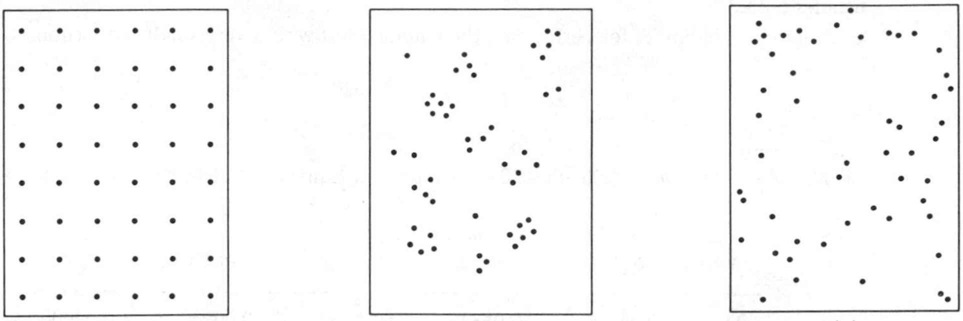

(a) (b) (c)

Figure 23.3 Distributions in two-dimensional space: (a) uniform, in which $\sigma^2 < \mu$; (b) contagious, or clustered, in which $\sigma^2 > \mu$; (c) random (Poisson), in which $\sigma^2 = \mu$.

Graphical testing of goodness of fit is sometimes encountered. The reader may consult Gart (1969b) for such considerations.

23.4 THE BINOMIAL TEST REVISITED

The binomial test was introduced (in Section 22.5) as a goodness of fit test for counts in two categories. If n is large, the binomial test is unwieldy and its normal approximation can be used. If p is very small, the normal approximation is not applicable, but inasmuch as the Poisson and binomial distributions converge for small p the following procedure may be employed (as in Traut, 1980).

One-Tailed Testing. Let us consider the following example. It is assumed (as from a very large body of previous information) that a certain type of genetic mutation naturally occurs in an insect population with a frequency of 0.0020 (i.e., in 20 out of 10,000 insects). On exposing a large number of these insects to a particular chemical, we wish to ask whether that chemical increases the rate of this mutation. Thus, we state $H_0: p \leq 0.0020$ and $H_A: p > 0.0020$. (The general one-tailed hypotheses of this sort would be H_0: $p \leq p_0$ and H_A: $p > p_0$.) If we had reason to ask whether some treatment reduced the natural rate of mutations, then the one-tailed test would have used H_0: $p \geq p_0$ and H_A: $p < p_0$.

Performing the experiment for the hypotheses H_0: $p \leq 0.0020$ and H_A: $p > 0.0020$ (Examples 23.3 and 23.4) yields 28 of the sought mutations in 8000 insects observed, for a sample mutation rate of $\hat{p} = X/n = 28/8000 = 0.0035$. The question, then, is whether the rate of 0.0035 is significantly greater than 0.0020. If we conclude that there is a low probability (i.e., $\leq \alpha$) of a sample rate being at least as large as 0.0035 when the sample is taken at random from a population having a rate of 0.0020, then H_0 is to be rejected. The hypotheses could also be stated H_0: $\mu \leq \mu_0$ and H_A: $\mu > \mu_0$; where $\mu_0 = p_0 n$ (which is 16 in Example 23.4).

Example 23.3

Poisson probabilities for performing the binomial test with a very small proportion.

$$p_0 = 0.0020$$

$$n = 8000$$

We substitute $p_0 n = (0.0020)(8000) = 16$ for μ in Equation 23.1 to compute the following:*

X	$P(X)$	*Cumulative* $P(X)$	X	$P(X)$	*Cumulative* $P(X)$
23	0.02156	0.05825	0	0.00000	0.00000
24	0.01437	0.03669	1	0.00000	0.00000
25	0.00920	0.02232	2	0.00001	0.00001
26	0.00566	0.01312	3	0.00008	0.00009
27	0.00335	0.00746	4	0.00031	0.00040
28	0.00192	0.00411	5	0.00098	0.00138
29	0.00106	0.00219	6	0.00262	0.00400
30	0.00056	0.00113	7	0.00599	0.00999
31	0.00029	0.00057	8	0.01199	0.02198
32	0.00015	0.00028	9	0.02131	0.04329
33	0.00007	0.00013	10	0.03410	0.07739
34	0.00003	0.00006			
35	0.00002	0.00003			
36	0.00001	0.00001			
37†	0.00000	0.00000			

*For example, $P(X = 28) = \frac{e^{-16}16^{28}}{28!} = 0.00192$; and $P(X = 29) = \frac{e^{-16}16^{29}}{29!} = 0.00106$.

†This series of computations terminates when we reach a $P(X)$ that is zero to the number of decimal places used.

Example 23.4

A one-tailed test for a proportion from a Poisson population.

H_0: $p \leq 0.0020$
H_A: $p > 0.0020$
$\alpha = 0.05$
$n = 8000$
$X = 28$
$p_0 n = (0.0020)(8000) = 16$

Therefore, we could state

H_0: $\mu \leq 16$
H_A: $\mu > 16$

From Example 23.3, we see that $P(X = 28) = 0.00192$ and

$$P(X \geq 28) = 0.00411.$$

Since $0.00411 < 0.05$, reject H_0.

By substituting $p_0 n$ for μ in Equation 23.1 (or Equation 23.2), we determine the probability of observing $X = 28$ mutations if our sample came from a population with $p_0 = 0.0020$. To test the hypothesis at hand we determine the probability of observing $X \geq 28$ mutations in a sample. (If the alternate hypothesis being considered were $H_A: p < p_0$, then we would compute the probability of mutations $\leq$ the number observed.) If the one-tailed probability is $\leq \alpha$, then H_0 is rejected at the α level of significance.

Two-Tailed Testing. If there is no reason, a priori, to hypothesize that a change in mutation rate would be in one specified direction (e.g., an increase) from the natural rate, then a two-tailed test is appropriate. The probability of the observed number of mutations is computed as shown in Example 23.3. Then we calculate and sum all the probabilities (in both tails) that are equal to or smaller than that of the observed. This is demonstrated in Example 23.5.

Example 23.5

A two-tailed test for a proportion from a Poisson population.

$$H_0: p = 0.0020$$
$$H_A: p \neq 0.0020$$
$$\alpha = 0.05$$
$$n = 8000$$
$$X = 28$$
$$p_0 n = (0.0020)(8000) = 16$$

Therefore, we can state $H_0: \mu = 16$
$H_A: \mu \neq 16$

From Example 23.3, we see that $P(X = 28) = 0.00192$.

The sum of the probabilities in one tail that are ≤ 0.00192 is 0.00411; the sum of the probabilities in the other tail that are ≤ 0.00192 is 0.00138.

Therefore, the probability of H_0 being true is $0.00411 + 0.00138 = 0.00549$.

Since $0.00549 < 0.05$, reject H_0.

Power of the Test. Recall that the power of a statistical test is the probability of that test rejecting a null hypothesis that is in fact a false statement about the population. We can determine the power of the above test when it is performed with a sample size of n at a significance level of α. For a one-tailed test, we first determine the critical value of X (i.e., the smallest X that delineates a proportion of the Poisson distribution $\leq \alpha$). Examining the distribution of Example 23.3, for example, for $\alpha = 0.05$, we see that the appropriate X is 24 [as $P(X \geq 24) = 0.037$, while $P(X \geq 23) = 0.058$]. We then examine the Poisson distribution having the sample X, replacing μ with that X in Equation 23.1 (or Equation 23.2), as shown in Example 23.6. The power of the test is $\geq$ the probability of an X at least as extreme as the critical value of X.*

*If the critical value delineates exactly α of the tail of the Poisson distribution, then the test's power is exactly what was calculated; if the critical value cuts off $< \alpha$ of the tail, then the power is $>$ that calculated.

Example 23.6

Determination of the power of the small-probability binominal tests of Examples 23.4 and 23.5, using $\alpha = 0.05$.

Substituting $X = 28$ for μ in Equation 23.1, we compute the following:*

X	$P(X)$	*Cumulative* $P(X)$	X	$P(X)$	*Cumulative* $P(X)$
24	0.060	0.798	0	0.000	0.000
25	0.067	0.738	1	0.000	0.000
26	0.072	0.671	2	0.000	0.000
27	0.075	0.599	3	0.000	0.000
28	0.075	0.524	4	0.000	0.000
29	0.073	0.449	5	0.000	0.000
30	0.068	0.376	6	0.000	0.000
31	0.061	0.308	7	0.000	0.000
32	0.054	0.247	8	0.000	0.000
33	0.045	0.193			
34	0.037	0.148			
35	0.030	0.111			
36	0.023	0.081			
37	0.018	0.058			
38	0.013	0.040			
39	0.009	0.027			
40	0.007	0.018			
41	0.004	0.011			
42	0.003	0.007			
43	0.002	0.004			
44	0.001	0.002			
45	0.001	0.001			
46	0.000	0.000			

*For example, $P(X = 24) = \dfrac{e^{-28}28^{24}}{24!} = 0.06010$; and $P(X = 25) = \dfrac{e^{-28}28^{25}}{25!}$.

The critical value for the one-tailed test of Example 23.3 is $X = 24$. The power of this test is $> P(X \geq 24)$ in the above distribution. That is, the power is > 0.798.

The critical values for the two-tailed test of Example 23.4 are 25 and 8. The power of this test is $> P(X \geq 25) + P(X \leq 8) = 0.738 + 0.000$. That is, the power is > 0.738.

For a two-tailed hypothesis, we identify one critical value of X as the smallest X that cuts off $\leq \alpha/2$ of the distribution in the upper tail and one as the largest X that cuts off $\leq \alpha/2$ of the lower tail. In Example 23.4, these two critical values for $\alpha = 0.05$ (i.e., $\alpha/2 = 0.025$) are $X = 25$ and $X = 8$ [as $P(X \geq 25) = 0.022$ and $P(X \geq 8) = 0.022$]. Then we examine the Poisson distribution having the sample X replace μ in Equation 23.1 (or Equation 23.2). As shown in Example 23.6, the power of the two-tailed test is at least as large as the probability of X in the latter Poisson distribution being more extreme than either of the critical values. That is, power $\geq P(X \geq$ upper critical value) $+ P(X \leq$ lower critical value).

23.5 COMPARING TWO POISSON COUNTS

If we have two counts, X_1 and X_2, each from a population with a Poisson distribution, we can ask whether they are likely to have come from the same population (or from populations with the same mean). The test of H_0: $\mu_1 = \mu_2$ (against H_A: $\mu_1 \neq \mu_2$) is related to the binomial test with $p = 0.5$ (Pearson and Hartley, 1966: 78–79), so that Table B.26 can be utilized, using $n = X_1 + X_2$. For the two-tailed test, H_0 is rejected if either X_1 or X_2 is $\leq$ the critical value, $C_{\alpha(2),n}$. This is demonstrated in Example 23.7.

Example 23.7

A two-sample test with Poisson data.

One fish is found to be infected with 13 parasites, and a second fish with 22. Test whether these two fish are likely to have come from the same population. (If the two are of different species, or sexes, then we can be asking whether the two species, or sexes, are equally infected.) The test is two-tailed for hypotheses H_0: $\mu_1 = \mu_2$ and H_A: $\mu_1 \neq \mu_2$.

Using Table B.26 for $n = X_1 + X_2 = 13 + 22 = 35$, we find a critical value of $C_{0.05(2),35} = 11$. Since neither X_1 nor X_2 is ≤ 11, H_0 is not rejected. Using the smaller of the two X's, we conclude that the probability is between 0.20 and 0.50 that a fish with 13 parasites and one with 22 parasites come from the same Poisson population (or from two Poisson populations having the same means).

Using the normal approximation of Equation 23.10,

$$Z = \frac{|X_1 - X_2|}{\sqrt{X_1 + X_2}} = \frac{|13 - 22|}{\sqrt{13 + 22}} = \frac{9}{5.916} = 1.521$$

$Z_{0.05(2)} = t_{0.05(2),\infty} = 1.960$

Therefore, do not reject H_0.

$0.10 < P < 0.20$

For a one-tailed test of H_0: $\mu_1 \leq \mu_2$ against H_A: $\mu_1 > \mu_2$, we reject H_0 if $X_1 > X_2$ and $X_2 \leq C_{\alpha(1),n}$, where $n = X_1 + X_2$. For H_0: $\mu_1 \geq \mu_2$ and H_A: $\mu_1 < \mu_2$, H_0 is rejected if $X_1 < X_2$ and $X_1 \leq C_{\alpha(1),n}$, where $n = X_1 + X_2$.

If n is at least 5, then a normal approximation can be used (Best, 1975). For the two-tailed test:

$$Z = \frac{|X_1 - X_2|}{\sqrt{X_1 + X_2}} \tag{23.10}$$

is considered a normal deviate (Ractliffe, 1964), so the critical value is $Z_{\alpha(2)}$ (which can be read as $t_{\alpha(2),\infty}$ at the end of Table B.3). This is demonstrated in Example 23.7.

For a one-tailed test,

$$Z = \frac{X_1 - X_2}{\sqrt{X_1 + X_2}}. \tag{23.11}$$

For H_0: $\mu_1 \leq \mu_2$ vs. H_A: $\mu_1 > \mu_2$, H_0: is rejected if $X_1 > X_2$ and $Z \geq Z_{\alpha(1)}$. For H_0: $\mu_1 \geq \mu_2$ vs. H_A: $\mu_1 < \mu_2$, H_0 is rejected if $X_1 < X_2$ and $Z \leq -Z_{\alpha(1)}$.

An alternative normal approximation, based on a square root transformation is

$$Z = \left| \sqrt{2X_1 + \frac{3}{4}} - \sqrt{2X_2 + \frac{3}{4}} \right|. \tag{23.12}$$

It may be used in place of Equation 23.10, and it has superior power when testing at $\alpha < 0.05$. Equation 23.12 is for a two-tailed test; for one-tailed testing, use

$$Z = \sqrt{2X_1 + \frac{3}{4}} - \sqrt{2X_2 + \frac{3}{4}}, \tag{23.13}$$

with the same procedure for concluding to reject H_0 as with Equation 23.11.

23.6 SERIAL RANDOMNESS OF NOMINAL SCALE CATEGORIES

Representatives of two different nominal scale categories may appear serially in space or time, and their randomness of occurrence may be assessed as in the following example. Members of two species of antelope are observed drinking along a river, and their linear order is as shown in Example 23.8. We may ask whether the order of occurrence of members of the two species is random (as opposed to the animals either forming groups with individuals of the same species or shunning members of the same species). A sequence of like elements, bounded on either side by either unlike elements or no elements, is termed a *run*. Thus, any of the following arrangements of 5 members of antelope species A and 7 members of species *B* would be considered to consist of five runs:

BAABBBAAABBB, or *BBAAAABBBBAB*, or *BABAAAABBBBB*, or *BAAABAABBBBB*, etc.

Example 23.8
The two-tailed runs test with elements of two kinds.

Members of two species of antelope (denoted as species *A* and *B*) are drinking along a river in the following order: *AABBAABBBBAAABBBBAABBB*.

H_0: The distribution of members of the two species along the river is random.
H_A: The sequential distribution of members of the two species is not random.

$n_1 = 9, \quad n_2 = 13, \quad u = 8$
$u_{0.05(2),9,13} = 6$ and 17 (from Table B.28)
Since u is neither ≤ 6 nor ≥ 17, do not reject H_0.
$P \approx 0.20$

To test the null hypothesis of randomness, we may use the *runs test*.* If we let n_1 be the total number of elements of the first category (in the present example, the number of antelopes of species *A*), n_2 the number if antelope of species *B*, and u the number of runs in the entire sequence, then the critical values, $u_{\alpha(2),n_1,n_2}$, can be read from Table B.28 for cases where both $n_1 \leq 30$ and $n_2 \leq 30$. The critical values in this table are given in pairs; if the u in the sample is $\leq$ the first member of the pair *or* $\geq$ the second, then H_0 is rejected.

*From its inception the runs test has also been considered to be a nonparametric test of whether two samples come from the same population (e.g., Wald and Wolfowitz, 1940), but as a two-sample test it has poor power and the Mann-Whitney test of Section 9.9 is preferable.

If either n_1 or n_2 is larger than 30, then Table B.28 cannot be employed; but for such large samples, the distribution of u approaches normality, with a mean of

$$\mu_u = \frac{2n_1n_2}{n_1 + n_2} + 1 \tag{23.14}$$

and a standard deviation of

$$\sigma_u = \sqrt{\frac{2n_1n_2(2n_1n_2 - n_1 - n_2)}{(n_1 + n_2)^2(n_1 + n_2 - 1)}} \tag{23.15}$$

(Brownlee, 1965: 266–30). And, the statistic

$$Z_c = \frac{|u - \mu_u| - 0.5}{\sigma_u} \tag{23.16}$$

may be considered a normal deviate with $Z_{\alpha(2)}$ being the critical value for the test.

Brownlee (1965:231) extends the preceding considerations to the case of three nominal categories, where to use Equation 23.16 we compute

$$\mu_u = \frac{2(n_1n_2 + n_1n_3 + n_2n_3)}{n_1 + n_2 + n_3} + 1 \tag{23.17}$$

and

$$\sigma_u = \sqrt{\frac{[2(n_1n_2 + n_1n_3 + n_2n_3)]^2}{(n_1 + n_2 + n_3)^2(n_1 + n_2 + n_3 - 1)} - \frac{2(n_1n_2 + n_1n_3 + n_2n_3) + 6n_1n_2n_3}{(n_1 + n_2 + n_3)(n_1 + n_2 + n_3 - 1)}}. \tag{23.18}$$

One-Tailed Testing. There are two ways in which a distribution of nominal scale categories can be nonrandom: (a) the distribution may have fewer runs than would occur at random, in which case the distribution is more clustered, or contagious, than random; (b) the distribution may have more runs than would occur at random, indicating a tendency toward a uniform distribution.

To test for the one-tailed situation of contagion, we state H_0: The elements in the population are not distributed contagiously, vs. H_A: The elements in the population are distributed contagiously; and H_0 would be rejected at the $\alpha(1)$ significance level if $u \leq$ the lower of the pair of critical values in Table B.28. Thus, had the animals in Example 23.8 been arranged *AAAAABBBBBBAAAABBBBBBB*, then $u = 4$ and the one-tailed 5% critical value would be the lower value of $u_{0.05(1),5,6}$, which is 7; as $4 < 7$, H_0 is rejected and the distribution is concluded to be clustered. In using the normal approximation, H_0 is rejected if $Z_c \geq Z_{\alpha(1)}$ and $u \leq \mu_u$.

To test for uniformity, we use H_0: The elements in the population are not uniformly distributed vs. H_A: The elements in the population are uniformly distributed. If $u \geq$ the upper critical value in Table B.28 for $\alpha(1)$, then H_0 is rejected. If the animals in Example 23.8 had been arranged as *ABABABBABABBABABBABBAB*, then $u = 18$, which is greater than the upper critical value of $u_{0.05(1),5,6} = 16$; therefore, H_0 is rejected. If the normal approximation were used, H_0 would be rejected if $Z_c \geq Z_{\alpha(1)}$ and $u \geq \mu_u$.

23.7 SERIAL RANDOMNESS OF MEASUREMENTS: PARAMETRIC TESTING

The biologist may encounter continuous data that have been collected serially in space or time. For example, rates of conduction might be measured at successive lengths along a nerve. A null hypothesis of no difference in conduction rate as one examines successive portions essentially is stating that all the measurements obtained are a random sample from a population of such measurements.

Example 23.9 presents data consisting of dissolved oxygen measurements of a water solution determined on the same instrument each 5 minutes. The desire is to conclude whether fluctuations in measurements are random or whether they indicate a nonrandom instability in the measuring device (or in the solution). The null hypothesis that the sequential variability among measurements is random may be subjected to the *mean square successive difference* test, a test that assumes normality in the underlying distribution. In this procedure, we calculate the sample variance, s^2, which is an estimate of the population variance, σ^2:

$$s^2 = \frac{\sum_{i=1}^{n}(X_i - \bar{X})^2}{n-1}, \tag{4.6}$$

or,

$$s^2 = \frac{\sum_{i=1}^{n} X_i^2 - \frac{\left(\sum_{i=1}^{n} X_i\right)^2}{n}}{n-1}, \tag{4.8}$$

Example 23.9

The mean square successive difference test.

An instrument for measuring dissolved oxygen is used to record a measurement each 5 minutes from a container of lake water. It is desired to know whether the differences in measurements are random or whether they are systematic. (If the latter, it could be due to the dissolved oxygen content in the water changing, or the instrument's response changing, or to both.) The data (in ppm) are as follows, recorded in the sequence in which they were obtained: 9.4, 9.3, 9.3, 9.2, 9.3, 9.2, 9.1, 9.3, 9.2, 9.1, 9.1.

H_0: Consecutive measurements obtained on the lake water with this instrument have random variability.

H_A: Consecutive measurements obtained on the lake water with this instrument have nonrandom variability and are serially correlated.

$$n = 11$$

$$s^2 = 0.01018 \text{ (ppm)}^2$$

$$s_*^2 = \frac{(9.3-9.4)^2 + (9.3-9.3)^2 + (9.2-9.3)^2 + \ldots + (9.1-9.1)^2}{2(11-1)}$$

$$= 0.00550$$

$$C = 1 - \frac{0.00550}{0.01018} = 1 - 0.540 = 0.460$$

$C_{0.05,11} = 0.452$

Therefore, reject H_0.

$0.025 < P < 0.05$

as introduced in Section 4.3. If the null hypothesis is true, then another estimate of σ^2 is

$$s_*^2 = \frac{\sum_{i=1}^{n-1} (X_{i+1} - X_i)^2}{2(n-1)} \tag{23.19}$$

(von Neumann et al., 1941). Therefore, the ratio s_*^2/s^2 should equal 1 when H_0 is true. Using Young's (1941) notation, the test statistic is

$$C = 1 - \frac{s_*^2}{s^2}, \tag{23.20}$$

and if this value equals or exceeds the critical value, $C_{\alpha,n}$, in Table B.29, we reject the null hypothesis of serial randomness.* The mean square successive difference test considers the one-tailed alternate hypothesis that measurements are serially correlated.

For n larger than those in Table B.29, the hypothesis may be tested by a normal approximation:

$$Z = \frac{C}{\sqrt{\dfrac{n-2}{n^2-1}}} \tag{23.22}$$

(von Neumann et al., 1941), with the value of the calculated Z being compared with the critical value of $Z_{\alpha(1)} = t_{\alpha(1),\infty}$. This approximation is very good for $\alpha = 0.05$, for n as small as 10; for $\alpha = 0.10$, 0.25, or 0.025, for n as small as 25; and for $\alpha = 0.01$ and 0.005, for n of at least 100.

23.8 SERIAL RANDOMNESS OF MEASUREMENTS: NONPARAMETRIC TESTING

If a researcher does not wish to assume that a sample of serially obtained measurements came from a normal population, then the procedure of Section 23.7 is inappropriate, and the nonparametric test of Section 23.6 may be employed. To apply the runs test (Section 23.6), we first determine the median of the sample of data (as explained in Section 3.2). Then we record each datum as being either above (+) or below (−) the median. (If a sample datum is equal to the median, it is discarded from the analysis.) We then determine the number of runs in which the n_1 +'s and the n_2 −'s are found and proceed as in the one-tailed test for contagion in Section 23.6. This is demonstrated in Example 23.10. For sample sizes larger than those in Table B.28, the normal approximation may be used to test this one-tailed hypothesis.

*Equations 4.6 and 23.19 may be combined so that C might be computed as

$$C = 1 - \frac{\sum_{i=1}^{n-1} (X_i - X_{i+1})^2}{2(\text{SS})}, \tag{23.21}$$

where SS is the numerator of either Equation 4.6 or 4.8.

Example 23.10

The test for runs above and below the median, demonstrated for the data of Example 23.9.

H_0: Consecutive measurements obtained on the lake water with this instrument have random variability.

H_A: Consecutive measurements obtained on the lake water with this instrument have nonrandom variability and are serially correlated.

The median of the sample of 11 measurements is determined by Equation 3.6; because data are tied at the middle value, the median is

$$9.2 + \left(\frac{(0.5)(11) - 3}{3}\right)(0.1) = 9.28 \text{ ppm}.$$

The sequence of data, indicating whether they are above (+) or below (−) the median, is $+++-+--+---$.

$n_1 = 5$, $n_2 = 6$, and $u = 6$

The critical value is the lower critical value for $u_{0.05(1),5,6}$, which is 3.

Since 6 is not ≤ 3, do not reject H_0.

$P > 0.25$.

EXERCISES

23.1. If, in a Poisson distribution, $\mu = 1.5$, what is $P(0)$ and $P(5)$?

23.2. A solution contains bacterial viruses in a concentration of 5×10^8 virus particles per milliliter. In the same solution are 2×10^8 bacteria per milliliter. Assume random distribution of virus among the bacteria. **(a)** What proportion of the bacteria will have no virus particles? **(b)** What proportion of the bacteria will have virus particles? **(c)** What proportion of the bacteria will have at least two virus particles? **(d)** What proportion of the bacteria will have three virus particles?

23.3. A raisin cake is divided into equal-sized slices. The distribution of raisins among these slices is as follows. Test the null hypothesis that the raisins are distributed at random throughout the cake.

X_i	f_i
0	8
1	17
2	18
3	11
4	3
5	0

23.4. We wish to compile a list of certain types of metabolic human infant diseases that occur in more than 0.01% of the population. A random sample of 25,000 infants reveals 5 with one of these diseases. Should that disease be placed in our list?

23.5. A biologist counts 112 diatoms in a milliliter of lake water, and 134 diatoms are counted in a milliliter of a second collection of lake water. Test the hypothesis that the two water collections came from the same lake (or from lakes with the same diatom concentrations).

23.6. The following data are the magnitudes of fish kills along a certain river (measured in kilograms of fish killed) over a period of years. Test the null hypothesis that the magnitudes of the fish kills are randomly distributed over time.

Year	*Kill* (kg)
1955	147.4
1956	159.8
1957	155.2
1958	161.3
1959	173.2
1960	191.5
1961	198.2
1962	166.0
1963	171.7
1964	184.9
1965	177.6
1966	162.8
1967	177.9
1968	189.6
1969	206.9
1970	221.5

23.7. An economic entomologist rates the annual incidence of damage by a certain beetle as mild (M) or heavy (H). For a 27-year period he records the following: $H\ M\ M\ M\ H\ H\ M\ M\ H\ M\ H\ H\ H\ M\ M\ H\ H\ H\ H\ M\ M\ H\ H\ M\ M\ M\ M$. Test the null hypothesis that the incidence of heavy damage occurs randomly over the years.

24 Circular Distributions: Descriptive Statistics

24.1 DATA ON A CIRCULAR SCALE

In Section 1.1, an interval scale of measurement was defined as a scale with equal intervals but with no true zero point. A special type of interval scale is a circular scale, where not only is there no true zero, but any designation of high or low values is arbitrary. A common example of a circular scale of measurement would be compass direction (Fig. 24.1a), where a circle is said to be divided into 360 equal intervals, called degrees, and for which the zero point is arbitrary. There is no physical justification for a direction of north to be designated 0 (or 360) degrees, and 90° cannot be said to be a "larger" direction than 60°.*

Another common circular scale is time of day (Fig. 24.1b), where a day is divided into 24 equal intervals, called hours, but where the designation of midnight as the zero or starting point is arbitrary. One hour of a day corresponds to 15° (i.e., 360°/24) of a circle, and 1° of a circle corresponds to 4 minutes of a day. Other time divisions, such as weeks and years (see Fig. 24.1c), also represent circular scales of measurement.

In general, we may convert X time units to an angular direction (a, in degrees), where X has been measured on a circular scale having k time units in the full cycle:

$$a = \frac{(360°)(X)}{k}. \tag{24.1}$$

*Occasionally one will encounter angular measurements expressed in radians rather than in degrees. A radian is the angle that is subtended by an arc of a circle equal in length to the radius of the circle. As a circle's circumference is 2π times the radius, a radian is $360°/2\pi = 180°/\pi =$ 57.29577951° (or 57 deg, 17 min, 44.8062 sec).

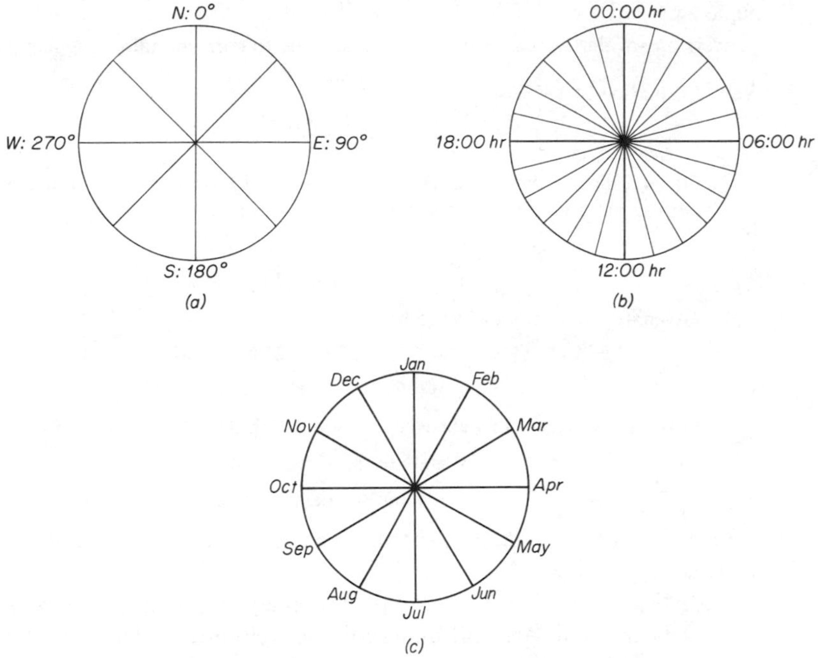

Figure 24.1 Common circular scales of measurement. (a) Compass directions. (b) Times of day. (c) Days of year (with the first day of each month shown).

For example, to convert a time of day (X, in hours) to an angular direction, $k = 24$ hr; to convert a day of the week to an angular direction, number the seven days from some arbitrary point (e.g., Sunday = day 1) and use Equation 24.1 with $k = 7$; to convert the Xth day of the year to an angular direction, $k = 365$ (or, $k = 366$ in a leap year); to convert a month of the year, $k = 12$; and so on.* Such conversions are demonstrated in Example 24.1. Also note that data from circular distributions generally may not be analyzed using the statistical methods presented earlier in this book.† This is so for theoretical reasons as well as for empirically obvious reasons stemming from the arbitrariness of the zero point on the circular scale. For example, consider three compass directions: 10°, 30°, and 350°, for which we wish to calculate an arithmetic mean. The arithmetic mean calculation of $(10° + 30° + 350°)/3 =$

*Equation 24.1 gives angular directions corresponding to the ends of time periods (e.g., the end of the Xth day of the year). If some other point in a time period is preferred, the equation can be adjusted accordingly. (For example, noon can be considered on the Xth day of the year by using $X - 0.5$ in place of X.) If the same point is used in each time period (e.g., always using either noon or midnight), then the statistical procedures of this and the following chapter will be unaffected by the choice of point. (However, graphical procedures, as in Section 24.2, will of course be affected, in the form of a rotation of the graph if Equation 24.1 is adjusted. If, for example, we considered noon on the Xth day of the year, the entire graph would be rotated about half a degree counterclockwise.)

†An exception is the case where the measurement scale is only a portion of a circle. For example, latitude on the earth's surface, even though measured in degrees, is constrained to a range of 0 to 90° on either side of the equator. Such data may be treated as ratio data measured on a linear scale.

Example 24.1

Conversions of times measured on a circular scale to corresponding angular directions.

We use Equation 24.1:

$$a = \frac{(360°)(X)}{k}.$$

1. Given a time of day of 06: 00 hr (which is one-fourth of the 24-hour clock and should correspond, therefore, to one-fourth of a circle):

$$X = 6 \text{ hr}, \quad k = 24 \text{ hr}, \quad \text{and}$$
$$a = (360°)(6 \text{ hr})/24 \text{ hr} = 90°.$$

2. Given a time of day of 06: 15 hr:

$$X = 6.25 \text{ hr}, \quad k = 24 \text{ hr}, \quad \text{and}$$
$$a = (360°)(6.25 \text{ hr})/24 \text{ hr} = 93.75°.$$

3. Given the 14th day of February, being the 45th day of the year:

$$X = 45 \text{ days}, \quad k = 365 \text{ days}, \quad \text{and}$$
$$a = (360°)(45 \text{ days})/365 \text{ days} = 44.38°.$$

$390°/3 = 130°$ is clearly absurd, for all data are northerly directions and the computed mean is southeasterly.

Statistical methods for describing and analyzing data from circular distributions are relatively new and are still undergoing development. This chapter will introduce some basic considerations useful in calculating descriptive statistics and Chapter 25 will discuss tests of hypotheses. Readers desiring a more through review are advised to consult the excellent monographs of Batschelet* (1965; 1972; 1981) and Mardia (1972) and the literature cited therein.

24.2 GRAPHICAL PRESENTATION OF DATA

Circular data are often presented as a scatter diagram. Figure 24.2 shows such a graph for the data of Example 24.2. If frequencies of data are too large to be plotted

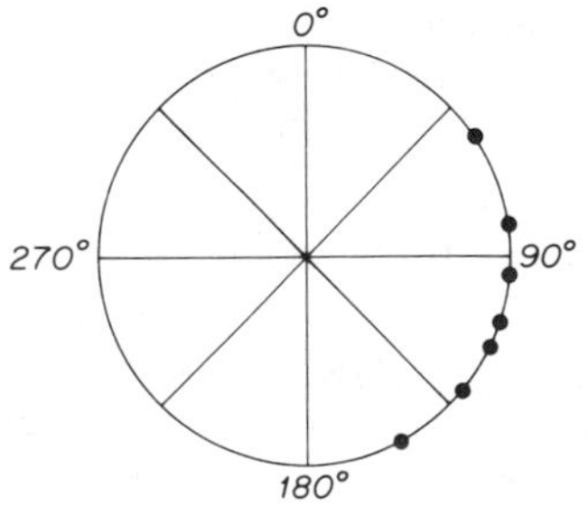

Figure 24.2 A circular scatter diagram for the data of Example 24.2.

Example 24.2

A sample of circular data. These data are plotted in Fig. 24.2.

Seven trees are found leaning in the following compass directions: 55°, 81°, 96°, 109°, 117°, 132°, 154°.

*Edward Batschelet (1914–1979), Swiss biomathematician, was one of the most influential writers in developing, explaining, and promulgating circular statistical methods.

conveniently on a scatter diagram, then a histogram may be drawn. This is demonstrated in Fig. 24.3, for the data presented in Example 24.3. Recall that in a histogram, the length, as well as the area, of each bar is an indication of the frequency observed

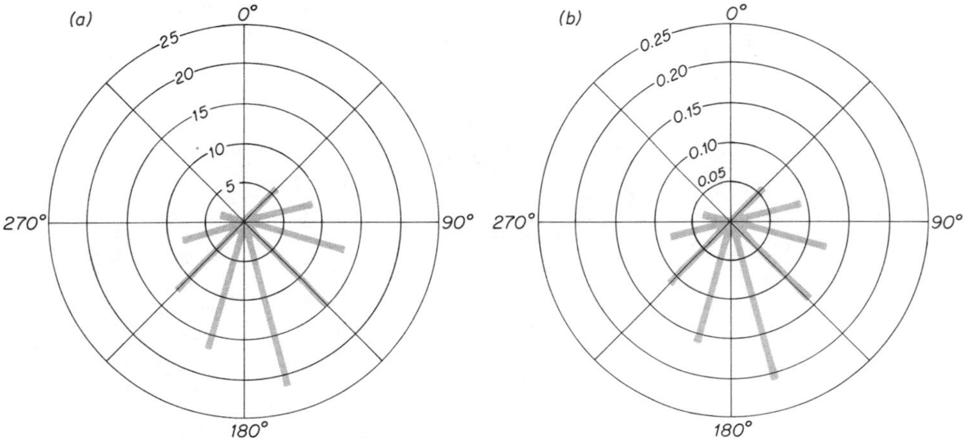

Figure 24.3 (a) Circular histogram for the data of Example 24.3 where the concentric circles represent frequency increments of 5. (b) A relative frequency histogram for the data of Example 24.3 with the concentric circles representing relative frequency increments of 0.05.

Example 24.3

A sample of circular data, presented as a frequency table, where a_i is an angle and f_i is the observed frequency of a_i. These data are plotted in Fig. 24.3.

a_i (deg)	f_i	*Relative* f_i
0–30	0	0.00
30–60	6	0.06
60–90	9	0.09
90–120	13	0.12
120–150	15	0.14
150–180	22	0.21
180–210	17	0.16
210–240	12	0.11
240–270	8	0.08
270–300	3	0.03
300–330	0	0.00
330–360	0	0.00
$n = 105$		Total = 1.00

at each plotted value of the variable (Section 1.3). Occasionally, as shown in Fig. 24.4, a histogram is seen presented with sectors, rather than bars, comprising the graph; this is sometimes called a *rose diagram*. Here, the radii forming the outer boundaries of the sectors are proportional to the frequencies being represented, but the areas of the sectors are not. Since it is likely that the areas will be judged by the eye to represent the frequencies, the reader of the graph is being deceived, and this type of graphical

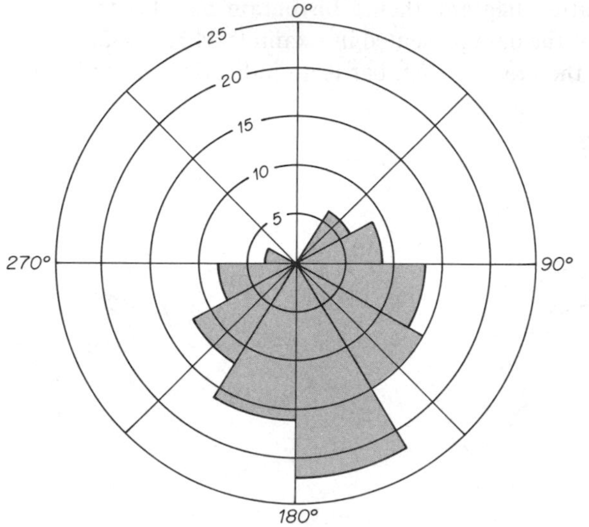

Figure 24.4 A graphical presentation of the data of Example 24.3, utilizing sectors instead of bars. This procedure is not recommended (see Section 22.2).

presentation is not recommended. (An equal-area rose diagram can be obtained by plotting the square roots of frequencies as radii.)

A histogram of circular data can also be plotted as a linear histogram (see Section 1.3), with degrees of the horizonal axis and frequencies (or relative frequencies) on the vertical. But the circular presentation of Fig. 24.3 is preferable.

24.3 SINES AND COSINES OF CIRCULAR DATA

A great many of the procedures that follow in this chapter and the next require the determination of two simple trigonometric functions. Let us consider that a circle (perhaps representing a compass face) is drawn on rectangular coordinates (as on common graph paper) with the center as the origin (i.e., zero) of a vertical X axis and a horizontal Y axis;* this is what is done in Fig. 24.5.

There are two methods that can be used to locate any point on a plane (such as a sheet of paper). One is to specify both the angle, a, with respect to some starting direction (say, clockwise from the top of the X axis, namely "north") and the straight-line distance, r, from some reference point (the center of the circle). This pair of numbers, a and r, is known as the "polar coordinates" of a point. Thus, for example, in Fig. 24.5, point 1 is uniquely identified by polar coordinates $a = 30°$ and $r = 1.00$, point 2 by $a = 120°$ and $r = 1.00$, and so on.†

*This is the opposite of the convention of having the X axis horizontal and the Y axis vertical. This is done for ease in obtaining trigonometric computations, for standard mathematical notation has angular measurement proceed counterclockwise from zero degrees on the right-hand portion of the horizontal axis (what we call "east" on the compass), rather than clockwise from the upper portion of the vertical axis (what we call "north" on the compass).

†If we specify that the radius of the circle is 1 unit, our figure is called a "unit circle."

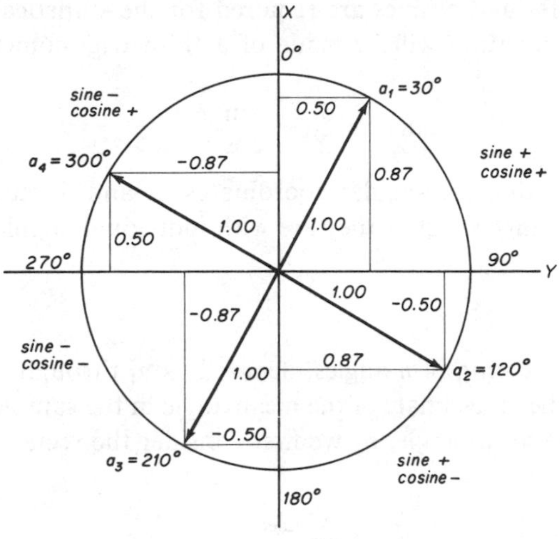

Figure 24.5 A unit circle, showing four points and their polar (a and r) and rectangular (X and Y) coordinates.

The second method of locating points on the graph is by referring to the X and Y axes, as we introduced when dealing with regression problems (Chapter 17). By this method, point 1 in Fig. 24.5 is located by the "rectangular coordinates"* $X = 0.87$ and $Y = 0.50$, point 2 by $X = -0.50$ and $Y = 0.87$, point 3 by $X = -0.87$ and $Y = -0.50$, and point 4 by $X = 0.50$ and $Y = -0.87$.

The *cosine* (abbreviated "cos") of an angle is defined as the ratio of the X and the r associated with the circular measurement:

$$\cos a = \frac{X}{r}, \tag{24.2}$$

while the *sine* (abbreviated "sin") of the angle is the ratio of the associated Y and r:

$$\sin a = \frac{Y}{r}. \tag{24.3}$$

Thus, for example, the sine of a_1 in Fig. 24.5 is $\sin 30° = 0.50/1.00 = 0.50$, and its cosine is $\cos 30° = 0.87/1.00 = 0.87$. Also, $\sin 120° = 0.87/1.00 = 0.87$, $\cos 120°$ $-0.50/1.00 = -0.50$, and so on. Sines and cosines (two of the most useful "trigonometric† functions") are readily available in published tables, and many electronic calculators give them (and sometimes convert between polar and rectangular coordinates as well). The sine of 0° and 180° is zero, angles between 0° and 180° have sines that are positive, and the sines are negative for $180° < a < 360°$. The cosine is zero for 90° and 270°, with positive cosines obtained for $0° < a < 90°$ and for $270° < a < 360°$, and negative cosines for angles between 90° and 270°.

*The familiar system of rectangular coordinates is also known as "Cartesian coordinates," after the French mathematician and philosopher, René Descartes (1596–1650), who wrote under the Latinized version of his name, Renatus Cartesius. His other noteworthy mathematical introductions were the use of exponents, the square root sign ($\surd$), and the use of X and Y to denote variables (Asimov, 1972: 106–107).

†"Trigonometry" refers, literally, to the measurement of triangles (such as the triangles that emanate from the center of the circle in Fig. 24.5).

At most, only sines and cosines are required for the statistical procedures that follow, although brief mention will be made of a third trigonometric function, the *tangent*:*

$$\tan a = \frac{Y}{X} = \frac{\sin a}{\cos a}. \qquad (24.5)$$

We shall see later that rectangular coordinates, X and Y may also be used in conjunction with mean angles just as they are with individual angular measurements.

24.4 THE MEAN ANGLE

If we have a sample consisting of n angles, denoted as a_1 through a_n, then the mean of these angles, $\bar{a}$, is to be an estimate of the mean angle in the sampled population μ_a. To compute the sample mean angle, $\bar{a}$, we first consider the rectangular coordinates of the mean angle:

$$X = \frac{\sum_{i=1}^{n} \cos a_i}{n} \qquad (24.6)$$

and

$$Y = \frac{\sum_{i=1}^{n} \sin a_i}{n}. \qquad (24.7)$$

Then, the quantity

$$r = \sqrt{X^2 + Y^2} \qquad (24.8)$$

is computed;† this is the length of the mean vector, which will be further discussed in Section 24.5. The value of $\bar{a}$ is determined as the angle having the following cosine and sine:

$$\cos \bar{a} = \frac{X}{r} \qquad (24.9)$$

and

$$\sin \bar{a} = \frac{Y}{r}. \qquad (24.10)$$

Example 24.4 demonstrates these calculations. It is also true that

$$\tan \bar{a} = \frac{Y}{X}, \qquad (24.11)$$

so we have a check on the calculation of the mean angle, $\bar{a}$. If $r = 0$, the mean angle is undefined and we conclude that there is no mean direction.

If we are dealing with data that are times instead of angles, then the mean time corresponding to the mean angle may be determined by inverting Equation 24.1:

$$\bar{X} = \frac{ka}{360°}. \qquad (24.12)$$

*The *cotangent* is the inverse of the tangent, namely

$$\cot a = \frac{X}{Y} = \frac{\cos a}{\sin a}. \qquad (24.4)$$

†This quantity, r, is not to be confused with a sample correlation coefficient (Section 19.1), with which it bears no relationship but which is denoted by the same symbol.

For example, to determine a mean time of day, $\bar{X}$, from a mean angle, $\bar{a}$: $\bar{X} = (24\text{ hr})(\bar{a})/360°$.

Example 24.4

Calculating the mean angle for the data of Example 24.2.

a_i (deg)	$\sin a_i$	$\cos a_i$
55	0.81915	0.57358
81	0.98769	0.15643
96	0.99452	−0.10453
109	0.94552	−0.32557
117	0.89101	−0.45399
132	0.74315	−0.66913
154	0.43837	−0.89879

$\sum \sin a_i = 5.81941$, $\quad \sum \cos a_i = -1.72200$

$n = 7 \qquad Y = \frac{\sum \sin a_i}{n} = 0.83134 \qquad X = \frac{\sum \cos a_i}{n} = -0.24600$

$$r = \sqrt{X^2 + Y^2} = \sqrt{(-0.24600)^2 + (0.83134)^2} = \sqrt{0.75164} = 0.86697$$

$$\cos \bar{a} = \frac{X}{r} = \frac{-0.24600}{0.86697} = -0.28375$$

$$\sin \bar{a} = \frac{Y}{r} = \frac{0.83134}{0.86697} = 0.95890$$

The angle with this cosine and sine is $\bar{a} = 106°$.

Grouped Data. Often circular data are recorded in a frequency table (as in Example 24.3). For such data, the following computations are convenient alternatives to Equations 24.6 and 24.7, respectively:

$$X = \frac{\sum f_i \cos a_i}{n} \tag{24.13}$$

$$Y = \frac{\sum f_i \sin a_i}{n} \tag{24.14}$$

(which are analogous to Equation 3.3 for linear data). In these equations, a_i is the midpoint of the measurement interval recorded (e.g., $a_2 = 45°$ in Example 24.3, which is the midpoint of the second recorded interval, 30–60°), and f_i is the frequency of occurrence of data within that interval (e.g., $f_2 = 6$ in that example).

There is a bias in computing r from grouped data, in that the result is too small. A correction for this is available (Batschelet, 1965: 16–17, 1981: 37–40; Mardia, 1972: 78–79). For data grouped into intervals of d degrees each,

$$r_c = cr, \tag{24.15}$$

where r_c is the corrected r, and c is a correction factor,

$$c = \frac{\frac{d\pi}{360°}}{\sin\left(\frac{d}{2}\right)}. \tag{24.16}$$

The correction becomes insignificant as the interval becomes smaller than 30°. This correction is for the quantity r, and for statistics calculated from it; but the mean angle, $\bar{a}$, requires no correction for grouping. The correction may be applied when the distribution is unimodal and does not deviate greatly from symmetry.

24.5 ANGULAR DISPERSION

When dealing with circular data, we wish to have a measure, analogous to those of Chapter 4 for a linear scale, to describe the disperson of the data.

We can define the *range* in a circular distribution of data as the smallest arc (i.e., the smallest portion of the circle's circumference) that contains all the data in the distribution. For example, in Fig. 24.6a, the range is zero; in Fig. 24.6b the shortest arc is from the data point at 38° to the datum at 60°, making the range 22°; in Fig.

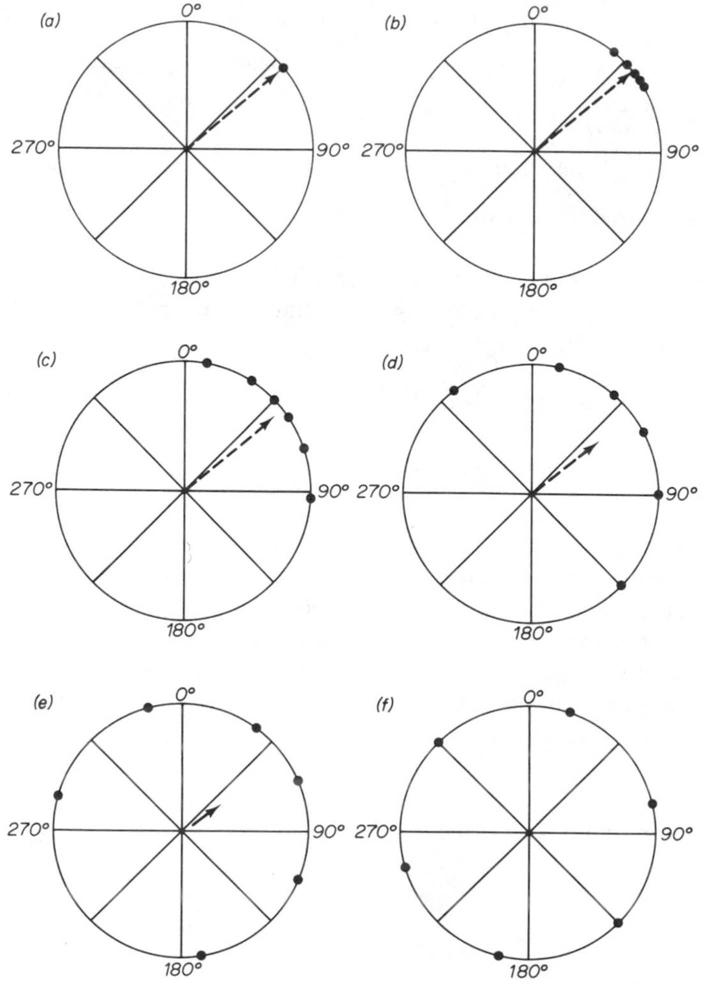

Figure 24.6 Circular distributions with various amounts of dispersion. The broken line indicates the mean angle, which is 50° in each case. Note that the value of r varies inversely with the amount of dispersion, and that the value of s varies directly with the amount of dispersion. (a) $r = 1.00$, $s = 0°$, $s' = 0°$. (b) $r = 0.99$, $s = 8.10°$, $s' = 8.12°$. (c) $r = 0.90$, $s = 25.62°$, $s' = 26.30°$. (d) $r = 0.60$, $s = 51.25°$, $s' = 57.91°$. (e) $r = 0.30$, $s = 67.79°$, $s' = 88.91$. (f) $r = 0.00$, $s = 81.03°$, $s' = \infty$.

24.6c, the data are found from 10° to 93°, with a range of 83°; in Fig. 24.6d, the data run from 322° to 135°, with a range of 173°; in Fig. 24.6e, the shortest arc containing all the data is that running clockwise from 285° to 171°, namely an arc of 246°; and in Fig. 24.6f the range is 300°. For the data of Example 24.2 the range is 99° (as the data run from 55° to 154°).

Another measure of dispersion is seen by examining Fig. 24.6; the value of r varies inversely with the amount of dispersion in the data. Therefore, r is a measure of concentration. It has no units and it may vary from 0 (when there is so much dispersion that a mean angle cannot be described) to 1.0 (when all the data are concentrated at the same direction).

In Section 3.1 the mean on a linear scale was noted to be the center of gravity of a group of data. Similarly, the length of the mean vector (i.e., the quantity r), in the direction of the mean angle ($\bar{a}$) is a center of gravity. (Consider that each circle in Fig. 24.6 is a disc of material of negligible weight, and each datum is a dot of unit weight. The disc, held parallel to the ground, would balance at the tip of the arrow in the figure. In Fig. 24.6f, $r = 0$ and the center of gravity is the center of the circle.)

Since r is a measure of concentration, $1 - r$ is a measure of dispersion. Lack of dispersion would be indicated by $1 - r = 0$, and maximum dispersion by $1 - r = 1.0$. A measure called "mean angular deviation," or simply the *angular deviation*, is

$$s = \frac{180°}{\pi}\sqrt{2(1 - r)}, \tag{24.17}$$

in degrees.* This ranges from a minimum of zero (e.g., Fig. 24.6a) to a maximum of 81.03° (e.g., Fig. 24.6f).†

Mardia (1972: 24, 74) defines *circular standard deviation* as

$$s' = \frac{180°}{\pi}\sqrt{-2 \ln r} \tag{24.18}$$

degrees; or, employing common, instead of natural, logarithms:

$$s' = \frac{180°}{\pi}\sqrt{-4.60517 \log r} \tag{24.19}$$

degrees. This is analogous to the standard deviation, s, on a linear scale (Section 4.4) in that it ranges from zero to infinity (see Fig. 24.6). For a given r, the values of s and s' differ by no more than 2 degrees for r as small as 0.80, by no more than 1 degree for r as small as 0.87, and by no more than 0.1 degree for r as small as 0.97. It is intuitively reasonable that a measure of angular dispersion should have a finite upper limit, so s is the measure preferred in this book. Appendix Tables B.30 and B.31 convert r to s and s', respectively. If the data are grouped, then s and s' are biased in being too high, so r_c (by Equation 24.15) can be used in place of r.

Measures of symmetry and kurtosis on a circular scale, analogous to the g_1 and g_2 that may be calculated on a linear scale (Section 7.1), are discussed byBatschelet (1965: 14–15, 1981: 43–44) and Mardia (1972: 36–38, 74–76).

*Simply delete the constant, $180°/\pi$, in this and in the following equations if the measurement is desired in radians rather than in degrees.

†This is a range of 0 to 1.41 radians.

24.6 THE MEDIAN AND MODAL ANGLES

In a fashion analogous to considerations for linear scales of measurement (Sections 3.2 and 3.3), one can determine the median and the mode of a set of data on a circular scale.

To find the *median angle*, we first determine that diameter of the circle which divides the data into two equal-sized groups. The median angle is the angle indicated by the radius on that diameter which is nearer to the majority of the data points. In Example 24.2, a diameter extending from 109° to 289° (as indicated by the dashed line in Fig. 24.2) divides the data into two groups of three data each. The data are concentrated around 109°, rather than 289°, so the sample median is 109°. If n is odd, the median will be one of the data points. If n is even, the median is midway between two of the data. Mardia (1972: Section 2.5.1) shows how the median is estimated, analogously to Equation 3.6, when the median is within a group of tied data. If the data distribution in question has the data equally spaced around the circle (as in Fig. 24.6f), then the median, as well as the mean, is undefined.

The *modal angle* is defined in the same way as is the mode for linear scale data (Section 3.3). Just as with linear data, there may be more than one mode.

24.7 CONFIDENCE LIMITS FOR THE MEAN AND MEDIAN ANGLES

The concept of confidence limits may be applied to the mean of angles by consulting Appendix Fig. B.2. This figure consists of two parts: a graph for 95% confidence and one for 99% confidence limits. For a given vector length, r, and sample size, n, the graph gives the quantity d, which defines the confidence interval for μ_a, as

$$\bar{a} \pm d. \tag{24.20}$$

(That is, the lower confidence limit is $L_1 = \bar{a} - d$ and the upper confidence limit is $L_2 = \bar{a} + d$.) For example, we may wish to estimate a population mean angle, μ_a, when from a sample of 30 data from the population we calculate $\bar{a} = 186°$ and $r = 0.80$. Using the graph of Fig. B.2 for $\alpha = 0.05$, d is found to be 14°, so the 95% confidence interval for μ_a is $186° \pm 14°$. (That is, we are 95% confident that 172° and 200° encompass the mean angle in the sampled population.) If the data are grouped, then (by Equation 24.15) we can use r_c instead of r.

Confidence limits for the median angle may be obtained by the procedure of Section 22.8. The data in the sample are numbered sequentially from 1 through n so that the median fits Equation 3.4 (if n is odd) or Equation 3.5 (if n is even).

24.8 DIAMETRICALLY BIMODAL DISTRIBUTIONS

Occasionally populations are encountered having data with two modes lying opposite each other on the diameter of the circle. (Such data are sometimes termed "axial.") For example, Fig. 24.7 shows a distribution having opposite modes at 45° and 225°.

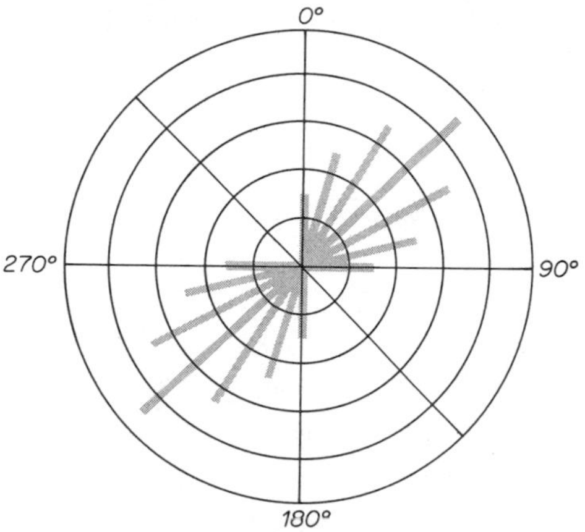

Figure 24.7 A bimodal circular distribution. For this distribution, $H_0 \colon \rho = 0$ would not be rejected.

If, as in this figure, the distribution is centrally symmetrical (i.e., each observation is matched by an observation 180° away), r computes to be zero and no mean angle can be determined. If the diametrically bimodal distribution is not centrally symmetrical, r will not be zero, but it may be so small as to have us conclude that there is no significant direction of orientation of the data (Section 25.2), and the calculated mean may be far from the diameter along which the bulk of the observations lie. However, we can engage in statistical analysis of such a distribution by a procedure involving the doubling of angles.

Example 24.5 shows the data that are graphed in Fig. 24.7. Each angle, a_i, is doubled; if the doubled angle is $< 360°$ it is recorded as $2a_i$, and if it is $\geq 360°$ then

Example 24.5

Descriptive statistics for the centrally symmetric distribution shown in Fig. 24.6.

a_i	f_i	$2a_i$	$\sin 2a_i$	$f_i \sin 2a_i$	$\cos 2a_i$	$f_i \cos 2a_i$
0°	15	0°	0.00000	0.00000	1.00000	15.00000
15°	25	30°	0.50000	12.50000	0.86603	21.65075
30°	35	60°	0.86603	30.31105	0.50000	17.50000
45°	45	90°	1.00000	45.00000	0.00000	0.00900
60°	35	120°	0.86603	30.31105	−0.50000	−17.50000
75°	25	150°	0.50000	12.50000	−0.86603	−21.65075
90°	15	180°	0.00000	0.00000	−1.00000	−15.00000
180°	15	0°	0.00000	0.00000	1.00000	15.00000
195°	25	30°	0.50000	12.50000	0.86603	21.65075
210°	35	60°	0.86603	30.31105	0.50000	17.50000
225°	45	90°	1.00000	45.00000	0.00000	0.00000
240°	35	120°	0.86603	30.31105	−0.50000	−17.50000
255°	25	150°	0.50000	12.50000	−0.86603	−21.65075
270°	15	180°	0.00000	0.00000	−1.00000	−15.00000

$n = 390$

$\sum f_i \sin 2a_i = 261.24420$

$$Y = \frac{261.24420}{390} = 0.66986$$

$\sum f_i \cos 2a_i = 0.00000$

$$X = \frac{0}{390} = 0$$

$$r = \sqrt{(0.66986)^2 + (0)^2} = 0.66986$$

$$\cos 2\bar{a} = \frac{0}{0.66986} = 0$$

$$\sin 2\bar{a} = \frac{0.66986}{0.66986} = 1.0000$$

Therefore, $2\bar{a} = 90°$ and $\bar{a} = 45°$, meaning that the bimodal distribution lies along a diameter line oriented at 45° (as can be seen by inspecting Fig. 24.7).

$$s \text{ for doubled angles} = \sqrt{2(1 - 0.66986)} = \sqrt{0.66028} = 0.81258$$

$$s = \frac{0.81258}{2} = 0.41$$

Inasmuch as the data are grouped (in intervals of 15°) Equation 24.15 may be used, in which case we would find $s = 0.40$.

360° is subtracted from it with the result being recorded as $2a_i$. (Also note in this example that Equations 24.13 and 24.14 are used to make the computations easier, since the data are grouped.) Note in Example 24.5 that the angular deviation is one-half the angular deviation of the doubled angles.

24.9 SECOND-ORDER ANALYSIS: THE MEAN OF MEAN ANGLES

If a mean is determined for each of several groups of angles, then we have a set of mean angles. Consider the data in Example 24.6. Here, a mean angle, $\bar{a}_j$, has been calculated for each of k samples of circular data, using procedures of Section 24.4. If, now, we desire to determine the grand mean of these several means, it is not appropriate to consider each of the sample means as an angle and employ the method of Section 24.4. To do so would be to assume that each mean had a vector length, r, of 1.0 (i.e., that an angular deviation, s, of zero was the case in each of the k samples), a most unlikely situation. Instead, we shall employ the procedure promulgated by Batschelet* (1978, 1981: Chapter 7), whereby the grand mean has rectangular coordinates

$$\bar{X} = \frac{\sum_{j=1}^{k} X_j}{k} \tag{24.21}$$

and

$$\bar{Y} = \frac{\sum_{j=1}^{k} Y_j}{k}, \tag{24.22}$$

where X_j and Y_j are the quantities X and Y, respectively, applying Equations 24.2 and 24.3 to sample j; k is the total number of samples. If we do not have X and Y

*Batschelet refers to the determination of the mean of a set of angles as a first-order analysis and the computation of the mean of a set of means as a second-order analysis.

Example 24.6

The mean of a set of mean angles.

Under particular light conditions, each of 7 butterflies is allowed to fly from the center of an experimental chamber 10 times. Using the procedures of Section 24.4, the values of $\bar{a}$ and r for each of the 7 samples of data are as follows.

$$k = 7; \quad n = 10$$

Sample (j)	$\bar{a}_j$	r_j	$X_j = r_j \cos \bar{a}_j$	$Y_j = r_j \sin \bar{a}_j$
1	160°	0.8954	−0.84140	0.30624
2	169	0.7747	−0.76047	0.14782
3	117	0.4696	−0.21319	0.41842
4	140	0.8794	−0.67366	0.56527
5	186	0.3922	−0.39005	−0.04100
6	134	0.6952	−0.48293	0.50009
7	171	0.3338	−0.32969	0.05222
			−3.69139	1.94906

$$\bar{X} = \frac{\sum r_j \cos \bar{a}_j}{k} = \frac{-3.69139}{7} = -0.52734$$

$$\bar{Y} = \frac{\sum r_j \sin \bar{a}_j}{k} = \frac{1.94906}{7} = 0.27844$$

$$r = \sqrt{\bar{X}^2 + \bar{Y}^2} = \sqrt{0.35562} = 0.59634$$

$$\cos \bar{a} = \frac{\bar{X}}{r} = \frac{-0.52734}{0.59634} = -0.88429$$

$$\sin \bar{a} = \frac{\bar{Y}}{r} = \frac{0.27844}{0.59634} = 0.46691$$

Therefore, $\bar{a} = 152°$.

for each sample, but we have $\bar{a}$ and r (polar coordinates) for each sample, then

$$\bar{X} = \frac{\sum_{j=1}^{k} r_j \cos \bar{a}_j}{k} \tag{24.23}$$

and

$$\bar{Y} = \frac{\sum_{j=1}^{k} r_j \sin \bar{a}_j}{k}. \tag{24.24}$$

Having obtained $\bar{X}$ and $\bar{Y}$, we may substitute them for X and Y, respectively, in Equations 24.8, 24.9, and 24.10 (and 24.11, if desired) in order to determine $\bar{a}$, which is the grand mean. For this calculation, all n_j's (sample sizes) should be equal, although a slight departure from this condition will not severely affect the results.

Figure 24.8 shows the individual means and the grand mean for Example 24.6. (By the hypothesis testing of Section 25.2 we would conclude that there is in this example no significant mean direction for samples 5 and 7. However, the data from these two samples should not be deleted from the present analysis.)

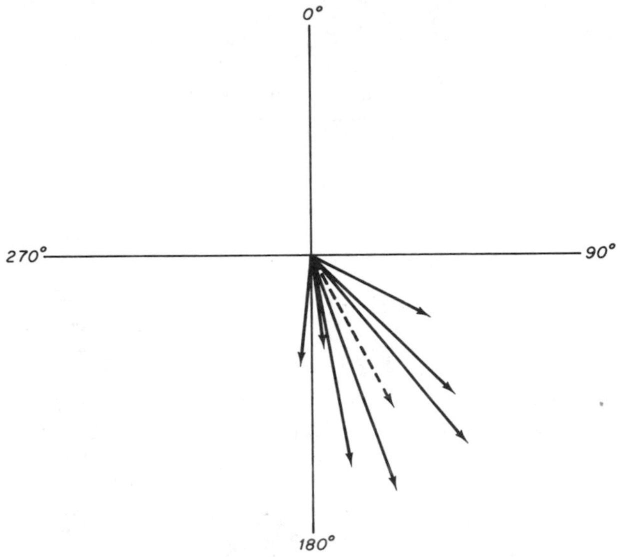

Figure 24.8 The data of Example 24.6. Each of the seven vectors in this sample is itself a mean vector. The mean of these seven means is indicated by the broken line.

24.10 CONFIDENCE LIMITS FOR THE SECOND-ORDER MEAN ANGLE

Section 24.9 explains how to obtain the mean of a set of mean angles. The mean thus computed is a sample estimate of a population mean, μ_a, and it is reasonable to ask how precise an estimate it is. The precision with which we estimate a population mean is typically expressed as a confidence interval for that parameter. For a first-order circular statistical analysis we may find confidence limits for μ_a by the procedure in Section 24.7. For a second-order analysis we may express confidence limits for μ_a if we first conclude (by the method of Section 25.7) that there is a significant directionality in the data.

Batschelet (1981: 144, 262–265) shows geometrically and analytically how the second-order confidence limits are obtained. Here we shall simply present the arithmetic employed:

$$A = \frac{k-1}{\sum x^2} \quad (24.25)$$

$$B = -\frac{(k-1)\sum xy}{\sum x^2 \sum y^2} \quad (24.26)$$

$$C = \frac{k-1}{\sum y^2} \quad (24.27)$$

$$D = \frac{2(k-1)\left[1 - \frac{(\sum xy)^2}{\sum x^2 \sum y^2}\right] F_{\alpha(1),2,k-2}}{k(k-2)} \quad (24.28)$$

$$H = AC - B^2 \quad (24.29)$$

$$G = A\bar{X}^2 + 2B\bar{X}\bar{Y} + C\bar{Y}^2 - D \tag{24.30}$$

$$U = H\bar{X}^2 - CD \tag{24.31}$$

$$V = \sqrt{DGH} \tag{24.32}$$

$$W = H\bar{X}\bar{Y} + BD \tag{24.33}$$

$$b_1 = \frac{W + V}{U} \tag{24.34}$$

$$b_2 = \frac{W - V}{U} \tag{24.35}$$

The quantities b_1 and b_2 are then examined separately, each yielding one of the confidence limits, as follows:

$$M = \sqrt{1 + b_i^2}, \tag{24.36}$$

after which we determine (from trigonometric tables or by calculator) that angle having

$$\text{sine} = \frac{b_i}{M} \tag{24.37}$$

and

$$\text{cosine} = \frac{1}{M}. \tag{24.38}$$

The confidence limit is either the angle thus determined or that angle + 180°, whichever is nearer the sample mean angle (and, if the angle + 180° is greater than 360°, simply subtract 360°). This procedure is demonstrated in Example 24.7. The confidence interval thus computed is a little conservative (i.e., the confidence coefficient is a little greater than the stated $1 - \alpha$), and the confidence limits are not necessarily symmetrical about the mean.

Example 24.7

Confidence limits for the mean of a set of mean angles, using the data of Example 24.6, for which $\bar{a} = 152°$.

j	X_j	X_j^2	Y_j	Y_j^2	X_jY_j
1	−0.84140	0.70795	0.30624	0.09378	−0.25767
2	−0.76047	0.57831	0.14782	0.02185	−0.11241
3	−0.21319	0.04545	0.41842	0.17508	−0.08920
4	−0.67366	0.45382	0.56527	0.31953	−0.38080
5	−0.39005	0.15214	−0.04100	0.00168	0.01599
6	−0.48293	0.23322	0.50009	0.25009	−0.24151
7	−0.32969	0.10870	0.05222	0.00273	−0.01722
	−3.69139	2.27959	1.94906	0.86474	−1.08282

For 95% confidence limits, $\alpha = 0.05$.

$k = 7$

$$\bar{X} = \frac{\sum X_j}{k} = \frac{-3.69139}{7} = -0.52734$$

$$\bar{Y} = \frac{\sum Y_j}{k} = \frac{1.94906}{7} = 0.27844$$

$$\sum x^2 = \sum X_j^2 - \frac{(\sum X_j)^2}{k} = 2.27959 - \frac{(-3.69139)^2}{7} = 0.33297$$

$$\sum y^2 = \sum Y_j^2 - \frac{(\sum Y_j)^2}{k} = 0.86474 - \frac{(1.94906)^2}{7} = 0.32205$$

$$\sum xy = \sum X_j Y_j - \frac{\sum X_j Y_j}{k} = -1.08282 - \frac{(-3.69139)(1.94906)}{7} = -0.05500$$

$$A = \frac{k-1}{\sum x^2} = \frac{7-1}{0.33297} = 18.01964$$

$$B = -\frac{(k-1)\sum xy}{\sum x^2 \sum y^2} = -\frac{(7-1)(-0.05500)}{(0.33297)(0.32205)} = 3.07741$$

$$C = \frac{k-1}{\sum y^2} = \frac{7-1}{0.32205} = 18.63065$$

$$F_{\alpha(1),2,k-2} = F_{0.05(1),2,5} = 5.79$$

$$D = \frac{2(k-1)\left[1 - \frac{(\sum xy)^2}{\sum x^2 \sum y^2}\right]F_{\alpha(1),2,k-2}}{k(k-2)}$$

$$= \frac{2(7-1)\left[1 - \frac{(-0.05500)^2}{(0.33297)(0.32205)}\right](5.79)}{7(7-2)} = 1.92914$$

$$H = AC - B^2 = (18.01964)(18.63065) - (3.07741)^2 = 326.24715$$

$$G = A\bar{X}^2 + 2B\bar{X}\bar{Y} + C\bar{Y}^2 - D$$

$$= (18.01964)(-0.52734)^2 + 2(3.07741)(-0.52734)(0.27844)$$

$$+ (18.63065)(0.27844)^2 - 1.92914 = 3.62258$$

$$U = H\bar{X}^2 - CD$$

$$= (326.24715)(-0.52734)^2 - (18.63065)(1.92914) = 54.78411$$

$$V = \sqrt{\mathrm{DGH}} = \sqrt{(1.92914)(3.62258)(326.24715)} = 47.748899$$

$$W = H\bar{X}\bar{Y} + BD$$

$$= (326.24715)(-0.52734)(0.27844) + (3.07741)(1.92914) = -41.96695$$

$$b_1 = \frac{W+V}{U} = \frac{-41.96695 + 47.74899}{54.78411} = 0.10554$$

$$b_2 = \frac{W-V}{U} = \frac{-41.96695 - 47.74899}{54.78411} = -1.63763$$

For b_1: $M = \sqrt{1 + b_1^2} = \sqrt{1 + (0.10554)^2} = 1.00555$

$$\text{sine} = \frac{b_1}{M} = \frac{0.10554}{1.00555} = 0.10496$$

$$\text{cosine} = \frac{1}{M} = \frac{1}{1.00555} = 0.99448$$

The angle with this sine and cosine is 6°, so one of the confidence limits is either 6° or 6° + 180° = 186°; of the two possibilities, 186° is closer to the mean (152°).

For b_2: $M = \sqrt{1 + b_2^2} = \sqrt{1 + (-1.63763)^2} = 1.91881$

$$\text{sine} = \frac{b_2}{M} = \frac{-1.63763}{1.91881} = -0.85346$$

$$\text{cosine} = \frac{1}{M} = \frac{1}{1.91881} = 0.52116$$

The angle with this sine and cosine is 301°, so the second confidence limit is either 301° or 301° + 180° = 481° = 121°; of the two possibilities, 121° is closer to the mean (152°).

EXERCISES

24.1. Twelve nests of a particular bird species were recorded facing outward from trees at the following directions:

Direction	*Frequency*
N: 0°	2
NE: 45	4
E: 90	3
SE: 135	1
S: 180	1
SW: 225	1
W: 270	0
NW: 315	0

(a) Compute the sample mean direction. **(b)** Compute the angular deviation for the data. **(c)** Determine 95% confidence limits for the population mean. **(d)** Determine the sample median direction.

24.2. A total of 15 human births occurred as follows:

1:15 AM	4:40 AM	5:30 AM	6:50 AM
2:00 AM	11:00 AM	4:20 AM	5:10 AM
4:30 AM	5:15 AM	10:30 PM	8:55 PM
6:10 AM	3:45 AM	3:10 AM	

(a) Compute the mean time of birth. **(b)** Compute the angular deviation for the data. **(c)** Determine 95% confidence limits for the population mean direction. **(d)** Determine the sample median direction.

24.3. Five samples of directional data were collected and were as follows. **(a)** Determine the mean of the five sample means. **(b)** Determine the 95% confidence limits for the second-order mean.

Sample	*Sample mean*	*Sample r*
1	230°	0.4542
2	245	0.6083
3	265	0.7862
4	210	0.5107
5	225	0.8639

25 Circular Distributions: Hypothesis Testing

Armed with the procedures in Chapter 24, and the information contained in the basic statistics of circular distributions (e.g., $\bar{a}$ and r), we can now examine a number of statistical methods for testing hypotheses about populations measured on a circular scale.

25.1 GOODNESS OF FIT TESTING

Either χ^2 or G may be used to test the goodness of fit of theoretical to an observed circular frequency distribution. (See Chapter 5 for general aspects of goodness of fit methods.) The procedure is to determine each expected frequency, $\hat{f}_i$, corresponding to each observed frequency, f_i, in each category, i. For the data of Example 24.3, for instance, we might hypothesize a uniform distribution of data among the 12 divisions of the data. The test of this hypothesis is presented in Example 25.1. Batschelet (1981: 72) recommends grouping the data so that no expected frequency is less than 4 in using chi-square.

Recall that goodness of fit testing by the chi-square, or G, statistic does not take into account the sequence of categories that occurs in the data distribution. In Sections 5.10 and 5.11 the Kolmogorov-Smirnov test was introduced as an improvement over chi-square when the categories of data are, in fact, ordered. Unfortunately, the Kolmogorov-Smirnov test yields different results for different starting points on a circular scale; however, a modification of this test by Kuiper (1960) provides a goodness of fit test, the results of which are unrelated to the starting point on a circle.

Example 25.1

Chi-square goodness of fit for the circular data of Example 24.3.

H_0: The data in the population are distributed uniformly around the circle.
H_A: The data in the population are not distributed uniformly around the circle.

a_i (deg)	f_i	$\hat{f}_i$
0–30	0	8.7500
30–60	6	8.7500
60–90	9	8.7500
90–120	13	8.7500
120–150	15	8.7500
150–180	22	8.7500
180–210	17	8.7500
210–240	12	8.7500
240–270	8	8.7500
270–300	3	8.7500
300–330	0	8.7500
330–360	0	8.7500

$k = 12 \qquad n = 105$
$\hat{f}_i = 105/12 = 8.7500$ for all i

$$\chi^2 = \frac{(0 - 8.7500)^2}{8.7500} + \frac{(6 - 8.7500)^2}{8.7500} + \frac{(9 - 8.7500)^2}{8.7500} + \cdots + \frac{(0 - 8.7500)^2}{8.7500}$$
$$= 8.7500 + 0.8643 + 0.0071 + \cdots + 8.7500$$
$$= 66.543$$

$\nu = k - 1 = 11$

$\chi^2_{0.05,11} = 19.675$

Reject H_0. $P \ll 0.001$.

The Kuiper test is discussed by Batschelet (1965: 26–27; 1981: 76–79) and Mardia (1972: Sect. 7.2.1).

Another goodness of fit test applicable to circular distributions is that of Watson (1962), often referred to as the *Watson one-sample U^2 test*, which is demonstrated in Example 25.2 To test the null hypothesis of uniformity, we first transform each angular measurement, a_i, by dividing it by 360°:

$$u_i = \frac{a_i}{360^\circ}. \tag{25.1}$$

Then the following quantities are obtained for the set of n values of u_i: $\sum u_i$, $\sum u_i^2$, $\bar{u}$, and $\sum iu_i$. The test statistic, called "Watson's U^2," is

$$U^2 = \sum u_i^2 - \frac{(\sum u_i)^2}{n} - \frac{2}{n}\sum iu_i + (n+1)\bar{u} + \frac{n}{12} \tag{25.2}$$

(Mardia, 1972: 182). Critical values for this test are $U^2_{\alpha,n,n}$ in Table B.35.

All the testing procedures for this section are nonparametric.

Example 25.2

Watson's goodness of fit testing using the data of Example 24.2.

H_0: The sample data come from a population uniformly distributed around the circle.
H_A: The sample data do not come from a population uniformly distributed around the circle.

i	a_i	u_i	u_i^2	iu_i
1	55°	0.1528	0.0233	0.1528
2	81°	0.2250	0.0506	0.4500
3	96°	0.2667	0.0711	0.8001
4	109°	0.3028	0.0917	1.2112
5	117°	0.3250	0.1056	1.6250
6	132°	0.3667	0.1345	2.2002
7	154°	0.4278	0.1830	2.9946
$n = 7$		$\sum u_i =$ 2.0668	$\sum u_i^2 =$ 0.6598	$\sum iu_i =$ 9.4339

$$\bar{u} = \frac{\sum u_i}{n} = \frac{2.0668}{7} = 0.2953$$

$$U^2 = \sum u_i^2 - \frac{(\sum u_i)^2}{n} - \frac{2}{n}\sum iu_i + (n+1)\bar{u} + \frac{n}{12}$$

$$= 0.6598 - \frac{(2.0668)^2}{7} - \frac{2}{7}(9.4339) + (7+1)(0.2953) + \frac{7}{12}$$

$$= 0.6598 - 0.6102 - 2.6954 + 2.3624 + 0.5833$$

$$= 0.2999$$

$U^2_{0.05,7,7} = 0.1986$

Therefore, reject H_0. $0.005 < P < 0.02$.

25.2 THE SIGNIFICANCE OF THE MEAN AND MEDIAN ANGLES

One can obviously place more confidence in $\bar{a}$ as an estimate of the population mean angle, μ_a, if s is small than if it is large. This is identical to stating that $\bar{a}$ is a better estimate of μ_a if r is large than if r is small. What is desired is a method of asking whether there is, in fact, a mean direction among the population of data which were sampled, for even if there is no mean direction (i.e., the circular distribution is uniform) in the population, a random sample might still display a calculable mean. The test we require is that concerning H_0: The sampled population is uniformly* distributed around a circle vs. H_A: The population is not a uniform circular distribution. This may be tested by the nonparametric *Rayleigh test*†. As circular uniformity implies no mean direction, the Rayleigh test may be said to test H_0: $\rho = 0$ vs. H_A: $\rho \neq 0$.

*In dealing with circular statistics, the terms "uniform distribution" and "random distribution" have been used synonymously in the literature.

†Named for Lord Rayleigh [John William Strutt, Third Baron Rayleigh (1842–1919)], a physicist and applied mathematician who gained his greatest fame for discovering and isolating the chemical element argon (winning him the Nobel Prize in physics in 1904), although some of his other contributions to physics were at least as important (Lindsay, 1976).

The Rayleigh test asks how large a sample r must be to indicate confidently a nonuniform population distribution. A quantity referred to as "Rayleigh's R" is obtainable as

$$R = nr, \tag{25.3}$$

and the so-called "Rayleigh's z" is utilized for testing the null hypothesis of no population mean direction:

$$z = \frac{R^2}{n} \quad \text{or} \quad z = nr^2. \tag{25.4}$$

Table B.32 presents critical values of $z_{\alpha,n}$. If the data are grouped, then we may substitute (by Equation 24.15) r_c for r in this test.

If H_0 is rejected by Rayleigh's test, we may conclude that there is a mean population direction (see Example 25.3), but only if the distribution is unimodal. If H_0 is not rejected, we may conclude the population distribution to be uniform around the circle only if we may assume that it does not have more than one mode. (For example, the data in Example 24.6 and Fig. 24.7 would result in a Rayleigh test failing to reject H_0. While these data do in fact have no mean direction, they are not uniform in their circular distribution. Section 24.8 explains how such data can be transformed into unimodal data, thereafter to be subjected to Rayleigh testing and other procedures requiring unimodality.)

Example 25.3

Rayleigh's test applied to the data of Example 24.2. These data are plotted in Fig. 24.2.

H_0: $\rho = 0$ (i.e., the population is uniformly distributed around the circle).
H_A: $\rho \neq 0$ (i.e., the population is not distributed uniformly around the circle).

The following were obtained in Example 24.4:

$$n = 7$$
$$r = 0.86697$$

Therefore,

$$R = nr = (7)(0.86697) = 6.06879$$

and

$$z = \frac{R^2}{n} = \frac{(6.06879)^2}{7} = 5.261$$

Using Table B.32, $z_{0.05,7} = 2.885$. Reject H_0. $0.001 < P < 0.002$.

A modification of the Rayleigh test (Greenwood and Durand, 1955; Durand and Greenwood, 1958) is available for use when the investigator has reason to expect, *in advance*, a specific mean direction. In Example 25.4a (and presented graphically in Fig. 25.1), 10 birds were released at a site directly west of their home. Therefore, the statistical hypotheses may include the expectation of the birds to tend to fly directly east (i.e., at an angle of 90°). The testing procedure considers H_0: The population angles are randomly distributed (i.e., H_0: $\rho = 0$) vs. H_A: The population angles are not randomly distributed (i.e., H_A: $\rho \neq 0$). By using additional information, namely the expected mean angle, this test is more powerful than Rayleigh's test (Batschelet, 1972; 1981: 60).

Example 25.4a

The V test.

H_0: The population is uniformly distributed around the circle (i.e., $H_0: \rho = 0$).
H_A: The population is not uniformly distributed around the circle (i.e., $H_A: \rho \neq 0$).

a_i (deg)	$\sin a_i$	$\cos a_i$
66	0.91355	0.40674
75	0.96593	0.25882
86	0.99756	0.06976
88	0.99939	0.03490
88	0.99939	0.03490
93	0.99863	−0.05234
97	0.99255	−0.12187
101	0.98163	−0.19081
118	0.88295	−0.46947
130	0.76604	−0.64279
$n = 10$	$\sum \sin a_i = 9.49762$	$\sum \cos a_i = -0.67216$

$$Y = \frac{9.49762}{10} = 0.94976 \qquad X = -\frac{0.67216}{10} = -0.06722$$

$$r = \sqrt{(-0.06722)^2 + (0.94976)^2} = 0.95213$$

$$\sin \bar{a} = \frac{Y}{r} = 0.99751$$

$$\cos \bar{a} = \frac{X}{r} = -0.07060$$

$$\bar{a} = 94^\circ.$$

$$R = (10)(0.95213) = 9.5213$$

$$\begin{aligned} V &= R \cos (94^\circ - 90^\circ) \\ &= 9.5213 \cos (4^\circ) \\ &= (9.5213)(0.99756) \\ &= 9.498 \end{aligned}$$

$$\begin{aligned} u &= V\sqrt{\frac{2}{n}} \\ &= (9.498)\sqrt{\frac{2}{10}} \\ &= 4.248 \end{aligned}$$

Using Table B.33, $u_{0.05,10} = 1.648$. Reject H_0. $P < 0.0005$.

The preceding hypotheses are tested by what we shall refer to as the V *test*, in which the test statistic is computed as

$$V = R \cos (\bar{a} - \mu_0), \tag{25.5}$$

where μ_0 is the mean angle predicted. Table B.33 gives critical values of $u_{\alpha,n}$, a statistic which, for large sample sizes, approaches a one-tailed normal deviate, Z.

$$u = V\sqrt{\frac{2}{n}}. \tag{25.6}$$

If the data are grouped, then R may be determined from r_c rather than r.

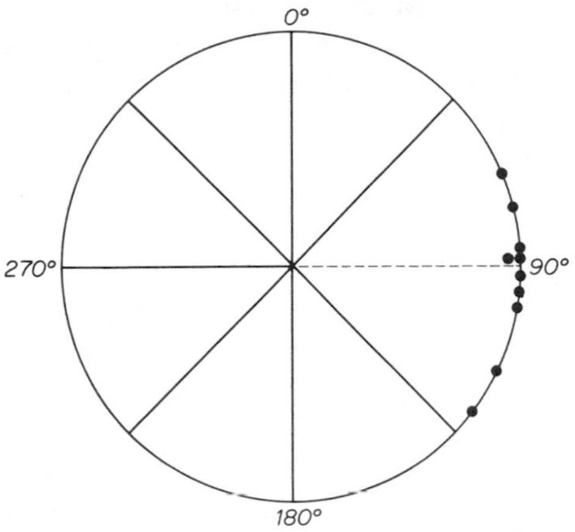

Figure 25.1 The data for the V test of Example 25.4. The broken line indicates the expected mean angle (94°).

One-Sample Test for the Mean Angle. The Rayleigh test and the V test are methods for testing for random distribution of a population of data around the circle. (See Batschelet, 1981: Chapter 4, for other tests of the null hypothesis of randomness.) If it is desired to test whether the population mean angle is equal to a specified value, say μ_0, then we have a one-sample test situation analogous to that of the one-sample t test for data on a linear scale (Section 8.1). The hypotheses are

$$H_0: \mu_a = \mu_0$$

and

$$H_A: \mu_a = \mu_0,$$

and H_0 is tested simply by observing whether μ_0 lies within the $1 - \alpha$ confidence interval for μ_a. If μ_0 lies outside the confidence interval, then H_0 is rejected. Section 24.7 describes the determination of confidence intervals for the population mean angle, and Example 25.4b demonstrates the hypothesis testing procedure.*

Example 25.4b

The one-sample test for the mean angle.

H_0: The population has a mean of 90° (i.e., $\mu_a = 90°$).
H_A: The population mean is not 90° (i.e., $\mu_a \neq 90°$).

The computation of the following is given in Example 25.4a:

$$r = 0.95$$
$$\bar{a} = 94°$$

Using appendix Fig. B.2, for $\alpha = 0.05$ and $n = 10$:

$d = 13°$; so the 95% confidence interval for μ_a is $94° \pm 13°$.

As this confidence interval does contain the hypothesized mean ($\mu_0 = 90°$), we do not reject H_0.

*For demonstration purposes (Examples 25.4a and 25.4b) we have applied the V test and the one-sample test for the mean angle to the same set of data. In practice this would not be done. Deciding which test to employ would depend, respectively, on whether the intention is to test for circular uniformity or to test whether the population mean angle is a specified value.

One-Sample Test for the Median Angle. We can perform a nonparametric test to determine whether the median angle equals a specified value simply by counting the number of observed angles on either side of a diameter through the hypothesized angle and subjecting these data to the binomial test of Section 22.5. This is demonstrated in Example 25.5.

Example 25.5

The significance of the median angle.

The sample data consist of the following directions: 97°, 104°, 121°, 136°, 159°, 164°, 172°, 195°, 213°. The median is 159°.

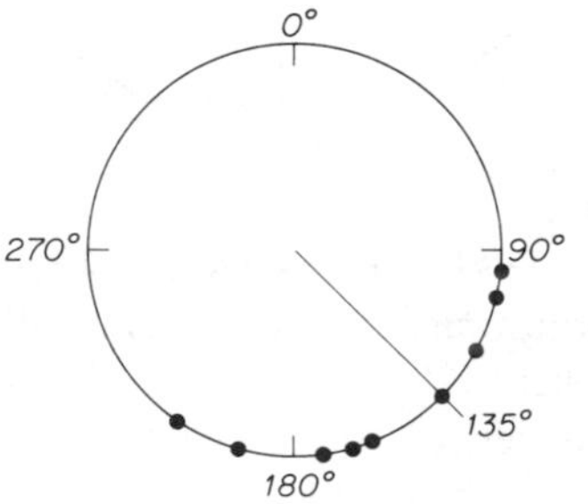

If we wish to test whether the population median is equal to some specific value—say, 135°—we can proceed as follows:

H_0: The population median angle is 135°.
H_A: The population median angle is not 135°.

Employing the two-tailed binomial test, we have $n = 9$ and $p = 0.5$. There are three sample directions $< 135°$ and six directions $> 135°$. It is found that $P = 0.508$. Do not reject H_0.

25.3 PARAMETRIC TWO-SAMPLE AND MULTISAMPLE TESTING OF ANGLES

It is common to consider the null hypothesis H_0: $\mu_1 = \mu_2$, where μ_1 and μ_2 are the mean angles for each of two circular distributions (see Example 25.6). Watson and Williams (1956, with an improvement by Stephens, 1972) proposed a test that utilizes the statistic

$$F = K\frac{(N-2)(R_1 + R_2 - R)}{N - R_1 - R_2}, \qquad (25.7)$$

where $N = n_1 + n_2$. In this equation, R is calculated by Equation 25.3 with the data from the two samples being combined; R_1 and R_2 are the values of Rayleigh's R for the two samples considered separately. K is a factor, obtained from Table B.34, that corrects for bias in the F calculation; in that table one uses for r the weighted mean of the two vector lengths:

$$r_w = \frac{n_1 r_1 + n_2 r_2}{N} = \frac{R_1 + R_2}{N}. \qquad (25.8)$$

Example 25.6

The Watson-Williams test for two samples.

H_0: $\mu_1 = \mu_2$
H_A: $\mu_1 \neq \mu_2$

	Sample 1			**Sample 2**	
a_i (deg)	$\sin a_i$	$\cos a_i$	a_i (deg)	$\sin a_i$	$\cos a_i$
94	0.99756	−0.06976	77	0.97437	0.22495
65	0.90631	0.42262	70	0.93969	0.34202
45	0.70711	0.70711	61	0.87462	0.48481
52	0.78801	0.61566	45	0.70711	0.70711
38	0.61566	0.78801	50	0.76604	0.64279
47	0.73135	0.68200	35	0.57358	0.81915
73	0.95630	0.29237	48	0.74314	0.66913
82	0.99027	0.13917	65	0.90631	0.42262
90	1.00000	0.00000	36	0.58779	0.80902
40	0.64279	0.76604	$n_2 = 9$	$\sum \sin a_i = 7.07265$	$\sum \cos a_i = 5.12160$
87	0.99863	0.05234			
$n_1 = 11$	$\sum \sin a_i = 9.33399$	$\sum \cos a_i = 4.39556$			

Sample 1:

$Y = 0.84854, \quad X = 0.39960$
$r_1 = 0.93792$
$\sin \bar{a}_1 = 0.90470$
$\cos \bar{a}_1 = 0.42605$
$\bar{a}_1 = 65^\circ$
$R_1 = 10.31712$

Sample 2:

$Y = 0.78585, \quad X = 0.56907$
$r_2 = 0.97026$
$\sin \bar{a}_2 = 0.80994$
$\cos \bar{a}_2 = 0.58651$
$\bar{a}_2 = 54^\circ$
$R_2 = 8.73234$

By combining the 20 data from both samples:

$$\sum \sin a_i = 9.33399 + 7.07265 = 16.40664$$
$$\sum \cos a_i = 4.39556 + 5.12160 = 9.51716$$
$$N = 11 + 9 = 20$$
$$Y = \frac{16.40664}{20} = 0.82033$$
$$X = \frac{9.51716}{20} = 0.47586$$
$$r = 0.94836$$
$$R = 18.96720$$
$$r_w = \frac{10.31712 + 8.73234}{20} = 0.952$$
$$F = K\frac{(N-2)(R_1 + R_2 - R)}{N - R_1 - R_2}$$
$$= (1.0351)\frac{(20-2)(10.31712 + 8.73234 - 18.96720)}{20 - 10.31712 - 8.73234}$$
$$= (1.0351)\frac{1.48068}{0.95054}$$
$$= 1.61$$
$$F_{0.05,1,18} = 4.41$$

Therefore, do not reject H_0.

$0.10 < P < 0.25$

Therefore, we conclude that the two sample means estimate the same population mean, and the best estimate of this population mean is obtained by:

$$\sin \bar{a} = \frac{Y}{r} = 0.86500$$

$$\cos \bar{a} = \frac{X}{r} = 0.50177$$

$$\bar{a} = 60^\circ$$

The critical value for this test is $F_{\alpha(1),1,N-2}$. Alternatively,

$$t = \sqrt{K\frac{(N-2)(R_1 + R_2 - R)}{N - R_1 - R_2}} \tag{25.9}$$

may be compared with $t_{\alpha(2),N-2}$. This test may be used for r_w as small as 0.75, if $5 \leq N/2 < 10$; or for r_w as low as 0.70, if $N/2 \geq 10$ (Batschelet, 1981: 97, 321; Mardia, 1972: 155). The data may be grouped as long as the grouping interval is $\leq 10^\circ$. See Batschelet (1972; 1981: Chapter 6) for a review of other two-sample testing procedures. Mardia (1972: 156–158) gives a procedure for an approximate confidence interval for $\mu_1 - \mu_2$.

Multisample Testing. The Watson-Williams test can be generalized to a multisample test for testing H_0: $\mu_1 = \mu_2 = \ldots = \mu_k$, a hypothesis reminiscent of analysis of variance considerations for linear data. In multisample tests (Example 25.7),

$$F = K\frac{(N-k)\left(\sum_{j=1}^{k} R_j - R\right)}{(k-1)\left(N - \sum_{j=1}^{k} R_j\right)}. \tag{25.10}$$

Here, k is the number of samples, R is the Rayleigh's R for all k samples combined, and $N = \sum_{j=1}^{n} n_j$. The correction factor, K, is obtained from Table B.34, using

$$r_w = \frac{\sum_{j=1}^{k} n_j r_j}{N} = \frac{\sum_{j=1}^{k} R_j}{N}. \tag{25.11}$$

Example 25.7

The Watson-Williams test for three samples.

H_0: All three samples are from populations with the same mean angle.
H_A: All three samples are not from populations with the same mean angle.

	Sample 1			Sample 2	
a_i (deg)	$\sin a_i$	$\cos a_i$	a_i (deg)	$\sin a_i$	$\cos a_i$
135	0.70711	−0.70711	150	0.50000	−0.86603
145	0.57358	−0.81915	130	0.76604	−0.64279
125	0.81915	−0.57358	175	0.08716	−0.99619
140	0.64279	−0.76604	190	−0.17365	−0.98481
165	0.25882	−0.96593	180	0.00000	−1.00000
170	0.17365	−0.98481	220	−0.64279	−0.76604
$n_1 = 6$	$\sum \sin a_i = 3.17510$	$\sum \cos a_i = -4.81662$	$n_2 = 6$	$\sum \sin a_i = 0.53676$	$\sum \cos a_i = -5.25586$
	$\bar{a}_1 = 147^\circ$			$\bar{a}_2 = 174^\circ$	
	$r_1 = 0.96150$			$r_2 = 0.88053$	
	$R_1 = 5.76894$			$R_2 = 5.28324$	

Sample 3

a_i (deg)	$\sin a_i$	$\cos a_i$
140	0.64279	−0.76604
165	0.25882	−0.96593
185	−0.08715	−0.99619
180	0.00000	−1.00000
125	0.81915	−0.57358
175	0.08716	−0.99619
140	0.64279	−0.76604
$n_3 = 7$	$\sum \sin a_i = 2.36356$	$\sum \cos a_i = -6.06397$

$\bar{a}_3 = 159°$

$r_3 = 0.92976$

$R_3 = 6.50832$

$k = 3$

$N = 6 + 6 + 7 = 19$

For all 19 data:

$\sum \sin a_i = 3.17510 + 0.53676 + 2.36356 = 6.07542$

$\sum \cos a_i = -4.81662 - 5.25586 - 6.06397 = -16.13645$

$Y = 0.31976$

$X = -0.84929$

$r = 0.90749$

$R = 17.24231$

$$r_w = \frac{5.76894 + 5.28324 + 6.50832}{19} = 0.924$$

$$F = K\frac{(N-k)(\sum R_j - R)}{(k-1)(N - \sum R_j)}$$

$$= (1.0546)\frac{(19-3)(5.76894 + 5.28324 + 6.50832 - 17.24231)}{(3-1)(19 - 5.76894 - 5.28324 - 6.50832)}$$

$$= (1.0546)\frac{5.09104}{2.87900}$$

$$= 1.86$$

$\nu_1 = k - 1 = 2$

$\nu_2 = N - k = 16$

$F_{0.05(1),2,16} = 3.63$

Therefore, do not reject H_0.

$0.10 < P < 0.25$

Therefore, we conclude that the three sample means estimate the same population mean, and the best estimate of that population mean is obtained by:

$$\sin \bar{a} = \frac{Y}{r} = 0.35236$$

$$\cos \bar{a} = \frac{X}{r} = -0.93587$$

$$\bar{a} = 159°$$

The critical value for this test is $F_{\alpha(1),k-1,N-k}$. Equation 25.11 (and, thus this test) may be used for r_w as small as 0.45, if $N/k > 10$; for r_w as small as 0.50, for $N/k > 6$; and for r_w as small as 0.55, for $N/k = 5$ or 6 (Mardia, 1972: 163; Batschelet (1981: 321). If the data are grouped, the grouping interval should be no larger than 10°. Upton (1976) presents an alternative to the Watson-Williams test that relies on χ^2, instead of F, but the Watson-Williams procedure is a little simpler to use.

The Watson-Williams test (for two or more samples) is parametric and assumes that each of the samples came from a population conforming to what is known as the von Mises, or circular normal, distribution.* In addition, the tests assume that the population dispersions are all the same. Fortunately, the tests are robust to departures from these assumptions. But if the underlying assumptions are known to be violated severely (as when the distributions are not unimodal), we should be wary of their use. In the two-sample case, the nonparametric test of Section 25.4 is preferable to the Watson-Williams test when the assumptions of the latter cannot be presumed to prevail.

Batschelet (1981: 122–126), Mardia (1972: 158–162, 165–166), and Stephens (1972) discuss testing of equality of population concentrations.

25.4 NONPARAMETRIC TWO-SAMPLE AND MULTISAMPLE TESTING OF ANGLES

If data are grouped—Batschelet (1981: 110) recommends a grouping interval larger than 10°—then contingency table analysis may be used (as introduced in Section 6.1) as a two-sample test. The runs test of Section 25.14 may be used as a two-sample test, but it is not as powerful for that purpose as is the procedure below, and it is best reserved for testing the hypothesis in Section 25.14.

Among the other nonparametric procedures applicable to two samples of circular data (e.g., see Batschelet, 1972; 1981: Chapter 6), is a powerful test developed by Watson (1962). It is recommended in place of the Watson-Williams two-sample test of Section 25.3 when at least one of the sampled populations is not unimodal or when there are other considerable departures from the assumptions of the latter test procedure. It may be used on grouped data if the grouping interval is no greater than 5° (Batschelet, 1981: 115).

The data in each sample are arranged in ascending order, as demonstrated in Example 25.8. For two sample sizes, n_1 and n_2, let us denote the ith observation in sample 1 as a_{1i} and the jth datum in sample 2 as a_{2j}. Then, for the data in Example 25.8, $a_{11} = 55°$, $a_{21} = 75°$, $a_{12} = 60°$, $a_{22} = 90°$, etc. The total number of data is $N = n_1 + n_2$. The cumulative relative frequencies for the observations in sample 1 are i/n_1, and those for sample 2 are j/n_2. As shown in the present example, we then define values of d_k (where k runs from 1 through N) as the differences between the two cumulative relative frequency distributions. The test statistic, called the Watson U^2, is computed as

$$U^2 = \frac{n_1 n_2}{N^2}\left[\sum_{k=1}^{N} d_k^2 - \frac{\left(\sum_{k=1}^{N} d_k\right)^2}{N}\right]. \tag{25.12}$$

Critical values of U^2_{α,n_1,n_2} are given in Table B.35 bearing in mind that $U^2_{\alpha,n_1,n_2} = U^2_{\alpha,n_2,n_1}$.

Watson's U^2 is especially useful for circular data because the starting point for determining the cumulative frequencies is immaterial. It may also be used in any

*Richard von Mises (1883–1953), an Austrian-born physicist and mathematician, introduced this distribution in 1918, and it was named "circular normal" by other authors in 1953 (Batschelet, 1965: 10).

Example 25.8

Watson's U^2 test for nonparametric two-sample testing.

H_0: The two samples came from the same population, or from two populations having the same directions.

H_A: The two samples did not come from the same population, or from two populations having the same directions.

i	a_{1i} (deg)	$\frac{i}{n_1}$	j	a_{2j} (deg)	$\frac{j}{n_2}$	$d_k = \frac{i}{n_1} - \frac{j}{n_2}$	d_k^2
1	55	0.1667			0.0000	0.1667	0.0278
2	60	0.3333			0.0000	0.3333	0.1111
3	70	0.5000			0.0000	0.5000	0.2500
		0.5000	1	75	0.1250	0.3750	0.1406
4	80	0.6667			0.1250	0.5417	0.2934
		0.6777	2	90	0.2500	0.4167	0.1736
5	95	0.8333			0.2500	0.5833	0.3402
		0.8333	3	100	0.3750	0.4583	0.2100
		0.8333	4	110	0.5000	0.3333	0.1111
6	120	1.0000			0.5000	0.5000	0.2500
		1.0000	5	135	0.6250	0.3750	0.1406
		1.0000	6	140	0.7500	0.2500	0.0625
		1.0000	7	150	0.8750	0.1250	0.0156
		1.0000	8	160	1.0000	0.0000	0.0000
	$n_1 = 6$			$n_2 = 8$		$\sum d_k = 4.9583$	$\sum d_k^2 = 2.1265$

$$N = n_1 + n_2 = 14$$

$$U^2 = \frac{n_1 n_2}{N^2}\left[\sum d_k^2 - \frac{(\sum d_k)^2}{N}\right]$$

$$= \frac{(6)(8)}{14^2}\left[2.1265 - \frac{(4.9583)^2}{14}\right]$$

$$= 0.0907$$

$$U^2_{0.05,6,8} = 0.1964$$

Do not reject H_0.

$0.20 < P < 0.50$

situation with linear data that are amenable to Mann-Whitney testing (Section 9.9), but it is generally not recommended as a substitute for the Mann-Whitney test; the latter is easier to perform and has access to more extensive tables of critical values, and the former may declare significance when the group dispersions are different.

Watson's Two-Sample Test with Ties. If there are some tied data (i.e., there are two or more observations having the same numerical value), then the Watson two-sample test is modified as demonstrated in Example 25.9. We define t_{1i} as the number of data in sample 1 with a value of a_{1i} and t_{2j} as the number of data in sample 2 that have a value of a_{2j}. Additionally, m_{1i} and m_{2j} are the cumulative number of data in samples 1 and 2, respectively; so the cumulative relative frequencies are m_{1i}/n_1 and m_{2j}/n_2, respectively. As above, d_k represents a difference between the two cumulative frequency distributions. The test statistic is

$$U^2 = \frac{n_1 n_2}{N^2}\left[\sum_{k=1}^{N} t_k d_k^2 - \frac{\left(\sum_{k=1}^{N} t_k d_k\right)^2}{N}\right]. \tag{25.13}$$

Example 25.9

Watson's U^2 test for data containing ties.

H_0: The two samples came from the same population, or from two populations having the same directions.

H_A: The two samples did not come from the same population, or from two populations having the same directions.

i	a_{1i}	t_{1i}	m_{1i}	$\frac{m_{1i}}{n_1}$	j	a_{2j}	t_{2j}	m_{2j}	$\frac{m_{2j}}{n_2}$	$d_k = \frac{m_{1i}}{n_1} - \frac{m_{2j}}{n_2}$	d_k^2	t_k
				0.0000	1	30°	1	1	0.1000	−0.1000	0.0100	1
				0.0000	2	35	1	2	0.2000	−0.2000	0.0400	1
1	40°	1	1	0.0833					0.2000	−0.1167	0.0136	1
2	45	1	2	0.1667					0.2000	−0.0333	0.0011	1
3	50	1	3	0.2500	3	50	1	3	0.3000	−0.0500	0.0025	2
4	55	1	4	0.3333					0.3000	0.0333	0.0011	1
				0.3333	4	60	1	4	0.4000	−0.0667	0.0044	1
				0.3333	5	65	2	6	0.6000	−0.2677	0.0711	2
5	70	1	5	0.4167					0.6000	−0.1833	0.0336	1
				0.4167	6	75	1	7	0.7000	−0.2833	0.0803	1
6	80	2	7	0.5833	7	80	1	8	0.8000	−0.2167	0.0470	3
				0.5833	8	90	1	9	0.9000	−0.3167	0.1003	1
7	95	1	8	0.6667					0.9000	−0.2333	0.0544	1
				0.6667	9	100	1	10	1.0000	−0.3333	0.1111	1
8	105	1	9	0.7500					1.0000	−0.2500	0.0625	1
9	110	2	11	0.9167					1.0000	−0.0833	0.0069	2
10	120	1	12	1.0000					1.0000	0.0000	0.0000	1
	$n_1 = 12$					$n_2 = 10$				$\sum t_k d_k =$ −3.5334	$\sum t_k d_k^2 =$ 0.8144	

$N = 12 + 10 = 22$

$$U^2 = \frac{n_1 n_2}{N^2}\left[\sum t_k d_k^2 - \frac{(\sum t_k d_k)^2}{N}\right]$$

$$= \frac{(12)(10)}{22^2}\left[0.8144 - \frac{(-3.5334)^2}{22}\right]$$

$$= 0.0612$$

$U^2_{0.05,10,12} = 0.2246$

Do not reject H_0.

$P > 0.50$

Multisample Testing. If the data are in groups with a grouping interval larger than 10°, then an $r \times c$ contingency table analysis may be performed, for r samples in c groups. See Section 6.1 for the computation procedure.

25.5 TWO-SAMPLE AND MULTISAMPLE TESTING OF ANGULAR DISTANCES

Angular distance is simply the shortest distance, in angles, between two points on a circle. For example, the angular distance between 95° and 120° is 25°, between 340° and 30° is 50°, and between 190° and 5° is 175°. In general, we shall refer to the angular distance between angles a_1 and a_2 as $d_{a_1-a_2}$. (Thus, $d_{95°-120°} = 25°$, and so on.)

Angular distances are useful in drawing inferences about departures of data from a specified direction. We may observed travel directions of animals trained to travel in a particular compass direction (perhaps "homeward"), or of animals faced with the odor of food coming from a specified direction. If dealing with times of day we might speak of the time of a physiological or behavioral activity in relation to the time of a particular stimulus.

If the specified angle (e.g., direction or time of day) is μ_0, we may ask whether the mean of a sample of data, $\bar{a}$, significantly departs from μ_0 by testing the one-sample hypothesis, H_0: $\mu_a = \mu_0$, as explained in Section 25.2. However, we may have two samples, sample 1 and sample 2, each of which has associated with it a specified angle of interest, μ_1 and μ_2, respectively (where μ_1 and μ_2 need not be the same). We may ask whether the angular distances for sample 1 ($d_{a_{1i}-\mu_1}$) are significantly different from those for sample 2 ($d_{a_{2i}-\mu_2}$). As shown in Example 25.10, we rank the angular distances of both samples combined and then perform a Mann-Whitney test (see Section 9.9). This was suggested by Wallraff (1979).

The procedure could be performed as a one-tailed, instead of a two-tailed test, if there were reason to be interested in whether the angular distances in one group were greater than those in the other.

Example 25.10

Two-Sample Testing of Angular Distances.

Birds of both sexes are transported away from their homes and released, with their directions of travel tabulated. The homeward direction for each sex is 135°.

H_0: Males and females orient equally well toward their homes.
H_A: Males and females do not orient equally well toward their homes.

Males			**Females**		
Direction travelled	*Angular distance*	*Rank*	*Direction travelled*	*Angular distance*	*Rank*
145°	10°	6	160°	25°	12.5
155	20	11	135	0	1
130	5	2.5	145	10	6
140	10	6	150	15	9.5
145	10	6	125	10	6
160	25	12.5	120	15	9.5
140	5	2.5			
		46.5			44.5

For the two-tailed Mann-Whitney test:

$n_1 = 7,\ R_1 = 46.5$

$n_2 = 6,\ R_2 = 44.5$

$$U = n_1 n_2 + \frac{n_1(n_1+1)}{2} - R_1 = (7)(6) + \frac{7(8)}{2} - 46.5 = 23.5$$

$U' = n_1 n_2 - U = (7)(6) - 23.5 = 18.5$

$U_{0.05(2),7,6} = U_{0.05(2),6,7} = 36$

Do not reject H_0.

$P > 0.20$

Multisample Testing. If more than two samples are involved, then the angular deviation of all of them are pooled and ranked, whereupon the Kruskal-Wallis test may be applied (Section 11.4), followed if necessary by nonparametric multiple comparison testing (Section 12.6).

25.6 TWO-SAMPLE AND MULTISAMPLE TESTING OF ANGULAR DISPERSION

The Wallraff (1979) procedure of analyzing angular distances (Section 25.5) may be applied to testing for dispersion. The angular distances of concern for sample 1 are $d_{a_{1i}-\bar{a}_1}$ and those for sample 2 are $d_{a_{2i}-\bar{a}_2}$. Thus, just as measures of dispersion for linear data may refer to deviations of the data from their mean (Sections 4.2 and 4.3), here we consider the deviations of circular data from their mean.

The angular distances of the two samples are pooled; then they are ranked for application of the Mann-Whitney test, which may be employed for either two-tailed (Example 25.11) or one-tailed testing.

Example 25.11

Two-sample testing for angular dispersion.

The times of day that males and females are born are tabulated. The mean time of day for each sex is determined (to the nearest 5 min) as in Section 24.4. (For males, $\bar{a}_1 = 7{:}55$ AM; for females, $\bar{a}_2 = 8{:}15$ AM.)

H_0: The times of day of male births are as variable as the times of day of female births.
H_A: The times of day of male births do not have the same variability as the times of day of female births.

Male			Female		
Time of day	*Angular distance*	*Rank*	*Time of day*	*Angular distance*	*Rank*
05:10 hr	2:45 hr	11	08:15 hr	0:00 hr	1
06:30	1:25	4	10:20	2:05	8.5
09:40	1:45	6	09:45	1:30	5
10:20	2:25	10	06:10	2:05	8.5
04:20	3:35	13	04:05	4:10	14
11:15	3:20	12	07:50	0:25	2
			09:00	0:45	3
			10:10	1:55	7
		$R_1 = 56$			$R_2 = 49$

For the two-tailed Mann-Whitney test:

$n_1 = 6,\ R_1 = 56$
$n_2 = 8,\ R_2 = 49$

$$U = n_1 n_2 + \frac{n_1(n_1+1)}{2} - R_1 = (6)(8) + \frac{6(7)}{2} - 56 = 13$$

$$U' = n_1 n_2 - U = (6)(8) - 13 = 35$$

$U_{0.05(2),6,8} = 40$
Do not reject H_0.
$P = 0.20$

Multisample Testing. If one wishes to compare the dispersions of more than two samples, then the above Mann-Whitney procedure may be expanded by using the Kruskal-Wallis test (Section 11.4), followed if necessary by nonparametric multiple comparisons (Section 12.6).

25.7 PARAMETRIC ONE-SAMPLE SECOND-ORDER ANALYSIS OF ANGLES

A set of n angles, a_i, has a mean angle, $\bar{a}$, and an associated mean vector length, r. This set of data may be referred to as a *first-order sample.* A set of k such means may be referred to as a *second-order sample.* Section 24.9 discussed the computation of the mean of a second-order sample, namely the mean of a set of means. We now wish to test the statistical significance of a mean of means.

For a second-order sample of k mean angles, we can obtain $\bar{X}$ with either Equation 24.21 or 24.23, and $\bar{Y}$ with either Equation 24.22 or 24.24. Assuming that the second-order sample comes from a bivariate normal distribution (i.e., a population in which the X_j's follow a normal distribution, and the Y_j's are normally distributed), a testing procedure due to Hotelling (1931) may be applied.

The sums of squares and crossproducts of the k means are

$$\sum x^2 = \sum X_j^2 - \frac{(\sum X_j)^2}{k}, \tag{25.14}$$

$$\sum y^2 = \sum Y_j^2 - \frac{(\sum Y_j)^2}{k}, \tag{25.15}$$

and

$$\sum xy = \sum X_j Y_j - \frac{\sum X_j \sum Y_j}{k}, \tag{25.16}$$

where $\sum$ in each instance refers to a summation over all k means (i.e., $\sum = \sum_{j=1}^{k}$).

Then, we can test the null hypothesis that there is no mean direction (i.e., H_0: $\rho = 0$) in the population from which the second-order sample came by using as a test statistic

$$F = \frac{k(k-2)}{2}\left[\frac{\bar{X}^2 \sum y^2 - 2\bar{X}\bar{Y} \sum xy + \bar{Y}^2 \sum x^2}{\sum x^2 \sum y^2 - (\sum xy)^2}\right], \tag{25.17}$$

with the critical value being the one-tailed F with degrees of freedom of 2 and $k - 2$ (Batschelet, 1978; 1981: 144-150). This test is demonstrated in Example 25.12, using the data from Example 24.7 (and shown in Fig. 24.7).

This test assumes the data are not grouped. The assumption of bivariate normality is a serious one. Although the test appears robust against departures due to kurtosis, the test may be badly affected by departures due to extreme skewness, rejecting a true H_0 far more often than indicated by the significance level, α (Everitt, 1979; Mardia, 1970).

Example 25.12

The second-order analysis for testing the significance of the mean of the sample means in Example 24.7.

H_0: There is no mean population direction (i.e., $\rho = 0$).
H_A: There is a mean population direction (i.e., $\rho \neq 0$).

$k = 7, \quad \bar{X} = -0.52734, \quad \bar{Y} = 0.27844$
$\sum X_j = -3.69139, \quad \sum X_j^2 = 2.27959, \quad \sum x^2 = 0.33297$ (by Equation 25.14)
$\sum Y_j = 1.94906, \quad \sum Y_j^2 = 0.86474, \quad \sum y^2 = 0.32205$ (by Equation 25.15)
$\sum X_j Y_j = -1.08282, \quad \sum xy = -0.05500$ (by Equation 25.16)

$$F = \frac{7(7-2)}{2}\left[\frac{(-0.52734)^2(0.32205) - 2(-0.52734)(0.27844)(-0.05500) + (0.27844)^2(0.33297)}{(0.33297)(0.32205) - (-0.05500)^2}\right]$$
$$= 16.66$$

$F_{0.05(1),2,5} = 5.79$
Reject H_0.
$0.005 < P < 0.01$
And, from Example 24.7, we see that the population mean angle is estimated to be 152°.

25.8 NONPARAMETRIC ONE-SAMPLE SECOND-ORDER ANALYSIS OF ANGLES

The Hotelling testing procedure of Section 25.6 requires that the k $\bar{X}$'s come from a normal distribution, as do the k $\bar{Y}$'s. Although we may assume the test to be robust to some departure from this bivariate normality, there may be considerable non-normality in a sample, in which case a nonparametric method is preferable.

Moore (1980) has provided a nonparametric modification of the Rayleigh test, which can be used to test a sample of mean angles; it is demonstrated in Example 25.13. We first rank the k vector lengths, so that r_1 is the smallest r_j and r_k is the largest. We will call the ranks i (where i ranges from 1 through k) and compute

$$X = \frac{\sum_{i=1}^{k} i \cos \bar{a}_i}{k} \tag{25.18}$$

$$Y = \frac{\sum_{i=1}^{k} i \sin \bar{a}_i}{k} \tag{25.19}$$

$$R' = \sqrt{\frac{X^2 + Y^2}{k}}. \tag{25.20}$$

The test statistic, R', is then compared to the appropriate critical value, $R'_{\alpha,n}$, in Table B.36.

Example 25.13

Nonparametric second-order analysis for significant direction in the sample of means of Example 24.7.

H_0: The population from which the sample of means came is uniformly distributed around the circle (i.e., $\rho = 0$).
H_A: The population of means is not uniformly distributed around the circle (i.e., $\rho \neq 0$).

Sample rank (i)	r_i	a_i	$i \cos \bar{a}_i$	$i \sin \bar{a}_i$
1	0.3338	171°	−0.98769	0.15643
2	0.3922	186	−1.98904	−0.20906
3	0.4696	117	−1.36197	2.67302
4	0.6962	134	−2.77863	2.87736
5	0.7747	169	−4.90814	0.95404
6	0.8794	140	−4.59627	3.85673
7	0.8954	160	−6.57785	2.39414
			−23.19959	12.70266

$$X = \frac{\sum i \cos \bar{a}_i}{k} = \frac{-23.19959}{7} = -3.31423$$

$$Y = \frac{\sum i \sin \bar{a}_i}{k} = \frac{12.70266}{7} = 1.81467$$

$$R' = \sqrt{\frac{X^2 + Y^2}{k}} = \sqrt{\frac{(-3.31423)^2 + (1.81467)^2}{7}} = \sqrt{2.03959} = 1.428$$

$R'_{0.05,7} = 1.150$

Therefore, reject H_0.

$P < 0.002$

Testing Weighted Angles. The Moore modification of the Rayleigh test can also be used when we have a sample of angles, each of which is weighted. We may then perform the ranking of the angles by the weights, instead of by the vector lengths, r_j. For example, the data of Example 24.2 could be ranked by the amount of leaning. Or, if we are recording the direction each of several birds flies from a release point, the weights could be the distances flown. (If the birds disappear at the horizon, then there are no weights and the Rayleigh test may be applied.)

25.9 PARAMETRIC TWO-SAMPLE SECOND-ORDER ANALYSIS OF ANGLES

Batschelet (1978, 1981: 150–154) explains how the Hotelling (1931) procedure of Section 25.7 can be extended to consider the hypothesis of equality of the means of two populations of means (assuming each population to be bivariate normal). We proceed as in Section 25.7, obtaining an $\bar{X}$ and $\bar{Y}$ for each of the two samples ($\bar{X}_1$ and $\bar{Y}_1$ for sample 1, and $\bar{X}_2$ and $\bar{Y}_2$ for sample 2). Then, we apply Equations 25.14, 25.15, and 25.16 to each of the two samples, obtaining $(\sum x^2)_1$, $(\sum xy)_1$, and $(\sum y^2)_1$ for sample 1, and $(\sum x^2)_2$, $(\sum xy)_2$, and $(\sum y^2)_2$ for sample 2. The two sample sizes, k_1 and k_2, do not have to be equal.

Then we calculate

$$(\textstyle\sum x^2)_c = (\sum x^2)_1 + (\sum x^2)_2; \qquad (25.21)$$

$$(\textstyle\sum y^2)_c = (\sum y^2)_1 + (\sum y^2)_2; \qquad (25.22)$$

$$(\textstyle\sum xy)_c = (\sum xy)_1 + (\sum xy)_2; \qquad (25.23)$$

and the null hypothesis of the two population mean angles being equal is tested by

$$F = \frac{N-3}{2\left(\frac{1}{k_1}+\frac{1}{k_2}\right)}\left[\frac{(\bar{X}_1-\bar{X}_2)^2(\sum y^2)_c - 2(\bar{X}_1-\bar{X}_2)(\bar{Y}_1-\bar{Y}_2)(\sum xy)_c + (\bar{Y}_1-\bar{Y}_2)^2(\sum x^2)_c}{(\sum x^2)_c(\sum y^2)_c - (\sum xy)_c^2}\right], \quad (25.24)$$

where $N = k_1 + k_2$, and F is one-tailed with 2 and $N - 3$ degrees of freedom. This test is shown in Example 25.14, using the data of Fig. 25.2.

The two-sample Hotelling test is robust to departures from the normality assumption (far more so than is the one-sample test of Section 25.7), the effect of nonnormality being slight conservatism (i.e., rejecting a false H_0 a little less frequently than indicated by the significance level, α) (Everitt, 1979). The two samples should all be of the same size, but departure from this assumption does not appear to have serious consequences (Batschelet, 1981: 202).

Example 25.14

Parametric two-sample second-order analysis for testing the difference between mean angles.

We have two samples, each consisting of mean directions and vector lengths, as shown in Fig. 25.2.

Sample 1 is the data from Examples 24.6 and 25.12, where

$k_1 = 7$; $\bar{X}_1 = -0.52734$; $\bar{Y}_1 = 0.27844$; $\bar{a}_1 = 152°$;
$(\sum x^2)_1 = 0.33297$; $(\sum y^2)_1 = 0.32205$; $(\sum xy)_1 = -0.05500$.

Sample 2 consists of the following 10 data:

j	$\bar{a}_j$	r_j
1	115°	0.9394
2	127	0.6403
3	143	0.3780
4	103	0.6671
5	130	0.8210
6	147	0.5534
7	107	0.8334
8	137	0.8139
9	127	0.2500
10	121	0.8746

Applying the procedures of Examples 24.7 and 25.12, we find:

$k_2 = 10$; $\sum r_j \cos a_j = -3.66655$; $\sum r_j \sin a_j = 5.47197$;
$\bar{X}_2 = -0.36660$; $\bar{Y}_2 = 0.54720$; $\bar{a}_2 = 124°$.
$(\Sigma x^2)_2 = 0.20897$; $(\Sigma y^2)_2 = 0.49793$; $(\Sigma xy)_2 = -0.05940$

Then, we can test

H_0: $\mu_1 = \mu_2$ (The means of the populations from which these two samples came are equal.)

H_A: $\mu_1 \neq \mu_2$ (The two population means are not equal.)

$N = 7 + 10$
$(\sum x^2)_c = 0.33297 + 0.20897 = 0.54194$
$(\sum y^2)_c = 0.32205 + 0.49793 = 0.81998$
$(\sum xy)_c = -0.05500 + (-0.05940) = -0.11440$

Using Equation 25.24:

$$F = \frac{(17-3)}{2\left(\frac{1}{7}+\frac{1}{10}\right)}\left[\frac{\begin{array}{r}[-0.52734-(-0.36660)]^2(0.81998)\\ -\,2[-0.52734-(-0.36660)](0.27844-0.54720)(-0.11440)\\ +\,(0.27844-0.54720)^2(0.54194)\end{array}}{(0.54194)(0.81998)-(-0.11440)^2}\right]$$

$= 4.69$

$F_{0.05(1),2,14} = 3.74$

Reject H_0.

$0.025 < P < 0.05$

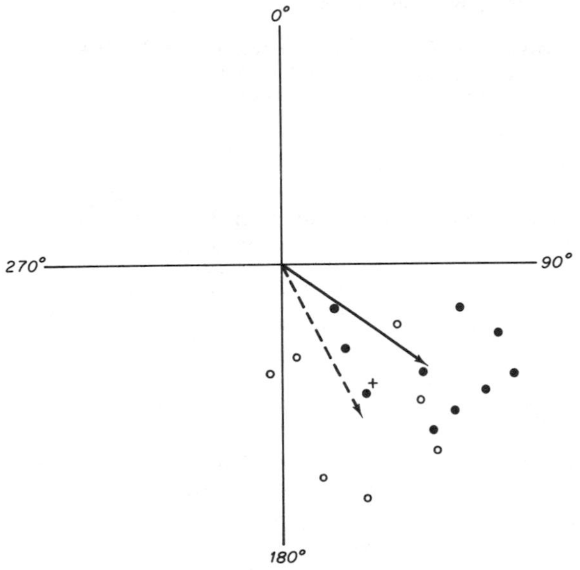

Figure 25.2 The data of Example 25.14. The open circles indicate the ends of the 7 mean vectors of sample 1 (also shown in Fig. 24.7), with the mean of these 7 indicated by the broken-line vector. The solid circles indicate the 10 data of sample 2, with their mean shown as a solid-line vector. (The "+" indicates the grand mean of all 17 data, which is used in Example 25.15.)

25.10 NONPARAMETRIC TWO-SAMPLE SECOND-ORDER ANALYSIS OF ANGLES

The parametric test of Section 25.9 is based on sampled populations being bivariate normal and the two populations having variances and covariances in common, unlikely assumptions to be strictly satisfied in practice. While the test is rather robust to these assumptions, one may be more comfortable in many cases employing a nonparametric test to assess whether two second-order populations have the same directional orientation.

Batschelet (1978; 1981: 154–156) presents the following nonparametric procedure (suggested by Mardia, 1967) as an alternative to the Hotelling test of Section 25.9. First we compute the grand mean vector, pooling all data from both samples. Then, the X coordinate of the grand mean is subtracted from the X coordinate of each of the data in both samples, and the Y of the grand mean is subtracted from the Y of each of the data. (This maneuver determines the direction of each datum from the grand mean.) As shown in Example 25.15, the resulting vectors are then tested by a nonparametric two-sample test (as in Section 25.4). This procedure requires that the data not be grouped.

Example 25.15

Nonparametric two-sample second-order analysis, using the data of Example 25.14.

H_0: The two samples came from the same population, or from two populations with the same directions.

H_A: The two samples did not come from the same population, nor from two populations with the same directions.

$$\text{Total number of vectors} = 7 + 10 = 17$$

To determine the grand mean vector (which is shown in Fig. 25.2):

$$\sum r_j \cos a_j = (-3.69139) + (-3.66655) = -7.35794$$
$$\sum r_j \sin a_j = 1.94906 + 5.47197 = 7.42103$$
$$\bar{X} = \frac{-7.35794}{17} = -0.43282$$
$$\bar{Y} = \frac{7.42103}{17} = 0.43653$$

$\bar{X}$ and $\bar{Y}$ are all that we need to define the grand mean; however, if we wish we can also determine the length and direction of the grand mean vector:

$$r = \sqrt{\bar{X}^2 + \bar{Y}^2} = \sqrt{(-0.43282)^2 + (0.43653)^2} = 0.61473$$
$$\cos \bar{a} = \frac{-0.43282}{0.61473} = -0.70408$$
$$\sin \bar{a} = \frac{0.43653}{0.61473} = 0.71012$$
$$\bar{a} = 135°$$

Returning to the hypothesis test, we subtract the above $\bar{X}$ from the X, and the $\bar{Y}$ from the Y, for each of the 17 data, arriving at 17 new vectors, as follows:

Sample 1

Datum	X	$X - \bar{X}$	Y	$Y - \bar{Y}$	*New a*
1	−0.84140	−0.40858	0.30624	−0.13029	184°
2	−0.76047	−0.32765	0.14782	−0.28871	210
3	−0.21319	0.21963	0.41842	−0.01811	20
4	−0.67366	−0.24084	0.56527	0.12874	137
5	−0.39005	0.04277	−0.04100	−0.47753	276
6	−0.48293	−0.05011	0.50009	0.06356	107
7	−0.32969	0.10313	0.05222	−0.38431	290

Sample 2

Datum	X	$X - \bar{X}$	Y	$Y - \bar{Y}$	*New a*
1	−0.39701	0.03581	0.85139	0.41485	86°
2	−0.38534	0.04748	0.51137	0.07484	75
3	−0.30188	0.13084	0.22749	−0.20904	320
4	−0.15006	0.28276	0.65000	0.21347	48
5	−0.52773	−0.09491	0.62892	0.19239	108
6	−0.46412	−0.03130	0.30140	−0.13513	230
7	−0.24366	0.18916	0.79698	0.36045	68
8	−0.59525	−0.16243	0.55508	0.11855	127
9	−0.15045	0.28237	0.19966	−0.23687	334
10	−0.45045	−0.01763	0.74968	0.31315	92

Now, using Watson's two-sample test (Section 25.4) on these new angles:

	Sample 1			Sample 2			
i	a_{1i}	i/n_1	j	a_{2j}	j/n_2	d_k	d_k^2
		0.0000	1	37°	0.1000	−0.1000	0.0100
		0.0000	2	58	0.2000	−0.2000	0.0400
		0.0000	3	62	0.3000	−0.3000	0.0900
		0.0000	4	85	0.4000	−0.4000	0.1600
		0.0000	5	93	0.5000	−0.5000	0.2500
		0.0000	6	116	0.6000	−0.6000	0.3600
1	128°	0.1429			0.6000	−0.4571	0.2089
		0.1429	7	144	0.7000	−0.5571	0.3104
2	152	0.2857			0.7000	−0.4143	0.1716
3	198	0.4286			0.7000	−0.2714	0.0737
4	221	0.5716			0.7000	−0.1284	0.0165
		0.5716	8	257	0.8000	−0.2284	0.0522
5	275	0.7143			0.8000	−0.0857	0.0073
6	285	0.8571			0.8000	0.0571	0.0033
		0.8571	9	302	0.9000	−0.0429	0.0018
		0.8571	10	320	1.0000	−0.1429	0.0204
7	355	1.0000			1.0000	0.0000	0.0000
	$n_1 = 7$			$n_2 = 10$		$\Sigma d_k =$ −4.3811	$\Sigma d_k^2 =$ 1.7761

$$U^2 = \frac{n_1 n_2}{N^2}\left[\sum d_k^2 - \frac{(\sum d_k)^2}{N}\right]$$

$$= \frac{(7)(10)}{17^2}\left[1.7761 - \frac{(-4.3811)^2}{17}\right]$$

$$= 0.1567$$

$U^2_{0.05,7,10} = 0.1866$

Do not reject H_0.

$0.05 < P < 0.10$

25.11 PARAMETRIC PAIRED-SAMPLE TESTING WITH ANGLES

The paired-sample experimental design was introduced in Chapter 10 for linear data, and Section 10.1 showed how the analysis of two samples having paired data could be reduced to a one-sample test employing the differences between members of pairs.

Circular data in two samples might also be paired, in which case the one-sample Hotelling test of Section 25.7 may be used after forming a single sample of data from the differences between the paired angles. If a_{ij} is the jth angle in the ith sample, then a_{1j} and a_{2j} are a pair of data. A single set of rectangular coordinates, X's and Y's, is formed by computing

$$X_j = \cos a_{2j} - \cos a_{1j} \qquad (25.25)$$

and

$$Y_j = \sin a_{2j} - \sin a_{1j}. \qquad (25.26)$$

Then the procedure of Section 25.7 may be applied, as shown in Example 25.16 (where k is the number of pairs).

Example 25.16

The Hotelling test for paired samples of angles.

Ten birds are marked for individual identification, and we record on which side of a tree each bird sits to rest in the morning and in the afternoon. We wish to test the following.

H_0: The side of a tree on which birds sit is the same in the morning and in the afternoon.
H_A: The side of a tree on which birds sit is not the same in the morning and in the afternoon.

	Morning			Afternoon			Difference	
Bird (j)	*Direction* (a_{1j})	$\sin a_{1j}$	$\cos a_{1j}$	*Direction* (a_{2j})	$\sin a_{2j}$	$\cos a_{2j}$	Y_j	X_j
1	105°	0.9659	−0.2588	205°	−0.4226	−0.9063	−1.3885	−0.6475
2	120	0.8660	−0.5000	210	−0.5000	−0.8660	−1.3660	−0.3660
3	135	0.7071	−0.7072	235	−0.8192	−0.5736	−1.5263	0.1336
4	95	0.9962	−0.0872	245	−0.9063	−0.4226	−1.9025	−0.3354
5	155	0.4226	−0.9063	260	−0.9848	−0.1736	−1.4074	0.7327
6	170	0.1736	−0.9848	255	−0.9659	−0.2588	−1.1395	0.7260
7	160	0.3420	−0.9397	240	−0.8660	−0.5000	−1.2080	0.4397
8	155	0.4226	−0.9063	245	−0.9063	−0.4226	−1.3289	0.4837
9	120	0.8660	−0.5000	210	−0.5000	−0.8660	−1.3660	−0.3660
10	115	0.9063	−0.4226	200	−0.3420	−0.9397	−1.2483	−0.5171

$k = 10$
$\bar{X} = 0.0284$
$\bar{Y} = -1.3881$

Using Equations 25.14 through 25.17:

$$\sum x^2 = 2.5761 - \frac{(0.2837)^2}{10} = 2.5681$$

$$\sum y^2 = 19.6717 - \frac{(-13.8814)^2}{10} = 0.4023$$

$$\sum xy = -0.0538 - \frac{(0.2837)(-13.8814)}{10} = 0.3400$$

$$F = \frac{10(10-2)}{2}\left[\frac{(0.0284)^2(0.4023) - 2(0.0284)(-1.3881)(-0.3400) + (-1.3881)^2(2.5681)}{(2.5681)(0.4023) - (-0.3400)^2}\right]$$

$= 217$

$F_{0.05(1),2,8} = 4.46$

Reject H_0.

$P \ll 0.0005$

Second-Order Data. If each member of a pair of data is a mean angle ($\bar{a}$) from a sample, with an associated vector length (r), then we are dealing with a second-order analysis. The above Hotelling test may be applied if the following computations are used in place of Equations 25.25 and 25.26, respectively:

$$X_j = r_{2j}\cos\bar{a}_{2j} - r_{1j}\cos\bar{a}_{1j} \quad (25.27)$$

$$Y_j = r_{2j}\sin\bar{a}_{2j} - r_{1j}\sin\bar{a}_{1j}. \quad (25.28)$$

25.12 NONPARAMETRIC PAIRED-SAMPLE TESTING WITH ANGLES

Circular data in a paired-sample experimental design may be tested nonparametrically by forming a single sample of the paired differences, which can then be subjected to the Moore test of Section 25.8. We calculate rectangular coordinates (X_j and Y_j) for each paired difference, as done in Equations 25.25 and 25.26. Then, for each of the j paired differences, we compute

$$r_j = \sqrt{X_j^2 + Y_j^2}, \tag{25.29}$$

$$\cos a_j = \frac{X_j}{r_j}, \tag{25.30}$$

$$\sin a_j = \frac{Y_j}{r_j}. \tag{25.31}$$

Then the values of r_j are ranked, with ranks (i) running from 1 through n, and we complete the analysis using Equations 25.18, 25.19, and 25.20, substituting n for k. The procedure is demonstrated in Example 25.17.

Second-Order Data. If each member of a pair of circular-scale data is a mean angle, a_j, with an associated vector length, r_j, then we modify the above analysis. Calculate X_j and Y_j by Equations 25.27 and 25.28, respectively, instead of by Equations 25.25 and 25.26, respectively. Then apply Equations 25.29 through 25.31 and Equations 25.18 through 25.20 to complete the analysis (using k as the number of paired means).

25.13 ANGULAR CORRELATION AND REGRESSION

The correlation of two variables, each measured on a linear scale, is discussed in Chapter 19, with linear-scale regression being introduced in Chapter 17. Correlation involving angular data may be of two kinds: Either both variables are measured on a circular scale (a situation sometimes termed "angular-angular," or "spherical," correlation), or one variable is on a circular scale with the other measured on a linear scale (sometimes called an "angular-linear," or "cylindrical," correlation). The study of biological rhythms deals essentially with the rhythmic dependence (i.e., regression) of a linear-scale variable (e.g., a measure of biological activity, such as body temperature) on a circular-scale variable (namely, time).

Numerous correlation measures have been proposed for dealing with angular data. Unfortunately, they have thus far each had drawbacks and none can be recommended enthusiastically. (For example, they may not distinguish between a positive and a negative relationship between variables, or the correlation coefficient may not be 1.0 when the correlation is perfect, or there may not be an adequate significance test for the correlation coefficient, or they may not be applicable to all sets of circular data.) This is a very young area in statistical method, and we can look forward to improvements in the future.

Example 25.17

The Moore test for paired data on a circular scale of measurement.

Ten birds are marked for individual identification, and we record on which side of a tree each bird sits to rest in the morning and in the afternoon. We wish to test the following:

H_0: The side of a tree on which birds sit is the same in the morning and in the afternoon.
H_A: The side of a tree on which birds sit is not the same in the morning and in the afternoon.

	Morning			Afternoon			Difference					
Bird (j)	Direction (a_{1j})	$\sin a_{1j}$	$\cos a_{1j}$	Direction (a_{2j})	$\sin a_{2j}$	$\cos a_{2j}$	Y_j	X_j	r_j	$\sin a_j$	$\cos a_j$	Rank of r_j (i)
1	105°	0.9659	−0.2588	205°	−0.4226	−0.9063	−1.3885	−0.6475	1.5321	−0.9063	−0.4226	7.5
2	120	0.8660	−0.5000	210	−0.5000	−0.8660	−1.3660	−0.3660	1.4142	−0.9659	−0.2588	4.5
3	135	0.7071	−0.7071	235	−0.8192	−0.5736	−1.5263	0.1335	1.5321	−0.9962	0.0871	7.5
4	95	0.9962	−0.0872	245	−0.9063	−0.4226	−1.9025	−0.3354	1.9318	−0.9848	−0.1736	10
5	155	0.4226	−0.9063	260	−0.9848	−0.1736	−1.4074	0.7327	1.5867	−0.8870	0.4618	9
6	170	0.1736	−0.9848	255	−0.9659	−0.2588	−1.1395	0.7260	1.3511	−0.8434	0.5373	2
7	160	0.3420	−0.9397	240	−0.8660	−0.5000	−1.2080	0.4397	1.2855	−0.9397	0.3420	1
8	155	0.4226	−0.9063	245	−0.9063	−0.4226	−1.3289	0.4837	1.4152	−0.9390	0.3418	6
9	120	0.8660	−0.5000	210	−0.5000	−0.8660	−1.3660	−0.3660	1.4142	−0.9659	−0.2588	4.5
10	150	0.9063	−0.4226	200	−0.3420	−0.9397	−1.2483	−0.5171	1.3512	−0.9238	−0.3827	3

$n = 10$

$$X = \frac{\sum_{i=1}^{n} i \cos a_i}{n} = \frac{1(0.3420) + 2(0.5373) + \cdots + 10(-0.1736)}{10} = -0.0106$$

$$Y = \frac{\sum_{i=1}^{n} i \sin a_i}{n} = \frac{1(-0.9397) + 2(-0.8434) + \cdots + 10(-0.9848)}{10} = -5.1825$$

$$R' = \sqrt{\frac{\bar{X}^2 + \bar{Y}^2}{n}} = 1.639$$

$R'_{0.05,10} = 1.144$

Reject H_0.

$P < 0.002$

Angular-Linear Correlation. One of the best methods available for correlating a circular with a linear variable is that of Mardia (1976), with Liddell and Ord (1978) providing an improved significance test. Mardia (1976) also presents a nonparametric coefficient of cylindrical correlation. Batschelet (1981: 191–196) reviews available methods.

Angular-Angular Correlation. Mardia and Puri (1978) describe a spherical correlation coefficient that is one of the best yet proposed; and Mardia (1975) presented a nonparametric coefficient for angular-angular correlation. A review of available methods is given by Batschelet (1981: 184–190).

Rhythmometry. The description of biological rhythms may be thought of as a regression (often called *periodic regression*) of the linear variable on time (a circular variable). Excellent discussion of such regression is provided by Batschelet (1972; 1974; 1981: Chapter 8), Bliss (1970: Chapter 17), Bloomfield (1976), and Nelson et al. (1979).

The *period*, or length, of the cycle,* is often stated in advance. Parameters to be estimated in the regression are the *amplitude* of the rhythm (which is the range from the minimum to the maximum value of the linear variable)† and the *phase angle*, or *acrophase*, of the cycle (which is the point on the circular time scale at which the linear variable is maximum). If the period is also a parameter to be estimated, then the situation is more complex and one may resort to the broad area of *time series analysis*. Some biological rhythms can be fitted, by least square regression, by a sine (or cosine) curve; but if the rhythm does not conform well to such a symmetrical functional relationship, then a "harmonic analysis" (also called a "Fourier analysis") may be employed.

25.14 SERIAL RANDOMNESS OF NOMINAL SCALE CATEGORIES ON A CIRCLE

When dealing with the occurrence of members of two nominal scale categories along a linear space or time, the runs test of Section 23.6 is appropriate. A runs test is also available for spatial or temporal measurements that are on a circular scale (Asano, 1965; Mardia, 1972: 203–204). This test may also be employed as a two-sample test, but the tests of Sections 25.3 and 25.4 are more powerful for that purpose; the circular runs test is best reserved for testing the hypothesis of random distribution of members of two categories around a circle.

We define a run on a circle as a sequence of like elements, bounded on each side

*A rhythm with one cycle every 24 hours is said to be "circadian" (from the Latin "circa," meaning "about" and "diem," meaning "day"); a rhythm with a 7-day period is said to be "circaseptan"; a rhythm with a 14-day period is "circadiseptan"; one with a period of 1 year is "circannual" (Halberg and Lee, 1974).

†The amplitude is often defined as half this range.

by unlike elements.* Similar to Section 23.6, we let n_1 be the total number of elements in the first category, n_2 the number of elements in the second category, and u' the number of runs in the entire sequence of elements. In Example 25.18 there are 7 members of species A and 10 members of species B, and there are 6 runs. The critical value, $u'_{\alpha(2),n_1,n_2}$, is found in Table B.37 for cases where n_1 and n_2 are no greater than 20. If u' is $\leq$ the first of the pair of critical values, or $\geq$ the second of the pair, then H_0 is rejected. If either n_1 or n_2 is larger than 20, this table cannot be used. For such large samples, the following statistic may be considered, approximately, to be from a normal distribution (Mardia, 1972: 203):

$$Z = \frac{u' + 1 - \dfrac{2n_1n_2 + N}{N}}{\sqrt{\dfrac{2n_1n_2(2n_1n_2 - N)}{N^2(N-1)}}}, \tag{25.32}$$

where $N = n_1 + n_2$.

Example 25.18
The two-tailed runs test on a circle.

Members of the two antelope species of Example 23.8 (referred to as species A and B) are observed drinking on the shore of a pond in the following sequence:

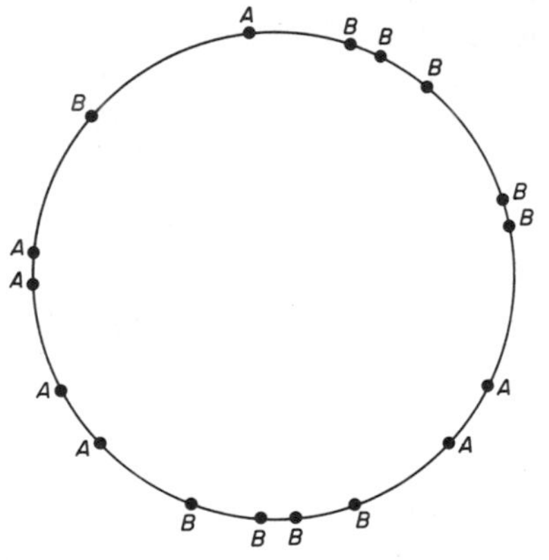

H_0: The distribution of members of the two species around the pond is random.
H_A: The distribution of members of the two species around the pond is not random.

$n_1 = 7, \quad n_2 = 10, \quad u' = 6$
$u'_{0.05(2),7,10} = 4$ and 14 (from Table B.37)
As 6 is neither ≤ 4 nor ≥ 14, do not reject H_0.
$P = 0.50$

*For the runs test on a linear scale (Section 23.6), the number of runs may be even or odd; however, on a circle the number of runs is always even.

One-Tailed Testing. For one-tailed testing we use the critical values of $u'_{\alpha(1),n_1,n_2}$ in Table B.37. We can test specifically whether the population is nonrandom due to clustering (also known as contagion) in the following manner. We state H_0: The elements in the population are not distributed contagiously and H_A: The elements in the population are distributed contagiously; and if $u' \leq$ the first member of the pair of critical values then H_0 is rejected. In using the normal approximation for these hypotheses, H_0 is rejected if $Z \leq -Z_{\alpha(1)}$.

If our interest is only in whether the two groups of elements are distributed uniformly in the population, then we hypothesize H_0: The elements in the population are not distributed uniformly, and H_A: The elements in the population are distributed uniformly; and if $u' \geq$ the second of the pair of one-tailed critical values then H_0 is rejected. In using the normal approximation, H_0 is rejected if $Z \geq Z_{\alpha(1)}$.

EXERCISES

25.1. Consider the data of Exercise 24.1. Test the null hypothesis that the population is distributed randomly around the circle (i.e., $\rho = 0$).

25.2. Consider the data of Exercise 24.2. Test the null hypothesis that time of birth is distributed randomly around the clock (i.e., $\rho = 0$).

25.3. Trees are planted in a circle to surround a cabin and protect it from prevailing west (i.e., 270°) winds. The trees suffering the greatest wind damage are the 11 at the following directions. **(a)** Using the V test, test the null hypothesis that tree damage is independent of wind direction. **(b)** Test H_0: $\rho = 0$ vs. H_A: $\rho \neq 0$ and $\mu_a = 270°$.

285°	295°	335°
240	275	260
280	310	300
255	260	

25.4. The direction of the spring flight of a certain bird species was recorded as follows in eight individuals released in full sunlight and seven individuals released under overcast skies:

Sunny	*Overcast*
350°	340°
340	305
315	255
10	270
20	305
355	320
345	335
360	

Using the Watson-Williams test, test the null hypothesis that the mean flight direction in this species is the same under both cloudy and sunny skies.

25.5. Using the data of Example 25.4, test nonparametrically the hypothesis that birds of the species under consideration fly in the same direction under sunny as well as under cloudy skies.

25.6. Times of arrival at a feeding station of members of three species of hummingbird were recorded as follows:

Species 1	*Species 2*	*Species 3*
05:40 hr	05:30 hr	05:35 hr
07:15	07:20	08:10
09:00	09:00	08:15
11:20	09:40	10:15
15:10	11:20	14:20
17:25	15:00	15:35
	17:05	16:05
	17:20	
	17:40	

Test the null hypothesis that members of all three species have the same mean time of visiting the feeding station.

25.7. For the data in Exercise 25.4, the birds were released at a site from which their home lies due north (i.e., in a compass direction of 0°). Test whether birds orient homeward better under sunny skies than under cloudy skies.

25.8. For the data in Exercise 25.4, test whether the variability in flight direction is the same under both sky conditions.

25.9. Consider the data of Exercise 24.3. Test for significance of the mean of the five sample means.

25.10. Nonparametrically test whether the data of Exercise 24.3 came from a population uniformly distributed around the circle.

25.11. Test for significant difference between the mean of the data of Exercise 24.3 and the mean of the data below.

j	a_j	r_j
1	265°	0.5283
2	270	0.4119
3	240	0.6086
4	250	0.8402
5	240	0.7436
6	235	0.7145
7	260	0.6459
8	270	0.7481

25.12. Perform the nonparametric analogue of the test in Exercise 25.11.

25.13. Eight men (M) and eight women (W) are asked to sit around a circular conference table; they did so in the following configuration (see figure). Test whether there is evidence that members of the same sex tend to sit next to each other.

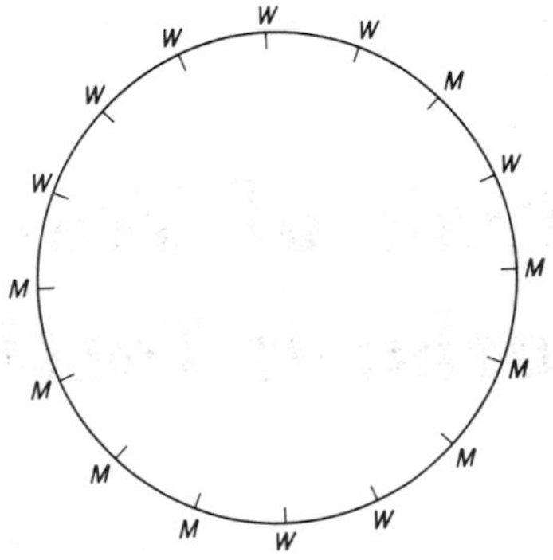

Appendix A

Analysis of Variance Hypothesis Testing

When an analysis of variance experimental design has more than one factor, then the appropriate F values (and degrees of freedom) depend upon which of the factors are fixed and which are random effects (see Section 13.1), and which factors (if any) are nested within which (see Chapter 16). Given below in Section A.1 is a procedure that enables us to determine the appropriate F's and DF's for a given experimental design. It is a simplification of the procedures outlined by Bennett and Franklin (1954: 413–415), Dunn and Clark (1974: 215–218), Hicks (1973: 177–180), Kirk (1968: 208–214), Scheffé (1959: 284–289), and Winer (1971: 371–375) and is the method used to produce the contents of Sections A.2 through A.4. If we have one of the experimental designs of Sections A.2–A.4, the F's and DF's are given therein; if the design is not found in those pages, then the procedures of Section A.1 may be used to determine the F's and DF's appropriate to the ANOVA.

A.1 DETERMINATION OF APPROPRIATE F'S AND DEGREES OF FREEDOM

We use as an example an analysis of variance design with three factors, A, B, and C, where factor A is a random effect, factor B is a fixed effect, and factor C is random and nested within combinations of factors A and B. The following steps are followed, and a table is prepared as indicated.

1. Assign to each factor a unique letter and subscript. If a factor is nested within one or more other factors, place the subscript(s) of the latter factor(s) in parentheses. For the present example, we would write A_i, B_j, $C_{l(ij)}$.

2. Prepare a table as follows.
   A. Row headings are the factors with their subscripts, the factor interactions (if any) with their subscripts, and a row for error (i.e., within cells) headed "e" with all factor subscripts in parentheses. Factors, interactions, and error will be referred to collectively as "effects."
   B. Column headings are the factor subscripts, with an indication of whether each is associated with a fixed effects factor or a random effects factor.

   For our example:

Effect	*Random* i	*Fixed* j	*Random* l
A_i			
B_j			
$C_{l(ij)}$			
AB_{ij}			
$e_{(ijl)}$			

3. The body of the table is filled in as follows:
   A. Examine each column corresponding to a fixed effects factor. (In our example, only column j is examined.) For each such column, enter a "0" in each row that the column subscript appears outside parentheses in the row subscript. (In our example, enter "0" in rows B_j and AB_{ij} of column j.)
   B. Enter "1" in every other position in the table.

   For our example:

Effect	*Random* i	*Fixed* j	*Random* l
A_i	1	1	1
B_j	1	0	1
$C_{l(ij)}$	1	1	1
AB_{ij}	1	0	1
$e_{(ijl)}$	1	1	1

4. For each row, list all effects that contain all the subscripts of the row heading. In our example:

Effect	*Random* i	*Fixed* j	*Random* l	*Effect list*
A_i	1	1	1	$A + C + AB + e$
B_j	1	0	1	$B + C + AB + e$
$C_{l(ij)}$	1	1	1	$C + e$
AB_{ij}	1	0	1	$AB + e$
$e_{(ijl)}$	1	1	1	e

5. For each row, examine the list of factors and interactions just prepared, as follows:
   A. Ignore those columns headed by the subscripts in the row heading.

B. Locate the table row corresponding to each factor or interaction in the list; if there is a zero in that row (ignoring the appropriate columns), then delete that factor or interaction from the list.

In our example, we examine row A_i by ignoring column i. Our factor and interaction list contains A, C, AB, and e. Rows A_i, C_i, and $e_{(ijl)}$ contain no zeroes, but row AB_{ij} has a zero. Therefore, we delete AB from our list for row A_i.

We examine row B_j by ignoring column j. Our list of factors and interactions contains B, C, AB, and e. None of the rows contain zeroes, so we retain the entire list.

We similarly examine row $C_{l(ij)}$, by ignoring columns i, j, and l; row AB_{ij}, by ignoring rows i and j; and row $e_{(ijl)}$, by ignoring columns i, j, and l. No items are deleted from the factor and interaction lists for these rows.

As a result we have the following:

Effect	*Effect list*
A_i	$A + C + e$
B_j	$B + C + AB + e$
$C_{l(ij)}$	$C + e$
AB_{ij}	$AB + e$
$e_{(ijl)}$	e

6. The appropriate F is determined as follows:
   A. The numerator for F is the mean square for the effect (factor or interaction) in question. The numerator degrees of freedom are the DF associated with the numerator MS.
   B. The denominator for F is the mean square for that effect having the same effect list as does the numerator effect, with the exception of not having the numerator effect in the effect list. The denominator degrees of freedom are the DF associated with the denominator MS.

To test for the significance of factor A in our example, MS_A would be the numerator. Since factor A has an effect list of A, C, e, we desire in the denominator the mean square for the effect having an effect list of C, e; therefore, the denominator is MS_C, and

$$F = \frac{\text{MS}_A}{\text{MS}_C}.$$

To test for the significance of factor C, place MS_C in the numerator of F. Since factor C has an effect list of C, e, the denominator should be the MS for the effect containing only e in its effect list. Therefore,

$$F = \frac{\text{MS}_C}{\text{MS}_e}.$$

Similarly, to test for the significance of factor interaction AB, we place MS_{AB} in the numerator of F; and, since AB has an effect list of AB, e, we use MS_e in the denominator. Therefore,

$$F = \frac{\text{MS}_{AB}}{\text{MS}_e}.$$

Occasionally, as in the present example, there is no single effect that has the effect list required in step 6B above. If this is the case, then a combination of effect lists may be considered as an approximate procedure (Satterthwaite, 1946). Thus, to test for the significance of factor B in our example, we place MS_B in the numerator of F and note that the associated effect list is B, C, AB, e, and we require a denominator MS associated with an effect list of C, AB, e. We note that if we add the effect lists for effect C (namely, C, e) and effect AB (AB, e) and then subtract that for error (e), we have the desired list, C, AB, e. Thus, we place in the denominator of F the combination of mean squares associated with the combinations of effect lists used, namely $MS_C + MS_{AB} - MS_e$. Therefore,

$$F = \frac{MS_B}{MS_C + MS_{AB} - MS_e}.$$

If a combination of mean squares is used as the denominator of F, then the denominator degrees of freedom are

$$\frac{(\text{denominator})^2}{\sum_i [(MS_i)^2/DF_i]},$$

where the summation takes place over all the mean squares in the denominator of F.

For our example, the numerator DF for testing the significance of factor B is DF_B, and the denominator DF is

$$\frac{(MS_C + MS_{AB} + MS_e)^2}{\dfrac{(MS_C)^2}{DF_C} + \dfrac{(MS_{AB})^2}{DF_{AB}} + \dfrac{(MS_e)^2}{DF_e}}.$$

In the following series of tables, mean squares for factors are denoted as MS_A, MS_B, etc.; mean squares for interaction between factors are indicated as MS_{AB}, MS_{AC}, MS_{ABC}, etc.; and the error mean square is denoted by MS_e. Degrees of freedom (DF) are indicated with the same subscripts as are the mean squares.

If an analysis of variance has only one replicate per cell, the calculation of MS_e is not possible, the quantity calculated as the "remainder" MS being the MS for the highest order interaction. If one assumes that the highest order interaction MS is insignificant, the remainder MS may be used in place of the error MS in the F calculations and other procedures where MS_e is called for.

A.2 TWO-FACTOR ANALYSIS OF VARIANCE

Factors *A* and *B* Both Fixed

Source of variation	F	ν_1	ν_2
A	MS_A/MS_e	DF_A	DF_e
B	MS_B/MS_e	DF_B	DF_e
AB	MS_{AB}/MS_e	DF_{AB}	DF_e

Factor A Fixed; Factor B Random

Source of variation	F	ν_1	ν_2
A	MS_A/MS_{AB}	DF_A	DF_{AB}
B	MS_B/MS_e	DF_B	DF_e
AB	MS_{AB}/MS_e	DF_{AB}	DF_e

Factors A and B Both Random

Source of variation	F	ν_1	ν_2
A	MS_A/MS_{AB}	DF_A	DF_{AB}
B	MS_B/MS_{AB}	DF_B	DF_{AB}
AB	MS_{AB}/MS_e	DF_{AB}	DF_e

A.3 THREE-FACTOR ANALYSIS OF VARIANCE

Factors A, B, and C All Fixed

Source of variation	F	ν_1	ν_2
A	MS_A/MS_e	DF_A	DF_e
B	MS_B/MS_e	DF_B	DF_e
C	MS_C/MS_e	DF_C	DF_e
AB	MS_{AB}/MS_e	DF_{AB}	DF_e
AC	MS_{AC}/MS_e	DF_{AC}	DF_e
BC	MS_{BC}/MS_e	DF_{BC}	DF_e
ABC	MS_{ABC}/MS_e	DF_{ABC}	DF_e

Factors A and B Fixed; Factor C Random

Source of variation	F	ν_1	ν_2
A	MS_A/MS_{AC}	DF_A	DF_{AC}
B	MS_B/MS_{BC}	DF_B	DF_{BC}
C	MS_C/MS_e	DF_C	DF_e
AB	MS_{AB}/MS_{ABC}	DF_{AB}	DF_{ABC}
AC	MS_{AC}/MS_e	DF_{AC}	DF_e
BC	MS_{BC}/MS_e	DF_{BC}	DF_e
ABC	MS_{ABC}/MS_e	DF_{ABC}	DF_e

A.4 NESTED ANALYSIS OF VARIANCE

Factor A Either Fixed or Random; Factor B Random and Nested within Factor A

Source of variation	F	ν_1	ν_2
A	MS_A/MS_B	DF_A	DF_B
B	MS_B/MS_e	DF_B	DF_e

Factor *A* Fixed; Factors *B* and *C* Random

Source of variation	*F*	ν_1	ν_2
A	$MS_A/(MS_{AB} + MS_{AC} - MS_{ABC})$	DF_A	$\dfrac{(MS_{AB} + MS_{AC} - MS_{ABC})^2}{(MS_{AB})^2/DF_{AB} + (MS_{AC})^2/DF_{AC} + (MS_{ABC})^2/DF_{ABC}}$
B	MS_B/MS_{BC}	DF_B	DF_{BC}
C	MS_C/MS_{BC}	DF_C	DF_{BC}
AB	MS_{AB}/MS_{ABC}	DF_{AB}	DF_{ABC}
AC	MS_{AC}/MS_{ABC}	DF_{AC}	DF_{ABC}
BC	MS_{BC}/MS_e	DF_{BC}	DF_e
ABC	MS_{ABC}/MS_e	DF_{ABC}	DF_e

Factors *A*, *B*, and *C* All Random

Source of variation	*F*	ν_1	ν_2
A	$MS_A/(MS_{AB} + MS_{AC} - MS_{ABC})$	DF_A	$\dfrac{(MS_{AB} + MS_{AC} - MS_{ABC})^2}{(MS_{AB})^2/DF_{AB} + (MS_{AC})^2/DF_{AC} + (MS_{ABC})^2/DF_{ABC}}$
B	$MS_B/(MS_{AB} + MS_{BC} - MS_{ABC})$	DF_B	$\dfrac{(MS_{AB} + MS_{BC} - MS_{ABC})^2}{(MS_{AB})^2/DF_{AB} + (MS_{BC})^2/DF_{BC} + (MS_{ABC})^2/DF_{ABC}}$
C	$MS_C/(MS_{AC} + MS_{BC} - MS_{ABC})$	DF_C	$\dfrac{(MS_{AC} + MS_{BC} - MS_{ABC})^2}{(MS_{AC})^2/DF_{AC} + (MS_{BC})^2/DF_{BC} + (MS_{ABC})^2/DF_{ABC}}$
AB	MS_{AB}/MS_{ABC}	DF_{AB}	DF_{ABC}
AC	MS_{AC}/MS_{ABC}	DF_{AC}	DF_{ABC}
BC	MS_{BC}/MS_{ABC}	DF_{BC}	DF_{ABC}
ABC	MS_{ABC}/MS_e	DF_{ABC}	DF_e

Factor *A* Either Fixed or Random; Factor *B* Random and Nested within Factor *A*; Factor *C* Random and Nested within Factor *B*

Source of variation	F	ν_1	ν_2
A	MS_A/MS_B	DF_A	DF_B
B	MS_B/MS_C	DF_B	DF_C
C	MS_C/MS_e	DF_C	DF_e

Factors *A* and *B* Fixed; Factor *C* Random and Nested within Factors *A* and *B*

Source of variation	F	ν_1	ν_2
A	MS_A/MS_C	DF_A	DF_C
B	MS_B/MS_C	DF_B	DF_C
AB	MS_{AB}/MS_C	DF_{AB}	DF_C
C	MS_C/MS_e	DF_C	DF_e

Factor *A* Fixed; Factor *B* Random; Factor *C* Random and Nested within Factors *A* and *B*

Source of variation	F	ν_1	ν_2
A	MS_A/MS_{AB}	DF_A	DF_{AB}
B	MS_B/MS_C	DF_B	DF_C
AB	MS_{AB}/MS_C	DF_{AB}	DF_C
C	MS_C/MS_e	DF_C	DF_e

Factors *A* and *B* Random; Factor *C* Random and Nested within Factors *A* and *B*

Source of variation	F	ν_1	ν_2
A	MS_A/MS_{AB}	DF_A	DF_{AB}
B	MS_B/MS_{AB}	DF_B	DF_{AB}
AB	MS_{AB}/MS_C	DF_{AB}	DF_C
C	MS_C/MS_e	DF_C	DF_e

Appendix B

Statistical Tables and Graphs

INTERPOLATION

In some of the statistical tables that follow (viz. Tables B.3, B.4, B.5, B.6, B.7, B.16) a critical value may be required for degrees of freedom not shown on the table, and interpolation may be used to compute an estimate of the required critical value.

Linear Interpolation. Let us say that we desire a critical value for ν degrees of freedom, where $a < \nu < b$; let us call it C_ν. We first determine the proportion $p = (\nu - a)/(b - a)$. Then, the required critical value is determined as $C_\nu = C_b + p(C_a - C_b)$.

For example, let us consider $F_{0.05(2),1,260}$, which lies between $F_{0.05(2),1,200} = 5.10$ and $F_{0.05(2),1,300} = 5.07$ in Table B.4, We calculate $p = (260 - 200)/(300 - 200) = 0.600$; then, $F_{0.05(2),1,260} = C_\nu = 5.07 + (0.600)(5.10 - 5.07) = 5.07 + 0.02 = 5.09$.

Linear interpolation cannot be used when $b = \infty$, but harmonic interpolation (below) can be.

Harmonic Interpolation. A more accurate interpolation procedure for critical value determination is one that uses the reciprocals of degrees of freedom. (Since reciprocals of large numbers are small numbers, it is good practice to use 100 times each reciprocal, and this is what will be demonstrated here.) Let us say we desire a critical value for ν degrees of freedom, where $a < \nu < b$; call it C_ν. We first determine $p = (100/a - 100/\nu)/(100/a - 100/b)$. Then, $C_\nu = C_b + (1 - p)(C_a - C_b)$.

For example, let us consider the above example of desiring $F_{0.05(2),1,260}$. We calculate $p = (100/200 - 100/260)/(100/200 - 100/300) = 0.692$; then, $F_{0.05(2),1,260} = C_v = 5.07 + (1 - 0.692)(5.10 - 5.07) = 5.07 + 0.01 = 5.08$.

Harmonic interpolation is especially useful when $b = \infty$. For example, to determine $t_{0.01(2),2800}$, which lies between $t_{0.01(2),1000} = 2.581$ and $t_{0.01(1),\infty} = 2.5758$ in Table B.3, we calculate $p = (100/1000 - 100/2800)/(100/1000 - 100/\infty) = 0.0643/0.1000 = 0.6430$. Then, $t_{0.01(2),2800} = C_v = 2.5758 + (1 - 0.6430)(2.581 - 2.5758) = 2.5758 + 0.0019 = 2.578$. (Note that $100/\infty = 0$.)

TABLE B.1 CRITICAL VALUES OF THE CHI-SQUARE DISTRIBUTION

ν | α =	0.999	0.995	0.99	0.975	0.95	0.90	0.75	0.50	0.25	0.10	0.05	0.025	0.01	0.005	0.001
1	0.000	0.000	0.000	0.001	0.004	0.016	0.102	0.455	1.323	2.706	3.841	5.024	6.635	7.879	10.828
2	0.002	0.010	0.020	0.051	0.103	0.211	0.575	1.386	2.773	4.605	5.991	7.378	9.210	10.597	13.816
3	0.024	0.072	0.115	0.216	0.352	0.584	1.213	2.366	4.108	6.251	7.815	9.348	11.345	12.838	16.266
4	0.091	0.207	0.297	0.484	0.711	1.064	1.923	3.357	5.385	7.779	9.488	11.143	13.277	14.860	18.467
5	0.210	0.412	0.554	0.831	1.145	1.610	2.675	4.351	6.626	9.236	11.070	12.833	15.086	16.750	20.515
6	0.381	0.676	0.872	1.237	1.635	2.204	3.455	5.348	7.841	10.645	12.592	14.449	16.812	18.548	22.458
7	0.599	0.989	1.239	1.690	2.167	2.833	4.255	6.346	9.037	12.017	14.067	16.013	18.475	20.278	24.322
8	0.857	1.344	1.646	2.180	2.733	3.490	5.071	7.344	10.219	13.362	15.507	17.535	20.090	21.955	26.124
9	1.152	1.735	2.088	2.700	3.325	4.168	5.899	8.343	11.389	14.684	16.919	19.023	21.666	23.589	27.877
10	1.479	2.156	2.558	3.247	3.940	4.865	6.737	9.342	12.549	15.987	18.307	20.483	23.209	25.188	29.588
11	1.834	2.603	3.053	3.816	4.575	5.578	7.584	10.341	13.701	17.275	19.675	21.920	24.725	26.757	31.264
12	2.214	3.074	3.571	4.404	5.226	6.304	8.438	11.340	14.845	18.549	21.026	23.337	26.217	28.300	32.909
13	2.617	3.565	4.107	5.009	5.892	7.042	9.299	12.340	15.984	19.812	22.362	24.736	27.688	29.819	34.528
14	3.041	4.075	4.660	5.629	6.571	7.790	10.165	13.339	17.117	21.064	23.685	26.119	29.141	31.319	36.123
15	3.483	4.601	5.229	6.262	7.261	8.547	11.037	14.339	18.245	22.307	24.996	27.488	30.578	32.801	37.697
16	3.942	5.142	5.812	6.908	7.962	9.312	11.912	15.338	19.369	23.542	26.296	28.845	32.000	34.267	39.252
17	4.416	5.697	6.408	7.564	8.672	10.085	12.792	16.338	20.489	24.769	27.587	30.191	33.409	35.718	40.790
18	4.905	6.265	7.015	8.231	9.390	10.865	13.675	17.338	21.605	25.989	28.869	31.526	34.805	37.156	42.312
19	5.407	6.844	7.633	8.907	10.117	11.651	14.562	18.338	22.718	27.204	30.144	32.852	36.191	38.582	43.820
20	5.921	7.434	8.260	9.591	10.851	12.443	15.452	19.337	23.828	28.412	31.410	34.170	37.566	39.997	45.315
21	6.447	8.034	8.897	10.283	11.591	13.240	16.344	20.337	24.935	29.615	32.671	35.479	38.932	41.401	46.797
22	6.983	8.643	9.542	10.982	12.338	14.041	17.240	21.337	26.039	30.813	33.924	36.781	40.289	42.796	48.268
23	7.529	9.260	10.196	11.689	13.091	14.848	18.137	22.337	27.141	32.007	35.172	38.076	41.638	44.181	49.728
24	8.085	9.886	10.856	12.401	13.848	15.659	19.037	23.337	28.241	33.196	36.415	39.364	42.980	45.559	51.179
25	8.649	10.520	11.524	13.120	14.611	16.473	19.939	24.337	29.339	34.382	37.652	40.646	44.314	46.928	52.620
26	9.222	11.160	12.198	13.844	15.379	17.292	20.843	25.336	30.435	35.563	38.885	41.923	45.642	48.290	54.052
27	9.803	11.808	12.879	14.573	16.151	18.114	21.749	26.336	31.528	36.741	40.113	43.195	46.963	49.645	55.476
28	10.391	12.461	13.565	15.308	16.928	18.939	22.657	27.336	32.620	37.916	41.337	44.461	48.278	50.993	56.892
29	10.986	13.121	14.256	16.047	17.708	19.768	23.567	28.336	33.711	39.087	42.557	45.722	49.588	52.336	58.301
30	11.588	13.787	14.953	16.791	18.493	20.599	24.478	29.336	34.800	40.256	43.773	46.979	50.892	53.672	59.703
31	12.196	14.458	15.655	17.539	19.281	21.434	25.390	30.336	35.887	41.422	44.985	48.232	52.191	55.003	61.098
32	12.811	15.134	16.362	18.291	20.072	22.271	26.304	31.336	36.973	42.585	46.194	49.480	53.486	56.328	62.487
33	13.431	15.815	17.074	19.047	20.867	23.110	27.219	32.336	38.058	43.745	47.400	50.725	54.776	57.648	63.870
34	14.057	16.501	17.789	19.806	21.664	23.952	28.136	33.336	39.141	44.903	48.602	51.966	56.061	58.964	65.247
35	14.688	17.192	18.509	20.569	22.465	24.797	29.054	34.336	40.223	46.059	49.802	53.203	57.342	60.275	66.619
36	15.324	17.887	19.233	21.336	23.269	25.643	29.973	35.336	41.304	47.212	50.998	54.437	58.619	61.581	67.985
37	15.965	18.586	19.960	22.106	24.075	26.492	30.893	36.336	42.383	48.363	52.192	55.668	59.893	62.883	69.346
38	16.611	19.289	20.691	22.878	24.884	27.343	31.815	37.335	43.462	49.513	53.384	56.896	61.162	64.181	70.703
39	17.262	19.996	21.426	23.654	25.695	28.196	32.737	38.335	44.539	50.660	54.572	58.120	62.428	65.476	72.055
40	17.916	20.707	22.164	24.433	26.509	29.051	33.660	39.335	45.616	51.805	55.758	59.342	63.691	66.766	73.402
41	18.576	21.421	22.906	25.215	27.326	29.907	34.585	40.335	46.692	52.949	56.942	60.561	64.950	68.053	74.745
42	19.239	22.138	23.650	25.999	28.144	30.765	35.510	41.335	47.766	54.090	58.124	61.777	66.206	69.336	76.084
43	19.906	22.859	24.398	26.785	28.965	31.625	36.436	42.335	48.840	55.230	59.304	62.990	67.459	70.616	77.419
44	20.576	23.584	25.148	27.575	29.787	32.487	37.363	43.335	49.913	56.369	60.481	64.201	68.710	71.893	78.750
45	21.251	24.311	25.901	28.366	30.612	33.350	38.291	44.335	50.985	57.505	61.656	65.410	69.957	73.166	80.077
46	21.929	25.041	26.657	29.160	31.439	34.215	39.220	45.335	52.056	58.641	62.830	66.617	71.201	74.437	81.400
47	22.610	25.775	27.416	29.956	32.268	35.081	40.149	46.335	53.127	59.774	64.001	67.821	72.443	75.704	82.720
48	23.295	26.511	28.177	30.755	33.098	35.949	41.079	47.335	54.196	60.907	65.171	69.023	73.683	76.969	84.037
49	23.983	27.249	28.941	31.555	33.930	36.818	42.010	48.335	55.265	62.038	66.339	70.222	74.919	78.231	85.351
50	24.674	27.991	29.707	32.357	34.764	37.689	42.942	49.335	56.334	63.167	67.505	71.420	76.154	79.490	86.661

TABLE B.1 (cont.) CRITICAL VALUES OF THE CHI-SQUARE DISTRIBUTION

ν	α = 0.999	0.995	0.99	0.975	0.95	0.90	0.75	0.50	0.25	0.10	0.05	0.025	0.01	0.005	0.001
51	25.368	28.735	30.475	33.162	35.600	38.560	43.874	50.335	57.401	64.295	68.669	72.616	77.386	80.747	87.968
52	26.065	29.481	31.246	33.968	36.437	39.433	44.808	51.335	58.468	65.422	69.832	73.810	78.616	82.001	89.272
53	26.765	30.230	32.019	34.776	37.276	40.308	45.741	52.335	59.534	66.548	70.993	75.002	79.843	83.253	90.573
54	27.468	30.981	32.793	35.586	38.116	41.183	46.676	53.335	60.600	67.673	72.153	76.192	81.069	84.502	91.872
55	28.173	31.735	33.570	36.398	38.958	42.060	47.610	54.335	61.665	68.796	73.311	77.380	82.292	85.749	93.168
56	28.881	32.491	34.350	37.212	39.801	42.937	48.546	55.335	62.729	69.919	74.468	78.567	83.513	86.994	94.461
57	29.592	33.248	35.131	38.027	40.646	43.816	49.482	56.335	63.793	71.040	75.624	79.752	84.733	88.236	95.751
58	30.305	34.008	35.913	38.844	41.492	44.696	50.419	57.335	64.857	72.160	76.778	80.936	85.950	89.477	97.039
59	31.021	34.770	36.698	39.662	42.339	45.577	51.356	58.335	65.919	73.279	77.931	82.117	87.166	90.715	98.324
60	31.738	35.535	37.485	40.482	43.188	46.459	52.294	59.335	66.981	74.397	79.082	83.298	88.379	91.952	99.607
61	32.459	36.301	38.273	41.303	44.038	47.342	53.232	60.335	68.043	75.514	80.232	84.476	89.591	93.186	100.888
62	33.181	37.068	39.063	42.126	44.889	48.226	54.171	61.335	69.104	76.630	81.381	85.654	90.802	94.419	102.166
63	33.906	37.838	39.855	42.950	45.741	49.111	55.110	62.335	70.165	77.745	82.529	86.830	92.010	95.649	103.442
64	34.633	38.610	40.649	43.776	46.595	49.996	56.050	63.335	71.225	78.860	83.675	88.004	93.217	96.878	104.716
65	35.362	39.383	41.444	44.603	47.450	50.883	56.990	64.335	72.285	79.973	84.821	89.177	94.422	98.105	105.988
66	36.093	40.158	42.240	45.431	48.305	51.770	57.931	65.335	73.344	81.085	85.965	90.349	95.626	99.330	107.258
67	36.826	40.935	43.038	46.261	49.162	52.659	58.872	66.335	74.403	82.197	87.108	91.519	96.828	100.554	108.526
68	37.561	41.713	43.838	47.092	50.020	53.548	59.814	67.335	75.461	83.308	88.250	92.689	98.028	101.776	109.791
69	38.298	42.494	44.639	47.924	50.879	54.438	60.756	68.334	76.519	84.418	89.391	93.856	99.228	102.996	111.055
70	39.036	43.275	45.442	48.758	51.739	55.329	61.698	69.334	77.577	85.527	90.531	95.023	100.425	104.215	112.317
71	39.777	44.058	46.246	49.592	52.600	56.221	62.641	70.334	78.634	86.635	91.670	96.189	101.621	105.432	113.577
72	40.520	44.843	47.051	50.428	53.462	57.113	63.585	71.334	79.690	87.743	92.808	97.353	102.816	106.648	114.835
73	41.264	45.629	47.858	51.265	54.325	58.006	64.528	72.334	80.747	88.850	93.945	98.516	104.010	107.862	116.092
74	42.010	46.417	48.666	52.103	55.189	58.900	65.472	73.334	81.803	89.956	95.081	99.678	105.202	109.074	117.346
75	42.757	47.206	49.475	52.942	56.054	59.795	66.417	74.334	82.858	91.061	96.217	100.839	106.393	110.286	118.599
76	43.507	47.997	50.286	53.782	56.920	60.690	67.362	75.334	83.913	92.166	97.351	101.999	107.583	111.495	119.850
77	44.258	48.788	51.097	54.623	57.786	61.586	68.307	76.334	84.968	93.270	98.484	103.158	108.771	112.704	121.100
78	45.010	49.582	51.910	55.466	58.654	62.483	69.252	77.334	86.022	94.374	99.617	104.316	109.958	113.911	122.348
79	45.764	50.376	52.725	56.309	59.522	63.380	70.198	78.334	87.077	95.476	100.749	105.473	111.144	115.117	123.594
80	46.520	51.172	53.540	57.153	60.391	64.278	71.145	79.334	88.130	96.578	101.879	106.629	112.329	116.321	124.839
81	47.277	51.969	54.357	57.998	61.261	65.176	72.091	80.334	89.184	97.680	103.010	107.783	113.512	117.524	126.083
82	48.036	52.767	55.174	58.845	62.132	66.076	73.038	81.334	90.237	98.780	104.139	108.937	114.695	118.726	127.324
83	48.796	53.567	55.993	59.692	63.004	66.976	73.985	82.334	91.289	99.880	105.267	110.090	115.876	119.927	128.565
84	49.557	54.368	56.813	60.540	63.876	67.876	74.933	83.334	92.342	100.980	106.395	111.242	117.057	121.126	129.804
85	50.320	55.170	57.634	61.389	64.749	68.777	75.881	84.334	93.394	102.079	107.522	112.393	118.236	122.325	131.041
86	51.085	55.973	58.456	62.239	65.623	69.679	76.829	85.334	94.446	103.177	108.648	113.544	119.414	123.522	132.277
87	51.850	56.777	59.279	63.089	66.498	70.581	77.777	86.334	95.497	104.275	109.773	114.693	120.591	124.718	133.512
88	52.617	57.582	60.103	63.941	67.373	71.484	78.726	87.334	96.548	105.372	110.898	115.841	121.767	125.913	134.745
89	53.386	58.389	60.928	64.793	68.249	72.387	79.675	88.334	97.599	106.469	112.022	116.989	122.942	127.106	135.978
90	54.155	59.196	61.754	65.647	69.126	73.291	80.625	89.334	98.650	107.565	113.145	118.136	124.116	128.299	137.208
91	54.926	60.005	62.581	66.501	70.003	74.196	81.574	90.334	99.700	108.661	114.268	119.282	125.289	129.491	138.438
92	55.698	60.815	63.409	67.356	70.882	75.100	82.524	91.334	100.750	109.756	115.390	120.427	126.462	130.681	139.666
93	56.472	61.625	64.238	68.211	71.760	76.006	83.474	92.334	101.800	110.850	116.511	121.571	127.633	131.871	140.893
94	57.246	62.437	65.068	69.068	72.640	76.912	84.425	93.334	102.850	111.944	117.632	122.715	128.803	133.059	142.119
95	58.022	63.250	65.898	69.925	73.520	77.818	85.376	94.334	103.899	113.038	118.752	123.858	129.973	134.247	143.344
96	58.799	64.063	66.730	70.783	74.401	78.725	86.327	95.334	104.948	114.131	119.871	125.000	131.141	135.433	144.567
97	59.577	64.878	67.562	71.642	75.282	79.633	87.278	96.334	105.997	115.223	120.990	126.141	132.309	136.619	145.789
98	60.356	65.694	68.396	72.501	76.164	80.541	88.229	97.334	107.045	116.315	122.108	127.282	133.476	137.803	147.010
99	61.137	66.510	69.230	73.361	77.046	81.449	89.181	98.334	108.093	117.407	123.225	128.422	134.642	138.987	148.230
100	61.918	67.328	70.065	74.222	77.929	82.358	90.133	99.334	109.141	118.498	124.342	129.561	135.807	140.169	149.449

TABLE B.1 (cont.) CRITICAL VALUES OF THE CHI-SQUARE DISTRIBUTION

ν	α = 0.999	0.995	0.99	0.975	0.95	0.90	0.75	0.50	0.25	0.10	0.05	0.025	0.01	0.005	0.001
101	62.701	68.146	70.901	75.083	78.813	83.267	91.085	100.334	110.189	119.589	125.458	130.700	136.971	141.351	150.667
102	63.484	68.965	71.737	75.946	79.697	84.177	92.038	101.334	111.236	120.679	126.574	131.838	138.134	142.532	151.884
103	64.269	69.785	72.575	76.809	80.582	85.088	92.991	102.334	112.284	121.769	127.689	132.975	139.297	143.712	153.099
104	65.055	70.606	73.413	77.672	81.468	85.998	93.944	103.334	113.331	122.858	128.804	134.111	140.459	144.891	154.314
105	65.841	71.428	74.252	78.536	82.354	86.909	94.897	104.334	114.378	123.947	129.918	135.247	141.620	146.070	155.528
106	66.629	72.251	75.092	79.401	83.240	87.821	95.850	105.334	115.424	125.035	131.031	136.382	142.780	147.247	156.740
107	67.418	73.075	75.932	80.267	84.127	88.733	96.804	106.334	116.471	126.123	132.144	137.517	143.940	148.424	157.952
108	68.207	73.899	76.774	81.133	85.015	89.645	97.758	107.334	117.517	127.211	133.257	138.651	145.099	149.593	159.162
109	68.998	74.724	77.616	82.000	85.903	90.558	98.712	108.334	118.563	128.298	134.369	139.784	146.257	150.774	160.372
110	69.790	75.550	78.458	82.867	86.792	91.471	99.666	109.334	119.608	129.385	135.480	140.917	147.414	151.948	161.581
111	70.582	76.377	79.302	83.735	87.681	92.385	100.620	110.334	120.654	130.472	136.591	142.049	148.571	153.122	162.788
112	71.376	77.204	80.146	84.604	88.570	93.299	101.575	111.334	121.699	131.558	137.701	143.180	149.727	154.294	163.995
113	72.170	78.033	80.991	85.473	89.461	94.213	102.530	112.334	122.744	132.643	138.811	144.311	150.882	155.466	165.201
114	72.965	78.862	81.836	86.342	90.351	95.128	103.485	113.334	123.789	133.729	139.921	145.441	152.037	156.637	166.406
115	73.761	79.692	82.682	87.213	91.242	96.043	104.440	114.334	124.834	134.813	141.030	146.571	153.191	157.808	167.610
116	74.558	80.522	83.529	88.084	92.134	96.958	105.396	115.334	125.878	135.898	142.138	147.700	154.344	158.977	168.813
117	75.356	81.353	84.377	88.955	93.026	97.874	106.352	116.334	126.923	136.982	143.246	148.829	155.496	160.146	170.016
118	76.155	82.185	85.225	89.827	93.918	98.790	107.307	117.334	127.967	138.066	144.354	149.957	156.648	161.314	171.217
119	76.955	83.018	86.074	90.700	94.811	99.707	108.263	118.334	129.011	139.149	145.461	151.084	157.800	162.481	172.418
120	77.755	83.852	86.923	91.573	95.705	100.624	109.220	119.334	130.055	140.233	146.567	152.211	158.950	163.648	173.617
121	78.557	84.686	87.773	92.446	96.598	101.541	110.176	120.334	131.098	141.315	147.674	153.338	160.100	164.814	174.816
122	79.359	85.521	88.624	93.320	97.493	102.458	111.133	121.334	132.142	142.398	148.779	154.464	161.250	165.980	176.014
123	80.162	86.356	89.475	94.195	98.387	103.376	112.089	122.334	133.185	143.480	149.885	155.589	162.398	167.144	177.212
124	80.965	87.192	90.327	95.070	99.283	104.295	113.046	123.334	134.228	144.562	150.989	156.714	163.546	168.308	178.408
125	81.770	88.029	91.180	95.946	100.178	105.213	114.004	124.334	135.271	145.643	152.094	157.839	164.694	169.471	179.604
126	82.575	88.866	92.033	96.822	101.074	106.132	114.961	125.334	136.313	146.724	153.198	158.962	165.841	170.634	180.799
127	83.381	89.704	92.887	97.698	101.971	107.051	115.918	126.334	137.356	147.805	154.302	160.086	166.987	171.796	181.993
128	84.188	90.543	93.741	98.576	102.867	107.971	116.876	127.334	138.398	148.885	155.405	161.209	168.133	172.957	183.186
129	84.996	91.383	94.596	99.453	103.765	108.891	117.834	128.334	139.440	149.965	156.508	162.331	169.278	174.118	184.379
130	85.804	92.223	95.451	100.331	104.662	109.811	118.792	129.334	140.482	151.045	157.610	163.453	170.423	175.278	185.571
131	86.613	93.063	96.307	101.210	105.560	110.732	119.750	130.334	141.524	152.125	158.712	164.575	171.567	176.438	186.762
132	87.423	93.904	97.163	102.089	106.459	111.652	120.708	131.334	142.566	153.204	159.814	165.696	172.711	177.597	187.953
133	88.233	94.746	98.021	102.968	107.357	112.573	121.667	132.334	143.608	154.283	160.915	166.816	173.854	178.755	189.142
134	89.044	95.588	98.878	103.848	108.257	113.495	122.625	133.334	144.649	155.361	162.016	167.936	174.996	179.913	190.331
135	89.856	96.431	99.736	104.729	109.156	114.417	123.584	134.334	145.690	156.440	163.116	169.056	176.138	181.070	191.520
136	90.669	97.275	100.595	105.609	110.056	115.338	124.543	135.334	146.731	157.518	164.216	170.175	177.280	182.226	192.707
137	91.482	98.119	101.454	106.491	110.956	116.261	125.502	136.334	147.772	158.595	165.316	171.294	178.421	183.382	193.894
138	92.296	98.964	102.314	107.372	111.857	117.183	126.461	137.334	148.813	159.673	166.415	172.412	179.561	184.538	195.080
139	93.111	99.809	103.174	108.254	112.758	118.106	127.421	138.334	149.854	160.750	167.514	173.530	180.701	185.693	196.266
140	93.926	100.655	104.034	109.137	113.659	119.029	128.380	139.334	150.894	161.827	168.613	174.648	181.840	186.847	197.451

Table B.1 was prepared using Equation 26.4.6 of Zelen and Severo (1964). The chi-square values were calculated to six decimal places and then rounded to three decimal places.

Examples:

$$\chi^2_{0.05,12} = 21.026 \quad \text{and} \quad \chi^2_{0.0,138} = 61.162.$$

For large degrees of freedom (ν), critical values of χ^2 can be approximated very well by

$$\chi^2_{\alpha,\nu} \cong \nu\left(1 - \frac{2}{9\nu} + Z_{\alpha(1)}\sqrt{\frac{2}{9\nu}}\right)^3$$

(Wilson and Hilferty, 1931). It is for this purpose that the values of $Z\alpha(1)$ are given below (from White, 1970).

$\alpha =$ 0.999	0.995	0.99	0.975	0.95	0.90	0.75
$Z_{\alpha(1)} = -3.09023$	-2.57583	-2.32635	-1.95996	-1.64485	-1.28155	-0.67449

$\alpha =$ 0.50	0.25	0.10	0.05	0.025	0.01	0.005	0.001
$Z_{\alpha(1)} = 0.00000$	0.67449	1.28155	1.64485	1.95996	2.32635	2.57583	3.09023

The percent error, i.e., (approximation − true value)/true value × 100%, resulting from the use of this approximation is as follows:

ν	$\alpha =$ 0.999	0.995	0.99	0.975	0.95	0.90	0.75	0.50	0.25	0.10	0.05	0.025	0.01	0.005	0.001
30	−0.7	−0.3	−0.2	−0.1	0.0*	0.0*	0.0*	0.0*	0.0*	0.0*	0.0*	0.0*	0.0*	0.1	0.2
100	−0.1	0.0*	0.0*	0.0*	0.0*	0.0*	0.0*	0.0*	0.0*	0.0*	0.0*	0.0*	0.0*	0.0*	0.0*
140	0.0*	0.0*	0.0*	0.0*	0.0*	0.0*	0.0*	0.0*	0.0*	0.0*	0.0*	0.0*	0.0*	0.0*	0.0*

where the asterisk indicates a percent error the absolute value of which is less than 0.05%. Zar (1978) discusses this and other approximations for $\chi^2_{\alpha,\nu}$.

TABLE B.2 PROPORTIONS OF THE NORMAL CURVE (ONE-TAILED)

This table gives the proportion of the normal curve that lies beyond (i.e., is more extreme than) a given normal deviate, (e.g., $Z = (X_i - \mu)/\sigma$ or $Z = (\bar{X} - \mu)/\sigma_{\bar{X}}$). For example, the proportion of a normal distribution for which $Z \geq 1.51$ is 0.0655.

Z	0	1	2	3	4	5	6	7	8	9	Z
0.0	0.5000	0.4960	0.4920	0.4880	0.4840	0.4801	0.4761	0.4721	0.4681	0.4641	0.0
0.1	0.4602	0.4562	0.4522	0.4483	0.4443	0.4404	0.4364	0.4325	0.4286	0.4247	0.1
0.2	0.4207	0.4168	0.4129	0.4090	0.4052	0.4013	0.3974	0.3936	0.3897	0.3859	0.2
0.3	0.3821	0.3783	0.3745	0.3707	0.3669	0.3632	0.3594	0.3557	0.3520	0.3483	0.3
0.4	0.3446	0.3409	0.3372	0.3336	0.3300	0.3264	0.3228	0.3192	0.3156	0.3121	0.4
0.5	0.3085	0.3050	0.3015	0.2981	0.2946	0.2912	0.2877	0.2843	0.2810	0.2776	0.5
0.6	0.2743	0.2709	0.2676	0.2643	0.2611	0.2578	0.2546	0.2514	0.2483	0.2451	0.6
0.7	0.2420	0.2389	0.2358	0.2327	0.2297	0.2266	0.2236	0.2207	0.2177	0.2148	0.7
0.8	0.2119	0.2090	0.2061	0.2033	0.2005	0.1977	0.1949	0.1922	0.1894	0.1867	0.8
0.9	0.1841	0.1814	0.1788	0.1762	0.1736	0.1711	0.1685	0.1660	0.1635	0.1611	0.9
1.0	0.1587	0.1562	0.1539	0.1515	0.1492	0.1469	0.1446	0.1423	0.1401	0.1379	1.0
1.1	0.1357	0.1335	0.1314	0.1292	0.1271	0.1251	0.1230	0.1210	0.1190	0.1170	1.1
1.2	0.1151	0.1131	0.1112	0.1093	0.1075	0.1056	0.1038	0.1020	0.1003	0.0985	1.2
1.3	0.0968	0.0951	0.0934	0.0918	0.0901	0.0885	0.0869	0.0853	0.0838	0.0823	1.3
1.4	0.0808	0.0793	0.0778	0.0764	0.0749	0.0735	0.0721	0.0708	0.0694	0.0681	1.4
1.5	0.0668	0.0655	0.0643	0.0630	0.0618	0.0606	0.0594	0.0582	0.0571	0.0559	1.5
1.6	0.0548	0.0537	0.0526	0.0516	0.0505	0.0495	0.0485	0.0475	0.0465	0.0455	1.6
1.7	0.0446	0.0436	0.0427	0.0418	0.0409	0.0401	0.0392	0.0384	0.0375	0.0367	1.7
1.8	0.0359	0.0351	0.0344	0.0336	0.0329	0.0322	0.0314	0.0307	0.0301	0.0294	1.8
1.9	0.0287	0.0281	0.0274	0.0268	0.0262	0.0256	0.0250	0.0244	0.0239	0.0233	1.9
2.0	0.0228	0.0222	0.0217	0.0212	0.0207	0.0202	0.0197	0.0192	0.0188	0.0183	2.0
2.1	0.0179	0.0174	0.0170	0.0166	0.0162	0.0158	0.0154	0.0150	0.0146	0.0143	2.1
2.2	0.0139	0.0136	0.0132	0.0129	0.0125	0.0122	0.0119	0.0116	0.0113	0.0110	2.2
2.3	0.0107	0.0104	0.0102	0.0099	0.0096	0.0094	0.0091	0.0089	0.0087	0.0084	2.3
2.4	0.0082	0.0080	0.0078	0.0075	0.0073	0.0071	0.0069	0.0068	0.0066	0.0064	2.4
2.5	0.0062	0.0060	0.0059	0.0057	0.0055	0.0054	0.0052	0.0051	0.0049	0.0048	2.5
2.6	0.0047	0.0045	0.0044	0.0043	0.0041	0.0040	0.0039	0.0038	0.0037	0.0036	2.6
2.7	0.0035	0.0034	0.0033	0.0032	0.0031	0.0030	0.0029	0.0028	0.0027	0.0026	2.7
2.8	0.0026	0.0025	0.0024	0.0023	0.0023	0.0022	0.0021	0.0021	0.0020	0.0019	2.8
2.9	0.0019	0.0018	0.0018	0.0017	0.0016	0.0016	0.0015	0.0015	0.0014	0.0014	2.9
3.0	0.0013	0.0013	0.0013	0.0012	0.0012	0.0011	0.0011	0.0011	0.0010	0.0010	3.0
3.1	0.0010	0.0009	0.0009	0.0009	0.0008	0.0008	0.0008	0.0008	0.0007	0.0007	3.1
3.2	0.0007	0.0007	0.0006	0.0006	0.0006	0.0006	0.0006	0.0005	0.0005	0.0005	3.2
3.3	0.0005	0.0005	0.0005	0.0004	0.0004	0.0004	0.0004	0.0004	0.0004	0.0003	3.3
3.4	0.0003	0.0003	0.0003	0.0003	0.0003	0.0003	0.0003	0.0003	0.0003	0.0002	3.4
3.5	0.0002	0.0002	0.0002	0.0002	0.0002	0.0002	0.0002	0.0002	0.0002	0.0002	3.5
3.6	0.0002	0.0002	0.0001	0.0001	0.0001	0.0001	0.0001	0.0001	0.0001	0.0001	3.6
3.7	0.0001	0.0001	0.0001	0.0001	0.0001	0.0001	0.0001	0.0001	0.0001	0.0001	3.7
3.8	0.0001	0.0001	0.0001	0.0001	0.0001	0.0001	0.0001	0.0001	0.0001	0.0001	3.8

Table B.2 was prepared using an algorithm of Hastings (1955: 187).
Probabilities for values of Z in between those shown in this table may be obtained by either linear or harmonic interpolation.

TABLE B.3 CRITICAL VALUES OF THE *t* DISTRIBUTION

ν	α(2): 0.50	0.20	0.10	0.05	0.02	0.01	0.005	0.002	0.001
	α(1): 0.25	0.10	0.05	0.025	0.01	0.005	0.0025	0.001	0.0005
1	1.000	3.078	6.314	12.706	31.821	63.657	127.321	318.309	636.619
2	0.816	1.886	2.920	4.303	6.965	9.925	14.089	22.327	31.599
3	0.765	1.638	2.353	3.182	4.541	5.841	7.453	10.215	12.924
4	0.741	1.533	2.132	2.776	3.747	4.604	5.598	7.173	8.610
5	0.727	1.476	2.015	2.571	3.365	4.032	4.773	5.893	6.869
6	0.718	1.440	1.943	2.447	3.143	3.707	4.317	5.208	5.959
7	0.711	1.415	1.895	2.365	2.998	3.499	4.029	4.785	5.408
8	0.706	1.397	1.860	2.306	2.896	3.355	3.833	4.501	5.041
9	0.703	1.383	1.833	2.262	2.821	3.250	3.690	4.297	4.781
10	0.700	1.372	1.812	2.228	2.764	3.169	3.581	4.144	4.587
11	0.697	1.363	1.796	2.201	2.718	3.106	3.497	4.025	4.437
12	0.695	1.356	1.782	2.179	2.681	3.055	3.428	3.930	4.318
13	0.694	1.350	1.771	2.160	2.650	3.012	3.372	3.852	4.221
14	0.692	1.345	1.761	2.145	2.624	2.977	3.326	3.787	4.140
15	0.691	1.341	1.753	2.131	2.602	2.947	3.286	3.733	4.073
16	0.690	1.337	1.746	2.120	2.583	2.921	3.252	3.686	4.015
17	0.689	1.333	1.740	2.110	2.567	2.898	3.222	3.646	3.965
18	0.688	1.330	1.734	2.101	2.552	2.878	3.197	3.610	3.922
19	0.688	1.328	1.729	2.093	2.539	2.861	3.174	3.579	3.883
20	0.687	1.325	1.725	2.086	2.528	2.845	3.153	3.552	3.850
21	0.686	1.323	1.721	2.080	2.518	2.831	3.135	3.527	3.819
22	0.686	1.321	1.717	2.074	2.508	2.819	3.119	3.505	3.792
23	0.685	1.319	1.714	2.069	2.500	2.807	3.104	3.485	3.768
24	0.685	1.318	1.711	2.064	2.492	2.797	3.091	3.467	3.745
25	0.684	1.316	1.708	2.060	2.485	2.787	3.078	3.450	3.725
26	0.684	1.315	1.706	2.056	2.479	2.779	3.067	3.435	3.707
27	0.684	1.314	1.703	2.052	2.473	2.771	3.057	3.421	3.690
28	0.683	1.313	1.701	2.048	2.467	2.763	3.047	3.408	3.674
29	0.683	1.311	1.699	2.045	2.462	2.756	3.038	3.396	3.659
30	0.683	1.310	1.697	2.042	2.457	2.750	3.030	3.385	3.646
31	0.682	1.309	1.696	2.040	2.453	2.744	3.022	3.375	3.633
32	0.682	1.309	1.694	2.037	2.449	2.738	3.015	3.365	3.622
33	0.682	1.308	1.692	2.035	2.445	2.733	3.008	3.356	3.611
34	0.682	1.307	1.691	2.032	2.441	2.728	3.002	3.348	3.601
35	0.682	1.306	1.690	2.030	2.438	2.724	2.996	3.340	3.591
36	0.681	1.306	1.688	2.028	2.434	2.719	2.990	3.333	3.582
37	0.681	1.305	1.687	2.026	2.431	2.715	2.985	3.326	3.574
38	0.681	1.304	1.686	2.024	2.429	2.712	2.980	3.319	3.566
39	0.681	1.304	1.685	2.023	2.426	2.708	2.976	3.313	3.558
40	0.681	1.303	1.684	2.021	2.423	2.704	2.971	3.307	3.551
41	0.681	1.303	1.683	2.020	2.421	2.701	2.967	3.301	3.544
42	0.680	1.302	1.682	2.018	2.418	2.698	2.963	3.296	3.538
43	0.680	1.302	1.681	2.017	2.416	2.695	2.959	3.291	3.532
44	0.680	1.301	1.680	2.015	2.414	2.692	2.956	3.286	3.526
45	0.680	1.301	1.679	2.014	2.412	2.690	2.952	3.281	3.520
46	0.680	1.300	1.679	2.013	2.410	2.687	2.949	3.277	3.515
47	0.680	1.300	1.678	2.012	2.408	2.685	2.946	3.273	3.510
48	0.680	1.299	1.677	2.011	2.407	2.682	2.943	3.269	3.505
49	0.680	1.299	1.677	2.010	2.405	2.680	2.940	3.265	3.500
50	0.679	1.299	1.676	2.009	2.403	2.678	2.937	3.261	3.496

TABLE B.3 (cont.) CRITICAL VALUES OF THE t DISTRIBUTION

ν	α(2): 0.50	0.20	0.10	0.05	0.02	0.01	0.005	0.002	0.001
	α(1): 0.25	0.10	0.05	0.025	0.01	0.005	0.0025	0.001	0.0005
52	0.679	1.298	1.675	2.007	2.400	2.674	2.932	3.255	3.488
54	0.679	1.297	1.674	2.005	2.397	2.670	2.927	3.248	3.480
56	0.679	1.297	1.673	2.003	2.395	2.667	2.923	3.242	3.473
58	0.679	1.296	1.672	2.002	2.392	2.663	2.918	3.237	3.466
60	0.679	1.296	1.671	2.000	2.390	2.660	2.915	3.232	3.460
62	0.678	1.295	1.670	1.999	2.388	2.657	2.911	3.227	3.454
64	0.678	1.295	1.669	1.998	2.386	2.655	2.908	3.223	3.449
66	0.678	1.295	1.668	1.997	2.384	2.652	2.904	3.218	3.444
68	0.678	1.294	1.668	1.995	2.382	2.650	2.902	3.214	3.439
70	0.678	1.294	1.667	1.994	2.381	2.648	2.899	3.211	3.435
72	0.678	1.293	1.666	1.993	2.379	2.646	2.896	3.207	3.431
74	0.678	1.293	1.666	1.993	2.378	2.644	2.894	3.204	3.427
76	0.678	1.293	1.665	1.992	2.376	2.642	2.891	3.201	3.423
78	0.678	1.292	1.665	1.991	2.375	2.640	2.889	3.198	3.420
80	0.678	1.292	1.664	1.990	2.374	2.639	2.887	3.195	3.416
82	0.677	1.292	1.664	1.989	2.373	2.637	2.885	3.193	3.413
84	0.677	1.292	1.663	1.989	2.372	2.636	2.883	3.190	3.410
86	0.677	1.291	1.663	1.988	2.370	2.634	2.881	3.188	3.407
88	0.677	1.291	1.662	1.987	2.369	2.633	2.880	3.185	3.405
90	0.677	1.291	1.662	1.987	2.368	2.632	2.878	3.183	3.402
92	0.677	1.291	1.662	1.986	2.368	2.630	2.876	3.181	3.399
94	0.677	1.291	1.661	1.986	2.367	2.629	2.875	3.179	3.397
96	0.677	1.290	1.661	1.985	2.366	2.628	2.873	3.177	3.395
98	0.677	1.290	1.661	1.984	2.365	2.627	2.872	3.175	3.393
100	0.677	1.290	1.660	1.984	2.364	2.626	2.871	3.174	3.390
105	0.677	1.290	1.659	1.983	2.362	2.623	2.868	3.170	3.386
110	0.677	1.289	1.659	1.982	2.361	2.621	2.865	3.166	3.381
115	0.677	1.289	1.658	1.981	2.359	2.619	2.862	3.163	3.377
120	0.677	1.289	1.658	1.980	2.358	2.617	2.860	3.160	3.373
125	0.676	1.288	1.657	1.979	2.357	2.616	2.858	3.157	3.370
130	0.676	1.288	1.657	1.978	2.355	2.614	2.856	3.154	3.367
135	0.676	1.288	1.656	1.978	2.354	2.613	2.854	3.152	3.364
140	0.676	1.288	1.656	1.977	2.353	2.611	2.852	3.149	3.361
145	0.676	1.287	1.655	1.976	2.352	2.610	2.851	3.147	3.359
150	0.676	1.287	1.655	1.976	2.351	2.609	2.849	3.145	3.357
160	0.676	1.287	1.654	1.975	2.350	2.607	2.846	3.142	3.352
170	0.676	1.287	1.654	1.974	2.348	2.605	2.844	3.139	3.349
180	0.676	1.286	1.653	1.973	2.347	2.603	2.842	3.136	3.345
190	0.676	1.286	1.653	1.973	2.346	2.602	2.840	3.134	3.342
200	0.676	1.286	1.653	1.972	2.345	2.601	2.839	3.131	3.340
250	0.675	1.285	1.651	1.969	2.341	2.596	2.832	3.123	3.330
300	0.675	1.284	1.650	1.968	2.339	2.592	2.828	3.118	3.323
350	0.675	1.284	1.649	1.967	2.337	2.590	2.825	3.114	3.319
400	0.675	1.284	1.649	1.966	2.336	2.588	2.823	3.111	3.315
450	0.675	1.283	1.648	1.965	2.335	2.587	2.821	3.108	3.312
500	0.675	1.283	1.648	1.965	2.334	2.586	2.820	3.107	3.310
600	0.675	1.283	1.647	1.964	2.333	2.584	2.817	3.104	3.307
700	0.675	1.283	1.647	1.963	2.332	2.583	2.816	3.102	3.304
800	0.675	1.283	1.647	1.963	2.331	2.582	2.815	3.100	3.303
900	0.675	1.282	1.647	1.963	2.330	2.581	2.814	3.099	3.301
1000	0.675	1.282	1.646	1.962	2.330	2.581	2.813	3.098	3.300
∞	0.6745	1.2816	1.6449	1.9600	2.3263	2.5758	2.8070	3.0902	3.2905

This table was prepared using Equations 26.7.3 and 26.7.4 of Zelen and Severo (1964), except for the values at infinity degrees of freedom, which are adapted from White (1970). Except for the values at infinity degrees of freedom, t was calculated to eight decimal places and then rounded to three decimal places.

Examples:

$$t_{0.05(2),13} = 2.160 \quad \text{and} \quad t_{0.01(1),19} = 2.539.$$

If a critical value is needed for degrees of freedom not on this table one may conservatively employ the next smaller ν that is on the table. Or, the needed critical value, for $\nu < 1000$, may be calculated by linear interpolation, with an error or no more than 0.001. If a little more accuracy is desired, or if the needed ν is > 1000, then harmonic interpolation should be used.

TABLE B.4 CRITICAL VALUES OF THE *F* DISTRIBUTION

Numerator DF = 1

α(2):	0.50	0.20	0.10	0.05	0.02	0.01	0.005	0.002	0.001
α(1):	0.25	0.10	0.05	0.025	0.01	0.005	0.0025	0.001	0.0005
Denom. DF									
1	5.83	39.9	161.	648.	4050.	16200.	64800.	405000.	1620000.
2	2.57	8.53	18.5	38.5	98.5	199.	399.	999.	2000.
3	2.02	5.54	10.1	17.4	34.1	55.6	89.6	167.	267.
4	1.81	4.54	7.71	12.2	21.2	31.3	45.7	74.1	106.
5	1.69	4.06	6.61	10.0	16.3	22.8	31.4	47.2	63.6
6	1.62	3.78	5.99	8.81	13.7	18.6	24.8	35.5	46.1
7	1.57	3.59	5.59	8.07	12.2	16.2	21.1	29.2	37.0
8	1.54	3.46	5.32	7.57	11.3	14.7	18.8	25.4	31.6
9	1.51	3.36	5.12	7.21	10.6	13.6	17.2	22.9	28.0
10	1.49	3.29	4.96	6.94	10.0	12.8	16.0	21.0	25.5
11	1.47	3.23	4.84	6.72	9.65	12.2	15.2	19.7	23.7
12	1.46	3.18	4.75	6.55	9.33	11.8	14.5	18.6	22.2
13	1.45	3.14	4.67	6.41	9.07	11.4	13.9	17.8	21.1
14	1.44	3.10	4.60	6.30	8.86	11.1	13.5	17.1	20.2
15	1.43	3.07	4.54	6.20	8.68	10.8	13.1	16.6	19.5
16	1.42	3.05	4.49	6.12	8.53	10.6	12.8	16.1	18.9
17	1.42	3.03	4.45	6.04	8.40	10.4	12.6	15.7	18.4
18	1.41	3.01	4.41	5.98	8.29	10.2	12.3	15.4	17.9
19	1.41	2.99	4.38	5.92	8.18	10.1	12.1	15.1	17.5
20	1.40	2.97	4.35	5.87	8.10	9.94	11.9	14.8	17.2
21	1.40	2.96	4.32	5.83	8.02	9.83	11.8	14.6	16.9
22	1.40	2.95	4.30	5.79	7.95	9.73	11.6	14.4	16.6
23	1.39	2.94	4.28	5.75	7.88	9.63	11.5	14.2	16.4
24	1.39	2.93	4.26	5.72	7.82	9.55	11.4	14.0	16.2
25	1.39	2.92	4.24	5.69	7.77	9.48	11.3	13.9	16.0
26	1.38	2.91	4.23	5.66	7.72	9.41	11.2	13.7	15.8
27	1.38	2.90	4.21	5.63	7.68	9.34	11.1	13.6	15.6
28	1.38	2.89	4.20	5.61	7.64	9.28	11.0	13.5	15.5
29	1.38	2.89	4.18	5.59	7.60	9.23	11.0	13.4	15.3
30	1.38	2.88	4.17	5.57	7.56	9.18	10.9	13.3	15.2
35	1.37	2.85	4.12	5.48	7.42	8.98	10.6	12.9	14.7
40	1.36	2.84	4.08	5.42	7.31	8.83	10.4	12.6	14.4
45	1.36	2.82	4.06	5.38	7.23	8.71	10.3	12.4	14.1
50	1.35	2.81	4.03	5.34	7.17	8.63	10.1	12.2	13.9
60	1.35	2.79	4.00	5.29	7.08	8.49	9.96	12.0	13.5
70	1.35	2.78	3.98	5.25	7.01	8.40	9.84	11.8	13.3
80	1.34	2.77	3.96	5.22	6.96	8.33	9.75	11.7	13.2
90	1.34	2.76	3.95	5.20	6.93	8.28	9.68	11.6	13.0
100	1.34	2.76	3.94	5.18	6.90	8.24	9.62	11.5	12.9
120	1.34	2.75	3.92	5.15	6.85	8.18	9.54	11.4	12.8
140	1.33	2.74	3.91	5.13	6.82	8.14	9.48	11.3	12.7
160	1.33	2.74	3.90	5.12	6.80	8.10	9.44	11.2	12.6
180	1.33	2.73	3.89	5.11	6.78	8.08	9.40	11.2	12.6
200	1.33	2.73	3.89	5.10	6.76	8.06	9.38	11.2	12.5
300	1.33	2.72	3.87	5.07	6.72	8.00	9.30	11.0	12.4
500	1.33	2.72	3.86	5.05	6.69	7.95	9.23	11.0	12.3
∞	1.32	2.71	3.84	5.02	6.64	7.88	9.14	10.8	12.1

TABLE B.4 (cont.) CRITICAL VALUES OF THE *F* DISTRIBUTION

Numerator DF = 2

	α(2): 0.50	0.20	0.10	0.05	0.02	0.01	0.005	0.002	0.001
Denom. DF	α(1): 0.25	0.10	0.05	0.025	0.01	0.005	0.0025	0.001	0.0005
1	7.50	49.5	200.	800.	5000.	20000.	80000.	500000.	2000000.
2	3.00	9.00	19.0	39.0	99.0	199.	399.	999.	2000.
3	2.28	5.46	9.55	16.0	30.8	49.8	79.9	149.	237.
4	2.00	4.32	6.94	10.6	18.0	26.3	38.0	61.2	87.4
5	1.85	3.78	5.79	8.43	13.3	18.3	25.0	37.1	49.8
6	1.76	3.46	5.14	7.26	10.9	14.5	19.1	27.0	34.8
7	1.70	3.26	4.74	6.54	9.55	12.4	15.9	21.7	27.2
8	1.66	3.11	4.46	6.06	8.65	11.0	13.9	18.5	22.7
9	1.62	3.01	4.26	5.71	8.02	10.1	12.5	16.4	19.9
10	1.60	2.92	4.10	5.46	7.56	9.43	11.6	14.9	17.9
11	1.58	2.86	3.98	5.26	7.21	8.91	10.8	13.8	16.4
12	1.56	2.81	3.89	5.10	6.93	8.51	10.3	13.0	15.3
13	1.55	2.76	3.81	4.97	6.70	8.19	9.84	12.3	14.4
14	1.53	2.73	3.74	4.86	6.51	7.92	9.47	11.8	13.7
15	1.52	2.70	3.68	4.77	6.36	7.70	9.17	11.3	13.2
16	1.51	2.67	3.63	4.69	6.23	7.51	8.92	11.0	12.7
17	1.51	2.64	3.59	4.62	6.11	7.35	8.70	10.7	12.3
18	1.50	2.62	3.55	4.56	6.01	7.21	8.51	10.4	11.9
19	1.49	2.61	3.52	4.51	5.93	7.09	8.35	10.2	11.6
20	1.49	2.59	3.49	4.46	5.85	6.99	8.21	9.95	11.4
21	1.48	2.57	3.47	4.42	5.78	6.89	8.08	9.77	11.2
22	1.48	2.56	3.44	4.38	5.72	6.81	7.96	9.61	11.0
23	1.47	2.55	3.42	4.35	5.66	6.73	7.86	9.47	10.8
24	1.47	2.54	3.40	4.32	5.61	6.66	7.77	9.34	10.6
25	1.47	2.53	3.39	4.29	5.57	6.60	7.69	9.22	10.5
26	1.46	2.52	3.37	4.27	5.53	6.54	7.61	9.12	10.3
27	1.46	2.51	3.35	4.24	5.49	6.49	7.54	9.02	10.2
28	1.46	2.50	3.34	4.22	5.45	6.44	7.48	8.93	10.1
29	1.45	2.50	3.33	4.20	5.42	6.40	7.42	8.85	9.99
30	1.45	2.49	3.32	4.18	5.39	6.35	7.36	8.77	9.90
35	1.44	2.46	3.27	4.11	5.27	6.19	7.14	8.47	9.52
40	1.44	2.44	3.23	4.05	5.18	6.07	6.99	8.25	9.25
45	1.43	2.42	3.20	4.01	5.11	5.97	6.86	8.09	9.04
50	1.43	2.41	3.18	3.97	5.06	5.90	6.77	7.96	8.88
60	1.42	2.39	3.15	3.93	4.98	5.79	6.63	7.77	8.65
70	1.41	2.38	3.13	3.89	4.92	5.72	6.53	7.64	8.49
80	1.41	2.37	3.11	3.86	4.88	5.67	6.46	7.54	8.37
90	1.41	2.36	3.10	3.84	4.85	5.62	6.41	7.47	8.28
100	1.41	2.36	3.09	3.83	4.82	5.59	6.37	7.41	8.21
120	1.40	2.35	3.07	3.80	4.79	5.54	6.30	7.32	8.10
140	1.40	2.34	3.06	3.79	4.76	5.50	6.26	7.26	8.03
160	1.40	2.34	3.05	3.78	4.74	5.48	6.22	7.21	7.97
180	1.40	2.33	3.05	3.77	4.73	5.46	6.20	7.18	7.93
200	1.40	2.33	3.04	3.76	4.71	5.44	6.17	7.15	7.90
300	1.39	2.32	3.03	3.73	4.68	5.39	6.11	7.07	7.80
500	1.39	2.31	3.01	3.72	4.65	5.35	6.06	7.00	7.72
∞	1.39	2.30	3.00	3.69	4.61	5.30	5.99	6.91	7.60

TABLE B.4 (cont.) CRITICAL VALUES OF THE *F* DISTRIBUTION

Numerator DF = 3

α(2):	0.50	0.20	0.10	0.05	0.02	0.01	0.005	0.002	0.001
α(1):	0.25	0.10	0.05	0.025	0.01	0.005	0.0025	0.001	0.0005
Denom. DF									
1	8.20	53.6	216.	864.	5400.	21600.	86500.	540000.	2160000.
2	3.15	9.16	19.2	39.2	99.2	199.	399.	999.	2000.
3	2.36	5.39	9.28	15.4	29.5	47.5	76.1	141.	225.
4	2.05	4.19	6.59	9.98	16.7	24.3	35.0	56.2	80.1
5	1.88	3.62	5.41	7.76	12.1	16.5	22.4	33.2	44.4
6	1.78	3.29	4.76	6.60	9.78	12.9	16.9	23.7	30.5
7	1.72	3.07	4.35	5.89	8.45	10.9	13.8	18.8	23.5
8	1.67	2.92	4.07	5.42	7.59	9.60	12.0	15.8	19.4
9	1.63	2.81	3.86	5.08	6.99	8.72	10.7	13.9	16.8
10	1.60	2.73	3.71	4.83	6.55	8.08	9.83	12.6	15.0
11	1.58	2.66	3.59	4.63	6.22	7.60	9.17	11.6	13.7
12	1.56	2.61	3.49	4.47	5.95	7.23	8.65	10.8	12.7
13	1.55	2.56	3.41	4.35	5.74	6.93	8.24	10.2	11.9
14	1.53	2.52	3.34	4.24	5.56	6.68	7.91	9.73	11.3
15	1.52	2.49	3.29	4.15	5.42	6.48	7.63	9.34	10.8
16	1.51	2.46	3.24	4.08	5.29	6.30	7.40	9.01	10.3
17	1.50	2.44	3.20	4.01	5.19	6.16	7.21	8.73	9.99
18	1.49	2.42	3.16	3.95	5.09	6.03	7.04	8.49	9.69
19	1.49	2.40	3.13	3.90	5.01	5.92	6.89	8.28	9.42
20	1.48	2.38	3.10	3.86	4.94	5.82	6.76	8.10	9.20
21	1.48	2.36	3.07	3.82	4.87	5.73	6.64	7.94	8.99
22	1.47	2.35	3.05	3.78	4.82	5.65	6.54	7.80	8.82
23	1.47	2.34	3.03	3.75	4.76	5.58	6.45	7.67	8.66
24	1.46	2.33	3.01	3.72	4.72	5.52	6.36	7.55	8.51
25	1.46	2.32	2.99	3.69	4.68	5.46	6.29	7.45	8.39
26	1.45	2.31	2.98	3.67	4.64	5.41	6.22	7.36	8.27
27	1.45	2.30	2.96	3.65	4.60	5.36	6.16	7.27	8.16
28	1.45	2.29	2.95	3.63	4.57	5.32	6.10	7.19	8.07
29	1.45	2.28	2.93	3.61	4.54	5.28	6.05	7.12	7.98
30	1.44	2.28	2.92	3.59	4.51	5.24	6.00	7.05	7.89
35	1.43	2.25	2.87	3.52	4.40	5.09	5.80	6.79	7.56
40	1.42	2.23	2.84	3.46	4.31	4.98	5.66	6.59	7.33
45	1.42	2.21	2.81	3.42	4.25	4.89	5.55	6.45	7.15
50	1.41	2.20	2.79	3.39	4.20	4.83	5.47	6.34	7.01
60	1.41	2.18	2.76	3.34	4.13	4.73	5.34	6.17	6.81
70	1.40	2.16	2.74	3.31	4.07	4.66	5.26	6.06	6.67
80	1.40	2.15	2.72	3.28	4.04	4.61	5.19	5.97	6.57
90	1.39	2.15	2.71	3.26	4.01	4.57	5.14	5.91	6.49
100	1.39	2.14	2.70	3.25	3.98	4.54	5.11	5.86	6.43
120	1.39	2.13	2.68	3.23	3.95	4.50	5.05	5.78	6.34
140	1.38	2.12	2.67	3.21	3.92	4.47	5.01	5.73	6.28
160	1.38	2.12	2.66	3.20	3.91	4.44	4.98	5.69	6.23
180	1.38	2.11	2.65	3.19	3.89	4.42	4.95	5.66	6.19
200	1.38	2.11	2.65	3.18	3.88	4.41	4/94	5.63	6.16
300	1.38	2.10	2.63	3.16	3.85	4.36	4.88	5.56	6.08
500	1.37	2.09	2.62	3.14	3.82	4.33	4.84	5.51	6.01
∞	1.37	2.08	2.61	3.12	3.78	4.28	4.77	5.42	5.91

TABLE B.4 (cont.) CRITICAL VALUES OF THE *F* DISTRIBUTION

Numerator DF = 4

α(2):	0.50	0.20	0.10	0.05	0.02	0.01	0.005	0.002	0.001
α(1):	0.25	0.10	0.05	0.025	0.01	0.005	0.0025	0.001	0.0005
Denom. DF									
1	8.58	55.8	225.	900.	5620.	22500.	90000.	562000.	2250000.
2	3.23	9.24	19.2	39.2	99.2	199.	399.	999.	2000.
3	2.39	5.34	9.12	15.1	28.7	46.2	73.9	137.	218.
4	2.06	4.11	6.39	9.60	16.0	23.2	33.3	53.4	76.1
5	1.89	3.52	5.19	7.39	11.4	15.6	21.0	31.1	41.5
6	1.79	3.18	4.53	6.23	9.15	12.0	15.7	21.9	28.1
7	1.72	2.96	4.12	5.52	7.85	10.1	12.7	17.2	21.4
8	1.66	2.81	3.84	5.05	7.01	8.81	10.9	14.4	17.6
9	1.63	2.69	3.63	4.72	6.42	7.96	9.74	12.6	15.1
10	1.59	2.61	3.48	4.47	5.99	7.34	8.89	11.3	13.4
11	1.57	2.54	3.36	4.28	5.67	6.88	8.25	10.3	12.2
12	1.55	2.48	3.26	4.12	5.41	6.52	7.76	9.63	11.2
13	1.53	2.43	3.18	4.00	5.21	6.23	7.37	9.07	10.5
14	1.52	2.39	3.11	3.89	5.04	6.00	7.06	8.62	9.95
15	1.51	2.36	3.06	3.80	4.89	5.80	6.80	8.25	9.48
16	1.50	2.33	3.01	3.73	4.77	5.64	6.58	7.94	9.08
17	1.49	2.31	2.96	3.66	4.67	5.50	6.39	7.68	8.75
18	1.48	2.29	2.93	3.61	4.58	5.37	6.23	7.46	8.47
19	1.47	2.27	2.90	3.56	4.50	5.27	6.09	7.27	8.23
20	1.47	2.25	2.87	3.51	4.43	5.17	5.97	7.10	8.02
21	1.46	2.23	2.84	3.48	4.37	5.09	5.86	6.95	7.83
22	1.45	2.22	2.82	3.44	4.31	5.02	5.76	6.81	7.67
23	1.45	2.21	2.80	3.41	4.26	4.95	5.67	6.70	7.52
24	1.44	2.19	2.78	3.38	4.22	4.89	5.60	6.59	7.39
25	1.44	2.18	2.76	3.35	4.18	4.84	5.53	6.49	7.27
26	1.44	2.17	2.74	3.33	4.14	4.79	5.46	6.41	7.16
27	1.43	2.17	2.73	3.31	4.11	4.74	5.40	6.33	7.06
28	1.43	2.16	2.71	3.29	4.07	4.70	5.35	6.25	6.97
29	1.43	2.15	2.70	3.27	4.04	4.66	5.30	6.19	6.89
30	1.42	2.14	2.69	3.25	4.02	4.62	5.25	6.12	6.82
35	1.41	2.11	2.64	3.18	3.91	4.48	5.07	5.88	6.51
40	1.40	2.09	2.61	3.13	3.83	4.37	4.93	5.70	6.30
45	1.40	2.07	2.58	3.09	3.77	4.29	4.83	5.56	6.13
50	1.39	2.06	2.56	3.05	3.72	4.23	4.75	5.46	6.01
60	1.38	2.04	2.53	3.01	3.65	4.14	4.64	5.31	5.82
70	1.38	2.03	2.50	2.97	3.60	4.08	4.56	5.20	5.70
80	1.38	2.02	2.49	2.95	3.56	4.03	4.50	5.12	5.60
90	1.37	2.01	2.47	2.93	3.53	3.99	4.45	5.06	5.53
100	1.37	2.00	2.46	2.92	3.51	3.96	4.42	5.02	5.48
120	1.37	1.99	2.45	2.89	3.48	3.92	4.36	4.95	5.39
140	1.36	1.99	2.44	2.88	3.46	3.89	4.32	4.90	5.33
160	1.36	1.98	2.43	2.87	3.44	3.87	4.30	4.86	5.29
180	1.36	1.98	2.42	2.86	3.43	3.85	4.27	4.83	5.26
200	1.36	1.97	2.42	2.85	3.41	3.84	4.26	4.81	5.23
300	1.35	1.96	2.40	2.83	3.38	3.80	4.21	4.75	5.15
500	1.35	1.96	2.39	2.81	3.36	3.76	4.17	4.69	5.09
∞	1.35	1.94	2.37	2.79	3.32	3.72	4.11	4.62	5.00

TABLE B.4 (cont.) CRITICAL VALUES OF THE *F* DISTRIBUTION

Numerator DF = 5

α(2):	0.50	0.20	0.10	0.05	0.02	0.01	0.005	0.002	0.001
α(1):	0.25	0.10	0.05	0.025	0.01	0.005	0.0025	0.001	0.0005
Denom. DF									
1	8.82	57.2	230.	922.	5760.	23100.	92200.	576000.	2310000.
2	3.28	9.29	19.3	39.3	99.3	199.	399.	999.	2000.
3	2.41	5.31	9.01	14.9	28.2	45.4	72.6	135.	214.
4	2.07	4.05	6.26	9.36	15.5	22.5	32.3	51.7	73.6
5	1.89	3.45	5.05	7.15	11.0	14.9	20.2	29.8	39.7
6	1.79	3.11	4.39	5.99	8.75	11.5	14.9	20.8	26.6
7	1.71	2.88	3.97	5.29	7.46	9.52	12.0	16.2	20.2
8	1.66	2.73	3.69	4.82	6.63	8.30	10.3	13.5	16.4
9	1.62	2.61	3.48	4.48	6.06	7.47	9.12	11.7	14.1
10	1.59	2.52	3.33	4.24	5.64	6.87	8.29	10.5	12.4
11	1.56	2.45	3.20	4.04	5.32	6.42	7.67	9.58	11.2
12	1.54	2.39	3.11	3.89	5.06	6.07	7.20	8.89	10.4
13	1.52	2.35	3.03	3.77	4.86	5.79	6.82	8.35	9.66
14	1.51	2.31	2.96	3.66	4.69	5.56	6.51	7.92	9.11
15	1.49	2.27	2.90	3.58	4.56	5.37	6.26	7.57	8.66
16	1.48	2.24	2.85	3.50	4.44	5.21	6.05	7.27	8.29
17	1.47	2.22	2.81	3.44	4.34	5.07	5.87	7.02	7.98
18	1.46	2.20	2.77	3.38	4.25	4.96	5.72	6.81	7.71
19	1.46	2.18	2.74	3.33	4.17	4.85	5.58	6.62	7.48
20	1.45	2.16	2.71	3.29	4.10	4.76	5.46	6.46	7.27
21	1.44	2.14	2.68	3.25	4.04	4.68	5.36	6.32	7.10
22	1.44	2.13	2.66	3.22	3.99	4.61	5.26	6.19	6.94
23	1.43	2.11	2.64	3.18	3.94	4.54	5.18	6.08	6.80
24	1.43	2.10	2.62	3.15	3.90	4.49	5.11	5.98	6.68
25	1.42	2.09	2.60	3.13	3.85	4.43	5.04	5.89	6.56
26	1.42	2.08	2.59	3.10	3.82	4.38	4.98	5.80	6.46
27	1.42	2.07	2.57	3.08	3.78	4.34	4.92	5.73	6.37
28	1.41	2.06	2.56	3.06	3.75	4.30	4.87	5.66	6.28
29	1.41	2.06	2.55	3.04	3.73	4.26	4.82	5.59	6.21
30	1.41	2.05	2.53	3.03	3.70	4.23	4.78	5.53	6.13
35	1.40	2.02	2.49	2.96	3.59	4.09	4.60	5.30	5.85
40	1.39	2.00	2.45	2.90	3.51	3.99	4.47	5.13	5.64
45	1.38	1.98	2.42	2.86	3.45	3.91	4.37	5.00	5.49
50	1.37	1.97	2.40	2.83	3.41	3.85	4.30	4.90	5.37
60	1.37	1.95	2.37	2.79	3.34	3.76	4.19	4.76	5.20
70	1.36	1.93	2.35	2.75	3.29	3.70	4.11	4.66	5.08
80	1.36	1.92	2.33	2.73	3.26	3.65	4.05	4.58	4.99
90	1.35	1.91	2.32	2.71	3.23	3.62	4.01	4.53	4.92
100	1.35	1.91	2.31	2.70	3.21	3.59	3.97	4.48	4.87
120	1.35	1.90	2.29	2.67	3.17	3.55	3.92	4.42	4.79
140	1.34	1.89	2.28	2.66	3.15	3.52	3.89	4.37	4.74
160	1.34	1.88	2.27	2.65	3.13	3.50	3.86	4.33	4.69
180	1.34	1.88	2.26	2.64	3.12	3.48	3.84	4.31	4.66
200	1.34	1.88	2.26	2.63	3.11	3.47	3.82	4.29	4.64
300	1.33	1.87	2.24	2.61	3.08	3.43	3.77	4.22	4.56
500	1.33	1.86	2.23	2.59	3.05	3.40	3.73	4.18	4.51
∞	1.33	1.85	2.21	2.57	3.02	3.35	3.68	4.10	4.42

TABLE B.4 (cont.) CRITICAL VALUES OF THE *F* DISTRIBUTION

Numerator DF = 6

Denom. DF	α(2): 0.50 α(1): 0.25	0.20 0.10	0.10 0.05	0.05 0.025	0.02 0.01	0.01 0.005	0.005 0.0025	0.002 0.001	0.001 0.0005
1	8.98	58.2	234.	937.	5860.	23400.	93700.	586000.	2340000.
2	3.31	9.33	19.3	39.3	99.3	199.	399.	999.	2000.
3	2.42	5.28	8.94	14.7	27.9	44.8	71.7	133.	211.
4	2.08	4.01	6.16	9.20	15.2	22.0	31.5	50.5	71.9
5	1.89	3.40	4.95	6.98	10.7	14.5	19.6	28.8	38.5
6	1.78	3.05	4.28	5.82	8.47	11.1	14.4	20.0	25.6
7	1.71	2.83	3.87	5.12	7.19	9.16	11.5	15.5	19.3
8	1.65	2.67	3.58	4.65	6.37	7.95	9.83	12.9	15.7
9	1.61	2.55	3.37	4.32	5.80	7.13	8.68	11.1	13.3
10	1.58	2.46	3.22	4.07	5.39	6.54	7.87	9.93	11.7
11	1.55	2.39	3.09	3.88	5.07	6.10	7.27	9.05	10.6
12	1.53	2.33	3.00	3.73	4.82	5.76	6.80	8.38	9.74
13	1.51	2.28	2.92	3.60	4.62	5.48	6.44	7.86	9.07
14	1.50	2.24	2.85	3.50	4.46	5.26	6.14	7.44	8.53
15	1.48	2.21	2.79	3.41	4.32	5.07	5.89	7.09	8.10
16	1.47	2.18	2.74	3.34	4.20	4.91	5.68	6.80	7.74
17	1.46	2.15	2.70	3.28	4.10	4.78	5.51	6.56	7.43
18	1.45	2.13	2.66	3.22	4.01	4.66	5.36	6.35	7.18
19	1.44	2.11	2.63	3.17	3.94	4.56	5.23	6.18	6.95
20	1.44	2.09	2.60	3.13	3.87	4.47	5.11	6.02	6.76
21	1.43	2.08	2.57	3.09	3.81	4.39	5.01	5.88	6.59
22	1.42	2.06	2.55	3.05	3.76	4.32	4.92	5.76	6.44
23	1.42	2.05	2.53	3.02	3.71	4.26	4.84	5.65	6.30
24	1.41	2.04	2.51	2.99	3.67	4.20	4.76	5.55	6.18
25	1.41	2.02	2.49	2.97	3.63	4.15	4.70	5.46	6.07
26	1.41	2.01	2.47	2.94	3.59	4.10	4.64	5.38	5.98
27	1.40	2.00	2.46	2.92	3.56	4.06	4.58	5.31	5.89
28	1.40	2.00	2.45	2.90	3.53	4.02	4.53	5.24	5.80
29	1.40	1.99	2.43	2.88	3.50	3.98	4.48	5.18	5.73
30	1.39	1.98	2.42	2.87	3.47	3.95	4.44	5.12	5.66
35	1.38	1.95	2.37	2.80	3.37	3.81	4.27	4.89	5.39
40	1.37	1.93	2.34	2.74	3.29	3.71	4.14	4.73	5.19
45	1.36	1.91	2.31	2.70	3.23	3.64	4.05	4.61	5.04
50	1.36	1.90	2.29	2.67	3.19	3.58	3.98	4.51	4.93
60	1.35	1.87	2.25	2.63	3.12	3.49	3.87	4.37	4.76
70	1.34	1.86	2.23	2.59	3.07	3.43	3.79	4.28	4.64
80	1.34	1.85	2.21	2.57	3.04	3.39	3.74	4.20	4.56
90	1.33	1.84	2.20	2.55	3.01	3.35	3.70	4.15	4.50
100	1.33	1.83	2.19	2.54	2.99	3.33	3.66	4.11	4.45
120	1.33	1.82	2.18	2.52	2.96	3.28	3.61	4.04	4.37
140	1.32	1.82	2.16	2.50	2.93	3.26	3.58	4.00	4.32
160	1.32	1.81	2.16	2.49	2.92	3.24	3.55	3.97	4.28
180	1.32	1.81	2.15	2.48	2.90	3.22	3.53	3.94	4.25
200	1.32	1.80	2.14	2.47	2.89	3.21	3.52	3.92	4.22
300	1.32	1.79	2.13	2.45	2.86	3.17	3.47	3.86	4.15
500	1.31	1.79	2.12	2.43	2.84	3.14	3.43	3.81	4.10
∞	1.31	1.77	2.10	2.41	2.80	3.09	3.37	3.74	4.02

TABLE B.4 (cont.) CRITICAL VALUES OF THE *F* DISTRIBUTION

Numerator DF = 7

$\alpha(2)$:	0.50	0.20	0.10	0.05	0.02	0.01	0.005	0.002	0.001
$\alpha(1)$:	0.25	0.10	0.05	0.025	0.01	0.005	0.0025	0.001	0.0005
Denom. DF									
1	9.10	58.9	237.	948.	5930.	23700.	94900.	593000.	2370000.
2	3.34	9.35	19.4	39.4	99.4	199.	399.	999.	2000.
3	2.43	5.27	8.89	14.6	27.7	44.4	71.0	132.	209.
4	2.08	3.98	6.09	9.07	15.0	21.6	31.0	49.7	70.7
5	1.89	3.37	4.88	6.85	10.5	14.2	19.1	28.2	37.6
6	1.78	3.01	4.21	5.70	8.26	10.8	14.0	19.5	24.9
7	1.70	2.78	3.79	4.99	6.99	8.89	11.2	15.0	18.7
8	1.64	2.62	3.50	4.53	6.18	7.69	9.49	12.4	15.1
9	1.60	2.51	3.29	4.20	5.61	6.88	8.36	10.7	12.8
10	1.57	2.41	3.14	3.95	5.20	6.30	7.56	9.52	11.2
11	1.54	2.34	3.01	3.76	4.89	5.86	6.97	8.66	10.1
12	1.52	2.28	2.91	3.61	4.64	5.52	6.51	8.00	9.28
13	1.50	2.23	2.83	3.48	4.44	5.25	6.15	7.49	8.63
14	1.49	2.19	2.76	3.38	4.28	5.03	5.86	7.08	8.11
15	1.47	2.16	2.71	3.29	4.14	4.85	5.62	6.74	7.68
16	1.46	2.13	2.66	3.22	4.03	4.69	5.41	6.46	7.33
17	1.45	2.10	2.61	3.16	3.93	4.56	5.24	6.22	7.04
18	1.44	2.08	2.58	3.10	3.84	4.44	5.09	6.02	6.78
19	1.43	2.06	2.54	3.05	3.77	4.34	4.96	5.85	6.57
20	1.43	2.04	2.51	3.01	3.70	4.26	4.85	5.69	6.38
21	1.42	2.02	2.49	2.97	3.64	4.18	4.75	5.56	6.21
22	1.41	2.01	2.46	2.93	3.59	4.11	4.66	5.44	6.07
23	1.41	1.99	2.44	2.90	3.54	4.05	4.58	5.33	5.94
24	1.40	1.98	2.42	2.87	3.50	3.99	4.51	5.23	5.82
25	1.40	1.97	2.40	2.85	3.46	3.94	4.44	5.15	5.71
26	1.39	1.96	2.39	2.82	3.42	3.89	4.38	5.07	5.62
27	1.39	1.95	2.37	2.80	3.39	3.85	4.33	5.00	5.53
28	1.39	1.94	2.36	2.78	3.36	3.81	4.28	4.93	5.45
29	1.38	1.93	2.35	2.76	3.33	3.77	4.24	4.87	5.38
30	1.38	1.93	2.33	2.75	3.30	3.74	4.19	4.82	5.31
35	1.37	1.90	2.29	2.68	3.20	3.61	4.02	4.59	5.04
40	1.36	1.87	2.25	2.62	3.12	3.51	3.90	4.44	4.85
45	1.35	1.85	2.22	2.58	3.07	3.43	3.81	4.32	4.71
50	1.34	1.84	2.20	2.55	3.02	3.38	3.74	4.22	4.60
60	1.33	1.82	2.17	2.51	2.95	3.29	3.63	4.09	4.44
70	1.33	1.80	2.14	2.47	2.91	3.23	3.56	3.99	4.32
80	1.32	1.79	2.13	2.45	2.87	3.19	3.50	3.92	4.24
90	1.32	1.78	2.11	2.43	2.84	3.15	3.46	3.87	4.18
100	1.32	1.78	2.10	2.42	2.82	3.13	3.43	3.83	4.13
120	1.31	1.77	2.09	2.39	2.79	3.09	3.38	3.77	4.06
140	1.31	1.76	2.08	2.38	2.77	3.06	3.35	3.72	4.01
160	1.31	1.75	2.07	2.37	2.75	3.04	3.32	3.69	3.97
180	1.31	1.75	2.06	2.36	2.74	3.02	3.30	3.67	3.94
200	1.30	1.75	2.06	2.35	2.73	3.01	3.29	3.65	3.92
300	1.30	1.74	2.04	2.33	2.70	2.97	3.24	3.59	3.85
500	1.30	1.73	2.03	2.31	2.68	2.94	3.20	3.54	3.80
∞	1.29	1.72	2.01	2.29	2.64	2.90	3.15	3.47	3.72

TABLE B.4 (cont.) CRITICAL VALUES OF THE *F* DISTRIBUTION

Numerator DF = 8

α(2):	0.50	0.20	0.10	0.05	0.02	0.01	0.005	0.002	0.001
α(1):	0.25	0.10	0.05	0.025	0.01	0.005	0.0025	0.001	0.0005
Denom. DF									
1	9.19	59.4	239.	957.	5980.	23900.	95700.	598000.	2390000.
2	3.35	9.37	19.4	39.4	99.4	199.	399.	999.	2000.
3	2.44	5.25	8.85	14.5	27.5	44.1	70.5	131.	208.
4	2.08	3.95	6.04	8.98	14.8	21.4	30.6	49.0	69.7
5	1.89	3.34	4.82	6.76	10.3	14.0	18.8	27.6	36.9
6	1.78	2.98	4.15	5.60	8.10	10.6	13.7	19.0	24.3
7	1.70	2.75	3.73	4.90	6.84	8.68	10.9	14.6	18.2
8	1.64	2.59	3.44	4.43	6.03	7.50	9.24	12.0	14.6
9	1.60	2.47	3.23	4.10	5.47	6.69	8.12	10.4	12.4
10	1.56	2.38	3.07	3.85	5.06	6.12	7.33	9.20	10.9
11	1.53	2.30	2.95	3.66	4.74	5.68	6.74	8.35	9.76
12	1.51	2.24	2.85	3.51	4.50	5.35	6.29	7.71	8.94
13	1.49	2.20	2.77	3.39	4.30	5.08	5.93	7.21	8.29
14	1.48	2.15	2.70	3.29	4.14	4.86	5.64	6.80	7.78
15	1.46	2.12	2.64	3.20	4.00	4.67	5.40	6.47	7.37
16	1.45	2.09	2.59	3.12	3.89	4.52	5.20	6.19	7.02
17	1.44	2.06	2.55	3.06	3.79	4.39	5.03	5.96	6.73
18	1.43	2.04	2.51	3.01	3.71	4.28	4.89	5.76	6.48
19	1.42	2.02	2.48	2.96	3.63	4.18	4.76	5.59	6.27
20	1.42	2.00	2.45	2.91	3.56	4.09	4.65	5.44	6.09
21	1.41	1.98	2.42	2.87	3.51	4.01	4.55	5.31	5.92
22	1.40	1.97	2.40	2.84	3.45	3.94	4.46	5.19	5.78
23	1.40	1.95	2.37	2.81	3.41	3.88	4.38	5.09	5.65
24	1.39	1.94	2.36	2.78	3.36	3.83	4.31	4.99	5.54
25	1.39	1.93	2.34	2.75	3.32	3.78	4.25	4.91	5.43
26	1.38	1.92	2.32	2.73	3.29	3.73	4.19	4.83	5.34
27	1.38	1.91	2.31	2.71	3.26	3.69	4.14	4.76	5.25
28	1.38	1.90	2.29	2.69	3.23	3.65	4.09	4.69	5.18
29	1.37	1.89	2.28	2.67	3.20	3.61	4.04	4.64	5.11
30	1.37	1.88	2.27	2.65	3.17	3.58	4.00	4.58	5.04
35	1.36	1.85	2.22	2.58	3.07	3.45	3.83	4.36	4.78
40	1.35	1.83	2.18	2.53	2.99	3.35	3.71	4.21	4.59
45	1.34	1.81	2.15	2.49	2.94	3.28	3.62	4.09	4.45
50	1.33	1.80	2.13	2.46	2.89	3.22	3.55	4.00	4.34
60	1.32	1.77	2.10	2.41	2.82	3.13	3.45	3.86	4.19
70	1.32	1.76	2.07	2.38	2.78	3.08	3.37	3.77	4.08
80	1.31	1.75	2.06	2.35	2.74	3.03	3.32	3.70	4.00
90	1.31	1.74	2.04	2.34	2.72	3.00	3.28	3.65	3.94
100	1.30	1.73	2.03	2.32	2.69	2.97	3.25	3.61	3.89
120	1.30	1.72	2.02	2.30	2.66	2.93	3.20	3.55	3.82
140	1.30	1.71	2.01	2.28	2.64	2.91	3.17	3.51	3.77
160	1.29	1.71	2.00	2.27	2.62	2.88	3.14	3.48	3.73
180	1.29	1.70	1.99	2.26	2.61	2.87	3.12	3.45	3.70
200	1.29	1.70	1.98	2.26	2.60	2.86	3.11	3.43	3.68
300	1.29	1.69	1.97	2.23	2.57	2.82	3.06	3.38	3.61
500	1.28	1.68	1.96	2.22	2.55	2.79	3.03	3.33	3.56
∞	1.28	1.67	1.94	2.19	2.51	2.74	2.97	3.27	3.48

TABLE B.4 (cont.) CRITICAL VALUES OF THE *F* DISTRIBUTION

Numerator DF = 9

α(2):	0.50	0.20	0.10	0.05	0.02	0.01	0.005	0.002	0.001
α(1):	0.25	0.10	0.05	0.025	0.01	0.005	0.0025	0.001	0.0005
Denom. DF									
1	9.26	59.9	241.	963.	6020.	24100.	96400.	602000.	2410000.
2	3.37	9.38	19.4	39.4	99.4	199.	399.	999.	2000.
3	2.44	5.24	8.81	14.5	27.3	43.9	70.1	130.	207.
4	2.08	3.94	6.00	8.90	14.7	21.1	30.3	48.5	69.0
5	1.89	3.32	4.77	6.68	10.2	13.8	18.5	27.2	36.3
6	1.77	2.96	4.10	5.52	7.98	10.4	13.4	18.7	23.9
7	1.69	2.72	3.68	4.82	6.72	8.51	10.7	14.3	17.8
8	1.63	2.56	3.39	4.36	5.91	7.34	9.03	11.8	14.3
9	1.59	2.44	3.18	4.03	5.35	6.54	7.92	10.1	12.1
10	1.56	2.35	3.02	3.78	4.94	5.97	7.14	8.96	10.6
11	1.53	2.27	2.90	3.59	4.63	5.54	6.56	8.12	9.48
12	1.51	2.21	2.80	3.44	4.39	5.20	6.11	7.48	8.66
13	1.49	2.16	2.71	3.31	4.19	4.94	5.76	6.98	8.03
14	1.47	2.12	2.65	3.21	4.03	4.72	5.47	6.58	7.52
15	1.46	2.09	2.59	3.12	3.89	4.54	5.23	6.26	7.11
16	1.44	2.06	2.54	3.05	3.78	4.38	5.04	5.98	6.77
17	1.43	2.03	2.49	2.98	3.68	4.25	4.87	5.75	6.49
18	1.42	2.00	2.46	2.93	3.60	4.14	4.72	5.56	6.24
19	1.41	1.98	2.42	2.88	3.52	4.04	4.60	5.39	6.03
20	1.41	1.96	2.39	2.84	3.46	3.96	4.49	5.24	5.85
21	1.40	1.95	2.37	2.80	3.40	3.88	4.39	5.11	5.69
22	1.39	1.93	2.34	2.76	3.35	3.81	4.30	4.99	5.55
23	1.39	1.92	2.32	2.73	3.30	3.75	4.22	4.89	5.43
24	1.38	1.91	2.30	2.70	3.26	3.69	4.15	4.80	5.31
25	1.38	1.89	2.28	2.68	3.22	3.64	4.09	4.71	5.21
26	1.37	1.88	2.27	2.65	3.18	3.60	4.03	4.64	5.12
27	1.37	1.87	2.25	2.63	3.15	3.56	3.98	4.57	5.04
28	1.37	1.87	2.24	2.61	3.12	2.52	3.93	4.50	4.96
29	1.36	1.86	2.22	2.59	3.09	3.48	3.89	4.45	4.89
30	1.36	1.85	2.21	2.57	3.07	3.45	3.85	4.39	4.82
35	1.35	1.82	2.16	2.50	2.96	3.32	3.68	4.18	4.57
40	1.34	1.79	2.12	2.45	2.89	3.22	3.56	4.02	4.38
45	1.33	1.77	2.10	2.41	2.83	3.15	3.47	3.91	4.25
50	1.32	1.76	2.07	2.38	2.78	3.09	3.40	3.82	4.14
60	1.31	1.74	2.04	2.33	2.72	3.01	3.30	3.69	3.98
70	1.31	1.72	2.02	2.30	2.67	2.95	3.23	3.60	3.88
80	1.30	1.71	2.00	2.28	2.64	2.91	3.17	3.53	3.80
90	1.30	1.70	1.99	2.26	2.61	2.87	3.13	3.48	3.74
100	1.29	1.69	1.97	2.24	2.59	2.85	3.10	3.44	3.69
120	1.29	1.68	1.96	2.22	2.56	2.81	3.06	3.38	3.62
140	1.29	1.68	1.95	2.21	2.54	2.78	3.02	3.34	3.57
160	1.28	1.67	1.94	2.19	2.52	2.76	3.00	3.31	3.54
180	1.28	1.67	1.93	2.19	2.51	2.74	2.98	3.28	3.51
200	1.28	1.66	1.93	2.18	2.50	2.73	2.96	3.26	3.49
300	1.27	1.65	1.91	2.16	2.47	2.69	2.92	3.21	3.42
500	1.27	1.64	1.90	2.14	2.44	2.66	2.88	3.16	3.37
∞	1.27	1.63	1.88	2.11	2.41	2.62	2.83	3.10	3.30

TABLE B.4 (cont.) CRITICAL VALUES OF THE *F* DISTRIBUTION

Numerator DF = 10

α(2):	0.50	0.20	0.10	0.05	0.02	0.01	0.005	0.002	0.001
α(1):	0.25	0.10	0.05	0.025	0.01	0.005	0.0025	0.001	0.0005
Denom. DF									
1	9.32	60.2	242.	969.	6060.	24200.	96900.	606000.	2420000.
2	3.38	9.39	19.4	39.4	99.4	199.	399.	999.	2000.
3	2.44	5.23	8.79	14.4	27.2	43.7	69.8	129.	206.
4	2.08	3.92	5.96	8.84	14.5	21.0	30.0	48.1	68.3
5	1.89	3.30	4.74	6.62	10.1	13.6	18.3	26.9	35.9
6	1.77	2.94	4.06	5.46	7.87	10.3	13.2	18.4	23.5
7	1.69	2.70	3.64	4.76	6.62	8.38	10.5	14.1	17.5
8	1.63	2.54	3.35	4.30	5.81	7.21	8.87	11.5	14.0
9	1.59	2.42	3.14	3.96	5.26	6.42	7.77	9.89	11.8
10	1.55	2.32	2.98	3.72	4.85	5.85	6.99	8.75	10.3
11	1.52	2.25	2.85	3.53	4.54	5.42	6.41	7.92	9.24
12	1.50	2.19	2.75	3.37	4.30	5.09	5.97	7.29	8.43
13	1.48	2.14	2.67	3.25	4.10	4.82	5.62	6.80	7.81
14	1.46	2.10	2.60	3.15	3.94	4.60	5.33	6.40	7.31
15	1.45	2.06	2.54	3.06	3.80	4.42	5.10	6.08	6.91
16	1.44	2.03	2.49	2.99	3.69	4.27	4.90	5.81	6.57
17	1.43	2.00	2.45	2.92	3.59	4.14	4.73	5.58	6.29
18	1.42	1.98	2.41	2.87	3.51	4.03	4.59	5.39	6.05
19	1.41	1.96	2.38	2.82	3.43	3.93	4.46	5.22	5.84
20	1.40	1.94	2.35	2.77	3.37	3.85	4.35	5.08	5.66
21	1.39	1.92	2.32	2.73	3.31	3.77	4.26	4.95	5.50
22	1.39	1.90	2.30	2.70	3.26	3.70	4.17	4.83	5.36
23	1.38	1.89	2.27	2.67	3.21	3.64	4.09	4.73	5.24
24	1.38	1.88	2.25	2.64	3.17	3.59	4.03	4.64	5.13
25	1.37	1.87	2.24	2.61	3.13	3.54	3.96	4.56	5.03
26	1.37	1.86	2.22	2.59	3.09	3.49	3.91	4.48	4.94
27	1.36	1.85	2.20	2.57	3.06	3.45	3.85	4.41	4.86
28	1.36	1.84	2.19	2.55	3.03	3.41	3.81	4.35	4.78
29	1.35	1.83	2.18	2.53	3.00	3.38	3.76	4.29	4.71
30	1.35	1.82	2.16	2.51	2.98	3.34	3.72	4.24	4.65
35	1.34	1.79	2.11	2.44	2.88	3.21	3.56	4.03	4.39
40	1.33	1.76	2.08	2.39	2.80	3.12	3.44	3.87	4.21
45	1.32	1.74	2.05	2.35	2.74	3.04	3.35	3.76	4.08
50	1.31	1.73	2.03	2.32	2.70	2.99	3.28	3.67	3.97
60	1.30	1.71	1.99	2.27	2.63	2.90	3.18	3.54	3.82
70	1.30	1.69	1.97	2.24	2.59	2.85	3.11	3.45	3.71
80	1.29	1.68	1.95	2.21	2.55	2.80	3.05	3.39	3.64
90	1.29	1.67	1.94	2.19	2.52	2.77	3.01	3.34	3.58
100	1.28	1.66	1.93	2.18	2.50	2.74	2.98	3.30	3.53
120	1.28	1.65	1.91	2.16	2.47	2.71	2.94	3.24	3.46
140	1.28	1.64	1.90	2.14	2.45	2.68	2.90	3.20	3.42
160	1.27	1.64	1.89	2.13	2.43	2.66	2.88	3.17	3.38
180	1.27	1.63	1.88	2.12	2.42	2.64	2.86	3.14	3.35
200	1.27	1.63	1.88	2.11	2.41	2.63	2.84	3.12	3.33
300	1.26	1.62	1.86	2.09	2.38	2.59	2.80	3.07	3.27
500	1.26	1.61	1.85	2.07	2.36	2.56	2.76	3.02	3.22
∞	1.25	1.60	1.83	2.05	2.32	2.52	2.71	2.96	3.14

TABLE B.4 (cont.) CRITICAL VALUES OF THE *F* DISTRIBUTION

Numerator DF = 11

α(2):	0.50	0.20	0.10	0.05	0.02	0.01	0.005	0.002	0.001
α(1):	0.25	0.10	0.05	0.025	0.01	0.005	0.0025	0.001	0.0005
Denom. DF									
1	9.37	60.5	243.	973.	6080.	24300.	97300.	608000.	2430000.
2	3.39	9.40	19.4	39.4	99.4	199.	399.	999.	2000.
3	2.45	5.22	8.76	14.4	27.1	43.5	69.5	129.	205.
4	2.08	3.91	5.94	8.79	14.5	20.8	29.8	47.7	67.8
5	1.89	3.28	4.70	6.57	9.96	13.5	18.1	26.6	35.5
6	1.77	2.92	4.03	5.41	7.79	10.1	13.1	18.2	23.2
7	1.69	2.68	3.60	4.71	6.54	8.27	10.4	13.9	17.2
8	1.63	2.52	3.31	4.24	5.73	7.10	8.73	11.4	13.8
9	1.58	2.40	3.10	3.91	5.18	6.31	7.63	9.72	11.6
10	1.55	2.30	2.94	3.66	4.77	5.75	6.86	8.59	10.1
11	1.52	2.23	2.82	3.47	4.46	5.32	6.29	7.76	9.05
12	1.49	2.17	2.72	3.32	4.22	4.99	5.85	7.14	8.25
13	1.47	2.12	2.63	3.20	4.02	4.72	5.50	6.65	7.63
14	1.46	2.07	2.57	3.09	3.86	4.51	5.21	6.26	7.13
15	1.44	2.04	2.51	3.01	3.73	4.33	4.98	5.94	6.73
16	1.43	2.01	2.46	2.93	3.62	4.18	4.79	5.67	6.40
17	1.42	1.98	2.41	2.87	3.52	4.05	4.62	5.44	6.12
18	1.41	1.95	2.37	2.81	3.43	3.94	4.48	5.25	5.89
19	1.40	1.93	2.34	2.76	3.36	3.84	4.35	5.08	5.68
20	1.39	1.91	2.31	2.72	3.29	3.76	4.24	4.94	5.50
21	1.39	1.90	2.28	2.68	3.24	3.68	4.15	4.81	5.35
22	1.38	1.88	2.26	2.65	3.18	3.61	4.06	4.70	5.21
23	1.37	1.87	2.24	2.62	3.14	3.55	3.99	4.60	5.09
24	1.37	1.85	2.22	2.59	3.09	3.50	3.92	4.51	4.98
25	1.36	1.84	2.20	2.56	3.06	3.45	3.85	4.42	4.88
26	1.36	1.83	2.18	2.54	3.02	3.40	3.80	4.35	4.79
27	1.35	1.82	2.17	2.51	2.99	3.36	3.75	4.28	4.71
28	1.35	1.81	2.15	2.49	2.96	3.32	3.70	4.22	4.63
29	1.35	1.80	2.14	2.48	2.93	3.29	3.66	4.16	4.56
30	1.34	1.79	2.13	2.46	2.91	3.25	3.61	4.11	4.50
35	1.33	1.76	2.07	2.39	2.80	3.12	3.45	3.90	4.25
40	1.32	1.74	2.04	2.33	2.73	3.03	3.33	3.75	4.07
45	1.31	1.72	2.01	2.29	2.67	2.96	3.25	3.64	3.94
50	1.30	1.70	1.99	2.26	2.63	2.90	3.18	3.55	3.83
60	1.29	1.68	1.95	2.22	2.56	2.82	3.08	3.42	3.68
70	1.29	1.66	1.93	2.18	2.51	2.76	3.00	3.33	3.58
80	1.28	1.65	1.91	2.16	2.48	2.72	2.95	3.27	3.50
90	1.28	1.64	1.90	2.14	2.45	2.68	2.91	3.22	3.44
100	1.27	1.64	1.89	2.12	2.43	2.66	2.88	3.18	3.40
120	1.27	1.63	1.87	2.10	2.40	2.62	2.83	3.12	3.33
140	1.27	1.62	1.86	2.09	2.38	2.59	2.80	3.08	3.28
160	1.26	1.61	1.85	2.07	2.36	2.57	2.78	3.05	3.25
180	1.26	1.61	1.84	2.07	2.35	2.56	2.76	3.02	3.22
200	1.26	1.60	1.84	2.06	2.34	2.54	2.74	3.00	3.20
300	1.26	1.59	1.82	2.04	2.31	2.51	2.70	2.95	3.14
500	1.25	1.58	1.81	2.02	2.28	2.48	2.66	2.91	3.09
∞	1.25	1.57	1.79	1.99	2.25	2.43	2.61	2.84	3.01

TABLE B.4 (cont.) CRITICAL VALUES OF THE *F* DISTRIBUTION

Numerator DF = 12

α(2):	0.50	0.20	0.10	0.05	0.02	0.01	0.005	0.002	0.001
α(1):	0.25	0.10	0.05	0.025	0.01	0.005	0.0025	0.001	0.0005
Denom. DF									
1	9.41	60.7	244.	977.	6110.	24400.	97700.	611000.	2440000.
2	3.39	9.41	19.4	39.4	99.4	199.	399.	999.	2000.
3	2.45	5.22	8.74	14.3	27.1	43.4	69.3	128.	204.
4	2.08	3.90	5.91	8.75	14.4	20.7	29.7	47.4	67.4
5	1.89	3.27	4.68	6.52	9.89	13.4	18.0	26.4	35.2
6	1.77	2.90	4.00	5.37	7.72	10.0	12.9	18.0	23.0
7	1.68	2.67	3.57	4.67	6.47	8.18	10.3	13.7	17.0
8	1.62	2.50	3.28	4.20	5.67	7.01	8.61	11.2	13.6
9	1.58	2.38	3.07	3.87	5.11	6.23	7.52	9.57	11.4
10	1.54	2.28	2.91	3.62	4.71	5.66	6.75	8.45	9.94
11	1.51	2.21	2.79	3.43	4.40	5.24	6.18	7.63	8.88
12	1.49	2.15	2.69	3.28	4.16	4.91	5.74	7.00	8.09
13	1.47	2.10	2.60	3.15	3.96	4.64	5.40	6.52	7.48
14	1.45	2.05	2.53	3.05	3.80	4.43	5.12	6.13	6.99
15	1.44	2.02	2.48	2.96	3.67	4.25	4.88	5.81	6.59
16	1.43	1.99	2.42	2.89	3.55	4.10	4.69	5.55	6.26
17	1.41	1.96	2.38	2.82	3.46	3.97	4.52	5.32	5.98
18	1.40	1.93	2.34	2.77	3.37	3.86	4.38	5.13	5.75
19	1.40	1.91	2.31	2.72	3.30	3.76	4.26	4.97	5.55
20	1.39	1.89	2.28	2.68	3.23	3.68	4.15	4.82	5.37
21	1.38	1.87	2.25	2.64	3.17	3.60	4.06	4.70	5.21
22	1.37	1.86	2.23	2.60	3.12	3.54	3.97	4.58	5.08
23	1.37	1.84	2.20	2.57	3.07	3.47	3.89	4.48	4.96
24	1.36	1.83	2.18	2.54	3.03	3.42	3.83	4.39	4.85
25	1.36	1.82	2.16	2.51	2.99	3.37	3.76	4.31	4.75
26	1.35	1.81	2.15	2.49	2.96	3.33	3.71	4.24	4.66
27	1.35	1.80	2.13	2.47	2.93	3.28	3.66	4.17	4.58
28	1.34	1.79	2.12	2.45	2.90	3.25	3.61	4.11	4.51
29	1.34	1.78	2.10	2.43	2.87	3.21	3.56	4.05	4.44
30	1.34	1.77	2.09	2.41	2.84	3.18	3.52	4.00	4.38
35	1.32	1.74	2.04	2.34	2.74	3.05	3.36	3.79	4.13
40	1.31	1.71	2.00	2.29	2.66	2.95	3.25	3.64	3.95
45	1.30	1.70	1.97	2.25	2.61	2.88	3.16	3.53	3.82
50	1.30	1.68	1.95	2.22	2.56	2.82	3.09	3.44	3.71
60	1.29	1.66	1.92	2.17	2.50	2.74	2.99	3.32	3.57
70	1.28	1.64	1.89	2.14	2.45	2.68	2.92	3.23	3.46
80	1.27	1.63	1.88	2.11	2.42	2.64	2.87	3.16	3.39
90	1.27	1.62	1.86	2.09	2.39	2.61	2.83	3.11	3.33
100	1.27	1.61	1.85	2.08	2.37	2.58	2.80	3.07	3.28
120	1.26	1.60	1.83	2.05	2.34	2.54	2.75	3.02	3.22
140	1.26	1.59	1.82	2.04	2.31	2.52	2.72	2.98	3.17
160	1.26	1.59	1.81	2.03	2.30	2.50	2.69	2.95	3.14
180	1.25	1.58	1.81	2.02	2.28	2.48	2.67	2.92	3.11
200	1.25	1.58	1.80	2.01	2.27	2.47	2.66	2.90	3.09
300	1.25	1.57	1.78	1.99	2.24	2.43	2.61	2.85	3.02
500	1.24	1.56	1.77	1.97	2.22	2.40	2.58	2.81	2.97
∞	1.24	1.55	1.75	1.94	2.18	2.36	2.53	2.74	2.90

TABLE B.4 (cont.) CRITICAL VALUES OF THE *F* DISTRIBUTION

Numerator DF = 13

α(2):	0.50	0.20	0.10	0.05	0.02	0.01	0.005	0.002	0.001
α(1):	0.25	0.10	0.05	0.025	0.01	0.005	0.0025	0.001	0.0005
Denom. DF									
1	9.44	60.9	245.	980.	6130.	24500.	98000.	613000.	2450000.
2	3.40	9.41	19.4	39.4	99.4	199.	399.	999.	2000.
3	2.45	5.21	8.73	14.3	27.0	43.3	69.1	128.	204.
4	2.08	3.89	5.89	8.71	14.3	20.6	29.5	47.2	67.1
5	1.89	3.26	4.66	6.49	9.82	13.3	17.9	26.2	34.9
6	1.77	2.89	3.98	5.33	7.66	9.95	12.8	17.8	22.7
7	1.68	2.65	3.55	4.63	6.41	8.10	10.1	13.6	16.8
8	1.62	2.49	3.26	4.16	5.61	6.94	8.51	11.1	13.4
9	1.58	2.36	3.05	3.83	5.05	6.15	7.43	9.44	11.3
10	1.54	2.27	2.89	3.58	4.65	5.59	6.66	8.32	9.80
11	1.51	2.19	2.76	3.39	4.34	5.16	6.09	7.51	8.74
12	1.49	2.13	2.66	3.24	4.10	4.84	5.66	6.89	7.96
13	1.47	2.08	2.58	3.12	3.91	4.57	5.31	6.41	7.35
14	1.45	2.04	2.51	3.01	3.75	4.36	5.03	6.02	6.86
15	1.43	2.00	2.45	2.92	3.61	4.18	4.80	5.71	6.47
16	1.42	1.97	2.40	2.85	3.50	4.03	4.61	5.44	6.14
17	1.41	1.94	2.35	2.79	3.40	3.90	4.44	5.22	5.86
18	1.40	1.92	2.31	2.73	3.32	3.79	4.30	5.03	5.63
19	1.39	1.89	2.28	2.68	3.24	3.70	4.18	4.87	5.43
20	1.38	1.87	2.25	2.64	3.18	3.61	4.07	4.72	5.25
21	1.37	1.86	2.22	2.60	3.12	3.54	3.98	4.60	5.10
22	1.37	1.84	2.20	2.56	3.07	3.47	3.89	4.49	4.97
23	1.36	1.83	2.18	2.53	3.02	3.41	3.82	4.39	4.84
24	1.36	1.81	2.15	2.50	2.98	3.35	3.75	4.30	4.74
25	1.35	1.80	2.14	2.48	2.94	3.30	3.69	4.22	4.64
26	1.35	1.79	2.12	2.45	2.90	3.26	3.63	4.14	4.55
27	1.34	1.78	2.10	2.43	2.87	3.22	3.58	4.08	4.47
28	1.34	1.77	2.09	2.41	2.84	3.18	3.53	4.01	4.40
29	1.33	1.76	2.08	2.39	2.81	3.15	3.49	3.96	4.33
30	1.33	1.75	2.06	2.37	2.79	3.11	3.45	3.91	4.27
35	1.32	1.72	2.01	2.30	2.69	2.98	3.29	3.70	4.02
40	1.31	1.70	1.97	2.25	2.61	2.89	3.17	3.55	3.85
45	1.30	1.68	1.94	2.21	2.55	2.82	3.08	3.44	3.71
50	1.29	1.66	1.92	2.18	2.51	2.76	3.01	3.35	3.61
60	1.28	1.64	1.89	2.13	2.44	2.68	2.91	3.23	3.46
70	1.27	1.62	1.86	2.10	2.40	2.62	2.84	3.14	3.36
80	1.27	1.61	1.84	2.07	2.36	2.58	2.79	3.07	3.29
90	1.26	1.60	1.83	2.05	2.33	2.54	2.75	3.02	3.23
100	1.26	1.59	1.82	2.04	2.31	2.52	2.72	2.99	3.19
120	1.26	1.58	1.80	2.01	2.28	2.48	2.67	2.93	3.12
140	1.25	1.57	1.79	2.00	2.26	2.45	2.64	2.89	3.07
160	1.25	1.57	1.78	1.99	2.24	2.43	2.62	2.86	3.04
180	1.25	1.56	1.77	1.98	2.23	2.42	2.60	2.83	3.01
200	1.24	1.56	1.77	1.97	2.22	2.40	2.58	2.82	2.99
300	1.24	1.55	1.75	1.95	2.19	2.37	2.54	2.76	2.93
500	1.24	1.54	1.74	1.93	2.17	2.34	2.50	2.72	2.88
∞	1.23	1.52	1.72	1.90	2.13	2.29	2.45	2.66	2.81

TABLE B.4 (cont.) CRITICAL VALUES OF THE *F* DISTRIBUTION

Numerator DF = 14

α(2):	0.50	0.20	0.10	0.05	0.02	0.01	0.005	0.002	0.001
α(1):	0.25	0.10	0.05	0.025	0.01	0.005	0.0025	0.001	0.0005
Denom. DF									
1	9.47	61.1	245.	983.	6140.	24600.	98300.	614000.	2460000.
2	3.41	9.42	19.4	39.4	99.4	199.	399.	999.	2000.
3	2.45	5.20	8.71	14.3	26.9	43.2	69.0	128.	203.
4	2.08	3.88	5.87	8.68	14.2	20.5	29.4	46.9	66.8
5	1.89	3.25	4.64	6.46	9.77	13.2	17.8	26.1	34.7
6	1.76	2.88	3.96	5.30	7.60	9.88	12.7	17.7	22.6
7	1.68	2.64	3.53	4.60	6.36	8.03	10.1	13.4	16.6
8	1.62	2.48	3.24	4.13	5.56	6.87	8.43	10.9	13.3
9	1.57	2.35	3.03	3.80	5.01	6.09	7.35	9.33	11.1
10	1.54	2.26	2.86	3.55	4.60	5.53	6.58	8.22	9.67
11	1.51	2.18	2.74	3.36	4.29	5.10	6.02	7.41	8.62
12	1.48	2.12	2.64	3.21	4.05	4.77	5.58	6.79	7.84
13	1.46	2.07	2.55	3.08	3.86	4.51	5.24	6.31	7.23
14	1.44	2.02	2.48	2.98	3.70	4.30	4.96	5.93	6.75
15	1.43	1.99	2.42	2.89	3.56	4.12	4.73	5.62	6.36
16	1.42	1.95	2.37	2.82	3.45	3.97	4.54	5.35	6.03
17	1.41	1.93	2.33	2.75	3.35	3.84	4.37	5.13	5.76
18	1.40	1.90	2.29	2.70	3.27	3.73	4.23	4.94	5.53
19	1.39	1.88	2.26	2.65	3.19	3.64	4.11	4.78	5.33
20	1.38	1.86	2.22	2.60	3.13	3.55	4.00	4.64	5.15
21	1.37	1.84	2.20	2.56	3.07	3.48	3.91	4.51	5.00
22	1.36	1.83	2.17	2.53	3.02	3.41	3.82	4.40	4.87
23	1.36	1.81	2.15	2.50	2.97	3.35	3.75	4.30	4.75
24	1.35	1.80	2.13	2.47	2.93	3.30	3.68	4.21	4.64
25	1.35	1.79	2.11	2.44	2.89	3.25	3.62	4.13	4.54
26	1.34	1.77	2.09	2.42	2.86	3.20	3.56	4.06	4.46
27	1.34	1.76	2.08	2.39	2.82	3.16	3.51	3.99	4.38
28	1.33	1.75	2.06	2.37	2.79	3.12	3.46	3.93	4.30
29	1.33	1.75	2.05	2.36	2.77	3.09	3.42	3.88	4.24
30	1.33	1.74	2.04	2.34	2.74	3.06	3.38	3.82	4.18
35	1.31	1.70	1.99	2.27	2.64	2.93	3.22	3.62	3.93
40	1.30	1.68	1.95	2.21	2.56	2.83	3.10	3.47	3.76
45	1.29	1.66	1.92	2.17	2.51	2.76	3.02	3.36	3.63
50	1.28	1.64	1.89	2.14	2.46	2.70	2.95	3.27	3.52
60	1.27	1.62	1.86	2.09	2.39	2.62	2.85	3.15	3.38
70	1.27	1.60	1.84	2.06	2.35	2.56	2.78	3.06	3.28
80	1.26	1.59	1.82	2.03	2.31	2.52	2.73	3.00	3.20
90	1.26	1.58	1.80	2.02	2.29	2.49	2.69	2.95	3.14
100	1.25	1.57	1.79	2.00	2.27	2.46	2.65	2.91	3.10
120	1.25	1.56	1.78	1.98	2.23	2.42	2.61	2.85	3.03
140	1.24	1.55	1.76	1.96	2.21	2.40	2.58	2.81	2.99
160	1.24	1.55	1.75	1.95	2.20	2.38	2.55	2.78	2.95
180	1.24	1.54	1.75	1.94	2.18	2.36	2.53	2.76	2.93
200	1.24	1.54	1.74	1.93	2.17	2.35	2.52	2.74	2.91
300	1.23	1.53	1.72	1.91	2.14	2.31	2.47	2.69	2.84
500	1.23	1.52	1.71	1.89	2.12	2.28	2.44	2.64	2.79
∞	1.22	1.50	1.69	1.87	2.08	2.24	2.39	2.58	2.72

TABLE B.4 (cont.) CRITICAL VALUES OF THE *F* DISTRIBUTION

Numerator DF = 15

α(2):	0.50	0.20	0.10	0.05	0.02	0.01	0.005	0.002	0.001
α(1):	0.25	0.10	0.05	0.025	0.01	0.005	0.0025	0.001	0.0005
Denom. DF									
1	9.49	61.2	246.	985.	6160.	24600.	98500.	616000.	2460000.
2	3.41	9.42	19.4	39.4	99.4	199.	399.	999.	2000.
3	2.46	5.20	8.70	14.3	26.9	43.1	68.8	127.	203.
4	2.08	3.87	5.86	8.66	14.2	20.4	29.3	46.8	66.5
5	1.89	3.24	4.62	6.43	9.72	13.1	17.7	25.9	34.5
6	1.76	2.87	3.94	5.27	7.56	9.81	12.7	17.6	22.4
7	1.68	2.63	3.51	4.57	6.31	7.97	9.98	13.3	16.5
8	1.62	2.46	3.22	4.10	5.52	6.81	8.35	10.8	13.1
9	1.57	2.34	3.01	3.77	4.96	6.03	7.28	9.24	11.0
10	1.53	2.24	2.85	3.52	4.56	5.47	6.51	8.13	9.56
11	1.50	2.17	2.72	3.33	4.25	5.05	5.95	7.32	8.52
12	1.48	2.10	2.62	3.18	4.01	4.72	5.52	6.71	7.74
13	1.46	2.05	2.53	3.05	3.82	4.46	5.17	6.23	7.13
14	1.44	2.01	2.46	2.95	3.66	4.25	4.89	5.85	6.65
15	1.43	1.97	2.40	2.86	3.52	4.07	4.67	5.54	6.26
16	1.41	1.94	2.35	2.79	3.41	3.92	4.47	5.27	5.94
17	1.40	1.91	2.31	2.72	3.31	3.79	4.31	5.05	5.67
18	1.39	1.89	2.27	2.67	3.23	3.68	4.17	4.87	5.44
19	1.38	1.86	2.23	2.62	3.15	3.59	4.05	4.70	5.24
20	1.37	1.84	2.20	2.57	3.09	3.50	3.94	4.56	5.07
21	1.37	1.83	2.18	2.53	3.03	3.43	3.85	4.44	4.92
22	1.36	1.81	2.15	2.50	2.98	3.36	3.76	4.33	4.78
23	1.35	1.80	2.13	2.47	2.93	3.30	3.69	4.23	4.66
24	1.35	1.78	2.11	2.44	2.89	3.25	3.62	4.14	4.56
25	1.34	1.77	2.09	2.41	2.85	3.20	3.56	4.06	4.46
26	1.34	1.76	2.07	2.39	2.81	3.15	3.50	3.99	4.37
27	1.33	1.75	2.06	2.36	2.78	3.11	3.45	3.92	4.29
28	1.33	1.74	2.04	2.34	2.75	3.07	3.40	3.86	4.22
29	1.32	1.73	2.03	2.32	2.73	3.04	3.36	3.80	4.15
30	1.32	1.72	2.01	2.31	2.70	3.01	3.32	3.75	4.09
35	1.31	1.69	1.96	2.23	2.60	2.88	3.16	3.55	3.85
40	1.30	1.66	1.92	2.18	2.52	2.78	3.04	3.40	3.68
45	1.29	1.64	1.89	2.14	2.46	2.71	2.96	3.29	3.55
50	1.28	1.63	1.87	2.11	2.42	2.65	2.89	3.20	3.45
60	1.27	1.60	1.84	2.06	2.35	2.57	2.79	3.08	3.30
70	1.26	1.59	1.81	2.03	2.31	2.51	2.72	2.99	3.20
80	1.26	1.57	1.79	2.00	2.27	2.47	2.67	2.93	3.12
90	1.25	1.56	1.78	1.98	2.24	2.44	2.63	2.88	3.07
100	1.25	1.56	1.77	1.97	2.22	2.41	2.60	2.84	3.02
120	1.24	1.55	1.75	1.94	2.19	2.37	2.55	2.78	2.96
140	1.24	1.54	1.74	1.93	2.17	2.35	2.52	2.74	2.91
160	1.24	1.53	1.73	1.92	2.15	2.33	2.49	2.71	2.88
180	1.23	1.53	1.72	1.91	2.14	2.31	2.48	2.69	2.85
200	1.23	1.52	1.72	1.90	2.13	2.30	2.46	2.67	2.83
300	1.23	1.51	1.70	1.88	2.10	2.26	2.42	2.62	2.77
500	1.22	1.50	1.69	1.86	2.07	2.23	2.38	2.58	2.72
∞	1.22	1.49	1.67	1.83	2.04	2.19	2.33	2.51	2.65

TABLE B.4 (cont.) CRITICAL VALUES OF THE *F* DISTRIBUTION

Numerator DF = 16

α(2):	0.50	0.20	0.10	0.05	0.02	0.01	0.005	0.002	0.001
α(1):	**0.25**	**0.10**	**0.05**	**0.025**	**0.01**	**0.005**	**0.0025**	**0.001**	**0.0005**
Denom. DF									
1	9.52	61.3	246.	987.	6170.	24700.	98700.	617000.	2470000.
2	3.41	9.43	19.4	39.4	99.4	199.	399.	999.	2000.
3	2.46	5.20	8.69	14.2	26.8	43.0	68.7	127.	202.
4	2.08	3.86	5.84	8.63	14.2	20.4	29.2	46.6	66.2
5	1.88	3.23	4.60	6.40	9.68	13.1	17.6	25.8	34.3
6	1.76	2.86	3.92	5.24	7.52	9.76	12.6	17.4	22.3
7	1.68	2.62	3.49	4.54	6.28	7.91	9.91	13.2	16.4
8	1.62	2.45	3.20	4.08	5.48	6.76	8.29	10.8	13.0
9	1.57	2.33	2.99	3.74	4.92	5.98	7.21	9.15	10.9
10	1.53	2.23	2.83	3.50	4.52	5.42	6.45	8.05	9.46
11	1.50	2.16	2.70	3.30	4.21	5.00	5.89	7.24	8.43
12	1.48	2.09	2.60	3.15	3.97	4.67	5.46	6.63	7.65
13	1.46	2.04	2.51	3.03	3.78	4.41	5.11	6.16	7.05
14	1.44	2.00	2.44	2.92	3.62	4.20	4.84	5.78	6.57
15	1.42	1.96	2.38	2.84	3.49	4.02	4.61	5.46	6.18
16	1.41	1.93	2.33	2.76	3.37	3.87	4.42	5.20	5.86
17	1.40	1.90	2.29	2.70	3.27	3.75	4.25	4.99	5.59
18	1.39	1.87	2.25	2.64	3.19	3.64	4.11	4.80	5.36
19	1.38	1.85	2.21	2.59	3.12	3.54	3.99	4.64	5.16
20	1.37	1.83	2.18	2.55	3.05	3.46	3.89	4.49	4.99
21	1.36	1.81	2.16	2.51	2.99	3.38	3.79	4.37	4.84
22	1.36	1.80	2.13	2.47	2.94	3.31	3.71	4.26	4.71
23	1.35	1.78	2.11	2.44	2.89	3.25	3.63	4.16	4.59
24	1.34	1.77	2.09	2.41	2.85	3.20	3.56	4.07	4.48
25	1.34	1.76	2.07	2.38	2.81	3.15	3.50	3.99	4.39
26	1.33	1.75	2.05	2.36	2.78	3.11	3.45	3.92	4.30
27	1.33	1.74	2.04	2.34	2.75	3.07	3.40	3.86	4.22
28	1.32	1.73	2.02	2.32	2.72	3.03	3.35	3.80	4.15
29	1.32	1.72	2.01	2.30	2.69	2.99	3.31	3.74	4.08
30	1.32	1.71	1.99	2.28	2.66	2.96	3.27	3.69	4.02
35	1.30	1.67	1.94	2.21	2.56	2.83	3.11	3.48	3.78
40	1.29	1.65	1.90	2.15	2.48	2.74	2.99	3.34	3.61
45	1.28	1.63	1.87	2.11	2.43	2.66	2.90	3.23	3.48
50	1.27	1.61	1.85	2.08	2.38	2.61	2.84	3.14	3.38
60	1.26	1.59	1.82	2.03	2.31	2.53	2.74	3.02	3.23
70	1.26	1.57	1.79	2.00	2.27	2.47	2.67	2.93	3.13
80	1.25	1.56	1.77	1.97	2.23	2.43	2.62	2.87	3.06
90	1.25	1.55	1.76	1.95	2.21	2.39	2.58	2.82	3.00
100	1.24	1.54	1.75	1.94	2.19	2.37	2.55	2.78	2.96
120	1.24	1.53	1.73	1.92	2.15	2.33	2.50	2.72	2.89
140	1.23	1.52	1.72	1.90	2.13	2.30	2.47	2.68	2.84
160	1.23	1.52	1.71	1.89	2.11	2.28	2.44	2.65	2.81
180	1.23	1.51	1.70	1.88	2.10	2.26	2.42	2.63	2.78
200	1.23	1.51	1.69	1.87	2.09	2.25	2.41	2.61	2.76
300	1.22	1.49	1.68	1.85	2.06	2.21	2.36	2.56	2.70
500	1.22	1.49	1.66	1.83	2.04	2.19	2.33	2.52	2.65
∞	1.21	1.47	1.64	1.80	2.00	2.14	2.28	2.45	2.58

TABLE B.4 (cont.) CRITICAL VALUES OF THE *F* DISTRIBUTION

Numerator DF = 17

α(2):	0.50	0.20	0.10	0.05	0.02	0.01	0.005	0.002	0.001
α(1):	0.25	0.10	0.05	0.025	0.01	0.005	0.0025	0.001	0.0005
Denom. DF									
1	9.53	61.5	247.	989.	6180.	24700.	98900.	618000.	2470000.
2	3.42	9.43	19.4	39.4	99.4	199.	399.	999.	2000.
3	2.46	5.19	8.68	14.2	26.8	42.9	68.6	127.	202.
4	2.08	3.86	5.83	8.61	14.1	20.3	29.1	46.5	66.0
5	1.88	3.22	4.59	6.38	9.64	13.0	17.5	25.7	34.2
6	1.76	2.85	3.91	5.22	7.48	9.71	12.5	17.4	22.1
7	1.67	2.61	3.48	4.52	6.24	7.87	9.85	13.1	16.3
8	1.61	2.45	3.19	4.05	5.44	6.72	8.23	10.7	12.9
9	1.57	2.32	2.97	3.72	4.89	5.94	7.16	9.08	10.8
10	1.53	2.22	2.81	3.47	4.49	5.38	6.40	7.98	9.38
11	1.50	2.15	2.69	3.28	4.18	4.96	5.84	7.17	8.34
12	1.47	2.08	2.58	3.13	3.94	4.63	5.40	6.57	7.57
13	1.45	2.03	2.50	3.00	3.75	4.37	5.06	6.09	6.97
14	1.44	1.99	2.43	2.90	3.59	4.16	4.79	5.71	6.49
15	1.42	1.95	2.37	2.81	3.45	3.98	4.56	5.40	6.11
16	1.41	1.92	2.32	2.74	3.34	3.83	4.37	5.14	5.79
17	1.39	1.89	2.27	2.67	3.24	3.71	4.21	4.92	5.52
18	1.38	1.86	2.23	2.62	3.16	3.60	4.07	4.74	5.29
19	1.37	1.84	2.20	2.57	3.08	3.50	3.94	4.58	5.09
20	1.37	1.82	2.17	2.52	3.02	3.42	3.84	4.44	4.92
21	1.36	1.80	2.14	2.48	2.96	3.34	3.74	4.31	4.77
22	1.35	1.79	2.11	2.45	2.91	3.27	3.66	4.20	4.64
23	1.35	1.77	2.09	2.42	2.86	3.21	3.58	4.10	4.52
24	1.34	1.76	2.07	2.39	2.82	3.16	3.52	4.02	4.41
25	1.33	1.75	2.05	2.36	2.78	3.11	3.46	3.94	4.32
26	1.33	1.73	2.03	2.34	2.75	3.07	3.40	3.86	4.23
27	1.33	1.72	2.02	2.31	2.71	3.03	3.35	3.80	4.15
28	1.32	1.71	2.00	2.29	2.68	2.99	3.30	3.74	4.08
29	1.32	1.71	1.99	2.27	2.66	2.95	3.26	3.68	4.02
30	1.31	1.70	1.98	2.26	2.63	2.92	3.22	3.63	3.96
35	1.30	1.66	1.92	2.18	2.53	2.79	3.06	3.43	3.72
40	1.29	1.64	1.89	2.13	2.45	2.70	2.95	3.28	3.54
45	1.28	1.62	1.86	2.09	2.39	2.62	2.86	3.17	3.41
50	1.27	1.60	1.83	2.06	2.35	2.57	2.79	3.09	3.31
60	1.26	1.58	1.80	2.01	2.28	2.49	2.69	2.96	3.17
70	1.25	1.56	1.77	1.97	2.23	2.43	2.62	2.88	3.07
80	1.25	1.55	1.75	1.95	2.20	2.39	2.57	2.81	3.00
90	1.24	1.54	1.74	1.93	2.17	2.35	2.53	2.76	2.94
100	1.24	1.53	1.73	1.91	2.15	2.33	2.50	2.73	2.89
120	1.23	1.52	1.71	1.89	2.12	2.29	2.45	2.67	2.83
140	1.23	1.51	1.70	1.87	2.10	2.26	2.42	2.63	2.78
160	1.23	1.50	1.69	1.86	2.08	2.24	2.40	2.60	2.75
180	1.22	1.50	1.68	1.85	2.07	2.22	2.38	2.58	2.72
200	1.22	1.49	1.67	1.84	2.06	2.21	2.36	2.56	2.70
300	1.22	1.48	1.66	1.82	2.03	2.17	2.32	2.50	2.64
500	1.21	1.47	1.64	1.80	2.00	2.14	2.28	2.46	2.59
∞	1.21	1.46	1.62	1.78	1.97	2.10	2.23	2.40	2.52

TABLE B.4 (cont.) CRITICAL VALUES OF THE *F* DISTRIBUTION

Numerator DF = 18

α(2):	0.50	0.20	0.10	0.05	0.02	0.01	0.005	0.002	0.001
α(1):	0.25	0.10	0.05	0.025	0.01	0.005	0.0025	0.001	0.0005
Denom. DF									
1	9.55	61.6	247.	990.	6190.	24800.	99100.	619000.	2480000.
2	3.42	9.44	19.4	39.4	99.4	199.	399.	999.	2000.
3	2.46	5.19	8.67	14.2	26.8	42.9	68.5	127.	202.
4	2.08	3.85	5.82	8.59	14.1	20.3	29.0	46.3	65.8
5	1.88	3.22	4.58	6.36	9.61	13.0	17.4	25.6	34.0
6	1.76	2.85	3.90	5.20	7.45	9.66	12.4	17.3	22.0
7	1.67	2.61	3.47	4.50	6.21	7.83	9.79	13.1	16.2
8	1.61	2.44	3.17	4.03	5.41	6.68	8.18	10.6	12.8
9	1.56	2.31	2.96	3.70	4.86	5.90	7.11	9.01	10.7
10	1.53	2.22	2.80	3.45	4.46	5.34	6.35	7.91	9.30
11	1.50	2.14	2.67	3.26	4.15	4.92	5.79	7.11	8.27
12	1.47	2.08	2.57	3.11	3.91	4.59	5.36	6.51	7.50
13	1.45	2.02	2.48	2.98	3.72	4.33	5.02	6.03	6.90
14	1.43	1.98	2.41	2.88	3.56	4.12	4.74	5.66	6.43
15	1.42	1.94	2.35	2.79	3.42	3.95	4.51	5.35	6.04
16	1.40	1.91	2.30	2.72	3.31	3.80	4.32	5.09	5.72
17	1.39	1.88	2.26	2.65	3.21	3.67	4.16	4.87	5.45
18	1.38	1.85	2.22	2.60	3.13	3.56	4.02	4.68	5.23
19	1.37	1.83	2.18	2.55	3.05	3.46	3.90	4.52	5.03
20	1.36	1.81	2.15	2.50	2.99	3.38	3.79	4.38	4.86
21	1.36	1.79	2.12	2.46	2.93	3.31	3.70	4.26	4.71
22	1.35	1.78	2.10	2.43	2.88	3.24	3.62	4.15	4.58
23	1.34	1.76	2.08	2.39	2.83	3.18	3.54	4.05	4.46
24	1.34	1.75	2.05	2.36	2.79	3.12	3.47	3.96	4.35
25	1.33	1.74	2.04	2.34	2.75	3.08	3.41	3.88	4.26
26	1.33	1.72	2.02	2.31	2.72	3.03	3.36	3.81	4.17
27	1.32	1.71	2.00	2.29	2.68	2.99	3.31	3.75	4.10
28	1.32	1.70	1.99	2.27	2.65	2.95	3.26	3.69	4.02
29	1.31	1.69	1.97	2.25	2.63	2.92	3.22	3.63	3.96
30	1.31	1.69	1.96	2.23	2.60	2.89	3.18	3.58	3.90
35	1.29	1.65	1.91	2.16	2.50	2.76	3.02	3.38	3.66
40	1.28	1.62	1.87	2.11	2.42	2.66	2.90	3.23	3.49
45	1.27	1.60	1.84	2.07	2.36	2.59	2.82	3.12	3.36
50	1.27	1.59	1.81	2.03	2.32	2.53	2.75	3.04	3.26
60	1.26	1.56	1.78	1.98	2.25	2.45	2.65	2.91	3.11
70	1.25	1.55	1.75	1.95	2.20	2.39	2.58	2.83	3.01
80	1.24	1.53	1.73	1.92	2.17	2.35	2.53	2.76	2.94
90	1.24	1.52	1.72	1.91	2.14	2.32	2.49	2.71	2.88
100	1.23	1.52	1.71	1.89	2.12	2.29	2.46	2.68	2.84
120	1.23	1.50	1.69	1.87	2.09	2.25	2.41	2.62	2.78
140	1.22	1.50	1.68	1.85	2.07	2.22	2.38	2.58	2.73
160	1.22	1.49	1.67	1.84	2.05	2.20	2.35	2.55	2.70
180	1.22	1.48	1.66	1.83	2.04	2.19	2.34	2.53	2.67
200	1.22	1.48	1.66	1.82	2.03	2.18	2.32	2.51	2.65
300	1.21	1.47	1.64	1.80	1.99	2.14	2.28	2.46	2.59
500	1.21	1.46	1.62	1.78	1.97	2.11	2.24	2.41	2.54
∞	1.20	1.44	1.60	1.75	1.93	2.06	2.19	2.35	2.47

TABLE B.4 (cont.) CRITICAL VALUES OF THE *F* DISTRIBUTION

Numerator DF = 19

α(2):	0.50	0.20	0.10	0.05	0.02	0.01	0.005	0.002	0.001
α(1):	0.25	0.10	0.05	0.025	0.01	0.005	0.0025	0.001	0.0005
Denom. DF									
1	9.57	61.7	248.	992.	6200.	24800.	99200.	620000.	2480000.
2	3.42	9.44	19.4	39.4	99.4	199.	399.	999.	2000.
3	2.46	5.19	8.67	14.2	26.7	42.8	68.4	127.	201.
4	2.08	3.85	5.81	8.58	14.0	20.2	28.9	46.2	65.7
5	1.88	3.21	4.57	6.34	9.58	12.9	17.4	25.5	33.9
6	1.76	2.84	3.88	5.18	7.42	9.62	12.4	17.2	21.9
7	1.67	2.60	3.46	4.48	6.18	7.79	9.74	13.0	16.1
8	1.61	2.43	3.16	4.02	5.38	6.64	8.13	10.5	12.8
9	1.56	2.30	2.95	3.68	4.83	5.86	7.06	8.95	10.7
10	1.53	2.21	2.79	3.44	4.43	5.31	6.31	7.86	9.23
11	1.49	2.13	2.66	3.24	4.12	4.89	5.75	7.06	8.20
12	1.47	2.07	2.56	3.09	3.88	4.56	5.32	6.45	7.43
13	1.45	2.01	2.47	2.96	3.69	4.30	4.98	5.98	6.84
14	1.43	1.97	2.40	2.86	3.53	4.09	4.70	5.60	6.37
15	1.41	1.93	2.34	2.77	3.40	3.91	4.47	5.29	5.98
16	1.40	1.90	2.29	2.70	3.28	3.76	4.28	5.04	5.66
17	1.39	1.87	2.24	2.63	3.19	3.64	4.12	4.82	5.40
18	1.38	1.84	2.20	2.58	3.10	3.53	3.98	4.63	5.17
19	1.37	1.82	2.17	2.53	3.03	3.43	3.86	4.47	4.97
20	1.36	1.80	2.14	2.48	2.96	3.35	3.76	4.33	4.80
21	1.35	1.78	2.11	2.44	2.90	3.27	3.66	4.21	4.65
22	1.35	1.77	2.08	2.41	2.85	3.21	3.58	4.10	4.52
23	1.34	1.75	2.06	2.37	2.80	3.15	3.50	4.00	4.41
24	1.33	1.74	2.04	2.35	2.76	3.09	3.44	3.92	4.30
25	1.33	1.73	2.02	2.32	2.72	3.04	3.38	3.84	4.21
26	1.32	1.71	2.00	2.29	2.69	3.00	3.32	3.77	4.12
27	1.32	1.70	1.99	2.27	2.66	2.96	3.27	3.70	4.04
28	1.31	1.69	1.97	2.25	2.63	2.92	3.22	3.64	3.97
29	1.31	1.68	1.96	2.23	2.60	2.88	3.18	3.59	3.91
30	1.31	1.68	1.95	2.21	2.57	2.85	3.14	3.53	3.85
35	1.29	1.64	1.89	2.14	2.47	2.72	2.98	3.33	3.61
40	1.28	1.61	1.85	2.09	2.39	2.63	2.87	3.19	3.44
45	1.27	1.59	1.82	2.04	2.34	2.56	2.78	3.08	3.31
50	1.26	1.58	1.80	2.01	2.29	2.50	2.71	2.99	3.21
60	1.25	1.55	1.76	1.96	2.22	2.42	2.61	2.87	3.06
70	1.24	1.54	1.74	1.93	2.18	2.36	2.54	2.78	2.96
80	1.24	1.52	1.72	1.90	2.14	2.32	2.49	2.72	2.89
90	1.23	1.51	1.70	1.88	2.11	2.28	2.45	2.67	2.83
100	1.23	1.50	1.69	1.87	2.09	2.26	2.42	2.63	2.79
120	1.22	1.49	1.67	1.84	2.06	2.22	2.37	2.58	2.73
140	1.22	1.48	1.66	1.83	2.04	2.19	2.34	2.54	2.68
160	1.22	1.48	1.65	1.82	2.02	2.17	2.32	2.51	2.65
180	1.21	1.47	1.64	1.81	2.01	2.15	2.30	2.48	2.62
200	1.21	1.47	1.64	1.80	2.00	2.14	2.28	2.46	2.60
300	1.21	1.46	1.62	1.77	1.97	2.10	2.24	2.41	2.54
500	1.20	1.45	1.61	1.76	1.94	2.07	2.20	2.37	2.49
∞	1.20	1.43	1.59	1.73	1.90	2.03	2.15	2.31	2.42

TABLE B.4 (cont.) CRITICAL VALUES OF THE *F* DISTRIBUTION

Numerator DF = 20

α(2):	0.50	0.20	0.10	0.05	0.02	0.01	0.005	0.002	0.001
α(1):	0.25	0.10	0.05	0.025	0.01	0.005	0.0025	0.001	0.0005
Denom. DF									
1	9.58	61.7	248.	993.	6210.	24800.	99300.	621000.	2480000.
2	3.43	9.44	19.4	39.4	99.4	199.	399.	999.	2000.
3	2.46	5.18	8.66	14.2	26.7	42.8	68.3	126.	201.
4	2.08	3.84	5.80	8.56	14.0	20.2	28.9	46.1	65.5
5	1.88	3.21	4.56	6.33	9.55	12.9	17.3	25.4	33.8
6	1.76	2.84	3.87	5.17	7.40	9.59	12.3	17.1	21.8
7	1.67	2.59	3.44	4.47	6.16	7.75	9.70	12.9	16.0
8	1.61	2.42	3.15	4.00	5.36	6.61	8.09	10.5	12.7
9	1.56	2.30	2.94	3.67	4.81	5.83	7.02	8.90	10.6
10	1.52	2.20	2.77	3.42	4.41	5.27	6.27	7.80	9.17
11	1.49	2.12	2.65	3.23	4.10	4.86	5.71	7.01	8.14
12	1.47	2.06	2.54	3.07	3.86	4.53	5.28	6.40	7.37
13	1.45	2.01	2.46	2.95	3.66	4.27	4.94	5.93	6.78
14	1.43	1.96	2.39	2.84	3.51	4.06	4.66	5.56	6.31
15	1.41	1.92	2.33	2.76	3.37	3.88	4.44	5.25	5.93
16	1.40	1.89	2.28	2.68	3.26	3.73	4.25	4.99	5.61
17	1.39	1.86	2.23	2.62	3.16	3.61	4.09	4.78	5.34
18	1.38	1.84	2.19	2.56	3.08	3.50	3.95	4.59	5.12
19	1.37	1.81	2.16	2.51	3.00	3.40	3.83	4.43	4.92
20	1.36	1.79	2.12	2.46	2.94	3.32	3.72	4.29	4.75
21	1.35	1.78	2.10	2.42	2.88	3.24	3.63	4.17	4.60
22	1.34	1.76	2.07	2.39	2.83	3.18	3.54	4.06	4.47
23	1.34	1.74	2.05	2.36	2.78	3.12	3.47	3.96	4.36
24	1.33	1.73	2.03	2.33	2.74	3.06	3.40	3.87	4.25
25	1.33	1.72	2.01	2.30	2.70	3.01	3.34	3.79	4.16
26	1.32	1.71	1.99	2.28	2.66	2.97	3.28	3.72	4.07
27	1.32	1.70	1.97	2.25	2.63	2.93	3.23	3.66	3.99
28	1.31	1.69	1.96	2.23	2.60	2.89	3.19	3.60	3.92
29	1.31	1.68	1.94	2.21	2.57	2.86	3.14	3.54	3.86
30	1.30	1.67	1.93	2.20	2.55	2.82	3.11	3.49	3.80
35	1.29	1.63	1.88	2.12	2.44	2.69	2.95	3.29	3.56
40	1.28	1.61	1.84	2.07	2.37	2.60	2.83	3.14	3.39
45	1.27	1.58	1.81	2.03	2.31	2.53	2.74	3.04	3.26
50	1.26	1.57	1.78	1.99	2.27	2.47	2.68	2.95	3.16
60	1.25	1.54	1.75	1.94	2.20	2.39	2.58	2.83	3.02
70	1.24	1.53	1.72	1.91	2.15	2.33	2.51	2.74	2.92
80	1.23	1.51	1.70	1.88	2.12	2.29	2.46	2.68	2.85
90	1.23	1.50	1.69	1.86	2.09	2.25	2.42	2.63	2.79
100	1.23	1.49	1.68	1.85	2.07	2.23	2.38	2.59	2.75
120	1.22	1.48	1.66	1.82	2.03	2.19	2.34	2.53	2.68
140	1.22	1.47	1.65	1.81	2.01	2.16	2.31	2.49	2.64
160	1.21	1.47	1.64	1.80	1.99	2.14	2.28	2.47	2.60
180	1.21	1.46	1.63	1.79	1.98	2.12	2.26	2.44	2.58
200	1.21	1.46	1.62	1.78	1.97	2.11	2.25	2.42	2.56
300	1.20	1.45	1.61	1.75	1.94	2.07	2.20	2.37	2.49
500	1.20	1.44	1.59	1.74	1.92	2.04	2.17	2.33	2.45
∞	1.19	1.42	1.57	1.71	1.88	2.00	2.12	2.27	2.37

TABLE B.4 (cont.) CRITICAL VALUES OF THE *F* DISTRIBUTION

Numerator DF = 22

	α(2): 0.50	0.20	0.10	0.05	0.02	0.01	0.005	0.002	0.001
	α(1): 0.25	0.10	0.05	0.025	0.01	0.005	0.0025	0.001	0.0005
Denom. DF									
1	9.61	61.9	249.	995.	6220.	24900.	99600.	622000.	2490000.
2	3.43	9.45	19.5	39.5	99.5	199.	399.	999.	2000.
3	2.46	5.18	8.65	14.1	26.6	42.7	68.2	126.	201.
4	2.08	3.84	5.79	8.53	14.0	20.1	28.7	45.9	65.3
5	1.88	3.20	4.54	6.30	9.51	12.8	17.2	25.3	33.6
6	1.76	2.83	3.86	5.14	7.35	9.53	12.3	17.0	21.7
7	1.67	2.58	3.43	4.44	6.11	7.69	9.62	12.8	15.9
8	1.61	2.41	3.13	3.97	5.32	6.55	8.02	10.4	12.6
9	1.56	2.29	2.92	3.64	4.77	5.78	6.95	8.80	10.5
10	1.52	2.19	2.75	3.39	4.36	5.22	6.20	7.71	9.06
11	1.49	2.11	2.63	3.20	4.06	4.80	5.64	6.92	8.04
12	1.46	2.05	2.52	3.04	3.82	4.48	5.21	6.32	7.27
13	1.44	1.99	2.44	2.92	3.62	4.22	4.87	5.85	6.68
14	1.42	1.95	2.37	2.81	3.46	4.01	4.60	5.48	6.21
15	1.41	1.91	2.31	2.73	3.33	3.83	4.37	5.17	5.83
16	1.39	1.88	2.25	2.65	3.22	3.68	4.18	4.91	5.52
17	1.38	1.85	2.21	2.59	3.12	3.56	4.02	4.70	5.25
18	1.37	1.82	2.17	2.53	3.03	3.45	3.88	4.51	5.03
19	1.36	1.80	2.13	2.48	2.96	3.35	3.76	4.35	4.83
20	1.35	1.78	2.10	2.43	2.90	3.27	3.66	4.21	4.67
21	1.35	1.76	2.07	2.39	2.84	3.19	3.56	4.09	4.52
22	1.34	1.74	2.05	2.36	2.78	3.12	3.48	3.98	4.39
23	1.33	1.73	2.02	2.33	2.74	3.06	3.41	3.89	4.27
24	1.33	1.71	2.00	2.30	2.70	3.01	3.34	3.80	4.17
25	1.32	1.70	1.98	2.27	2.66	2.96	3.28	3.72	4.07
26	1.32	1.69	1.97	2.24	2.62	2.92	3.22	3.65	3.99
27	1.31	1.68	1.95	2.22	2.59	2.88	3.17	3.58	3.91
28	1.31	1.67	1.93	2.20	2.56	2.84	3.13	3.52	3.84
29	1.30	1.66	1.92	2.18	2.53	2.80	3.08	3.47	3.77
30	1.30	1.65	1.91	2.16	2.51	2.77	3.04	3.42	3.71
35	1.28	1.62	1.85	2.09	2.40	2.64	2.89	3.22	3.48
40	1.27	1.59	1.81	2.03	2.33	2.55	2.77	3.07	3.31
45	1.26	1.57	1.78	1.99	2.27	2.47	2.68	2.96	3.18
50	1.25	1.55	1.76	1.96	2.22	2.42	2.62	2.88	3.08
60	1.24	1.53	1.72	1.91	2.15	2.33	2.52	2.75	2.94
70	1.23	1.51	1.70	1.88	2.11	2.28	2.45	2.67	2.84
80	1.23	1.49	1.68	1.85	2.07	2.23	2.39	2.61	2.77
90	1.22	1.48	1.66	1.83	2.04	2.20	2.35	2.56	2.71
100	1.22	1.48	1.65	1.81	2.02	2.17	2.32	2.52	2.67
120	1.21	1.46	1.63	1.79	1.99	2.13	2.28	2.46	2.60
140	1.21	1.45	1.62	1.77	1.97	2.11	2.24	2.42	2.56
160	1.21	1.45	1.61	1.76	1.95	2.09	2.22	2.39	2.52
180	1.20	1.44	1.60	1.75	1.94	2.07	2.20	2.37	2.50
200	1.20	1.44	1.60	1.74	1.93	2.06	2.19	2.35	2.48
300	1.20	1.43	1.58	1.72	1.89	2.02	2.14	2.30	2.41
500	1.19	1.42	1.56	1.70	1.87	1.99	2.11	2.26	2.37
∞	1.18	1.40	1.54	1.67	1.83	1.95	2.05	2.19	2.30

TABLE B.4 (cont.) CRITICAL VALUES OF THE *F* DISTRIBUTION

Numerator DF = 24

α(2):	0.50	0.20	0.10	0.05	0.02	0.01	0.005	0.002	0.001
α(1):	0.25	0.10	0.05	0.025	0.01	0.005	0.0025	0.001	0.0005
Denom. DF									
1	9.63	62.0	249.	997.	6230.	24900.	99800.	623000.	2490000.
2	3.43	9.45	19.5	39.5	99.5	199.	399.	999.	2000.
3	2.46	5.18	8.64	14.1	26.6	42.6	68.1	126.	200.
4	2.08	3.83	5.77	8.51	13.9	20.0	28.7	45.8	65.0
5	1.88	3.19	4.53	6.28	9.47	12.8	17.1	25.1	33.4
6	1.75	2.82	3.84	5.12	7.31	9.47	12.2	16.9	21.5
7	1.67	2.58	3.41	4.41	6.07	7.64	9.56	12.7	15.8
8	1.60	2.40	3.12	3.95	5.28	6.50	7.95	10.3	12.5
9	1.56	2.28	2.90	3.61	4.73	5.73	6.89	8.72	10.4
10	1.52	2.18	2.74	3.37	4.33	5.17	6.14	7.64	8.96
11	1.49	2.10	2.61	3.17	4.02	4.76	5.58	6.85	7.95
12	1.46	2.04	2.51	3.02	3.78	4.43	5.16	6.25	7.19
13	1.44	1.98	2.42	2.89	3.59	4.17	4.82	5.78	6.60
14	1.42	1.94	2.35	2.79	3.43	3.96	4.55	5.41	6.13
15	1.41	1.90	2.29	2.70	3.29	3.79	4.32	5.10	5.75
16	1.39	1.87	2.24	2.63	3.18	3.64	4.13	4.85	5.44
17	1.38	1.84	2.19	2.56	3.08	3.51	3.97	4.63	5.18
18	1.37	1.81	2.15	2.50	3.00	3.40	3.83	4.45	4.95
19	1.36	1.79	2.11	2.45	2.92	3.31	3.71	4.29	4.76
20	1.35	1.77	2.08	2.41	2.86	3.22	3.61	4.15	4.59
21	1.34	1.75	2.05	2.37	2.80	3.15	3.51	4.03	4.44
22	1.33	1.73	2.03	2.33	2.75	3.08	3.43	3.92	4.31
23	1.33	1.72	2.01	2.30	2.70	3.02	3.35	3.82	4.20
24	1.32	1.70	1.98	2.27	2.66	2.97	3.29	3.74	4.09
25	1.32	1.69	1.96	2.24	2.62	2.92	3.23	3.66	4.00
26	1.31	1.68	1.95	2.22	2.58	2.87	3.17	3.59	3.92
27	1.31	1.67	1.93	2.19	2.55	2.83	3.12	3.52	3.84
28	1.30	1.66	1.91	2.17	2.52	2.79	3.07	3.46	3.77
29	1.30	1.65	1.90	2.15	2.49	2.76	3.03	3.41	3.70
30	1.29	1.64	1.89	2.14	2.47	2.73	2.99	3.36	3.64
35	1.28	1.60	1.83	2.06	2.36	2.60	2.83	3.16	3.41
40	1.26	1.57	1.79	2.01	2.29	2.50	2.72	3.01	3.24
45	1.26	1.55	1.76	1.96	2.23	2.43	2.63	2.90	3.11
50	1.25	1.54	1.74	1.93	2.18	2.37	2.56	2.82	3.01
60	1.24	1.51	1.70	1.88	2.12	2.29	2.46	2.69	2.87
70	1.23	1.49	1.67	1.85	2.07	2.23	2.39	2.61	2.77
80	1.22	1.48	1.65	1.82	2.03	2.19	2.34	2.54	2.70
90	1.22	1.47	1.64	1.80	2.00	2.15	2.30	2.50	2.64
100	1.21	1.46	1.63	1.78	1.98	2.13	2.27	2.46	2.60
120	1.21	1.45	1.61	1.76	1.95	2.09	2.23	2.40	2.53
140	1.20	1.44	1.60	1.74	1.93	2.06	2.19	2.36	2.49
160	1.20	1.43	1.59	1.73	1.91	2.04	2.17	2.33	2.45
180	1.20	1.43	1.58	1.72	1.90	2.02	2.15	2.31	2.43
200	1.19	1.42	1.57	1.71	1.89	2.01	2.13	2.29	2.41
300	1.19	1.41	1.55	1.69	1.85	1.97	2.09	2.24	2.35
500	1.18	1.40	1.54	1.67	1.83	1.94	2.05	2.20	2.30
∞	1.18	1.38	1.52	1.64	1.79	1.90	2.00	2.13	2.23

TABLE B.4 (cont.) CRITICAL VALUES OF THE *F* DISTRIBUTION

Numerator DF = 26

α(2):	0.50	0.20	0.10	0.05	0.02	0.01	0.005	0.002	0.001
α(1):	0.25	0.10	0.05	0.025	0.01	0.005	0.0025	0.001	0.0005
Denom. DF									
1	9.64	62.1	249.	999.	6240.	25000.	99900.	624000.	2500000.
2	3.44	9.45	19.5	39.5	99.5	199.	399.	999.	2000.
3	2.46	5.17	8.63	14.1	26.6	42.6	68.0	126.	200.
4	2.08	3.83	5.76	8.49	13.9	20.0	28.6	45.6	64.9
5	1.88	3.18	4.52	6.26	9.43	12.7	17.1	25.0	33.3
6	1.75	2.81	3.83	5.10	7.28	9.43	12.1	16.8	21.4
7	1.67	2.57	3.40	4.39	6.04	7.60	9.50	12.7	15.7
8	1.60	2.40	3.10	3.93	5.25	6.46	7.90	10.2	12.4
9	1.55	2.27	2.89	3.59	4.70	5.69	6.84	8.66	10.3
10	1.52	2.17	2.72	3.34	4.30	5.13	6.09	7.57	8.89
11	1.48	2.09	2.59	3.15	3.99	4.72	5.54	6.78	7.87
12	1.46	2.03	2.49	3.00	3.75	4.39	5.11	6.19	7.12
13	1.44	1.97	2.41	2.87	3.56	4.13	4.77	5.72	6.53
14	1.42	1.93	2.33	2.77	3.40	3.92	4.50	5.35	6.07
15	1.40	1.89	2.27	2.68	3.26	3.75	4.27	5.04	5.69
16	1.39	1.86	2.22	2.60	3.15	3.60	4.09	4.79	5.37
17	1.38	1.83	2.17	2.54	3.05	3.47	3.92	4.57	5.11
18	1.36	1.80	2.13	2.48	2.97	3.36	3.79	4.39	4.89
19	1.35	1.78	2.10	2.43	2.89	3.27	3.67	4.23	4.70
20	1.35	1.76	2.07	2.39	2.83	3.18	3.56	4.09	4.53
21	1.34	1.74	2.04	2.34	2.77	3.11	3.47	3.97	4.38
22	1.33	1.72	2.01	2.31	2.72	3.04	3.38	3.86	4.25
23	1.32	1.70	1.99	2.28	2.67	2.98	3.31	3.77	4.14
24	1.32	1.69	1.97	2.25	2.63	2.93	3.24	3.68	4.03
25	1.31	1.68	1.95	2.22	2.59	2.88	3.18	3.60	3.94
26	1.31	1.67	1.93	2.19	2.55	2.84	3.13	3.53	3.85
27	1.30	1.65	1.91	2.17	2.52	2.79	3.08	3.47	3.78
28	1.30	1.64	1.90	2.15	2.49	2.76	3.03	3.41	3.71
29	1.29	1.63	1.88	2.13	2.46	2.72	2.99	3.35	3.64
30	1.29	1.63	1.87	2.11	2.44	2.69	2.95	3.30	3.58
35	1.27	1.59	1.82	2.04	2.33	2.56	2.79	3.10	3.35
40	1.26	1.56	1.77	1.98	2.26	2.46	2.67	2.96	3.18
45	1.25	1.54	1.74	1.94	2.20	2.39	2.59	2.85	3.05
50	1.24	1.52	1.72	1.91	2.15	2.33	2.52	2.76	2.95
60	1.23	1.50	1.68	1.86	2.08	2.25	2.42	2.64	2.81
70	1.22	1.48	1.65	1.82	2.03	2.19	2.35	2.56	2.71
80	1.22	1.47	1.63	1.79	2.00	2.15	2.30	2.49	2.64
90	1.21	1.45	1.62	1.77	1.97	2.12	2.26	2.44	2.58
100	1.21	1.45	1.61	1.76	1.95	2.09	2.23	2.41	2.54
120	1.20	1.43	1.59	1.73	1.92	2.05	2.18	2.35	2.48
140	1.20	1.42	1.57	1.72	1.89	2.02	2.15	2.31	2.43
160	1.19	1.42	1.57	1.70	1.88	2.00	2.12	2.28	2.40
180	1.19	1.41	1.56	1.69	1.86	1.98	2.10	2.26	2.37
200	1.19	1.41	1.55	1.68	1.85	1.97	2.09	2.24	2.35
300	1.18	1.39	1.53	1.66	1.82	1.93	2.04	2.18	2.29
500	1.18	1.38	1.52	1.64	1.79	1.90	2.01	2.14	2.24
∞	1.17	1.37	1.50	1.61	1.76	1.86	1.95	2.08	2.17

TABLE B.4 (cont.) CRITICAL VALUES OF THE *F* DISTRIBUTION

Numerator DF = 28

α(2):	0.50	0.20	0.10	0.05	0.02	0.01	0.005	0.002	0.001
α(1):	0.25	0.10	0.05	0.025	0.01	0.005	0.0025	0.001	0.0005
Denom. DF									
1	9.66	62.2	250.	1000.	6250.	25000.	100000.	625000.	2500000.
2	3.44	9.46	19.5	39.5	99.5	199.	399.	999.	2000.
3	2.46	5.17	8.62	14.1	26.5	42.5	67.9	126.	200.
4	2.08	3.82	5.75	8.48	13.9	19.9	28.5	45.5	64.7
5	1.88	3.18	4.50	6.24	9.40	12.7	17.0	24.9	33.2
6	1.75	2.81	3.82	5.08	7.25	9.39	12.1	16.7	21.3
7	1.66	2.56	3.39	4.38	6.02	7.57	9.45	12.6	15.6
8	1.60	2.39	3.09	3.91	5.22	6.43	7.86	10.2	12.3
9	1.55	2.26	2.87	3.58	4.67	5.65	6.80	8.60	10.2
10	1.51	2.16	2.71	3.33	4.27	5.10	6.05	7.52	8.82
11	1.48	2.08	2.58	3.13	3.96	4.68	5.49	6.73	7.81
12	1.46	2.02	2.48	2.98	3.72	4.36	5.07	6.14	7.05
13	1.43	1.96	2.39	2.85	3.53	4.10	4.73	5.67	6.47
14	1.42	1.92	2.32	2.75	3.37	3.89	4.46	5.30	6.01
15	1.40	1.88	2.26	2.66	3.24	3.72	4.23	4.99	5.63
16	1.39	1.85	2.21	2.58	3.12	3.57	4.05	4.74	5.32
17	1.37	1.82	2.16	2.52	3.03	3.44	3.89	4.53	5.05
18	1.36	1.79	2.12	2.46	2.94	3.33	3.75	4.34	4.83
19	1.35	1.77	2.08	2.41	2.87	3.24	3.63	4.18	4.64
20	1.34	1.75	2.05	2.37	2.80	3.15	3.52	4.05	4.47
21	1.33	1.73	2.02	2.33	2.74	3.08	3.43	3.93	4.33
22	1.33	1.71	2.00	2.29	2.69	3.01	3.35	3.82	4.20
23	1.32	1.69	1.97	2.26	2.64	2.95	3.27	3.72	4.08
24	1.31	1.68	1.95	2.23	2.60	2.90	3.20	3.63	3.98
25	1.31	1.67	1.93	2.20	2.56	2.85	3.14	3.56	3.89
26	1.30	1.66	1.91	2.17	2.53	2.80	3.09	3.49	3.80
27	1.30	1.64	1.90	2.15	2.49	2.76	3.04	3.42	3.72
28	1.29	1.63	1.88	2.13	2.46	2.72	2.99	3.36	3.65
29	1.29	1.62	1.87	2.11	2.44	2.69	2.95	3.31	3.59
30	1.29	1.62	1.85	2.09	2.41	2.66	2.91	3.26	3.53
35	1.27	1.58	1.80	2.02	2.30	2.53	2.75	3.06	3.30
40	1.26	1.55	1.76	1.96	2.23	2.43	2.64	2.91	3.13
45	1.25	1.53	1.73	1.92	2.17	2.36	2.55	2.80	3.00
50	1.24	1.51	1.70	1.89	2.12	2.30	2.48	2.72	2.90
60	1.23	1.49	1.66	1.83	2.05	2.22	2.38	2.60	2.76
70	1.22	1.47	1.64	1.80	2.01	2.16	2.31	2.51	2.66
80	1.21	1.45	1.62	1.77	1.97	2.11	2.26	2.45	2.59
90	1.21	1.44	1.60	1.75	1.94	2.08	2.22	2.40	2.53
100	1.20	1.43	1.59	1.74	1.92	2.05	2.19	2.36	2.49
120	1.20	1.42	1.57	1.71	1.89	2.01	2.14	2.30	2.42
140	1.19	1.41	1.56	1.69	1.86	1.99	2.11	2.26	2.38
160	1.19	1.40	1.55	1.68	1.85	1.97	2.08	2.23	2.35
180	1.19	1.40	1.54	1.67	1.83	1.95	2.06	2.21	2.32
200	1.18	1.39	1.53	1.66	1.82	1.94	2.05	2.19	2.30
300	1.18	1.38	1.51	1.64	1.79	1.90	2.00	2.14	2.24
500	1.17	1.37	1.50	1.62	1.76	1.87	1.97	2.10	2.19
∞	1.17	1.35	1.48	1.59	1.72	1.82	1.91	2.03	2.12

TABLE B.4 (cont.) CRITICAL VALUES OF THE *F* DISTRIBUTION

Numerator DF = 30

	α(2): 0.50	0.20	0.10	0.05	0.02	0.01	0.005	0.002	0.001
Denom. DF	α(1): 0.25	0.10	0.05	0.025	0.01	0.005	0.0025	0.001	0.0005
1	9.67	62.3	250.	1000.	6260.	25000.	100000.	626000.	2500000.
2	3.44	9.46	19.5	39.5	99.5	199.	399.	999.	2000.
3	2.47	5.17	8.62	14.1	26.5	42.5	67.8	125.	200.
4	2.08	3.82	5.75	8.46	13.8	19.9	28.5	45.4	64.6
5	1.88	3.17	4.50	6.23	9.38	12.7	17.0	24.9	33.1
6	1.75	2.80	3.81	5.07	7.23	9.36	12.0	16.7	21.2
7	1.66	2.56	3.38	4.36	5.99	7.53	9.41	12.5	15.5
8	1.60	2.38	3.08	3.89	5.20	6.40	7.82	10.1	12.2
9	1.55	2.25	2.86	3.56	4.65	5.62	6.76	8.55	10.2
10	1.51	2.16	2.70	3.31	4.25	5.07	6.01	7.47	8.76
11	1.48	2.08	2.57	3.12	3.94	4.65	5.46	6.68	7.75
12	1.45	2.01	2.47	2.96	3.70	4.33	5.03	6.09	7.00
13	1.43	1.96	2.38	2.84	3.51	4.07	4.70	5.63	6.42
14	1.41	1.91	2.31	2.73	3.35	3.86	4.42	5.25	5.95
15	1.40	1.87	2.25	2.64	3.21	3.69	4.20	4.95	5.58
16	1.38	1.84	2.19	2.57	3.10	3.54	4.01	4.70	5.27
17	1.37	1.81	2.15	2.50	3.00	3.41	3.85	4.48	5.01
18	1.36	1.78	2.11	2.44	2.92	3.30	3.71	4.30	4.78
19	1.35	1.76	2.07	2.39	2.84	3.21	3.59	4.14	4.59
20	1.34	1.74	2.04	2.35	2.78	3.12	3.49	4.00	4.42
21	1.33	1.72	2.01	2.31	2.72	3.05	3.40	3.88	4.28
22	1.32	1.70	1.98	2.27	2.67	2.98	3.31	3.78	4.15
23	1.32	1.69	1.96	2.24	2.62	2.92	3.24	3.68	4.03
24	1.31	1.67	1.94	2.21	2.58	2.87	3.17	3.59	3.93
25	1.31	1.66	1.92	2.18	2.54	2.82	3.11	3.52	3.84
26	1.30	1.65	1.90	2.16	2.50	2.77	3.06	3.44	3.75
27	1.30	1.64	1.88	2.13	2.47	2.73	3.00	3.38	3.68
28	1.29	1.63	1.87	2.11	2.44	2.69	2.96	3.32	3.61
29	1.29	1.62	1.85	2.09	2.41	2.66	2.92	3.27	3.54
30	1.28	1.61	1.84	2.07	2.39	2.63	2.88	3.22	3.49
35	1.27	1.57	1.79	2.00	2.28	2.50	2.72	3.02	3.25
40	1.25	1.54	1.74	1.94	2.20	2.40	2.60	2.87	3.08
45	1.24	1.52	1.71	1.90	2.14	2.33	2.51	2.76	2.96
50	1.23	1.50	1.69	1.87	2.10	2.27	2.45	2.68	2.86
60	1.22	1.48	1.65	1.82	2.03	2.19	2.35	2.55	2.71
70	1.21	1.46	1.62	1.78	1.98	2.13	2.28	2.47	2.62
80	1.21	1.44	1.60	1.75	1.94	2.08	2.22	2.41	2.54
90	1.20	1.43	1.59	1.73	1.92	2.05	2.18	2.36	2.49
100	1.20	1.42	1.57	1.71	1.89	2.02	2.15	2.32	2.44
120	1.19	1.41	1.55	1.69	1.86	1.98	2.11	2.26	2.38
140	1.19	1.40	1.54	1.67	1.84	1.96	2.07	2.22	2.33
160	1.18	1.39	1.53	1.66	1.82	1.93	2.05	2.19	2.30
180	1.18	1.39	1.52	1.65	1.81	1.92	2.03	2.17	2.27
200	1.18	1.38	1.52	1.64	1.79	1.91	2.01	2.15	2.25
300	1.17	1.37	1.50	1.62	1.76	1.87	1.97	2.10	2.19
500	1.17	1.36	1.48	1.60	1.74	1.84	1.93	2.05	2.14
∞	1.16	1.34	1.46	1.57	1.70	1.79	1.88	1.99	2.07

TABLE B.4 (cont.) CRITICAL VALUES OF THE *F* DISTRIBUTION

Numerator DF = 40

α(2):	0.50	0.20	0.10	0.05	0.02	0.01	0.005	0.002	0.001
α(1):	0.25	0.10	0.05	0.025	0.01	0.005	0.0025	0.001	0.0005
Denom. DF									
1	9.71	62.5	251.	1010.	6290.	25100.	101000.	629000.	2510000.
2	3.45	9.47	19.5	39.5	99.5	199.	399.	999.	2000.
3	2.47	5.16	8.59	14.0	26.4	42.3	67.5	125.	199.
4	2.08	3.80	5.72	8.41	13.7	19.8	28.2	45.1	64.1
5	1.88	3.16	4.46	6.18	9.29	12.5	16.8	24.6	32.7
6	1.75	2.78	3.77	5.01	7.14	9.24	11.9	16.4	21.0
7	1.66	2.54	3.34	4.31	5.91	7.42	9.26	12.3	15.2
8	1.59	2.36	3.04	3.84	5.12	6.29	7.68	9.92	12.0
9	1.54	2.23	2.83	3.51	4.57	5.52	6.62	8.37	9.94
10	1.51	2.13	2.66	3.26	4.17	4.97	5.88	7.30	8.55
11	1.47	2.05	2.53	3.06	3.86	4.55	5.33	6.52	7.55
12	1.45	1.99	2.43	2.91	3.62	4.23	4.91	5.93	6.81
13	1.42	1.93	2.34	2.78	3.43	3.97	4.57	5.47	6.23
14	1.41	1.89	2.27	2.67	3.27	3.76	4.30	5.10	5.77
15	1.39	1.85	2.20	2.59	3.13	3.58	4.08	4.80	5.40
16	1.37	1.81	2.15	2.51	3.02	3.44	3.89	4.54	5.09
17	1.36	1.78	2.10	2.44	2.92	3.31	3.73	4.33	4.83
18	1.35	1.75	2.06	2.38	2.84	3.20	3.59	4.15	4.61
19	1.34	1.73	2.03	2.33	2.76	3.11	3.47	3.99	4.42
20	1.33	1.71	1.99	2.29	2.69	3.02	3.37	3.86	4.25
21	1.32	1.69	1.96	2.25	2.64	2.95	3.27	3.74	4.11
22	1.31	1.67	1.94	2.21	2.58	2.88	3.19	3.63	3.98
23	1.31	1.66	1.91	2.18	2.54	2.82	3.12	3.53	3.87
24	1.30	1.64	1.89	2.15	2.49	2.77	3.05	3.45	3.76
25	1.29	1.63	1.87	2.12	2.45	2.72	2.99	3.37	3.67
26	1.29	1.61	1.85	2.09	2.42	2.67	2.93	3.30	3.59
27	1.28	1.60	1.84	2.07	2.38	2.63	2.88	3.23	3.51
28	1.28	1.59	1.82	2.05	2.35	2.59	2.84	3.18	3.44
29	1.27	1.58	1.81	2.03	2.33	2.56	2.79	3.12	3.38
30	1.27	1.57	1.79	2.01	2.30	2.52	2.76	3.07	3.32
35	1.25	1.53	1.74	1.93	2.19	2.39	2.60	2.87	3.09
40	1.24	1.51	1.69	1.88	2.11	2.30	2.48	2.73	2.92
45	1.23	1.48	1.66	1.83	2.05	2.22	2.39	2.62	2.79
50	1.22	1.46	1.63	1.80	2.01	2.16	2.32	2.53	2.69
60	1.21	1.44	1.59	1.74	1.94	2.08	2.22	2.41	2.55
70	1.20	1.42	1.57	1.71	1.89	2.02	2.15	2.32	2.45
80	1.19	1.40	1.54	1.68	1.85	1.97	2.10	2.26	2.38
90	1.19	1.39	1.53	1.66	1.82	1.94	2.06	2.21	2.32
100	1.18	1.38	1.52	1.64	1.80	1.91	2.02	2.17	2.28
120	1.18	1.37	1.50	1.61	1.76	1.87	1.98	2.11	2.21
140	1.17	1.36	1.48	1.60	1.74	1.84	1.94	2.07	2.17
160	1.17	1.35	1.47	1.58	1.72	1.82	1.92	2.04	2.13
180	1.16	1.34	1.46	1.57	1.71	1.80	1.90	2.02	2.11
200	1.16	1.34	1.46	1.56	1.69	1.79	1.88	2.00	2.09
300	1.15	1.32	1.43	1.54	1.66	1.75	1.84	1.94	2.02
500	1.15	1.31	1.42	1.52	1.63	1.72	1.80	1.90	1.98
∞	1.14	1.30	1.39	1.48	1.59	1.67	1.74	1.84	1.90

TABLE B.4 (cont.) CRITICAL VALUES OF THE *F* DISTRIBUTION

Numerator DF = 50

α(2):	0.50	0.20	0.10	0.05	0.02	0.01	0.005	0.002	0.001
α(1):	0.25	0.10	0.05	0.025	0.01	0.005	0.0025	0.001	0.0005
Denom. DF									
1	9.74	62.7	252.	1010.	6300.	25200.	101000.	630000.	2520000.
2	3.46	9.47	19.5	39.5	99.5	199.	399.	999.	2000.
3	2.47	5.15	8.58	14.0	26.4	42.2	67.4	125.	198.
4	2.08	3.80	5.70	8.38	13.7	19.7	28.1	44.9	63.8
5	1.88	3.15	4.44	6.14	9.24	12.5	16.7	24.4	32.5
6	1.75	2.77	3.75	4.98	7.09	9.17	11.8	16.3	20.8
7	1.66	2.52	3.32	4.28	5.86	7.35	9.17	12.2	15.1
8	1.59	2.35	3.02	3.81	5.07	6.22	7.59	9.80	11.8
9	1.54	2.22	2.80	3.47	4.52	5.45	6.54	8.26	9.81
10	1.50	2.12	2.64	3.22	4.12	4.90	5.80	7.19	8.43
11	1.47	2.04	2.51	3.03	3.81	4.49	5.25	6.42	7.43
12	1.44	1.97	2.40	2.87	3.57	4.17	4.83	5.83	6.69
13	1.42	1.92	2.31	2.74	3.38	3.91	4.50	5.37	6.11
14	1.40	1.87	2.24	2.64	3.22	3.70	4.23	5.00	5.66
15	1.38	1.83	2.18	2.55	3.08	3.52	4.00	4.70	5.29
16	1.37	1.79	2.12	2.47	2.97	3.37	3.81	4.45	4.98
17	1.36	1.76	2.08	2.41	2.87	3.25	3.65	4.24	4.72
18	1.34	1.74	2.04	2.35	2.78	3.14	3.52	4.06	4.50
19	1.33	1.71	2.00	2.30	2.71	3.04	3.40	3.90	4.31
20	1.32	1.69	1.97	2.25	2.64	2.96	3.29	3.77	4.15
21	1.32	1.67	1.94	2.21	2.58	2.88	3.20	3.64	4.00
22	1.31	1.65	1.91	2.17	2.53	2.82	3.12	3.54	3.88
23	1.30	1.64	1.88	2.14	2.48	2.76	3.04	3.44	3.76
24	1.29	1.62	1.86	2.11	2.44	2.70	2.98	3.36	3.66
25	1.29	1.61	1.84	2.08	2.40	2.65	2.91	3.28	3.57
26	1.28	1.59	1.82	2.05	2.36	2.61	2.86	3.21	3.49
27	1.28	1.58	1.81	2.03	2.33	2.57	2.81	3.14	3.41
28	1.27	1.57	1.79	2.01	2.30	2.53	2.76	3.09	3.34
29	1.27	1.56	1.77	1.99	2.27	2.49	2.72	3.03	3.28
30	1.26	1.55	1.76	1.97	2.25	2.46	2.68	2.98	3.22
35	1.24	1.51	1.70	1.89	2.14	2.33	2.52	2.78	2.98
40	1.23	1.48	1.66	1.83	2.06	2.23	2.40	2.64	2.82
45	1.22	1.46	1.63	1.79	2.00	2.16	2.31	2.53	2.69
50	1.21	1.44	1.60	1.75	1.95	2.10	2.24	2.44	2.59
60	1.20	1.41	1.56	1.70	1.88	2.01	2.14	2.32	2.45
70	1.19	1.39	1.53	1.66	1.83	1.95	2.07	2.23	2.35
80	1.18	1.38	1.51	1.63	1.79	1.90	2.02	2.16	2.28
90	1.18	1.36	1.49	1.61	1.76	1.87	1.98	2.11	2.22
100	1.17	1.35	1.48	1.59	1.74	1.84	1.94	2.08	2.18
120	1.16	1.34	1.46	1.56	1.70	1.80	1.89	2.02	2.11
140	1.16	1.33	1.44	1.55	1.67	1.77	1.86	1.98	2.06
160	1.15	1.32	1.43	1.53	1.66	1.75	1.83	1.95	2.03
180	1.15	1.32	1.42	1.52	1.64	1.73	1.81	1.92	2.00
200	1.15	1.31	1.41	1.51	1.63	1.71	1.80	1.90	1.98
300	1.14	1.29	1.39	1.48	1.59	1.67	1.75	1.85	1.92
500	1.14	1.28	1.38	1.46	1.57	1.64	1.71	1.80	1.87
∞	1.13	1.26	1.35	1.43	1.52	1.59	1.65	1.73	1.79

TABLE B.4 (cont.) CRITICAL VALUES OF THE *F* DISTRIBUTION

Numerator DF = 60

	α(2): 0.50	0.20	0.10	0.05	0.02	0.01	0.005	0.002	0.001
Denom. DF	α(1): 0.25	0.10	0.05	0.025	0.01	0.005	0.0025	0.001	0.0005
1	9.76	62.8	252.	1010.	6310.	25300.	101000.	631000.	2530000.
2	3.46	9.47	19.5	39.5	99.5	199.	399.	999.	2000.
3	2.47	5.15	8.57	14.0	26.3	42.1	67.3	124.	198.
4	2.08	3.79	5.69	8.36	13.7	19.6	28.0	44.7	63.6
5	1.87	3.14	4.43	6.12	9.20	12.4	16.6	24.3	32.4
6	1.74	2.76	3.74	4.96	7.06	9.12	11.7	16.2	20.7
7	1.65	2.51	3.30	4.25	5.82	7.31	9.12	12.1	15.0
8	1.59	2.34	3.01	3.78	5.03	6.18	7.54	9.73	11.8
9	1.54	2.21	2.79	3.45	4.48	5.41	6.49	8.19	9.72
10	1.50	2.11	2.62	3.20	4.08	4.86	5.75	7.12	8.34
11	1.47	2.03	2.49	3.00	3.78	4.45	5.20	6.35	7.35
12	1.44	1.96	2.38	2.85	3.54	4.12	4.78	5.76	6.61
13	1.42	1.90	2.30	2.72	3.34	3.87	4.44	5.30	6.04
14	1.40	1.86	2.22	2.61	3.18	3.66	4.17	4.94	5.58
15	1.38	1.82	2.16	2.52	3.05	3.48	3.95	4.64	5.21
16	1.36	1.78	2.11	2.45	2.93	3.33	3.76	4.39	4.91
17	1.35	1.75	2.06	2.38	2.83	3.21	3.60	4.18	4.65
18	1.34	1.72	2.02	2.32	2.75	3.10	3.47	4.00	4.43
19	1.33	1.70	1.98	2.27	2.67	3.00	3.35	3.84	4.24
20	1.32	1.68	1.95	2.22	2.61	2.92	3.24	3.70	4.08
21	1.31	1.66	1.92	2.18	2.55	2.84	3.15	3.58	3.93
22	1.30	1.64	1.89	2.14	2.50	2.77	3.07	3.48	3.81
23	1.30	1.62	1.86	2.11	2.45	2.71	2.99	3.38	3.69
24	1.29	1.61	1.84	2.08	2.40	2.66	2.92	3.29	3.59
25	1.28	1.59	1.82	2.05	2.36	2.61	2.86	3.22	3.50
26	1.28	1.58	1.80	2.03	2.33	2.56	2.81	3.15	3.42
27	1.27	1.57	1.79	2.00	2.29	2.52	2.76	3.08	3.34
28	1.27	1.56	1.77	1.98	2.26	2.48	2.71	3.02	3.27
29	1.26	1.55	1.75	1.96	2.23	2.45	2.67	2.97	3.21
30	1.26	1.54	1.74	1.94	2.21	2.42	2.63	2.92	3.15
35	1.24	1.50	1.68	1.86	2.10	2.28	2.47	2.72	2.91
40	1.22	1.47	1.64	1.80	2.02	2.18	2.35	2.57	2.75
45	1.21	1.44	1.60	1.76	1.96	2.11	2.26	2.46	2.62
50	1.20	1.42	1.58	1.72	1.91	2.05	2.19	2.38	2.52
60	1.19	1.40	1.53	1.67	1.84	1.96	2.09	2.25	2.38
70	1.18	1.37	1.50	1.63	1.78	1.90	2.01	2.16	2.28
80	1.17	1.36	1.48	1.60	1.75	1.85	1.96	2.10	2.20
90	1.17	1.35	1.46	1.58	1.72	1.82	1.92	2.05	2.15
100	1.16	1.34	1.45	1.56	1.69	1.79	1.89	2.01	2.10
120	1.16	1.32	1.43	1.53	1.66	1.75	1.84	1.95	2.04
140	1.15	1.31	1.41	1.51	1.63	1.72	1.80	1.91	1.99
160	1.15	1.30	1.40	1.50	1.61	1.69	1.77	1.88	1.95
180	1.14	1.29	1.39	1.48	1.60	1.68	1.75	1.85	1.93
200	1.14	1.29	1.39	1.47	1.58	1.66	1.74	1.83	1.90
300	1.13	1.27	1.36	1.45	1.55	1.62	1.69	1.78	1.84
500	1.13	1.26	1.35	1.42	1.52	1.58	1.65	1.73	1.79
∞	1.12	1.24	1.32	1.39	1.47	1.53	1.59	1.66	1.71

TABLE B.4 (cont.) CRITICAL VALUES OF THE *F* DISTRIBUTION

Numerator DF = 70

α(2):	0.50	0.20	0.10	0.05	0.02	0.01	0.005	0.002	0.001
α(1):	0.25	0.10	0.05	0.025	0.01	0.005	0.0025	0.001	0.0005
Denom. DF									
1	9.77	62.9	252.	1010.	6320.	25300.	101000.	632000.	2530000.
2	3.46	9.48	19.5	39.5	99.5	199.	399.	999.	2000.
3	2.47	5.15	8.57	14.0	26.3	42.1	67.2	124.	198.
4	2.08	3.79	5.68	8.35	13.6	19.6	28.0	44.6	63.4
5	1.87	3.14	4.42	6.11	9.18	12.4	16.6	24.3	32.3
6	1.74	2.76	3.73	4.94	7.03	9.09	11.7	16.1	20.6
7	1.65	2.51	3.29	4.24	5.80	7.28	9.07	12.1	14.9
8	1.59	2.33	2.99	3.77	5.01	6.15	7.49	9.67	11.7
9	1.54	2.20	2.78	3.43	4.46	5.38	6.45	8.13	9.65
10	1.50	2.10	2.61	3.18	4.06	4.83	5.71	7.07	8.28
11	1.46	2.02	2.48	2.99	3.75	4.41	5.16	6.30	7.29
12	1.44	1.95	2.37	2.83	3.51	4.09	4.74	5.71	6.55
13	1.41	1.90	2.28	2.70	3.32	3.84	4.41	5.26	5.98
14	1.39	1.85	2.21	2.60	3.16	3.62	4.14	4.89	5.53
15	1.38	1.81	2.15	2.51	3.02	3.45	3.91	4.59	5.16
16	1.36	1.77	2.09	2.43	2.91	3.30	3.73	4.34	4.85
17	1.35	1.74	2.05	2.36	2.81	3.18	3.57	4.13	4.60
18	1.34	1.71	2.00	2.30	2.72	3.07	3.43	3.95	4.38
19	1.33	1.69	1.97	2.25	2.65	2.97	3.31	3.79	4.19
20	1.32	1.67	1.93	2.20	2.58	2.88	3.20	3.66	4.03
21	1.31	1.65	1.90	2.16	2.52	2.81	3.11	3.54	3.88
22	1.30	1.63	1.88	2.13	2.47	2.74	3.03	3.43	3.76
23	1.29	1.61	1.85	2.09	2.42	2.68	2.95	3.34	3.64
24	1.28	1.60	1.83	2.06	2.38	2.63	2.89	3.25	3.54
25	1.28	1.58	1.81	2.03	2.34	2.58	2.83	3.17	3.45
26	1.27	1.57	1.79	2.01	2.30	2.53	2.77	3.10	3.37
27	1.27	1.56	1.77	1.98	2.27	2.49	2.72	3.04	3.29
28	1.26	1.55	1.75	1.96	2.24	2.45	2.67	2.98	3.22
29	1.26	1.54	1.74	1.94	2.21	2.42	2.63	2.92	3.16
30	1.25	1.53	1.72	1.92	2.18	2.38	2.59	2.87	3.10
35	1.23	1.49	1.66	1.84	2.07	2.25	2.43	2.67	2.86
40	1.22	1.46	1.62	1.78	1.99	2.15	2.31	2.53	2.70
45	1.21	1.43	1.59	1.74	1.93	2.08	2.22	2.42	2.57
50	1.20	1.41	1.56	1.70	1.88	2.02	2.15	2.33	2.47
60	1.19	1.38	1.52	1.64	1.81	1.93	2.05	2.21	2.33
70	1.18	1.36	1.49	1.60	1.75	1.86	1.97	2.12	2.22
80	1.17	1.34	1.46	1.57	1.71	1.82	1.92	2.05	2.15
90	1.16	1.33	1.44	1.55	1.68	1.78	1.88	2.00	2.09
100	1.16	1.32	1.43	1.53	1.66	1.75	1.84	1.96	2.05
120	1.15	1.31	1.41	1.50	1.62	1.71	1.79	1.90	1.98
140	1.14	1.29	1.39	1.48	1.60	1.68	1.76	1.86	1.93
160	1.14	1.29	1.38	1.47	1.58	1.65	1.73	1.83	1.90
180	1.14	1.28	1.37	1.46	1.56	1.64	1.71	1.80	1.87
200	1.13	1.27	1.36	1.45	1.55	1.62	1.69	1.78	1.85
300	1.13	1.26	1.34	1.42	1.51	1.58	1.64	1.72	1.78
500	1.12	1.24	1.32	1.39	1.48	1.54	1.60	1.68	1.73
∞	1.11	1.22	1.29	1.36	1.43	1.49	1.54	1.60	1.65

TABLE B.4 (cont.) CRITICAL VALUES OF THE *F* DISTRIBUTION

Numerator DF = 80

Denom. DF	α(2): 0.50 α(1): 0.25	0.20 0.10	0.10 0.05	0.05 0.025	0.02 0.01	0.01 0.005	0.005 0.0025	0.002 0.001	0.001 0.0005
1	9.78	62.9	253.	1010.	6330.	25300.	101000.	633000.	2530000.
2	3.46	9.48	19.5	39.5	99.5	199.	399.	999.	2000.
3	2.47	5.15	8.56	14.0	26.3	42.1	67.1	124.	198.
4	2.08	3.78	5.67	8.33	13.6	19.5	27.9	44.6	63.3
5	1.87	3.13	4.41	6.10	9.16	12.3	16.5	24.2	32.2
6	1.74	2.75	3.72	4.93	7.01	9.06	11.6	16.1	20.5
7	1.65	2.50	3.29	4.23	5.78	7.25	9.04	12.0	14.8
8	1.59	2.33	2.99	3.76	4.99	6.12	7.46	9.63	11.6
9	1.54	2.20	2.77	3.42	4.44	5.36	6.42	8.09	9.61
10	1.50	2.09	2.60	3.17	4.04	4.80	5.68	7.03	8.23
11	1.46	2.01	2.47	2.97	3.73	4.39	5.13	6.26	7.25
12	1.44	1.95	2.36	2.82	3.49	4.07	4.71	5.68	6.51
13	1.41	1.89	2.27	2.69	3.30	3.81	4.38	5.22	5.94
14	1.39	1.84	2.20	2.58	3.14	3.60	4.11	4.86	5.49
15	1.37	1.80	2.14	2.49	3.00	3.43	3.89	4.56	5.12
16	1.36	1.77	2.08	2.42	2.89	3.28	3.70	4.31	4.81
17	1.35	1.74	2.03	2.35	2.79	3.15	3.54	4.10	4.56
18	1.33	1.71	1.99	2.29	2.70	3.04	3.40	3.92	4.34
19	1.32	1.68	1.96	2.24	2.63	2.95	3.28	3.76	4.15
20	1.31	1.66	1.92	2.19	2.56	2.86	3.18	3.62	3.99
21	1.30	1.64	1.89	2.15	2.50	2.79	3.08	3.50	3.84
22	1.30	1.62	1.86	2.11	2.45	2.72	3.00	3.40	3.72
23	1.29	1.61	1.84	2.08	2.40	2.66	2.93	3.30	3.60
24	1.28	1.59	1.82	2.05	2.36	2.60	2.86	3.22	3.50
25	1.28	1.58	1.80	2.02	2.32	2.55	2.80	3.14	3.41
26	1.27	1.56	1.78	1.99	2.28	2.51	2.74	3.07	3.33
27	1.26	1.55	1.76	1.97	2.25	2.47	2.69	3.00	3.25
28	1.26	1.54	1.74	1.94	2.22	2.43	2.64	2.94	3.18
29	1.25	1.53	1.73	1.92	2.19	2.39	2.60	2.89	3.12
30	1.25	1.52	1.71	1.90	2.16	2.36	2.56	2.84	3.06
35	1.23	1.48	1.65	1.82	2.05	2.22	2.40	2.64	2.83
40	1.22	1.45	1.61	1.76	1.97	2.12	2.28	2.49	2.66
45	1.21	1.42	1.57	1.72	1.91	2.05	2.19	2.38	2.53
50	1.20	1.40	1.54	1.68	1.86	1.99	2.12	2.30	2.43
60	1.18	1.37	1.50	1.63	1.78	1.90	2.02	2.17	2.29
70	1.17	1.35	1.47	1.59	1.73	1.84	1.94	2.08	2.18
80	1.16	1.33	1.45	1.55	1.69	1.79	1.89	2.01	2.11
90	1.16	1.32	1.43	1.53	1.66	1.75	1.84	1.96	2.05
100	1.15	1.31	1.41	1.51	1.63	1.72	1.81	1.92	2.01
120	1.14	1.29	1.39	1.48	1.60	1.68	1.76	1.86	1.94
140	1.14	1.28	1.38	1.46	1.57	1.65	1.72	1.82	1.89
160	1.13	1.27	1.36	1.45	1.55	1.62	1.69	1.79	1.85
180	1.13	1.27	1.35	1.43	1.53	1.61	1.67	1.76	1.83
200	1.13	1.26	1.35	1.42	1.52	1.59	1.66	1.74	1.80
300	1.12	1.24	1.32	1.39	1.48	1.55	1.61	1.68	1.74
500	1.11	1.23	1.30	1.37	1.45	1.51	1.56	1.63	1.68
∞	1.10	1.21	1.27	1.33	1.40	1.45	1.50	1.56	1.60

TABLE B.4 (cont.) CRITICAL VALUES OF THE *F* DISTRIBUTION

Numerator DF = 90

α(2):	0.50	0.20	0.10	0.05	0.02	0.01	0.005	0.002	0.001
α(1):	0.25	0.10	0.05	0.025	0.01	0.005	0.0025	0.001	0.0005
Denom. DF									
1	9.79	63.0	253.	1010.	6330.	25300.	101000.	633000.	2530000.
2	3.46	9.48	19.5	39.5	99.5	199.	399.	999.	2000.
3	2.47	5.15	8.56	14.0	26.3	42.0	67.1	124.	197.
4	2.08	3.78	5.67	8.33	13.6	19.5	27.9	44.5	63.2
5	1.87	3.13	4.41	6.09	9.14	12.3	16.5	24.2	32.1
6	1.74	2.75	3.72	4.92	7.00	9.04	11.6	16.1	20.4
7	1.65	2.50	3.28	4.22	5.77	7.23	9.01	12.0	14.8
8	1.59	2.32	2.98	3.75	4.97	6.10	7.44	9.60	11.6
9	1.53	2.19	2.76	3.41	4.43	5.34	6.40	8.06	9.57
10	1.49	2.09	2.59	3.16	4.03	4.79	5.66	7.00	8.20
11	1.46	2.01	2.46	2.96	3.72	4.37	5.11	6.23	7.21
12	1.43	1.94	2.36	2.81	3.48	4.05	4.69	5.65	6.48
13	1.41	1.89	2.27	2.68	3.28	3.79	4.36	5.19	5.91
14	1.39	1.84	2.19	2.57	3.12	3.58	4.09	4.83	5.45
15	1.37	1.80	2.13	2.48	2.99	3.41	3.86	4.53	5.09
16	1.36	1.76	2.07	2.40	2.87	3.26	3.68	4.28	4.78
17	1.34	1.73	2.03	2.34	2.78	3.13	3.52	4.07	4.53
18	1.33	1.70	1.98	2.28	2.69	3.02	3.38	3.89	4.31
19	1.32	1.68	1.95	2.23	2.61	2.93	3.26	3.73	4.12
20	1.31	1.65	1.91	2.18	2.55	2.84	3.16	3.60	3.96
21	1.30	1.63	1.88	2.14	2.49	2.77	3.06	3.48	3.81
22	1.29	1.62	1.86	2.10	2.43	2.70	2.98	3.37	3.69
23	1.29	1.60	1.83	2.07	2.39	2.64	2.90	3.28	3.57
24	1.28	1.58	1.81	2.03	2.34	2.58	2.84	3.19	3.47
25	1.27	1.57	1.79	2.01	2.30	2.53	2.78	3.11	3.38
26	1.27	1.56	1.77	1.98	2.26	2.49	2.72	3.04	3.30
27	1.26	1.54	1.75	1.95	2.23	2.45	2.67	2.98	3.22
28	1.26	1.53	1.73	1.93	2.20	2.41	2.62	2.92	3.15
29	1.25	1.52	1.72	1.91	2.17	2.37	2.58	2.86	3.09
30	1.25	1.51	1.70	1.89	2.14	2.34	2.54	2.81	3.03
35	1.23	1.47	1.64	1.81	2.03	2.20	2.38	2.61	2.80
40	1.21	1.44	1.60	1.75	1.95	2.10	2.26	2.47	2.63
45	1.20	1.41	1.56	1.70	1.89	2.03	2.17	2.36	2.50
50	1.19	1.39	1.53	1.67	1.84	1.97	2.10	2.27	2.40
60	1.18	1.36	1.49	1.61	1.76	1.88	1.99	2.14	2.25
70	1.17	1.34	1.46	1.57	1.71	1.81	1.92	2.05	2.15
80	1.16	1.33	1.44	1.54	1.67	1.77	1.86	1.98	2.08
90	1.15	1.31	1.42	1.52	1.64	1.73	1.82	1.93	2.02
100	1.15	1.30	1.40	1.50	1.61	1.70	1.78	1.89	1.97
120	1.14	1.28	1.38	1.47	1.58	1.66	1.73	1.83	1.90
140	1.13	1.27	1.36	1.45	1.55	1.62	1.70	1.79	1.86
160	1.13	1.26	1.35	1.43	1.53	1.60	1.67	1.75	1.82
180	1.13	1.26	1.34	1.42	1.51	1.58	1.65	1.73	1.79
200	1.12	1.25	1.33	1.41	1.50	1.56	1.63	1.71	1.77
300	1.11	1.23	1.31	1.38	1.46	1.52	1.58	1.65	1.70
500	1.11	1.22	1.29	1.35	1.43	1.48	1.53	1.60	1.65
∞	1.10	1.20	1.26	1.31	1.38	1.43	1.47	1.52	1.56

TABLE B.4 (cont.) CRITICAL VALUES OF THE *F* DISTRIBUTION

Numerator DF = 100

α(2): α(1): Denom. DF	0.50 0.25	0.20 0.10	0.10 0.05	0.05 0.025	0.02 0.01	0.01 0.005	0.005 0.0025	0.002 0.001	0.001 0.0005
1	9.80	63.0	253.	1010.	6330.	25300.	101000.	633000.	2530000.
2	3.47	9.48	19.5	39.5	99.5	199.	399.	999.	2000.
3	2.47	5.14	8.55	14.0	26.2	42.0	67.1	124.	197.
4	2.08	3.78	5.66	8.32	13.6	19.5	27.9	44.5	63.2
5	1.87	3.13	4.41	6.08	9.13	12.3	16.5	24.1	32.1
6	1.74	2.75	3.71	4.92	6.99	9.03	11.6	16.0	20.4
7	1.65	2.50	3.27	4.21	5.75	7.22	8.99	12.0	14.8
8	1.58	2.32	2.97	3.74	4.96	6.09	7.42	9.57	11.6
9	1.53	2.19	2.76	3.40	4.41	5.32	6.38	8.04	9.54
10	1.49	2.09	2.59	3.15	4.01	4.77	5.64	6.98	8.17
11	1.46	2.01	2.46	2.96	3.71	4.36	5.09	6.21	7.19
12	1.43	1.94	2.35	2.80	3.47	4.04	4.67	5.63	6.45
13	1.41	1.88	2.26	2.67	3.27	3.78	4.34	5.17	5.88
14	1.39	1.83	2.19	2.56	3.11	3.57	4.07	4.81	5.43
15	1.37	1.79	2.12	2.47	2.98	3.39	3.85	4.51	5.06
16	1.36	1.76	2.07	2.40	2.86	3.25	3.66	4.26	4.76
17	1.34	1.73	2.02	2.33	2.76	3.12	3.50	4.05	4.50
18	1.33	1.70	1.98	2.27	2.68	3.01	3.36	3.87	4.28
19	1.32	1.67	1.94	2.22	2.60	2.91	3.24	3.71	4.10
20	1.31	1.65	1.91	2.17	2.54	2.83	3.14	3.58	3.93
21	1.30	1.63	1.88	2.13	2.48	2.75	3.04	3.46	3.79
22	1.29	1.61	1.85	2.09	2.42	2.69	2.96	3.35	3.66
23	1.29	1.59	1.82	2.06	2.37	2.62	2.89	3.25	3.55
24	1.28	1.58	1.80	2.02	2.33	2.57	2.82	3.17	3.45
25	1.27	1.56	1.78	2.00	2.29	2.52	2.76	3.09	3.36
26	1.27	1.55	1.76	1.97	2.25	2.47	2.70	3.02	3.27
27	1.26	1.54	1.74	1.94	2.22	2.43	2.65	2.96	3.20
28	1.25	1.53	1.73	1.92	2.19	2.39	2.60	2.90	3.13
29	1.25	1.52	1.71	1.90	2.16	2.36	2.56	2.84	3.06
30	1.25	1.51	1.70	1.88	2.13	2.32	2.52	2.79	3.01
35	1.23	1.47	1.63	1.80	2.02	2.19	2.36	2.59	2.77
40	1.21	1.43	1.59	1.74	1.94	2.09	2.24	2.44	2.60
45	1.20	1.41	1.55	1.69	1.88	2.01	2.15	2.33	2.47
50	1.19	1.39	1.52	1.66	1.82	1.95	2.08	2.25	2.37
60	1.18	1.36	1.48	1.60	1.75	1.86	1.97	2.12	2.23
70	1.16	1.34	1.45	1.56	1.70	1.80	1.90	2.03	2.13
80	1.16	1.32	1.43	1.53	1.65	1.75	1.84	1.96	2.05
90	1.15	1.30	1.41	1.50	1.62	1.71	1.80	1.91	1.99
100	1.14	1.29	1.39	1.48	1.60	1.68	1.76	1.87	1.95
120	1.14	1.28	1.37	1.45	1.56	1.64	1.71	1.81	1.88
140	1.13	1.26	1.35	1.43	1.53	1.60	1.67	1.76	1.83
160	1.13	1.26	1.34	1.42	1.51	1.58	1.64	1.73	1.79
180	1.12	1.25	1.33	1.40	1.49	1.56	1.62	1.70	1.76
200	1.12	1.24	1.32	1.39	1.48	1.54	1.60	1.68	1.74
300	1.11	1.22	1.30	1.36	1.44	1.50	1.55	1.62	1.67
500	1.10	1.21	1.28	1.34	1.41	1.46	1.51	1.57	1.62
∞	1.09	1.18	1.24	1.30	1.36	1.40	1.44	1.49	1.53

TABLE B.4 (cont.) CRITICAL VALUES OF THE *F* DISTRIBUTION

Numerator DF = 120

α(2):	0.50	0.20	0.10	0.05	0.02	0.01	0.005	0.002	0.001
α(1):	0.25	0.10	0.05	0.025	0.01	0.005	0.0025	0.001	0.0005
Denom. DF									
1	9.80	63.1	253.	1010.	6340.	25400.	101000.	634000.	2540000.
2	3.47	9.48	19.5	39.5	99.5	199.	399.	999.	2000.
3	2.47	5.14	8.55	13.9	26.2	42.0	67.0	124.	197.
4	2.08	3.78	5.66	8.31	13.6	19.5	27.8	44.4	63.1
5	1.87	3.12	4.40	6.07	9.11	12.3	16.4	24.1	32.0
6	1.74	2.74	3.70	4.90	6.97	9.00	11.6	16.0	20.3
7	1.65	2.49	3.27	4.20	5.74	7.19	8.96	11.9	14.7
8	1.58	2.32	2.97	3.73	4.95	6.06	7.39	9.53	11.5
9	1.53	2.18	2.75	3.39	4.40	5.30	6.35	8.00	9.49
10	1.49	2.08	2.58	3.14	4.00	4.75	5.61	6.94	8.12
11	1.46	2.00	2.45	2.94	3.69	4.34	5.07	6.18	7.14
12	1.43	1.93	2.34	2.79	3.45	4.01	4.65	5.59	6.41
13	1.41	1.88	2.25	2.66	3.25	3.76	4.31	5.14	5.84
14	1.39	1.83	2.18	2.55	3.09	3.55	4.04	4.77	5.39
15	1.37	1.79	2.11	2.46	2.96	3.37	3.82	4.47	5.02
16	1.35	1.75	2.06	2.38	2.84	3.22	3.63	4.23	4.72
17	1.34	1.72	2.01	2.32	2.75	3.10	3.47	4.02	4.46
18	1.33	1.69	1.97	2.26	2.66	2.99	3.34	3.84	4.25
19	1.32	1.67	1.93	2.20	2.58	2.89	3.22	3.68	4.06
20	1.31	1.64	1.90	2.16	2.52	2.81	3.11	3.54	3.90
21	1.30	1.62	1.87	2.11	2.46	2.73	3.02	3.42	3.75
22	1.29	1.60	1.84	2.08	2.40	2.66	2.93	3.32	3.62
23	1.28	1.59	1.81	2.04	2.35	2.60	2.86	3.22	3.51
24	1.28	1.57	1.79	2.01	2.31	2.55	2.79	3.14	3.41
25	1.27	1.56	1.77	1.98	2.27	2.50	2.73	3.06	3.32
26	1.26	1.54	1.75	1.95	2.23	2.45	2.68	2.99	3.24
27	1.26	1.53	1.73	1.93	2.20	2.41	2.62	2.92	3.16
28	1.25	1.52	1.71	1.91	2.17	2.37	2.58	2.86	3.09
29	1.25	1.51	1.70	1.89	2.14	2.33	2.53	2.81	3.03
30	1.24	1.50	1.68	1.87	2.11	2.30	2.49	2.76	2.97
35	1.22	1.46	1.62	1.79	2.00	2.16	2.33	2.56	2.73
40	1.21	1.42	1.58	1.72	1.92	2.06	2.21	2.41	2.56
45	1.20	1.40	1.54	1.68	1.85	1.99	2.12	2.30	2.44
50	1.19	1.38	1.51	1.64	1.80	1.93	2.05	2.21	2.34
60	1.17	1.35	1.47	1.58	1.73	1.83	1.94	2.08	2.19
70	1.16	1.32	1.44	1.54	1.67	1.77	1.87	1.99	2.09
80	1.15	1.31	1.41	1.51	1.63	1.72	1.81	1.92	2.01
90	1.15	1.29	1.39	1.48	1.60	1.68	1.76	1.87	1.95
100	1.14	1.28	1.38	1.46	1.57	1.65	1.73	1.83	1.90
120	1.13	1.26	1.35	1.43	1.53	1.61	1.68	1.77	1.83
140	1.13	1.25	1.33	1.41	1.50	1.57	1.64	1.72	1.78
160	1.12	1.24	1.32	1.39	1.48	1.55	1.61	1.69	1.75
180	1.12	1.23	1.31	1.38	1.47	1.53	1.59	1.66	1.72
200	1.11	1.23	1.30	1.37	1.45	1.51	1.57	1.64	1.69
300	1.10	1.21	1.28	1.34	1.41	1.46	1.51	1.58	1.62
500	1.10	1.19	1.26	1.31	1.38	1.42	1.47	1.53	1.57
∞	1.08	1.17	1.22	1.27	1.32	1.36	1.40	1.45	1.48

TABLE B.4 (cont.) CRITICAL VALUES OF THE *F* DISTRIBUTION

Numerator DF = 140

α(2):	0.50	0.20	0.10	0.05	0.02	0.01	0.005	0.002	0.001
α(1):	0.25	0.10	0.05	0.025	0.01	0.005	0.0025	0.001	0.0005
Denom. DF									
1	9.81	63.1	253.	1010.	6340.	25400.	101000.	634000.	2540000.
2	3.47	9.48	19.5	39.5	99.5	199.	399.	999.	2000.
3	2.47	5.14	8.55	13.9	26.2	42.0	67.0	124.	197.
4	2.08	3.77	5.65	8.30	13.5	19.4	27.8	44.4	63.0
5	1.87	3.12	4.39	6.06	9.10	12.3	16.4	24.0	31.9
6	1.74	2.74	3.70	4.90	6.96	8.98	11.5	15.9	20.3
7	1.65	2.49	3.26	4.19	5.72	7.18	8.94	11.9	14.7
8	1.58	2.31	2.96	3.72	4.93	6.05	7.37	9.50	11.5
9	1.53	2.18	2.74	3.38	4.39	5.28	6.33	7.97	9.46
10	1.49	2.08	2.57	3.13	3.98	4.73	5.59	6.92	8.09
11	1.46	2.00	2.44	2.94	3.68	4.32	5.05	6.15	7.11
12	1.43	1.93	2.33	2.78	3.44	4.00	4.63	5.57	6.38
13	1.41	1.87	2.25	2.65	3.24	3.74	4.29	5.11	5.81
14	1.39	1.82	2.17	2.54	3.08	3.53	4.02	4.75	5.36
15	1.37	1.78	2.11	2.45	2.95	3.36	3.80	4.45	5.00
16	1.35	1.75	2.05	2.37	2.83	3.21	3.61	4.20	4.69
17	1.34	1.71	2.00	2.31	2.73	3.08	3.45	3.99	4.44
18	1.33	1.69	1.96	2.25	2.65	2.97	3.32	3.81	4.22
19	1.32	1.66	1.92	2.19	2.57	2.87	3.20	3.66	4.03
20	1.31	1.64	1.89	2.15	2.50	2.79	3.09	3.52	3.87
21	1.30	1.62	1.86	2.10	2.44	2.71	3.00	3.40	3.73
22	1.29	1.60	1.83	2.07	2.39	2.65	2.92	3.29	3.60
23	1.28	1.58	1.81	2.03	2.34	2.59	2.84	3.20	3.49
24	1.27	1.57	1.78	2.00	2.30	2.53	2.77	3.11	3.38
25	1.27	1.55	1.76	1.97	2.26	2.48	2.71	3.03	3.29
26	1.26	1.54	1.74	1.94	2.22	2.43	2.66	2.96	3.21
27	1.26	1.53	1.72	1.92	2.18	2.39	2.60	2.90	3.13
28	1.25	1.51	1.71	1.90	2.15	2.35	2.56	2.84	3.06
29	1.24	1.50	1.69	1.88	2.12	2.32	2.51	2.79	3.00
30	1.24	1.49	1.68	1.86	2.10	2.28	2.47	2.74	2.94
35	1.22	1.45	1.61	1.77	1.98	2.15	2.31	2.53	2.71
40	1.21	1.42	1.57	1.71	1.90	2.05	2.19	2.39	2.54
45	1.19	1.39	1.53	1.66	1.84	1.97	2.10	2.27	2.41
50	1.18	1.37	1.50	1.63	1.79	1.91	2.03	2.19	2.31
60	1.17	1.34	1.46	1.57	1.71	1.81	1.92	2.06	2.16
70	1.16	1.32	1.42	1.53	1.65	1.75	1.84	1.96	2.06
80	1.15	1.30	1.40	1.49	1.61	1.70	1.79	1.90	1.98
90	1.14	1.28	1.38	1.47	1.58	1.66	1.74	1.84	1.92
100	1.14	1.27	1.36	1.45	1.55	1.63	1.70	1.80	1.87
120	1.13	1.26	1.34	1.42	1.51	1.58	1.65	1.74	1.80
140	1.12	1.24	1.32	1.39	1.48	1.55	1.61	1.69	1.75
160	1.12	1.23	1.31	1.38	1.46	1.52	1.58	1.66	1.71
180	1.11	1.22	1.30	1.36	1.45	1.50	1.56	1.63	1.68
200	1.11	1.22	1.29	1.35	1.43	1.49	1.54	1.61	1.66
300	1.10	1.20	1.26	1.32	1.39	1.44	1.49	1.55	1.59
500	1.09	1.18	1.24	1.29	1.35	1.40	1.44	1.49	1.53
∞	1.08	1.16	1.20	1.25	1.30	1.33	1.37	1.41	1.44

TABLE B.4 (cont.) CRITICAL VALUES OF THE *F* DISTRIBUTION

Numerator DF = 200

α(2):	0.50	0.20	0.10	0.05	0.02	0.01	0.005	0.002	0.001
α(1):	0.25	0.10	0.05	0.025	0.01	0.005	0.0025	0.001	0.0005
Denom. DF									
1	9.82	63.2	254.	1020.	6350.	25400.	102000.	635000.	2540000.
2	3.47	9.49	19.5	39.5	99.5	199.	399.	999.	2000.
3	2.47	5.14	8.54	13.9	26.2	41.9	66.9	124.	197.
4	2.08	3.77	5.65	8.29	13.5	19.4	27.7	44.3	62.9
5	1.87	3.12	4.39	6.05	9.08	12.2	16.4	24.0	31.8
6	1.74	2.73	3.69	4.88	6.93	8.95	11.5	15.9	20.2
7	1.65	2.48	3.25	4.18	5.70	7.15	8.90	11.8	14.6
8	1.58	2.31	2.95	3.70	4.91	6.02	7.33	9.45	11.4
9	1.53	2.17	2.73	3.37	4.36	5.26	6.29	7.93	9.40
10	1.49	2.07	2.56	3.12	3.96	4.71	5.56	6.87	8.04
11	1.46	1.99	2.43	2.92	3.66	4.29	5.01	6.10	7.06
12	1.43	1.92	2.32	2.76	3.41	3.97	4.59	5.52	6.33
13	1.40	1.86	2.23	2.63	3.22	3.71	4.26	5.07	5.76
14	1.38	1.82	2.16	2.53	3.06	3.50	3.99	4.71	5.31
15	1.37	1.77	2.10	2.44	2.92	3.33	3.77	4.41	4.95
16	1.35	1.74	2.04	2.36	2.81	3.18	3.58	4.16	4.64
17	1.34	1.71	1.99	2.29	2.71	3.05	3.42	3.95	4.39
18	1.32	1.68	1.95	2.23	2.62	2.94	3.28	3.77	4.17
19	1.31	1.65	1.91	2.18	2.55	2.85	3.16	3.61	3.98
20	1.30	1.63	1.88	2.13	2.48	2.76	3.06	3.48	3.82
21	1.29	1.61	1.84	2.09	2.42	2.68	2.96	3.36	3.68
22	1.28	1.59	1.82	2.05	2.36	2.62	2.88	3.25	3.55
23	1.28	1.57	1.79	2.01	2.32	2.56	2.81	3.16	3.44
24	1.27	1.56	1.77	1.98	2.27	2.50	2.74	3.07	3.34
25	1.26	1.54	1.75	1.95	2.23	2.45	2.68	2.99	3.24
26	1.26	1.53	1.73	1.92	2.19	2.40	2.62	2.92	3.16
27	1.25	1.52	1.71	1.90	2.16	2.36	2.57	2.86	3.09
28	1.25	1.50	1.69	1.88	2.13	2.32	2.52	2.80	3.02
29	1.24	1.49	1.67	1.86	2.10	2.29	2.48	2.74	2.95
30	1.24	1.48	1.66	1.84	2.07	2.25	2.44	2.69	2.89
35	1.22	1.44	1.60	1.75	1.96	2.11	2.27	2.49	2.66
40	1.20	1.41	1.55	1.69	1.87	2.01	2.15	2.34	2.49
45	1.19	1.38	1.51	1.64	1.81	1.93	2.06	2.23	2.36
50	1.18	1.36	1.48	1.60	1.76	1.87	1.99	2.14	2.26
60	1.16	1.33	1.44	1.54	1.68	1.78	1.88	2.01	2.11
70	1.15	1.30	1.40	1.50	1.62	1.71	1.80	1.92	2.00
80	1.14	1.28	1.38	1.47	1.58	1.66	1.74	1.85	1.93
90	1.13	1.27	1.36	1.44	1.55	1.62	1.70	1.79	1.86
100	1.13	1.26	1.34	1.42	1.52	1.59	1.66	1.75	1.82
120	1.12	1.24	1.32	1.39	1.48	1.54	1.60	1.68	1.74
140	1.11	1.22	1.30	1.36	1.45	1.51	1.56	1.64	1.69
160	1.11	1.21	1.28	1.35	1.42	1.48	1.53	1.60	1.65
180	1.10	1.21	1.27	1.33	1.41	1.46	1.51	1.57	1.62
200	1.10	1.20	1.26	1.32	1.39	1.44	1.49	1.55	1.60
300	1.09	1.18	1.23	1.28	1.35	1.39	1.43	1.48	1.52
500	1.08	1.16	1.21	1.25	1.31	1.35	1.38	1.43	1.46
∞	1.07	1.13	1.17	1.21	1.25	1.28	1.30	1.34	1.36

TABLE B.4 (cont.) CRITICAL VALUES OF THE *F* DISTRIBUTION

Numerator DF = ∞

α(2):	0.50	0.20	0.10	0.05	0.02	0.01	0.005	0.002	0.001
α(1):	0.25	0.10	0.05	0.025	0.01	0.005	0.0025	0.001	0.0005
Denom. DF									
1	9.85	63.3	254.	1020.	6370.	25500.	102000.	637000.	2550000.
2	3.48	9.49	19.5	39.5	99.5	199.	399.	999.	2000.
3	2.47	5.13	8.53	13.9	26.1	41.8	66.8	123.	196.
4	2.08	3.76	5.63	8.26	13.5	19.3	27.6	44.0	62.6
5	1.87	3.11	4.37	6.02	9.02	12.1	16.3	23.8	31.6
6	1.74	2.72	3.67	4.85	6.88	8.88	11.4	15.7	20.0
7	1.65	2.47	3.23	4.14	5.65	7.08	8.81	11.7	14.4
8	1.58	2.29	2.93	3.67	4.86	5.95	7.25	9.33	11.3
9	1.53	2.16	2.71	3.33	4.31	5.19	6.21	7.81	9.26
10	1.48	2.06	2.54	3.08	3.91	4.64	5.47	6.76	7.91
11	1.45	1.97	2.40	2.88	3.60	4.23	4.93	6.00	6.93
12	1.42	1.90	2.30	2.72	3.36	3.90	4.51	5.42	6.20
13	1.40	1.85	2.21	2.60	3.17	3.65	4.18	4.97	5.64
14	1.38	1.80	2.13	2.49	3.00	3.44	3.91	4.60	5.19
15	1.36	1.76	2.07	2.40	2.87	3.26	3.69	4.31	4.83
16	1.34	1.72	2.01	2.32	2.75	3.11	3.50	4.06	4.52
17	1.33	1.69	1.96	2.25	2.65	2.98	3.34	3.85	4.27
18	1.32	1.66	1.92	2.19	2.57	2.87	3.20	3.67	4.05
19	1.30	1.63	1.88	2.13	2.49	2.78	3.08	3.51	3.87
20	1.29	1.61	1.84	2.09	2.42	2.69	2.97	3.38	3.71
21	1.28	1.59	1.81	2.04	2.36	2.61	2.88	3.26	3.56
22	1.28	1.57	1.78	2.00	2.31	2.55	2.80	3.15	3.43
23	1.27	1.55	1.76	1.97	2.26	2.48	2.72	3.05	3.32
24	1.26	1.53	1.73	1.94	2.21	2.43	2.65	2.97	3.22
25	1.25	1.52	1.71	1.91	2.17	2.38	2.59	2.89	3.13
26	1.25	1.50	1.69	1.88	2.13	2.33	2.54	2.82	3.05
27	1.24	1.49	1.67	1.85	2.10	2.29	2.48	2.75	2.97
28	1.24	1.48	1.65	1.83	2.06	2.25	2.44	2.69	2.90
29	1.23	1.47	1.64	1.81	2.03	2.21	2.39	2.64	2.84
30	1.23	1.46	1.62	1.79	2.01	2.18	2.35	2.59	2.78
35	1.20	1.41	1.56	1.70	1.89	2.04	2.18	2.38	2.54
40	1.19	1.38	1.51	1.64	1.80	1.93	2.06	2.23	2.37
45	1.18	1.35	1.47	1.59	1.74	1.85	1.97	2.12	2.23
50	1.16	1.33	1.44	1.55	1.68	1.79	1.89	2.03	2.13
60	1.15	1.29	1.39	1.48	1.60	1.69	1.78	1.89	1.98
70	1.13	1.27	1.35	1.44	1.54	1.62	1.69	1.79	1.87
80	1.12	1.24	1.32	1.40	1.49	1.56	1.63	1.72	1.79
90	1.12	1.23	1.30	1.37	1.46	1.52	1.58	1.66	1.72
100	1.11	1.21	1.28	1.35	1.43	1.49	1.54	1.62	1.67
120	1.10	1.19	1.25	1.31	1.38	1.43	1.48	1.54	1.59
140	1.09	1.18	1.23	1.28	1.35	1.39	1.43	1.49	1.53
160	1.08	1.16	1.21	1.26	1.32	1.36	1.40	1.45	1.49
180	1.08	1.15	1.20	1.24	1.30	1.33	1.37	1.42	1.45
200	1.07	1.14	1.19	1.23	1.28	1.31	1.35	1.39	1.42
300	1.06	1.11	1.15	1.18	1.22	1.25	1.27	1.30	1.33
500	1.05	1.09	1.11	1.14	1.16	1.18	1.20	1.23	1.24
∞	1.00	1.00	1.00	1.00	1.00	1.00	1.00	1.00	1.00

Table B.4 was prepared using Equations 26.6.4, 26.6.5, 26.6.8, 26.6.11, 26.6.12, 26.4.6, and 26.4.14 of Zelen and Severo (1964). Values of F were calculated to a relative error $\leq 10^{-8}$ and then were rounded to three significant figures.

Examples:

$$F_{0.05(1),2,18} = 3.55, \quad F_{0.01(1),8,10} = 5.06 \quad \text{and} \quad F_{0.05(2),20,40} = 2.07.$$

If a critical value is needed for degrees of freedom not on this table, one may conservatively employ the next smaller degrees of freedom that are on the table. Or, the needed critical value may be obtained by linear interpolation, with an error no greater than 0.01 for $\alpha(1) \geq 0.01$ and no greater than 0.02 for $\alpha(1) < 0.01$. If a little more accuracy is desired, of if $\nu_1 > 200$ or $\nu_2 > 500$, harmonic interpolation should be used.

TABLE B.5 CRITICAL VALUES OF THE q DISTRIBUTION

$\alpha = 0.50$

v	k(or p) = 2	3	4	5	6	7	8	9	10
1	1.414	2.338	2.918	3.335	3.658	3.920	4.139	4.327	4.491
2	1.155	1.908	2.377	2.713	2.973	3.184	3.361	3.513	3.645
3	1.082	1.791	2.230	2.545	2.789	2.987	3.152	3.294	3.418
4	1.048	1.737	2.163	2.468	2.704	2.895	3.055	3.193	3.313
5	1.028	1.705	2.124	2.423	2.655	2.843	3.000	3.135	3.253
6	1.015	1.685	2.098	2.394	2.623	2.809	2.964	3.097	3.214
7	1.006	1.670	2.080	2.375	2.601	2.785	2.938	3.070	3.186
8	0.9990	1.659	2.067	2.359	2.584	2.767	2.920	3.051	3.165
9	0.9938	1.651	2.057	2.347	2.572	2.753	2.905	3.035	3.149
10	0.9897	1.645	2.049	2.338	2.561	2.742	2.893	3.023	3.137
11	0.9863	1.639	2.042	2.330	2.553	2.733	2.884	3.014	3.127
12	0.9836	1.635	2.037	2.324	2.546	2.726	2.876	3.005	3.118
13	0.9812	1.631	2.032	2.319	2.540	2.719	2.870	2.998	3.111
14	0.9792	1.628	2.028	2.314	2.535	2.714	2.864	2.993	3.105
15	0.9775	1.625	2.025	2.310	2.531	2.709	2.859	2.987	3.099
16	0.9760	1.623	2.022	2.307	2.527	2.705	2.855	2.983	3.095
17	0.9747	1.621	2.020	2.304	2.524	2.702	2.851	2.979	3.091
18	0.9735	1.619	2.018	2.301	2.521	2.699	2.848	2.976	3.087
19	0.9724	1.618	2.015	2.299	2.518	2.696	2.845	2.973	3.084
20	0.9715	1.616	2.013	2.297	2.516	2.693	2.842	2.970	3.081
24	0.9685	1.611	2.007	2.290	2.509	2.685	2.834	2.961	3.071
30	0.9656	1.606	2.001	2.284	2.501	2.678	2.825	2.952	3.062
40	0.9626	1.602	1.996	2:277	2.494	2.670	2.817	2.943	3.053
60	0.9597	1.597	1.990	2.270	2.486	2.662	2.808	2.934	3.043
120	0.9568	1.592	1.984	2.264	2.479	2.654	2.799	2.925	3.034
∞	0.9539	1.588	1.978	2.257	2.472	2.645	2.791	2.915	3.024

v	k(or p) = 1	12	13	14	15	16	17	18	19
1	4.637	4.767	4.885	4.992	5.091	5.182	5.266	5.345	5.420
2	3.762	3.867	3.963	4.049	4.129	4.203	4.271	4.335	4.395
3	3.528	3.626	3.715	3.797	3.871	3.940	4.004	4.064	4.120
4	3.419	3.515	3.601	3.680	3.752	3.819	3.881	3.939	3.993
5	3.357	3.451	3.535	3.613	3.684	3.749	3.811	3.867	3.920
6	3.317	3.409	3.493	3.569	3.639	3.704	3.764	3.820	3.873
7	3.288	3.380	3.463	3.538	3.608	3.672	3.732	3.788	3.840
8	3.267	3.358	3.440	3.515	3.585	3.648	3.708	3.763	3.815
9	3.250	3.341	3.423	3.498	3.567	3.630	3.689	3.744	3.796
10	3.237	3.328	3.410	3.484	3.552	3.616	3.674	3.729	3.781
11	3.227	3.317	3.398	3.472	3.540	3.604	3.662	3.717	3.769
12	3.219	3.308	3.389	3.463	3.531	3.594	3.652	3.706	3.757
13	3.211	3.300	3.381	3.455	3.523	3.585	3.643	3.698	3.749
14	3.204	3.293	3.375	3.448	3.515	3.578	3.636	3.690	3.741
15	3.199	3.288	3.369	3.442	3.509	3.572	3.630	3.684	3.735
16	3.194	3.283	3.364	3.436	3.504	3.567	3.624	3.678	3.729
17	3.190	3.278	3.359	3.432	3.499	3.562	3.620	3.673	3.724
18	3.186	3.274	3.355	3.428	3.495	3.558	3.615	3.669	3.719
19	3.183	3.271	3.352	3.424	3.491	3.554	3.611	3.665	3.715
20	3.179	3.268	3.348	3.421	3.488	3.550	3.608	3.661	3.712
24	3.170	3.258	3.338	3.410	3.477	3.539	3.596	3.650	3.700
30	3.160	3.248	3.327	3.400	3.466	3.528	3.585	3.639	3.688
40	3.150	3.238	3.317	3.389	3.456	3.517	3.574	3.627	3.677
60	3.141	3.228	3.306	3.378	3.444	3.505	3.562	3.615	3.665
120	3.131	3.217	3.296	3.367	3.433	3.494	3.551	3.603	3.653
∞	3.121	3.207	3.285	3.356	3.422	3.482	3.538	3.591	3.640

TABLE B.5 (cont.) CRITICAL VALUES OF THE q DISTRIBUTION

$\alpha = 0.50$

v	k(or p) = 20	22	24	26	28	30	32	34	36
1	5.489	5.616	5.731	5.835	5.930	6.017	6.098	6.173	6.244
2	4.451	4.554	4.646	4.730	4.807	4.878	4.943	5.004	5.061
3	4.172	4.269	4.356	4.434	4.507	4.573	4.634	4.691	4.745
4	4.044	4.138	4.222	4.298	4.368	4.432	4.492	4.547	4.599
5	3.970	4.062	4.145	4.220	4.288	4.352	4.410	4.464	4.515
6	3.922	4.014	4.095	4.169	4.237	4.299	4.357	4.411	4.461
7	3.889	3.979	4.060	4.133	4.200	4.262	4.319	4.372	4.422
8	3.864	3.953	4.033	4.106	4.174	4.234	4.291	4.344	4.394
9	3.844	3.933	4.013	4.086	4.152	4.214	4.270	4.323	4.372
10	3.829	3.917	3.997	4.069	4.135	4.196	4.253	4.305	4.354
11	3.816	3.904	3.984	4.056	4.122	4.183	4.239	4.291	4.340
12	3.806	3.894	3.973	4.045	4.110	4.171	4.227	4.279	4.328
13	3.797	3.884	3.964	4.035	4.101	4.162	4.218	4.269	4.318
14	3.789	3.877	3.956	4.027	4.092	4.153	4.209	4.261	4.309
15	3.783	3.870	3.949	4.020	4.085	4.145	4.201	4.253	4.302
16	3.777	3.864	3.942	4.014	4.079	4.139	4.195	4.246	4.295
17	3.772	3.859	3.937	4.009	4.073	4.133	4.189	4.241	4.289
18	3.767	3.854	3.932	4.003	4.069	4.128	4.184	4.236	4.284
19	3.763	3.850	3.928	3.999	4.064	4.124	4.180	4.231	4.279
20	3.759	3.846	3.995	3.995	4.060	4.120	4.175	4.227	4.275
24	3.748	3.834	3.912	3.983	4.048	4.107	4.162	4.214	4.262
30	3.736	3.822	3.899	3.970	4.035	4.094	4.149	4.200	4.248
40	3.724	3.809	3.887	3.957	4.021	4.080	4.135	4.186	4.234
60	3.711	3.797	3.874	3.944	4.008	4.067	4.121	4.172	4.220
120	3.699	3.784	3.861	3.930	3.994	4.053	4.107	4.157	4.204
∞	3.686	3.771	3.847	3.916	3.979	4.037	4.091	4.141	4.188

v	k(or p) = 38	40	50	60	70	80	90	100
1	6.310	6.372	6.637	6.847	7.021	7.169	7.297	7.411
2	5.115	5.165	5.379	5.550	5.690	5.810	5.914	6.006
3	4.795	4.842	5.043	5.202	5.335	5.447	5.544	5.630
4	4.647	4.693	4.888	5.043	5.171	5.280	5.374	5.458
5	4.563	4.608	4.799	4.951	5.077	5.184	5.277	5.359
6	4.500	4.552	4.741	4.891	5.016	5.121	5.213	5.294
7	4.469	4.513	4.700	4.850	4.973	5.078	5.169	5.249
8	4.440	4.484	4.671	4.819	4.941	5.045	5.136	5.215
9	4.418	4.462	4.647	4.794	4.916	5.020	5.110	5.189
10	4.400	4.444	4.629	4.775	4.897	5.000	5.090	5.169
11	4.386	4.429	4.613	4.760	4.881	4.984	5.073	5.152
12	4.374	4.417	4.600	4.747	4.867	4.970	5.059	5.138
13	4.364	4.406	4.590	4.736	4.856	4.959	5.048	5.126
14	4.355	4.397	4.581	4.726	4.846	4.949	5.037	5.116
15	4.347	4.390	4.573	4.718	4.838	4.940	5.029	5.107
16	4.340	4.383	4.566	4.710	4.831	4.932	5.021	5.099
17	4.334	4.377	4.559	4.704	4.824	4.926	5.014	5.092
18	4.329	4.372	4.554	4.698	4.818	4.920	5.008	5.086
19	4.324	4.367	4.549	4.693	4.813	4.914	5.003	5.081
20	4.320	4.363	4.545	4.689	4.808	4.910	4.998	5.076
24	4.307	4.349	4.530	4.674	4.793	4.894	4.982	5.060
30	4.293	4.335	4.515	4.659	4.778	4.878	4.966	5.044
40	4.279	4.321	4.500	4.644	4.762	4.862	4.950	5.027
60	4.264	4.306	4.485	4.627	4.745	4.845	4.932	5.009
120	4.249	4.290	4.469	4.610	4.727	4.827	4.914	4.990
∞	4.232	4.274	4.450	4.591	4.707	4.806	4.892	4.968

TABLE B.5 (cont.) CRITICAL VALUES OF THE q DISTRIBUTION

$\alpha = 0.20$

ν	k(or p) = 2	3	4	5	6	7	8	9	10
1	4.353	6.615	8.075	9.138	9.966	10.64	11.21	11.70	12.12
2	2.667	3.820	4.559	5.098	5.521	5.867	6.158	6.409	6.630
3	2.316	3.245	3.833	4.261	4.597	4.872	5.104	5.305	5.481
4	2.168	3.004	3.527	3.907	4.205	4.449	4.655	4.832	4.989
5	2.087	2.872	3.358	3.712	3.988	4.214	4.405	4.570	4.715
6	2.036	2.788	3.252	3.588	3.850	4.065	4.246	4.403	4.540
7	2.001	2.731	3.179	3.503	3.756	3.962	4.136	4.287	4.419
8	1.976	2.689	3.126	3.440	3.686	3.886	4.055	4.201	4.330
9	1.956	2.658	3.085	3.393	3.633	3.828	3.994	4.136	4.261
10	1.941	2.632	3.053	3.355	3.590	3.782	3.944	4.084	4.206
11	1.928	2.612	3.027	3.325	3.557	3.745	3.905	4.042	4.162
12	1.918	2.596	3.006	3.300	3.529	3.715	3.872	4.007	4.126
13	1.910	2.582	2.988	3.279	3.505	3.689	3.844	3.978	4.095
14	1.902	2.570	2.973	3.261	3.485	3.667	3.820	3.953	4.069
15	1.896	2.560	2.960	3.246	3.467	3.648	3.800	3.931	4.046
16	1.891	2.551	2.948	3.232	3.452	3.631	3.782	3.912	4.026
17	1.886	2.543	2.938	3.220	3.439	3.617	3.766	3.895	4.008
18	1.882	2.536	2.930	3.210	3.427	3.604	3.753	3.881	3.993
19	1.878	2.530	2.922	3.200	3.416	3.592	3.740	3.867	3.979
20	1.874	2.524	2.914	3.192	3.407	3.582	3.729	3.855	3.966
24	1.864	2.507	2.892	3.166	3.377	3.549	3.694	3.818	3.927
30	1.853	2.490	2.870	3.140	3.348	3.517	3.659	3.781	3.887
40	1.843	2.473	2.848	3.114	3.318	3.484	3.624	3.743	3.848
60	1.833	2.456	2.826	3.089	3.290	3.452	3.589	3.707	3.809
120	1.822	2.440	2.805	3.063	3.260	3.420	3.554	3.669	3.770
∞	1.812	2.424	2.784	3.037	3.232	3.389	3.520	3.632	3.730

ν	k(or p) = 11	12	13	14	15	16	17	18	19
1	12.50	12.84	13.14	13.43	13.68	13.93	14.14	14.35	14.54
2	6.826	7.002	7.162	7.308	7.442	7.566	7.682	7.790	7.891
3	5.637	5.778	5.906	6.023	6.131	6.230	6.323	6.410	6.491
4	5.128	5.253	5.367	5.471	5.566	5.655	5.738	5.815	5.888
5	4.844	4.960	5.066	5.162	5.251	5.334	5.411	5.482	5.550
6	4.663	4.773	4.873	4.965	5.049	5.128	5.201	5.269	5.333
7	4.537	4.643	4.739	4.827	4.908	4.984	5.054	5.120	5.181
8	4.444	4.547	4.640	4.726	4.805	4.877	4.945	5.009	5.069
9	4.372	4.473	4.564	4.647	4.724	4.796	4.862	4.924	4.982
10	4.316	4.414	4.503	4.585	4.660	4.730	4.795	4.856	4.913
11	4.270	4.366	4.454	4.534	4.608	4.677	4.741	4.801	4.857
12	4.231	4.327	4.413	4.492	4.565	4.633	4.696	4.755	4.810
13	4.199	4.293	4.379	4.457	4.529	4.596	4.658	4.716	4.770
14	4.172	4.265	4.349	4.426	4.498	4.564	4.625	4.683	4.737
15	4.148	4.240	4.324	4.400	4.471	4.536	4.597	4.654	4.707
16	4.127	4.218	4.301	4.377	4.447	4.512	4.572	4.628	4.681
17	4.109	4.199	4.282	4.357	4.426	4.490	4.550	4.606	4.659
18	4.093	4.182	4.264	4.339	4.407	4.471	4.531	4.586	4.638
19	4.078	4.167	4.248	4.323	4.391	4.454	4.513	4.569	4.620
20	4.065	4.154	4.234	4.308	4.376	4.439	4.498	4.552	4.604
24	4.024	4.111	4.190	4.262	4.329	4.391	4.448	4.502	4.552
30	3.982	4.068	4.145	4.216	4.281	4.342	4.398	4.451	4.500
40	3.941	4.025	4.101	4.170	4.234	4.293	4.348	4.399	4.447
60	3.900	3.982	4.056	4.124	4.186	4.244	4.297	4.347	4.395
120	3.859	3.938	4.011	4.077	4.138	4.194	4.246	4.295	4.341
∞	3.817	3.895	3.966	4.030	4.089	4.144	4.195	4.242	4.287

TABLE B.5 (cont.) CRITICAL VALUES OF THE q DISTRIBUTION

$\alpha = 0.20$

ν	k(or p) = 20	22	24	26	28	30	32	34	36
1	14.72	15.06	15.36	15.63	15.88	16.11	16.32	16.52	16.71
2	7.986	8.162	8.320	8.463	8.594	8.715	8.827	8.931	9.029
3	6.568	6.709	6.835	6.951	7.057	7.154	7.244	7.328	7.407
4	5.956	6.082	6.195	6.298	6.392	6.479	6.560	6.635	6.706
5	5.613	5.730	5.835	5.931	6.019	6.100	6.175	6.245	6.311
6	5.393	5.504	5.604	5.695	5.779	5.856	5.927	5.994	6.056
7	5.239	5.346	5.442	5.530	5.610	5.684	5.753	5.817	5.877
8	5.125	5.228	5.322	5.407	5.485	5.557	5.624	5.686	5.744
9	5.037	5.138	5.229	5.312	5.388	5.459	5.524	5.585	5.641
10	4.967	5.066	5.155	5.237	5.311	5.380	5.444	5.504	5.559
11	4.910	5.007	5.095	5.175	5.248	5.316	5.379	5.437	5.492
12	4.862	4.958	5.044	5.123	5.196	5.262	5.324	5.382	5.436
13	4.822	4.917	5.002	5.080	5.151	5.217	5.278	5.335	5.388
14	4.787	4.881	4.965	5.042	5.113	5.178	5.238	5.295	5.347
15	4.757	4.850	4.934	5.010	5.080	5.144	5.204	5.260	5.312
16	4.731	4.823	4.906	4.981	5.050	5.114	5.174	5.229	5.281
17	4.708	4.799	4.881	4.956	5.025	5.088	5.147	5.202	5.253
18	4.688	4.778	4.859	4.934	5.002	5.065	5.123	5.177	5.228
19	4.669	4.759	4.840	4.914	4.981	5.044	5.102	5.156	5.206
20	4.652	4.742	4.822	4.895	4.963	5.025	5.082	5.136	5.186
24	4.599	4.687	4.766	4.838	4.904	4.964	5.021	5.073	5.122
30	4.546	4.632	4.710	4.779	4.844	4.903	4.958	5.010	5.058
40	4.493	4.576	4.652	4.720	4.783	4.841	4.895	4.945	4.993
60	4.439	4.520	4.594	4.661	4.722	4.778	4.831	4.880	4.925
120	4.384	4.463	4.535	4.600	4.659	4.714	4.765	4.812	4.857
∞	4.329	4.405	4.475	4.537	4.595	4.648	4.697	4.743	4.786

ν	k(or p) = 38	40	50	60	70	80	90	100
1	16.88	17.05	17.74	18.30	18.76	19.15	19.49	19.79
2	9.121	9.207	9.576	9.869	10.11	10.32	10.50	10.66
3	7.481	7.551	7.849	8.086	8.283	8.450	8.596	8.725
4	6.771	6.834	7.100	7.313	7.489	7.639	7.769	7.885
5	6.372	6.430	6.678	6.877	7.041	7.181	7.303	7.411
6	6.115	6.170	6.406	6.595	6.751	6.885	7.001	7.103
7	5.934	5.987	6.214	6.397	6.548	6.676	6.788	6.887
8	5.799	5.851	6.072	6.249	6.395	6.520	6.629	6.725
9	5.695	5.745	5.961	6.134	6.277	6.399	6.506	6.600
10	5.612	5.661	5.873	6.042	6.182	6.302	6.407	6.499
11	5.544	5.592	5.800	5.967	6.105	6.223	6.326	6.416
12	5.487	5.535	5.740	5.904	6.040	6.156	6.257	6.347
13	5.438	5.486	5.689	5.850	5.985	6.100	6.200	6.288
14	5.397	5.444	5.644	5.804	5.937	6.051	6.150	6.237
15	5.361	5.407	5.606	5.764	5.896	6.008	6.106	6.193
16	5.329	5.375	5.572	5.729	5.859	5.971	6.068	6.154
17	5.301	5.347	5.542	5.698	5.827	5.938	6.034	6.119
18	5.276	5.321	5.515	5.670	5.798	5.908	6.004	6.089
19	5.254	5.299	5.491	5.645	5.772	5.881	5.976	6.061
20	5.233	5.278	5.469	5.622	5.749	5.857	5.951	6.035
24	5.169	5.212	5.400	5.549	5.674	5.780	5.872	5.954
30	5.103	5.146	5.329	5.475	5.597	5.701	5.791	5.871
40	5.037	5.078	5.257	5.399	5.518	5.619	5.708	5.786
60	4.969	5.009	5.183	5.321	5.437	5.535	5.621	5.697
120	4.899	4.938	5.106	5.240	5.352	5.447	5.530	5.603
∞	4.826	4.864	5.026	5.155	5.262	5.353	5.433	5.503

TABLE B.5 (cont.) CRITICAL VALUES OF THE q DISTRIBUTION

$\alpha = 0.10$

ν	k(or p) = 2	3	4	5	6	7	8	9	10
1	8.929	13.44	16.36	18.49	20.15	21.51	22.64	23.62	24.48
2	4.130	5.733	6.773	7.538	8.139	8.633	9.049	9.409	9.725
3	3.328	4.467	5.199	5.738	6.162	6.511	6.806	7.062	7.287
4	3.015	3.976	4.586	5.035	5.388	5.679	5.926	6.139	6.327
5	2.850	3.717	4.264	4.664	4.979	5.238	5.458	5.648	5.816
6	2.748	3.559	4.065	4.435	4.726	4.966	5.168	5.344	5.499
7	2.680	3.451	3.931	4.280	4.555	4.780	4.972	5.137	5.283
8	2.630	3.374	3.834	4.169	4.431	4.646	4.829	4.987	5.126
9	2.592	3.316	3.761	4.084	4.337	4.545	4.721	4.873	5.007
10	2.563	3.270	3.704	4.018	4.264	4.465	4.636	4.783	4.913
11	2.540	3.234	3.658	3.965	4.205	4.401	4.568	4.711	4.838
12	2.521	3.204	3.621	3.922	4.156	4.349	4.511	4.652	4.776
13	2.505	3.179	3.589	3.885	4.116	4.305	4.464	4.602	4.724
14	2.491	3.158	3.563	3.854	4.081	4.267	4.424	4.560	4.680
15	2.479	3.140	3.540	3.828	4.052	4.235	4.390	4.524	4.641
16	2.469	3.124	3.520	3.804	4.026	4.207	4.360	4.492	4.608
17	2.460	3.110	3.503	3.784	4.004	4.183	4.334	4.464	4.579
18	2.452	3.098	3.488	3.767	3.984	4.161	4.311	4.440	4.554
19	2.445	3.087	3.474	3.751	3.966	4.142	4.290	4.418	4.531
20	2.439	3.078	3.462	3.736	3.950	4.124	4.271	4.398	4.510
24	2.420	3.047	3.423	3.692	3.900	4.070	4.213	4.336	4.445
30	2.400	3.017	3.386	3.648	3.851	4.016	4.155	4.275	4.381
40	2.381	2.988	3.349	3.605	3.803	3.963	4.099	4.215	4.317
60	2.363	2.959	3.312	3.562	3.755	3.911	4.042	4.155	4.254
120	2.344	2.930	3.276	3.520	3.707	3.859	3.987	4.096	4.191
∞	2.326	2.902	3.240	3.478	3.661	3.808	3.931	4.037	4.129

ν	k(or p) = 11	12	13	14	15	16	17	18	19
1	25.24	25.92	26.54	27.10	27.62	28.10	28.54	28.96	29.35
2	10.01	10.26	10.49	10.70	10.89	11.07	11.24	11.39	11.54
3	7.487	7.667	7.832	7.982	8.120	8.249	8.368	8.479	8.584
4	6.495	6.645	6.783	6.909	7.025	7.133	7.233	7.327	7.414
5	5.966	6.101	6.223	6.336	6.440	6.536	6.626	6.710	6.789
6	5.637	5.762	5.875	5.979	6.075	6.164	6.247	6.325	6.398
7	5.413	5.530	5.637	5.735	5.826	5.910	5.988	6.061	6.130
8	5.250	5.362	5.464	5.558	5.644	5.724	5.799	5.869	5.935
9	5.127	5.234	5.333	5.423	5.506	5.583	5.655	5.723	5.786
10	5.029	5.134	5.229	5.317	5.397	5.472	5.542	5.607	5.668
11	4.951	5.053	5.146	5.231	5.309	5.382	5.450	5.514	5.573
12	4.886	4.986	5.077	5.160	5.236	5.308	5.374	5.436	5.495
13	4.832	4.930	5.019	5.100	5.176	5.245	5.311	5.372	5.429
14	4.786	4.882	4.970	5.050	5.124	5.192	5.256	5.316	5.373
15	4.746	4.841	4.927	5.006	5.079	5.147	5.209	5.269	5.324
16	4.712	4.805	4.890	4.968	5.040	5.107	5.169	5.227	5.282
17	4.682	4.774	4.858	4.935	5.005	5.071	5.133	5.190	5.244
18	4.655	4.746	4.829	4.905	4.975	5.040	5.101	5.158	5.211
19	4.631	4.721	4.803	4.879	4.948	5.012	5.073	5.129	5.182
20	4.609	4.699	4.780	4.855	4.924	4.987	5.047	5.103	5.155
24	4.541	4.628	4.708	4.780	4.847	4.909	4.966	5.021	5.071
30	4.474	4.559	4.635	4.706	4.770	4.830	4.886	4.939	4.988
40	4.408	4.490	4.564	4.632	4.695	4.752	4.807	4.857	4.905
60	4.342	4.421	4.493	4.558	4.619	4.675	4.727	4.775	4.821
120	4.276	4.353	4.422	4.485	4.543	4.597	4.647	4.694	4.738
∞	4.211	4.285	4.351	4.412	4.468	4.519	4.568	4.612	4.654

TABLE B.5 (cont.) CRITICAL VALUES OF THE q DISTRIBUTION

$\alpha = 0.10$

ν	k(or p) = 20	22	24	26	28	30	32	34	36
1	29.71	30.39	30.99	31.54	32.04	32.50	32.93	33.33	33.71
2	11.68	11.93	12.16	12.36	12.55	12.73	12.89	13.04	13.18
3	8.683	8.864	9.029	9.177	9.314	9.440	9.557	9.666	9.768
4	7.497	7.650	7.789	7.914	8.029	8.135	8.234	8.326	8.412
5	6.863	7.000	7.123	7.236	7.340	7.435	7.523	7.606	7.683
6	6.467	6.593	6.708	6.812	6.908	6.996	7.078	7.155	7.227
7	6.195	6.315	6.422	6.521	6.611	6.695	6.773	6.845	6.913
8	5.997	6.111	6.214	6.308	6.395	6.475	6.549	6.618	6.683
9	5.846	5.956	6.055	6.146	6.229	6.306	6.378	6.444	6.507
10	5.726	5.833	5.930	6.017	6.098	6.173	6.242	6.307	6.368
11	5.630	5.734	5.828	5.914	5.992	6.065	6.132	6.196	6.255
12	5.550	5.652	5.744	5.827	5.904	5.976	6.042	6.103	6.161
13	5.483	5.583	5.673	5.755	5.830	5.900	5.965	6.025	6.082
14	5.426	5.524	5.612	5.693	5.767	5.836	5.899	5.959	6.014
15	5.376	5.473	5.560	5.639	5.713	5.780	5.843	5.901	5.956
16	5.333	5.428	5.515	5.593	5.665	5.732	5.793	5.851	5.905
17	5.295	5.389	5.474	5.552	5.623	5.689	5.750	5.806	5.860
18	5.262	5.355	5.439	5.515	5.585	5.650	5.711	5.767	5.820
19	5.232	5.324	5.407	5.483	5.552	5.616	5.676	5.732	5.784
20	5.205	5.296	5.378	5.453	5.522	5.586	5.645	5.700	5.752
24	5.119	5.208	5.287	5.360	5.427	5.489	5.546	5.600	5.650
30	5.034	5.120	5.197	5.267	5.332	5.392	5.447	5.499	5.547
40	4.949	5.032	5.107	5.174	5.236	5.294	5.347	5.397	5.444
60	4.864	4.944	5.015	5.081	5.141	5.196	5.247	5.295	5.340
120	4.779	4.856	4.924	4.987	5.044	5.097	5.146	5.192	5.235
∞	4.694	4.767	4.832	4.892	4.947	4.997	5.044	5.087	5.128

ν	k(or p) = 38	40	50	60	70	80	90	100
1	34.06	34.38	35.79	36.91	37.83	38.62	39.30	39.91
2	13.31	13.44	13.97	14.40	14.75	15.05	15.31	15.54
3	9.864	9.954	10.34	10.65	10.91	11.12	11.31	11.48
4	8.493	8.569	8.896	9.156	9.373	9.557	9.718	9.860
5	7.756	7.825	8.118	8.353	8.548	8.715	8.859	8.988
6	7.294	7.358	7.630	7.848	8.029	8.184	8.319	8.438
7	6.976	7.036	7.294	7.500	7.672	7.818	7.946	8.059
8	6.744	6.801	7.048	7.245	7.409	7.550	7.672	7.780
9	6.566	6.621	6.859	7.050	7.208	7.343	7.461	7.566
10	6.425	6.479	6.709	6.895	7.048	7.180	7.295	7.396
11	6.310	6.363	6.588	6.768	6.918	7.047	7.158	7.258
12	6.215	6.267	6.487	6.663	6.810	6.936	7.045	7.142
13	6.135	6.186	6.402	6.575	6.719	6.842	6.949	7.045
14	6.067	6.116	6.329	6.499	6.641	6.762	6.868	6.961
15	6.008	6.057	6.266	6.433	6.573	6.692	6.796	6.888
16	5.956	6.004	6.210	6.376	6.513	6.631	6.734	6.825
17	5.910	5.958	6.162	6.325	6.461	6.577	6.679	6.769
18	5.870	5.917	6.118	6.280	6.414	6.529	6.630	6.719
19	5.833	5.880	6.079	6.239	6.372	6.486	6.585	6.674
20	5.801	5.847	6.044	6.203	6.335	6.447	6.546	6.633
24	5.697	5.741	5.933	6.086	6.214	6.324	6.419	6.503
30	5.593	5.636	5.821	5.969	6.093	6.198	6.291	6.372
40	5.488	5.529	5.708	5.850	5.969	6.071	6.160	6.238
60	5.382	5.422	5.593	5.730	5.844	5.941	6.026	6.102
120	5.275	5.313	5.476	5.606	6.715	5.808	5.888	5.960
∞	5.166	5.202	5.357	5.480	5.582	5.669	5.745	5.812

TABLE B.5 (cont.) CRITICAL VALUES OF THE q DISTRIBUTION

$\alpha = 0.05$

ν	k(or p) = 2	3	4	5	6	7	8	9	10
1	17.97	26.98	32.82	37.08	40.41	43.12	45.40	47.36	49.07
2	6.085	8.331	9.798	10.88	11.74	12.44	13.03	13.54	13.99
3	4.501	5.910	6.825	7.502	8.037	8.478	8.853	9.177	9.462
4	3.927	5.040	5.757	6.287	6.707	7.053	7.347	7.602	7.826
5	3.635	4.602	5.218	5.673	6.033	6.330	6.582	6.802	6.995
6	3.461	4.339	4.896	5.305	5.628	5.895	6.122	6.319	6.493
7	3.344	4.165	4.681	5.060	5.359	5.606	5.815	5.998	6.158
8	3.261	4.041	4.529	4.886	5.167	5.399	5.597	5.767	5.918
9	3.199	3.949	4.415	4.756	5.024	5.244	5.432	5.595	5.739
10	3.151	3.877	4.327	4.654	4.912	5.124	5.305	5.461	5.599
11	3.113	3.820	4.256	4.574	4.823	5.028	5.202	5.353	5.487
12	3.082	3.773	4.199	4.508	4.751	4.950	5.119	5.265	5.395
13	3.055	3.735	4.151	4.453	4.690	4.885	5.049	5.192	5.318
14	3.033	3.702	4.111	4.407	4.639	4.829	4.990	5.131	5.254
15	3.014	3.674	4.076	4.367	4.595	4.782	4.940	5.077	5.198
16	2.998	3.649	4.046	4.333	4.557	4.741	4.897	5.031	5.150
17	2.984	3.628	4.020	4.303	4.524	4.705	4.858	4.991	5.108
18	2.971	3.609	3.997	4.277	4.495	4.673	4.824	4.956	5.071
19	2.960	3.593	3.977	4.253	4.469	4.645	4.794	4.924	5.038
20	2.950	3.578	3.958	4.232	4.445	4.620	4.768	4.896	5.008
24	2.919	3.532	3.901	4.166	4.373	4.541	4.684	4.807	4.915
30	2.888	3.486	3.845	4.102	4.302	4.464	4.602	4.720	4.824
40	2.858	3.442	3.791	4.039	4.232	4.389	4.521	4.635	4.735
60	2.829	3.399	3.737	3.977	4.163	4.314	4.441	4.550	4.646
120	2.800	3.356	3.685	3.917	4.096	4.241	4.363	4.468	4.560
∞	2.772	3.314	3.633	3.858	4.030	4.170	4.286	4.387	4.474

ν	k(or p) = 11	12	13	14	15	16	17	18	19
1	50.59	51.96	53.20	54.33	55.36	56.32	57.22	58.04	58.83
2	14.39	14.75	15.08	15.38	15.65	15.91	16.14	16.37	16.57
3	9.717	9.946	10.15	10.35	10.53	10.69	10.84	10.98	11.11
4	8.027	8.208	8.373	8.525	8.664	8.794	8.914	9.028	9.134
5	7.168	7.324	7.466	7.596	7.717	7.828	7.932	8.030	8.122
6	6.649	6.789	6.917	7.034	7.143	7.244	7.338	7.426	7.508
7	6.302	6.431	6.550	6.658	6.759	6.852	6.939	7.020	7.097
8	6.054	6.175	6.287	6.389	6.483	6.571	6.653	6.729	6.802
9	5.867	5.983	6.089	6.186	6.276	6.359	6.437	6.510	6.579
10	5.722	5.833	5.935	6.028	6.114	6.194	6.269	6.339	6.405
11	5.605	5.713	5.811	5.901	5.984	6.062	6.134	6.202	6.265
12	5.511	5.615	5.710	5.798	5.878	5.953	6.023	6.089	6.151
13	5.431	5.533	5.625	5.711	5.789	5.862	5.931	5.995	6.055
14	5.364	5.463	5.554	5.637	5.714	5.786	5.852	5.915	5.974
15	5.306	5.404	5.493	5.574	5.649	5.720	5.785	5.846	5.904
16	5.256	5.352	5.439	5.520	5.593	5.662	5.727	5.786	5.843
17	5.212	5.307	5.392	5.471	5.544	5.612	5.675	5.734	5.790
18	5.174	5.267	5.352	5.429	5.501	5.568	5.630	5.688	5.743
19	5.140	5.231	5.315	5.391	5.462	5.528	5.589	5.647	5.701
20	5.108	5.199	5.282	5.357	5.427	5.493	5.553	5.610	5.663
24	5.012	5.099	5.179	5.251	5.319	5.381	5.439	5.494	5.545
30	4.917	5.001	5.077	5.147	5.211	5.271	5.327	5.379	5.429
40	4.824	4.904	4.977	5.044	5.106	5.163	5.216	5.266	5.313
60	4.732	4.808	4.878	4.942	5.001	5.056	5.107	5.154	5.199
120	4.641	4.714	4.781	4.842	4.898	4.950	4.998	5.044	5.086
∞	4.552	4.622	4.685	4.743	4.796	4.845	4.891	4.934	4.974

TABLE B.5 (cont.) CRITICAL VALUES OF THE q DISTRIBUTION

$\alpha = 0.05$

ν	k(or p) = 20	22	24	26	28	30	32	34	36
1	59.56	60.91	62.12	63.22	64.23	65.15	66.01	66.81	67.56
2	16.77	17.13	17.45	17.75	18.02	18.27	18.50	18.72	18.92
3	11.24	11.47	11.68	11.87	12.05	12.21	12.36	12.50	12.63
4	9.233	9.418	9.584	9.736	9.875	10.00	10.12	10.23	10.34
5	8.208	8.368	8.512	8.643	8.764	8.875	8.979	9.075	9.165
6	7.587	7.730	7.861	7.979	8.088	8.189	8.283	8.370	8.452
7	7.170	7.303	7.423	7.533	7.634	7.728	7.814	7.895	7.972
8	6.870	6.995	7.109	7.212	7.307	7.395	7.477	7.554	7.625
9	6.644	6.763	6.871	6.970	7.061	7.145	7.222	7.295	7.363
10	6.467	6.582	6.686	6.781	6.868	6.948	7.023	7.093	7.159
11	6.326	6.436	6.536	6.628	6.712	6.790	6.863	6.930	6.994
12	6.209	6.317	6.414	6.503	6.585	6.660	6.731	6.796	6.858
13	6.112	6.217	6.312	6.398	6.478	6.551	6.620	6.684	6.744
14	6.029	6.132	6.224	6.309	6.387	6.459	6.526	6.588	6.647
15	5.958	6.059	6.149	6.233	6.309	6.379	6.445	6.506	6.564
16	5.897	5.995	6.084	6.166	6.241	6.310	6.374	6.434	6.491
17	5.842	5.940	6.027	6.107	6.181	6.249	6.313	6.372	6.427
18	5.794	5.890	5.977	6.055	6.128	6.195	6.258	6.316	6.371
19	5.752	5.846	5.932	6.009	6.081	6.147	6.209	6.267	6.321
20	5.714	5.807	5.891	5.968	6.039	6.104	6.165	6.222	6.275
24	5.594	5.683	5.764	5.838	5.906	5.968	6.027	6.081	6.132
30	5.475	5.561	5.638	5.709	5.774	5.833	5.889	5.941	5.990
40	5.358	5.439	5.513	5.581	5.642	5.700	5.753	5.803	5.849
60	5.241	5.319	5.389	5.453	5.512	5.566	5.617	5.664	5.708
120	5.126	5.200	5.266	5.327	5.382	5.434	5.481	5.526	5.568
∞	5.012	5.081	5.144	5.201	5.253	5.301	5.346	5.388	5.427

ν	k(or p) = 38	40	50	60	70	80	90	100
1	68.26	68.92	71.73	73.97	75.82	77.40	78.77	79.98
2	19.11	19.28	20.05	20.66	21.16	21.59	21.96	22.29
3	12.75	12.87	13.36	13.76	14.08	14.36	14.61	14.82
4	10.44	10.53	10.93	11.24	11.51	11.73	11.92	12.09
5	9.250	9.330	9.674	9.949	10.18	10.38	10.54	10.69
6	8.529	8.601	8.913	9.163	9.370	9.548	9.702	9.839
7	8.043	8.110	8.400	8.632	8.824	8.989	9.133	9.261
8	7.693	7.756	8.029	8.248	8.430	8.586	8.722	8.843
9	7.428	7.488	7.749	7.958	8.132	8.281	8.410	8.526
10	7.220	7.279	7.529	7.730	7.897	8.041	8.166	8.276
11	7.053	7.110	7.352	7.546	7.708	7.847	7.968	8.075
12	6.916	6.970	7.205	7.394	7.552	7.687	7.804	7.909
13	6.800	6.854	7.083	7.267	7.421	7.552	7.667	7.769
14	6.702	6.754	6.979	7.159	7.309	7.438	7.550	7.650
15	6.618	6.669	6.888	7.065	7.212	7.339	7.449	7.546
16	6.544	6.594	6.810	6.984	7.128	7.252	7.360	7.457
17	6.479	6.529	6.741	6.912	7.054	7.176	7.283	7.377
18	6.422	6.471	6.680	6.848	6.989	7.109	7.213	7.307
19	6.371	6.419	6.626	6.792	6.930	7.048	7.152	7.244
20	6.325	6.373	6.576	6.740	6.877	6.994	7.097	7.187
24	6.181	6.226	6.421	6.579	6.710	6.822	6.920	7.008
30	6.037	6.080	6.267	6.417	6.543	6.650	6.744	6.827
40	5.893	5.934	6.112	6.255	6.375	6.477	6.566	6.645
60	5.750	5.789	5.958	6.093	6.206	6.303	6.387	6.462
120	5.607	5.644	5.802	5.929	6.035	6.126	6.205	6.275
∞	5.463	5.498	5.646	5.764	5.863	5.947	6.020	6.085

TABLE B.5 (cont.) CRITICAL VALUES OF THE q DISTRIBUTION

$\alpha = 0.025$

ν	k(or p) = 2	3	4	5	6	7	8	9	10
1	35.99	54.00	65.69	74.22	80.87	86.29	90.85	94.77	98.20
2	8.776	11.94	14.01	15.54	16.75	17.74	18.58	19.31	19.95
3	5.907	7.661	8.808	9.660	10.34	10.89	11.37	11.78	12.14
4	4.943	6.244	7.088	7.716	8.213	8.625	8.976	9.279	9.548
5	4.474	5.558	6.257	6.775	7.186	7.527	7.816	8.068	8.291
6	4.199	5.158	5.772	6.226	6.586	6.884	7.138	7.359	7.554
7	4.018	4.897	5.455	5.868	6.194	6.464	6.695	6.895	7.072
8	3.892	4.714	5.233	5.616	5.919	6.169	6.382	6.568	6.732
9	3.797	4.578	5.069	5.430	5.715	5.950	6.151	6.325	6.479
10	3.725	4.474	4.943	5.287	5.558	5.782	5.972	6.138	6.285
11	3.667	4.391	4.843	5.173	5.433	5.648	5.831	5.989	6.130
12	3.620	4.325	4.762	5.081	5.332	5.540	5.716	5.869	6.004
13	3.582	4.269	4.694	5.004	5.248	5.449	5.620	5.769	5.900
14	3.550	4.222	4.638	4.940	5.178	5.374	5.540	5.684	5.811
15	3.522	4.182	4.589	4.885	5.118	5.309	5.471	5.612	5.737
16	3.498	4.148	4.548	4.838	5.066	5.253	5.412	5.550	5.672
17	3.477	4.118	4.512	4.797	5.020	5.204	5.361	5.496	5.615
18	3.458	4.092	4.480	4.761	4.981	5.162	5.315	5.448	5.565
19	3.442	4.068	4.451	4.728	4.945	5.123	5.275	5.405	5.521
20	3.427	4.047	4.426	4.700	4.914	5.089	5.238	5.368	5.481
24	3.381	3.983	4.347	4.610	4.816	4.984	5.126	5.250	5.358
30	3.337	3.919	4.271	4.523	4.720	4.881	5.017	5.134	5.238
40	3.294	3.858	4.197	4.439	4.627	4.780	4.910	5.022	5.120
60	3.251	3.798	4.124	4.356	4.536	4.682	4.806	4.912	5.006
120	3.210	3.739	4.053	4.276	4.447	4.587	4.704	4.805	4.894
∞	3.170	3.682	3.984	4.197	4.361	4.494	4.605	4.700	4.784

ν	k(or p) = 11	12	13	14	15	16	17	18	19
1	101.3	104.0	106.5	108.8	110.8	112.7	114.5	116.2	117.7
2	20.52	21.03	21.49	21.91	22.30	22.67	23.01	23.32	23.62
3	12.46	12.75	13.01	13.26	13.48	13.69	13.88	14.06	14.23
4	9.788	10.01	10.20	10.39	10.55	10.71	10.85	10.99	11.11
5	8.490	8.670	8.834	8.984	9.124	9.253	9.374	9.486	9.593
6	7.729	7.887	8.031	8.163	8.286	8.399	8.506	8.605	8.698
7	7.230	7.373	7.504	7.624	7.735	7.839	7.935	8.025	8.111
8	6.879	7.011	7.132	7.244	7.347	7.443	7.532	7.616	7.695
9	6.617	6.742	6.856	6.961	7.058	7.148	7.232	7.311	7.385
10	6.416	6.534	6.643	6.742	6.834	6.920	7.000	7.075	7.146
11	6.256	6.369	6.473	6.568	6.657	6.739	6.815	6.887	6.955
12	6.125	6.235	6.335	6.427	6.512	6.591	6.665	6.734	6.799
13	6.017	6.123	6.220	6.309	6.392	6.468	6.539	6.607	6.670
14	5.926	6.029	6.123	6.210	6.290	6.364	6.434	6.499	6.560
15	5.848	5.949	6.041	6.125	6.203	6.276	6.344	6.407	6.467
16	5.781	5.879	5.969	6.052	6.128	6.199	6.265	6.328	6.386
17	5.722	5.818	5.907	5.987	6.062	6.132	6.197	6.258	6.315
18	5.670	5.765	5.852	5.931	6.004	6.073	6.137	6.197	6.253
19	5.624	5.718	5.803	5.881	5.954	6.020	6.083	6.142	6.198
20	5.583	5.675	5.759	5.836	5.907	5.974	6.036	6.093	6.148
24	5.455	5.543	5.623	5.697	5.764	5.827	5.886	5.941	5.994
30	5.330	5.414	5.490	5.560	5.624	5.684	5.740	5.792	5.841
40	5.208	5.288	5.360	5.426	5.487	5.544	5.597	5.646	5.693
60	5.089	5.164	5.232	5.295	5.352	5.406	5.456	5.503	5.546
120	4.972	5.043	5.107	5.166	5.221	5.271	5.318	5.362	5.403
∞	4.858	4.925	4.985	5.041	5.092	5.139	5.183	5.224	5.262

TABLE B.5 (cont.) CRITICAL VALUES OF THE *q* DISTRIBUTION

$\alpha = 0.025$

v	k(or p) = 20	22	24	26	28	30	32	34	36
1	119.2	121.9	124.3	126.5	128.6	130.4	132.1	133.7	135.2
2	23.89	24.41	24.87	25.29	25.67	26.03	26.35	26.66	26.95
3	14.39	14.69	14.95	15.19	15.41	15.62	15.81	15.99	16.15
4	11.23	11.46	11.66	11.84	12.00	12.16	12.30	12.44	12.56
5	9.693	9.878	10.04	10.20	10.34	10.47	10.59	10.70	10.80
6	8.787	8.949	9.097	9.231	9.355	9.469	9.575	9.674	9.767
7	8.191	8.339	8.473	8.595	8.708	8.812	8.909	8.999	9.084
8	7.769	7.907	8.031	8.145	8.250	8.346	8.436	8.520	8.599
9	7.455	7.585	7.702	7.809	7.908	7.999	8.084	8.163	8.237
10	7.212	7.335	7.447	7.549	7.643	7.729	7.810	7.885	7.956
11	7.019	7.137	7.244	7.341	7.431	7.514	7.592	7.664	7.732
12	6.861	6.974	7.078	7.172	7.258	7.338	7.413	7.483	7.548
13	6.730	6.840	6.939	7.031	7.115	7.192	7.265	7.332	7.396
14	6.619	6.726	6.823	6.911	6.993	7.069	7.139	7.204	7.266
15	6.523	6.628	6.723	6.809	6.889	6.962	7.031	7.095	7.155
16	6.441	6.543	6.636	6.721	6.799	6.870	6.938	7.000	7.059
17	6.370	6.469	6.560	6.644	6.720	6.790	6.856	6.917	6.975
18	6.306	6.404	6.493	6.575	6.650	6.720	6.784	6.844	6.900
19	6.250	6.347	6.434	6.514	6.588	6.656	6.719	6.779	6.835
20	6.200	6.295	6.381	6.460	6.532	6.600	6.662	6.720	6.775
24	6.043	6.133	6.215	6.290	6.359	6.423	6.482	6.538	6.589
30	5.888	5.974	6.052	6.123	6.188	6.248	6.305	6.357	6.406
40	5.737	5.818	5.891	5.958	6.020	6.077	6.130	6.179	6.226
60	5.588	5.664	5.733	5.797	5.854	5.908	5.958	6.004	6.048
120	5.442	5.513	5.578	5.637	5.691	5.741	5.788	5.831	5.872
∞	5.299	5.365	5.425	5.480	5.530	5.577	5.620	5.660	5.698

v	k(or p) = 38	40	50	60	70	80	90	100
1	136.6	137.9	143.6	148.1	151.8	154.9	157.7	160.0
2	27.22	27.47	28.55	29.42	30.13	30.74	31.27	31.74
3	16.31	16.46	17.08	17.59	18.00	18.36	18.67	18.95
4	12.68	12.79	13.27	13.65	13.96	14.23	14.47	14.68
5	10.91	11.00	11.40	11.72	11.99	12.21	12.41	12.59
6	9.855	9.938	10.30	10.58	10.81	11.02	11.19	11.35
7	9.164	9.239	9.563	9.822	10.04	10.23	10.38	10.53
8	8.673	8.743	9.044	9.286	9.487	9.660	9.810	9.944
9	8.307	8.373	8.657	8.885	9.076	9.238	9.381	9.507
10	8.023	8.086	8.356	8.574	8.755	8.911	9.046	9.167
11	7.796	7.856	8.116	8.325	8.499	8.648	8.779	8.894
12	7.610	7.668	7.919	8.120	8.289	8.433	8.559	8.671
13	7.455	7.512	7.755	7.950	8.113	8.253	8.375	8.484
14	7.324	7.379	7.615	7.806	7.965	8.101	8.220	8.325
15	7.212	7.265	7.496	7.682	7.837	7.970	8.086	8.189
16	7.115	7.167	7.393	7.574	7.726	7.856	7.969	8.070
17	7.030	7.081	7.302	7.480	7.628	7.756	7.868	7.966
18	6.954	7.005	7.221	7.396	7.543	7.667	7.777	7.874
19	6.887	6.936	7.150	7.322	7.465	7.589	7.696	7.792
20	6.827	6.876	7.086	7.255	7.397	7.518	7.624	7.718
24	6.639	6.685	6.885	7.046	7.180	7.296	7.397	7.486
30	6.453	6.497	6.686	6.839	6.965	7.075	7.171	7.255
40	6.270	6.311	6.489	6.633	6.753	6.855	6.945	7.025
60	6.089	6.127	6.295	6.429	6.540	6.636	6.720	6.795
120	5.910	5.946	6.101	6.225	6.329	6.418	6.495	6.564
∞	5.733	5.766	5.909	6.023	6.118	6.199	6.270	6.333

TABLE B.5 (cont.) CRITICAL VALUES OF THE q DISTRIBUTION

$\alpha = 0.01$

v	k(or p) = 2	3	4	5	6	7	8	9	10
1	90.03	135.0	164.3	185.6	202.2	215.8	227.2	237.0	245.6
2	14.04	19.02	22.29	24.72	26.63	28.20	29.53	30.68	31.69
3	8.261	10.62	12.17	13.33	14.24	15.00	15.64	16.20	16.69
4	6.512	8.120	9.173	9.958	10.58	11.10	11.55	11.93	12.27
5	5.702	6.976	7.804	8.421	8.913	9.321	9.669	9.972	10.24
6	5.243	6.331	7.033	7.556	7.973	8.318	8.613	8.869	9.097
7	4.949	5.919	6.543	7.005	7.373	7.679	7.939	8.166	8.368
8	4.746	5.635	6.204	6.625	6.960	7.237	7.474	7.681	7.863
9	4.596	5.428	5.957	6.348	6.658	6.915	7.134	7.325	7.495
10	4.482	5.270	5.769	6.136	6.428	6.669	6.875	7.055	7.213
11	4.392	5.146	5.621	5.970	6.247	6.476	6.672	6.842	6.992
12	4.320	5.046	5.502	5.836	6.101	6.321	6.507	6.670	6.814
13	4.260	4.964	5.404	5.727	5.981	6.192	6.372	6.528	6.667
14	4.210	4.895	5.322	5.634	5.881	6.085	6.258	6.409	6.543
15	4.168	4.836	5.252	5.556	5.796	5.994	6.162	6.309	6.439
16	4.131	4.786	5.192	5.489	5.722	5.915	6.079	6.222	6.349
17	4.099	4.742	5.140	5.430	5.659	5.847	6.007	6.147	6.270
18	4.071	4.703	5.094	5.379	5.603	5.788	5.944	6.081	6.201
19	4.046	4.670	5.054	5.334	5.554	5.735	5.889	6.022	6.141
20	4.024	4.639	5.018	5.294	5.510	5.688	5.839	5.970	6.087
24	3.956	4.546	4.907	5.168	5.374	5.542	5.685	5.809	5.919
30	3.889	4.455	4.799	5.048	5.242	5.401	5.536	5.653	5.756
40	3.825	4.367	4.696	4.931	5.114	5.265	5.392	5.502	5.559
60	3.762	4.282	4.595	4.818	4.991	5.133	5.253	5.356	5.447
120	3.702	4.200	4.497	4.709	4.872	5.005	5.118	5.214	5.299
∞	3.643	4.120	4.403	4.603	4.757	4.882	4.987	5.078	5.157

v	k(or p) = 11	12	13	14	15	16	17	18	19
1	253.2	260.0	266.2	271.8	277.0	281.8	286.3	290.4	294.3
2	32.59	33.40	34.13	34.81	35.43	36.00	36.53	37.03	37.50
3	17.13	17.53	17.89	18.22	18.52	18.81	19.07	19.32	19.55
4	12.57	12.84	13.09	13.32	13.53	13.73	13.91	14.08	14.24
5	10.48	10.70	10.89	11.08	11.24	11.40	11.55	11.68	11.81
6	9.301	9.485	9.653	9.808	9.951	10.08	10.21	10.32	10.43
7	8.548	8.711	8.860	8.997	9.124	9.242	9.353	9.456	9.554
8	8.027	8.176	8.312	8.436	8.552	8.659	8.760	8.854	8.943
9	7.647	7.784	7.910	8.025	8.132	8.232	8.325	8.412	8.495
10	7.356	7.485	7.603	7.712	7.812	7.906	7.993	8.076	8.153
11	7.128	7.250	7.362	7.465	7.560	7.649	7.732	7.809	7.883
12	6.943	7.060	7.167	7.265	7.356	7.441	7.520	7.594	7.665
13	6.791	6.903	7.006	7.101	7.188	7.269	7.345	7.417	7.485
14	6.664	6.772	6.871	6.962	7.047	7.126	7.199	7.268	7.333
15	6.555	6.660	6.757	6.845	6.927	7.003	7.074	7.142	7.204
16	6.462	6.564	6.658	6.744	6.823	6.898	6.967	7.032	7.093
17	6.381	6.480	6.572	6.656	6.734	6.806	6.873	6.937	6.997
18	6.310	6.407	6.497	6.579	6.655	6.725	6.792	6.854	6.912
19	6.247	6.342	6.430	6.510	6.585	6.654	6.719	6.780	6.837
20	6.191	6.285	6.371	6.450	6.523	6.591	6.654	6.714	6.771
24	6.017	6.106	6.186	6.261	6.330	6.394	6.453	6.510	6.563
30	5.849	5.932	6.008	6.078	6.143	6.203	6.259	6.311	6.361
40	5.686	5.764	5.835	5.900	5.961	6.017	6.069	6.119	6.165
60	5.528	5.601	5.667	5.728	5.785	5.837	5.886	5.931	5.974
120	5.375	5.443	5.505	5.562	5.614	5.662	5.708	5.750	5.790
∞	5.227	5.290	5.348	5.400	5.448	5.493	5.535	5.574	5.611

TABLE B.5 (cont.) CRITICAL VALUES OF THE q DISTRIBUTION

$\alpha = 0.01$

v	k(or p) = 20	22	24	26	28	30	32	34	36
1	298.0	304.7	310.8	316.3	321.3	326.0	330.3	334.3	338.0
2	37.95	38.76	39.49	40.15	40.76	41.32	41.84	42.33	42.78
3	19.77	20.17	20.53	20.86	21.16	21.44	21.70	21.95	22.17
4	14.40	14.68	14.93	15.16	15.37	15.57	15.75	15.92	16.08
5	11.93	12.16	12.36	12.54	12.71	12.87	13.02	13.15	13.28
6	10.54	10.73	10.91	11.06	11.21	11.34	11.47	11.58	11.69
7	9.646	9.815	9.970	10.11	10.24	10.36	10.47	10.58	10.67
8	9.027	9.182	9.322	9.450	9.569	9.678	9.779	9.874	9.964
9	8.573	8.717	8.847	8.966	9.075	9.177	9.271	9.360	9.443
10	8.226	8.361	8.483	8.595	8.698	8.794	8.883	8.966	9.044
11	7.952	8.080	8.196	8.303	8.400	8.491	8.575	8.654	8.728
12	7.731	7.853	7.964	8.066	8.159	8.246	8.327	8.402	8.473
13	7.548	7.665	7.772	7.870	7.960	8.043	8.121	8.193	8.262
14	7.395	7.508	7.611	7.705	7.792	7.873	7.948	8.018	8.084
15	7.264	7.374	7.474	7.566	7.650	7.728	7.800	7.869	7.932
16	7.152	7.258	7.356	7.445	7.527	7.602	7.673	7.739	7.802
17	7.053	7.158	7.253	7.340	7.420	7.493	7.563	7.627	7.687
18	6.968	7.070	7.163	7.247	7.325	7.398	7.465	7.528	7.587
19	6.891	6.992	7.082	7.166	7.242	7.313	7.379	7.440	7.498
20	6.823	6.922	7.011	7.092	7.168	7.237	7.302	7.362	7.419
24	6.612	6.705	6.789	6.865	6.936	7.001	7.062	7.119	7.173
30	6.407	6.494	6.572	6.644	6.710	6.772	6.828	6.881	6.932
40	6.209	6.289	6.362	6.429	6.490	6.547	6.600	6.650	6.697
60	6.015	6.090	6.158	6.220	6.277	6.330	6.378	6.424	6.467
120	5.827	5.897	5.959	6.016	6.069	6.117	6.162	6.204	6.244
∞	5.645	5.709	5.766	5.818	5.866	5.911	5.952	5.990	6.026

v	k(or p) = 38	40	50	60	70	80	90	100
1	341.5	344.8	358.9	370.1	379.4	387.3	394.1	400.1
2	43.21	43.61	45.33	46.70	47.83	48.80	49.64	50.38
3	22.39	22.59	23.45	24.13	24.71	25.19	25.62	25.99
4	16.23	16.37	16.98	17.46	17.86	18.02	18.50	18.77
5	13.40	13.52	14.00	14.39	14.72	14.99	15.23	15.45
6	11.80	11.90	12.31	12.65	12.92	13.16	13.37	13.55
7	10.77	10.85	11.23	11.52	11.77	11.99	12.17	12.34
8	10.05	10.13	10.47	10.75	10.97	11.17	11.34	11.49
9	9.521	9.594	9.912	10.17	10.38	10.57	10.73	10.87
10	9.117	9.187	9.486	9.726	9.927	10.10	10.25	10.39
11	8.798	8.864	9.148	9.377	9.568	9.732	9.875	10.00
12	8.539	8.603	8.875	9.094	9.277	9.434	9.571	9.693
13	8.326	8.387	8.648	8.859	9.035	9.187	9.318	9.436
14	8.146	8.204	8.457	8.661	8.832	8.978	9.106	9.219
15	7.992	8.049	8.295	8.492	8.658	8.800	8.924	9.035
16	7.860	7.916	8.154	8.347	8.507	8.646	8.767	8.874
17	7.745	7.799	8.031	8.219	8.377	8.511	8.630	8.735
18	7.643	7.696	7.924	8.107	8.261	8.393	8.508	8.611
19	7.553	7.605	7.828	8.008	8.159	8.288	8.401	8.502
20	7.473	7.523	7.742	7.919	8.067	8.194	8.305	8.404
24	7.223	7.270	7.476	7.642	7.780	7.900	8.004	8.097
30	6.978	7.023	7.215	7.370	7.500	7.611	7.709	7.796
40	6.740	6.782	6.960	7.104	7.225	7.328	7.419	7.500
60	6.507	6.546	6.710	6.843	6.954	7.050	7.133	7.207
120	6.281	6.316	6.467	6.588	6.689	6.776	6.852	6.919
∞	6.060	6.092	6.228	6.338	6.429	6.507	6.575	6.636

TABLE B.5 (cont.) CRITICAL VALUES OF THE q DISTRIBUTION

$\alpha = 0.005$

ν	k(or p) = 2	3	4	5	6	7	8	9	10
1	180.1	270.1	328.5	371.2	404.4	431.6	454.4	474.0	491.1
2	19.93	26.97	31.60	35.02	37.73	39.95	41.83	43.46	44.89
3	10.55	13.50	15.45	16.91	18.06	19.01	19.83	20.53	21.15
4	7.916	9.814	11.06	11.99	12.74	13.35	13.88	14.33	14.74
5	6.751	8.196	9.141	9.847	10.41	10.88	11.28	11.63	11.93
6	6.105	7.306	8.088	8.670	9.135	9.522	9.852	10.14	10.40
7	5.699	6.750	7.429	7.935	8.339	8.674	8.961	9.211	9.433
8	5.420	6.370	6.981	7.435	7.797	8.097	8.354	8.578	8.777
9	5.218	6.096	6.657	7.074	7.405	7.680	7.915	8.120	8.303
10	5.065	5.888	6.412	6.800	7.109	7.365	7.584	7.775	7.944
11	4.945	5.727	6.222	6.588	6.878	7.119	7.325	7.505	7.664
12	4.849	5.597	6.068	6.416	6.693	6.922	7.118	7.288	7.439
13	4.770	5.490	5.943	6.277	6.541	6.760	6.947	7.111	7.255
14	4.704	5.401	5.838	6.160	6.414	6.626	6.805	6.962	7.101
15	4.647	5.325	5.750	6.061	6.308	6.511	6.685	6.837	6.971
16	4.599	5.261	5.674	5.977	6.216	6.413	6.582	6.729	6.859
17	4.557	5.205	5.608	5.903	6.136	6.329	6.493	6.636	6.763
18	4.521	3.156	5.550	5.839	6.067	6.255	6.415	6.554	6.678
19	4.488	5.113	5.500	5.783	6.005	6.189	6.346	6.482	6.603
20	4.460	5.074	5.455	5.732	5.951	6.131	6.285	6.418	6.537
24	4.371	4.955	5.315	5.577	5.783	5.952	6.096	6.221	6.332
30	4.285	4.841	5.181	5.428	5.621	5.780	5.914	6.031	6.135
40	4.202	4.731	5.053	5.284	5.465	5.614	5.739	5.848	5.944
60	4.122	4.625	4.928	5.146	5.316	5.454	5.571	5.673	5.762
120	4.045	4.523	4.809	5.013	5.172	5.301	5.410	5.504	5.586
∞	3.970	4.424	4.694	4.886	5.033	5.154	5.255	5.341	5.418

ν	k(or p) = 11	12	13	14	15	16	17	18	19
1	506.3	520.0	532.4	543.6	554.0	563.6	572.5	580.9	588.7
2	46.16	47.31	48.35	49.30	50.17	50.99	51.74	52.45	53.12
3	21.70	22.20	22.66	23.08	23.46	23.82	24.15	24.46	24.76
4	15.10	15.42	15.72	15.99	16.24	16.48	16.70	16.90	17.09
5	12.21	12.46	12.69	12.90	13.09	13.27	13.44	13.60	13.75
6	10.63	10.83	11.02	11.20	11.36	11.51	11.65	11.78	11.90
7	9.632	9.812	9.977	10.13	10.27	10.40	10.52	10.64	10.75
8	8.955	9.117	9.265	9.401	9.527	9.644	9.754	9.857	9.953
9	8.466	8.614	8.749	8.874	8.990	9.097	9.198	9.292	9.381
10	8.096	8.234	8.360	8.476	8.583	8.683	8.777	8.865	8.947
11	7.807	7.937	8.055	8.164	8.265	8.359	8.447	8.530	8.608
12	7.575	7.697	7.810	7.914	8.009	8.099	8.183	8.261	8.335
13	7.384	7.502	7.609	7.708	7.800	7.886	7.965	8.040	8.111
14	7.225	7.338	7.442	7.537	7.625	7.707	7.784	7.856	7.924
15	7.091	7.200	7.300	7.392	7.477	7.556	7.630	7.699	7.765
16	6.976	7.081	7.178	7.267	7.349	7.426	7.498	7.566	7.629
17	6.876	6.979	7.072	7.159	7.239	7.314	7.384	7.449	7.511
18	6.788	6.888	6.980	7.064	7.142	7.215	7.283	7.347	7.407
19	6.711	6.809	6.898	6.981	7.057	7.128	7.195	7.257	7.316
20	6.642	6.738	6.826	6.907	6.981	7.051	7.116	7.177	7.235
24	6.431	6.520	6.602	6.677	6.747	6.812	6.872	6.930	6.983
30	6.227	6.310	6.387	6.456	6.521	6.581	6.638	6.691	6.741
40	6.030	6.108	6.179	6.244	6.304	6.360	6.412	6.461	6.507
60	5.841	5.913	5.979	6.039	6.094	6.146	6.194	6.239	6.281
120	5.660	5.726	5.786	5.842	5.893	5.940	5.984	6.025	6.064
∞	5.485	5.546	5.602	5.652	5.699	5.742	5.783	5.820	5.856

TABLE B.5 (cont.) CRITICAL VALUES OF THE q DISTRIBUTION

$\alpha = 0.005$

ν	k(or p) = 20	22	24	26	28	30	32	34	36
1	596.0	609.5	621.7	632.6	642.7	652.0	660.6	668.5	676.0
2	53.74	54.89	55.92	56.86	57.73	58.52	59.26	59.95	60.59
3	25.03	25.54	26.00	26.42	26.80	27.15	27.48	27.79	28.07
4	17.28	17.61	17.91	18.19	18.44	18.68	18.89	19.09	19.28
5	13.89	14.14	14.38	14.59	14.79	14.96	15.13	15.29	15.44
6	12.02	12.23	12.43	12.61	12.77	12.92	13.06	13.19	13.32
7	10.85	11.03	11.21	11.36	11.50	11.64	11.76	11.88	11.99
8	10.04	10.22	10.37	10.51	10.64	10.76	10.87	10.97	11.07
9	9.465	9.620	9.761	9.890	10.01	10.12	10.22	10.32	10.41
10	9.026	9.170	9.302	9.422	9.532	9.635	9.730	9.820	9.904
11	8.682	8.818	8.941	9.055	9.159	9.256	9.345	9.430	9.509
12	8.405	8.534	8.652	8.759	8.858	8.950	9.036	9.116	9.191
13	8.178	8.302	8.414	8.516	8.611	8.699	8.781	8.857	8.929
14	7.988	8.107	8.215	8.314	8.404	8.489	8.568	8.641	8.710
15	7.827	7.942	8.046	8.141	8.229	8.311	8.387	8.458	8.524
16	7.689	7.800	7.901	7.994	8.078	8.158	8.231	8.300	8.365
17	7.569	7.677	7.775	7.865	7.948	8.024	8.096	8.163	8.226
18	7.464	7.570	7.665	7.753	7.833	7.908	7.978	8.043	8.104
19	7.372	7.474	7.568	7.653	7.732	7.805	7.873	7.937	7.996
20	7.289	7.390	7.481	7.565	7.642	7.713	7.780	7.842	7.901
24	7.034	7.128	7.213	7.291	7.362	7.429	7.491	7.549	7.603
30	6.788	6.875	6.954	7.026	7.093	7.154	7.212	7.265	7.316
40	6.550	6.631	6.704	6.770	6.832	6.889	6.942	6.991	7.038
60	6.321	6.396	6.462	6.523	6.580	6.632	6.861	6.726	6.769
120	6.101	6.169	6.230	6.286	6.337	6.385	6.428	6.470	6.508
∞	5.889	5.951	6.006	6.057	6.103	6.146	6.186	6.223	6.258

ν	k(or p) = 38	40	50	60	70	80	90	100
1	683.0	689.6	717.8	740.2	758.8	774.5	788.2	800.3
2	61.19	61.76	64.19	66.13	67.74	69.10	70.29	71.35
3	28.34	28.60	29.68	30.55	31.27	31.88	32.42	32.90
4	19.46	19.63	20.36	20.93	21.42	21.83	22.18	22.50
5	15.58	15.71	16.27	16.72	17.09	17.41	17.69	17.94
6	13.43	13.54	14.02	14.40	14.71	14.98	15.21	15.43
7	12.09	12.18	12.60	12.93	13.21	13.44	13.65	13.84
8	11.16	11.25	11.63	11.93	12.18	12.39	12.58	12.75
9	10.49	10.58	10.92	11.20	11.43	11.63	11.80	11.96
10	9.983	10.06	10.38	10.64	10.86	11.04	11.20	11.35
11	9.583	9.654	9.957	10.20	10.41	10.59	10.74	10.88
12	9.262	9.328	9.617	9.850	10.04	10.21	10.36	10.49
13	8.997	9.061	9.337	9.560	9.747	9.907	10.05	10.17
14	8.775	8.837	9.103	9.317	9.497	9.652	9.787	9.907
15	8.587	8.647	8.904	9.111	9.285	9.434	9.565	9.680
16	8.425	8.483	8.733	8.933	9.102	9.247	9.373	9.486
17	8.285	8.341	8.583	8.779	8.943	9.084	9.206	9.316
18	8.162	8.217	8.452	8.643	8.803	8.940	9.061	9.167
19	8.053	8.106	8.337	8.523	8.679	8.813	8.931	9.036
20	7.956	8.008	8.234	8.416	8.569	8.700	8.815	8.917
24	7.655	7.704	7.914	8.083	8.226	8.348	8.455	8.551
30	7.364	7.409	7.603	7.760	7.893	8.006	8.105	8.193
40	7.082	7.123	7.302	7.447	7.568	7.672	7.763	7.845
60	6.808	6.846	7.010	7.143	7.252	7.347	7.431	7.504
120	6.545	6.580	6.728	6.846	6.946	7.032	7.107	7.173
∞	6.291	6.322	6.454	6.561	6.649	6.725	6.792	6.850

TABLE B.5 (cont.) CRITICAL VALUES OF THE q DISTRIBUTION

$\alpha = 0.001$

ν	k(or p) = 2	3	4	5	6	7	8	9	10
1	900.3	1351.	1643.	1856.	2022.	2158.	2272.	2370.	2455.
2	44.69	60.42	70.77	78.43	84.49	89.46	93.67	97.30	100.5
3	18.28	23.32	26.65	29.13	31.11	32.74	34.12	35.33	36.39
4	12.18	14.99	16.84	18.23	19.34	20.26	21.04	21.73	22.33
5	9.714	11.67	12.96	13.93	14.71	15.35	15.90	16.38	16.81
6	8.427	9.960	10.97	11.72	12.32	12.83	13.26	13.63	13.97
7	7.648	8.930	9.763	10.40	10.90	11.32	11.68	11.99	12.27
8	7.130	8.250	8.978	9.522	9.958	10.32	10.64	10.91	11.15
9	6.762	7.768	8.419	8.906	9.295	9.619	9.897	10.14	10.36
10	6.487	7.411	8.006	8.450	8.804	9.099	9.352	9.573	9.769
11	6.275	7.136	7.687	8.098	8.426	8.699	8.933	9.138	9.319
12	6.106	6.917	7.436	7.821	8.127	8.383	8.601	8.793	8.962
13	5.970	6.740	7.231	7.595	7.885	8.126	8.333	8.513	8.673
14	5.856	6.594	7.062	7.409	7.685	7.915	8.110	8.282	8.434
15	5.760	6.470	6.920	7.252	7.517	7.736	7.925	8.088	8.234
16	5.678	6.365	6.799	7.119	7.374	7.585	7.766	7.923	8.063
17	5.608	6.275	6.695	7.005	7.250	7.454	7.629	7.781	7.916
18	5.546	6.196	6.604	6.905	7.143	7.341	7.510	7.657	7.788
19	5.492	6.127	6.525	6.817	7.049	7.242	7.405	7.549	7.676
20	5.444	6.065	6.454	6.740	6.966	7.154	7.313	7.453	7.577
24	5.297	5.877	6.238	6.503	6.712	6.884	7.031	7.159	7.272
30	5.156	5.698	6.033	6.278	6.470	6.628	6.763	6.880	6.984
40	5.022	5.528	5.838	6.063	6.240	6.386	6.509	6.616	6.711
60	4.894	5.365	5.653	5.860	6.022	6.155	6.268	6.366	6.451
120	4.771	5.211	5.476	5.667	5.815	5.937	6.039	6.128	6.206
∞	4.654	5.063	5.309	5.484	5.619	5.730	5.823	5.903	5.973

ν	k(or p) = 11	12	13	14	15	16	17	18	19
1	2532.	2600.	2662.	2718.	2770.	2818.	2863.	2904.	2943.
2	103.3	105.9	108.2	110.4	112.3	114.2	115.9	117.4	118.9
3	37.34	38.20	38.98	39.69	40.35	40.97	41.54	42.07	42.58
4	22.87	23.36	23.81	24.21	24.59	24.94	25.27	25.58	25.87
5	17.18	17.53	17.85	18.13	18.41	18.66	18.89	19.10	19.31
6	14.27	14.54	14.79	15.01	15.22	15.42	15.60	15.78	15.94
7	12.52	12.74	12.95	13.14	13.32	13.48	13.64	13.78	13.92
8	11.36	11.56	11.74	11.91	12.06	12.21	12.34	12.47	12.59
9	10.55	10.73	10.89	11.03	11.18	11.30	11.42	11.54	11.64
10	9.946	10.11	10.25	10.39	10.52	10.64	10.75	10.85	10.95
11	9.482	9.630	9.766	9.892	10.01	10.12	10.22	10.31	10.41
12	9.115	9.254	9.381	9.498	9.606	9.707	9.802	9.891	9.975
13	8.817	8.948	9.068	9.178	9.281	9.376	9.466	9.550	9.629
14	8.571	8.696	8.809	8.914	9.012	9.103	9.188	9.267	9.343
15	8.365	8.483	8.592	8.693	8.786	8.872	8.954	9.030	9.102
16	8.189	8.303	8.407	8.504	8.593	8.676	8.755	8.828	8.897
17	8.037	8.148	8.248	8.342	8.427	8.508	8.583	8.654	8.720
18	7.906	8.012	8.110	8.199	8.283	8.361	8.434	8.502	8.567
19	7.790	7.893	7.988	8.075	8.156	8.232	8.303	8.369	8.432
20	7.688	7.788	7.880	7.966	8.044	8.118	8.186	8.251	8.312
24	7.374	7.467	7.551	7.629	7.701	7.768	7.831	7.890	7.946
30	7.077	7.162	7.239	7.310	7.375	7.437	7.494	7.548	7.599
40	6.796	6.872	6.942	7.007	7.067	7.122	7.174	7.223	7.269
60	6.528	6.598	6.661	6.720	6.774	6.824	6.871	6.914	6.956
120	6.276	6.339	6.396	6.448	6.496	6.542	6.583	6.623	6.660
∞	6.036	6.092	6.144	6.191	6.234	6.274	6.312	6.347	6.380

TABLE B.5 (cont.) CRITICAL VALUES OF THE q DISTRIBUTION

$\alpha = 0.001$

ν	k(or p)= 20	22	24	26	28	30	32	34	36
1	2980.	3047.	3108.	3163.	3213.	3260.	3303.	3343.	3380.
2	120.3	122.9	125.2	127.3	129.3	131.0	132.7	134.2	135.7
3	43.05	43.92	44.70	45.42	46.07	46.68	47.24	47.77	48.26
4	26.14	26.65	27.10	27.51	27.89	28.24	28.57	28.88	29.16
5	19.51	19.86	20.19	20.48	20.75	21.01	21.24	21.46	21.66
6	16.09	16.38	16.64	16.87	17.08	17.28	17.47	17.64	17.81
7	14.04	14.29	14.50	14.70	14.88	15.05	15.20	15.35	15.49
8	12.70	12.91	13.09	13.26	13.42	13.57	13.71	13.84	13.96
9	11.75	11.93	12.10	12.25	12.39	12.53	12.65	12.77	12.87
10	11.03	11.20	11.36	11.50	11.63	11.75	11.87	11.97	12.07
11	10.49	10.65	10.79	10.92	11.04	11.16	11.26	11.35	11.45
12	10.06	10.20	10.34	10.46	10.57	10.68	10.78	10.87	10.96
13	9.704	9.843	9.969	10.09	10.19	10.29	10.39	10.47	10.55
14	9.414	9.546	9.666	9.776	9.878	9.972	10.06	10.14	10.22
15	9.170	9.296	9.411	9.517	9.613	9.703	9.788	9.867	9.940
16	8.963	9.084	9.194	9.295	9.388	9.475	9.556	9.631	9.702
17	8.784	8.900	9.007	9.104	9.194	9.277	9.355	9.429	9.497
18	8.628	8.741	8.844	8.938	9.025	9.106	9.181	9.251	9.318
19	8.491	8.601	8.701	8.792	8.876	8.955	9.028	9.096	9.161
20	8.370	8.477	8.574	8.663	8.745	8.821	8.892	8.959	9.021
24	7.999	8.097	8.185	8.267	8.342	8.411	8.476	8.537	8.594
30	7.647	7.735	7.816	7.890	7.958	8.021	8.080	8.135	8.188
40	7.312	7.393	7.466	7.533	7.594	7.651	7.704	7.754	7.801
60	6.995	7.067	7.133	7.193	7.248	7.299	7.347	7.392	7.433
120	6.695	6.760	6.818	6.872	6.921	6.966	7.008	7.048	7.085
∞	6.411	6.469	6.520	6.568	6.611	6.651	6.689	6.723	6.756

ν	k(or p)= 38	40	50	60	70	80	90	100
1	3415.	3448.	3589.	3701.	3794.	3873.	3941.	4002.
2	137.0	138.3	143.7	148.0	151.6	154.7	157.4	159.7
3	48.72	49.16	51.02	52.51	53.75	54.81	55.72	56.53
4	29.43	29.68	30.78	31.65	32.37	32.98	33.52	34.00
5	21.86	22.03	22.82	23.45	23.97	24.41	24.80	25.15
6	17.96	18.10	18.73	19.22	19.64	20.00	20.31	20.58
7	15.62	15.74	16.27	16.69	17.04	17.35	17.61	17.85
8	14.07	14.18	14.64	15.01	15.32	15.59	15.82	16.02
9	12.97	13.07	13.49	13.82	14.10	14.34	14.55	14.74
10	12.16	12.25	12.63	12.94	13.20	13.42	13.61	13.78
11	11.53	11.62	11.97	12.25	12.49	12.70	12.88	13.04
12	11.03	11.11	11.44	11.71	11.94	12.13	12.29	12.45
13	10.63	10.70	11.01	11.27	11.48	11.66	11.82	11.97
14	10.30	10.37	10.66	10.91	11.11	11.28	11.43	11.57
15	10.01	10.08	10.37	10.59	10.79	10.96	11.10	11.23
16	9.769	9.833	10.11	10.34	10.52	10.68	10.82	10.95
17	9.562	9.623	9.888	10.10	10.29	10.44	10.58	10.70
18	9.381	9.440	9.696	9.904	10.08	10.23	10.36	10.48
19	9.221	9.279	9.528	9.730	9.899	10.04	10.17	10.29
20	9.081	9.137	9.379	9.575	9.740	9.881	10.01	10.12
24	8.648	8.700	8.921	9.100	9.250	9.380	9.494	9.596
30	8.237	8.283	8.484	8.647	8.783	8.901	9.004	9.096
40	7.845	7.887	8.067	8.214	8.337	8.442	8.535	8.618
60	7.473	7.510	7.671	7.802	7.911	8.005	8.088	8.161
120	7.121	7.153	7.296	7.411	7.507	7.590	7.662	7.726
∞	6.787	6.816	6.941	7.041	7.124	7.196	7.259	7.314

Table B.5 is reprinted, with permission of the author, from the more extensive table B.2 of H. L. Harter (1970).

Examples:

$$q_{0.05,24,3} = 5.532 \quad \text{and} \quad q_{0.01,20,5} = 5.294.$$

If a critical value is needed for degrees of freedom not on this table, one may conservatively use the next lower degrees of freedom in the table. The required critical value may be estimated by harmonic interpolation. See Harter (1960) for further considerations of interpolation.

TABLE B.6 CRITICAL VALUES OF q' FOR THE ONE-TAILED DUNNETT'S TEST

$\alpha = 0.05$

ν	p = 2	3	4	5	6	7	8	9	10
5	2.02	2.44	2.68	2.85	2.98	3.08	3.16	3.24	3.30
6	1.94	2.34	2.56	2.71	2.83	2.92	3.00	3.07	3.12
7	1.89	2.27	2.48	2.62	2.73	2.82	2.89	2.95	3.01
8	1.86	2.22	2.42	2.55	2.66	2.74	2.81	2.87	2.92
9	1.83	2.18	2.37	2.50	2.60	2.68	2.75	2.81	2.86
10	1.81	2.15	2.34	2.47	2.56	2.64	2.70	2.76	2.81
11	1.80	2.13	2.31	2.44	2.53	2.60	2.67	2.72	2.77
12	1.78	2.11	2.29	2.41	2.50	2.58	2.64	2.69	2.74
13	1.77	2.09	2.27	2.39	2.48	2.55	2.61	2.66	2.71
14	1.76	2.08	2.25	2.37	2.46	2.53	2.59	2.64	2.69
15	1.75	2.07	2.24	2.36	2.44	2.51	2.57	2.62	2.67
16	1.75	2.06	2.23	2.34	2.43	2.50	2.56	2.61	2.65
17	1.74	2.05	2.22	2.33	2.42	2.49	2.54	2.59	2.64
18	1.73	2.04	2.21	2.32	2.41	2.48	2.53	2.58	2.62
19	1.73	2.03	2.20	2.31	2.40	2.47	2.52	2.57	2.61
20	1.72	2.03	2.19	2.30	2.39	2.46	2.51	2.56	2.60
24	1.71	2.01	2.17	2.28	2.36	2.43	2.48	2.53	2.57
30	1.70	1.99	2.15	2.25	2.33	2.40	2.45	2.50	2.54
40	1.68	1.97	2.13	2.23	2.31	2.37	2.42	2.47	2.51
60	1.67	1.95	2.10	2.21	2.28	2.35	2.39	2.44	2.48
120	1.66	1.93	2.08	2.18	2.26	2.32	2.37	2.41	2.45
∞	1.64	1.92	2.06	2.16	2.23	2.29	2.34	2.38	2.42

$\alpha = 0.01$

ν	p = 2	3	4	5	6	7	8	9	10
5	3.37	3.90	4.21	4.43	4.60	4.73	4.85	4.94	5.03
6	3.14	3.61	3.88	4.07	4.21	4.33	4.43	4.51	4.59
7	3.00	3.42	3.66	3.83	3.96	4.07	4.15	4.23	4.30
8	2.90	3.29	3.51	3.67	3.79	3.88	3.96	4.03	4.09
9	2.82	3.19	3.40	3.55	3.66	3.75	3.82	3.89	3.94
10	2.76	3.11	3.31	3.45	3.56	3.64	3.71	3.78	3.83
11	2.72	3.06	3.25	3.38	3.48	3.56	3.63	3.69	3.74
12	2.68	3.01	3.19	3.32	3.42	3.50	3.56	3.62	3.67
13	2.65	2.97	3.15	3.27	3.37	3.44	3.51	3.56	3.61
14	2.62	2.94	3.11	3.23	3.32	3.40	3.46	3.51	3.56
15	2.60	2.91	3.08	3.20	3.29	3.36	3.42	3.47	3.52
16	2.58	2.88	3.05	3.17	3.26	3.33	3.39	3.44	3.48
17	2.57	2.86	3.03	3.14	3.23	3.30	3.36	3.41	3.45
18	2.55	2.84	3.01	3.12	3.21	3.27	3.33	3.38	3.42
19	2.54	2.83	2.99	3.10	3.18	3.25	3.31	3.36	3.40
20	2.53	2.81	2.97	3.08	3.17	3.23	3.29	3.34	3.38
24	2.49	2.77	2.92	3.03	3.11	3.17	3.22	3.27	3.31
30	2.46	2.72	2.87	2.97	3.05	3.11	3.16	3.21	3.24
40	2.42	2.68	2.82	2.92	2.99	3.05	3.10	3.14	3.18
60	2.39	2.64	2.78	2.87	2.94	3.00	3.04	3.08	3.12
120	2.36	2.60	2.73	2.82	2.89	2.94	2.99	3.03	3.06
∞	2.33	2.56	2.68	2.77	2.84	2.89	2.93	2.97	3.00

Values in Table B.6 are reprinted, with the permission of the author and publisher, from the tables of C. W. Dunnett (1955, J. Amer. Statist. Assoc. 50: 1096–1121).

Examples:

$$q'_{0.05(1),16,4} = 2.23 \quad \text{and} \quad q'_{0.01(1),24,3} = 2.77.$$

If a critical value is required for degrees of freedom not on this table, one may conservatively use the critical value with the next lower degrees of freedom. Or, the critical value may be estimated by harmonic interpolation.

TABLE B.7 CRITICAL VALUES OF q' FOR THE TWO-TAILED DUNNETT'S TEST

$\alpha = 0.05$

ν	$p=$ 2	3	4	5	6	7	8	9	10	11	12	13	16	21
5	2.57	3.03	3.29	3.48	3.62	3.73	3.82	3.90	3.97	4.03	4.09	4.14	4.26	4.42
6	2.45	2.86	3.10	3.26	3.39	3.49	3.57	3.64	3.71	3.76	3.81	3.86	3.97	4.11
7	2.36	2.75	2.97	3.12	3.24	3.33	3.41	3.47	3.53	3.58	3.63	3.67	3.78	3.91
8	2.31	2.67	2.88	3.02	3.13	3.22	3.29	3.35	3.41	3.46	3.50	3.54	3.64	3.76
9	2.26	2.61	2.81	2.95	3.05	3.14	3.20	3.26	3.32	3.36	3.40	3.44	3.53	3.65
10	2.23	2.57	2.76	2.89	2.99	3.07	3.14	3.19	3.24	3.29	3.33	3.36	3.45	3.57
11	2.20	2.53	2.72	2.84	2.94	3.02	3.08	3.14	3.19	3.23	3.27	3.30	3.39	3.50
12	2.18	2.50	2.68	2.81	2.90	2.98	3.04	3.09	3.14	3.18	3.22	3.25	3.34	3.45
13	2.16	2.48	2.65	2.78	2.87	2.94	3.00	3.06	3.10	3.14	3.18	3.21	3.29	3.40
14	2.14	2.46	2.63	2.75	2.84	2.91	2.97	3.02	3.07	3.11	3.14	3.18	3.26	3.36
15	2.13	2.44	2.61	2.73	2.82	2.89	2.95	3.00	3.04	3.08	3.12	3.15	3.23	3.33
16	2.12	2.42	2.59	2.71	2.80	2.87	2.92	2.97	3.02	3.06	3.09	3.12	3.20	3.30
17	2.11	2.41	2.58	2.69	2.78	2.85	2.90	2.95	3.00	3.03	3.07	3.10	3.18	3.27
18	2.10	2.40	2.56	2.68	2.76	2.83	2.89	2.94	2.98	3.01	3.05	3.08	3.16	3.25
19	2.09	2.39	2.55	2.66	2.75	2.81	2.87	2.92	2.96	3.00	3.03	3.06	3.14	3.23
20	2.09	2.38	2.54	2.65	2.73	2.80	2.86	2.90	2.95	2.98	3.02	3.05	3.12	3.22
24	2.06	2.35	2.51	2.61	2.70	2.76	2.81	2.86	2.90	2.94	2.97	3.00	3.07	3.16
30	2.04	2.32	2.47	2.58	2.66	2.72	2.77	2.82	2.86	2.89	2.92	2.95	3.02	3.11
40	2.02	2.29	2.44	2.54	2.62	2.68	2.73	2.77	2.81	2.85	2.87	2.90	2.97	3.06
60	2.00	2.27	2.41	2.51	2.58	2.64	2.69	2.73	2.77	2.80	2.83	2.86	2.92	3.00
120	1.98	2.24	2.38	2.47	2.55	2.60	2.65	2.69	2.73	2.76	2.79	2.81	2.87	2.95
∞	1.96	2.21	2.35	2.44	2.51	2.57	2.61	2.65	2.69	2.72	2.74	2.77	2.83	2.91

$\alpha = 0.01$

ν	$p=$ 2	3	4	5	6	7	8	9	10	11	12	13	16	21
5	4.03	4.63	4.98	5.22	5.41	5.56	5.69	5.80	5.89	5.98	6.05	6.12	6.30	6.52
6	3.71	4.21	4.51	4.71	4.87	5.00	5.10	5.20	5.28	5.35	5.41	5.47	5.62	5.81
7	3.50	3.95	4.21	4.39	4.53	4.64	4.74	4.82	4.89	4.95	5.01	5.06	5.19	5.36
8	3.36	3.77	4.00	4.17	4.29	4.40	4.48	4.56	4.62	4.68	4.73	4.78	4.90	5.05
9	3.25	3.63	3.85	4.01	4.12	4.22	4.30	4.37	4.43	4.48	4.53	4.57	4.68	4.82
10	3.17	3.53	3.74	3.88	3.99	4.08	4.16	4.22	4.28	4.33	4.37	4.42	4.52	4.65
11	3.11	3.45	3.65	3.79	3.89	3.98	4.05	4.11	4.16	4.21	4.25	4.29	4.30	4.52
12	3.05	3.39	3.58	3.71	3.81	3.89	3.96	4.02	4.07	4.12	4.16	4.19	4.29	4.41
13	3.01	3.33	3.52	3.65	3.74	3.82	3.89	3.94	3.99	4.04	4.08	4.11	4.20	4.32
14	2.98	3.29	3.47	3.59	3.69	3.76	3.83	3.88	3.93	3.97	4.01	4.05	4.13	4.24
15	2.95	3.25	3.43	3.55	3.64	3.71	3.78	3.83	3.88	3.92	3.95	3.99	4.07	4.18
16	2.92	3.22	3.39	3.51	3.60	3.67	3.73	3.78	3.83	3.87	3.91	3.94	4.02	4.13
17	2.90	3.19	3.36	3.47	3.56	3.63	3.69	3.74	3.79	3.83	3.86	3.90	3.98	4.08
18	2.88	3.17	3.33	3.44	3.53	3.60	3.66	3.71	3.75	3.79	3.83	3.86	3.94	4.04
19	2.86	3.15	3.31	3.42	3.50	3.57	3.63	3.68	3.72	3.76	3.79	3.83	3.90	4.00
20	2.85	3.13	3.29	3.40	3.48	3.55	3.60	3.65	3.69	3.73	3.77	3.80	3.87	3.97
24	2.80	3.07	3.22	3.32	3.40	3.47	3.52	3.57	3.61	3.64	3.68	3.70	3.78	3.87
30	2.75	3.01	3.15	3.25	3.33	3.39	3.44	3.49	3.52	3.56	3.59	3.62	3.69	3.78
40	2.70	2.95	3.09	3.19	3.26	3.32	3.37	3.41	3.44	3.48	3.51	3.53	3.60	3.68
60	2.66	2.90	3.03	3.12	3.19	3.25	3.29	3.33	3.37	3.40	3.42	3.45	3.51	3.59
120	2.62	2.85	2.97	3.06	3.12	3.18	3.22	3.26	3.29	3.32	3.35	3.37	3.43	3.51
∞	2.58	2.79	2.92	3.00	3.06	3.11	3.15	3.19	3.22	3.25	3.27	3.29	3.35	2.42

Values in Table B.7 are reprinted, with the permission of the author and editor, from the tables of C. W. Dunnett (1964, Biometrics 20: 482–491).

Examples:

$$q'_{0.05(2),30,5} = 2.58 \quad \text{and} \quad q'_{0.01(2),20,4} = 2.54.$$

If a critical value is required for degrees of freedom not on this table, one may conservatively use the critical value with the next lower degrees of freedom. Or, the critical value may be estimated by harmonic interpolation.

TABLE B.8 CRITICAL VALUES OF d_{max} FOR THE KOLMOGOROV-SMIRNOV GOODNESS OF FIT FOR DISCRETE OR GROUPED DATA

k	n	α < 0.50	0.20	0.10	0.05	0.02	0.01	0.005	0.002	0.001
3	3	2	2	2	3	3	3	3	3	3
3	6	2	3	3	3	4	4	4	5	5
3	9	2	3	4	4	4	5	5	5	6
3	12	3	4	4	4	5	5	6	6	7
3	15	3	4	4	5	6	6	6	7	7
3	18	3	4	5	5	6	6	7	7	8
3	21	3	4	5	6	6	7	7	8	8
3	24	3	5	5	6	7	7	8	8	9
3	27	3	5	6	6	7	8	8	9	9
3	30	4	5	6	7	8	8	9	9	10
3	33	4	5	6	7	8	8	9	10	10
3	35	4	5	6	7	8	9	9	10	11
3	39	4	6	7	7	8	9	10	11	11
3	42	4	6	7	8	9	9	10	11	12
3	45	4	6	7	8	9	10	10	11	12
3	48	4	6	7	8	9	10	11	12	12
3	51	4	6	7	8	10	10	11	12	13
3	54	4	6	8	9	10	11	11	12	13
3	57	5	7	8	9	10	11	12	13	13
3	60	5	7	8	9	10	11	12	13	14
3	63	5	7	8	9	10	11	12	13	14
3	66	5	7	8	9	10	11	12	13	14
3	69	5	7	8	9	11	12	13	14	14
3	72	5	7	8	9	11	12	13	14	15
3	75	4	7	8	10	11	12	13	14	15
3	78	5	7	9	10	11	12	13	14	15
3	81	5	7	9	10	11	13	13	15	16
3	84	5	7	9	10	12	13	14	15	16
3	87	4	7	9	10	12	13	14	15	16
3	90	4	7	9	10	12	13	14	15	16
3	93	4	7	9	11	12	13	14	15	16
3	95	4	7	9	10	12	13	14	15	16
3	99	4	7	9	10	12	13	14	15	16
4	4	2	2	3	3	3	3	4	4	4
4	3	2	3	4	4	4	5	5	5	5
4	12	3	4	4	5	5	6	6	6	7
4	15	3	4	5	5	6	6	7	7	8
4	20	3	4	5	6	6	7	7	8	8
4	24	4	5	6	6	7	8	8	9	9
4	28	4	5	6	7	7	8	9	9	10
4	32	4	5	6	7	8	9	9	10	10
4	36	4	6	7	7	8	9	10	10	11
4	40	4	6	7	8	9	9	10	11	12
4	44	5	6	7	8	9	10	11	11	12
4	48	5	6	7	8	10	10	11	12	13
4	52	5	7	8	9	10	11	11	12	13
4	56	5	7	8	9	10	11	12	13	13
4	60	5	7	8	9	10	11	12	13	14
4	64	5	7	8	9	11	12	13	14	14
4	63	5	7	9	10	11	12	13	14	15

TABLE B.8 (cont.) CRITICAL VALUES OF d_{max} FOR THE KOLMOGOROV-SMIRNOV GOODNESS OF FIT FOR DISCRETE OR GROUPED DATA

k	n	α < 0.50	0.20	0.10	0.05	0.02	0.01	0.005	0.002	0.001
4	72	5	7	9	10	11	12	13	14	15
4	76	5	8	9	10	11	12	13	14	15
4	80	5	8	9	10	11	12	13	15	15
4	84	5	7	9	10	12	13	14	15	16
4	88	5	7	9	10	12	13	14	15	16
4	92	5	7	9	10	12	13	14	16	16
4	96	5	7	9	10	12	13	14	16	17
4	100	5	8	9	11	12	13	14	16	17
5	5	2	3	3	3	4	4	4	4	4
5	10	3	3	4	4	5	5	5	6	6
5	15	3	4	5	5	6	6	7	7	7
5	20	4	5	5	6	7	7	7	8	8
5	25	4	5	6	6	7	8	8	9	9
5	30	4	5	6	7	8	8	9	10	10
5	35	4	6	7	7	8	9	10	10	11
5	40	5	6	7	8	9	10	10	11	12
5	45	5	6	7	8	9	10	11	12	12
5	50	5	7	8	9	10	11	11	12	13
5	55	5	7	8	9	10	11	12	13	14
5	60	5	7	8	9	11	12	12	13	14
5	65	5	7	9	10	11	12	13	14	14
5	70	6	8	9	10	11	12	13	14	15
5	75	6	8	9	10	12	13	14	15	15
5	80	5	8	9	11	12	13	14	15	16
5	85	5	8	9	11	12	13	14	15	16
5	90	6	8	10	11	12	13	14	15	16
5	95	6	8	9	11	12	13	14	16	17
5	100	5	8	9	11	12	14	15	16	17
6	6	2	3	3	4	4	4	4	5	5
6	12	3	4	4	5	5	6	6	6	7
6	18	3	4	5	6	6	7	7	8	8
6	24	4	5	6	6	7	8	8	9	9
6	30	4	6	6	7	8	9	9	10	10
6	36	5	6	7	8	9	9	10	11	11
6	42	5	6	7	8	9	10	11	11	12
6	48	5	7	8	9	10	11	11	12	13
6	54	5	7	8	9	10	11	12	13	14
6	60	6	7	9	10	11	12	13	13	14
6	66	6	8	9	10	11	12	13	14	15
6	72	6	8	9	10	12	13	13	14	15
6	78	6	8	9	11	12	13	14	15	16
6	84	6	8	9	11	12	13	14	15	16
6	90	5	8	10	11	13	14	15	16	16
6	96	6	8	10	11	13	14	15	16	17
7	7	3	3	4	4	4	5	5	5	5

TABLE B.8 (cont.) CRITICAL VALUES OF d_{max} FOR THE KOLMOGOROV-SMIRNOV GOODNESS OF FIT FOR DISCRETE OR GROUPED DATA

k	n	$\alpha < 0.50$	0.20	0.10	0.05	0.02	0.01	0.005	0.002	0.001
7	14	3	4	5	5	6	6	7	7	7
7	21	4	5	6	6	7	7	8	8	9
7	28	4	5	6	7	8	8	9	10	10
7	35	5	6	7	8	9	9	10	11	11
7	42	5	6	7	8	9	10	11	12	12
7	49	5	7	8	9	10	11	12	12	13
7	56	6	7	8	9	11	12	12	13	14
7	63	6	8	9	10	11	12	13	14	15
7	70	6	8	9	10	12	13	13	15	15
7	77	6	8	9	11	12	13	14	15	16
7	84	6	8	10	12	12	13	14	15	16
7	91	6	8	10	11	13	14	15	16	17
7	98	6	8	10	11	13	14	15	16	17
8	8	3	3	4	4	5	5	5	5	6
8	16	3	4	5	6	6	7	7	7	8
8	24	4	5	6	7	7	8	8	9	9
8	32	4	6	7	7	8	9	10	10	11
8	40	5	6	7	8	9	10	11	11	12
8	48	5	7	8	9	10	11	12	12	13
8	56	6	7	9	10	11	12	12	13	14
8	64	6	8	9	10	11	12	13	14	15
8	72	6	8	9	11	12	13	14	15	15
8	80	6	8	10	11	12	13	14	15	16
8	88	6	8	10	11	13	14	15	16	17
8	95	6	9	10	11	13	14	15	16	17
9	9	3	4	4	4	5	5	5	6	6
9	18	4	5	5	6	7	7	7	8	8
9	27	4	6	6	7	8	8	9	10	10
9	36	5	6	7	8	9	10	10	11	11
9	45	5	7	8	9	10	11	11	12	13
9	54	6	7	9	10	11	11	12	13	14
9	63	6	8	9	10	11	12	13	14	15
9	72	6	8	10	11	12	13	14	15	16
9	81	6	8	10	11	13	13	14	15	16
9	90	6	9	10	11	13	14	15	16	17
9	99	6	9	10	12	13	14	15	17	18
10	10	3	4	4	5	5	5	6	6	6
10	20	4	5	6	6	7	7	8	8	9
10	30	4	6	7	7	8	9	9	10	11
10	40	5	7	8	8	9	10	11	11	12
10	50	6	7	8	9	10	11	12	13	13
10	60	6	8	9	10	11	12	13	14	15
10	70	6	8	10	11	12	13	14	15	15
10	80	6	9	10	11	13	14	14	16	16
10	90	6	9	10	12	13	14	15	16	17
10	100	6	9	10	12	13	14	15	17	18
11	11	3	4	4	5	5	6	6	6	7
11	22	4	5	6	6	7	8	8	9	9
11	33	5	6	7	8	9	9	10	11	11
11	44	5	7	8	9	10	11	11	12	13

TABLE B.8 (cont.) CRITICAL VALUES OF d_{max} FOR THE KOLMOGOROV-SMIRNOV GOODNESS OF FIT FOR DISCRETE OR GROUPED DATA

k	n	$\alpha <$ 0.50	0.20	0.10	0.05	0.02	0.01	0.005	0.002	0.001
11	55	6	8	9	10	11	12	12	13	14
11	66	6	8	9	11	12	13	14	15	15
11	77	6	9	10	11	12	13	14	15	16
11	88	6	9	10	12	13	14	15	16	17
11	99	6	9	10	12	13	14	16	17	18
12	12	3	4	5	5	6	6	6	7	7
12	24	4	5	6	7	8	8	9	9	10
12	36	5	6	7	8	9	10	10	11	12
12	48	6	7	8	9	10	11	12	13	13
12	60	6	8	9	10	11	12	13	14	15
12	72	6	9	10	11	12	13	14	15	16
12	84	7	9	10	11	13	14	15	16	17
12	96	7	9	10	12	13	14	15	17	18
13	13	3	4	5	5	6	6	6	7	7
13	26	4	6	6	7	8	8	9	10	10
13	39	5	7	8	8	9	10	11	11	12
13	52	6	8	9	10	11	12	12	13	14
13	65	6	8	9	11	12	13	14	15	15
13	78	7	9	10	11	13	14	14	16	16
13	91	7	9	11	12	13	14	15	16	17
14	14	3	4	5	5	6	6	7	7	7
14	28	5	6	7	7	8	9	9	10	10
14	42	5	7	8	9	10	10	11	12	12
14	56	6	8	9	10	11	12	13	14	14
14	70	7	8	10	11	12	13	14	15	16
14	84	7	9	10	12	13	14	15	16	17
14	98	7	9	11	12	13	15	16	17	18
15	15	4	4	5	6	6	7	7	7	8
15	30	5	6	7	8	8	9	10	10	11
15	45	6	7	8	9	10	11	12	12	13
15	60	6	8	9	10	12	12	13	14	15
15	75	7	9	10	11	13	14	14	15	16
15	90	7	9	11	12	13	14	15	16	17
16	16	4	5	5	6	6	7	7	8	8
16	32	5	6	7	8	9	9	10	11	11
16	48	6	7	8	9	10	11	12	13	13
16	64	6	8	10	11	12	13	14	15	15
16	80	7	9	10	11	13	14	15	16	17
16	96	7	9	11	12	14	15	16	17	18
17	17	4	5	5	6	7	7	7	8	8
17	34	5	6	7	8	9	10	10	11	11
17	51	6	8	9	10	11	12	12	13	14
17	68	7	9	10	11	12	13	14	15	16
17	85	7	9	10	12	13	14	15	16	17
18	18	4	5	6	6	7	7	8	8	8
18	36	5	7	7	8	9	10	10	11	12
18	54	6	8	9	10	11	12	13	14	14
18	72	7	9	10	11	13	13	14	15	16
18	90	7	9	11	12	13	14	15	17	17
19	19	4	5	6	6	7	7	8	8	9

TABLE B.8 (cont.) CRITICAL VALUES OF d_{max} FOR THE KOLMOGOROV-SMIRNOV GOODNESS OF FIT FOR DISCRETE OR GROUPED DATA

k	n	$\alpha <$ 0.50	0.20	0.10	0.05	0.02	0.01	0.005	0.002	0.001
19	33	5	7	8	8	9	10	11	11	12
19	57	6	8	9	10	11	12	13	14	15
19	76	7	9	10	11	13	14	15	16	16
19	95	7	9	11	12	14	15	16	17	18
20	20	4	5	6	6	7	8	8	9	9
20	40	5	7	8	9	10	10	11	12	12
20	60	6	8	9	10	12	13	13	14	15
20	80	7	9	10	12	13	14	15	16	17
20	100	7	9	11	12	14	15	16	17	18
21	21	4	5	6	7	7	8	8	9	9
21	42	5	7	8	9	10	11	11	12	13
21	63	7	8	10	11	12	13	14	15	15
21	84	7	9	11	12	13	14	15	16	17
22	22	4	5	6	7	7	8	8	9	9
22	44	6	7	8	9	10	11	12	12	13
22	66	7	9	10	11	12	13	14	15	16
22	88	7	9	11	12	13	15	15	17	17
23	23	4	5	6	7	8	8	9	9	9
23	46	6	7	8	9	10	11	12	13	13
23	69	7	9	10	11	12	13	14	15	16
23	92	7	9	11	12	14	15	16	17	18
24	24	4	6	6	7	8	8	9	9	10
24	48	6	8	9	10	11	11	12	13	14
24	72	7	9	10	11	13	14	14	16	16
24	95	7	10	11	12	14	15	16	17	18
25	25	4	6	6	7	8	8	9	10	10
25	50	6	8	9	10	11	12	12	13	14
25	75	7	9	10	11	13	14	15	16	16
25	100	7	10	11	12	14	15	16	17	18
26	25	5	6	7	7	8	9	9	10	10
26	52	6	8	9	10	11	12	13	13	14
25	78	7	9	11	12	13	14	15	16	17
27	27	5	6	7	7	8	9	9	10	10
27	54	6	8	9	10	11	12	13	14	14
27	81	7	9	11	12	13	14	15	16	17
28	28	5	6	7	7	8	9	9	10	11
28	55	6	8	9	10	11	12	13	14	15
28	84	7	9	11	12	13	14	15	16	17
29	29	5	6	7	8	8	9	10	10	11
29	58	6	8	9	10	12	12	13	14	15
29	87	7	10	11	12	14	15	16	17	17
30	30	5	6	7	8	9	9	10	10	11
30	60	7	8	10	11	12	13	13	14	15
30	90	7	9	11	12	14	15	16	17	18
31	31	5	6	7	8	9	9	10	11	11
31	62	7	9	10	11	12	13	14	15	15
31	93	7	9	11	12	14	15	16	17	18
32	32	5	6	7	8	9	9	10	11	11
32	64	7	9	10	11	12	13	14	15	16
32	95	7	10	11	12	14	15	16	17	18

TABLE B.8 (cont.) CRITICAL VALUES OF d_{max} FOR THE KOLMOGOROV-SMIRNOV GOODNESS OF FIT FOR DISCRETE OR GROUPED DATA

k	n	$\alpha <$ 0.50	0.20	0.10	0.05	0.02	0.01	0.005	0.002	0.001
33	33	5	6	7	8	9	10	10	11	11
33	66	7	9	10	11	12	13	14	15	16
33	99	7	10	11	13	14	15	16	17	18
34	34	5	7	7	8	9	10	10	11	12
34	68	7	9	10	11	13	13	14	15	16
35	35	5	7	8	8	9	10	11	11	12
35	70	7	9	10	11	13	14	14	16	16
36	36	5	7	8	8	9	10	11	11	12
36	72	7	9	10	11	13	14	15	16	16
37	37	5	7	8	9	10	10	11	12	12
37	74	7	9	11	12	13	14	15	16	17
38	33	5	7	8	9	10	10	11	12	12
38	76	7	9	11	12	13	14	15	16	17
39	39	6	7	8	9	10	10	11	12	12
39	78	7	9	11	12	13	14	15	16	17
40	40	6	7	8	9	10	11	11	12	13
40	80	7	9	11	12	13	14	15	16	17
41	41	6	7	8	9	10	11	11	12	13
41	82	7	10	11	12	13	14	15	16	17
42	42	6	7	8	9	10	11	11	12	13
42	84	7	10	11	12	14	15	15	17	17
43	43	6	7	8	9	10	11	12	12	13
43	86	8	10	11	12	13	15	16	17	18
44	44	6	7	8	9	10	11	12	13	13
44	88	7	10	11	12	14	15	16	17	18
45	45	6	8	9	9	10	11	12	13	13
45	90	7	10	11	12	14	15	16	17	18
46	45	6	8	9	10	11	11	12	13	13
46	92	7	10	11	13	14	15	16	17	18
47	47	6	8	9	10	11	11	12	13	14
47	94	8	10	11	13	14	15	16	17	18
48	48	6	8	9	10	11	12	12	13	14
48	95	7	10	11	13	14	15	16	18	18
49	49	6	8	9	10	11	12	12	13	14
49	98	8	10	11	13	14	15	16	18	19
50	50	6	8	9	10	11	12	13	13	14
50	100	7	10	11	13	14	15	16	18	19

These values were determined by the method used in Pettitt and Stephens (1977) by modifying a computer program provided by Pettitt.

Examples:

$$(d_{max})_{0.05,30,90} = 11 \quad \text{and} \quad (d_{max})_{0.01,25,100} = 15.$$

The table above is applicable when all expected frequencies are equal; it also works well with expected frequencies that are slightly to moderately unequal (Pettitt and Stephens 1977).

Note: The values of d_{max} shown have probabilities slightly less than their column headings. Therefore, for example, $0.02 < P(d_{max} = 12$, for $k = 30$, and $n = 90) < 0.05$.

TABLE B.9 CRITICAL VALUES OF *D* FOR THE KOLMOGOROV-SMIRNOV GOODNESS OF FIT FOR CONTINUOUS DISTRIBUTIONS

n	α = 0.50	0.20	0.10	0.05	0.02	0.01	0.005	0.002	0.001
1	0.75000	0.90000	0.95000	0.97500	0.99000	0.99500	0.99750	0.99900	0.99950
2	0.50000	0.68377	0.77639	0.84189	0.90000	0.92929	0.95000	0.96838	0.97764
3	0.43529	0.56481	0.63604	0.70760	0.78456	0.82900	0.86428	0.90000	0.92063
4	0.38209	0.49265	0.56522	0.62394	0.68887	0.73424	0.77639	0.82217	0.85047
5	0.34319	0.44698	0.50945	0.56328	0.62718	0.66853	0.70543	0.75000	0.78137
6	0.31447	0.41037	0.46799	0.51926	0.57741	0.61661	0.65287	0.69571	0.72479
7	0.29312	0.38148	0.43607	0.48342	0.53844	0.57581	0.60975	0.65071	0.67930
8	0.27567	0.35831	0.40962	0.45427	0.50654	0.54179	0.57429	0.61368	0.64098
9	0.26082	0.33910	0.38746	0.43001	0.47960	0.51332	0.54443	0.58210	0.60846
10	0.24809	0.32260	0.36866	0.40925	0.45662	0.48893	0.51872	0.55500	0.58042
11	0.23709	0.30829	0.35242	0.39122	0.43670	0.46770	0.49639	0.53135	0.55588
12	0.22748	0.29577	0.33815	0.37543	0.41918	0.44905	0.47672	0.51047	0.53422
13	0.21901	0.28470	0.32549	0.36143	0.40362	0.43247	0.45921	0.49189	0.51490
14	0.21146	0.27481	0.31417	0.34890	0.38970	0.41762	0.44352	0.47520	0.49753
15	0.20465	0.26589	0.30397	0.33760	0.37713	0.40420	0.42934	0.46011	0.48182
16	0.19844	0.25778	0.29472	0.32733	0.36571	0.39201	0.41644	0.44637	0.46750
17	0.19277	0.25039	0.28627	0.31796	0.35528	0.38086	0.40464	0.43380	0.45440
18	0.18757	0.24360	0.27851	0.30936	0.34569	0.37062	0.39380	0.42224	0.44234
19	0.18277	0.23735	0.27136	0.30143	0.33685	0.36117	0.38379	0.41156	0.43119
20	0.17833	0.23156	0.26473	0.29408	0.32866	0.35241	0.37451	0.40165	0.42085
21	0.17421	0.22617	0.25858	0.28724	0.32104	0.34426	0.36588	0.39243	0.41122
22	0.17036	0.22115	0.25283	0.28087	0.31394	0.33666	0.35782	0.38382	0.40223
23	0.16676	0.21646	0.24746	0.27490	0.30728	0.32954	0.35027	0.37575	0.39380
24	0.16338	0.21205	0.24242	0.26931	0.30104	0.32286	0.34318	0.36817	0.38588
25	0.16021	0.20790	0.23768	0.26404	0.29516	0.31657	0.33651	0.36104	0.37843
26	0.15721	0.20399	0.23320	0.25908	0.28962	0.31063	0.33022	0.35431	0.37139
27	0.15437	0.20030	0.22898	0.25438	0.28438	0.30502	0.32426	0.34794	0.36473
28	0.15169	0.19680	0.22497	0.24993	0.27942	0.29971	0.31862	0.34190	0.35842
29	0.14914	0.19348	0.22117	0.24571	0.27471	0.29466	0.31327	0.33617	0.35242
30	0.14672	0.19032	0.21756	0.24170	0.27023	0.28986	0.30818	0.33072	0.34672
31	0.14442	0.18732	0.21412	0.23788	0.26596	0.28529	0.30333	0.32553	0.34129
32	0.14222	0.18445	0.21085	0.23424	0.26189	0.28094	0.29870	0.32058	0.33611
33	0.14012	0.18171	0.20771	0.23076	0.25801	0.27677	0.29428	0.31584	0.33115
34	0.13811	0.17909	0.20472	0.22743	0.25429	0.27279	0.29005	0.31131	0.32641
35	0.13618	0.17659	0.20185	0.22425	0.25073	0.26897	0.28600	0.30697	0.32187
36	0.13434	0.17418	0.19910	0.22119	0.24732	0.26532	0.28211	0.30281	0.31751
37	0.13257	0.17188	0.19646	0.21826	0.24404	0.26180	0.27838	0.29882	0.31333
38	0.13086	0.16966	0.19392	0.21544	0.24089	0.25843	0.27480	0.29498	0.30931
39	0.12923	0.16753	0.19148	0.21273	0.23786	0.25518	0.27135	0.29128	0.30544
40	0.12765	0.16547	0.18913	0.21012	0.23494	0.25205	0.26803	0.28772	0.30171
41	0.12613	0.16349	0.18687	0.20760	0.23213	0.24904	0.26482	0.28429	0.29811
42	0.12466	0.16158	0.18468	0.20517	0.22941	0.24613	0.26173	0.28097	0.29465
43	0.12325	0.15974	0.18257	0.20283	0.22679	0.24332	0.25875	0.27778	0.29130
44	0.12188	0.15796	0.18053	0.20056	0.22426	0.24060	0.25587	0.27468	0.28806
45	0.12056	0.15623	0.17856	0.19837	0.22181	0.23798	0.25308	0.27169	0.28493
46	0.11927	0.15457	0.17665	0.19625	0.21944	0.23544	0.25038	0.26880	0.28190
47	0.11803	0.15295	0.17481	0.19420	0.21715	0.23298	0.24776	0.26600	0.27896
48	0.11683	0.15139	0.17301	0.19221	0.21493	0.23059	0.24523	0.26328	0.27611
49	0.11567	0.14987	0.17128	0.19028	0.21277	0.22828	0.24277	0.26065	0.27335
50	0.11453	0.14840	0.16959	0.18841	0.21068	0.22604	0.24039	0.25809	0.27067

TABLE B.9 (cont.) CRITICAL VALUES OF *D* FOR THE KOLMOGOROV-SMIRNOV GOODNESS OF FIT FOR CONTINUOUS DISTRIBUTIONS

n	α = 0.50	0.20	0.10	0.05	0.02	0.01	0.005	0.002	0.001
51	0.11344	0.14697	0.16796	0.18659	0.20864	0.22386	0.23807	0.25561	0.26807
52	0.11237	0.14558	0.16637	0.18482	0.20667	0.22174	0.23582	0.25319	0.26555
53	0.11133	0.14423	0.16483	0.18311	0.20475	0.21968	0.23364	0.25085	0.26309
54	0.11032	0.14292	0.16332	0.18144	0.20289	0.21768	0.23151	0.24857	0.26070
55	0.10934	0.14164	0.16186	0.17981	0.20107	0.21574	0.22944	0.24635	0.25837
56	0.10839	0.14040	0.16044	0.17823	0.19930	0.21384	0.22742	0.24419	0.25611
57	0.10746	0.13919	0.15906	0.17669	0.19758	0.21199	0.22546	0.24208	0.25390
58	0.10655	0.13801	0.15771	0.17519	0.19590	0.21020	0.22355	0.24003	0.25175
59	0.10566	0.13686	0.15639	0.17373	0.19427	0.20844	0.22169	0.23803	0.24966
60	0.10480	0.13573	0.15511	0.17231	0.19267	0.20673	0.21987	0.23608	0.24761
61	0.10396	0.13464	0.15385	0.17091	0.19112	0.20506	0.21809	0.23418	0.24562
62	0.10314	0.13357	0.15263	0.16956	0.18960	0.20343	0.21636	0.23232	0.24367
63	0.10234	0.13253	0.15144	0.16823	0.18812	0.20184	0.21467	0.23051	0.24177
64	0.10155	0.13151	0.15027	0.16693	0.18667	0.20029	0.21302	0.22873	0.23991
65	0.10079	0.13052	0.14913	0.16567	0.18525	0.19877	0.21141	0.22700	0.23810
66	0.10004	0.12954	0.14802	0.16443	0.18387	0.19729	0.20983	0.22531	0.23633
67	0.09931	0.12859	0.14693	0.16322	0.18252	0.19584	0.20829	0.22365	0.23459
68	0.09859	0.12766	0.14587	0.16204	0.18119	0.19442	0.20678	0.22204	0.23289
69	0.09789	0.12675	0.14483	0.16088	0.17990	0.19303	0.20530	0.22045	0.23123
70	0.09721	0.12586	0.14381	0.15975	0.17863	0.19167	0.20386	0.21890	0.22961
71	0.09653	0.12499	0.14281	0.15864	0.17739	0.19034	0.20244	0.21738	0.22802
72	0.09588	0.12413	0.14183	0.15755	0.17618	0.18903	0.20105	0.21589	0.22646
73	0.09523	0.12329	0.14087	0.15649	0.17499	0.18776	0.19970	0.21444	0.22493
74	0.09460	0.12247	0.13993	0.15544	0.17382	0.18650	0.19837	0.21301	0.22343
75	0.09398	0.12167	0.13901	0.15442	0.17268	0.18528	0.19706	0.21161	0.22196
76	0.09338	0.12088	0.13811	0.15342	0.17155	0.18408	0.19578	0.21024	0.22053
77	0.09278	0.12011	0.13723	0.15244	0.17045	0.18290	0.19453	0.20889	0.21912
78	0.09220	0.11935	0.13636	0.15147	0.16938	0.18174	0.19330	0.20757	0.21773
79	0.09162	0.11860	0.13551	0.15052	0.16832	0.18060	0.19209	0.20628	0.21637
80	0.09106	0.11787	0.13467	0.14960	0.16728	0.17949	0.19091	0.20501	0.21504
81	0.09051	0.11716	0.13385	0.14868	0.16626	0.17840	0.18974	0.20376	0.21373
82	0.08997	0.11645	0.13305	0.14779	0.16526	0.17732	0.18860	0.20253	0.21245
83	0.08944	0.11576	0.13226	0.14691	0.16428	0.17627	0.18748	0.20133	0.21119
84	0.08891	0.11508	0.13148	0.14605	0.16331	0.17523	0.18638	0.20015	0.20995
85	0.08840	0.11442	0.13072	0.14520	0.16236	0.17421	0.18530	0.19898	0.20873
86	0.08790	0.11376	0.12997	0.14437	0.16143	0.17321	0.18423	0.19784	0.20753
87	0.08740	0.11311	0.12923	0.14355	0.16051	0.17223	0.18319	0.19672	0.20635
88	0.08691	0.11248	0.12850	0.14274	0.15961	0.17126	0.18216	0.19562	0.20520
89	0.08643	0.11186	0.12779	0.14195	0.15873	0.17031	0.18115	0.19453	0.20406
90	0.08596	0.11125	0.12709	0.14117	0.15786	0.16938	0.18016	0.19347	0.20294
91	0.08550	0.11064	0.12640	0.14040	0.15700	0.16846	0.17918	0.19242	0.20184
92	0.08504	0.11005	0.12572	0.13965	0.15616	0.16755	0.17822	0.19138	0.20076
93	0.08459	0.10947	0.12506	0.13891	0.15533	0.16666	0.17727	0.19037	0.19969
94	0.08415	0.10889	0.12440	0.13818	0.15451	0.16579	0.17634	0.18937	0.19865
95	0.08371	0.10833	0.12375	0.13746	0.15371	0.16493	0.17542	0.18838	0.19761
96	0.08328	0.10777	0.12312	0.13675	0.15291	0.16408	0.17452	0.18741	0.19660
97	0.08286	0.10722	0.12249	0.13606	0.15214	0.16324	0.17363	0.18646	0.19560
98	0.08245	0.10668	0.12187	0.13537	0.15137	0.16242	0.17275	0.18552	0.19461
99	0.08204	0.10615	0.12126	0.13469	0.15061	0.16162	0.17189	0.18460	0.19364
100	0.08163	0.10563	0.12067	0.13403	0.14987	0.16081	0.17104	0.18368	0.19268

TABLE B.9 (cont.) CRITICAL VALUES OF D FOR THE KOLMOGOROV-SMIRNOV GOODNESS OF FIT FOR CONTINUOUS DISTRIBUTIONS

n	α = 0.50	0.20	0.10	0.05	0.02	0.01	0.005	0.002	0.001
102	0.08084	0.10460	0.11949	0.13273	0.14841	0.15925	0.16938	0.18190	0.19081
104	0.08008	0.10361	0.11836	0.13146	0.14700	0.15773	0.16777	0.18017	0.18900
106	0.07933	0.10264	0.11725	0.13023	0.14562	0.15625	0.16620	0.17848	0.18723
108	0.07861	0.10170	0.11618	0.12904	0.14429	0.15482	0.16467	0.17685	0.18551
110	0.07790	0.10079	0.11513	0.12787	0.14299	0.15342	0.16319	0.17525	0.18384
112	0.07722	0.09990	0.11411	0.12674	0.14172	0.15207	0.16174	0.17370	0.18222
114	0.07655	0.09903	0.11312	0.12564	0.14049	0.15074	0.16034	0.17219	0.18063
116	0.07590	0.09818	0.11215	0.12457	0.13929	0.14945	0.15897	0.17072	0.17909
118	0.07527	0.09736	0.11121	0.12352	0.13812	0.14820	0.15763	0.16929	0.17759
120	0.07465	0.09656	0.11029	0.12250	0.13697	0.14697	0.15633	0.16789	0.17612
122	0.07404	0.09577	0.10940	0.12150	0.13586	0.14578	0.15506	0.16652	0.17469
124	0.07345	0.09501	0.10852	0.12053	0.13477	0.14461	0.15382	0.16519	0.17329
126	0.07288	0.09426	0.10767	0.11958	0.13371	0.14347	0.15261	0.16389	0.17193
128	0.07232	0.09353	0.10684	0.11866	0.13268	0.14236	0.15142	0.16262	0.17060
130	0.07177	0.09282	0.10602	0.11775	0.13166	0.14128	0.15027	0.16138	0.16930
132	0.07123	0.09213	0.10523	0.11687	0.13068	0.14021	0.14914	0.16017	0.16802
134	0.07071	0.09144	0.10445	0.11600	0.12971	0.13918	0.14804	0.15898	0.16678
136	0.07019	0.09078	0.10369	0.11516	0.12876	0.13816	0.14696	0.15782	0.16556
138	0.06969	0.09013	0.10294	0.11433	0.12784	0.13717	0.14590	0.15669	0.16437
140	0.06920	0.08949	0.10221	0.11352	0.12693	0.13620	0.14487	0.15558	0.16321
142	0.06872	0.08887	0.10150	0.11273	0.12604	0.13524	0.14385	0.15449	0.16207
144	0.06825	0.08826	0.10080	0.11195	0.12517	0.13431	0.14286	0.15343	0.16095
146	0.06778	0.08766	0.10012	0.11119	0.12432	0.13340	0.14189	0.15238	0.15986
148	0.06733	0.08707	0.09944	0.11044	0.12349	0.13250	0.14094	0.15136	0.15879
150	0.06689	0.08650	0.09879	0.10971	0.12267	0.13163	0.14001	0.15036	0.15774
152	0.06646	0.08593	0.09814	0.10900	0.12187	0.13077	0.13909	0.14938	0.15671
154	0.06603	0.08538	0.09751	0.10830	0.12109	0.12993	0.13820	0.14842	0.15570
156	0.06561	0.08484	0.09689	0.10761	0.12032	0.12910	0.13732	0.14747	0.15471
158	0.06520	0.08430	0.09628	0.10693	0.11956	0.12829	0.13645	0.14655	0.15374
160	0.06480	0.08378	0.09569	0.10627	0.11882	0.12749	0.13561	0.14564	0.15278
d_α	0.83255	1.07298	1.22387	1.35810	1.51743	1.62762	1.73082	1.85846	1.94947
A_α	−0.042554	0.002557	0.052556	0.112820	0.205662	0.284642	0.370673	0.494581	0.595698

Table B.9 was prepared using Equation 3.0 of Birnbaum and Tingey (1951). The values of $D_{\alpha,n}$ were computed to eight decimal places and then rounded to five decimal places.

Examples:

$$D_{0.05,12} = 0.37543 \quad \text{and} \quad D_{0.01,55} = 0.21574.$$

For large n, critical values of $D_{\alpha,n}$ can be approximated by

$$D_{\alpha,n} \cong \sqrt{\frac{-\ln(\alpha/2)}{2n}}$$

(Smirnov, 1939a) or, more accurately, by either

$$D_{\alpha,n} \cong \sqrt{\frac{-\ln(\alpha/2)}{2n}} - \frac{0.16693}{n}$$

or

$$D_{\alpha,n} \cong \sqrt{\frac{-\ln(\alpha/2)}{2n}} - \frac{0.16693}{n} - \frac{A_\alpha}{\sqrt{n^3}}$$

(Miller, 1956), where

$$A_\alpha = 0.09037\left[-\log\left(\frac{\alpha}{2}\right)\right]^{3/2} + 0.01515\left[\log\left(\frac{\alpha}{2}\right)\right]^2 + 0.08467\left(\frac{\alpha}{2}\right) - 0.11143.$$

For the significance levels, α, in Table B.9, the appropriate A_α's are given at the end of the table.

The accuracy of each of these three approximations is given as follows as percent error, where percent error $= |(\text{approximate } D_{\alpha,n} - \text{true } D_{\alpha,n})|/\text{true } D_{\alpha,n} \times 100\%$.

For the first approximating equation:

n	$\alpha = 0.50$	0.20	0.10	0.05	0.02	0.01	0.005	0.002	0.001
20	4.4%	3.6%	3.4%	3.3%	3.2%	3.3%	3.3%	3.5%	3.6%
50	2.8	2.3	2.1	1.9	1.9	1.8	1.8	1.8	1.9
100	2.0	1.6	1.4	1.3	1.2	1.2	1.2	1.2	1.2
160	1.6	1.3	1.1	1.0	1.0	0.9	0.9	0.9	0.9

For the second approximating equation:

n	$\alpha = 0.50$	0.20	0.10	0.05	0.02	0.01	0.005	0.002	0.001
10	0.6%	0.0*	0.5%	0.9%	1.4%	1.9%	2.3%	2.9%	3.3%
20	0.3	0.0*	0.2	0.4	0.7	0.9	1.1	1.4	1.6
50	0.1	0.0*	0.1	0.2	0.3	0.4	0.4	0.5	0.6
100	0.1	0.0*	0.0*	0.1	0.1	0.2	0.2	0.3	0.3
160	0.0*	0.0*	0.0*	0.1	0.1	0.1	0.1	0.2	0.2

For the third approximating equation:

n	$\alpha = 0.50$	0.20	0.10	0.05	0.02	0.01	0.005	0.002	0.001
5	0.1%	0.2%	0.0*	0.1%	0.1%	0.1%	0.3%	0.5%	0.5%
10	0.1	0.0*	0.0*	0.0*	0.0*	0.1	0.0*	0.1	0.1
20	0.0*	0.0*	0.0*	0.0*	0.0*	0.0*	0.0*	0.0*	0.0*

In the above, the asterisk indicates a percent error the absolute value of which is less than 0.05%.

The first approximation may also be written as

$$D_{\alpha,n} = \frac{d_\alpha}{\sqrt{n}},$$

where d_α is given at the end of Table B.9.

TABLE B.10 CRITICAL VALUES OF THE MANN-WHITNEY *U* DISTRIBUTION

n_1	n_2	α(2): 0.20 α(1): 0.10	α(2): 0.10 α(1): 0.05	α(2): 0.05 α(1): 0.025	α(2): 0.02 α(1): 0.01	α(2): 0.01 α(1): 0.005	α(2): 0.005 α(1): 0.0025	α(2): 0.002 α(1): 0.001	α(2): 0.001 α(1): 0.0005
1	1	--	--	--	--	--	--	--	--
	2	--	--	--	--	--	--	--	--
	3	--	--	--	--	--	--	--	--
	4	--	--	--	--	--	--	--	--
	5	--	--	--	--	--	--	--	--
	6	--	--	--	--	--	--	--	--
	7	--	--	--	--	--	--	--	--
	8	--	--	--	--	--	--	--	--
	9	9	--	--	--	--	--	--	--
	10	10	--	--	--	--	--	--	--
	11	11	--	--	--	--	--	--	--
	12	12	--	--	--	--	--	--	--
	13	13	--	--	--	--	--	--	--
	14	14	--	--	--	--	--	--	--
	15	15	--	--	--	--	--	--	--
	16	16	--	--	--	--	--	--	--
	17	17	--	--	--	--	--	--	--
	18	18	--	--	--	--	--	--	--
	19	18	19	--	--	--	--	--	--
	20	19	20	--	--	--	--	--	--
	21	20	21	--	--	--	--	--	--
	22	21	22	--	--	--	--	--	--
	23	22	23	--	--	--	--	--	--
	24	23	24	--	--	--	--	--	--
	25	24	25	--	--	--	--	--	--
	26	25	26	--	--	--	--	--	--
	27	26	27	--	--	--	--	--	--
	28	27	28	--	--	--	--	--	--
	29	27	29	--	--	--	--	--	--
	30	28	30	--	--	--	--	--	--
	31	29	31	--	--	--	--	--	--
	32	30	32	--	--	--	--	--	--
	33	31	33	--	--	--	--	--	--
	34	32	34	--	--	--	--	--	--
	35	33	35	--	--	--	--	--	--
	36	34	36	--	--	--	--	--	--
	37	35	37	--	--	--	--	--	--
	38	36	38	--	--	--	--	--	--
	39	36	38	39	--	--	--	--	--
1	40	37	39	40	--	--	--	--	--
2	2	--	--	--	--	--	--	--	--
	3	6	--	--	--	--	--	--	--
	4	8	--	--	--	--	--	--	--
	5	9	10	--	--	--	--	--	--
	6	11	12	--	--	--	--	--	--
	7	10	14	--	--	--	--	--	--
	8	14	15	16	--	--	--	--	--
	9	16	17	18	--	--	--	--	--
2	10	17	19	20	--	--	--	--	--

TABLE B.10 (cont.) CRITICAL VALUES OF THE MANN-WHITNEY *U* DISTRIBUTION

n_1	n_2	α(2): 0.20 α(1): 0.10	0.10 0.05	0.05 0.025	0.02 0.01	0.01 0.005	0.005 0.0025	0.002 0.001	0.001 0.0005
2	11	19	21	22	--	--	--	--	--
	12	20	22	23	--	--	--	--	--
	13	22	24	25	26	--	--	--	--
	14	23	25	27	28	--	--	--	--
	15	25	27	29	30	--	--	--	--
	16	27	29	31	32	--	--	--	--
	17	28	31	32	34	--	--	--	--
	18	30	32	34	36	--	--	--	--
	19	31	34	36	37	38	--	--	--
	20	33	36	38	39	40	--	--	--
	21	34	37	39	41	42	--	--	--
	22	36	39	41	43	44	--	--	--
	23	37	41	43	45	46	--	--	--
	24	39	42	45	47	48	--	--	--
	25	41	44	47	49	50	--	--	--
	26	42	46	48	51	52	--	--	--
	27	44	47	50	52	53	54	--	--
	28	45	49	52	54	55	56	--	--
	29	47	51	54	56	57	58	--	--
	30	48	53	55	58	59	60	--	--
	31	50	54	57	60	61	62	--	--
	32	51	56	59	62	63	64	--	--
	33	53	58	61	64	65	66	--	--
	34	55	59	63	65	67	68	--	--
	35	56	61	64	67	69	70	--	--
	36	58	63	66	69	71	72	--	--
	37	59	64	68	71	73	74	--	--
	38	61	66	70	73	75	76	--	--
	39	62	68	71	75	76	77	--	--
2	40	64	69	73	77	78	79	--	--
3	3	8	9	--	--	--	--	--	--
	4	11	12	--	--	--	--	--	--
	5	13	14	15	--	--	--	--	--
	6	15	16	17	--	--	--	--	--
	7	15	19	20	21	--	--	--	--
	8	19	21	22	24	--	--	--	--
	9	22	23	25	26	27	--	--	--
	10	24	26	27	29	30	--	--	--
	11	26	28	30	32	33	--	--	--
	12	28	31	32	34	35	36	--	--
	13	30	33	35	37	38	39	--	--
	14	32	35	37	40	41	42	--	--
	15	35	38	40	42	43	44	--	--
	16	37	40	42	45	46	47	--	--
	17	39	42	45	47	49	50	51	--
	18	41	45	47	50	52	53	54	--
	19	43	47	50	53	54	56	57	--
3	20	45	49	52	55	57	58	60	--

TABLE B.10 (cont.) CRITICAL VALUES OF THE MANN-WHITNEY U DISTRIBUTION

	$\alpha(2)$:	0.20	0.10	0.05	0.02	0.01	0.005	0.002	0.001
	$\alpha(1)$:	0.10	0.05	0.025	0.01	0.005	0.0025	0.001	0.0005
n_1	n_2								
3	21	48	52	55	58	60	61	62	63
	22	50	54	57	60	62	64	65	66
	23	52	56	60	63	65	67	68	69
	24	54	59	62	66	68	69	71	72
	25	56	61	65	68	70	72	74	75
	26	58	63	67	71	73	75	77	78
	27	60	66	70	74	76	78	79	80
	28	63	68	72	76	79	80	82	83
	29	65	70	74	79	81	83	85	86
	30	67	73	77	81	84	86	88	89
	31	69	75	79	84	87	89	91	92
	32	71	77	82	87	89	91	94	95
	33	73	80	84	89	92	94	96	98
	34	76	82	87	92	95	97	99	101
	35	78	84	89	94	97	100	102	103
	36	80	87	92	97	100	103	105	106
	37	82	89	94	100	103	105	108	109
	38	84	91	97	102	105	108	111	112
	39	86	94	99	105	108	111	113	115
3	40	89	96	102	107	111	114	116	118
4	4	13	15	16	--	--	--	--	--
	5	16	18	19	20	--	--	--	--
	6	19	21	22	23	24	--	--	--
	7	20	24	25	27	28	--	--	--
	8	25	27	28	30	31	32	--	--
	9	27	30	32	33	35	36	--	--
	10	30	33	35	37	38	39	40	--
	11	33	36	38	40	42	43	44	--
	12	36	39	41	43	45	46	48	--
	13	39	42	44	47	49	50	51	52
	14	41	45	47	50	52	53	55	56
	15	44	48	50	53	55	57	59	60
	16	47	50	53	57	59	60	62	63
	17	50	53	57	60	62	64	66	67
	18	52	56	60	63	66	67	69	71
	19	55	59	63	67	69	71	73	74
	20	58	62	66	70	72	75	77	78
	21	61	65	69	73	76	78	80	82
	22	63	68	72	77	79	82	84	85
	23	66	71	75	80	83	85	88	89
	24	69	74	79	83	86	89	91	93
	25	72	77	82	87	90	92	95	97
	26	74	80	85	90	93	96	98	100
	27	77	83	88	93	96	99	102	104
	28	80	86	91	96	100	103	106	108
	29	83	89	94	100	103	106	109	111
	30	85	92	97	103	107	110	113	115
4	31	88	95	100	106	110	113	117	119

TABLE B.10 (cont.) CRITICAL VALUES OF THE MANN-WHITNEY *U* DISTRIBUTION

	α(2):	0.20	0.10	0.05	0.02	0.01	0.005	0.002	0.001
	α(1):	0.10	0.05	0.025	0.01	0.005	0.0025	0.001	0.0005
n_1	n_2								
4	32	91	98	104	110	114	117	120	122
	33	94	101	107	113	117	120	124	126
	34	96	104	110	116	120	124	127	130
	35	99	107	113	120	124	127	131	133
	36	102	110	116	123	127	131	135	137
	37	105	113	119	126	131	134	138	141
	38	107	116	122	130	134	138	142	144
	39	110	118	125	133	137	141	145	148
4	40	113	121	129	136	141	145	149	152
5	5	20	21	23	24	25	--	--	--
	6	23	25	27	28	29	30	--	--
	7	24	29	30	32	34	35	--	--
	8	30	32	34	36	38	39	40	--
	9	33	36	38	40	42	43	44	45
	10	37	39	42	44	46	47	49	50
	11	40	43	46	48	50	52	53	54
	12	43	47	49	52	54	56	58	59
	13	47	50	53	56	58	60	62	63
	14	50	54	57	60	63	64	67	68
	15	53	57	61	64	67	69	71	72
	16	57	61	65	68	71	73	75	77
	17	60	65	68	72	75	77	80	81
	18	63	68	72	76	79	81	84	86
	19	67	72	76	80	83	86	88	90
	20	70	75	80	84	87	90	93	95
	21	73	79	83	88	91	94	97	99
	22	77	82	87	92	96	98	102	104
	23	80	86	91	96	100	103	106	108
	24	84	90	95	100	104	107	110	113
	25	87	93	98	104	108	111	115	117
	26	90	97	102	108	112	115	119	121
	27	94	100	106	112	119	120	123	126
	28	97	104	110	116	120	124	128	130
	29	100	107	113	120	124	128	132	135
	30	104	111	117	124	128	132	136	139
	31	107	115	121	128	133	136	141	144
	32	110	118	125	132	137	141	145	148
	33	114	122	128	136	141	145	150	153
	34	117	125	132	140	145	149	154	157
	35	120	129	136	144	149	153	158	161
	36	124	132	140	148	153	158	163	166
	37	127	136	144	152	157	162	167	170
	38	130	140	147	156	161	166	171	175
	39	134	143	151	160	165	170	176	179
5	40	137	147	155	164	169	174	180	184
6	6	27	29	31	33	34	35	--	--
	7	29	34	36	38	39	40	42	--
6	8	35	38	40	42	44	45	47	48

TABLE B.10 (cont.) CRITICAL VALUES OF THE MANN-WHITNEY U DISTRIBUTION

	α(2):	0.20	0.10	0.05	0.02	0.01	0.005	0.002	0.001
	α(1):	0.10	0.05	0.025	0.01	0.005	0.0025	0.001	0.0005
n_1	n_2								
6	9	39	42	44	47	49	50	52	53
	10	43	46	49	52	54	55	57	58
	11	47	50	53	57	59	60	62	64
	12	51	55	58	61	63	65	68	69
	13	55	59	62	66	68	70	73	74
	14	59	63	67	71	73	75	78	79
	15	63	67	71	75	78	80	83	85
	16	67	71	75	80	83	85	88	90
	17	71	76	80	84	87	90	93	95
	18	74	80	84	89	92	95	98	100
	19	78	84	89	94	97	100	103	106
	20	82	88	93	98	102	105	108	111
	21	86	92	97	103	107	110	114	116
	22	90	96	102	108	111	115	119	121
	23	94	101	106	112	116	120	124	126
	24	98	105	111	117	121	125	129	132
	25	102	109	115	121	126	130	134	137
	26	106	113	119	126	131	134	139	142
	27	110	117	124	131	135	139	144	147
	28	114	122	128	135	140	144	149	152
	29	118	126	132	140	145	149	154	157
	30	122	130	137	145	150	154	159	163
	31	125	134	141	149	154	159	164	168
	32	129	138	146	154	159	164	169	173
	33	133	142	150	158	164	169	174	178
	34	137	147	154	163	169	174	179	183
	35	141	151	159	168	173	179	185	188
	36	145	155	163	172	178	184	190	194
	37	149	159	167	177	183	188	195	199
	38	153	163	172	182	188	193	200	204
	39	157	167	176	186	193	198	205	209
6	40	161	172	181	191	197	203	210	214
7	7	36	38	41	43	45	46	48	49
	8	40	43	46	49	50	52	54	55
	9	45	48	51	54	56	58	60	61
	10	49	53	56	59	61	63	65	67
	11	54	58	61	65	67	69	71	73
	12	58	63	66	70	72	75	77	79
	13	63	67	71	75	78	80	83	85
	14	67	72	76	81	83	86	89	91
	15	72	77	81	86	89	92	95	97
	16	76	82	86	91	94	97	101	103
	17	81	86	91	96	100	103	106	109
	18	85	91	96	102	105	108	112	115
	19	90	96	101	107	111	114	118	120
	20	94	101	106	112	116	120	124	126
	21	99	106	111	117	122	125	129	132
7	22	103	110	116	123	127	131	135	138

TABLE B.10 (cont.) CRITICAL VALUES OF THE MANN-WHITNEY *U* DISTRIBUTION

	α(2):	0.20	0.10	0.05	0.02	0.01	0.005	0.002	0.001
	α(1):	0.10	0.05	0.025	0.01	0.005	0.0025	0.001	0.0005
n_1	n_2								
7	23	108	115	121	128	132	136	141	144
	24	112	120	126	133	138	142	147	150
	25	117	125	131	139	143	148	153	156
	26	121	129	136	144	149	153	158	162
	27	126	134	141	149	154	159	164	168
	28	130	139	146	154	160	164	170	174
	29	135	144	151	160	165	170	176	179
	30	139	149	156	165	170	176	181	185
	31	144	153	161	170	176	181	187	191
	32	148	158	166	175	181	187	193	197
	33	153	163	171	181	187	192	199	203
	34	157	168	176	186	192	198	204	209
	35	162	172	181	191	198	203	210	215
	36	166	177	186	196	203	209	216	221
	37	171	182	191	202	208	215	222	227
	38	175	187	196	207	214	220	227	232
	39	180	191	201	212	219	226	233	238
7	40	184	196	206	217	225	231	239	244
8	8	45	49	51	55	57	58	60	62
	9	50	54	57	61	63	65	67	68
	10	56	60	63	67	69	71	74	75
	11	61	65	69	73	75	77	80	82
	12	66	70	74	79	81	84	87	89
	13	71	76	80	84	87	90	93	95
	14	76	81	86	90	94	96	100	102
	15	81	87	91	96	100	103	106	109
	16	86	92	97	102	106	109	113	115
	17	91	97	102	108	112	115	119	122
	18	96	103	108	114	118	122	126	129
	19	101	108	114	120	124	128	132	135
	20	106	113	119	126	130	134	139	142
	21	112	119	125	132	136	140	145	148
	22	117	124	131	138	142	147	152	155
	23	122	130	136	144	149	153	158	162
	24	127	135	142	150	155	159	165	168
	25	132	140	147	155	161	165	171	175
	26	137	146	153	161	167	172	177	181
	27	142	151	159	167	173	178	184	188
	28	147	156	164	173	179	184	190	195
	29	152	162	170	179	185	190	197	201
	30	157	167	175	185	191	197	203	208
	31	162	172	181	191	197	203	210	214
	32	167	178	187	197	203	209	216	221
	33	172	183	192	203	209	215	223	227
	34	177	188	198	208	215	222	229	234
	35	182	194	203	214	221	228	235	241
	36	188	199	209	220	228	234	242	247
8	37	193	205	215	226	234	240	248	254

TABLE B.10 (cont.) CRITICAL VALUES OF THE MANN-WHITNEY *U* DISTRIBUTION

n_1	n_2	α(2): 0.20 α(1): 0.10	0.10 0.05	0.05 0.025	0.02 0.01	0.01 0.005	0.005 0.0025	0.002 0.001	0.001 0.0005
8	38	198	210	220	232	240	247	255	260
	39	203	215	226	238	246	253	261	267
8	40	208	221	231	244	252	259	268	273
9	9	56	60	64	67	70	72	74	76
	10	62	66	70	74	77	79	82	83
	11	68	72	76	81	83	86	89	91
	12	73	78	82	87	90	93	96	98
	13	79	84	89	94	97	100	103	106
	14	85	90	95	100	104	107	111	113
	15	90	96	101	107	111	114	118	120
	16	96	102	107	113	117	121	125	128
	17	101	108	114	120	124	128	132	135
	18	107	114	120	126	131	135	139	142
	19	113	120	126	133	138	142	146	150
	20	118	126	132	140	144	149	154	157
	21	124	132	139	146	151	155	161	164
	22	130	138	145	153	158	162	168	172
	23	135	144	151	159	164	169	175	179
	24	141	150	157	166	171	176	182	186
	25	147	156	163	172	178	183	189	193
	26	152	162	170	179	185	190	196	201
	27	158	168	176	185	191	197	203	208
	28	164	174	182	192	198	204	211	215
	29	169	179	188	198	205	211	218	222
	30	175	185	194	205	212	218	225	230
	31	180	191	201	211	218	224	232	237
	32	186	197	207	218	225	231	239	244
	33	192	203	213	224	232	238	246	251
	34	197	209	219	231	238	245	253	259
	35	203	215	226	237	245	252	260	266
	36	209	221	232	244	252	259	267	273
	37	214	227	238	250	258	266	275	280
	38	220	233	244	257	265	273	282	288
	39	225	239	250	263	272	280	289	295
9	40	231	245	257	270	279	286	296	302
10	10	68	73	77	81	84	87	90	92
	11	74	79	84	88	92	94	98	100
	12	81	86	91	96	99	102	106	108
	13	87	93	97	103	106	110	113	116
	14	93	99	104	110	114	117	121	124
	15	99	106	111	117	121	125	129	132
	16	106	112	118	124	129	133	137	140
	17	112	119	125	132	136	140	145	148
	18	118	125	132	139	143	148	153	156
	19	124	132	138	146	151	155	161	164
	20	130	138	145	153	158	163	168	172
10	21	137	145	152	160	166	170	176	180

TABLE B.10 (cont.) CRITICAL VALUES OF THE MANN-WHITNEY U DISTRIBUTION

	$\alpha(2)$:	0.20	0.10	0.05	0.02	0.01	0.005	0.002	0.001
	$\alpha(1)$:	0.10	0.05	0.025	0.01	0.005	0.0025	0.001	0.0005
n_1	n_2								
10	22	143	152	159	167	173	178	184	188
	23	149	158	166	175	180	186	192	196
	24	155	165	173	182	188	193	200	204
	25	161	171	179	189	195	201	207	212
	26	168	178	186	196	202	208	215	220
	27	174	184	193	203	210	216	223	228
	28	180	191	200	210	217	223	231	236
	29	186	197	207	217	224	231	238	244
	30	192	204	213	224	232	238	246	252
	31	199	210	220	232	239	246	254	259
	32	205	217	227	239	246	253	262	267
	33	211	223	234	246	254	261	269	275
	34	217	230	241	253	261	268	277	283
	35	223	236	247	260	268	276	285	291
	36	229	243	254	267	276	284	293	299
	37	236	249	261	274	283	291	300	307
	38	242	256	268	281	290	299	308	315
	39	248	262	275	289	298	306	316	323
10	40	254	269	281	296	305	314	324	331
11	11	81	87	91	96	100	103	106	109
	12	88	94	99	104	108	111	115	117
	13	95	101	106	112	116	119	123	126
	14	102	108	114	120	124	128	132	135
	15	108	115	121	128	132	136	141	144
	16	115	122	129	135	140	144	149	152
	17	122	130	136	143	148	152	158	161
	18	129	137	143	151	156	161	166	170
	19	136	144	151	159	164	169	175	178
	20	142	151	158	167	172	177	183	187
	21	149	158	166	174	180	185	191	196
	22	156	165	173	182	188	193	200	204
	23	163	172	180	190	196	202	208	213
	24	169	179	188	198	204	210	217	222
	25	176	186	195	205	212	218	225	230
	26	183	194	203	213	220	226	234	239
	27	190	201	210	221	228	234	242	247
	28	196	208	218	229	236	243	251	256
	29	203	215	225	236	244	251	259	265
	30	210	222	232	244	252	259	267	273
	31	217	229	240	252	260	267	276	282
	32	223	236	247	260	268	275	284	290
	33	230	243	255	267	276	283	293	299
	34	237	250	262	275	284	292	301	307
	35	244	257	269	283	292	300	309	316
	36	250	265	277	290	300	308	318	325
	37	257	272	284	298	308	316	326	333
	38	264	279	291	306	316	324	335	342
	39	271	286	299	314	323	332	343	350
11	40	277	293	306	321	331	341	351	359

TABLE B.10 (cont.) CRITICAL VALUES OF THE MANN-WHITNEY *U* DISTRIBUTION

n_1	n_2	α(2): 0.20 α(1): 0.10	0.10 0.05	0.05 0.025	0.02 0.01	0.01 0.005	0.005 0.0025	0.002 0.001	0.001 0.0005
12	12	95	102	107	113	117	120	124	127
	13	103	109	115	121	125	129	133	136
	14	110	117	123	130	134	138	143	146
	15	117	125	131	138	143	147	152	155
	16	125	132	139	146	151	156	161	165
	17	132	140	147	155	160	165	170	174
	18	139	148	155	163	169	173	179	183
	19	147	156	163	172	177	182	188	193
	20	154	163	171	180	186	191	198	202
	21	161	171	179	188	194	200	207	211
	22	169	179	187	197	203	209	216	220
	23	176	186	195	205	212	218	225	230
	24	183	194	203	213	220	227	234	239
	25	191	202	211	222	229	235	243	248
	26	198	209	219	230	238	244	252	258
	27	205	217	227	239	246	253	261	267
	28	213	225	235	247	255	262	270	276
	29	220	232	243	255	263	271	279	285
	30	227	240	251	264	272	279	288	295
	31	235	248	259	272	280	288	297	304
	32	242	256	267	280	289	297	307	313
	33	249	263	275	289	298	306	316	322
	34	257	271	283	297	306	315	325	332
	35	264	279	291	305	315	323	334	341
	36	271	286	299	314	323	332	343	350
	37	278	294	307	322	332	341	352	359
	38	286	302	315	330	340	350	361	368
	39	293	309	323	339	349	359	370	378
12	40	300	317	331	347	358	367	379	387
13	13	111	118	124	130	135	139	143	146
	14	119	126	132	139	144	148	153	157
	15	127	134	141	148	153	158	163	167
	16	134	143	149	157	163	167	173	177
	17	142	151	158	166	172	177	183	187
	18	150	159	167	175	181	186	192	197
	19	158	167	175	184	190	196	202	207
	20	166	176	184	193	200	205	212	217
	21	174	184	193	202	209	215	222	227
	22	182	192	201	211	218	224	232	237
	23	190	201	210	220	227	234	241	247
	24	198	209	218	229	237	243	251	256
	25	205	217	227	238	246	253	261	266
	26	213	225	236	247	255	262	270	276
	27	221	234	244	256	264	271	280	286
	28	229	242	253	265	273	281	290	296
	29	237	250	261	274	283	290	300	306
	30	245	258	270	283	292	300	309	316
13	31	253	267	278	292	301	309	319	326

TABLE B.10 (cont.) CRITICAL VALUES OF THE MANN-WHITNEY *U* DISTRIBUTION

	α(2):	0.20	0.10	0.05	0.02	0.01	0.005	0.002	0.001
	α(1):	0.10	0.05	0.025	0.01	0.005	0.0025	0.001	0.0005
n_1	n_2								
13	32	260	275	287	301	310	319	329	336
	33	268	283	296	310	319	328	338	346
	34	276	291	304	319	329	337	348	355
	35	284	299	313	328	338	347	358	365
	36	292	308	321	337	347	356	367	375
	37	300	316	330	346	356	366	377	385
	38	308	324	338	355	365	375	387	395
	39	315	332	347	363	374	385	397	405
13	40	323	341	355	372	384	394	406	415
14	14	127	135	141	149	154	161	164	167
	15	136	144	151	159	164	169	174	178
	16	144	153	160	168	174	179	185	189
	17	153	161	169	178	184	189	195	199
	18	161	170	178	187	194	199	206	210
	19	169	179	188	197	203	209	216	221
	20	178	188	197	207	213	219	226	231
	21	186	197	206	216	223	229	237	242
	22	195	206	215	226	233	240	247	253
	23	203	215	224	235	243	250	258	263
	24	212	223	234	245	253	260	268	274
	25	220	232	243	255	263	270	278	284
	26	228	241	252	264	272	280	289	295
	27	237	250	261	274	282	290	299	306
	28	245	259	270	283	292	300	309	316
	29	254	268	279	293	302	310	320	327
	30	262	276	289	302	312	320	330	337
	31	271	285	298	312	321	330	340	348
	32	279	294	307	321	331	340	351	358
	33	287	303	316	331	341	350	361	369
	34	296	312	325	341	351	360	371	379
	35	304	320	334	350	361	370	382	390
	36	313	329	343	360	370	380	392	400
	37	321	338	353	369	380	390	402	411
	38	329	347	362	379	390	400	413	421
	39	338	356	371	388	400	410	423	432
14	40	346	364	380	398	410	420	433	442
15	15	145	153	161	169	174	179	185	189
	16	154	163	170	179	185	190	197	201
	17	163	172	180	189	195	201	208	212
	18	172	182	190	200	206	212	219	224
	19	181	191	200	210	216	223	230	235
	20	190	200	210	220	227	233	241	246
	21	199	210	219	230	237	244	252	257
	22	208	219	229	240	248	255	263	269
	23	217	229	239	251	258	265	274	280
	24	226	238	249	261	269	276	285	291
15	25	235	247	258	271	279	287	296	302

TABLE B.10 (cont.) CRITICAL VALUES OF THE MANN-WHITNEY *U* DISTRIBUTION

	α(2):	0.20	0.10	0.05	0.02	0.01	0.005	0.002	0.001
	α(1):	0.10	0.05	0.025	0.01	0.005	0.0025	0.001	0.0005
n_1	n_2								
15	26	244	257	268	281	290	298	307	313
	27	253	266	278	291	300	308	318	325
	28	262	276	288	301	311	319	329	336
	29	271	285	297	312	321	330	340	347
	30	280	294	307	322	331	340	351	358
	31	288	304	317	332	342	351	362	369
	32	297	313	327	342	352	362	373	381
	33	306	323	336	352	363	372	384	392
	34	315	332	346	362	373	383	395	403
	35	324	341	356	372	383	394	406	414
	36	333	351	366	382	394	404	417	425
	37	342	360	375	393	404	415	428	436
	38	351	369	385	403	415	425	439	448
	39	360	379	395	413	425	436	449	459
15	40	369	388	404	423	435	447	460	470
16	16	163	173	181	190	196	202	208	213
	17	173	183	191	201	207	213	220	225
	18	182	193	202	212	218	224	232	237
	19	192	203	212	222	230	236	244	249
	20	201	213	222	233	241	247	255	261
	21	211	223	233	244	252	259	267	273
	22	221	233	243	255	263	270	279	285
	23	230	243	253	266	274	281	290	296
	24	240	253	264	276	285	293	302	308
	25	249	263	274	287	296	304	314	320
	26	259	273	284	298	307	315	325	332
	27	268	283	295	309	318	327	337	344
	28	278	292	305	319	329	338	348	356
	29	287	302	315	330	340	349	360	367
	30	297	312	326	341	351	360	372	379
	31	306	322	336	352	362	372	383	391
	32	316	332	346	362	373	383	395	403
	33	325	342	357	373	384	394	406	415
	34	335	352	367	384	395	406	418	427
	35	344	362	377	395	406	417	429	438
	36	354	372	388	405	417	428	441	450
	37	363	382	398	416	428	439	453	462
	38	373	392	408	427	439	451	464	474
	39	382	402	418	437	450	462	476	485
16	40	392	412	429	448	461	473	487	497
17	17	183	193	202	212	219	225	232	238
	18	193	204	213	224	231	237	245	250
	19	203	214	224	235	242	249	257	263
	20	213	225	235	247	254	261	270	275
	21	223	236	246	258	266	273	282	288
	22	233	246	257	269	278	285	294	300
	23	244	257	268	281	289	297	306	313
	24	254	267	279	292	301	309	319	325
17	25	264	278	290	303	313	321	331	338

TABLE B.10 (cont.) CRITICAL VALUES OF THE MANN-WHITNEY *U* DISTRIBUTION

	α(2):	0.20	0.10	0.05	0.02	0.01	0.005	0.002	0.001
	α(1):	0.10	0.05	0.025	0.01	0.005	0.0025	0.001	0.0005
n_1	n_2								
17	26	274	288	301	315	324	333	343	350
	27	284	299	312	326	336	345	355	363
	28	294	309	322	337	348	357	368	375
	29	304	320	333	349	359	369	380	388
	30	314	330	344	360	371	380	392	400
	31	324	341	355	371	382	392	404	413
	32	334	351	366	383	394	404	417	425
	33	344	362	377	394	406	416	429	438
	34	354	372	388	405	417	428	441	450
	35	365	383	399	417	429	440	453	462
	36	375	393	410	428	440	452	465	475
	37	385	404	420	439	452	464	478	487
	38	395	414	431	451	464	476	490	500
	39	405	425	442	462	475	487	502	512
17	40	415	435	453	473	487	499	514	525
18	18	204	215	225	236	243	250	258	263
	19	214	226	236	248	255	257	271	277
	20	225	237	248	260	268	275	284	287
	21	236	248	259	272	280	288	297	303
	22	246	260	271	284	292	300	310	316
	23	257	271	282	296	305	313	323	329
	24	268	282	294	308	317	325	336	343
	25	278	293	305	320	329	338	348	356
	26	289	304	317	332	341	350	361	369
	27	300	315	328	344	354	363	374	382
	28	310	326	340	355	366	376	387	395
	29	321	337	351	367	378	388	400	408
	30	331	348	363	379	390	401	413	421
	31	342	359	374	391	403	413	426	434
	32	353	370	386	403	415	426	438	447
	33	363	382	397	415	427	438	451	460
	34	374	393	409	427	439	451	464	473
	35	385	404	420	439	451	463	477	487
	36	395	415	432	451	464	475	490	500
	37	406	426	443	463	476	488	502	513
	38	416	437	454	475	488	500	515	526
	39	427	448	466	486	500	513	528	539
18	40	438	459	477	498	512	525	541	552
19	19	226	238	248	260	268	361	284	291
	20	237	250	261	273	281	289	298	304
	21	248	261	273	286	294	302	312	318
	22	259	273	285	298	307	315	325	332
	23	270	285	297	311	320	329	339	346
	24	282	296	309	323	333	342	352	360
	25	293	308	321	336	346	355	366	373
	26	304	320	333	348	359	368	379	387
	27	315	331	345	361	371	381	393	401
19	28	326	343	357	373	384	394	406	415

TABLE B.10 (cont.) CRITICAL VALUES OF THE MANN-WHITNEY U DISTRIBUTION

	α(2):	0.20	0.10	0.05	0.02	0.01	0.005	0.002	0.001
	α(1):	0.10	0.05	0.025	0.01	0.005	0.0025	0.001	0.0005
n_1	n_2								
19	29	338	355	369	386	397	407	420	428
	30	349	366	381	398	410	421	433	442
	31	360	378	393	411	423	434	447	456
	32	371	390	405	423	436	447	460	469
	33	382	401	417	436	448	460	474	483
	34	393	413	429	448	461	473	487	497
	35	405	424	441	461	474	486	500	511
	36	416	436	453	473	487	499	514	524
	37	427	448	465	486	500	512	527	538
	38	438	459	477	498	512	525	541	552
	39	449	471	489	511	525	538	554	565
19	40	--	482	502	523	538	551	568	579
20	20	249	262	273	286	295	303	312	319
	21	260	274	286	299	308	317	326	333
	22	272	276	299	313	322	330	341	348
	23	284	299	311	326	335	344	355	362
	24	296	311	324	339	349	358	369	377
	25	307	323	337	352	362	372	383	391
	26	319	335	349	365	376	386	397	405
	27	331	348	362	378	389	399	411	420
	28	343	360	374	391	403	413	425	434
	29	354	372	387	404	416	427	440	449
	30	366	384	400	418	430	440	454	463
	31	378	396	412	431	443	454	468	477
	32	389	409	425	444	456	468	482	492
	33	401	421	438	457	470	482	496	506
	34	413	433	450	470	483	495	510	520
	35	425	445	463	483	497	509	524	534
	36	436	457	475	496	510	523	538	549
	37	448	469	488	509	523	536	552	563
	38	--	482	501	522	537	550	566	577
	39	--	--	513	535	550	564	580	592
20	40	--	--	526	548	563	577	594	606

The preceding values were derived, with permission of the publisher, from the tables of Milton (1964, J. Amer. Statist. Assoc. 59: 925–934).

Examples:

$$U_{0.05(2),5,8} = 34 \quad \text{and} \quad U_{0.05(1),10,8} = U_{0.05(1),8,10} = 60.$$

For the Mann-Whitney test involving n_1 and/or n_2 larger than those in this table, the normal approximation (Section 9.9) may be used. This approximation is excellent for two-tailed testing at $\alpha = 0.10$ or 0.05 (or one-tailed testing at $\alpha = 0.05$ or 0.025, respectively), especially if n_1 and n_2 are similar in magnitude. The approximation becomes progressively poorer as we consider more extreme significance levels.

TABLE B.11 CRITICAL VALUES OF THE WILCOXON T DISTRIBUTION

n	$\alpha(2) = 0.50$ $\alpha(1) = 0.25$	0.20 0.10	0.10 0.05	0.05 0.025	0.02 0.01	0.01 0.005	0.005 0.0025	0.001 0.0005
4	2	0						
5	4	2	0					
6	6	3	2	0				
7	9	5	3	2	0			
8	12	8	5	3	1	0		
9	16	10	8	5	3	1	0	
10	20	14	10	8	5	3	1	
11	24	17	13	10	7	5	3	0
12	29	21	17	13	9	7	5	1
13	35	26	21	17	12	9	7	2
14	40	31	25	21	15	12	9	4
15	47	36	30	25	19	15	12	6
16	54	42	35	29	23	19	15	8
17	61	48	41	34	27	23	19	11
18	69	55	47	40	32	27	23	14
19	77	62	53	46	37	32	27	18
20	86	69	60	52	43	37	32	21
21	95	77	67	58	49	42	37	25
22	104	86	75	65	55	48	42	30
23	114	94	83	73	62	54	48	35
24	125	104	91	81	69	61	54	40
25	136	113	100	89	76	68	60	45
26	148	124	110	98	84	75	67	51
27	160	134	119	107	92	83	74	57
28	172	145	130	116	101	91	82	64
29	185	157	140	126	110	100	90	71
30	198	169	151	137	120	109	98	78
31	212	181	163	147	130	118	107	86
32	226	194	175	159	140	128	116	94
33	241	207	187	170	151	138	126	102
34	257	221	200	182	162	148	136	111
35	272	235	213	195	173	159	146	120
36	289	250	227	208	185	171	157	130
37	305	265	241	221	198	182	168	140
38	323	281	256	235	211	194	180	150
39	340	297	271	249	224	207	192	161
40	358	313	286	264	238	220	204	172
41	377	330	302	279	252	233	217	183
42	396	348	319	294	266	247	230	195
43	416	365	336	310	281	261	244	207
44	436	384	353	327	296	276	258	220
45	456	402	371	343	312	291	272	233
46	477	422	389	361	328	307	287	246
47	499	441	407	378	345	322	302	260
48	521	462	426	396	362	339	318	274
49	543	482	446	415	379	355	334	289
50	566	503	466	434	397	373	350	304
51	590	525	486	453	416	390	367	319
52	613	547	507	473	434	408	384	335
53	638	569	529	494	454	427	402	351
54	668	592	550	514	473	445	420	368
55	688	615	573	536	493	465	438	385
56	714	639	595	557	514	484	457	402
57	740	664	618	579	535	504	477	420
58	767	688	642	602	556	525	497	438
59	794	714	666	625	578	546	517	457
60	822	739	690	648	600	567	537	476

TABLE B.11 (cont.) CRITICAL VALUES OF THE WILCOXON T DISTRIBUTION

n	$\alpha(2) = 0.50$ $\alpha(1) = 0.25$	0.20 0.10	0.10 0.05	0.05 0.025	0.02 0.01	0.01 0.005	0.005 0.0025	0.001 0.0005
61	850	765	715	672	623	589	558	495
62	879	792	741	697	646	611	580	515
63	908	819	767	721	669	634	602	535
64	938	847	793	747	693	657	624	556
65	968	875	820	772	718	681	647	577
66	998	903	847	798	742	705	670	599
67	1029	932	875	825	768	729	694	621
68	1061	962	903	852	793	754	718	643
69	1093	992	931	879	819	779	742	666
70	1126	1022	960	907	846	805	767	689
71	1159	1053	990	936	873	831	792	712
72	1192	1084	1020	964	901	858	818	736
73	1226	1116	1050	994	928	884	844	761
74	1261	1148	1081	1023	957	912	871	786
75	1296	1181	1112	1053	986	940	898	811
76	1331	1214	1144	1084	1015	968	925	836
77	1367	1247	1176	1115	1044	997	953	862
78	1403	1282	1209	1147	1075	1026	981	889
79	1440	1316	1242	1179	1105	1056	1010	916
80	1478	1351	1276	1211	1136	1086	1039	943
81	1516	1387	1310	1244	1168	1116	1069	971
82	1554	1423	1345	1277	1200	1147	1099	999
83	1593	1459	1380	1311	1232	1178	1129	1028
84	1632	1496	1415	1345	1265	1210	1160	1057
85	1672	1533	1451	1380	1298	1242	1191	1086
86	1712	1571	1487	1415	1332	1275	1223	1116
87	1753	1609	1524	1451	1366	1308	1255	1146
88	1794	1648	1561	1487	1400	1342	1288	1177
89	1836	1688	1599	1523	1435	1376	1321	1208
90	1878	1727	1638	1560	1471	1410	1355	1240
91	1921	1767	1676	1597	1507	1445	1389	1271
92	1964	1808	1715	1635	1543	1480	1423	1304
93	2008	1849	1755	1674	1580	1516	1458	1337
94	2052	1891	1795	1712	1617	1552	1493	1370
95	2097	1933	1836	1752	1655	1589	1529	1404
96	2142	1976	1877	1791	1693	1626	1565	1438
97	2187	2019	1918	1832	1731	1664	1601	1472
98	2233	2062	1960	1872	1770	1702	1638	1507
99	2280	2106	2003	1913	1810	1740	1676	1543
100	2327	2151	2045	1955	1850	1779	1714	1578

Table B.11 is taken, with permission of the publisher, from the more extensive table of R. L. McCornack (1965, *J. Amer. Statist. Assoc.* 60: 864–871).

Examples:

$$T_{0.05(2),16} = 29 \quad \text{and} \quad T_{0.01(1),62} = 646.$$

For performing the Wilcoxon paired-sample test when $n > 100$, we may use the normal approximation (Section 10.4). The excellence of this approximation (without the correction for continuity) is shown as follows, as relative error = (true critical T − critical T from approximation)/true critical $T \times 100\%$. (Claypool and Holbert (1974) have determined that the approximation is improved by the correction for continuity—see Section 10.4—for $\alpha > 0.035$.)

$\alpha(2) =$ n	0.20 %	0.10 %	0.05 %	0.02 %	0.01 %	0.005 %	0.001 %
20	1.4	0.0	0.0	0.3	2.7	9.4	23.8
30	0.6	0.7	0.0	0.3	1.8	2.0	7.7
40	0.3	0.4	0.0	0.3	0.9	1.6	4.1
50	0.2	0.2	0.0	0.3	0.8	1.1	2.6
60	0.1	0.1	0.0	0.3	0.4	0.7	1.7
80	0.1	0.1	0.0	0.2	0.4	0.5	1.1
100	0.0	0.0	0.0	0.1	0.2	0.4	0.7

TABLE B.12 CRITICAL VALUES OF THE KRUSKAL-WALLIS *H* DISTRIBUTION

n_1	n_2	n_3	α = 0.10	0.05	0.02	0.01	0.005	0.002	0.001
2	2	2	4.571						
3	2	1	4.286						
3	2	2	4.500	4.714					
3	3	1	4.571	5.143					
3	3	2	4.556	5.361	6.250				
3	3	3	4.622	5.600	6.489	(7.200)	7.200		
4	2	1	4.500						
4	2	2	4.458	5.333	6.000				
4	3	1	4.056	5.208					
4	3	2	4.511	5.444	6.144	6.444	7.000		
4	3	3	4.709	5.791	6.564	6.745	7.318	8.018	
4	4	1	4.167	4.967	(6.667)	6.667			
4	4	2	4.555	5.455	6.600	7.036	7.282	7.855	
4	4	3	4.545	5.598	6.712	7.144	7.598	8.227	8.909
4	4	4	4.654	5.692	6.962	7.654	8.000	8.654	9.269
5	2	1	4.200	5.000					
5	2	2	4.373	5.160	6.000	6.533			
5	3	1	4.018	4.960	6.044				
5	3	2	4.651	5.251	6.124	6.909	7.182		
5	3	3	4.533	5.648	6.533	7.079	7.636	8.048	8.727
5	4	1	3.987	4.985	6.431	6.955	7.364		
5	4	2	4.541	5.273	6.505	7.205	7.573	8.114	8.591
5	4	3	4.549	5.656	6.676	7.445	7.927	8.481	8.795
5	4	4	4.619	5.657	6.953	7.760	8.189	8.868	9.168
5	5	1	4.109	5.127	6.145	7.309	8.182		
5	5	2	4.623	5.338	6.446	7.338	8.131	6.446	7.338
5	5	3	4.545	5.705	6.866	7.578	8.316	8.809	9.521
5	5	4	4.523	5.666	7.000	7.823	8.523	9.163	9.606
5	5	5	4.940	5.780	7.220	8.000	8.780	9.620	9.920
5	1	1	-----						
6	2	1	4.200	4.822					
6	2	2	4.545	5.345	6.182	6.982			
6	3	1	3.909	4.855	6.236				
6	3	2	4.682	5.348	6.227	6.970	7.515	8.182	
6	3	3	4.538	5.615	6.590	7.410	7.872	8.628	9.346
6	4	1	4.038	4.947	6.174	7.106	7.614		
6	4	2	4.494	5.340	6.571	7.340	7.846	8.494	8.827
6	4	3	4.604	5.610	6.725	7.500	8.033	8.918	9.170
6	4	4	4.595	5.681	6.900	7.795	8.381	9.167	9.861
6	5	1	4.128	4.990	6.138	7.182	8.077	8.515	
6	5	2	4.596	5.338	6.585	7.376	8.196	8.967	9.189
6	5	3	4.535	5.602	6.829	7.590	8.314	9.150	9.669
6	5	4	4.522	5.661	7.018	7.936	8.643	9.458	9.960
6	5	5	4.547	5.729	7.110	8.028	8.859	9.771	10.271
6	6	1	4.000	4.945	6.286	7.121	8.165	9.077	9.692
6	6	2	4.438	5.410	6.667	7.467	8.210	9.219	9.752
6	6	3	4.558	5.625	6.900	7.725	8.458	9.458	10.150
6	6	4	4.548	5.724	7.107	8.000	8.754	9.662	10.342
6	6	5	4.542	5.765	7.152	8.124	8.987	9.948	10.524
6	6	6	4.643	5.801	7.240	8.222	9.170	10.187	10.889

TABLE B.12 (cont.) CRITICAL VALUES OF THE KRUSKAL-WALLIS *H* DISTRIBUTION

n_1	n_2	n_3			α = 0.10	0.05	0.02	0.01	0.005	0.002	0.001
7	7	7			4.594	5.819	7.332	8.378	9.373	10.516	11.310
8	8	8			4.595	5.805	7.355	8.465	9.495	10.805	11.705
2	2	1	1		-----						
2	2	2	1		5.357	5.679					
2	2	2	2		5.667	6.167	(6.667)	6.667			
3	1	1	1		-----						
3	2	1	1		5.143						
3	2	2	1		5.556	5.833	6.500				
3	2	2	2		5.544	5.333	6.978	7.133	7.533		
3	3	1	1		5.333	6.333					
3	3	2	1		5.689	6.244	6.689	7.200	7.400		
3	3	2	2		5.745	6.527	7.182	7.636	7.873	8.018	8.455
3	3	3	1		5.655	6.600	7.109	7.400	8.055	8.345	
3	3	3	2		5.879	6.727	7.636	8.105	8.379	8.803	9.030
3	3	3	3		6.026	7.000	7.872	8.538	8.897	9.462	9.513
4	1	1	1		-----						
4	2	1	1		5.250	5.833					
4	2	2	1		5.533	6.133	6.667	7.000			
4	2	2	2		5.755	6.545	7.091	7.391	7.964	8.291	
4	3	1	1		5.067	6.178	6.711	7.067			
4	3	2	1		5.591	6.309	7.018	7.455	7.773	8.182	
4	3	2	2		5.750	6.621	7.530	7.871	8.273	8.689	8.909
4	3	3	1		5.539	6.545	7.485	7.758	8.212	8.697	9.182
4	3	3	2		5.872	6.795	7.763	8.333	8.718	9.167	8.455
4	3	3	3		6.016	6.984	7.995	8.659	9.253	9.709	10.016
4	4	1	1		5.182	5.945	7.091	7.909	7.909		
4	4	2	1		5.568	6.386	7.364	7.886	8.341	8.591	8.909
4	4	2	2		5.808	6.731	7.750	8.346	8.692	9.269	9.462
4	4	3	1		5.692	6.635	7.660	8.231	8.583	9.038	9.327
4	4	3	2		5.901	6.874	7.951	8.621	9.165	9.615	9.945
4	4	3	3		6.019	7.038	8.181	8.876	9.495	10.105	10.467
4	4	4	1		5.564	6.725	7.879	8.588	9.000	9.478	9.758
4	4	4	2		5.914	6.957	8.157	8.871	9.486	10.043	10.429
4	4	4	3		6.042	7.142	8.350	9.075	9.742	10.542	10.929
4	4	4	4		6.088	7.235	8.515	9.287	9.971	10.809	11.338
2	1	1	1	1	-----						
2	2	1	1	1	5.786						
2	2	2	1	1	6.250	6.750					
2	2	2	2	1	6.600	7.133	(7.533)	7.533			
2	2	2	2	2	6.982	7.418	8.073	8.291	(8.727)	8.727	
3	1	1	1	1	-----						
3	2	1	1	1	6.139	6.583					
3	2	2	1	1	6.511	6.800	7.400	7.600			
3	2	2	2	1	6.709	7.309	7.836	8.127	8.327	8.618	
3	2	2	2	2	6.955	7.682	8.303	8.682	8.985	9.273	9.364
3	3	1	1	1	6.311	7.111	7.467				
3	3	2	1	1	6.600	7.200	7.892	8.073	8.345		
3	3	2	2	1	6.788	7.591	8.258	8.576	8.924	9.167	9.303
3	3	2	2	2	7.026	7.910	8.667	9.115	9.474	9.769	10.026
3	3	3	1	1	6.788	7.576	8.242	8.424	8.848	(9.455)	9.455
3	3	3	2	1	6.910	7.759	8.590	9.051	9.410	9.769	9.974
3	3	3	2	2	7.121	8.044	9.011	9.505	9.890	10.330	10.637
3	3	3	3	1	7.077	8.000	8.879	9.451	9.846	10.286	10.549
3	3	3	3	2	7.210	8.200	9.267	9.876	10.333	10.838	11.171
3	3	3	3	3	7.333	8.333	9.467	10.200	10.733	10.267	11.667

The above values of *H* were determined from *Selected Tables in Mathematical Statistics*, Volume III, pp. 320–384, by permission of the American Mathematical Society. © 1975 by the American Mathematical Society (Iman, Quade, and Alexander 1975.)

Examples:

$$H_{0.05,4,3,2} = 5.444 \quad \text{and} \quad H_{0,01,4,4,5} = H_{0.01,5,4,4} = 7.760.$$

TABLE B.13 CRITICAL VALUES OF THE FRIEDMAN χ_r^2 DISTRIBUTION

a (n)	b (M) *	α = 0.50	0.20	0.10	0.05	0.02	0.01	0.005	0.002	0.001
3	2	3.000	4.000							
3	3	2.667	4.667	(6.000)	6.000					
3	4	2.000	4.500	6.000	6.500	(8.000)	(8.000)	8.000		
3	5	2.800	3.600	5.200	6.400	(8.400)	8.400	(10.000)	(10.000)	10.000
3	6	2.330	4.000	5.33	7.000	8.330	9.000	(10.330)	10.330	12.000
3	7	2.000	3.714	5.429	7.143	8.000	8.857	10.286	11.143	12.286
3	8	2.250	4.000	5.250	6.250	7.750	9.000	9.750	12.000	12.250
3	9	2.000	3.556	5.556	6.222	8.000	9.556	10.667	11.556	12.667
3	10	1.800	3.800	5.000	6.200	7.600	9.600	10.400	12.200	12.600
3	11	1.636	3.818	4.909	6.545	7.618	9.455	10.364	11.636	13.273
3	12	1.500	3.500	5.167	6.167	8.000	9.500	10.167	12.167	12.500
3	13	1.846	3.846	4.769	6.000	8.000	9.385	10.308	11.538	12.923
3	14	1.714	3.571	5.143	6.143	8.143	9.000	10.429	12.000	13.286
3	15	1.733	3.600	4.933	6.400	8.133	8.933	10.000	12.133	12.933
4	2	3.600	5.400	(6.000)	6.000					
4	3	3.400	5.400	6.600	7.400	8.200	(9.000)	(9.000)	9.000	
4	4	3.000	4.800	6.300	7.800	8.400	9.600	(10.200)	10.200	11.100
4	5	3.000	5.160	6.360	7.800	9.240	9.960	10.920	11.640	12.600
4	6	3.000	4.800	6.400	7.600	9.400	10.200	11.400	12.200	12.800
4	7	2.829	4.886	6.429	7.600	9.343	10.371	11.400	12.771	13.800
4	8	2.550	4.800	6.300	7.650	9.450	10.350	11.850	12.900	12.950

*For Kendall's coefficient of concordance (W), use the column headings in parentheses.

Examples:

$$(X_r^2)_{0.05,3,6} = 7.000 \quad \text{and} \quad W_{0.05,4,3} = 7.400.$$

TABLE B.14 CRITICAL VALUES OF *Q* FOR NONPARAMETRIC MULTIPLE COMPARISON TESTING

k	α : 0.50	0.20	0.10	0.05	0.02	0.01	0.005	0.002	0.001
2	0.674	1.282	1.645	1.960	2.327	2.576	2.807	3.091	3.291
3	1.383	1.834	2.128	2.394	2.713	2.936	3.144	3.403	3.588
4	1.732	2.128	2.394	2.639	2.936	3.144	3.342	3.588	3.765
5	1.960	2.327	2.576	2.807	3.091	3.291	3.481	3.719	3.891
6	2.128	2.475	2.713	2.936	3.209	3.403	3.588	3.820	3.986
7	2.261	2.593	2.823	3.038	3.304	3.494	3.675	3.902	4.067
8	2.369	2.690	2.914	3.124	3.384	3.570	3.748	3.972	4.134
9	2.461	2.773	2.992	3.197	3.453	3.635	3.810	4.031	4.191
10	2.540	2.845	3.059	3.261	3.512	3.692	3.865	4.083	4.241
11	2.609	2.908	3.119	3.317	3.565	3.743	3.914	4.129	4.286
12	2.671	2.965	3.172	3.368	3.613	3.789	3.957	4.171	4.326
13	2.726	3.016	3.220	3.414	3.656	3.830	3.997	4.209	4.363
14	2.777	3.062	3.264	3.456	3.695	3.868	4.034	4.244	4.397
15	2.823	3.105	3.304	3.494	3.731	3.902	4.067	4.276	4.428
16	2.866	3.144	3.342	3.529	3.765	3.935	4.098	4.305	4.456
17	2.905	3.181	3.376	3.562	3.796	3.965	4.127	4.333	4.483
18	2.942	3.215	3.409	3.593	3.825	3.993	4.154	4.359	4.508
19	2.976	3.246	3.439	3.622	3.852	4.019	4.179	4.383	4.532
20	3.008	3.276	3.467	3.649	3.878	4.044	4.203	4.406	4.554
21	3.038	3.304	3.494	3.675	3.902	4.067	4.226	4.428	4.575
22	3.067	3.331	3.519	3.699	3.925	4.089	4.247	4.448	4.595
23	3.094	3.356	3.543	3.722	3.947	4.110	4.268	4.468	4.614
24	3.120	3.380	3.566	3.744	3.968	4.130	4.287	4.486	4.632
25	3.144	3.403	3.588	3.765	3.988	4.149	4.305	4.504	4.649

This table was prepared using Equation 26.2.23 of Zelen and Severo (1964) which determines Q_α such that $P(Q_\alpha) \leq \alpha(1)/[k(k-1)]$, where Q_α is a normal deviate.

TABLE B.15 CRITICAL VALUES OF Q' FOR NONPARAMETRIC MULTIPLE COMPARISON TESTING WITH A CONTROL

k	α(2): 0.50	0.20	0.10	0.05	0.02	0.01	0.005	0.002	0.001
	α(1): 0.25	0.10	0.05	0.025	0.01	0.005	0.0025	0.001	0.0005
2	0.674	1.282	1.645	1.960	2.327	2.576	2.807	3.091	3.291
3	1.150	1.645	1.960	2.242	2.576	2.807	3.024	3.291	3.481
4	1.383	1.834	2.128	2.394	2.713	2.936	3.144	3.403	3.588
5	1.534	1.960	2.242	2.498	2.807	3.024	3.227	3.481	3.662
6	1.645	2.054	2.327	2.576	2.879	3.091	3.291	3.540	3.719
7	1.732	2.128	2.394	2.639	2.936	3.144	3.342	3.588	3.765
8	1.803	2.190	2.450	2.690	2.983	3.189	3.384	3.628	3.803
9	1.863	2.242	2.498	2.735	3.024	3.227	3.421	3.662	3.836
10	1.915	2.287	2.540	2.773	3.059	3.261	3.453	3.692	3.865
11	1.960	2.327	2.576	2.807	3.091	3.291	3.481	3.719	3.891
12	2.001	2.362	2.609	2.838	3.119	3.317	3.506	3.743	3.914
13	2.037	2.394	2.639	2.866	3.144	3.342	3.529	3.765	3.935
14	2.070	2.424	2.666	2.891	3.168	3.364	3.551	3.785	3.954
15	2.101	2.450	2.690	2.914	3.189	3.384	3.570	3.803	3.972
16	2.128	2.475	2.713	2.936	3.209	3.403	3.588	3.820	3.988
17	2.154	2.498	2.735	2.955	3.227	3.421	3.605	3.836	4.003
18	2.178	2.520	2.755	2.974	3.245	3.437	3.621	3.851	4.018
19	2.201	2.540	2.773	2.992	3.261	3.453	3.635	3.865	4.031
20	2.222	2.558	2.791	3.008	3.276	3.467	3.649	3.878	4.044
21	2.242	2.576	2.807	3.024	3.291	3.481	3.662	3.891	4.056
22	2.261	2.593	2.823	3.038	3.304	3.494	3.675	3.902	4.067
23	2.278	2.609	2.838	3.052	3.317	3.506	3.687	3.914	4.078
24	2.295	2.624	2.852	3.066	3.330	3.518	3.698	3.924	4.088
25	2.311	2.639	2.866	3.078	3.342	3.529	3.709	3.935	4.098

This table was prepared using Equation 26.2.23 of Zelen and Severo (1964), which determines Q'_α such that $P(Q'_\alpha) \leq \alpha(2)/[(k-1)] = \alpha(1)/(k-1)$, where Q'_α is a normal deviate.

TABLE B.16 CRITICAL VALUES OF THE CORRELATION COEFFICIENT, *r*

α(2):	0.50	0.20	0.10	0.05	0.02	0.01	0.005	0.002	0.001
α(1):	0.25	0.10	0.05	0.025	0.01	0.005	0.0025	0.001	0.0005
ν									
1	0.707	0.951	0.988	0.997	1.000	1.000	1.000	1.000	1.000
2	0.500	0.800	0.900	0.950	0.980	0.990	0.995	0.998	0.999
3	0.404	0.687	0.805	0.878	0.934	0.959	0.974	0.986	0.991
4	0.347	0.608	0.729	0.811	0.882	0.917	0.942	0.963	0.974
5	0.309	0.551	0.669	0.755	0.833	0.875	0.906	0.935	0.951
6	0.281	0.507	0.621	0.707	0.789	0.834	0.870	0.905	0.925
7	0.260	0.472	0.582	0.666	0.750	0.798	0.836	0.875	0.898
8	0.242	0.443	0.549	0.632	0.715	0.765	0.805	0.847	0.872
9	0.228	0.419	0.521	0.602	0.685	0.735	0.776	0.820	0.847
10	0.216	0.398	0.497	0.576	0.658	0.708	0.750	0.795	0.823
11	0.206	0.380	0.476	0.553	0.634	0.684	0.726	0.772	0.801
12	0.197	0.365	0.457	0.532	0.612	0.661	0.703	0.750	0.780
13	0.189	0.351	0.441	0.514	0.592	0.641	0.683	0.730	0.760
14	0.182	0.338	0.426	0.497	0.574	0.623	0.664	0.711	0.742
15	0.176	0.327	0.412	0.482	0.558	0.606	0.647	0.694	0.725
16	0.170	0.317	0.400	0.468	0.542	0.590	0.631	0.678	0.708
17	0.165	0.308	0.389	0.456	0.529	0.575	0.616	0.662	0.693
18	0.160	0.299	0.378	0.444	0.515	0.561	0.602	0.648	0.679
19	0.156	0.291	0.369	0.433	0.503	0.549	0.589	0.635	0.665
20	0.152	0.284	0.360	0.423	0.492	0.537	0.576	0.622	0.652
21	0.148	0.277	0.352	0.413	0.482	0.526	0.565	0.610	0.640
22	0.145	0.271	0.344	0.404	0.472	0.515	0.554	0.599	0.629
23	0.141	0.265	0.337	0.396	0.462	0.505	0.543	0.588	0.618
24	0.138	0.260	0.330	0.388	0.453	0.496	0.534	0.578	0.607
25	0.136	0.255	0.323	0.381	0.445	0.487	0.524	0.568	0.597
26	0.133	0.250	0.317	0.374	0.437	0.479	0.515	0.559	0.588
27	0.131	0.245	0.311	0.367	0.430	0.471	0.507	0.550	0.579
28	0.128	0.241	0.306	0.361	0.423	0.463	0.499	0.541	0.570
29	0.126	0.237	0.301	0.355	0.416	0.456	0.491	0.533	0.562
30	0.124	0.233	0.296	0.349	0.409	0.449	0.484	0.526	0.554
31	0.122	0.229	0.291	0.344	0.403	0.442	0.477	0.518	0.546
32	0.120	0.225	0.287	0.339	0.397	0.436	0.470	0.511	0.539
33	0.118	0.222	0.283	0.334	0.392	0.430	0.464	0.504	0.532
34	0.116	0.219	0.279	0.329	0.386	0.424	0.458	0.498	0.525
35	0.115	0.216	0.275	0.325	0.381	0.418	0.452	0.492	0.519
36	0.113	0.213	0.271	0.320	0.376	0.413	0.446	0.486	0.513
37	0.111	0.210	0.267	0.316	0.371	0.408	0.441	0.480	0.507
38	0.110	0.207	0.264	0.312	0.367	0.403	0.435	0.474	0.501
39	0.108	0.204	0.261	0.308	0.362	0.398	0.430	0.469	0.495
40	0.107	0.202	0.257	0.304	0.358	0.393	0.425	0.463	0.490
41	0.106	0.199	0.254	0.301	0.354	0.389	0.420	0.458	0.484
42	0.104	0.197	0.251	0.297	0.350	0.384	0.416	0.453	0.479
43	0.103	0.195	0.248	0.294	0.346	0.380	0.411	0.449	0.474
44	0.102	0.192	0.246	0.291	0.342	0.376	0.407	0.444	0.469
45	0.101	0.190	0.243	0.288	0.338	0.372	0.403	0.439	0.465
46	0.100	0.188	0.240	0.285	0.335	0.368	0.399	0.435	0.460
47	0.099	0.186	0.238	0.282	0.331	0.365	0.395	0.431	0.456
48	0.098	0.184	0.235	0.279	0.328	0.361	0.391	0.427	0.451
49	0.097	0.182	0.233	0.276	0.325	0.358	0.387	0.423	0.447
50	0.096	0.181	0.231	0.273	0.322	0.354	0.384	0.419	0.443

TABLE B.16 (cont.) CRITICAL VALUES OF THE CORRELATION COEFFICIENT, r

α(2):	0.50	0.20	0.10	0.05	0.02	0.01	0.005	0.002	0.001
α(1):	0.25	0.10	0.05	0.025	0.01	0.005	0.0025	0.001	0.0005
ν									
52	0.094	0.177	0.226	0.268	0.316	0.348	0.377	0.411	0.435
54	0.092	0.174	0.222	0.263	0.310	0.341	0.370	0.404	0.428
56	0.090	0.171	0.218	0.259	0.305	0.336	0.364	0.398	0.421
58	0.089	0.168	0.214	0.254	0.300	0.330	0.358	0.391	0.414
60	0.087	0.165	0.211	0.250	0.295	0.325	0.352	0.385	0.408
62	0.086	0.162	0.207	0.246	0.290	0.320	0.347	0.379	0.402
64	0.084	0.160	0.204	0.242	0.286	0.315	0.342	0.374	0.396
66	0.083	0.157	0.201	0.239	0.282	0.310	0.337	0.368	0.390
68	0.082	0.155	0.198	0.235	0.278	0.306	0.332	0.363	0.385
70	0.081	0.153	0.195	0.232	0.274	0.302	0.327	0.358	0.380
72	0.080	0.151	0.193	0.229	0.270	0.298	0.323	0.354	0.375
74	0.079	0.149	0.190	0.226	0.266	0.294	0.319	0.349	0.370
76	0.078	0.147	0.188	0.223	0.263	0.290	0.315	0.345	0.365
78	0.077	0.145	0.185	0.220	0.260	0.286	0.311	0.340	0.361
80	0.076	0.143	0.183	0.217	0.257	0.283	0.307	0.336	0.357
82	0.075	0.141	0.181	0.215	0.253	0.280	0.304	0.333	0.328
84	0.074	0.140	0.179	0.212	0.251	0.276	0.300	0.329	0.349
86	0.073	0.138	0.177	0.210	0.248	0.273	0.297	0.325	0.345
88	0.072	0.136	0.174	0.207	0.245	0.270	0.293	0.321	0.341
90	0.071	0.135	0.173	0.205	0.242	0.267	0.290	0.318	0.338
92	0.070	0.133	0.171	0.203	0.240	0.264	0.287	0.315	0.334
94	0.070	0.132	0.169	0.201	0.237	0.262	0.284	0.312	0.331
96	0.069	0.131	0.167	0.199	0.235	0.259	0.281	0.308	0.327
98	0.068	0.129	0.165	0.197	0.232	0.256	0.279	0.305	0.324
100	0.068	0.128	0.164	0.195	0.230	0.254	0.276	0.303	0.321
105	0.066	0.125	0.160	0.190	0.225	0.248	0.270	0.296	0.314
110	0.064	0.122	0.156	0.186	0.220	0.242	0.264	0.289	0.307
115	0.063	0.119	0.153	0.182	0.215	0.237	0.258	0.283	0.300
120	0.062	0.117	0.150	0.178	0.210	0.232	0.253	0.277	0.294
125	0.060	0.114	0.147	0.174	0.206	0.228	0.248	0.272	0.289
130	0.059	0.112	0.144	0.171	0.202	0.223	0.243	0.267	0.283
135	0.058	0.110	0.141	0.168	0.199	0.219	0.239	0.262	0.278
140	0.057	0.108	0.139	0.165	0.195	0.215	0.234	0.257	0.273
145	0.056	0.106	0.136	0.162	0.192	0.212	0.230	0.253	0.269
150	0.055	0.105	0.134	0.159	0.189	0.208	0.227	0.249	0.264
160	0.053	0.101	0.130	0.154	0.183	0.202	0.220	0.241	0.256
170	0.052	0.098	0.126	0.150	0.177	0.196	0.213	0.234	0.249
180	0.050	0.095	0.122	0.145	0.172	0.190	0.207	0.228	0.242
190	0.049	0.093	0.119	0.142	0.168	0.185	0.202	0.222	0.236
200	0.048	0.091	0.116	0.138	0.164	0.181	0.197	0.216	0.230
250	0.043	0.081	0.104	0.124	0.146	0.162	0.176	0.194	0.206
300	0.039	0.074	0.095	0.113	0.134	0.148	0.161	0.177	0.188
350	0.036	0.068	0.088	0.105	0.124	0.137	0.149	0.164	0.175
400	0.034	0.064	0.082	0.098	0.116	0.128	0.140	0.154	0.164
450	0.032	0.060	0.077	0.092	0.109	0.121	0.132	0.145	0.154
500	0.030	0.057	0.074	0.088	0.104	0.115	0.125	0.138	0.146
600	0.028	0.052	0.067	0.080	0.095	0.105	0.114	0.126	0.134
700	0.026	0.048	0.062	0.074	0.088	0.097	0.106	0.116	0.124
800	0.024	0.045	0.058	0.069	0.082	0.091	0.099	0.109	0.116
900	0.022	0.043	0.055	0.065	0.077	0.086	0.093	0.103	0.109
1000	0.021	0.041	0.052	0.062	0.073	0.081	0.089	0.098	0.104

The preceding values were computed using Equation 19.3 and Table B.3.

Examples:

$$r_{0.05(2),25} = 0.381 \quad \text{and} \quad r_{0.01(1),30} = 0.409.$$

If one requires a critical value for degrees of freedom not on this table, the critical value for the next lower degrees of freedom in the table may be conservatively used. Or, linear or harmonic interpolation may be used. If $\nu > 1000$, then use harmonic interpolation, setting the critical value equal to zero for $\nu = \infty$.

TABLE B.17 FISHER'S z TRANSFORMATION FOR CORRELATION COEFFICIENTS, r

r	0	1	2	3	4	5	6	7	8	9	r
0.000	0.0000	0.0010	0.0020	0.0030	0.0040	0.0050	0.0060	0.0070	0.0080	0.0090	0.000
0.010	0.0100	0.0110	0.0120	0.0130	0.0140	0.0150	0.0160	0.0170	0.0180	0.0190	0.010
0.020	0.0200	0.0210	0.0220	0.0230	0.0240	0.0250	0.0260	0.0270	0.0280	0.0290	0.020
0.030	0.0300	0.0310	0.0320	0.0330	0.0340	0.0350	0.0360	0.0370	0.0380	0.0390	0.030
0.040	0.0400	0.0410	0.0420	0.0430	0.0440	0.0450	0.0460	0.0470	0.0480	0.0490	0.040
0.050	0.0500	0.0510	0.0520	0.0530	0.0541	0.0551	0.0561	0.0571	0.0581	0.0591	0.050
0.060	0.0601	0.0611	0.0621	0.0631	0.0641	0.0651	0.0661	0.0671	0.0681	0.0691	0.060
0.070	0.0701	0.0711	0.0721	0.0731	0.0741	0.0751	0.0761	0.0772	0.0782	0.0792	0.070
0.080	0.0802	0.0812	0.0822	0.0832	0.0842	0.0852	0.0862	0.0872	0.0882	0.0892	0.080
0.090	0.0902	0.0913	0.0923	0.0933	0.0943	0.0953	0.0963	0.0973	0.0983	0.0993	0.090
0.100	0.1003	0.1013	0.1024	0.1034	0.1044	0.1054	0.1064	0.1074	0.1084	0.1094	0.100
0.110	0.1104	0.1115	0.1125	0.1135	0.1145	0.1155	0.1165	0.1175	0.1186	0.1196	0.110
0.120	0.1206	0.1216	0.1226	0.1236	0.1246	0.1257	0.1267	0.1277	0.1287	0.1297	0.120
0.130	0.1307	0.1318	0.1328	0.1338	0.1348	0.1358	0.1368	0.1379	0.1389	0.1399	0.130
0.140	0.1409	0.1419	0.1430	0.1440	0.1450	0.1460	0.1470	0.1481	0.1491	0.1501	0.140
0.150	0.1511	0.1522	0.1532	0.1542	0.1552	0.1563	0.1573	0.1583	0.1593	0.1604	0.150
0.160	0.1614	0.1624	0.1634	0.1645	0.1655	0.1665	0.1675	0.1686	0.1696	0.1706	0.160
0.170	0.1717	0.1727	0.1737	0.1748	0.1758	0.1768	0.1779	0.1789	0.1799	0.1809	0.170
0.180	0.1820	0.1830	0.1840	0.1851	0.1861	0.1872	0.1882	0.1892	0.1903	0.1913	0.180
0.190	0.1923	0.1934	0.1944	0.1955	0.1965	0.1975	0.1986	0.1996	0.2006	0.2017	0.190
0.200	0.2027	0.2038	0.2048	0.2059	0.2069	0.2079	0.2090	0.2100	0.2111	0.2121	0.200
0.210	0.2132	0.2142	0.2153	0.2163	0.2174	0.2184	0.2195	0.2205	0.2216	0.2226	0.210
0.220	0.2237	0.2247	0.2258	0.2268	0.2279	0.2289	0.2300	0.2310	0.2321	0.2331	0.220
0.230	0.2342	0.2352	0.2363	0.2374	0.2384	0.2395	0.2405	0.2416	0.2427	0.2437	0.230
0.240	0.2448	0.2458	0.2469	0.2480	0.2490	0.2501	0.2512	0.2522	0.2533	0.2543	0.240
0.250	0.2554	0.2565	0.2575	0.2586	0.2597	0.2608	0.2618	0.2629	0.2640	0.2650	0.250
0.260	0.2661	0.2672	0.2683	0.2693	0.2704	0.2715	0.2726	0.2736	0.2747	0.2758	0.260
0.270	0.2769	0.2779	0.2790	0.2801	0.2812	0.2823	0.2833	0.2844	0.2855	0.2866	0.270
0.280	0.2877	0.2888	0.2899	0.2909	0.2920	0.2931	0.2942	0.2953	0.2964	0.2975	0.280
0.290	0.2986	0.2997	0.3008	0.3018	0.3029	0.3040	0.3051	0.3062	0.3073	0.3084	0.290
0.300	0.3095	0.3106	0.3117	0.3128	0.3139	0.3150	0.3161	0.3172	0.3183	0.3194	0.300
0.310	0.3205	0.3217	0.3228	0.3239	0.3250	0.3261	0.3272	0.3283	0.3294	0.3305	0.310
0.320	0.3316	0.3328	0.3339	0.3350	0.3361	0.3372	0.3383	0.3395	0.3406	0.3417	0.320
0.330	0.3428	0.3440	0.3451	0.3462	0.3473	0.3484	0.3496	0.3507	0.3518	0.3530	0.330
0.340	0.3541	0.3552	0.3564	0.3575	0.3586	0.3598	0.3609	0.3620	0.3632	0.3643	0.340
0.350	0.3654	0.3666	0.3677	0.3689	0.3700	0.3712	0.3723	0.3734	0.3746	0.3757	0.350
0.360	0.3769	0.3780	0.3792	0.3803	0.3815	0.3826	0.3838	0.3850	0.3861	0.3873	0.360
0.370	0.3884	0.3896	0.3907	0.3919	0.3931	0.3942	0.3954	0.3966	0.3977	0.3989	0.370
0.380	0.4001	0.4012	0.4024	0.4036	0.4047	0.4059	0.4071	0.4083	0.4094	0.4106	0.380
0.390	0.4118	0.4130	0.4142	0.4153	0.4165	0.4177	0.4189	0.4201	0.4213	0.4225	0.390
0.400	0.4236	0.4248	0.4260	0.4272	0.4284	0.4296	0.4308	0.4320	0.4332	0.4344	0.400
0.410	0.4356	0.4368	0.4380	0.4392	0.4404	0.4416	0.4428	0.4441	0.4453	0.4465	0.410
0.420	0.4477	0.4489	0.4501	0.4513	0.4526	0.4538	0.4550	0.4562	0.4574	0.4587	0.420
0.430	0.4599	0.4611	0.4624	0.4636	0.4648	0.4660	0.4673	0.4685	0.4698	0.4710	0.430
0.440	0.4722	0.4735	0.4747	0.4760	0.4772	0.4784	0.4797	0.4809	0.4822	0.4834	0.440
0.450	0.4847	0.4860	0.4872	0.4885	0.4897	0.4910	0.4922	0.4935	0.4948	0.4960	0.450
0.460	0.4973	0.4986	0.4999	0.5011	0.5024	0.5037	0.5049	0.5062	0.5075	0.5088	0.460
0.470	0.5101	0.5114	0.5126	0.5139	0.5152	0.5165	0.5178	0.5191	0.5204	0.5217	0.470
0.480	0.5230	0.5243	0.5256	0.5269	0.5282	0.5295	0.5308	0.5321	0.5334	0.5347	0.480
0.490	0.5361	0.5374	0.5387	0.5400	0.5413	0.5427	0.5440	0.5453	0.5466	0.5480	0.490

TABLE B.17 (cont.) FISHER'S z TRANSFORMATION FOR CORRELATION COEFFICIENTS, r

r	0	1	2	3	4	5	6	7	8	9	r
0.500	0.5493	0.5506	0.5520	0.5533	0.5547	0.5560	0.5573	0.5587	0.5600	0.5614	0.500
0.510	0.5627	0.5641	0.5654	0.5668	0.5681	0.5695	0.5709	0.5722	0.5736	0.5750	0.510
0.520	0.5763	0.5777	0.5791	0.5805	0.5818	0.5832	0.5846	0.5860	0.5874	0.5888	0.520
0.530	0.5901	0.5915	0.5929	0.5943	0.5957	0.5971	0.5985	0.5999	0.6013	0.6027	0.530
0.540	0.6042	0.6056	0.6070	0.6084	0.6098	0.6112	0.6127	0.6141	0.6155	0.6169	0.540
0.550	0.6184	0.6198	0.6213	0.6227	0.6241	0.6256	0.6270	0.6285	0.6299	0.6314	0.550
0.560	0.6328	0.6343	0.6358	0.6372	0.6387	0.6401	0.6416	0.6431	0.6446	0.6460	0.560
0.570	0.6475	0.6490	0.6505	0.6520	0.6535	0.6550	0.6565	0.6580	0.6595	0.6610	0.570
0.580	0.6625	0.6640	0.6655	0.6670	0.6685	0.6700	0.6716	0.6731	0.6746	0.6761	0.580
0.590	0.6777	0.6792	0.6807	0.6823	0.6838	0.6854	0.6869	0.6885	0.6900	0.6916	0.590
0.600	0.6931	0.6947	0.6963	0.6978	0.6994	0.7010	0.7026	0.7042	0.7057	0.7073	0.600
0.610	0.7089	0.7105	0.7121	0.7137	0.7153	0.7169	0.7185	0.7201	0.7218	0.7234	0.610
0.620	0.7250	0.7266	0.7283	0.7299	0.7315	0.7332	0.7348	0.7365	0.7381	0.7398	0.620
0.630	0.7414	0.7431	0.7447	0.7464	0.7481	0.7497	0.7514	0.7531	0.7548	0.7565	0.630
0.640	0.7582	0.7599	0.7616	0.7633	0.7650	0.7667	0.7684	0.7701	0.7718	0.7736	0.640
0.650	0.7753	0.7770	0.7788	0.7805	0.7823	0.7840	0.7858	0.7875	0.7893	0.7910	0.650
0.660	0.7928	0.7946	0.7964	0.7981	0.7999	0.8017	0.8035	0.8053	0.8071	0.8089	0.660
0.670	0.8107	0.8126	0.8144	0.8162	0.8180	0.8199	0.8217	0.8236	0.8254	0.8273	0.670
0.680	0.8291	0.8310	0.8328	0.8347	0.8366	0.8385	0.8404	0.8422	0.8441	0.8460	0.680
0.690	0.8480	0.8499	0.8518	0.8537	0.8556	0.8576	0.8595	0.8614	0.8634	0.8653	0.690
0.700	0.8673	0.8693	0.8712	0.8732	0.8752	0.8772	0.8792	0.8812	0.8832	0.8852	0.700
0.710	0.8872	0.8892	0.8912	0.8933	0.8953	0.8973	0.8994	0.9014	0.9035	0.9056	0.710
0.720	0.9076	0.9097	0.9118	0.9139	0.9160	0.9181	0.9202	0.9223	0.9245	0.9266	0.720
0.730	0.9287	0.9309	0.9330	0.9352	0.9373	0.9395	0.9417	0.9439	0.9461	0.9483	0.730
0.740	0.9505	0.9527	0.9549	0.9571	0.9594	0.9616	0.9639	0.9661	0.9684	0.9707	0.740
0.750	0.9730	0.9752	0.9775	0.9798	0.9822	0.9845	0.9868	0.9891	0.9915	0.9938	0.750
0.760	0.9962	0.9986	1.0010	1.0034	1.0058	1.0082	1.0106	1.0130	1.0154	1.0179	0.760
0.770	1.0203	1.0228	1.0253	1.0277	1.0302	1.0327	1.0352	1.0378	1.0403	1.0428	0.770
0.780	1.0454	1.0479	1.0505	1.0531	1.0557	1.0583	1.0609	1.0635	1.0661	1.0688	0.780
0.790	1.0714	1.0741	1.0768	1.0795	1.0822	1.0849	1.0876	1.0903	1.0931	1.0958	0.790
0.800	1.0986	1.1014	1.1042	1.1070	1.1098	1.1127	1.1155	1.1184	1.1212	1.1241	0.800
0.810	1.1270	1.1299	1.1329	1.1358	1.1388	1.1417	1.1447	1.1477	1.1507	1.1538	0.810
0.820	1.1568	1.1599	1.1630	1.1660	1.1691	1.1723	1.1754	1.1786	1.1817	1.1849	0.820
0.830	1.1881	1.1914	1.1946	1.1979	1.2011	1.2044	1.2077	1.2111	1.2144	1.2178	0.830
0.840	1.2212	1.2246	1.2280	1.2314	1.2349	1.2384	1.2419	1.2454	1.2490	1.2526	0.840
0.850	1.2561	1.2598	1.2634	1.2671	1.2707	1.2744	1.2782	1.2819	1.2857	1.2895	0.850
0.860	1.2933	1.2972	1.3011	1.3050	1.3089	1.3129	1.3169	1.3209	1.3249	1.3290	0.860
0.870	1.3331	1.3372	1.3414	1.3456	1.3498	1.3540	1.3583	1.3626	1.3670	1.3713	0.870
0.880	1.3758	1.3802	1.3847	1.3892	1.3938	1.3984	1.4030	1.4077	1.4124	1.4171	0.880
0.890	1.4219	1.4268	1.4316	1.4365	1.4415	1.4465	1.4516	1.4566	1.4618	1.4670	0.890
0.900	1.4722	1.4775	1.4828	1.4882	1.4937	1.4992	1.5047	1.5103	1.5160	1.5217	0.900
0.910	1.5275	1.5334	1.5393	1.5453	1.5513	1.5574	1.5636	1.5698	1.5762	1.5825	0.910
0.920	1.5890	1.5956	1.6022	1.6089	1.6157	1.6226	1.6296	1.6366	1.6438	1.6510	0.920
0.930	1.6584	1.6658	1.6734	1.6811	1.6888	1.6967	1.7047	1.7129	1.7211	1.7295	0.930
0.940	1.7380	1.7467	1.7555	1.7645	1.7736	1.7828	1.7923	1.8019	1.8116	1.8216	0.940
0.950	1.8318	1.8421	1.8527	1.8635	1.8745	1.8857	1.8972	1.9090	1.9210	1.9333	0.950
0.960	1.9459	1.9588	1.9721	1.9856	1.9996	2.0139	2.0287	2.0439	2.0595	2.0756	0.960
0.970	2.0923	2.1095	2.1273	2.1457	2.1648	2.1847	2.2054	2.2269	2.2494	2.2729	0.970
0.980	2.2975	2.3234	2.3507	2.3795	2.4101	2.4426	2.4774	2.5147	2.5549	2.5987	0.980
0.990	2.6466	2.6995	2.7587	2.8257	2.9030	2.9944	3.1062	3.2502	3.4531	3.7997	0.990

TABLE B.18 CORRELATION COEFFICIENTS, *r*, CORRESPONDING TO FISHER'S *z* TRANSFORMATION

z	0	1	2	3	4	5	6	7	8	9	z
0.0	0.0000	0.0010	0.0020	0.0030	0.0040	0.0050	0.0060	0.0070	0.0080	0.0090	0.00
0.1	0.0100	0.0110	0.0120	0.0130	0.0140	0.0150	0.0160	0.0170	0.0180	0.0190	0.01
0.2	0.0200	0.0210	0.0220	0.0230	0.0240	0.0250	0.0260	0.0270	0.0280	0.0290	0.02
0.3	0.0300	0.0310	0.0320	0.0330	0.0340	0.0350	0.0360	0.0370	0.0380	0.0390	0.03
0.4	0.0400	0.0410	0.0420	0.0430	0.0440	0.0450	0.0460	0.0470	0.0480	0.0490	0.04
0.5	0.0500	0.0510	0.0520	0.0530	0.0539	0.0549	0.0559	0.0569	0.0579	0.0589	0.05
0.6	0.0599	0.0609	0.0619	0.0629	0.0639	0.0649	0.0659	0.0669	0.0679	0.0689	0.06
0.7	0.0699	0.0709	0.0719	0.0729	0.0739	0.0749	0.0759	0.0768	0.0778	0.0788	0.07
0.8	0.0798	0.0808	0.0818	0.0828	0.0838	0.0848	0.0858	0.0868	0.0878	0.0888	0.08
0.9	0.0898	0.0907	0.0917	0.0927	0.0937	0.0947	0.0957	0.0967	0.0977	0.0987	0.09
0.10	0.0997	0.1007	0.1016	0.1026	0.1036	0.1046	0.1056	0.1066	0.1076	0.1086	0.10
0.11	0.1096	0.1105	0.1115	0.1125	0.1135	0.1145	0.1155	0.1165	0.1175	0.1184	0.11
0.12	0.1194	0.1204	0.1214	0.1224	0.1234	0.1244	0.1253	0.1263	0.1273	0.1283	0.12
0.13	0.1293	0.1303	0.1312	0.1322	0.1332	0.1342	0.1352	0.1361	0.1371	0.1381	0.13
0.14	0.1391	0.1401	0.1411	0.1420	0.1430	0.1440	0.1450	0.1460	0.1469	0.1479	0.14
0.15	0.1489	0.1499	0.1508	0.1518	0.1528	0.1538	0.1547	0.1557	0.1567	0.1577	0.15
0.16	0.1586	0.1596	0.1606	0.1616	0.1625	0.1635	0.1645	0.1655	0.1664	0.1674	0.16
0.17	0.1684	0.1694	0.1703	0.1713	0.1723	0.1732	0.1742	0.1752	0.1761	0.1771	0.17
0.18	0.1781	0.1790	0.1800	0.1810	0.1820	0.1829	0.1839	0.1849	0.1858	0.1868	0.18
0.19	0.1877	0.1887	0.1897	0.1906	0.1916	0.1926	0.1935	0.1945	0.1955	0.1964	0.19
0.20	0.1974	0.1983	0.1993	0.2003	0.2012	0.2022	0.2031	0.2041	0.2051	0.2060	0.20
0.21	0.2070	0.2079	0.2089	0.2098	0.2108	0.2117	0.2127	0.2137	0.2146	0.2156	0.21
0.22	0.2165	0.2175	0.2184	0.2194	0.2203	0.2213	0.2222	0.2232	0.2241	0.2251	0.22
0.23	0.2260	0.2270	0.2279	0.2289	0.2298	0.2308	0.2317	0.2327	0.2336	0.2346	0.23
0.24	0.2355	0.2364	0.2374	0.2383	0.2393	0.2402	0.2412	0.2421	0.2430	0.2440	0.24
0.25	0.2449	0.2459	0.2468	0.2477	0.2487	0.2496	0.2506	0.2515	0.2524	0.2534	0.25
0.26	0.2543	0.2552	0.2562	0.2571	0.2580	0.2590	0.2599	0.2608	0.2618	0.2627	0.26
0.27	0.2636	0.2646	0.2655	0.2664	0.2673	0.2683	0.2692	0.2701	0.2711	0.2720	0.27
0.28	0.2729	0.2738	0.2748	0.2757	0.2766	0.2775	0.2784	0.2794	0.2803	0.2812	0.28
0.29	0.2821	0.2831	0.2840	0.2849	0.2858	0.2867	0.2876	0.2886	0.2895	0.2904	0.29
0.30	0.2913	0.2922	0.2931	0.2941	0.2950	0.2959	0.2968	0.2977	0.2986	0.2995	0.30
0.31	0.3004	0.3013	0.3023	0.3032	0.3041	0.3050	0.3059	0.3068	0.3077	0.3086	0.31
0.32	0.3095	0.3104	0.3113	0.3122	0.3131	0.3140	0.3149	0.3158	0.3167	0.3176	0.32
0.33	0.3185	0.3194	0.3203	0.3212	0.3221	0.3230	0.3239	0.3248	0.3257	0.3266	0.33
0.34	0.3275	0.3284	0.3293	0.3302	0.3310	0.3319	0.3328	0.3337	0.3346	0.3355	0.34
0.35	0.3364	0.3373	0.3381	0.3390	0.3399	0.3408	0.3417	0.3426	0.3435	0.3443	0.35
0.36	0.3452	0.3461	0.3470	0.3479	0.3487	0.3496	0.3505	0.3514	0.3522	0.3531	0.36
0.37	0.3540	0.3549	0.3557	0.3566	0.3575	0.3584	0.3592	0.3601	0.3610	0.3618	0.37
0.38	0.3627	0.3636	0.3644	0.3653	0.3662	0.3670	0.3679	0.3688	0.3696	0.3705	0.38
0.39	0.3714	0.3722	0.3731	0.3739	0.3748	0.3757	0.3765	0.3774	0.3782	0.3791	0.39
0.40	0.3799	0.3808	0.3817	0.3825	0.3834	0.3842	0.3851	0.3859	0.3868	0.3876	0.40
0.41	0.3885	0.3893	0.3902	0.3910	0.3919	0.3927	0.3936	0.3944	0.3952	0.3961	0.41
0.42	0.3969	0.3978	0.3986	0.3995	0.4003	0.4011	0.4020	0.4028	0.4036	0.4045	0.42
0.43	0.4053	0.4062	0.4070	0.4078	0.4087	0.4095	0.4103	0.4112	0.4120	0.4128	0.43
0.44	0.4136	0.4145	0.4153	0.4161	0.4170	0.4178	0.4186	0.4194	0.4203	0.4211	0.44
0.45	0.4219	0.4227	0.4235	0.4244	0.4252	0.4260	0.4268	0.4276	0.4285	0.4293	0.45
0.46	0.4301	0.4309	0.4317	0.4325	0.4333	0.4342	0.4350	0.4358	0.4366	0.4374	0.46
0.47	0.4382	0.4390	0.4398	0.4406	0.4414	0.4422	0.4430	0.4438	0.4446	0.4454	0.47
0.48	0.4462	0.4470	0.4478	0.4486	0.4494	0.4502	0.4510	0.4518	0.4526	0.4534	0.48
0.49	0.4542	0.4550	0.4558	0.4566	0.4574	0.4582	0.4590	0.4598	0.4605	0.4613	0.49

TABLE B.18 (cont.) CORRELATION COEFFICIENTS, r CORRESPONDING TO FISHER'S z TRANSFORMATION

z	0	1	2	3	4	5	6	7	8	9	z
0.50	0.4621	0.4629	0.4637	0.4645	0.4653	0.4660	0.4668	0.4676	0.4684	0.4692	0.50
0.51	0.4699	0.4707	0.4715	0.4723	0.4731	0.4738	0.4746	0.4754	0.4762	0.4769	0.51
0.52	0.4777	0.4785	0.4792	0.4800	0.4808	0.4815	0.4823	0.4831	0.4838	0.4846	0.52
0.53	0.4854	0.4861	0.4869	0.4877	0.4884	0.4892	0.4900	0.4907	0.4915	0.4922	0.53
0.54	0.4930	0.4937	0.4945	0.4953	0.4960	0.4968	0.4975	0.4983	0.4990	0.4998	0.54
0.55	0.5005	0.5013	0.5020	0.5028	0.5035	0.5043	0.5050	0.5057	0.5065	0.5072	0.55
0.56	0.5080	0.5087	0.5095	0.5102	0.5109	0.5117	0.5124	0.5132	0.5139	0.5146	0.56
0.57	0.5154	0.5161	0.5168	0.5176	0.5183	0.5190	0.5198	0.5205	0.5212	0.5219	0.57
0.58	0.5227	0.5234	0.5241	0.5248	0.5256	0.5263	0.5270	0.5277	0.5285	0.5292	0.58
0.59	0.5299	0.5306	0.5313	0.5320	0.5328	0.5335	0.5342	0.5349	0.5356	0.5363	0.59
0.60	0.5370	0.5378	0.5385	0.5392	0.5399	0.5406	0.5413	0.5420	0.5427	0.5434	0.60
0.61	0.5441	0.5448	0.5455	0.5462	0.5469	0.5476	0.5483	0.5490	0.5497	0.5504	0.61
0.62	0.5511	0.5518	0.5525	0.5532	0.5539	0.5546	0.5553	0.5560	0.5567	0.5574	0.62
0.63	0.5581	0.5587	0.5594	0.5601	0.5608	0.5615	0.5622	0.5629	0.5635	0.5642	0.63
0.64	0.5649	0.5656	0.5663	0.5669	0.5676	0.5683	0.5690	0.5696	0.5703	0.5710	0.64
0.65	0.5717	0.5723	0.5730	0.5737	0.5744	0.5750	0.5757	0.5764	0.5770	0.5777	0.65
0.66	0.5784	0.5790	0.5797	0.5804	0.5810	0.5817	0.5823	0.5830	0.5837	0.5843	0.66
0.67	0.5850	0.5856	0.5863	0.5869	0.5876	0.5883	0.5889	0.5896	0.5902	0.5909	0.67
0.68	0.5915	0.5922	0.5928	0.5935	0.5941	0.5948	0.5954	0.5961	0.5967	0.5973	0.68
0.69	0.5980	0.5986	0.5993	0.5999	0.6005	0.6012	0.6018	0.6025	0.6031	0.6037	0.69
0.70	0.6044	0.6050	0.6056	0.6063	0.6069	0.6075	0.6082	0.6088	0.6094	0.6100	0.70
0.71	0.6107	0.6113	0.6119	0.6126	0.6132	0.6138	0.6144	0.6150	0.6157	0.6163	0.71
0.72	0.6169	0.6175	0.6181	0.6188	0.6194	0.6200	0.6206	0.6212	0.6218	0.6225	0.72
0.73	0.6231	0.6237	0.6243	0.6249	0.6255	0.6261	0.6267	0.6273	0.6279	0.6285	0.73
0.74	0.6291	0.6297	0.6304	0.6310	0.6316	0.6322	0.6328	0.6334	0.6340	0.6346	0.74
0.75	0.6351	0.6357	0.6363	0.6369	0.6375	0.6381	0.6387	0.6393	0.6399	0.6405	0.75
0.76	0.6411	0.6417	0.6423	0.6428	0.6434	0.6440	0.6446	0.6452	0.6458	0.6463	0.76
0.77	0.6469	0.6475	0.6481	0.6487	0.6492	0.6498	0.6504	0.6510	0.6516	0.6521	0.77
0.78	0.6527	0.6533	0.6539	0.6544	0.6550	0.6556	0.6561	0.6567	0.6573	0.6578	0.78
0.79	0.6584	0.6590	0.6595	0.6601	0.6607	0.6612	0.6618	0.6624	0.6629	0.6635	0.79
0.80	0.6640	0.6646	0.6652	0.6657	0.6663	0.6668	0.6674	0.6679	0.6685	0.6690	0.80
0.81	0.6696	0.6701	0.6707	0.6712	0.6718	0.6723	0.6729	0.6734	0.6740	0.6745	0.81
0.82	0.6751	0.6756	0.6762	0.6767	0.6772	0.6778	0.6783	0.6789	0.6794	0.6799	0.82
0.83	0.6805	0.6810	0.6815	0.6821	0.6826	0.6832	0.6837	0.6842	0.6847	0.6853	0.83
0.84	0.6858	0.6863	0.6869	0.6874	0.6879	0.6884	0.6890	0.6895	0.6900	0.6905	0.84
0.85	0.6911	0.6916	0.6921	0.6926	0.6932	0.6937	0.6942	0.6947	0.6952	0.6957	0.85
0.86	0.6963	0.6968	0.6973	0.6978	0.6983	0.6988	0.6993	0.6998	0.7004	0.7009	0.86
0.87	0.7014	0.7019	0.7024	0.7029	0.7034	0.7039	0.7044	0.7049	0.7054	0.7059	0.87
0.88	0.7064	0.7069	0.7074	0.7079	0.7084	0.7089	0.7094	0.7099	0.7104	0.7109	0.88
0.89	0.7114	0.7119	0.7124	0.7129	0.7134	0.7139	0.7143	0.7148	0.7153	0.7158	0.89
0.90	0.7163	0.7168	0.7173	0.7178	0.7182	0.7187	0.7192	0.7197	0.7202	0.7207	0.90
0.91	0.7211	0.7216	0.7221	0.7226	0.7230	0.7235	0.7240	0.7245	0.7249	0.7254	0.91
0.92	0.7259	0.7264	0.7268	0.7273	0.7278	0.7283	0.7287	0.7292	0.7297	0.7301	0.92
0.93	0.7306	0.7311	0.7315	0.7320	0.7325	0.7329	0.7334	0.7338	0.7343	0.7348	0.93
0.94	0.7352	0.7357	0.7361	0.7366	0.7371	0.7375	0.7380	0.7384	0.7389	0.7393	0.94
0.95	0.7398	0.7402	0.7407	0.7411	0.7416	0.7420	0.7425	0.7429	0.7434	0.7438	0.95
0.96	0.7443	0.7447	0.7452	0.7456	0.7461	0.7465	0.7469	0.7474	0.7478	0.7483	0.96
0.97	0.7487	0.7491	0.7496	0.7500	0.7505	0.7509	0.7513	0.7518	0.7522	0.7526	0.97
0.98	0.7531	0.7535	0.7539	0.7544	0.7548	0.7552	0.7557	0.7561	0.7565	0.7569	0.98
0.99	0.7574	0.7578	0.7582	0.7586	0.7591	0.7595	0.7599	0.7603	0.7608	0.7612	0.99

TABLE B.18 (cont.) CORRELATION COEFFICIENTS, r CORRESPONDING TO FISHER'S z TRANSFORMATION

z	0	1	2	3	4	5	6	7	8	9	z
1.0	0.7616	0.7658	0.7699	0.7739	0.7779	0.7818	0.7857	0.7895	0.7932	0.7969	1.0
1.1	0.8005	0.8041	0.8076	0.8110	0.8144	0.8178	0.8210	0.8243	0.8274	0.8306	1.1
1.2	0.8337	0.8367	0.8397	0.8426	0.8455	0.8483	0.8511	0.8538	0.8565	0.8591	1.2
1.3	0.8617	0.8643	0.8668	0.8692	0.8717	0.8741	0.8764	0.8787	0.8809	0.8832	1.3
1.4	0.8854	0.8875	0.8896	0.8917	0.8937	0.8957	0.8977	0.8996	0.9015	0.9033	1.4
1.5	0.9051	0.9069	0.9087	0.9104	0.9121	0.9138	0.9154	0.9170	0.9186	0.9201	1.5
1.6	0.9217	0.9232	0.9246	0.9261	0.9275	0.9289	0.9302	0.9316	0.9329	0.9341	1.6
1.7	0.9354	0.9366	0.9379	0.9391	0.9402	0.9414	0.9425	0.9436	0.9447	0.9458	1.7
1.8	0.9468	0.9478	0.9488	0.9498	0.9508	0.9517	0.9527	0.9536	0.9545	0.9554	1.8
1.9	0.9562	0.9571	0.9579	0.9587	0.9595	0.9603	0.9611	0.9618	0.9626	0.9633	1.9
2.0	0.9640	0.9647	0.9654	0.9661	0.9667	0.9674	0.9680	0.9687	0.9693	0.9699	2.0
2.1	0.9705	0.9710	0.9716	0.9721	0.9727	0.9732	0.9737	0.9743	0.9748	0.9753	2.1
2.2	0.9757	0.9762	0.9767	0.9771	0.9776	0.9780	0.9785	0.9789	0.9793	0.9797	2.2
2.3	0.9801	0.9805	0.9809	0.9812	0.9816	0.9820	0.9823	0.9827	0.9830	0.9833	2.3
2.4	0.9837	0.9840	0.9843	0.9846	0.9849	0.9852	0.9855	0.9858	0.9861	0.9863	2.4
2.5	0.9866	0.9869	0.9871	0.9874	0.9876	0.9879	0.9881	0.9884	0.9886	0.9888	2.5
2.6	0.9890	0.9892	0.9895	0.9897	0.9899	0.9901	0.9903	0.9905	0.9906	0.9908	2.6
2.7	0.9910	0.9912	0.9914	0.9915	0.9917	0.9919	0.9920	0.9922	0.9923	0.9925	2.7
2.8	0.9926	0.9928	0.9929	0.9931	0.9932	0.9933	0.9935	0.9936	0.9937	0.9938	2.8
2.9	0.9940	0.9941	0.9942	0.9943	0.9944	0.9945	0.9946	0.9947	0.9949	0.9950	2.9
3.0	0.9951	0.9952	0.9952	0.9953	0.9954	0.9955	0.9956	0.9957	0.9958	0.9959	3.0
3.1	0.9959	0.9960	0.9961	0.9962	0.9963	0.9963	0.9964	0.9965	0.9965	0.9966	3.1
3.2	0.9967	0.9967	0.9968	0.9969	0.9969	0.9970	0.9971	0.9971	0.9972	0.9972	3.2
3.3	0.9973	0.9973	0.9974	0.9974	0.9975	0.9975	0.9976	0.9976	0.9977	0.9977	3.3
3.4	0.9978	0.9978	0.9979	0.9979	0.9979	0.9980	0.9980	0.9981	0.9981	0.9981	3.4
3.5	0.9982	0.9982	0.9982	0.9983	0.9983	0.9984	0.9984	0.9984	0.9984	0.9985	3.5
3.6	0.9985	0.9985	0.9986	0.9986	0.9986	0.9986	0.9987	0.9987	0.9987	0.9988	3.6
3.7	0.9988	0.9988	0.9988	0.9988	0.9989	0.9989	0.9989	0.9989	0.9990	0.9990	3.7
3.8	0.9990	0.9990	0.9990	0.9991	0.9991	0.9991	0.9991	0.9991	0.9991	0.9992	3.8
3.9	0.9992	0.9992	0.9992	0.9992	0.9992	0.9993	0.9993	0.9993	0.9993	0.9993	3.9
4.0	0.9993	0.9993	0.9994	0.9994	0.9994	0.9994	0.9994	0.9994	0.9994	0.9994	4.0
4.1	0.9995	0.9995	0.9995	0.9995	0.9995	0.9995	0.9995	0.9995	0.9995	0.9995	4.1
4.2	0.9996	0.9996	0.9996	0.9996	0.9996	0.9996	0.9996	0.9996	0.9996	0.9996	4.2
4.3	0.9996	0.9996	0.9996	0.9997	0.9997	0.9997	0.9997	0.9997	0.9997	0.9997	4.3
4.4	0.9997	0.9997	0.9997	0.9997	0.9997	0.9997	0.9997	0.9997	0.9997	0.9997	4.4
4.5	0.9998	0.9998	0.9998	0.9998	0.9998	0.9998	0.9998	0.9998	0.9998	0.9998	4.5
4.6	0.9998	0.9998	0.9998	0.9998	0.9998	0.9998	0.9998	0.9998	0.9998	0.9998	4.6
4.7	0.9998	0.9998	0.9998	0.9998	0.9998	0.9999	0.9999	0.9999	0.9999	0.9999	4.7
4.8	0.9999	0.9999	0.9999	0.9999	0.9999	0.9999	0.9999	0.9999	0.9999	0.9999	4.8
4.9	0.9999	0.9999	0.9999	0.9999	0.9999	0.9999	0.9999	0.9999	0.9999	0.9999	4.9

For example:

$$z = 2.42, \quad r = 0.9843.$$

TABLE B.19 CRITICAL VALUES OF THE SPEARMAN RANK CORRELATION COEFFICIENT, r_s

α(2):	0.50	0.20	0.10	0.05	0.02	0.01	0.005	0.002	0.001
α(1):	0.25	0.10	0.05	0.025	0.01	0.005	0.0025	0.001	0.0005
n									
4	0.600	1.000	1.000						
5	0.500	0.800	0.900	1.000	1.000				
6	0.371	0.657	0.829	0.886	0.943	1.000	1.000		
7	0.321	0.571	0.714	0.786	0.893	0.929	0.964	1.000	1.000
8	0.310	0.524	0.643	0.738	0.833	0.881	0.905	0.952	0.976
9	0.267	0.483	0.600	0.700	0.783	0.833	0.867	0.917	0.933
10	0.248	0.455	0.564	0.648	0.745	0.794	0.830	0.879	0.903
11	0.236	0.427	0.536	0.618	0.709	0.755	0.800	0.845	0.873
12	0.217	0.406	0.503	0.587	0.678	0.727	0.769	0.818	0.846
13	0.209	0.385	0.484	0.560	0.648	0.703	0.747	0.791	0.824
14	0.200	0.367	0.464	0.538	0.626	0.679	0.723	0.771	0.802
15	0.189	0.354	0.446	0.521	0.604	0.654	0.700	0.750	0.779
16	0.182	0.341	0.429	0.503	0.582	0.635	0.679	0.729	0.762
17	0.176	0.328	0.414	0.485	0.566	0.615	0.662	0.713	0.748
18	0.170	0.317	0.401	0.472	0.550	0.600	0.643	0.695	0.728
19	0.165	0.309	0.391	0.460	0.535	0.584	0.628	0.677	0.712
20	0.161	0.299	0.380	0.447	0.520	0.570	0.612	0.662	0.696
21	0.156	0.292	0.370	0.435	0.508	0.556	0.599	0.648	0.681
22	0.152	0.284	0.361	0.425	0.496	0.544	0.586	0.634	0.667
23	0.148	0.278	0.353	0.415	0.486	0.532	0.573	0.622	0.654
24	0.144	0.271	0.344	0.406	0.476	0.521	0.562	0.610	0.642
25	0.142	0.265	0.337	0.398	0.466	0.511	0.551	0.598	0.630
26	0.138	0.259	0.331	0.390	0.457	0.501	0.541	0.587	0.619
27	0.136	0.255	0.324	0.382	0.448	0.491	0.531	0.577	0.608
28	0.133	0.250	0.317	0.375	0.440	0.483	0.522	0.567	0.598
29	0.130	0.245	0.312	0.368	0.433	0.475	0.513	0.558	0.589
30	0.128	0.240	0.306	0.362	0.425	0.467	0.504	0.549	0.580
31	0.126	0.236	0.301	0.356	0.418	0.459	0.496	0.541	0.571
32	0.124	0.232	0.296	0.350	0.412	0.452	0.489	0.533	0.563
33	0.121	0.229	0.291	0.345	0.405	0.446	0.482	0.525	0.554
34	0.120	0.225	0.287	0.340	0.399	0.439	0.475	0.517	0.547
35	0.118	0.222	0.283	0.335	0.394	0.433	0.468	0.510	0.539
36	0.116	0.219	0.279	0.330	0.388	0.427	0.462	0.504	0.533
37	0.114	0.216	0.275	0.325	0.383	0.421	0.456	0.497	0.526
38	0.113	0.212	0.271	0.321	0.378	0.415	0.450	0.491	0.519
39	0.111	0.210	0.267	0.317	0.373	0.410	0.444	0.485	0.513
40	0.110	0.207	0.264	0.313	0.368	0.405	0.439	0.479	0.507
41	0.108	0.204	0.261	0.309	0.364	0.400	0.433	0.473	0.501
42	0.107	0.202	0.257	0.305	0.359	0.395	0.428	0.468	0.495
43	0.105	0.199	0.254	0.301	0.355	0.391	0.423	0.463	0.490
44	0.104	0.197	0.251	0.298	0.351	0.386	0.419	0.458	0.484
45	0.103	0.194	0.248	0.294	0.347	0.382	0.414	0.453	0.479
46	0.102	0.192	0.246	0.291	0.343	0.378	0.410	0.448	0.474
47	0.101	0.190	0.243	0.288	0.340	0.374	0.405	0.443	0.469
48	0.100	0.188	0.240	0.285	0.336	0.370	0.401	0.439	0.465
49	0.098	0.186	0.238	0.282	0.333	0.366	0.397	0.434	0.460
50	0.097	0.184	0.235	0.279	0.329	0.363	0.393	0.430	0.456

TABLE B.19 (cont.) CRITICAL VALUES OF THE SPEARMAN RANK CORRELATION COEFFICIENT, r_s

α(2):	0.50	0.20	0.10	0.05	0.02	0.01	0.005	0.002	0.001
α(1):	0.25	0.10	0.05	0.025	0.01	0.005	0.0025	0.001	0.0005
n									
51	0.096	0.182	0.233	0.276	0.326	0.359	0.390	0.426	0.451
52	0.095	0.180	0.231	0.274	0.323	0.356	0.386	0.422	0.447
53	0.095	0.179	0.228	0.271	0.320	0.352	0.382	0.418	0.443
54	0.094	0.177	0.226	0.268	0.317	0.349	0.379	0.414	0.439
55	0.093	0.175	0.224	0.266	0.314	0.346	0.375	0.411	0.435
56	0.092	0.174	0.222	0.264	0.311	0.343	0.372	0.407	0.432
57	0.091	0.172	0.220	0.261	0.308	0.340	0.369	0.404	0.428
58	0.090	0.171	0.218	0.259	0.306	0.337	0.366	0.400	0.424
59	0.089	0.169	0.216	0.257	0.303	0.334	0.363	0.397	0.421
60	0.089	0.168	0.214	0.255	0.300	0.331	0.360	0.394	0.418
61	0.088	0.166	0.213	0.252	0.298	0.329	0.357	0.391	0.414
62	0.087	0.165	0.211	0.250	0.296	0.326	0.354	0.388	0.411
63	0.086	0.163	0.209	0.248	0.293	0.323	0.351	0.385	0.408
64	0.086	0.162	0.207	0.246	0.291	0.321	0.348	0.382	0.405
65	0.085	0.161	0.206	0.244	0.289	0.318	0.346	0.379	0.402
66	0.084	0.160	0.204	0.243	0.287	0.316	0.343	0.376	0.399
67	0.084	0.158	0.203	0.241	0.284	0.314	0.341	0.373	0.396
68	0.083	0.157	0.201	0.239	0.282	0.311	0.338	0.370	0.393
69	0.082	0.156	0.200	0.237	0.280	0.309	0.336	0.368	0.390
70	0.082	0.155	0.198	0.235	0.278	0.307	0.333	0.365	0.388
71	0.081	0.154	0.197	0.234	0.276	0.305	0.331	0.363	0.385
72	0.081	0.153	0.195	0.232	0.274	0.303	0.329	0.360	0.382
73	0.080	0.152	0.194	0.230	0.272	0.301	0.327	0.358	0.380
74	0.080	0.151	0.193	0.229	0.271	0.299	0.324	0.355	0.377
75	0.079	0.150	0.191	0.227	0.269	0.297	0.322	0.353	0.375
76	0.078	0.149	0.190	0.226	0.267	0.295	0.320	0.351	0.372
77	0.078	0.148	0.189	0.224	0.265	0.293	0.318	0.349	0.370
78	0.077	0.147	0.188	0.223	0.264	0.291	0.316	0.346	0.368
79	0.077	0.146	0.186	0.221	0.262	0.289	0.314	0.344	0.365
80	0.076	0.145	0.185	0.220	0.260	0.287	0.312	0.342	0.363
81	0.076	0.144	0.184	0.219	0.259	0.285	0.310	0.340	0.361
82	0.075	0.143	0.183	0.217	0.257	0.284	0.308	0.338	0.359
83	0.075	0.142	0.182	0.216	0.255	0.282	0.306	0.336	0.357
84	0.074	0.141	0.181	0.215	0.254	0.280	0.305	0.334	0.355
85	0.074	0.140	0.180	0.213	0.252	0.279	0.303	0.332	0.353
86	0.074	0.139	0.179	0.212	0.251	0.277	0.301	0.330	0.351
87	0.073	0.139	0.177	0.211	0.250	0.276	0.299	0.328	0.349
88	0.073	0.138	0.176	0.210	0.248	0.274	0.298	0.327	0.347
89	0.072	0.137	0.175	0.209	0.247	0.272	0.296	0.325	0.345
90	0.072	0.136	0.174	0.207	0.245	0.271	0.294	0.323	0.343
91	0.072	0.135	0.173	0.206	0.244	0.269	0.293	0.321	0.341
92	0.071	0.135	0.173	0.205	0.243	0.268	0.291	0.319	0.339
93	0.071	0.134	0.172	0.204	0.241	0.267	0.290	0.318	0.338
94	0.070	0.133	0.171	0.203	0.240	0.265	0.288	0.316	0.336
95	0.070	0.133	0.170	0.202	0.239	0.264	0.287	0.314	0.334
96	0.070	0.132	0.169	0.201	0.238	0.262	0.285	0.313	0.332
97	0.069	0.131	0.168	0.200	0.236	0.261	0.284	0.311	0.331
98	0.069	0.130	0.167	0.199	0.235	0.260	0.282	0.310	0.329
99	0.068	0.130	0.166	0.198	0.234	0.258	0.281	0.308	0.327
100	0.068	0.129	0.165	0.197	0.233	0.257	0.279	0.307	0.326

For the table entries through $n = 11$, we used the exact distributions of $\sum d^2$ (Owen, 1962: 400–406). For $n = 12$, the exact distribution of de Jonge and van Montfort (1972) was used; and for n's of 13 through 16, we used the exact distributions of Otten (1973). For larger n, the Pearson curve approximations described by Olds (1938) were employed, the accuracy obtained being discussed elsewhere (Zar, 1972).

Examples:

$$(r_s)_{0.05(2),9} = 0.700 \quad \text{and} \quad (r_s)_{0.01(2),52} = 0.356.$$

For n larger than those in this table, we may utilize either Table B.16 or, equivalently, Equation 19.3. The accuracy of this procedure is discussed by Zar (1972).

TABLE B.20 CRITICAL VALUES OF g_1

$\alpha(2)$:	0.20	0.10	0.05	0.02	0.01	0.002
$\alpha(1)$:	0.10	0.05	0.025	0.01	0.005	0.001
n						
4	0.831	0.987	1.070	1.120	1.137	1.151
5	0.821	1.049	1.207	1.337	1.396	1.464
6	0.795	1.042	1.239	1.429	1.531	1.671
7	0.782	1.018	1.230	1.457	1.589	1.797
8	0.765	0.998	1.208	1.452	1.605	1.866
9	0.746	0.977	1.184	1.433	1.598	1.898
10	0.728	0.954	1.159	1.407	1.578	1.906
11	0.710	0.931	1.134	1.381	1.553	1.899
12	0.693	0.910	1.109	1.353	1.526	1.882
13	0.677	0.890	1.085	1.325	1.497	1.859
14	0.662	0.870	1.061	1.298	1.468	1.832
15	0.648	0.851	1.039	1.272	1.440	1.803
16	0.635	0.834	1.018	1.247	1.412	1.773
17	0.622	0.817	0.997	1.222	1.385	1.744
18	0.610	0.801	0.879	1.199	1.359	1.714
19	0.599	0.786	0.960	1.176	1.334	1.685
20	0.588	0.772	0.942	1.155	1.310	1.657
21	0.578	0.758	0.925	1.134	1.287	1.628
22	0.568	0.746	0.909	1.114	1.265	1.602
23	0.559	0.733	0.894	1.096	1.243	1.575
24	0.550	0.722	0.880	1.078	1.223	1.550
25	0.542	0.710	0.866	1.060	1.203	1.526
26	0.534	0.700	0.853	1.043	1.182	1.498
27	0.526	0.690	0.840	1.028	1.164	1.475
28	0.519	0.680	0.828	1.013	1.147	1.453
29	0.512	0.671	0.817	0.998	1.131	1.432
30	0.505	0.662	0.806	0.985	1.115	1.411
31	0.499	0.653	0.795	0.971	1.100	1.392
32	0.492	0.645	0.785	0.959	1.086	1.373
33	0.486	0.637	0.775	0.946	1.071	1.354
34	0.480	0.629	0.765	0.934	1.058	1.336
35	0.475	0.621	0.756	0.923	1.045	1.319
36	0.469	0.614	0.747	0.912	1.032	1.302
37	0.464	0.607	0.738	0.901	1.019	1.286
38	0.459	0.600	0.730	0.891	1.007	1.271
39	0.454	0.594	0.722	0.881	0.996	1.256
40	0.449	0.588	0.714	0.871	0.985	1.241
41	0.445	0.581	0.707	0.861	0.974	1.227
42	0.440	0.576	0.699	0.852	0.963	1.213
43	0.436	0.570	0.692	0.843	0.953	1.200
44	0.432	0.564	0.685	0.835	0.943	1.186
45	0.428	0.559	0.678	0.826	0.934	1.174
46	0.424	0.553	0.672	0.818	0.924	1.162
47	0.420	0.548	0.666	0.810	0.915	1.150
48	0.416	0.543	0.659	0.803	0.906	1.138
49	0.412	0.538	0.653	0.795	0.898	1.127
50	0.409	0.534	0.648	0.788	0.889	1.116

TABLE B.20 (cont.) CRITICAL VALUES OF g_1

α(2):	0.20	0.10	0.05	0.02	0.01	0.002
α(1):	0.10	0.05	0.025	0.01	0.005	0.001
n						
51	0.405	0.529	0.642	0.781	0.881	1.105
52	0.402	0.525	0.636	0.774	0.873	1.095
53	0.399	0.520	0.631	0.767	0.865	1.084
54	0.395	0.516	0.626	0.760	0.858	1.075
55	0.392	0.512	0.620	0.754	0.850	1.065
56	0.389	0.508	0.615	0.748	0.843	1.055
57	0.386	0.504	0.610	0.742	0.836	1.046
58	0.383	0.500	0.606	0.736	0.829	1.037
59	0.380	0.496	0.601	0.730	0.822	1.028
60	0.378	0.492	0.596	0.724	0.816	1.020
61	0.375	0.489	0.592	0.718	0.809	1.011
62	0.372	0.485	0.588	0.713	0.803	1.003
63	0.370	0.482	0.583	0.708	0.797	0.995
64	0.367	0.478	0.579	0.702	0.791	0.987
65	0.365	0.475	0.575	0.697	0.785	0.980
66	0.362	0.472	0.571	0.692	0.779	0.972
67	0.360	0.468	0.567	0.687	0.774	0.965
68	0.357	0.465	0.563	0.683	0.768	0.958
69	0.355	0.462	0.559	0.678	0.763	0.951
70	0.353	0.459	0.556	0.673	0.758	0.944
71	0.351	0.456	0.552	0.669	0.752	0.937
72	0.348	0.453	0.548	0.664	0.747	0.930
73	0.346	0.451	0.545	0.660	0.742	0.924
74	0.344	0.448	0.541	0.656	0.737	0.918
75	0.342	0.445	0.538	0.651	0.733	0.911
76	0.340	0.442	0.535	0.647	0.728	0.905
77	0.338	0.440	0.532	0.643	0.723	0.899
78	0.336	0.437	0.528	0.639	0.719	0.893
79	0.334	0.435	0.525	0.635	0.714	0.888
80	0.332	0.432	0.522	0.632	0.710	0.882
81	0.330	0.430	0.519	0.628	0.706	0.876
82	0.329	0.427	0.516	0.624	0.701	0.871
83	0.327	0.425	0.513	0.621	0.697	0.865
84	0.325	0.422	0.510	0.617	0.693	0.860
85	0.323	0.420	0.507	0.613	0.689	0.855
86	0.322	0.418	0.505	0.610	0.685	0.850
87	0.320	0.416	0.502	0.607	0.681	0.845
88	0.318	0.413	0.499	0.603	0.677	0.840
89	0.317	0.411	0.497	0.600	0.674	0.835
90	0.315	0.409	0.494	0.597	0.670	0.830
91	0.313	0.407	0.491	0.594	0.666	0.825
92	0.312	0.405	0.489	0.590	0.663	0.821
93	0.310	0.403	0.486	0.587	0.659	0.816
94	0.309	0.401	0.484	0.584	0.656	0.812
95	0.307	0.399	0.481	0.581	0.652	0.807

TABLE B.20 (cont.) CRITICAL VALUES OF g_1

$\alpha(2)$:	0.20	0.10	0.05	0.02	0.01	0.002
$\alpha(1)$:	0.10	0.05	0.025	0.01	0.005	0.001
n						
96	0.306	0.397	0.479	0.578	0.649	0.803
97	0.304	0.395	0.477	0.575	0.646	0.799
98	0.303	0.393	0.474	0.573	0.642	0.795
99	0.302	0.391	0.472	0.570	0.639	0.790
100	0.300	0.390	0.470	0.567	0.636	0.786
102	0.297	0.386	0.465	0.562	0.630	0.778
104	0.295	0.383	0.461	0.556	0.624	0.771
106	0.292	0.379	0.457	0.551	0.618	0.763
108	0.290	0.376	0.453	0.546	0.612	0.756
110	0.287	0.373	0.449	0.541	0.607	0.749
112	0.285	0.369	0.445	0.536	0.601	0.742
114	0.283	0.366	0.441	0.532	0.596	0.735
116	0.280	0.363	0.438	0.527	0.591	0.728
118	0.278	0.360	0.434	0.523	0.586	0.722
120	0.276	0.358	0.431	0.519	0.581	0.716
122	0.274	0.355	0.427	0.514	0.576	0.709
124	0.272	0.352	0.424	0.510	0.571	0.704
126	0.270	0.349	0.421	0.506	0.567	0.698
128	0.268	0.347	0.417	0.502	0.562	0.692
130	0.266	0.344	0.414	0.499	0.558	0.686
132	0.264	0.342	0.411	0.495	0.554	0.681
134	0.262	0.339	0.408	0.491	0.550	0.676
136	0.260	0.337	0.405	0.488	0.546	0.671
138	0.258	0.335	0.403	0.484	0.542	0.666
140	0.257	0.332	0.400	0.481	0.538	0.661
142	0.255	0.330	0.397	0.477	0.534	0.656
144	0.253	0.328	0.394	0.474	0.530	0.651
146	0.252	0.326	0.392	0.471	0.526	0.646
148	0.250	0.324	0.389	0.468	0.523	0.642
150	0.249	0.322	0.387	0.465	0.519	0.637
155	0.245	0.317	0.381	0.457	0.511	0.627
160	0.241	0.312	0.375	0.450	0.503	0.616
165	0.238	0.307	0.369	0.443	0.495	0.607
170	0.234	0.303	0.364	0.437	0.488	0.597
175	0.231	0.299	0.359	0.430	0.481	0.588
180	0.228	0.295	0.354	0.425	0.474	0.580
185	0.225	0.291	0.349	0.419	0.467	0.572
190	0.222	0.287	0.345	0.413	0.461	0.564
195	0.219	0.284	0.340	0.408	0.455	0.556
200	0.217	0.280	0.336	0.403	0.449	0.549
210	0.212	0.274	0.328	0.393	0.439	0.535
220	0.207	0.267	0.321	0.384	0.428	0.523
230	0.203	0.262	0.314	0.376	0.419	0.511
240	0.199	0.256	0.307	0.368	0.410	0.500
250	0.195	0.251	0.301	0.361	0.402	0.489

TABLE B.20 (cont.) CRITICAL VALUES OF g_1

α(2):	0.20	0.10	0.05	0.02	0.01	0.002
α(1):	0.10	0.05	0.025	0.01	0.005	0.001
n						
275	0.186	0.240	0.287	0.344	0.383	0.466
300	0.178	0.230	0.275	0.329	0.366	0.445
325	0.172	0.221	0.265	0.316	0.352	0.427
350	0.165	0.213	0.255	0.305	0.339	0.411
375	0.160	0.206	0.247	0.294	0.327	0.397
400	0.155	0.200	0.239	0.285	0.317	0.384
425	0.151	0.194	0.232	0.277	0.307	0.372
450	0.146	0.188	0.225	0.269	0.299	0.362
475	0.143	0.184	0.219	0.262	0.291	0.352
500	0.139	0.179	0.214	0.255	0.283	0.343
550	0.133	0.171	0.204	0.243	0.270	0.326
600	0.127	0.164	0.195	0.233	0.258	0.312
650	0.122	0.157	0.188	0.224	0.248	0.300
700	0.118	0.152	0.181	0.216	0.239	0.289
750	0.114	0.146	0.175	0.208	0.231	0.279
800	0.110	0.142	0.169	0.202	0.224	0.270
850	0.107	0.138	0.164	0.196	0.217	0.262
900	0.104	0.134	0.160	0.190	0.211	0.254
950	0.101	0.130	0.155	0.185	0.205	0.247
1000	0.099	0.127	0.152	0.180	0.200	0.241
1200	0.090	0.116	0.138	0.165	0.182	0.220
1400	0.084	0.107	0.128	0.152	0.169	0.203
1600	0.078	0.101	0.120	0.143	0.158	0.190
1800	0.074	0.095	0.113	0.134	0.149	0.179
2000	0.070	0.090	0.107	0.127	0.141	0.170
2500	0.063	0.080	0.096	0.114	0.126	0.152
3000	0.057	0.073	0.088	0.104	0.115	0.139
3500	0.053	0.068	0.081	0.096	0.107	0.128
4000	0.050	0.064	0.076	0.090	0.100	0.120
4500	0.047	0.060	0.072	0.085	0.094	0.113
5000	0.044	0.057	0.068	0.081	0.089	0.107
10000	0.031	0.040	0.048	0.057	0.063	0.076

Values in this table for $n < 25$ were taken with permission of the publisher from Table 2 of Mulholland (1977, Biometrika 64: 401–409). Other values were prepared using the approximating function described by D'Agostino (1970). Comparing these values to those for $\alpha(1) = 0.05$ and 0.01 of Pearson and Hartley (1966: 207), we find the discrepancy is never greater than 1 in the last decimal place (i.e., relative error is never > 0.14%).

Examples:

$$(g_1)_{0.05(2),110} = 0.449 \quad \text{and} \quad (g_1)_{0.01(2),600} = 0.258.$$

TABLE B.21 CRITICAL VALUES OF *D* FOR THE KOLMOGOROV-SMIRNOV GOODNESS OF FIT OF A NORMAL DISTRIBUTION

n	α = 0.15	0.10	0.05	0.025	0.01
1	0.421	0.445	0.486	0.519	0.562
2	0.386	0.408	0.446	0.476	0.516
3	0.350	0.370	0.404	0.432	0.468
4	0.321	0.339	0.371	0.395	0.429
5	0.297	0.314	0.343	0.366	0.397
6	0.278	0.294	0.321	0.343	0.371
7	0.262	0.277	0.303	0.323	0.350
8	0.248	0.263	0.287	0.306	0.332
9	0.237	0.250	0.273	0.292	0.316
10	0.227	0.239	0.262	0.279	0.303
11	0.218	0.230	0.251	0.268	0.290
12	0.209	0.221	0.242	0.258	0.280
13	0.202	0.214	0.234	0.249	0.270
14	0.196	0.207	0.226	0.241	0.261
15	0.190	0.201	0.219	0.234	0.254
16	0.184	0.195	0.213	0.227	0.246
17	0.179	0.190	0.207	0.221	0.240
18	0.175	0.185	0.202	0.215	0.233
19	0.171	0.180	0.197	0.210	0.228
20	0.167	0.176	0.192	0.205	0.222
21	0.163	0.172	0.188	0.201	0.218
22	0.159	0.168	0.184	0.196	0.213
23	0.156	0.165	0.180	0.192	0.209
24	0.153	0.162	0.177	0.189	0.204
25	0.150	0.159	0.173	0.185	0.201
26	0.147	0.156	0.170	0.182	0.197
27	0.145	0.153	0.167	0.179	0.193
28	0.142	0.150	0.164	0.175	0.190
29	0.140	0.148	0.162	0.173	0.187
30	0.138	0.146	0.159	0.170	0.184
31	0.136	0.143	0.157	0.167	0.181
32	0.134	0.141	0.154	0.165	0.179
33	0.132	0.139	0.152	0.162	0.176
34	0.130	0.137	0.150	0.160	0.173
35	0.128	0.135	0.148	0.158	0.171
36	0.126	0.134	0.146	0.156	0.169
37	0.125	0.132	0.144	0.154	0.167
38	0.123	0.130	0.142	0.152	0.164
39	0.122	0.129	0.140	0.150	0.162
40	0.120	0.127	0.139	0.148	0.160
41	0.119	0.126	0.137	0.146	0.159
42	0.117	0.124	0.136	0.145	0.157
43	0.116	0.123	0.134	0.143	0.155
44	0.115	0.121	0.133	0.141	0.153
45	0.114	0.120	0.131	0.140	0.152
46	0.112	0.119	0.130	0.138	0.150
47	0.111	0.118	0.128	0.137	0.149
48	0.110	0.116	0.127	0.136	0.147
49	0.109	0.115	0.126	0.134	0.146
50	0.108	0.114	0.125	0.133	0.144

TABLE B.21 (cont.) CRITICAL VALUES OF *D* FOR THE KOLMOGOROV-SMIRNOV GOODNESS OF FIT OF A NORMAL DISTRIBUTION

n	α = 0.15	0.10	0.05	0.025	0.01
52	0.106	0.112	0.122	0.130	0.141
54	0.104	0.110	0.120	0.128	0.139
56	0.102	0.108	0.118	0.126	0.136
58	0.100	0.106	0.116	0.124	0.134
60	0.099	0.104	0.114	0.122	0.132
62	0.097	0.103	0.112	0.120	0.130
64	0.096	0.101	0.111	0.118	0.128
66	0.094	0.100	0.109	0.116	0.126
68	0.093	0.098	0.107	0.115	0.124
70	0.092	0.097	0.106	0.113	0.122
72	0.090	0.096	0.104	0.111	0.121
74	0.089	0.094	0.103	0.110	0.119
76	0.088	0.093	0.102	0.108	0.118
78	0.087	0.092	0.100	0.107	0.116
80	0.086	0.091	0.099	0.106	0.115
82	0.085	0.090	0.098	0.104	0.113
84	0.084	0.089	0.097	0.103	0.112
86	0.083	0.088	0.096	0.102	0.111
88	0.082	0.087	0.095	0.101	0.109
90	0.081	0.086	0.094	0.100	0.108
92	0.080	0.085	0.093	0.099	0.107
94	0.079	0.084	0.092	0.098	0.106
96	0.078	0.083	0.091	0.097	0.105
98	0.078	0.082	0.090	0.096	0.104
100	0.077	0.081	0.089	0.095	0.103
105	0.075	0.079	0.087	0.093	0.100
110	0.073	0.078	0.085	0.090	0.098
115	0.072	0.076	0.083	0.088	0.096
120	0.070	0.074	0.081	0.087	0.094
125	0.069	0.073	0.080	0.085	0.092
130	0.068	0.071	0.078	0.083	0.090
135	0.066	0.070	0.077	0.082	0.089
140	0.065	0.069	0.075	0.080	0.087
145	0.064	0.068	0.074	0.079	0.086
150	0.063	0.067	0.073	0.078	0.084

These values were determined by the Stephens' (1974) equation for testing normality with D when the population mean and variance are unknown.

Example:

$$D_{0.05,70} = 0.106.$$

TABLE B.22 CRITICAL VALUES OF D'AGOSTINO'S D FOR NORMALITY TESTING

n	$\alpha = 0.20$	0.10	0.05	0.02	0.01
10	0.2632, 0.2835	0.2573, 0.2843	0.2513, 0.2849	0.2436, 0.2855	0.2379, 0.2857
12	0.2653, 0.2841	0.2598, 0.2849	0.2544, 0.2854	0.2473, 0.2859	0.2420, 0.2862
14	0.2669, 0.2846	0.2618, 0.2853	0.2568, 0.2858	0.2503, 0.2862	0.2455, 0.2865
16	0.2681, 0.2848	0.2634, 0.2855	0.2587, 0.2860	0.2527, 0.2865	0.2482, 0.2867
18	0.2690, 0.2850	0.2646, 0.2855	0.2603, 0.2862	0.2547, 0.2866	0.2505, 0.2868
20	0.2699, 0.2852	0.2657, 0.2857	0.2617, 0.2863	0.2564, 0.2867	0.2525, 0.2869
22	0.2705, 0.2853	0.2670, 0.2859	0.2629, 0.2864	0.2579, 0.2869	0.2542, 0.2870
24	0.2711, 0.2853	0.2675, 0.2860	0.2638, 0.2865	0.2591, 0.2870	0.2557, 0.2871
26	0.2717, 0.2854	0.2682, 0.2861	0.2647, 0.2866	0.2603, 0.2870	0.2570, 0.2872
28	0.2721, 0.2854	0.2688, 0.2861	0.2655, 0.2866	0.2612, 0.2870	0.2581, 0.2873
30	0.2725, 0.2854	0.2693, 0.2861	0.2662, 0.2866	0.2622, 0.2871	0.2592, 0.2872
32	0.2729, 0.2854	0.2698, 0.2862	0.2668, 0.2867	0.2630, 0.2871	0.2600, 0.2873
34	0.2732, 0.2854	0.2703, 0.2862	0.2674, 0.2867	0.2636, 0.2871	0.2609, 0.2873
36	0.2735, 0.2854	0.2707, 0.2862	0.2679, 0.2867	0.2643, 0.2871	0.2617, 0.2873
38	0.2738, 0.2854	0.2710, 0.2862	0.2683, 0.2867	0.2649, 0.2871	0.2623, 0.2873
40	0.2740, 0.2854	0.2714, 0.2862	0.2688, 0.2867	0.2655, 0.2871	0.2630, 0.2874
42	0.2743, 0.2854	0.2717, 0.2861	0.2691, 0.2867	0.2659, 0.2871	0.2636, 0.2874
44	0.2745, 0.2854	0.2720, 0.2861	0.2695, 0.2867	0.2664, 0.2871	0.2641, 0.2874
46	0.2747, 0.2854	0.2722, 0.2861	0.2698, 0.2866	0.2668, 0.2871	0.2646, 0.2874
48	0.2749, 0.2854	0.2725, 0.2861	0.2702, 0.2866	0.2672, 0.2871	0.2651, 0.2874
50	0.2751, 0.2853	0.2727, 0.2861	0.2705, 0.2866	0.2676, 0.2871	0.2655, 0.2874
60	0.2757, 0.2852	0.2737, 0.2860	0.2717, 0.2865	0.2692, 0.2870	0.2673, 0.2873
70	0.2763, 0.2851	0.2744, 0.2859	0.2726, 0.2864	0.2708, 0.2869	0.2687, 0.2872
80	0.2768, 0.2850	0.2750, 0.2857	0.2734, 0.2863	0.2713, 0.2868	0.2698, 0.2871
90	0.2771, 0.2849	0.2755, 0.2856	0.2740, 0.2862	0.2721, 0.2866	0.2707, 0.2870
100	0.2774, 0.2849	0.2759, 0.2855	0.2745, 0.2860	0.2727, 0.2865	0.2714, 0.2869
120	0.2779, 0.2847	0.2765, 0.2853	0.2752, 0.2858	0.2737, 0.2863	0.2725, 0.2866
140	0.2782, 0.2846	0.2770, 0.2852	0.2758, 0.2856	0.2744, 0.2862	0.2734, 0.2865
160	0.2785, 0.2845	0.2774, 0.2851	0.2763, 0.2855	0.2750, 0.2860	0.2741, 0.2863
180	0.2787, 0.2844	0.2777, 0.2850	0.2767, 0.2854	0.2755, 0.2859	0.2746, 0.2862
200	0.2789, 0.2843	0.2779, 0.2848	0.2770, 0.2853	0.2759, 0.2857	0.2751, 0.2860
250	0.2793, 0.2841	0.2784, 0.2846	0.2776, 0.2850	0.2767, 0.2855	0.2760, 0.2858
300	0.2796, 0.2840	0.2788, 0.2844	0.2781, 0.2848	0.2772, 0.2853	0.2766, 0.2855
350	0.2798, 0.2839	0.2791, 0.2843	0.2784, 0.2847	0.2776, 0.2851	0.2771, 0.2853
400	0.2799, 0.2838	0.2793, 0.2842	0.2787, 0.2845	0.2780, 0.2849	0.2775, 0.2852
450	0.2801, 0.2837	0.2795, 0.2841	0.2789, 0.2844	0.2782, 0.2848	0.2778, 0.2851
500	0.2802, 0.2836	0.2796, 0.2840	0.2791, 0.2843	0.2785, 0.2847	0.2780, 0.2849
600	0.2804, 0.2835	0.2799, 0.2839	0.2794, 0.2842	0.2788, 0.2845	0.2784, 0.2847
700	0.2805, 0.2834	0.2800, 0.2838	0.2796, 0.2840	0.2791, 0.2844	0.2787, 0.2846
800	0.2806, 0.2833	0.2802, 0.2837	0.2798, 0.2839	0.2793, 0.2842	0.2790, 0.2844
900	0.2807, 0.2833	0.2803, 0.2836	0.2799, 0.2838	0.2795, 0.2841	0.2792, 0.2843
1000	0.2808, 0.2832	0.2804, 0.2835	0.2800, 0.2838	0.2796, 0.2840	0.2793, 0.2842
1250	0.2809, 0.2831	0.2806, 0.2834	0.2803, 0.2836	0.2799, 0.2839	0.2797, 0.2840
1500	0.2810, 0.2830	0.2807, 0.2833	0.2805, 0.2835	0.2801, 0.2837	0.2799, 0.2839
1750	0.2811, 0.2830	0.2808, 0.2832	0.2806, 0.2834	0.2803, 0.2836	0.2801, 0.2838
2000	0.2812, 0.2829	0.2809, 0.2831	0.2807, 0.2833	0.2804, 0.2835	0.2802, 0.2837

For each significance level, α is given a pair of critical values. If the calculated D is $\leq$ the first member of the pair, or $\geq$ the second, then, the null hypothesis of population normality is rejected.

Table B.22 appears through the courtesy of its author, Ralph B. D'Agostino (1971b).

Example:

$$D_{0.05,120} = 0.2752 \quad \text{and} \quad 0.2858.$$

TABLE B.23 THE ARCSINE TRANSFORMATION

X	0	1	2	3	4	5	6	7	8	9	X
0.000	0.00	0.57	0.81	0.99	1.15	1.28	1.40	1.52	1.62	1.72	0.000
0.001	1.81	1.90	1.99	2.07	2.14	2.22	2.29	2.36	2.43	2.50	0.001
0.002	2.56	2.63	2.69	2.75	2.81	2.87	2.92	2.98	3.03	3.09	0.002
0.003	3.14	3.19	3.24	3.29	3.34	3.39	3.44	3.49	3.53	3.58	0.003
0.004	3.63	3.67	3.72	3.76	3.80	3.85	3.89	3.93	3.97	4.01	0.004
0.005	4.05	4.10	4.14	4.17	4.21	4.25	4.29	4.33	4.37	4.41	0.005
0.006	4.44	4.48	4.52	4.55	4.59	4.62	4.66	4.70	4.73	4.76	0.006
0.007	4.80	4.83	4.87	4.90	4.93	4.97	5.00	5.03	5.07	5.10	0.007
0.008	5.13	5.16	5.20	5.23	5.26	5.29	5.32	5.35	5.38	5.41	0.008
0.009	5.44	5.47	5.50	5.53	5.56	5.59	5.62	5.65	5.68	5.71	0.009
0.01	5.74	6.02	6.29	6.55	6.80	7.03	7.27	7.49	7.71	7.92	0.01
0.02	8.13	8.33	8.53	8.72	8.91	9.10	9.28	9.46	9.63	9.80	0.02
0.03	9.97	10.14	10.30	10.47	10.63	10.78	10.94	11.09	11.24	11.39	0.03
0.04	11.54	11.68	11.83	11.97	12.11	12.25	12.38	12.52	12.66	12.79	0.04
0.05	12.92	13.05	13.18	13.31	13.44	13.56	13.69	13.81	13.94	14.06	0.05
0.06	14.18	14.30	14.42	14.54	14.65	14.77	14.89	15.00	15.12	15.23	0.06
0.07	15.34	15.45	15.56	15.68	15.79	15.89	16.00	16.11	16.22	16.32	0.07
0.08	16.43	16.54	16.64	16.74	16.85	16.95	17.05	17.15	17.26	17.36	0.08
0.09	17.46	17.56	17.66	17.76	17.85	17.95	18.05	18.15	18.24	18.34	0.09
0.10	18.43	18.53	18.63	18 72	18.81	18.91	19.00	19.09	19.19	19.28	0.10
0.11	19.37	19.46	19.55	19.64	19.73	19.82	19.91	20.00	20.09	20.18	0.11
0.12	20.27	20.36	20.44	20.53	20.62	20.70	20.79	20.88	20.96	21.05	0.12
0.13	21.13	21.22	21.30	21.39	21.47	21.56	21.64	21.72	21.81	21.89	0.13
0.14	21.97	22.06	22.14	22.22	22.30	22.38	22.46	22.54	22.63	22.71	0.14
0.15	22.79	22.87	22.95	23.03	23.11	23.18	23.26	23.34	23.42	23.50	0.15
0.16	23.58	23.66	23.73	23.81	23.89	23.97	24.04	24.12	24.20	24.27	0.16
0.17	24.35	24.43	24.50	24.58	24.65	24.73	24.80	24.88	24.95	25.03	0.17
0.18	25.10	25.18	25.25	25.33	25.40	25.47	25.55	25.62	25.70	25.77	0.18
0.19	25.84	25.91	25.99	26.06	26.13	26.21	26.28	26.35	26.42	26.49	0.19
0.20	26.57	26.64	26.71	26.78	26.85	26.92	26.99	27.06	27.13	27.20	0.20
0.21	27.27	27.35	27.42	27.49	27.56	27.62	27.69	27.76	27.83	27.90	0.21
0.22	27.97	28.04	28.11	28.18	28.25	28.32	28.39	28.45	28.52	28.59	0.22
0.23	28.66	28.73	28.79	28.86	28.93	29.00	29.06	29.13	29.20	29.27	0.23
0.24	29.33	29.40	29.47	29.53	29.60	29.67	29.73	29.80	29.87	29.93	0.24
0.25	30.00	30.07	30.13	30.20	30.26	30.33	30.40	30.46	30.53	30.59	0.25
0.26	30.66	30.72	30.79	30.85	30.92	30.98	31.05	31.11	31.18	31.24	0.26
0.27	31.31	31.37	31.44	31.50	31.56	31.63	31.69	31.76	31.82	31.88	0.27
0.28	31.95	32.01	32.08	32.14	32.20	32.27	32.33	32.39	32.46	32.52	0.28
0.29	32.58	32.65	32.71	32.77	32.83	32.90	32.96	33.02	33.09	33.15	0.29
0.30	33.21	33.27	33.34	33.40	33.46	33.52	33.58	33.65	33.71	33.77	0.30

TABLE B.23 (cont.) THE ARCSINE TRANSFORMATION

X	0	1	2	3	4	5	6	7	8	9	X
0.31	33.83	33.90	33.96	34.02	34.08	34.14	34.20	34.27	34.33	34.39	0.31
0.32	34.45	34.51	34.57	34.63	34.70	34.76	34.82	34.88	34.94	35.00	0.32
0.33	35.06	35.12	35.18	35.24	35.30	35.37	35.43	35.49	35.55	35.61	0.33
0.34	35.67	35.73	35.79	35.85	35.91	35.97	36.03	36.09	36.15	36.21	0.34
0.35	36.27	36.33	36.39	36.45	36.51	36.57	36.63	36.69	36.75	36.81	0.35
0.36	36.87	36.93	36.99	37.05	37.11	37.17	37.23	37.29	37.35	37.41	0.36
0.37	37.46	37.52	37.58	37.64	37.70	37.76	37.82	37.88	37.94	38.00	0.37
0.38	38.06	38.12	38.17	38.23	38.29	38.35	38.41	38.47	38.53	38.59	0.38
0.39	38.65	38.70	38.76	38.82	38.88	38.94	39.00	39.06	39.11	39.17	0.39
0.40	39.23	39.29	39.35	39.41	39.47	39.52	39.58	39.64	39.70	39.76	0.40
0.41	39.82	39.87	39.93	39.99	40.05	40.11	40.16	40.22	40.28	40.34	0.41
0.42	40.40	40.45	40.51	40.57	40.63	40.69	40.74	40.80	40.86	40.92	0.42
0.43	40.98	41.03	41.09	41.15	41.21	41.27	41.32	41.38	41.44	41.50	0.43
0.44	41.55	41.61	41.67	41.73	41.78	41.84	41.90	41.96	42.02	42.07	0.44
0.45	42.13	42.19	42.25	42.30	42.36	42.42	42.48	42.53	42.59	42.65	0.45
0.46	42.71	42.76	42.82	42.88	42.94	42.99	43.05	43.11	43.17	43.22	0.46
0.47	43.28	43.34	43.39	43.45	43.51	43.57	43.62	43.68	43.74	43.80	0.47
0.48	43.85	43.91	43.97	44.03	44.08	44.14	44.20	44.26	44.31	44.37	0.48
0.49	44.43	44.48	44.54	44.60	44.66	44.71	44.77	44.83	44.89	44.94	0.49
0.50	45.00	45.06	45.11	45.17	45.23	45.29	45.34	45.40	45.46	45.52	0.50
0.51	45.57	45.63	45.69	45.74	45.80	45.86	45.92	45.97	46.03	46.09	0.51
0.52	46.15	46.20	46.26	46.32	46.38	46.43	46.49	46.55	46.61	46.66	0.52
0.53	46.72	46.78	46.83	46.89	46.95	47.01	47.06	47.12	47.18	47.24	0.53
0.54	47.29	47.35	47.41	47.47	47.52	47.58	47.64	47.70	47.75	47.81	0.54
0.55	47.87	47.93	47.98	48.04	48.10	48.16	48.22	48.27	48.33	48.39	0.55
0.56	48.45	48.50	48.56	48.62	48.68	48.73	48.79	48.85	48.91	48.97	0.56
0.57	49.02	49.08	49.14	49.20	49.26	49.31	49.37	49.43	49.49	49.55	0.57
0.58	49.60	49.66	49.72	49.78	49.84	49.89	49.95	50.01	50.07	50.13	0.58
0.59	50.18	50.24	50.30	50.36	50.42	50.48	50.53	50.59	50.65	50.71	0.59
0.60	50.77	50.83	50.89	50.94	51.00	51.06	51.12	51.18	51.24	51.30	0.60
0.61	51.35	51.41	51.47	51.53	51.59	51.65	51.71	51.77	51.83	51.88	0.61
0.62	51.94	52.00	52.06	52.12	52.18	52.24	52.30	52.36	52.42	52.48	0.62
0.63	52.54	52.59	52.65	52.71	52.77	52.83	52.89	52.95	53.01	53.07	0.63
0.64	53.13	53.19	53.25	53.31	53.37	53.43	53.49	53.55	53.61	53.67	0.64
0.65	53.73	53.79	53.85	53.91	53.97	54.03	54.09	54.15	54.21	54.27	0.65
0.66	54.33	54.39	54.45	54.51	54.57	54.63	54.70	54.76	54.82	54.88	0.66
0.67	54.94	55.00	55.06	55.12	55.18	55.24	55.30	55.37	55.43	55.49	0.67
0.68	55.55	55.61	55.67	55.73	55.80	55.86	55.92	55.98	56.04	56.10	0.68
0.69	56.17	56.23	56.29	56.35	56.42	56.48	56.54	56.60	56.66	56.73	0.69
0.70	56.79	56.85	56.91	56.98	57.04	57.10	57.17	57.23	57.29	57.35	0.70

TABLE B.23 (cont.) THE ARCSINE TRANSFORMATION

X	0	1	2	3	4	5	6	7	8	9	X
0.71	57.42	57.48	57.54	57.61	57.67	57.73	57.80	57.86	57.92	57.99	0.71
0.72	58.05	58.12	58.18	58.24	58.31	58.37	58.44	58.50	58.56	58.63	0.72
0.73	58.69	58.76	58.82	58.89	58.95	59.02	59.08	59.15	59.21	59.28	0.73
0.74	59.34	59.41	59.47	59.54	59.60	59.67	59.74	59.80	59.87	59.93	0.74
0.75	60.00	60.07	60.13	60.20	60.27	60.33	60.40	60.47	60.53	60.60	0.75
0.76	60.67	60.73	60.80	60.87	60.94	61.00	61.07	61.14	61.21	61.27	0.76
0.77	61.34	61.41	61.48	61.55	61.61	61.68	61.75	61.82	61.89	61.96	0.77
0.78	62.03	62.10	62.17	62.24	62.31	62.38	62.44	62.51	62.58	62.65	0.78
0.79	62.73	62.80	62.87	62.94	63.01	63.08	63.15	63.22	63.29	63.36	0.79
0.80	63.43	63.51	63.58	63.65	63.72	63.79	63.87	63.94	64.01	64.09	0.80
0.81	64.16	64.23	64.30	64.38	64.45	64.53	64.60	64.67	64.75	64.82	0.81
0.82	64.90	64.97	65.05	65.12	65.20	65.27	65.35	65.42	65.50	65.57	0.82
0.83	65.65	65.73	65.80	65.88	65.96	66.03	66.11	66.19	66.27	66.34	0.83
0.84	66.42	66.50	66.58	66.66	66.74	66.82	66.89	66.97	67.05	67.13	0.84
0.85	67.21	67.29	67.37	67.46	67.54	67.62	67.70	67.78	67.86	67.94	0.85
0.86	68.03	68.11	68.19	68.28	68.36	68.44	68.53	68.61	68.70	68.78	0.86
0.87	68.87	68.95	69.04	69.12	69.21	69.30	69.38	69.47	69.56	69.64	0.87
0.88	69.73	69.82	69.91	70.00	70.09	70.18	70.27	70.36	70.45	70.54	0.88
0.89	70.63	70.72	70.81	70.91	71.00	71.09	71.19	71.28	71.37	71.47	0.89
0.90	71.57	71.66	71.76	71.85	71.95	72.05	72.15	72.24	72.34	72.44	0.90
0.91	72.54	72.64	72.74	72.85	72.95	73.05	73.15	73.26	73.36	73.46	0.91
0.92	73.57	73.68	73.78	73.89	74.00	74.11	74.21	74.32	74.44	74.55	0.92
0.93	74.66	74.77	74.88	75.00	75.11	75.23	75.35	75.46	75.58	75.70	0.93
0.94	75.82	75.94	76.06	76.19	76.31	76.44	76.56	76.69	76.82	76.95	0.94
0.95	77.08	77.21	77.34	77.48	77.62	77.75	77.89	78.03	78.17	78.32	0.95
0.96	78.46	78.61	78.76	78.91	79.06	79.22	79.37	79.53	79.70	79.86	0.96
0.97	80.03	80.20	80.37	80.54	80.72	80.90	81.09	81.28	81.47	81.67	0.97
0.98	81.87	82.08	82.29	82.51	82.73	82.97	83.20	83.45	83.71	83.98	0.98
0.990	84.26	84.29	84.32	84.35	84.38	84.41	84.44	84.47	84.50	84.53	0.990
0.991	84.56	84.59	84.62	84.65	84.68	84.71	84.74	84.77	84.80	84.84	0.991
0.992	84.87	84.90	84.93	84.97	85.00	85.03	85.07	85.10	85.13	85.17	0.992
0.993	85.20	85.24	85.27	85.30	85.34	85.38	85.41	85.45	85.48	85.52	0.993
0.994	85.56	85.59	85.63	85.67	85.71	85.75	85.79	85.83	85.86	85.90	0.994
0.995	85.95	85.99	86.03	86.07	86.11	86.15	86.20	86.24	86.28	86.33	0.995
0.996	86.37	86.42	86.47	86.51	86.56	86.61	86.66	86.71	86.76	86.81	0.996
0.997	86.86	86.91	86.97	87.02	87.08	87.13	87.19	87.25	87.31	87.37	0.997
0.998	87.44	87.50	87.57	87.64	87.71	87.78	87.86	87.93	88.01	88.10	0.998
0.999	88.19	88.28	88.38	88.48	88.60	88.72	88.85	89.01	89.19	89.43	0.999
1.000	90.00										

Examples:

$X = 0.712,\ X' = 57.54$ and $X = 0.9921,\ X' = 84.90.$

TABLE B.24 PROPORTIONS CORRESPONDING TO ARCSINE TRANSFORMATIONS, X'

X'	0	1	2	3	4	5	6	7	8	9	X'
0.0	0.0000	0.0000	0.0000	0.0000	0.0000	0.0001	0.0001	0.0001	0.0002	0.0002	0.0
1.0	0.0003	0.0004	0.0004	0.0005	0.0006	0.0007	0.0008	0.0009	0.0010	0.0011	1.0
2.0	0.0012	0.0013	0.0015	0.0016	0.0018	0.0019	0.0021	0.0022	0.0024	0.0026	2.0
3.0	0.0027	0.0029	0.0031	0.0033	0.0035	0.0037	0.0039	0.0042	0.0044	0.0046	3.0
4.0	0.0049	0.0051	0.0054	0.0056	0.0059	0.0062	0.0064	0.0067	0.0070	0.0073	4.0
5.0	0.0076	0.0079	0.0082	0.0085	0.0089	0.0092	0.0095	0.0099	0.0102	0.0106	5.0
6.0	0.0109	0.0113	0.0117	0.0120	0.0124	0.0128	0.0132	0.0136	0.0140	0.0144	6.0
7.0	0.0149	0.0153	0.0157	0.0161	0.0166	0.0170	0.0175	0.0180	0.0184	0.0189	7.0
8.0	0.0194	0.0199	0.0203	0.0208	0.0213	0.0218	0.0224	0.0229	0.0234	0.0239	8.0
9.0	0.0245	0.0250	0.0256	0.0261	0.0267	0.0272	0.0278	0.0284	0.0290	0.0296	9.0
10.0	0.0302	0.0308	0.0314	0.0320	0.0326	0.0332	0.0338	0.0345	0.0351	0.0358	10.0
11.0	0.0364	0.0371	0.0377	0.0384	0.0391	0.0397	0.0404	0.0411	0.0418	0.0425	11.0
12.0	0.0432	0.0439	0.0447	0.0454	0.0461	0.0468	0.0476	0.0483	0.0491	0.0498	12.0
13.0	0.0506	0.0514	0.0521	0.0529	0.0537	0.0545	0.0553	0.0561	0.0569	0.0577	13.0
14.0	0.0585	0.0593	0.0602	0.0610	0.0618	0.0627	0.0635	0.0644	0.0653	0.0661	14.0
15.0	0.0670	0.0679	0.0687	0.0696	0.0705	0.0714	0.0723	0.0732	0.0741	0.0751	15.0
16.0	0.0760	0.0769	0.0778	0.0788	0.0797	0.0807	0.0816	0.0826	0.0835	0.0845	16.0
17.0	0.0855	0.0865	0.0874	0.0884	0.0894	0.0904	0.0914	0.0924	0.0934	0.0945	17.0
18.0	0.0955	0.0965	0.0976	0.0986	0.0996	0.1007	0.1017	0.1028	0.1039	0.1049	18.0
19.0	0.1060	0.1071	0.1082	0.1092	0.1103	0.1114	0.1125	0.1136	0.1147	0.1159	19.0
20.0	0.1170	0.1181	0.1192	0.1204	0.1215	0.1226	0.1238	0.1249	0.1261	0.1273	20.0
21.0	0.1284	0.1296	0.1308	0.1320	0.1331	0.1343	0.1355	0.1367	0.1379	0.1391	21.0
22.0	0.1403	0.1415	0.1428	0.1440	0.1452	0.1464	0.1477	0.1489	0.1502	0.1514	22.0
23.0	0.1527	0.1539	0.1552	0.1565	0.1577	0.1590	0.1603	0.1616	0.1628	0.1641	23.0
24.0	0.1654	0.1667	0.1680	0.1693	0.1707	0.1720	0.1733	0.1746	0.1759	0.1773	24.0
25.0	0.1786	0.1799	0.1813	0.1826	0.1840	0.1853	0.1867	0.1881	0.1894	0.1908	25.0
26.0	0.1922	0.1935	0.1949	0.1963	0.1977	0.1991	0.2005	0.2019	0.2033	0.2047	26.0
27.0	0.2061	0.2075	0.2089	0.2104	0.2118	0.2132	0.2146	0.2161	0.2175	0.2190	27.0
28.0	0.2204	0.2219	0.2233	0.2248	0.2262	0.2277	0.2291	0.2306	0.2321	0.2336	28.0
29.0	0.2350	0.2365	0.2380	0.2395	0.2410	0.2425	0.2440	0.2455	0.2470	0.2485	29.0
30.0	0.2500	0.2515	0.2530	0.2545	0.2561	0.2576	0.2591	0.2607	0.2622	0.2637	30.0
31.0	0.2653	0.2668	0.2684	0.2699	0.2715	0.2730	0.2746	0.2761	0.2777	0.2792	31.0
32.0	0.2808	0.2824	0.2840	0.2855	0.2871	0.2887	0.2903	0.2919	0.2934	0.2950	32.0
33.0	0.2966	0.2982	0.2998	0.3014	0.3030	0.3046	0.3062	0.3079	0.3095	0.3111	33.0
34.0	0.3127	0.3143	0.3159	0.3176	0.3192	0.3208	0.3224	0.3241	0.3257	0.3274	34.0
35.0	0.3290	0.3306	0.3323	0.3339	0.3356	0.3372	0.3389	0.3405	0.3422	0.3438	35.0
36.0	0.3455	0.3472	0.3488	0.3505	0.3521	0.3538	0.3555	0.3572	0.3588	0.3605	36.0
37.0	0.3622	0.3639	0.3655	0.3672	0.3689	0.3706	0.3723	0.3740	0.3757	0.3773	37.0
38.0	0.3790	0.3807	0.3824	0.3841	0.3858	0.3875	0.3892	0.3909	0.3926	0.3943	38.0
39.0	0.3960	0.3978	0.3995	0.4012	0.4029	0.4046	0.4063	0.4080	0.4097	0.4115	39.0
40.0	0.4132	0.4149	0.4166	0.4183	0.4201	0.4218	0.4235	0.4252	0.4270	0.4287	40.0
41.0	0.4304	0.4321	0.4339	0.4356	0.4373	0.4391	0.4408	0.4425	0.4443	0.4460	41.0
42.0	0.4477	0.4495	0.4512	0.4529	0.4547	0.4564	0.4582	0.4599	0.4616	0.4634	42.0
43.0	0.4651	0.4669	0.4686	0.4703	0.4721	0.4738	0.4756	0.4773	0.4791	0.4808	43.0
44.0	0.4826	0.4843	0.4860	0.4878	0.4895	0.4913	0.4930	0.4948	0.4965	0.4983	44.0

TABLE B.24 (cont.) PROPORTIONS CORRESPONDING TO ARCSINE TRANSFORMATIONS, X'

X'	0	1	2	3	4	5	6	7	8	9	X'
45.0	0.5000	0.5017	0.5035	0.5052	0.5070	0.5087	0.5105	0.5122	0.5140	0.5157	45.0
46.0	0.5174	0.5192	0.5209	0.5227	0.5244	0.5262	0.5279	0.5297	0.5314	0.5331	46.0
47.0	0.5349	0.5366	0.5384	0.5401	0.5418	0.5436	0.5453	0.5471	0.5488	0.5505	47.0
48.0	0.5523	0.5540	0.5557	0.5575	0.5592	0.5609	0.5627	0.5644	0.5661	0.5679	48.0
49.0	0.5696	0.5713	0.5730	0.5748	0.5765	0.5782	0.5799	0.5817	0.5834	0.5851	49.0
50.0	0.5868	0.5885	0.5903	0.5920	0.5937	0.5954	0.5971	0.5988	0.6005	0.6022	50.0
51.0	0.6040	0.6057	0.6074	0.6091	0.6108	0.6125	0.6142	0.6159	0.6176	0.6193	51.0
52.0	0.6210	0.6227	0.6243	0.6260	0.6277	0.6294	0.6311	0.6328	0.6345	0.6361	52.0
53.0	0.6378	0.6395	0.6412	0.6428	0.6445	0.6462	0.6479	0.6495	0.6512	0.6528	53.0
54.0	0.6545	0.6562	0.6578	0.6595	0.6611	0.6628	0.6644	0.6661	0.6677	0.6694	54.0
55.0	0.6710	0.6726	0.6743	0.6759	0.6776	0.6792	0.6808	0.6824	0.6841	0.6857	55.0
56.0	0.6873	0.6889	0.6905	0.6921	0.6938	0.6954	0.6970	0.6986	0.7002	0.7018	56.0
57.0	0.7034	0.7050	0.7066	0.7081	0.7097	0.7113	0.7129	0.7145	0.7160	0.7176	57.0
58.0	0.7192	0.7208	0.7223	0.7239	0.7254	0.7270	0.7285	0.7301	0.7316	0.7332	58.0
59.0	0.7347	0.7363	0.7378	0.7393	0.7409	0.7424	0.7439	0.7455	0.7470	0.7485	59.0
60.0	0.7500	0.7515	0.7530	0.7545	0.7560	0.7575	0.7590	0.7605	0.7620	0.7635	60.0
61.0	0.7650	0.7664	0.7679	0.7694	0.7709	0.7723	0.7738	0.7752	0.7767	0.7781	61.0
62.0	0.7796	0.7810	0.7825	0.7839	0.7854	0.7868	0.7882	0.7896	0.7911	0.7925	62.0
63.0	0.7939	0.7953	0.7967	0.7981	0.7995	0.8009	0.8023	0.8037	0.8051	0.8065	63.0
64.0	0.8078	0.8092	0.8106	0.8119	0.8133	0.8147	0.8160	0.8174	0.8187	0.8201	64.0
65.0	0.8214	0.8227	0.8241	0.8254	0.8267	0.8280	0.8293	0.8307	0.8320	0.8333	65.0
66.0	0.8346	0.8359	0.8372	0.8384	0.8397	0.8410	0.8423	0.8435	0.8448	0.8461	66.0
67.0	0.8473	0.8486	0.8498	0.8511	0.8523	0.8536	0.8548	0.8560	0.8572	0.8585	67.0
68.0	0.8597	0.8609	0.8621	0.8633	0.8645	0.8657	0.8669	0.8680	0.8692	0.8704	68.0
69.0	0.8716	0.8727	0.8739	0.8751	0.8762	0.8774	0.8785	0.8796	0.8808	0.8819	69.0
70.0	0.8830	0.8841	0.8853	0.8864	0.8875	0.8886	0.8897	0.8908	0.8918	0.8929	70.0
71.0	0.8940	0.8951	0.8961	0.8972	0.8983	0.8993	0.9004	0.9014	0.9024	0.9035	71.0
72.0	0.9045	0.9055	0.9066	0.9076	0.9086	0.9096	0.9106	0.9116	0.9126	0.9135	72.0
73.0	0.9145	0.9155	0.9165	0.9174	0.9184	0.9193	0.9203	0.9212	0.9222	0.9231	73.0
74.0	0.9240	0.9249	0.9259	0.9268	0.9277	0.9286	0.9295	0.9304	0.9313	0.9321	74.0
75.0	0.9330	0.9339	0.9347	0.9356	0.9365	0.9373	0.9382	0.9390	0.9398	0.9407	75.0
76.0	0.9415	0.9423	0.9431	0.9439	0.9447	0.9455	0.9463	0.9471	0.9479	0.9486	76.0
77.0	0.9494	0.9502	0.9509	0.9517	0.9524	0.9532	0.9539	0.9546	0.9553	0.9561	77.0
78.0	0.9568	0.9575	0.9582	0.9589	0.9596	0.9603	0.9609	0.9616	0.9623	0.9629	78.0
79.0	0.9636	0.9642	0.9649	0.9655	0.9662	0.9668	0.9674	0.9680	0.9686	0.9692	79.0
80.0	0.9698	0.9704	0.9710	0.9716	0.9722	0.9728	0.9733	0.9739	0.9744	0.9750	80.0
81.0	0.9755	0.9761	0.9766	0.9771	0.9776	0.9782	0.9787	0.9792	0.9797	0.9801	81.0
82.0	0.9806	0.9811	0.9816	0.9820	0.9825	0.9830	0.9834	0.9839	0.9843	0.9847	82.0
83.0	0.9851	0.9856	0.9860	0.9864	0.9868	0.9872	0.9876	0.9880	0.9883	0.9887	83.0
84.0	0.9891	0.9894	0.9898	0.9901	0.9905	0.9908	0.9911	0.9915	0.9918	0.9921	84.0
85.0	0.9924	0.9927	0.9930	0.9933	0.9936	0.9938	0.9941	0.9944	0.9946	0.9949	85.0
86.0	0.9951	0.9954	0.9956	0.9958	0.9961	0.9963	0.9965	0.9967	0.9969	0.9971	86.0
87.0	0.9973	0.9974	0.9976	0.9978	0.9979	0.9981	0.9982	0.9984	0.9985	0.9987	87.0
88.0	0.9988	0.9989	0.9990	0.9991	0.9992	0.9993	0.9994	0.9995	0.9996	0.9996	88.0
89.0	0.9997	0.9998	0.9998	0.9999	0.9999	0.9999	1.0000	1.0000	1.0000	1.0000	89.0
90.0	1.0000										

Examples:

$X' = 46.2$, $X = 0.5209$ and $X' = 85.3$, $X = 0.9933$.

TABLE B.25 PROPORTIONS OF THE BINOMIAL DISTRIBUTION FOR $p = q = 0.5$

X	n = 1	2	3	4	5	6	7	8	9	10	X
0	0.50000	0.25000	0.12500	0.06250	0.03125	0.01563	0.00781	0.00391	0.00195	0.00098	0
1	0.50000	0.50000	0.37500	0.25000	0.15625	0.09375	0.05469	0.03125	0.01758	0.00977	1
2		0.25000	0.37500	0.37500	0.31250	0.23438	0.16406	0.10938	0.07031	0.04395	2
3			0.12500	0.25000	0.31250	0.31250	0.27344	0.21875	0.16406	0.11719	3
4				0.06250	0.15625	0.23438	0.27344	0.27344	0.24609	0.20508	4
5					0.03125	0.09375	0.16406	0.21875	0.24609	0.24609	5
6						0.01563	0.05469	0.10938	0.16406	0.20508	6
7							0.00781	0.03125	0.07031	0.11719	7
								0.00391	0.01758	0.04395	8
9									0.00195	0.00977	9
10										0.00098	10

X	n = 11	12	13	14	15	16	17	18	19	20	X
0	0.00049	0.00024	0.00012	0.00006	0.00003	0.00002	0.00001	0.00000	0.00000	0.00000	0
1	0.00537	0.00293	0.00159	0.00085	0.00046	0.00024	0.00013	0.00007	0.00004	0.00002	1
2	0.02686	0.01611	0.00952	0.00555	0.00320	0.00183	0.00104	0.00058	0.00033	0.00018	2
3	0.08057	0.05371	0.03491	0.02222	0.01389	0.00854	0.00519	0.00311	0.00185	0.00109	3
4	0.16113	0.12085	0.08728	0.06110	0.04166	0.02777	0.01816	0.01167	0.00739	0.00462	4
5	0.22559	0.19336	0.15710	0.12219	0.09164	0.06665	0.04721	0.03268	0.02218	0.01479	5
6	0.22559	0.22559	0.20947	0.18329	0.15274	0.12219	0.09442	0.07082	0.05175	0.03696	6
7	0.16113	0.19336	0.20947	0.20947	0.19638	0.17456	0.14838	0.12140	0.09611	0.07393	7
8	0.08057	0.12085	0.15710	0.18329	0.19638	0.19638	0.18547	0.16692	0.14416	0.12013	8
9	0.02686	0.05371	0.08728	0.12219	0.15274	0.17456	0.18547	0.18547	0.17620	0.16018	9
10	0.00537	0.01611	0.03491	0.06110	0.09164	0.12219	0.14838	0.16692	0.17620	0.17620	10
11	0.00049	0.00293	0.00952	0.02222	0.04166	0.06665	0.09442	0.12140	0.14416	0.16018	11
12		0.00024	0.00159	0.00555	0.01389	0.02777	0.04721	0.07082	0.09611	0.12013	12
13			0.00012	0.00085	0.00320	0.00854	0.01816	0.03268	0.05175	0.07393	13
14				0.00006	0.00046	0.00183	0.00519	0.01167	0.02218	0.03696	14
15					0.00003	0.00024	0.00104	0.00311	0.00739	0.01479	15
16						0.00002	0.00013	0.00058	0.00185	0.00462	16
17							0.00001	0.00007	0.00033	0.00109	17
18								0.00000	0.00004	0.00018	18
19									0.00000	0.00002	19
20										0.00000	20

$$P(X) = \frac{n!}{X!(n-X)!} p^X (1-p)^{n-X}$$

TABLE B.26 CRITICAL VALUES OF C FOR THE SIGN TEST OR THE BINOMIAL TEST WITH $p = 0.5$

α(2):	.50	.20	.10	.05	.02	.01	.005	.002	.001
α(1):	.25	.10	.05	.025	.01	.005	.0025	.001	.0005
n									
2	---	---	---	---	---	---	---	---	---
3	0	---	---	---	---	---	---	---	---
4	0	0	---	---	---	---	---	---	---
5	1	0	0	---	---	---	---	---	---
6	1	0	0	0	---	---	---	---	---
7	2	1	0	0	0	---	---	---	---
8	2	1	1	0	0	0	---	---	---
9	2	2	1	1	0	0	0	---	---
10	3	2	1	1	0	0	0	0	---
11	3	2	2	1	1	0	0	0	0
12	4	3	2	2	1	1	0	0	0
13	4	3	3	2	1	1	1	0	0
14	5	4	3	2	2	1	1	1	0
15	5	4	3	3	2	2	1	1	1
16	6	4	4	3	2	2	2	1	1
17	6	5	4	4	3	2	2	1	1
18	7	5	5	4	3	3	2	2	1
19	7	6	5	4	4	3	3	2	2
20	7	6	5	5	4	3	3	2	2
21	8	7	6	5	4	4	3	3	2
22	8	7	6	5	5	4	4	3	3
23	9	7	7	6	5	4	4	3	3
24	9	8	7	6	5	5	4	4	3
25	10	8	7	7	6	5	5	4	4
26	10	9	8	7	6	6	5	4	4
27	11	9	8	7	7	6	5	5	4
28	11	10	9	8	7	6	6	5	5
29	12	10	9	8	7	7	6	5	5
30	12	10	10	9	8	7	6	6	5
31	13	11	10	9	8	7	7	6	6
32	13	11	10	9	8	8	7	6	6
33	14	12	11	10	9	8	8	7	6
34	14	12	11	10	9	9	8	7	7
35	15	13	12	11	10	9	8	8	7
36	15	13	12	11	10	9	9	8	7
37	15	14	13	12	10	10	9	8	8
38	16	14	13	12	11	10	9	9	8
39	16	15	13	12	11	11	10	9	8
40	17	15	14	13	12	11	10	9	9
41	17	15	14	13	12	11	11	10	9
42	18	16	15	14	13	12	11	10	10
43	18	16	15	14	13	12	11	11	10
44	19	17	16	15	13	13	12	11	10
45	19	17	16	15	14	13	12	11	11
46	20	18	16	15	14	13	13	12	11
47	20	18	17	16	15	14	13	12	11
48	21	19	17	16	15	14	13	12	12
49	21	19	18	17	15	15	14	13	12
50	22	19	18	17	16	15	14	13	13

α(2):	.50	.20	.10	.05	.02	.01	.005	.002	.001
α(1):	.25	.10	.05	.025	.01	.005	.0025	.001	.0005
n									
51	22	20	19	18	16	15	15	14	13
52	23	20	19	18	17	16	15	14	13
53	23	21	20	18	17	16	15	14	14
54	24	21	20	19	18	17	16	15	14
55	24	22	20	19	18	17	16	15	14
56	24	22	21	20	18	17	17	16	15
57	25	23	21	20	19	18	17	16	15
58	25	23	22	21	19	18	17	16	16
59	26	24	22	21	20	19	18	17	16
60	26	24	23	21	20	19	18	17	16
61	27	24	23	22	20	20	19	18	17
62	27	25	24	22	21	20	19	18	17
63	28	25	24	23	21	20	19	18	18
64	28	26	24	23	22	21	20	19	18
65	29	26	25	24	22	21	20	19	18
66	29	27	25	24	23	22	21	20	19
67	30	27	26	25	23	22	21	20	19
68	30	28	26	25	23	22	22	20	20
69	31	28	27	25	24	23	22	21	20
70	31	29	27	26	24	23	22	21	20
71	32	29	28	26	25	24	23	22	21
72	32	30	28	27	25	24	23	22	21
73	33	30	28	27	26	25	24	22	22
74	33	30	29	28	26	25	24	23	22
75	34	31	29	28	26	25	24	23	22
76	34	31	30	28	27	26	25	24	23
77	35	32	30	29	27	26	25	24	23
78	35	32	31	29	28	27	26	24	24
79	36	33	31	30	28	27	26	25	24
80	36	33	32	30	29	28	27	25	24
81	36	34	32	31	29	28	27	26	25
82	37	34	33	31	30	28	27	26	25
83	37	35	33	32	30	29	28	27	26
84	38	35	33	32	30	29	28	27	26
85	38	36	34	32	31	30	29	27	26
86	39	36	34	33	31	30	29	28	27
87	39	37	35	33	32	31	29	28	27
88	40	37	35	34	32	31	30	29	28
89	40	37	36	34	33	31	30	29	28
90	41	38	36	35	33	32	31	29	29
91	41	38	37	35	33	32	31	30	29
92	42	39	37	36	34	33	32	30	29
93	42	39	38	36	34	33	32	31	30
94	43	40	38	37	35	34	32	31	30
95	43	40	38	37	35	34	33	32	31
96	44	41	39	37	36	34	33	32	31
97	44	41	39	38	36	35	34	32	31
98	45	42	40	38	37	35	34	33	32
99	45	42	40	39	37	36	35	33	32
100	46	43	41	39	37	36	35	34	33

TABLE B.26 (cont.) CRITICAL VALUES OF *C* FOR THE SIGN TEST OR THE BINOMIAL TEST WITH $p = 0.5$

α(2):	.50	.20	.10	.05	.02	.01	.005	.002	.001	α(2):	.50	.20	.10	.05	.02	.01	.005	.002	.001
α(1):	.25	.10	.05	.025	.01	.005	.0025	.001	.0005	α(1):	.25	.10	.05	.025	.01	.005	.0025	.001	.0005
n										n									
101	46	43	41	40	38	37	35	34	33	151	70	67	64	62	60	59	57	56	54
102	47	44	42	40	38	37	36	35	34	152	71	67	65	63	61	59	58	56	55
103	47	44	42	41	39	37	36	35	34	153	71	68	65	63	61	60	58	56	55
104	48	44	43	41	39	38	37	35	34	154	72	68	66	64	62	60	59	57	56
105	48	45	43	41	40	38	37	36	35	155	72	69	66	64	62	61	59	57	56
106	49	45	44	42	40	39	38	36	35	156	73	69	67	65	63	61	60	58	57
107	49	46	44	42	41	39	38	37	36	157	73	69	67	65	63	61	60	58	57
108	49	46	44	43	41	40	38	37	36	158	74	70	68	66	63	62	60	59	57
109	50	47	45	43	41	40	39	37	36	159	74	70	68	66	64	62	61	59	58
110	50	47	45	44	42	41	39	38	37	160	75	71	69	67	64	63	61	60	58
111	51	48	46	44	42	41	40	38	37	161	75	71	69	67	65	63	62	60	59
112	51	48	46	45	43	41	40	39	38	162	76	72	70	68	65	64	62	60	59
113	52	49	47	45	43	42	41	39	38	163	76	72	70	68	66	64	63	61	60
114	52	49	47	46	44	42	41	40	39	164	77	73	70	68	66	65	63	61	60
115	53	50	48	46	44	43	42	40	39	165	77	73	71	69	67	65	64	62	60
116	53	50	48	46	45	43	42	40	39	166	78	74	71	69	67	65	64	62	61
117	54	51	49	47	45	44	42	41	40	167	78	74	72	70	68	66	64	63	61
118	54	51	49	47	45	44	43	41	40	168	79	75	72	70	68	66	65	63	62
119	55	52	50	48	46	45	43	42	41	169	79	75	73	71	68	67	65	63	62
120	55	52	50	48	46	45	44	42	41	170	80	76	73	71	69	67	66	64	63
121	56	52	50	49	47	45	44	43	42	171	80	76	74	72	69	68	66	64	63
122	56	53	51	49	47	46	45	43	42	172	81	77	74	72	70	68	67	65	64
123	57	53	51	50	48	46	45	43	42	173	81	77	75	73	70	69	67	65	64
124	57	54	52	50	48	47	45	44	43	174	82	78	75	73	71	69	68	66	64
125	58	54	52	51	49	47	46	44	43	175	82	78	76	74	71	70	68	66	65
126	58	55	53	51	49	48	46	45	44	176	83	79	76	74	72	70	68	67	65
127	59	55	53	51	49	48	47	45	44	177	83	79	77	74	72	70	69	67	66
128	59	56	54	52	50	48	47	46	45	178	83	79	77	75	73	71	69	67	66
129	60	56	54	52	50	49	48	46	45	179	84	80	78	75	73	71	70	68	67
130	60	57	55	53	51	49	48	46	45	180	84	80	78	76	73	72	70	68	67
131	61	57	55	53	51	50	49	47	46	181	85	81	78	76	74	72	71	69	67
132	61	58	56	54	52	50	49	47	46	182	85	81	79	77	74	73	71	69	68
133	62	58	56	54	52	51	49	48	47	183	86	82	79	77	75	73	72	70	68
134	62	59	56	55	53	51	50	48	47	184	86	82	80	78	75	74	72	70	69
135	63	59	57	55	53	52	50	49	48	185	87	83	80	78	76	74	72	71	69
136	63	60	57	56	53	52	51	49	48	186	87	83	81	79	76	74	73	71	70
137	64	60	58	56	54	52	51	50	48	187	88	84	81	79	77	75	73	71	70
138	64	60	58	57	54	53	52	50	49	188	88	84	82	80	77	75	74	72	71
139	65	61	59	57	55	53	52	50	49	189	89	85	82	80	78	76	74	72	71
140	65	61	59	57	55	54	52	51	50	190	89	85	83	81	78	76	75	73	71
141	65	62	60	58	56	54	53	51	50	191	90	86	83	81	78	77	75	73	72
142	66	62	60	58	56	55	53	52	51	192	90	86	84	81	79	77	76	74	72
143	66	63	61	59	57	55	54	52	51	193	91	87	84	82	79	78	76	74	73
144	67	63	61	59	57	56	54	53	51	194	91	87	85	82	80	78	77	75	73
145	67	64	62	60	58	56	55	53	52	195	92	88	85	83	80	79	77	75	74
146	68	64	62	60	58	56	55	53	52	196	92	88	85	83	81	79	77	75	74
147	68	65	63	61	58	57	56	54	53	197	93	89	86	84	81	79	78	76	75
148	69	65	63	61	59	57	56	54	53	198	93	89	86	84	82	80	78	76	75
149	69	66	63	62	59	58	56	55	54	199	94	89	87	85	82	80	79	77	75
150	70	66	64	62	60	58	57	55	54	200	94	90	87	85	83	81	79	77	76

TABLE B.26 (cont.) CRITICAL VALUES OF C FOR THE SIGN TEST OR THE BINOMIAL TEST WITH $p = 0.5$

α(2):	.50	.20	.10	.05	.02	.01	.005	.002	.001
α(1):	.25	.10	.05	.025	.01	.005	.0025	.001	.0005
n									
201	95	90	88	86	83	81	80	78	76
202	95	91	88	86	83	82	80	78	77
203	96	91	89	87	84	82	81	79	77
204	96	92	89	87	84	83	81	79	78
205	97	92	90	87	85	83	81	79	78
206	97	93	90	88	85	84	82	80	78
207	98	93	91	88	86	84	82	80	79
208	98	94	91	89	86	84	83	81	79
209	99	94	92	89	87	85	83	81	80
210	99	95	92	90	87	85	84	82	80
211	100	95	93	90	88	86	84	82	81
212	100	96	93	91	88	86	85	83	81
213	101	96	94	91	89	87	85	83	82
214	101	97	94	92	89	87	86	83	82
215	102	97	94	92	89	88	86	84	82
216	102	98	95	93	90	88	86	84	83
217	103	98	95	93	90	89	87	85	83
218	103	99	96	94	91	89	87	85	84
219	104	99	96	94	91	89	88	86	84
220	104	99	97	94	92	90	88	86	85
221	104	100	97	95	92	90	89	87	85
222	105	100	98	95	93	91	89	87	86
223	105	101	98	96	93	91	90	88	86
224	106	101	99	96	94	92	90	88	86
225	106	102	99	97	94	92	91	88	87
226	107	102	100	97	95	93	91	89	87
227	107	103	100	98	95	93	91	89	88
228	108	103	101	98	95	94	92	90	88
229	108	104	101	99	96	94	92	90	89
230	109	104	102	99	96	95	93	91	89
231	109	105	102	100	97	95	93	91	90
232	110	105	102	100	97	95	94	92	90
233	110	106	103	101	98	96	94	92	90
234	111	106	103	101	98	96	95	92	91
235	111	107	104	101	99	97	95	93	91
236	112	107	104	102	99	97	95	93	92
237	112	108	105	102	100	98	96	94	92
238	113	108	105	103	100	98	96	94	93
239	113	109	106	103	101	99	97	95	93
240	114	109	106	104	101	99	97	95	94
241	114	110	107	104	101	100	98	96	94
242	115	110	107	105	102	100	98	96	94
243	115	111	108	105	102	100	99	96	95
244	116	111	108	106	103	101	99	97	95
245	116	111	109	106	103	101	100	97	96
246	117	112	109	107	104	102	100	98	96
247	117	112	110	107	104	102	100	98	97
248	118	113	110	108	105	103	101	99	97
249	118	113	111	108	105	103	101	99	98
250	119	114	111	109	106	104	102	100	98

α(2):	.50	.20	.10	.05	.02	.01	.005	.002	.001
α(1):	.25	.10	.05	.025	.01	.005	.0025	.001	.0005
n									
251	119	114	111	109	106	104	102	100	99
252	120	115	112	109	107	105	103	101	99
253	120	115	112	110	107	105	103	101	99
254	121	116	113	110	107	106	104	101	100
255	121	116	113	111	108	106	104	102	100
256	122	117	114	111	108	106	105	102	101
257	122	117	114	112	109	107	105	103	101
258	123	118	115	112	109	107	106	103	102
259	123	118	115	113	110	108	106	104	102
260	124	119	116	113	110	108	106	104	103
261	124	119	116	114	111	109	107	105	103
262	125	120	117	114	111	109	107	105	103
263	125	120	117	115	112	110	108	106	104
264	126	121	118	115	112	110	108	106	104
265	126	121	118	116	113	111	109	106	105
266	126	122	119	116	113	111	109	107	105
267	127	122	119	117	114	111	110	107	106
268	127	123	120	117	114	112	110	108	106
269	128	123	120	117	114	112	111	108	107
270	128	123	120	118	115	113	111	109	107
271	129	124	121	118	115	113	111	109	107
272	129	124	121	119	116	114	112	110	108
273	130	125	122	119	116	114	112	110	108
274	130	125	122	120	117	115	113	110	109
275	131	126	123	120	117	115	113	111	109
276	131	126	123	121	118	116	114	111	110
277	132	127	124	121	118	116	114	112	110
278	132	127	124	122	119	117	115	112	111
279	133	128	125	122	119	117	115	113	111
280	133	128	125	123	120	117	116	113	112
281	134	129	126	123	120	118	116	114	112
282	134	129	126	124	120	118	116	114	112
283	135	130	127	124	121	119	117	115	113
284	135	130	127	124	121	119	117	115	113
285	136	131	128	125	122	120	118	115	114
286	136	131	128	125	122	120	118	116	114
287	137	132	129	126	123	121	119	116	115
288	137	132	129	126	123	121	119	117	115
289	138	133	130	127	124	122	120	117	116
290	138	133	130	127	124	122	120	118	116
291	139	134	130	128	125	123	121	118	117
292	139	134	131	128	125	123	121	119	117
293	140	135	131	129	126	123	122	119	117
294	140	135	132	129	126	124	122	120	118
295	141	135	132	130	127	124	122	120	118
296	141	136	133	130	127	125	123	120	119
297	142	136	133	131	127	125	123	121	119
298	142	137	134	131	128	126	124	121	120
299	143	137	134	132	128	126	124	122	120
300	143	138	135	132	129	127	125	122	121

TABLE B.26 (cont.) CRITICAL VALUES OF *C* FOR THE SIGN TEST OR THE BINOMIAL TEST WITH $p = 0.5$

α(2):	.50	.20	.10	.05	.02	.01	.005	.002	.001
α(1):	.25	.10	.05	.025	.01	.005	.0025	.001	.0005
n									
301	144	138	135	133	129	127	125	123	121
302	144	139	136	133	130	128	126	123	121
303	145	139	136	133	130	128	126	124	122
304	145	140	137	134	131	129	127	124	122
305	146	140	137	134	131	129	127	125	123
306	146	141	138	135	132	130	127	125	123
307	147	141	138	135	132	130	128	125	124
308	147	142	139	136	133	130	128	126	124
309	148	142	139	136	133	131	129	126	125
310	148	143	140	137	134	131	129	127	125
311	149	143	140	137	134	132	130	127	126
312	149	144	140	138	134	132	130	128	126
313	150	144	141	138	135	133	131	128	126
314	150	145	141	139	135	133	131	129	127
315	151	145	142	139	136	134	132	129	127
316	151	146	142	140	136	134	132	130	128
317	151	146	143	140	137	135	133	130	128
318	152	147	143	141	137	135	133	131	129
319	152	147	144	141	138	136	133	131	129
320	153	148	144	141	138	136	134	131	130
321	153	148	145	142	139	136	134	132	130
322	154	149	145	142	139	137	135	132	131
323	154	149	146	143	140	137	135	133	131
324	155	149	146	143	140	138	136	133	131
325	155	150	147	144	141	138	136	134	132
326	156	150	147	144	141	139	137	134	132
327	156	151	148	145	141	139	137	135	133
328	157	151	148	145	142	140	138	135	133
329	157	152	149	146	142	140	138	136	134
330	158	152	149	146	143	141	139	136	134
331	158	153	150	147	143	141	139	136	135
332	159	153	150	147	144	142	139	137	135
333	159	154	150	148	144	142	140	137	136
334	160	154	151	148	145	142	140	138	136
335	160	155	151	149	145	143	141	138	136
336	161	155	152	149	146	143	141	139	137
337	161	156	152	150	146	144	142	139	137
338	162	156	153	150	147	144	142	140	138
339	162	157	153	150	147	145	143	140	138
340	163	157	154	151	148	145	143	141	139
341	163	158	154	151	148	146	144	141	139
342	164	158	155	152	149	146	144	141	140
343	164	159	155	152	149	147	145	142	140
344	165	159	156	153	149	147	145	142	141
345	165	160	156	153	150	148	145	143	141
346	166	160	157	154	150	148	146	143	141
347	166	161	157	154	151	149	146	144	142
348	167	161	158	155	151	149	147	144	142
349	167	162	158	155	152	149	147	145	143
350	168	162	159	156	152	150	148	145	143

α(2):	.50	.20	.10	.05	.02	.01	.005	.002	.001
α(1):	.25	.10	.05	.025	.01	.005	.0025	.001	.0005
n									
351	168	162	159	156	153	150	148	146	144
352	169	163	160	157	153	151	149	146	144
353	169	163	160	157	154	151	149	147	145
354	170	164	161	158	154	152	150	147	145
355	170	164	161	158	155	152	150	147	146
356	171	165	161	159	155	153	151	148	146
357	171	165	162	159	156	153	151	148	146
358	172	166	162	159	156	154	151	149	147
359	172	166	163	160	156	154	152	149	147
360	173	167	163	160	157	155	152	150	148
361	173	167	164	161	157	155	153	150	148
362	174	168	164	161	158	156	153	151	149
363	174	168	165	162	158	156	154	151	149
364	175	169	165	162	159	156	154	152	150
365	175	169	166	163	159	157	155	152	150
366	176	170	166	163	160	157	155	152	151
367	176	170	167	164	160	158	156	153	151
368	177	171	167	164	161	158	156	153	152
369	177	171	168	165	161	159	157	154	152
370	178	172	168	165	162	159	157	154	152
371	178	172	169	166	162	160	158	155	153
372	178	173	169	166	163	160	158	155	153
373	179	173	170	167	163	161	158	156	154
374	179	174	170	167	164	161	159	156	154
375	180	174	171	168	164	162	159	157	155
376	180	175	171	168	164	162	160	157	155
377	181	175	172	168	165	163	160	158	156
378	181	176	172	169	165	163	161	158	156
379	182	176	172	169	166	163	161	158	157
380	182	177	173	170	166	164	162	159	157
381	183	177	173	170	167	164	162	159	157
382	183	177	174	171	167	165	163	160	158
383	184	178	174	171	168	165	163	160	158
384	184	178	175	172	168	166	164	161	159
385	185	179	175	172	169	166	164	161	159
386	185	179	176	173	169	167	164	162	160
387	186	180	176	173	170	167	165	162	160
388	186	180	177	174	170	168	165	163	161
389	187	181	177	174	171	168	166	163	161
390	187	181	178	175	171	169	166	164	162
391	188	182	178	175	172	169	167	164	162
392	188	182	179	176	172	170	167	164	162
393	189	183	179	176	172	170	168	165	163
394	189	183	180	177	173	170	168	165	163
395	190	184	180	177	173	171	169	166	164
396	190	184	181	178	174	171	169	166	164
397	191	185	181	178	174	172	170	167	165
398	191	185	182	178	175	172	170	167	165
399	192	186	182	179	175	173	171	168	166
400	192	186	183	179	176	173	171	168	166

TABLE B.26 (cont.) CRITICAL VALUES OF C FOR THE SIGN TEST OR THE BINOMIAL TEST WITH $p = 0.5$

$\alpha(2)$:	.50	.20	.10	.05	.02	.01	.005	.002	.001
$\alpha(1)$:	.25	.10	.05	.025	.01	.005	.0025	.001	.0005
n									
401	193	187	183	180	176	174	171	169	167
402	193	187	184	180	177	174	172	169	167
403	194	188	184	181	177	175	172	170	168
404	194	188	184	181	178	175	173	170	168
405	195	189	185	182	178	176	173	170	168
406	195	189	185	182	179	176	174	171	169
407	196	190	186	183	179	177	174	171	169
408	196	190	186	183	180	177	175	172	170
409	197	191	187	184	180	177	175	172	170
410	197	191	187	184	180	178	176	173	171
411	198	192	188	185	181	178	176	173	171
412	198	192	188	185	181	179	177	174	172
413	199	192	189	186	182	179	177	174	172
414	199	193	189	186	182	180	177	175	173
415	200	193	190	187	183	180	178	175	173
416	200	194	190	187	183	181	178	176	174
417	201	194	191	187	184	181	179	176	174
418	201	195	191	188	184	182	179	176	174
419	202	195	192	188	185	182	180	177	175
420	202	196	192	189	185	183	180	177	175
421	203	196	193	189	186	183	181	178	176
422	203	197	193	190	186	184	181	178	176
423	204	197	194	190	187	184	182	179	177
424	204	198	194	191	187	185	182	179	177
425	205	198	195	191	188	185	183	180	178
426	205	199	195	192	188	185	183	180	178
427	206	199	196	192	188	186	184	181	179
428	206	200	196	193	189	186	184	181	179
429	207	200	196	193	189	187	184	182	179
430	207	201	197	194	190	187	185	182	180
431	207	201	197	194	190	188	185	182	180
432	208	202	198	195	191	188	186	183	181
433	208	202	198	195	191	189	186	183	181
434	209	203	199	196	192	189	187	184	182
435	209	203	199	196	192	190	187	184	182
436	210	204	200	197	193	190	188	185	183
437	210	204	200	197	193	191	188	185	183
438	211	205	201	198	194	191	189	186	184
439	211	205	201	198	194	192	189	186	184
440	212	206	202	198	195	192	190	187	185
441	212	206	202	199	195	192	190	187	185
442	213	207	203	199	196	193	191	188	185
443	213	207	203	200	196	193	191	188	186
444	214	207	204	200	197	194	191	188	186
445	214	208	204	201	197	194	192	189	187
446	215	208	205	201	197	195	192	189	187
447	215	209	205	202	198	195	193	190	188
448	216	209	206	202	198	196	193	190	188
449	216	210	206	203	199	196	194	191	189
450	217	210	207	203	199	197	194	191	189

$\alpha(2)$:	.50	.20	.10	.05	.02	.01	.005	.002	.001
$\alpha(1)$:	.25	.10	.05	.025	.01	.005	.0025	.001	.0005
n									
451	217	211	207	204	200	197	195	192	190
452	218	211	208	204	200	198	195	192	190
453	218	212	208	205	201	198	196	193	191
454	219	212	208	205	201	199	196	193	191
455	219	213	209	206	202	199	197	194	191
456	220	213	209	206	202	200	197	194	192
457	220	214	210	207	203	200	198	195	192
458	221	214	210	207	203	200	198	195	193
459	221	215	211	208	204	201	198	195	193
460	222	215	211	208	204	201	199	196	194
461	222	216	212	208	205	202	199	196	194
462	223	216	212	209	205	202	200	197	195
463	223	217	213	209	205	203	200	197	195
464	224	217	213	210	206	203	201	198	196
465	224	218	214	210	206	204	201	198	196
466	225	218	214	211	207	204	202	199	197
467	225	219	215	211	207	205	202	199	197
468	226	219	215	212	208	205	203	200	197
469	226	220	216	212	208	206	203	200	198
470	227	220	216	213	209	206	204	201	198
471	227	221	217	213	209	207	204	201	199
472	228	221	217	214	210	207	205	201	199
473	228	222	218	214	210	208	205	202	200
474	229	222	218	215	211	208	205	202	200
475	229	223	219	215	211	208	206	203	201
476	230	223	219	216	212	209	206	203	201
477	230	224	220	216	212	209	207	204	202
478	231	224	220	217	213	210	207	204	202
479	231	224	221	217	213	210	208	205	203
480	232	225	221	218	214	211	208	205	203
481	232	225	221	218	214	211	209	206	203
482	233	226	222	218	214	212	209	206	204
483	233	226	222	219	215	212	210	207	204
484	234	227	223	219	215	213	210	207	205
485	234	227	223	220	216	213	211	208	205
486	235	228	224	220	216	214	211	208	206
487	235	228	224	221	217	214	212	208	206
488	236	229	225	221	217	215	212	209	207
489	236	229	225	222	218	215	212	209	207
490	237	230	226	222	218	216	213	210	208
491	237	230	226	223	219	216	213	210	208
492	238	231	227	223	219	216	214	211	209
493	238	231	227	224	220	217	214	211	209
494	239	232	228	224	220	217	215	212	209
495	239	232	228	225	221	218	215	212	210
496	239	233	229	225	221	218	216	213	210
497	240	233	229	226	222	219	216	213	211
498	240	234	230	226	222	219	217	214	211
499	241	234	230	227	223	220	217	214	212
500	241	235	231	227	223	220	218	214	212

TABLE B.26 (cont.) CRITICAL VALUES OF *C* FOR THE SIGN TEST OR THE BINOMIAL TEST WITH $p = 0.5$

α(2):	.50	.20	.10	.05	.02	.01	.005	.002	.001
α(1):	.25	.10	.05	.025	.01	.005	.0025	.001	.0005
n									
501	242	235	231	228	223	221	218	215	213
502	242	236	232	228	224	221	219	215	213
503	243	236	232	229	224	222	219	216	214
504	243	237	233	229	225	222	220	216	214
505	244	237	233	229	225	223	220	217	215
506	244	238	234	230	226	223	220	217	215
507	245	238	234	230	226	224	221	218	216
508	245	239	234	231	227	224	221	218	216
509	246	239	235	231	227	224	222	219	216
510	246	240	235	232	228	225	222	219	217
511	247	240	236	232	228	225	223	220	217
512	247	241	236	233	229	226	223	220	218
513	248	241	237	233	229	226	224	221	218
514	248	241	237	234	230	227	224	221	219
515	249	242	238	234	230	227	225	221	219
516	249	242	238	235	231	228	225	222	220
517	250	243	239	235	231	228	226	222	220
518	250	243	239	236	232	229	226	223	221
519	251	244	240	236	232	229	227	223	221
520	251	244	240	237	232	230	227	224	222
521	252	245	241	237	233	230	227	224	222
522	252	245	241	238	233	231	228	225	222
523	253	246	242	238	234	231	228	225	223
524	253	246	242	239	234	232	229	226	223
525	254	247	243	239	235	232	229	226	224
526	254	247	243	240	235	232	230	227	224
527	255	248	244	240	236	233	230	227	225
528	255	248	244	240	236	233	231	228	225
529	256	249	245	241	237	234	231	228	226
530	256	249	245	241	237	234	232	228	226
531	257	250	246	242	238	235	232	229	227
532	257	250	246	242	238	235	233	229	227
533	258	251	247	243	239	236	233	230	228
534	258	251	247	243	239	236	234	230	228
535	259	252	247	244	240	237	234	231	229
536	259	252	248	244	240	237	235	231	229
537	260	253	248	245	241	238	235	232	229
538	260	253	249	245	241	238	235	232	230
539	261	254	249	246	242	239	236	233	230
540	261	254	250	246	242	239	236	233	231
541	262	255	250	247	242	240	237	234	231
542	262	255	251	247	243	240	237	234	232
543	263	256	251	248	243	241	238	235	232
544	263	256	252	248	244	241	238	235	233
545	264	257	252	249	244	241	239	235	233
546	264	257	253	249	245	242	239	236	234
547	265	258	253	250	245	242	240	236	234
548	265	258	254	250	246	243	240	237	235
549	266	258	254	251	246	243	241	237	235
550	266	259	255	251	247	244	241	238	235

α(2):	.50	.20	.10	.05	.02	.01	.005	.002	.001
α(1):	.25	.10	.05	.025	.01	.005	.0025	.001	.0005
n									
551	267	259	255	252	247	244	242	238	236
552	267	260	256	252	248	245	242	239	236
553	268	260	256	252	248	245	243	239	237
554	268	261	257	253	249	246	243	240	237
555	269	261	257	253	249	246	243	240	238
556	269	262	258	254	250	247	244	241	238
557	270	262	258	254	250	247	244	241	239
558	270	263	259	255	251	248	245	242	239
559	271	263	259	255	251	248	245	242	240
560	271	264	260	256	251	249	246	242	240
561	272	264	260	256	252	249	246	243	241
562	272	265	261	257	252	249	247	243	241
563	272	265	261	257	253	250	247	244	242
564	273	266	261	258	253	250	248	244	242
565	273	266	262	258	254	251	248	245	242
566	274	267	262	259	254	251	249	245	243
567	274	267	263	259	255	252	249	246	243
568	275	268	263	260	255	252	250	246	244
569	275	268	264	260	256	253	250	247	244
570	276	269	264	261	256	253	251	247	245
571	276	269	265	261	257	254	251	248	245
572	277	270	265	262	257	254	251	248	246
573	277	270	266	262	258	255	252	249	246
574	278	271	266	263	258	255	252	249	247
575	278	271	267	263	259	256	253	249	247
576	279	272	267	263	259	256	253	250	248
577	279	272	268	264	260	257	254	250	248
578	280	273	268	264	260	257	254	251	249
579	280	273	269	265	261	258	255	251	249
580	281	274	269	265	261	258	255	252	249
581	281	274	270	266	261	258	256	252	250
582	282	275	270	266	262	259	256	253	250
583	282	275	271	267	262	259	257	253	251
584	283	276	271	267	263	260	257	254	251
585	283	276	272	268	263	260	258	254	252
586	284	276	272	268	264	261	258	255	252
587	284	277	273	269	264	261	259	255	253
588	285	277	273	269	265	262	259	256	253
589	285	278	274	270	265	262	259	256	254
590	286	278	274	270	266	263	260	257	254
591	286	279	275	271	266	263	260	257	255
592	287	279	275	271	267	264	261	257	255
593	287	280	275	272	267	264	261	258	255
594	288	280	276	272	268	265	262	258	256
595	288	281	276	273	268	265	262	259	256
596	289	281	277	273	269	266	263	259	257
597	289	282	277	274	269	266	263	260	257
598	290	282	278	274	270	267	264	260	258
599	290	283	278	275	270	267	264	261	258
600	291	283	279	275	271	267	265	261	259

TABLE B.26 (cont.) CRITICAL VALUES OF C FOR THE SIGN TEST OR THE BINOMIAL TEST WITH $p = 0.5$

α(2):	.50	.20	.10	.05	.02	.01	.005	.002	.001
α(1):	.25	.10	.05	.025	.01	.005	.0025	.001	.0005
n									
601	291	284	279	275	271	268	265	262	259
602	292	284	280	276	271	268	266	262	260
603	292	285	280	276	272	269	266	263	260
604	293	285	281	277	272	269	267	263	261
605	293	286	281	277	273	270	267	264	261
606	294	286	282	278	273	270	267	264	262
607	294	287	282	278	274	271	268	264	262
608	295	287	283	279	274	271	268	265	262
609	295	288	283	279	275	272	269	265	263
610	296	288	284	280	275	272	269	266	263
611	296	289	284	280	276	273	270	266	264
612	297	289	285	281	276	273	270	267	264
613	297	290	285	281	277	274	271	267	265
614	298	290	286	282	277	274	271	268	265
615	298	291	286	282	278	275	272	268	266
616	299	291	287	283	278	275	272	269	266
617	299	292	287	283	279	276	273	269	267
618	300	292	288	284	279	276	273	270	267
619	300	293	288	284	280	276	274	270	268
620	301	293	289	285	280	277	274	271	268
621	301	294	289	285	281	277	275	271	269
622	302	294	289	286	281	278	275	272	269
623	302	295	290	286	281	278	276	272	269
624	303	295	290	287	282	279	276	272	270
625	303	295	291	287	282	279	276	273	270
626	304	296	291	287	283	280	277	273	271
627	304	296	292	288	283	280	277	274	271
628	305	297	292	288	284	281	278	274	272
629	305	297	293	289	284	281	278	275	272
630	306	298	293	289	285	282	279	275	273
631	306	298	294	290	285	282	279	276	273
632	307	299	294	290	286	283	280	276	274
633	307	299	295	291	286	283	280	277	274
634	308	300	295	291	287	284	281	277	275
635	308	300	296	292	287	284	281	278	275
636	308	301	296	292	288	285	282	278	276
637	309	301	297	293	288	285	282	279	276
638	309	302	297	293	289	285	283	279	276
639	310	302	298	294	289	286	283	279	277
640	310	303	298	294	290	286	284	280	277
641	311	303	299	295	290	287	284	280	278
642	311	304	299	295	291	287	284	281	278
643	312	304	300	296	291	288	285	281	279
644	312	305	300	296	291	288	285	282	279
645	313	305	301	297	292	289	286	282	280
646	313	306	301	297	292	289	286	283	280
647	314	306	302	298	293	290	287	283	281
648	314	307	302	298	293	290	287	284	281
649	315	307	303	299	294	291	288	284	282
650	315	308	303	299	294	291	288	285	282

α(2):	.50	.20	.10	.05	.02	.01	.005	.002	.001
α(1):	.25	.10	.05	.025	.01	.005	.0025	.001	.0005
n									
651	316	308	304	300	295	292	289	285	283
652	316	309	304	300	295	292	289	286	283
653	317	309	304	300	296	293	290	286	284
654	317	310	305	301	296	293	290	287	284
655	318	310	305	301	297	294	291	287	284
656	318	311	306	302	297	294	291	287	285
657	319	311	306	302	298	295	292	288	285
658	319	312	307	303	298	295	292	288	286
659	320	312	307	303	299	295	293	289	286
660	320	313	308	304	299	296	293	289	287
661	321	313	308	304	300	296	293	290	287
662	321	314	309	305	300	297	294	290	288
663	322	314	309	305	301	297	294	291	288
664	322	314	310	306	301	298	295	291	289
665	323	315	310	306	302	298	295	292	289
666	323	315	311	307	302	299	296	292	290
667	324	316	311	307	302	299	296	293	290
668	324	316	312	308	303	300	297	293	291
669	325	317	312	308	303	300	297	294	291
670	325	317	313	309	304	301	298	294	291
671	326	318	313	309	304	301	298	295	292
672	326	318	314	310	305	302	299	295	292
673	327	319	314	310	305	302	299	295	293
674	327	319	315	311	306	303	300	296	293
675	328	320	315	311	306	303	300	296	294
676	328	320	316	312	307	304	301	297	294
677	329	321	316	312	307	304	301	297	295
678	329	321	317	312	308	304	301	298	295
679	330	322	317	313	308	305	302	298	296
680	330	322	318	313	309	305	302	299	296
681	331	323	318	314	309	306	303	299	297
682	331	323	319	314	310	306	303	300	297
683	332	324	319	315	310	307	304	300	298
684	332	324	319	315	311	307	304	301	298
685	333	325	320	316	311	308	305	301	298
686	333	325	320	316	312	308	305	302	299
687	334	326	321	317	312	309	306	302	299
688	334	326	321	317	313	309	306	303	300
689	335	327	322	318	313	310	307	303	300
690	335	327	322	318	313	310	307	303	301
691	336	328	323	319	314	311	308	304	301
692	336	328	323	319	314	311	308	304	302
693	337	329	324	320	315	312	309	305	302
694	337	329	324	320	315	312	309	305	303
695	338	330	325	321	316	313	310	306	303
696	338	330	325	321	316	313	310	306	304
697	339	331	326	322	317	314	310	307	304
698	339	331	326	322	317	314	311	307	305
699	340	332	327	323	318	314	311	308	305
700	340	332	327	323	318	315	312	308	306

TABLE B.26 (cont.) CRITICAL VALUES OF C FOR THE SIGN TEST OR THE BINOMIAL TEST WITH $p = 0.5$

$\alpha(2)$:	.50	.20	.10	.05	.02	.01	.005	.002	.001
$\alpha(1)$:	.25	.10	.05	.025	.01	.005	.0025	.001	.0005
n									
701	341	333	328	324	319	315	312	309	306
702	341	333	328	324	319	316	313	309	306
703	342	334	329	325	320	316	313	310	307
704	342	334	329	325	320	317	314	310	307
705	343	334	330	325	321	317	314	311	308
706	343	335	330	326	321	318	315	311	308
707	344	335	331	326	322	318	315	311	309
708	344	336	331	327	322	319	316	312	309
709	345	336	332	327	323	319	316	312	310
710	345	337	332	328	323	320	317	313	310
711	346	337	333	328	324	320	317	313	311
712	346	338	333	329	324	321	318	314	311
713	346	338	334	329	324	321	318	314	312
714	347	339	334	330	325	322	319	315	312
715	347	339	335	330	325	322	319	315	313
716	348	340	335	331	326	323	319	316	313
717	348	340	335	331	326	323	320	316	313
718	349	341	336	332	327	324	320	317	314
719	349	341	336	332	327	324	321	317	314
720	350	342	337	333	328	324	321	318	315
721	350	342	337	333	328	325	322	318	315
722	351	343	338	334	329	325	322	319	316
723	351	343	338	334	329	326	323	319	316
724	352	344	339	335	330	326	323	319	317
725	352	344	339	335	330	327	324	320	317
726	353	345	340	336	331	327	324	320	318
727	353	345	340	336	331	328	325	321	318
728	354	346	341	337	332	328	325	321	319
729	354	346	341	337	332	329	326	322	319
730	355	347	342	338	333	329	326	322	320
731	355	347	342	338	333	330	327	323	320
732	356	348	343	338	334	330	327	323	321
733	356	348	343	339	334	331	328	324	321
734	357	349	344	339	335	331	328	324	321
735	357	349	344	340	335	332	328	325	322
736	358	350	345	340	335	332	329	325	322
737	358	350	345	341	336	333	329	326	323
738	359	351	346	341	336	333	330	326	323
739	359	351	346	342	337	334	330	327	324
740	360	352	347	342	337	334	331	327	324
741	360	352	347	343	338	334	331	327	325
742	361	353	348	343	338	335	332	328	325
743	361	353	348	344	339	335	332	328	326
744	362	354	349	344	339	336	333	329	326
745	362	354	349	345	340	336	333	329	327
746	363	354	350	345	340	337	334	330	327
747	363	355	350	346	341	337	334	330	328
748	364	355	351	346	341	338	335	331	328
749	364	356	351	347	342	338	335	331	329
750	365	356	351	347	342	339	336	332	329

$\alpha(2)$:	.50	.20	.10	.05	.02	.01	.005	.002	.001
$\alpha(1)$:	.25	.10	.05	.025	.01	.005	.0025	.001	.0005
n									
751	365	357	352	348	343	339	336	332	329
752	366	357	352	348	343	340	337	333	330
753	366	358	353	349	344	340	337	333	330
754	367	358	353	349	344	341	337	334	331
755	367	359	354	350	345	341	338	334	331
756	368	359	354	350	345	342	338	335	332
757	368	360	355	351	346	342	339	335	332
758	369	360	355	351	346	343	339	336	333
759	369	361	356	352	346	343	340	336	333
760	370	361	356	352	347	344	340	336	334
761	370	362	357	352	347	344	341	337	334
762	371	362	357	353	348	344	341	337	335
763	371	363	358	353	348	345	342	338	335
764	372	363	358	354	349	345	342	338	336
765	372	364	359	354	349	346	343	339	336
766	373	364	359	355	350	346	343	339	337
767	373	365	360	355	350	347	344	340	337
768	374	365	360	356	351	347	344	340	337
769	374	366	361	356	351	348	345	341	338
770	375	366	361	357	352	348	345	341	338
771	375	367	362	357	352	349	346	342	339
772	376	367	362	358	353	349	346	342	339
773	376	368	363	358	353	350	347	343	340
774	377	368	363	359	354	350	347	343	340
775	377	369	364	359	354	351	347	344	341
776	378	369	364	360	355	351	348	344	341
777	378	370	365	360	355	352	348	344	342
778	379	370	365	361	356	352	349	345	342
779	379	371	366	361	356	353	349	345	343
780	380	371	366	362	357	353	350	346	343
781	380	372	367	362	357	354	350	346	344
782	381	372	367	363	357	354	351	347	344
783	381	373	367	363	358	354	351	347	345
784	382	373	368	364	358	355	352	348	345
785	382	374	368	364	359	355	352	348	345
786	383	374	369	365	359	356	353	349	346
787	383	375	369	365	360	356	353	349	346
788	384	375	370	365	360	357	354	350	347
789	384	376	370	366	361	357	354	350	347
790	385	376	371	366	361	358	355	351	348
791	385	376	371	367	362	358	355	351	348
792	386	377	372	367	362	359	356	352	349
793	386	377	372	368	363	359	356	352	349
794	386	378	373	368	363	360	356	353	350
795	387	378	373	369	364	360	357	353	350
796	387	379	374	369	364	361	357	353	351
797	388	379	374	370	365	361	358	354	351
798	388	380	375	370	365	362	358	354	352
799	389	380	375	371	366	362	359	355	352
800	389	381	376	371	366	363	359	355	353

TABLE B.26 (cont.) CRITICAL VALUES OF C FOR THE SIGN TEST OR THE BINOMIAL TEST WITH $p = 0.5$

α(2):	.50	.20	.10	.05	.02	.01	.005	.002	.001
α(1):	.25	.10	.05	.025	.01	.005	.0025	.001	.0005
n									
801	390	381	376	372	367	363	360	356	353
802	390	382	377	372	367	364	360	356	353
803	391	382	377	373	368	364	361	357	354
804	391	383	378	373	368	365	361	357	354
805	392	383	378	374	369	365	362	358	355
806	392	384	379	374	369	365	362	358	355
807	393	384	379	375	369	366	363	359	356
808	393	385	380	375	370	366	363	359	356
809	394	385	380	376	370	367	364	360	357
810	394	386	381	376	371	367	364	360	357
811	395	386	381	377	371	368	365	361	358
812	395	387	382	377	372	368	365	361	358
813	396	387	382	378	372	369	366	361	359
814	396	388	383	378	373	369	366	362	359
815	397	388	383	379	373	370	366	362	360
816	397	389	384	379	374	370	367	363	360
817	398	389	384	379	374	371	367	363	361
818	398	390	384	380	375	371	368	364	361
819	399	390	385	380	375	372	368	364	361
820	399	391	385	381	376	372	369	365	362
821	400	391	386	381	376	373	369	365	362
822	400	392	386	382	377	373	370	366	363
823	401	392	387	382	377	374	370	366	363
824	401	393	387	383	378	374	371	367	364
825	402	393	388	383	378	375	371	367	364
826	402	394	388	384	379	375	372	368	365
827	403	394	389	384	379	375	372	368	365
828	403	395	389	385	380	376	373	369	366
829	404	395	390	385	380	376	373	369	366
830	404	396	390	386	381	377	374	370	367
831	405	396	391	386	381	377	374	370	367
832	405	397	391	387	381	378	375	370	368
833	406	397	392	387	382	378	375	371	368
834	406	397	392	388	382	379	375	371	369
835	407	398	393	388	383	379	376	372	369
836	407	398	393	389	383	380	376	372	369
837	408	399	394	389	384	380	377	373	370
838	408	399	394	390	384	381	377	373	370
839	409	400	395	390	385	381	378	374	371
840	409	400	395	391	385	382	378	374	371
841	410	401	396	391	386	382	379	375	372
842	410	401	396	392	386	383	379	375	372
843	411	402	397	392	387	383	380	376	373
844	411	402	397	393	387	384	380	376	373
845	412	403	398	393	388	384	381	377	374
846	412	403	398	394	388	385	381	377	374
847	413	404	399	394	389	385	382	378	375
848	413	404	399	394	389	386	382	378	375
849	414	405	400	395	390	386	383	379	376
850	414	405	400	395	390	386	383	379	376

α(2):	.50	.20	.10	.05	.02	.01	.005	.002	.001
α(1):	.25	.10	.05	.025	.01	.005	.0025	.001	.0005
n									
851	415	406	401	396	391	387	384	379	377
852	415	406	401	396	391	387	384	380	377
853	416	407	401	397	392	388	385	380	377
854	416	407	402	397	392	388	385	381	378
855	417	408	402	398	393	389	385	381	378
856	417	408	403	398	393	389	386	382	379
857	418	409	403	399	393	390	386	382	379
858	418	409	404	399	394	390	387	383	380
859	419	410	404	400	394	391	387	383	380
860	419	410	405	400	395	391	388	384	381
861	420	411	405	401	395	392	388	384	381
862	420	411	406	401	396	392	389	385	382
863	421	412	406	402	396	393	389	385	382
864	421	412	407	402	397	393	390	386	383
865	422	413	407	403	397	394	390	386	383
866	422	413	408	403	398	394	391	387	384
867	423	414	408	404	398	395	391	387	384
868	423	414	409	404	399	395	392	388	385
869	424	415	409	405	399	396	392	388	385
870	424	415	410	405	400	396	393	388	386
871	425	416	410	406	400	397	393	389	386
872	425	416	411	406	401	397	394	389	386
873	426	417	411	407	401	397	394	390	387
874	426	417	412	407	402	398	395	390	387
875	427	418	412	408	402	398	395	391	388
876	427	418	413	408	403	399	395	391	388
877	428	419	413	408	403	399	396	392	389
878	428	419	414	409	404	400	396	392	389
879	429	420	414	409	404	400	397	393	390
880	429	420	415	410	405	401	397	393	390
881	429	420	415	410	405	401	398	394	391
882	430	421	416	411	405	402	398	394	391
883	430	421	416	411	406	402	399	395	392
884	431	422	417	412	406	403	399	395	392
885	431	422	417	412	407	403	400	396	393
886	432	423	418	413	407	404	400	396	393
887	432	423	418	413	408	404	401	397	394
888	433	424	418	414	408	405	401	397	394
889	433	424	419	414	409	405	402	397	394
890	434	425	419	415	409	406	402	398	395
891	434	425	420	415	410	406	403	398	395
892	435	426	420	416	410	407	403	399	396
893	435	426	421	416	411	407	404	399	396
894	436	427	421	417	411	408	404	400	397
895	436	427	422	417	412	408	405	400	397
896	437	428	422	418	412	408	405	401	398
897	437	428	423	418	413	409	405	401	398
898	438	429	423	419	413	409	406	402	399
899	438	429	424	419	414	410	406	402	399
900	439	430	424	420	414	410	407	403	400

TABLE B.26 (cont.) CRITICAL VALUES OF C FOR THE SIGN TEST OR THE BINOMIAL TEST WITH $p = 0.5$

α(2):	.50	.20	.10	.05	.02	.01	.005	.002	.001
α(1):	.25	.10	.05	.025	.01	.005	.0025	.001	.0005
n									
901	439	430	425	420	415	411	407	403	400
902	440	431	425	421	415	411	408	404	401
903	440	431	426	421	416	412	408	404	401
904	441	432	426	422	416	412	409	405	402
905	441	432	427	422	417	413	409	405	402
906	442	433	427	423	417	413	410	406	403
907	442	433	428	423	417	414	410	406	403
908	443	434	428	423	418	414	411	406	403
909	443	434	429	424	418	415	411	407	404
910	444	435	429	424	419	415	412	407	404
911	444	435	430	425	419	416	412	408	405
912	445	436	430	425	420	416	413	408	405
913	445	436	431	426	420	417	413	409	406
914	446	437	431	426	421	417	414	409	406
915	446	437	432	427	421	418	414	410	407
916	447	438	432	427	422	418	415	410	407
917	447	438	433	428	422	419	415	411	408
918	448	439	433	428	423	419	416	411	408
919	448	439	434	429	423	419	416	412	409
920	449	440	434	429	424	420	416	412	409
921	449	440	435	430	424	420	417	413	410
922	450	441	435	430	425	421	417	413	410
923	450	441	436	431	425	421	418	414	411
924	451	442	436	431	426	422	418	414	411
925	451	442	436	432	426	422	419	415	412
926	452	443	437	432	427	423	419	415	412
927	452	443	437	433	427	423	420	415	412
928	453	443	438	433	428	424	420	416	413
929	453	444	438	434	428	424	421	416	413
930	454	444	439	434	429	425	421	417	414
931	454	445	439	435	429	425	422	417	414
932	455	445	440	435	430	426	422	418	415
933	455	446	440	436	430	426	423	418	415
934	456	446	441	436	430	427	423	419	416
935	456	447	441	437	431	427	424	419	416
936	457	447	442	437	431	428	424	420	417
937	457	448	442	438	432	428	425	420	417
938	458	448	443	438	432	429	425	421	418
939	458	449	443	438	433	429	426	421	418
940	459	449	444	439	433	430	426	422	419
941	459	450	444	439	434	430	426	422	419
942	460	450	445	440	434	430	427	423	420
943	460	451	445	440	435	431	427	423	420
944	461	451	446	441	435	431	428	424	420
945	461	452	446	441	436	432	428	424	421
946	462	452	447	442	436	432	429	425	421
947	462	453	447	442	437	433	429	425	422
948	463	453	448	443	437	433	430	425	422
949	463	454	448	443	438	434	430	426	423
950	464	454	449	444	438	434	431	426	423

α(2):	.50	.20	.10	.05	.02	.01	.005	.002	.001
α(1):	.25	.10	.05	.025	.01	.005	.0025	.001	.0005
n									
951	464	455	449	444	439	435	431	427	424
952	465	455	450	445	439	435	432	427	424
953	465	456	450	445	440	436	432	428	425
954	466	456	451	446	440	436	433	428	425
955	466	457	451	446	441	437	433	429	426
956	467	457	452	447	441	437	434	429	426
957	467	458	452	447	442	438	434	430	427
958	468	458	453	448	442	438	435	430	427
959	468	459	453	448	442	439	435	431	428
960	469	459	454	449	443	439	436	431	428
961	469	460	454	449	443	440	436	432	429
962	470	460	454	450	444	440	436	432	429
963	470	461	455	450	444	441	437	433	429
964	471	461	455	451	445	441	437	433	430
965	471	462	456	451	445	442	438	434	430
966	472	462	456	452	446	442	438	434	431
967	472	463	457	452	446	442	439	434	431
968	473	463	457	453	447	443	439	435	432
969	473	464	458	453	447	443	440	435	432
970	473	464	458	453	448	444	440	436	433
971	474	465	459	454	448	444	441	436	433
972	474	465	459	454	449	445	441	437	434
973	475	466	460	455	449	445	442	437	434
974	475	466	460	455	450	446	442	438	435
975	476	466	461	456	450	446	443	438	435
976	476	467	461	456	451	447	443	439	436
977	477	467	462	457	451	447	444	439	436
978	477	468	462	457	452	448	444	440	437
979	478	468	463	458	452	448	445	440	437
980	478	469	463	458	453	449	445	441	438
981	479	469	464	459	453	449	446	441	438
982	479	470	464	459	454	450	446	442	438
983	480	470	465	460	454	450	447	442	439
984	480	471	465	460	455	451	447	443	439
985	481	471	466	461	455	451	447	443	440
986	481	472	466	461	455	452	448	444	440
987	482	472	467	462	456	452	448	444	441
988	482	473	467	462	456	453	449	444	441
989	483	473	468	463	457	453	449	445	442
990	483	474	468	463	457	453	450	445	442
991	484	474	469	464	458	454	450	446	443
992	484	475	469	464	458	454	451	446	443
993	485	475	470	465	459	455	451	447	444
994	485	476	470	465	459	455	452	447	444
995	486	476	471	466	460	456	452	448	445
996	486	477	471	466	460	456	453	448	445
997	487	477	472	467	461	457	453	449	446
998	487	478	472	467	461	457	454	449	446
999	488	478	473	468	462	458	454	450	447
1000	488	479	473	468	462	458	455	450	447

This table was prepared by considering binomial probabilities, such as those in Table B.25. These are lower critical values, $C_{\alpha,n}$; upper critical values are $n - C_{\alpha,n}$.

Example:

$C_{0.01,950} = 434$, with $950 - 434 = 516$ as the upper critical value.

TABLE B.27 CRITICAL VALUES FOR FISHER'S EXACT TEST

n	m_1	m_2	α = 0.50		0.20		0.10		0.05		0.02		0.01		0.005		0.002		0.001	
2	1	1	0, 1	-, -	-, -	-, -	-, -	-, -	-, -	-, -	-, -	-, -	-, -	-, -	-, -	-, -	-, -	-, -	-, -	-, -
3	1	1	-, 1	-, 1	-, -	-, -	-, -	-, -	-, -	-, -	-, -	-, -	-, -	-, -	-, -	-, -	-, -	-, -	-, -	-, -
4	1	1	-, 1	-, 1	-, -	-, -	-, -	-, -	-, -	-, -	-, -	-, -	-, -	-, -	-, -	-, -	-, -	-, -	-, -	-, -
4	1	2	0, 1	-, -	-, -	-, -	-, -	-, -	-, -	-, -	-, -	-, -	-, -	-, -	-, -	-, -	-, -	-, -	-, -	-, -
4	2	2	0, 2	0, 2	0, 2	-, -	-, -	-, -	-, -	-, -	-, -	-, -	-, -	-, -	-, -	-, -	-, -	-, -	-, -	-, -
5	1	1	-, 1	-, 1	-, 1	-, 1	-, -	-, -	-, -	-, -	-, -	-, -	-, -	-, -	-, -	-, -	-, -	-, -	-, -	-, -
5	1	2	-, 1	-, 1	-, -	-, -	-, -	-, -	-, -	-, -	-, -	-, -	-, -	-, -	-, -	-, -	-, -	-, -	-, -	-, -
5	2	2	0, 2	0, 2	-, 2	-, 2	-, 2	-, 2	-, -	-, -	-, -	-, -	-, -	-, -	-, -	-, -	-, -	-, -	-, -	-, -
6	1	1	-, 1	-, 1	-, 1	-, 1	-, -	-, -	-, -	-, -	-, -	-, -	-, -	-, -	-, -	-, -	-, -	-, -	-, -	-, -
6	1	2	-, 1	-, 1	-, -	-, -	-, -	-, -	-, -	-, -	-, -	-, -	-, -	-, -	-, -	-, -	-, -	-, -	-, -	-, -
6	1	3	0, 1	-, -	-, -	-, -	-, -	-, -	-, -	-, -	-, -	-, -	-, -	-, -	-, -	-, -	-, -	-, -	-, -	-, -
6	2	2	0, 2	0, 2	-, 2	-, 2	-, 2	-, 2	-, -	-, -	-, -	-, -	-, -	-, -	-, -	-, -	-, -	-, -	-, -	-, -
6	2	3	0, 2	0, 2	0, 2	-, -	-, -	-, -	-, -	-, -	-, -	-, -	-, -	-, -	-, -	-, -	-, -	-, -	-, -	-, -
6	3	3	1, 2	0, 3	0, 3	0, 3	0, 3	0, 3	0, 3	-, -	-, -	-, -	-, -	-, -	-, -	-, -	-, -	-, -	-, -	-, -
7	1	1	-, 1	-, 1	-, 1	-, 1	-, -	-, -	-, -	-, -	-, -	-, -	-, -	-, -	-, -	-, -	-, -	-, -	-, -	-, -
7	1	2	-, 1	-, 1	-, -	-, -	-, -	-, -	-, -	-, -	-, -	-, -	-, -	-, -	-, -	-, -	-, -	-, -	-, -	-, -
7	1	3	-, 1	-, 1	-, -	-, -	-, -	-, -	-, -	-, -	-, -	-, -	-, -	-, -	-, -	-, -	-, -	-, -	-, -	-, -
7	2	2	0, 2	-, 2	-, 2	-, 2	-, 2	-, 2	-, 2	-, 2	-, -	-, -	-, -	-, -	-, -	-, -	-, -	-, -	-, -	-, -
7	2	3	0, 2	0, 2	-, 2	-, 2	-, -	-, -	-, -	-, -	-, -	-, -	-, -	-, -	-, -	-, -	-, -	-, -	-, -	-, -
7	3	3	0, 2	0, 2	0, 3	0, 3	-, 3	-, 3	-, 3	-, 3	-, -	-, -	-, -	-, -	-, -	-, -	-, -	-, -	-, -	-, -
8	1	1	-, 1	-, 1	-, 1	-, 1	-, -	-, -	-, -	-, -	-, -	-, -	-, -	-, -	-, -	-, -	-, -	-, -	-, -	-, -
8	1	2	-, 1	-, 1	-, -	-, -	-, -	-, -	-, -	-, -	-, -	-, -	-, -	-, -	-, -	-, -	-, -	-, -	-, -	-, -
8	1	3	-, 1	-, 1	-, -	-, -	-, -	-, -	-, -	-, -	-, -	-, -	-, -	-, -	-, -	-, -	-, -	-, -	-, -	-, -
8	1	4	0, 1	-, -	-, -	-, -	-, -	-, -	-, -	-, -	-, -	-, -	-, -	-, -	-, -	-, -	-, -	-, -	-, -	-, -
8	2	2	-, 1	-, 1	-, 2	-, 2	-, 2	-, 2	-, 2	-, 2	-, -	-, -	-, -	-, -	-, -	-, -	-, -	-, -	-, -	-, -
8	2	3	0, 2	0, 2	-, 2	-, 2	-, -	-, -	-, -	-, -	-, -	-, -	-, -	-, -	-, -	-, -	-, -	-, -	-, -	-, -
8	2	4	0, 2	0, 2	-, -	-, -	-, -	-, -	-, -	-, -	-, -	-, -	-, -	-, -	-, -	-, -	-, -	-, -	-, -	-, -
8	3	3	0, 2	0, 2	0, 3	0, 3	-, 3	-, 3	-, 3	-, 3	-, 3	-, 3	-, -	-, -	-, -	-, -	-, -	-, -	-, -	-, -
8	3	4	1, 2	0, 3	0, 3	0, 3	0, 3	-, -	-, -	-, -	-, -	-, -	-, -	-, -	-, -	-, -	-, -	-, -	-, -	-, -
8	4	4	1, 3	1, 3	0, 4	0, 4	0, 4	0, 4	0, 4	0, 4	0, 4	-, -	-, -	-, -	-, -	-, -	-, -	-, -	-, -	-, -
9	1	1	-, 1	-, 1	-, 1	-, 1	-, -	-, -	-, -	-, -	-, -	-, -	-, -	-, -	-, -	-, -	-, -	-, -	-, -	-, -
9	1	2	-, 1	-, 1	-, -	-, -	-, -	-, -	-, -	-, -	-, -	-, -	-, -	-, -	-, -	-, -	-, -	-, -	-, -	-, -
9	1	3	-, 1	-, 1	-, -	-, -	-, -	-, -	-, -	-, -	-, -	-, -	-, -	-, -	-, -	-, -	-, -	-, -	-, -	-, -
9	1	4	-, 1	-, 1	-, -	-, -	-, -	-, -	-, -	-, -	-, -	-, -	-, -	-, -	-, -	-, -	-, -	-, -	-, -	-, -
9	2	2	-, 1	-, 1	-, 2	-, 2	-, 2	-, 2	-, 2	-, 2	-, -	-, -	-, -	-, -	-, -	-, -	-, -	-, -	-, -	-, -
9	2	3	0, 2	0, 2	-, 2	-, 2	-, 2	-, 2	-, -	-, -	-, -	-, -	-, -	-, -	-, -	-, -	-, -	-, -	-, -	-, -
9	2	4	0, 2	0, 2	-, 2	-, 2	-, -	-, -	-, -	-, -	-, -	-, -	-, -	-, -	-, -	-, -	-, -	-, -	-, -	-, -
9	3	3	0, 2	0, 2	-, 3	-, 3	-, 3	-, 3	-, 3	-, 3	-, 3	-, 3	-, -	-, -	-, -	-, -	-, -	-, -	-, -	-, -
9	3	4	0, 2	0, 3	0, 3	0, 3	-, 3	-, 3	-, 3	-, 3	-, -	-, -	-, -	-, -	-, -	-, -	-, -	-, -	-, -	-, -
9	4	4	1, 3	0, 3	0, 3	0, 4	0, 4	0, 4	0, 4	0, 4	-, 4	-, 4	-, 4	-, 4	-, -	-, -	-, -	-, -	-, -	-, -
10	1	1	-, 1	-, 1	-, 1	-, 1	-, 1	-, 1	-, -	-, -	-, -	-, -	-, -	-, -	-, -	-, -	-, -	-, -	-, -	-, -
10	1	2	-, 1	-, 1	-, 1	-, 1	-, -	-, -	-, -	-, -	-, -	-, -	-, -	-, -	-, -	-, -	-, -	-, -	-, -	-, -
10	1	3	-, 1	-, 1	-, -	-, -	-, -	-, -	-, -	-, -	-, -	-, -	-, -	-, -	-, -	-, -	-, -	-, -	-, -	-, -
10	1	4	-, 1	-, 1	-, -	-, -	-, -	-, -	-, -	-, -	-, -	-, -	-, -	-, -	-, -	-, -	-, -	-, -	-, -	-, -
10	1	5	0, 1	-, -	-, -	-, -	-, -	-, -	-, -	-, -	-, -	-, -	-, -	-, -	-, -	-, -	-, -	-, -	-, -	-, -

This table contains critical values of f, where m_1 = smallest of the four marginal totals (i.e., row and column totals); m_2 = smaller marginal total from the margin other than the margin in which m_1 is located; f = observed frequency contributing to both m_1 and m_2; n = total frequency in the table. Given are pairs of critical values; if f is less than or equal to the first, or greater than or equal to the second, member of the pair, then the null hypothesis is rejected at the indicated α. For each α, the first pair of f values refer to one-tailed hypotheses, and the second pair pertain to two-tailed hypotheses. See Section 22.9 for examples of the use of this table.

TABLE B.27 (cont.) CRITICAL VALUES FOR FISHER'S EXACT TEST

n	m_1	m_2	α = 0.50		0.20		0.10		0.05		0.02		0.01		0.005		0.002		0.001	
10	2	2	-, 1	-, 1	-, 2	-, 2	-, 2	-, 2	-, 2	-, 2	-, -	-, -	-, -	-, -	-, -	-, -	-, -	-, -	-, -	-, -
10	2	3	0, 2	-, 2	-, 2	-, 2	-, 2	-, 2	-, -	-, -	-, -	-, -	-, -	-, -	-, -	-, -	-, -	-, -	-, -	-, -
10	2	4	0, 2	0, 2	-, 2	-, 2	-, -	-, -	-, -	-, -	-, -	-, -	-, -	-, -	-, -	-, -	-, -	-, -	-, -	-, -
10	2	5	0, 2	0, 2	-, -	-, -	-, -	-, -	-, -	-, -	-, -	-, -	-, -	-, -	-, -	-, -	-, -	-, -	-, -	-, -
10	3	3	0, 2	0, 2	-, 2	-, 2	-, 3	-, 3	-, 3	-, 3	-, 3	-, 3	-, 3	-, 3	-, -	-, -	-, -	-, -	-, -	-, -
10	3	4	0, 2	0, 2	0, 3	0, 3	-, 3	-, 3	-, 3	-, 3	-, -	-, -	-, -	-, -	-, -	-, -	-, -	-, -	-, -	-, -
10	3	5	1, 2	0, 3	0, 3	0, 3	0, 3	-, -	-, -	-, -	-, -	-, -	-, -	-, -	-, -	-, -	-, -	-, -	-, -	-, -
10	4	4	1, 3	0, 3	0, 3	0, 3	0, 4	0, 4	-, 4	-, 4	-, 4	-, 4	-, 4	-, 4	-, 4	-, 4	-, -	-, -	-, -	-, -
10	4	5	1, 3	0, 4	0, 4	0, 4	0, 4	0, 4	0, 4	0, 4	-, -	-, -	-, -	-, -	-, -	-, -	-, -	-, -	-, -	-, -
10	5	5	2, 3	1, 4	1, 4	0, 5	0, 5	0, 5	0, 5	0, 5	0, 5	0, 5	0, 5	0, 5	0, 5	-, -	-, -	-, -	-, -	-, -
11	1	1	-, 1	-, 1	-, 1	-, 1	-, 1	-, 1	-, -	-, -	-, -	-, -	-, -	-, -	-, -	-, -	-, -	-, -	-, -	-, -
11	1	2	-, 1	-, 1	-, 1	-, 1	-, -	-, -	-, -	-, -	-, -	-, -	-, -	-, -	-, -	-, -	-, -	-, -	-, -	-, -
11	1	3	-, 1	-, 1	-, -	-, -	-, -	-, -	-, -	-, -	-, -	-, -	-, -	-, -	-, -	-, -	-, -	-, -	-, -	-, -
11	1	4	-, 1	-, 1	-, -	-, -	-, -	-, -	-, -	-, -	-, -	-, -	-, -	-, -	-, -	-, -	-, -	-, -	-, -	-, -
11	1	5	-, 1	-, 1	-, -	-, -	-, -	-, -	-, -	-, -	-, -	-, -	-, -	-, -	-, -	-, -	-, -	-, -	-, -	-, -
11	2	2	-, 1	-, 1	-, 2	-, 2	-, 2	-, 2	-, 2	-, 2	-, 2	-, 2	-, -	-, -	-, -	-, -	-, -	-, -	-, -	-, -
11	2	3	-, 1	-, 1	-, 2	-, 2	-, 2	-, 2	-, -	-, -	-, -	-, -	-, -	-, -	-, -	-, -	-, -	-, -	-, -	-, -
11	2	4	0, 2	0, 2	-, 2	-, 2	-, -	-, -	-, -	-, -	-, -	-, -	-, -	-, -	-, -	-, -	-, -	-, -	-, -	-, -
11	2	5	0, 2	0, 2	-, 2	-, 2	-, -	-, -	-, -	-, -	-, -	-, -	-, -	-, -	-, -	-, -	-, -	-, -	-, -	-, -
11	3	3	0, 2	0, 2	-, 2	-, 2	-, 3	-, 3	-, 3	-, 3	-, 3	-, 3	-, 3	-, 3	-, -	-, -	-, -	-, -	-, -	-, -
11	3	4	0, 2	0, 2	-, 3	-, 3	-, 3	-, 3	-, 3	-, 3	-, -	-, -	-, -	-, -	-, -	-, -	-, -	-, -	-, -	-, -
11	3	5	0, 2	0, 3	0, 3	0, 3	-, 3	-, 3	-, -	-, -	-, -	-, -	-, -	-, -	-, -	-, -	-, -	-, -	-, -	-, -
11	4	4	0, 2	0, 3	0, 3	0, 3	-, 3	-, 3	-, 4	-, 4	-, 4	-, 4	-, 4	-, 4	-, 4	-, 4	-, -	-, -	-, -	-, -
11	4	5	1, 3	0, 3	0, 3	0, 4	0, 4	0, 4	0, 4	-, 4	-, 4	-, 4	-, -	-, -	-, -	-, -	-, -	-, -	-, -	-, -
11	5	5	1, 3	1, 4	1, 4	0, 4	0, 4	0, 4	0, 5	0, 5	0, 5	0, 5	-, 5	-, 5	-, 5	-, 5	-, -	-, -	-, -	-, -
12	1	1	-, 1	-, 1	-, 1	-, 1	-, 1	-, 1	-, -	-, -	-, -	-, -	-, -	-, -	-, -	-, -	-, -	-, -	-, -	-, -
12	1	2	-, 1	-, 1	-, 1	-, 1	-, -	-, -	-, -	-, -	-, -	-, -	-, -	-, -	-, -	-, -	-, -	-, -	-, -	-, -
12	1	3	-, 1	-, 1	-, -	-, -	-, -	-, -	-, -	-, -	-, -	-, -	-, -	-, -	-, -	-, -	-, -	-, -	-, -	-, -
12	1	4	-, 1	-, 1	-, -	-, -	-, -	-, -	-, -	-, -	-, -	-, -	-, -	-, -	-, -	-, -	-, -	-, -	-, -	-, -
12	1	5	-, 1	-, 1	-, -	-, -	-, -	-, -	-, -	-, -	-, -	-, -	-, -	-, -	-, -	-, -	-, -	-, -	-, -	-, -
12	1	6	0, 1	-, -	-, -	-, -	-, -	-, -	-, -	-, -	-, -	-, -	-, -	-, -	-, -	-, -	-, -	-, -	-, -	-, -
12	2	2	-, 1	-, 1	-, 2	-, 2	-, 2	-, 2	-, 2	-, 2	-, 2	-, 2	-, -	-, -	-, -	-, -	-, -	-, -	-, -	-, -
12	2	3	-, 1	-, 1	-, 2	-, 2	-, 2	-, 2	-, 2	-, 2	-, -	-, -	-, -	-, -	-, -	-, -	-, -	-, -	-, -	-, -
12	2	4	0, 2	-, 2	-, 2	-, 2	-, 2	-, 2	-, -	-, -	-, -	-, -	-, -	-, -	-, -	-, -	-, -	-, -	-, -	-, -
12	2	5	0, 2	0, 2	-, 2	-, 2	-, -	-, -	-, -	-, -	-, -	-, -	-, -	-, -	-, -	-, -	-, -	-, -	-, -	-, -
12	2	6	0, 2	0, 2	-, -	-, -	-, -	-, -	-, -	-, -	-, -	-, -	-, -	-, -	-, -	-, -	-, -	-, -	-, -	-, -
12	3	3	0, 2	-, 2	-, 2	-, 2	-, 3	-, 3	-, 3	-, 3	-, 3	-, 3	-, 3	-, 3	-, 3	-, 3	-, -	-, -	-, -	-, -
12	3	4	0, 2	0, 2	-, 3	-, 3	-, 3	-, 3	-, 3	-, 3	-, 3	-, 3	-, -	-, -	-, -	-, -	-, -	-, -	-, -	-, -
12	3	5	0, 2	0, 3	0, 3	-, 3	-, 3	-, 3	-, 3	-, 3	-, -	-, -	-, -	-, -	-, -	-, -	-, -	-, -	-, -	-, -
12	3	6	1, 2	0, 3	0, 3	0, 3	0, 3	-, -	-, -	-, -	-, -	-, -	-, -	-, -	-, -	-, -	-, -	-, -	-, -	-, -
12	4	4	0, 2	0, 3	0, 3	-, 3	-, 3	-, 3	-, 4	-, 4	-, 4	-, 4	-, 4	-, 4	-, 4	-, 4	-, -	-, -	-, -	-, -
12	4	5	1, 3	0, 3	0, 3	0, 4	0, 4	0, 4	-, 4	-, 4	-, 4	-, 4	-, -	-, -	-, -	-, -	-, -	-, -	-, -	-, -
12	4	6	1, 3	0, 4	0, 4	0, 4	0, 4	0, 4	0, 4	-, -	-, -	-, -	-, -	-, -	-, -	-, -	-, -	-, -	-, -	-, -
12	5	5	1, 3	1, 4	0, 4	0, 4	0, 4	0, 4	0, 4	0, 5	-, 5	-, 5	-, 5	-, 5	-, 5	-, 5	-, 5	-, 5	-, -	-, -
12	5	6	2, 3	1, 4	1, 4	0, 5	0, 5	0, 5	0, 5	0, 5	0, 5	0, 5	0, 5	-, -	-, -	-, -	-, -	-, -	-, -	-, -
12	6	6	2, 4	1, 5	1, 5	1, 5	1, 5	1, 5	1, 5	0, 6	0, 6	0, 6	0, 6	0, 6	0, 6	0, 6	0, 6	-, -	-, -	-, -
13	1	1	-, 1	-, 1	-, 1	-, 1	-, 1	-, 1	-, -	-, -	-, -	-, -	-, -	-, -	-, -	-, -	-, -	-, -	-, -	-, -
13	1	2	-, 1	-, 1	-, 1	-, 1	-, -	-, -	-, -	-, -	-, -	-, -	-, -	-, -	-, -	-, -	-, -	-, -	-, -	-, -
13	1	3	-, 1	-, 1	-, -	-, -	-, -	-, -	-, -	-, -	-, -	-, -	-, -	-, -	-, -	-, -	-, -	-, -	-, -	-, -
13	1	4	-, 1	-, 1	-, -	-, -	-, -	-, -	-, -	-, -	-, -	-, -	-, -	-, -	-, -	-, -	-, -	-, -	-, -	-, -

TABLE B.27 (cont.) CRITICAL VALUES FOR FISHER'S EXACT TEST

n	m_1	m_2	α = 0.50	0.20	0.10	0.05	0.02	0.01	0.005	0.002	0.001
13	1	5	-, 1 -, 1	-, - -, -	-, - -, -	-, - -, -	-, - -, -	-, - -, -	-, - -, -	-, - -, -	-, - -, -
13	1	6	-, 1 -, 1	-, - -, -	-, - -, -	-, - -, -	-, - -, -	-, - -, -	-, - -, -	-, - -, -	-, - -, -
13	2	2	-, 1 -, 1	-, 2 -, 2	-, 2 -, 2	-, 2 -, 2	-, 2 -, 2	-, - -, -	-, - -, -	-, - -, -	-, - -, -
13	2	3	-, 1 -, 1	-, 2 -, 2	-, 2 -, 2	-, 2 -, 2	-, - -, -	-, - -, -	-, - -, -	-, - -, -	-, - -, -
13	2	4	0, 2 -, 2	-, 2 -, 2	-, 2 -, 2	-, - -, -	-, - -, -	-, - -, -	-, - -, -	-, - -, -	-, - -, -
13	2	5	0, 2 0, 2	-, 2 -, 2	-, - -, -	-, - -, -	-, - -, -	-, - -, -	-, - -, -	-, - -, -	-, - -, -
13	2	6	0, 2 0, 2	-, 2 -, 2	-, - -, -	-, - -, -	-, - -, -	-, - -, -	-, - -, -	-, - -, -	-, - -, -
13	3	3	0, 2 -, 2	-, 2 -, 2	-, 3 -, 3	-, 3 -, 3	-, 3 -, 3	-, 3 -, 3	-, 3 -, 3	-, - -, -	-, - -, -
13	3	4	0, 2 0, 2	-, 3 -, 3	-, 3 -, 3	-, 3 -, 3	-, 3 -, 3	-, - -, -	-, - -, -	-, - -, -	-, - -, -
13	3	5	0, 2 0, 3	0, 3 -, 3	-, 3 -, 3	-, 3 -, 3	-, - -, -	-, - -, -	-, - -, -	-, - -, -	-, - -, -
13	3	6	0, 2 0, 3	0, 3 0, 3	-, 3 -, 3	-, - -, -	-, - -, -	-, - -, -	-, - -, -	-, - -, -	-, - -, -
13	4	4	0, 2 0, 3	0, 3 -, 3	-, 3 -, 3	-, 4 -, 4	-, 4 -, 4	-, 4 -, 4	-, 4 -, 4	-, 4 -, 4	-, - -, -
13	4	5	1, 3 0, 3	0, 3 0, 4	0, 4 -, 4	-, 4 -, 4	-, 4 -, 4	-, 4 -, 4	-, - -, -	-, - -, -	-, - -, -
13	4	6	1, 3 0, 3	0, 4 0, 4	0, 4 0, 4	0, 4 -, 4	-, - -, -	-, - -, -	-, - -, -	-, - -, -	-, - -, -
13	5	5	1, 3 0, 3	0, 4 0, 4	0, 4 0, 4	0, 4 -, 4	-, 5 -, 5	-, 5 -, 5	-, 5 -, 5	-, 5 -, 5	-, 5 -, 5
13	5	6	1, 3 1, 4	1, 4 0, 4	0, 4 0, 5	0, 5 0, 5	0, 5 -, 5	-, 5 -, 5	-, 5 -, 5	-, - -, -	-, - -, -
13	6	6	2, 4 1, 4	1, 5 1, 5	1, 5 0, 5	0, 5 0, 5	0, 6 0, 6	0, 6 0, 6	0, 6 0, 6	-, 6 -, 6	-, 6 -, 6
14	1	1	-, 1 -, 1	-, 1 -, 1	-, 1 -, 1	-, - -, -	-, - -, -	-, - -, -	-, - -, -	-, - -, -	-, - -, -
14	1	2	-, 1 -, 1	-, 1 -, 1	-, - -, -	-, - -, -	-, - -, -	-, - -, -	-, - -, -	-, - -, -	-, - -, -
14	1	3	-, 1 -, 1	-, - -, -	-, - -, -	-, - -, -	-, - -, -	-, - -, -	-, - -, -	-, - -, -	-, - -, -
14	1	4	-, 1 -, 1	-, - -, -	-, - -, -	-, - -, -	-, - -, -	-, - -, -	-, - -, -	-, - -, -	-, - -, -
14	1	5	-, 1 -, 1	-, - -, -	-, - -, -	-, - -, -	-, - -, -	-, - -, -	-, - -, -	-, - -, -	-, - -, -
14	1	6	-, 1 -, 1	-, - -, -	-, - -, -	-, - -, -	-, - -, -	-, - -, -	-, - -, -	-, - -, -	-, - -, -
14	1	7	0, 1 -, -	-, - -, -	-, - -, -	-, - -, -	-, - -, -	-, - -, -	-, - -, -	-, - -, -	-, - -, -
14	2	2	-, 1 -, 1	-, 2 -, 2	-, 2 -, 2	-, 2 -, 2	-, 2 -, 2	-, - -, -	-, - -, -	-, - -, -	-, - -, -
14	2	3	-, 1 -, 1	-, 2 -, 2	-, 2 -, 2	-, 2 -, 2	-, - -, -	-, - -, -	-, - -, -	-, - -, -	-, - -, -
14	2	4	0, 1 -, 2	-, 2 -, 2	-, 2 -, 2	-, - -, -	-, - -, -	-, - -, -	-, - -, -	-, - -, -	-, - -, -
14	2	5	0, 2 -, 2	-, 2 -, 2	-, - -, -	-, - -, -	-, - -, -	-, - -, -	-, - -, -	-, - -, -	-, - -, -
14	2	6	0, 2 0, 2	-, 2 -, 2	-, - -, -	-, - -, -	-, - -, -	-, - -, -	-, - -, -	-, - -, -	-, - -, -
14	2	7	0, 2 0, 2	-, - -, -	-, - -, -	-, - -, -	-, - -, -	-, - -, -	-, - -, -	-, - -, -	-, - -, -
14	3	3	0, 2 -, 2	-, 2 -, 2	-, 2 -, 2	-, 3 -, 3	-, 3 -, 3	-, 3 -, 3	-, 3 -, 3	-, - -, -	-, - -, -
14	3	4	0, 2 -, 2	-, 2 -, 2	-, 3 -, 3	-, 3 -, 3	-, 3 -, 3	-, - -, -	-, - -, -	-, - -, -	-, - -, -
14	3	5	0, 2 0, 3	-, 3 -, 3	-, 3 -, 3	-, 3 -, 3	-, - -, -	-, - -, -	-, - -, -	-, - -, -	-, - -, -
14	3	6	0, 2 0, 3	0, 3 -, 3	-, 3 -, 3	-, - -, -	-, - -, -	-, - -, -	-, - -, -	-, - -, -	-, - -, -
14	3	7	1, 2 0, 3	0, 3 0, 3	0, 3 -, -	-, - -, -	-, - -, -	-, - -, -	-, - -, -	-, - -, -	-, - -, -
14	4	4	0, 2 0, 3	-, 3 -, 3	-, 3 -, 3	-, 3 -, 3	-, 4 -, 4	-, 4 -, 4	-, 4 -, 4	-, 4 -, 4	-, 4 -, 4
14	4	5	0, 2 0, 3	0, 3 -, 3	-, 3 -, 3	-, 4 -, 4	-, 4 -, 4	-, 4 -, 4	-, 4 -, 4	-, - -, -	-, - -, -
14	4	6	1, 3 0, 3	0, 3 0, 4	0, 4 0, 4	-, 4 -, 4	-, 4 -, 4	-, - -, -	-, - -, -	-, - -, -	-, - -, -
14	4	7	1, 3 0, 4	0, 4 0, 4	0, 4 0, 4	0, 4 -, -	-, - -, -	-, - -, -	-, - -, -	-, - -, -	-, - -, -
14	5	5	1, 3 0, 3	0, 4 0, 4	0, 4 0, 4	-, 4 -, 4	-, 5 -, 5	-, 5 -, 5	-, 5 -, 5	-, 5 -, 5	-, 5 -, 5
14	5	6	1, 3 1, 4	0, 4 0, 4	0, 4 0, 4	0, 5 0, 5	-, 5 -, 5	-, 5 -, 5	-, 5 -, 5	-, - -, -	-, - -, -
14	5	7	2, 3 1, 4	1, 4 0, 5	0, 5 0, 5	0, 5 0, 5	0, 5 -, -	-, - -, -	-, - -, -	-, - -, -	-, - -, -
14	6	6	2, 4 1, 4	1, 4 1, 5	0, 5 0, 5	0, 5 0, 5	0, 5 0, 6	0, 6 0, 6	-, 6 -, 6	-, 6 -, 6	-, 6 -, 6
14	6	7	2, 4 1, 5	1, 5 1, 5	1, 5 0, 6	0, 6 0, 6	0, 6 0, 6	0, 6 0, 6	0, 6 0, 6	-, - -, -	-, - -, -
14	7	7	3, 4 2, 5	2, 5 1, 6	1, 6 1, 6	1, 6 1, 6	1, 6 0, 7	0, 7 0, 7	0, 7 0, 7	0, 7 0, 7	0, 7 0, 7
15	1	1	-, 1 -, 1	-, 1 -, 1	-, 1 -, 1	-, - -, -	-, - -, -	-, - -, -	-, - -, -	-, - -, -	-, - -, -
15	1	2	-, 1 -, 1	-, 1 -, 1	-, - -, -	-, - -, -	-, - -, -	-, - -, -	-, - -, -	-, - -, -	-, - -, -
15	1	3	-, 1 -, 1	-, 1 -, 1	-, - -, -	-, - -, -	-, - -, -	-, - -, -	-, - -, -	-, - -, -	-, - -, -
15	1	4	-, 1 -, 1	-, - -, -	-, - -, -	-, - -, -	-, - -, -	-, - -, -	-, - -, -	-, - -, -	-, - -, -
15	1	5	-, 1 -, 1	-, - -, -	-, - -, -	-, - -, -	-, - -, -	-, - -, -	-, - -, -	-, - -, -	-, - -, -

TABLE B.27 (cont.) CRITICAL VALUES FOR FISHER'S EXACT TEST

n	m_1	m_2	α = 0.50	0.20	0.10	0.05	0.02	0.01	0.005	0.002	0.001
15	1	6	-, 1 -, 1	-, - -, -	-, - -, -	-, - -, -	-, - -, -	-, - -, -	-, - -, -	-, - -, -	-, - -, -
15	1	7	-, 1 -, 1	-, - -, -	-, - -, -	-, - -, -	-, - -, -	-, - -, -	-, - -, -	-, - -, -	-, - -, -
15	2	2	-, 1 -, 1	-, 2 -, 2	-, 2 -, 2	-, 2 -, 2	-, 2 -, 2	-, 2 -, 2	-, - -, -	-, - -, -	-, - -, -
15	2	3	-, 1 -, 1	-, 2 -, 2	-, 2 -, 2	-, 2 -, 2	-, - -, -	-, - -, -	-, - -, -	-, - -, -	-, - -, -
15	2	4	-, 1 -, 1	-, 2 -, 2	-, 2 -, 2	-, - -, -	-, - -, -	-, - -, -	-, - -, -	-, - -, -	-, - -, -
15	2	5	0, 2 -, 2	-, 2 -, 2	-, 2 -, 2	-, - -, -	-, - -, -	-, - -, -	-, - -, -	-, - -, -	-, - -, -
15	2	6	0, 2 0, 2	-, 2 -, 2	-, - -, -	-, - -, -	-, - -, -	-, - -, -	-, - -, -	-, - -, -	-, - -, -
15	2	7	0, 2 0, 2	-, 2 -, 2	-, - -, -	-, - -, -	-, - -, -	-, - -, -	-, - -, -	-, - -, -	-, - -, -
15	3	3	0, 1 -, 2	-, 2 -, 2	-, 2 -, 2	-, 3 -, 3	-, 3 -, 3	-, 3 -, 3	-, 3 -, 3	-, - -, -	-, - -, -
15	3	4	0, 2 -, 2	-, 2 -, 2	-, 3 -, 3	-, 3 -, 3	-, 3 -, 3	-, 3 -, 3	-, - -, -	-, - -, -	-, - -, -
15	3	5	0, 2 -, 2	-, 3 -, 3	-, 3 -, 3	-, 3 -, 3	-, - -, -	-, - -, -	-, - -, -	-, - -, -	-, - -, -
15	3	6	0, 2 0, 3	0, 3 -, 3	-, 3 -, 3	-, 3 -, 3	-, - -, -	-, - -, -	-, - -, -	-, - -, -	-, - -, -
15	3	7	0, 2 0, 3	0, 3 0, 3	-, 3 -, 3	-, - -, -	-, - -, -	-, - -, -	-, - -, -	-, - -, -	-, - -, -
15	4	4	0, 2 -, 3	-, 3 -, 3	-, 3 -, 3	-, 3 -, 3	-, 4 -, 4	-, 4 -, 4	-, 4 -, 4	-, 4 -, 4	-, 4 -, 4
15	4	5	0, 2 0, 3	0, 3 -, 3	-, 3 -, 3	-, 4 -, 4	-, 4 -, 4	-, 4 -, 4	-, 4 -, 4	-, - -, -	-, - -, -
15	4	6	1, 3 0, 3	0, 3 0, 4	0, 4 -, 4	-, 4 -, 4	-, 4 -, 4	-, - -, -	-, - -, -	-, - -, -	-, - -, -
15	4	7	1, 3 0, 3	0, 4 0, 4	0, 4 0, 4	-, 4 -, 4	-, - -, -	-, - -, -	-, - -, -	-, - -, -	-, - -, -
15	5	5	1, 3 0, 3	0, 3 0, 4	0, 4 -, 4	-, 4 -, 4	-, 4 -, 4	-, 5 -, 5	-, 5 -, 5	-, 5 -, 5	-, 5 -, 5
15	5	6	1, 3 0, 3	0, 4 0, 4	0, 4 0, 4	0, 4 0, 5	-, 5 -, 5	-, 5 -, 5	-, 5 -, 5	-, 5 -, 5	-, - -, -
15	5	7	1, 3 1, 4	1, 4 0, 4	0, 5 0, 5	0, 5 0, 5	0, 5 -, 5	-, 5 -, 5	-, - -, -	-, - -, -	-, - -, -
15	6	6	1, 4 1, 4	1, 4 0, 4	0, 5 0, 5	0, 5 0, 5	0, 5 -, 5	-, 6 -, 6	-, 6 -, 6	-, 6 -, 6	-, 6 -, 6
15	6	7	2, 4 1, 4	1, 5 1, 5	1, 5 0, 5	0, 5 0, 5	0, 6 0, 6	0, 6 0, 6	-, 6 -, 6	-, 6 -, 6	-, - -, -
15	7	7	2, 4 2, 5	1, 5 1, 5	1, 6 1, 6	1, 6 1, 6	0, 6 0, 6	0, 6 0, 7	0, 7 0, 7	0, 7 0, 7	-, 7 -, 7
16	1	1	-, 1 -, 1	-, 1 -, 1	-, 1 -, 1	-, - -, -	-, - -, -	-, - -, -	-, - -, -	-, - -, -	-, - -, -
16	1	2	-, 1 -, 1	-, 1 -, 1	-, - -, -	-, - -, -	-, - -, -	-, - -, -	-, - -, -	-, - -, -	-, - -, -
16	1	3	-, 1 -, 1	-, 1 -, 1	-, - -, -	-, - -, -	-, - -, -	-, - -, -	-, - -, -	-, - -, -	-, - -, -
16	1	4	-, 1 -, 1	-, - -, -	-, - -, -	-, - -, -	-, - -, -	-, - -, -	-, - -, -	-, - -, -	-, - -, -
16	1	5	-, 1 -, 1	-, - -, -	-, - -, -	-, - -, -	-, - -, -	-, - -, -	-, - -, -	-, - -, -	-, - -, -
16	1	6	-, 1 -, 1	-, - -, -	-, - -, -	-, - -, -	-, - -, -	-, - -, -	-, - -, -	-, - -, -	-, - -, -
16	1	7	-, 1 -, 1	-, - -, -	-, - -, -	-, - -, -	-, - -, -	-, - -, -	-, - -, -	-, - -, -	-, - -, -
16	1	8	0, 1 -, -	-, - -, -	-, - -, -	-, - -, -	-, - -, -	-, - -, -	-, - -, -	-, - -, -	-, - -, -
16	2	2	-, 1 -, 1	-, 2 -, 2	-, 2 -, 2	-, 2 -, 2	-, 2 -, 2	-, 2 -, 2	-, - -, -	-, - -, -	-, - -, -
16	2	3	-, 1 -, 1	-, 2 -, 2	-, 2 -, 2	-, 2 -, 2	-, - -, -	-, - -, -	-, - -, -	-, - -, -	-, - -, -
16	2	4	-, 1 -, 1	-, 2 -, 2	-, 2 -, 2	-, 2 -, 2	-, - -, -	-, - -, -	-, - -, -	-, - -, -	-, - -, -
16	2	5	0, 2 -, 2	-, 2 -, 2	-, 2 -, 2	-, - -, -	-, - -, -	-, - -, -	-, - -, -	-, - -, -	-, - -, -
16	2	6	0, 2 0, 2	-, 2 -, 2	-, - -, -	-, - -, -	-, - -, -	-, - -, -	-, - -, -	-, - -, -	-, - -, -
16	2	7	0, 2 0, 2	-, 2 -, 2	-, - -, -	-, - -, -	-, - -, -	-, - -, -	-, - -, -	-, - -, -	-, - -, -
16	2	8	0, 2 0, 2	-, - -, -	-, - -, -	-, - -, -	-, - -, -	-, - -, -	-, - -, -	-, - -, -	-, - -, -
16	3	3	-, 1 -, 1	-, 2 -, 2	-, 2 -, 2	-, 3 -, 3	-, 3 -, 3	-, 3 -, 3	-, 3 -, 3	-, 3 -, 3	-, - -, -
16	3	4	0, 2 -, 2	-, 2 -, 2	-, 3 -, 3	-, 3 -, 3	-, 3 -, 3	-, 3 -, 3	-, - -, -	-, - -, -	-, - -, -
16	3	5	0, 2 -, 2	-, 3 -, 3	-, 3 -, 3	-, 3 -, 3	-, 3 -, 3	-, - -, -	-, - -, -	-, - -, -	-, - -, -
16	3	6	0, 2 0, 3	-, 3 -, 3	-, 3 -, 3	-, 3 -, 3	-, - -, -	-, - -, -	-, - -, -	-, - -, -	-, - -, -
16	3	7	0, 2 0, 3	0, 3 -, 3	-, 3 -, 3	-, - -, -	-, - -, -	-, - -, -	-, - -, -	-, - -, -	-, - -, -
16	3	8	1, 2 0, 3	0, 3 0, 3	0, 3 -, -	-, - -, -	-, - -, -	-, - -, -	-, - -, -	-, - -, -	-, - -, -
16	4	4	0, 2 -, 2	-, 3 -, 3	-, 3 -, 3	-, 3 -, 3	-, 4 -, 4	-, 4 -, 4	-, 4 -, 4	-, 4 -, 4	-, 4 -, 4
16	4	5	0, 2 0, 3	0, 3 -, 3	-, 3 -, 3	-, 4 -, 4	-, 4 -, 4	-, 4 -, 4	-, 4 -, 4	-, - -, -	-, - -, -
16	4	6	0, 2 0, 3	0, 3 -, 3	-, 4 -, 4	-, 4 -, 4	-, 4 -, 4	-, 4 -, 4	-, - -, -	-, - -, -	-, - -, -
16	4	7	1, 3 0, 3	0, 3 0, 4	0, 4 0, 4	-, 4 -, 4	-, 4 -, 4	-, - -, -	-, - -, -	-, - -, -	-, - -, -
16	4	8	1, 3 0, 4	0, 4 0, 4	0, 4 0, 4	0, 4 -, -	-, - -, -	-, - -, -	-, - -, -	-, - -, -	-, - -, -
16	5	5	1, 3 0, 3	0, 3 0, 4	-, 4 -, 4	-, 4 -, 4	-, 4 -, 4	-, 5 -, 5	-, 5 -, 5	-, 5 -, 5	-, 5 -, 5

TABLE B.27 (cont.) CRITICAL VALUES FOR FISHER'S EXACT TEST

n	m_1	m_2	α = 0.50	0.20	0.10	0.05	0.02	0.01	0.005	0.002	0.001
16	5	6	1, 3 0, 3	0, 4 0, 4	0, 4 0, 4	-, 4 -, 4	-, 5 -, 5	-, 5 -, 5	-, 5 -, 5	-, 5 -, 5	-, - -, -
16	5	7	1, 3 1, 4	0, 4 0, 4	0, 4 0, 5	0, 5 0, 5	-, 5 -, 5	-, 5 -, 5	-, 5 -, 5	-, - -, -	-, - -, -
16	5	8	2, 3 1, 4	1, 4 0, 5	0, 5 0, 5	0, 5 0, 5	0, 5 -, -	-, - -, -	-, - -, -	-, - -, -	-, - -, -
16	6	6	1, 4 1, 4	0, 4 0, 4	0, 4 0, 5	0, 5 0, 5	-, 5 -, 5	-, 5 -, 5	-, 6 -, 6	-, 6 -, 6	-, 6 -, 6
16	6	7	2, 4 1, 4	1, 4 1, 5	0, 5 0, 5	0, 5 0, 5	0, 6 0, 6	-, 6 -, 6	-, 6 -, 6	-, 6 -, 6	-, 6 -, 6
16	6	8	2, 4 1, 5	1, 5 1, 5	1, 5 0, 6	0, 6 0, 6	0, 6 0, 6	0, 6 0, 6	0, 6 -, -	-, - -, -	-, - -, -
16	7	7	2, 4 2, 5	1, 5 1, 5	1, 5 1, 6	0, 6 0, 6	0, 6 0, 6	0, 6 0, 6	0, 7 0, 7	-, 7 -, 7	-, 7 -, 7
16	7	8	3, 4 2, 5	2, 5 1, 6	1, 6 1, 6	1, 6 1, 6	0, 7 0, 7	0, 7 0, 7	0, 7 0, 7	0, 7 0, 7	0, 7 -, -
16	8	8	3, 5 2, 6	2, 6 2, 6	2, 6 1, 7	1, 7 1, 7	1, 7 1, 7	1, 7 0, 8	0, 8 0, 8	0, 8 0, 8	0, 8 0, 8
17	1	1	-, 1 -, 1	-, 1 -, 1	-, 1 -, 1	-, - -, -	-, - -, -	-, - -, -	-, - -, -	-, - -, -	-, - -, -
17	1	2	-, 1 -, 1	-, 1 -, 1	-, - -, -	-, - -, -	-, - -, -	-, - -, -	-, - -, -	-, - -, -	-, - -, -
17	1	3	-, 1 -, 1	-, 1 -, 1	-, - -, -	-, - -, -	-, - -, -	-, - -, -	-, - -, -	-, - -, -	-, - -, -
17	1	4	-, 1 -, 1	-, - -, -	-, - -, -	-, - -, -	-, - -, -	-, - -, -	-, - -, -	-, - -, -	-, - -, -
17	1	5	-, 1 -, 1	-, - -, -	-, - -, -	-, - -, -	-, - -, -	-, - -, -	-, - -, -	-, - -, -	-, - -, -
17	1	6	-, 1 -, 1	-, - -, -	-, - -, -	-, - -, -	-, - -, -	-, - -, -	-, - -, -	-, - -, -	-, - -, -
17	1	7	-, 1 -, 1	-, - -, -	-, - -, -	-, - -, -	-, - -, -	-, - -, -	-, - -, -	-, - -, -	-, - -, -
17	1	8	-, 1 -, 1	-, - -, -	-, - -, -	-, - -, -	-, - -, -	-, - -, -	-, - -, -	-, - -, -	-, - -, -
17	2	2	-, 1 -, 1	-, 2 -, 2	-, 2 -, 2	-, 2 -, 2	-, 2 -, 2	-, 2 -, 2	-, - -, -	-, - -, -	-, - -, -
17	2	3	-, 1 -, 1	-, 2 -, 2	-, 2 -, 2	-, 2 -, 2	-, - -, -	-, - -, -	-, - -, -	-, - -, -	-, - -, -
17	2	4	-, 1 -, 1	-, 2 -, 2	-, 2 -, 2	-, 2 -, 2	-, - -, -	-, - -, -	-, - -, -	-, - -, -	-, - -, -
17	2	5	0, 1 -, 2	-, 2 -, 2	-, 2 -, 2	-, - -, -	-, - -, -	-, - -, -	-, - -, -	-, - -, -	-, - -, -
17	2	6	0, 2 -, 2	-, 2 -, 2	-, - -, -	-, - -, -	-, - -, -	-, - -, -	-, - -, -	-, - -, -	-, - -, -
17	2	7	0, 2 0, 2	-, 2 -, 2	-, - -, -	-, - -, -	-, - -, -	-, - -, -	-, - -, -	-, - -, -	-, - -, -
17	2	8	0, 2 0, 2	-, - -, -	-, - -, -	-, - -, -	-, - -, -	-, - -, -	-, - -, -	-, - -, -	-, - -, -
17	3	3	-, 1 -, 1	-, 2 -, 2	-, 2 -, 2	-, 3 -, 3	-, 3 -, 3	-, 3 -, 3	-, 3 -, 3	-, 3 -, 3	-, - -, -
17	3	4	0, 2 -, 2	-, 2 -, 2	-, 3 -, 3	-, 3 -, 3	-, 3 -, 3	-, 3 -, 3	-, - -, -	-, - -, -	-, - -, -
17	3	5	0, 2 -, 2	-, 2 -, 2	-, 3 -, 3	-, 3 -, 3	-, 3 -, 3	-, - -, -	-, - -, -	-, - -, -	-, - -, -
17	3	6	0, 2 -, 3	-, 3 -, 3	-, 3 -, 3	-, 3 -, 3	-, - -, -	-, - -, -	-, - -, -	-, - -, -	-, - -, -
17	3	7	0, 2 0, 3	0, 3 -, 3	-, 3 -, 3	-, - -, -	-, - -, -	-, - -, -	-, - -, -	-, - -, -	-, - -, -
17	3	8	0, 2 0, 3	0, 3 -, 3	-, 3 -, 3	-, - -, -	-, - -, -	-, - -, -	-, - -, -	-, - -, -	-, - -, -
17	4	4	0, 2 -, 2	-, 2 -, 3	-, 3 -, 3	-, 3 -, 3	-, 4 -, 4	-, 4 -, 4	-, 4 -, 4	-, 4 -, 4	-, 4 -, 4
17	4	5	0, 2 0, 3	-, 3 -, 3	-, 3 -, 3	-, 4 -, 4	-, 4 -, 4	-, 4 -, 4	-, 4 -, 4	-, - -, -	-, - -, -
17	4	6	0, 2 0, 3	0, 3 -, 3	-, 3 -, 3	-, 4 -, 4	-, 4 -, 4	-, 4 -, 4	-, - -, -	-, - -, -	-, - -, -
17	4	7	1, 3 0, 3	0, 3 0, 4	0, 4 -, 4	-, 4 -, 4	-, 4 -, 4	-, - -, -	-, - -, -	-, - -, -	-, - -, -
17	4	8	1, 3 0, 3	0, 4 0, 4	0, 4 0, 4	-, 4 -, 4	-, - -, -	-, - -, -	-, - -, -	-, - -, -	-, - -, -
17	5	5	0, 2 0, 3	0, 3 -, 3	-, 4 -, 4	-, 4 -, 4	-, 4 -, 4	-, 4 -, 4	-, 5 -, 5	-, 5 -, 5	-, 5 -, 5
17	5	6	1, 3 0, 3	0, 4 0, 4	0, 4 -, 4	-, 4 -, 4	-, 5 -, 5	-, 5 -, 5	-, 5 -, 5	-, 5 -, 5	-, 5 -, 5
17	5	7	1, 3 1, 4	0, 4 0, 4	0, 4 0, 5	0, 5 0, 5	-, 5 -, 5	-, 5 -, 5	-, 5 -, 5	-, - -, -	-, - -, -
17	5	8	1, 3 1, 4	1, 4 0, 4	0, 5 0, 5	0, 5 0, 5	-, 5 -, 5	-, 5 -, 5	-, - -, -	-, - -, -	-, - -, -
17	6	6	1, 4 1, 4	0, 4 0, 4	0, 4 0, 5	0, 5 0, 5	-, 5 -, 5	-, 5 -, 5	-, 6 -, 6	-, 6 -, 6	-, 6 -, 6
17	6	7	1, 3 1, 4	1, 4 0, 4	0, 5 0, 5	0, 5 0, 5	0, 5 -, 6	-, 6 -, 6	-, 6 -, 6	-, 6 -, 6	-, 6 -, 6
17	6	8	2, 4 1, 4	1, 5 1, 5	1, 5 0, 5	0, 5 0, 5	0, 6 0, 6	0, 6 0, 6	-, 6 -, 6	-, - -, -	-, - -, -
17	7	7	2, 4 1, 4	1, 5 1, 5	1, 5 0, 5	0, 6 0, 6	0, 6 0, 6	0, 6 0, 6	-, 6 -, 6	-, 7 -, 7	-, 7 -, 7
17	7	8	2, 4 2, 5	1, 5 1, 5	1, 6 1, 6	1, 6 1, 6	0, 6 0, 6	0, 7 0, 7	0, 7 0, 7	0, 7 -, 7	-, 7 -, 7
17	8	8	3, 5 2, 5	2, 6 2, 6	1, 6 1, 6	1, 6 1, 7	1, 7 1, 7	0, 7 0, 7	0, 7 0, 7	0, 8 0, 8	0, 8 0, 8
18	1	1	-, 1 -, 1	-, 1 -, 1	-, 1 -, 1	-, - -, -	-, - -, -	-, - -, -	-, - -, -	-, - -, -	-, - -, -
18	1	2	-, 1 -, 1	-, 1 -, 1	-, - -, -	-, - -, -	-, - -, -	-, - -, -	-, - -, -	-, - -, -	-, - -, -
18	1	3	-, 1 -, 1	-, 1 -, 1	-, - -, -	-, - -, -	-, - -, -	-, - -, -	-, - -, -	-, - -, -	-, - -, -
18	1	4	-, 1 -, 1	-, - -, -	-, - -, -	-, - -, -	-, - -, -	-, - -, -	-, - -, -	-, - -, -	-, - -, -
18	1	5	-, 1 -, 1	-, - -, -	-, - -, -	-, - -, -	-, - -, -	-, - -, -	-, - -, -	-, - -, -	-, - -, -

TABLE B.27 (cont.) CRITICAL VALUES FOR FISHER'S EXACT TEST

n	m_1	m_2	α = 0.50	0.20	0.10	0.05	0.02	0.01	0.005	0.002	0.001
18	1	6	-, 1 -, 1	-, - -, -	-, - -, -	-, - -, -	-, - -, -	-, - -, -	-, - -, -	-, - -, -	-, - -, -
18	1	7	-, 1 -, 1	-, - -, -	-, - -, -	-, - -, -	-, - -, -	-, - -, -	-, - -, -	-, - -, -	-, - -, -
18	1	8	-, 1 -, 1	-, - -, -	-, - -, -	-, - -, -	-, - -, -	-, - -, -	-, - -, -	-, - -, -	-, - -, -
18	1	9	0, 1 -, -	-, - -, -	-, - -, -	-, - -, -	-, - -, -	-, - -, -	-, - -, -	-, - -, -	-, - -, -
18	2	2	-, 1 -, 1	-, 2 -, 2	-, 2 -, 2	-, 2 -, 2	-, 2 -, 2	-, 2 -, 2	-, - -, -	-, - -, -	-, - -, -
18	2	3	-, 1 -, 1	-, 2 -, 2	-, 2 -, 2	-, 2 -, 2	-, 2 -, 2	-, - -, -	-, - -, -	-, - -, -	-, - -, -
18	2	4	-, 1 -, 1	-, 2 -, 2	-, 2 -, 2	-, 2 -, 2	-, - -, -	-, - -, -	-, - -, -	-, - -, -	-, - -, -
18	2	5	-, 1 -, 1	-, 2 -, 2	-, 2 -, 2	-, - -, -	-, - -, -	-, - -, -	-, - -, -	-, - -, -	-, - -, -
18	2	6	0, 2 -, 2	-, 2 -, 2	-, 2 -, 2	-, - -, -	-, - -, -	-, - -, -	-, - -, -	-, - -, -	-, - -, -
18	2	7	0, 2 0, 2	-, 2 -, 2	-, - -, -	-, - -, -	-, - -, -	-, - -, -	-, - -, -	-, - -, -	-, - -, -
18	2	8	0, 2 0, 2	-, 2 -, 2	-, - -, -	-, - -, -	-, - -, -	-, - -, -	-, - -, -	-, - -, -	-, - -, -
18	2	9	0, 2 0, 2	-, - -, -	-, - -, -	-, - -, -	-, - -, -	-, - -, -	-, - -, -	-, - -, -	-, - -, -
18	3	3	-, 1 -, 1	-, 2 -, 2	-, 2 -, 2	-, 3 -, 3	-, 3 -, 3	-, 3 -, 3	-, 3 -, 3	-, 3 -, 3	-, - -, -
18	3	4	0, 2 -, 2	-, 2 -, 2	-, 3 -, 3	-, 3 -, 3	-, 3 -, 3	-, 3 -, 3	-, 3 -, 3	-, - -, -	-, - -, -
18	3	5	0, 2 -, 2	-, 2 -, 2	-, 3 -, 3	-, 3 -, 3	-, 3 -, 3	-, - -, -	-, - -, -	-, - -, -	-, - -, -
18	3	6	0, 2 -, 2	-, 3 -, 3	-, 3 -, 3	-, 3 -, 3	-, - -, -	-, - -, -	-, - -, -	-, - -, -	-, - -, -
18	3	7	0, 2 0, 3	-, 3 -, 3	-, 3 -, 3	-, 3 -, 3	-, - -, -	-, - -, -	-, - -, -	-, - -, -	-, - -, -
18	3	8	0, 2 0, 3	0, 3 -, 3	-, 3 -, 3	-, - -, -	-, - -, -	-, - -, -	-, - -, -	-, - -, -	-, - -, -
18	3	9	1, 2 0, 3	0, 3 -, -	-, - -, -	-, - -, -	-, - -, -	-, - -, -	-, - -, -	-, - -, -	-, - -, -
18	4	4	0, 2 -, 2	-, 2 -, 2	-, 3 -, 3	-, 3 -, 3	-, 3 -, 3	-, 4 -, 4	-, 4 -, 4	-, 4 -, 4	-, 4 -, 4
18	4	5	0, 2 0, 3	-, 3 -, 3	-, 3 -, 3	-, 3 -, 3	-, 4 -, 4	-, 4 -, 4	-, 4 -, 4	-, 4 -, 4	-, - -, -
18	4	6	0, 2 0, 3	0, 3 -, 3	-, 3 -, 3	-, 4 -, 4	-, 4 -, 4	-, 4 -, 4	-, 4 -, 4	-, - -, -	-, - -, -
18	4	7	1, 3 0, 3	0, 3 0, 4	-, 4 -, 4	-, 4 -, 4	-, 4 -, 4	-, - -, -	-, - -, -	-, - -, -	-, - -, -
18	4	8	1, 3 0, 3	0, 4 0, 4	0, 4 0, 4	-, 4 -, 4	-, - -, -	-, - -, -	-, - -, -	-, - -, -	-, - -, -
18	4	9	1, 3 0, 4	0, 4 0, 4	0, 4 0, 4	0, 4 -, -	-, - -, -	-, - -, -	-, - -, -	-, - -, -	-, - -, -
18	5	5	0, 2 0, 3	0, 3 -, 3	-, 3 -, 3	-, 4 -, 4	-, 4 -, 4	-, 4 -, 4	-, 5 -, 5	-, 5 -, 5	-, 5 -, 5
18	5	6	1, 3 0, 3	0, 3 0, 4	0, 4 -, 4	-, 4 -, 4	-, 5 -, 5	-, 5 -, 5	-, 5 -, 5	-, 5 -, 5	-, 5 -, 5
18	5	7	1, 3 0, 3	0, 4 0, 4	0, 4 -, 4	-, 4 -, 4	-, 5 -, 5	-, 5 -, 5	-, 5 -, 5	-, - -, -	-, - -, -
18	5	8	1, 3 1, 4	0, 4 0, 4	0, 4 0, 5	0, 5 0, 5	-, 5 -, 5	-, 5 -, 5	-, - -, -	-, - -, -	-, - -, -
18	5	9	2, 3 1, 4	1, 4 0, 5	0, 5 0, 5	0, 5 0, 5	0, 5 -, -	-, - -, -	-, - -, -	-, - -, -	-, - -, -
18	6	6	1, 3 0, 3	0, 4 0, 4	0, 4 0, 5	0, 5 -, 5	-, 5 -, 5	-, 5 -, 5	-, 5 -, 5	-, 6 -, 6	-, 6 -, 6
18	6	7	1, 4 1, 4	1, 4 0, 4	0, 5 0, 5	0, 5 0, 5	-, 5 -, 5	-, 6 -, 6	-, 6 -, 6	-, 6 -, 6	-, 6 -, 6
18	6	8	2, 4 1, 4	1, 5 1, 5	0, 5 0, 5	0, 5 0, 5	0, 6 0, 6	-, 6 -, 6	-, 6 -, 6	-, 6 -, 6	-, - -, -
18	6	9	2, 4 1, 5	1, 5 1, 5	1, 5 0, 6	0, 6 0, 6	0, 6 0, 6	0, 6 0, 6	0, 6 -, -	-, - -, -	-, - -, -
18	7	7	2, 4 1, 4	1, 5 1 5	0, 5 0, 5	0, 5 0, 5	0, 6 0, 6	-, 6 -, 6	-, 6 -, 6	-, 7 -, 7	-, 7 -, 7
18	7	8	2, 4 2, 5	1, 5 1, 5	1, 5 1, 6	0, 6 0, 6	0, 6 0, 6	0, 6 0, 7	0, 7 0, 7	-, 7 -, 7	-, 7 -, 7
18	7	9	3, 4 2, 5	2, 5 1, 6	1, 6 1, 6	1, 6 1, 6	0, 7 0, 7	0, 7 0, 7	0, 7 0, 7	0, 7 -, -	-, - -, -
18	8	8	3, 5 2, 5	2, 5 2, 6	1, 6 1, 6	1, 6 1, 7	0, 7 0, 7	0, 7 0, 7	0, 7 0, 7	0, 7 0, 8	-, 8 -, 8
18	8	9	3, 5 2, 6	2, 6 2, 6	2, 6 1, 7	1, 7 1, 7	1, 7 1, 7	1, 7 0, 8	0, 8 0, 8	0, 8 0, 8	0, 8 0, 8
18	9	9	4, 5 3, 6	3, 6 2, 7	2, 7 2, 7	2, 7 1, 8	1, 8 1, 8	1, 8 1, 8	1, 8 1, 8	1, 8 0, 9	0, 9 0, 9
19	1	1	-, 1 -, 1	-, 1 -, 1	-, 1 -, 1	-, - -, -	-, - -, -	-, - -, -	-, - -, -	-, - -, -	-, - -, -
19	1	2	-, 1 -, 1	-, 1 -, 1	-, - -, -	-, - -, -	-, - -, -	-, - -, -	-, - -, -	-, - -, -	-, - -, -
19	1	3	-, 1 -, 1	-, 1 -, 1	-, - -, -	-, - -, -	-, - -, -	-, - -, -	-, - -, -	-, - -, -	-, - -, -
19	1	4	-, 1 -, 1	-, - -, -	-, - -, -	-, - -, -	-, - -, -	-, - -, -	-, - -, -	-, - -, -	-, - -, -
19	1	5	-, 1 -, 1	-, - -, -	-, - -, -	-, - -, -	-, - -, -	-, - -, -	-, - -, -	-, - -, -	-, - -, -
19	1	6	-, 1 -, 1	-, - -, -	-, - -, -	-, - -, -	-, - -, -	-, - -, -	-, - -, -	-, - -, -	-, - -, -
19	1	7	-, 1 -, 1	-, - -, -	-, - -, -	-, - -, -	-, - -, -	-, - -, -	-, - -, -	-, - -, -	-, - -, -
19	1	8	-, 1 -, 1	-, - -, -	-, - -, -	-, - -, -	-, - -, -	-, - -, -	-, - -, -	-, - -, -	-, - -, -
19	1	9	-, 1 -, 1	-, - -, -	-, - -, -	-, - -, -	-, - -, -	-, - -, -	-, - -, -	-, - -, -	-, - -, -
19	2	2	-, 1 -, 1	-, 1 -, 2	-, 2 -, 2	-, 2 -, 2	-, 2 -, 2	-, 2 -, 2	-, - -, -	-, - -, -	-, - -, -

TABLE B.27 (cont.) CRITICAL VALUES FOR FISHER'S EXACT TEST

n	m_1	m_2	α = 0.50	0.20	0.10	0.05	0.02	0.01	0.005	0.002	0.001
19	2	3	-, 1 -, 1	-, 2 -, 2	-, 2 -, 2	-, 2 -, 2	-, 2 -, 2	-, - -, -	-, - -, -	-, - -, -	-, - -, -
19	2	4	-, 1 -, 1	-, 2 -, 2	-, 2 -, 2	-, 2 -, 2	-, - -, -	-, - -, -	-, - -, -	-, - -, -	-, - -, -
19	2	5	-, 1 -, 1	-, 2 -, 2	-, 2 -, 2	-, - -, -	-, - -, -	-, - -, -	-, - -, -	-, - -, -	-, - -, -
19	2	6	0, 2 -, 2	-, 2 -, 2	-, 2 -, 2	-, - -, -	-, - -, -	-, - -, -	-, - -, -	-, - -, -	-, - -, -
19	2	7	0, 2 -, 2	-, 2 -, 2	-, - -, -	-, - -, -	-, - -, -	-, - -, -	-, - -, -	-, - -, -	-, - -, -
19	2	8	0, 2 0, 2	-, 2 -, 2	-, - -, -	-, - -, -	-, - -, -	-, - -, -	-, - -, -	-, - -, -	-, - -, -
19	2	9	0, 2 0, 2	-, - -, -	-, - -, -	-, - -, -	-, - -, -	-, - -, -	-, - -, -	-, - -, -	-, - -, -
19	3	3	-, 1 -, 1	-, 2 -, 2	-, 2 -, 2	-, 3 -, 3	-, 3 -, 3	-, 3 -, 3	-, 3 -, 3	-, 3 -, 3	-, - -, -
19	3	4	0, 1 -, 2	-, 2 -, 2	-, 2 -, 2	-, 3 -, 3	-, 3 -, 3	-, 3 -, 3	-, 3 -, 3	-, - -, -	-, - -, -
19	3	5	0, 2 -, 2	-, 2 -, 2	-, 3 -, 3	-, 3 -, 3	-, 3 -, 3	-, - -, -	-, - -, -	-, - -, -	-, - -, -
19	3	6	0, 2 -, 2	-, 3 -, 3	-, 3 -, 3	-, 3 -, 3	-, - -, -	-, - -, -	-, - -, -	-, - -, -	-, - -, -
19	3	7	0, 2 0, 3	-, 3 -, 3	-, 3 -, 3	-, 3 -, 3	-, - -, -	-, - -, -	-, - -, -	-, - -, -	-, - -, -
19	3	8	0, 2 0, 3	0, 3 -, 3	-, 3 -, 3	-, - -, -	-, - -, -	-, - -, -	-, - -, -	-, - -, -	-, - -, -
19	3	9	0, 2 0, 3	0, 3 -, 3	-, 3 -, 3	-, - -, -	-, - -, -	-, - -, -	-, - -, -	-, - -, -	-, - -, -
19	4	4	0, 2 -, 2	-, 2 -, 2	-, 3 -, 3	-, 3 -, 3	-, 3 -, 3	-, 4 -, 4	-, 4 -, 4	-, 4 -, 4	-, 4 -, 4
19	4	5	0, 2 -, 2	-, 3 -, 3	-, 3 -, 3	-, 3 -, 3	-, 4 -, 4	-, 4 -, 4	-, 4 -, 4	-, 4 -, 4	-, - -, -
19	4	6	0, 2 0, 3	0, 3 -, 3	-, 3 -, 3	-, 4 -, 4	-, 4 -, 4	-, 4 -, 4	-, 4 -, 4	-, - -, -	-, - -, -
19	4	7	0, 2 0, 3	0, 3 -, 3	-, 4 -, 4	-, 4 -, 4	-, 4 -, 4	-, 4 -, 4	-, - -, -	-, - -, -	-, - -, -
19	4	8	1, 3 0, 3	0, 3 0, 4	0, 4 -, 4	-, 4 -, 4	-, 4 -, 4	-, - -, -	-, - -, -	-, - -, -	-, - -, -
19	4	9	1, 3 0, 3	0, 4 0, 4	0, 4 0, 4	-, 4 -, 4	-, - -, -	-, - -, -	-, - -, -	-, - -, -	-, - -, -
19	5	5	0, 2 0, 3	0, 3 -, 3	-, 3 -, 3	-, 4 -, 4	-, 4 -, 4	-, 4 -, 4	-, 5 -, 5	-, 5 -, 5	-, 5 -, 5
19	5	6	1, 3 0, 3	0, 3 0, 4	-, 4 -, 4	-, 4 -, 4	-, 4 -, 4	-, 5 -, 5	-, 5 -, 5	-, 5 -, 5	-, 5 -, 5
19	5	7	1, 3 0, 3	0, 4 0, 4	0, 4 -, 4	-, 4 -, 4	-, 5 -, 5	-, 5 -, 5	-, 5 -, 5	-, 5 -, 5	-, - -, -
19	5	8	1, 3 1, 4	0, 4 0, 4	0, 4 0, 5	0, 5 0, 5	-, 5 -, 5	-, 5 -, 5	-, 5 -, 5	-, - -, -	-, - -, -
19	5	9	1, 3 1, 4	1, 4 0, 4	0, 5 0, 5	0, 5 0, 5	-, 5 -, 5	-, - -, -	-, - -, -	-, - -, -	-, - -, -
19	6	6	1, 3 0, 3	0, 4 0, 4	0, 4 -, 4	-, 4 -, 4	-, 5 -, 5	-, 5 -, 5	-, 5 -, 5	-, 6 -, 6	-, 6 -, 6
19	6	7	1, 4 1, 4	0, 4 0, 4	0, 4 0, 5	0, 5 0, 5	-, 5 -, 5	-, 5 -, 5	-, 6 -, 6	-, 6 -, 6	-, 6 -, 6
19	6	8	2, 4 1, 4	1, 4 1, 5	0, 5 0, 5	0, 5 0, 5	0, 6 0, 6	-, 6 -, 6	-, 6 -, 6	-, 6 -, 6	-, - -, -
19	6	9	2, 4 1, 4	1, 5 1, 5	1, 5 0, 5	0, 5 0, 6	0, 6 0, 6	0, 6 -, 6	-, 6 -, 6	-, - -, -	-, - -, -
19	7	7	2, 4 1, 4	1, 4 1, 5	0, 5 0, 5	0, 5 0, 5	0, 6 0, 6	-, 6 -, 6	-, 6 -, 6	-, 6 -, 6	-, 7 -, 7
19	7	8	2, 4 1, 4	1, 5 1, 5	1, 5 0, 5	0, 6 0, 6	0, 6 0, 6	0, 6 -, 6	-, 7 -, 7	-, 7 -, 7	-, 7 -, 7
19	7	9	2, 4 2, 5	1, 5 1, 5	1, 6 1, 6	1, 6 0, 6	0, 6 0, 6	0, 7 0, 7	0, 7 0, 7	-, 7 -, 7	-, 7 -, 7
19	8	8	2, 5 2, 5	1, 5 1, 5	1, 6 1, 6	1, 6 0, 6	0, 7 0, 7	0, 7 0, 7	0, 7 0, 7	-, 7 -, 7	-, 8 -, 8
19	8	9	3, 5 2, 5	2, 6 2, 6	1, 6 1, 6	1, 7 1, 7	1, 7 1, 7	0, 7 0, 7	0, 7 0, 8	0, 8 0, 8	0, 8 0, 8
19	9	9	3, 5 3, 6	2, 6 2, 6	2, 7 2, 7	1, 7 1, 7	1, 7 1, 8	1, 8 1, 8	1, 8 0, 8	0, 8 0, 8	0, 8 0, 9
20	1	1	-, 1 -, 1	-, 1 -, 1	-, 1 -, 1	-, 1 -, 1	-, - -, -	-, - -, -	-, - -, -	-, - -, -	-, - -, -
20	1	2	-, 1 -, 1	-, 1 -, 1	-, 1 -, 1	-, - -, -	-, - -, -	-, - -, -	-, - -, -	-, - -, -	-, - -, -
20	1	3	-, 1 -, 1	-, 1 -, 1	-, - -, -	-, - -, -	-, - -, -	-, - -, -	-, - -, -	-, - -, -	-, - -, -
20	1	4	-, 1 -, 1	-, 1 -, 1	-, - -, -	-, - -, -	-, - -, -	-, - -, -	-, - -, -	-, - -, -	-, - -, -
20	1	5	-, 1 -, 1	-, - -, -	-, - -, -	-, - -, -	-, - -, -	-, - -, -	-, - -, -	-, - -, -	-, - -, -
20	1	6	-, 1 -, 1	-, - -, -	-, - -, -	-, - -, -	-, - -, -	-, - -, -	-, - -, -	-, - -, -	-, - -, -
20	1	7	-, 1 -, 1	-, - -, -	-, - -, -	-, - -, -	-, - -, -	-, - -, -	-, - -, -	-, - -, -	-, - -, -
20	1	8	-, 1 -, 1	-, - -, -	-, - -, -	-, - -, -	-, - -, -	-, - -, -	-, - -, -	-, - -, -	-, - -, -
20	1	9	-, 1 -, 1	-, - -, -	-, - -, -	-, - -, -	-, - -, -	-, - -, -	-, - -, -	-, - -, -	-, - -, -
20	1	10	0, 1 -, -	-, - -, -	-, - -, -	-, - -, -	-, - -, -	-, - -, -	-, - -, -	-, - -, -	-, - -, -
20	2	2	-, 1 -, 1	-, 1 -, 1	-, 2 -, 2	-, 2 -, 2	-, 2 -, 2	-, 2 -, 2	-, - -, -	-, - -, -	-, - -, -
20	2	3	-, 1 -, 1	-, 2 -, 2	-, 2 -, 2	-, 2 -, 2	-, 2 -, 2	-, - -, -	-, - -, -	-, - -, -	-, - -, -
20	2	4	-, 1 -, 1	-, 2 -, 2	-, 2 -, 2	-, 2 -, 2	-, - -, -	-, - -, -	-, - -, -	-, - -, -	-, - -, -
20	2	5	-, 1 -, 1	-, 2 -, 2	-, 2 -, 2	-, - -, -	-, - -, -	-, - -, -	-, - -, -	-, - -, -	-, - -, -
20	2	6	0, 1 -, 2	-, 2 -, 2	-, 2 -, 2	-, - -, -	-, - -, -	-, - -, -	-, - -, -	-, - -, -	-, - -, -

TABLE B.27 (cont.) CRITICAL VALUES FOR FISHER'S EXACT TEST

n	m_1	m_2	α = 0.50		0.20		0.10		0.05		0.02		0.01		0.005		0.002		0.001	
20	2	7	0, 2	-, 2	-, 2	-, 2	-, -	-, -	-, -	-, -	-, -	-, -	-, -	-, -	-, -	-, -	-, -	-, -	-, -	-, -
20	2	8	0, 2	0, 2	-, 2	-, 2	-, -	-, -	-, -	-, -	-, -	-, -	-, -	-, -	-, -	-, -	-, -	-, -	-, -	-, -
20	2	9	0, 2	0, 2	-, 2	-, 2	-, -	-, -	-, -	-, -	-, -	-, -	-, -	-, -	-, -	-, -	-, -	-, -	-, -	-, -
20	2	10	0, 2	0, 2	-, -	-, -	-, -	-, -	-, -	-, -	-, -	-, -	-, -	-, -	-, -	-, -	-, -	-, -	-, -	-, -
20	3	3	-, 1	-, 1	-, 2	-, 2	-, 2	-, 2	-, 2	-, 2	-, 3	-, 3	-, 3	-, 3	-, 3	-, 3	-, 3	-, 3	-, 3	-, 3
20	3	4	0, 1	-, 2	-, 2	-, 2	-, 2	-, 2	-, 3	-, 3	-, 3	-, 3	-, 3	-, 3	-, 3	-, 3	-, -	-, -	-, -	-, -
20	3	5	0, 2	-, 2	-, 2	-, 2	-, 3	-, 3	-, 3	-, 3	-, 3	-, 3	-, 3	-, 3	-, -	-, -	-, -	-, -	-, -	-, -
20	3	6	0, 2	-, 2	-, 3	-, 3	-, 3	-, 3	-, 3	-, 3	-, 3	-, 3	-, -	-, -	-, -	-, -	-, -	-, -	-, -	-, -
20	3	7	0, 2	-, 2	-, 3	-, 3	-, 3	-, 3	-, 3	-, 3	-, -	-, -	-, -	-, -	-, -	-, -	-, -	-, -	-, -	-, -
20	3	8	0, 2	0, 3	0, 3	-, 3	-, 3	-, 3	-, 3	-, 3	-, -	-, -	-, -	-, -	-, -	-, -	-, -	-, -	-, -	-, -
20	3	9	0, 2	0, 3	0, 3	-, 3	-, 3	-, 3	-, -	-, -	-, -	-, -	-, -	-, -	-, -	-, -	-, -	-, -	-, -	-, -
20	3	10	1, 2	0, 3	0, 3	-, -	-, -	-, -	-, -	-, -	-, -	-, -	-, -	-, -	-, -	-, -	-, -	-, -	-, -	-, -
20	4	4	0, 2	-, 2	-, 2	-, 2	-, 3	-, 3	-, 3	-, 3	-, 3	-, 3	-, 4	-, 4	-, 4	-, 4	-, 4	-, 4	-, 4	-, 4
20	4	5	0, 2	-, 2	-, 3	-, 3	-, 3	-, 3	-, 3	-, 3	-, 4	-, 4	-, 4	-, 4	-, 4	-, 4	-, 4	-, 4	-, -	-, -
20	4	6	0, 2	0, 3	-, 3	-, 3	-, 3	-, 3	-, 4	-, 4	-, 4	-, 4	-, 4	-, 4	-, 4	-, 4	-, -	-, -	-, -	-, -
20	4	7	0, 2	0, 3	0, 3	-, 3	-, 4	-, 4	-, 4	-, 4	-, 4	-, 4	-, 4	-, 4	-, -	-, -	-, -	-, -	-, -	-, -
20	4	8	1, 3	0, 3	0, 3	0, 4	-, 4	-, 4	-, 4	-, 4	-, 4	-, 4	-, -	-, -	-, -	-, -	-, -	-, -	-, -	-, -
20	4	9	1, 3	0, 3	0, 4	0, 4	0, 4	0, 4	-, 4	-, 4	-, -	-, -	-, -	-, -	-, -	-, -	-, -	-, -	-, -	-, -
20	4	10	1, 3	0, 4	0, 4	0, 4	0, 4	0, 4	0, 4	-, -	-, -	-, -	-, -	-, -	-, -	-, -	-, -	-, -	-, -	-, -
20	5	5	0, 2	0, 3	0, 3	-, 3	-, 3	-, 3	-, 4	-, 4	-, 4	-, 4	-, 4	-, 4	-, 4	-, 4	-, 5	-, 5	-, 5	-, 5
20	5	6	0, 2	0, 3	0, 3	-, 3	-, 4	-, 4	-, 4	-, 4	-, 4	-, 4	-, 5	-, 5	-, 5	-, 5	-, 5	-, 5	-, 5	-, 5
20	5	7	1, 3	0, 3	0, 4	0, 4	0, 4	-, 4	-, 4	-, 4	-, 5	-, 5	-, 5	-, 5	-, 5	-, 5	-, 5	-, 5	-, -	-, -
20	5	8	1, 3	0, 3	0, 4	0, 4	0, 4	0, 5	-, 5	-, 5	-, 5	-, 5	-, 5	-, 5	-, 5	-, 5	-, -	-, -	-, -	-, -
20	5	9	1, 3	1, 4	0, 4	0, 4	0, 4	0, 5	0, 5	0, 5	-, 5	-, 5	-, 5	-, 5	-, -	-, -	-, -	-, -	-, -	-, -
20	5	10	2, 3	1, 4	1, 4	0, 5	0, 5	0, 5	0, 5	0, 5	0, 5	-, -	-, -	-, -	-, -	-, -	-, -	-, -	-, -	-, -
20	6	6	1, 3	0, 3	0, 4	0, 4	0, 4	-, 4	-, 4	-, 4	-, 5	-, 5	-, 5	-, 5	-, 5	-, 5	-, 6	-, 6	-, 6	-, 6
20	6	7	1, 4	1, 4	0, 4	0, 4	0, 4	0, 5	0, 5	-, 5	-, 5	-, 5	-, 5	-, 5	-, 6	-, 6	-, 6	-, 6	-, 6	-, 6
20	6	8	1, 4	1, 4	1, 4	0, 4	0, 5	0, 5	0, 5	0, 5	-, 5	-, 5	-, 6	-, 6	-, 6	-, 6	-, 6	-, 6	-, 6	-, 6
20	6	9	2, 4	1, 4	1, 5	1, 5	0, 5	0, 5	0, 5	0, 5	0, 6	0, 6	-, 6	-, 6	-, 6	-, 6	-, -	-, -	-, -	-, -
20	6	10	2, 4	1, 5	1, 5	1, 5	1, 5	0, 6	0, 6	0, 6	0, 6	0, 6	0, 6	-, -	-, -	-, -	-, -	-, -	-, -	-, -
20	7	7	1, 3	1, 4	1, 4	0, 4	0, 5	0, 5	0, 5	0, 5	-, 6	-, 6	-, 6	-, 6	-, 6	-, 6	-, 6	-, 6	-, 7	-, 7
20	7	8	2, 4	1, 4	1, 5	1, 5	0, 5	0, 5	0, 6	0, 6	0, 6	0, 6	-, 6	-, 6	-, 6	-, 6	-, 7	-, 7	-, 7	-, 7
20	7	9	2, 4	2, 5	1, 5	1, 5	1, 6	1, 6	0, 6	0, 6	0, 6	0, 6	0, 7	0, 7	0, 7	0, 7	-, 7	-, 7	-, 7	-, 7
20	7	10	3, 4	2, 5	2, 5	1, 6	1, 6	1, 6	1, 6	0, 7	0, 7	0, 7	0, 7	0, 7	0, 7	0, 7	0, 7	-, -	-, -	-, -
20	8	8	2, 5	2, 5	1, 5	1, 5	1, 6	1, 6	0, 6	0, 6	0, 6	0, 6	0, 7	0, 7	0, 7	0, 7	-, 7	-, 7	-, 7	-, 7
20	8	9	3, 5	2, 5	2, 6	2, 6	1, 6	1, 6	1, 6	1, 7	0, 7	0, 7	0, 7	0, 7	0, 7	0, 7	0, 8	0, 8	-, 8	-, 8
20	8	10	3, 5	2, 6	2, 6	2, 6	2, 6	1, 7	1, 7	1, 7	1, 7	1, 7	1, 7	0, 8	0, 8	0, 8	0, 8	0, 8	0, 8	0, 8
20	9	9	3, 5	3, 6	2, 6	2, 6	2, 6	2, 7	1, 7	1, 7	1, 7	1, 8	1, 8	1, 8	0, 8	0, 8	0, 8	0, 8	0, 8	0, 8
20	9	10	4, 5	3, 6	3, 6	2, 7	2, 7	2, 7	2, 7	1, 8	1, 8	1, 8	1, 8	1, 8	1, 8	0, 9	0, 9	0, 9	0, 9	0, 9
20	10	10	4, 6	3, 7	3, 7	3, 7	3, 7	2, 8	2, 8	2, 8	2, 8	1, 9	1, 9	1, 9	1, 9	1, 9	1, 9	1, 9	1, 9	0, 10
21	1	1	-, 1	-, 1	-, 1	-, 1	-, 1	-, 1	-, 1	-, 1	-, -	-, -	-, -	-, -	-, -	-, -	-, -	-, -	-, -	-, -
21	1	2	-, 1	-, 1	-, 1	-, 1	-, 1	-, 1	-, -	-, -	-, -	-, -	-, -	-, -	-, -	-, -	-, -	-, -	-, -	-, -
21	1	3	-, 1	-, 1	-, 1	-, 1	-, -	-, -	-, -	-, -	-, -	-, -	-, -	-, -	-, -	-, -	-, -	-, -	-, -	-, -
21	1	4	-, 1	-, 1	-, 1	-, 1	-, -	-, -	-, -	-, -	-, -	-, -	-, -	-, -	-, -	-, -	-, -	-, -	-, -	-, -
21	1	5	-, 1	-, 1	-, -	-, -	-, -	-, -	-, -	-, -	-, -	-, -	-, -	-, -	-, -	-, -	-, -	-, -	-, -	-, -
21	1	6	-, 1	-, 1	-, -	-, -	-, -	-, -	-, -	-, -	-, -	-, -	-, -	-, -	-, -	-, -	-, -	-, -	-, -	-, -
21	1	7	-, 1	-, 1	-, -	-, -	-, -	-, -	-, -	-, -	-, -	-, -	-, -	-, -	-, -	-, -	-, -	-, -	-, -	-, -
21	1	8	-, 1	-, 1	-, -	-, -	-, -	-, -	-, -	-, -	-, -	-, -	-, -	-, -	-, -	-, -	-, -	-, -	-, -	-, -
21	1	9	-, 1	-, 1	-, -	-, -	-, -	-, -	-, -	-, -	-, -	-, -	-, -	-, -	-, -	-, -	-, -	-, -	-, -	-, -
21	1	10	-, 1	-, 1	-, -	-, -	-, -	-, -	-, -	-, -	-, -	-, -	-, -	-, -	-, -	-, -	-, -	-, -	-, -	-, -

TABLE B.27 (cont.) CRITICAL VALUES FOR FISHER'S EXACT TEST

n	m_1	m_2	α = 0.50	0.20	0.10	0.05	0.02	0.01	0.005	0.002	0.001
21	2	2	-, 1 -, 1	-, 1 -, 1	-, 2 -, 2	-, 2 -, 2	-, 2 -, 2	-, 2 -, 2	-, 2 -, 2	-, - -, -	-, - -, -
21	2	3	-, 1 -, 1	-, 2 -, 2	-, 2 -, 2	-, 2 -, 2	-, 2 -, 2	-, - -, -	-, - -, -	-, - -, -	-, - -, -
21	2	4	-, 1 -, 1	-, 2 -, 2	-, 2 -, 2	-, 2 -, 2	-, - -, -	-, - -, -	-, - -, -	-, - -, -	-, - -, -
21	2	5	-, 1 -, 1	-, 2 -, 2	-, 2 -, 2	-, 2 -, 2	-, - -, -	-, - -, -	-, - -, -	-, - -, -	-, - -, -
21	2	6	0, 1 -, 1	-, 2 -, 2	-, 2 -, 2	-, - -, -	-, - -, -	-, - -, -	-, - -, -	-, - -, -	-, - -, -
21	2	7	0, 2 -, 2	-, 2 -, 2	-, 2 -, 2	-, - -, -	-, - -, -	-, - -, -	-, - -, -	-, - -, -	-, - -, -
21	2	8	0, 2 -, 2	-, 2 -, 2	-, - -, -	-, - -, -	-, - -, -	-, - -, -	-, - -, -	-, - -, -	-, - -, -
21	2	9	0, 2 0, 2	-, 2 -, 2	-, - -, -	-, - -, -	-, - -, -	-, - -, -	-, - -, -	-, - -, -	-, - -, -
21	2	10	0, 2 0, 2	-, - -, -	-, - -, -	-, - -, -	-, - -, -	-, - -, -	-, - -, -	-, - -, -	-, - -, -
21	3	3	-, 1 -, 1	-, 2 -, 2	-, 2 -, 2	-, 2 -, 2	-, 3 -, 3	-, 3 -, 3	-, 3 -, 3	-, 3 -, 3	-, 3 -, 3
21	3	4	-, 1 -, 1	-, 2 -, 2	-, 2 -, 2	-, 3 -, 3	-, 3 -, 3	-, 3 -, 3	-, 3 -, 3	-, - -, -	-, - -, -
21	3	5	0, 2 -, 2	-, 2 -, 2	-, 3 -, 3	-, 3 -, 3	-, 3 -, 3	-, 3 -, 3	-, - -, -	-, - -, -	-, - -, -
21	3	6	0, 2 -, 2	-, 2 -, 2	-, 3 -, 3	-, 3 -, 3	-, 3 -, 3	-, - -, -	-, - -, -	-, - -, -	-, - -, -
21	3	7	0, 2 -, 2	-, 3 -, 3	-, 3 -, 3	-, 3 -, 3	-, - -, -	-, - -, -	-, - -, -	-, - -, -	-, - -, -
21	3	8	0, 2 0, 3	-, 3 -, 3	-, 3 -, 3	-, 3 -, 3	-, - -, -	-, - -, -	-, - -, -	-, - -, -	-, - -, -
21	3	9	0, 2 0, 3	0, 3 -, 3	-, 3 -, 3	-, - -, -	-, - -, -	-, - -, -	-, - -, -	-, - -, -	-, - -, -
21	3	10	0, 2 0, 3	0, 3 -, 3	-, 3 -, 3	-, - -, -	-, - -, -	-, - -, -	-, - -, -	-, - -, -	-, - -, -
21	4	4	0, 2 -, 2	-, 2 -, 2	-, 3 -, 3	-, 3 -, 3	-, 3 -, 3	-, 4 -, 4	-, 4 -, 4	-, 4 -, 4	-, 4 -, 4
21	4	5	0, 2 -, 2	-, 3 -, 3	-, 3 -, 3	-, 3 -, 3	-, 4 -, 4	-, 4 -, 4	-, 4 -, 4	-, 4 -, 4	-, 4 -, 4
21	4	6	0, 2 0, 3	-, 3 -, 3	-, 3 -, 3	-, 4 -, 4	-, 4 -, 4	-, 4 -, 4	-, 4 -, 4	-, - -, -	-, - -, -
21	4	7	0, 2 0, 3	0, 3 -, 3	-, 3 -, 3	-, 4 -, 4	-, 4 -, 4	-, 4 -, 4	-, - -, -	-, - -, -	-, - -, -
21	4	8	0, 2 0, 3	0, 3 0, 4	-, 4 -, 4	-, 4 -, 4	-, 4 -, 4	-, - -, -	-, - -, -	-, - -, -	-, - -, -
21	4	9	1, 3 0, 3	0, 3 0, 4	0, 4 -, 4	-, 4 -, 4	-, - -, -	-, - -, -	-, - -, -	-, - -, -	-, - -, -
21	4	10	1, 3 0, 3	0, 4 0, 4	0, 4 0, 4	-, 4 -, 4	-, - -, -	-, - -, -	-, - -, -	-, - -, -	-, - -, -
21	5	5	0, 2 0, 3	-, 3 -, 3	-, 3 -, 3	-, 4 -, 4	-, 4 -, 4	-, 4 -, 4	-, 4 -, 4	-, 5 -, 5	-, 5 -, 5
21	5	6	0, 2 0, 3	0, 3 -, 3	-, 4 -, 4	-, 4 -, 4	-, 4 -, 4	-, 5 -, 5	-, 5 -, 5	-, 5 -, 5	-, 5 -, 5
21	5	7	1, 3 0, 3	0, 3 0, 4	0, 4 -, 4	-, 4 -, 4	-, 5 -, 5	-, 5 -, 5	-, 5 -, 5	-, 5 -, 5	-, - -, -
21	5	8	1, 3 0, 3	0, 4 0, 4	0, 4 -, 4	-, 4 -, 4	-, 5 -, 5	-, 5 -, 5	-, 5 -, 5	-, - -, -	-, - -, -
21	5	9	1, 3 1, 4	0, 4 0, 4	0, 4 0, 5	0, 5 0, 5	-, 5 -, 5	-, 5 -, 5	-, - -, -	-, - -, -	-, - -, -
21	5	10	1, 3 1, 4	1, 4 0, 4	0, 5 0, 5	0, 5 0, 5	-, 5 -, 5	-, - -, -	-, - -, -	-, - -, -	-, - -, -
21	6	6	1, 3 0, 3	0, 4 0, 4	0, 4 -, 4	-, 4 -, 4	-, 5 -, 5	-, 5 -, 5	-, 5 -, 5	-, 5 -, 5	-, 6 -, 6
21	6	7	1, 3 0, 3	0, 4 0, 4	0, 4 0, 5	-, 5 -, 5	-, 5 -, 5	-, 5 -, 5	-, 6 -, 6	-, 6 -, 6	-, 6 -, 6
21	6	8	1, 4 1, 4	0, 4 0, 4	0, 5 0, 5	0, 5 0, 5	-, 5 -, 5	-, 6 -, 6	-, 6 -, 6	-, 6 -, 6	-, 6 -, 6
21	6	9	2, 4 1, 4	1, 4 1, 5	0, 5 0, 5	0, 5 0, 5	0, 6 0, 6	-, 6 -, 6	-, 6 -, 6	-, 6 -, 6	-, - -, -
21	6	10	2, 4 1, 4	1, 5 1, 5	1, 5 0, 5	0, 6 0, 6	0, 6 0, 6	0, 6 -, 6	-, 6 -, 6	-, - -, -	-, - -, -
21	7	7	1, 4 1, 4	0, 4 0, 4	0, 5 0, 5	0, 5 0, 5	-, 5 -, 5	-, 6 -, 6	-, 6 -, 6	-, 6 -, 6	-, 6 -, 6
21	7	8	2, 4 1, 4	1, 5 1, 5	0, 5 0, 5	0, 5 0, 6	0, 6 0, 6	-, 6 -, 6	-, 6 -, 6	-, 7 -, 7	-, 7 -, 7
21	7	9	2, 4 1, 4	1, 5 1, 5	1, 5 0, 6	0, 6 0, 6	0, 6 0, 6	0, 6 0, 7	-, 7 -, 7	-, 7 -, 7	-, 7 -, 7
21	7	10	2, 4 2, 5	1, 5 1, 5	1, 6 1, 6	1, 6 0, 6	0, 7 0, 7	0, 7 0, 7	0, 7 0, 7	-, 7 -, 7	-, - -, -
21	8	8	2, 5 2, 5	1, 5 1, 5	1, 5 1, 6	0, 6 0, 6	0, 6 0, 6	0, 7 0, 7	-, 7 -, 7	-, 7 -, 7	-, 7 -, 7
21	8	9	2, 5 2, 5	1, 5 1, 6	1, 6 1, 6	1, 6 0, 6	0, 7 0, 7	0, 7 0, 7	0, 7 0, 7	-, 8 -, 8	-, 8 -, 8
21	8	10	3, 5 2, 5	2, 6 2, 6	1, 6 1, 6	1, 7 1, 7	1, 7 0, 7	0, 7 0, 7	0, 8 0, 8	0, 8 0, 8	0, 8 -, 8
21	9	9	3, 5 2, 5	2, 6 2, 6	1, 6 1, 6	1, 7 1, 7	1, 7 0, 7	0, 7 0, 7	0, 8 0, 8	0, 8 0, 8	0, 8 -, 8
21	9	10	3, 5 3, 6	2, 6 2, 6	2, 7 2, 7	1, 7 1, 7	1, 8 1, 8	1, 8 1, 8	0, 8 0, 8	0, 8 0, 8	0, 9 0, 9
21	10	10	4, 6 3, 6	3, 7 3, 7	2, 7 2, 7	2, 8 2, 8	1, 8 1, 8	1, 8 1, 8	1, 9 1, 9	1, 9 1, 9	0, 9 0, 9
22	1	1	-, 1 -, 1	-, 1 -, 1	-, 1 -, 1	-, 1 -, 1	-, - -, -	-, - -, -	-, - -, -	-, - -, -	-, - -, -
22	1	2	-, 1 -, 1	-, 1 -, 1	-, 1 -, 1	-, - -, -	-, - -, -	-, - -, -	-, - -, -	-, - -, -	-, - -, -
22	1	3	-, 1 -, 1	-, 1 -, 1	-, - -, -	-, - -, -	-, - -, -	-, - -, -	-, - -, -	-, - -, -	-, - -, -
22	1	4	-, 1 -, 1	-, 1 -, 1	-, - -, -	-, - -, -	-, - -, -	-, - -, -	-, - -, -	-, - -, -	-, - -, -
22	1	5	-, 1 -, 1	-, - -, -	-, - -, -	-, - -, -	-, - -, -	-, - -, -	-, - -, -	-, - -, -	-, - -, -

TABLE B.27 (cont.) CRITICAL VALUES FOR FISHER'S EXACT TEST

n	m_1	m_2	α = 0.50	0.20	0.10	0.05	0.02	0.01	0.005	0.002	0.001
22	1	6	-, 1 -, 1	-, - -, -	-, - -, -	-, - -, -	-, - -, -	-, - -, -	-, - -, -	-, - -, -	-, - -, -
22	1	7	-, 1 -, 1	-, - -, -	-, - -, -	-, - -, -	-, - -, -	-, - -, -	-, - -, -	-, - -, -	-, - -, -
22	1	8	-, 1 -, 1	-, - -, -	-, - -, -	-, - -, -	-, - -, -	-, - -, -	-, - -, -	-, - -, -	-, - -, -
22	1	9	-, 1 -, 1	-, - -, -	-, - -, -	-, - -, -	-, - -, -	-, - -, -	-, - -, -	-, - -, -	-, - -, -
22	1	10	-, 1 -, 1	-, - -, -	-, - -, -	-, - -, -	-, - -, -	-, - -, -	-, - -, -	-, - -, -	-, - -, -
22	1	11	0, 1 -, -	-, - -, -	-, - -, -	-, - -, -	-, - -, -	-, - -, -	-, - -, -	-, - -, -	-, - -, -
22	2	2	-, 1 -, 1	-, 1 -, 1	-, 2 -, 2	-, 2 -, 2	-, 2 -, 2	-, 2 -, 2	-, 2 -, 2	-, - -, -	-, - -, -
22	2	3	-, 1 -, 1	-, 2 -, 2	-, 2 -, 2	-, 2 -, 2	-, 2 -, 2	-, - -, -	-, - -, -	-, - -, -	-, - -, -
22	2	4	-, 1 -, 1	-, 2 -, 2	-, 2 -, 2	-, 2 -, 2	-, - -, -	-, - -, -	-, - -, -	-, - -, -	-, - -, -
22	2	5	-, 1 -, 1	-, 2 -, 2	-, 2 -, 2	-, 2 -, 2	-, - -, -	-, - -, -	-, - -, -	-, - -, -	-, - -, -
22	2	6	-, 1 -, 1	-, 2 -, 2	-, 2 -, 2	-, - -, -	-, - -, -	-, - -, -	-, - -, -	-, - -, -	-, - -, -
22	2	7	0, 2 -, 2	-, 2 -, 2	-, 2 -, 2	-, - -, -	-, - -, -	-, - -, -	-, - -, -	-, - -, -	-, - -, -
22	2	8	0, 2 -, 2	-, 2 -, 2	-, - -, -	-, - -, -	-, - -, -	-, - -, -	-, - -, -	-, - -, -	-, - -, -
22	2	9	0, 2 0, 2	-, 2 -, 2	-, - -, -	-, - -, -	-, - -, -	-, - -, -	-, - -, -	-, - -, -	-, - -, -
22	2	10	0, 2 0, 2	-, 2 -, 2	-, - -, -	-, - -, -	-, - -, -	-, - -, -	-, - -, -	-, - -, -	-, - -, -
22	2	11	0, 2 0, 2	-, - -, -	-, - -, -	-, - -, -	-, - -, -	-, - -, -	-, - -, -	-, - -, -	-, - -, -
22	3	3	-, 1 -, 1	-, 2 -, 2	-, 2 -, 2	-, 2 -, 2	-, 3 -, 3	-, 3 -, 3	-, 3 -, 3	-, 3 -, 3	-, 3 -, 3
22	3	4	-, 1 -, 1	-, 2 -, 2	-, 2 -, 2	-, 3 -, 3	-, 3 -, 3	-, 3 -, 3	-, 3 -, 3	-, - -, -	-, - -, -
22	3	5	0, 2 -, 2	-, 2 -, 2	-, 3 -, 3	-, 3 -, 3	-, 3 -, 3	-, 3 -, 3	-, - -, -	-, - -, -	-, - -, -
22	3	6	0, 2 -, 2	-, 2 -, 2	-, 3 -, 3	-, 3 -, 3	-, 3 -, 3	-, - -, -	-, - -, -	-, - -, -	-, - -, -
22	3	7	0, 2 -, 2	-, 3 -, 3	-, 3 -, 3	-, 3 -, 3	-, - -, -	-, - -, -	-, - -, -	-, - -, -	-, - -, -
22	3	8	0, 2 0, 3	-, 3 -, 3	-, 3 -, 3	-, 3 -, 3	-, - -, -	-, - -, -	-, - -, -	-, - -, -	-, - -, -
22	3	9	0, 2 0, 3	0, 3 -, 3	-, 3 -, 3	-, - -, -	-, - -, -	-, - -, -	-, - -, -	-, - -, -	-, - -, -
22	3	10	0, 2 0, 3	0, 3 -, 3	-, 3 -, 3	-, - -, -	-, - -, -	-, - -, -	-, - -, -	-, - -, -	-, - -, -
22	3	11	1, 2 0, 3	0, 3 -, -	-, - -, -	-, - -, -	-, - -, -	-, - -, -	-, - -, -	-, - -, -	-, - -, -
22	4	4	0, 2 -, 2	-, 2 -, 2	-, 3 -, 3	-, 3 -, 3	-, 3 -, 3	-, 3 -, 3	-, 4 -, 4	-, 4 -, 4	-, 4 -, 4
22	4	5	0, 2 -, 2	-, 2 -, 3	-, 3 -, 3	-, 3 -, 3	-, 4 -, 4	-, 4 -, 4	-, 4 -, 4	-, 4 -, 4	-, 4 -, 4
22	4	6	0, 2 -, 2	-, 3 -, 3	-, 3 -, 3	-, 3 -, 3	-, 4 -, 4	-, 4 -, 4	-, 4 -, 4	-, - -, -	-, - -, -
22	4	7	0, 2 0, 3	0, 3 -, 3	-, 3 -, 3	-, 4 -, 4	-, 4 -, 4	-, 4 -, 4	-, 4 -, 4	-, - -, -	-, - -, -
22	4	8	0, 2 0, 3	0, 3 -, 3	-, 4 -, 4	-, 4 -, 4	-, 4 -, 4	-, 4 -, 4	-, - -, -	-, - -, -	-, - -, -
22	4	9	1, 3 0, 3	0, 3 0, 4	0, 4 -, 4	-, 4 -, 4	-, 4 -, 4	-, - -, -	-, - -, -	-, - -, -	-, - -, -
22	4	10	1, 3 0, 3	0, 4 0, 4	0, 4 0, 4	-, 4 -, 4	-, - -, -	-, - -, -	-, - -, -	-, - -, -	-, - -, -
22	4	11	1, 3 0, 4	0, 4 0, 4	0, 4 0, 4	0, 4 -, -	-, - -, -	-, - -, -	-, - -, -	-, - -, -	-, - -, -
22	5	5	0, 2 0, 3	-, 3 -, 3	-, 3 -, 3	-, 4 -, 4	-, 4 -, 4	-, 4 -, 4	-, 4 -, 4	-, 5 -, 5	-, 5 -, 5
22	5	6	0, 2 0, 3	0, 3 -, 3	-, 4 -, 4	-, 4 -, 4	-, 4 -, 4	-, 4 -, 4	-, 5 -, 5	-, 5 -, 5	-, 5 -, 5
22	5	7	1, 3 0, 3	0, 3 0, 4	-, 4 -, 4	-, 4 -, 4	-, 5 -, 5	-, 5 -, 5	-, 5 -, 5	-, 5 -, 5	-, 5 -, 5
22	5	8	1, 3 0, 3	0, 4 0, 4	0, 4 -, 4	-, 4 -, 4	-, 5 -, 5	-, 5 -, 5	-, 5 -, 5	-, - -, -	-, - -, -
22	5	9	1, 3 1, 4	0, 4 0, 4	0, 4 0, 5	0, 5 -, 5	-, 5 -, 5	-, 5 -, 5	-, 5 -, 5	-, - -, -	-, - -, -
22	5	10	1, 3 1, 4	0, 4 0, 4	0, 5 0, 5	0, 5 0, 5	-, 5 -, 5	-, 5 -, 5	-, - -, -	-, - -, -	-, - -, -
22	5	11	2, 3 1, 4	1, 4 0, 5	0, 5 0, 5	0, 5 0, 5	0, 5 -, -	-, - -, -	-, - -, -	-, - -, -	-, - -, -
22	6	6	1, 3 0, 3	0, 4 0, 4	-, 4 -, 4	-, 4 -, 4	-, 5 -, 5	-, 5 -, 5	-, 5 -, 5	-, 5 -, 5	-, 6 -, 6
22	6	7	1, 3 0, 3	0, 4 0, 4	0, 4 -, 4	-, 5 -, 5	-, 5 -, 5	-, 5 -, 5	-, 5 -, 5	-, 6 -, 6	-, 6 -, 6
22	6	8	1, 4 1, 4	0, 4 0, 4	0, 4 0, 5	0, 5 -, 5	-, 5 -, 5	-, 6 -, 6	-, 6 -, 6	-, 6 -, 6	-, 6 -, 6
22	6	9	1, 4 1, 4	1, 4 0, 4	0, 5 0, 5	0, 5 0, 5	-, 6 -, 6	-, 6 -, 6	-, 6 -, 6	-, 6 -, 6	-, - -, -
22	6	10	2, 4 1, 4	1, 5 1, 5	0, 5 0, 5	0, 5 0, 6	0, 6 0, 6	-, 6 -, 6	-, 6 -, 6	-, - -, -	-, - -, -
22	6	11	2, 4 1, 5	1, 5 1, 5	1, 5 0, 6	0, 6 0, 6	0, 6 0, 6	0, 6 -, -	-, - -, -	-, - -, -	-, - -, -
22	7	7	1, 4 1, 4	0, 4 0, 4	0, 5 0, 5	0, 5 -, 5	-, 5 -, 5	-, 6 -, 6	-, 6 -, 6	-, 6 -, 6	-, 6 -, 6
22	7	8	2, 4 1, 4	1, 4 1, 5	0, 5 0, 5	0, 5 0, 6	-, 6 -, 6	-, 6 -, 6	-, 6 -, 6	-, 7 -, 7	-, 7 -, 7
22	7	9	2, 4 1, 4	1, 5 1, 5	0, 5 0, 5	0, 6 0, 6	0, 6 0, 6	-, 6 -, 6	-, 7 -, 7	-, 7 -, 7	-, 7 -, 7
22	7	10	2, 4 2, 5	1, 5 1, 5	1, 6 1, 6	0, 6 0, 6	0, 6 0, 7	0, 7 0, 7	0, 7 -, 7	-, 7 -, 7	-, 7 -, 7

TABLE B.27 (cont.) CRITICAL VALUES FOR FISHER'S EXACT TEST

n	m_1	m_2	α = 0.50	0.20	0.10	0.05	0.02	0.01	0.005	0.002	0.001
22	7	11	3, 4 2, 5	2, 5 1, 6	1, 6 1, 6	1, 6 0, 7	0, 7 0, 7	0, 7 0, 7	0, 7 0, 7	0, 7 -, -	-, - -, -
22	8	8	2, 4 1, 4	1, 5 1, 5	1, 5 0, 5	0, 6 0, 6	0, 6 0, 6	0, 6 -, 6	-, 7 -, 7	-, 7 -, 7	-, 7 -, 7
22	8	9	2, 5 2, 5	1, 5 1, 5	1, 6 1, 6	0, 6 0, 6	0, 7 0, 7	0, 7 0, 7	0, 7 -, 7	-, 7 -, 7	-, 8 -, 8
22	8	10	3, 5 2, 5	2, 6 1, 6	1, 6 1, 6	1, 6 1, 7	0, 7 0, 7	0, 7 0, 7	0, 7 0, 8	0, 8 0, 8	-, 8 -, 8
22	8	11	3, 5 2, 6	2, 6 2, 6	2, 6 1, 7	1, 7 1, 7	1, 7 0, 8	0, 8 0, 8	0, 8 0, 8	0, 8 0, 8	0, 8 -, -
22	9	9	3, 5 2, 5	2, 6 1, 6	1, 6 1, 6	1, 7 1, 7	0, 7 0, 7	0, 7 0, 7	0, 8 0, 8	0, 8 0, 8	-, 8 -, 8
22	9	10	3, 5 3, 6	2, 6 2, 6	2, 7 2, 7	1, 7 1, 7	1, 7 1, 8	0, 8 0, 8	0, 8 0, 8	0, 8 0, 8	0, 9 0, 9
22	9	11	4, 5 3, 6	3, 6 2, 7	2, 7 2, 7	2, 7 1, 8	1, 8 1, 8	1, 8 1, 8	1, 8 0, 9	0, 9 0, 9	0, 9 0, 9
22	10	10	4, 6 3, 6	3, 7 2, 7	2, 7 2, 7	2, 7 2, 8	1, 8 1, 8	1, 8 1, 8	1, 8 1, 9	0, 9 0, 9	0, 9 0, 9
22	10	11	4, 6 3, 7	3, 7 3, 7	3, 7 2, 8	2, 8 2, 8	2, 8 1, 9	1, 9 1, 9	1, 9 1, 9	1, 9 1, 9	1, 9 0,10
22	11	11	5, 6 4, 7	4, 7 3, 8	3, 8 3, 8	3, 8 2, 9	2, 9 2, 9	2, 9 2, 9	2, 9 1,10	1,10 1,10	1,10 1,10
23	1	1	-, 1 -, 1	-, 1 -, 1	-, 1 -, 1	-, 1 -, 1	-, - -, -	-, - -, -	-, - -, -	-, - -, -	-, - -, -
23	1	2	-, 1 -, 1	-, 1 -, 1	-, 1 -, 1	-, - -, -	-, - -, -	-, - -, -	-, - -, -	-, - -, -	-, - -, -
23	1	3	-, 1 -, 1	-, 1 -, 1	-, - -, -	-, - -, -	-, - -, -	-, - -, -	-, - -, -	-, - -, -	-, - -, -
23	1	4	-, 1 -, 1	-, 1 -, 1	-, - -, -	-, - -, -	-, - -, -	-, - -, -	-, - -, -	-, - -, -	-, - -, -
23	1	5	-, 1 -, 1	-, - -, -	-, - -, -	-, - -, -	-, - -, -	-, - -, -	-, - -, -	-, - -, -	-, - -, -
23	1	6	-, 1 -, 1	-, - -, -	-, - -, -	-, - -, -	-, - -, -	-, - -, -	-, - -, -	-, - -, -	-, - -, -
23	1	7	-, 1 -, 1	-, - -, -	-, - -, -	-, - -, -	-, - -, -	-, - -, -	-, - -, -	-, - -, -	-, - -, -
23	1	8	-, 1 -, 1	-, - -, -	-, - -, -	-, - -, -	-, - -, -	-, - -, -	-, - -, -	-, - -, -	-, - -, -
23	1	9	-, 1 -, 1	-, - -, -	-, - -, -	-, - -, -	-, - -, -	-, - -, -	-, - -, -	-, - -, -	-, - -, -
23	1	10	-, 1 -, 1	-, - -, -	-, - -, -	-, - -, -	-, - -, -	-, - -, -	-, - -, -	-, - -, -	-, - -, -
23	1	11	-, 1 -, 1	-, - -, -	-, - -, -	-, - -, -	-, - -, -	-, - -, -	-, - -, -	-, - -, -	-, - -, -
23	2	2	-, 1 -, 1	-, 1 -, 1	-, 2 -, 2	-, 2 -, 2	-, 2 -, 2	-, 2 -, 2	-, 2 -, 2	-, - -, -	-, - -, -
23	2	3	-, 1 -, 1	-, 2 -, 2	-, 2 -, 2	-, 2 -, 2	-, 2 -, 2	-, - -, -	-, - -, -	-, - -, -	-, - -, -
23	2	4	-, 1 -, 1	-, 2 -, 2	-, 2 -, 2	-, 2 -, 2	-, - -, -	-, - -, -	-, - -, -	-, - -, -	-, - -, -
23	2	5	-, 1 -, 1	-, 2 -, 2	-, 2 -, 2	-, 2 -, 2	-, - -, -	-, - -, -	-, - -, -	-, - -, -	-, - -, -
23	2	6	-, 1 -, 1	-, 2 -, 2	-, 2 -, 2	-, - -, -	-, - -, -	-, - -, -	-, - -, -	-, - -, -	-, - -, -
23	2	7	0, 1 -, 2	-, 2 -, 2	-, 2 -, 2	-, - -, -	-, - -, -	-, - -, -	-, - -, -	-, - -, -	-, - -, -
23	2	8	0, 2 -, 2	-, 2 -, 2	-, - -, -	-, - -, -	-, - -, -	-, - -, -	-, - -, -	-, - -, -	-, - -, -
23	2	9	0, 2 -, 2	-, 2 -, 2	-, - -, -	-, - -, -	-, - -, -	-, - -, -	-, - -, -	-, - -, -	-, - -, -
23	2	10	0, 2 0, 2	-, 2 -, 2	-, - -, -	-, - -, -	-, - -, -	-, - -, -	-, - -, -	-, - -, -	-, - -, -
23	2	11	0, 2 0, 2	-, - -, -	-, - -, -	-, - -, -	-, - -, -	-, - -, -	-, - -, -	-, - -, -	-, - -, -
23	3	3	-, 1 -, 1	-, 2 -, 2	-, 2 -, 2	-, 2 -, 2	-, 3 -, 3	-, 3 -, 3	-, 3 -, 3	-, 3 -, 3	-, 3 -, 3
23	3	4	-, 1 -, 1	-, 2 -, 2	-, 2 -, 2	-, 3 -, 3	-, 3 -, 3	-, 3 -, 3	-, 3 -, 3	-, - -, -	-, - -, -
23	3	5	0, 1 -, 2	-, 2 -, 2	-, 3 -, 3	-, 3 -, 3	-, 3 -, 3	-, 3 -, 3	-, - -, -	-, - -, -	-, - -, -
23	3	6	0, 2 -, 2	-, 2 -, 2	-, 3 -, 3	-, 3 -, 3	-, 3 -, 3	-, - -, -	-, - -, -	-, - -, -	-, - -, -
23	3	7	0, 2 -, 2	-, 3 -, 3	-, 3 -, 3	-, 3 -, 3	-, 3 -, 3	-, - -, -	-, - -, -	-, - -, -	-, - -, -
23	3	8	0, 2 -, 2	-, 3 -, 3	-, 3 -, 3	-, 3 -, 3	-, - -, -	-, - -, -	-, - -, -	-, - -, -	-, - -, -
23	3	9	0, 2 0, 3	-, 3 -, 3	-, 3 -, 3	-, 3 -, 3	-, - -, -	-, - -, -	-, - -, -	-, - -, -	-, - -, -
23	3	10	0, 2 0, 3	0, 3 -, 3	-, 3 -, 3	-, - -, -	-, - -, -	-, - -, -	-, - -, -	-, - -, -	-, - -, -
23	3	11	0, 2 0, 3	0, 3 -, 3	-, 3 -, 3	-, - -, -	-, - -, -	-, - -, -	-, - -, -	-, - -, -	-, - -, -
23	4	4	0, 2 -, 2	-, 2 -, 2	-, 3 -, 3	-, 3 -, 3	-, 3 -, 3	-, 3 -, 3	-, 4 -, 4	-, 4 -, 4	-, 4 -, 4
23	4	5	0, 2 -, 2	-, 2 -, 2	-, 3 -, 3	-, 3 -, 3	-, 4 -, 4	-, 4 -, 4	-, 4 -, 4	-, 4 -, 4	-, 4 -, 4
23	4	6	0, 2 -, 2	-, 3 -, 3	-, 3 -, 3	-, 3 -, 3	-, 4 -, 4	-, 4 -, 4	-, 4 -, 4	-, 4 -, 4	-, - -, -
23	4	7	0, 2 0, 3	-, 3 -, 3	-, 3 -, 3	-, 4 -, 4	-, 4 -, 4	-, 4 -, 4	-, 4 -, 4	-, - -, -	-, - -, -
23	4	8	0, 2 0, 3	0, 3 -, 3	-, 4 -, 4	-, 4 -, 4	-, 4 -, 4	-, 4 -, 4	-, - -, -	-, - -, -	-, - -, -
23	4	9	1, 3 0, 3	0, 3 0, 4	-, 4 -, 4	-, 4 -, 4	-, 4 -, 4	-, - -, -	-, - -, -	-, - -, -	-, - -, -
23	4	10	1, 3 0, 3	0, 3 0, 4	0, 4 -, 4	-, 4 -, 4	-, - -, -	-, - -, -	-, - -, -	-, - -, -	-, - -, -
23	4	11	1, 3 0, 3	0, 4 0, 4	0, 4 0, 4	-, 4 -, 4	-, - -, -	-, - -, -	-, - -, -	-, - -, -	-, - -, -
23	5	5	0, 2 -, 2	-, 3 -, 3	-, 3 -, 3	-, 3 -, 3	-, 4 -, 4	-, 4 -, 4	-, 4 -, 4	-, 5 -, 5	-, 5 -, 5

TABLE B.27 (cont.) CRITICAL VALUES FOR FISHER'S EXACT TEST

n	m_1	m_2	α = 0.50	0.20	0.10	0.05	0.02	0.01	0.005	0.002	0.001
23	5	6	0, 2 0, 3	0, 3 -, 3	-, 3 -, 3	-, 4 -, 4	-, 4 -, 4	-, 4 -, 4	-, 5 -, 5	-, 5 -, 5	-, 5 -, 5
23	5	7	0, 2 0, 3	0, 3 -, 3	-, 4 -, 4	-, 4 -, 4	-, 4 -, 4	-, 5 -, 5	-, 5 -, 5	-, 5 -, 5	-, 5 -, 5
23	5	8	1, 3 0, 3	0, 4 0, 4	0, 4 -, 4	-, 4 -, 4	-, 5 -, 5	-, 5 -, 5	-, 5 -, 5	-, 5 -, 5	-, - -, -
23	5	9	1, 3 0, 3	0, 4 0, 4	0, 4 -, 4	-, 5 -, 5	-, 5 -, 5	-, 5 -, 5	-, 5 -, 5	-, - -, -	-, - -, -
23	5	10	1, 3 1, 4	0, 4 0, 4	0, 4 0, 5	0, 5 0, 5	-, 5 -, 5	-, 5 -, 5	-, - -, -	-, - -, -	-, - -, -
23	5	11	1, 3 1, 4	1, 4 0, 4	0, 5 0, 5	0, 5 0, 5	-, 5 -, 5	-, - -, -	-, - -, -	-, - -, -	-, - -, -
23	6	6	1, 2 0, 3	0, 4 0, 4	-, 4 -, 4	-, 4 -, 4	-, 5 -, 5	-, 5 -, 5	-, 5 -, 5	-, 5 -, 5	-, 6 -, 6
23	6	7	1, 3 0, 3	0, 4 0, 4	0, 4 -, 4	-, 4 -, 4	-, 5 -, 5	-, 5 -, 5	-, 5 -, 5	-, 6 -, 6	-, 6 -, 6
23	6	8	1, 4 1, 4	0, 4 0, 4	0, 4 0, 5	0, 5 -, 5	-, 5 -, 5	-, 5 -, 5	-, 6 -, 6	-, 6 -, 6	-, 6 -, 6
23	6	9	1, 4 1, 4	0, 4 0, 4	0, 5 0, 5	0, 5 0, 5	-, 5 -, 5	-, 6 -, 6	-, 6 -, 6	-, 6 -, 6	-, 6 -, 6
23	6	10	2, 4 1, 4	1, 4 1, 5	0, 5 0, 5	0, 5 0, 6	0, 6 0, 6	-, 6 -, 6	-, 6 -, 6	-, - -, -	-, - -, -
23	6	11	2, 4 1, 4	1, 5 1, 5	1, 5 0, 5	0, 6 0, 6	0, 6 0, 6	0, 6 -, 6	-, 6 -, 6	-, - -, -	-, - -, -
23	7	7	1, 4 1, 4	0, 4 0, 4	0, 4 0, 5	0, 5 -, 5	-, 5 -, 5	-, 6 -, 6	-, 6 -, 6	-, 6 -, 6	-, 6 -, 6
23	7	8	1, 4 1, 4	1, 4 0, 4	0, 5 0, 5	0, 5 -, 5	-, 6 -, 6	-, 6 -, 6	-, 6 -, 6	-, 6 -, 6	-, 7 -, 7
23	7	9	2, 4 1, 4	1, 5 1, 5	0, 5 0, 5	0, 6 0, 6	0, 6 0, 6	-, 6 -, 6	-, 6 -, 6	-, 7 -, 7	-, 7 -, 7
23	7	10	2, 4 2, 5	1, 5 1, 5	1, 5 1, 6	0, 6 0, 6	0, 6 0, 6	0, 7 0, 7	-, 7 -, 7	-, 7 -, 7	-, 7 -, 7
23	7	11	2, 4 2, 5	1, 5 1, 5	1, 6 1, 6	1, 6 0, 6	0, 7 0, 7	0, 7 0, 7	0, 7 0, 7	-, 7 -, 7	-, - -, -
23	8	8	2, 4 1, 4	1, 5 1, 5	0, 5 0, 5	0, 6 0, 6	0, 6 0, 6	-, 6 -, 6	-, 7 -, 7	-, 7 -, 7	-, 7 -, 7
23	8	9	2, 5 2, 5	1, 5 1, 5	1, 6 1, 6	0, 6 0, 6	0, 6 0, 7	0, 7 0, 7	-, 7 -, 7	-, 7 -, 7	-, 8 -, 8
23	8	10	2, 5 2, 5	2, 5 1, 6	1, 6 1, 6	1, 6 0, 6	0, 7 0, 7	0, 7 0, 7	0, 7 0, 8	-, 8 -, 8	-, 8 -, 8
23	8	11	3, 5 2, 5	2, 6 2, 6	1, 6 1, 6	1, 7 1, 7	1, 7 0, 7	0, 7 0, 7	0, 8 0, 8	0, 8 0, 8	-, 8 -, 8
23	9	9	3, 5 2, 5	2, 5 1, 6	1, 6 1, 6	1, 6 1, 7	0, 7 0, 7	0, 7 0, 7	0, 7 0, 8	-, 8 -, 8	-, 8 -, 8
23	9	10	3, 5 2, 5	2, 6 2, 6	1, 6 1, 7	1, 7 1, 7	1, 7 0, 7	0, 8 0, 8	0, 8 0, 8	0, 8 0, 8	0, 8 -, 8
23	9	11	3, 5 3, 6	2, 6 2, 7	2, 7 2, 7	1, 7 1, 7	1, 8 1, 8	1, 8 1, 8	0, 8 0, 8	0, 9 0, 9	0, 9 0, 9
23	10	10	3, 6 3, 6	2, 6 2, 7	2, 7 2, 7	1, 7 1, 7	1, 8 1, 8	1, 8 1, 8	0, 8 0, 8	0, 9 0, 9	0, 9 0, 9
23	10	11	4, 6 3, 6	3, 7 2, 7	2, 7 2, 7	2, 8 2, 8	1, 8 1, 8	1, 9 1, 9	1, 9 1, 9	0, 9 0, 9	0, 9 0, 9
23	11	11	4, 6 4, 7	3, 7 3, 8	3, 8 3, 8	2, 8 2, 8	2, 9 2, 9	2, 9 1, 9	1, 9 1, 9	1,10 1,10	1,10 1,10
24	1	1	-, 1 -, 1	-, 1 -, 1	-, 1 -, 1	-, 1 -, 1	-, - -, -	-, - -, -	-, - -, -	-, - -, -	-, - -, -
24	1	2	-, 1 -, 1	-, 1 -, 1	-, 1 -, 1	-, - -, -	-, - -, -	-, - -, -	-, - -, -	-, - -, -	-, - -, -
24	1	3	-, 1 -, 1	-, 1 -, 1	-, - -, -	-, - -, -	-, - -, -	-, - -, -	-, - -, -	-, - -, -	-, - -, -
24	1	4	-, 1 -, 1	-, 1 -, 1	-, - -, -	-, - -, -	-, - -, -	-, - -, -	-, - -, -	-, - -, -	-, - -, -
24	1	5	-, 1 -, 1	-, - -, -	-, - -, -	-, - -, -	-, - -, -	-, - -, -	-, - -, -	-, - -, -	-, - -, -
24	1	6	-, 1 -, 1	-, - -, -	-, - -, -	-, - -, -	-, - -, -	-, - -, -	-, - -, -	-, - -, -	-, - -, -
24	1	7	-, 1 -, 1	-, - -, -	-, - -, -	-, - -, -	-, - -, -	-, - -, -	-, - -, -	-, - -, -	-, - -, -
24	1	8	-, 1 -, 1	-, - -, -	-, - -, -	-, - -, -	-, - -, -	-, - -, -	-, - -, -	-, - -, -	-, - -, -
24	1	9	-, 1 -, 1	-, - -, -	-, - -, -	-, - -, -	-, - -, -	-, - -, -	-, - -, -	-, - -, -	-, - -, -
24	1	10	-, 1 -, 1	-, - -, -	-, - -, -	-, - -, -	-, - -, -	-, - -, -	-, - -, -	-, - -, -	-, - -, -
24	1	11	-, 1 -, 1	-, - -, -	-, - -, -	-, - -, -	-, - -, -	-, - -, -	-, - -, -	-, - -, -	-, - -, -
24	1	12	0, 1 -, -	-, - -, -	-, - -, -	-, - -, -	-, - -, -	-, - -, -	-, - -, -	-, - -, -	-, - -, -
24	2	2	-, 1 -, 1	-, 1 -, 1	-, 2 -, 2	-, 2 -, 2	-, 2 -, 2	-, 2 -, 2	-, 2 -, 2	-, - -, -	-, - -, -
24	2	3	-, 1 -, 1	-, 2 -, 2	-, 2 -, 2	-, 2 -, 2	-, 2 -, 2	-, - -, -	-, - -, -	-, - -, -	-, - -, -
24	2	4	-, 1 -, 1	-, 2 -, 2	-, 2 -, 2	-, 2 -, 2	-, - -, -	-, - -, -	-, - -, -	-, - -, -	-, - -, -
24	2	5	-, 1 -, 1	-, 2 -, 2	-, 2 -, 2	-, 2 -, 2	-, - -, -	-, - -, -	-, - -, -	-, - -, -	-, - -, -
24	2	6	-, 1 -, 1	-, 2 -, 2	-, 2 -, 2	-, - -, -	-, - -, -	-, - -, -	-, - -, -	-, - -, -	-, - -, -
24	2	7	0, 1 -, 2	-, 2 -, 2	-, 2 -, 2	-, - -, -	-, - -, -	-, - -, -	-, - -, -	-, - -, -	-, - -, -
24	2	8	0, 2 -, 2	-, 2 -, 2	-, - -, -	-, - -, -	-, - -, -	-, - -, -	-, - -, -	-, - -, -	-, - -, -
24	2	9	0, 2 -, 2	-, 2 -, 2	-, - -, -	-, - -, -	-, - -, -	-, - -, -	-, - -, -	-, - -, -	-, - -, -
24	2	10	0, 2 0, 2	-, 2 -, 2	-, - -, -	-, - -, -	-, - -, -	-, - -, -	-, - -, -	-, - -, -	-, - -, -
24	2	11	0, 2 0, 2	-, 2 -, 2	-, - -, -	-, - -, -	-, - -, -	-, - -, -	-, - -, -	-, - -, -	-, - -, -
24	2	12	0, 2 0, 2	-, - -, -	-, - -, -	-, - -, -	-, - -, -	-, - -, -	-, - -, -	-, - -, -	-, - -, -

TABLE B.27 (cont.) CRITICAL VALUES FOR FISHER'S EXACT TEST

n	m_1	m_2	α = 0.50	0.20	0.10	0.05	0.02	0.01	0.005	0.002	0.001
24	3	3	-, 1 -, 1	-, 2 -, 2	-, 2 -, 2	-, 2 -, 2	-, 3 -, 3	-, 3 -, 3	-, 3 -, 3	-, 3 -, 3	-, 3 -, 3
24	3	4	-, 1 -, 1	-, 2 -, 2	-, 2 -, 2	-, 3 -, 3	-, 3 -, 3	-, 3 -, 3	-, 3 -, 3	-, 3 -, 3	-, - -, -
24	3	5	0, 1 -, 2	-, 2 -, 2	-, 2 -, 2	-, 3 -, 3	-, 3 -, 3	-, 3 -, 3	-, 3 -, 3	-, - -, -	-, - -, -
24	3	6	0, 2 -, 2	-, 2 -, 2	-, 3 -, 3	-, 3 -, 3	-, 3 -, 3	-, 3 -, 3	-, - -, -	-, - -, -	-, - -, -
24	3	7	0, 2 -, 2	-, 2 -, 2	-, 3 -, 3	-, 3 -, 3	-, 3 -, 3	-, - -, -	-, - -, -	-, - -, -	-, - -, -
24	3	8	0, 2 -, 2	-, 3 -, 3	-, 3 -, 3	-, 3 -, 3	-, - -, -	-, - -, -	-, - -, -	-, - -, -	-, - -, -
24	3	9	0, 2 0, 3	-, 3 -, 3	-, 3 -, 3	-, 3 -, 3	-, - -, -	-, - -, -	-, - -, -	-, - -, -	-, - -, -
24	3	10	0, 2 0, 3	0, 3 -, 3	-, 3 -, 3	-, - -, -	-, - -, -	-, - -, -	-, - -, -	-, - -, -	-, - -, -
24	3	11	0, 2 0, 3	0, 3 -, 3	-, 3 -, 3	-, - -, -	-, - -, -	-, - -, -	-, - -, -	-, - -, -	-, - -, -
24	3	12	1, 2 0, 3	0, 3. -, -	-, - -, -	-, - -, -	-, - -, -	-, - -, -	-, - -, -	-, - -, -	-, - -, -
24	4	4	0, 1 -, 2	-, 2 -, 2	-, 3 -, 3	-, 3 -, 3	-, 3 -, 3	-, 3 -, 3	-, 4 -, 4	-, 4 -, 4	-, 4 -, 4
24	4	5	0, 2 -, 2	-, 2 -, 2	-, 3 -, 3	-, 3 -, 3	-, 3 -, 3	-, 4 -, 4	-, 4 -, 4	-, 4 -, 4	-, 4 -, 4
24	4	6	0, 2 -, 2	-, 3 -, 3	-, 3 -, 3	-, 3 -, 3	-, 4 -, 4	-, 4 -, 4	-, 4 -, 4	-, 4 -, 4	-, - -, -
24	4	7	0, 2 0, 3	-, 3 -, 3	-, 3 -, 3	-, 4 -, 4	-, 4 -, 4	-, 4 -, 4	-, 4 -, 4	-, - -, -	-, - -, -
24	4	8	0, 2 0, 3	0, 3 -, 3	-, 3 -, 3	-, 4 -, 4	-, 4 -, 4	-, 4 -, 4	-, - -, -	-, - -, -	-, - -, -
24	4	9	0, 2 0, 3	0, 3 -, 3	-, 4 -, 4	-, 4 -, 4	-, 4 -, 4	-, - -, -	-, - -, -	-, - -, -	-, - -, -
24	4	10	1, 3 0, 3	0, 3 0, 4	0, 4 -, 4	-, 4 -, 4	-, 4 -, 4	-, - -, -	-, - -, -	-, - -, -	-, - -, -
24	4	11	1, 3 0, 3	0, 4 0, 4	0, 4 0, 4	-, 4 -, 4	-, - -, -	-, - -, -	-, - -, -	-, - -, -	-, - -, -
24	4	12	1, 3 0, 4	0, 4 0, 4	0, 4 0, 4	0, 4 -, -	-, - -, -	-, - -, -	-, - -, -	-, - -, -	-, - -, -
24	5	5	0, 2 -, 2	-, 3 -, 3	-, 3 -, 3	-, 3 -, 3	-, 4 -, 4	-, 4 -, 4	-, 4 -, 4	-, 5 -, 5	-, 5 -, 5
24	5	6	0, 2 0, 3	-, 3 -, 3	-, 3 -, 3	-, 4 -, 4	-, 4 -, 4	-, 4 -, 4	-, 5 -, 5	-, 5 -, 5	-, 5 -, 5
24	5	7	0, 2 0, 3	0, 3 -, 3	-, 4 -, 4	-, 4 -, 4	-, 4 -, 4	-, 5 -, 5	-, 5 -, 5	-, 5 -, 5	-, 5 -, 5
24	5	8	1, 3 0, 3	0, 3 0, 4	-, 4 -, 4	-, 4 -, 4	-, 5 -, 5	-, 5 -, 5	-, 5 -, 5	-, 5 -, 5	-, - -, -
24	5	9	1, 3 0, 3	0, 4 0, 4	0, 4 -, 4	-, 4 -, 4	-, 5 -, 5	-, 5 -, 5	-, 5 -, 5	-, - -, -	-, - -, -
24	5	10	1, 3 1, 4	0, 4 0, 4	0, 4 0, 5	0, 5 -, 5	-, 5 -, 5	-, 5 -, 5	-, - -, -	-, - -, -	-, - -, -
24	5	11	1, 3 1, 4	0, 4 0, 4	0, 5 0, 5	0, 5 0, 5	-, 5 -, 5	-, - -, -	-, - -, -	-, - -, -	-, - -, -
24	5	12	2, 3 1, 4	1, 4 0, 5	0, 5 0, 5	0, 5 0, 5	0, 5 -, -	-, - -, -	-, - -, -	-, - -, -	-, - -, -
24	6	6	0, 2 0, 3	0, 3 -, 3	-, 4 -, 4	-, 4 -, 4	-, 4 -, 4	-, 5 -, 5	-, 5 -, 5	-, 5 -, 5	-, 5 -, 5
24	6	7	1, 3 0, 3	0, 4 0, 4	0, 4 -, 4	-, 4 -, 4	-, 5 -, 5	-, 5 -, 5	-, 5 -, 5	-, 6 -, 6	-, 6 -, 6
24	6	8	1, 3 0, 3	0, 4 0, 4	0, 4 0, 5	-, 5 -, 5	-, 5 -, 5	-, 5 -, 5	-, 6 -, 6	-, 6 -, 6	-, 6 -, 6
24	6	9	1, 4 1, 4	0, 4 0, 4	0, 5 0, 5	0, 5 -, 5	-, 5 -, 5	-, 6 -, 6	-, 6 -, 6	-, 6 -, 6	-, 6 -, 6
24	6	10	1, 3 1, 4	1, 4 0, 4	0, 5 0, 5	0, 5 0, 6	-, 6 -, 6	-, 6 -, 6	-, 6 -, 6	-, 6 -, 6	-, - -, -
24	6	11	2, 4 1, 4	1, 5 1, 5	0, 5 0, 5	0, 5 0, 6	0, 6 0, 6	-, 6 -, 6	-, 6 -, 6	-, - -, -	-, - -, -
24	6	12	2, 4 1, 5	1, 5 1, 5	1, 5 0, 6	0, 6 0, 6	0, 6 0, 6	0, 6 -, -	-, - -, -	-, - -, -	-, - -, -
24	7	7	1, 3 0, 3	0, 4 0, 4	0, 4 0, 5	-, 5 -, 5	-, 5 -, 5	-, 5 -, 5	-, 6 -, 6	-, 6 -, 6	-, 6 -, 6
24	7	8	1, 4 1, 4	0, 4 0, 4	0, 5 0, 5	0, 5 -, 5	-, 6 -, 6	-, 6 -, 6	-, 6 -, 6	-, 6 -, 6	-, 7 -, 7
24	7	9	2, 4 1, 4	1, 5 1, 5	0, 5 0, 5	0, 5 0, 6	0, 6 -, 6	-, 6 -, 6	-, 6 -, 6	-, 7 -, 7	-, 7 -, 7
24	7	10	2, 4 1, 4	1, 5 1, 5	1, 5 0, 5	0, 6 0, 6	0, 6 0, 6	0, 6 -, 6	-, 7 -, 7	-, 7 -, 7	-, 7 -, 7
24	7	11	2, 4 2, 5	1, 5 1, 5	1, 6 1, 6	0, 6 0, 6	0, 6 0, 7	0, 7 0, 7	0, 7 -, 7	-, 7 -, 7	-, 7 -, 7
24	7	12	3, 4 2, 5	2, 5 1, 6	1, 6 1, 6	1, 6 0, 7	0, 7 0, 7	0, 7 0, 7	0, 7 0, 7	-, - -, -	-, - -, -
24	8	8	2, 4 1, 4	1, 5 1, 5	0, 5 0, 5	0, 5 0, 6	0, 6 -, 6	-, 6 -, 6	-, 6 -, 6	-, 7 -, 7	-, 7 -, 7
24	8	9	2, 4 1, 4	1, 5 1, 5	1, 5 0, 5	0, 6 0, 6	0, 6 0, 7	0, 7 0, 7	-, 7 -, 7	-, 7 -, 7	-, 7 -, 7
24	8	10	2, 5 2, 5	1, 5 1, 6	1, 6 1, 6	0, 6 0, 6	0, 7 0, 7	0, 7 0, 7	0, 7 -, 7	-, 8 -, 8	-, 8 -, 8
24	8	11	3, 5 2, 5	2, 6 1, 6	1, 6 1, 6	1, 7 1, 7	0, 7 0, 7	0, 7 0, 7	0, 8 0, 8	0, 8 0, 8	-, 8 -, 8
24	8	12	3, 5 2, 6	2, 6 2, 6	2, 6 1, 7	1, 7 1, 7	1, 7 0, 8	0, 8 0, 8	0, 8 0, 8	0, 8 0, 8	0, 8 -, -
24	9	9	2, 5 2, 5	1, 5 1, 6	1, 6 1, 6	1, 6 0, 6	0, 7 0, 7	0, 7 0, 7	0, 7 -, 7	-, 8 -, 8	-, 8 -, 8
24	9	10	3, 5 2, 5	2, 6 1, 6	1, 6 1, 6	1, 7 1, 7	0, 7 0, 7	0, 7 0, 8	0, 8 0, 8	0, 8 -, 8	-, 8 -, 8
24	9	11	3, 5 3, 6	2, 6 2, 7	2, 7 1, 7	1, 7 1, 7	1, 8 1, 8	0, 8 0, 8	0, 8 0, 8	0, 8 0, 9	0, 9 0, 9
24	9	12	4, 5 3, 6	2, 7 2, 7	2, 7 2, 7	2, 7 1, 8	1, 8 1, 8	1, 8 1, 8	1, 8 0, 9	0, 9 0, 9	0, 9 0, 9
24	10	10	3, 6 3, 6	2, 6 2, 7	2, 7 1, 7	1, 7 1, 7	1, 8 1, 8	0, 8 0, 8	0, 8 0, 8	0, 9 0, 9	0, 9 0, 9

TABLE B.27 (cont.) CRITICAL VALUES FOR FISHER'S EXACT TEST

n	m_1	m_2	α = 0.50		0.20		0.10		0.05		0.02		0.01		0.005		0.002		0.001	
24	10	11	4, 6	3, 6	3, 7	2, 7	2, 7	2, 7	2, 8	2, 8	1, 8	1, 8	1, 8	1, 9	1, 9	1, 9	0, 9	0, 9	0, 9	0, 9
24	10	12	4, 6	3, 7	3, 7	2, 8	2, 8	2, 8	2, 8	2, 8	2, 8	1, 9	1, 9	1, 9	1, 9	1, 9	1, 9	0,10	0,10	0,10
24	11	11	4, 6	4, 7	3, 7	3, 8	2, 8	2, 8	2, 8	2, 8	2, 9	2, 9	1, 9	1, 9	1, 9	1, 9	1, 9	1,10	0,10	0,10
24	11	12	5, 6	4, 7	3, 8	3, 8	3, 8	3, 8	3, 8	2, 9	2, 9	2, 9	2, 9	1,10	1,10	1,10	1,10	1,10	1,10	1,10
24	12	12	5, 7	4, 8	4, 8	3, 9	3, 9	3, 9	3, 9	3, 9	3, 9	2,10	2,10	2,10	2,10	2,10	2,10	1,11	1,11	1,11
25	1	1	-, 1	-, 1	-, 1	-, 1	-, 1	-, 1	-, 1	-, 1	-, -	-, -	-, -	-, -	-, -	-, -	-, -	-, -	-, -	-, -
25	1	2	-, 1	-, 1	-, 1	-, 1	-, 1	-, 1	-, -	-, -	-, -	-, -	-, -	-, -	-, -	-, -	-, -	-, -	-, -	-, -
25	1	3	-, 1	-, 1	-, 1	-, 1	-, -	-, -	-, -	-, -	-, -	-, -	-, -	-, -	-, -	-, -	-, -	-, -	-, -	-, -
25	1	4	-, 1	-, 1	-, 1	-, 1	-, -	-, -	-, -	-, -	-, -	-, -	-, -	-, -	-, -	-, -	-, -	-, -	-, -	-, -
25	1	5	-, 1	-, 1	-, 1	-, 1	-, -	-, -	-, -	-, -	-, -	-, -	-, -	-, -	-, -	-, -	-, -	-, -	-, -	-, -
25	1	6	-, 1	-, 1	-, -	-, -	-, -	-, -	-, -	-, -	-, -	-, -	-, -	-, -	-, -	-, -	-, -	-, -	-, -	-, -
25	1	7	-, 1	-, 1	-, -	-, -	-, -	-, -	-, -	-, -	-, -	-, -	-, -	-, -	-, -	-, -	-, -	-, -	-, -	-, -
25	1	8	-, 1	-, 1	-, -	-, -	-, -	-, -	-, -	-, -	-, -	-, -	-, -	-, -	-, -	-, -	-, -	-, -	-, -	-, -
25	1	9	-, 1	-, 1	-, -	-, -	-, -	-, -	-, -	-, -	-, -	-, -	-, -	-, -	-, -	-, -	-, -	-, -	-, -	-, -
25	1	10	-, 1	-, 1	-, -	-, -	-, -	-, -	-, -	-, -	-, -	-, -	-, -	-, -	-, -	-, -	-, -	-, -	-, -	-, -
25	1	11	-, 1	-, 1	-, -	-, -	-, -	-, -	-, -	-, -	-, -	-, -	-, -	-, -	-, -	-, -	-, -	-, -	-, -	-, -
25	1	12	-, 1	-, 1	-, -	-, -	-, -	-, -	-, -	-, -	-, -	-, -	-, -	-, -	-, -	-, -	-, -	-, -	-, -	-, -
25	2	2	-, 1	-, 1	-, 1	-, 1	-, 2	-, 2	-, 2	-, 2	-, 2	-, 2	-, 2	-, 2	-, 2	-, 2	-, -	-, -	-, -	-, -
25	2	3	-, 1	-, 1	-, 2	-, 2	-, 2	-, 2	-, 2	-, 2	-, 2	-, 2	-, 2	-, 2	-, -	-, -	-, -	-, -	-, -	-, -
25	2	4	-, 1	-, 1	-, 2	-, 2	-, 2	-, 2	-, 2	-, 2	-, 2	-, 2	-, -	-, -	-, -	-, -	-, -	-, -	-, -	-, -
25	2	5	-, 1	-, 1	-, 2	-, 2	-, 2	-, 2	-, 2	-, 2	-, -	-, -	-, -	-, -	-, -	-, -	-, -	-, -	-, -	-, -
25	2	6	-, 1	-, 1	-, 2	-, 2	-, 2	-, 2	-, 2	-, 2	-, -	-, -	-, -	-, -	-, -	-, -	-, -	-, -	-, -	-, -
25	2	7	-, 1	-, 1	-, 2	-, 2	-, 2	-, 2	-, -	-, -	-, -	-, -	-, -	-, -	-, -	-, -	-, -	-, -	-, -	-, -
25	2	8	0, 2	-, 2	-, 2	-, 2	-, 2	-, 2	-, -	-, -	-, -	-, -	-, -	-, -	-, -	-, -	-, -	-, -	-, -	-, -
25	2	9	0, 2	-, 2	-, 2	-, 2	-, -	-, -	-, -	-, -	-, -	-, -	-, -	-, -	-, -	-, -	-, -	-, -	-, -	-, -
25	2	10	0, 2	0, 2	-, 2	-, 2	-, -	-, -	-, -	-, -	-, -	-, -	-, -	-, -	-, -	-, -	-, -	-, -	-, -	-, -
25	2	11	0, 2	0, 2	-, 2	-, 2	-, -	-, -	-, -	-, -	-, -	-, -	-, -	-, -	-, -	-, -	-, -	-, -	-, -	-, -
25	2	12	0, 2	0, 2	-, -	-, -	-, -	-, -	-, -	-, -	-, -	-, -	-, -	-, -	-, -	-, -	-, -	-, -	-, -	-, -
25	3	3	-, 1	-, 1	-, 2	-, 2	-, 2	-, 2	-, 2	-, 2	-, 3	-, 3	-, 3	-, 3	-, 3	-, 3	-, 3	-, 3	-, 3	-, 3
25	3	4	-, 1	-, 1	-, 2	-, 2	-, 2	-, 2	-, 3	-, 3	-, 3	-, 3	-, 3	-, 3	-, 3	-, 3	-, 3	-, 3	-, -	-, -
25	3	5	0, 1	-, 2	-, 2	-, 2	-, 2	-, 2	-, 3	-, 3	-, 3	-, 3	-, 3	-, 3	-, 3	-, 3	-, -	-, -	-, -	-, -
25	3	6	0, 2	-, 2	-, 2	-, 2	-, 3	-, 3	-, 3	-, 3	-, 3	-, 3	-, 3	-, 3	-, -	-, -	-, -	-, -	-, -	-, -
25	3	7	0, 2	-, 2	-, 2	-, 2	-, 3	-, 3	-, 3	-, 3	-, 3	-, 3	-, -	-, -	-, -	-, -	-, -	-, -	-, -	-, -
25	3	8	0, 2	-, 2	-, 3	-, 3	-, 3	-, 3	-, 3	-, 3	-, -	-, -	-, -	-, -	-, -	-, -	-, -	-, -	-, -	-, -
25	3	9	0, 2	0, 3	-, 3	-, 3	-, 3	-, 3	-, 3	-, 3	-, -	-, -	-, -	-, -	-, -	-, -	-, -	-, -	-, -	-, -
25	3	10	0, 2	0, 3	0, 3	-, 3	-, 3	-, 3	-, -	-, -	-, -	-, -	-, -	-, -	-, -	-, -	-, -	-, -	-, -	-, -
25	3	11	0, 2	0, 3	0, 3	-, 3	-, 3	-, 3	-, -	-, -	-, -	-, -	-, -	-, -	-, -	-, -	-, -	-, -	-, -	-, -
25	3	12	0, 2	0, 3	0, 3	-, 3	-, 3	-, 3	-, -	-, -	-, -	-, -	-, -	-, -	-, -	-, -	-, -	-, -	-, -	-, -
25	4	4	0, 1	-, 2	-, 2	-, 2	-, 2	-, 3	-, 3	-, 3	-, 3	-, 3	-, 3	-, 3	-, 4	-, 4	-, 4	-, 4	-, 4	-, 4
25	4	5	0, 2	-, 2	-, 2	-, 2	-, 3	-, 3	-, 3	-, 3	-, 3	-, 3	-, 4	-, 4	-, 4	-, 4	-, 4	-, 4	-, 4	-, 4
25	4	6	0, 2	-, 2	-, 3	-, 3	-, 3	-, 3	-, 3	-, 3	-, 4	-, 4	-, 4	-, 4	-, 4	-, 4	-, 4	-, 4	-, -	-, -
25	4	7	0, 2	0, 3	-, 3	-, 3	-, 3	-, 3	-, 4	-, 4	-, 4	-, 4	-, 4	-, 4	-, 4	-, 4	-, -	-, -	-, -	-, -
25	4	8	0, 2	0, 3	0, 3	-, 3	-, 3	-, 3	-, 4	-, 4	-, 4	-, 4	-, 4	-, 4	-, -	-, -	-, -	-, -	-, -	-, -
25	4	9	0, 2	0, 3	0, 3	-, 3	-, 4	-, 4	-, 4	-, 4	-, 4	-, 4	-, 4	-, 4	-, -	-, -	-, -	-, -	-, -	-, -
25	4	10	1, 3	0, 3	0, 3	0, 4	-, 4	-, 4	-, 4	-, 4	-, 4	-, 4	-, -	-, -	-, -	-, -	-, -	-, -	-, -	-, -
25	4	11	1, 3	0, 3	0, 4	0, 4	0, 4	-, 4	-, 4	-, 4	-, -	-, -	-, -	-, -	-, -	-, -	-, -	-, -	-, -	-, -
25	4	12	1, 3	0, 3	0, 4	0, 4	0, 4	0, 4	-, 4	-, 4	-, -	-, -	-, -	-, -	-, -	-, -	-, -	-, -	-, -	-, -
25	5	5	0, 2	-, 2	-, 3	-, 3	-, 3	-, 3	-, 3	-, 3	-, 4	-, 4	-, 4	-, 4	-, 4	-, 4	-, 4	-, 4	-, 5	-, 5
25	5	6	0, 2	0, 3	-, 3	-, 3	-, 3	-, 3	-, 4	-, 4	-, 4	-, 4	-, 4	-, 4	-, 5	-, 5	-, 5	-, 5	-, 5	-, 5
25	5	7	0, 2	0, 3	0, 3	-, 3	-, 4	-, 4	-, 4	-, 4	-, 4	-, 4	-, 5	-, 5	-, 5	-, 5	-, 5	-, 5	-, 5	-, 5

TABLE B.27 (cont.) CRITICAL VALUES FOR FISHER'S EXACT TEST

n	m_1	m_2	α = 0.50	0.20	0.10	0.05	0.02	0.01	0.005	0.002	0.001
25	5	8	1, 3 0, 3	0, 3 0, 4	-, 4 -, 4	-, 4 -, 4	-, 5 -, 5	-, 5 -, 5	-, 5 -, 5	-, 5 -, 5	-, - -, -
25	5	9	1, 3 0, 3	0, 4 0, 4	0, 4 -, 4	-, 4 -, 4	-, 5 -, 5	-, 5 -, 5	-, 5 -, 5	-, - -, -	-, - -, -
25	5	10	1, 3 0, 3	0, 4 0, 4	0, 4 0, 5	-, 5 -, 5	-, 5 -, 5	-, 5 -, 5	-, 5 -, 5	-, - -, -	-, - -, -
25	5	11	1, 3 1, 4	0, 4 0, 4	0, 4 0, 5	0, 5 0, 5	-, 5 -, 5	-, 5 -, 5	-, - -, -	-, - -, -	-, - -, -
25	5	12	1, 3 1, 4	1, 4 0, 4	0, 5 0, 5	0, 5 0, 5	-, 5 -, 5	-, - -, -	-, - -, -	-, - -, -	-, - -, -
25	6	6	0, 2 0, 3	0, 3 -, 3	-, 4 -, 4	-, 4 -, 4	-, 4 -, 4	-, 5 -, 5	-, 5 -, 5	-, 5 -, 5	-, 5 -, 5
25	6	7	1, 3 0, 3	0, 4 0, 4	-, 4 -, 4	-, 4 -, 4	-, 5 -, 5	-, 5 -, 5	-, 5 -, 5	-, 6 -, 6	-, 6 -, 6
25	6	8	1, 3 0, 3	0, 4 0, 4	0, 4 -, 4	-, 5 -, 5	-, 5 -, 5	-, 5 -, 5	-, 6 -, 6	-, 6 -, 6	-, 6 -, 6
25	6	9	1, 4 1, 4	0, 4 0, 4	0, 4 0, 5	0, 5 -, 5	-, 5 -, 5	-, 6 -, 6	-, 6 -, 6	-, 6 -, 6	-, 6 -, 6
25	6	10	1, 4 1, 4	1, 4 0, 4	0, 5 0, 5	0, 5 -, 5	-, 6 -, 6	-, 6 -, 6	-, 6 -, 6	-, 6 -, 6	-, - -, -
25	6	11	2, 4 1, 4	1, 5 1, 5	0, 5 0, 5	0, 5 0, 6	0, 6 0, 6	-, 6 -, 6	-, 6 -, 6	-, - -, -	-, - -, -
25	6	12	2, 4 1, 4	1, 5 1, 5	1, 5 0, 5	0, 6 0, 6	0, 6 0, 6	0, 6 -, 6	-, - -, -	-, - -, -	-, - -, -
25	7	7	1, 3 0, 3	0, 4 0, 4	0, 4 -, 4	-, 5 -, 5	-, 5 -, 5	-, 5 -, 5	-, 6 -, 6	-, 6 -, 6	-, 6 -, 6
25	7	8	1, 4 1, 4	0, 4 0, 4	0, 5 0, 5	0, 5 -, 5	-, 5 -, 5	-, 6 -, 6	-, 6 -, 6	-, 6 -, 6	-, 7 -, 7
25	7	9	1, 3 1, 4	1, 4 0, 5	0, 5 0, 5	0, 5 0, 6	-, 6 -, 6	-, 6 -, 6	-, 6 -, 6	-, 7 -, 7	-, 7 -, 7
25	7	10	2, 4 1, 4	1, 5 1, 5	0, 5 0, 5	0, 6 0, 6	0, 6 -, 6	-, 6 -, 6	-, 7 -, 7	-, 7 -, 7	-, 7 -, 7
25	7	11	2, 4 2, 5	1, 5 1, 5	1, 6 1, 6	0, 6 0, 6	0, 6 0, 7	0, 7 0, 7	-, 7 -, 7	-, 7 -, 7	-, 7 -, 7
25	7	12	2, 4 2, 5	1, 5 1, 6	1, 6 1, 6	1, 6 0, 6	0, 7 0, 7	0, 7 0, 7	0, 7 -, 7	-, 7 -, 7	-, - -, -
25	8	8	2, 4 1, 4	1, 5 0, 5	0, 5 0, 5	0, 5 0, 6	-, 6 -, 6	-, 6 -, 6	-, 6 -, 6	-, 7 -, 7	-, 7 -, 7
25	8	9	2, 4 1, 4	1, 5 1, 5	0, 5 0, 5	0, 6 0, 6	0, 6 -, 6	-, 6 -, 6	-, 7 -, 7	-, 7 -, 7	-, 7 -, 7
25	8	10	2, 5 2, 5	1, 5 1, 5	1, 6 1, 6	0, 6 0, 6	0, 7 0, 7	0, 7 0, 7	-, 7 -, 7	-, 7 -, 7	-, 8 -, 8
25	8	11	3, 5 2, 5	2, 5 1, 6	1, 6 1, 6	1, 6 1, 7	0, 7 0, 7	0, 7 0, 7	0, 7 0, 8	-, 8 -, 8	-, 8 -, 8
25	8	12	3, 5 2, 5	2, 6 1, 6	1, 6 1, 6	1, 7 1, 7	0, 7 0, 7	0, 7 0, 8	0, 8 0, 8	0, 8 0, 8	-, 8 -, 8
25	9	9	2, 5 2, 5	1, 5 1, 5	1, 6 1, 6	0, 6 0, 6	0, 7 0, 7	0, 7 0, 7	-, 7 -, 7	-, 8 -, 8	-, 8 -, 8
25	9	10	3, 5 2, 5	2, 6 1, 6	1, 6 1, 6	1, 7 1, 7	0, 7 0, 7	0, 7 0, 7	0, 8 0, 8	-, 8 -, 8	-, 8 -, 8
25	9	11	3, 5 2, 5	2, 6 1, 6	1, 6 1, 7	1, 7 1, 7	1, 7 0, 7	0, 8 0, 8	0, 8 0, 8	0, 8 0, 9	0, 9 -, 9
25	9	12	3, 5 3, 6	2, 6 2, 7	2, 7 2, 7	1, 7 1, 7	1, 8 1, 8	1, 8 0, 8	0, 8 0, 8	0, 9 0, 9	0, 9 0, 9
25	10	10	3, 5 2, 5	2, 6 1, 6	1, 7 1, 7	1, 7 1, 7	1, 7 1, 8	0, 8 0, 8	0, 8 0, 8	0, 8 0, 9	0, 9 0, 9
25	10	11	3, 6 3, 6	2, 6 2, 7	2, 7 2, 7	1, 7 1, 7	1, 8 1, 8	1, 8 0, 8	0, 8 0, 9	0, 9 0, 9	0, 9 0, 9
25	10	12	4, 6 3, 6	3, 7 2, 7	2, 7 2, 8	2, 8 2, 8	1, 8 1, 8	1, 9 1, 9	1, 9 1, 9	0, 9 0, 9	0, 9 0, 9
25	11	11	4, 6 3, 6	3, 7 2, 7	2, 7 2, 8	2, 8 2, 8	1, 8 1, 8	1, 9 1, 9	1, 9 1, 9	0, 9 0, 9	0,10 0,10
25	11	12	4, 6 4, 7	3, 7 3, 8	3, 8 2, 8	2, 8 2, 8	2, 9 2, 9	1, 9 1, 9	1, 9 1, 9	1,10 1,10	1,10 1,10
25	12	12	5, 7 4, 7	4, 8 3, 8	3, 8 3, 9	3, 9 3, 9	2, 9 2, 9	2,10 2,10	2,10 2,10	1,10 1,10	1,11 1,11
26	1	1	-, 1 -, 1	-, 1 -, 1	-, 1 -, 1	-, 1 -, 1	-, - -, -	-, - -, -	-, - -, -	-, - -, -	-, - -, -
26	1	2	-, 1 -, 1	-, 1 -, 1	-, 1 -, 1	-, - -, -	-, - -, -	-, - -, -	-, - -, -	-, - -, -	-, - -, -
26	1	3	-, 1 -, 1	-, 1 -, 1	-, - -, -	-, - -, -	-, - -, -	-, - -, -	-, - -, -	-, - -, -	-, - -, -
26	1	4	-, 1 -, 1	-, 1 -, 1	-, - -, -	-, - -, -	-, - -, -	-, - -, -	-, - -, -	-, - -, -	-, - -, -
26	1	5	-, 1 -, 1	-, 1 -, 1	-, - -, -	-, - -, -	-, - -, -	-, - -, -	-, - -, -	-, - -, -	-, - -, -
26	1	6	-, 1 -, 1	-, - -, -	-, - -, -	-, - -, -	-, - -, -	-, - -, -	-, - -, -	-, - -, -	-, - -, -
26	1	7	-, 1 -, 1	-, - -, -	-, - -, -	-, - -, -	-, - -, -	-, - -, -	-, - -, -	-, - -, -	-, - -, -
26	1	8	-, 1 -, 1	-, - -, -	-, - -, -	-, - -, -	-, - -, -	-, - -, -	-, - -, -	-, - -, -	-, - -, -
26	1	9	-, 1 -, 1	-, - -, -	-, - -, -	-, - -, -	-, - -, -	-, - -, -	-, - -, -	-, - -, -	-, - -, -
26	1	10	-, 1 -, 1	-, - -, -	-, - -, -	-, - -, -	-, - -, -	-, - -, -	-, - -, -	-, - -, -	-, - -, -
26	1	11	-, 1 -, 1	-, - -, -	-, - -, -	-, - -, -	-, - -, -	-, - -, -	-, - -, -	-, - -, -	-, - -, -
26	1	12	-, 1 -, 1	-, - -, -	-, - -, -	-, - -, -	-, - -, -	-, - -, -	-, - -, -	-, - -, -	-, - -, -
26	1	13	0, 1 -, -	-, - -, -	-, - -, -	-, - -, -	-, - -, -	-, - -, -	-, - -, -	-, - -, -	-, - -, -
26	2	2	-, 1 -, 1	-, 1 -, 1	-, 2 -, 2	-, 2 -, 2	-, 2 -, 2	-, 2 -, 2	-, 2 -, 2	-, - -, -	-, - -, -
26	2	3	-, 1 -, 1	-, 2 -, 2	-, 2 -, 2	-, 2 -, 2	-, 2 -, 2	-, 2 -, 2	-, - -, -	-, - -, -	-, - -, -
26	2	4	-, 1 -, 1	-, 2 -, 2	-, 2 -, 2	-, 2 -, 2	-, 2 -, 2	-, - -, -	-, - -, -	-, - -, -	-, - -, -
26	2	5	-, 1 -, 1	-, 2 -, 2	-, 2 -, 2	-, 2 -, 2	-, - -, -	-, - -, -	-, - -, -	-, - -, -	-, - -, -

TABLE B.27 (cont.) CRITICAL VALUES FOR FISHER'S EXACT TEST

n	m_1	m_2	α = 0.50	0.20	0.10	0.05	0.02	0.01	0.005	0.002	0.001
26	2	6	-, 1 -, 1	-, 2 -, 2	-, 2 -, 2	-, 2 -, 2	-, - -, -	-, - -, -	-, - -, -	-, - -, -	-, - -, -
26	2	7	-, 1 -, 1	-, 2 -, 2	-, 2 -, 2	-, - -, -	-, - -, -	-, - -, -	-, - -, -	-, - -, -	-, - -, -
26	2	8	0, 1 -, 2	-, 2 -, 2	-, 2 -, 2	-, - -, -	-, - -, -	-, - -, -	-, - -, -	-, - -, -	-, - -, -
26	2	9	0, 2 -, 2	-, 2 -, 2	-, - -, -	-, - -, -	-, - -, -	-, - -, -	-, - -, -	-, - -, -	-, - -, -
26	2	10	0, 2 -, 2	-, 2 -, 2	-, - -, -	-, - -, -	-, - -, -	-, - -, -	-, - -, -	-, - -, -	-, - -, -
26	2	11	0, 2 0, 2	-, 2 -, 2	-, - -, -	-, - -, -	-, - -, -	-, - -, -	-, - -, -	-, - -, -	-, - -, -
26	2	12	0, 2 0, 2	-, - -, -	-, - -, -	-, - -, -	-, - -, -	-, - -, -	-, - -, -	-, - -, -	-, - -, -
26	2	13	0, 2 0, 2	-, - -, -	-, - -, -	-, - -, -	-, - -, -	-, - -, -	-, - -, -	-, - -, -	-, - -, -
26	3	3	-, 1 -, 1	-, 2 -, 2	-, 2 -, 2	-, 2 -, 2	-, 3 -, 3	-, 3 -, 3	-, 3 -, 3	-, 3 -, 3	-, 3 -, 3
26	3	4	-, 1 -, 1	-, 2 -, 2	-, 2 -, 2	-, 3 -, 3	-, 3 -, 3	-, 3 -, 3	-, 3 -, 3	-, 3 -, 3	-, - -, -
26	3	5	-, 1 -, 1	-, 2 -, 2	-, 2 -, 2	-, 3 -, 3	-, 3 -, 3	-, 3 -, 3	-, 3 -, 3	-, - -, -	-, - -, -
26	3	6	0, 2 -, 2	-, 2 -, 2	-, 3 -, 3	-, 3 -, 3	-, 3 -, 3	-, 3 -, 3	-, - -, -	-, - -, -	-, - -, -
26	3	7	0, 2 -, 2	-, 2 -, 2	-, 3 -, 3	-, 3 -, 3	-, 3 -, 3	-, - -, -	-, - -, -	-, - -, -	-, - -, -
26	3	8	0, 2 -, 2	-, 3 -, 3	-, 3 -, 3	-, 3 -, 3	-, - -, -	-, - -, -	-, - -, -	-, - -, -	-, - -, -
26	3	9	0, 2 -, 2	-, 3 -, 3	-, 3 -, 3	-, 3 -, 3	-, - -, -	-, - -, -	-, - -, -	-, - -, -	-, - -, -
26	3	10	0, 2 0, 3	-, 3 -, 3	-, 3 -, 3	-, 3 -, 3	-, - -, -	-, - -, -	-, - -, -	-, - -, -	-, - -, -
26	3	11	0, 2 0, 3	0, 3 -, 3	-, 3 -, 3	-, - -, -	-, - -, -	-, - -, -	-, - -, -	-, - -, -	-, - -, -
26	3	12	0, 2 0, 3	0, 3 -, 3	-, 3 -, 3	-, - -, -	-, - -, -	-, - -, -	-, - -, -	-, - -, -	-, - -, -
26	3	13	1, 2 0, 3	0, 3 -, -	-, - -, -	-, - -, -	-, - -, -	-, - -, -	-, - -, -	-, - -, -	-, - -, -
26	4	4	0, 1 -, 2	-, 2 -, 2	-, 2 -, 2	-, 3 -, 3	-, 3 -, 3	-, 3 -, 3	-, 4 -, 4	-, 4 -, 4	-, 4 -, 4
26	4	5	0, 2 -, 2	-, 2 -, 2	-, 3 -, 3	-, 3 -, 3	-, 3 -, 3	-, 4 -, 4	-, 4 -, 4	-, 4 -, 4	-, 4 -, 4
26	4	6	0, 2 -, 2	-, 2 -, 3	-, 3 -, 3	-, 3 -, 3	-, 4 -, 4	-, 4 -, 4	-, 4 -, 4	-, 4 -, 4	-, - -, -
26	4	7	0, 2 -, 2	-, 3 -, 3	-, 3 -, 3	-, 3 -, 3	-, 4 -, 4	-, 4 -, 4	-, 4 -, 4	-, - -, -	-, - -, -
26	4	8	0, 2 0, 3	-, 3 -, 3	-, 3 -, 3	-, 4 -, 4	-, 4 -, 4	-, 4 -, 4	-, 4 -, 4	-, - -, -	-, - -, -
26	4	9	0, 2 0, 3	0, 3 -, 3	-, 4 -, 4	-, 4 -, 4	-, 4 -, 4	-, 4 -, 4	-, - -, -	-, - -, -	-, - -, -
26	4	10	1, 2 0, 3	0, 3 0, 4	-, 4 -, 4	-, 4 -, 4	-, 4 -, 4	-, - -, -	-, - -, -	-, - -, -	-, - -, -
26	4	11	1, 3 0, 3	0, 3 0, 4	0, 4 -, 4	-, 4 -, 4	-, - -, -	-, - -, -	-, - -, -	-, - -, -	-, - -, -
26	4	12	1, 3 0, 3	0, 4 0, 4	0, 4 -, 4	-, 4 -, 4	-, - -, -	-, - -, -	-, - -, -	-, - -, -	-, - -, -
26	4	13	1, 3 0, 4	0, 4 0, 4	0, 4 0, 4	0, 4 -, -	-, - -, -	-, - -, -	-, - -, -	-, - -, -	-, - -, -
26	5	5	0, 2 -, 2	-, 3 -, 3	-, 3 -, 3	-, 3 -, 3	-, 4 -, 4	-, 4 -, 4	-, 4 -, 4	-, 4 -, 4	-, 5 -, 5
26	5	6	0, 2 0, 3	-, 3 -, 3	-, 3 -, 3	-, 4 -, 4	-, 4 -, 4	-, 4 -, 4	-, 4 -, 4	-, 5 -, 5	-, 5 -, 5
26	5	7	0, 2 0, 3	0, 3 -, 3	-, 4 -, 4	-, 4 -, 4	-, 4 -, 4	-, 5 -, 5	-, 5 -, 5	-, 5 -, 5	-, 5 -, 5
26	5	8	0, 2 0, 3	0, 3 -, 4	-, 4 -, 4	-, 4 -, 4	-, 5 -, 5	-, 5 -, 5	-, 5 -, 5	-, 5 -, 5	-, 5 -, 5
26	5	9	1, 3 0, 3	0, 4 0, 4	0, 4 -, 4	-, 4 -, 4	-, 5 -, 5	-, 5 -, 5	-, 5 -, 5	-, 5 -, 5	-, - -, -
26	5	10	1, 3 0, 3	0, 4 0, 4	0, 4 -, 4	-, 5 -, 5	-, 5 -, 5	-, 5 -, 5	-, 5 -, 5	-, - -, -	-, - -, -
26	5	11	1, 3 1, 4	0, 4 0, 4	0, 4 0, 5	0, 5 -, 5	-, 5 -, 5	-, 5 -, 5	-, - -, -	-, - -, -	-, - -, -
26	5	12	1, 3 1, 4	0, 4 0, 4	0, 5 0, 5	0, 5 0, 5	-, 5 -, 5	-, - -, -	-, - -, -	-, - -, -	-, - -, -
26	5	13	2, 3 1, 4	1, 4 0, 5	0, 5 0, 5	0, 5 0, 5	0, 5 -, -	-, - -, -	-, - -, -	-, - -, -	-, - -, -
26	6	6	0, 2 0, 3	0, 3 -, 3	-, 3 -, 4	-, 4 -, 4	-, 4 -, 4	-, 5 -, 5	-, 5 -, 5	-, 5 -, 5	-, 5 -, 5
26	6	7	1, 3 0, 3	0, 4 0, 4	-, 4 -, 4	-, 4 -, 4	-, 5 -, 5	-, 5 -, 5	-, 5 -, 5	-, 5 -, 5	-, 6 -, 6
26	6	8	1, 3 0, 3	0, 4 0, 4	0, 4 -, 4	-, 5 -, 5	-, 5 -, 5	-, 5 -, 5	-, 5 -, 5	-, 6 -, 6	-, 6 -, 6
26	6	9	1, 4 1, 4	0, 4 0, 4	0, 4 0, 5	-, 5 -, 5	-, 5 -, 5	-, 5 -, 5	-, 6 -, 6	-, 6 -, 6	-, 6 -, 6
26	6	10	1, 4 1, 4	0, 4 0, 4	0, 5 0, 5	0, 5 -, 5	-, 5 -, 5	-, 6 -, 6	-, 6 -, 6	-, 6 -, 6	-, 6 -, 6
26	6	11	2, 4 1, 4	1, 4 1, 5	0, 5 0, 5	0, 5 0, 6	-, 6 -, 6	-, 6 -, 6	-, 6 -, 6	-, - -, -	-, - -, -
26	6	12	2, 4 1, 4	1, 5 1, 5	0, 5 0, 5	0, 6 0, 6	0, 6 0, 6	-, 6 -, 6	-, 6 -, 6	-, - -, -	-, - -, -
26	6	13	2, 4 1, 5	1, 5 1, 5	1, 5 0, 6	0, 6 0, 6	0, 6 0, 6	0, 6 -, -	-, - -, -	-, - -, -	-, - -, -
26	7	7	1, 3 0, 3	0, 4 0, 4	0, 4 -, 4	-, 5 -, 5	-, 5 -, 5	-, 5 -, 5	-, 6 -, 6	-, 6 -, 6	-, 6 -, 6
26	7	8	1, 4 1, 4	0, 4 0, 4	0, 5 0, 5	0, 5 -, 5	-, 5 -, 5	-, 6 -, 6	-, 6 -, 6	-, 6 -, 6	-, 6 -, 6
26	7	9	1, 4 1, 4	1, 4 0, 4	0, 5 0, 5	0, 5 -, 5	-, 6 -, 6	-, 6 -, 6	-, 6 -, 6	-, 7 -, 7	-, 7 -, 7
26	7	10	2, 4 1, 4	1, 5 1, 5	0, 5 0, 5	0, 6 0, 6	0, 6 -, 6	-, 6 -, 6	-, 7 -, 7	-, 7 -, 7	-, 7 -, 7

TABLE B.27 (cont.) CRITICAL VALUES FOR FISHER'S EXACT TEST

n	m_1	m_2	α = 0.50		0.20		0.10		0.05		0.02		0.01		0.005		0.002		0.001	
26	7	11	2, 4	1, 4	1, 5	1, 5	1, 5	0, 5	0, 6	0, 6	0, 6	0, 7	0, 7	-, 7	-, 7	-, 7	-, 7	-, 7	-, 7	-, 7
26	7	12	2, 4	2, 5	1, 5	1, 5	1, 6	1, 6	0, 6	0, 6	0, 7	0, 7	0, 7	0, 7	-, 7	-, 7	-, 7	-, 7	-, -	-, -
26	7	13	3, 4	2, 5	2, 5	1, 6	1, 6	1, 6	1, 6	0, 7	0, 7	0, 7	0, 7	0, 7	0, 7	-, -	-, -	-, -	-, -	-, -
26	8	8	1, 3	1, 4	1, 4	0, 4	0, 5	0, 5	0, 5	0, 6	-, 6	-, 6	-, 6	-, 6	-, 6	-, 6	-, 7	-, 7	-, 7	-, 7
26	8	9	2, 4	1, 4	1, 5	1, 5	0, 5	0, 5	0, 6	0, 6	0, 6	-, 6	-, 6	-, 6	-, 7	-, 7	-, 7	-, 7	-, 7	-, 7
26	8	10	2, 5	2, 5	1, 5	1, 5	1, 6	1, 6	0, 6	0, 6	0, 6	0, 7	0, 7	0, 7	-, 7	-, 7	-, 7	-, 7	-, 8	-, 8
26	8	11	2, 5	2, 5	1, 5	1, 6	1, 6	1, 6	1, 6	0, 6	0, 7	0, 7	0, 7	0, 7	0, 7	-, 7	-, 8	-, 8	-, 8	-, 8
26	8	12	3, 5	2, 5	2, 6	1, 6	1, 6	1, 6	1, 7	1, 7	0, 7	0, 7	0, 7	0, 7	0, 8	0, 8	0, 8	-, 8	-, 8	-, 8
26	8	13	3, 5	2, 6	2, 6	1, 7	1, 7	1, 7	1, 7	1, 7	1, 7	0, 8	0, 8	0, 8	0, 8	0, 8	0, 8	0, 8	0, 8	-, -
26	9	9	2, 5	2, 5	1, 5	1, 5	1, 6	1, 6	0, 6	0, 6	0, 6	0, 7	0, 7	0, 7	-, 7	-, 7	-, 7	-, 7	-, 8	-, 8
26	9	10	2, 5	2, 5	1, 5	1, 6	1, 6	1, 6	1, 6	0, 6	0, 7	0, 7	0, 7	0, 7	0, 7	0, 8	-, 8	-, 8	-, 8	-, 8
26	9	11	3, 5	2, 5	2, 6	1, 6	1, 6	1, 7	1, 7	1, 7	0, 7	0, 7	0, 8	0, 8	0, 8	0, 8	0, 8	-, 8	-, 8	-, 8
26	9	12	3, 5	3, 6	2, 6	2, 7	2, 7	1, 7	1, 7	1, 7	1, 8	1, 8	0, 8	0, 8	0, 8	0, 8	0, 9	0, 9	0, 9	0, 9
26	9	13	4, 5	3, 6	2, 7	2, 7	2, 7	2, 7	2, 7	1, 8	1, 8	1, 8	1, 8	0, 9	0, 9	0, 9	0, 9	0, 9	0, 9	0, 9
26	10	10	3, 5	2, 5	2, 6	1, 6	1, 6	1, 7	1, 7	1, 7	0, 7	0, 7	0, 8	0, 8	0, 8	0, 8	0, 8	-, 8	-, 9	-, 9
26	10	11	3, 6	3, 6	2, 6	2, 7	2, 7	1, 7	1, 7	1, 7	1, 8	1, 8	0, 8	0, 8	0, 8	0, 8	0, 9	0, 9	0, 9	0, 9
26	10	12	4, 6	3, 6	3, 7	2, 7	2, 7	2, 8	2, 8	1, 8	1, 8	1, 8	1, 8	1, 9	1, 9	0, 9	0, 9	0, 9	0, 9	0, 9
26	10	13	4, 6	3, 7	3, 7	2, 8	2, 8	2, 8	2, 8	2, 8	1, 9	1, 9	1, 9	1, 9	1, 9	1, 9	1, 9	0,10	0,10	0,10
26	11	11	4, 6	3, 6	3, 7	2, 7	2, 7	2, 8	2, 8	1, 8	1, 8	1, 8	1, 9	1, 9	1, 9	0, 9	0, 9	0, 9	0, 9	0, 9
26	11	12	4, 6	4, 7	3, 7	3, 8	2, 8	2, 8	2, 8	2, 8	2, 9	1, 9	1, 9	1, 9	1, 9	1, 9	1,10	1,10	0,10	0,10
26	11	13	5, 6	4, 7	3, 8	3, 8	3, 8	2, 9	2, 9	2, 9	2, 9	2, 9	2, 9	1,10	1,10	1,10	1,10	1,10	1,10	1,10
26	12	12	5, 7	4, 7	3, 8	3, 8	3, 8	3, 9	2, 9	2, 9	2, 9	2, 9	2, 9	2,10	1,10	1,10	1,10	1,10	1,10	1,11
26	12	13	5, 7	4, 8	4, 8	3, 9	3, 9	3, 9	3, 9	3, 9	2,10	2,10	2,10	2,10	2,10	2,10	1,11	1,11	1,11	1,11
26	13	13	6, 7	5, 8	4, 9	4, 9	4, 9	3,10	3,10	3,10	3,10	3,10	3,10	2,11	2,11	2,11	2,11	2,11	2,11	1,12
27	1	1	-, 1	-, 1	-, 1	-, 1	-, 1	-, 1	-, 1	-, 1	-, -	-, -	-, -	-, -	-, -	-, -	-, -	-, -	-, -	-, -
27	1	2	-, 1	-, 1	-, 1	-, 1	-, 1	-, 1	-, -	-, -	-, -	-, -	-, -	-, -	-, -	-, -	-, -	-, -	-, -	-, -
27	1	3	-, 1	-, 1	-, 1	-, 1	-, -	-, -	-, -	-, -	-, -	-, -	-, -	-, -	-, -	-, -	-, -	-, -	-, -	-, -
27	1	4	-, 1	-, 1	-, 1	-, 1	-, -	-, -	-, -	-, -	-, -	-, -	-, -	-, -	-, -	-, -	-, -	-, -	-, -	-, -
27	1	5	-, 1	-, 1	-, 1	-, 1	-, -	-, -	-, -	-, -	-, -	-, -	-, -	-, -	-, -	-, -	-, -	-, -	-, -	-, -
27	1	6	-, 1	-, 1	-, -	-, -	-, -	-, -	-, -	-, -	-, -	-, -	-, -	-, -	-, -	-, -	-, -	-, -	-, -	-, -
27	1	7	-, 1	-, 1	-, -	-, -	-, -	-, -	-, -	-, -	-, -	-, -	-, -	-, -	-, -	-, -	-, -	-, -	-, -	-, -
27	1	8	-, 1	-, 1	-, -	-, -	-, -	-, -	-, -	-, -	-, -	-, -	-, -	-, -	-, -	-, -	-, -	-, -	-, -	-, -
27	1	9	-, 1	-, 1	-, -	-, -	-, -	-, -	-, -	-, -	-, -	-, -	-, -	-, -	-, -	-, -	-, -	-, -	-, -	-, -
27	1	10	-, 1	-, 1	-, -	-, -	-, -	-, -	-, -	-, -	-, -	-, -	-, -	-, -	-, -	-, -	-, -	-, -	-, -	-, -
27	1	11	-, 1	-, 1	-, -	-, -	-, -	-, -	-, -	-, -	-, -	-, -	-, -	-, -	-, -	-, -	-, -	-, -	-, -	-, -
27	1	12	-, 1	-, 1	-, -	-, -	-, -	-, -	-, -	-, -	-, -	-, -	-, -	-, -	-, -	-, -	-, -	-, -	-, -	-, -
27	1	13	-, 1	-, 1	-, -	-, -	-, -	-, -	-, -	-, -	-, -	-, -	-, -	-, -	-, -	-, -	-, -	-, -	-, -	-, -
27	2	2	-, 1	-, 1	-, 1	-, 1	-, 2	-, 2	-, 2	-, 2	-, 2	-, 2	-, 2	-, 2	-, 2	-, 2	-, -	-, -	-, -	-, -
27	2	3	-, 1	-, 1	-, 2	-, 2	-, 2	-, 2	-, 2	-, 2	-, 2	-, 2	-, 2	-, 2	-, -	-, -	-, -	-, -	-, -	-, -
27	2	4	-, 1	-, 1	-, 2	-, 2	-, 2	-, 2	-, 2	-, 2	-, 2	-, 2	-, -	-, -	-, -	-, -	-, -	-, -	-, -	-, -
27	2	5	-, 1	-, 1	-, 2	-, 2	-, 2	-, 2	-, 2	-, 2	-, -	-, -	-, -	-, -	-, -	-, -	-, -	-, -	-, -	-, -
27	2	6	-, 1	-, 1	-, 2	-, 2	-, 2	-, 2	-, 2	-, 2	-, -	-, -	-, -	-, -	-, -	-, -	-, -	-, -	-, -	-, -
27	2	7	-, 1	-, 1	-, 2	-, 2	-, 2	-, 2	-, -	-, -	-, -	-, -	-, -	-, -	-, -	-, -	-, -	-, -	-, -	-, -
27	2	8	0, 1	-, 2	-, 2	-, 2	-, 2	-, 2	-, -	-, -	-, -	-, -	-, -	-, -	-, -	-, -	-, -	-, -	-, -	-, -
27	2	9	0, 2	-, 2	-, 2	-, 2	-, -	-, -	-, -	-, -	-, -	-, -	-, -	-, -	-, -	-, -	-, -	-, -	-, -	-, -
27	2	10	0, 2	-, 2	-, 2	-, 2	-, -	-, -	-, -	-, -	-, -	-, -	-, -	-, -	-, -	-, -	-, -	-, -	-, -	-, -
27	2	11	0, 2	0, 2	-, 2	-, 2	-, -	-, -	-, -	-, -	-, -	-, -	-, -	-, -	-, -	-, -	-, -	-, -	-, -	-, -
27	2	12	0, 2	0, 2	-, 2	-, 2	-, -	-, -	-, -	-, -	-, -	-, -	-, -	-, -	-, -	-, -	-, -	-, -	-, -	-, -
27	2	13	0, 2	0, 2	-, -	-, -	-, -	-, -	-, -	-, -	-, -	-, -	-, -	-, -	-, -	-, -	-, -	-, -	-, -	-, -
27	3	3	-, 1	-, 1	-, 2	-, 2	-, 2	-, 2	-, 2	-, 2	-, 3	-, 3	-, 3	-, 3	-, 3	-, 3	-, 3	-, 3	-, 3	-, 3

TABLE B.27 (cont.) CRITICAL VALUES FOR FISHER'S EXACT TEST

n	m_1	m_2	α = 0.50		0.20		0.10		0.05		0.02		0.01		0.005		0.002		0.001	
27	3	4	-, 1	-, 1	-, 2	-, 2	-, 2	-, 2	-, 2	-, 2	-, 3	-, 3	-, 3	-, 3	-, 3	-, 3	-, 3	-, 3	-, -	-, -
27	3	5	-, 1	-, 1	-, 2	-, 2	-, 2	-, 2	-, 3	-, 3	-, 3	-, 3	-, 3	-, 3	-, 3	-, 3	-, -	-, -	-, -	-, -
27	3	6	0, 1	-, 2	-, 2	-, 2	-, 3	-, 3	-, 3	-, 3	-, 3	-, 3	-, 3	-, 3	-, -	-, -	-, -	-, -	-, -	-, -
27	3	7	0, 2	-, 2	-, 2	-, 2	-, 3	-, 3	-, 3	-, 3	-, 3	-, 3	-, -	-, -	-, -	-, -	-, -	-, -	-, -	-, -
27	3	8	0, 2	-, 2	-, 3	-, 3	-, 3	-, 3	-, 3	-, 3	-, 3	-, 3	-, -	-, -	-, -	-, -	-, -	-, -	-, -	-, -
27	3	9	0, 2	-, 2	-, 3	-, 3	-, 3	-, 3	-, 3	-, 3	-, -	-, -	-, -	-, -	-, -	-, -	-, -	-, -	-, -	-, -
27	3	10	0, 2	0, 3	-, 3	-, 3	-, 3	-, 3	-, 3	-, 3	-, -	-, -	-, -	-, -	-, -	-, -	-, -	-, -	-, -	-, -
27	3	11	0, 2	0, 3	0, 3	-, 3	-, 3	-, 3	-, -	-, -	-, -	-, -	-, -	-, -	-, -	-, -	-, -	-, -	-, -	-, -
27	3	12	0, 2	0, 3	0, 3	-, 3	-, 3	-, 3	-, -	-, -	-, -	-, -	-, -	-, -	-, -	-, -	-, -	-, -	-, -	-, -
27	3	13	0, 2	0, 3	0, 3	-, 3	-, 3	-, 3	-, -	-, -	-, -	-, -	-, -	-, -	-, -	-, -	-, -	-, -	-, -	-, -
27	4	4	-, 1	-, 1	-, 2	-, 2	-, 2	-, 2	-, 3	-, 3	-, 3	-, 3	-, 3	-, 3	-, 4	-, 4	-, 4	-, 4	-, 4	-, 4
27	4	5	0, 2	-, 2	-, 2	-, 2	-, 3	-, 3	-, 3	-, 3	-, 3	-, 3	-, 4	-, 4	-, 4	-, 4	-, 4	-, 4	-, 4	-, 4
27	4	6	0, 2	-, 2	-, 2	-, 3	-, 3	-, 3	-, 3	-, 3	-, 4	-, 4	-, 4	-, 4	-, 4	-, 4	-, 4	-, 4	-, 4	-, 4
27	4	7	0, 2	-, 2	-, 3	-, 3	-, 3	-, 3	-, 3	-, 3	-, 4	-, 4	-, 4	-, 4	-, 4	-, 4	-, 4	-, 4	-, -	-, -
27	4	8	0, 2	0, 3	-, 3	-, 3	-, 3	-, 3	-, 4	-, 4	-, 4	-, 4	-, 4	-, 4	-, 4	-, 4	-, -	-, -	-, -	-, -
27	4	9	0, 2	0, 3	0, 3	-, 3	-, 3	-, 3	-, 4	-, 4	-, 4	-, 4	-, 4	-, 4	-, -	-, -	-, -	-, -	-, -	-, -
27	4	10	0, 2	0, 3	0, 3	-, 3	-, 4	-, 4	-, 4	-, 4	-, 4	-, 4	-, -	-, -	-, -	-, -	-, -	-, -	-, -	-, -
27	4	11	1, 3	0, 3	0, 3	0, 4	-, 4	-, 4	-, 4	-, 4	-, 4	-, 4	-, -	-, -	-, -	-, -	-, -	-, -	-, -	-, -
27	4	12	1, 3	0, 3	0, 4	0, 4	0, 4	-, 4	-, 4	-, 4	-, -	-, -	-, -	-, -	-, -	-, -	-, -	-, -	-, -	-, -
27	4	13	1, 3	0, 3	0, 4	0, 4	0, 4	0, 4	-, 4	-, 4	-, -	-, -	-, -	-, -	-, -	-, -	-, -	-, -	-, -	-, -
27	5	5	0, 2	-, 2	-, 2	-, 3	-, 3	-, 3	-, 3	-, 3	-, 4	-, 4	-, 4	-, 4	-, 4	-, 4	-, 4	-, 4	-, 5	-, 5
27	5	6	0, 2	-, 2	-, 3	-, 3	-, 3	-, 3	-, 4	-, 4	-, 4	-, 4	-, 4	-, 4	-, 4	-, 4	-, 5	-, 5	-, 5	-, 5
27	5	7	0, 2	0, 3	0, 3	-, 3	-, 3	-, 3	-, 4	-, 4	-, 4	-, 4	-, 4	-, 4	-, 5	-, 5	-, 5	-, 5	-, 5	-, 5
27	5	8	0, 2	0, 3	0, 3	-, 3	-, 4	-, 4	-, 4	-, 4	-, 4	-, 4	-, 5	-, 5	-, 5	-, 5	-, 5	-, 5	-, 5	-, 5
27	5	9	1, 3	0, 3	0, 3	0, 4	-, 4	-, 4	-, 4	-, 4	-, 5	-, 5	-, 5	-, 5	-, 5	-, 5	-, 5	-, 5	-, -	-, -
27	5	10	1, 3	0, 3	0, 4	0, 4	0, 4	-, 4	-, 4	-, 4	-, 5	-, 5	-, 5	-, 5	-, 5	-, 5	-, -	-, -	-, -	-, -
27	5	11	1, 3	0, 3	0, 4	0, 4	0, 4	0, 5	-, 5	-, 5	-, 5	-, 5	-, 5	-, 5	-, -	-, -	-, -	-, -	-, -	-, -
27	5	12	1, 3	1, 4	0, 4	0, 4	0, 5	0, 5	0, 5	0, 5	-, 5	-, 5	-, 5	-, 5	-, -	-, -	-, -	-, -	-, -	-, -
27	5	13	1, 3	1, 4	1, 4	0, 4	0, 5	0, 5	0, 5	0, 5	-, 5	-, 5	-, -	-, -	-, -	-, -	-, -	-, -	-, -	-, -
27	6	6	0, 2	0, 3	0, 3	-, 3	-, 3	-, 4	-, 4	-, 4	-, 4	-, 4	-, 5	-, 5	-, 5	-, 5	-, 5	-, 5	-, 5	-, 5
27	6	7	1, 2	0, 3	0, 4	0, 4	-, 4	-, 4	-, 4	-, 4	-, 5	-, 5	-, 5	-, 5	-, 5	-, 5	-, 5	-, 5	-, 6	-, 6
27	6	8	1, 3	0, 3	0, 4	0, 4	0, 4	-, 4	-, 4	-, 4	-, 5	-, 5	-, 5	-, 5	-, 5	-, 5	-, 6	-, 6	-, 6	-, 6
27	6	9	1, 3	0, 3	0, 4	0, 4	0, 4	0, 5	-, 5	-, 5	-, 5	-, 5	-, 5	-, 5	-, 6	-, 6	-, 6	-, 6	-, 6	-, 6
27	6	10	1, 4	1, 4	0, 4	0, 4	0, 5	0, 5	0, 5	-, 5	-, 5	-, 5	-, 6	-, 6	-, 6	-, 6	-, 6	-, 6	-, 6	-, 6
27	6	11	1, 4	1, 4	1, 4	0, 4	0, 5	0, 5	0, 5	-, 5	-, 6	-, 6	-, 6	-, 6	-, 6	-, 6	-, 6	-, 6	-, -	-, -
27	6	12	2, 4	1, 4	1, 5	1, 5	0, 5	0, 5	0, 5	0, 6	0, 6	-, 6	-, 6	-, 6	-, 6	-, 6	-, -	-, -	-, -	-, -
27	6	13	2, 4	1, 4	1, 5	1, 5	1, 5	0, 5	0, 6	0, 6	0, 6	0, 6	-, 6	-, 6	-, -	-, -	-, -	-, -	-, -	-, -
27	7	7	1, 3	0, 3	0, 4	0, 4	0, 4	-, 4	-, 4	-, 4	-, 5	-, 5	-, 5	-, 5	-, 5	-, 5	-, 6	-, 6	-, 6	-, 6
27	7	8	1, 4	0, 4	0, 4	0, 4	0, 4	0, 5	-, 5	-, 5	-, 5	-, 5	-, 6	-, 6	-, 6	-, 6	-, 6	-, 6	-, 6	-, 6
27	7	9	1, 4	1, 4	0, 4	0, 4	0, 5	0, 5	0, 5	-, 5	-, 6	-, 6	-, 6	-, 6	-, 6	-, 6	-, 6	-, 6	-, 7	-, 7
27	7	10	2, 4	1, 4	1, 5	0, 5	0, 5	0, 5	0, 5	0, 6	-, 6	-, 6	-, 6	-, 6	-, 6	-, 6	-, 7	-, 7	-, 7	-, 7
27	7	11	2, 4	1, 4	1, 5	1, 5	0, 5	0, 5	0, 6	0, 6	0, 6	-, 6	-, 6	-, 6	-, 7	-, 7	-, 7	-, 7	-, 7	-, 7
27	7	12	2, 4	2, 5	1, 5	1, 5	1, 6	1, 6	0, 6	0, 6	0, 6	0, 7	0, 7	0, 7	-, 7	-, 7	-, 7	-, 7	-, 7	-, 7
27	7	13	2, 4	2, 5	1, 5	1, 6	1, 6	1, 6	1, 6	0, 6	0, 7	0, 7	0, 7	0, 7	0, 7	-, 7	-, 7	-, 7	-, -	-, -
27	8	8	1, 4	1, 4	0, 4	0, 4	0, 5	0, 5	0, 5	-, 5	-, 6	-, 6	-, 6	-, 6	-, 6	-, 6	-, 7	-, 7	-, 7	-, 7
27	8	9	2, 4	1, 4	1, 5	0, 5	0, 5	0, 5	0, 6	0, 6	0, 6	-, 6	-, 6	-, 6	-, 7	-, 7	-, 7	-, 7	-, 7	-, 7
27	8	10	2, 4	1, 4	1, 5	1, 5	1, 5	0, 6	0, 6	0, 6	0, 6	0, 7	-, 7	-, 7	-, 7	-, 7	-, 7	-, 7	-, 7	-, 7
27	8	11	2, 5	2, 5	1, 5	1, 6	1, 6	1, 6	0, 6	0, 6	0, 7	0, 7	0, 7	0, 7	-, 7	-, 7	-, 8	-, 8	-, 8	-, 8
27	8	12	3, 5	2, 5	2, 6	1, 6	1, 6	1, 6	1, 6	1, 7	0, 7	0, 7	0, 7	0, 7	0, 8	0, 8	-, 8	-, 8	-, 8	-, 8
27	8	13	3, 5	2, 5	2, 6	1, 6	1, 6	1, 7	1, 7	1, 7	0, 7	0, 7	0, 8	0, 8	0, 8	0, 8	0, 8	0, 8	-, 8	-, 8

TABLE B.27 (cont.) CRITICAL VALUES FOR FISHER'S EXACT TEST

n	m_1	m_2	α = 0.50	0.20	0.10	0.05	0.02	0.01	0.005	0.002	0.001
27	9	9	2, 4 1, 4	1, 5 1, 5	1, 5 0, 6	0, 6 0, 6	0, 6 0, 7	-, 7 -, 7	-, 7 -, 7	-, 7 -, 7	-, 8 -, 8
27	9	10	2, 5 2, 5	1, 5 1, 6	1, 6 1, 6	0, 6 0, 6	0, 7 0, 7	0, 7 0, 7	-, 7 -, 7	-, 8 -, 8	-, 8 -, 8
27	9	11	3, 5 2, 5	2, 6 1, 6	1, 6 1, 6	1, 7 1, 7	0, 7 0, 7	0, 7 0, 8	0, 8 0, 8	-, 8 -, 8	-, 8 -, 8
27	9	12	3, 5 2, 5	2, 6 1, 6	1, 7 1, 7	1, 7 1, 7	1, 7 1, 8	0, 8 0, 8	0, 8 0, 8	0, 8 0, 9	-, 9 -, 9
27	9	13	3, 5 3, 6	2, 6 2, 7	2, 7 1, 7	1, 7 1, 7	1, 8 1, 8	1, 8 0, 8	0, 8 0, 8	0, 9 0, 9	0, 9 0, 9
27	10	10	3, 5 2, 5	2, 6 1, 6	1, 6 1, 7	1, 7 1, 7	0, 7 0, 7	0, 8 0, 8	0, 8 0, 8	-, 8 -, 8	-, 8 -, 8
27	10	11	3, 6 3, 6	2, 6 2, 7	2, 7 1, 7	1, 7 1, 7	1, 8 1, 8	0, 8 0, 8	0, 8 0, 8	0, 9 0, 9	0, 9 -, 9
27	10	12	3, 6 3, 6	2, 6 2, 7	2, 7 1, 7	1, 7 1, 8	1, 8 1, 8	1, 8 0, 8	0, 9 0, 9	0, 9 0, 9	0, 9 0, 9
27	10	13	4, 6 3, 6	3, 7 2, 7	2, 7 2, 8	2, 8 2, 8	1, 8 1, 8	1, 9 1, 9	1, 9 1, 9	0, 9 0, 9	0,10 0,10
27	11	11	3, 6 3, 6	2, 7 2, 7	2, 7 1, 7	1, 8 1, 8	1, 8 1, 8	1, 8 1, 9	0, 9 0, 9	0, 9 0, 9	0, 9 0, 9
27	11	12	4, 6 3, 6	3, 7 2, 7	2, 8 2, 8	2, 8 2, 8	1, 8 1, 9	1, 9 1, 9	1, 9 1, 9	0, 9 0, 9	0,10 0,10
27	11	13	4, 6 4, 7	3, 7 3, 8	3, 8 2, 8	2, 8 2, 9	2, 9 2, 9	1, 9 1, 9	1,10 1,10	1,10 1,10	0,10 0,10
27	12	12	4, 7 4, 7	3, 7 3, 8	3, 8 2, 8	2, 8 2, 9	2, 9 2, 9	1, 9 1, 9	1,10 1,10	1,10 1,10	1,10 0,10
27	12	13	5, 7 4, 7	4, 8 3, 8	3, 8 3, 9	3, 9 2, 9	2, 9 2,10	2,10 2,10	2,10 1,10	1,10 1,10	1,11 1,11
27	13	13	5, 7 5, 8	4, 8 4, 9	4, 9 3, 9	3, 9 3,10	3,10 2,10	2,10 2,10	2,11 2,11	2,11 2,11	1,11 1,11
28	1	1	-, 1 -, 1	-, 1 -, 1	-, 1 -, 1	-, 1 -, 1	-, - -, -	-, - -, -	-, - -, -	-, - -, -	-, - -, -
28	1	2	-, 1 -, 1	-, 1 -, 1	-, 1 -, 1	-, - -, -	-, - -, -	-, - -, -	-, - -, -	-, - -, -	-, - -, -
28	1	3	-, 1 -, 1	-, 1 -, 1	-, - -, -	-, - -, -	-, - -, -	-, - -, -	-, - -, -	-, - -, -	-, - -, -
28	1	4	-, 1 -, 1	-, 1 -, 1	-, - -, -	-, - -, -	-, - -, -	-, - -, -	-, - -, -	-, - -, -	-, - -, -
28	1	5	-, 1 -, 1	-, 1 -, 1	-, - -, -	-, - -, -	-, - -, -	-, - -, -	-, - -, -	-, - -, -	-, - -, -
28	1	6	-, 1 -, 1	-, - -, -	-, - -, -	-, - -, -	-, - -, -	-, - -, -	-, - -, -	-, - -, -	-, - -, -
28	1	7	-, 1 -, 1	-, - -, -	-, - -, -	-, - -, -	-, - -, -	-, - -, -	-, - -, -	-, - -, -	-, - -, -
28	1	8	-, 1 -, 1	-, - -, -	-, - -, -	-, - -, -	-, - -, -	-, - -, -	-, - -, -	-, - -, -	-, - -, -
28	1	9	-, 1 -, 1	-, - -, -	-, - -, -	-, - -, -	-, - -, -	-, - -, -	-, - -, -	-, - -, -	-, - -, -
28	1	10	-, 1 -, 1	-, - -, -	-, - -, -	-, - -, -	-, - -, -	-, - -, -	-, - -, -	-, - -, -	-, - -, -
28	1	11	-, 1 -, 1	-, - -, -	-, - -, -	-, - -, -	-, - -, -	-, - -, -	-, - -, -	-, - -, -	-, - -, -
28	1	12	-, 1 -, 1	-, - -, -	-, - -, -	-, - -, -	-, - -, -	-, - -, -	-, - -, -	-, - -, -	-, - -, -
28	1	13	-, 1 -, 1	-, - -, -	-, - -, -	-, - -, -	-, - -, -	-, - -, -	-, - -, -	-, - -, -	-, - -, -
28	1	14	0, 1 -, -	-, - -, -	-, - -, -	-, - -, -	-, - -, -	-, - -, -	-, - -, -	-, - -, -	-, - -, -
28	2	2	-, 1 -, 1	-, 1 -, 1	-, 2 -, 2	-, 2 -, 2	-, 2 -, 2	-, 2 -, 2	-, 2 -, 2	-, - -, -	-, - -, -
28	2	3	-, 1 -, 1	-, 1 -, 2	-, 2 -, 2	-, 2 -, 2	-, 2 -, 2	-, 2 -, 2	-, - -, -	-, - -, -	-, - -, -
28	2	4	-, 1 -, 1	-, 2 -, 2	-, 2 -, 2	-, 2 -, 2	-, 2 -, 2	-, - -, -	-, - -, -	-, - -, -	-, - -, -
28	2	5	-, 1 -, 1	-, 2 -, 2	-, 2 -, 2	-, 2 -, 2	-, - -, -	-, - -, -	-, - -, -	-, - -, -	-, - -, -
28	2	6	-, 1 -, 1	-, 2 -, 2	-, 2 -, 2	-, 2 -, 2	-, - -, -	-, - -, -	-, - -, -	-, - -, -	-, - -, -
28	2	7	-, 1 -, 1	-, 2 -, 2	-, 2 -, 2	-, - -, -	-, - -, -	-, - -, -	-, - -, -	-, - -, -	-, - -, -
28	2	8	-, 1 -, 1	-, 2 -, 2	-, 2 -, 2	-, - -, -	-, - -, -	-, - -, -	-, - -, -	-, - -, -	-, - -, -
28	2	9	0, 2 -, 2	-, 2 -, 2	-, 2 -, 2	-, - -, -	-, - -, -	-, - -, -	-, - -, -	-, - -, -	-, - -, -
28	2	10	0, 2 -, 2	-, 2 -, 2	-, - -, -	-, - -, -	-, - -, -	-, - -, -	-, - -, -	-, - -, -	-, - -, -
28	2	11	0, 2 -, 2	-, 2 -, 2	-, - -, -	-, - -, -	-, - -, -	-, - -, -	-, - -, -	-, - -, -	-, - -, -
28	2	12	0, 2 0, 2	-, 2 -, 2	-, - -, -	-, - -, -	-, - -, -	-, - -, -	-, - -, -	-, - -, -	-, - -, -
28	2	13	0, 2 0, 2	-, - -, -	-, - -, -	-, - -, -	-, - -, -	-, - -, -	-, - -, -	-, - -, -	-, - -, -
28	2	14	0, 2 0, 2	-, - -, -	-, - -, -	-, - -, -	-, - -, -	-, - -, -	-, - -, -	-, - -, -	-, - -, -
28	3	3	-, 1 -, 1	-, 2 -, 2	-, 2 -, 2	-, 2 -, 2	-, 3 -, 3	-, 3 -, 3	-, 3 -, 3	-, 3 -, 3	-, 3 -, 3
28	3	4	-, 1 -, 1	-, 2 -, 2	-, 2 -, 2	-, 2 -, 2	-, 3 -, 3	-, 3 -, 3	-, 3 -, 3	-, 3 -, 3	-, - -, -
28	3	5	-, 1 -, 1	-, 2 -, 2	-, 2 -, 2	-, 3 -, 3	-, 3 -, 3	-, 3 -, 3	-, 3 -, 3	-, - -, -	-, - -, -
28	3	6	0, 1 -, 2	-, 2 -, 2	-, 3 -, 3	-, 3 -, 3	-, 3 -, 3	-, 3 -, 3	-, - -, -	-, - -, -	-, - -, -
28	3	7	0, 2 -, 2	-, 2 -, 2	-, 3 -, 3	-, 3 -, 3	-, 3 -, 3	-, - -, -	-, - -, -	-, - -, -	-, - -, -
28	3	8	0, 2 -, 2	-, 2 -, 2	-, 3 -, 3	-, 3 -, 3	-, 3 -, 3	-, - -, -	-, - -, -	-, - -, -	-, - -, -
28	3	9	0, 2 -, 2	-, 3 -, 3	-, 3 -, 3	-, 3 -, 3	-, - -, -	-, - -, -	-, - -, -	-, - -, -	-, - -, -
28	3	10	0, 2 -, 2	-, 3 -, 3	-, 3 -, 3	-, 3 -, 3	-, - -, -	-, - -, -	-, - -, -	-, - -, -	-, - -, -

TABLE B.27 (cont.) CRITICAL VALUES FOR FISHER'S EXACT TEST

n	m_1	m_2	α = 0.50	0.20	0.10	0.05	0.02	0.01	0.005	0.002	0.001
28	3	11	0, 2 0, 3	-, 3 -, 3	-, 3 -, 3	-, - -, -	-, - -, -	-, - -, -	-, - -, -	-, - -, -	-, - -, -
28	3	12	0, 2 0, 3	0, 3 -, 3	-, 3 -, 3	-, - -, -	-, - -, -	-, - -, -	-, - -, -	-, - -, -	-, - -, -
28	3	13	0, 2 0, 3	0, 3 -, 3	-, 3 -, 3	-, - -, -	-, - -, -	-, - -, -	-, - -, -	-, - -, -	-, - -, -
28	3	14	1, 2 0, 3	0, 3 -, -	-, - -, -	-, - -, -	-, - -, -	-, - -, -	-, - -, -	-, - -, -	-, - -, -
28	4	4	-, 1 -, 1	-, 2 -, 2	-, 2 -, 2	-, 3 -, 3	-, 3 -, 3	-, 3 -, 3	-, 3 -, 3	-, 4 -, 4	-, 4 -, 4
28	4	5	0, 2 -, 2	-, 2 -, 2	-, 3 -, 3	-, 3 -, 3	-, 3 -, 3	-, 4 -, 4	-, 4 -, 4	-, 4 -, 4	-, 4 -, 4
28	4	6	0, 2 -, 2	-, 2 -, 2	-, 3 -, 3	-, 3 -, 3	-, 4 -, 4	-, 4 -, 4	-, 4 -, 4	-, 4 -, 4	-, 4 -, 4
28	4	7	0, 2 -, 2	-, 3 -, 3	-, 3 -, 3	-, 3 -, 3	-, 4 -, 4	-, 4 -, 4	-, 4 -, 4	-, 4 -, 4	-, - -, -
28	4	8	0, 2 0, 3	-, 3 -, 3	-, 3 -, 3	-, 4 -, 4	-, 4 -, 4	-, 4 -, 4	-, 4 -, 4	-, - -, -	-, - -, -
28	4	9	0, 2 0, 3	0, 3 -, 3	-, 3 -, 3	-, 4 -, 4	-, 4 -, 4	-, 4 -, 4	-, - -, -	-, - -, -	-, - -, -
28	4	10	0, 2 0, 3	0, 3 -, 3	-, 4 -, 4	-, 4 -, 4	-, 4 -, 4	-, - -, -	-, - -, -	-, - -, -	-, - -, -
28	4	11	1, 3 0, 3	0, 3 0, 4	-, 4 -, 4	-, 4 -, 4	-, 4 -, 4	-, - -, -	-, - -, -	-, - -, -	-, - -, -
28	4	12	1, 3 0, 3	0, 3 0, 4	0, 4 -, 4	-, 4 -, 4	-, - -, -	-, - -, -	-, - -, -	-, - -, -	-, - -, -
28	4	13	1, 3 0, 3	0, 4 0, 4	0, 4 -, 4	-, 4 -, 4	-, - -, -	-, - -, -	-, - -, -	-, - -, -	-, - -, -
28	4	14	1, 3 0, 4	0, 4 0, 4	0, 4 0, 4	0, 4 -, -	-, - -, -	-, - -, -	-, - -, -	-, - -, -	-, - -, -
28	5	5	0, 2 -, 2	-, 2 -, 3	-, 3 -, 3	-, 3 -, 3	-, 4 -, 4	-, 4 -, 4	-, 4 -, 4	-, 4 -, 4	-, 5 -, 5
28	5	6	0, 2 -, 2	-, 3 -, 3	-, 3 -, 3	-, 4 -, 4	-, 4 -, 4	-, 4 -, 4	-, 4 -, 4	-, 5 -, 5	-, 5 -, 5
28	5	7	0, 2 0, 3	-, 3 -, 3	-, 3 -, 3	-, 4 -, 4	-, 4 -, 4	-, 4 -, 4	-, 5 -, 5	-, 5 -, 5	-, 5 -, 5
28	5	8	0, 2 0, 3	0, 3 -, 3	-, 4 -, 4	-, 4 -, 4	-, 4 -, 4	-, 5 -, 5	-, 5 -, 5	-, 5 -, 5	-, 5 -, 5
28	5	9	1, 3 0, 3	0, 3 0, 4	-, 4 -, 4	-, 4 -, 4	-, 5 -, 5	-, 5 -, 5	-, 5 -, 5	-, 5 -, 5	-, - -, -
28	5	10	1, 3 0, 3	0, 4 0, 4	0, 4 -, 4	-, 4 -, 4	-, 5 -, 5	-, 5 -, 5	-, 5 -, 5	-, - -, -	-, - -, -
28	5	11	1, 3 0, 3	0, 4 0, 4	0, 4 -, 4	-, 5 -, 5	-, 5 -, 5	-, 5 -, 5	-, 5 -, 5	-, - -, -	-, - -, -
28	5	12	1, 3 1, 4	0, 4 0, 4	0, 4 0, 5	0, 5 -, 5	-, 5 -, 5	-, 5 -, 5	-, - -, -	-, - -, -	-, - -, -
28	5	13	1, 3 1, 4	0, 4 0, 4	0, 5 0, 5	0, 5 0, 5	-, 5 -, 5	-, - -, -	-, - -, -	-, - -, -	-, - -, -
28	5	14	2, 3 1, 4	1, 4 0, 5	0, 5 0, 5	0, 5 0, 5	-, - -, -	-, - -, -	-, - -, -	-, - -, -	-, - -, -
28	6	6	0, 2 0, 3	0, 3 -, 3	-, 3 -, 3	-, 4 -, 4	-, 4 -, 4	-, 4 -, 4	-, 5 -, 5	-, 5 -, 5	-, 5 -, 5
28	6	7	0, 2 0, 3	0, 3 -, 3	-, 4 -, 4	-, 4 -, 4	-, 5 -, 5	-, 5 -, 5	-, 5 -, 5	-, 5 -, 5	-, 6 -, 6
28	6	8	1, 3 0, 3	0, 4 0, 4	-, 4 -, 4	-, 4 -, 4	-, 5 -, 5	-, 5 -, 5	-, 5 -, 5	-, 6 -, 6	-, 6 -, 6
28	6	9	1, 3 0, 3	0, 4 0, 4	0, 4 -, 4	-, 5 -, 5	-, 5 -, 5	-, 5 -, 5	-, 6 -, 6	-, 6 -, 6	-, 6 -, 6
28	6	10	1, 4 1, 4	0, 4 0, 4	0, 4 0, 5	0, 5 -, 5	-, 5 -, 5	-, 6 -, 6	-, 6 -, 6	-, 6 -, 6	-, 6 -, 6
28	6	11	1, 4 1, 4	0, 4 0, 4	0, 5 0, 5	0, 5 -, 5	-, 6 -, 6	-, 6 -, 6	-, 6 -, 6	-, 6 -, 6	-, - -, -
28	6	12	2, 4 1, 4	1, 4 1, 5	0, 5 0, 5	0, 5 0, 6	-, 6 -, 6	-, 6 -, 6	-, 6 -, 6	-, - -, -	-, - -, -
28	6	13	2, 4 1, 4	1, 5 1, 5	0, 5 0, 5	0, 6 0, 6	0, 6 0, 6	-, 6 -, 6	-, 6 -, 6	-, - -, -	-, - -, -
28	6	14	2, 4 1, 5	1, 5 1, 5	1, 5 0, 6	0, 6 0, 6	0, 6 0, 6	0, 6 -, -	-, - -, -	-, - -, -	-, - -, -
28	7	7	1, 3 0, 3	0, 4 0, 4	0, 4 -, 4	-, 4 -, 4	-, 5 -, 5	-, 5 -, 5	-, 5 -, 5	-, 6 -, 6	-, 6 -, 6
28	7	8	1, 3 0, 3	0, 4 0, 4	0, 4 0, 5	-, 5 -, 5	-, 5 -, 5	-, 5 -, 5	-, 6 -, 6	-, 6 -, 6	-, 6 -, 6
28	7	9	1, 4 1, 4	0, 4 0, 4	0, 5 0, 5	0, 5 -, 5	-, 5 -, 5	-, 6 -, 6	-, 6 -, 6	-, 6 -, 6	-, 7 -, 7
28	7	10	1, 3 1, 4	1, 4 0, 5	0, 5 0, 5	0, 5 0, 6	-, 6 -, 6	-, 6 -, 6	-, 6 -, 6	-, 7 -, 7	-, 7 -, 7
28	7	11	2, 4 1, 4	1, 5 1, 5	0, 5 0, 5	0, 6 0, 6	0, 6 -, 6	-, 6 -, 6	-, 7 -, 7	-, 7 -, 7	-, 7 -, 7
28	7	12	2, 4 1, 4	1, 5 1, 5	1, 5 0, 6	0, 6 0, 6	0, 6 0, 7	0, 7 -, 7	-, 7 -, 7	-, 7 -, 7	-, 7 -, 7
28	7	13	2, 4 2, 5	1, 5 1, 5	1, 6 1, 6	0, 6 0, 6	0, 7 0, 7	0, 7 0, 7	-, 7 -, 7	-, 7 -, 7	-, - -, -
28	7	14	3, 4 2, 5	2, 5 1, 6	1, 6 1, 6	1, 6 0, 7	0, 7 0, 7	0, 7 0, 7	0, 7 -, -	-, - -, -	-, - -, -
28	8	8	1, 4 1, 4	0, 4 0, 4	0, 5 0, 5	0, 5 -, 5	-, 6 -, 6	-, 6 -, 6	-, 6 -, 6	-, 6 -, 6	-, 7 -, 7
28	8	9	2, 4 1, 4	1, 5 0, 5	0, 5 0, 5	0, 5 0, 6	-, 6 -, 6	-, 6 -, 6	-, 6 -, 6	-, 7 -, 7	-, 7 -, 7
28	8	10	2, 4 1, 4	1, 5 1, 5	0, 5 0, 5	0, 6 0, 6	0, 6 -, 6	-, 7 -, 7	-, 7 -, 7	-, 7 -, 7	-, 7 -, 7
28	8	11	2, 5 2, 5	1, 5 1, 5	1, 6 1, 6	0, 6 0, 6	0, 7 0, 7	0, 7 0, 7	-, 7 -, 7	-, 7 -, 7	-, 8 -, 8
28	8	12	2, 5 2, 5	1, 5 1, 6	1, 6 1, 6	1, 6 0, 6	0, 7 0, 7	0, 7 0, 7	0, 7 -, 7	-, 8 -, 8	-, 8 -, 8
28	8	13	3, 5 2, 5	2, 6 1, 6	1, 6 1, 6	1, 7 1, 7	0, 7 0, 7	0, 7 0, 8	0, 8 0, 8	-, 8 -, 8	-, 8 -, 8
28	8	14	3, 5 2, 6	2, 6 1, 7	1, 7 1, 7	1, 7 1, 7	1, 7 0, 8	0, 8 0, 8	0, 8 0, 8	0, 8 0, 8	0, 8 -, -
28	9	9	2, 4 1, 4	1, 5 1, 5	0, 5 0, 5	0, 6 0, 6	0, 6 -, 6	-, 7 -, 7	-, 7 -, 7	-, 7 -, 7	-, 7 -, 7

TABLE B.27 (cont.) CRITICAL VALUES FOR FISHER'S EXACT TEST

n	m_1	m_2	α = 0.50	0.20	0.10	0.05	0.02	0.01	0.005	0.002	0.001
28	9	10	2, 5 2, 5	1, 5 1, 6	1, 6 1, 6	0, 6 0, 6	0, 7 0, 7	0, 7 0, 7	-, 7 -, 7	-, 8 -, 8	-, 8 -, 8
28	9	11	3, 5 2, 5	2, 6 1, 6	1, 6 1, 6	1, 7 1, 7	0, 7 0, 7	0, 7 0, 8	0, 8 0, 8	-, 8 -, 8	-, 8 -, 8
28	9	12	3, 5 2, 5	2, 6 1, 6	1, 6 1, 7	1, 7 1, 7	0, 7 0, 7	0, 8 0, 8	0, 8 0, 8	0, 8 -, 8	-, 9 -, 9
28	9	13	3, 5 3, 6	2, 6 2, 7	2, 7 1, 7	1, 7 1, 7	1, 8 1, 8	0, 8 0, 8	0, 8 0, 8	0, 9 0, 9	0, 9 0, 9
28	9	14	4, 5 3, 6	2, 7 2, 7	2, 7 1, 8	1, 8 1, 8	1, 8 1, 8	1, 8 0, 9	0, 9 0, 9	0, 9 0, 9	0, 9 0, 9
28	10	10	3, 5 2, 5	2, 6 1, 6	1, 6 1, 6	1, 7 1, 7	0, 7 0, 7	0, 7 0, 8	0, 8 0, 8	-, 8 -, 8	-, 8 -, 8
28	10	11	3, 5 2, 5	2, 6 1, 6	1, 7 1, 7	1, 7 1, 7	0, 7 0, 8	0, 8 0, 8	0, 8 0, 8	0, 8 0, 9	-, 9 -, 9
28	10	12	3, 6 3, 6	2, 6 2, 7	2, 7 1, 7	1, 7 1, 7	1, 8 1, 8	0, 8 0, 8	0, 8 0, 9	0, 9 0, 9	0, 9 0, 9
28	10	13	4, 6 3, 6	3, 7 2, 7	2, 7 2, 8	2, 8 1, 8	1, 8 1, 8	1, 9 1, 9	0, 9 0, 9	0, 9 0, 9	0, 9 0,10
28	10	14	4, 6 3, 7	3, 7 2, 8	2, 8 2, 8	2, 8 2, 8	1, 9 1, 9	1, 9 1, 9	1, 9 1, 9	0,10 0,10	0,10 0,10
28	11	11	3, 6 3, 6	2, 6 2, 7	2, 7 1, 7	1, 7 1, 8	1, 8 1, 8	0, 8 0, 8	0, 9 0, 9	0, 9 0, 9	0, 9 0, 9
28	11	12	4, 6 3, 6	3, 7 2, 7	2, 7 2, 8	2, 8 1, 8	1, 8 1, 8	1, 9 1, 9	1, 9 0, 9	0, 9 0, 9	0,10 0,10
28	11	13	4, 6 4, 7	3, 7 3, 8	2, 8 2, 8	2, 8 2, 9	1, 9 1, 9	1, 9 1, 9	1, 9 1,10	1,10 0,10	0,10 0,10
28	11	14	5, 6 4, 7	3, 8 3, 8	3, 8 2, 9	2, 9 2, 9	2, 9 2, 9	2, 9 1,10	1,10 1,10	1,10 1,10	1,10 0,11
28	12	12	4, 7 4, 7	3, 7 3, 8	2, 8 2, 8	2, 8 2, 9	2, 9 1, 9	1, 9 1, 9	1, 9 1,10	1,10 0,10	0,10 0,10
28	12	13	5, 7 4, 7	3, 8 3, 8	3, 8 3, 9	2, 9 2, 9	2, 9 2,10	2,10 2,10	1,10 1,10	1,10 1,10	1,11 1,11
28	12	14	5, 7 4, 8	4, 8 3, 9	3, 9 3, 9	3, 9 2,10	2,10 2,10	2,10 2,10	2,10 1,11	1,11 1,11	1,11 1,11
28	13	13	5, 7 5, 8	4, 8 4, 9	3, 9 3, 9	3, 9 3,10	2,10 2,10	2,10 2,10	2,10 2,11	1,11 1,11	1,11 1,11
28	13	14	6, 7 5, 8	4, 9 4, 9	4, 9 3,10	3,10 3,10	3,10 2,11	2,11 2,11	2,11 2,11	2,11 2,11	2,11 1,12
28	14	14	6, 8 5, 9	5, 9 4,10	4,10 4,10	4,10 3,11	3,11 3,11	3,11 3,11	3,11 2,12	2,12 2,12	2,12 2,12
29	1	1	-, 1 -, 1	-, 1 -, 1	-, 1 -, 1	-, 1 -, 1	-, - -, -	-, - -, -	-, - -, -	-, - -, -	-, - -, -
29	1	2	-, 1 -, 1	-, 1 -, 1	-, 1 -, 1	-, - -, -	-, - -, -	-, - -, -	-, - -, -	-, - -, -	-, - -, -
29	1	3	-, 1 -, 1	-, 1 -, 1	-, - -, -	-, - -, -	-, - -, -	-, - -, -	-, - -, -	-, - -, -	-, - -, -
29	1	4	-, 1 -, 1	-, 1 -, 1	-, - -, -	-, - -, -	-, - -, -	-, - -, -	-, - -, -	-, - -, -	-, - -, -
29	1	5	-, 1 -, 1	-, 1 -, 1	-, - -, -	-, - -, -	-, - -, -	-, - -, -	-, - -, -	-, - -, -	-, - -, -
29	1	6	-, 1 -, 1	-, - -, -	-, - -, -	-, - -, -	-, - -, -	-, - -, -	-, - -, -	-, - -, -	-, - -, -
29	1	7	-, 1 -, 1	-, - -, -	-, - -, -	-, - -, -	-, - -, -	-, - -, -	-, - -, -	-, - -, -	-, - -, -
29	1	8	-, 1 -, 1	-, - -, -	-, - -, -	-, - -, -	-, - -, -	-, - -, -	-, - -, -	-, - -, -	-, - -, -
29	1	9	-, 1 -, 1	-, - -, -	-, - -, -	-, - -, -	-, - -, -	-, - -, -	-, - -, -	-, - -, -	-, - -, -
29	1	10	-, 1 -, 1	-, - -, -	-, - -, -	-, - -, -	-, - -, -	-, - -, -	-, - -, -	-, - -, -	-, - -, -
29	1	11	-, 1 -, 1	-, - -, -	-, - -, -	-, - -, -	-, - -, -	-, - -, -	-, - -, -	-, - -, -	-, - -, -
29	1	12	-, 1 -, 1	-, - -, -	-, - -, -	-, - -, -	-, - -, -	-, - -, -	-, - -, -	-, - -, -	-, - -, -
29	1	13	-, 1 -, 1	-, - -, -	-, - -, -	-, - -, -	-, - -, -	-, - -, -	-, - -, -	-, - -, -	-, - -, -
29	1	14	-, 1 -, 1	-, - -, -	-, - -, -	-, - -, -	-, - -, -	-, - -, -	-, - -, -	-, - -, -	-, - -, -
29	2	2	-, 1 -, 1	-, 1 -, 1	-, 2 -, 2	-, 2 -, 2	-, 2 -, 2	-, 2 -, 2	-, 2 -, 2	-, - -, -	-, - -, -
29	2	3	-, 1 -, 1	-, 1 -, 1	-, 2 -, 2	-, 2 -, 2	-, 2 -, 2	-, 2 -, 2	-, - -, -	-, - -, -	-, - -, -
29	2	4	-, 1 -, 1	-, 2 -, 2	-, 2 -, 2	-, 2 -, 2	-, 2 -, 2	-, - -, -	-, - -, -	-, - -, -	-, - -, -
29	2	5	-, 1 -, 1	-, 2 -, 2	-, 2 -, 2	-, 2 -, 2	-, - -, -	-, - -, -	-, - -, -	-, - -, -	-, - -, -
29	2	6	-, 1 -, 1	-, 2 -, 2	-, 2 -, 2	-, 2 -, 2	-, - -, -	-, - -, -	-, - -, -	-, - -, -	-, - -, -
29	2	7	-, 1 -, 1	-, 2 -, 2	-, 2 -, 2	-, - -, -	-, - -, -	-, - -, -	-, - -, -	-, - -, -	-, - -, -
29	2	8	-, 1 -, 1	-, 2 -, 2	-, 2 -, 2	-, - -, -	-, - -, -	-, - -, -	-, - -, -	-, - -, -	-, - -, -
29	2	9	0, 1 -, 2	-, 2 -, 2	-, 2 -, 2	-, - -, -	-, - -, -	-, - -, -	-, - -, -	-, - -, -	-, - -, -
29	2	10	0, 2 -, 2	-, 2 -, 2	-, - -, -	-, - -, -	-, - -, -	-, - -, -	-, - -, -	-, - -, -	-, - -, -
29	2	11	0, 2 -, 2	-, 2 -, 2	-, - -, -	-, - -, -	-, - -, -	-, - -, -	-, - -, -	-, - -, -	-, - -, -
29	2	12	0, 2 0, 2	-, 2 -, 2	-, - -, -	-, - -, -	-, - -, -	-, - -, -	-, - -, -	-, - -, -	-, - -, -
29	2	13	0, 2 0, 2	-, 2 -, 2	-, - -, -	-, - -, -	-, - -, -	-, - -, -	-, - -, -	-, - -, -	-, - -, -
29	2	14	0, 2 0, 2	-, - -, -	-, - -, -	-, - -, -	-, - -, -	-, - -, -	-, - -, -	-, - -, -	-, - -, -
29	3	3	-, 1 -, 1	-, 2 -, 2	-, 2 -, 2	-, 2 -, 2	-, 3 -, 3	-, 3 -, 3	-, 3 -, 3	-, 3 -, 3	-, 3 -, 3
29	3	4	-, 1 -, 1	-, 2 -, 2	-, 2 -, 2	-, 2 -, 2	-, 3 -, 3	-, 3 -, 3	-, 3 -, 3	-, 3 -, 3	-, - -, -
29	3	5	-, 1 -, 1	-, 2 -, 2	-, 2 -, 2	-, 3 -, 3	-, 3 -, 3	-, 3 -, 3	-, 3 -, 3	-, - -, -	-, - -, -

TABLE B.27 (cont.) CRITICAL VALUES FOR FISHER'S EXACT TEST

n	m_1	m_2	α = 0.50	0.20	0.10	0.05	0.02	0.01	0.005	0.002	0.001
29	3	6	0, 1 -, 2	-, 2 -, 2	-, 2 -, 2	-, 3 -, 3	-, 3 -, 3	-, 3 -, 3	-, - -, -	-, - -, -	-, - -, -
29	3	7	0, 2 -, 2	-, 2 -, 2	-, 3 -, 3	-, 3 -, 3	-, 3 -, 3	-, 3 -, 3	-, - -, -	-, - -, -	-, - -, -
29	3	8	0, 2 -, 2	-, 2 -, 2	-, 3 -, 3	-, 3 -, 3	-, 3 -, 3	-, - -, -	-, - -, -	-, - -, -	-, - -, -
29	3	9	0, 2 -, 2	-, 3 -, 3	-, 3 -, 3	-, 3 -, 3	-, - -, -	-, - -, -	-, - -, -	-, - -, -	-, - -, -
29	3	10	0, 2 -, 2	-, 3 -, 3	-, 3 -, 3	-, 3 -, 3	-, - -, -	-, - -, -	-, - -, -	-, - -, -	-, - -, -
29	3	11	0, 2 0, 3	-, 3 -, 3	-, 3 -, 3	-, 3 -, 3	-, - -, -	-, - -, -	-, - -, -	-, - -, -	-, - -, -
29	3	12	0, 2 0, 3	0, 3 -, 3	-, 3 -, 3	-, - -, -	-, - -, -	-, - -, -	-, - -, -	-, - -, -	-, - -, -
29	3	13	0, 2 0, 3	0, 3 -, 3	-, 3 -, 3	-, - -, -	-, - -, -	-, - -, -	-, - -, -	-, - -, -	-, - -, -
29	3	14	0, 2 0, 3	0, 3 -, 3	-, 3 -, 3	-, - -, -	-, - -, -	-, - -, -	-, - -, -	-, - -, -	-, - -, -
29	4	4	-, 1 -, 1	-, 2 -, 2	-, 2 -, 2	-, 3 -, 3	-, 3 -, 3	-, 3 -, 3	-, 3 -, 3	-, 4 -, 4	-, 4 -, 4
29	4	5	0, 1 -, 2	-, 2 -, 2	-, 3 -, 3	-, 3 -, 3	-, 3 -, 3	-, 4 -, 4	-, 4 -, 4	-, 4 -, 4	-, 4 -, 4
29	4	6	0, 2 -, 2	-, 2 -, 2	-, 3 -, 3	-, 3 -, 3	-, 3 -, 3	-, 4 -, 4	-, 4 -, 4	-, 4 -, 4	-, 4 -, 4
29	4	7	0, 2 -, 2	-, 3 -, 3	-, 3 -, 3	-, 3 -, 3	-, 4 -, 4	-, 4 -, 4	-, 4 -, 4	-, 4 -, 4	-, - -, -
29	4	8	0, 2 -, 2	-, 3 -, 3	-, 3 -, 3	-, 4 -, 4	-, 4 -, 4	-, 4 -, 4	-, 4 -, 4	-, - -, -	-, - -, -
29	4	9	0, 2 0, 3	-, 3 -, 3	-, 3 -, 3	-, 4 -, 4	-, 4 -, 4	-, 4 -, 4	-, - -, -	-, - -, -	-, - -, -
29	4	10	0, 2 0, 3	0, 3 -, 3	-, 4 -, 4	-, 4 -, 4	-, 4 -, 4	-, 4 -, 4	-, - -, -	-, - -, -	-, - -, -
29	4	11	0, 2 0, 3	0, 3 -, 3	-, 4 -, 4	-, 4 -, 4	-, 4 -, 4	-, - -, -	-, - -, -	-, - -, -	-, - -, -
29	4	12	1, 3 0, 3	0, 3 0, 4	-, 4 -, 4	-, 4 -, 4	-, - -, -	-, - -, -	-, - -, -	-, - -, -	-, - -, -
29	4	13	1, 3 0, 3	0, 4 0, 4	0, 4 -, 4	-, 4 -, 4	-, - -, -	-, - -, -	-, - -, -	-, - -, -	-, - -, -
29	4	14	1, 3 0, 3	0, 4 0, 4	0, 4 0, 4	-, 4 -, 4	-, - -, -	-, - -, -	-, - -, -	-, - -, -	-, - -, -
29	5	5	0, 2 -, 2	-, 2 -, 2	-, 3 -, 3	-, 3 -, 3	-, 4 -, 4	-, 4 -, 4	-, 4 -, 4	-, 4 -, 4	-, 5 -, 5
29	5	6	0, 2 -, 2	-, 3 -, 3	-, 3 -, 3	-, 3 -, 3	-, 4 -, 4	-, 4 -, 4	-, 4 -, 4	-, 5 -, 5	-, 5 -, 5
29	5	7	0, 2 0, 3	-, 3 -, 3	-, 3 -, 3	-, 4 -, 4	-, 4 -, 4	-, 4 -, 4	-, 5 -, 5	-, 5 -, 5	-, 5 -, 5
29	5	8	0, 2 0, 3	0, 3 -, 3	-, 4 -, 4	-, 4 -, 4	-, 4 -, 4	-, 5 -, 5	-, 5 -, 5	-, 5 -, 5	-, 5 -, 5
29	5	9	1, 2 0, 3	0, 3 0, 4	-, 4 -, 4	-, 4 -, 4	-, 5 -, 5	-, 5 -, 5	-, 5 -, 5	-, 5 -, 5	-, - -, -
29	5	10	1, 3 0, 3	0, 4 0, 4	0, 4 -, 4	-, 4 -, 4	-, 5 -, 5	-, 5 -, 5	-, 5 -, 5	-, - -, -	-, - -, -
29	5	11	1, 3 0, 3	0, 4 0, 4	0, 4 -, 4	-, 5 -, 5	-, 5 -, 5	-, 5 -, 5	-, 5 -, 5	-, - -, -	-, - -, -
29	5	12	1, 3 1, 4	0, 4 0, 4	0, 4 0, 5	-, 5 -, 5	-, 5 -, 5	-, 5 -, 5	-, - -, -	-, - -, -	-, - -, -
29	5	13	1, 3 1, 4	0, 4 0, 4	0, 5 0, 5	0, 5 0, 5	-, 5 -, 5	-, - -, -	-, - -, -	-, - -, -	-, - -, -
29	5	14	1, 3 1, 4	1, 4 0, 4	0, 5 0, 5	0, 5 0, 5	-, 5 -, 5	-, - -, -	-, - -, -	-, - -, -	-, - -, -
29	6	6	0, 2 0, 3	-, 3 -, 3	-, 3 -, 3	-, 4 -, 4	-, 4 -, 4	-, 4 -, 4	-, 5 -, 5	-, 5 -, 5	-, 5 -, 5
29	6	7	0, 2 0, 3	0, 3 -, 3	-, 4 -, 4	-, 4 -, 4	-, 4 -, 4	-, 5 -, 5	-, 5 -, 5	-, 5 -, 5	-, 5 -, 5
29	6	8	1, 3 0, 3	0, 4 0, 4	-, 4 -, 4	-, 4 -, 4	-, 5 -, 5	-, 5 -, 5	-, 5 -, 5	-, 6 -, 6	-, 6 -, 6
29	6	9	1, 3 0, 3	0, 4 0, 4	0, 4 -, 4	-, 5 -, 5	-, 5 -, 5	-, 5 -, 5	-, 6 -, 6	-, 6 -, 6	-, 6 -, 6
29	6	10	1, 4 0, 4	0, 4 0, 4	0, 4 0, 5	-, 5 -, 5	-, 5 -, 5	-, 6 -, 6	-, 6 -, 6	-, 6 -, 6	-, 6 -, 6
29	6	11	1, 4 1, 4	0, 4 0, 4	0, 5 0, 5	0, 5 -, 5	-, 5 -, 5	-, 6 -, 6	-, 6 -, 6	-, 6 -, 6	-, 6 -, 6
29	6	12	1, 3 1, 4	1, 4 0, 4	0, 5 0, 5	0, 5 0, 6	-, 6 -, 6	-, 6 -, 6	-, 6 -, 6	-, 6 -, 6	-, - -, -
29	6	13	2, 4 1, 4	1, 5 1, 5	0, 5 0, 5	0, 5 0, 6	0, 6 -, 6	-, 6 -, 6	-, 6 -, 6	-, - -, -	-, - -, -
29	6	14	2, 4 1, 4	1, 5 1, 5	1, 5 0, 5	0, 6 0, 6	0, 6 0, 6	-, 6 -, 6	-, - -, -	-, - -, -	-, - -, -
29	7	7	1, 3 0, 3	0, 4 0, 4	-, 4 -, 4	-, 4 -, 4	-, 5 -, 5	-, 5 -, 5	-, 5 -, 5	-, 6 -, 6	-, 6 -, 6
29	7	8	1, 3 0, 3	0, 4 0, 4	0, 4 -, 4	-, 5 -, 5	-, 5 -, 5	-, 5 -, 5	-, 6 -, 6	-, 6 -, 6	-, 6 -, 6
29	7	9	1, 4 1, 4	0, 4 0, 4	0, 5 0, 5	0, 5 -, 5	-, 5 -, 5	-, 6 -, 6	-, 6 -, 6	-, 6 -, 6	-, 7 -, 7
29	7	10	1, 4 1, 4	0, 4 0, 4	0, 5 0, 5	0, 5 -, 5	-, 6 -, 6	-, 6 -, 6	-, 6 -, 6	-, 7 -, 7	-, 7 -, 7
29	7	11	2, 4 1, 4	1, 5 0, 5	0, 5 0, 5	0, 6 0, 6	-, 6 -, 6	-, 6 -, 6	-, 7 -, 7	-, 7 -, 7	-, 7 -, 7
29	7	12	2, 4 1, 4	1, 5 1, 5	0, 5 0, 5	0, 6 0, 6	0, 6 -, 6	-, 7 -, 7	-, 7 -, 7	-, 7 -, 7	-, 7 -, 7
29	7	13	2, 4 2, 5	1, 5 1, 5	1, 6 1, 6	0, 6 0, 6	0, 6 0, 7	0, 7 0, 7	-, 7 -, 7	-, 7 -, 7	-, - -, -
29	7	14	2, 4 2, 5	1, 5 1, 6	1, 6 1, 6	1, 6 0, 6	0, 7 0, 7	0, 7 0, 7	0, 7 -, 7	-, - -, -	-, - -, -
29	8	8	1, 4 1, 4	0, 4 0, 4	0, 5 0, 5	0, 5 -, 5	-, 5 -, 5	-, 6 -, 6	-, 6 -, 6	-, 6 -, 6	-, 7 -, 7
29	8	9	1, 3 1, 4	1, 4 0, 5	0, 5 0, 5	0, 5 0, 6	-, 6 -, 6	-, 6 -, 6	-, 6 -, 6	-, 7 -, 7	-, 7 -, 7
29	8	10	2, 4 1, 4	1, 5 0, 5	0, 5 0, 5	0, 6 0, 6	0, 6 -, 6	-, 6 -, 6	-, 7 -, 7	-, 7 -, 7	-, 7 -, 7

TABLE B.27 (cont.) CRITICAL VALUES FOR FISHER'S EXACT TEST

n	m_1	m_2	α = 0.50	0.20	0.10	0.05	0.02	0.01	0.005	0.002	0.001
29	8	11	2, 4 1, 4	1, 5 1, 5	1, 6 0, 6	0, 6 0, 6	0, 6 0, 7	-, 7 -, 7	-, 7 -, 7	-, 7 -, 7	-, 8 -, 8
29	8	12	2, 5 2, 5	1, 5 1, 6	1, 6 1, 6	0, 6 0, 6	0, 7 0, 7	0, 7 0, 7	-, 7 -, 7	-, 8 -, 8	-, 8 -, 8
29	8	13	3, 5 2, 5	2, 6 1, 6	1, 6 1, 6	1, 7 1, 7	0, 7 0, 7	0, 7 0, 7	0, 8 0, 8	-, 8 -, 8	-, 8 -, 8
29	8	14	3, 5 2, 5	2, 6 1, 6	1, 6 1, 7	1, 7 1, 7	0, 7 0, 7	0, 8 0, 8	0, 8 0, 8	0, 8 -, 8	-, 8 -, 8
29	9	9	2, 4 1, 4	1, 5 0, 5	0, 5 0, 5	0, 6 0, 6	0, 6 -, 6	-, 7 -, 7	-, 7 -, 7	-, 7 -, 7	-, 7 -, 7
29	9	10	2, 5 2, 5	1, 5 1, 6	1, 6 0, 6	0, 6 0, 6	0, 7 0, 7	0, 7 -, 7	-, 7 -, 7	-, 8 -, 8	-, 8 -, 8
29	9	11	2, 5 2, 5	1, 5 1, 6	1, 6 1, 6	0, 6 0, 6	0, 7 0, 7	0, 7 0, 8	0, 8 -, 8	-, 8 -, 8	-, 8 -, 8
29	9	12	3, 5 2, 5	2, 6 1, 6	1, 6 1, 7	1, 7 1, 7	0, 7 0, 7	0, 8 0, 8	0, 8 0, 8	-, 8 -, 8	-, 8 -, 8
29	9	13	3, 5 3, 6	2, 6 2, 7	1, 7 1, 7	1, 7 1, 7	1, 8 1, 8	0, 8 0, 8	0, 8 0, 8	0, 9 0, 9	-, 9 -, 9
29	9	14	3, 5 3, 6	2, 6 2, 7	2, 7 1, 7	1, 7 1, 8	1, 8 1, 8	1, 8 0, 8	0, 8 0, 9	0, 9 0, 9	0, 9 0, 9
29	10	10	2, 5 2, 5	1, 5 1, 6	1, 6 1, 6	0, 6 0, 7	0, 7 0, 7	0, 7 0, 8	0, 8 -, 8	-, 8 -, 8	-, 8 -, 8
29	10	11	3, 5 2, 5	2, 6 1, 6	1, 6 1, 7	1, 7 1, 7	0, 7 0, 7	0, 8 0, 8	0, 8 0, 8	-, 8 -, 8	-, 9 -, 9
29	10	12	3, 6 3, 6	2, 6 2, 7	2, 7 1, 7	1, 7 1, 7	1, 8 1, 8	0, 8 0, 8	0, 8 0, 8	0, 9 0, 9	0, 9 -, 9
29	10	13	3, 6 3, 6	2, 7 2, 7	2, 7 1, 7	1, 8 1, 8	1, 8 1, 8	1, 8 1, 9	0, 9 0, 9	0, 9 0, 9	0, 9 0, 9
29	10	14	4, 6 3, 6	3, 7 2, 7	2, 7 2, 8	2, 8 1, 8	1, 8 1, 9	1, 9 1, 9	1, 9 0, 9	0, 9 0, 9	0,10 0,10
29	11	11	3, 6 3, 6	2, 6 2, 7	2, 7 1, 7	1, 7 1, 7	1, 8 1, 8	0, 8 0, 8	0, 8 0, 9	0, 9 0, 9	0, 9 -, 9
29	11	12	4, 6 3, 6	2, 7 2, 7	2, 7 2, 8	1, 8 1, 8	1, 8 1, 8	1, 9 1, 9	0, 9 0, 9	0, 9 0, 9	0, 9 0,10
29	11	13	4, 6 3, 6	3, 7 2, 7	2, 8 2, 8	2, 8 1, 8	1, 9 1, 9	1, 9 1, 9	1, 9 0, 9	0,10 0,10	0,10 0,10
29	11	14	4, 6 4, 7	3, 7 3, 8	3, 8 2, 8	2, 8 2, 9	2, 9 1, 9	1, 9 1, 9	1,10 1,10	1,10 1,10	0,10 0,10
29	12	12	4, 6 3, 6	3, 7 2, 7	2, 8 2, 8	2, 8 1, 8	1, 9 1, 9	1, 9 1, 9	1, 9 1,10	0,10 0,10	0,10 0,10
29	12	13	4, 7 4, 7	3, 7 3, 8	3, 8 2, 8	2, 9 2, 9	2, 9 1, 9	1, 9 1, 9	1,10 1,10	1,10 1,10	0,10 0,10
29	12	14	5, 7 4, 7	4, 8 3, 8	3, 8 3, 9	3, 9 2, 9	2, 9 2,10	2,10 2,10	1,10 1,10	1,11 1,11	1,11 1,11
29	13	13	5, 7 4, 7	4, 8 3, 8	3, 9 3, 9	3, 9 2, 9	2,10 2,10	2,10 2,10	1,10 1,10	1,11 1,11	1,11 1,11
29	13	14	5, 7 5, 8	4, 8 4, 9	4, 9 3, 9	3, 9 3,10	3,10 2,10	2,10 2,10	2,11 2,11	2,11 1,11	1,11 1,11
29	14	14	6, 8 5, 8	5, 9 4, 9	4, 9 4,10	4,10 3,10	3,10 3,11	3,11 3,11	2,11 2,11	2,12 2,12	2,12 2,12
30	1	1	-, 1 -, 1	-, 1 -, 1	-, 1 -, 1	-, 1 -, 1	-, - -, -	-, - -, -	-, - -, -	-, - -, -	-, - -, -
30	1	2	-, 1 -, 1	-, 1 -, 1	-, 1 -, 1	-, - -, -	-, - -, -	-, - -, -	-, - -, -	-, - -, -	-, - -, -
30	1	3	-, 1 -, 1	-, 1 -, 1	-, 1 -, 1	-, - -, -	-, - -, -	-, - -, -	-, - -, -	-, - -, -	-, - -, -
30	1	4	-, 1 -, 1	-, 1 -, 1	-, - -, -	-, - -, -	-, - -, -	-, - -, -	-, - -, -	-, - -, -	-, - -, -
30	1	5	-, 1 -, 1	-, 1 -, 1	-, - -, -	-, - -, -	-, - -, -	-, - -, -	-, - -, -	-, - -, -	-, - -, -
30	1	6	-, 1 -, 1	-, 1 -, 1	-, - -, -	-, - -, -	-, - -, -	-, - -, -	-, - -, -	-, - -, -	-, - -, -
30	1	7	-, 1 -, 1	-, - -, -	-, - -, -	-, - -, -	-, - -, -	-, - -, -	-, - -, -	-, - -, -	-, - -, -
30	1	8	-, 1 -, 1	-, - -, -	-, - -, -	-, - -, -	-, - -, -	-, - -, -	-, - -, -	-, - -, -	-, - -, -
30	1	9	-, 1 -, 1	-, - -, -	-, - -, -	-, - -, -	-, - -, -	-, - -, -	-, - -, -	-, - -, -	-, - -, -
30	1	10	-, 1 -, 1	-, - -, -	-, - -, -	-, - -, -	-, - -, -	-, - -, -	-, - -, -	-, - -, -	-, - -, -
30	1	11	-, 1 -, 1	-, - -, -	-, - -, -	-, - -, -	-, - -, -	-, - -, -	-, - -, -	-, - -, -	-, - -, -
30	1	12	-, 1 -, 1	-, - -, -	-, - -, -	-, - -, -	-, - -, -	-, - -, -	-, - -, -	-, - -, -	-, - -, -
30	1	13	-, 1 -, 1	-, - -, -	-, - -, -	-, - -, -	-, - -, -	-, - -, -	-, - -, -	-, - -, -	-, - -, -
30	1	14	-, 1 -, 1	-, - -, -	-, - -, -	-, - -, -	-, - -, -	-, - -, -	-, - -, -	-, - -, -	-, - -, -
30	1	15	0, 1 -, -	-, - -, -	-, - -, -	-, - -, -	-, - -, -	-, - -, -	-, - -, -	-, - -, -	-, - -, -
30	2	2	-, 1 -, 1	-, 1 -, 1	-, 2 -, 2	-, 2 -, 2	-, 2 -, 2	-, 2 -, 2	-, 2 -, 2	-, - -, -	-, - -, -
30	2	3	-, 1 -, 1	-, 1 -, 1	-, 2 -, 2	-, 2 -, 2	-, 2 -, 2	-, 2 -, 2	-, - -, -	-, - -, -	-, - -, -
30	2	4	-, 1 -, 1	-, 2 -, 2	-, 2 -, 2	-, 2 -, 2	-, 2 -, 2	-, - -, -	-, - -, -	-, - -, -	-, - -, -
30	2	5	-, 1 -, 1	-, 2 -, 2	-, 2 -, 2	-, 2 -, 2	-, - -, -	-, - -, -	-, - -, -	-, - -, -	-, - -, -
30	2	6	-, 1 -, 1	-, 2 -, 2	-, 2 -, 2	-, 2 -, 2	-, - -, -	-, - -, -	-, - -, -	-, - -, -	-, - -, -
30	2	7	-, 1 -, 1	-, 2 -, 2	-, 2 -, 2	-, 2 -, 2	-, - -, -	-, - -, -	-, - -, -	-, - -, -	-, - -, -
30	2	8	-, 1 -, 1	-, 2 -, 2	-, 2 -, 2	-, - -, -	-, - -, -	-, - -, -	-, - -, -	-, - -, -	-, - -, -
30	2	9	0, 1 -, 2	-, 2 -, 2	-, 2 -, 2	-, - -, -	-, - -, -	-, - -, -	-, - -, -	-, - -, -	-, - -, -
30	2	10	0, 2 -, 2	-, 2 -, 2	-, - -, -	-, - -, -	-, - -, -	-, - -, -	-, - -, -	-, - -, -	-, - -, -
30	2	11	0, 2 -, 2	-, 2 -, 2	-, - -, -	-, - -, -	-, - -, -	-, - -, -	-, - -, -	-, - -, -	-, - -, -

TABLE B.27 (cont.) CRITICAL VALUES FOR FISHER'S EXACT TEST

n	m_1	m_2	α = 0.50	0.20	0.10	0.05	0.02	0.01	0.005	0.002	0.001
30	2	12	0, 2 -, 2	-, 2 -, 2	-, - -, -	-, - -, -	-, - -, -	-, - -, -	-, - -, -	-, - -, -	-, - -, -
30	2	13	0, 2 0, 2	-, 2 -, 2	-, - -, -	-, - -, -	-, - -, -	-, - -, -	-, - -, -	-, - -, -	-, - -, -
30	2	14	0, 2 0, 2	-, - -, -	-, - -, -	-, - -, -	-, - -, -	-, - -, -	-, - -, -	-, - -, -	-, - -, -
30	2	15	0, 2 0, 2	-, - -, -	-, - -, -	-, - -, -	-, - -, -	-, - -, -	-, - -, -	-, - -, -	-, - -, -
30	3	3	-, 1 -, 1	-, 2 -, 2	-, 2 -, 2	-, 2 -, 2	-, 3 -, 3	-, 3 -, 3	-, 3 -, 3	-, 3 -, 3	-, 3 -, 3
30	3	4	-, 1 -, 1	-, 2 -, 2	-, 2 -, 2	-, 2 -, 2	-, 3 -, 3	-, 3 -, 3	-, 3 -, 3	-, 3 -, 3	-, 3 -, 3
30	3	5	-, 1 -, 1	-, 2 -, 2	-, 2 -, 2	-, 3 -, 3	-, 3 -, 3	-, 3 -, 3	-, 3 -, 3	-, - -, -	-, - -, -
30	3	6	0, 1 -, 2	-, 2 -, 2	-, 2 -, 2	-, 3 -, 3	-, 3 -, 3	-, 3 -, 3	-, 3 -, 3	-, - -, -	-, - -, -
30	3	7	0, 2 -, 2	-, 2 -, 2	-, 3 -, 3	-, 3 -, 3	-, 3 -, 3	-, 3 -, 3	-, - -, -	-, - -, -	-, - -, -
30	3	8	0, 2 -, 2	-, 2 -, 2	-, 3 -, 3	-, 3 -, 3	-, 3 -, 3	-, - -, -	-, - -, -	-, - -, -	-, - -, -
30	3	9	0, 2 -, 2	-, 3 -, 3	-, 3 -, 3	-, 3 -, 3	-, - -, -	-, - -, -	-, - -, -	-, - -, -	-, - -, -
30	3	10	0, 2 -, 2	-, 3 -, 3	-, 3 -, 3	-, 3 -, 3	-, - -, -	-, - -, -	-, - -, -	-, - -, -	-, - -, -
30	3	11	0, 2 0, 3	-, 3 -, 3	-, 3 -, 3	-, 3 -, 3	-, - -, -	-, - -, -	-, - -, -	-, - -, -	-, - -, -
30	3	12	0, 2 0, 3	-, 3 -, 3	-, 3 -, 3	-, - -, -	-, - -, -	-, - -, -	-, - -, -	-, - -, -	-, - -, -
30	3	13	0, 2 0, 3	0, 3 -, 3	-, 3 -, 3	-, - -, -	-, - -, -	-, - -, -	-, - -, -	-, - -, -	-, - -, -
30	3	14	0, 2 0, 3	0, 3 -, 3	-, 3 -, 3	-, - -, -	-, - -, -	-, - -, -	-, - -, -	-, - -, -	-, - -, -
30	3	15	1, 2 0, 3	0, 3 -, -	-, - -, -	-, - -, -	-, - -, -	-, - -, -	-, - -, -	-, - -, -	-, - -, -
30	4	4	-, 1 -, 1	-, 2 -, 2	-, 2 -, 2	-, 3 -, 3	-, 3 -, 3	-, 3 -, 3	-, 3 -, 3	-, 4 -, 4	-, 4 -, 4
30	4	5	0, 1 -, 2	-, 2 -, 2	-, 3 -, 3	-, 3 -, 3	-, 3 -, 3	-, 3 -, 3	-, 4 -, 4	-, 4 -, 4	-, 4 -, 4
30	4	6	0, 2 -, 2	-, 2 -, 2	-, 3 -, 3	-, 3 -, 3	-, 3 -, 3	-, 4 -, 4	-, 4 -, 4	-, 4 -, 4	-, 4 -, 4
30	4	7	0, 2 -, 2	-, 2 -, 3	-, 3 -, 3	-, 3 -, 3	-, 4 -, 4	-, 4 -, 4	-, 4 -, 4	-, 4 -, 4	-, - -, -
30	4	8	0, 2 -, 2	-, 3 -, 3	-, 3 -, 3	-, 3 -, 3	-, 4 -, 4	-, 4 -, 4	-, 4 -, 4	-, - -, -	-, - -, -
30	4	9	0, 2 0, 3	-, 3 -, 3	-, 3 -, 3	-, 4 -, 4	-, 4 -, 4	-, 4 -, 4	-, 4 -, 4	-, - -, -	-, - -, -
30	4	10	0, 2 0, 3	0, 3 -, 3	-, 3 -, 3	-, 4 -, 4	-, 4 -, 4	-, 4 -, 4	-, - -, -	-, - -, -	-, - -, -
30	4	11	0, 2 0, 3	0, 3 -, 3	-, 4 -, 4	-, 4 -, 4	-, 4 -, 4	-, - -, -	-, - -, -	-, - -, -	-, - -, -
30	4	12	1, 3 0, 3	0, 3 0, 4	-, 4 -, 4	-, 4 -, 4	-, 4 -, 4	-, - -, -	-, - -, -	-, - -, -	-, - -, -
30	4	13	1, 3 0, 3	0, 4 0, 4	0, 4 -, 4	-, 4 -, 4	-, - -, -	-, - -, -	-, - -, -	-, - -, -	-, - -, -
30	4	14	1, 3 0, 3	0, 4 0, 4	0, 4 -, 4	-, 4 -, 4	-, - -, -	-, - -, -	-, - -, -	-, - -, -	-, - -, -
30	4	15	1, 3 0, 4	0, 4 0, 4	0, 4 0, 4	0, 4 -, -	-, - -, -	-, - -, -	-, - -, -	-, - -, -	-, - -, -
30	5	5	0, 2 -, 2	-, 2 -, 2	-, 3 -, 3	-, 3 -, 3	-, 4 -, 4	-, 4 -, 4	-, 4 -, 4	-, 4 -, 4	-, 4 -, 4
30	5	6	0, 2 -, 2	-, 3 -, 3	-, 3 -, 3	-, 3 -, 3	-, 4 -, 4	-, 4 -, 4	-, 4 -, 4	-, 5 -, 5	-, 5 -, 5
30	5	7	0, 2 0, 3	-, 3 -, 3	-, 3 -, 3	-, 4 -, 4	-, 4 -, 4	-, 4 -, 4	-, 5 -, 5	-, 5 -, 5	-, 5 -, 5
30	5	8	0, 2 0, 3	0, 3 -, 3	-, 4 -, 4	-, 4 -, 4	-, 4 -, 4	-, 5 -, 5	-, 5 -, 5	-, 5 -, 5	-, 5 -, 5
30	5	9	0, 2 0, 3	0, 3 -, 3	-, 4 -, 4	-, 4 -, 4	-, 4 -, 4	-, 5 -, 5	-, 5 -, 5	-, 5 -, 5	-, 5 -, 5
30	5	10	1, 3 0, 3	0, 3 0, 4	-, 4 -, 4	-, 4 -, 4	-, 5 -, 5	-, 5 -, 5	-, 5 -, 5	-, 5 -, 5	-, - -, -
30	5	11	1, 3 0, 3	0, 4 0, 4	0, 4 -, 4	-, 4 -, 4	-, 5 -, 5	-, 5 -, 5	-, 5 -, 5	-, - -, -	-, - -, -
30	5	12	1, 3 0, 3	0, 4 0, 4	0, 4 0, 5	-, 5 -, 5	-, 5 -, 5	-, 5 -, 5	-, - -, -	-, - -, -	-, - -, -
30	5	13	1, 3 1, 4	0, 4 0, 4	0, 4 0, 5	0, 5 -, 5	-, 5 -, 5	-, 5 -, 5	-, - -, -	-, - -, -	-, - -, -
30	5	14	1, 3 1, 4	0, 4 0, 4	0, 5 0, 5	0, 5 0, 5	-, 5 -, 5	-, - -, -	-, - -, -	-, - -, -	-, - -, -
30	5	15	2, 3 1, 4	1, 4 0, 5	0, 5 0, 5	0, 5 0, 5	-, - -, -	-, - -, -	-, - -, -	-, - -, -	-, - -, -
30	6	6	0, 2 0, 3	-, 3 -, 3	-, 3 -, 3	-, 4 -, 4	-, 4 -, 4	-, 4 -, 4	-, 5 -, 5	-, 5 -, 5	-, 5 -, 5
30	6	7	0, 2 0, 3	0, 3 -, 3	-, 4 -, 4	-, 4 -, 4	-, 4 -, 4	-, 5 -, 5	-, 5 -, 5	-, 5 -, 5	-, 5 -, 5
30	6	8	1, 2 0, 3	0, 4 0, 4	-, 4 -, 4	-, 4 -, 4	-, 5 -, 5	-, 5 -, 5	-, 5 -, 5	-, 6 -, 6	-, 6 -, 6
30	6	9	1, 3 0, 3	0, 4 0, 4	0, 4 -, 4	-, 4 -, 4	-, 5 -, 5	-, 5 -, 5	-, 5 -, 5	-, 6 -, 6	-, 6 -, 6
30	6	10	1, 3 0, 3	0, 4 0, 4	0, 4 0, 5	-, 5 -, 5	-, 5 -, 5	-, 5 -, 5	-, 6 -, 6	-, 6 -, 6	-, 6 -, 6
30	6	11	1, 4 1, 4	0, 4 0, 4	0, 5 0, 5	0, 5 -, 5	-, 5 -, 5	-, 6 -, 6	-, 6 -, 6	-, 6 -, 6	-, 6 -, 6
30	6	12	1, 4 1, 4	0, 4 0, 4	0, 5 0, 5	0, 5 -, 5	-, 6 -, 6	-, 6 -, 6	-, 6 -, 6	-, 6 -, 6	-, - -, -
30	6	13	2, 4 1, 4	1, 5 1, 5	0, 5 0, 5	0, 5 0, 6	-, 6 -, 6	-, 6 -, 6	-, 6 -, 6	-, - -, -	-, - -, -
30	6	14	2, 4 1, 4	1, 5 1, 5	0, 5 0, 5	0, 6 0, 6	0, 6 0, 6	-, 6 -, 6	-, - -, -	-, - -, -	-, - -, -
30	6	15	2, 4 1, 5	1, 5 1, 5	1, 5 0, 6	0, 6 0, 6	0, 6 0, 6	0, 6 -, -	-, - -, -	-, - -, -	-, - -, -

TABLE B.27 (cont.) CRITICAL VALUES FOR FISHER'S EXACT TEST

n	m_1	m_2	α = 0.50	0.20	0.10	0.05	0.02	0.01	0.005	0.002	0.001
30	7	7	1, 3 0, 3	0, 4 0, 4	-, 4 -, 4	-, 4 -, 4	-, 5 -, 5	-, 5 -, 5	-, 5 -, 5	-, 6 -, 6	-, 6 -, 6
30	7	8	1, 3 0, 3	0, 4 0, 4	0, 4 -, 4	-, 5 -, 5	-, 5 -, 5	-, 5 -, 5	-, 6 -, 6	-, 6 -, 6	-, 6 -, 6
30	7	9	1, 4 1, 4	0, 4 0, 4	0, 4 0, 5	-, 5 -, 5	-, 5 -, 5	-, 6 -, 6	-, 6 -, 6	-, 6 -, 6	-, 6 -, 6
30	7	10	1, 4 1, 4	0, 4 0, 4	0, 5 0, 5	0, 5 -, 5	-, 6 -, 6	-, 6 -, 6	-, 6 -, 6	-, 7 -, 7	-, 7 -, 7
30	7	11	2, 4 1, 4	1, 5 0, 5	0, 5 0, 5	0, 5 0, 6	-, 6 -, 6	-, 6 -, 6	-, 6 -, 6	-, 7 -, 7	-, 7 -, 7
30	7	12	2, 4 1, 4	1, 5 1, 5	0, 5 0, 5	0, 6 0, 6	0, 6 -, 6	-, 6 -, 6	-, 7 -, 7	-, 7 -, 7	-, 7 -, 7
30	7	13	2, 4 2, 5	1, 5 1, 5	1, 6 0, 6	0, 6 0, 6	0, 6 0, 7	0, 7 -, 7	-, 7 -, 7	-, 7 -, 7	-, 7 -, 7
30	7	14	2, 4 2, 5	1, 5 1, 6	1, 6 1, 6	0, 6 0, 6	0, 7 0, 7	0, 7 0, 7	-, 7 -, 7	-, 7 -, 7	-, - -, -
30	7	15	3, 4 2, 5	2, 5 1, 6	1, 6 1, 6	1, 6 0, 7	0, 7 0, 7	0, 7 0, 7	0, 7 -, -	-, - -, -	-, - -, -
30	8	8	1, 4 1, 4	0, 4 0, 4	0, 5 0, 5	-, 5 -, 5	-, 5 -, 5	-, 6 -, 6	-, 6 -, 6	-, 6 -, 6	-, 7 -, 7
30	8	9	1, 4 1, 4	0, 4 0, 4	0, 5 0, 5	0, 5 -, 5	-, 6 -, 6	-, 6 -, 6	-, 6 -, 6	-, 7 -, 7	-, 7 -, 7
30	8	10	2, 4 1, 4	1, 5 0, 5	0, 5 0, 5	0, 6 0, 6	-, 6 -, 6	-, 6 -, 6	-, 7 -, 7	-, 7 -, 7	-, 7 -, 7
30	8	11	2, 4 1, 4	1, 5 1, 5	0, 5 0, 6	0, 6 0, 6	0, 6 0, 7	-, 7 -, 7	-, 7 -, 7	-, 7 -, 7	-, 8 -, 8
30	8	12	2, 5 2, 5	1, 5 1, 6	1, 6 1, 6	0, 6 0, 6	0, 7 0, 7	0, 7 0, 7	-, 7 -, 7	-, 8 -, 8	-, 8 -, 8
30	8	13	2, 5 2, 5	1, 5 1, 6	1, 6 1, 6	1, 6 0, 6	0, 7 0, 7	0, 7 0, 7	0, 8 0, 8	-, 8 -, 8	-, 8 -, 8
30	8	14	3, 5 2, 5	2, 6 1, 6	1, 6 1, 7	1, 7 1, 7	0, 7 0, 7	0, 7 0, 8	0, 8 0, 8	-, 8 -, 8	-, 8 -, 8
30	8	15	3, 5 2, 6	2, 6 1, 7	1, 7 1, 7	1, 7 1, 7	1, 7 0, 8	0, 8 0, 8	0, 8 0, 8	0, 8 -, -	-, - -, -
30	9	9	2, 4 1, 4	1, 5 0, 5	0, 5 0, 5	0, 6 0, 6	-, 6 -, 6	-, 6 -, 6	-, 7 -, 7	-, 7 -, 7	-, 7 -, 7
30	9	10	2, 4 1, 4	1, 5 0, 5	1, 6 0, 6	0, 6 0, 6	0, 6 0, 7	-, 7 -, 7	-, 7 -, 7	-, 7 -, 7	-, 8 -, 8
30	9	11	2, 5 2, 5	1, 5 1, 6	1, 6 0, 6	0, 6 0, 6	0, 7 0, 7	0, 7 -, 7	-, 7 -, 7	-, 8 -, 8	-, 8 -, 8
30	9	12	3, 5 2, 5	2, 6 1, 6	1, 6 1, 7	1, 7 1, 7	0, 7 0, 7	0, 7 0, 8	0, 8 0, 8	-, 8 -, 8	-, 8 -, 8
30	9	13	3, 5 2, 5	2, 6 1, 6	1, 6 1, 7	1, 7 1, 7	0, 7 0, 7	0, 8 0, 8	0, 8 0, 8	0, 8 -, 8	-, 9 -, 9
30	9	14	3, 5 3, 6	2, 6 2, 7	2, 7 1, 7	1, 7 1, 7	1, 8 1, 8	0, 8 0, 8	0, 8 0, 8	0, 9 0, 9	0, 9 0, 9
30	9	15	4, 5 3, 6	2, 7 2, 7	2, 7 1, 8	1, 8 1, 8	1, 8 1, 8	1, 8 0, 9	0, 9 0, 9	0, 9 0, 9	0, 9 0, 9
30	10	10	2, 5 2, 5	1, 5 1, 6	1, 6 0, 6	0, 6 0, 6	0, 7 0, 7	0, 7 -, 7	-, 7 -, 7	-, 8 -, 8	-, 8 -, 8
30	10	11	3, 5 2, 5	2, 6 1, 6	1, 6 1, 7	1, 7 1, 7	0, 7 0, 7	0, 8 0, 8	0, 8 0, 8	-, 8 -, 8	-, 8 -, 8
30	10	12	3, 5 2, 5	2, 6 1, 6	1, 7 1, 7	1, 7 1, 7	0, 8 0, 8	0, 8 0, 8	0, 8 0, 8	0, 9 0, 9	-, 9 -, 9
30	10	13	3, 6 3, 6	2, 6 2, 7	2, 7 1, 7	1, 7 1, 8	1, 8 1, 8	0, 8 0, 8	0, 9 0, 9	0, 9 0, 9	0, 9 -, 9
30	10	14	4, 6 3, 6	3, 7 2, 7	2, 7 2, 8	2, 8 1, 8	1, 8 1, 8	1, 9 1, 9	0, 9 0, 9	0, 9 0, 9	0,10 0,10
30	10	15	4, 6 3, 7	3, 7 2, 8	2, 8 2, 8	2, 8 1, 9	1, 9 1, 9	1, 9 1, 9	1, 9 0,10	0,10 0,10	0,10 0,10
30	11	11	3, 6 3, 6	2, 6 2, 7	1, 7 1, 7	1, 7 1, 7	1, 8 0, 8	0, 8 0, 8	0, 8 0, 8	0, 9 0, 9	-, 9 -, 9
30	11	12	3, 6 3, 6	2, 6 2, 7	2, 7 1, 7	1, 8 1, 8	1, 8 1, 8	0, 8 0, 8	0, 9 0, 9	0, 9 0, 9	0, 9 0,10
30	11	13	4, 6 3, 6	3, 7 2, 7	2, 7 2, 8	2, 8 1, 8	1, 8 1, 9	1, 9 1, 9	1, 9 0, 9	0, 9 0,10	0,10 0,10
30	11	14	4, 6 4, 7	3, 7 3, 8	2, 8 2, 8	2, 8 2, 9	1, 9 1, 9	1, 9 1, 9	1, 9 1,10	0,10 0,10	0,10 0,10
30	11	15	5, 6 4, 7	3, 8 3, 8	3, 8 2, 9	2, 9 2, 9	2, 9 1,10	1,10 1,10	1,10 1,10	1,10 1,10	1,10 0,11
30	12	12	4, 6 3, 6	3, 7 2, 7	2, 7 2, 8	2, 8 1, 8	1, 8 1, 9	1, 9 1, 9	1, 9 0, 9	0,10 0,10	0,10 0,10
30	12	13	4, 7 4, 7	3, 7 3, 8	3, 8 2, 8	2, 8 2, 9	2, 9 1, 9	1, 9 1, 9	1,10 1,10	1,10 0,10	0,10 0,10
30	12	14	5, 7 4, 7	3, 8 3, 8	3, 8 3, 9	2, 9 2, 9	2, 9 2,10	2,10 1,10	1,10 1,10	1,10 1,10	1,11 1,11
30	12	15	5, 7 4, 8	4, 8 3, 9	3, 9 3, 9	3, 9 2,10	2,10 2,10	2,10 2,10	2,10 1,11	1,11 1,11	1,11 1,11
30	13	13	5, 7 4, 7	3, 8 3, 8	3, 8 3, 9	2, 9 2, 9	2, 9 2,10	2,10 1,10	1,10 1,10	1,10 1,11	1,11 1,11
30	13	14	5, 7 5, 8	4, 8 4, 9	3, 9 3, 9	3, 9 3,10	2,10 2,10	2,10 2,10	2,11 2,11	1,11 1,11	1,11 1,11
30	13	15	6, 7 5, 8	4, 9 4, 9	4, 9 3,10	3,10 3,10	3,10 2,11	2,11 2,11	2,11 2,11	2,11 1,12	1,12 1,12
30	14	14	6, 8 5, 8	4, 9 4, 9	4, 9 4,10	3,10 3,10	3,10 3,11	2,11 2,11	2,11 2,11	2,11 2,12	1,12 1,12
30	14	15	6, 8 5, 9	5, 9 4,10	4,10 4,10	4,10 3,11	3,11 3,11	3,11 3,11	3,11 2,12	2,12 2,12	2,12 2,12
30	15	15	7, 8 6, 9	5,10 5,10	5,10 4,11	4,11 4,11	4,11 3,12	3,12 3,12	3,12 3,12	3,12 2,13	2,13 2,13

TABLE B.28 CRITICAL VALUES FOR THE RUNS TEST

		α(2): 0.50	0.20	0.10	0.05	0.02	0.01	0.005	0.002	0.001
n_1	n_2	α(1): 0.25	0.10	0.05	0.025	0.01	0.005	0.0025	0.001	0.0005
2	3	2, 4	-, 5	-, -	-, -	-, -	-, -	-, -	-, -	-, -
	4	2, 5	-, -	-, -	-, -	-, -	-, -	-, -	-, -	-, -
	5	2, -	2, -	-, -	-, -	-, -	-, -	-, -	-, -	-, -
	6	2, -	2, -	-, -	-, -	-, -	-, -	-, -	-, -	-, -
	7	3, -	2, -	-, -	-, -	-, -	-, -	-, -	-, -	-, -
	8	3, -	2, -	2, -	-, -	-, -	-, -	-, -	-, -	-, -
	9	3, -	2, -	2, -	-, -	-, -	-, -	-, -	-, -	-, -
	10	3, -	2, -	2, -	-, -	-, -	-, -	-, -	-, -	-, -
	11	3, -	2, -	2, -	-, -	-, -	-, -	-, -	-, -	-, -
	12	3, -	2, -	2, -	2, -	-, -	-, -	-, -	-, -	-, -
	13	3, -	2, -	2, -	2, -	-, -	-, -	-, -	-, -	-, -
	14	3, -	2, -	2, -	2, -	-, -	-, -	-, -	-, -	-, -
	15	3, -	2, -	2, -	2, -	-, -	-, -	-, -	-, -	-, -
	16	3, -	2, -	2, -	2, -	-, -	-, -	-, -	-, -	-, -
	17	3, -	2, -	2, -	2, -	-, -	-, -	-, -	-, -	-, -
	18	3, -	2, -	2, -	2, -	-, -	-, -	-, -	-, -	-, -
	19	3, -	3, -	2, -	2, -	2, -	-, -	-, -	-, -	-, -
	20	3, -	3, -	2, -	2, -	2, -	-, -	-, -	-, -	-, -
	21	4, -	3, -	2, -	2, -	2, -	-, -	-, -	-, -	-, -
	22	4, -	3, -	2, -	2, -	2, -	-, -	-, -	-, -	-, -
	23	4, -	3, -	2, -	2, -	2, -	-, -	-, -	-, -	-, -
	24	4, -	3, -	2, -	2, -	2, -	-, -	-, -	-, -	-, -
	25	4, -	3, -	2, -	2, -	2, -	-, -	-, -	-, -	-, -
	26	4, -	3, -	2, -	2, -	2, -	-, -	-, -	-, -	-, -
	27	4, -	3, -	2, -	2, -	2, -	2, -	-, -	-, -	-, -
	28	4, -	3, -	2, -	2, -	2, -	2, -	-, -	-, -	-, -
	29	4, -	3, -	2, -	2, -	2, -	2, -	-, -	-, -	-, -
2	30	4, -	3, -	2, -	2, -	2, -	2, -	-, -	-, -	-, -
3	3	2, 6	2, 6	-, -	-, -	-, -	-, -	-, -	-, -	-, -
	4	3, 6	2, 7	-, 7	-, -	-, -	-, -	-, -	-, -	-, -
	5	3, 7	2, 7	2, -	-, -	-, -	-, -	-, -	-, -	-, -
	6	3, 7	2, -	2, -	2, -	-, -	-, -	-, -	-, -	-, -
	7	3, 7	3, -	2, -	2, -	-, -	-, -	-, -	-, -	-, -
	8	4, 7	3, -	2, -	2, -	-, -	-, -	-, -	-, -	-, -
	9	4, -	3, -	2, -	2, -	2, -	-, -	-, -	-, -	-, -
	10	4, -	3, -	3, -	2, -	2, -	-, -	-, -	-, -	-, -
	11	4, -	3, -	3, -	2, -	2, -	-, -	-, -	-, -	-, -
	12	4, -	3, -	3, -	2, -	2, -	2, -	-, -	-, -	-, -
	13	4, -	3, -	3, -	2, -	2, -	2, -	-, -	-, -	-, -
	14	4, -	3, -	3, -	3, -	2, -	2, -	-, -	-, -	-, -
	15	4, -	4, -	3, -	3, -	2, -	2, -	2, -	-, -	-, -
	16	4, -	4, -	3, -	3, -	2, -	2, -	2, -	-, -	-, -
	17	4, -	4, -	3, -	3, -	2, -	2, -	2, -	-, -	-, -
	18	4, -	4, -	3, -	3, -	2, -	2, -	2, -	-, -	-, -
	19	4, -	4, -	3, -	3, -	2, -	2, -	2, -	-, -	-, -
	20	4, -	4, -	3, -	3, -	2, -	2, -	2, -	-, -	-, -
	21	5, -	4, -	3, -	3, -	2, -	2, -	2, -	2, -	-, -
	22	5, -	4, -	4, -	3, -	2, -	2, -	2, -	2, -	-, -
	23	5, -	4, -	4, -	3, -	3, -	2, -	2, -	2, -	-, -
3	24	5, -	4, -	4, -	3, -	3, -	2, -	2, -	2, -	-, -
3	25	5, -	4, -	4, -	3, -	3, -	2, -	2, -	2, -	-, -
	26	5, -	4, -	4, -	3, -	3, -	2, -	2, -	2, -	-, -
	27	5, -	4, -	4, -	3, -	3, -	2, -	2, -	2, -	2, -
	28	5, -	4, -	4, -	3, -	3, -	2, -	2, -	2, -	2, -
	29	5, -	4, -	4, -	3, -	3, -	2, -	2, -	2, -	2, -
3	30	5, -	4, -	4, -	3, -	3, -	2, -	2, -	2, -	2, -
4	4	3, 7	2, 8	2, 8	-, -	-, -	-, -	-, -	-, -	-, -
	5	3, 6	3, 8	2, 9	2, 9	-, 9	-, -	-, -	-, -	-, -
	6	4, 8	3, 9	3, 9	2, 9	2, -	-, -	-, -	-, -	-, -
	7	4, 8	3, 9	3, 9	2, -	2, -	-, -	-, -	-, -	-, -
	8	4, 8	3, 9	3, -	3, -	2, -	2, -	-, -	-, -	-, -
	9	5, 9	4, 9	3, -	3, -	2, -	2, -	-, -	-, -	-, -
	10	5, 9	4, -	3, -	3, -	2, -	2, -	2, -	-, -	-, -
	11	5, 9	4, -	3, -	3, -	2, -	2, -	2, -	-, -	-, -
	12	5, 9	4, -	4, -	3, -	3, -	2, -	2, -	-, -	-, -
	13	5, 9	4, -	4, -	3, -	3, -	2, -	2, -	2, -	-, -

TABLE B.28 (cont.) CRITICAL VALUES FOR THE RUNS TEST

n_1	n_2	$\alpha(2)$: 0.50 $\alpha(1)$: 0.25	0.20 0.10	0.10 0.05	0.05 0.025	0.02 0.01	0.01 0.005	0.005 0.0025	0.002 0.001	0.001 0.0005
	14	5, 9	4, -	4, -	3, -	3, -	2, -	2, -	2, -	-, -
	15	6, -	4, -	4, -	3, -	3, -	3, -	2, -	2, -	-, -
	16	6, -	5, -	4, -	4, -	3, -	3, -	2, -	2, -	2, -
	17	6, -	5, -	4, -	4, -	3, -	3, -	2, -	2, -	2, -
	18	6, -	5, -	4, -	4, -	3, -	3, -	2, -	2, -	2, -
	19	6, -	5, -	4, -	4, -	3, -	3, -	2, -	2, -	2, -
	20	6, -	5, -	4, -	4, -	3, -	3, -	3, -	2, -	2, -
	21	6, -	5, -	4, -	4, -	3, -	3, -	3, -	2, -	2, -
	22	6, -	5, -	4, -	4, -	3, -	3, -	3, -	2, -	2, -
	23	6, -	5, -	4, -	4, -	4, -	3, -	3, -	2, -	2, -
	24	6, -	5, -	5, -	4, -	4, -	3, -	3, -	2, -	2, -
	25	6, -	5, -	5, -	4, -	4, -	3, -	3, -	2, -	2, -
	26	6, -	5, -	5, -	4, -	4, -	3, -	3, -	2, -	2, -
	27	6, -	5, -	5, -	4, -	4, -	3, -	3, -	3, -	2, -
	28	6, -	6, -	5, -	4, -	4, -	3, -	3, -	3, -	2, -
	29	6, -	6, -	5, -	4, -	4, -	4, -	3, -	3, -	2, -
4	30	6, -	6, -	5, -	4, -	4, -	4, -	3, -	3, -	2, -
5	5	4, 8	3, 9	3, 9	2,10	2,10	-, -	-, -	-, -	-, -
	6	4, 9	3, 9	3,10	3,10	2,11	2,11	-,11	-, -	-, -
	7	5, 9	4,10	3,10	3,11	2,11	2, -	-, -	-, -	-, -
	8	5, 9	4,10	3,11	3,11	2, -	2, -	2, -	-, -	-, -
	9	5,10	4,10	4,11	3, -	3, -	2, -	2, -	2, -	-, -
	10	6,10	5,11	4,11	3, -	3, -	3, -	2, -	2, -	-, -
	11	6,10	5,11	4, -	4, -	3, -	3, -	2, -	2, -	2, -
	12	6,10	5,11	4, -	4, -	3, -	3, -	2, -	2, -	2, -
	13	6,10	5,11	4, -	4, -	3, -	3, -	3, -	2, -	2, -
	14	6,10	5, -	5, -	4, -	3, -	3, -	3, -	2, -	2, -
	15	6,11	5, -	5, -	4, -	4, -	3, -	3, -	2, -	2, -
	16	7,11	6, -	5, -	4, -	4, -	3, -	3, -	2, -	2, -
	17	7,11	6, -	5, -	4, -	4, -	3, -	3, -	3, -	2, -
	18	7,11	6, -	5, -	5, -	4, -	4, -	3, -	3, -	2, -
	19	7,11	6, -	5, -	5, -	4, -	4, -	3, -	3, -	2, -
	20	7,11	6, -	5, -	5, -	4, -	4, -	3, -	3, -	3, -
	21	7,11	6, -	5, -	5, -	4, -	4, -	3, -	3, -	3, -
5	22	7, -	6, -	6, -	5, -	4, -	4, -	4, -	3, -	3, -
5	23	7, -	6, -	6, -	5, -	4, -	4, -	4, -	3, -	3, -
	24	7, -	6, -	6, -	5, -	4, -	4, -	4, -	3, -	3, -
	25	8, -	6, -	6, -	5, -	4, -	4, -	4, -	3, -	3, -
	26	8, -	6, -	6, -	5, -	5, -	4, -	4, -	3, -	3, -
	27	8, -	6, -	6, -	5, -	5, -	4, -	4, -	3, -	3, -
	28	8, -	6, -	6, -	5, -	5, -	4, -	4, -	3, -	3, -
	29	8, -	7, -	6, -	6, -	5, -	4, -	4, -	4, -	3, -
5	30	8, -	7, -	6, -	6, -	5, -	4, -	4, -	4, -	3, -
6	6	5, 9	4,10	3,11	3,11	2,12	2,12	2,12	-, -	-, -
	7	5, 8	4,11	4,11	3,12	3,12	2,13	2,13	-,13	-, -
	8	6,10	5,11	4,12	3,12	3,13	3,13	2,13	2, -	-, -
	9	6,10	5,11	4,12	4,13	3,13	3, -	2, -	2, -	2, -
	10	6,11	5,12	5,12	4,13	3, -	3, -	3, -	2, -	2, -
	11	7,11	5,12	5,13	4,13	4, -	3, -	3, -	2, -	2, -
	12	7,11	6,12	5,13	4,13	4, -	3, -	3, -	3, -	2, -
	13	7,12	6,12	5,13	5, -	4, -	3, -	3, -	3, -	2, -
	14	7,12	6,13	5,13	5, -	4, -	4, -	3, -	3, -	2, -
	15	7,12	6,13	6, -	5, -	4, -	4, -	3, -	3, -	3, -
	16	8,12	6,13	6, -	5, -	4, -	4, -	4, -	3, -	3, -
	17	8,12	6,13	6, -	5, -	5, -	4, -	4, -	3, -	3, -
	18	8,12	7,13	6, -	5, -	5, -	4, -	4, -	3, -	3, -
	19	8,12	7, -	6, -	6, -	5, -	4, -	4, -	3, -	3, -
	20	8,12	7, -	6, -	6, -	5, -	4, -	4, -	4, -	3, -
	21	8,12	7, -	6, -	6, -	5, -	5, -	4, -	4, -	3, -
	22	8,13	7, -	6, -	6, -	5, -	5, -	4, -	4, -	3, -
	23	8,13	7, -	6, -	6, -	5, -	5, -	4, -	4, -	3, -
	24	8,13	7, -	7, -	6, -	5, -	5, -	4, -	4, -	4, -
	25	8,13	8, -	7, -	6, -	5, -	5, -	4, -	4, -	4, -
	26	9,13	8, -	7, -	6, -	6, -	5, -	5, -	4, -	4, -
	27	9,13	8, -	7, -	6, -	6, -	5, -	5, -	4, -	4, -

TABLE B.28 (cont.) CRITICAL VALUES FOR THE RUNS TEST

n_1	n_2	$\alpha(2)$: 0.50 $\alpha(1)$: 0.25	0.20 0.10	0.10 0.05	0.05 0.025	0.02 0.01	0.01 0.005	0.005 0.0025	0.002 0.001	0.001 0.0005
	28	9,13	8, -	7, -	6, -	6, -	5, -	5, -	4, -	4, -
	29	9,13	8, -	7, -	6, -	6, -	5, -	5, -	4, -	4, -
6	30	9,13	8, -	7, -	6, -	6, -	5, -	5, -	4, -	4, -
7	7	6,10	5,11	4,12	3,13	3,13	3,13	2,14	2,14	-, -
	8	6,11	5,12	4,13	4,13	3,14	3,14	3,14	2,15	2,15
	9	7,11	5,12	5,13	4,14	4,14	3,15	3,15	2,15	2, -
	10	7,12	6,13	5,13	5,14	4,15	3,15	3,15	3, -	2, -
	11	7,12	6,13	5,14	5,14	4,15	4,15	3, -	3, -	2, -
	12	8,12	6,13	6,14	5,14	4,15	4, -	3, -	3, -	3, -
	13	8,12	7,14	6,14	5,15	5, -	4, -	4, -	3, -	3, -
	14	8,13	7,14	6,14	5,15	5, -	4, -	4, -	3, -	3, -
	15	8,13	7,14	6,15	6,15	5, -	4, -	4, -	3, -	3, -
	16	8,13	7,14	6,15	6, -	5, -	5, -	4, -	4, -	3, -
	17	9,13	7,14	7,15	6, -	5, -	5, -	4, -	4, -	3, -
	18	9,14	8,14	7,15	6, -	5, -	5, -	4, -	4, -	4, -
	19	9,14	8,15	7,15	6, -	6, -	5, -	5, -	4, -	4, -
	20	9,14	8,15	7, -	6, -	6, -	5, -	5, -	4, -	4, -
	21	9,14	8,15	7, -	7, -	6, -	5, -	5, -	4, -	4, -
	22	9,14	8,15	7, -	7, -	6, -	5, -	5, -	4, -	4, -
7	23	10,14	8.15	8, -	7, -	6, -	6, -	5, -	5, -	4. -
7	24	10,14	8,15	8, -	7, -	6, -	6, -	5, -	5, -	4, -
	25	10,14	8, -	8, -	7, -	6, -	6, -	5, -	5, -	4, -
	26	10,14	8, -	8, -	7, -	6, -	6, -	5, -	5, -	4, -
	27	10,14	9, -	8, -	7, -	6, -	6, -	6, -	5, -	5, -
	28	10,14	9, -	8, -	7, -	7, -	6, -	6, -	5, -	5, -
	29	10,14	9, -	8, -	8, -	7, -	6, -	6, -	5, -	5, -
7	30	10,14	9, -	8, -	8, -	7, -	6, -	6, -	5, -	5, -
8	8	7,11	5,13	5,13	4,14	4,14	3,15	3,15	2,16	2,16
	9	7,10	6,13	5,14	5,14	4,15	3,15	3,16	3,16	2,17
	10	7,12	6,13	6,14	5,15	4,15	4,16	3,16	3,17	3,17
	11	8,13	7,14	6,15	5,15	5,16	4,16	4,17	3,17	3, -
	12	8,13	7,14	6,15	6,16	5,16	4,17	4,17	3, -	3, -
	13	8,13	7,15	6,15	6,16	5,17	5,17	4,17	4, -	3, -
	14	9,14	7,15	7,16	6,16	5,17	5,17	4, -	4, -	3, -
	15	9,14	8,15	7,16	6,16	5,17	5, -	5, -	4, -	4, -
	16	9,14	8,15	7,16	6,17	6,17	5, -	5, -	4, -	4, -
	17	9,14	8,16	7,16	7,17	6, -	5, -	5, -	4, -	4, -
	18	10,14	8,16	8,16	7,17	6, -	6, -	5, -	4, -	4, -
	20	10,15	9,16	8,17	7,17	6, -	6, -	5, -	5, -	4, -
	21	10,15	9,16	8,17	7, -	7, -	6, -	6, -	5, -	5, -
	22	10,15	9,16	8,17	8, -	7, -	6, -	6, -	5, -	5, -
	23	10,15	9,16	8,17	8, -	7, -	6, -	6, -	5, -	5, -
	24	11,16	9,16	8,17	8, -	7, -	6, -	6, -	5, -	5, -
	25	11,16	9,17	9, -	8, -	7, -	7, -	6, -	5, -	5, -
	26	11,16	10,17	9, -	8, -	7, -	7, -	6, -	6, -	5, -
	27	11,16	10,17	9, -	8, -	7, -	7, -	6, -	6, -	5, -
	28	11,16	10,17	9, -	8, -	8, -	7, -	6, -	6, -	5, -
	29	11,16	10,17	9, -	8, -	8, -	7, -	6, -	6, -	5, -
8	30	11,16	10,17	9, -	8, -	8, -	7, -	7, -	6, -	6, -
9	9	8,12	6,14	6,14	5,15	4,16	4,16	3,17	3,17	3,17
	10	8,13	7,14	6,15	5,16	5,16	4,17	4,17	3,18	3,18
	11	8,13	7,15	6,15	6,16	5,17	5,17	4,18	3,18	3,19
	12	9,14	7,15	7,16	6,16	5,17	5,18	4,18	4,19	3,19
	13	9,14	8,15	7,16	6,17	6,18	5,18	5,18	4,19	4,19
	14	9,14	8,16	7,17	7,17	6,18	5,18	5,19	4,19	4, -
	15	10,15	8,16	8,17	7,18	6,18	6,19	5,19	4, -	4, -
	16	10,15	9,16	8,17	7,18	6,18	6,19	5,19	5, -	4, -
	17	10,15	9,17	8,17	7,18	7,19	6,19	5, -	5, -	4, -
	18	10,16	9,17	8,18	8,18	7,19	6, -	6, -	5, -	5, -
	19	11,16	9,17	8,18	8,18	7,19	6, -	6, -	5, -	5, -
	20	11,16	10,17	9,18	8,18	7,19	7, -	6, -	5, -	5, -
	21	11,16	10,18	9,18	8,19	7, -	7, -	6, -	6, -	5, -
	22	11,16	10,18	9,18	8,19	7, -	7, -	6, -	6, -	5, -
	23	12,16	10,18	9,18	8,19	8, -	7, -	6, -	6, -	5, -

TABLE B.28 (cont.) CRITICAL VALUES FOR THE RUNS TEST

n_1	n_2	$\alpha(2)$: 0.50 $\alpha(1)$: 0.25	0.20 0.10	0.10 0.05	0.05 0.025	0.02 0.01	0.01 0.005	0.005 0.0025	0.002 0.001	0.001 0.0005
	24	12,17	10,18	9,18	9,19	8, -	7, -	7, -	6, -	6, -
	25	12,17	10,18	10,19	9,19	8, -	7, -	7, -	6, -	6, -
	26	12,17	10,18	10,19	9, -	8, -	7, -	7, -	6, -	6, -
	27	12,17	11,18	10,19	9, -	8, -	8, -	7, -	6, -	6, -
	28	12,17	11,18	10,19	9, -	8, -	8, -	7, -	6, -	6, -
	29	12,17	11,18	10,19	9, -	8, -	8, -	7, -	7, -	6, -
9	30	12,18	11,18	10,19	9, -	8, -	8, -	7, -	7, -	6, -
10	10	9,13	7,15	6,16	6,16	5,17	5,17	4,18	4,18	3,19
	11	9,12	8,15	7,16	6,17	5,18	5,18	4,19	4,19	3,19
	12	9,14	8,16	7,17	7,17	6,18	5,19	5,19	4,20	4,20
	13	10,15	8,16	8,17	7,18	6,19	5,19	5,20	4,20	4,20
	14	10,15	9,17	8,17	7,18	6,19	6,19	5,20	5,20	4,21
	15	10,16	9,17	8,18	7,18	7,19	6,20	6,20	5,21	5,21
	16	11,16	9,17	8,18	8,19	7,20	6,20	6,20	5,21	5, -
	17	11,16	10,18	9,18	8,19	7,20	6,20	6,20	5,21	5, -
	18	11,16	10,18	9,19	8,19	7,20	7,21	6,21	6, -	5, -
	19	12,17	10,18	9,19	8,20	8,20	7,21	6,21	6, -	5, -
	20	12,17	10,18	9,19	9,20	8,20	7,21	7, -	6, -	6, -
	21	12,17	10,18	10,19	9,20	8,21	7,21	7, -	6, -	6, -
	22	12,17	11,19	10,20	9,20	8,21	8, -	7, -	6, -	6, -
	23	12,18	11,19	10,20	9,20	8,21	8, -	7, -	6, -	6, -
	24	12,18	11,19	10,20	9,20	8,21	8, -	7, -	7, -	6, -
	25	13,18	11,19	10,20	10,20	9, -	8, -	7, -	7, -	6, -
	26	13,18	11,20	10,20	10,21	9, -	8, -	8, -	7, -	6, -
	27	13,18	12,20	11,20	10,21	9, -	8, -	8, -	7, -	7, -
	28	13,18	12,20	11,20	10,21	9, -	8, -	8, -	7, -	7, -
	29	13,18	12,20	11,20	10,21	9, -	9, -	8, -	7, -	7, -
10	30	14,18	12,20	11,20	10,21	9, -	9, -	8, -	8, -	7, -
11	11	9,15	8,16	7,17	7,17	6,18	5,19	5,19	4,20	4,20
	12	10,15	9,16	8,17	7,18	6,19	6,19	5,20	5,20	4,21
	13	10,16	9,17	8,18	7,19	6,19	6,20	5,20	5,21	4,21
	14	11,16	9,17	8,18	8,19	7,20	6,20	6,21	5,21	5,22
	15	11,16	10,18	9,19	8,19	7,20	7,21	6,21	5,22	5,22
	16	11,17	10,18	9,19	8,20	7,21	7,21	6,22	6,22	5,23
	17	12,17	10,18	9,19	9,20	8,21	7,22	7,22	6,22	5,23
	18	12,17	10,19	10,20	9,20	8,21	7,22	7,22	6,23	6,23
	19	12,18	11,19	10,20	9,21	8,22	8,22	7,22	6,23	6, -
	20	12,18	11,19	10,20	9,21	8,22	8,22	7,23	7,23	6, -
	21	13,18	11,20	10,20	10,21	9,22	8,22	7,23	7, -	6, -
	22	13,18	11,20	10,21	10,22	9,22	8,23	8,23	7, -	6, -
	23	13,19	12,20	11,21	10,22	9,22	8,23	8,23	7, -	7, -
	24	13,19	12,20	11,21	10,22	9,22	9,23	8, -	7, -	7, -
	25	14,19	12,20	11,21	10,22	9,23	9,23	8, -	7, -	7, -
	26	14,19	12,21	11,22	10,22	10,23	9, -	8, -	8, -	7, -
	27	14,19	12,21	11,22	11,22	10,23	9, -	8, -	8, -	7, -
	28	14,20	13,21	12,22	11,22	10,23	9, -	9, -	8, -	7, -
	29	14,20	13,21	12,22	11,22	10,23	9, -	9, -	8, -	8, -
11	30	14,20	13,21	12,22	11,22	10, -	10, -	9, -	8, -	8, -
12	12	10,16	9,17	8,18	7,19	7,19	6,20	5,21	5,21	4,22
	13	11,14	9,18	9,18	8,19	7,20	6,21	6,21	5,22	5,22
	14	11,17	10,18	9,19	8,20	7,21	7,21	6,22	5,22	5,23
	15	12,17	10,19	9,19	8,20	8,21	7,22	6,22	6,23	5,23
	16	12,17	10,19	10,20	9,21	8,22	7,22	7,23	6,23	6,24
	17	12,18	11,19	10,20	9,21	8,22	8,22	7,23	6,24	6,24
	18	13,18	11,20	10,21	9,21	8,22	8,23	7,23	7,24	6,24
	19	13,18	11,20	10,21	10,22	9,23	8,23	7,24	7,24	6,25
12	20	13,19	12,20	11,21	10,22	9,23	8,23	8,24	7,24	7,25
12	21	14,19	12,21	11,22	10,22	9,23	9,24	8,24	7,25	7,25
	22	14,19	12,21	11,22	10,22	9,23	9,24	8,24	7,25	7, -
	23	14,20	12,21	11,22	11,23	10,24	9,24	8,24	8,25	7, -
	24	14,20	13,21	12,22	11,23	10,24	9,24	9,25	8, -	7, -
	25	14,20	13,22	12,22	11,23	10,24	9,24	9,25	8, -	8, -
	26	15,20	13,22	12,23	11,23	10,24	10,25	9,25	8, -	8, -
	27	15,20	13,22	12,23	11,24	10,24	10,25	9,25	8, -	8, -
	28	15,21	13,22	12,23	12,24	11,24	10,25	9, -	9, -	8, -
	29	15,21	14,22	13,23	12,24	11,24	10,25	10, -	9, -	8, -
12	30	15,21	14,22	13,23	12,24	11,25	10,25	10, -	9, -	8, -

TABLE B.28 (cont.) CRITICAL VALUES FOR THE RUNS TEST

n_1	n_2	$\alpha(2)$: 0.50	0.20	0.10	0.05	0.02	0.01	0.005	0.002	0.001
		$\alpha(1)$: 0.25	0.10	0.05	0.025	0.01	0.005	0.0025	0.001	0.0005
13	13	11,17	10,18	9,19	8,20	7,21	7,21	6,22	5,23	5,23
	14	12,17	10,19	9,20	9,20	8,21	7,22	7,22	6,23	5,24
	15	12,18	11,19	10,20	9,21	8,22	7,22	7,23	6,24	6,24
	16	13,18	11,20	10,21	9,21	8,22	8,23	7,23	6,24	6,25
	17	13,19	11,20	10,21	10,22	9,23	8,23	7,24	7,25	6,25
	18	13,19	12,20	11,21	10,22	9,23	8,24	8,24	7,25	7,25
	19	14,19	12,21	11,22	10,23	9,24	9,24	8,25	7,25	7,26
	20	14,20	12,21	11,22	10,23	10,24	9,24	8,25	8,26	7,26
	21	14,20	13,22	12,22	11,23	10,24	9,25	9,25	8,26	7,26
	22	15,20	13,22	12,23	11,24	10,24	9,25	9,26	8,26	7,27
	23	15,20	13,22	12,23	11,24	10,25	10,25	9,26	8,26	8,27
	24	15,21	13,22	12,23	11,24	10,25	10,26	9,26	8,27	8,27
	25	15,21	14,23	13,24	12,24	11,25	10,26	9,26	9,27	8,27
	26	16,21	14,23	13,24	12,24	11,26	10,26	10,26	9,27	8, -
	27	16,21	14,23	13,24	12,25	11,26	10,26	10,26	9,27	9, -
	28	16,22	14,23	13,24	12,25	11,26	11,26	10,27	9, -	9, -
	29	16,22	14,24	13,24	13,25	12,26	11,26	10,27	9, -	9, -
13	30	16,22	15,24	14,24	13,25	12,26	11,26	10,27	10, -	9, -
14	14	12,18	11,19	10,20	9,21	8,22	7,23	7,23	6,24	6,24
	15	13,16	11,20	10,21	9,22	8,23	8,23	7,24	7,24	6,25
	16	13,19	11,20	11,21	10,22	9,23	8,24	8,24	7,25	6,25
	17	14,19	12,21	11,22	10,23	9,24	8,24	8,25	7,25	7,26
	18	14,20	12,21	11,22	10,23	9,24	9,25	8,25	7,26	7,26
	19	14,20	13,22	12,23	11,23	10,24	9,25	8,26	8,26	7,27
	20	15,20	13,22	12,23	11,24	10,25	9,25	9,26	8,27	7,27
	21	15,21	13,22	12,23	11,24	10,25	10,26	9,26	8,27	8,28
	22	15,21	14,23	12,24	12,24	11,26	10,26	9,27	9,27	8,28
	23	16,21	14,23	13,24	12,25	11,26	10,26	10,27	9,28	8,28
	24	16,22	14,23	13,24	12,25	11,26	10,27	10,27	9,28	8,28
	25	16,22	14,24	13,24	12,25	11,26	11,27	10,28	9,28	9,28
	26	16,22	15,24	14,25	13,26	12,26	11,27	10,28	9,28	9,29
	27	17,22	15,24	14,25	13,26	12,27	11,27	10,28	10,28	9,29
	28	17,23	15,24	14,25	13,26	12,27	11,28	11,28	10,29	9,29
	29	17,23	15,24	14,26	13,26	12,27	12,28	11,28	10,29	9, -
14	30	17,23	15,25	14,26	13,26	12,27	12,28	11,28	10,29	10, -
15	15	13,19	12,20	11,21	10,22	9,23	8,24	8,24	7,25	6,26
	16	14,19	12,21	11,22	10,23	9,24	9,24	8,25	7,26	7,26
	17	14,20	12,21	11,22	11,23	10,24	9,25	8,26	8,26	7,27
	18	14,20	13,22	12,23	11,24	10,25	9,25	9,26	8,27	7,27
15	19	15,21	13,22	12,23	11,24	10,25	10,26	9,27	8,27	8,28
15	20	15,21	13,23	12,24	12,25	11,26	10,26	9,27	8,28	8,28
	21	16,21	14,23	13,24	12,25	11,26	10,27	10,27	9,28	8,29
	22	16,22	14,24	13,25	12,25	11,26	10,27	10,28	9,28	8,29
	23	16,22	14,24	13,25	12,26	11,27	11,27	10,28	9,29	9,29
	24	16,22	15,24	14,25	13,26	12,27	11,28	10,28	10,29	9,30
	25	17,23	15,24	14,26	13,26	12,27	11,28	11,29	10,29	9,30
	26	17,23	15,25	14,26	13,27	12,28	11,28	11,29	10,30	9,30
	27	17,23	16,25	14,26	14,27	12,28	12,28	11,29	10,30	10,30
	28	18,24	16,25	15,26	14,27	13,28	12,29	11,29	10,30	10,30
	29	18,24	16,26	15,26	14,27	13,28	12,29	11,30	11,30	10,31
15	30	18,24	16,26	15,27	14,28	13,28	12,29	12,30	11,30	10,31
16	16	14,20	12,22	11,23	11,23	10,24	9,25	8,26	8,26	7,27
	17	15,18	13,22	12,23	11,24	10,25	9,26	9,26	8,27	7,28
	18	15,21	13,23	12,24	11,25	10,26	10,26	9,27	8,28	8,28
	19	15,21	14,23	13,24	12,25	11,26	10,27	9,27	9,28	8,29
	20	16,22	14,24	13,25	12,25	11,26	10,27	10,28	9,29	8,29
	21	16,22	14,24	13,25	12,26	11,27	11,28	10,28	9,29	9,30
	22	17,23	15,24	14,25	13,26	12,27	11,28	10,29	9,29	9,30
	23	17,23	15,25	14,26	13,27	12,28	11,28	11,29	10,30	9,30
	24	17,23	15,25	14,26	13,27	12,28	12,29	11,29	10,30	9,31
	25	17,24	16,25	15,26	14,27	13,28	12,29	11,30	10,30	10,31
	26	18,24	16,26	15,27	14,28	13,29	12,29	11,30	11,31	10,31
	27	18,24	16,26	15,27	14,28	13,29	12,30	12,30	11,31	10,32
	28	18,24	16,26	15,27	14,28	13,29	13,30	12,30	11,31	10,32
	29	19,25	17,26	16,28	15,28	14,30	13,30	12,31	11,32	11,32
16	30	19,25	17,27	16,28	15,29	14,30	13,30	12,31	11,32	11,32

TABLE B.28 (cont.) CRITICAL VALUES FOR THE RUNS TEST

n_1	n_2	$\alpha(2)$: 0.50 $\alpha(1)$: 0.25	0.20 0.10	0.10 0.05	0.05 0.025	0.02 0.01	0.01 0.005	0.005 0.0025	0.002 0.001	0.001 0.0005
17	17	15,21	13,23	12,24	11,25	10,26	10,26	9,27	8,28	8,28
	18	16,21	14,23	13,24	12,25	11,26	10,27	9,27	9,28	8,29
	19	16,22	14,24	13,25	12,26	11,27	10,27	10,28	9,29	8,29
	20	16,22	15,24	13,25	13,26	11,27	11,28	10,29	9,29	9,30
	21	17,23	15,25	14,26	13,27	12,28	11,28	10,29	10,30	9,30
	22	17,23	15,25	14,26	13,27	12,28	11,29	11,30	10,30	9,31
	23	17,24	16,25	15,27	14,27	13,29	12,29	11,30	10,31	10,31
	24	18,24	16,26	15,27	14,28	13,29	12,30	11,30	11,31	10,32
	25	18,24	16,26	15,27	14,28	13,29	12,30	12,31	11,31	10,32
	26	18,25	17,26	15,28	14,29	13,30	13,30	12,31	11,32	10,32
	27	19,25	17,27	16,28	15,29	14,30	13,31	12,31	11,32	11,33
	28	19,25	17,27	16,28	15,29	14,30	13,31	12,32	12,32	11,33
	29	19,26	17,27	16,28	15,29	14,30	13,31	13,32	12,33	11,33
17	30	20,26	18,28	17,29	16,30	14,31	14,32	13,32	12,33	11,33
18	18	16,22	14,24	13,25	12,26	11,27	11,27	10,28	9,29	9,29
	19	16,20	15,24	14,25	13,26	12,27	11,28	10,29	9,30	9,30
	20	17,23	15,25	14,26	13,27	12,28	11,29	11,29	10,30	9,31
	21	17,23	15,25	14,26	13,27	12,28	12,29	11,30	10,31	10,31
	22	18,24	16,26	15,27	14,28	13,29	12,30	11,30	10,31	10,32
	23	18,24	16,26	15,27	14,28	13,29	12,30	12,31	11,32	10,32
	24	18,25	17,27	15,28	14,29	13,30	13,30	12,31	11,32	10,33
	25	19,25	17,27	16,28	15,29	14,30	13,31	12,32	11,32	11,33
	26	19,25	17,27	16,28	15,29	14,30	13,31	12,32	12,33	11,33
18	27	19,26	18,28	16,29	15,30	14,31	13,32	13,32	12,33	11,34
18	28	20,26	18,28	17,29	16,30	14,31	14,32	13,33	12,33	12,34
	29	20,26	18,28	17,29	16,30	15,32	14,32	13,33	12,34	12,34
18	30	20,27	18,29	17,30	16,31	15,32	14,32	14,33	13,34	12,34
19	19	17,23	15,25	14,26	13,27	12,28	11,29	11,29	10,30	9,31
	20	17,24	16,25	14,27	13,27	12,29	12,29	11,30	10,31	10,31
	21	18,24	16,26	15,27	14,28	13,29	12,30	11,31	11,31	10,32
	22	18,25	16,26	15,28	14,29	13,30	12,30	12,31	11,32	10,32
	23	19,25	17,27	16,28	15,29	13,30	13,31	12,32	11,32	11,33
	24	19,25	17,27	16,28	15,29	14,31	13,31	12,32	11,33	11,33
	25	19,26	17,28	16,29	15,30	14,31	13,32	13,32	12,33	11,34
	26	20,26	18,28	17,29	16,30	14,31	14,32	13,33	12,34	11,34
	27	20,26	18,28	17,30	16,31	15,32	14,32	13,33	12,34	12,35
	28	20,27	18,29	17,30	16,31	15,32	14,33	14,34	13,34	12,35
	29	21,27	19,29	18,30	17,31	15,32	15,33	14,34	13,35	12,35
19	30	21,28	19,29	18,31	17,32	16,33	15,34	14,34	13,35	13,36
20	20	18,24	16,26	15,27	14,28	13,29	12,30	11,31	11,31	10,32
	21	18,22	16,27	15,28	14,29	13,30	12,31	12,31	11,32	10,33
	22	19,25	17,27	16,28	15,29	14,30	13,31	12,32	11,33	11,33
	23	19,26	17,28	16,29	15,30	14,31	13,32	12,32	12,33	11,34
	24	20,26	18,28	16,29	15,30	14,31	14,32	13,33	12,34	11,34
	25	20,26	18,28	17,30	16,31	15,32	14,33	13,33	12,34	12,35
	26	20,27	18,29	17,30	16,31	15,32	14,33	13,34	13,35	12,35
	27	21,27	19,29	18,30	17,31	15,33	15,33	14,34	13,35	12,36
	28	21,28	19,30	18,31	17,32	16,33	15,34	14,34	13,35	13,36
	29	21,28	19,30	18,31	17,32	16,33	15,34	14,35	13,36	13,36
20	30	22,28	20,30	18,32	17,32	16,34	15,34	15,35	14,36	13,37
21	21	19,25	17,27	16,28	15,29	14,30	13,31	12,32	11,33	11,33
	22	19,26	17,28	16,29	15,30	14,31	13,32	13,32	12,33	11,34
	23	20,26	18,28	17,29	16,30	14,31	14,32	13,33	12,34	11,35
	24	20,27	18,29	17,30	16,31	15,32	14,33	13,34	12,34	12,35
	25	21,27	19,29	17,30	16,31	15,32	14,33	14,34	13,35	12,36
	26	21,27	19,30	18,31	17,32	15,33	15,34	14,34	13,35	12,36
	27	21,28	19,30	18,31	17,32	16,33	15,34	14,35	13,36	13,36
	28	22,28	20,30	18,32	17,33	16,34	15,35	15,35	14,36	13,37
	29	22,29	20,31	19,32	18,33	16,34	16,35	15,36	14,37	13,37
21	30	22,29	20,31	19,32	18,33	17,35	16,35	15,36	14,37	13,38
22	22	20,26	18,28	17,29	16,30	14,32	14,32	13,33	12,34	11,35
	23	20,24	18,29	17,30	16,31	15,32	14,33	13,34	12,35	12,35
	24	21,27	19,29	17,30	16,31	15,33	14,33	14,34	13,35	12,36
	25	21,28	19,30	18,31	17,32	16,33	15,34	14,35	13,36	13,36
	26	22,28	19,30	18,31	17,32	16,34	15,34	14,35	13,36	13,37

TABLE B.28 (cont.) CRITICAL VALUES FOR THE RUNS TEST

n_1	n_2	$\alpha(2)$: 0.50 $\alpha(1)$: 0.25	0.20 0.10	0.10 0.05	0.05 0.025	0.02 0.01	0.01 0.005	0.005 0.0025	0.002 0.001	0.001 0.0005
	27	22,29	20,31	19,32	18,33	16,34	15,35	15,36	14,37	13,37
	28	22,29	20,31	19,32	18,33	17,35	16,35	15,36	14,37	13,38
	29	23,29	21,31	19,33	18,34	17,35	16,36	15,37	14,38	14,38
22	30	23,30	21,32	20,33	19,34	17,35	16,36	16,37	15,38	14,39
23	23	21,27	19,29	17,31	16,32	15,33	14,34	14,34	13,35	12,36
	24	21,28	19,30	18,31	17,32	16,33	15,34	14,35	13,36	13,36
	25	22,28	20,30	18,32	17,33	16,34	15,35	14,35	14,36	13,37
	26	22,29	20,31	19,32	18,33	16,34	16,35	15,36	14,37	13,38
23	27	23,29	20,31	19,33	18,34	17,35	16,36	15,36	14,37	14,38
23	28	23,30	21,32	20,33	18,34	17,35	16,36	16,37	15,38	14,39
	29	23,30	21,32	20,33	19,35	17,36	17,37	16,37	15,38	14,39
23	30	24,30	21,33	20,34	19,35	18,36	17,37	16,38	15,39	15,39
24	24	22,28	20,30	18,32	17,33	16,34	15,35	15,35	14,36	13,37
	25	22,26	20,31	19,32	18,33	17,34	16,35	15,36	14,37	13,38
	26	23,29	20,31	19,33	18,34	17,35	16,36	15,37	14,38	14,38
	27	23,30	21,32	20,33	19,34	17,36	16,36	16,37	15,38	14,39
	28	23,30	21,32	20,34	19,35	18,36	17,37	16,38	15,39	14,39
	29	24,31	22,33	20,34	19,35	18,36	17,37	16,38	15,39	15,40
24	30	24,31	22,33	21,35	20,36	18,37	17,38	17,39	16,40	15,40
25	25	23,29	21,31	19,33	18,34	17,35	16,36	15,37	14,38	14,38
	26	23,30	21,32	20,33	19,34	17,36	16,37	16,37	15,38	14,39
	27	24,30	21,33	20,34	19,35	18,36	17,37	16,38	15,39	14,39
	28	24,31	22,33	21,34	19,35	18,37	17,38	16,38	15,39	15,40
	29	24,31	22,33	21,35	20,36	18,37	18,38	17,39	16,40	15,41
25	30	25,32	23,34	21,35	20,36	19,38	18,39	17,39	16,40	15,41
26	26	24,30	21,33	20,34	19,35	18,36	17,37	16,38	15,39	14,40
	27	24,28	22,33	21,34	19,36	18,37	17,38	16,39	15,39	15,40
	28	25,31	22,34	21,35	20,36	19,37	18,38	17,39	16,40	15,41
	29	25,32	23,34	21,35	20,37	19,38	18,39	17,40	16,41	16,41
26	30	25,32	23,35	22,36	21,37	19,38	18,39	18,40	17,41	16,42
27	27	25,31	22,34	21,35	20,36	19,37	18,38	17,39	16,40	15,41
	28	25,32	23,34	21,36	20,37	19,38	18,39	17,40	16,41	16,41
	29	25,32	23,35	22,36	21,37	19,39	19,39	18,40	17,41	16,42
27	30	26,33	24,35	22,37	21,38	20,39	19,40	18,41	17,42	16,43
28	28	25,33	23,35	22,36	21,37	19,39	19,39	18,40	17,41	16,42
	29	26,30	24,35	22,37	21,38	20,39	19,40	18,41	17,42	16,43
28	30	26,34	24,36	23,37	22,38	20,40	19,41	18,41	17,42	17,43
29	29	26,34	24,36	23,37	22,38	20,40	19,41	19,41	17,43	17,43
29	30	27,34	25,36	23,38	22,39	21,40	20,41	19,42	18,43	17,44
30	30	27,35	25,37	24,38	23,39	21,41	20,42	19,43	18,44	18,44

This table was prepared using the procedure described by Brownlee (1965: 225–226) and Swed and Eisenhart (1943).

Example:

$$u_{0.05(2),24,30} = 20 \text{ and } 36.$$

TABLE B.29 CRITICAL VALUES OF *C* FOR THE MEAN SQUARE SUCCESSIVE DIFFERENCE TEST

n	α = 0.25	0.10	0.05	0.025	0.01	0.005	0.0025	0.001	0.0005
8	0.223	0.409	0.509	0.587	0.668	0.716	0.756	0.799	0.825
9	0.212	0.391	0.488	0.565	0.645	0.694	0.735	0.779	0.807
10	0.203	0.374	0.469	0.544	0.624	0.673	0.714	0.760	0.788
11	0.194	0.360	0.452	0.526	0.604	0.653	0.694	0.740	0.770
12	0.187	0.347	0.436	0.509	0.586	0.634	0.676	0.722	0.752
13	0.180	0.335	0.422	0.493	0.569	0.617	0.658	0.705	0.735
14	0.174	0.325	0.409	0.478	0.553	0.601	0.642	0.689	0.719
15	0.169	0.315	0.397	0.465	0.539	0.586	0.627	0.673	0.704
16	0.164	0.306	0.386	0.453	0.525	0.572	0.612	0.658	0.689
17	0.159	0.298	0.376	0.441	0.513	0.558	0.599	0.645	0.675
18	0.155	0.290	0.367	0.431	0.501	0.546	0.586	0.632	0.662
19	0.151	0.283	0.358	0.421	0.490	0.535	0.574	0.619	0.650
20	0.148	0.276	0.350	0.412	0.480	0.524	0.563	0.608	0.638
21	0.144	0.270	0.343	0.403	0.470	0.513	0.552	0.597	0.627
22	0.141	0.264	0.335	0.395	0.461	0.504	0.542	0.586	0.616
23	0.138	0.259	0.329	0.387	0.452	0.494	0.532	0.576	0.606
24	0.135	0.254	0.322	0.380	0.444	0.486	0.523	0.566	0.596
25	0.133	0.249	0.316	0.373	0.436	0.477	0.514	0.557	0.587
26	0.130	0.245	0.311	0.366	0.429	0.469	0.506	0.549	0.578
27	0.128	0.240	0.305	0.360	0.422	0.462	0.498	0.540	0.569
28	0.126	0.236	0.300	0.354	0.415	0.455	0.490	0.532	0.561
29	0.123	0.232	0.295	0.349	0.409	0.448	0.483	0.525	0.553
30	0.121	0.229	0.291	0.343	0.402	0.441	0.476	0.517	0.546
31	0.120	0.225	0.286	0.338	0.397	0.435	0.470	0.510	0.538
32	0.118	0.222	0.282	0.333	0.391	0.429	0.463	0.504	0.531
33	0.116	0.218	0.278	0.329	0.386	0.423	0.457	0.497	0.525
34	0.114	0.215	0.274	0.324	0.380	0.418	0.451	0.491	0.518
35	0.113	0.212	0.271	0.320	0.376	0.412	0.446	0.485	0.512
36	0.111	0.210	0.267	0.316	0.371	0.407	0.440	0.479	0.506
37	0.110	0.207	0.264	0.312	0.366	0.402	0.435	0.474	0.500
38	0.108	0.204	0.260	0.308	0.362	0.397	0.430	0.468	0.495
39	0.107	0.202	0.257	0.304	0.357	0.393	0.425	0.463	0.489
40	0.106	0.199	0.254	0.301	0.353	0.388	0.420	0.458	0.484
41	0.104	0.197	0.251	0.297	0.349	0.384	0.415	0.453	0.479
42	0.103	0.195	0.248	0.294	0.345	0.380	0.411	0.448	0.474
43	0.102	0.192	0.245	0.290	0.342	0.376	0.407	0.444	0.469
44	0.101	0.190	0.243	0.287	0.338	0.372	0.402	0.439	0.464
45	0.100	0.188	0.240	0.284	0.335	0.368	0.398	0.435	0.460
46	0.099	0.186	0.238	0.281	0.331	0.364	0.394	0.431	0.455
47	0.098	0.184	0.235	0.279	0.328	0.361	0.391	0.426	0.451
48	0.097	0.182	0.233	0.276	0.325	0.357	0.387	0.422	0.447
49	0.096	0.181	0.230	0.273	0.322	0.354	0.383	0.419	0.443
50	0.095	0.179	0.228	0.270	0.319	0.351	0.380	0.415	0.439

TABLE B.29 (cont.) CRITICAL VALUES OF C FOR THE MEAN SQUARE SUCCESSIVE DIFFERENCE TEST

n	α = 0.25	0.10	0.05	0.025	0.01	0.005	0.0025	0.001	0.0005
52	0.093	0.175	0.224	0.265	0.313	0.344	0.373	0.408	0.431
54	0.091	0.172	0.220	0.261	0.307	0.338	0.367	0.401	0.424
56	0.090	0.169	0.216	0.256	0.302	0.333	0.361	0.394	0.417
58	0.088	0.166	0.212	0.252	0.297	0.327	0.355	0.388	0.411
60	0.087	0.164	0.209	0.248	0.292	0.322	0.349	0.382	0.405
62	0.085	0.161	0.206	0.244	0.288	0.317	0.344	0.376	0.399
64	0.084	0.158	0.203	0.240	0.284	0.313	0.339	0.371	0.393
66	0.083	0.156	0.200	0.237	0.279	0.308	0.334	0.366	0.387
68	0.081	0.154	0.197	0.233	0.276	0.304	0.330	0.361	0.382
70	0.080	0.152	0.194	0.230	0.272	0.300	0.325	0.356	0.377
72	0.079	0.150	0.191	0.227	0.268	0.296	0.321	0.351	0.372
74	0.078	0.148	0.189	0.224	0.265	0.292	0.317	0.347	0.368
76	0.077	0.146	0.186	0.221	0.261	0.288	0.313	0.342	0.363
78	0.076	0.144	0.184	0.218	0.258	0.285	0.309	0.338	0.359
80	0.075	0.142	0.182	0.216	0.255	0.281	0.305	0.334	0.355
82	0.074	0.140	0.180	0.213	0.252	0.278	0.302	0.331	0.351
84	0.073	0.139	0.177	0.211	0.249	0.275	0.298	0.327	0.347
86	0.072	0.137	0.175	0.208	0.246	0.272	0.295	0.323	0.343
88	0.071	0.136	0.173	0.206	0.244	0.269	0.292	0.320	0.339
90	0.070	0.134	0.172	0.204	0.241	0.266	0.289	0.316	0.336
92	0.069	0.133	0.170	0.202	0.238	0.263	0.286	0.313	0.332
94	0.068	0.131	0.168	0.200	0.236	0.260	0.283	0.310	0.329
96	0.067	0.130	0.166	0.198	0.233	0.258	0.280	0.307	0.326
98	0.065	0.129	0.165	0.196	0.231	0.255	0.277	0.304	0.323
100	0.064	0.127	0.163	0.194	0.229	0.253	0.275	0.301	0.320
105		0.124	0.159	0.189	0.224	0.247	0.268	0.294	0.312
110		0.121	0.156	0.185	0.219	0.241	0.262	0.288	0.305
115		0.119	0.152	0.181	0.214	0.236	0.257	0.282	0.299
120		0.116	0.149	0.177	0.210	0.231	0.252	0.276	0.293
125		0.114	0.146	0.174	0.205	0.227	0.247	0.271	0.287
130		0.111	0.143	0.170	0.202	0.223	0.242	0.266	0.282
135		0.109	0.141	0.167	0.198	0.219	0.238	0.261	0.277
140		0.107	0.138	0.164	0.194	0.215	0.234	0.256	0.272
145		0.105	0.136	0.161	0.191	0.211	0.230	0.252	0.268
150		0.102	0.133	0.159	0.188	0.208	0.226	0.248	0.263

Table B.29 was prepared by the method outlined in Young (1941).
Examples:

$$C_{0.05,60} = 0.209 \quad \text{and} \quad C_{0.01,68} = 0.276.$$

For n greater than shown in the table, the normal approximation (Equation 23.22) may be utilized. This approximation is excellent, especially for α near 0.05, as shown in the following tabulation. The following table considers the absolute difference between the exact critical value of C and the value of C calculated from the normal approximation. Given in the table are the minimum sample sizes necessary to achieve such absolute differences of various specified magnitudes.

Absolute Difference	α = 0.25	0.10	0.05	0.025	0.01	0.005	0.0025	0.001	0.0005
≤ 0.002	30	35	8	35	70	90	120		
≤ 0.005	20	20	8	20	45	60	80	100	120
≤ 0.010	15	10	8	15	30	40	50	60	80
≤ 0.020	8	8	8	8	20	25	30	40	45

TABLE B.30 ANGULAR DEVIATION, *s*, AS A FUNCTION OF VECTOR LENGTH, *r*

r	0	1	2	3	4	5	6	7	8	9
0.00	81.0285	80.9879	80.9474	80.9068	80.8662	80.8256	80.7850	80.7444	80.7037	80.6630
0.01	80.6223	80.5815	80.5408	80.5000	80.4593	80.4185	80.3776	80.3368	80.2959	80.2550
0.02	80.2141	80.1731	80.1322	80.0912	80.0502	80.0092	79.9682	79.9271	79.8860	79.8449
0.03	79.8038	79.7626	79.7215	79.6803	79.6391	79.5978	79.5566	79.5153	79.4740	79.4327
0.04	79.3914	79.3500	79.3086	79.2672	79.2258	79.1843	79.1429	79.1014	79.0599	79.0183
0.05	78.9768	78.9352	78.8936	78.8520	78.8103	78.7687	78.7270	78.6853	78.6435	78.6018
0.06	78.5600	78.5182	78.4764	78.4345	78.3927	78.3508	78.3089	78.2670	78.2250	78.1830
0.07	78.1410	78.0990	78.0569	78.0149	77.9728	77.9307	77.8885	77.8464	77.8042	77.7620
0.08	77.7198	77.6775	77.6353	77.5930	77.5506	77.5083	77.4659	77.4235	77.3811	77.3387
0.09	77.2962	77.2537	77.2112	77.1687	77.1262	77.0836	77.0410	76.9984	76.9557	76.9130
0.10	76.8703	76.8276	76.7849	76.7421	76.6993	76.6565	76.6137	76.5708	76.5279	76.4850
0.11	76.4421	76.3991	76.3562	76.3132	76.2701	76.2271	76.1840	76.1409	76.0978	76.0546
0.12	76.0114	75.9582	75.9250	75.8818	75.8385	75.7952	75.7519	75.7085	75.6651	75.6217
0.13	75.5783	75.5349	75.4914	75.4479	75.4044	75.3608	75.3172	75.2737	75.2300	75.1864
0.14	75.1427	75.0990	75.0553	75.0115	74.9678	74.9240	74.8801	74.8363	74.7924	74.7485
0.15	74.7045	74.6606	74.6166	74.5726	74.5286	74.4845	74.4404	74.3963	74.3522	74.3080
0.15	74.2638	74.2195	74.1754	74.1311	74.0868	74.0425	73.9981	73.9537	73.9093	73.8649
0.17	73.8204	73.7760	73.7314	73.6869	73.6423	73.5978	73.5531	73.5085	73.4638	73.4191
0.18	73.3744	73.3296	73.2849	73.2401	73.1952	73.1503	73.1055	73.0605	73.0156	72.9706
0.19	72.9256	72.8305	72.8355	72.7904	72.7453	72.7002	72.6550	72.6098	72.5646	72.5193
0.20	72.4741	72.4287	72.3834	72.3380	72.2926	72.2472	72.2018	72.1563	72.1108	72.0652
0.21	72.0197	71.9741	71.9285	71.8828	71.8371	71.7914	71.7457	71.6999	71.6541	71.6083
0.22	71.5624	71.5165	71.4706	71.4247	71.3787	71.3327	71.2866	71.2406	71.1945	71.1483
0.23	71.1022	71.0560	71.0098	70.9635	70.9173	70.8710	70.8246	70.7783	70.7319	70.6854
0.24	70.6390	70.5925	70.5460	70.4994	70.4528	70.4062	70.3596	70.3129	70.2662	70.2195
0.25	70.1727	70.1259	70.0791	70.0322	69.9853	69.9384	69.8914	69.8445	69.7975	69.7504
0.26	69.7033	69.6562	59.5091	69.5619	69.5147	69.4674	69.4202	69.3729	69.3255	69.2782
0.27	69.2307	69.1833	69.1358	69.0883	69.0408	68.9933	68.9456	68.8980	68.8504	68.8027
0.28	68.7549	53.7072	68.6594	68.6115	68.5637	68.5158	68.4678	68.4199	68.3719	68.3239
0.29	68.2758	68.2277	68.1796	68.1314	68.0832	68.0350	67.9867	67.9384	67.8901	67.8417
0.30	67.7933	67.7448	67.6964	67.6478	67.5993	67.5507	67.5021	67.4535	67.4048	67.3560
0.31	67.3073	67.2585	67.2097	67.1608	67.1119	67.0630	67.0140	66.9650	66.9160	66.8669
0.32	66.8178	66.7686	66.7195	66.6702	66.6210	66.5717	66.5223	66.4730	66.4236	66.3741
0.33	66.3246	66.2751	66.2256	66.1760	66.1264	66.0767	66.0270	65.9773	65.9275	65.8777
0.34	65.8278	65.7779	65.7280	65.6781	65.6281	65.5780	65.5279	65.4778	65.4277	65.3775
0.35	65.3272	65.2770	65.2267	65.1763	65.1259	65.0755	65.0250	64.9745	64.9240	64.8734
0.35	54.8228	64.7721	64.7214	64.6707	64.6199	64.5691	64.5182	64.4673	64.4164	64.3654
0.37	64.3143	64.2533	64.2122	64.1610	64.1098	64.0586	64.0074	63.9561	63.9047	63.8533
0.38	63.8019	63.7504	63.6989	63.6473	63.5957	63.5441	63.4924	63.4407	63.3889	63.3371
0.39	63.2852	63.2334	63.1814	63.1294	63.0774	63.0253	62.9732	62.9211	62.8689	62.8167
0.40	62.7644	62.7121	62.6597	62.6073	62.5548	62.5023	62.4498	62.3972	62.3445	62.2919
0.41	62.2391	62.1354	62.1336	62.0807	62.0278	61.9749	61.9219	61.8688	61.8158	61.7626
0.42	61.7094	61.6562	61.6030	61.5496	61.4963	61.4429	61.3894	61.3359	61.2824	61.2288
0.43	61.1751	61.1215	61.0677	61.0139	60.9601	60.9063	60.8523	60.7984	60.7443	60.6903
0.44	60.6361	60.5320	50.5278	60.4735	60.4192	60.3648	60.3104	60.2560	60.2015	60.1469
0.45	60.0923	60.0377	59.9830	59.9282	59.8734	59.8185	59.7636	59.7087	59.6537	59.5986
0.46	59.5435	59.4884	59.4332	59.3779	59.3226	59.2672	59.2118	59.1563	59.1008	59.0452
0.47	58.9896	58.9339	58.8782	58.8224	58.7666	58.7107	58.6548	58.5988	58.5427	58.4866
0.48	58.4305	58.3743	58.3180	58.2617	58.2053	58.1489	58.0924	58.0358	57.9792	57.9226
0.49	57.8659	57.8091	57.7523	57.6954	57.6385	57.5815	57.5245	57.4674	57.4102	57.3530

TABLE B.30 (cont.) ANGULAR DEVIATION, s, AS A FUNCTION OF VECTOR LENGTH, r

r	0	1	2	3	4	5	6	7	8	9
0.50	57.2958	57.2384	57.1811	57.1236	57.0661	57.0086	56.9510	56.8933	56.8356	56.7778
0.51	56.7199	56.6520	56.5040	56.5460	56.4879	56.4298	56.3716	56.3133	56.2550	56.1966
0.52	56.1382	56.0797	56.0211	55.9625	55.9038	55.8450	55.7862	55.7273	55.6684	55.6094
0.53	55.5503	55.4912	55.4320	55.3727	55.3134	55.2540	55.1946	55.1351	55.0755	55.0159
0.54	54.9562	54.8964	54.8366	54.7767	54.7167	54.6567	54.5966	54.5364	54.4762	54.4159
0.55	54.3555	54.2951	54.2346	54.1741	54.1134	54.0527	53.9920	53.9311	53.8702	53.8092
0.56	53.7482	53.6371	53.6259	53.5647	53.5033	53.4419	53.3805	53.3189	53.2573	53.1957
0.57	53.1339	53.0721	53.0102	52.9482	52.8862	52.8241	52.7619	52.6997	52.6373	52.5749
0.58	52.5124	52.4499	52.3873	52.3246	52.2618	52.1989	52.1360	52.0730	52.0099	51.9468
0.59	51.8835	51.8202	51.7568	51.6934	51.6298	51.5662	51.5025	51.4387	51.3749	51.3109
0.60	51.2469	51.1828	51.1186	51.0544	50.9900	50.9256	50.8611	50.7965	50.7318	50.6671
0.61	50.6023	50.5373	50.4723	50.4073	50.3421	50.2768	50.2115	50.1461	50.0806	50.0150
0.62	49.9493	49.8835	49.8177	49.7517	49.6857	49.6196	49.5534	49.4871	49.4207	49.3542
0.63	49.2877	49.2210	49.1543	49.0875	49.0205	48.9535	48.8864	48.8192	48.7519	48.6846
0.64	48.6171	48.5495	48.4818	48.4141	48.3462	48.2783	48.2102	48.1421	48.0739	48.0055
0.65	47.9371	47.8685	47.7999	47.7312	47.6624	47.5934	47.5244	47.4553	47.3861	47.3167
0.66	47.2473	47.1778	47.1081	47.0384	46.9686	46.8986	46.8286	46.7584	46.6882	46.6178
0.67	46.5473	46.4767	46.4060	46.3353	46.2643	46.1933	46.1222	46.0510	45.9796	45.9082
0.68	45.8366	45.7549	45.6932	45.6213	45.5492	45.4771	45.4049	45.3325	45.2600	45.1875
0.69	45.1147	45.0419	44.9690	44.8959	44.8227	44.7494	44.6760	44.6025	44.5288	44.4550
0.70	44.3811	44.3071	44.2329	44.1586	44.0842	44.0097	43.9351	43.8603	43.7854	43.7103
0.71	43.6352	43.5599	43.4844	43.4089	43.3332	43.2574	43.1814	43.1053	43.0291	42.9527
0.72	42.8762	42.7996	42.7228	42.6459	42.5689	42.4917	42.4144	42.3369	42.2593	42.1815
0.73	42.1036	42.0256	41.9474	41.8691	41.7906	41.7120	41.6332	41.5543	41.4752	41.3960
0.74	41.3166	41.2370	41.1573	41.0775	40.9975	40.9174	40.8371	40.7566	40.6760	40.5952
0.75	40.5142	40.4331	40.3519	40.2704	40.1888	40.1070	40.0251	39.9430	39.8607	39.7783
0.76	39.6957	39.6129	39.5299	39.4468	39.3635	39.2800	39.1963	39.1125	39.0285	38.9443
0.77	38.8599	38.7753	38.6906	38.6056	38.5205	38.4352	38.3497	38.2640	38.1781	38.0920
0.78	38.0057	37.9192	37.8326	37.7457	37.6586	37.5714	37.4839	37.3962	37.3083	37.2202
0.79	37.1319	37.0434	36.9547	36.8657	36.7766	36.6872	36.5976	36.5078	36.4178	36.3275
0.80	36.2370	36.1463	36.0554	35.9642	35.8728	35.7812	35.6893	35.5972	35.5049	35.4123
0.81	35.3195	35.2254	35.1331	35.0396	34.9457	34.8517	34.7573	34.6628	34.5679	34.4728
0.82	34.3775	34.2818	34.1859	34.0898	33.9933	33.8966	33.7997	33.7024	33.6048	33.5070
0.83	33.4089	33.3105	33.2118	33.1128	33.0135	32.9139	32.8140	32.7138	32.6133	32.5125
0.84	32.4114	32.3099	32.2082	32.1061	32.0037	31.9009	31.7979	31.6945	31.5907	31.4866
0.85	31.3822	31.2774	31.1723	31.0668	30.9609	30.8547	30.7481	30.6412	30.5339	30.4262
0.86	30.3181	30.2096	30.1007	29.9915	29.8818	29.7718	29.6613	29.5504	29.4391	29.3274
0.87	29.2152	29.1026	28.9896	28.8762	28.7623	28.6479	28.5331	28.4178	28.3020	28.1858
0.88	28.0691	27.9519	27.8342	27.7160	27.5973	27.4781	27.3584	27.2381	27.1173	26.9960
0.89	26.8741	26.7517	26.6287	26.5051	26.3810	26.2562	26.1309	26.0050	25.8784	25.7513
0.90	25.6234	25.4950	25.3659	25.2362	25.1058	24.9746	24.8428	24.7104	24.5771	24.4432
0.91	24.3085	24.1731	24.0369	23.9000	23.7622	23.6237	23.4843	23.3441	23.2030	23.0611
0.92	22.9183	22.7746	22.6300	22.4845	22.3380	22.1906	22.0421	21.8927	21.7422	21.5907
0.93	21.4381	21.2844	21.1296	20.9737	20.8166	20.6583	20.4988	20.3380	20.1759	20.0126
0.94	19.8478	19.6817	19.5142	19.3453	19.1748	19.0029	18.8293	18.6542	18.4773	18.2988
0.95	18.1185	17.9364	17.7524	17.5666	17.3787	17.1887	16.9967	16.8024	16.6059	16.4070
0.96	16.2057	16.0018	15.7954	15.5861	15.3741	15.1590	14.9409	14.7196	14.4948	14.2666
0.97	14.0345	13.7987	13.5587	13.3144	13.0655	12.8117	12.5529	12.2886	12.0185	11.7422
0.98	11.4592	11.1590	10.8711	10.5648	10.2494	9.9239	9.5874	9.2387	8.8763	8.4984
0.99	8.1029	7.6371	7.2474	6.7794	6.2765	5.7296	5.1247	4.4382	3.6238	2.5625
1.00	0.0000									

TABLE B.31 CIRCULAR STANDARD DEVIATION, s', AS A FUNCTION OF r

r	0	1	2	3	4	5	6	7	8	9
0.00	∞	212.9639	201.9968	195.2961	190.3990	186.5119	183.2748	180.4925	178.0473	175.8622
0.01	173.8843	172.0755	170.4075	168.8585	167.4115	166.0531	164.7723	163.5600	162.4087	161.3121
0.02	160.2649	159.2623	158.3005	157.3760	156.4857	155.6270	154.7974	153.9950	153.2178	152.4640
0.03	151.7323	151.0212	150.3295	149.6560	148.9998	148.3597	147.7351	147.1250	146.5287	145.9456
0.04	145.3750	144.8163	144.2690	143.7326	143.2066	142.6905	142.1839	141.6865	141.1979	140.7177
0.05	140.2456	139.7813	139.3245	138.8749	138.4324	137.9966	137.5672	137.1442	136.7273	136.3162
0.06	135.9109	135.5110	135.1165	134.7272	134.3430	133.9636	133.5889	133.2189	132.8533	132.4920
0.07	132.1350	131.7822	131.4333	131.0883	130.7472	130.4097	130.0759	129.7455	129.4186	129.0951
0.08	128.7748	128.4578	128.1438	127.8329	127.5250	127.2200	126.9178	126.6184	126.3218	126.0278
0.09	125.7364	125.4476	125.1612	124.8774	124.5959	124.3167	124.0399	123.7654	123.4930	123.2228
0.10	122.9548	122.6888	122.4249	122.1630	121.9031	121.6452	121.3891	121.1349	120.8825	120.6320
0.11	120.3832	120.1361	119.8908	119.6472	119.4052	119.1648	118.9261	118.6889	118.4533	118.2192
0.12	117.9866	117.7554	117.5258	117.2975	117.0707	116.8452	116.6211	116.3984	116.1770	115.9569
0.13	115.7381	115.5205	115.3042	115.0891	114.8753	114.6626	114.4511	114.2408	114.0316	113.8236
0.14	113.6166	113.4108	113.2060	113.0023	112.7997	112.5981	112.3976	112.1980	111.9995	111.8019
0.15	111.6054	111.4097	111.2151	111.0213	110.8285	110.6367	110.4457	110.2556	110.0664	109.8780
0.16	109.6906	109.5039	109.3182	109.1332	108.9491	108.7657	108.5832	108.4015	108.2205	108.0404
0.17	107.8609	107.6823	107.5044	107.3272	107.1508	106.9751	106.8000	106.6258	106.4522	106.2793
0.18	106.1070	105.9355	105.7646	105.5944	105.4248	105.2559	105.0877	104.9200	104.7530	104.5866
0.19	104.4209	104.2557	104.0911	103.9272	103.7638	103.6010	103.4388	103.2772	103.1161	102.9556
0.20	102.7957	102.6362	102.4774	102.3191	102.1613	102.0040	101.8473	101.6911	101.5354	101.3802
0.21	101.2255	101.0714	100.9177	100.7645	100.6118	100.4595	100.3078	100.1565	100.0057	99.8553
0.22	99.7054	99.5560	99.4070	99.2585	99.1104	98.9628	98.8155	98.6688	98.5224	98.3765
0.23	98.2310	98.0859	97.9412	97.7969	97.6531	97.5096	97.3665	97.2239	97.0816	96.9397
0.24	96.7982	96.6571	96.5164	96.3760	96.2360	96.0964	95.9571	95.8182	95.6797	95.5415
0.25	95.4037	95.2663	95.1292	94.9924	94.8560	94.7199	94.5841	94.4487	94.3136	94.1789
0.26	94.0445	93.9104	93.7766	93.6432	93.5100	93.3772	93.2447	93.1125	92.9806	92.8490
0.27	92.7177	92.5867	92.4560	92.3257	92.1956	92.0657	91.9362	91.8070	91.6781	91.5494
0.28	91.4210	91.2929	91.1651	91.0375	90.9102	90.7832	90.6565	90.5300	90.4038	90.2778
0.29	90.1521	90.0267	89.9015	89.7766	89.6519	89.5275	89.4033	89.2794	89.1557	89.0322
0.30	88.9090	88.7861	88.6634	88.5409	88.4186	88.2966	88.1748	88.0533	87.9320	87.8109
0.31	87.6900	87.5693	87.4489	87.3287	87.2087	87.0889	86.9694	86.8500	86.7309	86.6120
0.32	86.4933	86.3748	86.2565	86.1384	86.0205	85.9028	85.7853	85.6680	85.5509	85.4340
0.33	85.3173	85.2008	85.0845	84.9684	84.8525	84.7367	84.6212	84.5058	84.3906	84.2757
0.34	84.1608	84.0462	83.9317	83.8175	83.7034	83.5895	83.4757	83.3621	83.2487	83.1355
0.35	83.0224	82.9095	82.7968	82.6843	82.5719	82.4597	82.3476	82.2357	82.1240	82.0124
0.36	81.9010	81.7897	81.6786	81.5676	81.4568	81.3462	81.2357	81.1254	81.0152	80.9052
0.37	80.7953	80.6855	80.5759	80.4665	80.3572	80.2480	80.1390	80.0301	79.9214	79.8128
0.38	79.7043	79.5960	79.4878	79.3798	79.2719	79.1641	79.0565	78.9490	78.8416	78.7343
0.39	78.6272	78.5202	78.4133	78.3066	78.2000	78.0935	77.9872	77.8809	77.7748	77.6688
0.40	77.5629	77.4572	77.3516	77.2460	77.1407	77.0354	76.9302	76.8252	76.7202	76.6154
0.41	76.5107	76.4061	76.3016	76.1973	76.0930	75.9888	75.8848	75.7809	75.6770	75.5733
0.42	75.4697	75.3662	75.2628	75.1594	75.0562	74.9531	74.8501	74.7472	74.6444	74.5417
0.43	74.4391	74.3366	74.2342	74.1319	74.0296	73.9275	73.8255	73.7235	73.6217	73.5199
0.44	73.4183	73.3167	73.2152	73.1138	73.0125	72.9113	72.8101	72.7091	72.6081	72.5072
0.45	72.4064	72.3057	72.2051	72.1046	72.0041	71.9037	71.8034	71.7032	71.6030	71.5030
0.46	71.4030	71.3031	71.2033	71.1035	71.0038	70.9042	70.8047	70.7052	70.6058	70.5065
0.47	70.4073	70.3081	70.2090	70.1100	70.0110	69.9121	69.8133	69.7146	69.6159	69.5173
0.48	69.4187	69.3202	69.2218	69.1234	69.0251	68.9269	68.8287	68.7306	68.6326	68.5346
0.49	68.4367	68.3388	68.2410	68.1433	68.0456	67.9479	67.8504	67.7528	67.6554	67.5580

TABLE B.31 (cont.) CIRCULAR STANDARD DEVIATION, s', AS A FUNCTION OF r

r	0	1	2	3	4	5	6	7	8	9
0.50	67.4606	67.3633	67.2661	67.1689	67.0718	66.9747	66.8776	66.7806	66.6837	66.5868
0.51	66.4900	66.3932	66.2965	66.1998	66.1031	66.0065	65.9100	65.8135	65.7170	65.6206
0.52	65.5243	65.4279	65.3316	65.2354	65.1392	65.0431	64.9469	64.8509	64.7548	64.6588
0.53	64.5629	64.4670	64.3711	64.2752	64.1794	64.0837	63.9879	63.8922	63.7966	63.7009
0.54	63.6053	63.5098	63.4143	63.3188	63.2233	63.1279	63.0325	62.9371	62.8417	62.7464
0.55	62.6511	62.5559	62.4607	62.3655	62.2703	62.1751	62.0800	61.9849	61.8899	61.7948
0.56	61.6998	61.6048	61.5098	61.4149	61.3199	61.2250	61.1301	61.0353	60.9404	60.8456
0.57	60.7508	60.6560	60.5612	60.4664	60.3717	60.2770	60.1823	60.0876	59.9929	59.8982
0.58	59.8036	59.7089	59.6143	59.5197	59.4251	59.3305	59.2359	59.1414	59.0468	58.9523
0.59	58.8577	58.7632	58.6687	58.5742	58.4797	58.3852	58.2907	58.1962	58.1017	58.0072
0.60	57.9127	57.8182	57.7238	57.6293	57.5348	57.4404	57.3459	57.2514	57.1570	57.0625
0.61	56.9680	56.8736	56.7791	56.6846	56.5902	56.4957	56.4012	56.3067	56.2122	56.1177
0.62	56.0232	55.9287	55.8342	55.7396	55.6451	55.5505	55.4560	55.3614	55.2668	55.1722
0.63	55.0776	54.9830	54.8884	54.7938	54.6991	54.6044	54.5097	54.4150	54.3203	54.2256
0.64	54.1308	54.0361	53.9413	53.8465	53.7517	53.6568	53.5619	53.4671	53.3722	53.2772
0.65	53.1823	53.0873	52.9923	52.8973	52.8022	52.7071	52.6120	52.5169	52.4217	52.3266
0.66	52.2313	52.1361	52.0408	51.9455	51.8502	51.7548	51.6594	51.5640	51.4685	51.3730
0.67	51.2775	51.1819	51.0863	50.9907	50.8950	50.7993	50.7035	50.6077	50.5119	50.4160
0.68	50.3201	50.2241	50.1281	50.0321	49.9360	49.8398	49.7437	49.6474	49.5512	49.4549
0.69	49.3585	49.2621	49.1656	49.0691	48.9725	48.8759	48.7792	48.6825	48.5858	48.4889
0.70	48.3920	48.2951	48.1981	48.1011	48.0039	47.9068	47.8095	47.7123	47.6149	47.5175
0.71	47.4200	47.3225	47.2249	47.1272	47.0295	46.9317	46.8338	46.7359	46.6379	46.5398
0.72	46.4417	46.3435	46.2452	46.1468	46.0484	45.9499	45.8513	45.7527	45.6539	45.5551
0.73	45.4562	45.3573	45.2582	45.1591	45.0599	44.9606	44.8612	44.7617	44.6622	44.5625
0.74	44.4628	44.3630	44.2631	44.1631	44.0630	43.9628	43.8625	43.7621	43.6617	43.5611
0.75	43.4604	43.3597	43.2588	43.1578	43.0568	42.9556	42.8543	42.7529	42.6515	42.5499
0.76	42.4482	42.3463	42.2444	42.1424	42.0402	41.9380	41.8356	41.7331	41.6305	41.5277
0.77	41.4249	41.3219	41.2188	41.1156	41.0122	40.9087	40.8051	40.7014	40.5975	40.4935
0.78	40.3894	40.2851	40.1807	40.0761	39.9715	39.8666	39.7617	39.6565	39.5513	39.4459
0.79	39.3403	39.2346	39.1288	39.0228	38.9166	38.8103	38.7038	38.5972	38.4904	38.3834
0.80	38.2763	38.1690	38.0615	37.9539	37.8461	37.7381	37.6300	37.5217	37.4132	37.3045
0.81	37.1956	37.0865	36.9773	36.8679	36.7583	36.6484	36.5384	36.4282	36.3178	36.2072
0.82	36.0964	35.9854	35.8742	35.7628	35.6511	35.5393	35.4272	35.3149	35.2024	35.0896
0.83	34.9767	34.8635	34.7501	34.6364	34.5225	34.4084	34.2940	34.1793	34.0645	33.9493
0.84	33.8340	33.7183	33.6024	33.4863	33.3698	33.2531	33.1362	33.0189	32.9014	32.7836
0.85	32.6655	32.5471	32.4285	32.3095	32.1902	32.0707	31.9508	31.8306	31.7101	31.5893
0.86	31.4682	31.3467	31.2249	31.1028	30.9803	30.8575	30.7343	30.6108	30.4869	30.3627
0.87	30.2381	30.1131	29.9877	29.8620	29.7359	29.6094	29.4825	29.3552	29.2274	29.0993
0.88	28.9708	28.8418	28.7124	28.5825	28.4522	28.3215	28.1903	28.0586	27.9265	27.7938
0.89	27.6607	27.5271	27.3930	27.2584	27.1233	26.9877	26.8515	26.7148	26.5775	26.4397
0.90	26.3013	26.1623	26.0227	25.8826	25.7418	25.6004	25.4584	25.3158	25.1725	25.0285
0.91	24.8839	24.7386	24.5925	24.4458	24.2984	24.1502	24.0012	23.8515	23.7011	23.5498
0.92	23.3977	23.2448	23.0910	22.9364	22.7808	22.6244	22.4671	22.3088	22.1496	21.9894
0.93	21.8282	21.6660	21.5027	21.3384	21.1729	21.0063	20.8386	20.6697	20.4996	20.3283
0.94	20.1556	19.9817	19.8064	19.6298	19.4517	19.2722	19.0912	18.9087	18.7246	18.5388
0.95	18.3513	18.1622	17.9712	17.7784	17.5837	17.3870	17.1882	16.9874	16.7843	16.5790
0.96	16.3714	16.1612	15.9486	15.7333	15.5152	15.2943	15.0703	14.8432	14.6128	14.3790
0.97	14.1415	13.9003	13.6550	13.4055	13.1516	12.8929	12.6292	12.3601	12.0854	11.8045
0.98	11.5171	11.2226	10.9205	10.6101	10.2907	9.9614	9.6212	9.2689	8.9030	8.5218
0.99	8.1232	7.7044	7.2620	6.7912	6.2859	5.7368	5.1298	4.4414	3.6255	2.5630
1.00	0.0000									

TABLE B.32 CRITICAL VALUES OF RAYLEIGH'S z

n	α: 0.50	0.20	0.10	0.05	0.02	0.01	0.005	0.002	0.001
6	0.734	1.639	2.274	2.865	3.576	4.058	4.491	4.985	5.297
7	0.727	1.634	2.278	2.885	3.627	4.143	4.617	5.181	5.556
8	0.723	1.631	2.281	2.899	3.665	4.205	4.710	5.322	5.743
9	0.719	1.628	2.283	2.910	3.694	4.252	4.780	5.430	5.885
10	0.717	1.626	2.285	2.919	3.716	4.289	4.835	5.514	5.996
11	0.715	1.625	2.287	2.926	3.735	4.319	4.879	5.582	6.085
12	0.713	1.623	2.288	2.932	3.750	4.344	4.916	5.638	6.158
13	0.711	1.622	2.289	2.937	3.763	4.365	4.947	5.685	6.219
14	0.710	1.621	2.290	2.941	3.774	4.383	4.973	5.725	6.271
15	0.709	1.620	2.291	2.945	3.784	4.398	4.996	5.759	6.316
16	0.708	1.620	2.292	2.948	3.792	4.412	5.015	5.789	6.354
17	0.707	1.619	2.292	2.951	3.799	4.423	5.033	5.815	6.388
18	0.706	1.619	2.293	2.954	3.806	4.434	5.048	5.838	6.418
19	0.705	1.618	2.293	2.956	3.811	4.443	5.061	5.858	6.445
20	0.705	1.618	2.294	2.958	3.816	4.451	5.074	5.877	6.469
21	0.704	1.617	2.294	2.960	3.821	4.459	5.085	5.893	6.491
22	0.704	1.617	2.295	2.961	3.825	4.466	5.095	5.908	6.510
23	0.703	1.616	2.295	2.963	3.829	4.472	5.104	5.922	6.528
24	0.703	1.616	2.295	2.964	3.833	4.478	5.112	5.935	6.544
25	0.702	1.616	2.296	2.966	3.836	4.483	5.120	5.946	6.559
26	0.702	1.616	2.296	2.967	3.839	4.488	5.127	5.957	6.573
27	0.702	1.615	2.296	2.968	3.842	4.492	5.133	5.966	6.586
28	0.701	1.615	2.296	2.969	3.844	4.496	5.139	5.975	6.598
29	0.701	1.615	2.297	2.970	3.847	4.500	5.145	5.984	6.609
30	0.701	1.615	2.297	2.971	3.849	4.504	5.150	5.992	6.619
32	0.700	1.614	2.297	2.972	3.853	4.510	5.159	6.006	6.637
34	0.700	1.614	2.297	2.974	3.856	4.516	5.168	6.018	6.654
36	0.700	1.614	2.298	2.975	3.859	4.521	5.175	6.030	6.668
38	0.699	1.614	2.298	2.976	3.862	4.525	5.182	6.039	6.681
40	0.699	1.613	2.298	2.977	3.865	4.529	5.188	6.048	6.692
42	0.699	1.613	2.298	2.978	3.867	4.533	5.193	6.056	6.703
44	0.698	1.613	2.299	2.979	3.869	4.536	5.198	6.064	6.712
46	0.698	1.613	2.299	2.979	3.871	4.539	5.202	6.070	6.721
48	0.698	1.613	2.299	2.980	3.873	4.542	5.206	6.076	6.729
50	0.698	1.613	2.299	2.981	3.874	4.545	5.210	6.082	6.736
55	0.697	1.612	2.299	2.982	3.878	4.550	5.218	6.094	6.752
60	0.697	1.612	2.300	2.983	3.881	4.555	5.225	6.104	6.765
65	0.697	1.612	2.300	2.984	3.883	4.559	5.231	6.113	6.776
70	0.696	1.612	2.300	2.985	3.885	4.562	5.235	6.120	6.786
75	0.696	1.612	2.300	2.986	3.887	4.565	5.240	6.127	6.794
80	0.696	1.611	2.300	2.986	3.889	4.567	5.243	6.132	6.801
90	0.696	1.611	2.301	2.987	3.891	4.572	5.249	6.141	6.813
100	0.695	1.611	2.301	2.988	3.893	4.575	5.254	6.149	6.822
120	0.695	1.611	2.301	2.990	3.896	4.580	5.262	6.160	6.837
140	0.695	1.611	2.301	2.990	3.899	4.584	5.267	6.168	6.847
160	0.695	1.610	2.301	2.991	3.900	4.586	5.271	6.174	6.855
180	0.694	1.610	2.302	2.992	3.902	4.588	5.274	6.178	6.861
200	0.694	1.610	2.302	2.992	3.903	4.590	5.276	6.182	6.865
300	0.694	1.610	2.302	2.993	3.906	4.595	5.284	6.193	6.879
500	0.694	1.610	2.302	2.994	3.908	4.599	5.290	6.201	6.891
∞	0.6931	1.6094	2.3026	2.9957	3.9120	4.6052	5.2983	6.2146	6.9078

The preceding values were computed using Durand and Greenwood's Equation 6 (1958). This procedure was found to give slightly more accurate results than Durand and Greenwood's Equation 4, distinctly better results than the Pearson curve approximation (Stephens, 1969), and very much better results than the chi-square approximation (Stephens, 1969). By examining the exact critical values of Greenwood and Durand (1955), we see that the preceding tabled values for $\alpha = 0.05$ are accurate to the third decimal place for n as small as 8, and for $\alpha = 0.01$ for n as small as 10. For n as small as 6, none of the tabled values for $\alpha = 0.05$ or 0.01 has a relative error greater than 0.3%.

Examples:

$$z_{0.05,80} = 2.986 \quad \text{and} \quad z_{0.01,32} = 4.510.$$

TABLE B.33 CRITICAL VALUES OF u FOR THE V TEST OF CIRCULAR UNIFORMITY

n	α: 0.25	0.10	0.05	0.025	0.01	0.005	0.0025	0.001	0.0005
8	0.688	1.296	1.649	1.947	2.280	2.498	2.691	2.916	3.066
9	0.687	1.294	1.649	1.948	2.286	2.507	2.705	2.937	3.094
10	0.685	1.293	1.648	1.950	2.290	2.514	2.716	2.954	3.115
11	0.684	1.292	1.648	1.950	2.293	2.520	2.725	2.967	3.133
12	0.684	1.291	1.648	1.951	2.296	2.525	2.732	2.978	3.147
13	0.683	1.290	1.647	1.952	2.299	2.529	2.738	2.987	3.159
14	0.682	1.290	1.647	1.953	2.301	2.532	2.743	2.995	3.169
15	0.682	1.289	1.647	1.953	2.302	2.535	2.748	3.002	3.177
16	0.681	1.289	1.647	1.953	2.304	2.538	2.751	3.008	3.185
17	0.681	1.288	1.647	1.954	2.305	2.540	2.755	3.013	3.191
18	0.681	1.288	1.647	1.954	2.306	2.542	2.758	3.017	3.197
19	0.680	1.287	1.647	1.954	2.308	2.544	2.761	3.021	3.202
20	0.680	1.287	1.646	1.955	2.308	2.546	2.763	3.025	3.207
21	0.680	1.287	1.646	1.955	2.309	2.547	2.765	3.028	3.211
22	0.679	1.287	1.646	1.955	2.310	2.549	2.767	3.031	3.215
23	0.679	1.286	1.646	1.955	2.311	2.550	2.769	3.034	3.218
24	0.679	1.286	1.646	1.956	2.311	2.551	2.770	3.036	3.221
25	0.679	1.286	1.646	1.956	2.312	2.552	2.772	3.038	3.224
26	0.679	1.286	1.646	1.956	2.313	2.553	2.773	3.040	3.227
27	0.678	1.286	1.646	1.956	2.313	2.554	2.775	3.042	3.229
28	0.678	1.285	1.646	1.956	2.314	2.555	2.776	3.044	3.231
29	0.678	1.285	1.646	1.956	2.314	2.555	2.777	3.046	3.233
30	0.678	1.285	1.646	1.957	2.315	2.556	2.778	3.047	3.235
32	0.678	1.285	1.646	1.957	2.315	2.557	2.780	3.050	3.239
34	0.678	1.285	1.646	1.957	2.316	2.558	2.781	3.052	3.242
36	0.677	1.285	1.646	1.957	2.316	2.559	2.783	3.054	3.245
38	0.677	1.284	1.646	1.957	2.317	2.560	2.784	3.056	3.247
40	0.677	1.284	1.646	1.957	2.317	2.561	2.785	3.058	3.249
42	0.677	1.284	1.646	1.958	2.318	2.562	2.786	3.060	3.251
44	0.677	1.284	1.646	1.958	2.318	2.562	2.787	3.061	3.253
46	0.677	1.284	1.646	1.958	2.319	2.563	2.788	3.062	3.255
48	0.677	1.284	1.645	1.958	2.319	2.564	2.789	3.063	3.256
50	0.677	1.284	1.645	1.958	2.319	2.564	2.790	3.065	3.258
55	0.676	1.284	1.645	1.958	2.320	2.565	2.791	3.067	3.261
60	0.676	1.283	1.645	1.958	2.320	2.566	2.793	3.069	3.263
65	0.676	1.283	1.645	1.958	2.321	2.567	2.794	3.071	3.265
70	0.676	1.283	1.645	1.958	2.321	2.567	2.795	3.072	3.267
75	0.676	1.283	1.645	1.959	2.322	2.568	2.796	3.073	3.269
80	0.676	1.283	1.645	1.959	2.322	2.568	2.796	3.074	3.270
90	0.676	1.283	1.645	1.959	2.322	2.569	2.797	3.076	3.272
100	0.676	1.283	1.645	1.959	2.323	2.570	2.798	3.077	3.274
120	0.675	1.282	1.645	1.959	2.323	2.571	2.800	3.080	3.277
140	0.675	1.282	1.645	1.959	2.324	2.572	2.801	3.081	3.279
160	0.675	1.282	1.645	1.959	2.324	2.572	2.802	3.082	3.280
180	0.675	1.282	1.645	1.959	2.324	2.573	2.802	3.083	3.282
200	0.675	1.282	1.645	1.959	2.325	2.573	2.803	3.084	3.282
300	0.675	1.282	1.645	1.960	2.325	2.574	2.804	3.086	3.285
∞	0.6747	1.2818	1.6449	1.9598	2.3256	2.5747	2.8053	3.0877	3.2873

The preceding values were computed using Durand and Greenwood's Equation 7 (1958).
Examples:

$$u_{0.05,25} = 1.646 \quad \text{and} \quad u_{0.01,20} = 2.308.$$

TABLE B.34 CORRECTION FACTOR, *K*, FOR THE WATSON AND WILLIAMS TEST

r	0	1	2	3	4	5	6	7	8	9
0.00		188.4989	94.7472	63.5015	47.8749	38.4992	32.2498	27.7851	24.4367	21.8325
0.01	19.7489	18.0444	16.6239	15.4219	14.3916	13.4986	12.7173	12.0278	11.4150	10.8667
0.02	10.3731	9.9266	9.5206	9.1500	8.8103	8.4976	8.2091	7.9419	7.6938	7.4628
0.03	7.2472	7.0455	6.8564	6.6787	6.5115	6.3539	6.2050	6.0641	5.9306	5.8040
0.04	5.6837	5.5693	5.4603	5.3564	5.2572	5.1625	5.0718	4.9850	4.9017	4.8219
0.05	4.7453	4.6717	4.6009	4.5328	4.4672	4.4039	4.3430	4.2841	4.2273	4.1724
0.06	4.1194	4.0680	4.0184	3.9703	3.9237	3.8785	3.8347	3.7922	3.7510	3.7109
0.07	3.6720	3.6342	3.5974	3.5616	3.5268	3.4930	3.4600	3.4278	3.3965	3.3660
0.08	3.3362	3.3072	3.2789	3.2512	3.2243	3.1979	3.1722	3.1470	3.1224	3.0984
0.09	3.0749	3.0519	3.0294	3.0074	2.9858	2.9648	2.9441	2.9239	2.9041	2.8846
0.10	2.8656	2.8469	2.8286	2.8107	2.7931	2.7758	2.7589	2.7423	2.7259	2.7099
0.11	2.6942	2.6787	2.6636	2.6487	2.6340	2.6196	2.6055	2.5915	2.5779	2.5644
0.12	2.5512	2.5382	2.5254	2.5128	2.5004	2.4882	2.4762	2.4644	2.4528	2.4413
0.13	2.4301	2.4189	2.4080	2.3972	2.3866	2.3762	2.3658	2.3557	2.3457	2.3358
0.14	2.3261	2.3165	2.3070	2.2977	2.2885	2.2794	2.2705	2.2616	2.2529	2.2443
0.15	2.2358	2.2275	2.2192	2.2110	2.2030	2.1950	2.1872	2.1794	2.1718	2.1642
0.16	2.1567	2.1494	2.1421	2.1349	2.1278	2.1208	2.1138	2.1070	2.1002	2.0935
0.17	2.0868	2.0803	2.0738	2.0674	2.0611	2.0549	2.0487	2.0426	2.0365	2.0305
0.18	2.0246	2.0188	2.0130	2.0072	2.0016	1.9960	1.9904	1.9849	1.9795	1.9741
0.19	1.9688	1.9635	1.9583	1.9532	1.9481	1.9430	1.9380	1.9331	1.9282	1.9233
0.20	1.9185	1.9137	1.9090	1.9043	1.8997	1.8951	1.8906	1.8861	1.8817	1.8772
0.21	1.8729	1.8685	1.8643	1.8600	1.8558	1.8516	1.8475	1.8434	1.8393	1.8353
0.22	1.8313	1.8274	1.8234	1.8195	1.8157	1.8119	1.8081	1.8043	1.8006	1.7969
0.23	1.7933	1.7896	1.7860	1.7825	1.7789	1.7754	1.7719	1.7685	1.7651	1.7617
0.24	1.7583	1.7550	1.7516	1.7484	1.7451	1.7419	1.7386	1.7355	1.7323	1.7292
0.25	1.7261	1.7230	1.7199	1.7169	1.7138	1.7108	1.7079	1.7049	1.7020	1.6991
0.26	1.6962	1.6933	1.6905	1.6877	1.6849	1.6821	1.6793	1.6766	1.6739	1.6712
0.27	1.6685	1.6658	1.6632	1.6606	1.6579	1.6554	1.6528	1.6502	1.6477	1.6452
0.28	1.6427	1.6402	1.6377	1.6353	1.6328	1.6304	1.6280	1.6256	1.6233	1.6209
0.29	1.6186	1.6162	1.6139	1.6116	1.6094	1.6071	1.6048	1.6026	1.6004	1.5982
0.30	1.5960	1.5938	1.5916	1.5895	1.5873	1.5852	1.5831	1.5810	1.5789	1.5768
0.31	1.5748	1.5727	1.5707	1.5687	1.5667	1.5647	1.5627	1.5607	1.5587	1.5568
0.32	1.5548	1.5529	1.5510	1.5491	1.5472	1.5453	1.5434	1.5416	1.5397	1.5379
0.33	1.5360	1.5342	1.5324	1.5306	1.5288	1.5270	1.5253	1.5235	1.5217	1.5200
0.34	1.5183	1.5165	1.5148	1.5131	1.5114	1.5097	1.5081	1.5064	1.5047	1.5031
0.35	1.5014	1.4998	1.4982	1.4966	1.4950	1.4934	1.4918	1.4902	1.4886	1.4871
0.36	1.4855	1.4839	1.4824	1.4809	1.4793	1.4778	1.4763	1.4748	1.4733	1.4718
0.37	1.4703	1.4689	1.4674	1.4659	1.4645	1.4630	1.4616	1.4602	1.4587	1.4573
0.38	1.4559	1.4545	1.4531	1.4517	1.4503	1.4490	1.4476	1.4462	1.4449	1.4435
0.39	1.4422	1.4408	1.4395	1.4382	1.4368	1.4355	1.4342	1.4329	1.4316	1.4303
0.40	1.4290	1.4277	1.4265	1.4252	1.4239	1.4227	1.4214	1.4202	1.4189	1.4177
0.41	1.4165	1.4152	1.4140	1.4128	1.4116	1.4104	1.4092	1.4080	1.4068	1.4056
0.42	1.4044	1.4033	1.4021	1.4009	1.3998	1.3986	1.3975	1.3963	1.3952	1.3940
0.43	1.3929	1.3918	1.3907	1.3895	1.3884	1.3873	1.3862	1.3851	1.3840	1.3829
0.44	1.3818	1.3808	1.3797	1.3786	1.3775	1.3765	1.3754	1.3744	1.3733	1.3723
0.45	1.3712	1.3702	1.3691	1.3681	1.3671	1.3660	1.3650	1.3640	1.3630	1.3620
0.46	1.3610	1.3600	1.3590	1.3580	1.3570	1.3560	1.3550	1.3540	1.3530	1.3521
0.47	1.3511	1.3501	1.3492	1.3482	1.3472	1.3463	1.3453	1.3444	1.3434	1.3425
0.48	1.3416	1.3406	1.3397	1.3388	1.3378	1.3369	1.3360	1.3351	1.3342	1.3333
0.49	1.3324	1.3315	1.3306	1.3297	1.3288	1.3279	1.3270	1.3261	1.3252	1.3243

TABLE B.34 (cont.) CORRECTION FACTOR, *K*, FOR THE WATSON AND WILLIAMS TEST

r	0	1	2	3	4	5	6	7	8	9
0.50	1.3235	1.3226	1.3217	1.3209	1.3200	1.3191	1.3183	1.3174	1.3166	1.3157
0.51	1.3148	1.3140	1.3132	1.3123	1.3115	1.3106	1.3098	1.3090	1.3081	1.3073
0.52	1.3065	1.3057	1.3049	1.3040	1.3032	1.3024	1.3016	1.3008	1.3000	1.2992
0.53	1.2984	1.2976	1.2968	1.2960	1.2952	1.2944	1.2936	1.2929	1.2921	1.2913
0.54	1.2905	1.2897	1.2890	1.2882	1.2874	1.2867	1.2859	1.2851	1.2844	1.2836
0.55	1.2829	1.2821	1.2814	1.2806	1.2799	1.2791	1.2784	1.2776	1.2769	1.2762
0.56	1.2754	1.2747	1.2740	1.2732	1.2725	1.2718	1.2710	1.2703	1.2696	1.2689
0.57	1.2682	1.2674	1.2667	1.2660	1.2653	1.2646	1.2639	1.2632	1.2625	1.2618
0.58	1.2611	1.2604	1.2597	1.2590	1.2583	1.2576	1.2569	1.2562	1.2555	1.2548
0.59	1.2542	1.2535	1.2528	1.2521	1.2514	1.2508	1.2501	1.2494	1.2487	1.2481
0.60	1.2474	1.2467	1.2461	1.2454	1.2447	1.2441	1.2434	1.2428	1.2421	1.2414
0.61	1.2408	1.2401	1.2395	1.2388	1.2382	1.2375	1.2369	1.2362	1.2356	1.2350
0.62	1.2343	1.2337	1.2330	1.2324	1.2318	1.2311	1.2305	1.2298	1.2292	1.2286
0.63	1.2280	1.2273	1.2267	1.2261	1.2254	1.2248	1.2242	1.2236	1.2230	1.2223
0.64	1.2217	1.2211	1.2205	1.2199	1.2193	1.2186	1.2180	1.2174	1.2168	1.2162
0.65	1.2156	1.2150	1.2144	1.2138	1.2132	1.2126	1.2120	1.2114	1.2108	1.2102
0.66	1.2096	1.2090	1.2084	1.2078	1.2072	1.2066	1.2060	1.2054	1.2048	1.2042
0.67	1.2036	1.2030	1.2024	1.2018	1.2013	1.2007	1.2001	1.1995	1.1989	1.1983
0.68	1.1977	1.1972	1.1966	1.1960	1.1954	1.1948	1.1943	1.1937	1.1931	1.1925
0.69	1.1920	1.1914	1.1908	1.1902	1.1897	1.1891	1.1885	1.1879	1.1874	1.1868
0.70	1.1862	1.1857	1.1851	1.1845	1.1840	1.1834	1.1828	1.1823	1.1817	1.1811
0.71	1.1806	1.1800	1.1794	1.1789	1.1783	1.1777	1.1772	1.1766	1.1761	1.1755
0.72	1.1749	1.1744	1.1738	1.1733	1.1727	1.1721	1.1716	1.1710	1.1705	1.1699
0.73	1.1694	1.1688	1.1682	1.1677	1.1671	1.1666	1.1660	1.1655	1.1649	1.1644
0.74	1.1638	1.1633	1.1627	1.1621	1.1616	1.1610	1.1605	1.1599	1.1594	1.1588
0.75	1.1583	1.1577	1.1572	1.1566	1.1561	1.1555	1.1550	1.1544	1.1539	1.1533
0.76	1.1528	1.1522	1.1517	1.1511	1.1505	1.1500	1.1494	1.1489	1.1483	1.1478
0.77	1.1472	1.1467	1.1461	1.1456	1.1450	1.1445	1.1439	1.1434	1.1428	1.1423
0.78	1.1417	1.1412	1.1406	1.1401	1.1395	1.1389	1.1384	1.1378	1.1373	1.1367
0.79	1.1362	1.1356	1.1351	1.1345	1.1340	1.1334	1.1328	1.1323	1.1317	1.1312
0.80	1.1306	1.1300	1.1295	1.1289	1.1284	1.1278	1.1272	1.1267	1.1261	1.1256
0.81	1.1250	1.1244	1.1239	1.1233	1.1227	1.1222	1.1216	1.1210	1.1205	1.1199
0.82	1.1193	1.1188	1.1182	1.1176	1.1170	1.1165	1.1159	1.1153	1.1147	1.1142
0.83	1.1136	1.1130	1.1124	1.1119	1.1113	1.1107	1.1101	1.1095	1.1090	1.1084
0.84	1.1078	1.1072	1.1066	1.1060	1.1054	1.1049	1.1043	1.1037	1.1031	1.1025
0.85	1.1019	1.1013	1.1007	1.1001	1.0995	1.0989	1.0983	1.0977	1.0971	1.0965
0.86	1.0959	1.0953	1.0947	1.0941	1.0935	1.0928	1.0922	1.0916	1.0910	1.0904
0.87	1.0898	1.0892	1.0885	1.0879	1.0873	1.0867	1.0861	1.0854	1.0848	1.0842
0.88	1.0835	1.0829	1.0823	1.0816	1.0810	1.0804	1.0797	1.0791	1.0785	1.0778
0.89	1.0772	1.0765	1.0759	1.0752	1.0746	1.0740	1.0733	1.0727	1.0720	1.0713
0.90	1.0707	1.0700	1.0694	1.0687	1.0681	1.0674	1.0667	1.0661	1.0654	1.0647
0.91	1.0641	1.0634	1.0627	1.0621	1.0614	1.0607	1.0601	1.0594	1.0587	1.0580
0.92	1.0573	1.0567	1.0560	1.0553	1.0546	1.0539	1.0533	1.0526	1.0519	1.0512
0.93	1.0505	1.0498	1.0491	1.0484	1.0477	1.0470	1.0463	1.0456	1.0449	1.0443
0.94	1.0436	1.0429	1.0422	1.0414	1.0407	1.0400	1.0393	1.0386	1.0379	1.0372
0.95	1.0365	1.0358	1.0351	1.0344	1.0337	1.0330	1.0322	1.0315	1.0308	1.0301
0.96	1.0294	1.0287	1.0279	1.0272	1.0265	1.0258	1.0251	1.0243	1.0236	1.0229
0.97	1.0222	1.0214	1.0207	1.0200	1.0192	1.0185	1.0178	1.0170	1.0163	1.0156
0.98	1.0148	1.0141	1.0134	1.0126	1.0119	1.0112	1.0104	1.0097	1.0089	1.0082
0.99	1.0075	1.0067	1.0060	1.0052	1.0045	1.0037	1.0030	1.0022	1.000*	1.000*

*No correction needed.

Values of K were determined as $1 + 3/8k$, where k was obtained using Equation 6.3.14 of Mardia (1972: 155), and Equations 9.8.1–9.8.4 of Olver (1964).

Examples:

$$r = 0.743,\ K = 1.1621 \quad \text{and} \quad r = 0.814,\ K = 1.227.$$

TABLE B.35 CRITICAL VALUES OF WATSON'S U^2

n_1	n_2	α = 0.50	0.20	0.10	0.05	0.02	0.01	0.005	0.002	0.001
4	4	0.1172	0.1875	------	------	------	------	------	------	------
4	5	0.0815	0.2037	0.2037	------	------	------	------	------	------
4	6	0.0875	0.1333	0.2167	0.2167	------	------	------	------	------
4	7	0.0844	0.1299	0.1688	0.2273	------	------	------	------	------
4	8	0.0903	0.1319	0.1632	0.2361	------	------	------	------	------
4	9	0.0855	0.1282	0.1752	0.2436	0.2436	------	------	------	------
4	10	0.0804	0.1232	0.1571	0.2018	0.2500	------	------	------	------
4	11	0.0828	0.1253	0.1556	0.1949	0.2556	------	------	------	------
4	12	0.0781	0.1302	0.1563	0.2031	0.2604	0.2604	------	------	------
4	13	0.0792	0.1244	0.1538	0.1855	0.2647	0.2647	------	------	------
4	14	0.0780	0.1227	0.1534	0.1931	0.2298	0.2685	------	------	------
4	15	0.0789	0.1228	0.1561	0.1807	0.2228	0.2719	0.2719	------	------
4	16	0.0781	0.1250	0.1531	0.1836	0.2281	0.2750	0.2750	------	------
4	17	0.0775	0.1223	0.1531	0.1839	0.2330	0.2778	0.2778	------	------
4	18	0.0764	0.1212	0.1490	0.1818	0.2197	0.2481	0.2803	------	------
4	19	0.0755	0.1213	0.1533	0.1796	0.2220	0.2517	0.2826	------	------
4	20	0.0764	0.1201	0.1535	0.1842	0.2264	0.2451	0.2847	------	------
4	21	0.0752	0.1200	0.1514	0.1819	0.2143	0.2486	0.2867	0.2867	------
4	22	0.0756	0.1211	0.1508	0.1823	0.2185	0.2517	0.2885	0.2885	------
4	23	0.0751	0.1194	0.1508	0.1814	0.2177	0.2394	0.2636	0.2901	------
4	24	0.0755	0.1202	0.1499	0.1797	0.2184	0.2411	0.2660	0.2917	------
4	25	0.0752	0.1200	0.1497	0.1814	0.2152	0.2441	0.2600	0.2931	------
4	26	0.0752	0.1191	0.1486	0.1816	0.2175	0.2396	0.2624	0.2944	------
4	27	0.0753	0.1189	0.1505	0.1786	0.2151	0.2360	0.2646	0.2957	0.2957
4	28	0.0748	0.1203	0.1496	0.1775	0.2165	0.2388	0.2667	0.2969	0.2969
4	29	0.0749	0.1198	0.1491	0.1794	0.2165	0.2369	0.2557	0.2980	0.2980
4	30	0.0745	0.1196	0.1493	0.1797	0.2140	0.2395	0.2578	0.2990	0.2990
5	5	0.0890	0.1610	0.2250	0.2250	------	------	------	------	------
5	6	0.0848	0.1333	0.1818	0.2424	------	------	------	------	------
5	7	0.0855	0.1284	0.1712	0.1998	0.2569	------	------	------	------
5	8	0.0846	0.1308	0.1654	0.2154	0.2692	------	------	------	------
5	9	0.0798	0.1242	0.1591	0.1909	0.2798	0.2798	------	------	------
5	10	0.0836	0.1236	0.1609	0.1956	0.2409	0.2889	0.2889	------	------
5	11	0.0810	0.1241	0.1560	0.1901	0.2287	0.2969	0.2969	------	------
5	12	0.0784	0.1235	0.1549	0.1863	0.2255	0.2608	0.3039	------	------
5	13	0.0777	0.1256	0.1563	0.1837	0.2298	0.2692	0.3102	------	------
5	14	0.0782	0.1218	0.1534	0.1820	0.2211	0.2571	0.2767	0.3158	------
5	15	0.0782	0.1235	0.1515	0.1835	0.2248	0.2515	0.2835	0.3208	------
5	15	0.0766	0.1206	0.1552	0.1825	0.2230	0.2552	0.2897	0.3254	------
5	17	0.0761	0.1199	0.1520	0.1820	0.2205	0.2472	0.2782	0.3295	0.3295
5	18	0.0763	0.1208	0.1536	0.1797	0.2164	0.2464	0.2715	0.3333	0.3333
5	19	0.0754	0.1201	0.1517	0.1824	0.2193	0.2526	0.2745	0.3052	0.3368
5	20	0.0760	0.1216	0.1520	0.1824	0.2200	0.2416	0.2664	0.3096	0.3400
5	21	0.0755	0.1195	0.1510	0.1810	0.2206	0.2448	0.2712	0.2990	0.3429
5	22	0.0756	0.1201	0.1524	0.1820	0.2191	0.2426	0.2689	0.3033	0.3457
5	23	0.0755	0.1196	0.1513	0.1811	0.2178	0.2451	0.2737	0.2960	0.3209
5	24	0.0747	0.1195	0.1511	0.1810	0.2190	0.2437	0.2736	0.2983	0.3241
5	25	0.0754	0.1197	0.1517	0.1810	0.2168	0.2461	0.2674	0.3021	0.3272
5	26	0.0749	0.1186	0.1514	0.1806	0.2189	0.2447	0.2675	0.2943	0.3176
5	27	0.0748	0.1193	0.1508	0.1804	0.2165	0.2443	0.2674	0.2975	0.3207

TABLE B.35 (cont.) CRITICAL VALUES OF WATSON'S U^2

n_1	n_2	α = 0.50	0.20	0.10	0.05	0.02	0.01	0.005	0.002	0.001
5	28	0.0746	0.1188	0.1512	0.1802	0.2170	0.2417	0.2694	0.2937	0.3136
5	29	0.0743	0.1189	0.1510	0.1802	0.2171	0.2443	0.2666	0.2970	0.3153
5	30	0.0743	0.1189	0.1512	0.1802	0.2160	0.2419	0.2678	0.2979	0.3181
6	6	0.0880	0.1319	0.1713	0.2060	0.2639	------	------	------	------
6	7	0.0806	0.1209	0.1538	0.1941	0.2821	0.2821	------	------	------
6	8	0.0833	0.1265	0.1607	0.1964	0.2455	0.2976	0.2976	------	------
6	9	0.0815	0.1259	0.1556	0.1926	0.2321	0.2617	0.3111	------	------
6	10	0.0771	0.1260	0.1563	0.1896	0.2313	0.2479	0.3229	0.3229	------
6	11	0.0784	0.1212	0.1569	0.1872	0.2246	0.2620	0.2888	0.3333	------
6	12	0.0802	0.1242	0.1551	0.1829	0.2261	0.2593	0.2747	0.3426	0.3426
6	13	0.0769	0.1215	0.1538	0.1849	0.2213	0.2497	0.2780	0.3509	0.3509
6	14	0.0768	0.1220	0.1536	0.1839	0.2250	0.2506	0.2821	0.3196	0.3583
6	15	0.0762	0.1217	0.1524	0.1852	0.2201	0.2487	0.2730	0.3058	0.3651
6	16	0.0758	0.1212	0.1534	0.1823	0.2235	0.2500	0.2789	0.3073	0.3357
6	17	0.0750	0.1211	0.1526	0.1833	0.2199	0.2472	0.2745	0.3129	0.3427
6	18	0.0760	0.1211	0.1535	0.1840	0.2199	0.2461	0.2739	0.2998	0.3295
6	19	0.0751	0.1200	0.1523	0.1832	0.2204	0.2498	0.2744	0.3060	0.3298
6	20	0.0747	0.1196	0.1526	0.1824	0.2196	0.2490	0.2734	0.3077	0.3333
6	21	0.0758	0.1205	0.1523	0.1834	0.2205	0.2475	0.2734	0.3057	0.3369
6	22	0.0749	0.1204	0.1518	0.1824	0.2202	0.2473	0.2752	0.3036	0.3260
6	23	0.0745	0.1194	0.1514	0.1824	0.2194	0.2469	0.2729	0.3073	0.3273
6	24	0.0743	0.1194	0.1519	0.1826	0.2206	0.2484	0.2715	0.3056	0.3289
6	25	0.0744	0.1191	0.1514	0.1819	0.2202	0.2473	0.2731	0.3015	0.3277
6	26	0.0739	0.1188	0.1510	0.1815	0.2198	0.2464	0.2710	0.3047	0.3265
6	27	0.0741	0.1193	0.1515	0.1822	0.2200	0.2469	0.2731	0.3053	0.3281
6	28	0.0737	0.1190	0.1507	0.1821	0.2201	0.2467	0.2731	0.3039	0.3270
6	29	0.0736	0.1189	0.1511	0.1816	0.2200	0.2473	0.2719	0.3038	0.3258
6	30	0.0736	0.1193	0.1509	0.1823	0.2194	0.2471	0.2725	0.3045	0.3262
7	7	0.0791	0.1345	0.1578	0.1986	0.2511	0.3036	0.3036	------	------
7	8	0.0794	0.1198	0.1556	0.1817	0.2246	0.2722	0.3222	------	------
7	9	0.0786	0.1223	0.1560	0.1818	0.2215	0.2552	0.2909	0.3385	------
7	10	0.0773	0.1227	0.1546	0.1866	0.2269	0.2622	0.2773	0.3529	0.3529
7	11	0.0771	0.1219	0.1551	0.1839	0.2214	0.2532	0.2806	0.3225	0.3657
7	12	0.0764	0.1216	0.1541	0.1855	0.2256	0.2519	0.2757	0.3083	0.3772
7	13	0.0765	0.1216	0.1545	0.1842	0.2227	0.2523	0.2776	0.3150	0.3479
7	14	0.0761	0.1228	0.1568	0.1840	0.2248	0.2530	0.2744	0.3210	0.3337
7	15	0.0754	0.1213	0.1525	0.1845	0.2235	0.2503	0.2780	0.3118	0.3378
7	16	0.0753	0.1203	0.1530	0.1848	0.2236	0.2508	0.2772	0.3113	0.3432
7	17	0.0749	0.1204	0.1526	0.1827	0.2227	0.2500	0.2752	0.3109	0.3340
7	18	0.0749	0.1200	0.1524	0.1841	0.2235	0.2502	0.2768	0.3117	0.3346
7	20	0.0743	0.1198	0.1526	0.1832	0.2219	0.2499	0.2780	0.3081	0.3330
7	21	0.0751	0.1203	0.1534	0.1840	0.2224	0.2496	0.2782	0.3123	0.3336
7	22	0.0743	0.1196	0.1518	0.1832	0.2221	0.2512	0.2763	0.3090	0.3341
7	23	0.0739	0.1194	0.1522	0.1832	0.2226	0.2499	0.2780	0.3103	0.3327
8	8	0.0781	0.1250	0.1563	0.1836	0.2256	0.2500	0.2959	0.3438	------
8	9	0.0784	0.1225	0.1552	0.1863	0.2255	0.2582	0.2827	0.3627	0.3627
8	10	0.0775	0.1220	0.1546	0.1852	0.2220	0.2491	0.2796	0.3359	0.3796
8	11	0.0766	0.1220	0.1543	0.1842	0.2249	0.2524	0.2799	0.3194	0.3529
8	12	0.0766	0.1208	0.1557	0.1854	0.2229	0.2521	0.2807	0.3167	0.3396
8	13	0.0754	0.1212	0.1532	0.1853	0.2237	0.2531	0.2778	0.3135	0.3446

TABLE B.35 (cont.) CRITICAL VALUES OF WATSON'S U^2

n_1	n_2	α = 0.50	0.20	0.10	0.05	0.02	0.01	0.005	0.002	0.001
8	14	0.0751	0.1205	0.1530	0.1855	0.2224	0.2516	0.2796	0.3137	0.3381
8	15	0.0746	0.1210	0.1536	0.1855	0.2232	0.2507	0.2783	0.3130	0.3341
8	16	0.0761	0.1220	0.1542	0.1854	0.2222	0.2531	0.2795	0.3156	0.3417
8	17	0.0747	0.1200	0.1529	0.1841	0.2241	0.2524	0.2782	0.3124	0.3388
8	18	0.0748	0.1199	0.1528	0.1840	0.2244	0.2513	0.2813	0.3152	0.3397
8	19	0.0742	0.1196	0.1527	0.1839	0.2243	0.2526	0.2799	0.3145	0.3384
8	20	0.0741	0.1196	0.1527	0.1839	0.2239	0.2527	0.2795	0.3134	0.3393
9	9	0.0770	0.1250	0.1552	0.1867	0.2251	0.2663	0.2855	0.3404	0.3843
9	10	0.0760	0.1216	0.1544	0.1860	0.2257	0.2538	0.2865	0.3205	0.3614
9	11	0.0764	0.1208	0.1542	0.1845	0.2249	0.2552	0.2814	0.3168	0.3410
9	12	0.0767	0.1217	0.1543	0.1852	0.2257	0.2540	0.2804	0.3157	0.3395
9	13	0.0755	0.1205	0.1532	0.1850	0.2247	0.2526	0.2798	0.3187	0.3389
9	14	0.0752	0.1201	0.1532	0.1843	0.2243	0.2526	0.2809	0.3168	0.3409
9	15	0.0757	0.1201	0.1535	0.1850	0.2245	0.2541	0.2831	0.3152	0.3393
9	16	0.0744	0.1200	0.1533	0.1850	0.2244	0.2539	0.2822	0.3172	0.3439
10	10	0.0750	0.1225	0.1545	0.1850	0.2250	0.2545	0.2825	0.3170	0.3450
10	11	0.0756	0.1215	0.1544	0.1856	0.2237	0.2548	0.2791	0.3172	0.3405
10	12	0.0758	0.1212	0.1534	0.1848	0.2246	0.2545	0.2818	0.3155	0.3409
10	13	0.0749	0.1204	0.1532	0.1853	0.2254	0.2542	0.2816	0.3184	0.3452
10	14	0.0749	0.1201	0.1535	0.1847	0.2252	0.2550	0.2823	0.3181	0.3439
10	15	0.0747	0.1211	0.1536	0.1856	0.2256	0.2549	0.2837	0.3189	0.3440
11	11	0.0760	0.1211	0.1541	0.1857	0.2262	0.2540	0.2826	0.3194	0.3442
11	12	0.0751	0.1206	0.1535	0.1851	0.2253	0.2543	0.2839	0.3182	0.3439
11	13	0.0746	0.1206	0.1532	0.1853	0.2255	0.2546	0.2838	0.3193	0.3461
12	12	0.0752	0.1215	0.1528	0.1863	0.2266	0.2558	0.2844	0.3192	0.3438
14	14	0.070	0.117	0.151	0.183	0.226	0.258	0.289	0.330	0.361
16	16	0.070	0.117	0.151	0.184	0.227	0.259	0.291	0.332	0.364
18	18	0.070	0.117	0.151	0.184	0.228	0.260	0.292	0.334	0.366
20	20	0.069	0.117	0.151	0.185	0.228	0.261	0.293	0.335	0.367
25	25	0.069	0.117	0.152	0.185	0.229	0.262	0.295	0.338	0.370
30	30	0.069	0.117	0.152	0.186	0.230	0.263	0.296	0.339	0.372
35	35	0.069	0.117	0.152	0.186	0.231	0.264	0.297	0.340	0.373
40	40	0.069	0.117	0.152	0.186	0.231	0.264	0.298	0.341	0.374
50	50	0.069	0.117	0.152	0.187	0.231	0.265	0.299	0.343	0.376
60	60	0.069	0.117	0.152	0.187	0.232	0.266	0.299	0.343	0.377
80	80	0.069	0.117	0.152	0.187	0.232	0.266	0.300	0.344	0.378
100	100	0.069	0.117	0.152	0.187	0.233	0.267	0.300	0.345	0.378
∞	∞	0.0710	0.1167	0.1518	0.1869	0.2333	0.2684	0.3035	0.3500	0.3851

The four-decimal-place critical values in this table (except for sample sizes of infinity) were obtained from distributions of U^2 calculated using the method described by Burr (1964). The three-decimal-place critical values shown were computed by the approximation of Tiku (1965), using the computer algorithms of Best and Roberts (1975), Bhattacharjee (1970), International Business Machines (1968: 362), and Odeh and Evans (1974). Comparing these values to those of Stephens (1974), for $\alpha = 0.005$ through 0.10, shows agreement to the third decimal place. The critical values for sample sizes of infinity were computed as

$$U^2_{\alpha,\infty,\infty} = -\left(\frac{1}{2\pi^2}\right)\left[\ln\left(\frac{\alpha}{2}\right) - \ln\left\{1 + \left(\frac{\alpha}{2}\right)^3\right\}\right]$$

(Watson 1962) and should be used if sample sizes are greater than 100.

Examples:

$$U^2_{0.05,6,8} = 0.1964 \quad \text{and} \quad U^2_{0.01,10,12} = 0.2545.$$

TABLE B.36 CRITICAL VALUES OF R′ FOR THE MOORE TEST OF CIRCULAR UNIFORMITY

n	α = 0.20	0.10	0.05	0.02	0.01	0.002
2	1.049	1.053	1.060	1.061	1.061	1.061
3	1.039	1.095	1.124	1.143	1.149	1.154
4	1.008	1.090	1.146	1.192	1.212	1.238
5	0.988	1.084	1.152	1.216	1.250	1.298
6	0.972	1.074	1.152	1.230	1.275	1.345
7	0.959	1.065	1.150	1.238	1.291	1.373
8	0.949	1.059	1.148	1.242	1.300	1.397
9	0.940	1.053	1.145	1.245	1.307	1.416
10	0.934	1.048	1.144	1.248	1.313	1.432
12	0.926	1.042	1.140	1.252	1.322	1.456
14	0.920	1.037	1.136	1.252	1.325	1.470
16	0.914	1.031	1.132	1.250	1.327	1.480
18	0.910	1.027	1.129	1.248	1.328	1.487
20	0.906	1.024	1.127	1.247	1.329	1.492
22	0.903	1.022	1.126	1.246	1.330	1.496
24	0.901	1.021	1.125	1.246	1.331	1.499
26	0.899	1.019	1.124	1.246	1.332	1.501
28	0.897	1.018	1.124	1.246	1.333	1.502
30	0.896	1.016	1.123	1.245	1.334	1.502
40	0.891	1.012	1.119	1.243	1.332	1.504
50	0.887	1.007	1.115	1.241	1.329	1.506
80	0.883	1.005	1.113	1.240	1.329	1.508
100	0.881	1.004	1.112	1.240	1.329	1.509
∞	0.876	0.999	1.109	1.239	1.329	1.517

Values in this table were taken, with permission of the publisher, from Table 1 of Moore (1980, Biometrika 80: 175–180).

Examples:

$$R'_{0.05,24} = 1.125 \quad \text{and} \quad R'_{0.10,30} = 1.016.$$

TABLE B.37 CRITICAL VALUES FOR THE RUNS TEST ON A CIRCLE

n_1	n_2	$\alpha(2) = 0.50$	0.20	0.10	0.05	0.02	0.01	0.005	0.002	0.001
		$\alpha(1) = 0.25$	0.10	0.05	0.025	0.01	0.005	0.0025	0.001	0.0005
2	6	2, -								
	7	2, -								
	8	2, -								
	9	2, -								
	10	2, -								
	11	2, -								
	12	2, -								
	13	2, -								
	14	2, -								
	15	2, -								
	16	2, -								
	17	2, -								
	18	2, -	2, -							
	19	2, -	2, -							
2	20	2, -	2, -							
3	3	2, 6								
	4	2, 6								
	5	2, -								
	6	2, -	2, -							
	7	2, -	2, -							
	8	2, -	2, -							
	9	2, -	2, -							
	10	2, -	2, -	2, -						
	11	2, -	2, -	2, -						
	12	2, -	2, -	2, -						
	13	2, -	2, -	2, -						
	14	2, -	2, -	2, -	2, -					
	15	2, -	2, -	2, -	2, -					
	16	2, -	2, -	2, -	2, -					
	17	2, -	2, -	2, -	2, -					
	18	2, -	2, -	2, -	2, -					
	19	2, -	2, -	2, -	2, -					
3	20	2, -	2, -	2, -	2, -					
4	4	2, 8	2, 8							
	5	2, 8	2, 8							
	6	2, 8	2, 8	2, -						
	7	2, 8	2, -	2, -						
	8	2, 8	2, -	2, -	2, -					
	9	4, -	2, -	2, -	2, -					
	10	4, -	2, -	2, -	2, -					
	11	4, -	2, -	2, -	2, -					
	12	4, -	2, -	2, -	2, -	2, -				
	13	4, -	2, -	2, -	2, -	2, -				
	14	4, -	2, -	2, -	2, -	2, -				
	15	4, -	2, -	2, -	2, -	2, -	2, -			
	16	4, -	4, -	2, -	2, -	2, -	2, -			
	17	4, -	4, -	2, -	2, -	2, -	2, -			
4	18	4, -	4, -	2, -	2, -	2, -	2, -			

TABLE B.37 (cont.) CRITICAL VALUES FOR THE RUNS TEST ON A CIRCLE

n_1	n_2	$\alpha(2) = 0.50$	0.20	0.10	0.05	0.02	0.01	0.005	0.002	0.001
		$\alpha(1) = 0.25$	0.10	0.05	0.025	0.01	0.005	0.0025	0.001	0.0005
4	19	4, -	4, -	2, -	2, -	2, -	2, -			
4	20	4, -	4, -	2, -	2, -	2, -	2, -	2, -		
5	5	2, 8	2,10	2,10						
	5	2,10	2,10	2,10	2,10					
	7	4,10	2,10	2,10	2, -					
	8	4,10	2,10	2, -	2, -					
	9	4,10	2,10	2, -	2, -	2, -				
	10	4,10	4, -	2, -	2, -	2, -	2, -			
		4,10	4, -	2, -	2, -	2, -	2, -			
	11	4,10	4, -	2, -	2, -	2, -	2, -			
	12	4,10	4, -	2, -	2, -	2, -	2, -			
	13	4,10	4, -	2, -	2, -	2, -	2, -	2, -		
	14	4,10	4, -	4, -	2, -	2, -	2, -	2, -		
	15	4, -	4, -	4, -	2, -	2, -	2, -	2, -		
	16	6, -	4, -	4, -	2, -	2, -	2, -	2, -		
	17	6, -	4, -	4, -	2, -	2, -	2, -	2, -	2, -	
	18	6, -	4, -	4, -	4, -	2, -	2, -	2, -	2, -	
	19	6, -	4, -	4, -	4, -	2, -	2, -	2, -	2, -	
5	20	6, -	4, -	4, -	4, -	2, -	2, -	2, -	2, -	2, -
6	6	4,10	2,10	2,12	2,12					
	7	4,10	4,12	2,12	2,12	2,12				
	8	4,10	4,12	2,12	2,12	2, -	2, -			
	9	4,10	4,12	2,12	2, -	2, -	2, -			
	10	4,12	4,12	4,12	2, -	2, -	2, -	2, -		
	11	4,12	4,12	4, -	2, -	2, -	2, -	2, -		
	12	4,12	4,12	4, -	2, -	2, -	2, -	2, -	2, -	
	13	6,12	4, -	4, -	4, -	2, -	2, -	2, -	2, -	
	14	6,12	4, -	4, -	4, -	2, -	2, -	2, -	2, -	
	15	6,12	4, -	4, -	4, -	2, -	2, -	2, -	2, -	2, -
	16	6,12	4, -	4, -	4, -	2, -	2, -	2, -	2, -	2, -
	17	6,12	4, -	4, -	4, -	4, -	2, -	2, -	2, -	2, -
	18	6,12	6, -	4, -	4, -	4, -	2, -	2, -	2, -	2, -
	19	6,12	6, -	4, -	4, -	4, -	2, -	2, -	2, -	2, -
6	20	6,12	6, -	4, -	4, -	4, -	2, -	2, -	2, -	2, -
7	7	4,10	4,12	2,12	2,14	2,14	2,14			
	8	4,12	4,12	2,14	2,14	2,14	2,14	2,14		
	9	4,12	4,12	4,12	2,14	2,14	2, -	2, -		
	10	6,12	4,14	4,14	4,14	2, -	2, -	2, -	2, -	
	11	6,12	4,14	4,14	4,14	2, -	2, -	2, -	2, -	
	12	6,12	4,14	4,14	4, -	2, -	2, -	2, -	2, -	2, -
	13	6,12	4,14	4,14	4, -	4, -	2, -	2, -	2, -	2, -
	14	6,14	6,14	4,14	4, -	4, -	2, -	2, -	2, -	2, -
	15	6,14	6,14	4, -	4, -	4, -	2, -	2, -	2, -	2, -
	16	6,14	6,14	4, -	4, -	4, -	4, -	2, -	2, -	2, -
	17	8,14	6,14	6, -	4, -	4, -	4, -	2, -	2, -	2, -
	18	8,14	6,14	6, -	4, -	4, -	4, -	2, -	2, -	2, -
	19	8,14	5, -	6, -	4, -	4, -	4, -	2, -	2, -	2, -
7	20	8,14	6, -	6, -	4, -	4, -	4, -	4, -	2, -	2, -

TABLE B.37 (cont.) CRITICAL VALUES FOR THE RUNS TEST ON A CIRCLE

n_1	n_2	$\alpha(2) = 0.50$ $\alpha(1) = 0.25$	0.20 0.10	0.10 0.05	0.05 0.025	0.02 0.01	0.01 0.005	0.005 0.0025	0.002 0.001	0.001 0.0005
8	8	6,12	4,14	4,14	2,14	2,14	2,16	2,16		
	9	6,12	4,14	4,14	4,14	2,16	2,16	2,16	2,16	
	10	6,12	4,14	4,14	4,16	2,16	2,16	2,16	2, -	2, -
	11	6,14	6,14	4,16	4,16	4,16	2,16	2, -	2, -	2, -
	12	6,14	6,14	4,16	4,16	4,16	2, -	2, -	2, -	2, -
	13	6,14	6,14	4,16	4,16	4, -	4, -	2, -	2, -	2, -
	14	8,14	6,16	6,18	4,18	4,18	4,18	2, -	2, -	2, -
	15	8,16	6,16	6,18	4,18	4,18	4, -	4, -	2, -	2, -
	16	8,16	6,16	6,18	4,18	4,18	4, -	4, -	2, -	2, -
	17	8,14	6,16	6,16	6, -	4, -	4, -	4, -	2, -	2, -
	18	8,14	6,16	6,16	6, -	4, -	4, -	4, -	2, -	2, -
	19	8,16	8,16	6,16	6, -	4, -	4, -	4, -	4, -	2, -
8	20	8,16	8,16	6, -	6, -	4, -	4, -	4, -	4, -	2, -
9	9	6,12	4,14	4,14	4,16	2,16	2,16	2,18	2,18	2,18
	10	6,14	6,14	4,16	4,16	4,16	2,18	2,18	2,18	2,18
	11	6,14	6,16	4,16	4,16	4,18	4,18	2,18	2,18	2, -
	12	8,14	6,16	6,16	4,16	4,18	4,18	2,18	2, -	2, -
	13	8,14	6,16	6,16	4,18	4,18	4,18	4,18	2, -	2, -
	14	8,14	6,16	6,18	6,18	4,18	4,18	4, -	2, -	2, -
	15	8,16	6,16	6,18	6,18	4,18	4, -	4, -	2, -	2, -
	16	8,16	8,18	6,18	6,18	4,18	4, -	4, -	4, -	2, -
	17	8,16	8,16	6,18	6,18	6, -	4, -	4, -	4, -	4, -
	18	8,16	8,18	6,18	6,18	6, -	4, -	4, -	4, -	4, -
	19	8,16	6,18	6,18	6,18	6, -	4, -	4, -	4, -	4, -
9	20	10,16	8,18	8,18	6,18	6, -	6, -	4, -	4, -	4, -
10	10	8,14	6,16	4,16	4,16	4,18	4,18	2,18	2,18	2,20
	11	8,14	6,16	6,16	4,18	4,18	4,18	2,20	2,20	2,20
	12	8,14	6,16	6,18	6,18	4,18	4,20	4,20	2,20	2,20
	13	8,16	6,16	6,18	6,18	4,20	4,20	4,20	2,20	2,20
	14	8,16	8,18	6,18	6,18	4,20	4,20	4,20	4,20	2, -
	15	8,16	8,18	6,18	6,18	6,20	4,20	4,20	4, -	4, -
	16	10,16	8,18	6,18	6,20	6,20	4,20	4,20	4, -	4, -
	17	10,16	8,18	8,18	6,20	6,20	6,20	4, -	4, -	4, -
	18	10,16	8,18	8,20	6,20	6,20	6, -	4, -	4, -	4, -
	19	10,18	8,20	8,20	6,20	6,20	6, -	4, -	4, -	4, -
10	20	10,18	8,20	8,20	8,20	6,20	6, -	4, -	4, -	4, -
11	11	8,16	6,18	6,18	6,18	4,18	4,20	4,20	2,20	2,20
	12	8,16	8,18	6,18	6,20	4,20	4,22	4,22	4,22	2,22
	13	8,16	8,18	6,18	6,20	4,20	4,20	4,20	4,22	2,22
	14	10,16	8,18	6,18	6,20	6,20	4,20	4,22	4,22	4,22
	15	10,16	8,18	8,20	6,20	6,20	6,22	4,22	4,22	4,22
	16	10,18	8,20	8,20	6,20	6,22	6,22	4,22	4,22	4, -
	17	10,18	8,18	8,20	8,20	6,22	6,22	6,22	4,22	4, -
	18	10,18	8,20	8,20	8,20	6,22	6,22	6,22	4, -	4, -
	19	10,18	10,20	8,20	8,22	6,22	6,22	6,22	4, -	4, -
11	20	10,18	10,20	8,20	8,22	6,22	6,22	6, -	6, -	4, -
12	12	8,16	8,18	6,18	6,20	6,20	4,20	4,22	4,22	4,22
12	13	10,16	8,18	8,18	6,20	6,20	4,22	4,22	4,22	4,22

TABLE B.37 (cont.) CRITICAL VALUES FOR THE RUNS TEST ON A CIRCLE

n_1	n_2	$\alpha(2) = 0.50$	0.20	0.10	0.05	0.02	0.01	0.005	0.002	0.001
		$\alpha(1) = 0.25$	0.10	0.05	0.025	0.01	0.005	0.0025	0.001	0.0005
12	14	10,18	8,18	8,20	6,20	6,22	6,22	4,22	4,22	4,24
	15	10,18	8,20	8,20	6,20	6,22	6,22	4,22	4,24	4,24
	16	10,18	8,20	8,20	8,22	6,22	6,22	6,24	4,24	4,24
	17	10,18	10,20	8,20	8,22	6,22	6,22	6,24	4,24	4,24
	18	12,18	10,20	8,22	8,22	6,22	6,24	6,24	6,24	4,24
	19	12,18	10,20	8,22	8,22	8,24	6,24	6,24	6,24	4, –
12	20	12,20	10,20	10,22	8,22	8,24	6,24	6,24	6,24	6, –
13	13	10,18	8,18	8,20	6,20	6,22	6,22	4,22	4, –	4, –
	14	10,18	8,20	8,20	8,20	6,22	6,22	6,22	4,24	4,24
	15	10,18	10,20	8,20	8,22	6,22	6,22	6,24	4,24	4,24
	16	12,18	10,20	8,22	8,22	6,22	6,24	6,24	4,24	4, –
	17	12,20	10,20	8,22	8,22	8,24	6,24	6,24	6,26	4,26
	18	12,20	10,20	10,22	8,22	8,24	6,24	6,24	6,26	6,26
	19	12,20	10,22	10,22	8,24	8,24	8,24	6,26	6,26	6,26
13	20	12,20	10,22	10,22	8,24	8,24	8,24	6,26	6,26	6,26
14	14	10,18	10,20	8,20	8,22	6,22	6,24	6,24	4,24	4,24
	15	12,18	10,20	8,22	8,22	6,24	6,24	6,24	6,24	4,26
	16	12,20	10,20	10,22	8,22	8,24	6,24	6,24	6,26	4,26
	17	12,20	10,22	10,22	8,24	8,24	6,24	6,26	6,26	6,26
	18	12,20	10,22	10,22	8,24	8,24	8,26	6,26	6,26	6,26
	19	12,20	12,22	10,24	10,24	8,24	8,26	6,26	6,26	6,28
14	20	14,20	12,22	10,24	10,24	8,24	8,26	8,26	6,28	6,28
15	15	12,20	10,20	10,22	8,22	8,24	6,24	6,24	6,26	4,26
	16	12,20	10,22	10,22	8,24	8,24	8,24	6,26	6,26	6,26
	17	12,20	10,22	10,22	10,24	8,24	8,26	6,26	6,26	6,28
	18	12,20	12,22	10,24	10,24	8,26	8,26	8,26	6,28	6,28
	19	14,22	12,22	10,24	10,24	8,26	8,26	8,28	6,28	6,28
15	20	14,22	12,24	10,24	10,26	8,26	8,26	8,28	6,28	6,28
16	16	12,20	10,22	10,24	10,24	8,24	8,26	6,26	6,26	6,28
	17	14,20	12,22	10,24	10,24	8,26	8,26	8,26	8,28	8,28
	18	14,22	12,24	10,24	10,26	10,26	8,26	8,28	8,28	6,28
	19	14,22	12,24	12,24	10,26	10,26	8,28	8,28	8,28	6,30
15	20	14,22	12,24	12,26	10,26	10,26	8,28	8,28	8,30	6,30
17	17	14,22	12,24	10,24	10,26	8,26	8,26	8,28	6,28	6,28
	18	14,22	12,24	12,24	10,26	10,26	8,28	8,28	8,28	6,30
	19	14,22	12,24	12,26	10,26	10,28	8,28	8,28	8,30	8,30
17	20	14,22	14,24	12,26	12,26	10,28	10,28	8,30	8,30	8,30
18	18	14,22	12,24	12,26	10,26	10,28	10,28	10,28	8,30	8,30
	19	16,24	14,26	12,28	12,28	10,30	10,30	10,32	8,32	8,32
18	20	16,24	14,26	12,26	12,28	10,28	10,30	10,30	8,30	8,32
19	19	16,24	14,26	12,26	12,28	10,28	10,30	10,30	8,30	8,32
19	20	16,24	14,26	12,28	12,28	10,30	10,30	10,30	8,32	8,32
20	20	16,24	14,26	14,28	12,28	12,30	12,30	10,32	10,32	8,32

These critical values were derived, with permission of the publisher, from the distribution of runs given in Asano (1965).

Example:

$$U'_{0.05(2),12,16} = 8 \text{ and } 22.$$

TABLE B.38 COMMON LOGARITHMS OF FACTORIALS

X	0	1	2	3	4	5	6	7	8	9
0	0.00000	0.00000	0.30103	0.77815	1.38021	2.07918	2.85733	3.70243	4.60552	5.55976
10	6.55976	7.60116	8.68034	9.79428	10.94041	12.11650	13.32062	14.55107	15.80634	17.08509
20	18.38612	19.70834	21.05077	22.41249	23.79271	25.19065	26.60562	28.03698	29.48414	30.94654
30	32.42366	33.91502	35.42017	36.93869	38.47016	40.01423	41.57054	43.13874	44.71852	46.30959
40	47.91165	49.52443	51.14768	52.78115	54.42460	56.07781	57.74057	59.41267	61.09391	62.78410
50	64.48307	66.19065	67.90665	69.63092	71.36332	73.10368	74.85187	76.60774	78.37117	80.14202
60	81.92017	83.70550	85.49790	87.29724	89.10342	90.91633	92.73587	94.56195	96.39446	98.23331
70	100.07841	101.92966	103.78700	105.65032	107.51955	109.39461	111.27543	113.16192	115.05401	116.95164
80	118.85473	120.76321	122.67703	124.53610	126.52038	128.44980	130.38430	132.32382	134.26830	136.21769
90	138.17194	140.13098	142.09477	144.06325	146.03638	148.01410	149.99637	151.98314	153.97437	155.97000
100	157.97000	159.97433	161.98293	163.99576	166.01280	168.03399	170.05929	172.08867	174.12210	176.15952
110	178.20092	180.24624	182.29546	184.34854	186.40544	188.46614	190.53060	192.59878	194.67067	196.74621
120	198.82539	200.90818	202.99454	205.08444	207.17787	209.27478	211.37515	213.47895	215.58616	217.69675
130	219.81069	221.92796	224.04854	226.17239	228.29949	230.42983	232.56337	234.70009	236.83997	238.98298
140	241.12911	243.27833	245.43062	247.58595	249.74432	251.90568	254.07004	256.23735	258.40762	260.58080
150	262.75689	264.93587	267.11771	269.30241	271.48993	273.68026	275.87338	278.06928	280.26794	282.46934
160	284.67346	286.88028	289.08980	291.30198	293.51683	295.73431	297.95442	300.17714	302.40245	304.63033
170	306.86078	309.09378	311.32931	313.56735	315.80790	318.05094	320.29645	322.54443	324.79485	327.04770
180	329.30297	331.56065	333.82072	336.08317	338.34799	340.61516	342.88467	345.15652	347.43067	349.70714
190	351.98589	354.26692	356.55022	358.83578	361.12358	363.41362	365.70587	368.00034	370.29701	372.59586
200	374.89689	377.20008	379.50544	381.81293	384.12256	386.43432	388.74818	391.06415	393.38222	395.70236
210	398.02458	400.34887	402.67520	405.00358	407.33399	409.66643	412.00089	414.33735	416.67580	419.01625
220	421.35867	423.70306	426.04941	428.39772	430.74797	433.10015	435.45426	437.81028	440.16822	442.52805
230	444.88978	447.25339	449.61888	451.98624	454.35545	456.72652	459.09943	461.47418	463.85076	466.22916
240	468.60937	470.99139	473.37520	475.76081	478.14820	480.53736	482.92830	485.32100	487.71545	490.11165
250	492.50959	494.90926	497.31066	499.71378	502.11861	504.52516	506.93340	509.34333	511.75495	514.16825
260	516.58322	518.99986	521.41816	523.83812	526.25972	528.68297	531.10785	533.53436	535.96250	538.39225
270	540.82361	543.25658	545.69115	548.12731	550.56506	553.00440	555.44531	557.88779	560.33183	562.77743
280	565.22459	567.67330	570.12355	572.57533	575.02865	577.48350	579.93986	582.39775	584.85714	587.31804
290	589.78043	592.24433	594.70971	597.17658	599.64492	602.11475	604.58604	607.05879	609.53301	612.00868
300	614.48580	616.96437	619.44438	621.92582	624.40869	626.89299	629.37871	631.86585	634.35440	636.84436
310	639.33572	641.82848	644.32264	646.81818	649.31511	651.81342	654.31311	656.81417	659.31660	661.82039
320	664.32554	666.83204	669.33990	671.84910	674.35965	676.87153	679.38475	681.89929	684.41517	686.93236
330	689.45088	691.97071	694.49184	697.01429	699.53803	702.06308	704.58942	707.11705	709.64597	712.17616
340	714.70764	717.24040	719.77442	722.30972	724.84628	727.38410	729.92317	732.46350	735.00508	737.54791
350	740.09197	742.63728	745.18382	747.73160	750.28060	752.83083	755.38228	757.93495	760.48883	763.04393
360	765.60023	768.15774	770.71644	773.27635	775.83745	778.39975	780.96323	783.52789	786.09374	788.66077
370	791.22897	793.79834	796.36889	798.94059	801.51347	804.08750	806.66268	809.23903	811.81652	814.39516
380	816.97494	819.55587	822.13793	824.72113	827.30546	829.89092	832.47751	835.06522	837.65405	840.24400
390	842.83506	845.42724	848.02053	850.61492	853.21042	855.80701	858.40471	861.00350	863.60338	866.20435
400	868.80641	871.40956	874.01378	876.61909	879.22547	881.83293	884.44145	887.05105	889.66171	892.27343
410	894.88621	897.50006	900.11495	902.73090	905.34790	907.96595	910.58504	913.20518	915.82636	918.44857
420	921.07182	923.69610	926.32141	928.94776	931.57512	934.20351	936.83292	939.46335	942.09479	944.72725
430	947.36072	949.99519	952.63068	955.26717	957.90466	960.54314	963.18263	965.82311	968.46459	971.10705
440	973.75050	976.39494	979.04037	981.68677	984.33415	986.98251	989.63185	992.28215	994.93343	997.58568
450	1000.23889	1002.89307	1005.54821	1008.20430	1010.86136	1013.51937	1016.17834	1018.83825	1021.49912	1024.16093
460	1026.82369	1029.48739	1032.15203	1034.81761	1037.48413	1040.15158	1042.81997	1045.48929	1048.15953	1050.83070
470	1053.50280	1056.17582	1058.84977	1061.52463	1064.20040	1066.87710	1069.55471	1072.23322	1074.91265	1077.59299
480	1080.27423	1082.95637	1085.63942	1088.32337	1091.00821	1093.69395	1096.38059	1099.06812	1101.75654	1104.44585
490	1107.13604	1109.82713	1112.51909	1115.21194	1117.90567	1120.60027	1123.29575	1125.99211	1128.68934	1131.38744

If $\log X!$ is needed for $X > 499$, consult Lloyd, Zar, and Karr (1968) or Pearson and Hartley (1966: Table 51), or note that "Stirling's approximation" is excellent:

$$\log X! = (X + 0.5) \log X - 0.434294\, X + 0.39909.$$

TABLE B.39 TEN THOUSAND RANDOM DIGITS

	00-04	05-09	10-14	15-19	20-24	25-29	30-34	35-39	40-44	45-49
00	22808	04391	45529	53968	57136	98228	85485	13801	68194	56382
01	49305	36965	44849	64987	59501	35141	50159	57369	76913	75739
02	81934	19920	73316	69243	69605	17022	53264	83417	55193	92929
03	10840	13508	48120	22467	54505	70536	91206	81038	22418	34800
04	99555	73289	59605	37105	24621	44100	72832	12268	97089	68112
05	32677	45709	62337	35132	45128	96761	08745	53388	98353	46724
06	09401	75407	27704	11569	52842	83543	44750	03177	50511	15301
07	73424	31711	65519	74869	56744	40864	75315	89866	96563	75142
08	37075	81378	59472	71858	86903	66860	03757	32723	54273	45477
09	02060	37158	55244	44812	45369	78939	08048	28036	40946	03898
10	94719	43565	40028	79866	43137	28063	52513	66405	71511	66135
11	70234	48272	59621	88778	16536	36505	41724	24776	63971	01685
12	07972	71752	92745	86465	01845	27416	50519	48458	68460	63113
13	58521	64882	26993	48104	61307	73933	17214	44827	88306	78177
14	32580	45202	21148	09684	39411	04892	02055	75276	51831	85686
15	88796	30829	35009	22695	23694	11220	71006	26720	39476	60538
16	31525	82746	78935	82980	61236	28940	96341	13790	66247	33839
17	02747	35989	70387	89571	34570	17002	79223	96817	31681	15207
18	46651	28987	20625	61347	63981	41085	67412	29053	00724	14841
19	43598	14436	33521	55637	39789	26560	66404	71802	18763	80560
20	30596	92319	11474	64546	60030	73795	60809	24016	29166	36059
21	56198	64370	85771	62633	78240	05766	32419	35769	14057	80674
22	68266	67544	06464	84956	18431	04015	89049	15098	12018	89338
23	31107	28597	65102	75599	17496	87590	68848	33021	69855	54015
24	37555	05069	38680	87274	55152	21792	77219	48732	03377	01160
25	90463	27249	43845	94391	12145	36882	48906	52336	00780	74407
26	99189	88731	93531	52638	54989	04237	32978	59902	05463	09245
27	37631	74016	89072	59598	55356	27346	80856	80875	52850	36548
28	73829	21651	50141	76142	72303	06694	61697	76662	23745	96282
29	15634	89428	47090	12094	42134	62381	87236	90118	53463	46969
30	00571	45172	78532	63863	98597	15742	41967	11821	91389	07476
31	83374	10184	56384	27050	77700	13875	96607	76479	80535	17454
32	78666	85645	13181	08700	08289	62956	64439	39150	95690	18555
33	47890	88197	21368	65254	35917	54035	83028	84636	38186	50581
34	56238	13559	79344	83198	94642	35165	40188	21456	67024	62771
35	36369	32234	38129	59963	99237	72648	66504	99065	61161	16186
36	42934	34578	28968	74028	42164	56647	76806	61023	33099	48293
37	09010	15226	43474	30174	26727	39317	48508	55438	85336	40762
38	83897	90073	72941	85613	85569	24183	08247	15946	02957	68504
39	82206	01230	93252	89045	25141	91943	75531	87420	99012	80751
40	14175	32992	49046	41272	94040	44929	98531	27712	05106	35242
41	58968	88367	70927	74765	18635	85122	27722	95388	61523	91745
42	62601	04595	76926	11007	67631	64641	07994	04639	39314	83126
43	97030	71165	47032	85021	65554	66774	21560	04121	57297	85415
44	89074	31587	21360	41673	71192	85795	82757	52928	62586	02179
45	07806	81312	81215	99858	26762	28993	74951	64680	50934	32011
46	91540	86466	13229	76624	44092	96604	08590	89705	03424	48033
47	99279	27334	33804	77988	93592	90708	56780	70097	39907	51006
48	63224	05074	83941	25034	43516	22840	35230	66048	80754	46302
49	98361	97513	27529	66419	35328	19738	82366	38573	50967	72754

TABLE B.39 (cont.) TEN THOUSAND RANDOM DIGITS

	00-04	05-09	10-14	15-19	20-24	25-29	30-34	35-39	40-44	45-49
50	27791	82504	33523	27623	16597	32089	81596	78429	14111	68245
51	33147	46058	92388	10150	63224	26003	56427	29945	44546	50233
52	67243	10454	40269	44324	46013	00061	21622	68213	47749	76398
53	78176	70368	95523	09134	31178	33857	26171	07063	41984	99310
54	70199	70547	94431	45423	48695	01370	68065	61982	20200	27066
55	19840	01143	18606	07622	77282	68422	70767	33026	15135	91212
56	32970	28267	17695	20571	50227	69447	45535	16845	68283	15919
57	43233	53872	68520	70013	31395	60361	39034	59444	17066	07418
58	08514	23921	16685	89184	71512	82239	72947	69523	75618	79826
59	28595	51196	96108	84384	80359	02346	60581	01488	63177	47496
60	83334	81552	88223	29934	68663	23726	18429	84855	26897	94782
61	66112	95787	84997	91207	67576	27496	01603	22395	41546	68178
62	25245	14749	30653	42355	88625	37412	87384	09392	11273	28116
63	21861	22185	41576	15238	92294	50643	69848	48020	19785	41518
64	74506	40569	90770	40812	57730	84150	91500	53850	52104	37988
65	23271	39549	33042	10661	37312	50914	73027	21010	76788	64037
66	08548	16021	64715	08275	50987	67327	11431	31492	86970	47335
67	14236	80869	90798	85659	10079	28535	35938	10710	67046	74021
68	55270	49583	86467	40633	27952	27187	35058	66628	94372	75665
69	02301	05524	91801	23647	51330	35677	05972	90729	26650	81684
70	72843	03767	62590	92077	91552	76853	45812	15503	93138	87788
71	49248	43346	29503	22494	08051	09035	75802	63967	74257	00046
72	62598	99092	87806	42727	30659	10118	83000	96198	47155	00361
73	27510	69457	98616	62172	07056	61015	22159	65590	51082	34912
74	84167	66640	69100	22944	19833	23961	80834	37418	42284	12951
75	14722	88488	54999	55244	03301	37344	01053	79305	94771	95215
76	46696	05477	32442	18738	43021	72933	14995	30408	64043	67834
77	13938	09867	28949	94761	38419	38695	90165	82841	75399	09932
78	48778	56434	42495	07050	35250	09660	56192	34793	36146	96806
79	00571	71281	01563	66448	94560	55920	31580	26640	91262	30863
80	96050	57641	21798	14917	21836	15053	33566	51177	91786	12610
81	30870	81575	14019	07831	81840	25506	29358	88668	42742	62048
82	59153	29135	00712	73025	14263	17253	95662	75535	26170	95240
83	78283	70379	54969	05821	26485	28990	40207	00434	38863	61892
84	12175	95800	41106	93962	06245	00883	65337	75506	66294	62241
85	14192	39242	17961	29448	84078	14545	39417	83649	26495	41672
86	69060	38669	00849	24991	84252	41611	62773	63024	57079	59283
87	46154	11705	29355	71523	21377	36745	00766	21549	51796	81340
88	93419	54353	41269	07014	28352	77594	57293	59219	26098	63041
89	13201	04017	68889	81388	60829	46231	46161	01360	25839	52380
90	62264	99963	98226	29972	95169	07546	01574	94986	06123	52804
91	58030	30054	27479	70354	12351	33761	94357	81081	74418	74297
92	81242	26739	92304	81425	29052	37708	49370	46749	59613	50749
93	16372	70531	92036	54496	50521	83872	30064	67555	40354	23671
94	54191	04574	58634	91370	40041	77649	42030	42547	47593	07435
95	15933	92602	19496	18703	63380	58017	14665	88867	84807	44672
96	21518	77770	53826	97114	82062	34592	87400	64938	75540	54751
97	34524	64627	92997	21198	14976	07071	91566	44335	83237	24335
98	46557	67780	59432	23250	63352	43890	07109	07911	85956	62699
99	31929	13996	05126	83561	03244	33635	26952	01638	22788	26393

TABLE B.39 (cont.) TEN THOUSAND RANDOM DIGITS

	50-54	55-59	60-64	65-69	70-74	75-79	80-84	85-89	90-94	95-99
00	53330	26487	85005	06384	13822	83736	95876	71355	31226	56063
01	96990	62825	97110	73006	32661	63408	03893	10333	41902	69175
02	30385	16588	63609	09132	53081	14478	50813	22887	03746	10289
03	75252	66905	60536	13408	25158	35825	10447	47375	89249	91238
04	52615	66504	78496	90443	84414	31981	88768	49629	15174	99795
05	39992	51082	74547	31022	71980	40900	84729	34286	96944	49502
06	51788	87155	13272	92461	06466	25392	22330	17336	42528	78628
07	88569	35645	50602	94043	35316	66344	78064	89651	89025	12722
08	14513	34794	44976	71244	60548	03041	03300	46389	25340	23804
09	50257	53477	24546	01377	20292	85097	00660	39561	62367	61424
10	35170	69025	46214	27085	83416	48597	19494	49380	28469	77549
11	22225	83437	43912	30337	75784	77689	60425	85588	93438	61343
12	90103	12542	97828	85859	85859	64101	00924	89012	17889	01154
13	68240	89649	85705	18937	30114	89827	89460	01998	81745	31281
14	01589	18335	24024	39498	82052	07868	49486	25155	61730	08946
15	36375	61694	90654	16475	92703	59561	45517	90922	93357	00207
16	11237	60921	51162	74153	94774	84150	39274	10089	45020	09624
17	48667	68353	40567	79819	48551	26789	07281	14669	00576	17435
18	99286	42806	02956	73762	04419	21676	67533	50553	21115	26742
19	44651	48349	13003	39656	99757	74964	00141	21387	66777	68533
20	83251	70164	05732	66842	77717	25305	36218	85600	23736	06629
21	41551	54630	88759	10085	48806	08724	50685	95638	20829	37264
22	68990	51280	51368	73661	21764	71552	69654	17776	51935	53169
23	63393	76820	33106	23322	16783	35630	50938	90047	97577	27699
24	93317	87564	32371	04190	27608	40658	11517	19646	82335	60088
25	48546	41090	69890	58014	04093	39286	12253	55859	83853	15023
26	31435	57566	99741	77250	43165	31150	20735	57406	85891	04806
27	56405	29392	76998	66849	29175	11641	85284	89978	73169	62140
28	70102	50882	85960	85955	03828	69417	55854	63173	60485	00327
29	92746	32004	52242	94763	32955	39848	09724	30029	45196	67606
30	67737	34389	57920	47081	60714	04935	48278	90687	99290	18554
31	35606	76646	14813	51114	52492	46778	08156	22372	59999	43938
32	64836	28649	45759	45788	43183	25275	25300	21548	33941	66314
33	86319	92367	37873	48993	71443	22768	69124	65611	79267	49709
34	90632	32314	24446	60301	31376	13575	99663	81929	39343	17648
35	83752	51966	43895	03129	37539	72989	52393	45542	70344	96712
36	56755	21142	86355	33569	63096	66780	97539	75150	25718	33724
37	14100	28857	60648	86304	97397	97210	74842	87483	51558	52883
38	69227	24872	48057	29318	74385	02097	63266	26950	73173	53025
39	77718	56967	36560	87155	26021	70903	32086	11722	32053	63723
40	09550	38799	88929	80877	87779	99905	17122	25985	16866	76005
41	12404	42453	88609	89148	85892	96045	10310	45021	62023	70061
42	07985	27418	92734	80000	58969	99011	73815	49705	68076	69605
43	58124	53830	08705	20916	46048	30342	86530	72608	93074	80937
44	46173	77223	75661	57691	24055	27568	41227	58542	73196	44886
45	13476	72301	85793	80516	59479	66985	24801	84009	71317	87321
46	82472	98647	17053	94591	36790	42275	51154	77765	01115	09331
47	55370	63433	80653	30739	68821	46854	41939	38962	20703	69424
48	89274	74795	82231	69384	53605	67860	01309	27273	76316	54253
49	55242	74511	62992	17981	17323	79325	35238	21393	13114	70084

TABLE B.39 (cont.) TEN THOUSAND RANDOM DIGITS

	50-54	55-59	60-64	65-69	70-74	75-79	80-84	85-89	90-94	95-99
50	03674	36059	46810	58367	82676	15051	57977	49410	02971	05797
51	26136	80623	96505	91089	02309	54743	15831	45538	96456	87272
52	61716	80405	84735	12997	86386	61606	75091	84996	76070	54923
53	67051	63246	99547	81223	52485	90333	24697	06266	07388	70389
54	17284	60347	87314	30218	87983	45426	84153	10569	64042	95618
55	12543	23999	95777	28105	66073	35174	67706	05181	35176	85558
56	45494	93037	29209	70724	86438	65354	71209	27969	85321	10216
57	39262	15415	93940	41615	43605	95675	53916	29580	07048	95838
58	29094	58703	92144	14287	50165	85661	95749	61118	36668	96852
59	77988	03222	57805	00725	91543	80021	16442	63360	33620	39324
60	02758	86823	52423	32355	96707	47448	06453	59430	43952	16775
61	46702	37467	66803	49344	59519	92717	97110	82087	36785	00880
62	61759	95153	80090	60626	55917	92812	63544	82295	50729	20116
63	82316	11402	28078	75325	43963	63105	99294	30285	61473	53613
64	92754	74241	14315	49697	61979	66711	61707	81589	53936	82115
65	37907	24080	31741	86653	81460	32304	99590	56644	41521	91172
66	16619	75264	12279	18996	16716	81959	65722	10058	91522	65410
67	66640	06195	84416	32836	53178	93810	36766	59778	26612	69017
68	45208	58525	07714	77126	67986	73140	12026	75550	84912	64691
69	00910	40237	91035	29125	03534	47246	64698	00608	39537	71755
70	19965	46945	59357	15551	20335	03145	21519	37882	99146	70161
71	37538	05747	54982	00494	51866	86172	82679	04152	56369	20356
72	38571	69663	03287	28101	46753	55715	93527	30508	19722	02072
73	76711	02864	00880	85518	25834	52317	48070	51582	03374	19540
74	07128	44400	48015	41449	21109	38948	21816	52089	64529	21510
75	00882	89357	80906	76476	58420	95793	34043	00991	38937	39859
76	96160	18580	40549	46562	45106	53768	76097	60504	85273	63076
77	13443	22235	46210	47755	05802	00311	15171	23818	89870	47578
78	99494	35395	71411	48281	92151	84465	63651	15969	61345	13324
79	90647	11809	96365	52409	17977	05971	35835	03889	43733	66100
80	33050	48785	92200	59319	36977	41111	28002	51580	10573	21763
81	21257	15066	72630	23206	03106	53140	50292	64012	83184	81304
82	45362	94324	81800	83980	97244	09691	08435	66723	06150	54972
83	93322	58684	95695	19096	98108	47678	98061	87193	99992	82870
84	20374	61803	62508	83696	54449	53649	86447	66115	90857	69114
85	00715	13209	17080	06890	38022	76469	27696	30778	31836	96676
86	85519	93677	90186	09579	98760	50320	98077	46048	79700	81431
87	71948	15871	84502	41330	46675	51342	93431	55566	90819	68923
88	43427	95500	02004	51802	59668	17806	87605	33010	20991	76269
89	64854	28815	74959	03531	77051	51807	89005	18898	23716	45862
90	62195	29095	23982	75883	41561	25897	43595	92703	86676	32038
91	61186	54041	60984	61602	18482	57941	59657	35924	21738	30646
92	88585	40218	69965	74354	62274	38948	44813	31558	40625	22477
93	15598	21389	79016	92151	21926	49901	16835	88055	30545	60306
94	27097	89653	21558	72731	66694	36703	92172	46129	32660	91356
95	40537	85697	78182	39711	59270	21934	78647	94801	78832	37287
96	74828	06544	13078	59528	31100	11132	91256	85899	72492	18200
97	43297	83195	66218	65838	63255	72093	38976	44892	96861	97848
98	32663	58127	73258	09220	49701	92357	43700	37214	56844	02048
99	45551	31330	08152	23712	23963	58274	94583	03761	73429	47328

Table B.39 was prepared using an algorithm for the IBM 360 computer (International Business Machines, 1968: 77).

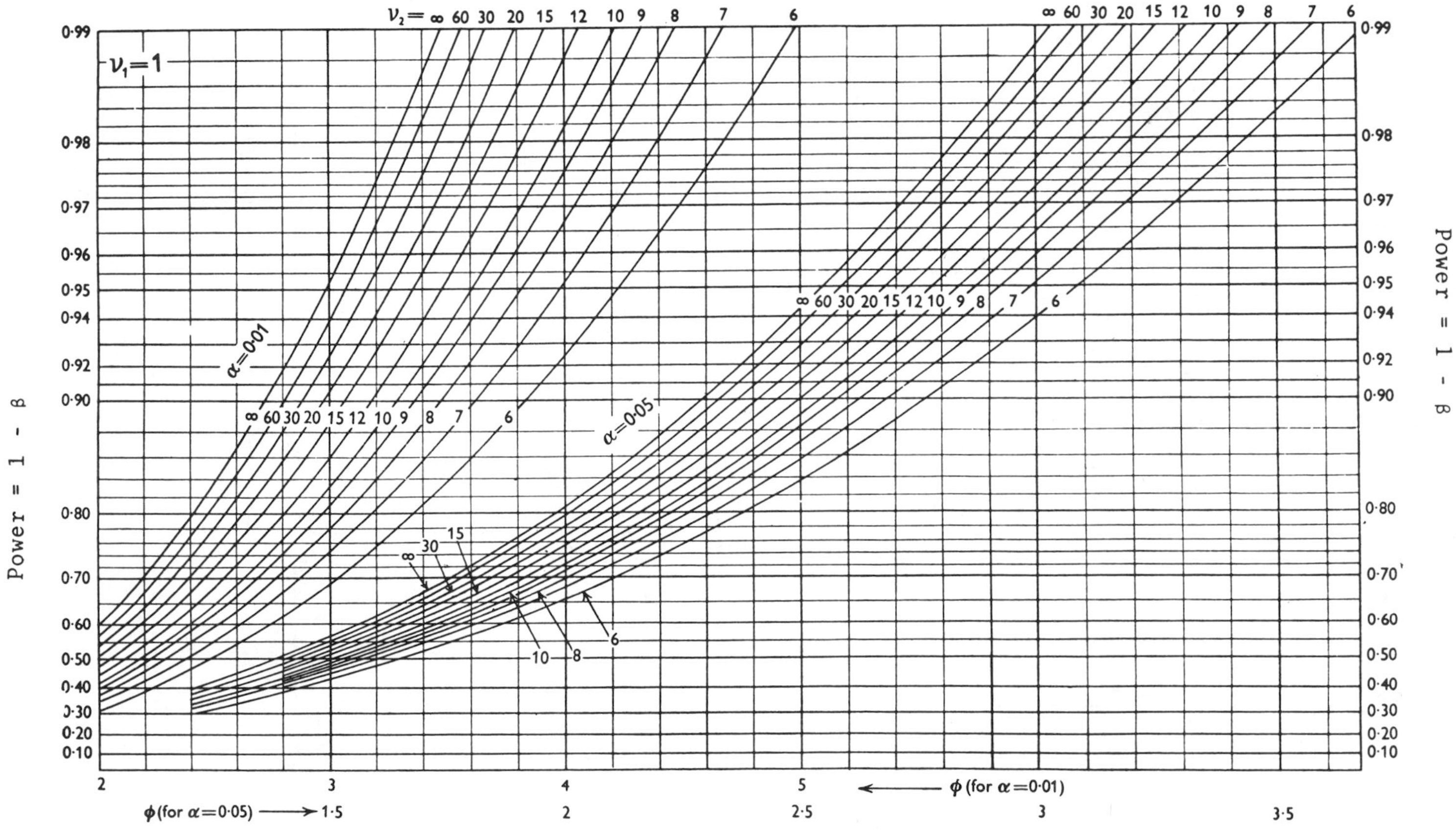

Figure B.1a. Power and sample size in analysis of variance: $\nu_1 = 1$.

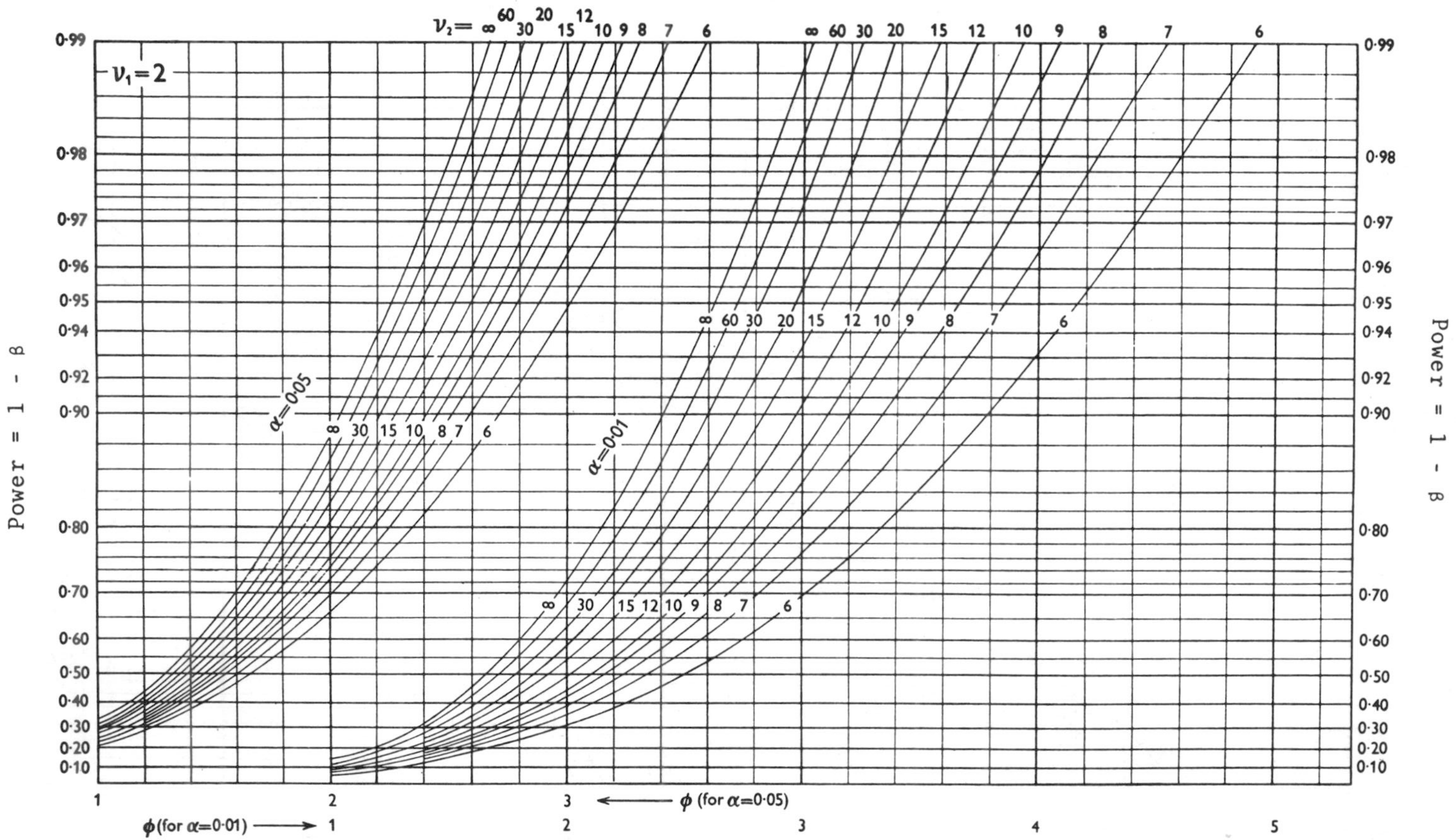

Figure B.1b. Power and sample size in analysis of variance: $\nu_1 = 2$.

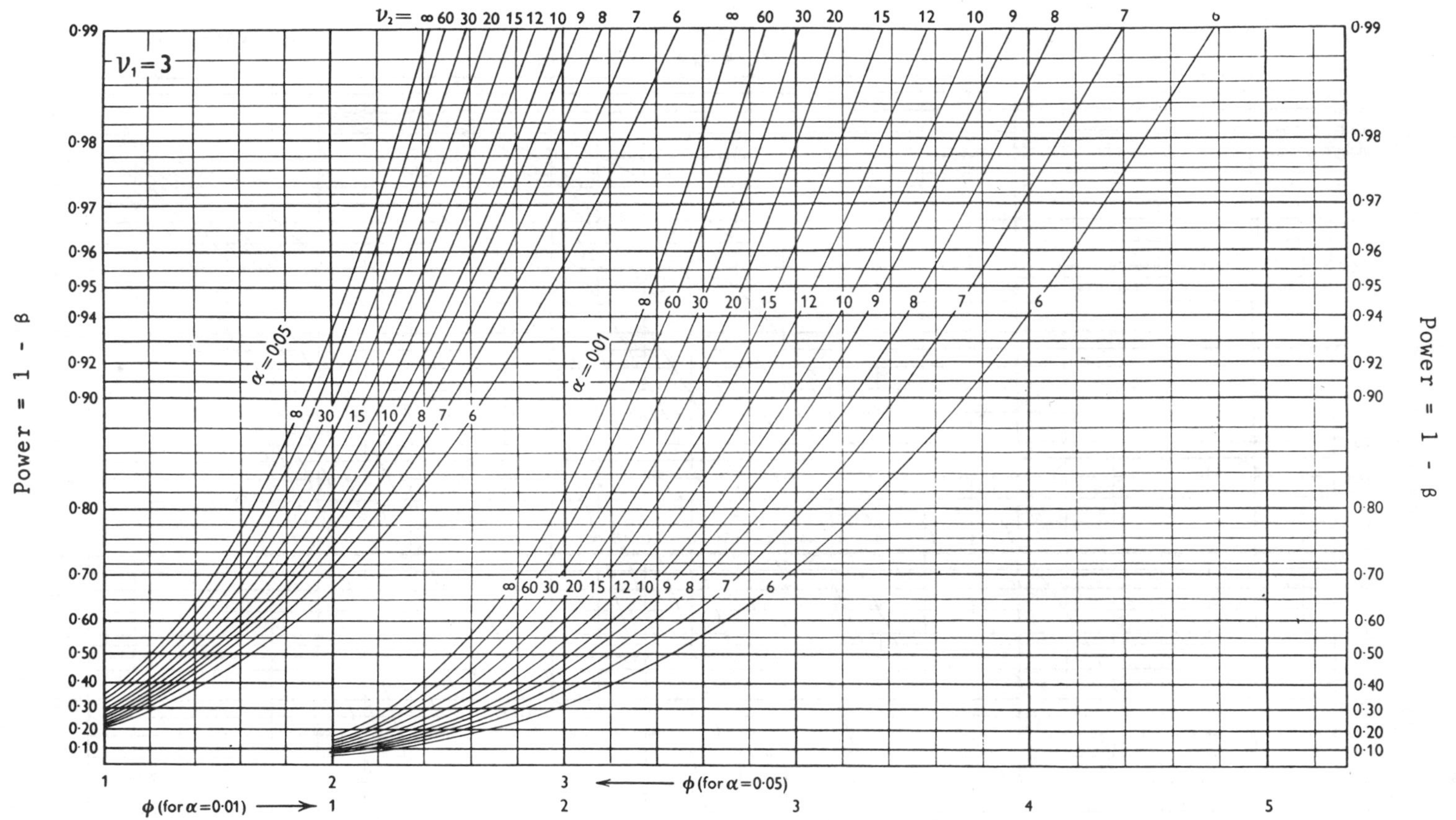

Figure B.1c. Power and sample size in analysis of variance: $\nu_1 = 3$.

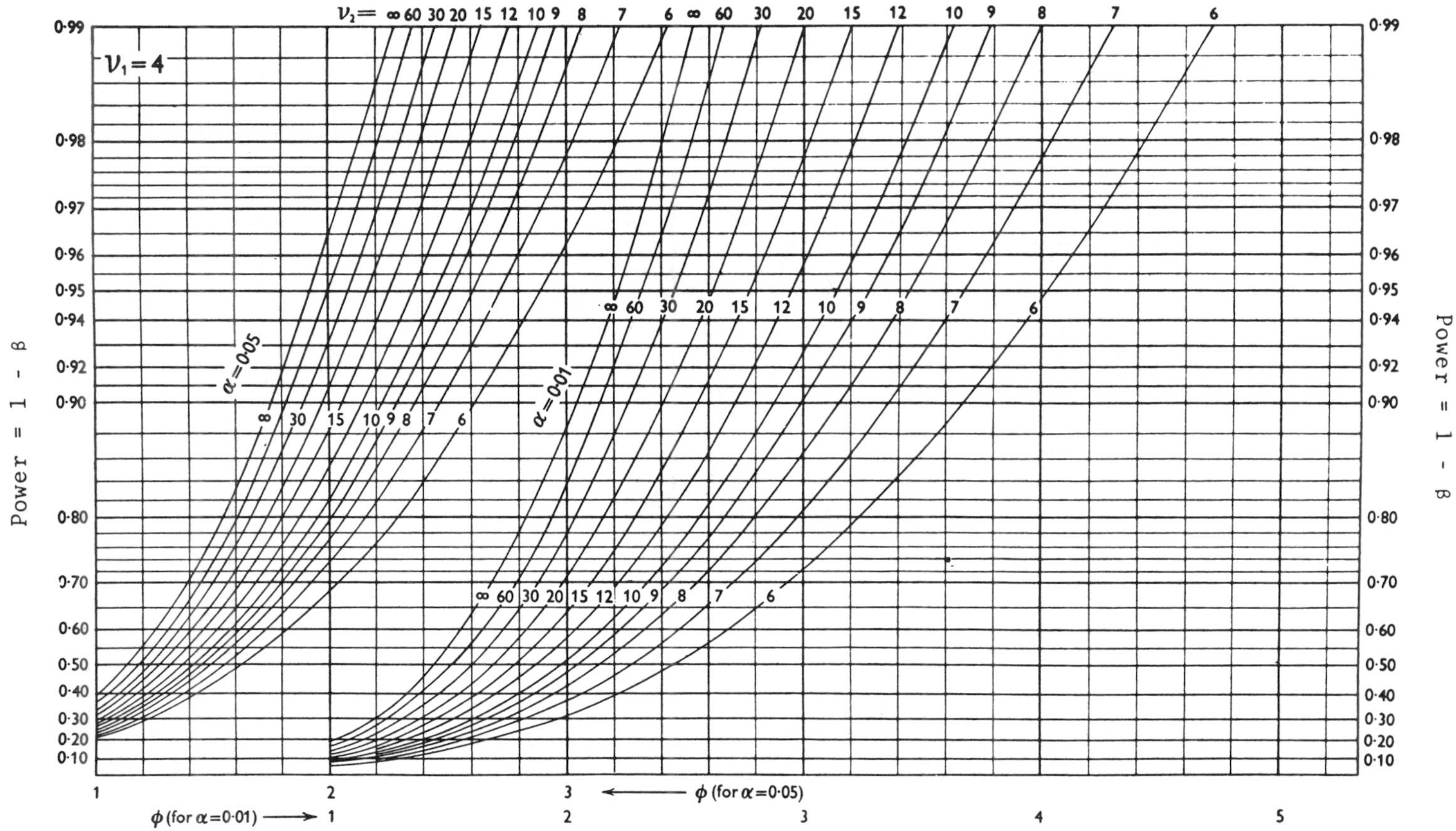

Figure B.1d. Power and sample size in analysis of variance: $\nu_1 = 4$.

Power = 1 - β

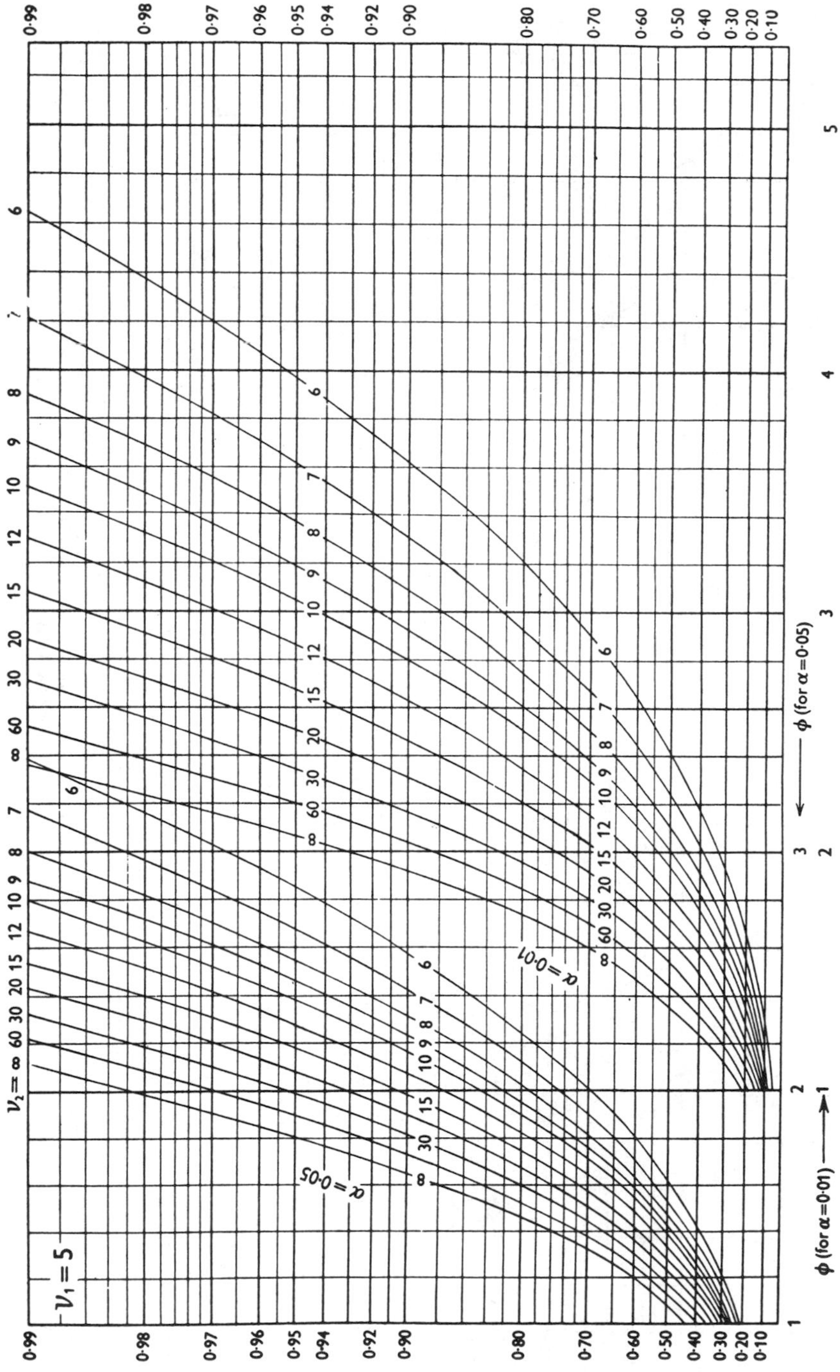

Power = 1 - β

Figure B.1e. Power and sample size in analysis of variance: $\nu_1 = 5$.

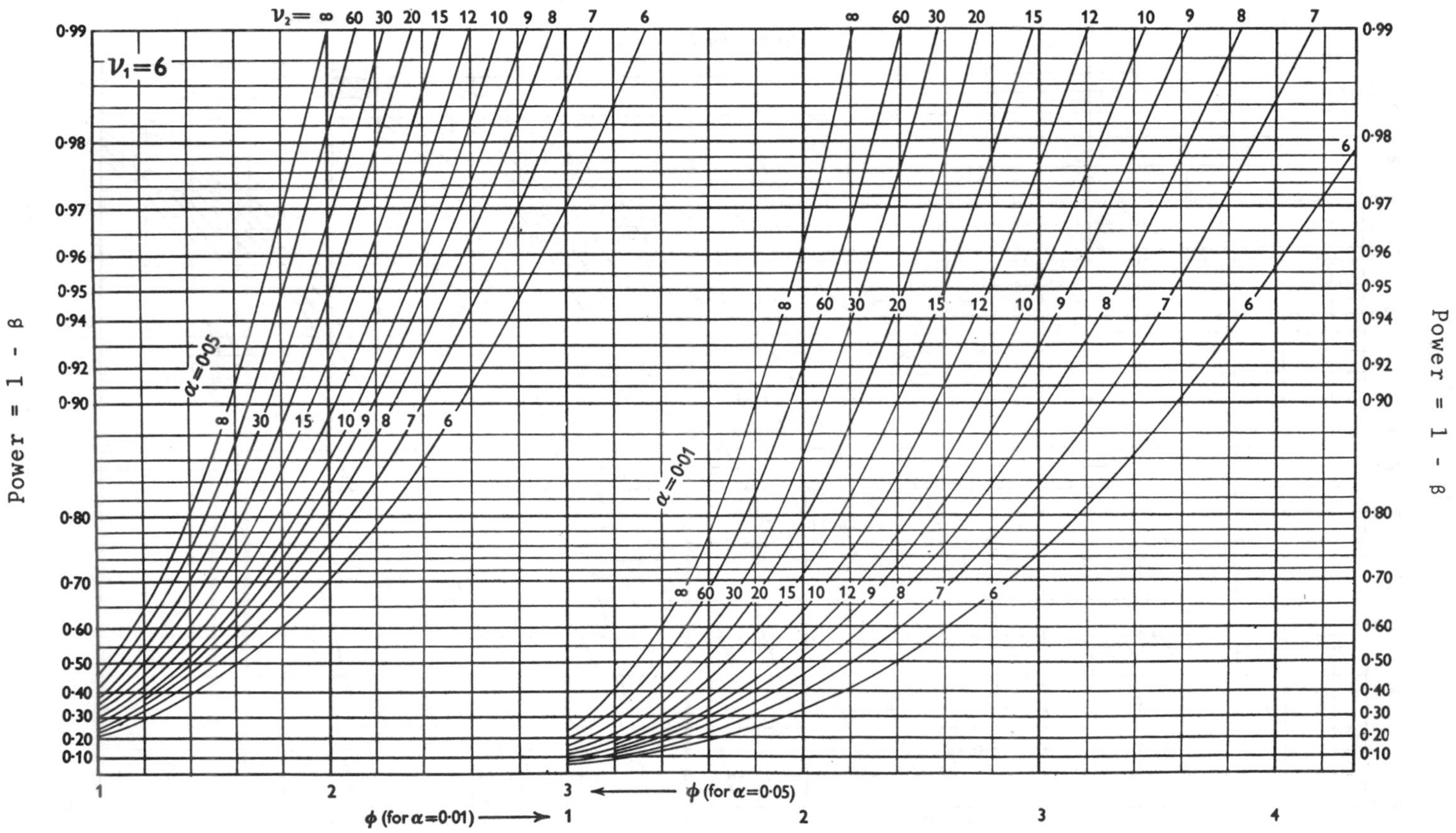

Figure B.1f. Power and sample size in analysis of variance: $\nu_1 = 6$.

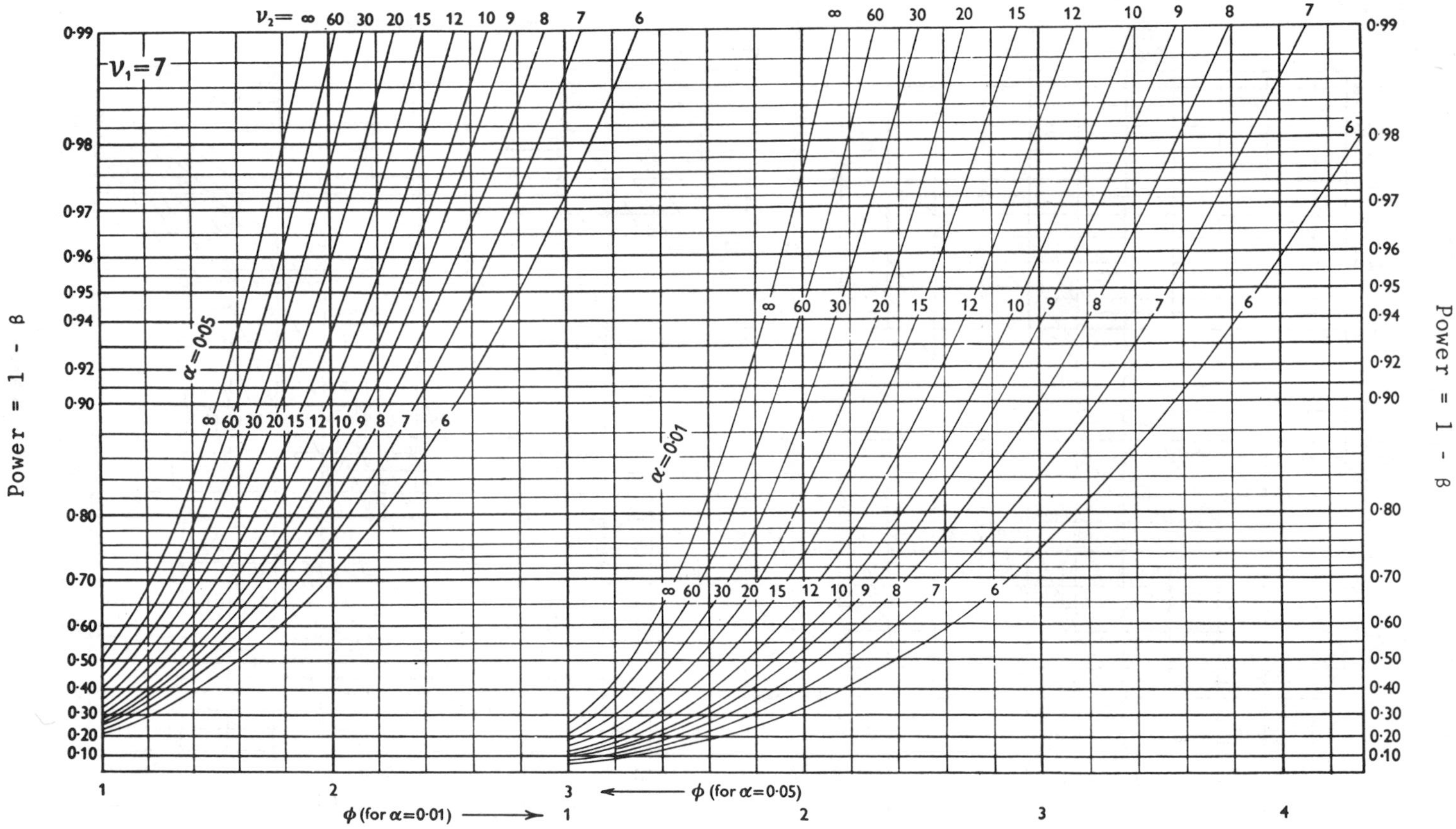

Figure B.1g. Power and sample size in analysis of variance: $\nu_1 = 7$.

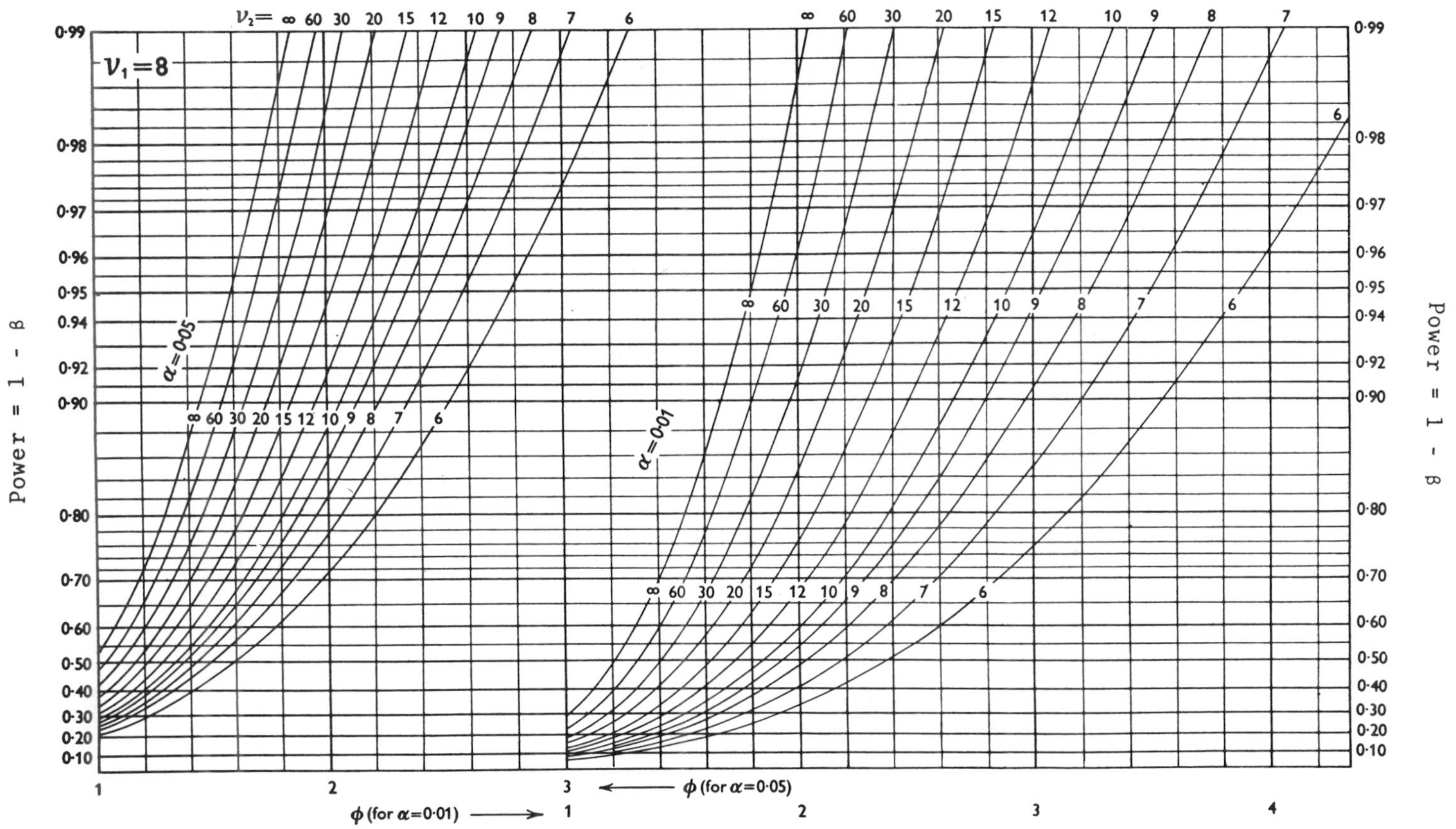

Figure B.1h. Power and sample size in analysis of variance: $\nu_1 = 8$.

These graphs were taken from Pearson and Hartley (1951, Biometrika 38: 112–130) with permission of the Biometrika Trustees.

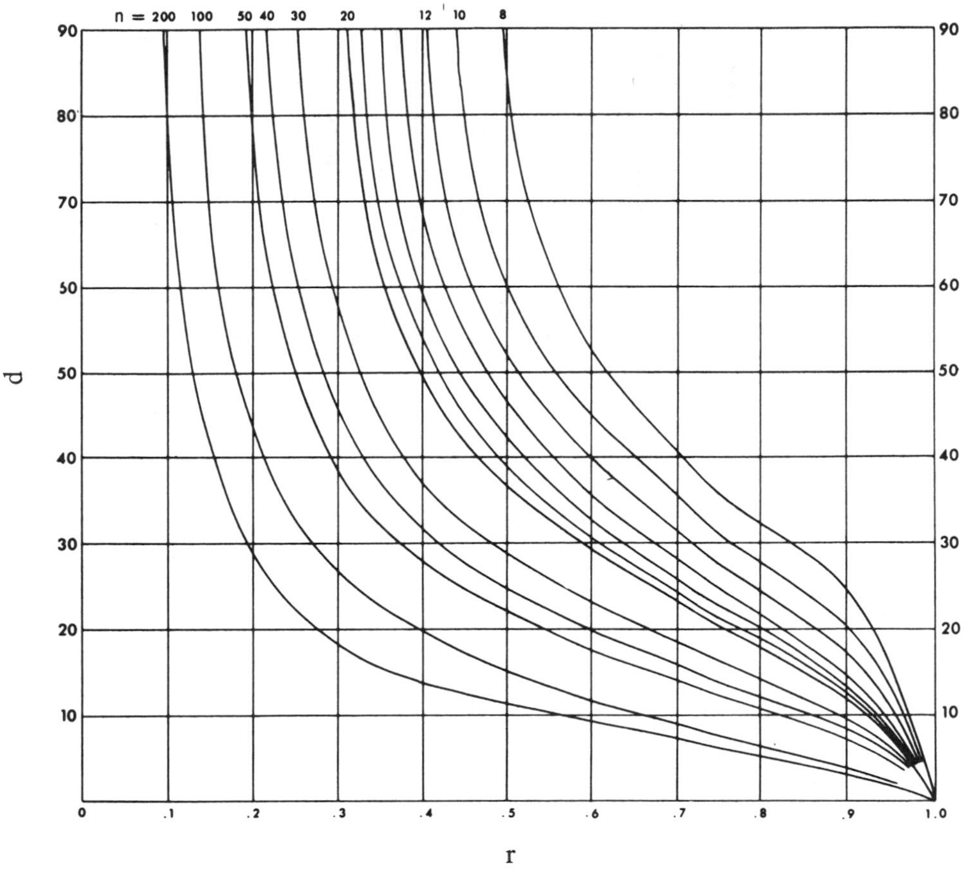

Figure B.2a. Confidence limits for the mean angle: 95% confidence.

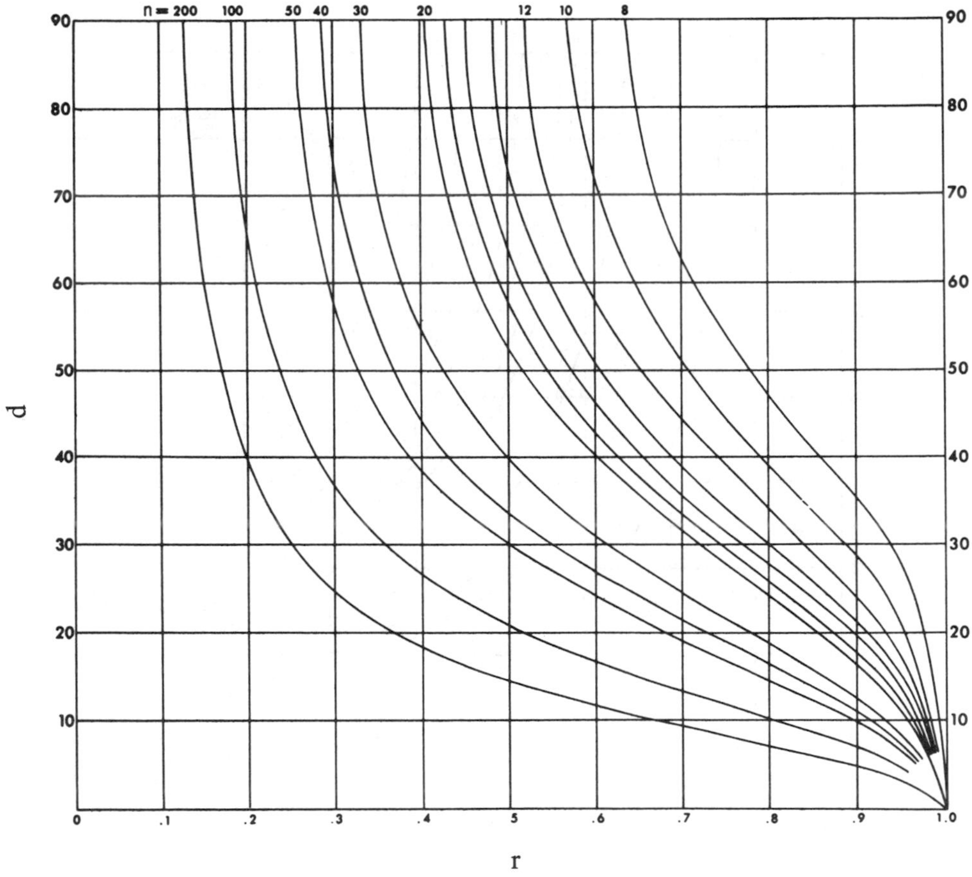

Figure B.2b. Confidence limits for the mean angle: 99% confidence.

Figures B.2a and B.2b were taken from Batschelet (1972) with permission of the publisher.

Answers to Exercises

CHAPTER 3

3.1. **(a)** 13.8 kg; **(b)** 10.7 kg; **(c)** 17.8 kg; **(d)** 17.8 kg.
3.2. **(a)** 3.56 kg; **(b)** 3.6 kg.
3.3. **(a)** 46.63 yr; **(b)** 46.3 yr; **(c)** 44.58 yr; **(d)** 46.3 yr.
3.4. **(a)** 2.33 g; **(b)** 2.33 g; **(c)** 2.4 g; **(d)** 2.358 g; **(e)** 2.4 g.

CHAPTER 4

4.1. **(a)** SS $= 156.028$ g^2, $s^2 = 39.007$ g^2; **(b)** same as (a).
4.2. **(a)** Range $= 236.4$ mg/100 ml to 244.8 mg/100 ml $= 8.4$ mg/100 ml;
(b) SS $= 46.1886$ (mg/100 ml)^2; **(c)** $s^2 = 7.6981$ (mg/100 ml)^2; **(d)** $s = 2.77$ mg/100 ml;
(e) $V = 0.0115 = 1.15\%$.
4.3. $k = 6$, $n = 97$; **(a)** $H' = 0.595$; **(b)** $H'_{max} = 0.778$; **(c)** $J' = 0.76$.
4.4. $k = 6$, $n = 97$; **(a)** $H = 0.554$; **(b)** $c = 16$, $d = 0.1667$, $H_{max} = 0.741$; **(c)** $J = 0.75$.

CHAPTER 5

5.1. **(a)** $P(\chi^2 \geq 3.452)$ is between 0.10 and 0.25; **(b)** $0.10 < (\chi^2 \geq 8.668) < 0.25$;
(c) $\chi^2_{0.05,4} = 9.488$; **(d)** $\chi^2_{0.01,8} = 20.090$.
5.2. **(a)** $\chi^2 = 16.000$, $\nu = 5$, $0.005 < P < 0.01$. As $P < 0.05$, reject H_0 of equal food item preference. **(b)** By grouping food items 1, 4, and 6: $n = 41$, and for H_0: equal food preference, $\chi^2 = 0.049$, $\nu = 2$, $0.975 < P < 0.99$; as $P > 0.05$, H_0 is not rejected. By grouping food items 2, 3, and 5: $n = 85$, and for H_0: equal food preference, $\chi^2 = 0.447$, $\nu = 2$, $0.75 <$

$P < 0.90$; as $P > 0.05$, H_0 is not rejected. By considering food items 1, 4, and 6 as one group and items 2, 3, and 5 as a second group, and H_0: equal preference for the two groups, $\chi_c^2 = 14.675$, $\nu = 1$, $P < 0.001$; H_0 is rejected.

5.3. $\chi_c^2 = 0.827$, $\nu = 1$, $0.25 < P < 0.50$; since $P > 0.05$, do not reject H_0: The population consists in equal numbers of males and females.

5.4.

Location	*Males*	*Females*	χ^2	ν
1	44	54	1.020	1
2	31	40	1.141	1
3	12	18	1.200	1
4	15	16	0.032	1
		Total	3.393	4
Total	102	128	2.939	1
Heterogeneity chi-square			0.454	3
				$0.90 < P < 0.95$

Since P(heterogeneity χ^2) > 0.05, the four samples may be pooled with the following results: $\chi_c^2 = 2.717$, $\nu = 1$, $0.05 < P < 0.10$; since $P > 0.05$, do not reject H_0: Equal numbers of males and females in the population.

5.5. $G = 15.881$, $\nu = 7$, $0.025 < P < 0.05$; since $P < 0.05$, reject H_0: No difference in nest orientation.

5.6. $d_{\max} = 1$; $(d_{\max})_{0.05,6,18} = 5$; do not reject H_0: The feeders are equally acceptable to the birds; $P > 0.50$.

5.7. $\max D_i = 0.2519$; $\max D'_i = 0.2197$; $D = 0.2519$; $D_{0.05,31} = 0.23788$; reject H_0: There is a uniform distribution of these animals from the water's edge to a distance of 10 meters upland; $0.02 < P < 0.05$.

5.8. $D_{0.05,27} = 0.25438$ and $D_{0.05,28} = 0.24993$, so a sample size of at least 28 is called for.

CHAPTER 6

6.1. $F_{11} = 157.1026$, $F_{12} = 133.7580$, $F_{13} = 70.0337$, $F_{14} = 51.1057$, $F_{21} = 91.8974$, $F_{22} = 78.2420$, $F_{23} = 40.9663$, $F_{24} = 29.8943$, $R_1 = 412$, $R_2 = 241$, $C_1 = 249$, $C_2 = 212$, $C_3 = 111$, $C_4 = 81$, $n = 653$; $\chi^2 = 0.2214 + 0.0115 + 0.0133 + 1.2856 + 0.3785 + 0.0197 + 0.0228 + 2.1978 = 4.151$; $\nu = (2-1)(4-1) = 3$; $\chi^2_{0.05,3} = 7.815$, $0.10 < P(\chi^2 \geq 4.156) < 0.25$; since $P > 0.05$, do not reject H_0.

6.2. $f_{11} = 14, f_{12} = 29, f_{21} = 12, f_{22} = 38$, $R_1 = 43$, $R_2 = 50$, $C_1 = 26$, $C_2 = 67$, $n = 93$; $\chi_c^2 = 0.469$, $\nu = 1$; $\chi^2_{0.05,1} = 3.841$, $0.25 < P(\chi^2 \geq 0.469) < 0.50$; as $P > 0.05$, do not reject H_0. $\hat{f} = 12.0215$, $f = 14$, $d = 1.9785$, $f < 2\hat{f}$ so $D = 1.5$, $\chi_c^2{}' = 0.483$, with conclusions as above.

6.3. $G = 4.032$, $\nu = 3$, $\chi^2_{0.05,3} = 7.815$, $0.25 < P(\chi^2 \geq 4.032) < 0.50$; since $P > 0.05$, do not reject H_0.

6.4. $G = 0.468$, $\nu = 1$, $\chi^2_{0.05,1} = 3.841$, $0.25 < P(\chi^2 \geq 0.468) < 0.50$; since $P > 0.05$, do not reject H_0.

6.5. H_0: Sex, area, and occurrence of rabies are mutually independent; $\chi^2 = 32.031$; $\nu = 4$; $\chi^2_{0.05,4} = 9.488$; reject H_0; $P < 0.001$. H_0: Area is independent of sex and rabies; $\chi^2 = 23.515$; $\nu = 3$; $\chi^2_{0.05,3} = 7.815$; reject H_0; $P < 0.001$. H_0: Sex is independent of area and rabies; $\chi^2 = 16.723$; $\nu = 3$; reject H_0; $P < 0.001$. H_0: Rabies is independent of area and sex; $\chi^2 = 32.170$; $\nu = 3$; reject H_0; $P < 0.001$.

CHAPTER 7

7.1. **(a)** $Z = (78.0\text{ g} - 63.5\text{ g})/12.2\text{ g} = 1.19$, $P(X \geq 78.0\text{ g}) = P(Z \geq 1.19) = 0.1170$; **(b)** $P(X \leq 78.0\text{ g}) = 1.0000 - P(X \geq 78.0\text{ g}) = 1.0000 - 0.1170 = 0.8830$; **(c)** $(0.1170)(1000) = 117$; (d) $Z = (41.0\text{ g} - 63.5\text{ g})/12.2\text{ g} = -1.84$, $P(X \leq 41.0\text{ g}) = P(Z \leq -1.84) = 0.0329$

7.2. **(a)** $P(X \leq 60.0\text{ g}) = P(Z \leq -0.29) = 0.3859$, $P(X \geq 70.0\text{ g}) = P(Z \geq 0.53) = 0.2981$, $P(60.0\text{ g} \leq X \leq 70.0\text{ g}) = 1.0000 - 0.3859 - 0.2981 = 0.3160$; **(b)** $P(X \leq 60.0\text{ g}) = P(Z \leq -0.29) = 0.3859$, $P(X \leq 50.0\text{ g}) = P(Z \leq -1.11) = 0.1335$, $P(50.0\text{ g} \leq X \leq 60.0\text{ g}) = P(-1.11 \leq Z \leq -0.29) = 0.3859 - 0.1335 = 0.2524$.

7.3. **(a)** $s_{\bar{X}} = s/\sqrt{n} = 12.2\text{ g}/\sqrt{10} = 3.86\text{ g}$; **(b)** $Z = (65.0\text{ g} - 63.5)/3.86\text{ g} = 0.39$, $P(\bar{X} \geq 65.0\text{ g}) = P(Z \geq 0.39) = 0.3483$; **(c)** $P(\bar{X} \leq 62.0\text{ g}) = P(Z \leq -0.39) = 0.3483$, $P(\bar{X} \leq 60.0\text{ g}) = P(Z \leq -0.91) = 0.1814$, $P(60.0\text{ g} \leq \bar{X} \leq 62.0\text{ g}) = 0.3483 - 0.1814 = 0.1669$.

CHAPTER 8

8.1. H_0: $\mu = 29.5$ days, H_A: $\mu \neq 29.5$ days, $\bar{X} = 27.7$ days, $s_{\bar{X}} = 0.708$ days, $n = 15$, $t = 2.542$, $\nu = 15 - 1 = 14$, $t_{0.05(2),14} = 2.145$, $0.02 < P(|t| \geq 2.542) < 0.05$; therefore, reject H_0 and conclude that the sample came from a population with a mean that is not 29.5 days.

8.2. H_0: $\mu \geq 26.2°\text{C}$, H_A: $\mu < 26.2°\text{C}$, $\bar{X} = 25.04°\text{C}$, $s_{\bar{X}} = 0.24°\text{C}$, $n = 8$, $t = -4.833$, $\nu = 7$, $t_{0.05(1),7} = 1.895$, $0.0005 < P(t \leq -4.833) < 0.001$; therefore, reject H_0 and conclude that the data came from a population with a mean body temperature less than 26.2°C.

8.3. Graph, which includes three 95% confidence intervals: 0.458 ± 0.057 kcal/g; 0.413 ± 0.059 kcal/g; 0.327 ± 0.038 kcal/g.

8.4. **(a)** 13.55 ± 1.26 cm; **(b)** $n = 28$; **(c)** $n = 9$; **(d)** $n = 15$.

8.5. **(a)** $n = 30$; **(b)** $n = 41$; **(c)** $n = 42$; **(d)** $d = 2.2$ cm; **(e)** $t_{\beta(1),49} = 1.432$; $0.05 < \beta < 0.10$, so $0.90 <$ power < 0.95; or, by normal approximation, $\beta = 0.08$ and power $= 0.92$.

8.6. **(a)** $N = 200$, $n = 50$, $s^2 = 97.8121\text{ yr}^2$, $t_{0.05(2),49} = 2.009$; $s_{\bar{X}} = 1.2143$ yr, 95% confidence interval $= 53.87 \pm 2.44$ yr; **(b)** $t_{0.05(2),99} = 1.984$; $s_{\bar{X}} = 0.7011$ yr, 95% confidence interval $= 53.87 \pm 1.39$ yr.

CHAPTER 9

9.1. H_0: $\sigma_1^2 = \sigma_2^2$, H_A: $\sigma_1^2 \neq \sigma_2^2$, $F = 14.62\text{ cm}^2/8.45\text{ cm}^2 = 1.73$, $\nu_1 = 28$, $\nu_2 = 24$, $F_{0.05(2),28,24} = 2.23$; therefore, do not reject H_0; $0.10 < P < 0.20$.

9.2. H_0: $\sigma_1^2 \leq \sigma_2^2$, H_A: $\sigma_1^2 > \sigma_2^2$, $F = 324.46\text{ sec}^2/158.95\text{ sec}^2 = 2.04$, $\nu_1 = 40$, $\nu_2 = 35$, $F_{0.05(1),40,35} = 1.74$; therefore, reject H_0; $0.01 < P < 0.025$.

9.3. H_0: $\mu_1 = \mu_2$, H_A: $\mu_1 \neq \mu_2$, $n_1 = 7$, $SS_1 = 108.6171\text{ (mg/100 ml)}^2$, $\bar{X}_1 = 224.24$ mg/100 ml, $\nu_1 = 6$, $n_2 = 6$, $SS_2 = 74.7533\text{ (mg/100 ml)}^2$, $\bar{X}_2 = 225.67$ mg/100 ml, $\nu_2 = 5$, $s_p^2 = 16.6700\text{ (mg/100 ml)}^2$, $s_{\bar{X}_1 - \bar{X}_2} = 2.27$ mg/100 ml, $t = -0.630$, $t_{0.05(2),11} = 2.201$; therefore, do not reject H_0; $P > 0.50$.

9.4. H_0: $\mu_1 \geq \mu_2$, H_A: $\mu_1 < \mu_2$, $n_1 = 7$, $SS_1 = 98.86\text{ mm}^2$, $\nu_1 = 6$, $\bar{X}_1 = 117.9$ mm, $n_2 = 8$, $SS_2 = 62.88\text{ mm}^2$, $\nu_2 = 7$, $\bar{X}_2 = 118.1$ mm, $s_p^2 = 12.44\text{ mm}^2$, $s_{\bar{X}_1 - \bar{X}_2} = 1.82$ mm, $t = -0.11$, $t_{0.05(1),13} = 1.771$; therefore, do not reject H_0; $P > 0.25$.

9.5. H_0: $\mu_1 \geq \mu_2$, H_A: $\mu_1 < \mu_2$, $\bar{X}_1 = 4.6$ kg, $s_1^2 = 3.88$ kg^2, $n_1 = 18$, $\nu_1 = 17$, $\bar{X}_2 = 6.0$ kg, $s_2^2 = 4.35$ kg^2, $n_2 = 26$, $\nu_2 = 25$, $s_p^2 = 4.16$ kg^2, $s_{\bar{X}_1 - \bar{X}_2} = 0.62$ kg, $t = -2.26$, $t_{0.05(1),42} = 1.682$; therefore, reject H_0; $0.01 < P < 0.025$.

9.6. H_0: $\mu_2 - \mu_1 \leq 10$ g, H_A: $\mu_2 - \mu_1 > 10$ g, $\bar{X}_1 = 334.6$ g, $\text{SS}_1 = 364.34$ g^2, $n_1 = 19$, $\nu_1 = 18$, $\bar{X}_2 = 349.8$ g, $\text{SS}_2 = 286.78$ g^2, $n_2 = 24$, $\nu_2 = 23$, $s_p^2 = 15.88$ g^2, $s_{\bar{X}_1 - \bar{X}_2} = 1.22$ g, $t = 4.26$, $t_{0.05(1),41} = 1.683$; therefore, reject H_0 and conclude that μ_2 is at least 10 g greater than μ_1; $P < 0.0005$.

9.7. H_0 is not rejected; $\bar{X}_p = [(7)(224.24 \text{ mg/100 ml}) + (6)(225.67 \text{ mg/100 ml})]/(7 + 6) = 224.90$ mg/100 ml, $s_p^2 = 16.6700$ (mg/100 ml)^2, $t_{0.05(2),11} = 2.201$; 95% confidence interval for $\mu = 224.9 \text{ mg/100 ml} \pm 2.201\sqrt{16.67 \text{ (mg/100 ml)}^2/13} = 224.9$ mg/100 ml $\pm$ 2.5 mg/100 ml; $L_1 = 222.4$ mg/100 ml, $L_2 = 227.4$ mg/100 ml.

9.8. $s_p^2 = (244.66 + 289.18)/(13 + 13) = 20.53$ (km/hr)^2; $d = 2.0$ km/hr. If we guess $n = 50$, then $\nu = 2(50 - 1) = 98$, $t_{0.05(2),98} = 1.984$, and $n = 40.4$. Then, guess $n = 41$; $\nu = 80$, $t_{0.05(2),80} = 1.990$, and $n = 40.6$. So, the desired $n = 41$.

9.9. **(a)** If we guess $n = 25$, then $\nu = 2(24) = 48$. $t_{0.05(2),24} = 2.064$, $t_{0.10(1),24} = 1.299$, and $n = 18.5$. Then, guess $n = 19$; $\nu = 36$, $t_{0.05(2),36} = 2.028$, $t_{0.10(1),36} = 1.306$, and $n = 18.3$. So, the desired sample size is $n = 19$. **(b)** $n = 20.95$, $\nu = 40$, $t_{0.05(2),40} = 2.021$, $t_{0.10(1),40} = 1.303$, and $d = 4.65$ km/hr. **(c)** $n = 50$, $\nu = 98$, $t_{0.05(2),98} = 1.984$, and $t_{\beta(1),98} = 0.223$; $\beta > 0.25$, so power < 0.75 (or, by the normal approximation, $\beta = 0.41$, so power $= 0.59$).

9.10. H_0: Male and female turtles have the same serum cholesterol concentrations; H_A: Male and female turtles do not have the same serum cholesterol concentrations.

Male ranks	*Female ranks*
2	5
1	3
11	12
10	8
4	6
7	13
9	

$R_1 = 44$, $n_1 = 7$, $R_2 = 47$, $n_2 = 6$; $U = 26$; $U' = (7)(6) - 26 = 16$; $U_{0.05(2),7,6} = U_{0.05(2),6,7} = 36$; therefore, do not reject H_0; $P > 0.20$.

9.11. H_0: Northern birds do not have shorter wings than southern birds; H_A: Northern birds have shorter wings than southern birds.

Northern ranks	*Southern ranks*
11.5	5
1	7
15	13
8.5	2.5
5	5
2.5	8.5
10	14
	11.5

$R_1 = 53.5$, $n_1 = 7$, $n_2 = 8$; $U = 30.5$; $U' = 25.5$; $U_{0.05(1),7,8} = 43$; therefore, do not reject H_0; $P > 0.10$.

9.12. H_0: Intersex cells have 1.5 times the volume of normal cells; H_A: Intersex cells do not have 1.5 times the volume of normal cells.

Normal × 1.5	*Rank*	*Intersex*	*Rank*
372	4	380	9
354	1	391	13
403.5	16	377	8
381	10	392	14
373.5	5	398	15
376.5	7	374	6
390	12		
367.5	3		
358.5	2		
382.5	11		

$R_1 = 71$, $n_1 = 10$, $n_2 = 6$; $U = 44$, $U' = 16$; $U_{0.05(2),10,6} = 49$; therefore, do not reject H_0; $0.10 < P < 0.20$.

9.13. H_0: Both species have the same food preference; H_A: Both species do not have the same food preference. By methods of Chapter 6, $\chi^2 = 9.175$ and $\nu = 3$, or $G = 9.375$; $\chi^2_{0.05,3} = 7.815$; therefore, reject H_0.

CHAPTER 10

10.1. H_0: $\mu_d = 0$, H_A: $\mu_d \neq 0$; $\bar{d} = -2.09\ \mu\text{g/m}^3$, $s_{\bar{d}} = 1.29\ \mu\text{g/m}^3$. **(a)** $t = 1.62$, $n = 11$, $\nu = 10$, $t_{0.05(2),10} = 2.228$; therefore, do not reject H_0: $0.10 < P < 0.20$. **(b)** 95% confidence interval for $\mu_d = -2.09 \pm (2.228)(1.29) = -2.09 \pm 2.87$; $L_1 = -4.96\ \mu\text{g/m}^3$, $L_2 = 0.78\ \mu\text{g/m}^3$.

10.2.

d_i	*Signed rank*
−4	−5.5
−2	−3
−5	−7.5
6	9
−5	−7.5
1	1.5
−7	−10.5
−4	−5.5
−7	−10.5
1	1.5
3	4

$T = 9 + 1.5 + 1.5 + 4 = 16$; $T_{0.05(2),11} = 10$; since T is not ≤ 10, do not reject H_0; $0.10 < P < 0.20$.

10.3. H_0: There is no difference in frequency of occurrence of varicose veins between overweight and normal weight men; H_A: There is a difference in frequency of occurrence of varicose veins between overweight men and normal weight men; $f_{11} = 19$, $f_{12} = 5$, $f_{21} = 12$, $f_{22} = 86$, $n = 122$; $\chi^2_c = 2.118$; $\chi^2_{0.05,1} = 3.841$; do not reject H_0; $0.10 < P < 0.25$.

CHAPTER 11

11.1. H_0: $\mu_1 = \mu_2 = \mu_3 = \mu_4$; H_A: The mean food consumption is not the same for all four months; $F = 0.7688/0.0348 = 22.1$; $F_{0.05(1),3,18} = 3.16$; reject H_0; $P < 0.0005$.

11.2. $k = 5, \nu_1 = 4, n = 12, \nu_2 = 55, \sigma^2 = 1.54\ (°C)^2, \delta = 2.0°C$; $\phi = 1.77$; from Fig. B.1d we find that the power is about 0.89.

11.3. $n = 16$, for which $\nu_2 = 80$ and $\phi = 2.04$. (The power is a little greater than 0.95; for $n = 15$ the power is about 0.94.)

11.4. $\nu_2 = 45$, power $= 0.95$, $\phi = 2.05$; minimum detectable difference is about 2.5°C.

11.5. H_0: The amount of food consumed is the same during all four months; H_A: The amount of food consumed is not the same during all four months; $n_1 = 5, n_2 = 6, n_3 = 6, n_4 = 5$; $R_1 = 69.5$, $R_2 = 23.5$, $R_3 = 61.5$, $R_4 = 98.5$; $N = 22$; $H = 17.08$; $\chi^2_{0.05,3} = 7.815$; reject H_0; $P \ll 0.001$. H_c (i.e., H corrected for ties) would be obtained as $\sum T = 120$, $C = 0.9887$, $H_c = 17.28$.

11.6. $H_0: \sigma_1^2 = \sigma_2^2 = \sigma_3^2$; H_A: The three population variances are not all equal; $B = 5.94517$, $C = 1.0889$, $B_c = B/C = 5.460$; $\chi^2_{0.05,2} = 5.991$; do not reject H_0; $0.05 < P < 0.10$.

CHAPTER 12

12.1. Ranked sample means: $\underline{14.8 \quad 16.2} \quad \underline{20.2}$; $k = 3$, $n = 8$, $\nu = 0.05$, $s^2 = 8.46$, $\nu = 21$ (which is not in Table B.5, so use $\nu = 20$, which is in the table); reject H_0: $\mu_2 = \mu_1$; reject H_0: $\mu_2 = \mu_3$; accept H_0: $\mu_3 = \mu_1$. Therefore, the overall conclusion is $\mu_1 = \mu_3 \neq \mu_2$. (c) $\bar{X}_p = \bar{X}_{1,3} = 15.5$, $t_{0.05(2),21} = 2.080$, $n_1 + n_2 = 16$, 95% CI for $\mu_{1,3} = 15.5 \pm 1.5$; 95% CI for $\mu_2 = 20.2 \pm 2.1$; $\bar{X}_{1,3} - \bar{X}_2 = -4.7$, SE $= 1.03$, 95% CI for $\mu_{1,3} - \mu_2 = -4.7 \pm 3.7$.

12.2. Ranked sample means: 133.36, 152.44, 189.72, 200.78; sample sizes of 5, 5, 5, and 4, respectively; $k = 4$, $\nu = 15$, $\alpha = 0.05$, $s^2 = 41.49$; control group is group 1; reject H_0: $\mu_4 = \mu_1$, reject H_0: $\mu_3 = \mu_1$, reject H_0: $\mu_2 = \mu_1$. Overall conclusion: The mean of the control population is different from the mean of each other population.

12.3. Ranked sample means: 133.36, 152.44, 189.72, 220.78; sample sizes of 5, 5, 5, and 4, respectively; $k = 4$, $\nu = 15$, $\alpha = 0.05$, $s^2 = 41.49$; critical value of S is $F_{0.05(1),3,15} = 3.29$; for H_0: $(\mu_1 + \mu_4)/2 - (\mu_2 + \mu_3)/2 = 0$, $S = 2.02$, accept H_0; for H_0: $(\mu_2 + \mu_4)/2 - \mu_3 = 0$, $S = 0.86$, accept H_0.

12.4. $R_1 = 26$, $R_2 = 30$, $R_3 = 64$. Overall conclusion: The variable being measured is the same magnitude in populations 1 and 2. The variable is of different magnitude in population 3.

CHAPTER 13

13.1. **(a)** H_0: There is no difference in mean blood calcium among the three hormone treatments; H_A: There is difference in mean blood calcium among the three hormone treatments; $F = 572.6139/3.1714 = 180.55$; $F_{0.05(1),2,24} = 3.40$; reject H_0; $P \ll 0.0005$. **(b)** H_0: There is no difference in mean blood calcium between males and females; H_A: There is a difference in mean blood calcium between males and females; $F = 3.9169/3.1714 = 1.24$; $F_{0.05(1),1,24} = 4.26$; do not reject H_0; $P > 0.25$. **(c)** H_0: There is no interaction between hormone treatment and sex in affecting blood calcium concentration; H_A: There is interaction between hormone treatment and sex in affecting blood calcium concentration; $F = 1.9123/3.1714$; since $F < 1.0$, do not reject H_0 (i.e., $P > 0.25$).

13.2. Hypotheses as in Exercise 13.1, but without the word "mean." **(a)** $H = 1751.2000/$

$77.5000 = 22.596$; $\chi^2_{0.05,2} = 5.991$; reject H_0; $P < 0.001$; **(b)** $H = 86.7000/77.5000 = 1.119$; $\chi^2_{0.05,1} = 3.841$; do not reject H_0; $0.25 < P < 0.50$; **(c)** $H = 160.0000/77.5000 = 2.065$; $\chi^2_{0.05,2} = 5.991$; do not reject H_0; $0.25 < P < 0.50$.

13.3. H_0: All four plant varieties reach the same mean height (i.e., H_0: $\mu_1 = \mu_2 = \mu_3 = \mu_4$); H_A: All four plant varieties do not reach the same mean height; $F = 62.8461/0.4351 = 144$; $F_{0.05(1),3,15} = 3.29$; reject H_0; $P \ll 0.0005$.

13.4. H_0: Adult bird weight is the same on all three diets; H_A: Adult bird weight is not the same on all three diets; $R_1 = 10$, $R_2 = 11$, $R_3 = 21$; $\chi^2_r = 10.57$; $\nu = 2$; $\chi^2_{0.05,2} = 5.991$; reject H_0; $0.005 < P < 0.01$.

13.5. H_0: There is no difference in potential acceptance among the three textbooks; H_A: The three textbooks do not have the same potential acceptance; $a = 4$, $b = 13$ (blocks 4 and 11 are deleted from the analysis); $Q = 3.69$; $\nu = 3$; $\chi^2_{0.05,3} = 7.815$; do not reject H_0; $0.25 < P < 0.50$.

CHAPTER 14

14.1. $\bar{X}' = 0.68339$, $s'_{\bar{X}} = 0.00363$; $L'_1 = 0.67481$, $L'_2 = 0.69197$; $L_1 = 3.73$ ml, $L_2 = 3.92$ ml.

14.2. $\bar{X}' = 61.48$, $s'_{\bar{X}} = 0.76$; $L'_1 = 59.53$, $L'_2 = 63.43$; $L_1 = 0.739$, $L_2 = 0.780$.

14.3. $\bar{X}' = 2.4280$, $s'_{\bar{X}} = 0.2329$; $L'_1 = 1.8292$, $L'_2 = 3.0268$; $L_1 = 2.85$, $L_2 = 8.66$.

CHAPTER 15

15.1. H_0: No effect of factor A; H_A: Factor A has an effect; $F = 10.07890/0.08370 = 120.4$; since $F_{0.05(1),3,72} \cong 2.74$, H_0 is rejected; $P \ll 0.0005$. H_0: No effect of factor B; $F = 3.77918/0.08370 = 45.2$; since $F_{0.05(1),2,72} \cong 3.13$, H_0 is rejected; $P \ll 0.0005$. H_0: No effect of factor C; $F = 4.25924/0.08370 = 50.9$; since $F_{0.05(1),1,72} \cong 3.98$, H_0 is rejected; $P \ll 0.0005$. H_0: No interaction between factors A and B; H_A: There is $A \times B$ interaction; $F = 0.10932/0.08370 = 1.31$; since $F_{0.05(1),6,72} \cong 2.23$, H_0 is not rejected; $P > 0.25$. H_0: No interaction between factors A and C; $F = 0.60984/0.08370 = 7.29$; since $F_{0.05(1),3,72} \cong 2.74$, H_0 is rejected; $P < 0.0005$. H_0: No interaction between factors B and C; $F = 0.00164/0.08370 = 0.020$; since $F_{0.05(1),2,72} \cong 1.41$, H_0 is not rejected; $P > 0.25$. H_0: No interaction between factors A, B, and C; $F = 0.12785/0.08370 = 1.53$; since $F_{0.05(1),6,72} \cong 2.23$, H_0 is not rejected; $0.10 < P < 0.25$.

15.2. H_0: No effect of factor A; H_A: Factor A has an effect; $F = 56.00347/0.03198 = 1751$; reject H_0; $P \ll 0.0005$. H_0: No effect of factor B; $F = 4.65125/0.03198 = 145.4$; reject H_0; $P \ll 0.0005$. H_0: No effect of factor C; $F = 8.61125/0.03198 = 269.3$; reject H_0; $P \ll 0.0005$. H_0: No effect of factor D; $F = 2.17056/0.03198 = 67.9$; reject H_0; $P \ll 0.0005$. H_0: No interaction between factors A and B; $F = 2.45681/0.03198 = 76.8$; reject H_0; $P \ll 0.0005$. H_0: No interaction between factors A and C; $F = 0.05014/0.03198 = 1.57$; do not reject H_0; $0.10 < P < 0.25$, H_0: No interaction between factors A and D; $F = 0.06889/0.03198 = 2.15$; do not reject H_0; $0.10 < P < 0.25$. H_0: No interaction between factors B and C; $F = 0.01681/0.03198 = 0.53$; do not reject H_0; $P > 0.25$. H_0: No interaction between factors B and D; $F = 0.15167/0.03198 = 4.74$; reject H_0; $0.01 < P < 0.025$. H_0: No interaction between factors C and D; $F = 0.26000/0.03198 = 8.13$; reject H_0; $0.0005 < P < 0.001$. H_0: No interaction among factors A, B, and C; $F = 0.00125/0.03198 = 0.04$; do not reject H_0; $P > 0.25$. H_0: No interaction among factors A, B, and D; $F = 0.14222/0.03198 = 2.11$; do not reject H_0; $0.10 < P < 0.25$. H_0: No interaction among factors B, C, and D; $F = 0.00222/0.03198 = 0.07$; do not reject H_0; $P > 0.25$. H_0: No interaction among factors A, B, C, and D; $F = 0.01167/0.03198 = 0.36$; do not reject H_0; $P > 0.25$.

15.3. H_0: No effect of factor A; H_A: There is an effect of factor A; $F = 239.39048/2.10954 = 113.5$; since $F_{0.05(1),1,22} = 4.30$, reject H_0; $P \ll 0.0005$. H_0: No effect of factor B; $F = 8.59013/2.10954 = 4.07$; since $F_{0.05(1),1,22} = 4.30$; do not reject H_0; $0.05 < P < 0.10$. H_0: No interaction between factors A and B; $F = 0.10440/2.10954 = 0.05$; since $F_{0.05(1),1,22} = 4.30$, do not reject H_0; $P > 0.25$.

CHAPTER 16

16.1. H_0: The mean fluoride concentrations are the same for all three samples at a given location; H_A: The mean fluoride concentrations are not the same for all three samples at a given location; $F = 0.008333/0.01778 = 0.469$; since $F < 1.0$, do not reject H_0 (i.e., $P > 0.25$). H_0: The mean fluoride concentration is the same at all three locations; H_A: The mean fluoride concentration is not the same at all three locations; $F = 1.1850/0.008333 = 142.2$; $F_{0.05(1),2,6} = 5.14$; reject H_0; $P \ll 0.0005$.

CHAPTER 17

17.1. **(a)** $b = -0.0878$ ml/g/hr/°C, $a = 3.471$ ml/g/hr. **(b)** H_0: $\beta = 0$, H_A: $\beta \neq 0$; $F = 309$; reject H_0; $P \ll 0.0005$. **(c)** H_0: $\beta = 0$, H_A: $\beta \neq 0$; $t = -17.6$; reject H_0: $P \ll 0.0001$. **(d)** $s_{Y \cdot X} = 0.17$ ml/g/hr; **(e)** $r^2 = 0.98$; (f) 95% confidence interval for $\beta = -0.0878 \pm 0.0122$; $L_1 = -0.1000$ ml/g/hr/°C, $L_2 = -0.0756$ ml/g/hr/°C.

17.2. **(a)** $\hat{Y} = 3.47 - (0.878)(15) = 2.15$ ml/g/hr. **(b)** $s_{\hat{Y}} = 0.1021$ ml/g/hr; $L_1 = 1.90$ ml/g/hr, $L_2 = 2.40$ ml/g/hr. **(c)** $\hat{Y} = 2.15$ ml/g/hr. **(d)** $s_{\hat{Y}} = 0.1960$ ml/g/hr; $L_1 = 1.67$ ml/g/hr, $L_2 = 2.63$ ml/g/hr.

17.3. **(a)** $b = 9.73$ impulses/sec/°C, $a = 44.2$ impulses/sec. **(b)** H_0: $\beta = 0$, H_A: $\beta \neq 0$; $F = 311.0$; reject H_0; $P \ll 0.0005$. **(c)** $s_{Y \cdot X} = 8.33$ impuses/sec. **(d)** $r^2 = 0.94$. (e) H_0: The population regression is linear; H_A: The population regression is not linear; $F = 1.78$, do not reject H_0; $0.10 < P < 0.25$.

CHAPTER 18

18.1. **(a)** H_0: $\beta_1 = \beta_2$; $b_1 = 0.488$, $b_2 = 0.537$; $s_{b_1 - b_2} = 0.202$; $t = -0.243$; since $t_{0.05(2),54} = 2.005$, do not reject H_0; $P > 0.50$. **(b)** H_0: The elevations of the two population regressions are the same; H_A: The elevations of the two population regressions are not the same; $b_c = 0.516$; $t = 10.7$; since $t_{0.05(2),55} \cong 2.004$, reject H_0; $P < 0.001$.

18.2. **(a)** H_0: $\beta_1 = \beta_2 = \beta_3$; H_A: All three β's are not equal; $F = 0.84$; since $F_{0.05(1),2,90} = 3.10$, do not reject H_0; $P > 0.25$; $b_c = 3.16$. **(b)** H_0: The three population regression lines have the same elevation; H_A: The three lines do not all have the same elevation; $F = 4.61$; since $F_{0.05(1),2,90} = 3.10$, reject H_0; $0.01 < P < 0.025$.

CHAPTER 19

19.1. **(a)** $r = 0.86$. **(b)** $r^2 = 0.73$. **(c)** H_0: $\rho = 0$; H_A: $\rho \neq 0$; $s_r = 0.16$; $t = 5.38$; since $t_{0.05(2),10} = 2.228$; reject H_0; $P < 0.001$. Or: $r = 0.86$, $r_{0.05(2),10} = 0.576$; reject H_0; $P < 0.001$. Or: $F = 13.29$, $F_{0.05(2),10,10} = 3.72$; reject H_0; $P < 0.001$. (d) $L_1 = 0.56$, $L_2 = 0.96$.

19.2. **(a)** H_0: $\rho \leq 0$; H_A: $\rho > 0$; $r = 0.86$; $t = 5.38$; $t_{0.05(1),10} = 1.812$; reject H_0; $P <$

0.0005. Or: $r_{0.05(1),10} = 0.407$; reject H_0; $P < 0.0005$. Or: $F = 13.29$; $F_{0.05(1),10,10} = 2.98$; reject H_0; $P < 0.0005$. **(b)** H_0: $\rho = 0.50$; H_A: $\rho \neq 0.50$; $r = 0.857$; $z = 1.2933$; $z_c = 0.5493$; $\sigma_z = 0.3333$; $Z = 2.232$; $Z_{0.05(2)} = 1.960$; reject H_0; $0.02 < P < 0.05$.

19.3. (a) H_0: $\rho_1 = \rho_2$; H_A: $\rho_1 \neq \rho_2$; $z_1 = -0.4722$. $z_2 = -0.4236$; $z_1 - z_2 = 0.2910$; $Z = -0.167$; $Z_{0.05(2)} = 1.960$; do not reject H_0; $P > 0.50$. **(b)** $z_w = -0.4449$; $r_w = -0.42$.

19.4. H_0: $\rho_1 \geq \rho_2$; H_A: $\rho_1 < \rho_2$; $z_1 = 0.4847$, $z_2 = 0.6328$; $\sigma_{z_1 - z_2} = 0.3789$; $Z = -0.3909$; $Z_{0.05(1)} = 1.645$; do not reject H_0; $P > 0.25$.

19.5. (a) H_0: $\rho_1 = \rho_2 = \rho_3$; H_A: The three population correlation coefficients are not all the same; $\chi^2 = 111.6607 - (92.9071)^2/78 = 0.998$; $\chi^2_{0.05,2} = 5.991$; do not reject H_0; $0.50 < P < 0.75$. **(b)** $z_w = 92.9071/78 = 1.1911$; $r_w = 0.83$.

19.6. (a) $\sum d_i^2 = 88.00$, $r_s = 0.69$; **(b)** H_0: $\rho_s = 0$; H_A: $\rho_s \neq 0$; since $(r_s)_{0.05(2),12} = 0.587$, reject H_0; $0.01 < P < 0.02$.

19.7. (a) $r_n = (16 - 7)/(16 + 7) = 0.39$. **(b)** H_0: There is no correlation between the type of institution a college president heads and the type of institution he attended as an undergraduate; H_A: There is a correlation between the type of school headed and the type attended. By Fisher exact test (using Table B.27): $n = 23$, $m_1 = 9$, $m_2 = 11$, $f = 2$, critical $f_{0.05(2)} = 1$ and 7; since f is not ≤ 1 and is not ≥ 7, do not reject H_0.

19.8. (a) $r_I = (0.000946 - 0.00213)/(0.000946 + 0.001213) = -0.12$. **(b)** H_0: There is no correlation between corticosterone determinations from the same laboratory (i.e., $\rho_I = 0$); H_A: There is correlation between corticosterone determinations from the same laboratory (i.e., $\rho_I \neq 0$); $F = 0.000946/0.001213 = 0.78$; since $F_{0.05(1),3,4} = 6.59$, do not reject H_0; $P > 0.25$.

CHAPTER 20

20.1. (a) $\hat{Y} = -30.14 + 2.07X_1 + 2.58X_2 + 0.64X_3 + 1.11X_4$. **(b)** H_0: No population regression; H_A: There is a population regression; $F = 90.2$, reject H_0, $P \ll 0.0005$. **(c)** H_0: $\beta_i = 0$, H_A: $\beta_i \neq 0$; $t_{0.05(2),9} = 2.262$; "*" below denotes significance:

i	b_i	s_{b_i}	$t = \frac{b_i}{s_{b_i}}$	*Conclusion*
1	2.07	0.46	4.50*	Reject H_0.
2	2.58	0.74	3.49*	Reject H_0.
3	0.64	0.46	1.39	Do not reject H_0.
4	1.11	0.76	1.46	Do not reject H_0.

(d) $s_{Y\cdot1,2,3,4} = 3.11$ g; $R^2 = 0.9757$. **(e)** $\hat{Y} = 61.73$ g. **(f)** $s_{\hat{Y}} = 2.9549$ g, $L_1 = 55.0$ g, $L_2 = 68.4$ g. **(g)** H_0: $\mu_Y \leq 50.0$ g, H_A: $\mu_Y > 50.0$ g, $t = 3.970$; reject H_0; $0.001 < P < 0.0025$.

20.2. (1) With X_1, X_2, X_3, and X_4 in the model, see Exercise 20.1c.
(2) Delete X_3. With X_1, X_2, and X_4 in the model,

i	b_i	t
1	1.48	9.15*
2	1.73	4.02*
4	2.14	0.50

$a = 16.83$

(3) Delete X_4. With X_1 and X_2 in the model,

i	b_i	t
1	1.48	9.47*
2	1.53	13.19

$a = 24.96$

(4) Therefore, the final equation is $\hat{Y} = 24.96 + 1.48X_1 + 1.53X_2$.

20.3. **(a)** $R = 0.9878$. **(b)** $F = 90.2$, reject H_0: There is no population correlation among the five variables; $P \ll 0.0005$. **(c)** Partial correlation coefficients:

	1	2	3	4	5
1	1.0000				
2	−0.9092*	1.0000			
3	−0.8203*	−0.8089*	1.0000		
4	−0.7578*	−0.9094*	−0.8724*	1.0000	
5	0.8342*	0.7583*	0.4183	0.4342	1.0000

(d) From Table B.16, $r_{0.05(2),9} = 0.602$, and the significant partial correlation coefficients are indicated with asterisks in part (c).

20.4. H_0: Each of the three sample regressions estimates the same population regression; H_A: Each of the three sample regressions does not estimate the same population regression; $F = 0.915$; since $F_{0.05(1),8,72} \cong 2.07$, do not reject H_0; $P > 0.25$.

20.5. **(a)** $W = 0.675$. **(b)** H_0: There is no agreement among the four faculty reviewers; H_A: There is agreement among the four faculty reviewers; $\chi_r^2 = 10.800$; $(\chi_r^2)_{0.05,4,5} = 7.800$; reject H_0; $0.005 < P < 0.01$. **(c)** $W = 0.713$. **(d)** H_0: There is no agreement among the four student reviewers; H_A: There is agreement among the four student reviewers; $\chi_r^2 = 11.408$; $(\chi_r^2)_{0.05,4,5} = 7.800$; reject H_0; $0.002 < P < 0.005$. **(e)** $\mathcal{W} = 0.606$, H_0: There is not agreement both within and between the two groups of reviewers; H_A: There is agreement both within and between the two groups of reviewers, $Z = 4.848$; $Z_{0.05(1)} = 1.6449$; reject H_0; $P < 0.0005$.

CHAPTER 21

21.1. In each step, H_0: $\beta_i = 0$ vs. H_A: $\beta_i \neq 0$ is tested, where i is the highest term in the polynomial expression. An asterisk indicates H_0 is rejected.

(1) Linear regression: $\hat{Y} = 8.8074 - 0.18646X$; $t = 7.136$*; $t_{0.05(2),13} = 2.160$.

(2) Quadratic regression: $\hat{Y} = -14.495 + 1.6595X - 0.036133X^2$; $t = 5.298$*; $t_{0.05(2),12} = 2.179$.

(3) Cubic regression: $\hat{Y} = -33.810 + 3.9550X - 0.12649X^2 + 0.0011781X^3$; $t = 0.374$; $t_{0.05(2),11} = 2.201$.

(4) Quartic regression: $\hat{Y} = 525.30 - 84.708X + 5.1223X^2 - 0.13630X^3 + 0.0013443X^4$; $t = 0.911$; $t_{0.05(2),10} = 2.28$.

Therefore, the quadratic expression is concluded to be the "best."

21.2. **(a)** $\hat{Y} = 1.00 + 0.851X - 0.0259X^2$. **(b)** H_0: $\beta_2 = 0$; H_A: $\beta_2 \neq 0$; $F = 69.4$; $F_{0.05(1),1,4} = 7.71$; reject H_0; $0.001 < P < 0.0025$. **(c)** $\hat{Y} = 6.92$ eggs/cm²; $s_{\hat{Y}} = 0.26$ eggs/cm²; 95% confidence interval $= 6.92 \pm 0.72$ eggs/cm². **(d)** $\hat{X}_0 = 16.43$°C; $\hat{Y}_0 = 7.99$ eggs/cm². **(e)** For $\hat{X}_0$: 95% confidence interval $= 16.47 \pm 0.65$°C; for $\hat{Y}_0$: 95% confidence interval $= 7.99 \pm 0.86$ eggs/cm².

CHAPTER 22

22.1. $P(X = 2) = 0.32413$. **22.2.** $P(X = 4) = 0.00914$.

22.3. H_0: The sampled population is binomial with $p = 0.25$; H_A: The sampled population is not binomial with $p = 0.25$; $\sum f_i = 126$; $F_1 = (0.31641)(126) = 39.868$, $F_2 = 53.157$, $F_3 = 26.578$, $F_4 = 5.907$, $F_5 = 0.493$; combine F_4 and F_5 and combine f_4 and f_5; $\chi^2 = 11.524$, $\nu = k - 1 = 3$, $\chi^2_{0.05,3} = 7.815$; reject H_0; $0.005 < P < 0.01$.

22.4. H_0: The sampled population is binomial; H_A: The sampled population is not not binomial; $\hat{p} = \frac{156}{109}/4 = 0.3578$; $\chi^2 = 3.186$, $\nu = k - 2 = 3$, $\chi^2_{0.05,3} = 7.815$; do not reject H_0; $0.25 < P < 0.50$.

22.5. $H_0: p = 0.5$; $H_A: p \neq 0.5$; $n = 20$; $P(X \leq 6 \text{ or } X \geq 14) = 0.11532$; since this probability is greater than 0.05, do not reject H_0.

22.6. $H_0: p = 0.5$; $H_A: p \neq 0.5$; $\hat{p} = \frac{197}{412} = 0.4782$; $Z = -0.888$; $Z_c = 0.838$; $Z_{0.05(2)} = t_{0.05(2),\infty} = 1.960$; therefore, do not reject H_0; $P = 0.37$.

22.7. $H_0: p = 0.5$; $H_A: p \neq 0.5$; $\hat{p} = \frac{44}{98} = 0.4490$; $Z = -1.015$; $Z_c = 0.914$; $Z_{0.05(2)} = t_{0.05(2),\infty} = 1.960$; do not reject H_0; $0.20 < P < 0.50$.

22.8. $H_0: p = 0.5$; $H_A: p \neq 0.5$; number of positive differences = 7; for $n = 10$ and $p = 0.5$: $P(X \leq 3 \text{ or } X \geq 7) = 0.34378$; since this probability is greater than 0.05, do not reject H_0.

22.9. $X = 62$, $n = 1215$; $\hat{p} = 0.0510$; $P(0.0394 \leq p \leq 0.0649) = 0.95$.

22.10. $n = 20$, $p = 0.50$; critical values are 5 and 15; $\hat{p} = 6/20 = 0.30$, power $= 0.00080 + 0.00684 + 0.02785 + 0.07160 + 0.13042 + 0.17886 + 0.00004 + 0.00001 = 0.42$.

22.11. $p_0 = 0.50$, $p = 0.4782$, $n = 412$; power $= P(Z < -1.08) + P(Z > 2.84) = 0.1401 + 0.0023 = 0.14$.

22.12. H_0: The frequency of wind damage in mature spruce is the same as in mature pine; H_A: The frequency of wind damage is not the same in mature spruce as in mature pine.

	Damage	No Damage	
Spruce	4	4	8
Pine	1	8	9
	5	12	17

$$P = \frac{\dfrac{8!\,9!\,5!\,12!}{17!}}{4!\,4!\,1!\,8!} = 0.10181$$

5	3	8
0	9	9
5	12	17

$P = 0.00905$

For the Opposite Tail:

0	8	8
5	4	9
5	12	17

$P = 0.02036$

1	7	8
4	5	9
5	12	17

$P = 0.16290$

Total $P = 0.10181 + 0.00905 + 0.02036 = 0.13122$; since this probability is greater than 0.05, do not reject H_0. Using Table B.27: $n = 17$, $m_1 = 5$, and $m_2 = 8$; the critical f values are 0 and 5; and the observed f associated with m_1 and m_2 is 4; therefore, since 4 is not ≥ 5 and is not ≤ 0, do not reject H_0.

22.13. H_0: $p_1 = p_2$, H_A: $p_1 \neq p_2$, $X_1 = 31$, $X_2 = 43$, $n_1 = 87$, $n_2 = 158$; $\hat{p}_1 = 0.356$, $\hat{p}_2 = 0.272$, $\bar{p} = 0.302$; $Z = 1.370$, $z_{0.05(2)} = 1.960$, do not reject H_0, $0.05 < P < 0.10$.

22.14. $\alpha = 0.05$, $p_1 = 0.333$, $p_2 = 0.250$, $n_1 = n_2 = 300$; $\bar{p} = 0.292$, power $=$ $P(Z < -4.22) + P(Z > -0.28) = 0.00 + 0.61 = 0.61$.

22.15. $\alpha = 0.05$, $\beta = 0.10$, $p_1 = 0.333$, $p_2 = 0.250$, $\bar{p} = 0.292$; $A = 4.3287$, $n = 652.2$, so use sample sizes of at least 653.

22.16. H_0: $p_1 = p_2 = p_3 = p_4$, H_A: All four population proportions are not equal; $X_1 = 163$, $X_2 = 135$, $X_3 = 71$, $X_4 = 43$, $n_1 = 249$, $n_2 = 212$, $n_3 = 111$, $n_4 = 81$; $\hat{p}_1 = 0.655$, $\hat{p}_2 = 0.637$, $\hat{p}_3 = 0.640$, $\hat{p}_4 = 0.531$; $\bar{p} = 412/653 = 0.631$, $\chi^2 = 0.5966 + 0.0305 + 0.0356 + 3.4883 = 4.151$, $\chi^2_{0.05,3} = 7.815$, do not reject H_0, $0.10 < P < 0.25$.

22.17. H_0: $p_1 = p_2 = p_3 = p_4 = 0.5$, $p_0 = 0.5$; $\chi^2 = 36.3742 + 15.8679 + 8.6577 + 0.3086 = 61.208$, $\chi^2_{0.05,4} = 9.488$, reject H_0, $P \ll 0.001$.

22.18. Ranked proportions: 0.161, 0.261, 0.448; ranked transformed proportions: 23.76, 30.78, 42.06; $k = 3$, $q_{0.05,\infty,3} = 3.314$; reject $H_0: p_A = p_B$ for all three comparisons; overall conclusion: $p_1 \neq p_2 \neq p_3$.

CHAPTER 23

23.1. If $\mu = 1.5$, $P(X = 0) = 0.2231$ and $P(X = 5) = 0.0141$.

23.2. $\mu = \frac{5}{2} = 2.5$ viruses per bacterium. **(a)** $P(X = 0) = 0.0821$. **(b)** $P(X > 0) = 1.0000 - P(X = 0) = 1.0000 - 0.0821 = 0.9197$. **(c)** $P(X \geq 2) = 1.0000 - P(X = 0) - P(X = 1) = 1.0000 - 0.0821 - 0.2052 = 0.7127$. **(d)** $P(X = 3) = 0.2138$.

23.3. H_0: Raisins are distributed randomly throughout the cake; H_A: Raisins are not distributed randomly throughout the cake; $\bar{X} = \Sigma f_i X_i / \Sigma f_i = 98/57 = 1.7193$; $\chi^2 = 3.060$, $\nu = 6 - 2 = 4$, $\chi^2_{0.05,4} = 7.815$; do not reject H_0; $0.25 < P < 0.50$.

23.4. H_0: $p \leq 0.00010$; H_A: $p > 0.00010$; $p_0 = 0.00010$; $n = 25{,}000$; $p_0 n = 2.5$; $X = 5$; $P(X \geq 5) = 0.1087$; do not reject H_0; do not include this disease on the list.

23.5. H_0: $\mu_1 = \mu_2$; H_A: $\mu_1 \neq \mu_2$; $X_1 = 112$, $X_2 = 134$; $Z = 1.40$; $Z_{0.05(2)} = 1.9600$; do not reject H_0; $0.10 < P < 0.20$.

23.6. H_0: The magnitude of fish kills is randomly distributed over time; H_A: The magnitude of fish kills is not randomly distributed over time; $n = 16$, $s^2 = 400.25$, $s^2_* = 3549.17/30 = 118.31$; $C = 0.704$, $C_{0.05,16} = 0.386$; reject H_0, $P < 0.0005$.

23.7. H_0: The incidence of heavy damage is random over the years; H_A: The incidence of heavy damage is not random over the years; $n_1 = 14$, $n_2 = 13$, $u = 12$, $u_{0.05,14,13} = 9$ and 20. As 12 is neither ≤ 9 nor ≥ 20, do not reject H_0; $P = 0.50$.

CHAPTER 24

24.1. $n = 12$, $Y = 0.48570$, $X = 0.20118$, $r = 0.52572$ ($c = 1.02617$, $r_c = 0.53948$). **(a)** $\bar{a} = 68°$. **(b)** $s = 56°$ (using correction for grouping, $s = 55°$), $s' = 65°$ (using correction for grouping, $s' = 64°$). **(c)** $68° \pm 47°$ (using correction for grouping, $68° \pm 46°$). **(d)** median $= 67.5°$.

24.2. $n = 15$, $Y = 0.76319$, $X = 0.12614$, $r = 0.77354$. **(a)** $a = $ 5:22 AM. **(b)** $s = $ 2:34 hr, or $s' = $ 2:36 hr. **(c)** 5:22 hr $\pm$ 1:32 hr. **(d)** median $=$ 5:10 AM.

24.3. **(a)** $k = 5$, $\sum X_j = -1.67070$, $\sum Y_j = -2.54868$, $\bar{X} = -0.33414$, $\bar{Y} = -0.50974$, $r = 0.60949$, $\cos\bar{a} = -0.54823$, $\sin\bar{a} = -0.83634$, $\bar{a} = 237°$. **(b)** $\sum x^2 = 0.16655$, $\sum y^2 = 0.17763$, $\sum xy = -0.06854$; $A = 24.01739$, $B = 9.26709$, $C = 22.51812$, $F_{0.05(1),2,3} = 9.55$, $D = 4.28455$; $b_1 = -5.19495$ and one confidence limit is 281°; $b_2 = 0.06449$ and the other confidence limit is 184°.

CHAPTER 25

25.1. H_0: $\rho = 0$; H_A: $\rho \neq 0$; $r = 0.526$; $R = 6.309$; $z = 3.317$, $z_{0.05,12} = 2.932$; reject H_0; $0.02 < P < 0.05$.

25.2. H_0: $\rho = 0$, H_A: $\rho \neq 0$; $r = 0.774$; $R = 11.603$; $z = 8.975$, $z_{0.05,15} = 2.945$; reject H_0; $P < 0\,001$.

25.3. H_0: $\rho = 0$; H_A: $\rho \neq 0$; $n = 11$, $Y = -0.88268$, $X = 0.17138$, $r = 0.89917$, $\bar{a} = 281°$, $R = 9.891$, $\mu_0 = 270°$. **(a)** $V = 9.709$, $u = 4.140$, $u_{0.05,11} = 1.648$; reject H_0; $P < 0.0005$. **(b)** H_0: $\mu_a = 270°$, H_A: $\mu_a \neq 270°$, 95% confidence interval for $\mu_a = 281° \pm 19°$, so do not reject H_0.

25.4. H_0: Mean flight direction is the same under the two sky conditions; H_A: Mean flight direction is not the same under the two sky conditions; $n_1 = 8$, $n_2 = 7$, $R_1 = 7.5916$, $R_2 = 6.1130$, $\bar{a}_1 = 352°$, $\bar{a}_2 = 305°$, $N = 15$, $r_w = 0.914$, $R = 12.5774$; $F = 12.01$, $F_{0.05(1),1,13} = 4.67$; reject H_0; $0.0025 < P < 0.0005$.

25.5. H_0: The flight direction is the same under the two sky conditions; H_A: The flight direction is not the same under the two sky conditions; $n_1 = 8$, $n_2 = 7$, $N = 15$; $\sum d_k = -2.96429$, $\sum d_k^2 = 1.40243$, $U^2 = 0.2032$, $U^2_{0.05,8,7} = 0.1817$; do not reject H_0; $0.02 < P < 0.05$.

25.6. H_0: Members of all three hummingbird species have the same mean time of feeding at the feeding station; H_A: Members of all three species do not have the same mean time of feeding at the feeding station; $n_1 = 6$, $n_2 = 9$, $n_3 = 7$, $N = 22$; $R_1 = 2.965$, $R_2 = 3.938$, $R_3 = 3.868$; $\bar{a}_1 = 10{:}30$ hr, $\bar{a}_2 = 11{:}45$ hr, $\bar{a}_3 = 11{:}10$ hr; $r_w = 0.490$, $F = 0.206$, $F_{0.05(1),2,19} = 3.54$; do not reject H_0; $P > 0.25$. Therefore, all three $\bar{a}_i$'s estimate the same μ_a, the best estimate of which is 11:25 hr.

25.7. H_0: Birds do not orient better when skies are sunny than when cloudy; H_A: Birds do orient better when skies are sunny than when cloudy. Angular distances for group 1 (sunny): 10, 20, 45, 10, 20, 5, 15, and 0°; for group 2 (cloudy): 20, 55, 105, 90, 55, 40, and 25°. For the one-tailed Mann-Whitney test: $n_1 = 8$, $n_2 = 7$, $R_1 = 40$, $U = 52$, $U_{0.05(1),8,7} = 43$, reject H_0, $P = 0.0025$.

25.8. H_0: Variability in flight direction is the same under both sky conditions; H_A: Variability in flight direction is not the same under both sky conditions; $\bar{a}_1 = 352°$, $\bar{a}_2 = 305°$; angular distances for group 1 (sunny): 2, 12, 37, 18, 28, 3, 7, and 8°, and for group 2 (cloudy): 35, 0, 50, 35, 0, 15, and 30°; for the two-tailed Mann-Whitney test: $R_1 = 58$, $U = 34$, $U' = 22$, $U_{0.05(2),8,7} = 46$, reject H_0, $P < 0.001$.

25.9. $\sum x^2 = 0.16655$, $\sum y^2 = 0.17763$, $\sum xy = -0.06854$; $F = 26.1$; $F_{0.05(1),2,3} = 9.55$; reject H_0; $0.01 < P < 0.025$.

25.10. $X = -7.52684/5 = -1.50537$, $Y = -12.00527/5 = -2.40105$, $R' = 1.267$, $R'_{0.05,5} = 1.152$, reject H_0, $0.002 < P < 0.01$.

25.11. $\bar{X}_1 = -0.33414$, $\bar{X}_2 = -0.19144$, $\bar{Y}_1 = -0.50974$, $\bar{Y}_2 = -0.60853$, $(\sum x^2)_c = 0.36943$, $(\sum y^2)_c = 0.28449$, $(\sum xy)_c = -0.03836$; $F = 1.52$; $F_{0.05(1),2,10} = 4.10$; do not reject H_0; $P > 0.25$.

25.12. Grand mean: $\bar{X} = -0.24632$, $\bar{Y} = -0.57053$; $U^2 = 0.0775$; $U^2_{0.05,5,8} = 0.2154$; do not reject H_0; $P > 0.50$.

25.13. H_0: The distribution is not contagious; H_A: The distribution is contagious; $u' = 6$; $u'_{0.05(1),8,8} = 4$; do not reject H_0; $P = 0.25$.

Literature Cited

ACTON, F. S. 1966. *Analysis of Straight Line Data.* Dover, New York. 267 pp.

ANDERSON, N. H. 1961. Scales and statistics: parametric and non-parametric. *Psychol. Bull.* 58: 305–316.

ANDREWS, F. C. 1954. Asymptotic behavior of some rank tests for analysis of variance. *Ann. Math. Statist.* 25: 724–735.

ANDREWS, H. P., R. D. SNEE, and M. H. SARNER. 1980. Graphical display of means. *Amer. Statist.* 34(4): 195–199.

ANSCOMBE, F. J. 1948. The transformation of Poisson, binomial, and negative binomial data. *Biometrika* 35: 246–254.

ARBUTHNOTT, J. 1710. An argument for Divine Providence taken from the constant regularity in the births of both sexes. *Philos. Trans.* 27: 186–190.

ASANO, C. 1965. Runs test for a circular distribution and a table of probabilities. *Ann. Instit. Statist. Math.* 17: 331–346.

ASIMOV, I. 1972. *Asimov's Biographical Encyclopedia of Science and Technology.* Doubleday, Garden City, N.Y. 805 pp.

BANCROFT, T. A. 1968. *Topics in Intermediate Statistical Methods*, Vol. 1. Iowa State University, Ames, Iowa. 129 pp.

BARTLETT, M. S. 1936. The square root transformation in analysis of variance. *J. Royal Statist. Soc. Suppl.* 3: 68–78.

BARTLETT, M. S. 1937. Some examples of statistical methods of research in agriculture and applied biology. *J. Royal Statist. Soc. Suppl.* 4: 137–170.

BARTLETT, M. S. 1947. The use of transformations. *Biometrics* 3: 39–52.

BASHARIN, G. P. 1959. On a statistical estimate for the entropy of a sequence of independent variables. *Theory Prob. Appl.* 4: 333–336.

BATSCHELET, E. 1965. *Statistical Methods for the Analysis of Problems in Animal Orientation and Certain Biological Rhythms*. American Institute of Biological Sciences, Washington, D.C. 57 pp.

BATSCHELET, E. 1972. Recent statistical methods for orientation data, pp. 61–91. (With discussion.) *In* S. R. Galler, K. Schmidt-Koenig, G. J. Jacobs, and R. E. Belleville (eds.), *Animal Orientation and Navigation*. National Aeronautics and Space Administration, Washington, D.C.

BATSCHELET, E. 1974. Statistical rhythm evaluations, pp. 25–35. *In* M. Ferin, F. Halberg, R. M. Richart, and R. L. Vande Wiele (eds.), *Biorhythms and Human Reproduction*. John Wiley, New York.

BATSCHELET, E. 1976. *Mathematics for Life Scientists*. 2nd ed. Springer-Verlag, New York. 643 pp.

BATSCHELET, E. 1978. Second-order statistical analysis of directions, pp. 1–24. *In* K. Schmidt-Koenig and W. T. Keeton (eds.), *Animal Migration, Navigation, and Homing*. Springer-Verlag, Berlin.

BATSCHELET, E. 1981. *Circular Statistics in Biology*. Academic Press, New York. 371 pp.

BEALL, G. 1940. The transformation of data from entomological field experiments. *Can. Entomol.* 72: 168.

BEALL, G. 1942. The transformation of data from entomological field experiments so that the analysis of variance becomes applicable. *Biometrika* 32: 243–262.

BECKMAN, P. 1971. *A History of* π. 2nd ed. Golem Press, Boulder, Colorado. 196 pp.

BENNETT, C. A. and N. L. FRANKLIN. 1954. *Statistical Analysis in Chemistry and the Chemical Industry*. John Wiley, New York. 724 pp.

BERNHARDSON, C. S. 1975. Type I error rates when multiple comparison procedures follow a significant F test of ANOVA. *Biometrics* 31: 229–232.

BEST, D. J. 1975. The difference between two Poisson expectations. *Austral. J. Statist.* 17: 29–33.

BEST, D. J. and D. E. ROBERTS. 1975. The percentage points of the χ^2 distribution. *J. Royal Statist. Soc. Ser. C Appl. Statist.* 24: 385–388.

BHATTACHARJEE, G. P. 1970. The incomplete gamma interval. *J. Royal Statist. Soc. Ser. C Appl. Statist.* 19: 285–287.

BIRNBAUM, Z. W., and F. H. TINGEY. 1951. One-sided confidence contours for probability distribution functions. *Ann. Math. Statist.* 22: 592–596.

BISHOP, Y. M. M., S. E. FIENBERG, and P. W. HOLLAND. 1975. *Discrete Multivariate Analysis: Theory and Practice*. MIT Press, Cambridge, Mass. 557 pp.

BLISS, C. I. 1967. *Statistics in Biology*, Vol. 1. McGraw-Hill, New York. 558 pp.

BLISS, C. I. 1970. *Statistics in Biology*, Vol. 2. McGraw-Hill, New York. 639 pp.

BLOOMFIELD, P. 1976. *Fourier Analysis of Time Series: An Introduction*. John Wiley, New York. 258 pp.

BONEAU, C. A. 1960. The effects of violations of assumptions underlying the t test. *Psychol. Bull.* 57: 49–64.

BOWKER, A. H. 1948. A test for symmetry in contingency tables. *J. Amer. Statist. Assoc.* 43: 572–574.

BOWMAN, K. O., K. HUTCHESON, E. P. ODUM, and L. R. SHENTON. 1971. Comments on the distribution of indices of diversity, pp. 315–366. *In* G. P. Patil, E. C. Pielou, and W. E.

Waters (eds.), *Statistical Ecology*, Vol. 3. Many Species Populations, Ecosystems, and Systems Analysis. Pennsylvania State University Press, University Park.

Box, G. E. P. 1953. Non-normality and tests on variances. *Biometrika* 40: 318–335.

Box, G. E. P. 1954. Some theorems on quadratic forms applied in the study of analysis of variance problems, I. Effect of inequality of variance in the one-way classification. *Ann. Math. Statist.* 25: 290–302.

Box, G. E. P., and S. L. Anderson. 1955. Permutation theory in the derivation of robust criteria and the study of departures from assumption. *J. Royal Statist. Soc.* B17: 1–34.

Box, G. E. P., and D. R. Cox. 1964. An analysis of transformations. *J. Royal Statist. Soc.* B26: 211–243.

Brillouin, L. 1962. *Science and Information Theory*. Academic Press, New York. 351 pp.

Brower, J. E. and J. H. Zar. 1977. *Field and Laboratory Methods for General Ecology*. Wm. C. Brown, Dubuque, Iowa. 194 pp.

Brown, M. B. and A. B. Forsythe. 1974. Robust tests for the equality of variance. *J. Amer. Statist. Assoc.* 69: 364–367.

Browne, R. H. 1979. On visual assessment of the significance of a mean difference. *Biometrics* 35: 657–665.

Brownlee, K. A. 1965. *Statistical Theory and Methodology in Science and Engineering*, 2nd ed. John Wiley, New York. 590 pp.

Buckle, N., C. Kraft, and C. van Eeden. 1969. An approximation to the Wilcoxon-Mann-Whitney distribution. *J. Amer. Statist. Assoc.* 64: 591–599.

Burr, E. J. 1964. Small-sample distributions of the two-sample Cramér-von Mises' W^2 and Watson's U^2. *Ann. Math. Statist. Assoc.* 64: 1091–1098.

Burstein, H. 1971. *Attribute Sampling. Tables and Explanations.* McGraw-Hill, New York. 464 pp.

Burstein, H. 1975. Finite population correction for binomial confidence limits. *J. Amer. Statist. Assoc.* 70: 67–69.

Cacoullos, T. 1965. A relation between the *t* and *F* distributions. *J. Amer. Statist. Assoc.* 60: 528–531.

Cajori, F. 1954. Binomial formula, pp. 588–589. In *The Encyclopaedia Britannica*, Vol. 3. Encyclopaedia Britannica Inc., New York.

Casagrande, J. T., M. C. Pike, and P. G. Smith. 1978. An improved approximate formula for calculating sample sizes for comparing binomial distributions. *Biometrics* 34: 483–486.

Chow, B., J. E. Miller, and P. C. Dickinson. 1974. Extensions of Monte-Carlo comparison of some properties of two rank correlation coefficients in small samples. *J. Statist. Comput. Simul.* 3: 189–195.

Claypool, P. L. and D. Holbert. 1974. Accuracy of normal and Edgeworth approximations to the distribution of the Wilcoxon signed rank statistic. *J. Amer. Statist. Assoc.* 69: 255–258.

Cochran, W. G. 1942. The 2 × 2 correction for continuity. *Iowa State Coll. J. Sci.* 16: 421–436. [Cited in Haber (1980).]

Cochran, W. G. 1947. Some consequences when the assumptions for analysis of variance are not satisfied. *Biometrics* 3: 22–38.

Cochran, W. G. 1950. The comparison of percentages in matched samples. *Biometrika* 37: 256–266.

COCHRAN, W. G. 1952. The χ^2 test for goodness of fit. *Ann. Math. Statist.* 23: 315–345.

COCHRAN, W. G. 1954. Some methods for strengthening the common χ^2 tests. *Biometrics* 10: 417–451.

COCHRAN, W. G. 1964. Approximate significance levels of the Behrens-Fisher test. *Biometrics* 20: 191–195.

COCHRAN, W. G. 1977. *Sampling Techniques.* 3rd. ed. John Wiley, New York. 428 pp.

COCHRAN, W. G. and G. M. COX 1957. *Experimental Designs.* 2nd ed. John Wiley, New York. 617 pp.

COHEN, J. 1977. *Statistical Power Analysis for the Behavioral Sciences.* Academic Press, New York. 474 pp.

CONOVER, W. J. 1973. On methods of handling ties in the Wilcoxon signed-rank test. *J. Amer. Statist. Assoc.* 68: 985–988.

CONOVER, W. J. 1974. Some reasons for not using the Yates continuity correction on 2 × 2 contingency tables. *J. Amer. Statist. Assoc.* 69: 374–376. *Also*: Comment, by C. F. Starmer, J. E. Grizzle, and P. K. Sen, *ibid.* 69: 376–378; Comment and a suggestion, by N. Mantel, *ibid.* 69: 378–380; Comment, by O. S. Miettinen, *ibid.* 69: 380–382; Rejoinder, by W., J. Conover, *ibid.* 69: 382.

CONOVER, W. J. 1980. *Practical Nonparametric Statistics.* 2nd ed. John Wiley, New York. 494 pp.

COX, D. R. 1958. *Planning of Experiments.* John Wiley, New York. 308 pp.

CRAMER, E. M. 1972. Significance test and tests of models in multiple regression. *Amer. Statist.* 26(4): 26–30.

CRAMÉR, H. 1946. Mathematical Methods of Statistics. *Princeton University Press*, Princeton, N. J. 575 pp.

CROW, E. I. 1952. Some cases in which Yates' correction should not be applied. *J. Amer. Statist. Assoc.* 47: 303–304.

CROXTON, F. E., D. J. COWDEN, and S. KLEIN. 1967. *Applied General Statistics.* 3rd ed. Prentice-Hall, Englewood Cliffs, N. J. 754 pp.

CURETON, E. E. 1967. The normal approximation to the signed-rank sampling distribution when zero differences are present. *J. Amer. Statist. Assoc.* 62: 1068–1069.

D'AGOSTINO, R. B. 1970. Transformation to normality of the null distribution of g_1. *Biometrika* 57: 679–681.

D'AGOSTINO, R. B. 1971a. An omnibus test of normality for moderate and large samples. *Biometrika* 58: 341–348.

D'AGOSTINO, R. B. 1971b. Tables for the *D* test of normality. Department of Mathematics, Boston University. Research Report. April 1971.

D'AGOSTINO, R. B. 1972. Small sample probability points for the *D* test of normality. *Biometrika* 59: 219–221.

D'AGOSTINO, R. B. and G. E. NOETHER. 1973. On the evaluation of the Kolmogorov statistic. *Amer. Statist.* 27: 81–82.

D'AGOSTINO, R. B. and G. L. TIETJEN. 1971. Simulation probability points of b_2 for small samples. *Biometrika* 58: 669–672.

DANIEL, W. W. 1978. *Applied Nonparametric Statistics.* Houghton Mifflin Co., Boston. 510 pp.

DANIEL, C., and F. S. WOOD. 1971. *Fitting Equations to Data.* John Wiley, New York. 342 pp.

DAPSON, R. W. 1980. Guidelines for statistical usage in age-estimation technics. *J. Wildlife Manage.* 44: 541–548.

DAVENPORT, J. M. and J. T. WEBSTER. 1975. The Behrens-Fisher problem, and old solution revisited. *Metrika* 22: 47–54.

DE JONGE, C., and M. A. J. VAN MONTFORT. 1972. The null distribution of Spearman's S when $n = 12$. *Statist. Neerland.* 26: 15–17.

DAVID, F. N. 1962. *Games, Gods and Gambling.* Hafner Press, New York. 275 pp.

DIXON, W. J. (ed.) 1975. *BMDP Biomedical Computer Programs.* University of California Press, Berkeley. 792 pp.

DIXON, W. J. and M. B. BROWN (ed.) 1979. *BMDP-79. Biomedical Computer Program P-Series.* University of California Press, Berkeley. 880 pp.

DIXON, W. J., and F. J. MASSEY, JR. 1969. *Introduction to Statistical Analysis.* 3rd ed. McGraw-Hill, New York. 638 pp.

DOANE, D. P. 1976. Aesthetic frequency classifications. *Amer. Statist.* 30: 181–183.

DRAPER, N. R., and H. SMITH. 1981. *Applied Regression Analysis.* 2nd ed. John Wiley, New York. 709 pp.

DUNCAN, D. B. 1951. A significance test for difference between ranked treatments in an analysis of variance. *Virginia J. Sci.* 2: 171–189.

DUNCAN, D. B. 1955. Multiple range and multiple F tests. *Biometrics* 11: 1–42.

DUNN, O. J. 1964. Multiple contrasts using rank sums. *Technometrics* 6: 241–252.

DUNN, O. J. and V. A. CLARK. 1974. *Applied Statistics: Analysis of Variance and Regression.* John Wiley, New York. 349 pp.

DUNNETT, C. W. 1955. A multiple comparison procedure for comparing several treatments with a control. *J. Amer. Statist. Assoc.* 50: 1096–1121.

DUNNETT, C. W. 1964. New tables for multiple comparisons with a control. *Biometrics* 20: 482–491.

DUNNETT, C. W. 1970. Multiple comparison tests. *Biometrics* 26: 139–141.

DUNNETT, C. W. 1980a. Pairwise multiple comparisons in the homogeneous variance, equal sample size case. *J. Amer. Statist. Assoc.* 75: 789–795.

DUNNETT, C. W. 1980b. Pairwise multiple comparisons in the unequal variance case. *J. Amer. Statist. Assoc.* 75: 796–800.

DUNNETT, C. W. and M. GENT. 1977. Significance testing to establish equivalence between treatments, with special reference to data in the form of 2×2 tables. *Biometrics* 33: 593–602.

DURAND, D., and J. A. GREENWOOD. 1958. Modifications of the Rayleigh test for uniformity in analysis of two-dimensional orientation data. *J. Geol.* 66: 229–238.

EASON, G., C. W. COLES, and G. GETTINBY. 1980. *Mathematics and Statistics for the Bio-Sciences.* Ellis Horwood Ltd., Chichester, England. 578 pp.

EBERHARDT, K. R. and M. A. FLIGNER. 1977. A comparison of two tests for equality of proportions. *Amer. Statist.* 31: 151–155.

EFROYMSON, M. A. 1960. Multiple regression analysis, pp. 191–203. *In* A. Ralston and H. S. Wilf, *Mathematical Methods for Digital Computers.* John Wiley, New York.

EINOT, I. and K. R. GABRIEL. 1975. A survey of the powers of several methods of multiple comparisons. *J. Amer. Statist. Assoc.* 70: 574–583.

EISENHART, C. 1947. The assumptions underlying the analysis of variance. *Biometrics* 3: 1–21.

EISENHART, C. 1968. Expression of the uncertainties of final results. *Science* 160: 1201–1204.

EVERITT, B. S. 1977. *The Analysis of Contingency Tables.* Halsted Press, New York. 128 pp.

EVERITT, B. S. 1979. A Monte Carlo investigation of the robustness of Hotelling's one- and two-sample tests. *J. Amer. Statist. Assoc.* 74: 48–51.

EZEKIAL, M., and K. A. FOX. 1959. *Methods of Correlation and Regression Analysis.* 3rd ed. John Wiley, New York. 548 pp.

FEDERER, W. T. 1955. *Experimental Design.* Macmillan, New York. 591 pp.

FELDMAN, S. E., and E. KLUGER. 1963. Short cut calculation of the Fisher-Yates "exact test." *Psychometrika* 28: 289–291.

FIELLER, E. C., H. O. HARTLEY, and E. S. PEARSON. 1957. Tests for rank correlation coefficients. I. *Biometrika* 44: 470–481.

FIELLER, E. C., H. O. HARTLEY, and E. S. PEARSON. 1961. Tests for rank correlation coefficients. II. *Biometrika* 48: 29–40.

FIENBERG, S. E. 1970. The analysis of multidimensional contingency tables. *Ecology* 51: 419–433.

FIENBERG, S. E. 1972. The analysis of incomplete multi-way contingency tables. *Biometrics* 28: 177–202.

FIENBERG, S. E. 1980. *The Analysis of Cross-Classified Categorical Data.* 2nd ed. MIT Press, Cambridge, Mass. 198 pp.

FISHER, R. A. 1915. Frequency distributions of the values of the correlation coefficient in samples from an indefinitely large population. *Biometrika* 10: 507–521.

FISHER, R. A. 1921. On the "probable error" of a coefficient of correlation deduced from a small sample. *Metron* 1: 3–32.

FISHER, R. A. 1922. On the interpretation of χ^2 from contingency tables and the calculation of *P. J. Royal Statist. Soc.* 85: 87–94.

FISHER, R. A. 1925. Application of "Student's" distribution. *Metron* 5: 90–104.

FISHER, R. A. 1958. *Statistical Methods for Research Workers.* 13th ed. Hafner, New York. 356 pp.

FISHER, R. A., and F. YATES. 1963. *Statistical Tables for Biological, Agricultural, and Medical Research.* 6th ed. Hafner, New York. 146 pp.

FISZ, M. 1963. *Probability Theory and Mathematics Statistics.* 3rd ed. John Wiley, New York. 677 pp.

FIX, E., and J. L. HODGES, JR. 1955. Significance probabilities of the Wilcoxon test. *Ann. Math. Statist.* 26: 301–312.

FLEISHMAN, A. I. 1977. A program for calculating the exact probability along with explorations of M by N contingency tables. *Educ. Psychol. Meas.* 37: 799–803.

FLEISS, J. L., A. TYTUN, and H. K. URY. 1980. A simple approximation for calculating sample sizes for comparing independent proportions. *Biometrics* 36: 343–346.

FOX, M. 1956. Charts for the power of the *F*-test. *Ann. Math. Statist.* 27: 484–497.

FRAWLEY, W. H., and W. R. SCHUCANY. 1972. Tables of the distribution of the $\mathcal{W}$ concordance statistic. Tech. Rep. No. 116. Department of Statistics, Southern Methodist University, Dallas, Texas.

FREEMAN, M. F., and J. W. TUKEY. 1950. Transformations related to the angular and the square root. *Ann. Math. Statist.* 21: 607–611.

FREUND, R. J. 1963. A warning of roundoff errors in regression. *Amer. Statist.* 17(5): 13–15.

FREUND, R. J. 1971. Some observations on regressions with grouped data. *Amer. Statist.* 25(3): 29–30.

FRIEDMAN, M. 1937. The use of ranks to avoid the assumption of normality implicit in the analysis of variance. *J. Amer. Statist. Assoc.* 32: 675–701.

FRIEDMAN, M. 1940. A comparison of alternate tests of significance for the problem of *m* rankings. *Ann. Math. Statist.* 11: 86–92.

GAITO, J. 1959. Non-parametric methods in psychological research. *Psychol. Rep.* 5: 115–125.

GAITO, J. 1960. Scale classification and statistics. *Psychol. Bull.* 67: 277–278.

GAMES, P. A. and J. F. HOWELL. 1976. Pairwise multiple comparison procedures with unequal *n*'s and/or variances: A Monte Carlo study. *J. Educ. Statist.* 1: 113–125.

GAMES, P. A., H. B. WINKLER, and D. A. PROBERT. 1972. Robust tests for homogeneity of variance. *Educ. Psychol. Meas.* 32: 887–909.

GART, J. J. 1969a. An exact test for comparing matched proportions in crossover designs. *Biometrika* 56: 75–80.

GART, J. 1969b. Graphically oriented tests of the Poisson distribution. *Bull. Intern. Statist. Inst.*, 37th Session, pp. 119–121.

GARTSIDE, P. S. 1972. A study of methods for comparing several variances. *J. Amer. Statist. Assoc.* 67: 342–346.

GEARY, R. C., and C. E. V. LESER. 1968. Significance tests in multiple regression. *Amer. Statist.* 22(1): 20–21.

GHENT, A. W. 1972. A method for exact testing of 2×2, 2×3, 3×3, and other contingency tables, employing binomial coefficients. *Amer. Midland Natur.* 88: 15–27.

GHOSH, B. K. 1979. A comparison of some approximate confidence intervals for the binomial parameter. *J. Amer. Statist. Assoc.* 74: 894–900.

GIBBONS, J. D. 1976. *Nonparametric Methods for Quantitative Analysis.* Holt, Rinehart, and Winston, New York. 463 pp.

GILL, J. L. 1971. Analysis of data with heterogeneous data: A review. *J. Dairy Sci.* 54: 369–373.

GLASS, G. V., P. D. PECKHAM, and J. R. SANDERS. 1972. Consequences of failure to meet assumptions underlying the fixed effects analysis of variance and covariance. *Rev. Educ. Res.* 42: 239–288.

GLEJSER, H. 1969. A new test for heteroscedasticity. *J. Amer. Statist. Assoc.* 64: 316–323.

GLEN, W. A. and C. Y. KRAMER. 1958. Analysis of variance of a randomized block design with missing observations. *Appl. Statist.* 7: 173–185.

GOODMAN, L. A. 1970. The multivariate analysis of qualitative data: Interactions among multiple classifications. *J. Amer. Statist. Assoc.* 65: 226–256.

GREENWOOD, J. A., and D. DURAND. 1955. The distribution of length and components of the sum of *n* random unit vectors. *Ann. Math. Statist.* 26: 233–246.

GUENTHER, W. C. 1964. *Analysis of Variance.* Prentice-Hall, Englewood Cliffs, N. J. 199 pp.

GUENTHER, W. C. 1977. Some probabilities available from desk calculators and their relation to tables. *Amer. Statist.* 31: 41–45.

GURLAND, J., and R. C. TRIPATHI. 1971. A simple approximation for unbiased estimation of the standard deviation. *Amer. Statist.* 25(4): 30–32.

HABER, M. 1980. A comparison of some continuity corrections for the chi-squared test on 2×2 tables. *J. Amer. Statist. Assoc.* 75: 510–515.

HAIGHT, F. A. 1967. *Handbook of the Poisson Distribution.* John Wiley, New York. 168 pp.

HALBERG, F. and J.-K. LEE. 1974. Glossary of selected chronobiologic terms, pp. XXXVII–L. *In* L. E. Scheving, F. Halberg, and J. E. Pauly (eds.), *Chronobiology*. Igaku Shoin, Tokyo.

HAMAKER, H. C. 1962. On multiple regression analysis. *Statist. Neerland.* 16: 31–56.

HARRIS, M., D. G. HORVITZ, and A. M. MOOD. 1948. On the determination of sample sizes in designing experiments. *J. Amer. Statist. Assoc.* 43: 391–402.

HARTER, H. L. 1957. Error rates and sample sizes for range tests in multiple comparisons. *Biometrics* 13: 511–536.

HARTER, H. L. 1960. Tables of range and studentized range. *Ann. Math. Statist.* 31: 1122–1147.

HARTER, H. L. 1970. *Order Statistics and Their Use in Testing and Estimation*, Vol. 1. Tests Based on Range and Studentized Range of Samples from a Normal Population. U. S. Government Printing Office, Washington, D. C. 761 pp.

HASTINGS, C., JR. 1955. *Approximations for Digital Computers.* Princeton University Press, Princeton, N. J. 201 pp.

HAWKINS, D. M. 1980. A note on fitting a regression without an intercept term. *Amer. Statist.* 34(4): 233.

HAYS, W. L. 1963. *Statistics for Psychologists.* Holt, Rinehart & Winston, New York. 719 pp.

HEALY, M. J. R. 1963. Programming multiple regression. *Computer J.* 6: 57–61.

HICKS, C. R. 1973. *Fundamental Concepts in Design of Experiments.* Holt, Rinehart, & Winston, New York. 349 pp.

HODGES, J. L., and E. L. LEHMAN. 1956. The efficiency of some nonparametric competitors of the *t*-test. *Ann. Math. Statist.* 27: 324–335.

HOLLANDER, M., and D. A. WOLFE. 1973. *Nonparamtric Statistical Methods.* John Wiley, New York. 503 pp.

HOTELLING, H. 1931. The generalization of Student's ratio. *Ann. Math. Statist.* 2: 360–378.

HOTELLING, H., and M. R. PABST. 1936. Rank correlation and tests of significance involving no assumption of normality. *Ann. Math. Statist.* 7: 29–43.

HOWELL, J. F., and P. A. GAMES. 1974. The effects of variance heterogeneity on simultaneous multiple-comparison procedures with equal sample size. *Brit. J. Math. Statist. Psychol.* 27: 72–81.

HUCK, S. W., and B. H. LAYNE. 1974. Checking for proportional *n*'s in factorial ANOVA's. *Educ. Psychol. Meas.* 34: 281–287.

HUFF, D. 1954. *How to Lie with Statistics.* W. W. Norton, New York. 142 pp.

HUITEMA, B. E. 1974. Three multiple comparison procedures for contrasts among correlation coefficients. *Proc. Soc. Statist. Sect., Amer. Statist. Assoc.*, 1974, pp. 336–339.

HUTCHESON, K. 1970. A test for comparing diversities based on the Shannon formula. *J. Theoret. Biol.* 29: 151–154.

IMAN, R. L., D. QUADE, and D. A. ALEXANDER. 1975. Exact probability levels for the Kruskal-Wallis test, pp. 329–384. *In* H. L. Harter and D. B. Owen, *Selected Tables in Mathematical Statistics*, Volume III. American Mathematical Society, Providence, R. I.

INTERNATIONAL BUSINESS MACHINES CORPORATION. 1968. *System/360 Scientific Subroutine Package(360A-CM-03X) Version III. Programmer's Manual.* 4th ed. White Plains, N. Y. 454 pp.

IVES, K. H., and J. D. GIBBONS. 1967. A correlation measure for nominal data. *Amer. Statist.* 21(5): 16–17.

JACQUES, J. A., and M. NORUSIS. 1973. Sampling requirements on the estimation of parameters in heteroscedastic linear regression. *Biometrics* 29: 771–780.

JENRICH, R. I. 1977. Stepwise regression, pp. 58–75. *In* K. Enslein, A. Ralston, and H. S. Wilf (eds.), *Statistical Methods for Digital Computers.* John Wiley, New York.

KENDALL, M. G. 1962. *Rank Correlation Methods.* 3rd ed. Charles Griffin, London. 199 pp.

KENDALL, M. G., and B. BABINGTON-SMITH. 1939. The problem of *m* rankings. *Ann. Math. Statist.* 10: 275–287.

KENDALL, M. G., and A. STUART. 1966. *The Advanced Theory of Statistics,* Vol. 3. Hafner. New York. 552 pp.

KESELMAN, H. J. 1976. A power investigation of the Tukey multiple comparison statistic. *Educ. Psychol. Meas.* 36: 97–104.

KESELMAN, H. J., P. A. GAMES, and J. J. CLINCH. 1979. Tests for homogeneity of variance. *Communic. Statist. Simulat. Computat.* 88: 113–129.

KESELMAN, H. J., R. MURRAY, and J. C. ROGAN. 1976. Effect of very unequal group sizes on Tukey's multiple comparison test. *Educ. Psychol. Meas.* 36: 263–270.

KEULS, M. 1952. The use of the "studentized range" in connection with an analysis of variance. *Euphytica* 1: 112–122.

KIHLBERG, J. K., J. H. HERSON, and W. E. SCHUTZ. 1972. Square root transformation revisited. *J. Royal Statist. Soc. Ser. C. Appl. Statist.* 21: 76–81.

KIRK, R. E. 1968. *Experimental Design: Procedures for the Behavioral Sciences.* Brooks/Cole, Monterey, Calif. 577 pp.

KLEINBAUM, D. G., and L. L. KUPPER. 1978. *Applied Regression Analysis and Other Multivariate Methods.* Duxbury Press, North Scituate, Mass. 556 pp.

KOHR, R. L. and P. A. GAMES. 1974. Robustness of the analysis of variance, the Welch procedure, and a Box procedure to heterogeneous variances. *J. Exper. Educ.* 43: 61–69.

KOLMOGOROV, A. 1933. Sulla determinazione empirica di una legge di distribuzione. *Giornalle dell Instituto Italiano degli Attuari* 4: 1–11.

KRAMER, C. Y. 1956. Extension of multiple range tests to group means with unequal numbers of replications. *Biometrics* 12: 307–310.

KRUSKAL, W. H. 1957. Historical notes on the Wilcoxon unpaired two-sample test. *J. Amer. Statist. Assoc.* 52: 356–360.

KRUSKAL, W. H., and W. A. WALLIS. 1952. Use of ranks in one-criterion analysis of variance. *J. Amer. Statist. Assoc.* 47: 583–621.

KUIPER, N. H. 1960. Tests concerning random points on a circle. *Ned. Akad. Wetensch. Proc. Ser. A* 63: 38–47.

LANCASTER, H. O. 1969. *The Chi-Squared Distribution.* John Wiley, New York. 356 pp.

LEE, A. F. S., and J. GURLAND. 1975. Size and power of tests for equality of means of two normal populations with unequal variances. *J. Amer. Statist. Assoc.* 70: 933–941.

LESLIE, P. H. 1955. A simple method of calculating the exact probability in 2×2 contingency tables with small marginal totals. *Biometrika* 42: 522–523.

LEVY, K. J. 1975a. Some multiple range tests for variances. *Educ. Psychol. Meas.* 35: 599–604.

LEVY, K. J. 1975b. Comparing variances of several treatments with a control. *Educ. Psychol. Meas.* 35: 793–396.

LEVY, K. J. 1975c. An empirical comparison of several multiple range tests for variances. *J. Amer. Statist. Assoc.* 70: 180–183.

Levy, K. J. 1976. A multiple range procedure for independent correlations. *Educ. Psychol. Meas.* 36: 27–31.

Levy, K. J. 1979. Pairwise comparisons associated with the *K* independent sample median test. *Amer. Statist.* 33: 138–139.

Lewontin, R. C. 1966. On the measurement of relative variability. *Systematic Zool.* 15: 141–142.

Leyton, M. K. 1968. Rapid calculation of exact probabilities for 2 × 3 contingency tables. *Biometrics* 24: 714–717.

Li, C. C. 1964. *Introduction to Experimental Statistics.* McGraw-Hill, New York. 460 pp.

Li, J. C. R. 1964. *Statistical Inference*, II. Edward Brothers, Ann Arbor, Mich. 575 pp.

Liddell, I. G., and J. K. Ord. 1978. Linear-angular correlation coefficients: Some further results. *Biometrika* 65: 448–450.

Light, R. J., and B. H. Margolin. 1971. Analysis of variance for categorical data. *J. Amer. Statist. Assoc.* 66: 534–544.

Lindsay, R. B. 1976. John William Strutt, Third Baron Rayleigh, pp. 100–107. *In* C. C. Gillispie (ed.), *Dictionary of Scientific Biography*, Vol. XIII. Scribner's, New York.

Ling, R. F. 1974. Comparison of several algorithms for computing sample means and variances. *J. Amer. Statist. Assoc.* 69: 859–866.

Lloyd, M., J. H. Zar, and J. R. Karr. 1968. On the calculation of information-theoretical measures of diversity. *Amer. Midland Natur.* 79: 257–272.

Longley, J. W. 1967. An appraisal of least squares programs for the electronic computer from the point of view of the user. *J. Amer. Statist. Assoc.* 62: 819–841.

MacKinnon, W. J. 1964. Table for both the sign test and distribution-free confidence intervals of the median for sample sizes to 1,000. *J. Amer. Statist. Assoc.* 59: 935–956.

McCornack, R. L. 1965. Extended tables of the Wilcoxon matched pair signed rank statistic. *J. Amer. Statist. Assoc.* 60: 864–871.

McGill, R., J. W. Tukey, and W. A. Larsen. 1978. Variations of box plots. *Amer. Statist.* 32: 12–16.

McNemar, Q. 1947. Note on the sampling error of the difference between correlated proportions or percentages. *Psychometrica* 12: 153–157.

Madow, W. G. 1940. Notes on tests of departure from normality. *J. Amer. Statist. Assoc.* 35: 515–517.

Mann, H. B., and D. R. Whitney. 1947. On a test of whether one of two random variables is stochastically larger than the other. *Ann. Math. Statist.* 18: 50–60.

Mantel, N. 1976. The continuity correction. *Amer. Statist.* 30: 103–104.

Mantel, N. 1970. Why stepdown procedures in variable selection. *Technometrics* 12: 621–625.

Marascuilo, L. A. 1971. *Statistical Methods for Behavioral Science Research.* McGraw-Hill, New York. 578 pp.

Marascuilo, L. A., and M. McSweeney. 1967. Nonparametric post hoc comparisons for trend. *Psychol. Bull.* 67: 401–412.

Marascuilo, L. A., and M. McSweeney. 1977. *Nonparametric and Distribution-free Methods for the Social Sciences.* Brooks/Cole, Monterey, Calif. 556 pp.

Mardia, K. V. 1967. A non-parametric test for the bivariate two-sample location problem. *J. Royal Statist. Soc.* B29: 320–342.

MARDIA, K. V. 1970. Measures of multivariate skewness and kurtosis with applications. *Biometrika* 57: 519–530.

MARDIA, K. V. 1972. *Statistics of Directional Data.* Academic Press, New York. 357 pp.

MARDIA, K. V. 1975. Statistics of directional data. (With discussion.) *J. Royal Statist. Soc.* B37: 349–393.

MARDIA, K. V. 1976. Linear-angular correlation coefficients and rhythmometry. *Biometrika* 63: 403–405.

MARDIA, K. V., and M. L. PURI. 1978. A spherical correlation coefficient robust against scale. *Biometrika* 65: 391–395.

MASSEY, F. J., JR. 1951. The Kolmogorov-Smirnov test for goodness of fit. *J. Amer. Statist. Assoc.* 46: 68–78.

MAXWELL, A. E. 1970. Comparing the classification of subjects by two independent judges. *Brit. J. Psychiatry* 116: 651–655.

MEHTA, J. S., and R. SRINIVASAN. 1970. On the Behrens-Fisher problem. *Biometrics* 57: 649–655.

MENDEL, G. 1933. *Experiments in Plant Hybridization.* Harvard University Press, Cambridge, Mass. 353 pp.

MILLER, L. H. 1956. Table of percentage points of Kolmogorov statistics. *J. Amer. Statist. Assoc.* 51: 111–121.

MILLER, R. G., JR. 1966. *Simultaneous Statistical Inference.* McGraw-Hill, New York. 272 pp.

MILLER, R. G., JR. 1977. Developments in multiple comparisons, 1966–1977. *J. Amer. Statist. Assoc.* 72: 779–788.

MILTON, R. C. 1964. An extended table of critical values for the Mann-Whitney (Wilcoxon) two-sample statistic. *J. Amer. Statist. Assoc.* 59: 925–934.

MOOD, A. M. 1950. *Introduction to the Theory of Statistics.* McGraw-Hill, New York. 433 pp.

MOOD, A. M. 1954. On the asymptotic efficiency of certain non-parametric two-sample tests. *Ann. Math. Statist.* 25: 514–522.

MOORE, B. R. 1980. A modification of the Rayleigh test for vector data. *Biometrika* 67: 175–180.

MULHOLLAND, H. P. 1977. On the null distribution of $\sqrt{b_1}$ for samples of size at most 25, with tables. *Biometrika* 64: 401–409.

NELSON, W., Y. L. TONG, J.-K. LEE, and F. HALBERG. 1979. Methods for cosinor-rhythmometry. *Chronobiologia* 6: 305–323.

NEMENYI, P. 1963. *Distribution-Free Multiple Comparisons.* State University of New York, Downstate Medical Center. [Cited in Wilcoxon and Wilcox (1964).]

NETER, J., and W. WASSERMAN. 1974. *Applied Linear Statistical Models.* Richard D. Irwin, Homewood, Ill. 842 pp.

NEWMAN, D. 1939. The distribution of range in samples from a normal population, expressed in terms of an independent estimate of standard deviation. *Biometrika* 31: 20–30.

NIE, N. H., C. H. HULL, J. G. JENKINS, K. STEINBRENNER, and D. H. BENT. 1975. *SPSS. Statistical Package for the Social Sciences.* 2nd ed. McGraw Hill, New York. 675 pp.

NORRIS, R. C., and H. F. HJELM. 1961. Non-normality and product moment correlation. *J. Exp. Educ.* 29: 261–270.

ODEH, R. E. and J. O. EVANS. 1974. The percentage points of the normal distribution. *J. Royal Statist. Soc. Ser. C Appl. Statist.* 23: 98–99.

ODEH, R. E. and M. FOX. 1975. *Sample Size Choice: Charts for Experiments with Linear Models.* Marcel Dekker, New York. 190 pp.

OLDS, E. G. 1938. Distributions of sums of squares of rank differences for small numbers of individuals. *Ann. Math. Statist.* 9: 133–148.

OLVER, F. W. J. 1964. Bessel functions of integer order, pp. 355–433. *In* M. Abramowitz and I. Stegun (eds.), *Handbook of Mathematical Functions*, National Bureau of Standards, Washington, D.C. (Also, Dover, New York, 1965.)

OSTLE, B., and R. W. MENSING. 1975. *Statistics in Research.* 3rd ed. Iowa State University Press, Iowa. 596 pp.

OTTEN, A. 1973. The null distribution of Spearman's S when $n = 13(1)16$. *Statist. Neerland.* 27: 19–20.

OWEN, D. B. 1962. *Handbook of Statistical Tables.* Addison-Wesley Reading, Mass. 580 pp.

PATIL, K. D. 1975. Cochran's Q test: Exact distribution. *J. Amer. Statist. Assoc.* 70: 186–189.

PATNIAK, A. B. 1949. The non-central χ^2 and F-distributions and their applications. *Biometrika* 36: 202–232.

PAULL, A. E. 1950. On a preliminary test for pooling mean squares in the analysis of variance. *Ann. Math. Statist.* 21: 539–556.

PAZER, H. L., and L. A. SWANSON. 1972. *Modern Methods for Statistical Analysis.* Intext Educational Publishers, Scranton, Penna. 483 pp.

PEARSON, E. S. 1967. Studies in the history of probability and statistics. XVII. Some reflexions on continuity in the development of mathematical statistics, 1885–1920. *Biometrika* 54: 341–355.

PEARSON, E. S. 1971. Lectures on the Early History of Biometry. Tech. Rep. No. 3. Department of Statistics, University of California, Riverside. 20 pp.

PEARSON, E. S. and H. O. HARTLEY. 1951. Charts for the power function for analysis of variance tests, derived from the non-central F-distribution. *Biometrika* 38: 112–130.

PEARSON, E. S., and H. O. HARTLEY. 1966. *Biometrika Tables for Statisticians*, Vol. 1. 3rd ed. Cambridge University Press, Cambridge, England. 264 pp.

PEARSON, K. 1900. On a criterion that a given system of deviations from the probable in the case of a correlated system of variables is such that it can be reasonably supposed to have arisen in random sampling. *Phil. Mag. Ser. 5* 50: 157–175.

PEARSON, K. 1904. Mathematical contributions to the theory of evolution. XIII. On the theory of contingency and its relation to association and normal correlation. Draper's Co. Res. Mem., Biometric Ser. 1. 35 p. [Cited in Lancaster (1969).]

PEARSON, K. 1920. Notes on the history of correlation. *Biometrika* 13: 25–45.

PETTITT, A. N., and M. A. STEPHENS. 1977. The Kolmogorov-Smirnov goodness-of-fit statistic with discrete and grouped data. *Technometrics* 19: 205–210.

PIELOU, E. C. 1966. The measurement of diversity in different types of biological collections. *J. Theoret. Biol.* 13: 131–144.

PRATT, J. W. 1959. Remarks on zeroes and ties in the Wilcoxon signed rank procedures. *J. Amer. Statist. Assoc.* 54: 655–667.

QUENOUILLE, M. H. 1950. *Introductory Statistics.* Butterworth-Springer, London. [Cited in Thöni (1967: 18).]

RACTLIFFE, J. F. 1964. The significance of the difference between two Poisson variables: An experimental investigation. *J. Royal Statist. Soc. Ser. C. Appl. Statist.* 13: 84–86.

RACTLIFFE, J. F. 1968. The effect on the t-distribution of non-normality in the sampled population. *J. Royal Statist. Soc. Ser. C. Appl. Statist.* 17: 42–48.

RADLOW, R., and E. F. ALF, JR. 1975. An alternate multinomial assessment of the accuracy of the χ^2 test of goodness of fit. *J. Amer. Statist. Assoc.* 70: 811–813.

RAHE, A. J. 1974. Table of critical values for the Pratt matched pair signed rank statistic. *J. Amer. Statist. Assoc.* 69: 368–373.

RAMSEY, P. H. 1978. Power differences between pairwise multiple comparisons. *J. Amer. Statist. Assoc.* 73: 479–485.

RAO, C. R. 1973. *Linear Statistical Inference and Its Applications.* 2nd. ed. John Wiley, New York. 625 pp.

RAO, C. R., and I. M. CHAKRAVARTI. 1956. Some small sample tests of significance for a Poisson distribution. *Biometrics* 12: 264–282.

RAO, C. R., S. K. MITRA, and R. A. MATTHAI. 1966. *Formulae and Tables for Statistical Work.* Statistical Publishing Society, Calcutta. 233 pp.

ROSCOE, J. T., and J. A. BYARS. 1971. Sample size restraints commonly imposed on the use of the chi-square statistic. *J. Amer. Statist. Assoc.* 66: 755–759.

ROTHERY, P. 1979. A nonparametric measure of intraclass correlation. *Biometrika* 66: 629–639.

SAKODA, J. M., and B. H. COHEN. 1957. Exact probabilities for contingency tables using binomial coefficients. *Psychometrika* 22: 83–86.

SAS INSTITUTE. 1979. *SAS Users Guide.* SAS Institute, Raleigh, N. C. 494 pp.

SATTERTHWAITE, F. E. 1946. An approximate distribution of estimates of variance components. *Biometrics Bull.* 2: 110–114.

SAVAGE, I. R. 1957. Non-parametric statistics. *J. Amer. Statist. Assoc.* 52: 331–334.

SCHEFFÉ, H. 1953. A method of judging all contrasts in the analysis of variance. *Biometrika* 40: 87–104.

SCHEFFÉ, H. 1959. *The Analysis of Variance.* John Wiley, New York. 477 pp.

SCHEFFÉ, H. 1970. Practical solutions of the Behrens-Fisher problem. *J. Amer. Statist. Assoc.* 65: 1501–1508.

SCHEIRER, C. J., W. S. RAY, and N. HARE. 1976. The analysis of ranked data derived from completely randomized factorial designs. *Biometrics* 32: 429–434.

SCHUCANY, W. R., and W. H. FRAWLEY. 1973. A rank test for two group concordance. *Psychometrika* 38: 249–258.

SEARLE, S. R. 1966. *Matrix Algebra for the Biological Sciences (Including Applications in Statistics).* John Wiley, New York. 296 pp.

SEBER, G. A. F. 1977. *Linear Regression Analysis.* John Wiley, New York. 465 pp.

SENDERS, V. L. 1958. *Measurement and Statistics.* Oxford University Press, New York. 594 pp.

SHAFFER, J. P. 1977. Multiple comparison emphasizing selected contrasts: An extension and generalization of Dunnett's procedure. *Biometrics* 33: 293–303.

SHANNON, C. E. 1948. A mathematical theory of communication. *Bell System Tech. J.* 27: 379–423, 623–656.

SHAPIRO, S. S., and M. B. WILK. 1965. An analysis of variance test for normality (complete samples). *Biometrika* 52: 591–611.

SHAPIRO, S. S., M. B. WILK, and H. J. CHEN. 1968. A comparative study of various tests for normality. *J. Amer. Statist. Assoc.* 63: 1343–1372.

SHEARER, P. R. 1973. Missing data in quantitative designs. *J. Royal Statist. Soc. Ser. C. Appl. Statist.* 22: 135–140.

SIEGEL, S. 1956. *Nonparametric Statistics for the Behavioral Sciences.* McGraw-Hill, New York. 312 pp.

SIMPSON, G. G., A. ROE, and R. C. LEWONTIN. 1960. *Quantitative Zoology.* Harcourt, Brace, Jovanovich, New York. 440 pp.

SMIRNOV, N. V. 1939a. Sur les écarts de la courbe de distribution empirique. *Recueil Mathématique N. S.* 6: 3–26.

SMIRNOV, N. V. 1939b. On the estimation of the discrepancy between empirical curves of distribution for two independent samples. (In Russian.) *Bull. Moscow Univ. Intern. Ser.* (*Math.*) 2: 3–16. [Cited in Massey (1951).]

SMITH, H. 1936. The problem of comparing the results of two experiments with unequal means. *J. Council Sci. Indust. Res.* 9: 211–212. [Cited in Davenport and Webster (1975).]

SMITH, R. A. 1971. The effect of unequal group size on Tukey's HSD procedure. *Psychometrika* 36: 31–34.

SNEDECOR, G. W. 1934. *Calculation and Interpretation of Analysis of Variance and Covariance.* Collegiate Press, Ames, Iowa. 96 pp.

SNEDECOR, G. W., and W. G. COCHRAN. 1980. *Statistical Methods.* 7th ed. Iowa State University Press, Ames, Iowa. 507 pp.

SNEE, R. D. 1974. Graphical display of two-way contingency tables. *Amer. Statist.* 28: 9–12.

SOKAL, R. R., and F. J. ROHLF. 1981. *Biometry.* 2nd ed., W. H. Freeman & Co., San Francisco. 859 pp.

SPEARMAN, C. 1904. The proof and measurement of association between two things. *Amer. J. Psychol.* 15: 72–101.

SRIVASTAVA, A. B. L. 1958. Effect of non-normality on the power function of the *t*-test. *Biometrika* 45: 421–429.

SRIVASTAVA, A. B. L. 1959. Effects of non-normality on the power of the analysis of variance test. *Biometrika* 46: 114–122.

STEEL, R. G. D., and J. H. TORRIE. 1980. *Principles and Procedures of Statistics.* 2nd ed. McGraw-Hill, New York. 633 pp.

STEPHENS, M. A. 1969. Tests for randomness of directions against two circular alternatives. *J. Amer. Statist. Assoc.* 64: 280–289.

STEPHENS, M. A. 1972. Multisample tests for the von Mises distribution. *J. Amer. Statist. Assoc.* 67: 456–461.

STEPHENS, M. A. 1974. EDF statistics for goodness of fit and some comparisons. *J. Amer. Statist. Assoc.* 69: 730–737.

STEVENS, S. S. 1946. On the theory of scales of measurement. *Science* 103: 677–680.

STEVENS, S. S. 1968. Measurement, statistics, and the schemapiric view. *Science* 161: 849–856.

STRUIK, D. J. 1967. *A Concise History of Mathematics.* 3rd ed. Dover, New York. 299 pp.

"STUDENT." 1908. The probable error of a mean. *Biometrika* 6: 1–25.

STUDIER, E. H., R. W. DAPSON, and R. E. BIGELOW. 1975. Analysis of polynomial functions for determining maximum or minimum conditions in biological systems. *Comp. Biochem. Physiol.* 52A: 19–20.

SWED, F. S., and C. EISENHART, 1943. Tables for testing randomness of grouping in a sequence of alternatives. *Ann. Math. Statist.* 14: 66–87.

TANG, P. C. 1938. The power function of the analysis of variance tests with tables and illustrations of their use. *Statist. Res. Mem.* 2: 126–157.

TATE, M. W., and S. M. BROWN. 1964. *Tables for Comparing Related-Sample Percentages and for the Median Test.* Graduate School of Education, University of Pennsylvania, Philadelphia.

TATE, M. W., and S. M. BROWN. 1970. Note on the Cochran Q test. *J. Amer. Statist. Assoc.* 65: 155–160.

THÖNI, H. 1967. Transformation of Variables Used in the Analysis of Experimental and Observational Data. A Review. Tech. Rep. No. 7, Statistical Laboratory, Iowa State University, Ames. 61 pp.

THORNBY, J. I. 1972. A robust test for linear regression. *Biometrics* 28: 533–543.

TIKU, M. L. 1965. Chi-square approximations for the distributions of goodness-of-fit statistics U_N^2 and W_N^2. *Biometrika* 52: 630–633.

TIKU, M. L. 1967. Tables of the power of the F-test. *J. Amer. Statist. Assoc.* 62: 525–539.

TIKU, M. L. 1971. Power function of F-test under non-normal situations. *J. Amer. Statist. Assoc.* 66: 913–916.

TIKU, M. L. 1972. More tables of the power of the F-test. *J. Amer. Statist. Assoc.* 67: 709–710.

TOLMAN, H. 1971. A simple method for obtaining unbiased estimates of population standard deviations. *Amer. Statist.* 25(1): 60.

TRAUT, H. 1980. A method for determining the statistical significance of mutation frequencies. *Biomet. J.* 22: 73–78.

TUKEY, J. W. 1953. The problem of multiple comparisons. Department of Statistics, Princeton University. (unpublished)

TWAIN, M. (S. L. CLEMENS). 1950. *Life on the Mississippi.* Harper & Row, New York. 526 pp.

UPTON, G. J. G. 1976. More multisample tests for the von Mises distribution. *J. Amer. Statist. Assoc.* 71: 675–678.

UPTON, G. J. G. 1978. *The Analysis of Cross-Tabulated Data.* John Wiley, New York. 148 pp.

VAN ELTEREN, P., and G. E. NOETHER. 1959. The asymptotic efficiency of the χ^2-test for a balanced incomplete block design. *Biometrika* 46: 475–477.

VON NEUMANN, J., R. H. KENT, H. R. BELLINSON, and B. I. HART. 1941. The mean square successive difference. *Ann. Math. Statist.* 12: 153–162.

WALD, A., and J. WOLFOWITZ. 1940. On a test whether two samples are from the same population. *Ann. Math. Statist.* 11: 112.

WALDO, D. R. 1976. An evaluation of multiple comparison procedures. *J. Animal Sci.* 42: 539–544.

WALLIS, W. A. 1939. The correlation ratio for ranked data. *J. Amer. Statist. Assoc.* 34: 533–538.

WALLRAFF, H. G. 1979. Goal-oriented and compass-oriented movements of displaced homing pigeons after confinement in differentially shielded aviaries. *Behav. Ecol. Sociobiol.* 5: 201–225.

WAMPLER, R. H. 1970. A report on the accuracy of some widely used least squares computer programs. *J. Amer. Statist. Assoc.* 65: 549–565.

WANG, Y. Y. 1971. Probabilities of the Type I errors of the Welch tests for the Behrens-Fisher problem. *J. Amer. Statist. Assoc.* 66: 605–608.

WATSON, G. S. 1957. The χ^2 goodness of fit test for normal distributions. *Biometrika* 44: 336–348.

WATSON, G. S. 1962. Goodness of fit tests on a circle. II. *Biometrika* 49: 57–63.

WATSON, G. S., and E. J. WILLIAMS. 1956. On the construction of significance tests on the circle and the sphere. *Biometrika* 43: 344–352.

WELCH, B. L. 1951. On the comparison of several mean values: An alternate approach. *Biometrika* 38: 330–336.

WHITE, J. S. 1970. Tables of normal percentile points. *J. Amer. Statist. Assoc.* 65: 635–638.

WILCOXON, F. 1945. Individual comparisons by ranking methods. *Biometrics Bull.* 1: 80–83.

WILCOXON, F., and R. A. WILCOX. 1964. *Some Rapid Approximate Statistical Procedures.* Lederle Laboratories, Pearl River, N. Y. 59 pp.

WILKINSON, L., and G. E. DALLAL. 1977. Accuracy of sample moments calculations among widely used statistical programs. *Amer. Statist.* 31: 128–131.

WILKS, S. S. 1935. The likelihood test of independence in contingency tables. *Ann. Math. Statist.* 6: 190–196.

WILLIAMS, E. J. 1959. *Regression Analysis.* John Wiley, New York. 214 pp.

WILLIAMS, K. 1976. The failure of Pearson's goodness of fit statistic. *Statistician* 25: 49.

WILSON, E. B. and M. M. HILFERTY. 1931. The distribution of chi-square. *Proc. Nat. Acad. Sci.*, Washington, D. C. 17: 684–688.

WINDSOR, C. P. 1948. Factorial analysis of a multiple dichotomy. *Human Biol.* 20: 195–204.

WINER, B. J. 1971. *Statistical Concepts in Experimental Design.* McGraw-Hill, New York. 907 pp.

WOOLF, C. M. 1968. *Principles of Biometry*, D. Van Nostrand, Princeton, N. J. 359 pp.

YATES, F. 1934. Contingency tables involving small numbers and the χ^2 test. *J. Royal Statist. Soc. Suppl.* 1: 217–235.

YOUNG, L. C. 1941. On randomness in ordered sequences. *Ann. Math. Statist.* 12: 293–300.

YULE, G. U. 1900. On the association of attributes in statistics. *Phil. Trans. Royal Soc.* Ser. A 94: 257.

YULE, G. U. 1912. On the methods of measuring the association between two attributes. *J. Royal Statist. Soc.* 75: 579–642.

YULE, G. U. 1917. *An Introduction to the Theory of Statistics.* Griffin, London. 382 pp.

ZAR, J. H. 1967. The effect of changes in units of measurement on least squares regression lines. *BioScience* 17: 818–819.

ZAR, J. H. 1968a. The effect of the choice of temperature scale on simple linear regression equations. *Ecology* 49: 1161.

ZAR, J. H. 1968b. A FORTRAN IV program for polynomial curve fitting by least squares. *Behav. Sci.* 13: 428–429.

ZAR, J. H. 1969. FORTRAN IV program for least squares curve fitting of certain exponential models. *Behav. Sci.* 14: 80.

ZAR, J. H. 1972. Significance testing of the Spearman rank correlation coefficient. *J. Amer. Statist. Assoc.* 67: 578–580.

ZAR, J. H. 1978. Approximations for the percentage points of the chi-squared distribution. *J. Royal Statist. Soc. Ser. C Appl. Statist.* 27: 280–290.

ZELEN, M., and N. C. SEVERO. 1964. Probability functions, pp. 925–995. *In* M. Abramowitz and I. Stegun (eds.), *Handbook of Mathematical Functions*, National Bureau of Standards, Washington, D. C. (Also, Dover, New York, 1965.)

Index

A

a: angle, 422; Y intercept, 266; $\bar{a}$: mean angle, 428; a_c: common Y intercept, 298
Abramowitz, M., 691, 696
Abscissa, defined, 93
Accuracy, 4–5
Acrophase, 465
Acton, F. S., 286, 680
Additivity, 236
Agreement (*see* Concordance; Correlation)
Alexander, D. A., 567, 687
Alf, E. F., Jr., 49, 692
α (alpha): significance level, 43, 99, 102; as regression Y intercept, 263–66
Amplitude of rhythm, 465
Analysis of covariance, 300–302
Analysis of variance (ANOVA):
  components of variance model, 168
  confidence limits, 170–171, 213, 251, 260
  crossed factors, defined, 206–7, 253
  DF, determination of, 473
  F, determination of, 470–73
  factorial, 206, 244–51
    confidence limits, 251
    minimum detectable difference, 251
    multiple comparisons, 251
    nonparametric, 249, 250–51
    power and sample size, 251
    with nesting, 258–59
    with unequal replication, 250–51
  fixed effects, defined, 168
  hierarchical (nested), 253–60
    confidence intervals, 260
    with crossed factors, 258
    multiple comparisons, 259–60
    power and sample size, 260
    with unequal replication, 257
  for intraclass correlation, 323–25
  Latin square, 248
  Model I, defined, 168
    two factors, 211–13
    > two factors, 244–49, Appendix A
  Model II, defined, 168
    two factors, 212, 213
  Model III, defined, 212
    two factors, 212–13
    > two factors, 248–49, Appendix A
  multiple comparisons (*see* Multiple comparisons)
    for multiple correlation, 335–38
    for multiple regression, 335–38
  multiway (*see* Analysis of Variance, factorial)
  nested (*see* Analysis of Variance, hierarchical)
  nonparametric, 176–79
    factorial, 219–22, 249, 250–51
  one factor, 163–79
    confidence limits, 170–71
    maximum testable groups, 176
    nonparametric, 176–79
    power and sample size, 171–75
  power and sample size, 171–75
    graphs, 657–64
    one factor, 171–74
    > one factor, 227–28, 251
  random effects, defined, 168
  randomized block, 222–25
    confidence intervals, 226
    multiple comparisons, 226
  by ranks (*see* Analysis of Variance, nonparametric)
  for regression, 269–71, 272
    nonlinear, 351
  with repeated measures, 222

Note: Reference to a symbol is to its first appearance in the book.

Analysis of variance *(cont.)*
sample size (*see* Analysis of Variance, power)
Latin square, 248
of treatments by subjects, 222
two factors, 206–33
additivity, 237
confidence limits, 213–26
equal replication, 206–14
multiplicativity, 237
nonparametric, 219–22
power and sample size, 227–28
unequal replication, 214–17
> two factors, 245–51
confidence limits, 251
Latin square, 248
nonparametric, 249
Anderson, N. H., 138, 680
Anderson, S. L., 170, 183, 682
Andrews, F. C., 176, 680
Andrews, H. P., 108, 680
Angle (*see also* Angular; Circular distribution)
cosine (cos), defined, 426–28
phase, 465
sine (sin), defined, 426–28
tangent (tan), defined, 248
Angular correlation, 463, 465
Angular deviation, defined, 431
table of, 636–37
Angular dispersion (*see* Circular distribution, dispersion)
Angular distance, comparing two or more groups, 452–54
Angular regression, 463, 465
Angular transformation, 239–41 (*see also* Arcsine transformation)
ANOCOVA (*see* Analysis of covariance)
ANOV or ANOVA (*see* Analysis of variance)
Anscombe, F. J., 240, 241, 242, 402, 680
AOV (*see* Analysis of variance)
A posteriori test, defined, 186
Approximation (*see also* Normal approximation; Uniform approximation)
to chi–square distribution, 482
to Kolmogorov-Smirnov distribution, 548–49
Arbuthnott, J., 386, 680
Arc, defined, 430
Arcsine transformation, 239–41
in regression, 286
table of, 586–90
Arithmetic mean (*see* Mean)
Asano, C., 465, 651, 680
Asimov, I., 427, 680
Association (*see* Coefficient of Association; Concordance; Contingency table; Correlation)
Asymmetry (*see* Symmetry)
Asymptotic regression, 350
Attribute (*see* Nominal scale data)
Average (*see* Central tendency)
Axial distribution (*see* Bimodal)

B

B for Bartlett test, 181
b: regression coefficient (slope), 204; b_c: common regr. coeff., 295; b_i: partial regr. coeff., 334; b_i': standardized partial regr. coeff., 338
Babington-Smith, B., 229, 352
Balanced experimental design, 207
Bancroft, T. A., 186, 680
Bar graph, 6–11 (*see also* Histogram)
of circular data, 424–26
Bartlett, M. S., 181, 237, 238, 240, 241, 242, 680
Bartlett test, 170, 181–83
Basharin, G. P., 146, 680
Batschelet, E., 424, 429, 431, 434, 436, 440, 441, 443, 445, 448, 449, 450, 455, 457, 458, 459, 465, 666, 680, 681
Beall, G., 242, 681
Beckman, P., 79, 681
Behrens-Fisher problem, 130
in multiple comparisons, 186
Belleville, R. E., 681
Bellinson, H. R., 694
Bennett, C. A., 82, 470, 681
Bent, D. H., 690
Bernoulli, J., 370
Bernoulli distribution (*see* Binomial distribution)
Best, D. J., 646, 681
Best fit regression, defined, 263
β (beta): probability of Type II error, 43; regression coefficient (slope), 263–65; β_0: hypothesized β, 271; β_i, partial regression coefficient, 333
Bhattacharjee, G. P., 646, 681
Bias, 17 (*see also* Continuity; Fisher's *z*)
Bigelow, R. E., 693
Bimodal distribution, 432–34
circular, 432–34
Binomial coefficient, 371, 391
Binomial distribution, 369–403, 407–8, 410, 411–14
and CI for median, 113–14
of counts: mean standard deviation, symmetry, variance, 374–75
and data transformation, 239–41
goodness of fit, 380–90
in regression, 286
mean, 374
negative, 242
probabilities (proportions), 370–75
confidence limits of, 378–80
standard error and variance, 376–77
table of, 591
sampling from, 375–77
standard deviation, 375, 377
standard error, 375, 377
symmetry measure, 375
variance, 375–77
Binomial expansion, 371–72
Binomial test, 383–90, 415
for correlation, 323
critical values, 592–601
for Ives-Gibbons correlation, 323
for McNemar test, 157
for median, 114
for median angle, 446
normal approximation, 385–86
minimum detectable difference, 390
power and sample size, 389–90
power, 387–89
Biometry, defined, 1
Biostatistics, defined, 1
Birnbaum, Z. W., 548, 681
Bishop, Y. M. M., 72, 681
Bivariate normal, 306, 455
Bliss, C. I., 31, 70, 82, 95, 329, 367, 378, 465, 681
Block, defined, 222
in Latin square, 248
Bloomfield, P., 465, 681
Boneau, C. A., 130, 681
Bowker, A. H., 158, 681

Bowman, K. O., 33, 681
Box, G. E. P., 130, 170, 183, 237, 682
Briggsian logarithm (*see* Logarithm)
Brillouin, L. 34, 682
Brillouin diversity (*see* Diversity index)
Brower, J. E., 33, 34, 36, 682
Brown, M. B., 183, 329, 682, 684
Brown, S. M., 233, 694
Browne, R. H., 108, 682
Brownlee, K. A., 213, 378, 417, 633, 682
Buckle, N., 143, 682
Burr, E. J., 646, 682
Burstein, H., 378, 379, 380, 682
Byars, J. A., 49, 692

C

C: for mean square successive difference test, 419; critical values, 634-35; contingency table column total, 63, 73; for sign or binomial test, 113, 384-85; critical values, 592-601
${}_nC_X$: combinations, 371
c: number of contingency table columns, 61, 73; c_i, in multiple contrasts, 197; c_{ii}, c_{ij}: matrix elements, 332-33
Cacoullos, T., 309, 682
Cajori, F., 370, 372, 682
Casagrande, J. T., 399, 682
Cartesius, R., 427
Causation, 278
Cell in ANOVA, 210, 245
Cell size (*see* Replication)
Celsius, 290
Center of gravity (*see* Mean)
Centile, 23
Central limit theorem, 86
Central tendency, 16, 18-25 (*see also* Mean; Median; Mode)
  and coding, 24-25, 36-38
Chakravarti, I. M., 409, 692
Chen, H. J., 95, 119, 693
Chi-square (χ^2)
  approximation to, 482
  and Bartlett test, 181
  bias (*see* Continuity)
  for CI for Poisson parameter, 408-9
  for CI for SD or variance, 115
  Cochran correction (*see* Continuity)
  and Cochran test, 233
  for coefficients of association, 322-23
  for comparing correlations, 315-16
  in concordance, 355
  for contingency coefficients, 322-23
  for contingency table (*see* Contingency table)
  continuity (*see* Continuity correction)
  critical values, 42, 479-81
    approximations, 482
  and Friedman test, 229 (*see also* χ_r^2)
  for Gart test, 159-60
  goodness of fit (*see* Goodness of fit)
  for heterogeneity testing:
    for binomial distribution, 381
    contingency table, 67-68
    goodness of fit, 49-52
  for Kruskal-Wallis test, 178
  and log-likelihood testing, 5-53, 71-72
  for McNemar test, 157-58
  and Mann-Whitney test, 178
  for median test, 145, 179-80
  in multisample testing, 181
    for correlation coefficients, 315-16
    for mean angles, 449
    for proportions, 400-401
  for normality, 88-92
  for variance testing, 115–18
  Yates correction (*see* Continuity)
χ_r^2, defined, 229, 354
  critical values, 567
Chow, B., 320, 682
CI (*see* Confidence interval)
Circadian, Circadiseptan, circannual, Circaseptan rhythm, 465
Circle, unit, 426
Circular distribution (*see also* Angle, Angular)
  angular deviation, 431
    table of, 636-37
  concentration (*see* Dispersion)
  data, 3, 422-23
  descriptive statistics, 422-38
  dispersion, 430-31
    multisample text, 465
    two-sample test, 464
  goodness of fit testing, 440-42
  graphing, 424-26
  hypothesis testing, 440-67
  kurtosis, 431
  mean, 428-29
    comparing two, 446-48
    comparing > two, 448-50
    confidence limits, 432
      graphs, 665-66
    for grouped data, 429-30
    of means, 434-36
      comparing two, 457-59
      comparing > two, 463
      confidence limits of, 436-38
      significance of, 455-57
    significance of, 442-45
  mean angular deviation (*see* Angular deviation)
  median, 432
    confidence limits, 432
    significance of, 446
  mode, 432
  multisample test, 452
  paired-sample test, 461, 462
    nonparametric, 463-64
  range, 430-31
  standard deviation, 431
    table of, 638-39
  symmetry, 431
  two-sample test, 446-48
    nonparametric, 450-52
Circular normal distribution, 450
Circular standard deviation, 431
  table of, 638-39
Clark, V. A., 329, 470, 684
Claypool, P. L., 155, 564, 682
Clemens, S. L. 694
Clinch, J. J., 183
Clustered distribution, 410-11, 417
  around a circle, 466-67
Cochran, W. G., 49, 66, 70, 91, 107, 112, 119, 130, 131, 134, 163, 232, 248, 316, 350, 351, 377, 379, 380, 381, 385, 409, 682, 683, 693
Cochran correction for continuity, 66, 68
Cochran Q test, 232–33
Coding data:
  in ANOVA, 180

Coding data *(cont.)*
and central tendency, 24–25, 36–38
in correlation, 325, 340–41
multiple, 340–41
and dispersion, 36–38
in Kruskal–Wallis test, 180
in Mann–Whitney or median test, 146
in multiple correlation or in regression, 340–41
in multisample tests, 180
in one-sample tests, 120
in polynomial regression, 365–66
in regression, 289–90
multiple, 340–41
polynomial, 365–66
in two–sample tests, 146
in variance ratio test, 146
Coefficient:
of association, 322–23
Cramér, 322
Ives-Gibbons, r_n, 323
phi, 322
Yule, Q, 322
binomial, 371, 391
of concordance, 230, 320, 352–59
multigroup, 357–59
contingency, 322
of correlation (*see* Correlation)
of determination, 271, 308, 336
and coding data, 290, 325, 341, 366
in correlation, 308
multiple, 336
in regression, 271
multiple, 336
nonlinear, 351
polynomial, 366
of nondetermination, 271
phi, 322
regression (*see* Regression)
of variation, 32
and coding, 36–38
difference between, 125–26
Cohen, B. H., 391, 692
Cohen, J., 137, 171, 284, 302, 312, 314, 315, 389, 683
Coincidental regression, 304
Coles, C. W., 684
Collinearity (*see* Multicollinearity)
Combinations, 249, 371–72
in multiple comparisons, 186
Common logarithm, 181, 203
Comparisonwise error rate, 188
Compass direction, 422 (*see also* Circular distribution)
Completely randomized ANOVA design, 164
Components of variance ANOVA, 168
Computer calculation (*see* Multidimensional contingency table; Multiple regression; Multiway ANOVA; Nonlinear regression)
Computer rounding error, 340
Concentration of data (*see* Dispersion)
Concordance, 320, 352–59
between groups, 357–59
Confidence bands, 273
Confidence coefficient, 104
Confidence interval (*see* Confidence limits)
Confidence level, 104
Confidence limits:
in ANOVA, 170–71
factorial, 213–14, 251
hierarchical, 260
two factors, 213, 226
for correlation coefficient, 311
intraclass, 325
Spearman rank, 320
and data transformation, 239–40, 242
for estimated Y in regression, 273–75
in regression through origin, 285
for kurtosis measure, 119
for mean (*see also* ANOVA), 103–4, 112, 131, 132, 170, 171
difference between, 132, 191–93, 195, 198
reporting, 104–8
for mean angle, 432
difference between, 448
graphs, 665–66
for mean difference, 152–53
for median, 113–14, 390
for median angle, 432
for median difference, 156
in multiple comparisons, 132, 191–93, 198
with a control, 195
in multiple contrasts, 198
for Poisson parameter, 408–9
for predicted X in regression, 276–77
for predicted Y in regression, 273–75
for proportion, 378–80
difference between, 396
with finite population sampling, 379
normal approximation, 379–80
for regression coefficient (slope), 272–73
difference between, 293, 303
with control, 303
through origin, 285
partial, 337
for residual MS, 276
for standard deviation, 115
for symmetry measure, 119
for variance, 115
for variance ratio, 125
for Y intercept, 275
Conover, W. J., 65, 155, 322, 683
Conservatism, 56, 65, 211
Consistency, 17
Constant (*see* Coding data)
Contagious distribution, 410–11, 417
on a circle, 466–67
Contingency coefficient, 322
Contingency table, 61–77
bias, 70–71
chi-square, 62–71
for correlation, 322–23
for circular data, 450, 452
continuity correction, 64–66, 68, 72
in Gart test, 159–60
heterogeneity, 67–68
log-likelihood ratio, 71–72, 77
log-linear models, 77
for median test, 179–80
multidimensional, 77
for multisample test, 181
subdividing, 69–70
three-dimensional, 72–77
2×2, 64–68, 390–95
Continuity correction:
in binomial test, 386
for comparing proportions, 396
in contingency table, 64–66, 68, 72
in goodness of fit, 48–49
in heterogeneity testing, 51, 67

in Wilcoxon paired-sample test, 155–56
Continuous variable, defined, 4
Control group (*see also* Dunnett test), 194–95
Coordinates: Cartesian, polar, rectangular (*see also* Regression), 426–27
Corrected SS and crossproducts matrix, 331
Correction for continuity (*see* Continuity)
"Correction term," 209, 221, 255, 331
Correlation (and coefficient), 262–63, 306–20 (*see also* Multiple and Partial correlation)
angular, 463, 465
coding in, 325
common, 313
comparing two, 313–15
power, 314–15
comparing > two, 315–18
confidence limits, 311
critical values, 570–71
cylindrical, 463, 465
data transformation in, 306, 311
Fisher's z transformation, 310–18
tables of, 572–76
intraclass, 323–25
Ives-Gibbons, 323
Kendall rank, 318
matrix, 331–32
partial, 339
multiple (*see* Multiple correlation)
"best," 342–43
rank, 352–59
multiple comparisons, 316–18
with control, 317–18
Newman-Keuls test, 317
for nominal scale data, 321–23
nonparametric, 318–20, 356–57
partial (*see* Partial correlation)
product-moment, 307
rank, 318–20, 356–57
multiple, 352–59
vs. regression, 198–199
significance, 309–11
power and sample size, 312
critical values, 570–71
Spearman rank, 318–20, 356–57
confidence limits, 320
critical values, 570–71
Fisher's z transformation, 320
multiple comparisons, 320
power, 320
spherical, 463, 465
standard error, 309
testing, 309–11
power and sample size, 312
Tukey test, 316–17
weighted, 313
Correlation index, 308
Correlation ratio, 352
Cosine (cos), 426–28
curve, 465
Cotangent (cot), 428
Counts (*see* Poisson distribution)
Covariance (*see* Analysis of covariance)
Cowden, D. J., 24, 683
Cox, D. R., 163, 237, 682, 683
Cox, G. M., 131, 134, 163, 682
Cramer, E. M., 338, 683
Cramér, H., 322, 683
Cramér coefficient, 322
Critical region, defined, 99
Critical value, 43, 99, 479f (*see also* names of test statistics)
Crossed factors, defined, 206–7, 253
Crossproducts (*see* Sum of crossproducts)
Crow, E. I., 64, 683
Croxton, F. E., 24, 683
Cubic equation, 362
Cumulative frequency, 12–13 (*see also* Kolmogorov-Smirnov test)
Cureton, E. E., 156, 683
Curve fitting (*see* Regression; Polynomial)
Cycle, 465

D

D: for D'Agostino test, 95, 683; critical values, 585; for Kolmogorov-Smirnov test, 55; critical values, 583–84
d_i: difference, 54; d_{ii}, d_{ij}: matrix element, 332; $\bar{d}$: mean difference, 151; d_{max}: maximum d, 54; critical values, 54
D'Agostino, R. B., 56, 95, 96, 119, 582, 585
D'Agostino test for normality, 95–96
critical values, 585
Dallal, G. E., 31, 695
Daniel, C., 329
Daniel, W. W., 124, 268, 320, 683
Dapson, R. W., 271, 684, 693
Data, defined, 1–4 (*see also* Variable)
coding (*see* Coding data)
dichotomous (*see* Binomial distribution)
missing (*see* Missing data)
transformation (*see* Transformation)
Davenport, J. M., 131, 684
David, F. N., 406, 684
Decay, exponential, 350
Decile, 23
Degrees (direction) (*see* Circular distribution; Temperature)
Degrees of freedom (DF), 45 (*see also* ANOVA; ANOCOVA; Chi-square; Log-likelihood; t)
and variance, 29
Δ(delta): difference, 58
δ(delta): population difference, 111, 135, 173, 175; a regression parameter, 350
de Jonge, C., 578, 684
de Moivre, A., 79, 406
Dependence (*see* Contingency table; Regression)
Dependent variable, (Y), 262
predicting, 266–68
Derivative, partial, 351
Descartes, R., 427
Descriptive model, 351–52
Descriptive statistics (*see* Central tendency; Dispersion; Kurtosis; Symmetry)
for circular data, 422–38
Design (*see* Experimental design)
Detectable difference (*see* Minimum detectable difference)
Deviation:
angular, defined, 431
from mean (*see* SS; MS; Variance)
mean angular, defined, 431
from normality, 88–96
standard (*see* Standard deviation)

DF (*see* Degrees of freedom)
Dichotomous data (*see* Binomial distribution)
Dickinson, P. C., 320, 682
Difference:
  minimum detectable (*see* Minimum detectable difference)
  for paired data, 151
Digital computer (*see* Computer)
Direction (*see* Angle; Circular distribution)
Discontinuous or discrete variable, 4
  bar graph, 5–6
  frequency table, 5–6
Dispersion, 16, 27–38
  angular, 430–31
    testing between, 464–65
  relative (*see* Coefficient of variation)
Distance, angular, 452–54
Distribution (*see also* Binomial; Bivariate normal; Chi-square; Circular; Circular normal; Contagious; Cumulative frequency; *F*; Frequency; *G*; Hypergeometric; Lognormal; Multivariate normal; Negative binomial; Normal; Poisson; *q*; Random; Symmetrical; *T*; *t*; *U*; U^2; *u*; Uniform; *V*)
  of data, 6–13
  of means, 86–88
  sampling, defined, 86
Distribution-free tests, introduced, 138
Diversity index, 32–36
  Brillouin, 34–36
  and coding, 36–38
  relative, 34
  Shannon, 146–48
    difference between two, 146–48
  variance of, 146
Dixon, W. J., 31, 36, 131, 329, 351, 362, 683
Doane, D. P., 684
Dominance, 34
Dose, lethal, 23
Double precision, 331, 340
Draper, W. R., 288, 329, 342, 351, 684
Dummy variable, 346–47
Duncan, D. B., 185, 684
Duncan test, 185, 188
  robustness, 186
Dunn, O. J., 200, 201, 329, 470, 684
Dunnett, C. W., 65, 186, 189, 194, 195, 538, 539, 684
Dunnett test:
  for ANOVA, 194–95
    factorial, 213
    randomized block, 226
      nonparametric, 231
    two factors, 226
  and confidence limits, 195
  for correlation coefficients, 317–18
  critical values, 538–39
  for elevations in regression, 303–4
  for means (*see* for ANOVA)
  nonparametric, 201, 231
  for proportions, 403
  for regression coefficients (slopes), 303
  with unequal sample sizes, 194
  for variances, 204–5
Dunn test, 200
Durand, D., 443, 640, 641, 684, 686

E

e, logarithmic base, 80
Eason, G., 332, 684
Eberhardt, K. R., 396, 684
Eeden, C. van (*see* van Eeden, C.)
Efficiency, 17
Efroymson, M. A., 329, 684
Einot, I., 191, 684
Eisenhart, C., 87, 168, 632, 684, 685, 694
Elevation in regression, defined, 295
  comparing two, 295–98
  comparing > two, 302
  Dunnett test, 304
  multiple comparisons, 303–4
    with control, 304
  Scheffé test, 304
  Tukey test, 304
Enslein, K., 688
Enumeration data (*see* Nominal scale)
Enumeration methods (*see* Contingency table; Goodness of fit; Poisson)
ε (epsilon): as a residual, 350
Equation (*see* Regression)
Error (*see also* ANOVA; Regression)
  comparisonwise, 188
  experimentwise, 188
  of the first kind, 43–44
  in multiple comparisons, 188
  rounding, 340
    in polynomial regression, 365–66
    probable, 106
  of the second kind, 43–45 (*see also* Power)
  standard (*see* Standard error)
  statistical, 43–45
  Types I and II, 43–45 (*see also* Power)
Estimate (*see* Statistic)
Estimate, standard error of, 270, 337
Estimation (*see* Prediction)
Estimator (*see* Statistic)
η_r^2 (eta): correlation ratio, 352
Evans, J. O., 646, 690
Evenness, 34, 36
Everitt, B. S., 72, 77, 322, 455, 458, 685
Exact test (*see* Fisher exact test)
Expected frequency (*see* Contingency table; Goodness of fit)
Experimental design (*see* ANOVA; Balance; Multisample test; One—sample test Paired–sample test)
Experimentwise error rate, 188
Exponential:
  decay and growth, 350
  notation, 330
Extrapolation in regression, 267, 298
Ezekial, M., 329

F

F (distribution), 122 (*see also* ANOVA; ANOCOVA)

and CI for proportion, 378–79
for coefficients of variation comparison, 125–26
in correlation, 309–10
intraclass, 323–25
critical values, 486–521
and Friedman test, 229
for Hotelling test, 455, 458, 462
for means comparison (*see* ANOVA)
for angles, 446–50
second-order analysis, 455–56
noncentral, 171
partial, 337
in regression:
for elevations comparison, 302
partial, 337
polynomial, 365
for variance ratio test, 122–25
for residual MS's, 293
for Watson-Williams test, 446–50
F_i: cumulative observed frequency, 54; $\hat{F}_i$: cumulative expected frequency, 54
f_i, f_{ij}, f_{ijl}: observed frequency, 31, 41; $\hat{f}_i$: $\hat{f}_{ij}, \hat{f}_{ijl}$: expected frequency, 41, 63, 74–75
Factor, defined, 163
crossed, 206–7, 253
fixed effect, 168
in regression, 262
random effect, 168
in regression, 261–62
Factorial (!), 34, 371
logarithm of, table, 652
Fahrenheit, 290
Federer, W. T., 186, 684
Feldman, S. E., 392, 395, 685
Ferin, M., 681
Fiducial interval and limits, 104
Fieller, E. C., 320, 685
Fienberg, S. E., 72, 77, 681, 685
Figures (*see* Graph; Significant figures)
Finite population, 112
binomial distribution, 377
confidence limits, 379
correction, 112, 377
First-order analysis, defined, 434
First-order sample, defined, 455
Fisher, R. A., 40, 62, 97, 122, 131, 163, 248, 310, 316, 325, 378, 390, 391, 685
Fisher exact test, 390–95
for association, 323
critical values, 602–26
for Ives-Gibbons correlation, 323
power, 395
Fisher's z transformation, 310–18
bias, 316, 325
and coding, 325
for correlation coefficients, 309–18
comparing two or more, 313–17
multiple comparisons, 316–17
confidence limits, 311
intraclass, 325
partial, 340
power and sample size, 314–15, 320
Spearman rank, 320
standard error of, 310
table of, 572–76
Fisz, M., 56, 685
Fit (*see* Goodness of fit; Regression)
Fitting curves (*see* Polynomial)
Fix, E., 143, 685
Fixed effects ANOVA (*see* ANOVA, Model I)
Fixed effects factor, defined, 168
in regression, 262
Fleishman, A. I., 391, 685
Fleiss, J. L., 399, 684
Fligner, M. A., 396
Forsythe, A. B., 183, 682
Fourfold table (*see* Contingency table, 2 × 2)
Fourier analysis, 465
Fox, K. A., 329, 685
Fox, M. 171, 685, 690
Franklin, N. L., 82, 470, 681
Frawley, W. H., 358, 359, 685, 692
Freeman, M. F., 240, 241, 402, 685
Frequency:
cumulative (*see* Cumulative frequency)
expected (*see* Expected frequency)
observed (*see* Chi-square)
relative, 10ff
Frequency distribution (*see also* Distribution; specific name of distribution), of data, 6–13
Frequency polygon, 9–11
relative, 12–13
Frequency table (*see* Frequency distribution)
Freund, R. J., 278, 340, 685
Friedman, M., 228, 229, 686
Friedman test, 228–29, 354
critical values, 567
and multiple comparisons, 230–31

G

G: log-likelihood ratio, 53, 71
for association, 323
for contingency table, 71–2
for goodness of fit, 52–53
for binomial distribution, 381, 383
for normal distribution, 91
for Poisson distribution, 409–10
for median test, 145, 179
g_1: symmetry measure, 81; critical values, 579–582; g_2, kurtosis measure, 82
Gabriel, K. R., 191, 684
Gaito, J., 138, 686
Galler, S. R., 681
Galton, F., 307
Galton function, 307
Games, P. A., 130, 170, 180, 183, 686, 687, 688
γ (gamma): a regression parameter, 350; γ_1: symmetry measure, 81; for binomial distribution, 374; γ_2: kurtosis measure, 82
Gart, J., 159, 410, 686
Gartside, P. S., 183, 686
Gauss, K. F., 79
Gaussian distribution, 79 (*see also* Normal distribution)
Geary, R. C., 338, 686
Geometric mean, 24
General linear model, 250
Gent, M., 65, 684
Gettinby, G., 684
Ghent, A. W., 391, 686
Ghosh, B. K., 379, 686

Gibbons, J. D., 56, 322, 323, 686, 687
Gill, J. L., 131, 170, 686
Gillispie, C. C., 689
Glass, G. V., 130, 170, 686
Glejser, H., 288, 686
Glen, W. A., 225, 686
Goodman, L. A., 72
Goodness of fit, 40–58
  for binomial distribution, 380–83
  by binomial test, 383–90
  by chi-square, 40–42
    bias in, 49
    for binomial distribution, 380–83
    for circular distribution, 440–42
    heterogeneity, 49–52
    interaction (*see* Goodness of fit, by chi–square, heterogeneity)
  for circular data, 440–42
  by *G*, 71–72
    for binomial distribution, 381, 383
    for circular distribution, 440
  by Kolmogorov-Smirnov test (*see* Kolmogorov-Smirnov goodness of fit)
  by Kuiper test, 440–41
  by log-likelihood ratio (*see G*)
  for normal distribution, 88–92
  for ordinal data, 53–58
  for Poisson distribution, 409–11
  subdividing, 46–48
  by Watson test, 92, 441–42
    for normal distribution, 92
Gosset, W. S., 97, 406 (*see also* "Student")
Grand mean, 164
  of angles, 459
Graph:
  bar, 6–11
    of circular data, 424–26
  of circular data, 424–26
  of frequency distribution, 6–13
  frequency polygon, 9–13
    cumulative, 13
    histogram, 9
  for mean and variability, 106–8
  misleading, 11
  of normal distribution, 80, 82–85, 94
  for normality testing, 93–95
  ogive, 13
  polar coordinates, 426–27
  presenting statistics, 106–8
  rectangular coordinates, 427 (*see also* Regression)
  in regression, 262ff
    line, simple, 266–67
  rose diagram, 425–26
  scatter plot (*see also* Correlation; Regression)
    of circular data, 424
    in correlation, 307
    sector, 425–26
  of *t* distribution, 98–99
Greenwood, J. A., 443, 640, 641, 684, 686
Grizzle, J. E., 683
Growth: exponential and logistic, 350
Guenther, W. C., 43, 100, 123, 163, 171, 172, 173, 227, 686
Gurland, J., 31, 131, 686, 688

H

H: for Kruskal–Wallis test, 177, 221; critical values, 565–66; as Brillouin diversity index, 36; *H′*: as Shannon diversity index, 33; H_0: null hypothesis, 41; H_A: alternate hypothesis
Haber, M., 66, 682, 686
Haight, F. A., 406, 686
Halberg, F., 465, 681, 687, 689
Hamaker, H. C., 342, 687
Hare, N., 219, 692
Harmonic analysis, 465
Harmonic mean, 24
Harris, M., 109, 133, 687
Hart, B. I., 694
Harter, H. L., 186, 537, 687
Hartley, H. O., 119, 171, 320, 408, 415, 582, 652, 664, 685, 691
Hastings, C., Jr., 483, 687
Hawkins, D. M., 349, 687
Hays, W. L., 179, 687
Healy, M. J. R., 340, 687
Herson, J. H., 241, 688
Heterogeneity (*see also* Homogeneity)
  chi–square (*see* Chi-square, heterogeneity)
  and diversity, 4
Heteroscedasticity, 181 (*see also* Homogeneity of variance)
Hicks, C.R., 470, 687
Hierarchical ANOVA, 253–60 (*see also* ANOVA, hierarchical)
Hilferty, M. M., 482, 695
Histogram, 9
  of circular data, 424–26
Hjelm, H. G., 306, 690
Hodges, J. L., Jr., 141, 143, 685, 687
Holbert, D., 155, 564, 682
Holland, P. W., 72, 681
Hollander, M., 113, 200, 201, 687
Homogeneity (*see also* Multisample test)
  chi-square (*see* Chi-square, heterogeneity)
  and diversity, 34
  of regression coefficients (slopes), 300–302
  of residual MS, 300
  of variance, 122–24, 170, 181–83 (*see also* Transformation)
    and ANOVA, 170
    and multiple comparisons, 186
    in regression, 268, 285–89
    and *t* test, 130
Homoscedasticity, 181 (*see also* Homogeneity of variance)
Honestly significant difference, 186
Horvitz, D. G., 109, 133, 687
Hotelling, H., 320, 455, 457, 687
Hotelling test,
  one sample, 455–56
  paired samples, 461–62
  two samples, 457–59
Howell, J. F., 186, 686, 687
Huck, H. W., 215, 687
Huff, D., 11, 687
Huitema, B. E., 317, 687
Hull, C. H., 690
Hutchinson, K., 146, 681, 687
Hyperbolic tangent, inverse, 310
Hypergeometric distribution, 377
Hypothesis (*see also* Circular distribution; Contingency table; Correlation; Goodness of fit; Multiple comparisons; Multisample test; Normality; One-sample test; Paired-sample test; Randomness;

Regression; Two-sample test)
alternate and null, 41, 98ff
one-tailed, introduced, 101–3
two-tailed, introduced, 97–101

I

Iman, R. L., 567, 687
Independence (*see* Contingency table; Regression)
in ANOVA, 211
conditional, mutual, partial, 74
Independent variable (X), 262
Index:
correlation, 308
diversity (*see* Diversity index)
Inference, 14 (*see also* Hypothesis)
Inferential statistics, defined, 2
Information theory, 32–33
Integers, sum of consecutive, 221
Interaction:
in ANOVA, 211
chi-square (*see* Chi-square, heterogeneity)
Intercorrelation, 338–44 (*see also* Multicollinearity)
International Business Machines Corp., 646, 656, 687
Interpolation, harmonic and linear, 477–78
Interquartile range, 36
Interval (*see* Confidence interval; Prediction interval)
Interval scale, 2–4 (*see also* Circular distribution)
Intraclass correlation, 323–25
Inverse hyperbolic sine, 242
Inverse hyperbolic tangent, 310
Inverse of matrix, 332–33, 348
Inverse prediction, 276–77
regression through origin, 285
Inverse sine (*see* Arcsine)
Irrational number, 79, 80
Iteration, 109, 110, 116–17, 133, 135, 174, 217, 225, 250, 351
Ives, K. H., 323, 687
Ives-Gibbons correlation, 323

J

J and J', evenness measures, 36
Jacobs, G. J., 681
Jacques, J. A., 268, 688
Jenkins, J. G., 690
Jenrich, R. I., 329, 688
Jones, W., 79
Jonge, C. de (*see* de Jonge)

K

K: Watson-Williams correction factor, 446, 448; table of, 642–643
k_i: moment about the mean, 81–82
κ_i (kappa): moment about the mean, 81–82
Karr, J. R., 33, 34, 146, 652
Keeton, W. T., 681
Kendall, M. G., 229, 230, 237, 318, 320, 352, 688
Kendall coefficient:
of concordance, 230, 320, 352–59
of correlation (*see* Rank correlation)
Kent, R. H., 694
Keselman, H. J., 183, 186, 189, 688
Keuls, M., 185, 688
Kihlberg, J. K., 241, 688
Kirk, R. E., 137, 172, 173, 174, 227, 470, 688
Klein, S., 24, 683
Kleinbaum, D. G., 332, 688
Kluger, E., 392, 395, 685
Kohr, R. L., 130, 170, 688
Kolmogorov, A. N., 54, 688
Kolmogorov-Smirnov goodness of fit, 53–58
for binomial distribution, 381, 383
for continuous data, 55–58
critical values, 546–48
approximation, 548–49
grouped, 56–58
for discrete data, 53–55
critical values, 540–44
sample size, 58
for grouped data,
critical values, 540–44
for normal distribution, 91–92
critical values, 583–84
for Poisson distribution, 410
Kraft, C., 143, 681
Kramer, C. Y., 189, 225, 686, 688
Kruskal, W. H., 139, 176, 178, 688
Kruskal-Wallis test, 176–79
for angular dispersion, 455
for angular distance, 454
critical values of H, 565–66
for multiple contrasts, 201–2
for nonparametric ANOVA, factorial, 249
power, 176
Kuiper, N. H., 440, 688
Kupper, L. L., 332, 688
Kurtosis, 82–83
of circular distribution, 431
confidence limits for, 119
testing, 119

L

L_1 and L_2: confidence limits defined, 103
Lancaster, H. O., 40, 72, 688, 691
Laplace, P.–S. M. de, 79
Larsen, W. A., 106, 689
Latin square ANOVA, 248
Layne, B. H., 215, 687
LD_{50}, 23
Least significant difference test, 186
Least squares, defined, 263
Lee, A. F. S., 131, 688
Lee, J.-K., 465, 687, 690
Lehman, E. L., 141, 687
Leptokurtosis, 82, 93 (*see also* Normality)
Leser, C. E. V., 338, 686
Leslie, P. H., 391, 688
Lethal dose, 23
Level, defined, 163
Level of significance, 43
Levels, maximum testable in ANOVA, 176
hierarchical, 260
two-factor, 228
Levy, K. J., 202, 203, 204, 316, 401, 688, 689
Lewontin, R. C., 32, 70, 108, 125, 410, 689, 693
Leyton, M. K., 391, 689
Li, C. C., 225, 226, 248, 689
Li, J. C. R., 329, 332, 689
Liddell, I. G., 465, 689
Light, R. J., 181, 689
Lindsay, R. B., 442, 689
Linear correlation (*see* Correlation)
Linear equation (*see* Regression)

Linearity of regression, 278–84, 288–89 (*see also* Additivity)
Linear model, general, 250
Linear regression (*see* Regression)
Linear transformation (*see* Coding of data)
Ling, R. F., 689
Lloyd, M., 33, 34, 146, 652, 689
ln: natural logarithm, defined, 181, 203
Location (*see* Central tendency)
log: common logarithm, defined, 181, 203
Logarithm, 181, 203
  Briggsian, or common, 181, 203
  conversion among bases, 33
  of factorials, tables, 652
  Naperian, or natural, 80, 181, 203
Logarithmic transformation, 238–39
  in regression, 286–89
Logistic growth, 350
Log-likelihood ratio (*see* *G*)
Log-linear model, 77
Lognormal distribution, 239
Longley, J. W., 340, 689
LSD, defined, 186

M

M: number of variables in multiple regression or correlation, 329 (*see also* Mu)
m: number of independent variables in a multiple regression, 334; m_1, m_2, 2×2 table marginal frequencies, 393, 602
Machine formula (*see also* ANOVA)
  for analysis of variance, 166–67, 209–10, 212
  for sum of squares, 30–31
McCornack, R. L., 564, 689
McGill, R., 106, 689
MacKinnon, W. J., 39, 689
McNemar, Q., 156, 689
McNemar test, 156–58
McSweeney, M., 230, 231, 233, 289, 298, 689
Madow, W. G., 119, 689
Major mode, 23
Mann, H. B., 139, 689
Mann-Whitney test, 138–46
  for angular dispersion, 454
  for angular distance, 453
  and coding data, 146
  critical values, 550–62
  normal approximation, 142–43
  for ordinal data, 143–44
  power, 141
  uniform approximation, 143
Mantel, N., 65, 343, 683, 689
Marascuilo, L. A., 230, 231, 233, 318, 389, 398, 402, 689
Mardia, K. V., 424, 429, 431, 432, 441, 448, 449, 450, 455, 459, 465, 466, 643, 689, 690
Margolin, B. H., 181, 689
Massey, F. J., Jr., 31, 36, 131, 410, 684, 690, 693
Matched pairs test (*see* Paired–sample test)
Matthai, R. A., 325, 692
Matrix, 329, 356
  corrected SS and crossproducts, 331
  correlation coefficients, 339–341
    partial, 339
  half, 331
  inverse, 332–33
  in multiple regression, 329–33
  pseudocorrelation, 349
  residual SS and crossproducts, 331
  SS and crossproducts, 329, 331
  symmetry of, 331
Maxima in quadratic regression, 367
Maximum testable groups in ANOVA (*see* Levels, maximum testable)
Maxwell, A. E., 158, 690
Mean, 18–20, 24
  angle, 428–29 (*see also* Circular distribution, mean)
  angular deviation (*see* Angular deviation)
  arithmetic (*see* Mean)
  binomial (*see* Binomial distribution)
  and coding data, 428–30
  comparing two, 126–31, 132–38
    for angles, 446–48
  confidence limits for, 112, 131, 132, 170–71
    for angles, 432, 436–38
    for difference between two, 132
    for angles, 448
    with a control group (Dunnett test), 195
  distribution of, 86–88
  of distribution of runs, 417
  geometric, 24
  grand, 164
    of angles, 459
  harmonic, 24
  of mean angles, 434–36 (*see also* Circular distribution, mean)
  pooled, 132
  reporting variability, 104
  sample size for estimation, 108–10
  second-order (*see* Mean of mean angles)
  significance of, 97–103
  standard deviation and standard error of, 86
  variance of, 86
Mean absolute deviation, 29
Mean deviation, 29
  and coding, 36–38
Mean difference, 150–52
  confidence limits for, 152–53
Mean square (MS), 29, 166 (*see also* ANOVA; ANOCOVA; Variance)
  pooling, 213
  testing residual, 276
Mean square successive difference test, 418–19
  critical values, 634–35
  normal approximation, 419
Mean squared deviation (*see* Mean square)
Measurement (*see* Data)
Median, 20–22
  angle, 432
    confidence limits, 432
    significance, 446
  and coding data, 38
  comparing two, 145–46
  comparing > two, 179–80
  confidence limits for, 113–14, 390
  and multiple comparisons, 202–3
  significance, 97–103, 114
Median difference, CI for, 156
Median test, 145–46
  and coding data, 146

multisample, 179–80
power, 145–46
using Wilcoxon test, 154–55
Mehta, J. S., 131, 690
Mendel, G., 49, 690
Mensing, R. W., 27, 691
Meristic (*see* Discontinuous variable)
Mesokurtosis, 82
Midpoint (*see* Midrange)
Midrange, 23–24
Miettinen, O. S., 683
Miller, J. E., 682
Miller, L. H., 548, 690
Miller, R. G., Jr., 186, 191, 690
Milton, R. C., 562, 690
Minima in quadratic regression, 367
Minimum detectable difference:
in ANOVA, 175
factorial, 251
hierarchical, 260
two factors, 228
binomial test, 390
in testing between two means, 135–37
in testing mean, 111
in testing mean difference, 153
Minor mode, 23
Missing data in ANOVA:
with randomized blocks, 224–25
with two-factors, 216–17, 224–25
Mitra, S. K., 325, 692
Mixed model ANOVA, 212–13
factorial, 251
Mode, 23–24
of angles, 432
and coding, 38
major and minor, 23
Model (*see also* ANOVA; Regression)
descriptive, 351–52
general linear, 250
log-linear, 77
predictive, 351–52
Moivre, A. de (*see* de Moivre)
Moments, 81
Montfort, M. A. J. van (*see* van Montfort, M. A. J.)
Mood, A. M., 109, 133, 141, 145, 146, 153, 179, 687, 690
Moore, B. R., 456, 647, 690
Moore test, 456–57
critical values, 647
for paired data, 463–64
for weighted angles, 457
MS (*see* Mean square)
MSE, defined, 167
M (mu): median, 113; M_0: hypothesized M, 114
μ (mu): mean, 18, 416; μ_0: hypothesized, 99; μ_a: mean angle, 428; μ_d: mean difference, 150; μ_T: mean of T distribution, 155; μ_U: mean of U distribution; μ_u: mean of distribution of runs, 417; μ_X: mean binomial frequency, 374; $\mu_{Y \cdot X}$: mean in regression, 299
Mulholland, H. P., 582, 690
Multicollinearity, 338, 344
Multiple comparisons (*see also* Dunnett, Newman-Keuls, Tukey tests; Multiple contrasts)
for ANOVA, 185–205
with a control (*see* Dunnett test)
factorial, 213, 251
hierarchical, 259–60
randomized block, 226
nominal scale data, 233
nonparametric, 230–31
two factors, 226
with a control (Dunnett test), 226
nonparametric, test, 233
and confidence limits, 191–93
with a control (Dunnett test), 195
with a control (*see* Dunnett test)
for correlation coefficients, 316–18
with a control (Dunnett test), 317–18
Duncan test (*see* Duncan test)
Dunnett test (*see* Dunnett test)
Dunn test, 200
nonparametric, 200
for elevations in regression, 303–4
with control (Dunnett test), 304
least significant difference test, 185–86
for means (*see* for ANOVA)
for medians, 202–3
Newman-Keuls test (*see* Newman-Keuls test)
nonparametric, 199–202
for Cochran Q test, 233
Dunnett test, 201
Dunn test, 200
randomized block ANOVA, 230–31
power, 186
for proportions, 401–3
with control (Dunnett test), 402–3
for regression coefficients (slopes), 302–3
Tukey test (*see* Tukey test)
with unequal sample sizes, 189
with a control (Dunnett test), 204–5
Multiple contrasts, 197–98
in ANOVA, 186, 197–98
factorial, 213
randomized block, 226
nonparametric, 231
two factors, 213, 226
and confidence limits, 198
for correlation coefficients, 318
for elevations in regression, 304
nonparametric, 201–2
for randomized block ANOVA, 231
for proportions, 403
for regression coefficients (slopes), 303
robustness, 186
Multiple correlation, 328, 336
coding data in, 340–41
coefficient, 336
nonparametric (rank), 318–20, 356–57
significance, 336–37
Multiple range testing, 190 (*see also* Duncan test; Dunnett test: Newman–Keuls test)
Multiple regression, 328–29 (*see also* Partial regression)
coding data in, 340–41
comparing equations, 347–49
computations, 329–33
dummy variables, 346–47
equation, 333–35
"best," 342–43
nonlinear (*see* Nonlinear regression)
through origin, 349
parallel, 348–49
plane, 348

Multiple regression *(cont.)*
polynomial (*see* Polynomial regression)
predicting Y, 344–46
significance, 336
stepwise, 341–44
Y intercept, 345
Multiple t tests, 162–63, 185
Multiplicativity, 236
Multisample testing, 162–63
for angles, 448–50, 463
for angular dispersion, 455
for angular distance, 454
for circular concentrations, 450
for correlation coefficients, 315–18
Spearman rank, 320
circular concentration, 450
for elevations in regression, 303–4
for means (*see* ANOVA)
of angles, 448–50, 463
with a control (*see* Dunnett test)
for medians, 179–80
for nominal scale data, 181
nonparametric, 176–81
for regression coefficients (slopes), 302–3
for variances, 181–83
Multivariate normal distribution, 328
Multiway ANOVA (*see* ANOVA, factorial)
Murray, R., 189

N

N: population size, 18; combined sample size, 139
n: sample size or number of replicates, 19, 207
${}_nC_X$: combinations, 371; ${}_nP_X$: permutations, 371
Naperian, or natural, logarithm, 80, 181, 203
Negative binomial distribution, 242
Nelson, W., 465, 690
Nemenyi, P., 199, 690
Nested ANOVA (*see* ANOVA, heirarchical)
Neter, J., 329, 338, 690
Newman, D., 185
Newman, I., 370, 690
Newman-Keuls test:
for ANOVA, 185, 190–91
factorial, 213
randomized blocks, 226
two factors, 226
and confidence limits, 193
for correlation coefficients, 317
for means (*see* for ANOVA)
nonparametric, 200
for proportions, 401
robustness, 186
for variances, 204
Nie, N. H., 329, 362, 690
Noether, G. E., 56, 228, 683
Nominal scale data, 3–4 (*see also* Contingency table; Goodness of fit)
correlation of, 321–23
with two categories (*see* Binomial distribution)
Noncentrality, 171
Nonlinear regression, 351
logarithmic transformation, 350
Nonparametric testing, introduced, 138
Normal approximation:
for binomial test, 385–86, 389–90
minimum detectable difference, 390
power and sample size, 389–90
for CI for median, 113–14
for CI for proportions, 379–80
for Mann-Whitney test, 142–43
for mean square successive difference test, 419
for paired-sample t test, 153
Wilcoxon test, 155–56
for runs test, 417
for sign test, 387
for two-sample testing, 142–43
for Poisson counts, 415–16
for proportions (*see* Binomial test)
for Wilcoxon paired-sample test, 155–56
Normal curve (*see also* Normal distribution)
table of proportions, 483
Normal deviate, 83 (*see also* Normal approximation)
and binomial test, 386
for concordance, multigroup, 358
and correlation coefficient, 310–12
difference between, 314
for mean, 86
for normal approximation (*see* Normal approximation)
proportions associated with, 483
in two-sample testing:
for correlation coefficients, 313–15
for proportions, 396–97
and V test, 44
Normal distribution, 79–96, 98
bivariate, 306
circular, 450
cumulative, 91–95
departures from, 88–96, 119
graphed, 80, 82–85, 94
kurtosis, 81–83
multivariate, 38
proportions of, 83–86
table of, 483
standardized, 80–81
symmetry, 81–82
table of, 483
Normality (*see also* Transformation)
and ANOVA, 169–71
and Bartlett test, 183
and correlation, 306, 311, 318
and multiple comparisons, 186
and regression, 26, 285–89
and t test, 100, 130
testing for, 88–96, 119
D'Agostino D, 95–96
goodness of fit, 88–93
graphical, 93–95
Shapiro and Wilk W, 95
Normalization, 83, 340
Normal probability scale, 93
Norris, R. C., 306, 690
Norusis, M., 268, 688
Notation, exponential or scientific, 5, 330
ν (nu) (*see* Degrees of freedom)
Null hypothesis, 41, 98ff (*see also* Contingency table; Correlation; Goodness of fit; Multisample test; One-sample test; Randomness testing; Regression; Two-sample test)

O

Observed frequency (*see* Chi-square)

Odeh, R. E., 171, 646, 690
Odum, E. P., 681
Ogive, 13
Olds, E. G., 578
Olver, F. W. J., 643, 691
One-sample test (*see also* Goodness of fit)
of angles, 445–46
second-order, 455–56
nonparametric, 456
weighted, 457
binomial test, 383–86, 411–13
for correlation coefficient, 309–12
for kurtosis, 119
for mean, 97–103
of angles, 445
second-order, 455–56
for median, 114
of angles, 446
using Wilcoxon test, 154–55
for proportion, 383–86, 411–14
for regression coefficient, 268–72
power, 413–14
for standard deviation (*see* One–sample test, for standard deviation, for variance)
for symmetry, 118–19
for variance, 115–18
One-sided (*see* One-tailed)
One-tailed hypothesis, introduced, 101–3
Ord, J. K., 465, 689
Ordinal scale data, 3
bar graph and frequency table, 6–7
Ordinate, defined, 93
Origin (of graph), 284, 426–27
regression through, 284–86
in multiple regression, 349
in polynomial regression, 365
Ostle, B., 277, 691
Otten, A., 578
Owen, D. B., 567, 578, 691

P

P: probability, 43; ${}_nP_X$: permutations, 371
p: range of means, 239; p, $\hat{p}$: binomial proportion, 369, 375; p_i: proportion, 33, 190; $\bar{p}$: pooled proportion, 396
${}_nP_x$: permutations, 371
Pabst, M. R., 320, 687
Paired-sample test, 150–53
with angles, 461–62
nonparametric, 463–64
second-order, 462
nonparametric, 463
McNemar test, 156–58
for mean difference, 150–53
Moore test, 456–57
nonparametric (*see* McNemar test; Moore test; Wilcoxon test; sign test)
for nominal scale data, 156–60
considering treatment order, 159–60
sign test, 386–90
normal approximation, 387
minimal detectable difference, 390
power and sample size, 387–90
power, 389
Wilcoxon test, 153–56
Parabola (*see* Quadratic regression)
Parallel regressions, 295, 348
multiple, 348
Parameter, defined, 16–17
Partial correlation (and coefficient), 339–40
coding data in, 341
standard error of, 339
Partial *F*, 337
Partial derivative, 351
Partial regression, 328, 334
coding data in, 340–41
comparing two coefficients, 346
standard error, 346
confidence limits, 337
in nonlinear regression, 351
significance, 337–38
standard error, 337
standardized, 338
Pascal, B., 372
Pascal's triangle, 372
Patil, G. P., 681
Patil, K. D., 233, 691
Patniak, A. B., 171, 691
Paull, A. E., 213, 691
Pauly, J. E., 687
Pazer, H. L., 400, 691
PE (*see* Probable error)
Pearson, E. S., 31, 79, 119, 171, 307, 320, 328, 408, 415, 582, 652, 664, 685, 691
Pearson, K., 31, 32, 40, 62, 79, 307, 328, 691
Pearson correlation (*see* Rank correlation)
Peckham, P. D., 130, 170, 686
Percentages:
data transformation of, 239–41
in regression, 286
percentile, 23
Per-comparison error (*see* Comparison-wise error)
Per-experiment error (*see* Experimentwise error)
Period of cycle, 465
Periodic regression, 465
Permutations, 371
Pettitt, A. N., 55, 545, 691
Phase angle, 465
ϕ (phi): in determining power, 136, 137, 172–74, 227; ϕ_1, ϕ_1^2, ϕ_2: association coefficients, 322
Phi coefficient, 322
Π (pi): taking a product, 24, 34
π (pi): defined, 79, 422
Pielou, E. C., 34, 681, 691
Pike, M. C., 399, 682
Plane, regression, 348
Platykurtosis, 83, 93 (*see also* Normality)
Poisson, S., 406
Poisson distribution, 406–11, 415–16
goodness of fit, 409–11
mean, 407
confidence limits, 408–9
probabilities, 406–8
in regression, 286
standard deviation and CI, 408–9
two-sample testing, 415–16
variance and CI, 408–9
Polygon, frequency, 9–13
Polynomial regression, 361–67
coding data in, 365–66
through origin, 365
stepwise, 364–65
Pooled variance, 124, 128, 167, 181
Pooling in chi-square (*see* Chi-square heterogeneity)
Pooling MS, 283
in heirarchical ANOVA, 257
Population, 14–15

Population (*cont.*)
finite, 112
parameter, 16–17
Power, 43–45
of ANOVA, 171–74
factorial, 251
graphs, 657–64
hierarchical, 260
nonparametric, 176–228
one factor, 171–74
two factors, 227–28
of binomial test, 387–89
normal approximation, 389
of Fisher exact test, 395
of Friedman test, 228
of Kruskal-Wallis test, 176
of Mann-Whitney test, 141
of median test, 145–46
of one-sample test:
for correlation coefficient, 312
Spearman rank, 320
for mean, 111–12
for proportions, 413–14
for regression, 283
for variance, 117–18
of paired-sample test, 153
nonparametric, 153
of sign test, 387–89
of Spearman rank correlation, 320
of two-sample test:
for correlation coefficients, 314–15
for means, 136–38
nonparametric, 141
for proportions, 397–98
for regressions, 302
of Wilcoxon paired-sample test, 153
Pratt, J. W., 155, 156, 691
Precision:
of correlation coefficient, 309
double, 331, 340
of mean, 87–88
of measurement, 4
of regression coefficient, 272
partial, 335
of standard deviation, 87–88
of statistic, 17
of variance, 88
Prediction in regression, 266–68, 275–76, 351–52
confidence limits, 273–75
hypothesis testing, 275–76
inverse, 276–77
with regression through origin, 285
multiple, 344–46
standard error of, 344–46
Prediction interval, 275
Predictive model, 351–52
Probability:
of a test statistic, 42–44
of two events, 370
Probable error, 106
Probert, D. A., 183, 686
Probit, 83
Product-moment correlation, 307
Proportion (*see also* Binomial Distribution; Cochran test; Contingency table)
of binomial distribution, 591
comparing two, 395–400
confidence limits, 396
minimum detectable difference, 400
normal approximation, 396–97
power and sample size, 397–99
comparing > two, 400–403 (*see also* Contingency table; Fisher exact test)
confidence limits, 378–80
corresponding to arcsine, 589–90
of normal distribution, 83–86
table of, 483
of Poisson distribution, 406–8
pooled, $\bar{p}$, $\bar{q}$, 396, 400
in regression, 286
transformation of, 239–41, 401–3
Pseudocorrelation matrix, 349
Puri, M. L., 465, 690

Q

Q: for multiple comparison tests, 20; for Cochran test, 233; critical values, 568; Yule coefficient, 323; Q': for multiple comparison tests, 201; critical values, 569
q: distribution, 188, 369, (*see also* Dunnett, Newman-Keuls, Tukey tests); critical values, 522–37
q, $\hat{q}$: binomial proportion, 369, 375; q': for Dunnett test, 194; critical values, 538–39; $\bar{q}$, pooled proportion, 396
Q test, 232–33; critical values, 568
Quade, D., 567, 687
Quadratic regression, 288, 362–67
maxima and minima in, 367
Quantile, 20–23
Quartic regression, 362–65
Quartile, 23
Quenouille, M. H., 239, 692
Quetelet, L., 323
Quintic regression, 363–64
Quintile, 23

R

R, R^2: in multiple correlation, 336; R: contingency table row total, 63, 73; rank sum, 139–40; Rayleigh's, 443, R': for Moore test, 456; critical values, 647
r: number of contingency table rows, 61, 73; correlation coefficient, 307; critical values, 570–71; polar coordinate, 426, 428; vector length, 428; r^2: coefficient of determination, 271; r_c: vector length for grouped data, 431; r_I: for intraclass correlation, 323; r_{ik}: correlation coefficient, 331–32; $r_{ij\ldots}$: for partial correlation, 339; r_n: for Ives-Gibbons correlation, 323; r_s: for Spearman rank correlation, 319; critical values, 577–78; r_w: for weighted correlation, 446; weighted mean vector length, 446
Ractliffe, J. F., 100, 415
Radian, defined, 239, 422
Radlow, R., 49, 692
Rahe, A. J., 155, 692
Ralston, A., 684, 688
Ramsey, P. H., 191, 692
Random distribution (*see* Randomness)
Random effect ANOVA, 168 (*see also* ANOVA, Model II)
Random effect factor:
in ANOVA, 168

in regression, 261–62
Randomized block ANOVA, 222–25
confidence limits, 226
Friedman test, 228–31
multiple comparisons, 230–31
multiple comparisons, 226
with nominal scale data, 231–33
multiple comparisons, 233
nonparametric, 228–31
multiple comparisons, 230–31
Randomized experimental design, 164
Randomness (*see also* Poisson distribution; Uniformity)
on a circle, 442, 465–67
serial, 416–20
on a circle, 465–67
nominal scale, 416–17
non-nominal scale, 418–20
nonparametric test, 419–20
parametric test, 418–19
Random numbers, 15–16
table of random digits, 653–56
Random sampling, 15–16
Range, 27–28
circular, 430–31
and coding data, 36–38
interquartile, 36
midpoint, 23–24
reporting, 104, 106
semiquartile, 36
Rank correlation, 318–20
Kendall, 318, 320
multiple (*see* Coefficient of concordance)
Spearman, 318–20 (*see also* Correlation, Spearman rank)
critical values, 577–78
Ranking data, introduction, 139–40
tied ranks, introduction, 141
Rank sum, introduced, 140
Rank testing (*see* Nonparametric testing)
Rao, C. R., 314, 325, 409, 692
Ratio scale data, 2, 4
Ray, W. S., 219, 692
Rayleigh, Lord, 442
Rayleigh *R* and *z*, 443
Rayleigh test, 442–43
critical values, 640
modification, 443–44, 456–57
Reciprocal transformation, 242
Regression, 261–90 (*see also* Multiple and Partial regression)
assumptions in, 268
asymptotic, 350
coding data in, 289–90
coefficient (*see* Regression, quartic slope)
coincidental, 304
common, 294, 296, 300, 302
comparing points, 299
comparing two elevations, 295–98
comparing > two elevations, 302
multiple comparisons, 303–4
comparing two lines, 292–99
confidence limits, 293
comparing > two lines, 300–304
multiple comparisons, 302–4
comparing two regression coefficients (slopes), 292–95
confidence limits, 303
power and sample size, 302
comparing > two regression coefficients (slopes), 300–302
multiple comparisons, 302–3
confidence limits:
for estimated *Y*, 273–75
for regression coefficient (slope), 272–73
for *Y* intercept, 273–75
vs. correlation, 198–99
cosine, 465
curvilinear, 287–88
descriptive, 351–52
double coefficient, 328
elevation, 295 (*see also* comparing)
equation, 263
exponential, 350
graph, 262ff
interpretation of, 278
intersection of lines, 293–94
line, 263–66
linearity, 278–84, 288–89
linear model, 349–50
logarithmic transformation, 286–89, 350
logistic, 350
multiple (*see* Multiple regression)
multiple *Y* values, 278–84
predicting *X*, 277, 285
net coefficient, 328
with nominal scale variable, 346–47
nonlinear models, 349–51 (*see also* Nonlinear regression)
through origin, 284–85
parallel, 295, 348
partial (*see* Partial regression)
periodic, 465
plane, 348
polynomial (*see* Polynomial regression)
pooled, 300–302
power and sample size, 283
precision, 272
predicting *X*, 276–77, 285
predicting *Y*, 266–68, 273–75, 284–85
predictive, 351–52
quadratic, 288, 362–66
maxima and minima in, 367
quartic, 362–64
quintic, 363–64
significance, 268–72
sine, 465
slope, 264–65
standardized partial (*see* Partial regression)
transformation of data, 286–89
weighted, 294
Y intercept, 265–66
Relative dispersion, 32
Relative diversity, 34
Relative frequency, 10ff
of circular data, 425
polygon, 12–13
Relative variability, 32
Reliability, 17
Replication (*see also* Sample size)
disproportional, 216
equal, 216
in Friedman test, 230, 231
lack of, 217–19
proportional, 215–16
in regression, 278–84
two factors, 215
Residuals, 350
examination of, 288–89
SS and crossproducts matrix, 331
Response surface, 334
Reversion, 307
ρ(rho): correlation coefficient, 309; vector length, 442; ρ_0: hypothesized ρ, 310; ρ_I: for intraclass correlation, 323; $\rho_{ij...}$: for partial

ρ(rho) *(cont.)*
correlation, 339; ρ_s: for Spearman rank correlation, 319; ρ^2, $\hat{\rho}^2$: coefficient of determination, 336
Rhythms and rhythmometry, 465
Richart, R. M., 681
Roberts, D. E., 646, 681
Robustness, 100, 236–37
of ANOVA, 170
of Bartlett test, 183
of multiple comparisons, 186
of *t* test, 100, 130
of Tukey test, 186
Roe, A., 32, 70, 108, 410, 693
Rogan, J. C., 189, 688
Rohlf, F. J., 170, 257, 693
Root mean square deviation, 31
Roscoe, J. T., 49, 692
Rose diagram, 425–26
Rothery, P., 325, 692
Rounding error with computer, 340
Runs test, 416–17, 419–20
circular, 250, 465–67
critical values, 648–51
critical values, 627–33
normal approximation, 417

S

S: for Scheffé test, 197
S test, 196–98 (*see also* Multiple contrasts)
s: standard deviation, 31; angular deviation, 431; table of, 636–37; s_*^2: 29, 419 (*see also* Variance); s': circular standard deviation, 431, table of, 638–39; s_a: SE of *Y* intercept, 345; s_b: SE of *b*, 271–72, 284; s_{bi}: SE of b_i, 337; s_{bi-bj}: SE of difference between *b*'s or b_i's, 292, 346; s_d: SE of differences, 151; $s_{\bar{d}}$: SE of $\bar{d}$, 151; s_{g_1}: SE of g_1, 118; s_{g_2}: SE of g_2, 119; $s_{H'}$: SE of H', 146; s_p^2: pooled variance, 124; $s_{\hat{p}}$, $s_{\hat{q}}$: SE of binomial $\hat{p}$ and $\hat{q}$, 377; s_r: SE of correlation coefficient, 309; $s_{\bar{X}}$: SE of mean, 87; $s_{\bar{X}_1-\bar{X}_2}$: SE of difference between means, 127; $s_{\hat{Y}}$: SE of $\hat{Y}$, 273–75, 344–46; s_Y: SE of binomial *Y*, 377; $s^2_{Y\cdot X}$ and $s_{Y\cdot X}$: residual MS and SE of estimate, 270, 284, 337
Sakoda, J. M., 391, 692
Sample, defined, 15
random, 15–16
Sample size:
for ANOVA, 174–75
factorial, 251
graphs for, 657–66
hierarchical, 260
two factors, 228
for binomial test, 389–90
for confidence limits:
for mean, 108–10
for proportion, 380
for correlation coefficient, 312
for difference between means, 132–34
with a control, 195
in multisample testing, 193
for Kolmogorov-Smirnov test, 58
for mean, 110–12
for mean difference, 153
for regression, 283
and statistical errors, 44
for two-sample test:
for correlations, 314–15
for means, 134–35
for proportions, 399
for regressions, 302
for variance, 116–17
Sampling:
distribution, defined, 86
finite population, 112
Sampling fraction, 112, 377
Sanders, J. R., 130, 170, 692
Sarner, M. H., 108, 680
SAS Institute, 329, 692
Satterthwaite, F. E., 131, 473
Savage, I. R., 138
Scale (*see* Data)
of measurement (*see also* Circular, Interval, Nominal, Ordinal, Ratio Scale; Coding of data)
normal probability, 93–95
temperature (*see* Temperature)
Scatter plot, 424 (*see also* Correlation; Regression)
of circular data, 424
Scheffé, H., 131, 163, 186, 196, 470, 692
Scheffé test, 196–98 (*see also* Multiple contrasts)
Scheirer, C. J., 219
Scheving, L. E., 687
Schmidt-Koenig, K., 681
Schucany, W. R., 358, 359, 685, 692
Schutz, W. E., 241, 688
Scientific notation, 5, 330
SD (*see* Standard deviation)
SE (*see* Standard error)
Searle, S. R., 332, 692
Seber, G. A. F., 183, 277, 285, 692
Second-order analysis:
for mean of mean angles, 434–36
significance, 455–57
with paired data, 462–63
Sector graph, 425–26
SEM (*see* Standard error of mean)
Semiquartile range, 36
Sen, P. K., 683
Senders, V. L., 2, 692
Serial randomness (*see* Randomness, serial)
Severo, N. C., 482, 485, 521, 568, 569, 696
Shaffer, J. P., 198, 692
Shannon, C. E., 33, 692
Shannon diversity (*see* Diversity index)
Shannon-Weaver (*see* Shannon diversity)
Shannon-Wiener (*see* Shannon diversity)
Shapiro, S. S., 95, 119, 692, 693
Shapiro and Wilk test, 95
Shearer, P. R., 216, 250, 692
Shenton, L. R., 681
Siegel, S., 2, 320, 692
Σ (sigma): as summation, 19, 329ff; ΣT: correction for ties, 143, 320; Σx^2, Σxy, Σy^2, defined, 264, 280, 455
σ (sigma): standard deviation, 31; σ^2: 29 (*see also* Variance); $\sigma_{\hat{p}}$, $\sigma_{\hat{q}}$: SE of binomial proportion, 376–77; σ_T: SE of *T*, 155; σ_U: SE of *U*, 143; σ_u: SE of distribution of runs; $\sigma_{\bar{X}}$: SE of $\bar{X}$, 86; $\sigma_{\bar{X}_1-\bar{X}_2}$: SE of difference between means, 127; $\sigma^2_{\hat{Y}-X}$:

residual MS, 276; $\sigma_{Y \cdot X}$: SE of estimate, 27; σ_z: SE of Fisher z, 310; σ_{z_1}: SE of z_I, 325; $\sigma_{z_1 - z_2}$: SE of $z_1 - z_2$, 313
Sigmoid curve, 93
Signed rank testing, 153
Significance, statistical, 41–43
Significance level, 43, 98
Significant figures, 4–5
on computer, 31, 340
Sign test, 386–90
critical values, 592–601
for Ives-Gibbons correlation, 323
normal approximation, 387
minimum detectable difference, 390
power and sample size, 389–90
power, 387–89
Simple correlation (*see* Correlation)
Simple regression (*see* Regression)
Simpson, G. G., 32, 70, 108, 410, 693
$\sin^{-1}$ (*see* Arcsine)
Sine (sin), defined, 426–28
curve, 465
inverse, 239
Size (*see* Sample size)
Skewness, 93–95
Slope, defined, 364–65 (*see also* Regression)
Smallest detectable difference (*see* Minimum detectable difference)
Smirnov, N. V., 54, 548, 693
Smith, H., 131, 288, 329, 342, 351, 684, 693
Smith, P. G., 399, 682
Smith, R. A., 189, 693
Snedecor, G. W., 70, 119, 122, 163, 248, 316, 350, 351, 693
Snee, R. D., 70, 108, 680, 693
SNK (*see* Newman-Keuls test)
Sokal, R. R., 170, 257, 693
Spearman, S., 318, 693
Spearman rank correlation, 318–20 (*see also* Correlation, Spearman rank)
Square root transformation, 241–42, 286
in regression, 286
in two-sample test for Poisson data, 415–16
Srinivasan, R., 131, 690
Srivastava, A. B. L., 130, 170, 693
SS (*see* Sum of squares)
Standard deviation, 31–32 (*see also* Standard error)
of binomial frequencies and proportions, 376–77
circular, 431
table of, 638–39
and coding data, 36–38
confidence limits, 115
of difference between pairs, 151
of distribution of runs, 417
of mean, 86
for proportions, 377
reporting, 104–8
Standard error, 86 (*see also* Standard deviation)
of binomial distribution, 375, 377
of correlation coefficient, 309
partial, 339
of difference between pairs, 151
of estimate, 270, 337
of Fisher's z, 310
for intraclass correlation, 325
of kurtosis measure, 119
of mean, 86
difference between, 128
of predicted $\hat{Y}$, 272–76
in multiple regression, 344–46
for proportions, 377
of regression, 270
of regression coefficient, 272
difference between, 292
partial, 337
difference between, 346
reporting, 104–8
of symmetry measure, 118
of Y intercept, 275
in multiple regression, 345
Standardization, 83
Standardized normal distribution, 80–81
Standardized score, 340
Standard partial regression coefficient, 338
Starmer, C. F., 683
Statistic, defined, 16–17
test, 43
Statistical errors, 43–45
Statistical hypotheses (*see* Hypothesis testing)
Statistical significance, introduced, 44
Statistical test (*see* Hypothesis testing)
Statistics, defined, 1
descriptive and inferential, defined, 2
Steel, R. G. D., 186, 193, 248, 410, 693
Stegun, I., 691, 696
Steinbrenner, K., 690
Stephens, M. A., 55, 92, 446, 450, 545, 584, 640, 691, 693
Stepwise regression, 341–44
Stevens, S. S., 2, 138, 693
Stirling's approximation, 34, 652
Struik, D. J., 372, 406, 693
Strutt, J. W., 442
Stuart, A., 237, 688
"Student," 97, 406, 693
Studentized range, 188 (*see also* q distribution)
Student-Newman-Keuls test (*see* Newman-Keuls test)
Student t (*see* t)
Studier, E. H., 367, 693
Subgroups, subsamples (*see* ANOVA, hierarchical)
Sum of crossproducts, defined, 264
corrected, 331
matrix, 329, 331
raw, 329
Sum of squares, 29–31 (*see also* ANOVA)
and coding data, 36–38
corrected, 331
matrix, 329, 331
raw, 329
Summation, 19
Surface, response, 334
Swanson, L. A., 400, 691
Swed, F. S., 632, 694
Symmetry:
of circular data, 431
confidence limits for, 119
critical values, 579–82
of distribution, defined, 20–21
of matrix, 331
measure of, 81–82
testing of, 118–19

T

T: contingency table tier sum, 74–75; T, T': for Wilcoxon test, 153–54
t: contingency table tiers, 73; t_i: tied ranks defined, 143
t: distribution, 97–98
and coding data, 120

t: distribution *(cont.)*
for confidence limits (*see* Confidence limits)
for correlation, 309–10
partial, 339, 340, 343
critical values, 484–85
for difference between means, 167
general hypothesis, 118, 271
grouped, 98–99
multiple *t* tests, 162–63, 185
one-sample test:
for kurtosis, 119
for mean, 97–103
for symmetry, 118–19
for paired-sample test, 150–53
for regression coefficient (slope), 271–72
partial, 337–38, 351
polynomial, 365
for symmetry measure, 118–19
in two-sample test:
for diversity indices, 146–48
for elevations in regression, 295–98
for means, 126–38
partial, 346
for regression coefficients (slopes), 292–95
for Watson-Williams test, 448
Welch's, 131
Table:
contingency (*see* Contingency table)
of critical values, Appendix B
fourfold (*see* Contingency table, 2 × 2)
frequency (*see* Frequency distribution)
for circular data, 425
presenting mean and variability, 104–6
Tang, P. C., 171
Tangent (tan), 428
inverse hyperbolic, 310
$\tanh^{-1}$ (*see* Inverse hyperbolic tangent)
Tate, M. W., 233
τ (tau): Kendall rank correlation coefficient, 320
Temperature (*see also* Interval scale)
data transformation in regression, 289–90
scales, 2–3
Testing hypotheses (*see* Hypothesis testing)
Test statistic, defined, 43
Thöni, H., 238, 241, 242, 694
Thornby, J. I., 278, 69
Tied data, 22, 141, 353–55
in Friedman test, 230
in Kruskal-Wallis test, 178–79
in Mann-Whitney test, 141
in Spearman rank correlation, 320
in Watson test, 451–52
in Wilcoxon paired sample test, 155–56
Tietjen, G. L., 119, 683
Tiku, M. L., 170, 171, 646, 694
Time, in periodic regression, 465
Time series, 465
Times of day, week, or year, 422–23 (*see also* Circular distribution)
Tingey, F. H., 548, 681
Tolman, H., 31, 64
Torrie, J. H., 186, 193, 248, 410, 693
Transformation, 236–42
angular (*see* Transformation, arcsine)
arcsine, 239–241, 401–3
in regression, 286
table of, 586–90
in correlation, 306, 311
multiple, 328
Fisher's *z*, 310–18
inverse hyperbolic sine, 242
inverse sine (*see* Arcsine)
linear, as coding, 24–25, 36–38
logarithmic, 238–39
in regression, 286–89, 350
of percentages, 239–41
of proportions, 239–41
reciprocal, 242
in regression, 285–89
multiple, 328
square root, 241–42
in regression, 286
of temperature data, 289–90
Traut, H., 411, 694
Trigonometric functions, 426–28
Tripathi, R. C., 31, 686
Tukey, J. W., 106, 185, 189, 193, 241, 402, 685, 689, 694
Tukey test:
for ANOVA, 185–90
factorial, 213
randomized block, 226
nonparametric, 230–31
two factors, 226
and confidence limits, 191–93
for correlation coefficients, 316–17
for elevations in regression, 303–4
for medians, 202–3
nonparametric, 199
randomized block ANOVA, 230–31
for proportions, 401–2
for regression coefficients, 302–3
robustness, 186
for variances, 203–4
Twain, M., 267, 694
2 × 2 table (*see* Contingency table, 2 × 2)
Two-sample testing (*see also* Contingency table, 2 × 2)
for angles, 446–48
nonparametric, 450–52
second-order, 459–61
for angular dispersion, 454
for angular distance, 452–53
for circular concentration, 450
and coding data, 146
for coefficients of variation, 125–26
for correlation coefficients, 313–15
for diversity indices, 146–48
of independence (*see* Association; Contingency table: Correlation)
for means, 126–38, 167
of angles, 446–48
second-order, 457–59;
for medians, 145–46
for nominal scale data, 61–72, 146
nonparametric, 138–46
for Poisson counts, 415–16
normal approximation, 415–16
for proportions, 396–400
minimum detectable difference, 400
power and sample size, 397–99
for regression coefficients (slopes), 300–302
runs test, 416–17, 419–20
normal approximation, 417
for variances, 122–25
Two-sided (*see* Two–tailed)

Two-tailed hypotheses, introduced, 97–101
Type I error, 43–45
Type II error, 43–45 (*see also* Power)
Tytun, A., 399, 685

U

U and U': for Mann-Whitney test, 139–40; critical values, 550–62; U^2: for Watson test, 441, 450; critical values, 644–46
u: number of runs, 416; u: for V test, 444; critical values, 641; u': number of runs on a circle, 466; u_i: transformation of angle, 441; $\bar{u}$: mean of u_i, 441
Unbiasedness, 17
Uncertainty and diversity, 32–33
Uniform approximation, 143
Uniform distribution, 410–11, 417
 around a circle, 467
  testing, 441–46
Universe (*see* Population)
Upton, G. J. G., 72, 694
Ury, H. K., 399, 685

V

V: coefficient of variation, 32; V: for circular data, 444
V test, 444
 critical values, 641
Vande Wiele, R. L., 681
van Eeden, C., 143, 682
van Elteren, P., 228, 694
van Montfort, M. A. J., 578, 684
Variability (*see* Dispersion)
 coefficient of, 32
 relative, 32
Variable, defined, 2 (*see also* Data)
 circular scale, 3, 422–33 (*see also* Circular distribution)
 continuous, 4
 criterion, 334
 dependent, 261, 328, 334
 discontinuous or discrete, 4
 dummy, 346–47
 ill-conditioned, 338
 independent, 261, 328, 334
 interval scale, 2–3
 meristic, 4
 nominal scale, 3–4
 ordinal scale, 3
 predictor, 334
 ratio scale, 2–3
 regressor, 334
 response, 334
 selection in multiple regression, 341–44
 types of, 2–4
Variance, 29–31 (*see also* ANOVA; Mean square)
 analysis of (*see* ANOVA)
 and coding data, 36–38
 common (*see* Variance, pooled)
 comparing two, 122–25
 comparing > two, 181–83
 confidence limits for, 115
 equality (*see* Variance, homogeneity)
 heterogeneity (*see* Variance, homogeneity)
 homogeneity, 122–25, 170, 181–83 (*see also* Transformation)
  and ANOVA, 170
  and multiple comparisons, 186
  and t test, 130
 of mean, 86
 multiple comparisons, 203–5
  with a control, 204–5
 pooled, 124, 165, 167, 181–82
 testing, 115–18
Variance ratio:
 and coding data, 36–38
 for coefficients of variation, 125–26
 confidence limits for, 125
 test, 122–25, 167ff
Variate (*see* Variable)
Vector length, testing, 442–44
von Mises, R., 450
von Mises distribution, 450
von Montfort, M. A. J., 578
von Neumann, J., 419, 694

W

W: for Shapiro–Wilk test, 95; Kendall coefficient of concordance, 320, 354; $\mathcal{W}$: multisample coefficient of concordance, 358
Wald, A., 416, 694
Waldo, D. R., 186, 694
Wallis, W. A., 176, 178, 352, 688, 694
Wallraff, H. G., 453, 454, 694
Wampler, R. H., 340, 695
Wang, Y. Y., 131, 695
Wasserman, W., 329, 338, 690
Waters, W. E., 682
Watson, G. S., 91, 646, 695
Watson test:
 goodness of fit, 92, 441–42
 for two samples, 450–52
  for second-order analysis, 459–61
Watson U^2, defined, 441, 446, 450
 critical values, 644–46
Watson-Williams test:
 correction factors, 642–43
 for two mean angles, 446–48
 for > two mean angles, 448–50
Webster, J. T., 131, 684
Weighted correlation coefficient, 313
Weighted regression coefficient, 24
Welch, B. L., 170, 695
Welch's approximation, 131
Welch t, 131
White, J. S., 482, 485, 695
Whitney, D. R., 139, 689
Wholly significant difference, 186
Wilcox, R. A., 153, 199, 201, 695
Wilcoxon, F., 138, 153, 199, 201, 695
Wilcoxon-Mann-Whitney test (*see* Mann-Whitney test)
Wilcoxon T (*see* T)
 critical values, 563–64
  approximation, 564
  normal approximation, 155–56
Wilcoxon test:
 for median, 114
 for median test, 154–55
 for paired samples, 153–56
  continuity correction, 156
  normal approximation, 155–56
  power, 153
Wilf, H. S., 684, 688
Wilk, M. B., 95, 119, 692, 693
Wilkinson, L., 31, 695
Wilks, S. S., 52, 71, 695
Williams, E. J., 329, 446, 695
Williams, K., 53, 72, 695

Wilson, E. B., 482, 695
Windsor, C. P., 181, 695
Winer, B., 470, 695
Winkler, H. B., 183, 683
Wolfe, D. A., 113, 200, 201, 687
Wolfowitz, J., 416, 694
Wood, F. S., 329, 683
Woolf, C. M., 248, 695
Working formula (*see* Machine formula)

X

X: a datum, 18; binomial frequency, 371; rectangular coordinate, 427, 428 (*see also* Independent variable); $\bar{X}$: mean of X, 19; $\hat{X}$: predicted X, 276, 277, 285; X_I: X at regression intersection, 293; $\bar{X}_p$: pooled mean, 132; $\hat{X}_0$: X at regression maximum or minimum, 367
x_i (*see* Σx^2; Σxy)

Y

Y: rectangular coordinate, 427, 428 (*see also* Dependent variable); $\bar{Y}$: mean of Y, 264; Y_I: Y at regression intersection, 294; $\bar{Y}_p$: common mean of Y, 298; $\hat{Y}_0$: Y at regression maximum or minimum, 367
Y intercept, 265–66, 334 (*see also* Elevation)
  confidence limits, 275
  in multiple regression, 345
  standard error, 275
    in multiple regression, 345
Yates, F., 48, 131, 248, 378, 685, 695
Yates correction for continuity, 48–49, 64, 68, 72, 157
y_i (*see* Σxy; Σy^2)
Young, L. C., 419, 635, 695
Yule, G. U., 322, 323, 328, 695

Z

Z: normal deviate, 83, 86 (*see also* Normal deviate)
z: Rayleigh's, 443; critical values, 640; z: Fisher transformation, 310ff; z_I: Fisher transformation of r_I, 325; z_w: weighted Fisher z, 313
Zar, J. H., 33, 34, 36, 146, 290, 351, 362, 482, 578, 652, 682, 689, 695, 696
Zelen, M., 482, 485, 521, 568, 569, 696
Zero point (*see* Variable, types of)
ζ (zeta): Fisher transformation, 310; ζ_0: Fisher transformation of ρ_0, 310